DATE DUE

			PRINTED IN U.S.A.

HENDERSON'S DICTIONARY OF

Biological Terms

Eleventh Edition

Eleanor Lawrence

John Wiley & Sons, Inc.

John Wiley & Sons, Inc.
605 Third Avenue
New York, NY 10158–0012, USA

© Oliver and Boyd Limited 1963

© Longman Group UK Limited 1989

© Longman Group Limited 1975, 1979, 1995

Originally published by Oliver and Boyd under the title
A Dictionary of Scientific Terms
First edition 1920
Seventh edition 1960
First published under the title
A Dictionary of Biological Terms
Eighth edition 1963
Reprinted by Longman Group Limited 1975, 1976
Ninth edition 1979
Reprinted 1982
Reprinted in paperback 1985, 1986, 1987
Tenth edition 1989
Reprinted in paperback 1990
Eleventh edition 1995

Library of Congress Cataloging-in-Publication Data
A catalogue entry for this title is available
from the Library of Congress
ISBN 0–470–23507–1

Produced by Longman Singapore Publishers (Pte) Ltd.

Printed in Singapore

CONTENTS

PREFACE

In the six years that have elapsed since the publication of the Tenth Edition of *Henderson's Dictionary of Biological Terms* many new words have entered the biological vocabulary, and old ones have acquired new and more precise meanings. Progress in cell biology in particular continues to be rapid, driven by the use of recombinant DNA technology and advances in experimental cell biology. During the thorough revision undertaken for this edition we have also extended and updated the coverage of ecology and animal behaviour.

The classification of the living world followed in this edition is that of the 'five kingdoms' scheme. Entries in the body of the dictionary are given for all the main phyla, divisions and classes of plants, fungi, animals, protists and prokaryotes, with some orders being included for groups such as the insects, birds, mammals and flowering plants. Entries for many common names of organisms are also included. The appendices at the back give a fuller outline of the various kingdoms. An appendix covering viruses has been added for this edition.

Within an entry, different meanings of a term are separated by semicolons, which are otherwise not used in routine punctuation. The abbreviations (*bot.*), (*zool.*), (*mycol.*), etc. have been used in some cases to indicate more clearly which in a long list of meanings refer to plants, which to animals, and so on. We aim in this dictionary to define as a matter of course any technical or unfamiliar term that may be used in a definition. To avoid complicating the text with excessive cross-referencing, therefore, words that are defined elsewhere in the dictionary are not necessarily indicated by (*q.v.*). Common abbreviations and acronyms are gathered together in alphabetical order at the front of each letter section.

Common suffixes and prefixes derived from Latin and Greek are entered in the body of the dictionary, along with their usual meanings, and Appendix 9 gives etymological origins of some common word elements.

I should like to thank the staff of Longman for their help and encouragement throughout the project. Comments concerning errors or omissions in this edition will be greatly appreciated, so that they may be rectified in future reprints or editions.

Eleanor Lawrence
London, 1995

ABBREVIATIONS

a. adjective
adv. adverb
alt. alternative (synonm)
appl. applies or applied to
ca. circa (approximately)
cf. compare
EC Enzyme Commission number
  (1978)
e.g. for example
esp. especially
et al. and others
etc. and so forth
Gk. Greek
i.e. that is
L. Latin

n. noun
n haploid no. of chromosomes
pert. pertaining to
plu. plural
p.p.m. parts per million
q.v. see
r.n. Enzyme Commission
  recommended name,
  where it differs from that
  used as the headword
sing. singular
sp. species (*sing.*)
spp. species (*plu.*)
v. verb

UNITS AND CONVERSIONS

acre	4046.86 m^2 (4840 sq. yd)
ångström unit*, Å	10^{-10}m
atmosphere, standard, atm	101325 Pa (14.72 p.s.i.)
bar	10^5 Pa
British thermal unit, Btu	1.055 kJ
British thermal unit/hour, Btu/h	0.293 W
bushel, bu	0.0364 m^3
bushel (US), bu	0.0352 m^3
calorie, thermochemical	4.184 J
centimetre, cm	10^{-2} m (0.394 in)
cubic foot, ft^3	0.0283 m^3
cubic inch, in^3	16.387 cm^3
cubic yard, yd^3	0.7645 m^3
degree Celsius (centigrade), °C†	(9/5) °F
degree Fahrenheit, °F†	(5/9) °C
dram (avoirdupois), dr	1.772 g
fathom	1.829 m (6 ft)
fluid ounce, fl oz	28.413 cm^3
foot, ft	0.3048 m
gallon, gal	4.546 dm^3
gallon (US), US gal	3.785 dm^3
grain, gr	64.799 mg

hectare, ha	10^4 m^2 (2.471 acres)
hour, h	3600 s
hundredweight, cwt	50.802 kg
inch, in	25.4 mm
joule, J (SI unit of energy)	kg m^2 s^{-2} (0.239 cal)
kilocalorie/hour, kcal/h	1.163 W
kilogram, kg (SI unit of mass)	2.20 lb
litre, l	dm^3 (1.76 pt)
metre, m (SI unit of length)	39.37 in
micron‡, μm	10^{-6} m
mile	1.6093 km
millibar, mbar	100 Pa
millimetre, mm	10^{-3} m (0.039 in)
millimetre of mercury, mmHg	133.332 Pa
millimetre of water	9.807 Pa
minute (time), min	60 s
molar, mol/l, m	1 mol dm^{-3}
nanometre, nm	10^{-9} m (10Å)
newton, N (SI unit of force)	kg m s^{-2}
ounce (avoirdupois), oz	0.0283 kg
pascal, Pa (SI unit of pressure)	kg m^{-1} s^{-2}
pint, pt	0.568 dm^3
pound (avoirdupois)	0.4536 kg
square foot, ft^2	0.0929 m^2
square inch, in^2	645.16 mm^2
square mile, sq. mile	2.590 km^2 (640 acres)
square yard, yd^2	0.836 m^2
ton (long)	1016.05 kg (2240 lb)
tonne (metric ton), t	1 Mg (0.984 tons)
watt, W (SI unit of power)	kg m^2 s^{-3}
yard, yd	0.9144 m

* The use of this unit is being discouraged. It is being replaced by the nanometre (nm, 10^{-9} m).

† To convert temperature in °C to °F multiply by 9/5 and add 32; to convert °F to °C subtract 32, then multiply by 5/9.

‡ The term 'micron' is no longer recommended; 'micrometre' is preferred. The symbols μ and mμ should not be used; they should give way to μm (micrometre) and nm (nanometre) respectively. See p. ix for a full list of SI prefixes.

SI PREFIXES

The following prefixes may be used to construct decimal multiples of units.

Multiple	Prefix	Symbol	Multiple	Prefix	Symbol
10^{-1}	deci	d	10	deca	da
10^{-2}	centi	c	10^{2}	hecto	h
10^{-3}	milli	m	10^{3}	kilo	k
10^{-6}	micro	μ	10^{6}	mega	M
10^{-9}	nano	n	10^{9}	giga	G
10^{-12}	pico	p	10^{12}	tera	T
10^{-15}	femto	f			
10^{-18}	atto	a			

GREEK ALPHABET

Name	Greek letter	English equivalent	Name	Greek letter	English equivalent
alpha	A α	a	nu	N ν	n
beta	B β	b	xi	Ξ ξ	x
gamma	Γ γ	g	omicron	O o	o
delta	Δ δ	d	pi	Π π	p
epsilon	E ε	e	rho	P ρ	r
zeta	Z ζ	z	sigma	Σ σ	s
eta	H η	e	tau	T τ	t
theta	Θ θ	th	upsilon	Y υ	u
iota	I ι	i	phi	Φ ϕ	ph
kappa	K κ	k	chi	X χ	ch
lambda	Λ λ	l	psi	Ψ ψ	ps
mu	M μ	m	omega	Ω ω	o

COMMON LATIN AND GREEK NOUN ENDINGS

sing.	plu.
-a	-ae (L.)
-a	-ata (Gk.)
-is	-es (L.)
-on	-a (Gk.)
-um	-a (L.)
-us	-i (L.)

A

α heavy chain class corresponding to IgA.

α- *for all headwords with prefix* α- *or alpha refer also to headword itself.*

A absorbance *q.v.*; adenine *q.v.*; alanine *q.v.*

Å Ångström *q.v.*

AAV adeno-associated virus *q.v.*

Ab antibody *q.v.*

ABA abscisic acid *q.v.*

abl *v-abl*, oncogene carried by the RNA tumour virus Abelson leukaemia virus, specifies a tyrosine protein kinase. *c-abl*, corresponding proto-oncogene found in normal cells. The chromosome translocation typical of chronic myelogenous leukaemia appears to disrupt the function of the *c-abl* gene.

Ab-MLV Abelson murine leukaemia virus, an RNA tumour virus, carries oncogene *v-abl*.

ABO system of human blood groups *see* blood groups.

ACh acetylcholine *q.v.*

AChE acetylcholinesterase *q.v.*

AChR acetylcholine receptor *q.v.*; **mAChR** muscarinic acetylcholine receptor; **nAChR** nicotinic acetycholine receptor.

ACP acyl carrier protein *q.v.*

ACTH adrenocorticotropic hormone *q.v.*

ADA adenosine deaminase.

ADCC antibody-dependent cell-mediated cytotoxicity *q.v.*

ADH alcohol dehydrogenase *q.v.*; antidiuretic hormone, vasopressin *q.v.*

ADI aerobic dependence index *q.v.*

ADMR average daily metabolic rate *q.v.*

A-DNA *see* DNA.

ADP adenosine diphosphate *q.v.*

ADPR adenosine diphosphate ribosyl *see* ADP-ribosylation.

AEC adenylate energy charge, *see* energy charge.

AET actual evapotranspiration *q.v.*

AEV avian erythroblastosis virus, an RNA tumour virus.

AFDW ash-free dry weight.

aFGF acidic fibroblast growth factor, one of the two forms of this polypeptide factor.

AFP α-foetoprotein *q.v.*

Ag antigen *q.v.*

AI, AID artificial insemination by donor, widely used in livestock breeding.

AIDS acquired immunodeficiency syndrome *q.v.*

Ala alanine *q.v.*

Alu *see* Alu sequences.

AL-SV avian leukosis-sarcoma virus, an RNA tumour virus.

ALV avian leukosis virus, a retrovirus.

AMP adenosine monophosphate *q.v.*

AMV avian myeloblastosis virus, an RNA tumour virus, carries oncogene *v-myb*.

ANA antinuclear antibody, autoantibodies against various components of the cell nucleus.

ANF atrial natriuretic factor *q.v.*

ANOVAR analysis of variance *q.v.*

ANP atrial natriuretic polypeptide *see* factor.

ANS autonomic nervous system *q.v.*

AP aminopurine *q.v.*

4-AP 4-amino pyridine *q.v.*

AP-1 a transcriptional regulatory protein in mammalian cells, the product of the proto-oncogene c-*jun*, and which is believed to be involved in mediating a cell's response to agents that stimulate mitosis and cell division.

Arg arginine *q.v.*

Asn asparagine *q.v.*

Asp aspartic acid *q.v.*

ASV avian sarcoma virus, an RNA tumour virus, carries oncogene *v-fps* (*v-fes*).

Asx aspartic acid or asparagine.

ATCase aspartate transcarbamoylase *q.v.*

ATP adenosine triphosphate *q.v.*

ATPase adenosinetriphosphatase *q.v.*

azaC 5-azacytidine *q.v.*

AZT the deoxyribonucleoside analogue 3′-azido-3′-deoxythymidine, which inhibits the enzyme reverse transcriptase of retroviruses and is used in the chemotherapy of HIV infection. *alt.* zidovudine.

a- prefix denoting lacking, without, derived from Gk. *a*, not.

ab- prefix derived from Gk. *ab*, from.

abactinal *a.* situated on the part of echinoderm body not bearing tube-feet, and in which the madreporite is usually included. *alt.* abambulacral.

abambulacral abactinal *q.v.*

A-band dark band seen in longitudinal sections of striated muscle myofibrils, repeating alternately with light bands (I-bands), and containing thick filaments interdigitating with thin filaments at their ends.

abapical *a. pert.* or situated at lower pole; away from the apex.

abaxial *a. pert.* that surface of any structure which is furthest from or turned away from the axis.

abaxile *a. appl.* plant embryo whose axis has not the same direction as axis of seed.

abbreviated *a.* shortened; curtailed.

abcauline *a.* outwards from or not close to the stem.

abdomen *n.* in vertebrates, the lower part of the body cavity, containing the digestive organs, reproductive organs, kidneys and liver, in mammals separated from the thorax by the diaphragm; in arthropods and certain polychaete worms, the posterior part of the body, behind the head and thoracic regions; in tunicates, the section of the body containing stomach and intestines. *a.* abdominal.

abdominal histoblast nests *see* histoblast nests.

abdominal pores single or paired openings leading from coelom to exterior, in certain fishes.

abdominal reflex contraction of abdominal wall muscles when skin over side of abdomen is stimulated.

abdominal regions 9 regions into which the human abdomen is divided by 2 horizontal and 2 vertical imaginary lines, comprising hypochondriac (2), lumbar (2), inguinal (2), epigastric, umbilical, hypogastric.

abdominal ribs ossifications occurring in fibrous tissue in abdominal region between skin and muscles in certain reptiles.

abdominal ring one of 2 openings in fasciae of abdominal muscles through which passes spermatic cord in male, round ligament in female.

abducens *n.* the 6th cranial nerve, supplying the rectus externus muscle of the eyeball.

abductin *n.* elastic protein forming the inner hinge ligament of shells of bivalve molluscs.

abduction *n.* movement away from the median axis. *cf.* adduction.

abductor *n.* a muscle that draws a limb or part outwards.

abequose *n.* a 3,6-dideoxyhexose sugar found in the lipopolysaccharide outer membrane of some enteric bacteria.

aberrant *a.* with characteristics not in accordance with type, *appl.* species, etc.

abhymenial *a.* on or *pert.* the side of the gill opposite that of the hymenium in agaric fungi.

abience *n.* retraction from stimulus; avoiding reaction. *a.* abient.

abiocoen *n.* the non-living parts of the environment in total.

abiogenesis *n.* the production of living from non-living matter, as in the origin of life. Also sometimes refers to the theory of spontaneous generation, held in the 19th century and before, which stated that microorganisms or higher organisms could arise from non-living material.

abiology *n.* the study of non-living things in a biological context.

abioseston *n.* non-living material floating in the plankton.

abiosis *n.* apparent suspension of life.

abiotic *a.* non-living.

abiotic environment that part of an organism's environment consisting of non-biological factors such as topography, geology, climate, and inorganic nutrients.

abjection *n.* the shedding of spores, as from sporophores, usually with some force.

abjunction *n.* the delimitation of spores by septa at tip of hypha.

ablactation *n.* the cessation of milk secretion; weaning.

ablation *n.* the destruction or removal of a particular structure, piece of tissue or individual cell.

abomasum *n.* in ruminants, the fourth chamber of the stomach, into which acid and digestive enzymes are secreted, and in which the final stages of digestion take place.

aboospore *n.* a spore developed from an unfertilized female gamete.

aboral *a.* away from, or opposite to, the mouth.

abortion *n.* premature birth of a dead foetus, technically in humans expulsion of a foetus from the time of fertilization to 3 months gestation; arrest of development of an organ (in plants and animals). *v.* to abort.

abortive infection of a virus, an infection in which no new infectious viral particles are produced.

abranchiate *n.* without gills.

abrin *n.* lectin isolated from *Abrus precatorius*, specific for D-galactose.

abrupt *a.* appearing as if broken, or cut off, at extremity, *appl.* leaves, etc.

abrupt speciation the formation of a species as a result of a sudden change in chromosome number or constitution.

abruptly-acuminate *a.* having a broad extremity from which a point arises, *appl.* leaf.

abscise *v.* to become separated; to fall off, as leaves, fruit, etc.

abscisic acid (ABA) sesquiterpene plant hormone which promotes senescence, leaf fall, and dormancy in buds, and antagonizes the effect of the growth-promoting hormones.

abscisin abscisic acid *q.v.*

abscission *n.* the separation of a part from the rest of the plant.

abscission zone the region at the base of a leaf, flower, fruit, or other part of the plant, consisting of an abscission layer of weak cells whose breakdown separates the part from the rest of the plant, and a protective layer of corky cells formed over the wound when the part falls.

absenteeism *n.* the practice of certain animals of nesting away from their offspring and visiting them from time to time to provide them with food and a minimum of care.

absolute age the age of a rock, fossil or archaeological specimen in years before present (BP), which can be determined by radioisotope dating methods, or for wooden artefacts sometimes by dendrochronology.

absolute configuration for molecules that contain a centre of asymmetry and can occur in mirror-image structural isomers, as e.g. many sugars and amino acids, the handedness of a particular isomer is given the arbitrary designation D- or L-, which is based on the atomic configuration around the asymmetric carbon atom.

absorbance (A) *n.* a spectrophotometric measurement of the absorption of light at a particular wavelength by a substance in solution, which can be used, e.g., to determine the concentration of a substance in solution and to follow conversion of substrate to product in enzyme reactions. *alt.* optical density, extinction.

absorption *n.* uptake of fluid and solutes by living cells and tissues; passage of nutritive material through living cells; of light, when neither reflected nor transmitted.

absorption spectrum the pattern of absorption of light at different wavelengths shown by many organic substances, usually displayed as a plot of absorbance against wavelength, giving a characteristic curve for any particular substance.

abstriction *n.* the process of detaching spores or conidia from a hypha by rounding off and constriction of the tip.

abterminal *a.* going from the end inwards.

abundance *n.* of mRNA, the average number of molecules of a particular mRNA per cell.

abundant mRNA the class of cellular mRNA consisting of large numbers of copies of relatively few different sequences.

abyssal *a. appl.* or *pert.* the deep sea, below 2000 m; *appl.* the depths of a lake where light does not penetrate.

abyssobenthic *a. pert.* or found on the ocean floor at the depths of the ocean, in the abyssal zone.

abyssopelagic *a. pert.* or inhabiting the ocean depths of the abyssal zone, but floating, not on the ocean floor.

abzyme *n.* a protein with both antibody and enzymatic activity, which may occur naturally but is generally produced by protein engineering.

acanaceous *a.* prickly; bearing prickles, as leaves.

acantha *n.* a spine or similar structure.

acanthaceous *a.* bearing thorns or prickles.

acanthion *n.* the most prominent point on the nasal spine.

acanthocarpous *a.* having fruit covered in spines or prickles.

Acanthocephala *n.* a phylum of pseudocoelomate animals, commonly called thorny-headed worms, that as adults are intestinal parasites of vertebrates and as larvae have an arthropod host.

acanthocephalous *a.* with a hooked proboscis.

acanthocladous *a.* having spiny branches.

acanthocyst *n.* a sac containing lateral or reserve stylets in nemertine worms.

Acanthodii *n.* extinct group of cartilaginous fishes, first present in the Silurian, and which were the first known jawed vertebrates. *alt.* spiny sharks.

acanthoid *a.* resembling a spine or prickle.

Acanthopterygii, acanthopterygians *n., n.plu.* large group of teleost fishes, having spiny fins and spiny scales, and including perch, mackerel and plaice.

acanthozooid *n.* the tail part of proscolex of cestodes (tapeworms).

acapnia *n.* condition of low carbon dioxide content in blood.

Acari, Acarina *n.* a very large and varied order of arachnids, commonly called mites and ticks, the adults usually having a rounded body carrying the four pairs of legs. Ticks are relatively large and parasitic, living as ectoparasites on mammals and sucking their blood, and are the vectors of several serious diseases, including Rocky Mountain spotted fever and tickborne encephalitis. Mites are smaller, inhabiting both plants and animals and are common in soil. They include both parasites and non-parasites, and are the vectors of scrub typhus in humans.

acaricide *n.* chemical compound used to kill ticks and mites (Acarina).

acarocecidium *n.* a gall caused by gall mites.

acarology *n.* the study of mites and ticks.

acarophily *n.* symbiosis of plants and mites.

acarpous *a.* not fruiting.

acaryote, acaryotic akaryote, akaryotic *q.v.*

acaudate *a.* lacking a tail.

acaulescent *a.* having a shortened stem.

acauline, acaulous *a.* having no stem or stipe.

accelerator *n.* muscle or nerve that increases rate of activity.

accelerin Factor V *q.v.*

acceptor *n.* a substance that receives and unites with another substance, as in oxidation-reduction processes where the oxygen acceptor is the substance oxidized, the hydrogen acceptor the substance reduced.

accessorius *n.* a muscle aiding in the action of another; spinal accessory or 11th cranial nerve.

accessory bud an additional axillary bud; a bud formed on a leaf.

accessory cells in immune system, macrophages and other nonlymphoid cells of lymphoid organs, involved in trapping and presentation of foreign antigens to initiate immune response.

accessory chromosomes supernumerary chromosomes *q.v.*

accessory food factor vitamin *q.v.*

accessory nerve 11th cranial nerve, supplying muscle of soft palate and pharynx, and the sternomastoid and trapezius muscles.

accessory pigment pigments other than chlorophyll present in many photosynthetic organisms.

accessory pulsatory organs sac-like structures in insects, variously situated, pulsating independently of the heart.

accessory species in plant ecology, a species that is found in a quarter to a half of the area of a stand. *cf.* accidental species.

accidental species in plant ecology, a species that is found in less than a quarter of a stand. *cf.* accessory species.

accipiters *n.plu.* the hawks, medium-sized birds of prey with rounded wings and long tails, part of the family Accipitridae.

acclimation, acclimatization *n.* the gradual habituation of an organism to a different climate or environment; adaptation to slowly changing new conditions.

accommodation *n.* the adjustment that occurs in the eye to receive clear images of objects at different distances, by changing the shape of the lens and thus its focal length; adaptation of receptors to a different stimulus; capacity of a plant to adapt to new conditions if these are introduced gradually.

accrescence *n.* growth through addition of similar tissues; continued growth after flowering. *a.* accrescent.

accrete *a.* grown or joined together; formed by accretion.

accretion *n.* growth by external addition of new matter.

accumbent *a. appl.* embryo having cotyledons with edges turned towards radicle, as in dicots of the family Cruciferae.

accumulators *n.plu.* plants which accumulate relatively high concentrations of certain chemical elements, such as heavy metals, in their tissues.

A-cells α-cells of islets of Langerhans in pancreas, secreting glucagon.

acellular *a.* not divided into cells.

acellular slime moulds a group of simple eukaryotic soil microorganisms (Myxomycetes), sometimes considered as fungi, sometimes as protists, whose vegetative phase consists of a plasmodium - a naked, creeping, multinucleate mass of protoplasm sometimes covering up to several square metres.

acelomate acoelomate *q.v.*

acelous acoelous *q.v.*

acentric *a. appl.* chromosome or chromosome fragment lacking a centromere and which therefore does not segregate at mitosis or meiosis.

acentrous *a.* with no vertebral centra, but persistent notochord, as some fishes.

acephalocyst *n.* hydatid stage of some tapeworms.

acephalous *a.* having no structure comparable to a head, *appl.* some molluscs, *appl.* larvae of certain Diptera, *appl.* ovary without terminal stigma.

Acerales Sapindales *q.v.*

acerate *a.* needle-shaped; pointed at one end.

acerose *a.* narrow and slender, with sharp point, as leaf of pine.

acerous *a.* hornless; without antennae; without tentacles.

acervate *a.* heaped together; clustered.

acervuline *a.* irregularly heaped together, *appl.* shape of foraminiferal tests.

acervulus *n.* small cluster of spore-bearing hyphae. *a.* acervulate.

Acetabularia genus of large unicellular green algae of which *A. mediterranea* has been used for experiments in developmental biology.

acetabulum *n.* cup-shaped socket in pelvic girdle for head of femur, forming the hip joint in tetrapod vertebrates; in insects, thoracic cavity in which leg is inserted; socket of coxa in arachnids; sucker used for attachment to host in flukes, tapeworms and leeches; sucker on arm of cephalopod. *a.* acetabular.

acetic acid bacteria bacteria (e.g. *Acetobacter, Acetomonas*) that partially oxidize ethyl alcohol to produce acetic acid, a reaction used in manufacturing vinegar. *alt.* acetobacters.

acetoacetate *n.* a ketone body, produced in liver, important energy source esp. in heart muscle and renal cortex, and of brain during starvation.

acetobacters acetic acid bacteria *q.v.*

acetyl CoA acetyl coenzyme A *q.v.*

acetyl coenzyme A (acetyl CoA) an acetyl thioester of coenzyme A, produced during the breakdown of carbohydrates as in the tricarboxylic acid cycle where it is formed from pyruvate and coenzyme A and the acetyl group subsequently oxidized to CO_2. Also formed during the breakdown of some amino acids and esp. in fatty acid oxidation. Is an important carrier of activated acetyl groups in certain metabolic reactions.

acetyl group chemical group, $-COCH_3$, formed by removal of $-OH$ from acetic acid.

acetylase *see* acetyltransferase.

acetylation *n.* addition of an acetyl group to a molecule.

acetylcholine (ACh) *n.* a neurotransmitter, the acetyl ester of choline (*see* Appendix 1 (57)), which mediates the transmission of signals between motor nerve fibres and skeletal muscle, and which also acts as a neurotransmitter at some synapses in all branches of the peripheral nervous system and within the brain, neurons secreting acetylcholine being known as cholinergic. It acts at several different types of receptor.

acetylcholine receptor cell-surface protein which binds and is activated by the neurotransmitter acetylcholine. There are two main types, each present as several different subtypes. The nicotinic receptor (nAChR) is an ion channel and is found on the postsynaptic terminals of neurones and on skeletal muscle motor end plates (neuromuscular junctions). The muscarinic acetylcholine receptor (mAChR) is a G protein-coupled receptor and is mainly present on neurones, heart muscle and in the gastrointestinal tract.

acetylcholinesterase (AChE) cholinesterase $q.v.$

N-acetylgalactosamine (GalNAc) $n.$ acetyl derivative of the amino sugar galactosamine, with the acetyl group carried on the amino N; a common constituent of glycoproteins, various heteropolysaccharides (e.g. chondroitin) and glycolipids. *see* Appendix 1 (10).

N-acetylglucosamine (GlcNAc) $n.$ acetyl derivative of the amino sugar glucosamine, with the acetyl group carried on the amino N, a common constituent of glycoproteins, various heteropolysaccharides (e.g. hyaluronic acid) and glycolipids.

N-acetylmuramic acid (NAM) monosaccharide present in the peptidoglycans of bacterial cell walls, comprising N-acetylglucosamine condensed with lactic acid at carbon 3 of the sugar.

N-acetylneuraminic acid (NAN) a sialic acid, a 9-carbon acid sugar, a component of gangliosides and of the carbohydrate side chains of some glycoproteins.

acetylsalicylic acid aspirin, a compound with analgesic and anti-inflammatory properties, the latter being due to its inhibition of prostaglandin synthesis.

acetyltransferase $n.$ any of a group of enzymes catalysing the transfer of acetyl groups, usually from acetyl CoA, (included in EC 2.3.1). *alt.* transacetylase.

achaetous $a.$ without chaetae.

acheilary $a.$ having undeveloped labellum,

as some orchids.

achelate $a.$ without claws or chelae.

achene $n.$ a one-seeded dry, indehiscent fruit formed from one carpel, usually with one seed which is not fused to the fruit wall.

achiasmate, achiasmatic $a.$ lacking chiasma in meiosis, as some Diptera (e.g. in spermatogenesis in male *Drosophila*).

Achilles tendon the tendon of the heel, the united strong tendon of the gastrocnemius and solaeus muscles. *alt.* tendo calcaneus.

achlamydate $a.$ lacking a mantle, as some gastropods.

achlamydeous $a.$ without calyx or corolla. *alt.* gymnanthous.

acholeplasmas $n.plu.$ group of mycoplasm-like microorganisms.

achondroplasia $n.$ a dominantly inherited form of dwarfism characterized by disturbance of ossification of the long bones of the limbs and of certain facial bones during development.

ACh receptors *see* acetylcholine receptors.

achroglobin $n.$ colourless respiratory pigment found in some molluscs and tunicates.

achroic $a.$ colourless.

achromasie $n.$ emission of chromatin from nucleus.

achromatic $a.$ colourless; *appl.* threshold, the minimal stimulus inducing sensation of luminosity or brightness.

achromatous $a.$ colourless.

achromic $a.$ colourless, unpigmented.

A chromosomes the normal chromosomes of a diploid set, as opposed to the B or supernumerary chromosomes.

aciculate $a.$ having acicles or aciculae.

aciculilignosa $n.$ evergreen forest and bush made up of needle-leaved coniferous trees and shrubs.

aciculum $n.$ thick central bristle (chaeta) in tuft of chaetae on parapodia of polychaete worms.

acid $n.$ a substance that releases an H^+ ion (proton) in solution.

acid-base balance the correct ratio of acids to bases in blood maintaining a suitable pH.

acid deposition rain (acid rain) or other form of precipitation, or dry deposition, which contains acids and acid-forming compounds and has a pH less than pH 5.6. It can cause acidification of lakes, with harmful effects on the aquatic flora and fauna, and damage to terrestrial vegetation. Acid deposition is caused mainly by atmospheric sulphur dioxide produced by the burning of coal and other fossil fuels, which is precipitated as sulphuric acid and sulphates, and by nitrogen oxides emitted from fossil fuel burning and vehicle exhausts, which form nitric acid and nitrogen dioxide.

acid-fast *a.* remaining stained with aniline dyes on treatment with acid, *appl.* e.g. to mycobacteria, whose waxy membrane retains the dye.

acid gland acid-secreting gland of ants, bees and wasps; acid-secreting oxyntic cells of mammalian stomach.

acid hydrolase any of a class of hydrolytic enzymes active at acid pH (around pH 5), which includes some proteinases, phosphatases, nucleases, glycosidases, lipases, phospholipases and sulphatases, found esp. in lysosomes.

acid phosphatase a hydrolytic enzyme, found especially in lysosomes, that catalyses the hydrolysis of an orthophosphate monoester to an alcohol and orthophosphate at acid pH. EC 3.1.3.2.

acid protease any of a class of proteolytic enzymes that act in relatively acid conditions.

acid rain *see* acid deposition.

acid tide transient increase in acidity of body fluids that follows the decrease in acidity (alkaline tide) after taking food.

acidic *a.* having the properties of an acid; *appl.* stains such as eosin that react with basic components of protoplasm such as cytoplasm and collagen.

acidophil eosinophil *q.v.*

acidophile *n.* plant that grows best on acid soils; microorganism that thrives in acidic (< pH 5) conditions, and can be isolated on acidic media. *a.* acidophilic.

acidosis *n.* condition in which pH of blood becomes abnormally acid e.g. pH 7 rather than the normal pH 7.4.

aciduria *n.* condition in which pH of urine is lowered.

aciduric *a.* tolerating acid conditions. *see* acidophile. *alt.* acidophilic.

aciform *a.* needle-shaped.

acinaciform *a.* shaped like a sabre or scimitar, *appl.* leaf.

acinar *a. pert.* acinus.

acinar cell pancreatic secretory cell, characterized by sac-like terminations.

acinarious *a.* having globose vesicles, as some algae.

aciniform *a.* grape- or berry-shaped.

acinus *n.* a cluster of cells forming the inner secretory region of a gland, usually a branched or compound gland. *alt.* alveolus. *plu.* acini; (*bot.*) drupel *q.v.*

acleidian *a.* with clavicles vestigial or absent.

acme *n.* the highest point of evolutionary development reached by a group; the prime of an individual.

acoelomate *a. appl.* animals not having a true coelom (i.e. sponges, sea anemones and corals, nematodes, rotifers, platyhelminths, and nemertean worms). *alt.* acelomate, acoelous.

acoelous *a.* acoelomate *q.v.; appl.* vertebrae with flattened centra. *alt.* acelous.

acondylous *a.* without nodes or joints.

acone *a. appl.* insect compound eye without crystalline or liquid secretion from cone cells.

aconitase *n. see* aconitate, *r.n.* aconitate hydratase, EC 4.2.1.3.

aconitate (aconitic acid) *n.* as *cis*-aconitate a 6-carbon intermediate in the tricarboxylic acid cycle, formed from citrate and converted into isocitrate by the enzyme aconitase.

acont akont *q.v.*

acontium *n.* thread-like process armed with stinging cells, borne on the mesenteries of some sea anemones. *plu.* acontia.

acorn worms Enteropneusta *q.v.*

acotyledonous *a.* lacking cotyledons.

acoustic *a. pert.* organs or sense of hearing, *appl.* nerve, etc.

acoustic reflex adjustment of the muscles that regulate the positions of the ossicles in the mammalian ear, a protective mechanism against damage by too loud a noise.

acoustico-lateralis system sensory system concerned with movement detection, avoidance of obstacles etc., in fish and amphibians consisting of vibration-sensitive hair cells located in the ear and on the external surface (the lateral line system in fishes), which detect vibrations in the surrounding liquid.

acquired behaviour behaviour brought on by conditioning or learning.

acquired character modification or permanent structural change brought on during the lifetime of an individual by use or disuse of a particular organ, disease, trauma or other environmental influence, and which is not heritable.

acquired immmunity immunity to infection produced by previous exposure to a pathogen, or by immunization against it. *cf*. innate immunity; passive immunity.

acquired immune deficiency syndrome (AIDS) fatal T-cell deficiency disease caused by infection with the human immunodeficiency virus (HIV), a retrovirus, which results in severe depletion of a particular class of T lymphocytes and consequent immunodeficiency. The immunodeficient patient is susceptible to infection by opportunistic pathogens (e.g. *Candida*, *Pneumocystis*) and to the development of unusual cancers (Kaposi's sarcoma). HIV is transmitted sexually or by infected blood and blood products, and can be transmitted from mother to child at birth or by breast feeding.

acral *a. pert*. extremities.

acrandrous *n*. having antheridia borne at the tips, in bryophytes. *n*. acrandry.

Acrania *n*. a group of chordates, sometimes considered as a subphylum, which includes the urochordates and the cephalochordates, and excludes the craniates.

Acraniata *n*. invertebrates *q.v*.

acranthous *a*. having the inflorescence borne at the tip of the main axis.

Acrasiomycota *n*. cellular slime moulds (*q.v*.), when classified as a phylum of protists. *alt*. Acrasiomycotina, Acrasiomycetes.

Acrididae *n*. a family of grasshoppers (order Orthoptera) with antennae shorter than their bodies, and which includes the locust, *Locusta migratoria*. *alt*. short-horned grasshoppers.

acridines a group of dyes which are mutagenic because they intercalate into DNA causing addition or deletion of a single base during replication and producing frameshift mutations.

acroblast *n*. body in spermatid which gives rise to the acrosome.

acrocarpic, acrocarpous *a*. bearing fructifications at the tips of main stem or branches.

acrocentric *a. appl*. chromosomes with the centromere very near one end and which appear rod shaped during segregation in mitosis or meiosis; *n*. an acrocentric chromosome.

acrochordal *a. appl*. a chondrocranial unpaired frontal cartilage in birds.

acrocoracoid *n*. a process at the dorsal end of the coracoid bone in birds.

acrodont *a. appl*. teeth attached to the summit of a parapet of bone, as in lizards.

acrodromous *a. appl*. leaf with veins converging at its point.

acrogenous *a*. increasing in growth at summit or tip.

acromio-clavicular *a. appl*. ligaments covering joint between acromion and clavicle (shoulder blade).

acromion *n*. the ventral prolongation of the spine of the scapula (shoulder blade). *a*. acromial.

acron *n*. pre-oral region of insects.

acronematic *a. appl*. smooth whip-like flagella.

acropetal *a*. ascending; *appl*. leaves, flowers or roots developing successively along an axis so that the youngest are at the apex. *cf*. basipetal.

acropodium *n*. the digits - fingers or toes.

acrorhagus *n*. a tubercle near the margin of certain sea anemones, containing specialized nematocysts (stinging cells).

acrosarc *n*. pulpy berry arising from the union of ovary and calyx.

acrosomal process long actin-containing process that projects from the head of some invertebrate sperm on fertilization and which helps penetrate the egg.

acrosomal reaction the release of hydrolytic enzymes from the acrosome of sperm when it contacts the egg.

acrosome *n*. organelle at apex of sperm, containing hydrolytic enzymes that digest coating of egg enabling sperm to penetrate. *alt*. acrosomal vesicle.

acrospire *n*. the first shoot, being spiral, at the end of germinating seed.

acrospore *n*. the spore at the apex of a sporophore or hypha.

acrostichoid *a. appl.* fern sporangia produced all over surface of frond and not grouped in sori over a vein.

acroteric *a. pert.* outermost points, as tips of digits, ears, nose, etc.

acrotonic, acrotonous *a.* having anther united at its apex with rostellum.

acrotroch *n.* circlet of cilia at the extreme anterior end of the trochophore larva of some marine invertebrates.

acrotrophic *a. appl.* ovariole having nutritive cells at apex joined to oocytes by nutritive cords. *alt.* telotrophic.

actin *n.* a globular protein, G-actin, found in all eukaryotic cells but not present in prokaryotes, which can polymerize end-to-end into a fibrous form, F-actin, which forms the thin filaments of muscle and the microfilaments of the cytoskeleton. F-actin consists of two chains of actin monomers wound round each other in a helix, and, with myosin, can form a contractile complex, actomyosin.

actin-binding proteins large and disparate group of proteins that bind to the cytoskeletal protein actin, having diverse effects on actin filaments (microfilaments) which include bundling of filaments into fibres and severing of filaments into actin subunits.

actin filament *see* actin, microfilament.

actinal *a. appl.* area of echinoderm body bearing tube feet, *alt.* ambulacral; star-shaped.

actinians sea anemones *q.v.*

actinic *a. pert.* radiation of wavelength between that of visible violet and X-rays.

actiniform *a.* star-shaped.

α-actinin *n.* protein present in Z-lines of muscle fibrils, and at the site of non-muscle cell contact with substrate.

Actinobacteria Actinomycetes *q.v.*

actinoblast *n.* mother cell from which a spicule develops, as in sponges.

actinocarpous, actinocarpic *a.* having flowers and fruit radially arranged.

actinodromous *a.* palmately veined.

actinoid *a.* rayed; star-shaped.

actinology *n.* formerly used for the study of the action of radiation; the homology of successive regions or parts radiating from a common central region.

actinomorphic, actinomorphous *a.* radially symmetrical; regular.

Actinomycetes, actinomycetes *n., n.plu.* group of Gram-positive prokaryotic microorganisms found in soil, river muds and lake bottoms, which grow as slender branched filaments (hyphae), and which include many species (e.g. *Streptomyces*) producing antibiotics.

actinomycins *n.plu.* antibiotics produced by species of the actinomycete *Streptomyces*, which block transcription in both bacterial and eukaryotic cells by binding to DNA. Actinomycin C (cactinomycin) and actinomycin D (dactinomycin) are used as anticancer drugs.

actinophage *n.* bacteriophage that infects actinomycetes.

Actinopoda, actinopods *n., n.plu.* phylum of non-photosynthetic protists characterized by long slender cytoplasmic projections (axopodia) that are stiffened by a bundle of microtubules. They include the marine radiolarians and the mainly freshwater heliozoans. The radiolarians are divided into three groups: acantharians, with a radially symmetrical skeleton of rods of strontium sulphate, and polycystids and phaeodarians with silica spicules.

Actinopodea, actinopods *n.* in older classifications, the class of protozoans including the radiolarians and heliozoans. *see* Actinopoda.

Actinopterygii, actinopterygians *n., n.plu.* a subclass of bony fishes (Osteichthyes), the ray-finned fishes, including all extant bony fishes except the lungfishes (Dipnoi) and the coelacanth. *see also* Acanthopterygii.

actinorrhiza *n.* nitrogen-fixing nodule-like structure formed on the roots of some non-legumes (e.g. alders, *Alnus* spp.) by infection with nitrogen-fixing actinomycetes of the genus *Frankia. alt.* actinorhiza.

actinost *n.* the basal bone of fin-rays in teleost fishes.

actinostele *n.* column of vascular tissue in plant stem lacking pith, the xylem being star-shaped in cross section.

actinostome *n.* mouth of sea anemone, starfish.

actinotrichia *n.plu.* unjointed horny rays at edge of fins in many fishes.

actinotrocha *n.* free-swimming larval form

of Phoronida (*q.v.*).

Actinozoa Anthozoa *q.v.*

actinula *a.* larval stage in some hydrozoans.

action potential a potential difference produced across the plasma membrane of nerve or muscle cells when they are stimulated, reversing the resting potential from about −70 millivolts (mV) to about +30 mV, and being an easily measured manifestation of a nerve impulse. *cf.* resting potential.

action spectrum the range of wavelengths of light or other electromagnetic radiation within which a certain process takes place, often *pert.* to photosynthetic processes.

activated sludge process a sewage treatment process in which a mixture of protozoa and bacteria (activated sludge) is added to aerated sewage to break down the organic matter. As the microorganisms use the organic matter for food they multiply, producing more activated sludge.

activation energy free energy of activation *q.v.*

activation hormone hormone secreted by the brain in insects and which stimulates the prothoracic gland to produce the steroid hormone ecdysone, which triggers moulting.

activator *n.* any substance that stimulates a given process; of genes, a protein that acts as a positive regulator of transcription, *cf.* repressor.

active site (centre) region of an enzyme molecule at which substrates are bound and activation and chemical reaction of substrate takes place; on an antibody molecule, the antigen-binding site.

active space of a pheromone or other chemical signal, the space within which the chemical is above the threshold concentration for its detection by another individual.

active transport the movement of substances across biological membranes into cells or organelles other than by passive diffusion or passive transport, often occurring against concentration gradients. It involves carrier proteins and requires energy.

activin *n.* protein growth factor involved in mesoderm induction in early amphibian development.

activity *n.* of enzymes, the rate at which they catalyse a chemical reaction; in ecology,

the total flow of energy through a system in unit time.

actomyosin *n.* complex of myosin and actin, threads of which contract on addition of ATP, K^+, Mg^2+, formed in solution and in myofilaments.

actual evapotranspiration (AET) the total water loss over a given area with time by the removal of water from soil and other surfaces by evaporation and from plants by transpiration.

actual vegetation in ecology, the vegetation existing at the time of observation.

aculeate *a.* having prickles, sharp points or a sting.

aculeiform *a.* prickle-shaped.

aculeus *n.* a prickle growing from bark, as in a rose, a sting, a hair-like projection; sting of an insect.

acuminiferous *a.* having pointed tubercles.

acuminulate *a.* having a very sharp tapering point.

acute *a.* ending in a sharp point; temporarily severe, not chronic.

acute phase proteins proteins synthesized by the liver, which are present in blood of healthy people in small amounts but whose levels are greatly increased after infection and other traumas.

acyclic *a.* flowers with floral leaves arranged in a spiral.

acyclovir *n.* antiviral drug active against herpes viruses 1 and 2 and which is used topically to treat cold sores, etc. *alt.* acycloguanosine, Zovirax.

acyl carrier protein (ACP) small protein with a pantothenate prosthetic group that carries acyl groups in metabolic cycles concerned with fatty acid synthesis.

acyl CoA synthetase widely distributed enzyme catalysing formation of activated acyl CoA compounds, involved in fatty acid activation prior to oxidation in mitochondria. EC 6.2.1.3. *alt.* fatty acid thiokinase.

acyl coenzyme A (acyl CoA) any acyl thioester of coenzyme A, such as succinyl CoA.

acyl group a chemical group formed by the removal of -OH from a carboxylic acid.

acylation *n.* the addition of an acyl group (—RCO^-, where R is any alkyl or aryl group, derived from a carboxylic acid) to a molecule.

acyltransferase *n.* any of a group of enzymes catalysing the transfer of acyl groups: includes acetyltransferases, succinyltransferases etc. *alt.* transacylase.

adamantoblast enamel cell, ameloblast *q.v.*

adambulacral *a. appl.* structures adjacent to ambulacral areas in echinoderms.

adaptation *n.* evolutionary process by which an organism becomes fitted to its environment; a structure or habit fitted for some special environment or activity; process by which an organ or organism becomes habituated to a particular level of stimulus and ceases to respond to it, a more intense stimulus then being needed to produce a response; in the eye, increasing sensitivity of retina to the available light.

adaptive *a.* capable of fitting different conditions; adjustable; inducible, *appl.* enzymes formed only when their specific substrates are available, *cf.* constitutive enzymes; *appl.* control of metabolism, changes in rates of synthesis and degradation of enzymes in response to the organism's requirements; *appl.* any trait that confers some advantage on an organism and thus is maintained in a population by natural selection. Traits can only be defined as adaptive with reference to the environment pertaining at the time, as a change in environment can render a previously adaptive trait non-adaptive, and vice versa.

adaptive immunity long-lasting and specific protection against subsequent infection that follows infection with a pathogen or immunization against it, and which is one of the consequences of the adaptive immune response against the pathogen. *cf.* innate immunity, passive immunity.

adaptive radiation evolutionary process in which species descended from a common ancestor multiply and diverge to occupy different ecological niches.

adaxial *a.* the surface nearest to the axis (of leaves etc.).

adcauline *a.* towards or nearest the stem.

ad-digital *n.* a primary wing quill connected with phalanx of 3rd digit.

additive alleles alleles that interact in such a way that the phenotype of the heterozygote is exactly the average of the phenotypes of the two corresponding homozygotes.

additive variance in quantitative genetics, that part of the genotypic variance that can be attributed to the additive effects of genes.

addressins cell-surface proteins that enable cells such as lymphocytes to home to particular sites in the body.

adduction *n.* movement towards the median axis. *cf.* abduction.

adductor *n.* a muscle that brings one part towards another.

adeciduate *a.* not falling or coming away.

adecticous *a.* without functional mandibles to escape from puparium or cocoon, *appl.* pupa of some insects.

adelocodonic *a. appl.* undetached medusa of certain hydrozoans, which degenerates after discharging ripe sexual cells.

adelomorphic, adelomorphous *a.* indefinite in form.

adelophycean *a. appl.* stage or generation of many seaweeds when they appear as prostrate thalli.

adelphogamy *n.* brother sister mating as in certain ants; union of mother cell and one of the daughter cells formed from it by mitosis.

adelphous *a.* joined together in bundles, as filaments of stamens.

adendritic *a.* without dendrites or cellular processes, *appl.* cells.

adendroglia *n.* a type of neuroglia lacking cellular processes.

adenine (A) *n.* a purine base, constituent of DNA and RNA, and is a constituent of the nucleoside adenosine, the nucleotide ATP, the nicotinamide cofactors NAD and NADP, and the flavin nucleotide coenzymes FAD and FMN. *see* Appendix 1 (21) for chemical structure.

adenine deaminase enzyme that catalyses the deamination of adenine with the formation of hypoxanthine and ammonia. EC 3.5.4.2.

adeno-associated viruses (AAV) group of DNA viruses of the Parvoviridae, whose multiplication is dependent on the presence of adenovirus.

adenoblast *n.* embryonic glandular cell.

adenocarcinoma *n.* a tumour of glandular epithelium.

adenocheiri *n.plu.* elaborate accessory copulatory organs, outgrowths of atrial wall in turbellarians. *alt.* adenodactyli.

adenocyte *n.* secretory cell of gland.

adenodactyli adenocheiri *q.v.*

adenohypophysis *n.* the glandular, non-neural portion of the pituitary body, from which many pituitary hormones are secreted. *cf.* neurohypophysis.

adenoid *a. pert.* or resembling a gland or lymphoid tissue; *n.* lymphoid gland in nasopharynx.

adenophore *n.* stalk of nectar gland.

adenophyllous *a.* bearing glands on leaves.

adenopodous *a.* bearing glands on petiole or peduncle.

adenose *a.* glandular.

adenosine *n.* a nucleoside made up of the purine base adenine linked to ribose.

adenosine deaminase enzyme (EC 3.5.4.4) involved in nucleic acid metabolism, which catalyses the conversion of adenosine to inosine and whose genetic deficiency leads to a build-up of toxic nucleosides, causing the death and non-development of lymphocytes, especially T cells, and severe immunodeficiency.

adenosine diphosphate (ADP) a nucleotide made of adenosine linked to 2 phosphate groups in series, important in all living cells in energy transfer reactions where it is converted to ATP (e.g. during oxidative phosphorylation and photosynthesis) or formed from ATP.

adenosine monophosphate (AMP) sugar (ribose)-phosphate ester (nucleotide) of the nucleoside adenosine, the phosphate group being carried at C atom 5 (5′) on the ribose ring, one of the four types of nucleotide subunit in RNA. *alt.* adenylic acid, adenylate, adenosine 5′-phosphoric acid. *see also* cyclic AMP, deoxyadenylic acid (dAMP).

adenosine triphosphatase (ATPase) enzyme activity that catalyses the hydrolysis of ATP to ADP and orthophosphate (P_i, phosphate(V)) with the release of free energy, which is used to drive mechanical work (as in muscle), enzyme reactions, ion transport across membranes, etc. Many proteins possess ATPase activity, including motor proteins such as myosin and dynein associated with the cytoskeleton,

and transport proteins such as the sodium and hydrogen pumps in cellular membranes. An ATPase is also part of mitochondrial and chloroplast coupling factors, in which it catalyses the reverse reaction, $ADP + P_i = ATP$ and is often known as ATP synthase.

adenosine triphosphate (ATP) nucleotide made of the purine base adenine linked to the pentose sugar ribose which carries 3 phosphate groups linked in series, an energy-rich molecule important as a source of energy and phosphate groups for metabolic reactions in all living cells when it is converted to ADP with release of energy and phosphate or to AMP with release of energy and pyrophosphate (PP_i), regenerated chiefly from ADP during photosynthesis and oxidative phosphorylation. *see* Appendix 1 (23) for chemical structure.

Adenoviridae, adenoviruses *n., n.plu.* family of double-stranded DNA viruses including adenovirus type 2, which causes a mild respiratory infection in various mammals including humans, but which is tumorigenic in newborn hamsters.

adenylate (adenylic acid) adenosine monophosphate *q.v.*

adenylate cyclase membrane-associated enzyme that converts AMP into the second messenger and regulatory molecule cyclic AMP. One of the first steps in the actions of many hormones and other chemical messengers on cells is to stimulate or inhibit the activity of adenylate cyclase. *see* G-protein coupled receptors. EC 4.6.1.1. *alt.* adenylyl cyclase.

adenylate energy charge *see* energy charge.

adenylate kinase enzyme catalysing the interconversion of adenosine tri-, di-, and monophosphates. EC 2.7.4.3. *alt.* (formerly) myokinase.

adenyl cyclase adenylate cyclase *q.v.*

adenylyl cyclase adenylate cyclase *q.v.*

adenylyltransferase *see* nucleotidyltransferase.

adequate *a. appl.* stimulus acting on a given receptor and producing the appropriate sensation.

adesmic *a. appl.* scales (as of some extinct fish) that grow from margin outwards, and which are made up of separate small tooth-like units.

adesmy *n.* a break or division in an organ usually entire.

adhaerens junction intermediate junction, belt desmosome. *see* desmosome.

adherent *a.* touching but not growing together; attached to substratum.

adhesion *n.* condition of touching but not growing together of parts normally separate; the specific binding together of cells, esp. animal cells, by means of intercellular junctions and/or cell-surface molecules, to form organized tissues.

adhesion molecule, adhesion receptor any of a large variety of different cell-surface proteins and glycoproteins that mediate contact and adhesion of cells to one another and to the extracellular matrix. Some types of adhesion molecule bind to an identical molecule on the other cell whereas others bind to different receptor molecules. *see* cadherin, cell adhesion molecule, integrin, selectin.

adhesion plaque area of a cell contacting substrate and at which it makes connections with extracellular matrix through membrane proteins.

adiabatic *a.* without losing or gaining heat.

adience *n.* urge, or advance towards stimulus; approaching reaction. *a.* adient.

adient *a.* approaching the source of a stimulus.

adipocellulose *n.* cellulose with a large amount of suberin, as in cork tissue.

adipocyte *n.* an animal cell specialized for fat (lipid) storage, containing large globules of triglycerides in the cytoplasm. *alt.* adipose cell, fat cell.

adipoleukocyte *n.* blood cell containing fat or wax droplets, in insects.

adipolysis lipolysis *q.v.*

adipose *a.* fatty, *pert.* animal fat.

adipose body fat body *q.v.*

adipose cell adipocyte *q.v.*

adipose fin modified rayless posterior dorsal fin in some fishes.

adipose tissue type of connective tissue in animals made up of cells (adipocytes) filled with fat droplets. *see* brown adipose tissue, white adipose tissue.

aditus *n.* anatomical structure forming approach or entrance to a part, e.g. to larynx.

adjacent-1 and adjacent-2 segregation types of segregation at meiosis of the members of a quadrivalent involving a heterozygous reciprocal translocation, resulting in unbalanced gametes (i.e. gametes bearing duplications and deficiencies of the translocated regions). *cf.* alternate segregation.

adjuvant *n.* substance that stimulates and prolongs antibody synthesis when injected together with an antigen.

adlacrimal *n.* the lacrimal bone of reptiles, not homologous with that of mammals.

admedial *a.* near the middle; near the median plane.

adminiculum *n.* a spine used for locomotion in some insect pupae.

adnasal *n.* small bone in front of each nasal bone in some fishes.

adnate *a.* joined to another organ of a different kind; of leaves and stipules, closely attached to petiole or stalk; of anther, attached throughout its length to filament; of gills of agarics, fused with the stem for the whole of their width. *alt.* conjoined.

adnexa *n.plu.* structures or parts closely associated with an organ; extraembryonic membranes, as foetal membranes, placenta.

adnexed *a.* reaching to the stem only; *appl.* gills of agarics which are fused to stem for only part of their width.

adolescence *n.* stage in human and animal development from the onset of puberty to full sexual maturity. *a.* adolescent.

adoptive transfer transfer of cell-mediated immunity from an immunized to an unimmunized animal by the transfer of lymphocytes.

adoral *a.* near or *pert.* mouth.

adpressed appressed *q.v.*

ADP-ribosylation the addition of a ribosyl group derived from the ADP moiety of NAD to a protein. The effects of diphtheria toxin and cholera toxin are due to ADP-ribosylation and consequent inactivation of protein synthesis elongation factor 2 and the GTPase activity of G_s protein respectively.

adradius *n.* in coelenterates, the radius midway between perradius and interradius, a radius of the 3rd order.

adrectal *a.* near to or closely connected with rectum.

adrenal *a.* situated near kidneys; *pert.* the adrenal glands *q.v.*

adrenal glands, adrenals paired bodies adjacent to kidneys in mammals, consisting of an inner medulla secreting adrenaline and noradrenaline, and an outer cortex secreting various steroid hormones (corticosteroids). In some vertebrates the two types of secretory tissue are segregated into separate glands.

adrenalin(e) *n.* a catecholamine secreted by adrenal medulla and by nerve endings in the sympathetic nervous system. It acts via specific adrenergic receptors on a wide variety of tissues and has many effects, e.g. speeding up of heartbeat, breakdown of glycogen to glucose in muscle and liver. It is also a neurotransmitter. *alt.* epinephrine. *see* Appendix 1(32) for structure.

adrenergic *a. appl.* nerve fibres, of the sympathetic system, that liberate adrenaline or noradrenaline from their terminals.

adrenergic receptor any of several types of specific receptor for adrenaline and noradrenaline, on the surface of a variety of cells. β-adrenergic receptors and α-adrenergic receptors are the two main classes and they mediate their effects through different intracellular second messenger pathways.

adrenoceptor adrenergic receptor *q.v.*

adrenocortical *a.* produced by the adrenal cortex, *appl.* hormones.

adrenocorticotropic hormone, adrenocorticotropin (ACTH) polypeptide hormone synthesized by anterior pituitary and which acts on adrenal cortex, stimulating growth and synthesis and release of adrenocortical steroids. Also has effects on other tissues, stimulating lipid breakdown and release of fatty acids from fat cells.

adrenodoxin *n.* non-haem iron protein component of the cytochrome P_{450} system important in detoxification of foreign compounds.

adrenogenital syndrome form of pseudohermaphroditism in genetically female (XX) humans caused by a hereditary defect in the adrenal glands which results in accumulation of the hormone progesterone and its breakdown to aldosterone, which has similar but weaker effects to testosterone, with consequent partial virilization.

adrostral *a.* near to or closely connected with beak.

adsere *n.* in plant ecology, that stage in a plant succession that precedes its development into another at any time before the climax stage is reached.

adsorption *n.* the adhesion of molecules to a surface; formation of a thin layer of a substance at the surface of a solid or liquid.

adtidal *a. appl.* organisms living just below the low-tide mark.

adult neurogenesis the generation of new neurones in the brains of adults, occurring in fish and other cold-blooded vertebrates, and in birds, but not to any appreciable extent in mammals.

adultoid *a. appl.* nymph having imaginal characters differentiated further than in normal nymph.

aduncate *a.* crooked, bent in the form of a hook.

adustous *a.* browned, as if scorched.

advanced *a.* of more recent evolutionary origin. *cf.* primitive.

advehent *a.* carrying to an organ, afferent.

adventitia *n.* external connective tissue layer of blood vessels.

adventitious *a.* accidental; found in an unusual place; *appl.* tissues and organs arising in abnormal positions; secondary, *appl.* dentine.

adventive *a.* not native; *appl.* organism in a new habitat but not completely established there.

aecidiospore *n.* spore produced in aecidium.

aecidium *n.* in rust fungi a cup-shaped structure containing chains of spores (aecidiospores). *plu.* aecidia.

aecium aecidium *q.v.*

aedeagus *n.* the copulatory organ of male insects. *alt.* intromittent organ.

aegithognathous *a. appl.* type of palate characteristic of passerine birds (e.g. sparrow), with maxillopalatines separate, vomers forming a wedge in front and diverging behind.

aeolian *a.* wind-borne, *appl.* deposits.

Aepyornithiformes *n.plu.* order of very large Pleistocene birds of the subclass Neornithes from Madagascar, known as elephant birds.

aequorin *n.* a calcium-binding protein from the jellyfish *Aequorea*, which emits a flash of light when it binds Ca^{2+}.

aerenchyma *n.* parenchyma with large intercellular spaces; air-storing tissue in cortex of some aquatic plants.

aerial *a. appl.* roots growing above ground; *appl.* small bulbs appearing in leaf axils.

aerial plankton spores, bacteria and other microorganisms floating in the air.

aeroallergen *n.* an antigen present in the air, such as pollen, which causes an allergy in those susceptible to it when inhaled.

aero-aquatic *a. appl.* fungi living in water and liberating spores into the air.

aerobe *n.* any organism capable of living in the presence of oxygen, obligate aerobes being unable to live without oxygen. All animals and higher plants are obligate aerobes, many bacteria and some fungi are anaerobes or facultative anaerobes. *cf.* anaerobe. *a.* aerobic.

aerobic dependence index (ADI) a measure of the relative use of aerobic and anaerobic metabolism in muscle during exercise in animals, which is calculated using the lactate content of resting muscles after exercise as a measure of anaerobic metabolism and maximum oxygen consumption as a measure of aerobic metabolism. Higher ADIs tend to indicate more active animals.

aerobic respiration respiration occurring in the presence of oxygen. *see also* oxidative phosphorylation.

aerobiology *n.* study of air-borne organisms and their distribution.

aerobiosis *n.* existence in the presence of oxygen.

aerobiotic *a.* living mainly in the air.

aerocyst *n.* air vesicle of algae.

Aeroendospora *n.* name sometimes given to the aerobic endospore-forming Gram-positive bacteria (e.g. *Bacillus* spp.).

aerogenic *a.* gas-producing, *appl.* certain bacteria.

aerolae *n.plu.* large depressed box-like structures in the walls of diatoms.

aerophora *n.* aerating outgrowth or pneumatophore in certain ferns.

aerophyte *n.* epiphyte attached to the aerial portion of another plant.

aeroplankton *n.* spores, pollen, microorganisms, etc. drifting in the air. *alt.* aerial plankton.

aerostat *n.* air sac in insect body or bird bone.

aerostatic *a.* containing air spaces.

aerotaxis *n.* movement towards or away from oxygen. *a.* aerotactic.

aerotropism *n.* reaction to gases, generally to oxygen, particularly the growth curvature of roots and other parts of plants in response to changes in oxygen tension. *a.* aerotropic.

aesthesis *n.* sensibility; sense perception.

aestidurilignosa *n.* mixed evergreen and deciduous broadleaf forest.

aestilignosa *n.* broadleaf deciduous forest which loses its leaves in winter.

aestival *a.* produced in, or *pert.* summer; *pert.* early summer.

aestivation *n.* the mode in which different parts of a flower are arranged in the bud; torpor during heat and drought during summer in some animals.

aethalium *n.* reproductive stage of some plasmodial slime moulds in which the plasmodium forms a spherical sporangium in which the spores are formed.

aetiological *a.* causal.

aetiology *n.* cause (of a disease).

affectional *a. appl.* behaviour concerned with social relationships, as in monkeys, important in development and maintenance of social cohesion and organization.

afferent *a.* bringing towards, *appl.* sensory nerves bringing impulses towards the central nervous system, *appl.* vessels carrying blood to an organ or set of organs. *cf.* efferent.

affinity *n.* the strength of binding of two molecules, such as an antigen and its antibody or a hormone and its receptor, which is defined for antibodies and other molecules with more than one binding site as the strength of binding of the ligand at one specified binding site. *cf.* avidity.

affinity chromatography technique for purifying specific proteins (or other macromolecules) by their affinity for particular chemical groups, by passage through a column containing those groups to which the protein binds.

affinity constant

affinity constant measure of the strength of binding between two molecules, being the concentration of the bound form divided by the product of the concentrations of the free forms of each molecule at equilibrium. *alt.* equilibrium constant.

affinity maturation in an immune response, the increase in affinity (binding strength) of the antibodies for the immunizing antigen as the response progresses, which is the result of selection of antibody-producing cells producing antibodies of higher affinity. Affinity maturation is most marked in secondary immune responses.

afforestation *n.* the production of forest over an area, either by planting or by allowing natural regeneration or colonization.

aflagellate *a.* lacking flagella.

aflatoxin *n.* a mycotoxin formed by the mould *Aspergillus* and thought to cause liver damage.

African subkingdom subdivision of the Palaeotropical Floral Realm, comprising all of the African continent, Madagascar, Ascension Island and St Helena.

afterbirth *n.* placenta and foetal membranes expelled after offspring's birth.

afterpotential *n.* slow change in membrane potential in nerve cells following trains of impulses.

after-ripening period after dispersal in which a seed cannot yet germinate, even if conditions are favourable.

after-sensation persistent sensation due to continued activity in sense receptor after cessation of external stimulation.

aftershaft *n.* small tuft of down near an opening in the base of a feather.

agameon *n.* species comprising only apomictic individuals.

agamete *n.* young form or gamete which develops directly into an adult without fusing with another gamete.

agametoblast *n.* one of the cells into which an agamont or schizont divides by multiple fission, and which gives rise to merozoites.

agamic, agamous *a.* asexual; parthenogenetic.

agamic complexes group of apomictic plants that are usually allopolyploids and consist of many different biotypes, forming a taxonomically difficult group.

agammaglobulinaemia *n.* lack of immunoglobulin in the blood.

agamodeme *n.* small assemblage of closely related individuals consisting predominantly of apomictic plants or asexual organisms.

agamogenesis *n.* any reproduction without participation of a male gamete; asexual reproduction. *a.* agamogenetic.

agamogony *n.* any reproduction without the sexual process.

agamohermaphrodite *a.* with neuter and hermaphrodite flowers on the same plant, usually in the same inflorescence.

agamont *n.* in some protozoans, esp. sporozoans, the stage following the trophozoite which reproduces in the host by multiple fission; that stage in protozoan life cycle that gives rise to agametes.

agamospecies *n.* species that only reproduces non-sexually.

agamospermy *n.* any form of apomixis in which embryos and seeds are produced asexually.

agamotropic *a. appl.* flowers which having opened do not close again.

agar *n.* gelatinous substrate for bacterial cultures and constituent of some gels used for electrophoresis, prepared from agar-agar, a gelatinous polysaccharide extracted from red algae.

agarics *n.plu.* common name for a large group of basidiomycete fungi, the order Agaricales, typified by mushrooms and toadstools, with conspicuous fruiting bodies, usually composed of a stalk and cap, with the spores borne on the surface of gills on the underside of the cap. In some classifications the order Agaricales covers all gilled fungi, in others, it includes only certain families, but the term agarics is often used for the gilled fungi as a whole.

agarose *n.* polysaccharide prepared from agar and which is used to make gels for electrophoresis.

agarose gel *see* agarose.

age distribution *see* age structure.

age polyethism in social animals, the changing of roles by members of a society as they age.

age-specific reproductive value an index of the extent to which the members of

16

a given age group contribute to the next generation between now and the time they die.

age structure of a population, the percentage of the population at each age level, or the number of individuals of each sex at each age level. *alt.* age distribution.

ageotropism *n*. not responding to gravity; negative geotropism.

agglomerate *a*. clustered, as a head of flowers.

agglutinate *v*. to cause or undergo agglutination; *n*. the mass formed by agglutination; *a*. stuck together.

agglutination *n*. the formation of clumps or floccules of cells, esp. pollen, bacteria, red blood cells, spermatozoa and some protozoans, either spontaneously or after treatment with a specific antibody or other agent.

agglutinin *n*. any substance causing agglutination of cells (e.g. specific antibodies, lectins, viral haemagglutinins).

agglutinogen *n*. cell-surface antigens such as the ABO antigens of human red cells that interact with specific antibodies (agglutinins) resulting in agglutination of the red cells.

aggregate *a*. formed in a cluster; *appl*. fruit formed from apocarpous gynaecium of a single flower, as raspberry; *appl*. fruit formed from several flowers, as pineapple.

aggregation *n*. a group of individuals of the same species gathered in the same place but not socially organized or engaged in cooperative behaviour.

aggregation centre central collecting point to which myxamoebae of cellular slime moulds are attracted by cyclic AMP signals to form the "slug".

aggregation chimaera an animal (usually a mouse) that has developed from an embryo formed by the mixing *in vitro* of cells from two genetically different embryos.

aggression *n*. an act or threat of action by one individual that limits the freedom of action of another individual, often shown by animals to each other in ritualized trials of strength to gain mates, territory, etc.

aglomerular *a*. devoid of glomeruli, as kidney in certain fishes.

aglossate *a*. having no tongue.

aglutone *n*. non-sugar residue produced on

hydrolysis of glycosides.

aglycone *n*. the non-carbohydrate portion of a glycoside, produced from the glycoside by hydrolysis.

Agnatha, agnathans *n*., *n.plu.* subphylum, class or superclass of primitive jawless vertebrates, including the lampreys (Monorhina), hagfishes (Diplorhina) and their extinct relatives. *see also* heterostracans, osteostracans, anaspids.

agnathous, agnathostomatous *a*. having mouth without jaws, as lamprey.

agonist *n*. substance responsible for triggering a response in a cell, such as a hormone, neurotransmitter etc.; a prime mover or muscle directly responsible for a change in position of a part.

agonistic *a*. in animal behaviour, *pert*. any activity related to fighting, whether aggressive or conciliatory; *pert*. an agonist.

agranular *a*. lacking granules; without a conspicuous layer of granular cells, *appl*. motor areas in brain.

agrestal *a*. *appl*. uncultivated plants growing on arable land.

Agricultural Revolution the gradual shift from a hunter–gatherer way of life to settled agriculture, in which animals and wild plants were domesticated, which began between 10,000 and 12,000 years ago.

Agrobacterium tumefaciens a plant pathogenic bacterium, the cause of crown gall in numerous dicotyledonous plants. It carries a plasmid, the Ti plasmid, which becomes integrated into the chromosomes of infected tissue. *Agrobacterium* and its plasmid have been extensively modified by genetic engineering and are widely used to introduce novel genes into plant cells to produce transgenic plants.

agrocoenosis *n*. an ecosystem of cultivated land.

agroecosystem *n*. an ecosystem that develops on farmed land, and which includes the indigenous microorganisms, plants and animals, and the crop species.

agroinfection, agroinoculation *n*. the infection or inoculation of plants with genetically engineered *Agrobacterium* spp. as a means of introducing novel genes.

agrostology *n*. that part of botany dealing with grasses.

A-horizon the upper, or leached soil layers.

alt. eluvial layer.

air bladder the swim bladder in fishes; hollow dilation of thallus in bladderwrack.

air cells thin-walled cavities in ethmoid bone; numerous cavities in mastoid bone; air spaces in plant tissue.

air chamber gas-filled compartment of *Nautilus* shell, previously occupied by the animal; accessory respiratory organ or respiratory sac in some air-breathing teleost fishes.

air pollution any gaseous or particulate matter in the air that is not a normal constituent of air or not normally present in such large amounts. It may be the result of human activity, as sulphur dioxide from burning of coal, and carbon monoxide and nitrogen oxides from exhaust emissions, or can result from natural causes, as desert dust, methane and hydrogen sulphide from microbial activity in bogs, and volcanic debris in the atmosphere.

air quality the level of pollution in the air, which may be judged by a variety of criteria such as chemical and physical analysis, medical symptoms, damage to plants, damage to buildings, etc. Air quality is deemed to be high when pollution is low.

air sacs spaces filled with air and connected with lungs in birds; dilatations of tracheae in many insects; sacs representing tracheal system in some insect larvae and having hydrostatic function.

air sinuses cavities in various facial bones with passages to nasal cavities.

aitiogenic *a. appl.* reactions, movements, etc. caused by an external agent.

aitionastic *a. appl.* curvature of part of a plant induced by a diffuse stimulus.

akaryote *n.* a cell lacking a nucleus. *a.* akaryotic, non-nucleated.

akinesis *n.* absence or arrest of motion.

akinete *n.* resting cell in some algae, which will later reproduce.

ala *n.* any wing-like projection or structure; lateral petal of papilionaceous flower; membranous extension on some fruits or seeds for wind dispersal; outgrowth from petiole of decurrent leaf; wing-like projection on bone. *plu.* alae.

ala spuria bastard wing *q.v.*

alanine (Ala, A) α-amino propionic acid, simple amino acid with a methyl side chain, constituent of protein, non-essential in the human diet as it can be synthesized in the body by the reaction of pyruvate and glutamate giving alanine and α-ketoglutarate.

alar *a.* wing-like; *pert.* wings or alae; axillary.

alarm behaviour diverse types of behaviour shown by animals when disturbed and which is intended to distract predators or to hide the animal from view.

alarm pheromone chemical substance released in minute amounts by an animal which induces a fright response in other members of the species.

alary *a.* wing-like; *pert.* wings.

alate *a.* winged; having a wing-like extension, as of petiole or stem; broad-lipped, *appl.* shells.

albedo *n.* the ratio of the amount of light reflected by a surface to the amount of incident light; (*bot.*) mesocarp, the white tissue of rind of orange, lemon, etc.

albescent *a.* growing whitish.

albinism *n.* genetically determined or environmentally induced absence of pigmentation in animals normally pigmented, leading e.g. to lack of pigmentation in hair, skin and eyes. In humans, albinism is generally an autosomal recessive trait in which there is a deficiency of the enzyme tyrosinase (tyrosine 3-monooxygenase); (*bot.*) absence of green or other colour in plants due to genetically determined lack or non-development of chloroplasts or other chromoplasts. Organisms showing albinism are known as albinos.

albino *n. see* albinism.

albomaculus *a. appl.* variegation in plants consisting of irregular green and white patches as a result of mitotic segregation of chloroplasts or genes directing their development and function into certain cells, with their absence from others, during development.

albuginea *n.* dense white connective tissue surrounding testis, ovary, spleen or eye.

albumen *n.* white of egg, containing several proteins such as ovalbumin, ovomucoid and conalbumin; endosperm *q.v.*

albuminous cell *see* phloem.

albumins *n.plu.* general name for a class of small globular proteins found in blood (serum albumin), synovial fluid, milk and other mammalian secretions, and as storage proteins in plant seeds, characterized originally as proteins that are soluble in water and dilute buffers at neutral pH.

alburnum *n.* sapwood, splintwood, the young wood of dicotyledons.

alcaptonuria, alkaptonuria inherited enzyme deficiency in which homogentisate oxidase is absent leading to accumulation of homogentisate (*q.v.*) in urine, which turns dark on exposure to air.

alcohol dehydrogenase (ADH) any one of several enzymes catalysing the conversion of acetaldehydes or ketones to alcohols (e.g. in alcoholic fermentation in yeast) and the reverse reaction. E.C.1.1.1.1–2.

alcoholic fermentation the production in anaerobic conditions of ethyl alcohol from sugars by yeasts and other microorganisms.

Alcyonaria *n.* soft corals, sea pens and sea fans, a class of colonial coelenterates in which the lower parts of the polyps fuse to form a soft mass.

aldehyde group -CHO.

alder flies *see* Neuroptera.

aldolase *n.* enzyme that catalyses the conversion of fructose 1,6-bisphosphate into dihydroxyacetone phosphate and glyceraldehyde 3-phosphate in glycolysis and the reverse reaction in photosynthesis, EC 4.1.2.13, *r.n.* fructose-bisphosphate aldolase; used in a general sense for any enzyme that catalyses aldol condensations involving aldose and ketose monosaccharides. *alt.* aldehyde-lyases, EC 4.1.2.

aldose *n.* any monosaccharide containing an aldehyde (CHO) group. *cf.* ketose.

aldosterone *n.* steroid hormone produced by adrenal cortex, involved in regulation of mineral and water balance in body by its action on kidneys.

alecithal *a.* with little or no yolk.

alepidote *a.* without scales.

aleuriospore *n.* fungal spore cut off from the apex of a hypha at an early stage, borne by many fungal parasites of skin and other hyphomycetes. *alt.* microconidium.

aleurone grains storage granules in endosperm and cotyledon of seeds, consisting of proteins, phytin and hydrolytic

enzymes, forming aleurone layer in seeds such as cereals.

aleurone layer in cereal seeds, a protein-rich layer of cells found immediately under the testa, containing aleurone grains.

aleuroplast *n.* protein storage body in plant cells.

alexin(e) *n.* formerly used for complement *q.v.* Now virtually obsolete.

alfalfa *n.* lucerne, *Medicago sativa*, leguminous plant widely grown for forage and green manure.

alfalfa mosaic virus group plant virus group containing a single member, alfalfa mosaic virus, which is a rod-shaped single-stranded RNA virus causing mosaic symptoms (mottling of leaves). It is a multicomponent virus in which the four separate genomic RNAs are encapsidated in different virus particles.

algae *n.plu.* a general term with no taxonomic status for a heterogeneous group of unicellular, colonial and multicellular eukaryotic photosynthetic organisms of simple structure. Traditionally included in the plant kingdom, the different groups of algae are now often classified as divisions of the kingdom Protoctista (Protista). They are aquatic or live in damp habitats on land and include unicellular organisms such as *Chlamydomonas* and diatoms, colonial forms such as *Volvox*, the multicellular green, red and brown seaweeds, and fresh-water multicellular algae such as *Spirogyra*. The algal body is known as a thallus and in multicellular forms is generally filamentous or flattened into a thin sheet or ribbon. *sing.* alga. *see* Bacillariophyceae (Bacillariophyta), Charophyceae (Charophyta), Chlorophyta, Chrysophyta (Bacillariophyceae, Chrysophyceae, Xanthophyceae), dinoflagellates, Euglenophyta, Phaeophyta, Pyrrophyta, Rhodophyta, zooxanthellae.

algal bloom an exceptional growth of algae or cyanobacteria in lakes, rivers or oceans, which may occur in particular climatic conditions or as a result of excess nutrients in the water. In some cases, the microorganisms produce toxic compounds.

algesis *n.* the sense of pain.

algicolous *a.* living on algae.

alginate *n.* carbohydrate polymer derived

from algin, a gel-like polysaccharide found in the cell walls of brown algae, and which is used as a food stabilizing and texturing agent and in dental moulding materials.

algoid *a. pert.*, resembling or of the nature of an alga.

Algonkian *a. pert.* the late Proterozoic era.

alien *n.* plant species thought to have been introduced by man but now more or less naturalized.

aliform *a.* wing-shaped.

alignment *n.* of nucleotide or amino acid sequences, the process of matching two sequences for comparative purposes, e.g. in molecular evolutionary studies, in which gaps may be introduced into one or other sequence to produce maximum homology.

alima *n.* a larval stage of certain crustaceans.

alimentary *a. pert.* nutritive functions.

alimentary canal the tube running from mouth to anus into which food is ingested and in which it is digested, in vertebrates comprising the oesophagus, stomach and intestines.

alimentation *n.* the process of nourishing or being nourished.

Alismales *n.* order of herbaceous monocots, aquatic or partly aquatic, and including the families Alismaceae (water plantain), Butomaceae (flowering rush) and Limnocharitaceae.

Alismatales Alismales *q.v.*

alisphenoid *n.* wing-like portion of the sphenoid bone forming part of the cranium.

alkali soils soils with a very high surface content of mineral salts such as sodium chloride, sodium sulphate, sodium carbonate and borax, which are formed in dry regions where evaporation is much greater than rainfall. *see* solonchaks, solonetz.

alkaline gland Dufour's gland *q.v.*

alkaline soils in general agriculture or horticulture, temperate-region soils rich in calcium compounds, with a pH > 7.5 and up to 8 or 9, which develop over chalk or limestone. *alt.* calcareous soils. *cf.* alkali soils.

alkaline tide transient decrease in acidity of body fluids after taking food.

alkaloid *n.* any of a group of nitrogenous organic bases found in plants, with toxic or medicinal properties, such as caffeine, morphine, nicotine, strychnine.

alkane *n.* member of a group of saturated

hydrocarbons thought to be chemical fossils indicating the existence of life, which have been found in Precambrian rocks.

alkaptonuria alcaptonuria *q.v.*

alkylating agents highly reactive chemicals that attach to bases in DNA causing errors in replication and thus are potent mutagens, as well as causing widespread damage to tissues, e.g. mustard gas.

all-or-none principle principle that a response to a stimulus is either completely effected or is absent, first observed in heart muscle. *alt.* Bowditch's Law.

allaesthetic *a. appl.* characters effective when perceived by another organism.

allantoate (allantoic acid) hydration product of allantoin, nitrogenous excretion product of teleost fish, also found in other animals and plants.

allantochorion *n.* foetal membrane formed of outer wall of allantois and the primitive chorion.

allantoicase enzyme catalysing the hydrolytic breakdown of allantoate to glyoxylic acid and urea, which is found in amphibia, certain fishes and invertebrates. EC 3.5.3.4.

allantoin *n.* a derivative of urate (uric acid), an end-product of purine and pyrimidine metabolism, occurring in allantoic fluid and urine of mammals other than primates, some gastropods and insects, and in plants.

allantoinase enzyme catalysing the conversion of allantoin to allantoate. EC 3.5. 2.5.

allantois *n.* sac-like outgrowth from posterior part of alimentary canal in embryos of reptiles, amphibians, birds and mammals, which acts as an organ of respiration and/or nutrition and/or excretion. *a.* allantoic.

allatectomy *n.* excision or removal of corpora allata.

allele *n.* one of a number of alternative forms of a gene that can occupy a given genetic locus on a chromosome.

allelic *a. pert.*, or the state of being, alleles; *appl.* two or more mutations mapping to the same area and which do not complement each other in the heterozygous state, showing that they are affecting the same genetic locus or cistron.

allelic complementation *see* complementation.

allelic exclusion the situation in any particular antibody-producing cell that antibody synthesis is specified by the genes on only one of the relevant pair of homologous chromosomes.

allelomimetic *a. appl.* animal behaviour involving imitation of another animal, usually of the same species.

allelomorph *n.* the characteristic specified by an allele; formerly also used for allele *q.v.*

allelopathic *a. pert.* the influence or effects (sometimes inhibitory or harmful) of a liv ing plant on other nearby plants or microorganisms. *n.* allelopathy.

allelotype *n.* the frequency of different alleles in a population.

Allen's rule rule that in a widely distributed species of endothermic animal (animals that generate their own body heat), the extremities (e.g. ears, feet, tail) tend to be smaller in the colder regions of the species range than in the warmer regions.

allergen *n.* a substance to which an individual is hypersensitive and which causes an immune response often characterized by local inflammatory reactions but sometimes by severe shock symptoms (anaphylactic shock).

allergy *n.* type of hypersensitive immune reaction exhibited by certain individuals on exposure to an otherwise innocuous antigen (e.g. pollen, food, drugs, dust-mites, etc.). Some allergies (e.g. hay-fever) involve the production of IgE antibodies, which cause the release of inflammatory products such as histamine from mast cells, causing the characteristic symptoms. Other allergies, such as some contact sensitivities, are caused by T-cell mediated reactions. *a.* allergic. *see also* anaphylaxis, delayed hypersensitivity.

alliaceous *a. pert.* or like garlic or onion.

alloantibody *n.* alloantiserum *q.v.*; antibody produced in another member of the same species (refers to antibodies then used for treatment or immunization of an individual).

alloantigens *n. plu.* antigens, e.g. histocompatibility antigens, which are present in different forms in different individuals of the same species. An alloantigen from one individual therefore provokes an immune response in a genetically dissimilar individual of the same species.

alloantiserum *n.* antiserum raised in one animal against the antigens of a genetically dissimilar animal of the same species.

allobiosis *n.* the changed reactivity of an organism in a changed internal or external environment.

allocarpy *n.* production of fruit after cross-fertilization.

allocheiral *a.* having right and left sides reversed; *pert.* reversed symmetry.

allochroic *a.* able to change colour; with colour variation.

allochronic *a.* not contemporary, *appl.* species in evolutionary time; *appl.* species or populations that have non-overlapping breeding seasons or flowering periods.

allochthonous *a. appl.* material or species that has originated elsewhere; exotic; not aboriginal.

allocortex *n.* the primitive cortical areas or cortex of olfactory brain.

alloenzyme allozyme *q.v.*

allogamy *n.* cross-fertilization *q.v. a.* allogamous.

allogeneic *a.* genetically different, when *appl.* animals of the same species.

allogenic *a. appl.* plant successions, caused by external factors, such as fire or grazing.

allogenous *a. appl.* floras persisting from an earlier environment.

allograft *n.* a graft of tissue or an organ from one individual to another genetically non-identical individual of the same species.

allogrooming *n.* grooming directed at another individual of the same species. *alt.* social grooming.

alloheteroploid *n.* heteroploid derived from genomes of different species.

allokinesis *n.* reflex or passive movement; involuntary movement.

allokinetic *a.* moving passively; drifting, as plankton.

allolactose *n.* form of lactose that is the natural inducer for the *lac* operon.

allometric *a.* differing in growth rate; *pert.* allometry.

allometry *n.* study of relative growth; change of proportions with increase in size; growth rate of a part differing from standard growth rate or from growth rate as a whole.

allomixis *n.* cross-fertilization *q.v. a.* allomictic.

allomone *n.* chemical secreted by one individual which causes an individual of another species to react favourably to it, such as scent given out by flowers to attract pollinating insects.

allomorphosis *n.* evolution with rapid increase in specialization.

alloparent *n.* individual that assists the parent in the care of the young.

alloparental care assistance in the care of the young by individuals other than the parents.

allopatric *a.* having separate and mutually exclusive areas of geographical distribution.

allophenic *a.* chimaeric (*appl.* animals); *appl.* a phenotype not due to a mutation in the actual cells showing the characteristic, but due to the influence of other cells.

allophore *n.* cell or chromatophore containing red pigment in skin of fishes, amphibians and reptiles.

allophycocyanin *n.* red protein pigment in the phycobilisomes of red algae and cyanobacteria which acts as a light-harvesting pigment for photosynthesis.

allopolyploid, alloploid *n.* a polyploid produced from a hybrid between two or more different species and therefore possessing two or more unlike sets of chromosomes. *alt.* amphiploid.

alloreactive *a. appl.* antibodies, T cells, etc. that react against antigens or cells from a genetically dissimilar individual of the same species.

allosematic *a.* having markings or coloration imitating warning signs in other, usually dangerous, species. *cf.* aposematic.

allosomal *n. pert.* inheritance of characters controlled by genes carried on an allosome.

allosome *n.* a chromosome other than an autosome, such as a sex chromosome.

allosteric *a. appl.* proteins showing allostery.

allostery *n.* the property displayed by many proteins, that the binding of a small molecule at one site induces a change in the properties of another, distant, site.

allosynapsis, allosyndesis *n.* pairing of homologous chromosomes from opposite parents in a polyploid.

allotetraploid *n.* an allopolyploid produced when a hybrid between two species doubles its chromosome number. *alt.* amphidiploid.

allotherm *n.* organism with body temperature dependent on environmental temperature.

allotopic *a. appl.* sympatric populations occupying different habitats within the same geographical range of distribution.

allotopic gene expression expression of a gene in a cell or organelle in which it is not normally expressed.

allotriploid *n.* an organism whose somatic cells contain three sets of chromosomes, one of which differs from the others.

allotrophic heterotrophic *q.v.*

allotropous *a. appl.* insects not limited to or adapted to visiting special kinds of flowers; *appl.* flowers whose nectar is available to all kinds of insects.

allotropy *n.* tendency of certain cells or structures to approach each other; mutual attraction, as between gametes.

allotype *n.* antigenic determinant that characterizes allelic differences in light or heavy chains of immunoglobulins, and which is therefore inherited in Mendelian fashion. *cf.* isotype, idiotype.

allozygous *a. appl.* alleles at the same locus that are different, or, if identical, of different origins, i.e. their identity is not due to common descent.

allozyme *n.* one of a number of forms of the same enzyme having different electrophoretic mobilities. *alt.* alloenzyme.

alluvial *a. pert.* soils composed of sediment transported and deposited by flowing water.

alpestrine *a.* growing high on mountains but not above the tree line.

alpha- *for headwords with prefix alpha- refer also to headword itself, e.g. for alpha-globin look under globin.*

alpha *n.* the highest-ranking individual within a dominance hierarchy.

alpha diversity biological diversity resulting from competition between species that reduces the variation within species as each species becomes more finely adapted to the niche it occupies. *alt.* niche diversification. *cf.* beta diversity.

alphaviruses *n. plu.* group of viruses in the Togaviridae, including Semliki Forest virus.

alpine *a. appl.* the part of a mountain above the tree line and below the permanent snow line, and to species mainly restricted to this zone.

alpine grassland grassland found above the tree line on high mountains.

alpine tundra the zone of tundra-like vegetation found on high mountains above the alpine grasslands and below the permanent snow line.

alternate *a.* not opposite, *appl.* leaves, branches etc. occurring at different levels successively on opposite sides of stem.

alternate segregation a type of segregation at meiosis of the members of a quadrivalent involving a heterozygous reciprocal translocation, resulting in balanced gametes.

alternation of generations the alternation of haploid and diploid stages which occurs in the life cycle of sexually-reproducing eukaryotic organisms. In some organisms, e.g. mosses, the haploid phase is predominant, in others, e.g. flowering plants and many animals, the diploid phase is dominant and the haploid phase is represented only by the gametes, and in others e.g. hydrozoan coelenterates, diploid and haploid organisms alternate.

alternation of parts general rule that leaves of different whorls of flower alternate in position with each other, sepals with petals, stamens with petals.

alternative pathway of complement activation, pathway triggered by cell surfaces possessing certain properties, and endotoxins of Gram-negative bacteria, involving complement component C3 and serum factors B, D and P (properdin), resulting in formation of C3b which then follows the classical pathway. *see also* complement system.

alterne *n.* vegetation exhibiting disturbed zonation due to abrupt change in environment, or to interference with normal plant succession.

alternipinnate *a. appl.* leaflets or pinnae arising alternately on each side of midrib.

altrices *n.plu.* birds whose young are hatched in a very immature condition.

altricial *a.* requiring care or nursing after hatching or birth.

altruism *n.* any act or behaviour which re-

sults in an individual increasing the genetic fitness of another at the expense of its own, e.g. by devoting large amounts of time and resources to caring for another individual's offspring at the expense of producing its own. *a.* altruistic. *see also* reciprocal altruism.

Alu sequences a family of repetitive interspersed DNA sequences of around 300 bp present in up to 1 million copies in the human genome (named after the restriction enzyme AluI which is used to identify them in restriction maps). Similar sequences are present in other mammals.

alula *n.* a small lobe of a wing.

alveolar *a. pert.* an alveolus; *pert.* tooth socket, *appl.* nerve, artery etc. in connection with jaw bone.

alveolate *a.* deeply pitted or honeycombed.

alveolus *n.* air cavity in lungs; small pit or depression; a cavity; tooth socket.

alveus *n.* layer of white matter on ventricular surface of hippocampus (in brain); utricle of ear; dilation of thoracic duct.

Alzheimer's disease a neurological disease relatively common in elderly people, with atrophy of neurones in certain parts of the forebrain and consequent disturbance of brain function. Symptoms include profound confusion, memory loss and often changes in personality. Formerly called senile dementia. A rare inherited form of the disease, familial Alzheimer's disease (FAD), causes symptoms at a much earlier age than usual.

amacrine *a.* having no conspicuous axon, *appl.* type of neurone in retina which forms a layer with bipolar and horizontal cells and makes lateral connections.

α-amanitin *n.* cyclic octapeptide toxin from *Amanita phalloides,* a potent inhibitor of RNA polymerase II and III.

Amastigomycota *n.* in modern classifications, a major division of the Fungi including the subdivisions Zygomycotina, Ascomycotina, Basidiomycotina, and the form subdivision Deuteromycotina (Fungi Imperfecti). They are terrestrial fungi, usually with a well-developed mycelium, and do not have motile flagellate zoospores or gametes. They include the familiar moulds (e.g. *Mucor, Penicillium*) and the mushrooms and toadstools.

amb *n.* ambulacral area, in echinoderms.

amber *n.* translucent yellow or brown material, known from the Cretaceous onwards, which is the fossilized resin of coniferous trees.

amber the UAG termination (nonsense) codon.

ambergris *n.* secretion of the sperm whale, formerly used as a musk fragrance in perfumery, now superseded by synthetic compounds.

amber mutation a mutation generating an amber codon and resulting in premature termination of synthesis of the protein product of the mutated gene.

amber suppressor a mutant gene producing a tRNA which overcomes the effects of an amber mutation by inserting an amino acid at UAG.

ambiens *n.* thigh muscle in some birds, whose action causes the toes to maintain grasp on perch.

ambient *a.* surrounding.

ambilateral *a. pert.* both sides.

ambiparous *a. appl.* buds containing the beginnings of both flowers and leaves.

ambiquitous *a. appl.* enzymes for which the degree to which they are associated with subcellular particulate structures is dependent on metabolic activity.

ambisexual *a. pert.* both sexes; monoecious *q.v.*

ambitus *n.* the outer edge or margin; outline of echinoid shell viewed from apical pole.

ambon *n.* ring of fibrous cartilage surrounding the socket of a joint.

ambosexual *a.* common to, or *pert.*, both sexes; activated by both male and female hormones.

ambrosial *a. appl.* a class of odours typified by ambergris and musk.

ambulacra *n.plu.* region containing the tube-feet of echinoderms; the bands of tube-feet themselves. *sing.* ambulacrum.

ambulacral *a. pert.* or used for walking, *appl.* to legs of arthropods; *pert.* ambulacra.

ameba *alt.* spelling of amoeba *q.v.*

ameiosis *n.* occurrence of only one division in meiosis instead of two; the absence of pairing of chromosomes in meiosis.

ameiotic *a. appl.* parthenogenesis in which meiosis is suppressed.

amelification *n.* formation of tooth enamel.

ameloblast *n.* columnar or hexagonal epithelial cell that secretes enamel, and is part of enamel organ in tooth. *alt.* enamel cell.

amensalism *n.* a form of competition between two species in which one is inhibited and the other is not. *alt.* antagonism.

amentaceous, amentiferous *a.* bearing catkins.

amentum *n.* catkin. *plu.* amenta.

ameristic *a.* not divided into parts; unsegmented; undifferentiated or undeveloped.

Ames test simple *in vitro* test devised by the American biochemist Bruce Ames to screen compounds for potential mutagens and carcinogens by their ability to cause mutations in bacteria.

Ametabola the ametabolous (*q.v.*) insects.

ametabolous *a.* not changing form, *appl.* the orders of primitive wingless insects (the Ametabola) in which the young hatch from the egg resembling young adults, and comprising the Diplura (two-pronged bristletails), Thysanura (three-pronged bristletails), Collembola (springtails) and Protura (bark-lice).

amethopterin *n.* a folate analogue which blocks regeneration of tetrahydrofolate and dTMP synthesis and is used as an anticancer drug to inhibit rapidly dividing cells. *alt.* methotrexate.

ametoecious *a.* parasitic on one host during one life cycle. *alt.* autoecious.

amicronucleate *a. appl.* fragments of certain protozoans in which there is no micronucleus.

amictic *a. appl.* eggs that cannot be fertilized and which develop parthenogenetically into females; *appl.* females producing such eggs.

amidase *n.* an enzyme catalysing the hydrolysis of a monocarboxylic acid amide to a monocarboxylic acid and ammonia. EC 3.5.1.4; any of a group of enzymes hydrolysing non-peptide C–N linkages of amides and including asparaginase, urease and glutaminase. EC 3.5.1–2.

amide *n.* a compound which contains the group –CO.NH$_2$, biological amides being derived from carboxylic acids and amino acids by replacement of the –OH of the carboxyl group with $-NH_2$.

amidinase *n*. any of a group of enzymes hydrolysing non-peptide C–N linkages in amidines, and including arginase. EC 3.5.3–4.

amidine *n*. a compound which contains the group -CNH.NH$_2$.

amine n. chemically, any compound containing the functional group -NH$_2$,

amino acid *n*. any of a class of compounds of the general formula $RCH(NH_2)COOH$ (α-amino acids) where R is a distinctive side chain, which can occur as optically active D- and L-isomers, of which only L-isomers are found in proteins. Around 20 different amino acids are present in proteins, all of which can be synthesized by autotrophs but which in heterotrophs are chiefly obtained by breakdown of dietary protein. Amino acids are also biosynthetic precursors of many important molecules such as purines, pyrimidines, histamine, thyroxine, adrenaline, melanin, serotonin, the nicotinamide ring and porphyrins amongst others. *see* Appendix 1 (29) and table.

D-amino acid type of amino acid atomic configuration, found in peptides in bacterial cell walls and a few other instances but never in proteins.

amino acid analysis determination of the amino acid composition of a protein.

amino acid neurotransmitters the amino acids glycine and glutamate, which act respectively as inhibitory and excitatory neurotransmitters in the central nervous system.

amino acid notation the abbreviations for the amino acids found in proteins, which are used in displaying protein sequences and which may be of one or three letters. *see* Appendix 1.

amino acid racemization the conversion of L-amino acids to D-amino acids, which occurs at a very slow rate in nature, and which can be used to date certain fossils by measuring the amount of D-amino acids in the sample and calculating the time taken for them to form from the original L-amino acids (only L-amino acids are found in living tissue to any appreciable extent), this method being applicable to fossils between 15,000 and 100,000 years old.

amino acid sequence order of amino acid subunits in a polypeptide or protein.

4-amino pyridine compound that selectively blocks potassium conductance channels in neurones, thereby blocking the generation of nerve impulses.

amino sugar any monosaccharide in which a hydroxyl group has been replaced by an amino group, e.g. galactosamine and glucosamine.

aminoacylase *n*. enzyme catalysing the transfer of an acyl group from an acyl-amino acid during fatty acid synthesis. EC 3.5.1.14.

aminoacyltransferase *n*. any of a group of enzymes that transfer aminoacyl groups and including peptidyltransferase. EC 2.3.2.

aminoacyl-tRNA *n*. tRNA carrying a specific activated amino acid covalently attached to its 3′ end (e.g. methionyl-tRNA, seryl-tRNA, tyrosyl-tRNA).

aminoacyl-tRNA synthetase *n*. any of a large group of enzymes catalysing the attachment of an amino acid to tRNA, each aminoacyl-tRNA synthetase being specific for a particular amino acid and one or more acceptor tRNAs.

γ-aminobutyric acid (GABA) an amino acid which acts as an inhibitory neurotransmitter in the central nervous system.

aminopeptidase *n*. any of a group of enzymes that remove amino-terminal amino acid residues from a protein or peptide. EC 3.4.11.

aminopterin *n*. a folate analogue which blocks the regeneration of tetrahydrofolate and dTMP synthesis and is used as an anticancer drug to inhibit rapidly dividing cells.

aminopurine (2-aminopurine) *n*. an analogue of adenine, pairs with cytosine rather than thymine and therefore mutagenic.

amino-terminus the end of a protein chain that bears the free α-amino group, often abbreviated N-terminus. A protein is synthesized starting at the amino-terminus.

aminotransferase *n*. any of a class of enzymes that catalyse transfer of an α-amino group, usually from an α-amino acid to a α-keto acid. EC 2.6.1. *alt*.

transaminase.

amitosis *n.* division of the nucleus by constriction without the condensation of mitotic chromosomes or formation of a spindle and without breakdown of the nuclear membrane, e.g. in the macronucleus of ciliates.

amixia *n.* cross-sterility between members of the same species as a result of morphological, geographical, or physiological isolating mechanisms.

amixis *n.* absence of fertilization; sometimes used for the absence of gonads; apomixis in haploid organisms.

ammocoete *n.* larval lamprey.

ammonia, ammonium ion *n.* NH_3, NH_4^+, produced in the biological nitrogen cycle by biological nitrogen fixation, by the action of soil microorganisms that break down protein in dead organic matter, and as an excretion product of e.g. teleost fishes. It is the form in which atmospheric nitrogen enters the nitrogen cycle and is taken up by plants. *see also* ammonification, ammonotelic.

ammonification *n.* the production of ammonia (as ammonium ion) from organic nitrogenous compounds, carried out by a variety of heterotrophic microorganisms (ammonifiers).

ammonifier *n.* any of diverse group of heterotrophic microorganisms (including bacteria and fungi) that can produce ammonia (ammonium ion) from
organic matter breakdown, esp. in soil.

ammonites *n.plu.* an extinct group of cephalopods, familiar from their coiled shells, similar to the nautiloids but probably had a calcareous larval shell.

ammonitiferous *a.* carrying fossil remains of ammonites.

ammonotelic *a.* excreting nitrogen mainly as ammonia, as most aquatic invertebrates, tadpoles and some teleost fish.

amniocentesis *n.* procedure in which cells from the amniotic fluid surrounding the foetus are withdrawn for prenatal diagnosis of hereditary defects.

amnion *n.* the innermost of the embryonic membranes derived from the blastula and surrounding the developing embryo of reptiles, birds and mammals; the inner embryonic membrane of insects; a membrane like

an amnion found in other invertebrates.

amniotes *n.plu.* vertebrates which have an amnion around the developing embryo, as reptiles, birds and mammals, considered as a superclass Amniota in some classifications.

amniotic *a. pert.* amnion, *appl.* folds, sac, cavity, fluid.

amniotic egg egg of birds, reptiles and prototherians (egg-laying mammals) within which extraembyronic membranes are formed during embryonic development.

amoeba *n.* unicellular non-photosynthetic wall-less protist whose shape is subject to constant change due to formation and retraction of pseudopodia. The amoebas are classified in the protist phylum Rhizopoda, and were formerly classified in the protozoan class Rhizopodea; myxamoeba *q.v.. plu.* amoebae, amoebas. *alt.* ameba.

amoebic *a. pert.*, or caused by, amoebae, *appl.* dysentery (amoebiasis) caused by the parasitic amoeba *Entamoeba histolytica*.

amoebiform amoeboid *q.v.*

amoebism *n.* amoeboid form or behaviour, as of some cells.

amoebocyte *n.* any cell having the shape or properties of an amoeba; a cell in coelomic fluid of echinoderms.

amoeboid *a.* resembling an amoeba in shape, in properties or mode of movement.

amoebula *n.* the amoeboid swarm spore of various protists.

amorph *n.* a mutation in which no active gene product is formed, a null mutation.

amorphous *a.* of indeterminate or irregular form; with no visible differentiation in structure.

ampheclexis sexual selection *q.v.*

ampherotoky amphitoky *q.v.*

amphetamine *n.* sympathomimetic drug chemically related to adrenaline and which is a powerful stimulant of the CNS.

amphi- Gk. prefix denoting both, on both sides.

amphiapomict *n.* group of genetically identical individuals reproduced from facultatively sexual forms.

amphiarthrosis *n.* a slightly moveable joint.

amphiaster *n.* a sponge spicule star-shaped

at both ends.

Amphibia, amphibians *n.*, *n.plu.* vertebrate class including the extant subclass Lissamphibia, comprising the frogs and toads (order Anura), newts and salamanders (order Urodela) and the worm-like caecilians (order Apoda). There are also a number of extinct subclasses dating from the Devonian onwards, of which the ichthyostegalians are the earliest fossils found. Amphibians are anamniote tetrapod vertebrates which typically return to the water for reproduction and pass through an aquatic larval stage with gills. Adults generally have lungs and may be at least partly terrestrial, and modern amphibians have a skin without scales. *see also* anthracosaurs, ichthyostegalians, lepospondyls, temnospondyls.

amphibian *a.* adapted for life either in water or on land.

amphibivalent *n.* a ring of chromosomes arising in the metaphase and anaphase of the 1st meiotic division as a result of the reciprocal translocation of chromosome segments between two chromosomes.

amphiblastic *a. appl.* telolecithal ova with complete but unequal segmentation.

amphibolic *a.* capable of turning backwards or forwards, as outer toe of some birds.

amphicarpous *a.* producing two kinds of fruit.

amphicoelous, amphicelous *a.* concave on both surfaces, *appl.* biconcave vertebral centra.

amphicondylous *a.* having two occipital condyles.

amphicribral *a.* with the phloem surrounding the xylem, *appl.* some concentric vascular bundles. *alt.* amphiphloic. *cf.* amphivasal.

amphid *n.* one of a pair of anterior sense organs in nematodes, possibly detecting chemical stimuli.

amphidetic *a.* extending behind and in front of umbo, *appl.* hinge ligaments of some bivalve shells.

amphidiploid allotetraploid *q.v.*

amphidromous *a.* going in both directions, *appl.* animal migration.

amphigenous *a.* borne or growing on both sides of a structure, as of leaf; borne or growing on all sides of an organism or structure, *alt.* perigenous.

amphigony *n.* sexual reproduction involving 2 individuals.

amphigynous *a. appl.* antheridium surrounding the base of oogonium, as in some Peronosporales.

amphihaploid *n.* a haploid arising from an amphidiploid species.

amphimict *n.* a group of individuals resulting from sexual reproduction; an obligate sexual organism.

amphimixis *n.* reproducing by seed produced by normal sexual fusion. *a.* amphimictic. *cf.* apomixis.

Amphineura *n.* class of marine molluscs, commonly called chitons, having an elongated body and a mantle bearing calcareous plates.

amphioxus *n.* lancelet, a cephalochordate of the genus *Branchiostoma*.

amphipathic *a.* possessing both a hydrophobic and a hydrophilic (polar) portion in the same molecule, *appl.* proteins or parts of proteins, and to molecules such as phospholipids. The amphipathic nature of phospholipids allows them to orient in aqueous solution into micelles with hydrophobic groups together at the centre of the sphere and the polar groups on the surface, and to form bilayered biological membranes with the hydrophobic groups pointing towards the centre.

amphiphloic *a.* with phloem both external and internal to xylem, *appl.* stems, vascular bundles.

amphiphyte *n.* an amphibious plant, one that can live on land or in water.

amphiplatyan *a.* flat on both ends, *appl.* vertebral centra.

amphiploid allopolyploid *q.v.*

amphipneustic *a.* having both gills and lungs throughout life history; with only anterior and posterior pairs of spiracles functioning, as in most dipteran larvae.

amphipod *n.* member of the Amphipoda, an order of terrestrial, marine and freshwater malacostracan crustaceans, having a laterally compressed body, elongated abdomen and no carapace, e.g. sand-hopper.

amphirhinal *a.* having, or *pert.* two nostrils.

amphisarca *n.* a superior, many-seeded fruit with pulpy interior and woody exterior.

Amphisbaenia, amphisbaenians,

amphisbaenids *n., n. plu., n. plu.* group of worm-like, burrowing, generally limbless reptiles with inconspicuous eyes and a rounded tail.

amphispermous *a.* having seed closely surrounded by pericarp.

amphisporangiate *a.* having sporophylls bearing both megasporangia and microsporangia; hermaphrodite, *appl.* flowers.

amphispore *n.* a reproductive spore which functions as a resting spore in certain algae; a uredospore modified to withstand dry conditions.

amphisternous *a. appl.* type of sternum structure in some sea urchins.

amphistomatous *a.* having stomata on both surfaces, *appl.* some leaves.

amphistomous *a.* having a sucker at each end of body, as leeches.

amphistylic *a.* having jaw arch connected with skull by both hyoid and quadrate, or by both hyoid and palatoquadrate.

amphitelic *a. appl.* orientation of chromosomes on the spindle equator at metaphase of mitosis, with centromeres exactly equidistant from each pole.

amphithecium *n.* in bryophytes, peripheral layer of cells in sporogonium.

amphitoky *n.* parthenogenetic reproduction of both males and females.

amphitrichous *a.* with a flagellum at each pole, *appl.* bacteria.

amphitroph *n.* normally autotrophic organism which adapts itself to heterotrophic nutrition if placed in the dark for long periods.

amphitropous *a.* having the ovule inverted, with hilum in middle of one side.

amphiumas *n.plu.* small family of wholly aquatic eel-like amphibians from southeast USA, comprising a single genus with three species.

amphivasal *a.* with the xylem surrounding the phloem, *appl.* some concentric vascular bundles. *alt.* amphixylic.

amphogenic *a.* producing offspring consisting of males and females.

amphoteric *a.* possessing both acidic and basic properties, e.g. amino acids; with opposite characters.

ampicillin *n.* a semi-synthetic aminophenylacetyl penicillin.

amplectant *a.* clasping or winding tightly round a support, as tendrils.

amplexicaul *n.* clasping or surrounding the stem, as base of sessile leaf.

amplexus *n.* mating embrace in frogs and toads when eggs are shed into the water and fertilized.

ampliate *a.* having outer edge of wing prominent, as in some insects.

amplicon *n.* a stretch of DNA that has become copied many times to form an array of repeated sequences.

amplification *n.* changes towards increased structural or functional complexity in ontogeny or phylogeny; of genes or DNA, multiplication of a gene or DNA sequence to produce numerous copies within the chromosomes.

ampulla *n.* a membranous vesicle; dilated portion at one end of each semicircular canal of ear; dilated portion of various ducts and tubules; internal reservoir on ring canal of water vascular system in echinoderms; terminal vesicle of sensory canals in elasmobranch fishes; (*bot.*) submerged bladder of bladderwort, *Utricularia*.

ampullaceous *a.* flask-shaped.

ampullae of Lorenzini jelly-filled tubes in the head of selachian fish (sharks, rays, etc.) opening to the exterior and terminating in sensory cells that detect changes in temperature or salinity of water or changes in electrical potential in the tissue.

amygdala *n.* almond, or almond-shaped structure; subcortical structure within the temporal lobes of cerebral hemispheres, thought to be involved in memory.

amygdalin *n.* cyanogenic glycoside found in kernels of bitter almond, peach, cherry.

amylase *n.* α-amylase (EC 3.2.1.1), which randomly hydrolyses α-1,4 linkages in starch, glycogen and other glucose polysaccharides, or β-amylase (EC 3.2.1.2) which successively removes maltose units, or a mixture of these two enzymes, and formerly called diastase.

amyliferous *a.* containing or producing starch.

amylogenesis *n.* starch formation.

amyloid *a.* starch-like; *n.* complex proteinaceous fibrillar material deposited in heart, liver, spleen, and other organs in various forms of amyloidosis. Amyloid deposits in brain are characteristic of Alzheimer's disease.

amylolytic *a*. starch-digesting.

amylopectin *n*. a branch-chained polysaccharide found in starch with a structure similar to glycogen.

amyloplast *n*. colourless starch-forming granule in plants.

amylose *n*. a straight chain polysaccharide found in starch. *see* Appendix 1 (14) for structure.

anabiosis *n*. a condition of apparent death or suspended animation produced in certain organisms by, e.g., desiccation, and from which they can be revived to normal metabolism.

anabolism *n*. the constructive biochemical processes in living organisms involving the formation of complex molecules from simpler ones and the storage of energy. *a*. anabolic.

anabolite *n*. any substance involved in anabolism.

anacanthous *a*. without spines or thorns.

anachoresis *n*. the phenomenon of living in holes or crevices.

anadromous *a*. *appl*. fishes which migrate from salt to fresh water annually.

anaemia *n*. blood disorder characterized by a lack of red cells, which may be due to a variety of causes.

anaerobe *n*. any organism that can live in the absence of oxygen, obligate anaerobes being unable to live in even low oxygen concentrations, facultative anaerobes being able to live in low or normal oxygen concentrations as well. *cf*. aerobe. *a*. anaerobic.

anaerobic decomposition the breakdown of organic matter under anaerobic conditions by microorganisms.

anaerobic respiration respiration occurring in the absence of oxygen. *alt*. glycolysis, fermentation.

anaerobiotic *a*. lacking or depleted in oxygen, *appl*. habitats.

anaerogenic *a*. not producing gas during fermentation, *appl*. microorganisms.

anagenesis *n*. progressive evolution within a lineage, *cf*. cladogenesis; regeneration of tissues.

anagenetic *a*. *appl*. evolution occurring by

the gradual change of one type into another. *n*. anagenesis.

anal *a*. *pert*. or situated at or near the anus; *appl*. posterior median ventral fin of fishes.

analgesic *a*. reducing or abolishing pain; *n*. pain-killing drug.

analogous *a*. *appl*. structures that are similar in function but not in structure and developmental and evolutionary origin, e.g. the wings of insects and birds.

analogue *n*. any organ or part similar in function to one in a different plant or animal, but of unlike origin; any compound chemically related to but not identical with another and which in the case of analogues of natural metabolites compete with them for binding sites on enzymes, receptors etc., often blocking the normal reaction.

analogy *n*. resemblance in function though not in structure or origin. *a*. analogous.

analysis of variance (ANOVAR) a statistical method by which the variance of a set of data can be apportioned to different causes.

anamestic *a*. *appl*. small variable bones filling spaces between larger bones of more fixed position.

anamnestic *a*. in immunology, *appl*. secondary immune responses.

anamniotes *n.plu*. fishes, amphibians and the Agnatha (lampreys and hagfishes), characterized by the absence of an amnion around the embryo.

anamorpha *n.plu*. larvae hatched with an incomplete number of segments.

anamorphosis *n*. evolution from one type to another through a series of gradual changes; excessive or abnormal formation of plant origin.

anandrous *a*. without anthers.

anangian *a*. without a vascular system.

ananthous *a*. not flowering; without an inflorescence.

anaphase *n*. stage of mitosis or meiosis which follows metaphase. In mitosis duplicated chromosomes split lengthways, each chromatid moving to opposite poles of the mitotic spindle. In meiosis, homologous chromosomes move to opposite poles of the spindle in anaphase of 1st meiotic division, and sister chromatids separate and move to opposite poles in anaphase of 2nd meiotic division.

anaphylactic shock severe and sometimes fatal shock symptoms produced by exposure to an antigen to which an individual is hypersensitive (allergic).

anaphylaxis *n.* rapid hypersensitive response arising on second exposure to a foreign antigen after a first or sensitizing dose.

anaphysis *n.* an outgrowth; sterigma-like filament in apothecium of certain lichens.

anaplasia *n.* reversion to a less differentiated structure.

anaplast(id) leucoplast *q.v.*

anaplerotic reaction a replenishing reaction in intermediary metabolism, such as the carboxylation of pyruvate to oxaloacetate to replenish oxaloacetate in the tricarboxylic acid cycle after its withdrawal for amino acid synthesis.

anapleurite *n.* upper thoracic pleurite, as in some bristletails.

anapophysis *n.* small dorsal projection rising near transverse process in lumbar vertebrae.

anapsid *a.* with skull completely roofed over, with no temporal fenestrae, the only gaps in the dorsal surface being the nares, eye orbits and the parietal foramen.

anapsids *n.plu.* tortoises and turtles and extinct members of the reptilian subclass Anapsida, characterized by a sprawling gait and a skull with no temporal opening.

anarthous *a.* having no distinct joints.

anaschistic *a. appl.* tetrads which divide twice longitudinally in meiosis.

anastomosis *n.* formation of network or meshwork, as union of fine ramifications of leaf veins, union of blood vessels arising from a single trunk, union of nerves.

anastral *a.* appl. type of mitosis without aster formation.

anatomical *a. pert.* the structure of a plant or animal.

anatomy *n.* study of the structure of plants and animals as determined by dissection.

anatoxin toxoid *q.v.*

anatriaene *n.* a trident-shaped spicule with backwardly directed branches.

anatropous *a.* inverted, *appl.* ovule bent over so that hilum and micropyle are close together and chalaza is at other end.

anautogenous *a. appl.* adult female insect that must feed if her eggs are to mature.

anaxial *a.* having no distinct axis; asymmetrical.

anchorage-dependent growth the requirement of many mammalian cells, such as fibroblasts, for a suitable surface, such as glass or plastic, on which to grow and divide in culture.

anchorage-independent growth the property shown by some transformed mammalian cells in culture, which can divide in semi-solid agar, no longer requiring a surface such as glass or plastic on which to grow.

ancient forest, ancient woodland native forest or woodland, which may be virgin forest or old secondary forest developed by secondary succession, that has been continuously present on a site, often with management but without extensive clear-felling, for hundreds of years, and which can be recognized by its characteristic flora. *cf.* old-growth forest, secondary forest.

ancipital *a.* flattened and having 2 edges.

anconeal *a. pert.* the elbow.

anconeus *n.* small extensor muscle situated over elbow.

Andreaeidae granite mosses *q.v.*

andric *a.* male.

androconia *n.plu.* modified wing scales producing a sexually attractive scent in certain male butterflies.

androdioecious *a.* having male and hermaphrodite flowers on different plants.

androecium *n.* male reproductive organs of a plant; stamens collectively.

androgamone *n.* any substance produced by a male gamete which acts on a female gamete.

androgen *n.* any of various male steroid sex hormones concerned with development of male reproductive system and production and maintenance of secondary sexual characteristics, and secreted chiefly by testis, e.g. androsterone and testosterone.

androgenesis *n.* development in which the embryo contains paternal chromosomes only, due to the failure of the nucleus of the female gamete to participate in fertilization; development from a male gamete, i.e. male parthenogenesis. *a.* androgenetic.

androgenic *a.* stimulating male characters, masculinizing, *appl.* hormones; *appl.* tissue capable of making androgenic hormones.

androgenous *a.* producing only male offspring.

androgonidia *n.plu.* male sexual individuals produced after repeated divisions of reproductive individuals of the colonial protistan, *Volvox*.

androgonium *n.* cell in antheridium (male reproductive organ of cryptogams) which gives rise to antherozoid mother cell.

androgynary *a.* having flowers with stamens and pistils developed into petals.

androgyne *a.* hermaphrodite *q.v.*

androgynism hermaphroditism *q.v.*

androgynous *a.* hermaphrodite; bearing both staminate and pistillate flowers in the same inflorescence; with antheridium and oogonium on the same hypha.

andromerogony *n.* development of an egg fragment with only paternal chromosomes.

andromonoecious *a.* having male and hermaphrodite flowers on the same plant.

andromorphic *a.* having a morphological resemblance to males.

andropetalous *a.* having petaloid stamens.

androphore *n.* stalk, hypha etc. carrying male reproductive organs (e.g. antheridia or androecium).

androsporangium *n.* sporangium containing androspores.

androspore *n.* an asexual zoospore which gives rise to a male dwarf plant.

androstenedione *n.* steroid sex hormone, an androgen, synthesized in gonads and adrenals.

androsterone *n.* a male steroid sex hormone, produced chiefly by testis, and which is less active than testosterone.

androtype *n.* type specimen of the male of a species.

anellus *n.* a small ring-shaped or triangular plate supported by valves and vinculum, in Lepidoptera.

anelytrous *a.* without elytra.

anemo- prefix derived from Gk. *anemos*, wind.

anemochorous *a.* dispersed by wind; having seeds so dispersed. *n.* anemochory, wind dispersal.

anemophily *n.* wind pollination or any other type of fertilization brought about by wind. *a.* anemophilous.

anemoplankton *n.* wind-borne microorganisms, spores, pollen etc.

anemosporic *a.* having spores or seeds dispersed by air currents.

anemotaxis *n.* movement in response to air currents.

anemotropism *n.* orientation of body, or plant curvature, in response to air currents.

anencephaly *n.* condition of having no brain. *a.* anencephalous.

anenterous, anenteric *a.* lacking a gut.

aner *n.* insect male, especially of ants.

anergy *n.* state of non-responsiveness of a lymphocyte to antigen.

aneuploid *a.* having more or fewer than an exact multiple of the haploid number of chromosomes or haploid gene dosage; *appl.* chromosomal abnormalities that disrupt relative gene dosage, such as deletions, *alt.* unbalanced. *n.* aneuploidy.

aneuronic *a.* without innervation, *appl.* chromatophores controlled by hormones.

aneusomic *a. appl.* organisms whose cells have varying numbers of chromosomes.

angienchyma *n.* vascular tissue.

angioblast *n.* cell from which endothelial lining of blood vessels is derived.

angiocarpic, angiocarpous *a.* having fruit enclosed in a covering; having spores enclosed in some kind of receptacle. *n.* angiocarpy.

angiogenesis *n.* the formation of new capillaries by sprouting from pre-existing small blood vessels.

angiogenic *a.* stimulating the formation of new blood vessels.

angiology *n.* anatomy of blood and lymph systems.

Angiospermae, angiosperms Anthophyta *q.v.*

angiosporous *a.* having spores contained in a theca or capsule.

angiostomatous *a.* narrow-mouthed, *appl.* molluscs and snakes with non-distensible mouth.

angiotensin *n.* either of the 2 short polypeptides angiotensin I (inactive) or II (active hormone) released in the blood by action of renin on angiotensinogen, angiotensin II being formed from angiotensin I and found only in people with high blood pressure and acting on blood vessels, causing constriction and a rise in blood pressure. It also causes contraction of the uterus and stimulates aldosterone secretion from the adrenal cortex.

angiotensinogen *n*. protein formed in liver and released into the blood where it may be split by the enzyme renin to produce angiotensin I, the inactive precursor of the vasopressor angiotensin II.

Ångström (Å) *n*. unit of ultramicroscopic measurement, 10^{-10} m, 0.1 nm.

angular *a*. having or *pert.* an angle; *appl.* leaf originating at forking of stem, as in many ferns; *n*. membrane bone in lower jaw of most vertebrates.

angulosplenial *n*. bone forming most of lower and inner part of mandible in amphibians.

angustifoliate *a*. with narrow leaves.

angustirostral *a*. with narrow beak or snout.

angustiseptate *a*. having a silicula laterally compressed with a narrow septum.

anholocyclic *a. pert.* alternation of generations with suppression of sexual part of cycle; permanently parthenogenetic.

animal pole in yolky eggs, that part free of yolk and which cleaves more rapidly than the opposite, vegetal, pole; the end of a blastula at which the smaller cleavage products (micromeres) collect.

animal traps name given to (1) a group of zygomycete fungi (Zoopagales *q.v.*) that capture and parasitize small soil animals, and (2) to deuteromycete fungi (e.g. *Arthrobotrys*) that form hyphal loops and branches to which soil nematodes adhere and become entangled.

animal viruses viruses that infect animals.

Animalia *n*. the animal kingdom. In most modern classifications comprising multicellular eukaryotic organisms with wall-less, non-photosynthetic cells. Heterotrophic unicellular Protista – e.g. protozoans such as *Amoeba* – are also often included.

anion *n*. negatively charged ion (e.g. Cl^-) which moves towards the anode, the positive electrode. *cf.* cation.

anion channel protein spanning cell membrane and allowing the passive transport of anions across the membrane.

aniso- prefix from Gk. *anisos*, unequal.

anisocarpous *a*. having number of carpels less than that of other floral whorls.

anisocercal *a*. with lobes of tail-fin unequal.

anisocytic *a. appl.* stomata of a type in which three subsidiary cells, one distinctly smaller than the other two, surround the stoma. Formerly called cruciferous.

anisodactylous *a*. having unequal toes, 3 toes forward, 1 backward.

anisodont heterodont *q.v.*

anisogamete *n*. one of two conjugating gametes differing in form or size. *alt.* heterogamete.

anisogamy *n*. the union of morphologically unlike gametes. *a*. anisogamous. *alt.* heterogamy. *cf.* isogamy.

anisognathous *a*. with jaws of unequal width; having teeth in upper and lower jaws unlike.

anisomerous *a*. having unequal numbers of parts in floral whorls. *n*. anisomery.

anisomorphic *a*. differing in size, shape or structure.

anisophylly heterophylly *q.v.*

anisopleural *a*. asymmetrical bilaterally.

anisoploid *a*. with an odd number of chromosome sets in somatic cells.

anisopterans *n.plu.* dragonflies, members of the suborder Anisoptera of the order Odonata.

anisopterous *a*. unequally winged, of seeds.

anisospore *n*. anisogamete *q.v.*

anisostemonous *a*. having number of stamens unequal to number of parts in other floral whorls; having stamens of unequal size.

anisotropic *a*. doubly refracting, *appl.* dark bands of striated muscle fibres; *appl.* eggs with predetermined axis or axes.

ankistroid *a*. like a barb; barbed.

ankyloblastic *a*. with a curved germ band.

ankylosis *n*. union of 2 or more bones or hard parts to form one, e.g. bone to bone, tooth to bone. *a*. ankylosing.

ankyrin *n*. protein on cytoplasmic face of membrane in red blood cells that interacts with spectrin and band III protein.

ankyroid *a*. hook-shaped.

anlage *n*. the first structure or cell group indicating development of a part or organ. *alt.* primordium.

annealing *n*. reconstitution of a double-stranded nucleic acid from single strands.

annectant *a*. linking, *appl*. intermediate species or genera.

Annelida, annelids *n., n.plu.* phylum of segmented coelomate worms, commonly called ringed worms, having a soft elongated body with a muscular body wall, divided into many similar segments, usually separated by septa, and the body covered with a thin, flexible collagenous cuticle. They possess a blood system, nephridia and a central nervous system. The Annelida contains three main classes, Polychaeta (ragworms, lugworms), Oligochaeta (e.g. earthworms) and Hirudinea (leeches).

annexin *n*. any of a structurally related group of proteins with calcium-binding and phospholipid-binding properties and a wide range of biochemical functions including inhibition of blood coagulation.

annidation *n*. situation in which a mutant organism survives in a population because an ecological niche exists which the normal individual cannot use.

Annonales Magnoliales *q.v.*

annotinous *a. appl.* growth during the previous year.

annual *a. appl.* structures or growth features that are marked off or completed yearly; living for a year only; completing life cycle in a year from germination; *n*. plant that completes its life cycle in a year.

annual ring growth ring *q.v.*

annular *a*. ring-like; *appl*. certain ligaments in wrist and ankle; *appl*. orbicular ligament encircling head of radius and attached to radial notch of ulna; (*bot.*) *appl*. certain vessels in xylem having ring-like thickenings in their interior; *appl*. bands formed on inner surface of cell wall.

Annulata, annulates *n., n.plu.* a group of invertebrates including the annelid worms, arthropods and some related forms, having bilateral symetry and true metameric segmentation.

annulate *a*. ring-shaped; composed of ring-like segments; with ring-like constrictions; having colour arranged in ring-like bands or annuli.

annulate lamellae stacks of membranes containing structures like nuclear pore complexes, seen in cytoplasm of some eukaryotic cells.

annulus *n*. any ring-like structure, as ring or segment of annelid; remains of veil forming ring around stalk in mushrooms and toadstools; growth ring of fish scale; 4th digit of hand. *plu.* annuli.

anococcygeal *a. pert.* region between coccus and anus, *appl*. body of fibrous or muscular tissue.

anoestrus *n*. the non-breeding period; period of absence of sexual receptiveness in females.

anomaly *n*. any departure from type characteristics.

anomer *n*. either of two isomers of a monosaccharide differing only in the arrangement of atoms around the carbonyl carbon atom, as α-D-glucopyranose and β-D-glucopyranose.

anomocytic *a. appl.* stomata of a type in which no subsidiary cells are associated with the guard cells. Formerly called ranunculaceous.

anomophyllous *a*. with irregularly placed leaves.

anopheline *a. appl.* mosquitoes of the genus *Anopheles*, vectors of malaria and some other diseases.

Anoplura *n*. order of insects, the sucking or body lice, which are ectoparasites of mammals. *alt.* Siphunculata.

anorthogenesis *n*. evolution showing changes in direction of adaptations.

anorthospiral paranemic *q.v.*

anosmic *a*. having no sense of smell. *n*. anosmia.

anoxia *n*. lack of oxygen.

anoxic *a*. devoid of molecular oxygen, *appl*. habitats.

Anoxyphotobacteria *n*. class of bacteria including photosynthetic bacteria other than the cyanobacteria, i.e. those that do not produce oxygen as a byproduct of photosynthesis, and including the green and purple photosynthetic sulphur bacteria.

ansa *n*. loop, as of certain nerves.

Anseriformes *n*. a large order of birds, the waterfowl, including ducks, geese and swans. *a*. anseriform.

anserine *n*. a dipeptide, methylcarnosine, present in muscle of birds, reptiles and fishes, and in mammalian urine; *a. pert.* a goose.

ansiform *a*. loop-shaped, or looped.

antagonism *n*. the effect of a hormone, etc.

that counteracts the effects of another; the inhibitory action of one species on another, such as the action of certain substances secreted by plant roots which inhibit other plants nearby, *alt.* amensalism. *a.* antagonistic.

antagonist *n.* a muscle working against the action of another; any substance that counteracts the effects of a hormone, neurotransmitter, drug, etc.

Antarctic kingdom phytogeographical area comprising the Antarctic, New Zealand and the southern tip of South America.

ante- prefix from the L. *ante*, before, in front of.

anteclypeus *n.* the anterior portion of clypeus when it is differentiated by a suture.

antecosta *n.* internal ridge of tergum for attachment of intersegmental muscles in insects, extended to phragma in segments that bear wings.

antecubital *a.* in front of the elbow, *appl.* fossa.

antedorsal *a.* situated in front of the dorsal fin in fishes.

antefrons *n.* the portion of frons anterior to base of antenna in certain insects.

antefurca *n.* forked process of anterior thoracic segment in some insects.

antelabrum *n.* the anterior portion of insect labrum when differentiated.

antemarginal *a. appl.* sori of ferns when they lie within margin of frond.

antenatal *a.* before birth, *appl.* tests for genetic defects performed on the foetus in the womb.

antenna *n.* one of a pair of jointed feelers on head of various arthropods; feeler of rotifers; in some fish, a modified flap on dorsal fin which attracts prey; group of chlorophyll and other pigment molecules involved in light capture in photosynthesis. *plu.* antennae. *a.* antennal, antennary.

Antennapedia complex (ANT) cluster of homeotic genes in the fruitfly *Drosophila melanogaster*, which control the specification of particular segments in the embryo.

antennary *a. pert.* antenna, *appl.* nerve, artery etc.

antennation *n.* touching with the antenna, serving as tactile communication signal or an exploratory probing.

antennifer *n.* socket of antenna.

antennule *n.* small antenna or feeler, especially the 1st pair of antennae in Crustacea.

anterior *a.* nearer head end; ventral in human anatomy; facing outwards from axis; previous. *cf.* posterior.

anterior commissure tract of fibres connecting the two cerebral hemispheres, anterior and ventral to the anterior end of corpus callosum.

anterior horn of the spinal cord, that part of grey matter containing cell bodies of motor neurones. *cf.* posterior horn.

anterior lobe of pituitary gland adenohypophysis (*q.v.*) excluding the pars intermedia.

anterograde *a. appl.* transport of movement of material in axons of neurones away from cell body.

antero-posterior *a. appl.* axis, from head to tail of the animal body.

antesternite *n.* anterior sternal sclerite of insects. *alt.* basisternum.

anthela *n.* the cymose inflorescence of the rush family, Juncaceae.

anther *n.* terminal part of stamen, which produces the pollen. *see* Fig. 4.

antheridiophore *n.* structure bearing antheridia.

antheridium *n.* organ or receptacle in which male gametes are produced in many cryptogams (ferns, mosses, etc.) and fungi. *plu.* antheridia.

antherophore *n.* stalk of a stamen bearing many anthers, in male cone of some gymnosperms.

antherozoid *n.* motile male gamete produced from antherozoid mother cells in antheridia of algae, mosses and ferns and other lower plants. *alt.* sperm, spermatozoid.

anthesis *n.* stage or period at which flower bud opens; flowering.

anthoblast *n.* young sessile polyp of a stony coral.

anthocarp *n.* a collective, composite or aggregated fruit formed from an entire inflorescence, as pineapple, fig.

Anthoceratopsida, Anthocerotae, Anthocerotales Anthocerophyta *q.v.*

Anthocerophyta *n.* a group of small sporebearing non-vascular green plants with a thalloid gametophyte and rosette-like habit of growth, commonly called hornworts, and with the liverworts and mosses, known as bryophytes (*q.v.*). The cells typically contain a single large chloroplast with a pyrenoid. Hornworts often carry symbiotic photosynthetic cyanobacteria in the intercellular spaces. The sporophyte is typically an upright elongated sporangium on a stalk (the foot) growing from the gametophyte. *alt.* Anthoceratopsida, Anthocerotae, Anthocerotales. *see* also Appendix 2.

anthoclore *n.* yellow pigment dissolved in cell sap of corolla, as of primrose.

anthocodia *n.* distal portion of polyp of soft corals (Alcyonaria) bearing mouth and tentacles.

anthocyan *n.* any of various water-soluble red, blue or purple flavonoid pigments found in the cell vacuole of plants and comprising the anthocyanidins and the anthocyanins.

anthocyanidin *n.* any of a group of red, purple or blue flavonoid pigments whose glycosides are anthocyanins.

anthocyanin *n.* any of a group of important plant pigments found in flowers, fruits, leaves and stems, which are sap-soluble flavonoid glycosides giving scarlet, purple and blue colours, also found in some insects, absorbed with plant food.

anthodium *n.* head of florets, as in Compositae.

anthogenesis *n.* in some aphids, production of both males and females by asexual forms.

anthophilous *a.* attracted by flowers; feeding on flowers.

anthophore *n.* elongation of receptacle between calyx and corolla.

Anthophyta *n.* the flowering plants, one of the five main divisions of extant seed-bearing plants. Reproductive organs (stamens and ovary) are carried in flowers, in which the sporophylls (stamens and carpels) are typically surrounded by sterile leaves (petals and sepals). After pollination and fertilization the closed ovary containing the seeds develops into a fruit. The gametophyte generation is much reduced, being restricted to the male and female gametes. *alt.* angiosperms, Magnoliophyta.

anthostrobilus *n.* strobilus (cone) of certain cycads.

anthotaxis *n.* arrangement of flowers on an axis.

anthoxanthins *n.plu.* sap-soluble flavone flower pigments giving colours from ivory to deep yellow, also found in insects, having been absorbed from the plant on which the insect feeds.

Anthozoa, anthozoans *n., n.plu.* class of coelenterates of the phylum Cnidaria, comprising the soft corals, sea pens and sea fans (subclass Alcyonaria), and the sea anemones and stony corals (subclass Zoantharia). The soft corals, sea fans and stony corals are generally colonial, with individual polyps connected by living tissue and, in the case of the stony corals, embedded in a calcium carbonate matrix. Sea anemones are generally solitary.

anthracobiontic *a.* growing on burned soil or scorched material.

Anthracosauria, anthracosaurs *n., n.plu.* order of Carboniferous to Permian labyrinthodont amphibians, among whose members were the ancestors of the reptiles.

anthraquinone *n.* any of a class of orange or red pigments found in lichens, fungi, higher plants and insects such as the cochineal beetle and lac insect.

anthropic zone that area of the Earth's surface that is under the influence of humans.

anthropocentrism *n.* an exclusively human-centred view that human activities are paramount and need take no account of non-human species.

anthropochory *n.* accidental dispersal by man (via spores, pollen etc.).

anthropogenesis *n.* the evolutionary descent of humans. *a.* anthropogenetic.

anthropogenic *a.* produced or caused by man.

anthropoid *a.* resembling or related to humans, as the anthropoid apes (family Pongidae): orang utan, chimpanzee and gorilla. Gibbons (family Hylobatidae) are sometimes also included.

Anthropoidea *n.* the suborder of primates consisting of monkeys, apes and man.

anthropology *n.* the scientific study of human beings and human societies, especially differences in social organization, racial differences, physiological differences and social and religious development.

anthropometry *n.* study of proportional

measurements of parts of the human body.

anthropomorphism *n*. ascribing human emotions to animals.

anthropomorphous *a*. resembling man.

anti- prefix derived from Gk. *anti*, against or L. *ante*, before.

antiae *n.plu.* feathers at base of bill in some birds.

antiauxin *n*. any compound that regulates or inhibits growth stimulation by auxins.

antibiosis *n*. antagonistic association of organisms in which one produces compounds, antibiotics, harmful to the other(s).

antibiotic *n*. any of a diverse group of organic compounds produced by microorganisms which selectively inhibit the growth of or kill other microorganisms, many antibiotics being used therapeutically against bacterial and fungal infections in man and animals; *a*. killing or inhibiting growth.

antibiotic resistance *see* drug-resistance factors, R plasmid, transposon.

antibiotin avidin *q.v.*

antiblastic *a. appl.* immunity due to factors that inhibit growth of invading organisms.

antibody (Ab) *n*. protein of the immunoglobulin class which is produced by plasma cells (derived from B lymphocytes) on exposure to an antigen and which specifically recognizes that antigen, binding selectively to it and thus aiding its elimination by other components of the immune system. The body can make an almost unlimited variety of different antibodies, each B lymphocyte being genetically programmed early in its development to produce antibody of a single specificity. *see also* immunoglobulin, IgA, IgD, IgE, IgG, IgM.

antibody combining site the site on an antibody that binds and is in contact with the antigen.

antibody-dependent cell-mediated cytotoxicity (ADCC) cell killing by various types of white blood cell, especially natural killer cells and eosinophils, which requires the target cell to be coated with specific antibody.

antibody diversity the production of an almost unlimited repertoire of antibodies of different specificities by an individual immune system. Different antibodies are pro-

duced by different B cells. Each B cell is programmed, by gene rearrangment at the immunoglobulin loci early in its development, to produce an antibody of a single specificity. *see* immunoglobulin genes.

antibody engineering the production by genetic engineering of antibodies with new properties (*e.g.* abzymes), or hybrid antibodies in which sequences from two different species (*e.g.* mouse and human) have been combined. One product of antibody engineering is human antibodies bearing specific antigen-binding sites derived from mouse antibodies. In this way, antibodies against a particular antigen, which can be obtained more easily in mice, are rendered non-immunogenic to humans and thus less likely to be inactivated than the unmodified animal antibody.

antibody genes *see* immunoglobulin genes.

antiboreal *a. pert.* cool or temperate regions of the Southern Hemisphere.

antibrachial *a. pert.* the forearm or corresponding portion of a forelimb.

antical *a. appl.* the upper or front surface of a thallus, leaf, or stem, esp. in liverworts.

anticlinal *a. appl.* plane of cell division at right angles to surface of apex of a growing point; in quadrupeds, *appl.* one of the lower thoracic vertebrae with upright spine towards which those on either side incline.

anticoagulant *n*. any substance which prevents coagulation or clotting of blood, such as dicoumarol, warfarin, heparin.

anticoding strand sense strand *q.v.*

anticodon *n*. group of 3 consecutive bases in tRNA complementary to a codon on mRNA.

anticryptic *a. appl.* protective coloration facilitating attack.

antidiuretic *a*. reducing the volume of urine; *appl.* hormone (vasopressin) that controls water reabsorption by kidney tubules.

antidiuretic hormone vasopressin *q.v.*

antidromic *a*. contrary to normal direction; (*bot.*) *appl.* stipules with fused outer margins.

antifertilizin *n*. a protein in cytoplasm of spermatozoa which reacts with fertilizin produced by ovum.

antifreeze compounds compounds such as glycerol, sorbitol and mannitol, which lower the freezing point of body fluids and protect against freezing, found in the haemolymph of some insects; glycoproteins found in the blood of some polar fish, and which depress the freezing point of the blood by enveloping small ice crystals that would otherwise form ice nuclei and cause the blood to freeze.

antigen (Ag) *n.* any substance capable of specific binding to an antibody or a T-cell receptor. An antigen may be unable to induce a specific immune response when administered on its own, but will do so if attached to a suitable carrier. *a.* antigenic. *cf.* immunogen.

antigen–antibody complex complex of antibody with its specific antigen noncovalently bound to the antigen-binding site, which forms when antigen and antibody come together, and is the form in which foriegn antigens are most effectively scavenged by phagocytic cells and thus removed from the body. Some antigen-antibody complexes may form a precipitate, which is the basis for the precipitin reaction formerly used to detect and quantify antibodies in serum, and for immunodiffusion assays.

antigen-binding site the part of an immunoglobulin molecule that specifically binds an antigen, composed of the variable regions of a light and a heavy chain, each antigen-binding site recognizing a single antigenic determinant. Each antibody molecule has two identical antigen-binding sites.

antigen presentation the process by which a foreign antigen, esp. *appl.* a protein antigen, is recognized by the immune system. Antigen-presenting cells (e.g. macrophages, dendritic cells) take up incoming protein antigens, break them down into peptide fragments intracellularly, form complexes between the peptides and MHC molecules, and display the complexes on the cell surface. In this form the antigen is recognized and responded to by T lymphocytes specific for the peptide–MHC combination, and various types of immune response are initiated. *see also* helper T lymphocyte, T-cell receptor, major histocompatibility antigens.

antigen-presenting cell *see* antigen presentation.

antigen processing the uptake and partial breakdown of foreign protein antigens by phagocytic cells of the immune system. The resulting peptide fragments are then displayed on the cell surface in combination with MHC molecules.

antigen receptor cell-surface protein on B cells (where it is an immunoglobulin M) and T cells, which binds a specific antigen. Each B or T cell carries antigen receptors of a single specificity.

antigenic determinant particular site on antigen molecule eliciting the formation of a specific antibody or activating a specific T cell, and against which the antibody or T cell activity is directed. *alt.* epitope.

antigenic drift a gradual change in the antigens carried by some viruses, esp. influenza viruses, as a result of small genetic changes.

antigenic shift a substantial change in the antigens carried by some viruses, esp. influenza viruses, caused by the recombination between two virus strains, which manifests itself as the sudden appearance of a new virus type.

antigenic variation the ability of African trypanosomes (e.g. *Trypanosoma brucei*) and some other microorganisms to change the cell-surface antigens that they synthesize in succeeding generations. In trypanosomes, this is due to DNA rearrangements which bring a different gene into a position where it is expressed.

antigenicity *n.* the property possessed by a substance that can bind specifically to the antigen receptors on B or T lymphocytes, and thus, in principle, is able to stimulate a specific immune response. *see* antigen. *cf.* immunogenicity.

antigiberellin *n.* any compound (e.g. phosphon, maleic hydrazide) with action on plant growth opposite to that of giberellins, causing plants to grow with short thick stems.

antihaemophilia factor factor VIII *q.v.*

antihaemorrhagic *a.* appl. agents that stop bleeding, *appl.* vitamin: vitamin K *q.v.*

antihelix *n.* the curved prominence in front of helix of ear.

antihelminthic *n.* drug effective against parasitic flatworms or roundworms.

antihormone *n.* any substance that prevents the action of a hormone.

anti-immunoglobulin an antibody against an immunoglobulin.

anti-idiotypic antibody antibody specific for an antigenic determinant located in the variable region of another antibody.

anti-idiotypic networks *see* network theory.

antilysin *n.* any substance that counteracts a lysin or the process of lysis.

antimeres *n.plu.* corresponding parts, as left and right limbs, of a bilaterally symmetrical animal; a series of equal radial parts, or actinomeres, of a radially symmetrical animal.

antimetabolite *n.* any substance that blocks a metabolic reaction, e.g., by competing with the natural substrate for enzyme active sites.

antimitotic *a.* inhibiting or preventing mitosis.

antimorph *n.* mutant allele that has an opposite effect to the normal allele, competing with the normal allele when in the heterozygous state.

antimutagen *n.* any substance or other agent that slows down the mutation rate or reverses the action of a mutagen.

antimycin *n.* compound used experimentally as an inhibitor of cellular respiration.

antineuritic *a. appl.* vitamin: thiamine *q.v.*, lack of which causes polyneuritis.

antinociceptive *a. appl.* any agent that can lessen or prevent the generation or transmission of a painful or injurious stimulus.

anti-oncogene *n.* a gene that counteracts the activity of an oncogene.

antiparallel *a.* describes two similar structures arranged in opposite orientations (e.g. the two strands of the DNA double helix).

antipepsin *n.* stomach secretion that prevents the action of pepsin.

antiperistalsis *n.* peristalsis in the posterior-anterior direction.

antipetalous *a.* inserted opposite the insertion of the petals.

antipodal *a.* in plant embryo, *appl.* group of 3 cells at chalazal end of embryo sac.

antiport *n.* membrane protein that transports a solute across the membrane, transport depending on the simultaneous or sequential transport of another solute in the opposite direction.

antiprostate n. bulbo-urethral gland *q.v.*

antipyretic *a. appl.* drugs that lower body temperature.

antirachitic *a.* preventing rickets, *appl.* vitamin: vitamin D.

anti-reductionism *see* reductionism.

Antirrhinum majus snapdragon, an ornamental dicot plant widely used in experimental plant genetics.

antiscorbutic *a.* preventing or counteracting scurvy, *appl.* vitamin: vitamin C (ascorbic acid).

antisense RNA RNA complementary to the normal RNA transcript of a gene, and which can block its expression by hybridizing to the RNA transcript and preventing its translation.

antisense strand the DNA strand in a duplex complementary to the sense strand. *alt.* coding strand.

antisepalous *a.* inserted opposite the insertion of the sepals.

antiseptic *n.* substance which destroys harmful microorganisms; *a.* preventing putrefaction.

antiserum *n.* blood serum containing specific antibodies, obtained after immunization or natural infection.

antisocial factor any selection pressure that tends to reverse social evolution.

antispadix *n.* a group of four modified tentacles in internal lateral lobes of *Nautilus*.

antitermination *n.* the continuation of transcription by RNA polymerase past the usual termination point in a gene, caused by the interaction of specific protein factors (antitermination factors) with the enzyme.

antithesis, principle of *see* principle of antithesis.

antithrombin *n.* former name for heparin *q.v.*; antithrombin III now used for a plasma protein which specifically inactivates thrombin and other activated blood clotting factors.

antitoxin *n.* substance which neutralizes a toxin by combining with it.

antitragus *n.* prominence opposite tragus of external ear.

antitrochanter *n.* in birds, an articular surface on ilium, against which trochanter of femur moves.

antitrope *n.* any structure that forms a bilaterally symmetrical pair with another.

antitropic *a.* turned or arranged in opposite directions; arranged to form bilaterally symmetrical pairs.

antitropous *a.* inverted; *appl.* plant embryos with radicle directed away from hilum.

α₁-antitrypsin deficiency familial emphysema, an inherited defect leading to emphysema (overinflation and distension of air sacs in the lungs causing shortness of breath, etc.), caused by a genetic defect resulting in the production of inactive antitrypsin.

antitrypsins *n.plu.* protein inhibitors of the enzyme trypsin, produced in various animals and plants, some of which are serpins (e.g. α₁-antitrypsin). The genes for some plant antitrypsins have been transferred to crop species that lack them in order to make them resistant to insect pests, which cannot digest the plant material as a result of the antitrypsins it contains and so starve to death.

antitype *n.* a specimen of the same type as that chosen for designation of a species, and gathered at the same time and place.

antiviral *a. appl.* antibodies, drugs etc. that destroy or neutralize a virus.

antlia *n.* the spiral sucking proboscis of Lepidoptera.

antlers *n.plu.* paired bony growths, projections from the skull, on the heads of members of the deer family, which are often branched, are shed annually and are usually confined to males.

ant lions a group of insects in the order Neuroptera *q.v.*

antral follicle immature fluid-filled ovarian follicle.

antrorse *a.* directed forwards or upwards.

antrum *n.* a cavity or sinus; fluid-filled cavity in developing ovarian follicle. *a.* antral.

ants *n.plu.* social insects of the superfamily Formicoidea of the order Hymenoptera, which live in colonies composed of a queen, with male, worker and in some cases, soldier, castes.

anucleate *a.* without a nucleus.

anucleolate *a.* without a nucleolus.

Anura, anurans *n., n.plu.* one of the three orders of extant amphibians, comprising the frogs and toads. In some classifications called the Salientia.

anural, anurous *a.* tailless.

anus *n.* the opening of the alimentary canal (usually posterior) through which undigested food is voided. *a.* anal.

aorta *n.* in mammals the great trunk artery that carries blood from the heart to the arterial system of the body; in other animals, major blood vessels carrying oxygenated blood. *see* dorsal aorta, ventral aorta.

aortic *a. pert.* aorta, *appl.* hiatus, isthmus, lymph glands, semilunar valves, etc.

aortic arches paired arteries in vertebrate embryos, which connect dorsal and ventral arteries, running between gill slits on either side.

aortic bodies two small masses of chromaffin cells in a capillary plexus, one on each side of foetal abdominal aorta, being part of system for controlling the oxygen content and acidity of blood. *alt.* Zuckerkandl's bodies.

apandrous *a.* without functional male sex organs; without antheridia; parthenogenetic, as oospores in certain oomycete fungi.

apatetic *a. appl.* misleading coloration.

AP endonuclease any of a group of endonucleases that make a single-stranded incision in DNA to the 5¼ side of a nucleotide from which the purine or pyrimidine base has been removed (as in certain types of DNA repair).

aperispermic *a. appl.* seeds without nutritive tissue.

apertura piriformis anterior nasal aperture of skull.

apes *see* Primates.

apetalous *a.* without petals.

apex *n.* tip or summit, as of wing, heart, lung, root, shoot. *plu.* apices.

Aphaniptera Siphonaptera *q.v.*

aphanipterous *a.* apparently without wings.

Aphasmidia *n.* class of nematode worms with no phasmids, and whose amphids open on to the posterior part of the head capsule.

apheliotropism *n.* tendency to turn away from light, strictly from the sun.

aphid *n.* insects of the family Aphididae

(Aphidae) of the Hemiptera with mouthparts adapted for piercing and sucking plants, which are of economic importance as vectors of virus diseases, and which have a parthenogenetic and a sexual reproductive phase.

aphidicolin *n.* fungal antibiotic that inhibits DNA replication in eukaryotes and DNA polymerase α *in vitro*.

aphins *n.plu.* red and yellow fat-soluble pigments extracted from various aphids, probably arising after death from protaphin.

aphlebia *n.* lateral outgrowth from base of frond stalk in certain ferns.

aphodus *n.* short tube leading from internal chamber lined with flagellate cells to the excurrent canal system in sponges. *a.* aphodal.

aphotic *a. pert.* absence of light, *appl.* zone of deep sea where daylight fails to penetrate.

aphototropism *n.* tendency to turn away from light.

Aphragmabacteria *n.* in some classifications the name for the group of prokaryotes comprising the mycoplasmas (*q.v.*) and similar organisms, small prokaryotes lacking the typical bacterial cell wall, and bounded by a triple-layered lipid membrane. They live mostly as obligate intracellular parasites of plants and animals. *see also* Appendix 6.

aphthous *a.* producing blisters.

aphthoviruses *n. plu.* group of picornaviruses including foot-and-mouth disease virus.

aphyllous *a.* without foliage leaves. *n.* aphylly.

aphytic *a.* without plant life, *appl.* zone of coastal waters below approx. 100 m, the bottoms of deep lakes, etc.

Apiales Cornales *q.v.*

apical *a.* at the tip of any cell, structure or organ; *pert.* distal end; *appl.* cell at tip of growing point; (*bot.*) *appl.* style arising from summit of ovary.

apical dominance phenomenon common in plant development in which the bud at the tip of the shoot, the apical bud, suppresses the development of lateral buds which have formed further down the stem. If the apical bud is removed, the lateral buds then develop. The apical bud is thought to produce a growth-inhibitory hormone that is carried back down the stem.

apical membrane in an epithelium lining an internal cavity (e.g. gut, lungs, glands), the face of an epithelial cell which is adjacent to the cavity.

apical meristems dividing tissue at tip of developing shoot and young root, at which growth occurs.

apical placentation in plant ovary, placentation where ovule is at the apex of ovary.

apical ridge ridge of thickened ectoderm on vertebrate limb and whose presence is required for a complete limb to develop.

apices *plu.* of apex.

Apicomplexa *n.* phylum of non-photosynthetic heterotrophic protists parasitic in animals, comprising the sporozoan protozoans, e.g. gregarines, coccidians, *Plasmodium*, and piroplasms. They are transmitted from host to host in the form of "spores", small infective bodies produced by schizogony. *alt.* Sporozoa.

apiculate *a.* forming abruptly to a small tip, as leaf.

apiculus *n.* a small apical termination, as in some protozoans or spores; the reflected portion of the club-end of antenna in some Lepidoptera.

apilary *a.* having upper lip missing or suppressed in corolla.

apileate *a.* having no pileus.

apitoxin *n.* main toxic fraction of bee venom.

apivorous *a.* feeding on bees.

aplacental *a.* having no placenta, as monotremes.

aplanetic *a.* non-motile, *appl.* spores.

aplanetism *n.* absence of motile spores or gametes.

aplanetogametangium *n.* gametangium in which non-motile gametes are formed.

aplanogamete *n.* a non-motile gamete.

aplanosporangium *n.* sporangium in which non-motile spores are formed.

aplanospore *n.* non-motile resting spore.

aplasia *n.* arrested development; non-development; defective development.

aplastic *a. pert.* aplasia; without change in development or structure.

aplerotic *a.* not entirely filling a space.

aploperistomatous *a*. having a peristome with one row of teeth, as mosses.

aplostemonous *a*. with a single row of stamens.

Aplysia the opisthobranch mollusc *Aplysia californica*, the sea hare, used as an experimental animal in neurobiology.

apneustic *a*. with spiracles closed or absent, *appl.* aquatic larvae of certain insects.

apocarp *n*. individual carpel of a composite fruit.

apocarpous *a*. having separate or partially united carpels. *n*. apocarpy.

apocentric *a*. diverging or differing from the original type.

apocratic *a*. opportunistic, *appl.* species.

apocrine *a. appl.* glands whose secretion accumulates beneath the surface and is released by breaking away of the outer part of the cells, e.g. mammary glands.

apocytium *n*. multinucleate mass of naked cytoplasm.

Apoda *n*. an order of limbless burrowing amphibians, commonly known as caecilians, having a reduced or absent larval stage and minute calcified scales in the skin. In some classifications called the Gymnophiona; the name has also been given to orders of parasitic barnacles (crustaceans) and burrowing sea cucumbers (echinoderms).

apodal *a*. having no feet; having no ventral fin, of fishes; stemless.

apodeme *n*. an internal skeletal projection in arthropods.

apoderma *n*. enveloping membrane secreted during resting stage between instars by certain ticks and mites.

Apodiformes *n*. an order of birds including the swifts.

apoenzyme *n*. inactive protein part of an enzyme remaining after removal of the prosthetic group. *cf.* holoenzyme.

apogamy *n*. a type of apomixis in which the embryo is produced from the unfertilized female gamete or from an associated cell. *see also* generative apogamy.

apogeotropism ageotropism *q.v.*

apogynous *a*. lacking functional female reproductive organs.

apoinducer *n*. regulatory protein that activates a gene by binding to control regions in the DNA and allowing transcription to take place.

apolegamic *a. appl.* mating associated with sexual selection.

apolipoprotein *n*. the protein component of a lipoprotein, esp. of the lipoproteins that transport lipids in the blood.

apomeiosis *n*. sporogenesis without meiosis. *a*. apomeiotic.

apomict *n*. an organism reproducing by apomixis.

apomixis *n*. reproductive process without fertilization in plants, akin to parthenogenesis but including development from cells other than ovules, as apogamy and apospory *q.v.*; vegetative apomixis *q.v. a*. apomictic. *cf.* amphimixis.

apomorphous *a*. in cladistic phylogenetics, *appl.* novel character evolved from a pre-existing character. The two form a homologous pair of characters, termed an evolutionary transformation series. *see also* synapomorphy.

aponeurosis *n*. the flattened tendon for insertion of, or membrane investing, certain muscles.

apopetalous polypetalous *q.v.*

apophyllous *a*. having free perianth leaves.

apophysis *n*. (*zool.*) a projecting process on bone or other skeletal material, usually for muscle attachment; (*bot. & mycol.*) various small protuberances, as on hyphae, capsule of mosses, etc.; small protuberance at base of seed-bearing scales in pine cones.

apoplasmodial *a*. not forming a typical plasmodium.

apoplast *n*. the cell walls collectively of a tissue or a complete plant. *a*. apoplastic.

apoplastic pathway in plant tissue the movement of ions and other solutes across stems or roots via cell walls.

apoplastid *n*. a plastid lacking chromatophores.

apoprotein *n*. protein lacking its prosthetic group. *see also* apoenzyme.

apoptosis *n*. cell death as a result of induction of an internal "suicide" programme, and which is a normal and essential event in many developmental stages. *cf.* necrosis.

apopyle *n*. exhalent pore of sponges.

aporogamy *n*. entry of pollen tube into ovule by some method other than through the micropyle.

aporrhysa *n.plu.* exhalent canals in sponges.

aposematic *a. appl.* warning coloration or markings which signal to a predator that an organism is toxic, dangerous or distasteful. *cf.* epidematic, parasematic.

aposporogony *n.* absence of sporogony.

apospory *n.* a type of apomixis in which a diploid gamete is produced from the sporophyte without spore formation.

apostasis *n.* condition of abnormal growth of axis that causes separation of perianth whorls from one another.

apostatic *a.* differing markedly from the normal.

apostatic selection type of frequency-dependent selection in which a predator selects the most common morph in the population.

apostaxis *n.* abnormal or excessive exudation.

apostrophe *n.* arrangement of chloroplasts along lateral walls of leaf cells in bright light.

apothecium *n.* cup-shaped fruiting body of Discomycetes (cup fungi, morels and truffles) bearing asci on the inner surface; fruiting body of some lichens. *plu.* apothecia.

apothelium *n.* a secondary tissue derived from a primary epithelium.

apotracheal *a.* with xylem parenchyma independent of vessels, or dispersed, *appl.* wood.

apotropous *a. appl.* an anatropous ovule with a ventrally situated line of fusion with funicle.

apotypic *a.* diverging from a type.

apparent free space that part of a tissue lying outside the plasmalemmas of its constituent cells, i.e. intercellular spaces, and in plant tissues, the cell wall as well.

apparent mortality a measure of mortality in a population at a given developmental stage (e.g. age groups) expressed as the percentage of the number alive at the beginning of the stage. *alt.* percentage successive mortality.

appeasement *n. appl.* behaviour which ends the attack of one animal on another of the same species by the loser adopting a submissive posture or gesture.

appendage *n.* organ or part attached to a trunk, as limb, branch etc.

appendical *a. pert.* appendix.

appendices *plu.* of appendix *q.v.*

appendicular *a. pert.* appendages, *appl.* skeleton of limbs; *pert.* appendix.

appendiculate *a.* having a small appendage, as a stamen or filament; having an appendiculum.

appendiculum *n.* remains of the partial veil on rim of cap of some agaric fungi.

appendix *a.* an outgrowth, esp. the vermiform appendix of human intestine.

appendix colli the hanging tuft of hairs on neck of goat, sheep, pig, etc.

appetitive *a. appl.* behaviour at the beginning of a fixed behaviour pattern, which can be very variable, from unoriented wanderings to apparently purposeful behaviour.

applanate *a.* flattened.

application factor a factor used to determine the maximum safe concentration of a substance for an organism, and which is the ratio of the concentration of the substance that produces a certain long-term response in the organism to the concentration causing death in 50% of the population within a given time period.

apposition *n.* laying down of material on a preformed surface, as in growth of cell wall, bone, etc.

appressed *a.* pressed together without being united.

appressorium *n.* adhesive disc, as of sucker or haustorium; modified hyphal tip which may form haustorium or penetrate substrate, as of parasitic fungi.

aproterodont *a.* having no premaxillary teeth.

apteria *n.plu.* naked or down-covered surfaces between feather tracts on bird skin.

apterous *a.* (*zool.*) wingless; (*bot.*) having no wing-like extensions on stems or petioles.

apterygial *a.* wingless; without fins.

Apterygiformes *n.* an order of flightless birds including the kiwis.

apterygote, apterygotous *a. appl.* a group of insects, the subclass Apterygota, that have no wings, little or no metamorphosis, and abdominal appendages in the adult, and that comprises the orders Thysanura, Diplura, Protura and Collembola. *see individual entries.*

apurinic *a. appl.* nucleotide in DNA lacking its purine base.

apyrene *a.* seedless, *appl.* certain cultivated fruits.

apyrimidinic *a.* appl. nucleotide in DNA lacking its pyrimidine base.

aquaculture *n.* raising of algae, fish and shellfish for human use in artificial or natural freshwater ponds, lakes, irrigated fields and irrigation ditches, and, for marine organisms, in enclosures in coastal inlets and estuaries.

aquatic *a.* living in or near water.

aquatic ecosystems any ecosystem of which the principal component is water, such as ponds, lakes, rivers, streams and oceans. *cf.* wetlands.

aqueduct *n.* fluid-filled channel or passage, as that of cochlea, and of vestibule of ear.

aqueduct of Sylvius channel running through midbrain, connecting 3rd and 4th ventricles. *alt.* aqueduct, cerebral aqueduct.

aqueous *a.* watery, *appl.* humor: fluid between lens and cornea.

Arabidopsis thaliana small annual plant of the Cruciferae, widely used as a model organism in plant molecular genetic and developmental research because of its ease of culture and small simple genome.

arabinan *n.* any of a group of polysaccharides composed predominantly of arabinose units, found in some plant cell walls.

arabinogalactan *n.* polysaccharide found in plant cell walls, composed of arabinose and galactose sugar units.

arabinose *n.* a 5-carbon aldose sugar found esp. in plant gums, pectins and cell wall polysaccharides.

arable *a. appl.* land that is cultivated, and to crops grown on cultivated land, except trees. *cf.* pasture, rangeland.

arachidonic acid a fatty acid, a precursor of prostaglandins and prostacyclins.

Arachnida, arachnids *n., n.plu.* class of mainly terrestrial, carnivorous arthropods, included in the subphylum Chelicerata, comprising spiders (order Araneae), scorpions (Scorpiones), mites and ticks (Acari), false scorpions (order Pseudoscorpiones), palpigrades (order Palpigrada), solifugids (order Solifugae), and harvestmen (Opiliones). They have a body usually divided into a prosoma of eight fused segments and a posterior opisthosoma of 13 fused segments. The prosoma is not differentiated into a head and thorax and bears the clawed and prehensile chelicerae, the pedipalps, and four pairs of walking legs.

arachnidium *n.* the spinning apparatus of a spider, including spinning glands and spinnerets.

arachniform *a.* arachnoid, stellate.

arachnoid *a. pert.* or resembling a spider; like a cobweb; consisting of fine entangled hairs.

arachnoid *n.* the middle of the three membranes surrounding the brain and spinal cord.

arachnoidal *a. pert.* the arachnoid membrane.

arachnoidal granulations Pacchionian bodies *q.v.*

aragonite *n.* a crystalline form of calcium carbonite, one of the constituents of mollusc shells.

Arales *n.* an order of herbaceous monocots comprising the families Araceae (arum) and Lemnaceae (duckweed).

Araliales Cornales *q.v.*

Araneida spiders *q.v.*

araneose, araneous *a.* covered with or consisting of fine entangled filaments.

arbacioid *a.* of sea urchins, having one primary pore plate, with a secondary on either side.

arboreal *a.* living in trees; *pert.* trees.

arborescence arborization *q.v.*

arborescent *a.* branched like a tree.

arboretum *n.* a collection of species of trees.

arboriculture *n.* the cultivation of trees.

arborization *n.* tree-like branching, as of dendrites and axon on nerve cells.

arboroid *a.* tree-like.

arbovirus *n.* virus that replicates in an arthropod as its intermediate host and in a vertebrate as its definitive host, e.g. yellow fever, transmitted by mosquitoes, and the tick-borne encephalitis complex. Previously known as arbor viruses, arborviruses.

arbuscular *a.* shrub-like; *appl.* complex branched hyphal systems of vesicular arbuscular mycorrhizae.

arbuscule *n.* small tree-like shrub or dwarf tree.

arbutoid *a. appl.* endomycorrhizas formed on members of the tribe Arbutoideae (family Ericaceae), with a well-defined fungal sheath and Hartig net and extensive penetration of the cells of the root cortex.

arcade *n*. an arched channel or passage; a bony arch.

arch-, arche- prefix derived from Gk. *arch*, beginning.

archae- prefix derived from Gk. *archaios*, primitive.

Archaea *n*. organisms found in the oldest Precambrian rocks, i.e. rocks of the Archaean era.

Archaean *a. appl.* the earlier eon of the Precambrian, ending at around 2500 million years ago.

Archaebacteria *n.plu.* class of prokaryotic microorganisms forming the division Mendosicutes, including extreme thermophiles, extreme halophiles and methanogens (methane producers). Typically found in extreme environments, e.g. hot springs, salt lakes. They differ in many ways from other bacteria, e.g. in the structure of their cell walls and membrane lipids and possession of introns in some genes, and are thought to represent a quite separate group from the true bacteria or eubacteria.

Archaeopteryx and ***Archaeornis*** genera of fossil birds from the Jurassic, which show many reptilian features.

Archaeornithes *n*. a subclass of primitive birds containing the fossils *Archaeopteryx* and *Archaeornis*.

archaeostomatous *a*. having the blastopore persistent and forming the mouth.

Archaeozoic *a. pert.* earliest geological era, the lower division of the Precambrian, the time of Archaean rocks and solely unicellular life.

arch-centra *n.plu.* vertebral centra formed by fusion of basal growths of primary dorsal and ventral cartilaginous outgrowths (arcualia) from centrum external to chordal sheath. *a*. archecentrous, archicentrous, archocentrous.

arche- *see* archi-, arch-.

Archean Archaean *q.v.*

archecentric *a*. conforming more or less with the original type.

archedictyon *n*. intervein network in the wings of some primitive insects.

archegoniophore *n*. branches of bryophytes, or parts of fern prothallus, that bear archegonia.

archegonium *n*. female sex organ in liverworts, mosses, ferns and related plants and in most gymnosperms, consisting of a multicellular, flask-shaped structure containing one ovum (oosphere) in the base (venter). *plu*. archegonia.

archencephalon *n*. primitive forebrain or cerebrum.

archenteron *n*. the cavity formed at gastrulation which develops into gut of embryo.

archeo- archaeo-.

archespore sporoblast *q.v.*

archesporium *n*. a cell or mass of cells dividing to form spore mother cells, in anthers generating the pollen mother cells.

archi- prefix derived from Gk. *archi*, first. *see also* arche-, arch-.

archibenthic *a. pert.* sea bottom from edge of continental shelf to upper limit of abyssal zone, at depths of *ca.* 200–1000 m.

archiblastic *a. appl.* eggs which develop into a blastula by total and equal segmentation.

archicarp *n*. spirally coiled region of thallus or stalk bearing female sex organ in certain fungi; cell which gives rise to a fruiting body.

archichlamydeous *a*. having no petals, or having petals entirely separate from one another.

archinephridium *n*. excretory organ of certain larval invertebrates, usually a solenocyte.

archipallium *n*. the olfactory region of cerebral hemispheres, comprising olfactory bulbs and tubercles, pyriform lobes, hippocampus and fornix.

archipterygium *n*. type of fin in which skeleton consists of elongated segmental central axis and two rows of jointed rays.

archisternum *n*. cartilaginous elements in ligaments joining muscle blocks in ventral region of thorax, as in tailed amphibians.

archistriatum *n*. region of forebrain in birds involved in organization of motor function.

architomy *n*. reproduction by fission with subsequent regeneration, in certain annelids.

architype *n*. an original type from which

others may be derived. *alt.* archetype.

Archosauria, archosaurs *n., n.plu.* subclass of reptiles, the "ruling reptiles", that included the dinosaurs, mainly extinct but including the living crocodilians, with specializations of the skeleton showing trend towards bipedalism.

arciform *n.* shaped like an arch or bow.

arcocentrous *a. appl.* vertebral column with inconspicuous chordal sheath and centra mainly derived from arch tissue.

arcocentrum *n.* vertebral centrum formed from parts of neural and haemal arches.

Arctic Circle latitude 66°30′N, to the north of which there is at least one period of 24 hours in summer in which the Sun does not set, and at least one period of 24 hours in winter in which the Sun does not rise.

Arctic Floral Region the region of the Holarctic Realm that extends from the far north, south to central Alaska, Labrador, central Scandinavia and northern Siberia.

Arctogaea *n.* zoogeographical area comprising Holarctic, Ethiopian and Oriental regions. *alt.* Arctogea.

Arctogea Arctogaea *q.v.*

arcualia *n.* small cartilaginous pieces, dorsal and ventral, fused or free, on vertebral column of fishes.

arcuate *a.* shaped like an arch or bow.

arculus *n.* arc formed by the two wing veins of certain insects.

ardellae *n.* small apothecia of certain lichens, having appearance of dust.

area centralis area corresponding to the fovea in some animal eyes.

Arecales *n.* order of tree-like or climbing monocots with feather- or fan-like leaves, comprising the family Arecaceae (Palmae) (palms, yuccas, etc.).

arena *n.* area used for communal courtship displays.

arenaceous *a.* having properties or appearance of sand.

Arenaviridae, arenaviruses *n., n.plu.* family of RNA viruses with single-stranded genomes in two parts, and which include Lassa fever and lymphocytic choriomeningitis virus.

arenicolous *a.* living or growing in sand.

areola *n.* small coloured circle round nipple; part of iris bordering pupil of eye; a small pit. *a.* areolar.

areolar glands sebaceous glands on areola of nipple.

areolar tissue type of connective tissue consisting of cells (macrophages, fibroblasts and mast cells) embedded in a matrix of glycoproteins and proteoglycans in which are embedded collagen and elastin fibres.

areolate *a.* divided into small areas by cracks or other margins.

areole *n.* areola *q.v.*; (*bot.*) space occupied by a group of hairs or spines, as in cacti; small area of mesophyll in leaf delimited by intersecting veins.

arescent *a.* becoming dry.

argentaffin *a.* staining with silver salts, *appl.* certain cells in gastric glands and crypts of Lieberk uhn which secrete digestive enzymes.

argenteal *a. appl.* layer of eye as in fish, having a silvery appearance.

argenteous *a.* like silver.

argenteum *n.* dermal silvery reflecting tissue layer of iridocytes, without chromatophores, in fishes.

Argentinian Floral Region part of the Austral Realm that includes Argentina, Paraguay, southern Chile and the offshore islands including the Falkands (Malvinas).

argentophil *a.* staining with silver salts.

argillaceous *a.* having clay-sized particles, *appl.* soil; having the properties of clay.

arginase *n.* enzyme hydrolysing the amino acid arginine to urea and ornithine, important in the urea cycle and also found in some plants. EC 3.5.3.1.

arginine (Arg, R) *n.* basic amino acid, positively charged at physiological pH, essential in human diet, constituent of protein, hydrolysed to ornithine and urea in urea cycle.

arginine-urea cycle urea cycle *q.v.*

arginosuccinate compound formed from arginine and succinate, intermediate in the urea cycle of vertebrates.

argyrophil *a.* staining with silver salts.

ariboflavinosis *n.* condition of skin cracking and lesions caused by a deficiency of the vitamin riboflavin.

arid *a. appl.* climate or habitat with less than 250 mm annual rainfall, very high evaporation and sparse vegetation.

arid zone regions extending from latitudes

15 to 30° in both hemispheres in which rainfall is very low and either evaporates in the high daytime temperatures or drains away rapidly so that it is unavailable to vegetation. This zone contains most of the world's deserts. Parts of the zone support vegetation and some cultivation but are subject to overgrazing and overcultivation in times of drought, which can lead to desertification.

aridity index a measure of the aridity of an area which takes into account both rainfall and evaporation.

aril *n*. additional covering formed on some seeds after fertilization, and which may be spongy, fleshy (as the red aril of yew berries), or a tuft of hairs.

arillate *a*. having an aril.

arillode *n*. a false aril arising from region of micropyle as an expansion of the opening in outer wall of ovule.

arista *n*. (*bot*.) awn, long-pointed process as in many grasses; (*zool*.) bristle borne on antenna of some dipteran flies.

aristate *a*. with awns, or with a well-developed bristle.

Aristolochiales *n*. an order of dicots of the subclass Magnoliidae, containing one family, the birthworts (Aristolochiaceae), which comprise herbs and climbing plants.

Aristotle's lantern calcareous structure around mouth of sea-urchin supporting 5 long teeth.

aristulate *a*. having a short awn or bristle.

arithmetic growth linear growth *q.v.*

arm-palisade palisade tissue in which the chloroplast-bearing surface is enlarged by infolding of cell walls beneath the epidermis.

armature *n*. any structure that serves as a defence, e.g. hairs, prickles, thorns, spines, stings.

armilla *n*. bracelet-like fringe; superior ring of certain fungi.

armillate *a*. fringed.

arms race the sequence of evolutionary changes seen in e.g. a predator and its prey, as each advantageous adaptation in one organism is countered by a further adaptation in the other.

arolium *n*. soft hairy pad at extreme tip of insect leg.

aromatic amino acids amino acids with an aromatic side chain: phenylalanine, tryptophan, tyrosine. *see* table in Appendix 1.

aromorphosis *n*. evolution with an increase in degree of organization without much increase in specialization.

arousal *n*. level of responsiveness to a stimulus in an animal.

array *n*. arrangement in order of magnitude.

arrect *a*. upright; erect.

Arrhenius principle the principle that a relatively small percentage change in the average kinetic energy of a population of molecules may result in a relatively large change in the fraction of molecules having energy greater than the activation energy.

arrhenogenic *a*. producing offspring preponderantly or entirely male. *n*. arrhenogeny.

arrhenotoky *n*. type of parthenogenesis where males are formed from unfertilized eggs and are haploid.

arrhizal *a*. without true roots, as some parasitic plants.

arrow worms Chaetognatha *q.v.*

arsenate *n*. metabolic analogue of phosphate, toxic, used in experimental biochemistry to study oxidative phosphorylation, where it uncouples oxidation and phosphorylation.

artefact, artifact *n*. an apparent structure or experimental result obtained due to method of preparing specimen or experimental conditions; a man-made object.

artenkreis *n*. complex of species which replace one another geographically.

arterial *a. pert*. an artery, or to the system of vessels which carry blood from the heart to the rest of the body.

arteriole *n*. a very small artery.

artery *n*. vessel that conveys blood from heart to rest of the body.

arthral, arthritic *a. pert*. or at joints.

arthrobranchiae *n.plu*. joint gills, arising at junction of thoracic appendage with trunk, in some arthropods.

arthrocyte *n*. a large resorptive cell in nephridium of bryozoans; type of coelomocyte in nematodes.

arthrocytosis *n*. the capacity of cells to selectively absorb and retain solid particles in suspension, as dyes.

arthrodia *n.* a joint admitting only of gliding movements.

arthrodial *a. appl.* articular membranes connecting appendages with thorax in arthropods.

arthrogenous *a.* formed as a separate joint, as spores developed from separated portions of a plant.

arthromere *n.* an arthropod body segment or somite.

Arthropoda, arthropods *n., n.plu.* very large phylum of segmented invertebrate animals with heads, jointed appendages (feelers, mouthparts and legs), and a thickened chitinous cuticle forming an exoskeleton. The main body cavity is a haemocoel. The phylum is generally divided into several different groups, most commonly the Chelicerata, Atelocerata, Crustacea and the extinct Trilobita. In this classification, the Chelicerata includes the spiders, ticks, mites, scorpions, pycnogonids, horseshoe crabs and the extinct eurypterids, the Atelocerata (sometimes known as the Uniramia) includes the insects and myriapods (centipedes and millipedes), and the Crustacea includes the crustaceans (e.g. crabs, shrimps, barnacles). The velvet worms (Onychophora) are sometimes placed in a separate phylum. *see* Appendix 4.

arthropterus *a.* having jointed fin-rays, as fishes.

arthrosis *n.* articulation, as of joints.

arthrospore *n.* in some cyanobacteria, thick-walled resting cell formed by segmentation of filament; a cell formed by fragmentation of hyphae in fungi, as ooidia.

arthrostraceous *a.* having a segmented shell.

arthrous articulate *q.v.*

articulamentum *n.* in chitons, the lower part of each of the body plates.

articular *a. pert.* or situated at a joint, *appl.* cartilage, surface, capsule etc.

articular(e) *n.* bone of skull articulating with quadrate to constitute the upper part of hyoid arch from which lower jaw is suspended.

Articulata *n.* class of brachiopods with shells joined by a hinge joint in which two teeth on one shell move in sockets on the other. *cf.* Inarticulata.

articulate *a.* jointed; separating easily at certain points.

articulation *n.* joint between bones or between segments of a stem or fruit.

artifact *alt.* spelling of artefact *q.v.*

artificial chromosomes small chromosome-like structures constructed by genetic engineering and which include a centromere, origin of replication and telomeres, and which are thus replicated and segregated like chromosomes when introduced into a eukaryotic cell. Large pieces of additional DNA can be introduced into these chromosomes. *see also* YAC.

artificial classification or key classification that groups organisms or objects together on the basis of a few convenient characteristics rather than on the basis of evolutionary relationships. *cf.* natural classification.

artificial formation a pattern of vegetation caused by human activity.

artificial insemination the artificial introduction of sperm collected from a male into the female reproductive tract in mammals.

artificial selection the selection of particular forms as a result of environmental pressures deliberately imposed, either in whole plant or animal breeding or in *in vitro* cultures.

artiodactyl *a.* having an even number of digits.

Artiodactyla, artiodactyls *n., n.plu.* even-toed ungulates, including pigs, sheep, cattle and camels, which have a complex stomach for dealing with plant food and in which the 3rd and 4th digit of the limb forms a cloven hoof.

arytaenoid *a.* pitcher-like, *appl.* two cartilages at back of larynx.

ascending *a.* curving or sloping upwards.

ascending aestivation arrangement of petals, where each petal overlaps the edge of the one posterior to it.

Aschelminthes, aschelminths *n.plu.* group of pseudocoelomate, mainly worm-like animals, in some classifications considered as a phylum, including the Gastrotricha, Kinorhyncha, Nematoda, Nematomorpha and Rotifera (all usually considered as separate phyla).

asci *plu.* of ascus.

Ascidiacea, ascidians *n., n.plu.* class of marine tunicates (urochordates), commonly called sea squirts, in which the adults are generally colonial and fixed to a substrate.

ascidium *n.* a pitcher-leaf or part of leaf as in *Nepenthes* and other pitcher plants.

ascigerous *a.* the sexual, ascus-bearing, reproductive phase in the life history of ascomycete fungi.

ascites *n.* accumulation of watery fluid and cells in abdominal cavity. *a.* ascitic.

ascocarp *n.* the sexual fruiting body of ascomycete fungi, containing asci surrounded by a protective covering, and which may be an apothecium, cleistocarp, or perithecium.

ascogenous *a.* producing asci, *appl.* hyphae.

ascogonium *n.* specialized hyphal branch which gives rise to asci.

Ascolichenes *n.* lichens in which the fungal partner is an ascomycete.

ascoma *n.* disc-shaped ascocarp in certain fungi. *plu.* ascomata.

Ascomycota, ascomycotina, ascomycetes *n., n., n. plu.* large group of terrestrial fungi, commonly called sac fungi, which have a septate mycelium and develop their spores in sac-like structures called asci. They include the yeasts, leaf-curl fungi, black and green moulds, powdery mildews, cup fungi, morels and truffles. *see* Appendix 3.

ascophore *n.* a hypha producing asci in an ascocarp.

ascorbic acid vitamin C *q.v.*

ascospore *n.* haploid spore of ascomycete fungi, produced in an ascus.

ascostroma *n.* type of ascocarp characteristic of the Loculoascomycetidae, in which the asci arise in locules of the stroma. *plu.* ascostromata.

ascus *n.* generally club-shaped or cylindrical sac-like structure containing ascospores in ascomycete fungi. The ascus and the (usually) eight ascospores it contains are formed from a single multinucleate cell.

ascyphous *a.* without a cup-shaped expansion of the thallus bearing archegonia, as in some lichens.

-ase suffix denoting an enzyme, usually joined to a root denoting the substance acted on or the type of reaction, e.g. proteinase, lipase, glucosidase, asparaginase, hydrolase, oxidase.

asemic *a.* without markings.

asepsis *n.* sterile conditions. *see* aseptic.

aseptate *a.* without any septum.

aseptic *a.* sterile; *appl.* certain infectious diseases in which no bacterial agent can be isolated, as aseptic meningitis, and which may be due to viruses or other infectious agents.

asexual reproduction reproduction which does not involve formation and fusion of gametes, and which may be by binary fission, budding, asexual spore formation or vegetative propagation, resulting in progeny with an identical genetic constitution to the parent and to each other.

asialoglycoprotein *n.* glycoprotein that has lost the terminal sialic acid residues from its carbohydrate side chains. Such damaged glycoproteins are removed from the circulation by the liver.

Asiatic subregion subdivision of the Boreal phytogeographical kingdom, comprising central Asia between latitudes 50° and 70° N.

A-site site on the ribosome at which the next codon on mRNA is exposed and incoming aminoacyl-tRNAs attach.

asparaginase enzyme hydrolysing asparagine via aspartate to glutamate and fumarate. E.C.3.5.1.1.

asparagine (Asn,N) amino acid, uncharged derivative of aspartic acid, constituent of proteins, first discovered in asparagus, important in nitrogen metabolism in plants. Required in human diet.

aspartase *n.* enzyme that hydrolyses aspartic acid to yield fumaric acid and ammonia, present in some bacteria and higher plants. EC 4.3.1.1. *r.n.* aspartate ammonia-lyase.

aspartate, aspartic acid (Asp,D) amino acid, constituent of proteins, negatively charged form of amino succinic acid, important in transamination reactions. Required in human diet.

aspartate transcarbamoylase enzyme involved in control of pyrimidine nucleotide biosynthesis. EC 2.1.3.2. *r.n.* aspartate carbamoyltransferase.

aspect *n*. direction in which a surface faces; appearance or look; seasonal appearance.

asperate *a*. having a rough surface.

aspergilliform *a*. tufted like a brush.

Aspergillus nidulans ascomycete fungus much used for genetic studies.

asperity *n*. roughness, as on a leaf.

asperous asperate *q.v.*

aspirin *see* acetylsalicylic acid.

asplanchnic *a*. without an alimentary canal.

asporocystid *a*. *appl*. oocyst of certain sporozoan protozoans when zygote divides into sporozoites without sporocyst formation.

asporogenous *a*. not originating from spores; not producing spores.

asporous *a*. having no spores.

assay *n*. a procedure for measurement or identification.

assembly *n*. the smallest community unit of plants or animals, e.g. a colony of aphids on a stem.

assimilate *n*. any of the first organic compounds produced during assimilation in autotrophs. *v. see* assimilation.

assimilate stream the movement of sugars out of the leaves where they are manufactured during photosynthesis to other parts of the plant via the phloem.

assimilated energy in an animal's energy budget, assimilated energy (A) = consumption (C) − faeces (F) and is equivalent to production (P) + energy lost as heat during respiration (R) + energy lost by small metabolites voided in urine (U).

assimilation *n*. in autotrophic organisms, the uptake of elements and simple inorganic compounds such as CO_2, N_2, H_2O from the environment and their incorporation into complex organic compounds; in heterotrophs, the conversion of digested food material into complex biomolecules. *v.* assimilate.

assimilation efficiency in animal physiology and ecophysiology, a measure of the efficiency of utilization of food. It is expressed as a ratio of assimilated energy (A) divided by energy consumed (C) and is mainly influenced by the nature of the food consumed, carnivores having a greater assimilation efficiency than herbivores.

assimilative *a*. *pert*. or used for assimilation; *appl*. growth preceding reproduction.

association *n*. *see* associative learning; (*bot*.) plant community forming a division of a larger unit of vegetation and characterized by a dominant species.

association centres, association cortex areas of cerebral cortex where different aspects and types of incoming sensory information are believed to be integrated.

association constant affinity constant *q.v.*

association fibres nerve fibres connecting the white matter of interior of brain with cortex.

associative learning learning by associating a stimulus (the cause) with a particular outcome (the effect).

associative nitrogen fixation non-symbiotic nitrogen fixation by bacteria (e.g. *Azospirillum*, *Azotobacter*) associated with the rhizosphere of certain grasses and cereals.

associes *n*. an association representing a stage in the process of succession.

assortative mating non-random mating within a population where individuals tend to mate with individuals resembling themselves. In human populations for example, mating tends to be random for certain characteristics such as blood groups and assortative for others such as height, ethnic group, etc.

astely *n*. condition of stem in which apical meristem gives rise to a number of strands, each composed of one vascular bundle, and there is no central stele or vascular cylinder.

aster *n*. star-shaped system of microtubules radiating from the centriole, seen in many cells during cell division at either end of the spindle but not found in plants.

Asteraceae in some classifications the name for the Compositae *q.v.*

Asterales *n*. an order of herbaceous dicots, rarely trees or woody climbers, with flowers usually crowded into closely packed heads and comprising the family Asteraceae (Compositae) (daisy, etc.).

asterigmate *a*. not borne on sterigma, *appl*. spores.

asternal ribs false ribs *q.v.*

asteroid *a*. star-shaped; *pert*. starfish.

Asteroidea, asteroids *n.*, *n.plu.* class of echinoderms, commonly called starfish or sea stars, having a star-shaped body with five radiating arms not sharply marked off from the central disc.

asterophysis *n*. a rayed cystidium-like hair in hymenium of certain fungi.

asterospondylous *a. appl.* vertebra with centrum of radiating calcified cartilage.

asthenic *a*. weak; tall and slender, *appl.* physical constitutional type.

astichous *a*. not set in a row or rows.

astigmatous *a*. (*bot.*) without a stigma; (*zool.*) without spiracles.

astipulate exstipulate *q.v.*

astogeny *n*. the process by which a siphonophore "colony" develops from a single egg by a budding off of the individual zooids from a larva that develops from the zygote.

astomatous *a*. not having a mouth; without a cytostome; (*bot.*) without stomata.

astomous *a*. without a stomium or line of dehiscence; bursting irregularly.

astragalus *n*. one of the bones of the vertebrate ankle (tarsus).

astrocyte *n*. type of neuroglial cell forming large part of the supporting non-neuronal tissue in the central nervous system, fibrous astrocytes being found mainly in white matter, bearing fine branched processes some of which abut on blood vessels, protoplasmic astrocytes being found in grey matter and bearing thick branched processes similar to pseudopodia. Astrocytes are thought to have numerous functions, one of which is to regulate the ionic environment of the neurones.

astroglia *n*. astrocytes collectively.

astropodia *n.plu.* fine unbranched radiating pseudopodia, as in some protozoans.

astropyle *n*. chief aperture of central capsule in some radiolarian protozoans.

astrosclereid *n*. a multiradiate sclereid or stone cell.

asymmetrical *a. pert.* lack of symmetry; having 2 sides unlike or disproportionate; *appl.* structures that cannot be divided into similar halves by any plane.

asynapsis *n*. absence of pairing of homologous chromosomes in meiosis.

asyndesis asynapsis *q.v.*

atactostele *n*. a complex stele, having vascular bundles scattered in the ground tissue, as in monocotyledons.

atavism *n*. presence of an ancestral characteristic not observed in more recent progenitors.

atavistic *a. pert.*, marked by or tending to atavism.

ateleosis *n*. dwarfism where individual is a miniature adult.

atelia *n*. the apparent uselessness of a character of unknown biological significance; incomplete development. *a.* atelic.

Atelocerata *n*. in some classifications the name of the group of arthropods that includes the insects and myriapods. *see* Appendix 4.

Atherinomorpha *n*. small group of advanced teleost fish including the toothcarps (guppies and swordtails).

athermopause *n*. dormancy of animals due to lack of water or food.

atherosclerosis *n*. deposition of plaques of cholesterol esters in blood vessels, narrowing vessel lumen and restricting blood flow.

Atlantic North American Floral Region part of the Holarctic floral realm comprising North America east to the Rockies, south to the Gulf of Mexico and north to the Arctic Circle.

atlanto-occipital occipito-atlantal *q.v.*

atlas *n*. the 1st cervical vertebra.

atoll *n*. coral reef surrounding a central lagoon.

atopy *n*. an idiosyncratic sensitivity to a particular compound or antigen. *a.* atopic.

ATP synthase enzyme that synthesizes ATP from ADP and inorganic phosphate in mitochondria and chloroplasts and in bacterial membranes using energy derived from a proton (H^+) gradient across the membrane. Such enzymes can also perform the reverse reaction of hydrolysing ATP and pumping protons across the membrane. *alt.* F-type ATPase.

atractoid *a*. spindle-shaped.

atretic *a*. having no opening; imperforate; *appl.* vesicles resulting from degeneration of Graafian follicles, spurious corpora lutea.

atrial *a. pert.* atrium.

atrial natriuretic factor or polypeptide (ANF(P)) one or all of several polypeptides (α, β, γ) isolated from atrium of heart and having natriuretic-diuretic activity, regulating extracellular fluid volume and electrolyte balance. Also present in brain.

atrichous *a.* having no flagella or cilia.

atriopeptin atrial natriuretic factor *q.v.*

atriopore *n.* opening to exterior from atrial cavity in cephalochordates; spiracle in tadpole.

atrioventricular *a. pert.* atrium and ventricle of heart; *appl.* node, mass of tissue in the wall of the right auricle.

atrioventricular bundle His' bundle *q.v.*

atrium *n.* the main chamber of the auricle of the heart, sometimes used to indicate the whole auricle; various chambers or cavities, e.g. the tympanic cavity in the ear, chamber from which tracheae extend into body in insects, chamber surrounding pharynx in tunicates and cephalochordates.

atrochal *a. appl.* trochophore larvae in which the preoral circlet of cilia is absent and surface is uniformly ciliated.

atropal atropous *q.v.*

atrophy *n.* diminution in size and function.

atropine *n.* alkaloid obtained from the deadly nightshade, *Atropa bella-donna* and other plants of the Solanaceae, used medically as a muscle relaxant.

atropous *a. appl.* ovule which is not inverted.

attached chromosome isochromosome *q.v.*

attachment site (*att* **site)** in bacterial genetics, the sites on phage and bacterial chromosomes at which phage DNA is integrated into and excised from the bacterial chromosome, in *E. coli* designated e.g. *att*λ (for phage λ).

attenuate(d) *a.* gradually tapering to a point; reduced in density, strength or pathogenicity; *appl.* vaccines prepared from live strains of virus that have become non-pathogenic by mutation during longterm growth in culture.

attenuation *n.* a regulatory process occurring in some bacterial biosynthetic operons in which translation of a leader sequence in mRNA terminates transcription at a site called the attenuator; of pathogens, loss of virulence.

attic *n.* the recess in bone above the eardrum.

auditory *a. pert.* sense of hearing; *pert.* hearing apparatus, as auditory organ; auditory canal, auditory meatus, auditory ossicle,

auditory capsule *see* ear; *appl.* nerve: 8th cranial nerve, connecting inner ear with hindbrain, carrying sensations of sound and pitch from cochlea for relay to auditory area of cerebral cortex, and postural information from semicircular canals to cerebellum.

Auerbach's plexus myenteric plexus *q.v.*

augmentation *n.* in plants, increase in the number of whorls of floral leaves.

aulostomatous *a.* having a tubular mouth or snout.

aural *a. pert.* ear or hearing.

auricle *n.* the external ear; the anterior chamber of the heart; any ear-shaped appendage.

auricula auricle *q.v.*

auricular *a. pert.* an auricle.

auricular *n.* feathers covering apertures of ears in birds.

auricularis *n.* superior, anterior, posterior, extrinsic muscles of the external ear.

auriculate *a.* eared, having auricles; *appl.* leaf with expanded bases surrounding stem; *appl.* leaf with lobes separate from rest of blade.

auriculotemporal *a. pert.* external ear and temples (temporal regions); *appl.* nerve: a branch of the mandibular nerve.

auriculoventricular *a. pert.* or connecting auricle and ventricle of heart.

auriform *a.* ear-shaped.

aurones *n.plu.* a group of yellow flavonoid plant pigments.

Austral Realm the floristic area that includes the southern part of South America, Australasia, the southern tip of Africa and the southern oceanic islands. It comprises five floral regions, *see* Argentinian Floral Region, Australian Floral Region, New Zealand Floral Region, South African Floral Region, South Oceanic Floral Region. *alt.* Southern Realm.

Australasian *a. appl.* or *pert.* zoogeographical region including Papua-New Guinea, Australia, New Zealand and Pacific Islands.

Australian Floral Region part of the Austral Realm that includes Australia and Tasmania.

Australian Region zoogeographical region including Australia, New Guinea, Sulawesi and other islands south-east of Wallace's line.

australopithecines, _Australopithecus_
n.plu, n. a genus of fossil hominids, at present believed to have lived from at least 4 million until 1 million years ago, found in southern and eastern Africa. They include the "gracile" australopithecines (_Australopithecus africanus_) at around 2.5 million years, the "robust" australopithecines (_A. robustus_) at around 1.5 million years, and an older species, _A. afarensis_ at around 3.5 million years, found in Kenya and Ethiopia. They had an upright posture and a relatively small brain compared to _Homo_. Fossils now accepted as australopithecine were also formerly known as _Paranthropus_ and _Zinjanthropus_. The evolutionary position of the australopithecines in relation to _Homo_ is still unclear, but _A. afarensis_ is thought by some palaeoanthropologists to be directly ancestral to the human line. The later _A. africanus_ and _A. robustus_ are offshoots from the direct line of human evolution and _A. robustus_ is contemporaneous with early species of _Homo_.

autacoid _n._ an internal secretion having a physiological effect such as a hormone, growth factor, etc.

autapses _n.plu._ synapses formed between one part of a nerve cell and another part of the same cell, typically between axon and dendrite or axon and soma.

autarchic _a. appl._ genes in a mosaic organism which are not inhibited from expressing their effect by the presence of a different neighbouring genotype.

autecious, autecism _alt._ spelling of autoecious, autoecism _q.v._

autecology _n._ the biological relations between a single species and its environment; the ecology of a single organism. _cf._ synecology.

auto- prefix derived from Gk. _autos_, self.

autoagglutination _n._ the clumping of an individual's cells by its own serum.

autoantibody _n._ antibody against an antigen of one's own body. Autoantibodies are produced in many autoimmune diseases and are in many cases responsible for causing damage to tissues.

autoantigen _n._ any molecule of an individual's own body which is recognized by the individual's own immune system, pro-voking an autoimmune response.

autobasidium _n._ basidium having sterigmata bearing spores laterally; a non-septate basidium.

autoblast _n._ a free-living unicellular organism.

autocarp _n._ fruit resulting from self-fertilization.

autocatalysis _n._ catalysis of a reaction by one of its own products.

autochory _n._ self-dispersal of spores or seeds by an explosive mechanism.

autochthonous _a._ aboriginal; indigenous; _appl._ indigenous soil microflora which is normally active; inherited or hereditary; _appl._ characteristics originating within an organ, as the pulsating of an excised heart; formed where found.

autoclave _n._ equipment used for sterilization of glassware and media by steam heat.

autocoprophagy refection _q.v._

autocrine _a. appl._ effects of a substance secreted by a cell on that cell itself.

autocyst _n._ thick membrane formed by some sporozoan parasites separating them from host tissues.

autodeliquescent _a._ becoming liquid as the result of self-digestion, as the cap and gills of fungi of the genus _Coprinus_ (inkcaps).

autodont _a._ designating or _pert._ teeth not directly attached to jaws, as in cartilaginous fishes.

autoecious _a._ parasitic on one host only during a complete life cycle.

autogamy _n._ self-fertilization in which the two nuclei that fuse are products of different meioses during gametogenesis; conjugation of nuclei within a single cell; conjugation of two protozoans originating from division of the same cell. _a._ autogamous. _cf._ automixis.

autogenesis _n._ origin, production or reproduction within the same organism.

autogenic _a. appl._ plant successions, caused by interactions between the members of the community itself.

autogenous _a._ produced in same organism, _cf._ exogenous; _appl._ adult female insect that does not need to feed for her eggs to mature.

autogenous control case where the expression of a gene is controlled by its own product, either at the level of transcription or translation.

autogeny, autogony autogenesis *q.v.*

autograft *n.* graft, as of skin, transplanted to a different site on the same individual.

autoheteroploid *a.* heteroploid derived from a single genome.

autoimmune *a. appl.* immune responses against own cells. *n.* autoimmunity.

autoimmunity *n.* condition where the body mounts an immune response against its own components, leading e.g. to chronic inflammation, and tissue destruction.

autoinfection *n.* reinfection from a host's own parasites or body flora.

autointoxication *n.* intoxication by poisonous substances formed within the body.

autokinetic *a.* moving by its own action.

autologous *a. appl.* tissue graft from one part of body to another part in same individual; *appl.* immunization of animal of one species with antigen from another individual of the same species.

autolysin *n.* any enzyme that causes autolysis.

autolysis *n.* self-digestion of a cell by its own hydrolytic enzymes.

autolytic *a.* causing or *pert.* autolysis, *appl.* enzymes.

automimicry *n.* imitation of a communication signal used by a particular sex or age-group in a species, by a member of the opposite sex or another age-group; intraspecific mimicry as when some members of a species are unpalatable, and palatable members of the same species mimic them.

automixis *n.* self-fertilization in which the two nuclei that fuse to form the zygote are the products of the same meiosis during gametogenesis. *a.* automictic.

autonarcosis *n.* the state of being poisoned, rendered dormant or arrested in growth, owing to self-produced carbon dioxide.

autonomic *a.* autonomous, self-governing, spontaneous; *appl.* nervous system: the involuntary nervous system as a whole comprising sympathetic and parasympathetic systems supplying smooth and cardiac muscles and glands, and whose actions are not under conscious control.

autonomously replicating sequence (ARS) DNA sequence that enables any

DNA containing it to replicate in the yeast *Saccharomyces cerevisiae*, and which may represent a yeast DNA origin of replication.

autopalatine *n.* in some fish, an ossification at anterior end of pterygoquadrate.

autoparasite *n.* parasite subsisting on another parasite.

autoparthenogenesis *n.* development from unfertilized eggs activated by a chemical or physical stimulus.

autophagic *a.* involved in or *pert.* autophagy.

autophagic vacuole type of lysosome containing intracellular membranes and organelles, presumably being digested.

autophagous *a. appl.* birds capable of running about and securing food for themselves when newly hatched.

autophagy *n.* self-digestion, sometimes seen in cells, apparently mediated by lysosomes; subsistence by self-absorption of products of cellular metabolism, as consumption of their own glycogen by yeasts.

autophilous *a.* self-pollinating.

autophosphorylation *n.* self-phosphorylation by a protein with kinase activity. Autophosphorylation by receptors with tyrosine kinase activity is one of the first events after their activation by ligand binding and is an essential part of the signal transduction mechanism of such receptors.

autophyllogeny *n.* growth of one leaf on or out of another.

autophyte *n.* an autotrophic plant. *a.* autophytic.

autoploid autopolyploid *q.v.*

autopodium *n.* hand or foot.

autopolyploid, autoploid *a.* organism having more than 2 sets of homologous chromosomes; polyploid in which chromosome sets are all derived from a single species.

autoradiography *n.* technique by which large molecules, cell components or body organs are radioactively labelled and their image recorded on photographic film, producing an autoradiograph or autoradigram. *alt.* (formerly) radioautography.

autoreceptor *n.* receptor that is activated by a compound secreted by the cell that bears it. In the nervous system e.g., autoreceptors for neurotransmitter on the presynaptic neuron can regulate the release of that transmitter.

autoregulation *n.* the regulation of initiation of transcription of a gene by its own protein product.

autoshaping *n.* a Pavlovian conditioned response which may be induced in learning experiments in which animals are exposed to repeated pairings of stimulus and reward and begin to exhibit the behaviour appropriate to the reward on presentation of the stimulus without any training or reinforcement.

autoskeleton *n.* a true skeleton formed from elements secreted by the animal itself.

autosomal *a. appl.* to a gene carried on an autosome, i.e. any chromosome other than a sex chromosome.

autosomal dominant inheritance pattern of inheritance characteristic of a dominant allele carried on an autosome. The associated phenotypic trait is displayed by individuals who carry only one copy of the allele so that each offspring of a heterozygous individual has a 50% chance of inheriting the allele and showing the trait. *cf.* autosomal recessive inheritance; X-linked inheritance.

autosomal recessive inheritance typical pattern of inheritance of a simple recessive Mendelian genetic trait carried on an autosome. The offspring of parents both heterozygous for the recessive allele will each have a 1 in 4 chance of being homozygous for the allele and displaying the trait.

autosome *n.* any chromosome other than the sex chromosomes.

autospore *n.* a non-motile spore resembling the parent cell; the protoplast formed by longitudinal division of a diatom, and which forms new valves.

autostoses cartilage bones *q.v.*

autostyly *n.* condition of having the mandibular arch self-supporting, articulating directly with skull. *a.* autostylic.

autosynapsis autosyndesis *q.v.*

autosyndesis *n.* in a polyploid organism, pairing of chromosomes from the same parent at meiosis.

autotetraploid *n.* organism whose nuclei contain 4 sets of chromosomes of identi-

cal origin.

autotomy *n.* shedding of a part, as in some worms, arthropods and lizards.

autotransplantation *n.* grafting of tissue from one part of the body to another in the same individual.

autotriploid *n.* organism with nuclei containing 3 sets of chromosomes of identical origin.

autotroph *n.* any organism able to utilize inorganic sources of carbon (as carbon dioxide), nitrogen (as nitrates, ammonium salts) etc. as starting materials for biosynthesis, using either sunlight (photoautotroph) or chemical energy sources (chemoautotroph).

autotropism *n.* tendency to grow in a straight line, *appl.* plants unaffected by external influences; tendency of organs to resume original form, after bending or straightening due to external factors.

autoxenous autoecious *q.v.*

autozooid *n.* fully formed polyp of an alcyonarian colony, which can feed and digest nutrients.

autozygote *n.* homozygote in which the 2 homologous genes have a common origin.

autozygous *a. appl.* alleles at the same locus that are identical by virtue of common descent.

auxesis *n.* growth owing to increase in cell size.

auxetic *a. appl.* growth due to increase in cell size rather than proliferation of cells.

auxiliaries *n.plu.* female social insects that associate with other females of the same generation and become workers.

auxiliary cells two or more modified epidermal cells adjoining guard cells, or surrounding stomata.

auxins *n.plu.* various related plant growth hormones, of which indoleacetic acid is most common in nature. They are involved in cellular elongation and differentiation, in root growth, the development of vascular tissue, phototropism, the development of fruits and the normal suppression of the growth of lateral buds. They are responsible for the curvature of plant shoots towards the light by causing a differential elongation of cells on the side away from light where a greater concentration of auxin accumulates. Auxin produced by the tip of the shoot also normally suppresses the development of lateral buds.

auxoautotroph *n.* any organism that can synthesize all the growth substances needed for its development.

auxoheterotroph *n.* any organism that cannot itself synthesize all the growth substances needed for its development.

auxospore *n.* in certain diatoms, zygote formed by fusion of gametes produced by small individuals, which then expands to full size before forming silica valves.

auxotonic *a.* (*bot.*) induced by growth, *appl.* movements of immature plants; (*zool.*) *appl.* contractions against an increasing resistance in muscle.

auxotroph *n.* an organism that is auxotrophic for one or more substances. *see* also auxoheterotroph.

auxotrophic *a. appl.* any organism (esp. a microorganism) that has a nutritional requirement for some specific substance (e.g. vitamins, amino acids, nucleotides, etc.) because it is unable to synthesize it; *appl.* esp. to mutant bacteria or fungi that have lost the ability to synthesize enzymes present in the parental strain and which therefore require a specific nutritional supplement.

average daily metabolic rate (ADMR) a measure of the daily energy requirements of an animal engaging in its normal activities, and which is calculated from the energy content of the food eaten in a day.

average heterozygosity a measure of genetic variation in a population, being the average frequency of heterozygotes among a group of genetic loci.

average life expectancy the number of years a newborn individual can be expected to live, averaged over the population.

aversive *a. appl.* stimuli which decrease the strength of a response if applied several times, and evoke fear and avoidance behaviour.

Aves *n.plu.* birds, a class of bipedal homoiothermic vertebrates having the body clothed in feathers and front limbs modified as wings, and the skin of the jaw forming a horny bill (beak). They are descended from the extinct archosaurian reptiles. The earliest known fossil bird is *Archaeopteryx* (subclass Archaeornithes) from the upper

Jurassic. Modern birds belong to the subclass Neornithes. *see* Appendix 4.

avian *a. pert.* birds.

avicularium *n.* a type of zooid in certain bryozoans, modified in the form of a claw.

avidin *n.* egg white protein which leads to biotin deficiency if given in large quantities by binding to biotin in the intestinal tract. *alt.* antibiotin.

avidity *n.* the total binding strength of a multivalent ligand bound to a receptor with more than one binding site, e.g. an antibody binding to a repeated antigenic determinant on the surface of a pathogenic microorganism. *cf.* affinity.

avifauna *n.* all the bird species of a region or period.

avirulence gene gene found in some bacterial and plant pathogens that determines their ability to cause disease on a host plant containing a corresponding resistance gene. *see* gene-for-gene resistance.

avirulent *a. appl.* a strain of bacterium, virus or other potential pathogen that does not cause disease.

avitaminosis *n.* vitamin deficiency.

avoidance behaviour a wide range of defensive behaviour in animals (e.g. freezing posture, running for cover, giving warning signals) by which they minimize exposure to apparently harmful situations, and which may be innate or learned.

avoidance reaction movement away from stimulus.

awn *n.* stiff bristle-like projection from the tip or back of the lemma or glumes in grasses, or from a fruit, or from tip of leaf.

axenic *a. appl.* pure cultures of microorganisms *in vitro* in which the organism is grown in the absence of any other contaminating microorganism or, in the case of parasites or symbionts, its plant or animal host.

axial *a. pert.* axis or stem; *appl.* filaments and other structures running longitudinally along stem, axon of nerve cell, etc.

axial sinus a nearly vertical canal of the water vascular system in echinoderms, opening into internal division of oral ring sinus, and communicating with stone canal.

axial skeleton skeleton of head and trunk.

axil *n.* the angle between leaf or branch and the axis from which it springs.

axile *a. pert.*, situated in, or belonging to the axis; *appl.* placentation in which ovules are situated in middle of ovary in the angles formed by the meeting of the septa.

axilla *n.* armpit.

axillary *a.* (*bot.*), *pert.* axil; growing in axil, as buds; (*zool.*) *pert.* armpit.

axipetal *a.* travelling in direction of cell body, *appl.* nerve impulses.

axis *n.* the central line of a structure; in embryos, one of usually three (in bilaterally symmetrical animals) axes of polarity: the antero-posterior axis, the dorso-ventral axis, and the medio-lateral axis, which are established very early in development; the main stem of plant or central cylinder of plant stem; the 2nd cervical vertebra.

axis cylinder the axon of a myelinated nerve fibre.

axoaxonic *a. appl.* synapse between axon terminal and axon of another neurone.

axodendritic *a. appl.* synapse in which the axon terminal contacts a dendrite.

axon *n.* cytoplasmic process which carries impulses away from the cell body of neurone, may be up to several meters long in the case of motor neurones of large animals. *alt.* nerve fibre.

axon hillock small clear area of nerve cell at point of exit of axon, at which conducted impulses often start.

axon terminal the (usually) branched tip of an axon, each branch making a synapse on the dendrites or cell body of another nerve cell and secreting neurotransmitter in response to the arrival of a nerve impulse down the axon.

axonal *a. pert.* axon or axons.

axonal transport active transport of material in small cytoplasmic vesicles between cell body and axon terminus in nerve cells. Transport may be anterograde (from cell body to axon terminus) or retrograde (in the other direction) and is believed to occur along tracks formed from microtubules.

axoneme *n.* central core of eukaryotic cilium or flagellum, composed of a regular array of microtubules.

axoplast *n.* filament extending from kinetoplast to end of body in some trypanosomes.

axopod *n.* spike-like, usually permanent pseudopodium with a strengthening axial filament, in heliozoan protozoans. *alt.* axopodium.

axosomatic *a. appl.* synapses in which axon terminal contacts nerve cell body.

axospermous *a.* with axile placentation, of carpel.

axostyle *n.* supporting axial filament in body or flagella of many flagellate protozoans.

axotomy *n.* cutting of an axon.

5-azacytidine nucleotide analogue which is incorporated into DNA in place of cytidine and which, unlike cytidine, cannot be methylated, and which causes various changes in cells some of which appear to be due to derepression of previously represssed genes.

azoic *a.* uninhabited; without remains of organisms or their products.

azonal *a. appl.* soils without definite horizons.

azurophilic *a.* staining readily with blue aniline dyes.

azygoid *a.* haploid, *appl.* parthenogenesis.

azygomatous *a.* without a cheek bone arch.

azygomelous *a.* having unpaired appendages.

azygospore *n.* spore developed directly from a gamete without conjugation.

azygote *n.* organism resulting from haploid parthenogenesis.

azygous *a.* unpaired.

azymic *a.* devoid of enzymes.

B

β- *for all headwords with prefix β- or beta-refer to headword itself.*

B either asparagine or aspartic acid in the single-letter code for amino acids.

BAT brown adipose tissue, brown fat *q.v.*

BCG Bacille Calmette-Guéerin, a modified variant of a bovine strain of *Mycobacterium tuberculosis* used as a vaccine against human tuberculosis.

B-DNA *see* DNA *q.v.*

bFGF basic fibroblast growth factor, one of the two forms of this polypeptide factor.

BFU-E erythrocytic burst-forming unit *q.v.*

BMD Becker muscular dystrophy *q.v.*

BOD biochemical oxygen demand *q.v.*

bp base pair (in a nucleic acid). *q.v.*; a unit of length in DNA.

BP, b.p. before present.

Bq becquerel *q.v.*

BrdU, BrU bromouracil *q.v.*

BSA bovine serum albumin.

BSF-1 B-cell stimulatory factor *q.v.*

by abbreviation for a billion (10^9) years.

bacca *n.* berry, esp. if formed from an inferior ovary.

baccate *a.* pulpy, fleshy, as a berry; bearing berries.

bacciferous *a.* berry-producing or bearing.

bacciform *a.* berry-shaped.

Bacillariophyceae, Bacillariophyta *n.* the diatoms, a class (or division) of unicellular photosynthetic protists, which have a silicified wall in two halves, chlorophyll and carotenoid pigments and store oils and leucosin instead of starch. There are two main groups: centric diatoms, which are radially symmetrical, and pennate diatoms, which have bilateral symmetry.

bacillary *a.* rod-like; *appl.* layer of rods and cones of retina; *pert.* or caused by bacilli.

bacillus *n.* formerly much used, esp. in medical bacteriology, for any rod-shaped bacterium; more specifically any member of the genus *Bacillus*, aerobic spore-forming rods, widely distributed in the soil. *plu.* bacilli.

Bacillus thuringiensis soil bacterium that produces endotoxins active against the larvae of many insect pests, and which is used as a biological insecticide.

back cross a cross between the heterozygous F_1 generation and (usually) the homozygous recessive parent, which allows the different genotypes present in the F_1 to be distinguished.

back mutation a mutation which reverses the effects of a previous mutation in the same gene, either by an exact reversal of the original mutation or by a compensatory mutation elsewhere in the gene (also known as intragenic suppression). *alt.* reverse mutation, reversion.

background *a. appl.* the rate at which spontaneous mutations occur in any particular organism.

bacteria *n.plu.* a group of extremely metabolically diverse, prokaryotic, unicellular microorganisms usually possessing cell walls, sometimes forming filaments, which are found in soil, water, or parasitic or saprophytic on plants and animals, the parasitic forms causing many familiar infectious diseases. They reproduce by binary fission or asexual spores and also transfer genetic material by sexual processes (conjugation) and by virus (bacteriophage)-mediated transfer (transduction). They comprise the "true" bacteria (eubacteria), the "archaebacteria", the actinomycetes and related organisms, mycoplasmas, rickettsias and rickettsia-like organisms, and spirochaetes. *see individual entries. sing.* bacterium. *a.* bacterial.

bactericidal, bacteriocidal *a.* causing death of bacteria.

bactericidin *n.* a substance that kills bacteria.

bacteriochlorophyll *n.* various photosynthetic pigments related to chlorophyll, which act as light-collecting pigments in photosynthetic bacteria.

bacteriocidal bactericidal *q.v.*

bacteriocin *n.* any of a class of proteins produced by bacteria that kill bacteria of another strain or species.

Bacteriological Code International Code of Nomenclature of Bacteria.

bacteriology *n.* the study of bacteria. *a.* bacteriological.

bacteriolysin *n.* substance that causes the dissolution of bacteria.

bacteriolysis *n.* disintegration or dissolution of bacteria.

bacteriophage *n.* a virus infecting bacteria, such as lambda, T2, T4 (infecting *E. coli*). *alt.* phage. *see also* lysogeny, lytic infection.

bacteriophagous *a.* feeding on bacteria. *alt.* bacterivorous.

bacteriorhodopsin *n.* a purple protein found in the purple membrane of extreme halobacteria (salt-loving bacteria), which acts as a light-driven transmembrane proton pump.

bacteriostatic *a.* inhibiting the growth of, but not killing, bacteria.

bacterium *sing.* of bacteria *q.v.*

bacterivorous *a.* feeding on bacteria. *alt.* bacteriophagous.

bacteroid *n.* irregularly shaped rhizobial cell, the form in which rhizobia are found in root nodules of legumes.

bacteroidal *a. appl.* rod-shaped uric acid crystals, in certain annelids.

baculiform *a.* rod-shaped.

Baculoviridae, baculoviruses *n., n.plu.* family of enveloped double-stranded rod-shaped DNA viruses infecting arthropods, some of which are being developed as vectors for expressing genes and manufacturing proteins in cultures of insect cells.

Benson-Calvin-Bassham cycle Calvin cycle *q.v.*

benthic *a. pert.*, or living on, the bottom of sea, lake, river, etc. *alt.* benthal.

baeocyte *n.* cell showing gliding motility, produced by multiple fission of the fibrous-coated cells of pleurocapsalean cyanobacteria.

bag cell secretory neurone in *Aplysia*, involved in regulation of egg-laying.

balanced lethal the existence of two non-allelic recessive lethal genes in different homologous chromosomes, so that the double heterozygote is viable. The lethal genes are maintained in the population since on interbreeding half the offspring are homozygous for one or the other gene and die, but the other half are heterozygous.

balanced polymorphism the stable coexistence of 2 or more distinct types of individual, forms of a character or different alleles of a gene in a population, the proportions of each being maintained by selection.

balanced translocation a mutual exchange of material between two non-homologous chromosomes in which none is lost, and in which two monocentric chromosomes are produced that segregate normally at mitosis, so preserving the normal diploid number of chromosomes in the somatic cells. *alt.* balanced reciprocal translocation.

balancers *n.plu.* halteres *q.v.*; paired larval head appendages functioning as props until forelegs are developed in some salamanders.

balance trials a methodology for determining quantitative nutritional requirements of animals, in which total intake and total loss of a particular nutrient are measured over a long period.

balancing selection selection that maintains a balanced polymorphism.

balanic *a. pert.* glans penis or glans clitoridis.

balanoid *a.* acorn-shaped; *pert.* barnacles.

Balanopales *n.* an order of dicot trees and shrubs comprising the family Balanopaceae, sometimes placed in the Fagales.

balausta *n.* many-celled, many-seeded, indehiscent fruit with tough outer rind, such as a pomegranate.

Balbiani rings type of very large puff seen on certain dipteran polytene chromosomes, esp. in the gnat *Chironomus tentans* in which they were first discovered by E.G. Balbiani in 1881.

baleen *n*. whalebone, horny plates attached to upper jaw in some whales, used to filter plankton etc. from water.

baler scaphognathite *q.v.*

ball-and socket *appl.* joints in which the hemispherical end of one bone fits into a socket on another allowing movement in several planes, as in shoulder and hip joints.

ballast *n*. elements present in plants and which are not apparently essential for growth, such as aluminium or silicon.

ballistic *a. appl.* fruits which explode when ripe, forcibly discharging seeds.

ballistospore *n*. asexual spore produced by the yeast-like Sporobolomycetaceae, which is borne on a sterigma and violently discharged.

balsam *n*. any of various complex fragrant substances found in some plants, consisting of resin acids, esters and terpenes, often exuded from wounds.

balsamic *a. appl.* class of odours typified by vanilla, heliotrope, resins etc.

balsamiferous *a*. producing balsam.

band *n*. in gel electrophoresis, the region of a gel containing molecules of a particular size class, visualized by staining, autoradiography, etc.; in mitotic chromosomes, densely staining or otherwise distinguishable regions which make up an invariant pattern for each chromosome; in polytene chromosomes, a densely staining region of variable size, alternating with lighter staining regions (interbands) along the length of the chromosome (each chromosome having a characteristic and invariant banding pattern).

band III protein protein spanning the red blood cell membrane, forming an anion channel through which HCO_3^- is exchanged for Cl^- in the lungs as red cells dispose of carbon dioxide.

banner standard *q.v.*

baraesthesia *n*. sensation of pressure.

barb *n*. one of delicate thread-like structures extending from the shaft of a feather and forming the vane; a hooked hair-like bristle; a type of fish.

barbate *a*. bearded; tufted.

barbel *n*. tactile process arising on the lower jaw in some fishes.

barbellate *a*. with stiff hooked hair-like bristles.

Barbeyales *n*. order of dicot trees with one family Barbeyaceae including one genus *Barbeya*.

barbicels *n.plu.* tiny hooked processes on barbules of feather which interlock to hold the barbs together.

barbule *n*. lateral projection from barb of feather, serving to hold barbs together to form an unbroken vane; barbel *q.v.*

baresthesia baraesthesia *q.v.*

bark *n*. in strict botanical terms the layer of tissue external to the vascular cambium in woody plants, comprising the secondary phloem, cortex and periderm, the periderm being the layer commonly known as the bark.

bark lice common name for some members of the Psocoptera, small wingless insects with a globular abdomen and incomplete metamorphosis, inhabiting the bark of trees.

barnacle *n*. common name for a member of the Cirripedia *q.v.*

baroceptor, baroreceptor *n*. sensory receptor responding to stretch in the wall of large blood vessels, signalling changes in blood pressure.

barognosis *n*. capacity to detect changes in pressure.

barotaxis *n*. directed movement or orientation in response to a pressure stimulus.

Barr body densely staining body seen in the nuclei of somatic cells from female mammals and which represents the inactivated X chromosome, individuals of abnormal genetic constitution, e.g. 3X, 4X, having two and three such bodies respectively. *alt.* sex-chromatin body.

barrier forest forest in the mountains that holds back snow from the lower slopes.

Bartholin's duct the larger duct of the sublingual gland.

Bartholin's glands glands secreting vaginal lubricant, situated on either side of the vagina.

basad *adv*. towards the base.

basal *a. pert.*, at, or near the base.

basal body cylindrical structure at base of axoneme in eukaryotic flagella and cilia, composed of microtubules. It organizes the assembly and arrangement of the microtubules of the axoneme. *alt.* basal granule.

basal cell cell in lowest layer of a stratified tissue, such as epidermis and other epithelia, and from which that tissue is renewed; (*mycol.*) uninucleate cell which supports the dome and tip of a hyphal crozier; (*zool.*) contractile epithelial cell, as in coelenterates.

basal disc in corals, the area of ectoderm that secretes the calcareous skeleton; in hydra, lower end of body by which it attaches to substratum.

basal ganglia masses of grey matter in cerebral hemispheres which connect with other brain centres, involved in motor control.

basal knobs swellings or granules at points of emergence of cilia in ciliated epithelial cells.

basal lamina thin collagenous layer underlying many epithelia, forming part of the basement membrane separating the epithelial layer from underlying tissues; layer of extracellular matrix lying between nerve terminal and muscle membrane and surrounding muscle and nerve terminals.

basal leaf one of the leaves produced near base of stem, a radical leaf.

basal metabolic rate minimum metabolic rate required for survival, measured in humans at complete rest in a thermally neutral environment after fasting for 12 hours.

basal metabolism normal state of metabolic activity of organism at rest.

basal placentation condition where ovules are situated at the base of the ovary.

basal plates fused parachordal plates in developing skull; of placentae, outer wall of intervillous space; certain plates in echinoderms, situated at top of stalk in crinoids, in echinoids forming part of the apical disc.

basal ridge a ridge around base of crown in a tooth.

basalar *a. appl.* sclerites below the base of wing in insects.

basale *n.* bone of variable structure supporting fish fins.

basapophysis *n.* a transverse process arising from the ventrolateral side of a vertebra.

base *n.* a substance that accepts a H^+ ion (proton) in solution; in biochemistry often refers to the nitrogenous bases, the purine and pyrimidine constituents of nucleotides.

base analogue a substance chemically similar to one of the normal nucleotide bases and which is incorporated into DNA, often causing mutations.

base exchange capacity the extent to which exchangeable cations can be held in a soil. *alt.* cation exchange capacity.

base pair (bp) a single pair of complementary nucleotides from opposite strands of the DNA double helix. The number of base pairs is used as a measure of length of a double-stranded DNA.

base pairing weak bonding between purine and pyrimidine bases within nucleic acids, adenine pairing with thymine (in DNA) or uracil (in RNA) and cytosine with guanine (DNA and RNA).

base ratio the ratio of the bases (A+T)/(C+G) in DNA, which varies widely from species to species.

base-rich *a.* soils containing a relatively large amount of free basic ions such as magnesium or calcium.

base sequence *see* nucleotide sequence.

base sequencing *see* DNA sequencing.

base substitution replacement of one nucleotide with another in DNA.

basement membrane layer separating many types of epithelia from underlying tissues, consisting of basal lamina, mucopolysaccharides and a fine fibrous meshwork.

basibranchial *n.* central ventral or basal skeletal portion of branchial arch.

basibranchiostegal urohyal *q.v.*

basic *a.* having the properties of a base; *appl.* stains which act in general on the nuclear contents of the cell; *appl.* number, (i) the minimum haploid chromosome number occurring in a series of euploid species in a genus, (ii) chromosome number in gametes of diploid ancestor of a polyploid organism; of soils, rich in alkaline minerals.

basicranial *a.* situated or relating to base of skull.

basidia *plu.* of basidium *q.v.*

basidiocarp *n.* the fruiting body of basidiomycete fungi, which bears the basidia.

Basidiolichenes *n*. lichens in which the fungal partner is a basidiomycete.

basidiolum *n*. an undeveloped basidium.

Basidiomycota, basidiomycotina, basidiomycetes *n*., *n.plu.* a large group of fungi that have septate hyphae and bear their spores on the outside of spore-producing bodies (basidia) which are often borne on or in conspicuous fruiting structures. Basidiomycetes include the rusts, smuts, jelly fungi, mushrooms and toadstools, puffballs, stinkhorns, bracket fungi and bird's nest fungi.

basidiophore *n*. a hypha that carries basidia.

basidiospore *n*. spore formed by meiosis in basidiomycetes, borne on outside of a basidium.

basidium *n*. in basidiomycete fungi, a usually club-shaped cell on the surface of which basidiospores, usually 4 in number, are formed by meiosis. *plu*. basidia. *a*. basidial.

basifugal *a*. growing away from base; acropetal *q.v.*

basifuge *n*. a plant unable to tolerate basic soils. *alt*. calcifuge.

basigamic, basigamous *a*. having ovum and synergids at the far end of the embryo sac, away from the micropyle.

basihyal *n*. the basal or ventral portion of hyoid arch.

basilabium *n*. sclerite formed by fusion of labiostipites in insects.

basilar *a. pert.*, near, or growing from, base.

basilar membrane basal membrane of organ of Corti in ear.

basilemma basement membrane *q.v.*

basilic *a. appl.* large vein on inner side of biceps of arm.

basilingual *a. appl.* broad cartilaginous plate, the body of the hyoid, in crocodiles, turtles and amphibians.

basimandibula *n*. a small sclerite on insect head, at base of mandible.

basimaxilla *n*. a small sclerite on insect head, at base of maxilla.

basinym *n*. the name on which new names of species, etc. have been based. *alt*. basionym.

basion *n*. the middle of the anterior margin of the foramen magnum.

basiophthalmite *n*. basal joint of eye-stalk in crustaceans.

basiotic mesotic *q.v.*

basipetal *a*. descending; developing from apex to base, *appl*. leaves, roots or flowers.

basipharynx *n*. in insects, epipharynx and hypopharynx united.

basiphil basophil *q.v.*

basipodite *n*. 2nd joint of certain limbs of crustaceans.

basipodium *n*. wrist or ankle.

basiproboscis *n*. the membranous portion of proboscis of some insects.

basipterygium *n*. bone or cartilage in pelvic fin of fishes.

basipterygoid *n*. a process of the basisphenoid in some birds.

basirostral *a*. situated at, or *pert.*, the base of beak or rostrum.

basisphenoid *n*. cranial bone in middle of base of skull.

basistyle *n*. base of clasper in male mosquitoes.

basitarsus *n*. 1st segment of tarsus, usually the largest.

basitonic *a*. having anther united at its base with rostellum.

basivertebral *a. appl.* veins within bodies of vertebrae and communicating with vertebral plexuses.

basket cell one of the myoepithelial cells surrounding the base of certain glands; type of interneurone found in cerebellum with cell body situated just above Purkinje cell layer, acts on Purkinje cells.

basolateral *a. pert.* sides and base of any cell, structure or organ.

basophil *n*. type of white blood cell, classed as a granulocyte, which stains strongly with basic dyes, releases histamine and serotonin in certain immune reactions; secretory cells of the anterior lobe of pituitary gland that stain with basic dyes. *a*. staining strongly with basic dyes.

bass (*bot*.) bast *q.v.*; (*zool*.) type of fish.

bassorin *n*. a mucilaginous carbohydrate food store in orchids, used to make saloop.

bast *n*. an inner fibrous layer of some trees; phloem fibres.

bastard merogony activation of an enucleated egg fragment by a spermatozoon of a different species.

bastard wing group of 3 quill feathers on the 1st digit of bird's wing.

Batesian mimicry resemblance of one animal (the mimic) to another (the model) to the benefit of the mimic, as when the model is dangerous or unpalatable, described by the English naturalist H.W. Bates.

bathyaesthesia *n.* sensation of stimuli within the body.

bathyal *a. appl.* or *pert.* zone of seabed between the edge of the continental shelf and the abyssal zone at a water depth of 2000 m.

bathylimnetic *a.* living or growing in the depths of lakes or marshes.

bathymetric *a. pert.* vertical distribution of organisms in space.

bathypelagic *a.* inhabiting the deep sea (1000–3000 m).

bathyplankton *n.* plankton that undergo a daily migration, moving up towards the surface at dusk, and down to lower depths at dawn.

bathysmal *a. pert.* deepest depths of the sea.

batrachians *n.plu.* frogs and toads.

batrachosaurs *n.plu.* a group of labyrinthodonts of the Carboniferous and Permian, which may include the ancestors of reptiles.

batrachotoxin *n.* alkaloid poison from skin of certain frogs that acts on the nervous system.

bauplan *n.* generalized, idealized, archetypal body plan of a particular group of animals.

B cells B lymphocytes *q.v.*; β-cells of pancreas; β-cells of pituitary.

B-cell receptor the antigen receptor on B lymphocytes, which is a membrane-bound immunoglobulin.

B-cell stimulatory factor protein factor produced by T cells that acts on nonactivated B cells to enhance expression of class II MHC cell-surface molecules, and also stimulates antibody secretion by activated B cells, and also has stimulatory effects on some T cells and mast cells. *alt.* interleukin-4 (IL-4).

B chromosomes extra heterochromatic chromosomes present in some organisms above the normal number for the species.

bdelloid *a.* having the appearance of a leech.

beak *n.* bill (*q.v.*) of birds; elongated jaws or mandibles of other animals, as the elongated jaw of a dolphin; long angled projections on certain fruits, as those of cranesbills (Geraniales).

beard *n.* (*bot.*) barbed or bristly outgrowths on grain; awn *q.v.*

beard worms Pogonophora *q.v.*

Becker muscular dystrophy (BMD) a mild heritable form of muscular dystrophy, caused by a defect of the DMD (Duchenne muscular dystrophy) locus.

becquerel (Bq) derived SI unit for expressing the activity of a radionuclide, which is equal to 2.7×10^{-11} curies (Ci).

bedeguar *n.* moss-like outgrowth produced on rosebushes by gall wasps.

bees *n.plu.* insects of the superfamily Apoidea of the order Hymenoptera, some of which are social and some solitary, and which include the honey bees (*Apis*), bumblebees (*Bombus*) and flower bees (*Anthophora*). They feed themselves and their young on pollen and nectar gathered from flowers, and are important plant pollinators. The social bees form colonies with a single queen, males (drones) and workers.

beet sugar sucrose *q.v.*

beetles common name for the Coleoptera *q.v.*

Begoniales *n.* an order of mostly succulent dicot herbs, but also shrub-like herbs and some large trees, comprising the families Begoniaceae (begonia) and Datiscaceae.

behavioural ecology the relationship of animals to their environment and to other animals extended to take in the effects of their behaviour and the way it may be modified by environmental factors and by interactions with members of their own species.

behavioural genetics the study of the genetic basis of behaviour, which includes the study of the contribution of environment and nurture and heritable traits to behaviour.

behavioural silence the condition in which an animal is thought to learn even though no change in behaviour is observed.

behaviourism *n.* a mechanistic psychological theory postulating in its most extreme form that animal and human behaviour may be explained purely in terms of muscular and other physiological (e.g. hormonal) reactions to external and internal stimuli.

belemnoid *a.* shaped like a dart.

Bellini's ducts tubes opening at apex of kidney papilla, formed by union of smaller collecting tubules.

belonoid *a.* shaped like a needle.

belt desmosome intermediate junction *q.v.*

Beltian or Belt's bodies small nutritive organs containing oils and proteins, borne at the tips of leaves of swollen-thorn acacias and which provide food for ants that live on the plant and help it to survive by protecting against attack by other insect pests and by damaging neighbouring seedlings.

Bence Jones protein free immunoglobulin light chains found in urine of patients with multiple myeloma.

β-bend β-turn *q.v.*

Benedict's solution solution containing sodium citrate, sodium carbonate and copper sulphate, which forms a rust-brown cuprous oxide precipitate on boiling with reducing sugars.

benefit *n.* in animal behaviour, the quantity that is maximized by the behavioural choices made. *alt.* negative cost. *cf.* cost.

benign *a. appl.* neoplasms that do not become invasive and spread to other sites in the body.

Bennettitales *see* Cycadeoidophyta (*q.v.*)

benthic *a. pert.*,or living on, the bottom of sea, lake, river, etc. *alt.* benthal.

benthophyte *n.* a bottom-living plant.

benthos *n.* flora and fauna of sea or lake bottom from high water mark down to the deepest levels.

Benzedrine a trade name for amphetamine *q.v.*

benzo[a]pyrene, benzpyrene *n.* a carcinogenic polycylic aromatic hydrocarbon present in coal tar.

benzodiazepines *n.plu.* a class of compounds used as minor tranquillizers, and which are thought to exert their effects by inhibiting the action of the excitatory amino acid glycine in the central nervous system.

benzyladenine *n.* a synthetic plant growth hormone.

Bergmann's rule the idea that geographically variable warm-blooded animal species have smaller body sizes in the warmer parts of their range than in the colder.

beriberi *n.* disease resulting in degeneration of the nerves or polyneuritis, caused by a deficiency of the vitamin thiamine (vitamin B_1) in the diet, or to an inability to absorb thiamine.

berry *n.* a several-seeded indehiscent fruit with a fleshy covering and without a stony layer surrounding the seeds; dark knob-like structure on bill of swan.

Bertin's columns renal columns *q.v.*

beta- *for headwords with prefix beta- or β- refer also to headword itself, e.g. for β-globin look under globin.*

beta diversity biological diversity resulting from competition between species producing a finer adaptation of a species to the complete habitat, thus narrowing the range of tolerance to other environmental factors. *alt.* habitat diversification. *cf.* alpha diversity.

betacyanins *n.plu.* complex flavonoid pigments which give a reddish colour to some flowers.

betaine *n.* glycine derivative, intermediate in choline synthesis.

Betulales *n.* in some classifications an order of dicot trees including the families Betulaceae (birch) and Corylaceae (hazel).

Betz cells giant pyramidal cells in motor area of cerebral cortex.

B-horizon layer of deposition and accumulation below the topmost layer (the A-horizon) in soils. *alt.* illuvial layer.

bi- prefix from L. *bis*, twice, often indicating having 2 of.

biacuminate *a.* having 2 tapering points.

biarticulate *a.* two-jointed.

biaxial *a.* with 2 axes; allowing movement in 2 planes, as condyloid and ellipsoid joints.

bicapsular *a.* having 2 capsules; having a capsule with 2 chambers.

bicarinate *a.* with 2 keel-like processes.

bicarpellary, bicarpellate *a.* with 2 carpels.

bicaudal, bicaudate *a.* possessing 2 tail-like processes.

bicellular *a.* composed of 2 cells.

bicentric *a. pert.* 2 centres, *appl.* discontinuous distribution of species, etc.

biceps *n.* muscle with 2 heads or origins, esp. the large muscle in upper arm or leg.

biciliate *a.* having 2 cilia.

bicipital *a. pert.* biceps; *pert.* groove, the

intertubercular sulcus, on upper part of humerus; *appl*. ridges, the crests of the greater and lesser tubercles of the humerus; *appl*. a rib with dorsal tuberculum and ventral capitulum; divided into 2 parts at one end.

bicollateral *a*. having 2 sides similar.

biconjugate *a*. with 2 similar sets of pairs.

Bicornes Ericales *q.v.*

bicornute *a*. with 2 horn-like processes.

bicrenate *a*. doubly crenate, as crenate leaves with notched toothed margins; having 2 rounded teeth.

bicuspid *a*. having two longitudinal ridges or ribs, as leaf; having two cusps or points, *appl*. premolar teeth; *appl*. valve: mitral valve of heart between left auricle and ventricle, *alt*. bicuspidate.

bicyclic *a*. arranged in 2 whorls.

bidentate *a*. having 2 teeth or tooth-like processes.

bidenticulate *a*. with 2 small teeth or tooth-like processes, as some scales.

bidirectional replication DNA replication which proceeds in both directions from the origin of replication, as in the *E. coli* chromosome.

bidiscoidal *a*. consisting of 2 disc-shaped parts, *appl*. a type of placenta.

biennial *n*. plant living for two years and fruiting only in the second. *a*. biennial.

bifacial *a*. flattened, and having upper and lower surface of distinctly different structure.

bifarious *a*. arranged in 2 rows, one on each side of axis.

bifid *a*. forked; divided nearly to middle line.

biflagellate *a*. having 2 flagella.

biflex *a*. twice curved.

biflorate *a*. having 2 flowers.

biflorous *a*. flowering in both spring and summer.

bifoliar *a*. having 2 leaves.

bifoliate *a*. *appl*. palmate compound leaves with 2 leaflets; having 2 leaves.

biforate *a*. having 2 foramina or pores.

bifurcate *a*. forked; having 2 joints, the distal V-shaped and attached by its middle to the proximal.

bigeminal *a*. with structures arranged in double pairs; *pert*. optic lobes of vertebrate brain.

bigeminate *a*. twin-forked.

bigeneric *a*. *appl*. hybrids between 2 different genera.

bijugate *a*. with 2 pairs of leaflets.

bilabiate *a*. two-lipped.

bilamellar *a*. formed of 2 plates or scales.

bilaminar, bilaminate *a*. having 2 layers.

bilateral *a*. *pert*. or having 2 sides.

bilateral symmetry having two sides symmetrical about one median axis only, so that one side is a mirror image of the other.

bilayer *n*. bimolecular layer (double layer) formed by amphipathic molecules such as phospholipids in an aqueous environment, each molecule oriented with the hydrophilic group on the outside and the hydrophobic group to the interior of the layer, and which is the basic structure of the membranes of a cell. *alt*. lipid bilayer, bimolecular sheet.

bile *n*. secretion of liver cells which collects in gall bladder and passes via the bile duct to the duodenum, and which contains emulsifying bile salts, bile pigments (derived from breakdown of haemoglobin), cholesterol and lecithin, and some other substances. *alt*. gall.

bile salts sodium glycocholate and sodium taurocholate found in bile and which aid the emulsification of fats during digestion.

bilharzia *n*. schistosomiasis. *see* schistosome.

biliary *a*. *pert*. or conveying bile.

bilicyanin *n*. blue pigment resulting from oxidation of biliverdin and bilirubin.

bilin *n*. the chromophore in biliproteins.

biliproteins *n.plu*. protein pigments present in some groups of algae, including phycoerythrin and phycocyanin.

bilirubin *n*. red pigment, reduction product of biliverdin, breakdown product of the haem group of haemoglobin, found in spleen and seen in bruising, excreted in bile.

biliverdin *n*. green pigment, breakdown product of the haem group of haemoglobin, found in spleen and seen in bruising, excreted in bile.

bill *n*. the beak of a bird, formed from outgrowth of cornified skin at the corners of the jaws.

bilobate, bilobed *a*. having 2 lobes.

bilobular *a*. having 2 lobules.

bilocellate *a*. divided into 2 compartments;

having 2 locelli.

bilocular *a.* having 2 cavities or compartments.

bilophodont *a. appl.* molar teeth of tapir, which have ridges joining the anterior and posterior cusps.

bimaculate *a.* having 2 spots.

bimanous *a.* having 2 hands, *appl.* certain primates.

bimolecular sheet *see* bilayer.

bimuscular *a.* having 2 muscles.

binary *a.* composed of 2 units; *appl.* compounds of only 2 chemical elements.

binary fission in prokaryotic organisms, the chief mode of division, in which a cell divides into 2 equal daughter cells, each containing a copy of the "chromosome".

binary vector a genetic engineering vector composed of two separate plasmids, one of which carries the foreign DNA and the other some function required for its transfer or maintenance.

binate *a.* growing in pairs; *appl.* leaf composed of 2 leaflets.

binaural *a. pert.* both ears.

bindin *n.* protein isolated from sea urchin sperm which is responsible for the species-specific adherence of sperm to egg.

binemic *a.* two-stranded.

binervate *a.* having 2 veins, *appl.* insect wing, leaf.

binocular *a. pert.* both eyes; stereoscopic, *appl.* vision.

binodal *a.* having 2 nodes, as stem of plant.

binomial *a.* consisting of 2 names, *appl.* nomenclature, the system of double Latin names given to plants and animals, consisting of a generic name followed by a specific name, e.g. *Felis* (genus) *tigris* (species).

binotic binaural *q.v.*

binovular *a. pert.* 2 ova, *appl.* twins arising from 2 eggs.

binuclear, binucleate *a.* having 2 nuclei.

-bio- word element derived from Gk. *bios*, life, indicating living, *pert.* living organisms, etc.

bioaccumulation *n.* the increasing concentration of a compound, usually *appl.* fat-soluble pesticides such as DDT, in the bodies of living organisms at successively higher levels in the food chain. *alt.* biological amplification, biomagnification.

bioaccumulator *n.* plant or animal species that accumulates heavy metals or other environmental contaminants (e.g. fat-soluble pesticides) in its tissues, and can be used as an indicator of the presence of chronic pollution by these compounds, especially where amounts of pollutant in the environment are too low to be easily detectable.

bioassay *n.* any biological assay; use of a living organism or tissue for assay purposes; a quantitative biological analysis.

biocenosis biocoenosis *q.v.*

biochemical evolution *see* molecular evolution.

biochemical oxygen demand (BOD) measurement of the amount of organic pollution in water, measured as the amount of oxygen taken up from a sample containing a known amount of dissolved oxygen kept at 20 °C for 5 days, a low BOD indicating little pollution, a high BOD indicating increased activity of heterotrophic microorganisms and thus heavy pollution. *alt.* biological oxygen demand.

biochemistry *n.* the chemistry of living organisms and its study.

biochore *n.* boundary of a floral or faunal region; climatic boundary of a floral region.

biochrome *n.* any biological pigment.

biocide *n.* any agent that kills living organisms.

bioclimatology *n.* the study of the relationship of living organisms and climate.

biocoen *n.* the living parts of an environment; biosphere *q.v.*

biocoenosis *n.* community of organisms inhabiting a particular biotope.

bioconcentration *n.* uptake of a heavy metal or chemical compound such as a pesticide from the environment and its accumulation in the cells of living organisms, *cf.* bioaccumulation, biological amplification; accumulation of such compounds in a particular part of a plant or animal body.

biocontrol biological control *q.v.*

biocycle *n.* one of the 3 main divisions of the biosphere: marine, freshwater or terrestrial habitat.

biodegradable *a. appl.* compounds and materials that can be broken down by microorganisms.

biodegradation *n*. the breakdown of materials by living organisms, mainly bacteria.

biodemography *n*. the science dealing with the integration of ecology and population genetics.

biodiversity, biological diversity in different contexts may denote: the number of different species present in a given environment (species diversity); the genetic diversity within a species (genetic diversity); the number of different ecosystems present in a given environment (ecological diversity).

bioelectric *a. appl.* electric currents produced in living organisms.

bioenergetics *n*. the energy flow in an ecosystem; study of energy transformation in living organisms.

bioengineering *n*. use of artifical replacements for body organs; use of technology in the biosynthesis of economically important compounds. *see also* genetic engineering.

bioflavonoids *n.plu.* group of flavonoids present in citrus and other fruits such as paprika, which have activity in animals due to their reducing and chelating properties, e.g. citrin.

biofuel *n*. gas such as methane or liquid fuel such as ethanol (ethyl alcohol) made from organic waste material, usually by microbial action.

biogas *n*. a combustible gas produced by microbial activity, usually referring to methane produced by microbial fermentation of organic wastes.

biogenesis *n*. the theory that living organisms must be generated from living organisms, as opposed to the 19th century theory of spontaneous generation.

biogenic, biogenetic *a*. originating from living organisms, *appl.* deposits such as coal, oil and chalk.

biogenic amines 5-hydroxytryptophan (serotonin), adrenaline, noradrenaline and dopamine.

biogeochemistry *n*. the study of the distribution and movement of elements present in living organisms in relation to their geographical environment, and the movement of elements between living organisms and their non-living environment

(biogeochemical cycles).

biogeocoenosis *n*. a community of organisms in relation to its special habitat.

biogeographical province an area of the Earth's surface defined by the endemic species it contains.

biogeographical realms, biogeographical kingdoms the major geographical divisions of the terrestrial environment characterized by their overall flora and fauna, *see* floral realm, zoogeographical kingdom.

biogeography *n*. that part of biology dealing with the geographical distribution of plants (phytogeography) and animals (zoogeography).

biolistics *n*. technique of introducing DNA into a cell by firing minute (*ca.* 1.6 μm) DNA-coated particles (e.g. of gold) into the cell using a device powered by pressurized helium – the biolistic gun or "gene gun".

biological *a. pert.* to living organisms or to their study.

biological amplification *see* bioaccumulation.

biological clocks hypothetical mechanisms underlying the regular metabolic and behavioural rhythms seen in many cells and organisms, the biochemical basis for which is as yet unknown.

biological containment in genetic engineering, the use of noninfectious, en feebled and exceptionally nutritionally fastidious strains of microorganism, which cannot survive outside the laboratory, as vehicles in which to clone recombinant DNA in order to minimize risk in case of accident.

biological control control of pests and weeds by other living organisms, usually other insects, bacteria or viruses, or by biological products such as hormones.

biological indicator *see* indicator species.

biological oxygen demand biochemical oxygen demand *q.v.*

biological races strains of a species which are alike morphologically but differ in some physiological way, such as a parasite or saprophyte with particular host requirement, or a free-living organism with a food or habitat preference.

biological rhythm intrinsic (endogenous) periodic, diurnal or seasonal behaviour or metabolic change in living organisms, which will continue, for some time at least, even when the environmental rhythm (e.g. cycle of light and dark) to which they are entrained is absent. *see* circadian, circannual, diurnal. *alt.* biorhythm.

biological species population of individuals that can interbreed. i.e. a true species.

biology *n.* the science dealing with living organisms, a term coined by J.B. de Lamarck in 1802.

bioluminescence *n.* the production of light by living organisms, which is the result of an enzyme-catalysed biochemical reaction in which an inactive precursor is converted into a light-emitting chemical.

biomagnification bioaccumulation *q.v.*

biomanipulation *n.* the deliberate manipulation of the species composition of an ecosystem, e.g. to try to regenerate a hypereutrophic lake after the organic pollution itself has been ameliorated.

biomass *n.* total weight, volume or energy equivalent of organisms in a given area; plant materials and animal wastes used as a source of fuel or other industrial products; in biotechnology, the microbial matter in the system.

biome *n.* a climatically controlled group of plants and animals of a characteristic composition and distributed over a wide area, such as tropical rainforest, tundra, temperate grassland, desert, savanna, mountain habitats, taiga and northern coniferous forest, etc.

biometeorology *n.* the study of the effects of the weather on plants and animals.

biometrics, biometry *n.* statistical study of living organisms and their variations.

biomineralization *n.* the production of partly or wholly mineralized internal or external structures by living organisms.

biophage *n.* an organism feeding upon other living organisms. *cf.* saprophage.

biophysics *n.* study of biological processes in terms of their underlying physical principles; physics as applied to biology.

biophyte *n.* a parasitic plant.

bioplastic *see* biopolymer.

biopolymer *n.* biodegradable polymer produced by living organisms, e.g. polysaccharide gums (xanthans) produced by the bacterium *Xanthomonas* and other spp.,

and bioplastics such as poly-β-hydroxybutyric acid and other poly-β-hydroxyalkanoates, produced by the bacterium *Alcaligenes eutrophus*.

biopsy *n.* examination of living tissue.

biopterin *n.* a colourless pteridine, a possible intermediate in the formation of the eye pigment drosopterin in certain insects.

bioregion *n.* a unique area with distinctive soils, landforms, climates and indigenous plants and animals.

biorhythm *n. see* biological rhythm.

bios *n.* living organisms.

biosensor *n.* any organism, microorganism, enzyme system or other biological structure used as an assay or indicator.

bioseries *n.* a succession of changes of any single heritable character.

biospecies biological species *q.v.*

biospeleology *n.* the biology of cave-dwelling organisms and its study.

biosphere *n.* the part of the planet containing living organisms, the living world.

biostasis *n.* the ability of living organisms to withstand environmental changes without being changed themselves.

biostatics *n.* study of the structure of living organisms in relation to function.

biostratigraphic zone a stratum of rock containing a characteristic assemblage of fossils.

biosynthesis *n.* formation of organic compounds by living organisms.

biosystem ecosystem *q.v.*

biosystematics *n.* the study of the variation and evolution of taxa.

biota *n.* the total fauna and flora of a region; the population of living organisms in general.

biotechnology *n.* the use of living cells or microorganisms (e.g. bacteria) in industry and technology to manufacture drugs and chemicals, break down waste, etc. In recent years esp. refers to the use of genetically modified cells and microorganisms.

biotic *a. pert.* life and living organisms.

biotic climax a plant community that is maintained in the climax state by some biotic factor such as grazing.

biotic community a community of plants and animals as a whole.

biotic environment the part of an organism's environment produced by its interaction with other organisms.

biotic factors the influence on the environment of living organisms.

biotic index a measure of the ecological quality of the environment, generally with regard to organic polllution, using assessments of the number and abundance of key indicator species present, e.g. for rivers, biotic indices based on the type of invertebrate community present are used. A high biotic index, representing high diversity and the presence of pollution-sensitive species, indicates a clean river, a low biotic index, reflecting the presence of only a few pollution-tolerant species, indicates heavy organic pollution. *cf.* diversity index.

biotic potential highest possible rate of population increase (r_{max}), resulting from maximum rate of reproduction and minimum mortality.

biotic province a major ecological region of a continent.

biotic pyramid ecological pyramid *q.v.*

biotic succession the part of a succession that is controlled by the activities and interactions of the species present rather than by the physical environment.

biotin *n.* vitamin B_4, a water-soluble vitamin which is required as the prosthetic group of enzymes involved in the incorporation of carbon dioxide into organic compounds. Liver, egg yolk and yeast are rich sources. It is bound by avidin, a protein from egg white, and a similar protein, streptavidin, from bacteria, and this has been used in the development of biotin–streptavidin systems for detecting nucleic acids and proteins labelled with biotin.

biotope *n.* an area or habitat of a particular type, defined by the organisms (plants, animals, microorganisms) that typically inhabit it, e.g. grassland, woodland, etc., or on a smaller scale a microhabitat.

biotroph *n.* any organism that feeds on other living organisms, e.g. parasitic or symbiotic bacteria and fungi, carnivores, herbivores, etc. *a.* biotrophic.

biotype *n.* group of organisms of similar genetic constitution.

bipaleolate *a.* furnished with 2 small bracts (paleae).

bipalmate *a.* lobed with the lobes again lobed.

biparietal *a.* connected with the 2 parietal lobes of brain.

biparous *a.* bearing 2 young at a time.

bipectinate *a. appl.* structures having the 2 edges furnished with teeth like a comb.

biped *n.* a 2-footed animal. *a.* bipedal.

bipeltate *a.* having, or consisting of, 2 shield-like structures.

bipenniform *a.* shaped like a feather, with sides of vane of equal size.

bipetalous *a.* with 2 petals.

bipinnaria *n.* starfish larva with 2 bands of cilia.

bipinnate *a. appl.* compound pinnate leaf in which leaflets grow in pairs on paired stems.

bipinnatifid *a. appl.* pinnate leaf whose segments are again divided.

bipinnatipartite *a.* bipinnatifid, but with divisions extending nearly to midrib.

bipinnatisect *a.* bipinnatifid, but with divisions extending completely to midrib.

biplicate *a.* having 2 folds, having 2 distinct wavelengths, *appl.* flagellar movement in certain bacteria.

bipolar *a.* having, located at or *pert.* 2 ends or poles.

bipolar cell type of nerve cell with one process leaving cell body at either end, present in retina between photoreceptors and ganglion cells.

bipolarity *n.* the condition of having 2 distinct ends or poles.

biradial *a.* symmetrical both radially and bilaterally, as some coelenterates.

biradiate *a.* two-rayed.

biramous, biramose *a.* dividing into 2 branches.

bird lice common name for an order of insects, the Mallophaga or biting lice, which are ectoparasites of birds.

bird's nest fungi common name for the Nidulariales, an order of gasteromycete fungi which have fruiting bodies resembling minute bird's nests (the peridium) full of eggs (the peridioles containing the spores).

birds Aves *q.v. see also* Appendix 4.

Birnaviridae *n.* family of non-enveloped RNA viruses infecting fish and birds.

birostrate *a.* having 2 beak-like processes.

birth pore uterine pore of trematodes and

cestodes; birth opening of redia of trematodes.

birth rate the number of births within a population over a set period, usually a year. The crude birth rate is calculated for human populations as the annual number of live births per 1000 population in a given geographical area, with the population number usually taken at the midpoint of the year in question.

biscotiform *a.* biscuit-shaped, *appl.* spores.

biscuspid *a. appl.* tooth: premolar *q.v.*

bisect *n.* a stratum transect chart with root system as well as shoot included; *v.* to divide into 2 equal halves.

bisegmental *a.* involving or *pert.* two segments.

biseptate *a.* with 2 partitions.

biserial, biseriate *a.* arranged in 2 rows or series.

biserrate *a.* having marginal teeth which are themselves notched.

bisexual *a.* having both male and female organs.

bisporangiate *a.* having both micro- and megasporangia; *appl.* strobilus consisting of both micro- and megasporophylls.

bispore *n.* a paired spore, as of certain red algae.

bisporic bisporous *q.v.*

bistephanic *a.* joining 2 points where coronal suture crosses superior temporal ridges.

bistipulate *a.* having 2 stipules.

bistrate *a.* having 2 layers.

bistratose *a.* with cells arranged in 2 layers.

bisulcate *a.* having 2 grooves; cloven-hoofed.

biternate *a.* divided into 3 with each part again divided into 3.

bithorax complex cluster of homeotic genes in the fruitfly *Drosophila melanogaster* that control the identity of the segments along the antero-posterior body axis.

biting lice *see* bird lice.

bitubercular *a.* with 2 tubercles or cusps, *appl.* biscuspid premolar teeth.

bitunicate *a. appl.* asci in which the inner wall is elastic and expands beyond the outer wall at the time of spore liberation.

Biuret reaction simple test for the presence of proteins and peptides which is based on a reaction with the peptide bond. Solutions of copper sulphate and sodium hydroxide are added to the test sample, and a purple copper complex is formed if a compound containing a peptide bond is present.

bivalent *n.* may either refer to a chromosome which has duplicated to form two sister chromatids still held together at the centromere, or may refer to a pair of duplicated homologous chromosomes held together by chiasmata at meiosis (*alt.* tetrad); *a. appl.* antibody with 2 antigen-binding sites.

bivalve *a.* consisting of 2 plates or valves, as shell of brachiopods and bivalve molluscs; (*bot.*) *appl.* seed capsule of similar structure.

Bivalvia, bivalves *n., n.plu.* class of bilaterally symmetrical molluscs which are laterally flattened and have a shell made of 2 hinged valves, e.g. clams, mussels, scallops, cockles.

biventer cervicalis the spinalis capitis muscle, or central part of the spinalis, a muscle of the neck, consisting of 2 fleshy ends with a narrow portion of tendon in the middle.

biventral *a. appl.* muscles of the biventer type.

bivittate *a.* with 2 oil receptacles; with 2 stripes.

bivoltine *a.* having 2 broods in a year.

black corals common name for an order of stony corals which are colonial and have a black or brown skeleton.

black earth chernozem *q.v.*

bladder cell (*mycol.*) a globular modified hyphal cell in integument or stalk of fruit body in fungi; (*zool.*) in tunicates, a large vacuolated cell in outer layer of tunic.

bladder *n.* membranous sac filled with air or fluid.

bladderworm *n.* larval stage of some Cestoda (tapeworms) in the intermediate host, which is in the form of a bladder containing an inverted scolex. *see* cysticercoid.

blade *n.* flat part of leaf, bone, etc.

Blandin's glands anterior lingual glands.

blanket bog, blanket mire type of acid peat bog covering large stretches of country which develops in very wet climates on upland watersheds where drainage is impeded and the soil is acid.

-blast, -blastic word elements from Gk.

blastos, bud, signifying a cell or structure that can produce new cells.

blast cell, blast lymphocyte lymphocyte that has been stimulated by an antigen, has increased in size and will go on to divide.

blastema *n.* mass of undifferentiated cells that develops on end of amputated limb of e.g. some amphibians, reptiles, insects, etc. and from which limb is regenerated.

blastic *a. pert.* or stimulating growth by cell division.

blastocarpous *a.* developing while still surrounded by pericarp, *appl.* fruits.

blastocele blastocoel *q.v.*

blastocoel *n.* hollow fluid-filled cavity inside the blastula.

blastocyst *n.* a hollow sphere of cells developing from the morula in mammalian embryogenesis and which is the stage of implantation in the uterine wall.

blastocyte *n.* any undifferentiated embryonic cell.

blastoderm *n.* in insect eggs the layer of syncytial protoplasm that forms at periphery of egg and which develops into a layer of cells, the cellular blastoderm.

blastodisc *n.* in large yolky eggs as of birds and reptiles, a disc-shaped superficial layer of cells formed by cleavage and which will form the embryo. *alt.* germinal disc.

blastogenesis *n.* reproduction by budding.

blastokinesis *n.* movement of embryo in the egg, as in some insects and cephalopods.

blastomere *n.* any one of the cells formed by the first divisions of a fertilized egg.

blastopore *n.* indentation on surface of blastula in some animals at which invagination of gut cavity begins at the commencement of gastrulation.

blastospore *n.* spore developed by budding and itself capable of budding, as in yeast cells.

blastostyle *n.* specialized structure in which medusae develop in some colonial hydrozoans.

blastozooid *n.* an individual or zooid formed by budding.

blastula *n.* hollow ball of cells formed from the morula in embryogenesis in many animals.

blastulation *n.* development of a blastula.

blending inheritance the idea, now known to be mistaken, but widely current in the 19th century before the advent of Mendelian genetics, that the intermediate and novel characteristics seen in hybrids were due to the physical blending of hypothetical carriers of parental characteristics, rather than, as Mendel showed, to the inheritance of particular "factors" (genes) in different combinations, but which themselves remain unchanged.

blennogenous *a.* mucus-producing.

blennoid *a.* resembling mucus.

blephara *n.* a peristome tooth in mosses.

blepharal *a. pert.* eyelids.

blepharoplast *n.* basal body of flagella in flagellate protists.

blight *n.* insect or fungus disease of plants and also the agent causing it.

blind pit a pit in plant cell wall that is not backed by a complementary pit in opposing cell wall.

blind spot region of retina where optic nerve leaves and which is devoid of rods and cones.

blocky *a. appl.* soil crumbs of squarish or rectangular shape.

blood *n.* fluid circulating in the vascular system of animals, distributing nutrients, oxygen, hormones, etc., and taking up waste products for excretion. In vertebrates it contains cells specialized for oxygen transport (red blood cells or erythrocytes) and white blood cells (leukocytes) which are concerned with protection against infection and scavenging dead cells.

blood-brain barrier the structural and physiological barriers which prevent movement of most blood components into brain tissue or cerebrospinal fluid.

blood cell any cell forming a normal part of the blood, *see* erythrocyte, leukocyte, lymphocyte, haemocyte.

blood clotting a response to any wound that causes bleeding, in which liquid blood is converted into a solid clot that plugs the wound. Damaged blood vessels release local chemical mediators that cause the narrowing of blood vessels and adherence of platelets to vessel walls. Activation of plasma proteins (coagulation factors) results in the conversion of soluble plasma fibrinogen to insoluble fibrin, which forms a meshwork with platelets, red blood cells and other plasma proteins, forming a permanent clot. A genetically determined deficiency of coagulation factor VIII is the cause of classical haemophilia, in which the blood cannot clot.

blood corpuscle red or white blood cell, *see* erythrocyte, leukocyte.

blood gills delicate blood-filled sacs functioning in uptake of salts, in certain insects.

blood groups classification of the different types of blood in human populations, according to various antigenic systems, the main one being the ABO system. This is based on the presence or absence of either or both of two antigens (A or B) on the red blood cells and the presence of naturally occurring antibodies against the absent antigen(s) in the serum (e.g. α or anti-A in B individuals and β or anti-B in A individuals) which results in agglutination of red blood cells and "transfusion shock" if blood is transfused to an individual of a different blood group. AB individuals possess A and B antigens and are therefore universal recipients, being able to receive A, B, AB or O blood. O individuals lack either antigen on their red cells and therefore can only receive O blood, but are universal donors. Other blood group systems in humans include the Rh system (Rh positive and Rh negative), and the MN system.

blood islands groups of mesodermal cells in embryo in which primitive erythroblasts are enclosed by peripheral cells that develop into endothelium.

blood plasma the fluid part of the blood including the soluble blood proteins, which is left as a clear yellow fluid after whole vertebrate blood is centrifuged to remove the red and white cells. *cf.* blood serum.

blood platelet *see* platelet.

blood serum clear fluid which can be expressed from clotted blood or clotted blood plasma, and which comprises the fluid portion of the blood from which fibrin has been removed. It contains many plasma proteins, including the immunoglobulins (antibodies). *see also* antiserum, serum.

blood sugar glucose *q.v.*

blood vessel any vessel or space in which blood circulates (e.g. artery, vein or capillary), strictly used only in regard to special vessels with well-defined walls.

bloodworm *n.* reddish threadlike aquatic larva of the chironomid midge, *Chironomus riparius*, which is tolerant of heavy organic pollution; reddish oligochaete worm living in river mud; a red bristleworm found on muddy shores.

bloom *n.* waxy layer on surface of certain fruits, as grapes, some berries, etc.; blossom or flower; seasonal dense growth of algae or phytoplankton.

Bloom syndrome inherited condition characterized by low birth weight, short stature, extreme sensitivity to sunlight, and an enhanced frequency of sister-chromatid exchange during mitosis.

blubber *n.* insulating layer of fat in whales, seals, etc. lying between skin and muscle layer.

blue coral common name for corals of the genus *Heliopora*, having a solid calcareous skeleton with vertical tubular cavities containing polyps.

blue-green algae common name for the cyanobacteria *q.v.*

blue light receptor receptor in plants for light in the blue region of the spectrum (wavelength 400–500 nm) which is involved in phototropism, inhibition of stem growth, promotion of leaf expansion and induction of gene expression. The nature of the receptor(s) is still unknown but it is thought to be a flavoprotein. *alt.* cryptochrome.

blunt-end ligation technique used in the construction of recombinant DNAs in which any two DNA molecules may be joined.

B lymphocytes (B cells) small lymphocytes originating (in mammals) in bone marrow and found in large numbers in lymph nodes, spleen and other secondary lymphoid tissues, and in the blood. During its time in the bone marrow an individual B cell undergoes rearrangement of its immunoglobulin genes to produce genes encoding antibody of a single antigen-specificity. After encounter with antigen in secondary lymphoid tissues, B cells proliferate and after several cell generations differentiate into antibody-producing plasma cells.

body cavity the internal cavity in many animals in which various organs are suspended and which is bounded by the body wall.

body cell somatic cell as opposed to a germ or reproductive cell.

body fluids the fluid components of the animal body and fluids secreted or excreted by the body, including blood, lymph, urine, semen, sweat and secretions from mucous membranes.

body lice common name for members of the insect order Anoplura which are ectoparasitic on mammals (e.g. body louse, bed bug).

body stalk band of mesodermal tissue connecting tail end of embryo and chorion.

bog *n*. characteristic plant community developing on wet, very acid peat, containing e.g. sundews (*Drosera* spp.), sphagnum moss and bog myrtle (*Myrica gale*). *see* blanket bog, raised bog; sometimes also refers to alkaline bogs developing in valleys. *cf.* fen.

Bohr effect decrease in the affinity of haemoglobin for oxygen which occurs when pH is lowered and carbon dioxide concentration is increased, resulting in the release of oxygen from oxyhaemoglobin, as in metabolically active tissues.

boletes *n.plu*. basidomycete fungi of the genus *Boletus* and related genera, similar in general form to agarics, but in which the hymenium lines pores and not gills on the underside of the cap.

boletiform *a*. shaped like an elliptic spindle, *appl*. spores of some boletes.

boll *n*. a capsule or spherical pericarp as in cotton plant.

bolus *n*. a rounded mass; lump of chewed food.

bombesin *n*. peptide found in brain and gut, first isolated from skin of frog *Bombina bombina*.

bone *n*. calcified connective tissue forming the skeleton of some fishes, reptiles, amphibians, birds and mammals. It consists of an organic phase, mostly the protein collagen, and an inorganic phase composed of calcium phosphate as hydroxyapatite and other minerals. The organic matrix is secreted by osteoblasts which persist in mature bone as non-dividing osteocytes. Bone may be formed directly or by calcification of cartilage. *see also* compact bone, Haversian system, membrane bone, ossification, osteoblast, osteoclast, osteocyte.

bone-beds deposits formed largely by remains of bones of fishes and reptiles, as Liassic bone-beds.

bone cell *see* osteoblast, osteocyte.

bone marrow connective tissue filling up the cylindrical cavities of the long bones, the site of production of new blood cells from persistent pluripotent stem cells.

bonitation *n*. evaluation of the numerical distribution of a species in a particular locality or season, esp. in relation to agricultural, veterinary, or medical implications.

bony fishes common name for the Osteichthyes, a class of fishes with bony skeletons, usually possessing a swimbladder or lung, and gills covered by an operculum, and which include teleosts, lungfishes and crossopterygians.

book lice common name for some members of the Psocoptera, small wingless insects with a globular abdomen and incomplete metamorphosis, often living on paper.

book lung lung book *q.v.*

booted *a*. equipped with raised horny plates of skin, as feet of some birds.

bordered pit form of pit developed on walls of xylem vessels with overarching border of secondary cell wall.

boreal *a*. *pert*. or *appl*. northern biogeographical region; *appl*. the northern coniferous forest growing in that region; *pert*. post-glacial age with a continental type of climate.

Boreal kingdom phytogeographical kingdom comprising all of the Northern Hemisphere north of latitude 30°.

bosset *n*. the beginning of antler formation in deer in 1st year.

Botallo's duct ductus arteriosus, a small blood vessel representing the 6th gill arch and connecting pulmonary with systemic arch.

botany *n*. the branch of biology dealing with plants.

bothrenchyma *n*. plant tissue formed of pitted ducts.

bothridium *n*. muscular cup-shaped outgrowth from scolex of some tapeworms, used for attachment to host.

bothrionic *a*. *appl*. seta arising from the bottom of a pit in the integument.

bothrium *n.* a sucker.

botryoid(al) *a.* in the form of a bunch of grapes.

botryose racemose *q.v.*

bottleneck *n.* a sudden decrease in population density with a resulting decrease in genetic variability within a population.

botuliform *a.* sausage-shaped.

botulinum toxin protein produced by the bacterium *Clostridium botulinum* under anaerobic conditions, e.g. in inadequately sterilized canned or bottled food, a powerful poison affecting the nervous system and causing botulism.

bouquet *n.* bunch of muscles and ligaments connected with the styloid process of the temporal bone.

bourrelet *n.* poison gland associated with sting in ants.

bouton *n.* the enlarged terminal of an axon branch where it forms a synapse.

Bovidae, bovids *n.*, *n.plu.* mammalian family of the Artiodactyla comprising cattle, sheep, goats and bison, which have four-chambered stomachs and horns that are not shed.

Bowditch's Law all-or-none principle *q.v.*

Bowman's capsule the cup-like dilated end of a vertebrate kidney tubule, surrounding a glomerulus.

Bowman's glands glands in nasal mucous membranes secreting a watery fluid to wash over olfactory epithelium.

box-jellies a class of Cnidaria, the Cubozoa, with free-swimming cuboidal medusae with venomous tentacles at each corner, which can deliver a sting fatal to humans. *alt.* cubozoans, sea wasps.

braccate *a.* having additional feathers on legs or feet, *appl.* birds.

brachia *n.plu.* arms; two spirally coiled structures, one at each side of mouth in brachiopods. *sing.* brachium.

brachial *a. pert.* arms, arm-like.

brachialis *n.* a flexor muscle of the forearm.

brachiate *a.* branched; having arms; having opposite, widely spread, paired branches on alternate sides.

brachiating *n.* moving along by swinging the arms from one hold to another, as in the gibbon. *v.* to brachiate.

brachiation *n.* movement of forelimbs out

to the side, away from the median longitudinal plane; the act of brachiating *q.v.*

brachidia *n.plu.* calcareous skeleton supporting brachia in some brachiopods.

brachiocephalic *pert.* arm and head, *appl.* artery, vein.

brachiocubital *a. pert.* arm and forearm.

brachiolaria *n.* larval stage in some starfish.

Brachiopoda, brachiopods *n.*, *n.plu.* a small phylum of coelomate animals, the lamp shells, superficially resembling the bivalve molluscs but different in symmetry of the shell and internal structure. A characteristic structure is the lophophore, consisting of coiled tentacles (brachia) surrounding the mouth.

brachioradialis *n.* the supinator longus muscle in forearm, one of the muscles used in turning the palm of hand upwards.

brachium *n.* arm, or branching structure; forelimb of vertebrate; bundle of nerve fibres connecting cerebellum to cerebrum or pons.

brachy- prefix from Gk. *brachys*, short.

brachyblast brachyplast *q.v.*

brachyblastic *a.* with a short germ band.

brachycephalic *a.* short-headed.

Brachycera *n.* the short-horned flies, a suborder of Diptera with short stout antenna, which include the blood-sucking horse-flies and gadflies, the metallic-coloured soldier-flies, the slender, long-legged snipe-flies, the bee-flies, which have a strong superficial resemblance to bees and hover and feed in flowers, and the robber-flies, and other families.

brachycerous *a.* short-horned; with short antennae.

brachydactyly *n.* condition where fingers or toes are abnormally short.

brachydont, brachyodont *a. appl.* molar teeth with low crowns. *cf.* hypsodont.

brachyelytrous *a.* having short elytra.

brachyism *n.* dwarfism in plants caused by shortening of internodes.

brachyplast *n.* a short spur bearing tufts of leaves, occurring with normal branches on the same plant.

brachypleural *a.* with short pleura or side plates.

brachypodous *a.* with short legs, or stalk.

brachypterous *a.* with short wings, *appl.* insects.

brachysclereid(e) stone cell *q.v.*

brachystomatous *a*. with a short proboscis, *appl*. certain insects.

brachytic *a*. dwarfish, *appl*. plants in which dwarfism is caused by shortening of the internodes.

brachyural, brachyurous *a*.having short abdomen usually tucked in below thorax, *appl*. certain crabs.

brachyuric *a*. short-tailed.

brackish *a*. *appl*. water that contains dissolved salt (sodium chloride) in the range 0.5–30 parts per thousand, which is less than that in seawater (average 35 parts per thousand).

bract *n*. modified leaf in whose axil an inflorescence or flower arises; a floral leaf; a leaf-like structure.

bract scales small scales developed directly on axis of cones in conifers, not bearing ovules.

bracteate *a*. having bracts.

bracteiform *a*. like a bract.

bracteolate *a*. possessing bracteoles.

bracteole, bractlet *n*. secondary bract at the base of an individual flower.

bracteose *a*. having many bracts.

bractlet bracteole *q.v.*

bradyauxesis *n*. relatively slow growth; growth of a part at a slower rate that that of the whole.

bradycardia *n*. slowing of heart (and pulse) rate.

bradykinins *n.plu*. group of polypeptides found in blood and causing dilation of blood vessels and contraction of smooth muscle. *alt*. kinins.

bradytelic *a*. evolving at a rate slower than the standard rate.

brain *n*. the coordinating centre of the nervous system, found as an enlargement of the central nervous system of many animals, and most highly developed in vertebrates, where it forms an anterior continuation of the spinal cord and is enclosed in the bony cranium. The vertebrate brain is a highly organized mass of billions of interconnected nerve cells and supporting tissues, and is divided into three main regions: the forebrain, the midbrain and the hindbrain. The brain analyses and integrates incoming sensory information and generates output which is sent to muscles and glands. In invertebrates lacking a distinct brain the supraoesophageal or suprapharyngeal ganglia are the main coordinating centres.

brain stem part of brain at its base, before it becomes spinal cord, and which consists of the midbrain, pons and medulla oblongata.

brain vesicles three dilations at the anterior end of the embryonic neural tube which give rise to the forebrain, midbrain and hindbrain.

branch *n*. a taxonomic group used in different ways by different specialists but usually referring to a level between subphylum and class.

branch gaps gaps in the vascular cylinder of a main plant stem, subtending branch traces (vascular strands leading into lateral branches).

branch trace vascular bundle extending from stem into base of lateral bud.

branchia *sing*. of branchiae.

branchiae *n.plu*. gills (*q.v.*) of aquatic animals.

branchial *a*. *pert*. gills.

branchial basket framework of cartilaginous bars around the gill region in lampreys and cartilaginous fish.

branchial clefts gill slits *q.v.*

branchial grooves outer pharyngeal grooves or visceral clefts *q.v.*

branchial siphon in molluscs, the siphon through which water is drawn in over the gills.

branchiate *a*. having gills.

branchicolous *a*. parasitic on fish gills.

branchiform *a*. gill-like.

branchihyal *n*. an element of a gill arch *q.v.*

branching enzyme enzyme that catalyses the transfer of a segment of 1,4-α D-glucan chain to a primary hydroxyl group in a similar chain, forming a branchpoint in a polysaccharide chain. EC 2.4.1.18.

branchiocardiac *a*. *pert*. gills and heart; *appl*. a vessel leading off ventrally from ascidian heart; *appl*. vessels conveying blood from gills to pericardial cavity in certain crustaceans.

branchiomeric *a*. *appl*. muscles derived from gill arches.

branchiopallial *a*. *pert*. gill and mantle of molluscs.

branchiopneustic *a*. *appl*. insects having

spiracles replaced functionally by gills.

Branchiopoda, branchiopods *n., n.plu.* the water fleas, brine shrimps and their allies, a subclass of mainly freshwater crustaceans whose carapace, if present, forms a dorsal shield or bivalve shell, and which have broad lobed trunk appendages fringed with hairs.

branchiostegal *a.* with, or *pert.* a gill cover, *appl.* membrane, rays.

branchiostegal ray one of a group of dermal bones ventral to gill cover in bony fishes.

branchiostegite *n.* expanded lateral portion of carapace forming gill cover in some crustaceans.

Branchiura *n.* class of crustaceans, commonly called fish lice, that are ectoparasites on fish and amphibians.

brand *n.* a burnt appearance of leaves, caused by rust and smut fungi.

brand fungi a common name for the rust and smut fungi *q.v.*

brand spore teleutospore *q.v.*

breakage and reunion generally accepted model for crossing-over and recombination during meiosis, in which chromatids break and rejoin in a way that differs from the original, resulting in a visible chiasma.

breeding season a period each year, for animals that do not breed all the year round, in which animals court, mate and rear their young.

breeding size the number of individuals in a population actually involved in reproduction, in a generation.

breeding system the extent and mode of interbreeding within a species or group of closely related species.

bregma *n.* that part of skull where frontals and parietals meet; intersection of sagittal and coronal sutures.

brephic *a. appl.* a larval form preceding the adult form.

brevi- prefix from L. *brevis*, short.

brevicaudate *a.* with a short tail.

brevilingual *a.* with a short tongue.

brevipennate *a.* with short wings, *appl.* birds.

brevirostrate *a.* with a short beak.

brevissimus oculi the obliquus inferior, the shortest of the eye muscles.

bridge *n.* a chromosomal arrangement seen

at meiosis produced from a dicentric chromosome in which the two centromeres are segregating to opposite poles, and which may lead to breakage of the chromosome.

brigalow forest acacia forest covering large areas in Australia, in which the dominant species is the brigalow (*Acacia harpophylla*).

bright-field microscopy technique of optical microscopy in which a living cell is viewed by direct transmission of light through it.

brille *n.* transparent covering over the eyes of snakes.

brine shrimp small marine crustacean of the subclass Branchiopoda, usually refers to *Artemia*.

bristle worms common name for the Polychaeta, a class of annelid worms.

bristletails common name for certain insects of the orders Thysanura and Diplura, small wingless insects characterized by a single or two-pronged bristle at the tail end.

brittle stars common name for the Ophiuroidea, a class of echinoderms with five slender arms clearly marked off from the central disc.

broadleaf, broadleaved *a. appl.* trees, name commonly given to the angiosperm trees of temperate climates, characterized by thin flat leaves (as opposed to the needle-bearing conifers).

Broca's area region of left frontal cerebral cortex involved in speech production. *alt.* area 44.

Brodman areas small areas into which the human cerebral cortex is conventionally divided, each being numbered. Areas 17 and 18, for example, comprise the primary visual cortex. The division is based on the cytology and detailed morphology of the cortex.

bromatium *n.* a hyphal swelling on a fungus cultivated by ants, and which they use as food.

bromelain *n.* proteolytic and milk clotting enzyme found in pineapple. EC 3.4.22.4. *alt.* bromelin.

Bromeliales, bromeliads *n., n.plu.* order of terrestrial and epiphytic monocots with reduced stem and rosette of fleshy water-storing leaves and comprising the family Bromeliaceae (pineapple).

bromodeoxyuridine, bromouracil (BrdU, BrU) thymidine analogue in which the methyl group of thymine has been replaced by a bromine atom, and which causes mutations when incorporated into DNA because it can pair with guanine as well as adenine. Used as an anticancer drug to preferentially kill rapidly dividing cancer cells.

bromovirus group plant virus group named after the type member, brome mosaic virus, which is a small isometric single-stranded RNA virus, causing mosaic symptoms (mottling of leaves). They are multicomponent viruses in which four genomic RNAs are encapsidated in three different virus particles.

bronchi *n.plu.* tubes connecting the trachea (windpipe) and the lungs. *sing.* bronchus.

bronchia *n.plu.* the subdivisions or branches of each bronchus.

bronchial *a.pert.* bronchi.

bronchiole *n.* small terminal branch of a bronchus.

bronchopulmonary *a. pert.* bronchi and lungs.

bronchus *sing.* of bronchi.

brood *n.* the offspring of a single birth or clutch of eggs; any young animals being cared for by adults.

brood cell chamber built to house immature stages of insects.

brood parasite an animal that lays its eggs in the nest of another member of the same species (intraspecific brood parasitism) or of a different species (interspecific brood parasitism), who then rears them.

brood parasitism in birds, the placing of eggs of one species into the nest of another, which then raises the brood as if it were their own.

brood pouch sac-like cavity in which eggs or embryos are placed.

brown algae common name for the Phaeophyceae, mainly marine algae (seaweeds) containing the brown pigment fucoxanthin.

brown earths, brown forest soils dark brown friable soils associated with areas of the Earth's land surface originally covered with deciduous forest.

Brownian movement movement of small particles such as pollen grains, bacteria, etc. when suspended in a colloidal solution, due to their bombardment by molecules of the solution.

brown fat highly vascularized adipose tissue rich in mitochondria, the cytochromes of which help to give it a brown colour, and which is involved in thermoregulation in hibernators and in young mammals generally. It typically occurs around the shoulder blades, neck, heart, large blood vessels and lungs. It is specialized for heat generation as a result of uncoupling of fatty acid oxidation and electron transfer in mitochondria from ATP synthesis. *alt.* brown adipose tissue (BAT).

brown podzolic soil acid forest soil with a layer of litter over a greyish-brown organic and mineral layer and a pale leached layer below.

brown soils soils similar to chernozems (*q.v.*) but found in warmer and drier areas and supporting short grassland.

Bruch's membrane a thin basal membrane forming the inner layer of the choroid.

Brunner's glands much branched glands in the submucosa of the duodenum which open into the crypts of Lieberk ühn and which with the crypts secrete digestive juice containing various digestive enzymes and mucus. *alt.* duodenal glands.

brush border dense covering of minute fingerlike projections (microvilli) on the surface of epithelial cells lining the lumen of the intestine, and on renal epithelium.

Bryidae true mosses *q.v.*

bryocole *n.* animal living among moss.

bryology *n.* the branch of botany that deals with mosses and liverworts.

Bryophyta *n.* the mosses, a division of small spore-bearing, non-vascular green plants comprising the classes Sphagnidae (sphagnum or peat mosses), Andreaeidae (granite or rock mosses) and Bryidae (true mosses). With the liverworts and hornworts they comprise the bryophytes (*q.v.*). They are characterized by a leafy upright or creeping gametophyte with multicellular rhizoids, a ring of teeth (peristome) around the mouth of the spore capsule, no elaters in the sporangia, and spores germinating to produce a filamentous young gametophyte (a protonema) consisting of a branched chain of cells. The sporophyte, in the form of a stalked capsule containing the spores, is generally dependent on the gametophyte. Specialized conducting tissue is often present. *see* granite mosses, sphagnum mosses, true mosses. *alt.* Bryopsida, Muscopsida. *see also* Anthocerophyta, Hepatophyta. *see also* Appendix 2.

bryophytes *n.plu.* group of simple non-vascular spore-bearing green plants containing the mosses (Bryophyta *q.v.*), liverworts (Hepatophyta *q.v.*) and hornworts (Anthocerophyta *q.v.*). They all have well-marked alternation of generations, with an independent leafy or thalloid photosynthetic gametophyte and a sporophyte completely or partly dependent on it. The vegetative plant body is divided into a leafy or thalloid shoot and hair-like rhizoids which anchor the plant to the ground.

Bryopsida Bryophyta *q.v.*

Bryozoa, bryozoans *n., n. plu.* moss animals. *see* Ectoprocta.

buccae *n.plu.* the cheeks.

buccal *a. pert.* the cheek or mouth.

buccal cavity part of the alimentary tract between mouth and pharynx.

buccinator *n.* a broad thin muscle of the cheek.

buccolabial *a. pert.* mouth cavity and lips.

buccolingual *a. pert.* cheeks and tongue.

bucconasal *a. pert.* cheek and nose; *appl.* membrane closing posterior end of nasal pit.

buccopharyngeal *a. pert.* mouth and pharynx.

bud *n.* structure from which shoot, leaf or flower develops; incipient outgrowth, as limb buds in embryo, from which limbs develop.

budding *n.* the production of buds; (*zool.*) method of asexual reproduction common in sponges, coelenterates and some other invertebrates, in which new individuals develop as outgrowths of the parent organism, and may eventually be set free; (*bot.*) artificial vegetative propagation by insertion of a bud within the bark of another plant; (*mycol.*) cell division by the outgrowth of a new cell from the parent cell; (*virol.*) release of certain animal viruses from the cell by their envelopment in a piece of plasma membrane which subse-

quently pinches off from the cell.

budding yeast *Saccharomyces* and related species, which multiply by budding a small new daughter cell off the parent cell. *cf.* fission yeast.

budget *n. see* energy budget, time–energy budget.

buffer *n.* salt solution which minimizes changes in pH when an acid or alkali is added; any factor that reduces the impact of external changes on a system.

bufonin *n.* a toxin present in toads.

bufotoxin *n.* a toxin present in toads.

bugs common name for insects of the order Hemiptera *q.v.*

bulb *n.* (*bot.*) specialized underground reproductive organ consisting of a short stem bearing a number of swollen fleshy leaf bases or scale leaves, the whole enclosing next year's bud; (*general*) any part or structure resembling a bulb, a bulb-like swelling.

bulbar *a. pert.* bulb or a bulb-like part.

bulbiferous *a.* bearing bulbs or bulbils.

bulbil *n.* a fleshy axillary bud which may fall and produce a new plant, as in some lilies; any small bulb-shaped structure or swelling.

bulbonuclear *a. pert.* medulla oblongata and nuclei of cranial nerves.

bulbo-urethral *a. appl.* two branching glands opening into the bulb of the male urethra. *alt.* Cowper's or Mery's glands, antiprostate.

bulbous *a.* like a bulb; developing from a bulb; having bulbs.

bulbus *n.* knob-like or bulb-like structure or swelling.

bulla *n.* a rounded prominence, such as the bony projection of skull encasing the middle ear in many tetrapod vertebrates; various other bubble-shaped structures.

bullate *a.* appearing blistered; puckered like a savoy cabbage leaf.

bulliform *a.* bubble-shaped; *appl.* thin-walled cells which cause rolling, folding or opening of leaves by changes in turgor.

bundle scar traces of the vascular bundle remaining on a leaf scar after leaf fall.

bundle sheath one or more layers of large parenchyma or sclerenchyma cells surrounding a vascular bundle.

bundle sheath cells in some tropical

plants, the cells in photosynthetic tissues in which carbon dioxide incorporated into aspartate and malate in the C4 pathway is released and enters the Calvin cycle.

α-bungarotoxin poison obtained from the venom of snakes of the genus *Bungarus*, binds specifically to acetylcholine receptors.

bunodont *a. appl.* molar and premolar teeth with low crowns and cusps, as those of e.g. pigs, monkeys and humans.

bunoid *a. appl.* cusps of molar teeth, low and conical.

bunolophodont *a.* between bunodont and lophodont (having transverse ridges) in structure, *appl.* molar teeth.

bunoselenodont *a. appl.* molar teeth having internal cusps bunoid, external cusps crescent-shaped.

bunt fungi smut fungi *q.v.*

Bunyaviridae *n.* enveloped, spherical single-stranded, segmented RNA viruses, including the Bunyamwera virus and Rift Valley fever viruses.

burden *n.* of parasites, the total number or mass of parasites infecting an individual.

Burkitt's lymphoma a childhood lymphoma, rare except in certain regions of Africa, where Epstein-Barr virus infection and malarial infection are together thought to precipitate the disease.

bursa *n.* pouch or sac-like cavity; sac containing a viscid fluid that prevents friction at joints; (*imm.*) bursa of Fabricius *q.v. a.* bursal.

bursa copulatrix region of female genitalia in insects that receives the adeagus and sperm during copulation; genital pouch of various animals.

bursa of Fabricius a pouch of lymphoid tissue opening into the cloaca in birds, degenerating during adolescence, a site of development of B lymphocytes of the immune system.

bursa seminalis fertilization chamber of female genital ducts, as in turbellarians.

bursicle *n.* in orchid flowers, a flap- or purse-like structure surrounding the sticky disc at the base of stalk of pollinium, and containing a sticky liquid to prevent the disc drying up.

bursicule *n.* a small sac.

burster neurone neurone that typically produces rythmic bursts of impulses when activated.

bush *n.* small shrub; vegetation cover composed of grassland and shrubs.

bush layer the horizontal ecological stratum of a plant community composed of shrubs, which is higher than the field or herb layer and lower than the tree layer. *alt.* shrub layer.

butterflies common name for members of the Lepidoptera (*q.v.*) which have clubbed antennae.

butterfly bone sphenoid *q.v.*

buttress roots branch roots given off above ground, arching away from stem before entering the soil, forming additional support for trunk.

butyrin(e) *n.* one of the three glycerides of butyric acid, esp. tributyrin.

byssaceous byssoid *q.v.*

byssal *a. pert.* the byssus.

byssoid *a.* formed of fine threads, resembling a byssus.

byssus *n.* tuft of strong filaments secreted by the byssogenous gland of certain bivalve molluscs, by which they become attached to substrate; the stalk of certain fungi.

C

C symbol for the element carbon; cysteine *q.v.*; cytosine *q.v.*; Calorie (= 1000 cal).

Ca symbol for the element calcium.

C$_H$ constant region of antibody heavy chain. *see* C region.

C$_L$ constant region of antibody light chain. *see* C region.

C1 complement component, comprising 3 proteins, C1q, C1r and C1s, which binds to antibody-antigen complexes on cell surfaces to initiate the "classical pathway" of complement action.

C1(q, r, and s), C2, C3(a and b), C4, C5, C6, C7, C8, C9 components of the complement system *q.v.*

C3a complement component, derived from C3 by enzymatic cleavage, causes local blood vessel dilation. and attracts polymorphonuclear leukocytes.

C5a complement component, derived from C5 by enzymatic cleavage, causes blood vessel dilation and attracts polymorphonuclear leukocytes.

C6b, C6, C7 complement components that in free form attract polymorphonuclear leukocytes.

C9 complement component that forms pores in cell membrane leading to lysis.

Ca^{2+} ATPase protein in plasma membrane (and other membranes) of eukaryotic cells that actively transports calcium ions across the membrane.

cal calorie *q.v.*

CAM cell adhesion molecule *q.v.*; crassulacean acid metabolism *q.v.*

CaMV cauliflower mosaic virus, a plant DNA virus, infecting brassicas and a few other families of dicotyledonous plants.

cAMP cyclic AMP *q.v.*

CAP catabolite gene activator protein *q.v.*

CAT chloramphenicol acetyltransferase.

CCK cholecystokinin *q.v.*

C-DNA form of DNA occurring *in vitro* in the presence of lithium ions, with fewer base pairs per turn than B-DNA.

cDNA complementary DNA *q.v.*

CDP cytidine diphosphate *q.v.*

CDV canine distemper virus, a paramyxovirus affecting dogs and other canids.

CD1, CD2, CD3, CD4 etc. *see* CD antigens, cluster of differentiation.

CD25 the interleukin-2 receptor.

CEA carcinoembryonic antigen *q.v.*

CF complement fixation *q.v.*; cystic fibrosis *q.v.*

CF$_1$, CF$_0$ chloroplast factor *q.v.*

CFU colony-forming unit *q.v.*

CFU-E erythrocyte colony-forming unit *q.v.*

cGMP cyclic GMP *q.v.*

Ci curie *q.v.*

CML chronic myelogenous leukaemia.

CMP cytidine monophosphate *q.v.*

CMV cowpea mosaic virus.

CNS central nervous system *q.v.*

CoA coenzyme A *q.v.*

COD chemical oxygen demand *q.v.*

Con A concanavalin A *q.v.*

c-*onc* general designation for cellular oncogenes (also known as protoncogenes). *see also* v-*onc*, oncogene.

CoQ coenzyme Q, ubiquinone *q.v.*

CPE cytopathic effect *q.v.*

Cot a measure of DNA reassociation, being the product of single-stranded DNA concentration at time 0 (C$_0$) and the time of incubation (t), Cot being the value when reassociation is half complete. Given a standard amount of starting DNA, the larger and more complex the genome the greater the Cot value due to the longer period of incubation needed for reassociation of similar sequences to occur.

CRE cyclic AMP-responsive element, a control site in certain mammalian genes that is responsible for their activation in response to stimulation of cyclic AMP production within the cell.

CRF corticotropin-releasing factor *q.v.*

CRP cyclic AMP receptor protein, catabolite gene activator protein *q.v.*

CS conditioned stimulus *q.v.*

CSF cerebrospinal fluid *q.v.*

CSF-1 macrophage colony-stimulating factor, stimulates production of macrophages from bipotential macrophage-granulocyte precursor cells.

CTL cytotoxic T lymphocyte *q.v.*

CTP cytidine triphosphate *q.v.*

Cys cysteine *q.v.*

cyt cytochrome *q.v.*

$^{12}C/^{13}C$ ratio a ratio of the proportion of ^{12}C to ^{13}C in a geological deposit, etc. which determines whether or not it has originated from living matter. The difference arises because photosynthetic enzymes discriminate between the two isotopes, preferentially incorporating ^{13}C into living matter from carbon dioxide. The two isotopic forms of carbon exist in atmospheric carbon dioxide and other inorganic carbon compounds in a stable ratio.

CAAT box a conserved sequence in the promoter region of some eukaryotic genes about 70–80 base pairs upstream from the startpoint of transcription, and which is involved in control of initiation of transcription.

cacao *n.* plant (*Cacao theobroma*) whose beans provide the raw material for cocoa and chocolate manufacture.

cachectin *n.* a protein also known as tumour necrosis factor. Its entirely separate role as cachectin involves inhibiting enzymes of lipid utilization, disturbing lipid metabolism and causing cachexia (wasting), as in patients with some cancers.

cacogenesis *n.* inability to hybridize.

Cactales *n.* the cacti, an order of succulent dicots, often bearing clusters of spines, leaves usually absent or reduced. One family, the Cactaceae.

cacuminous *a.* with a pointed top, *appl.* trees.

cadavericole *n.* animal that feeds on carrion. *alt.* carrion feeder.

cadaverine *n.* a foul-smelling toxic polyamine, $H_2N(CH_2)_5NH_2$, formed by decarboxylation of lysine and produced during bacterial breakdown of protein, e.g. in putrefying meat.

caddis flies common name for the Trichoptera, an order of insects somewhat resembling moths, with weak flight and mouth parts adapted for licking, with aquatic larvae, caddis worms, that construct protective cases.

cadherin *n.* any member of a structurally related family of cell-surface proteins in animals that cause a strong calcium-dependent adhesion of one cell to another by cadherin molecules on each cell binding to each other.

cadophore *n.* a dorsal outgrowth bearing buds in certain tunicates.

caducous *a. pert.* parts that fall off early.

caecal *a.* ending without outlet, *appl.* to stomach with cardiac part prolonged into blind sac; *pert.* caecum.

caecilians *see* Apoda.

caecum *n.* a blind-ended diverticulum or pouch from part of the alimentary canal, or other hollow organ. *plu.* caeca.

Caenorhabditis elegans a species of small soil nematode, much used as an experimental subject in developmental biology.

Caenozoic Cenozoic *q.v.*

caespitose, cespitose *a.* having low, closely matted stems; growing densely in tufts.

caffeine *n.* 1,3,7-trimethylxanthine, a purine with a bitter taste, found in coffee, tea, matè and kola nuts, which is a stimulant of the central nervous system and a diuretic.

caino- *alt.* caeno-, ceno-, kaino-.

Cainozooic Cenozoic *q.v.*

caisson *n.* box-like arrangement of longitudinal muscle fibres in earthworms.

calactin *n.* cardiac glycoside present in the milkweeds (Asclepiadaceae).

calamistrum *n.* a comb-like structure on metatarsus of certain spiders.

calamus *n.* hollow reed-like stem without nodes; the quill of a feather.

calcaneus *n.* the heel or heel-bone.

calcar *n.* (*bot.*) a hollow prolongation or tube at base of petal or sepal; (*zool.*) spur-like process on leg or wing of birds; bony process on heel-bone that supports web between wing and tail in bats.

calcar avis a protuberance in posterior part of lateral ventricle of brain, the hippocampus minor.

calcarate *a*. spurred, *appl*. petal, corolla.

calcareous *a*. composed chiefly of calcium carbonate (lime); growing on limestone or chalky soil; *pert*. limestone.

calcareous sponges sponges of the class Calcarea, with a skeleton of 1-, 3- or 4-rayed spicules composed chiefly of calcite (calcium carbonate).

calcariform *a*. spur-shaped.

calcarine *a. appl*. a fissure extending to the hippocampal gyrus, on medial surface of cerebral hemisphere.

calceolate *a*. slipper-shaped, *appl*. corolla.

calcicole *n*. plant that thrives in soil rich in lime or other calcium salts. *a*. calcicolous, *appl*. grassland.

calciferol *n*. an unsaturated alcohol, vitamin D_2, which can be made by ultraviolet irradiation of ergosterol, and which is found in fish liver oils, egg yolk, milk, etc. It controls levels of calcium and phosphorus in the body and prevents rickets.

calciferous *a*. containing or producing calcium salts.

calcification *n*. deposition of calcium salts in tissue; accumulation of calcium salts in soil. *a*. calcified.

calcifuge *n*. plant that thrives only in soils poor in lime and usually acid. *a*. calcifugous, *appl*. grassland.

calcigerous calciferous *q.v.*

calciphile *n*. calcicole *q.v.*

calciphobe calcifuge *q.v.*

calciphyte calcicole *q.v.*

calcite *n*. a crystalline form of calcium carbonate, one of the constituents of mollusc shells and the skeletons of calcareous sponges.

calcitonin *n*. peptide hormone secreted by the thyroid and/or parathyroid gland in mammals and by the ultimobranchial bodies in other vertebrates, which lowers the level of calcium in the blood, opposing the activity of parathyroid hormone, by reducing release of calcium from bone.

calcium (Ca) *n*. an essential macronutrient. As the ion Ca^{2+}, an important regulatory ion and second messenger in living cells, involved e.g. in regulation of enzyme activity, stimulation of muscle contraction and control of secretion. As calcium salts (e.g. calcium phosphates), a major constituent of bone.

calcium-binding protein protein that binds Ca^{2+}, which usually results in a conformational change in the protein, either activating or inactivating it. Calcium-binding proteins are the main agents mediating the effects of calcium in living cells.

calcium channel ion channel in a biological membrane which allows the passage of calcium ions.

calcium cycle the movement of calcium from inorganic sources in the soil and water, from which it is taken up by plants and microorganisms, esp. into shells and bone, its passage through the food chain, and its return to the inorganic environment.

calcium pump Ca^{2+}-ATPase *q.v.*

caldesmon *n*. calcium-binding protein abundant in smooth muscle and associated with actin filaments in other cells, and which may be involved in the calcium-dependent control of contraction in smooth muscle.

calice, calicle calycle *q.v.*

Caliciviridae, caliciviruses *n., n.plu*. family of icosahedral, single-stranded RNA viruses including vesicular exanthema of swine.

caligate *a*. sheathed; veiled.

callosal *a. pert*. corpus callosum *q.v.*

callose *a*. having hardened thickened areas on skin or bark; *n*. amorphous polysaccharide of glucose, usually found on sieve plates in phloem but also in parenchyma cells after injury.

callow workers in colonies of social insects, newly emerged adult workers, having exoskeletons that are still rather soft and lightly pigmented.

callunetum *n*. plant community dominated by ling (*Calluna vulgaris*).

callus *a*. small hard outgrowth or swelling; mass of hard tissue that forms over cut or damaged plant surface; mass of undifferentiated cells that initially arises from plant cell or tissue in artificial culture.

calmodulin *n*. ubiquitous calcium-binding protein abundant in eukaryotic cells, forming the regulatory subunit of glycogen phosphorylase. Undergoes a conformational change on binding Ca^2+ and associates with many other proteins, mediating regulation of their activity by Ca^2+.

calobiosis *n.* in social insects, when one species lives in the nest of another and at its expense.

calorie (cal) *n.* unit of quantity of heat: amount of heat required to raise temperature of 1 g water by 1 °C from 14.5 to 15.5 °C. 1 kilocalorie (kcal, Calorie, C) = 1000 cal, and is the unit commonly used to denote energy value of food and daily energy requirement of a person. Becoming replaced by the SI unit the joule: 1 kcal = 4184 joules.

calorific *a.* heat-producing.

calorigenic *a.* promoting oxygen consumption and heat production.

calorimetry *n.* the measurement of heat, in animal physiology used to measure heat production and thus metabolic rate.

calotte *n.* an external cell group or polar cap in Dicyemidae, for adhesion to kidney of cephalopods; lid of an ascus.

calsequestrin *n.* calcium-binding protein from sarcoplasmic reticulum of muscle.

calvaria *n.* the dome of the skull.

Calvin cycle the cycle of reactions in the stroma of chloroplasts in which ATP and NADPH produced during the light reaction of photosynthesis provide energy and reducing power for the incorporation of carbon dioxide into carbohydrate. The first reaction is that of ribulose-1,5-bisphosphate with carbon dioxide to form 3-phosphoglycerate. This is converted in several stages to reform ribulose-1, 5-bisphosphate, producing in the process glyceraldehyde-3-phosphate, which is the precursor of starch, amino acids, fatty acids and sucrose.

Calycerales *n.* an order of herbaceous dicots comprising the family Calyceraceae.

calyces *plu.* of calyx.

calyciflorous *a. appl.* flowers in which stamens and petals are adnate to the calyx.

calyciform, calycoid *a.* calyx-like in shape.

calycine *a. pert.* a calyx; cup-like.

calycle *n.* a cup-shaped cavity in a coral; a theca in a hydroid; the calyx of a flower.

calycoid *a.* like a calyx in shape.

calyculus *n.* cup-shaped or bud-shaped structure; calycle *q.v.*

calypter *n.* a modified wing sheath covering haltere in certain flies.

calyptobranchiate *a.* with gills not visible from outside.

calyptopsis *n.* a larva of some crustaceans.

calyptra *n.* the enlarged archegonial wall surrounding the developing sporophyte in bryophytes, which in some cases persists as a protective covering to the spore capsule of the sporophyte.

calyptrate *a. appl.* a calyx that falls off, separating from its lower portion or thalamus; having a lid, as capsules, seed pods, etc.; (*zool.*) *appl.* certain flies which have halteres hidden by scales.

calyptrogen *n.* meristem giving rise to the root cap independently of other initials in an apical meristem. *cf.* calyptra.

calyptron calypter *q.v.*

calyx *n.* the sepals collectively, forming the outer whorl of the flower; various structures resembling the calyx of a flower, as cup-like body of crinoids.

cambial *a. pert.* cambium.

cambiform *a.* similar to cambial cells.

cambiogenetic *a. appl.* cells that produce cambium.

cambium *n.* the meristematic tissue from which secondary growth occurs in roots and shoots, producing xylem from one face and phloem from the other.

Cambrian *a. pert.* or *appl.* geological period lasting from about 590 to 505 million years ago and during which many phyla of multicellular animals first arose.

cameration *n.* division into a large number of separate chambers.

campaniform *a.* bell- or dome-shaped.

Campanulales *n.* an order of mainly herbaceous dicots, often with latex vessels, and including the families Campanulaceae (bellflower), Goodeniaceae and Lobeliaceae (lobelia).

Campanulatae Campanulales *q.v.*

campanulate *a.* bell-shaped, *appl.* corolla.

campodeiform *a.* flattened and elongated with well-developed legs and antennae, *appl.* lacewing and certain beetle larvae.

camptodactyly *n.* crookedness of little finger due to a congenital shortness of tendon, often due to inheritance of a simple

Mendelian dominant allele.

camptodrome *a. pert.* leaf venation in which secondary veins bend forward and anastomose before reaching end of leaf.

camptotrichia *n.plu.* jointed dermal fin-rays in some primitive fishes.

campylodrome *a. appl.* leaf with veins converging at its point.

campylospermous *a. appl.* seeds with groove along inner face.

campylotropous *a. appl.* ovules bent so that the funicle appears attached to the side halfway between the chalaza and micropyle.

canaliculus *n.* a minute channel, e.g. containing bile in liver; small channel for passage of nerves through bone. *plu.* canaliculi. *a.* canalicular.

canaliform *a.* canal-like.

canalizing selection selection for phenotypic characters which is largely unaffected by environmental fluctuations and genetic variability.

canavanine *n.* amino acid found in jack bean (*Canavalia*).

cancellated, cancellous *a.* consisting of slender fibres and lamellae, which join to form a meshwork, *appl.* inner spongy portion of bone.

cancer *n.* malignant, ill-regulated proliferation of cells, causing either a solid tumour or other abnormal conditions. Usually fatal if untreated. Cancer cells are abnormal in many ways, esp. in their ability to multiply indefinitely, to invade underlying tissue and to migrate to other sites in the body and multiply there (metastasis).

cancerous *a. appl.* cells that have undergone certain changes which enable them to divide indefinitely and invade underlying tissue. *alt.* malignant.

cancrisocial *a. appl.* commensals with crabs.

Candida genus of yeasts containing the normal component of the body flora and opportunistic pathogen *C. albicans*, which causes candidiasis.

candidiasis *n.* an infection caused by the yeast-like fungus *Candida albicans*, a normal inhabitant of the body. It commonly takes the form of thrush (infection of mucous membranes of either mouth or vagina), or, in immunosuppressed or

otherwise debilitated patients, a more serious systemic infection of the lungs.

cane sugar sucrose *q.v.*

canid *n.* member of the mammalian family Canidae, including the dogs, wolves, foxes, jackals and coyotes.

canine *a. pert.* to a dog, or the genus *Canis*; *n.* the tooth next to the incisors.

caninus *n.* muscle from canine fossa to angle of mouth, which lifts the corner of the mouth.

cannabis *n.* Indian hemp, *Cannabis sativa*, from which the hallucinogenic drug cannabis or marijuana is extracted, also cultivated for its stem fibres (hemp) used to make rope.

cannibalistic *a.* eating the flesh of one's own species.

cannon bone bone supporting limb from hock to fetlock.

canoids *n.plu.* mammals of the dog, hyena, bear, panda and related families.

canonical sequence consensus sequence *q.v.*

canopy *n.* the cover formed by the branches and leaves of trees in a wood or forest.

canthal *a. pert.* the angle at which upper and lower eyelids meet, *appl.* a scale in certain reptiles.

cantharidin *n.* toxic irritant and blister-causing terpene derivative produced from the Spanish fly and other beetles of the family Meloidae, blister beetles.

canthus *n.* angle where upper and lower eyelids meet. *a.* canthal.

cap *n.* in eukaryotic mRNA, a structure found at the 5' end of the molecule, comprising a terminal methylguanosine residue and sometimes methylations at other sites, and which is formed after transcription.

capacitation *n.* little-understood process whereby mammalian sperm undergo a final stage of maturation in contact with the secretions of the female genital tract, without which they are not capable of fertilization.

cap binding proteins accessory proteins needed for initiation of translation in eukaryotic (but not bacterial) cells and which are involved in ribosome binding to the cap region of mRNA (the first step in translation of eukaryotic mRNA).

capillary *a.* hair-like; *appl.* moisture held in

and around particles of soil; *appl.* spontaneous creeping movement of water in very fine tubes; *n.* one of the minute thin-walled vessels which form networks in various parts of the body, e.g. blood, lymph or biliary capillaries.

capillitium *n.* sterile, thread-like structure present among the spores in fruiting bodies of many Myxomycetes and Gasteromycetes. *plu.* capillitia.

capillovirus group plant virus group composed of filamentous single-stranded RNA viruses, type member apple stem grooving virus.

capitate *a.* enlarged or swollen at tip; gathered into a mass at apex, *appl.* inflorescence.

capitatum *n.* the third carpal bone.

capitellum capitulum *q.v.*

capitular *a. pert.* a capitulum.

capitulum *n.* (*zool.*) knob-like swelling on end of a bone, e.g. on humerus; part of cirriped body enclosed in mantle; swollen end of hair or tentacle; enlarged end of insect proboscis or antenna; (*bot.*) inflorescence forming a head of sessile florets crowded together on a receptacle and often surrounded by an involucre, as of dandelion and certain other Compositae.

Capparales *n.* order of dicot herbs, shrubs, small trees and lianas, including the families Brassicaceae (Cruciferae) (mustard, etc.), Capparaceae (caper) and Resedaceae (mignonette).

capping *n.* energy-requiring process in which membrane proteins induced to cluster by treatment with lectins or specific antibody eventually collect at one end of the cell to form a "cap"; RNA capping *see* cap.

capreolate *a.* having tendrils.

caprification *n.* the pollination of flowers of fig trees by chalcid wasps.

Caprimulgiformes *n.* an order of birds including the nightjars.

capsaicin *n.* a pungent compound found in chilli peppers.

capsid *n.* the external protein coat of a virus particle; common name for a bug of the family Capsidae.

capsomere *n.* one of the protein units of which a virus capsid is made.

capsula glomeruli Bowman's capsule *q.v.*

capsular *a.* like or *pert.* a capsule; *appl.* dry dehiscent many-seeded fruits.

capsule *n.* a sac-like membrane enclosing an organ; any closed box-like vessel containing spores, seeds or fruits; membrane surrounding the nerve cells of sympathetic ganglia; (*bot.*) a one- or more celled, many-seeded dehiscent fruit; in bryophytes, the portion of the sporogonium containing the spores; (*bact.*) thick slime layer surrounding certain bacteria, composed of polysaccharides, or more rarely, polypeptides.

capsuliferous, capsuligerous, capsulogenous *a.* with, or forming, a capsule.

captacula *n.plu.* exsertile filamentous tactile organs with sucker-like ends near mouth of tusk-shells (Scaphopoda).

Captorhinida, captorhinids *n., n.plu.* the earliest and most primitive order of extinct reptiles known, which evolved from the amphibians in the Late Carboniferous and is found until the end of the Triassic. *alt.* cotylosaurs.

caput *n.* head; knob-like swelling at apex of a structure.

carapace *n.* bony plates beneath the horny shell of tortoises and other chelonians; chitinous covering in crustaceans starting behind the head and covering the whole or part of the trunk.

carbohydrates *n.plu.* compounds of carbon, oxygen and hydrogen, of general formula $C_x(H_2O)_y$, including sugars (monosaccharides and disaccharides) and their derivatives, and polysaccharides such as starch and cellulose.

carbon-14 *see* radiocarbon.

carbon cycle the various biological processes by which carbon from atmospheric carbon dioxide, CO_2, enters the biosphere by photosynthetic fixation of carbon into organic compounds, circulates within it as organic carbon, and is eventually returned to the atmosphere as CO_2 chiefly by respiration of living organisms, but also by burning of wood and fossil fuels.

carbon dating *see* radiocarbon.

carbon dioxide compensation concentration *see* carbon dixoide compensation point.

carbon dioxide compensation point the ambient concentration of carbon dioxide at which, when light is not limiting, photosynthesis just compensates for respiration. At 25 °C and 21% O_2 this value is about 45 ppm for C3 plants. *see also* compensation point.

carbon dioxide fixation the pathway incorporating carbon dioxide into carbohydrates which occurs in the stroma of chloroplasts; any reaction in which carbon dioxide is incorporated into organic compounds.

carbon isotope ratio *see* $^{12}C/^{13}C$ ratio.

carbon monoxide (CO) poisonous gas which when inhaled in large amounts binds tightly to the haemoglobin of red blood cells, making it unable to bind and transport oxygen.

carbonic anhydrase enzyme of red blood cells that catalyses the formation of bicarbonate from carbon dioxide in the reaction $CO_2 + H_2O \rightarrow H^+ + HCO_3^-$. EC 4.2.1.1. *r.n.* carbonic dehydratase.

carbonicole, carbonicolous *a.* living on burnt soils or burnt wood.

Carboniferous *a. pert.* period of late Paleozoic era, lasting from approx. 350 to 285 million years ago, and during which the coal measures were formed.

carboxyhaemoglobin, carbomonoxyhaemoglobin (HbCO) a compound of carbon monoxide and haemoglobin formed in the blood following carbon monoxide poisoning.

carboxypeptidase *n.* any of a group of proteolytic enzymes which remove the C-terminal amino acid from a peptide, carboxypeptidases A and B (EC 3.4.17.1 and 2) being mammalian digestive enzymes synthesized in the pancreas as inactive precursors and enzymatically activated in the small intestine.

carboxylase *n.* any of a group of enzymes, mostly containing a biotin prosthetic group, that catalyse the addition of a carboxyl (–COOH, –COO⁻) group derived from CO_2 to a compound. EC 6.4.

carboxylic acids organic acids containing one or more carboxyl (–COOH, –COO⁻) groups.

carboxy-terminus the end of a protein chain that bears the free carboxyl group, often abbreviated C-terminus.

carcerulus *n.* a superior, dry, capsular fruit composed of several cells, each containing a single seed and which split off as separate nutlets when ripe (e.g. fruit of dead-nettle).

carcinoembryonic antigens (CEA) group of antigens found on surface of tumour cells derived from gastrointestinal tract, and also on embryonic gastrointestinal and derived tissues.

carcinogen *n.* any agent capable of causing cancer in humans or animals. *a.* carcinogenic.

carcinogenesis the development of a cancer; the process by which a cancerous cell arises from a normal cell.

carcinology *n.* the study of crustaceans.

carcinoma *n.* malignant tumour of epithelial tissues. *cf.* sarcoma.

cardenolides *n.plu.* cardiac glycosides found in milkweeds, which cause vertebrates feeding on the plant to vomit, and so learn to avoid it.

cardia *n.* the opening between oesophagus and stomach; in sucking insects the enlarged anterior part of the digestive chamber in front of the stomach.

cardiac *a. pert.*, near or supplying the heart; *pert.* anterior part of the stomach.

cardiac glycosides plant glycosides such as digitalin that have stimulatory effects on the vertebrate (more particularly human) heart, and are therefore often highly toxic.

cardiac muscle specialized muscle tissue of vertebrate heart, formed of muscle fibres made up of cylindrical muscle cells joined end to end.

cardiac sphincter thick ring of muscle around opening between oesophagus and stomach.

cardiac valve in insects, a valve at the junction of the fore-gut and mid-gut, probably serving to prevent or reduce regurgitation of food.

cardinal *a. pert.* that upon which something depends or hinges; *pert.* hinge of bivalve shell; *appl.* points for plant growth: maximum, minimum and optimal temperatures or temperature ranges.

cardinal sinuses and veins veins uniting in Cuvier's duct, persistent in most fishes, embryonic in other vertebrates.

cardines *plu.* of cardo *q.v.*

cardiobranchial *a. appl.* enlarged posterior

basibranchial cartilage below the heart in elasmobranch fishes.

cardiogenic *a*. arising in the heart.

cardiolipin *n*. a diphosphatidyl glycerol found in heart tissue, possesses antigenic activity and is used in the detection of antispirochaete antibodies in the Wassermann reaction syphilis because of its coincidental similarity to a syphilis spirochaete antigen.

cardiovascular *a*. *pert*. heart and blood vessels.

cardo *n*. basal segment of maxilla or secondary jaw in insects; hinge of bivalve shell.

carina *n*. (*zool*.) keel-like ridge on certain bones, as on breast bone of birds; median dorsal plate of a barnacle; (*bot*.) the 2 joined lower petals (the keel) of a leguminous flower; ridge on bracts of certain grasses.

Caribbean Floral Region Central American Floral Region *q.v.*

carinate *a*. having a ridge or keel.

cariniform *a*. keel-shaped.

carlavirus group plant virus group composed of rod-shaped single-stranded RNA viruses, type member, carnation latent virus.

carmovirus group plant virus group composed of isometric single-stranded RNA viruses, type member, carnation mottle virus.

carnassial *a*. *pert*. cutting teeth of carnivora, 4th premolar above and 1st molar below.

carneous *a*. flesh-coloured.

carnitine *n*. methyl-substituted amino acid found in muscle which carries long-chain fatty acids across the inner mitochondrial membrane for oxidation.

Carnivora *n*. order of flesh-eating mammals containing the sub-orders Fissipedia, comprising the terrestrial carnivores, and Pinnipedia, the aquatic carnivores (seals, walruses and sea-lions).

carnivore *n*. animal that feeds on other animals, esp. the flesh-eating mammals (Carnivora) such as dogs, cats, bears and seals, which feed on meat or fish.

carnivorous *a*. flesh-eating; (*bot*.) *appl*. certain plants that trap and digest insects and other small animals.

carnose *a*. fleshy or pulpy.

carnosine *n*. β-alanyl L-histidine, dipeptide found in muscle which has a histamine-like activity and produces a fall in blood pressure.

carotene *n*. any of several orange or yellow pigments, hydrocarbons with the formula $C_{40}H_{56}$, synthesized in plants, accessory pigment in photosynthesis (β-carotene), precursor or provitamin of the A vitamins such as retinol, to which they are converted in animals. *see* Appendix 1 (61) for structure of β-carotene.

carotenoid *n*. any of a group of widely distributed orange, yellow, red or brown fat-soluble pigments, synthesized in plants, involved in photosynthesis as accessory pigments and also found in flowers and fruits, consisting of 2 groups: carotenes, and xanthophylls (e.g. lutein, violaxanthin, fucoxanthin).

carotid *a*. *pert*. chief arteries in the neck.

carotid body one of two small masses of tissue associated with the carotid sinus in neck, involved in control of oxygen content and acidity of blood, composed of sensory cells and chromaffin cells secreting catecholamines.

carotid sinus small dilation inside carotid artery containing baroceptors that sense changes in arterial pressure and are involved in regulating heart rate and vasodilation.

carpal *n*. wrist bone.

carpel *n*. female reproductive structure in angiosperm flowers, containing one or more ovules, the carpels together making up the pistil or gynaecium.

carpellate *a*. *appl*. a flower containing carpels but not stamens.

carpellate *a*. having carpels.

carpocerite *n*. the 5th antennal joint in certain crustaceans.

carpogenic, carpogenous *a*. *appl*. those cells in red algae that form the carpogonium.

carpogenous *a*. growing in or on fruit, *appl*. fungi.

carpogonium *n*. female sex organ of red algae whose basal portion functions as the female gamete.

carpolith *n*. a fossil fruit.

carpometacarpus *n*. bone in wing of birds corresponding to three metacarpals and some carpals fused together.

carpophagous *a.* feeding on fruit.

carpophore *n.* part of flower axis to which carpels are attached; the stalk of a fruit body containing spores.

carpopodite *n.* the 3rd joint of endopodite in certain crustaceans; the patella in spiders.

carposporangium *n.* sporangium formed after fertilization of carpogonium in red algae, which produces spores (carpospores) from which the sporophyte arises.

carpotropism *n.* movements of fruit stalk, esp. after fertilization to place fruit in a good position for ripening and dispersal.

carpus *n.* the wrist; region of forelimb between forearm and metatarsus.

carr *n.* fen woodland, usually dominated by alder or willow.

carrier *n.* individual infected with a transmissible pathogen, who does not suffer from the disease but can transmit the pathogen to others; (*genet.*) individual heterozygous for a recessive allele, esp. one responsible for a genetic disease, who shows no symptoms of disease but can pass the allele on to their offspring; (*imm.*) protein to which a hapten is attached to render it immunogenic.

carrier protein any membrane protein that binds molecules and transports them across membranes, either by facilitated diffusion or by active transport.

carrier-mediated transport transport of ions and other solutes across cell membranes with the aid of a carrier protein in the membrane.

carrying capacity (*K*) the maximum number of individuals of a particular species that can be supported indefinitely by a given part of the environment; the number of grazing animals a piece of land can support without deterioration; the level of use an environment or resource can sustain without being destroyed or suffering an unacceptable deterioration.

cartilage *n.* firm elastic skeletal tissue in which the cells, chondrocytes, are embedded in a matrix of the proteoglycan chondrin and collagen fibres, and which is of three chief types, smooth bluish-white hyaline cartilage, fibrous cartilage and yellow elastic cartilage, containing elastin fibres. Hyaline cartilage is the main skeletal material of the cartilaginous fishes such as sharks and dogfish. In other vertebrates it forms the embryonic skeleton, to be partly replaced later in development by bone.

cartilage bone bones formed by the ossification of cartilage.

cartilaginous *a.* gristly, consisting of or *pert.* cartilage; resembling consistency of cartilage.

cartilaginous fishes common name for fishes of the class Selachii having a cartilaginous skeleton, a spiral valve in the gut and no lungs or air bladder, and which include the sharks and their allies.

caruncle *n.* naked fleshy excrescence.

Caryoblastea *n.* phylum of protists containing one species, the giant multinucleate amoeba *Pelomyxa palustris*, which lacks mitochondria and other organelles characteristic of eukaryotic cells and whose nuclei divide without mitosis.

caryophyllaceous *a. pert.* flowers of the family Caryophyllaceae, the pinks, campions, etc. whose flowers have long clawed petals. Also *appl.* any flower with long clawed petals.

Caryophyllales *n.* an order of herbaceous dicots, rarely shrubs or trees, containing betalain pigments and including the families Amaranthaceae (amaranth), Caryophyllaceae (pink), Chenopodiaceae (goosefoot), Phytolaccaceae (pokeweed) and others.

caryopsis *n.* achene with pericarp and tests inseparably fused, as in grasses.

cascade *n.* series of enzymatic reactions in which the activated form of one enzyme catalyses the activation of the next, greatly amplifying the initial response, as in the complement (*q.v.*) cascade and in blood clot formation.

casein *n.* mixture of milk proteins precipitated by the action of rennin; protein found in milk, synthesized in mammary glands in response to the hormone prolactin.

Casparian strip zone of corky cells in the endodermis of plant roots.

casque *n.* helmet-like structure in animals, as the horny outgrowth of beak in hornbill.

cassava *n. Manihot esculenta*, tropical plant of the family Euphorbiaceae, whose starchy roots are used as food (tapioca) in parts of the world, and which have to be processed before being turned into flour to remove toxic compounds.

cassette model description of the DNA rearrangements underlying switching of mating type in the yeast *Saccharomyces cerevisiae* in which the mating type locus MAT can be occupied by either of 2 genes (a or α) transposed from sites on either side of the locus.

cassideous *a*. helmet-like.

caste *n*. one of the distinct forms found among social insects, e.g. worker, drone, queen.

caste polyethism the division of labour between different castes in social insects.

castoreum *n*. material secreted by scent glands of beaver.

castrate *a*. animal from which gonads, esp. male gonad has been removed.

casual *n*. plant which has been introduced but has not yet become established as a wild plant, although occurring uncultivated.

casual society a temporary and highly unstable group formed by individuals within a society, as for play, feeding, etc.

Casuariformes *n*. an order of flightless birds including the cassowaries and emus.

Casuarinales *n*. an order of dicot trees or shrubs with whorls of branches and which comprises the family Casuarinaceae (she-oak).

catabolism *n*. the breaking down of complex molecules by living organisms with release of energy. *a*. catabolic.

catabolite *n*. any substance that is the product of catabolism.

catabolite gene activator protein (CAP, CRP) a protein in bacteria which when complexed with cyclic AMP can activate some catabolic operons.

catabolite repression phenomenon seen in bacteria when glucose is available as a substrate, when the genes involved in the uptake and metabolism of alternative energy sources are all repressed.

catacorolla *a*. a scondary corolla.

catadromous *a*. tending downward; *appl.* fishes which migrate from fresh to salt water for spawning.

catalase *n*. iron-containing enzyme found in all aerobic tissues which catalyses the breakdown of hydrogen peroxide into water and molecular oxygen. EC 1.11.1.6.

catalepsis *n*. a shamming dead reflex, as in spiders.

catalysis *n*. the acceleration (or, rarer usage, the retardation) of a reaction due to the presence of a substance (catalyst) which can be recovered unchanged after the reaction.

catalyst *n*. a substance which accelerates (or, rarer usage, retards) a chemical reaction, usually by forming temporary complexes with intermediates and reducing the free energy of activation, but which can be recovered unchanged at the end of the reaction, e.g. an enzyme.

catapetalous *a*. having the petals united with the base of stamens.

cataphyll *n*. simple form of leaf on lower part of plant, as cotyledon, bud scale, bulb scale, scale leaf.

cataphyllary *a. appl.* rudimentary scale-like leaves which act as covering of buds.

cataplasis *n*. regression or decline after maturity.

cataplasmic *a. appl.* irregular plant galls caused by parasites or other factors.

catapleurite *n*. a thoracic pleurite in certain insects.

cataplexis *n*. condition of an animal feigning death; maintenance of a postural reflex induced by restraint or shock.

catarrhines *n.plu*. the Old World monkeys, the apes and humans.

catastrophism *n*. the idea, held in the 18th and 19th centuries, that the fossil record represented a series of discrete creations, each terminated by a catastrophic mass extinction. *see* mass extinction for the modern view of this aspect of the fossil record.

catch muscles muscles such as those that close the shells of bivalve molluscs, which can remain in the contracted state for long periods with little expenditure of energy.

catechins *n.plu*. a group of colourless flavonoids, constituents of wines, etc.

catecholamines *n.plu*. a group of chemicals acting as neurotransmitters or hormones, which are amine derivatives of catechol (2-hydroxyphenol) and include adrenaline, noradrenaline and dopamine.

catena *n*. a sequence of soil types which is

repeated in a corresponding sequence of topographical sites, as between ridges and valleys of a region.

catenane *n.* 2 interlocked circles of duplex DNA, produced by the action of DNA topoisomerases.

catenation *n.* the production of a catenane *q.v.*

catenoid *a.* chain-like, *appl.* certain protozoan colonies.

catenular *a.* chain-like, *appl.* colonies of butterflies, colour markings on butterfly wings, shells, etc.

caterpillar *n.* fleshy thin-skinned larva, esp. of Lepidoptera, having segmented body, true legs and also prolegs on abdomen and no cerci.

catfishes *n.plu.* a group of mainly tropical, mainly freshwater bony fish (the order Siluriformes) often with long whisker-like barbels from which they take their name. (The marine fish *Anarhichas*, commonly called the catfish in English, is not a member of this group.)

cathepsins *n.plu.* a group of intralysosomal proteolytic enzymes present in many animal tissues such as liver, spleen, kidney, which are thought to be concerned in autolysis in some diseases and after death.

cation *n.* positively charged ion which moves towards cathode or negative pole, e.g. K^+, Na^+, Ca^{2+}. *cf.* anion.

cation exchange capacity base exchange capacity *q.v.*

catkin *n.* inflorescence consisting of a hanging spike of small unisexual flowers interspersed with bracts, as in willows, poplars, hazel (male catkins).

cauda *n.* tail or tail-like appendage.

caudad *a.* towards the tail or posterior region.

caudal *a.* of or *pert.* a tail, e.g. caudal fin of fishes; towards the tail-end of the body.

Caudata Urodela *q.v.*

caudate *a.* having a tail; *appl.* to a lobe of the liver.

caudate nucleus nucleus in brain lying on either side of the lateral ventricles.

caudatolenticular *a. appl.* caudate and lenticular (lentiform) nucleus of corpus striatum in brain.

caudex *n.* the axis or stem of a woody plant, as of tree ferns, palms, etc.

caudicle *n.* stalk of pollinium in orchids.

Caudofoveata *n.* class of shell-less worm-like molluscs.

caudostyle *n.* a terminal structure in some parasitic amoebae.

caudotibialis *n.* a muscle connecting caudal vertebrae and tibia, as in some seals.

caul *n.* amnion *q.v.*; an enclosing membrane.

caulescent *a.* with leaf-bearing stem above ground.

caulicle *n.* a small or rudimentary stem; axis of a young seedling.

caulicolous *a.* growing on the stem of another plant, usually *appl.* fungi.

cauliflory *n.* condition of having flowers arising from axillary buds on the main stem or older branches.

cauliform *a.* stem-like.

cauligenous *a.* borne on the stem.

caulimovirus group plant virus group containing isometric double-stranded DNA viruses, type member cauliflower mosaic virus.

caulimoviruses *n. plu.* group of double-stranded, non-enveloped DNA plant viruses which includes cauliflower mosaic virus.

cauline *a. pert.* stem; *appl.* leaves growing on upper portion of a stem; *appl.* vascular bundles not passing into leaves.

caulis *n.* the stem, esp. in herbaceous plants.

caulocarpous *a.* with fruit-bearing stems; fruiting repeatedly.

caulome *n.* the shoot structure of a plant as a whole.

caulonema *n.* a profusely branched portion of protonema with relatively few chloroplasts, present in some genera of mosses.

caulotrichome *n.* hair-like or filamentous outgrowths on a stem.

cavernicolous *a.* cave-dwelling.

cavernosus *a.* hollow or honeycombed, *appl.* tissue.

cavicorn *a.* hollow-horned, *appl.* certain ruminants.

cavitation *n.* formation of a cavity by enlargement of intercellular space within a cell mass.

cavum *n.* hollow or chamber; in helical shells, the lower division of the internal cavity caused by origin of the helix.

C-banding technique for staining constitutive heterochromatin which generates dense staining at centromeric regions.

CD antigens cell-surface antigens on white blood cells that are detected by particular sets of monoclonal antibodies, and which are designated by CD followed by a number, e.g. CD1, CD2, CD3, etc. *see* cluster of differentiation.

cDNA clone a DNA clone derived from a complementary DNA (cDNA) transcript of an mRNA.

Ceboidea *n.* a superfamily of primates comprising the New World monkeys.

cecal, cecum *see* caecal, caecum.

cecidium *see* gall.

Celastrales *n.* order of dicot trees, shrubs or vines, rarely herbs and including the families Aquifoliaceae (holly), Celastraceae (staff tree), and others.

celiac coeliac *q.v.*

cell *n.* the basic structural building block of living organisms, consisting of protoplasm delimited by a cell membrane, and in plants, bacteria and fungi also surrounded by a non-living rigid cell wall. Some organisms consist of a single cell (bacteria, protozoans and some algae), others of cells of a few different types, and the more complex animals and plants of billions of cells of many different types. Bacterial (prokaryotic) cells have a relatively simple internal structure in which the DNA is not enclosed in a discrete nucleus and the cytoplasm is not differentiated into specialized organelles, Cells from all other living organisms (eukaryotic cells) typically are comprised of a nucleus enclosing the DNA which is organized into chromosomes, and a cytoplasm containing a cytoskeleton of fine protein tubules and filaments and spe-

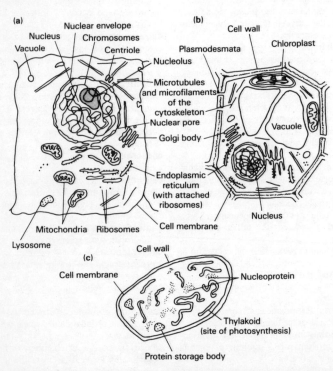

Fig. 1 Schematic diagram of (a) a generalized animal cell, (b) a generalized plant cell and (c) a prokaryotic cell (the example used is a cyanobacterium).

cell-mediated immune response

cialized membrane-bounded organelles such as mitochondria and (in photosynthetic plants) chloroplasts; a small cavity or hollow; space between veins of insect wing. *see* Fig. 1.

α-cell glucagon-secreting cell in islets of Langerhans in pancreas; oxyphilic cell in adenohypophysis of pituitary. *alt.* A cell.

β-cell cell in islets of Langerhans of pancreas that secretes insulin; secretory basophil cell in the anterior pituitary. *alt.* B cell.

cell adhesion molecule (CAM) any of a large and heterogenous group of cell-surface glycoproteins that can promote adhesion between animal cells, by binding to each other or to other receptor molecules. Cell adhesion molecules fall into several different families on the basis of structure: molecules of the immunoglobulin superfamily (e.g. the neural cell adhesion molecule, NCAM, which promotes aggregation of neurones and binding of axons together into nerve bundles), the cadherins, the integrins, and the selectins.

cell-autonomous *a. appl.* development of a blastomere or embryonic cell entirely determined by factors within the cell itself, and not by interactions with its neighbours.

cell body in neurone, that part containing the nucleus and most organelles. *alt.* soma.

cell centre centrosome *q.v.*

cell coat carbohydrate-rich peripheral zone at the surface of most eukaryotic cells, composed of the carbohydrate side-chains of membrane glycoproteins, and often including secreted and adsorbed carbohydrates and proteoglycans. *alt.* glycocalyx.

cell culture cells growing outside the organism in nutrient medium in the laboratory, *see* also primary cell culture, secondary cell culture; the process of growing cells in culture outside a living organism.

cell cycle collectively the changes that take place in a cell in the period between its formation as one of the products of cell division and its own subsequent division, and which in all cells include replication of the DNA. In eukaryotic cells the cell cycle is divided into phases termed G1, S, G2, and M. G1 is the period immediately after mitosis and cell division when the newly formed cell is in the diploid state, S

is the phase of DNA synthesis which is followed by G2 when the cell is in a tetraploid state, mitosis (M) follows to restore the diploid state, accompanied by cell division.

cell cycle genes a group of genes identified in yeast and mammalian cells, designated *cdc*, which are involved in controlling the progression of a cell through the cell cycle, from the point of its formation to its entry into mitosis.

cell division splitting of a cell into two complete new cells: by binary fission in bacteria and other prokaryotes, and by divison of both nucleus and cytoplasm in eukaryotic cells.

cell envelope in bacteria, all outer layers of the cell, including plasma membrane cell wall and outer membrane if any.

cell fractionation process of breaking cells up and then subjecting them to centrifugation to separate out the different cellular components.

cell-free system any mixture of cell components reconstituted *in vitro* in which processes such as translation, transcription, DNA replication, etc. can be studied.

cell fusion the fusion of the plasma membranes of two cells with the mingling of their cytoplasms, sometimes, but not necessarily, accompanied by fusion of the two cell nuclei. *see* cell hybrid.

cell hybrid cell produced by induced fusion *in vitro* of two somatic cells of different genetic constitution (often from different species) and in which, e.g., the linkage of various genetic markers can be determined. Plant cell hybrids can sometimes be regenerated into whole plants of novel genetic constitution.

cell junctions points at which the plasma membrane of one cell is closely apposed to that of its neighbouring cell. *see* desmosome, gap junction, intermediate junction, septate junction, tight junction.

cell line a line of cells that can be propagated indefinitely in culture.

cell lineage the family tree of cells in a tissue or part of a developing embryo, tracing their mitotic line of descent from a particular blastomere.

cell-mediated immune response or immunity *see* immune response.

cell membrane plasma membrane *q.v.*

cell memory refers to the fact that differentiated eukaryotic cells give rise only to cells of the same type; refers also to the fact that in eukaryotic development, determined cells give rise to similarly determined cells, even though the extracellular influences that resulted in determination no longer are acting. *see also* determination.

cell motility all movement within cells (e.g. muscle contraction) and by cells (cell locomotion).

cell of Boettcher granular epithelial cell found between cells of Claudius and the basal membrane in the ear.

cell of Claudius columnar or cuboid epithelial cell found in the lining of the endolymphatic space in the ear.

cell plate material laid down across middle of plant cell undergoing division from which new partition wall arises.

cell sap the fluid in vacuoles in plant cells, being a solution of small organic molecules in water; sometimes used for the more fluid part of the cytoplasm, especially in protozoans.

cell theory the idea that plant and animal bodies are made up of cells, and that the cell is the unit structure of an organism, first proposed by Schlieden for plants and Schwann for animals in 1838–40. *alt.* cell doctrine.

cell wall non-living rigid structure surrounding the plasma membrane of algal, plant, fungal and bacterial cells. It is predominantly composed of cellulose fibrils in green plants, and also contains many other polysaccharides. Fungal cell walls are composed largely of chitin, those of bacteria of peptidoglycans.

cellobiase *n.* enzyme which hydrolyses cellobiose to glucose. EC 3.2.1.21, *r.n.* b-D-glucosidase.

cellobiose *n.* disaccharide produced on partial hydrolysis of cellulose, made up of two glucose units joined by a β,1,4 linkage.

cellular *a. pert.* or consisting of cells.

cellular immunity or immune response *see* immune response.

cellular slime moulds a group of simple eukaryotic soil microorganisms (Acrasiomycetes) typified by *Dictyostelium discoideum* and considered either as fungi or as protists. Free-living unicellular amoebae (myxamoebae) aggregate to form a multicellular fruiting body differentiated into stalk and sporehead (a compound sporangium). *cf.* acellular or plasmodial slime moulds.

cellulase *n.* enzyme catalysing the degradation of cellulose by hydrolysis of internal β,1,4 linkages, found in some invertebrates, plants, fungi and bacteria but lacking in most animals. EC 3.2.1.4.

cellulolytic *a.* able to degrade cellulose.

cellulose *n.* a linear polysaccharide made up of glucose residues joined by β,1,4 linkages, the most abundant organic compound in the biosphere, comprising the bulk of plant and algal cell walls where it occurs as cellulose microfibrils, and also found in certain tunicates. *see* Appendix 1 (16).

cellulosic *a.* composed of cellulose.

celo- *alt.* spelling of coelo-.

cement, cementum bone-like material covering part of tooth; material that sticks cells together or anchors organism or part of organism to its substrate.

cementoblast, cementocyte *n.* cell secreting the bone-like cementum that covers the root of a tooth.

cen- *alt.* spelling of coen-.

Cenozoic *a. pert.* geological era following the Mesozoic, commencing *ca.* 65 million years ago. In some usages it is followed by the Quaternary era whereas in others it lasts until the present and is subdivided into the Tertiary and Quaternary epochs (periods).

censer mechanism method of seed dispersal by which seeds are shaken out of fruit by wind action.

census *n.* a complete counting of a whole population with respect to the variable under study.

centimorgan *n.* map unit, equal to 1% recombination between two gene loci on the same chromosome.

centipedes *n.plu.* common name for the Chilopoda, a group of arthropods having numerous and similar body segments each with one pair of walking legs, except the 1st segment which bears a pair of poison claws.

central *a.* situated in, or *pert.* the centre; *pert.* a vertebral centrum; *n.* a bone, situated between proximal and distal rows in wrist or ankle.

Central American Floral Region part of the Neotropical Realm comprising Central America, the southern tips of California and Florida, the islands of the Caribbean and the northern part of South America. *alt.* Caribbean Floral Region.

central cylinder stele *q.v.*

central dogma the principle that the transfer of genetic information from DNA to RNA by transcription and from RNA to protein by translation is irreversible, now modified to take into account the transfer of information from RNA to DNA by reverse transcription carried out by some viruses.

central nervous system (CNS) that part of the nervous sytem that serves an integrating and coordinating function. In vertebrates it consists of the brain and spinal cord. In those invertebrates with a central nervous system it consists of (sometimes) a brain, cerebral ganglia and a nerve cord which may be dorsal or ventral, single or double. *cf.* peripheral nervous system.

central sulcus the transverse middle fissure of cerebral hemisphere.

central vacuole large fluid-filled cavity that takes up most of the volume of many plant and algal cells, and which maintains the turgor of plant cells, and also contains digestive enzymes.

centrarch *a. appl.* protostele with central protoxylem.

centres of diversity regions identified originally by the Russian botanist N. Vavilov, in which large numbers of different strains or races of a particular cultivated plant occur. In some cases this diversity may indicate the geographical place of origin of the crop plant.

centric *a. appl.* leaves that are cylindrical or nearly so; *appl.* chromosomes having a centromere; *appl.* diatoms radially symmetrical when viewed face on.

centrifugal *a.* turning or turned away from centre or axis; (*bot.*) *appl.* radicle, *appl.* compact cymose inflorescences having youngest flowers towards outside, *appl.* xylem differentiating from centre towards edge of stem or root, *appl.* thickening of cell wall when material is deposited on

outside of wall, as in pollen grains; (*zool.*) away from centre, *appl.* transmission of nerve impulses from a nerve centre towards parts supplied by nerve.

centrifuge *see* ultracentrifuge.

centriole *n.* one of a pair of organelles with a characteristic arrangement of 27 microtubules, found just outside the nuclear envelope in animal cells but apparently absent from plant cells, and which separate to opposite poles at prophase and organize the microtubules of the mitotic and meiotic spindle.

centripetal *a.* turning or turned towards centre or axis; (*bot.*) *appl.* radicle, *appl.* racemose inflorescences having youngest flowers at apex, *appl.* xylem differentiating from edge towards centre of stem or root, *appl.* thickening of cell wall where material is deposited on the inside of wall, as most cells; (*zool.*) towards centre, *appl.* transmission of nerve impulses from periphery to nerve centres.

centrogenous *a. appl.* skeleton of spicules which meet in a common centre and grow outwards.

centrolecithal *a. appl.* ovum with yolk aggregated in centre.

centromere *n.* constricted region at which sister chromatids are attached in mitotic chromosomes and which becomes aligned on the spindle equator at mitosis, becoming visibly doubled and dividing late in mitosis.

centroplast *n.* extranuclear spherical body forming division centre of mitosis, as in some radiolarians.

centrosome *n.* organelle in plant and animal cells, situated near the nucleus, and in animal cells containing the centrioles, and from which the spindle develops in mitosis and meiosis.

Centrospermae Caryophyllales *q.v.*

centrum *n.* main body of a vertebra, from which neural and haemal arches arise; (*mycol.*) all structures enclosed by an ascocarp wall.

cephalad *a.* towards head region or anterior end.

cephalanthium *n.* a flower-head of closely-packed florets as in Compositae.

cephalic *a. pert.* head; in the head region.

cephalic index one hundred times maximum breadth of skull divided by maximum length.

cephalin *n*. phosphatidyl ethanolamine *q.v.*; in parasitology, an epimerite bearing trophozooites.

cephalization *n*. increasing differentiation and importance of anterior end in animal development.

cephalo-, -cephalic word elements derived from Gk. *kephal*, head.

Cephalocarida *n*. class of minute marine crustaceans living in sand and having a horseshoe-shaped carapace.

Cephalochordata, cephalochordates *n.*, *n.plu*. subphylum of small cigar-shaped chordates commonly called lancelets and including amphioxus (*Branchiostoma*), with a persistent notochord in adult and a large sac-like pharynx with gill slits for food collection and respiration.

cephalodia *n.plu*. gall-like outgrowths, usually brown, developing on, or occasionally within, certain lichens, and containing a different alga from that characteristic of the lichen.

cephalogenesis *n*. development of the head region.

cephalon *n*. the head of some arthropods.

cephalopedal *a. appl*. the haemocoel cavity in the head and foot of snails and their relatives.

Cephalopoda, cephalopods *n.*, *n.plu*. class of marine molluscs including octopus, squid, and nautiloids, with a well-developed head, large brain, and eyes resembling those of vertebrates in structure. The head is surrounded by prehensile tentacles and the animals can move very rapidly by jet propulsion, squirting water out of the large mantle cavity that communicates with the exterior by a siphon or funnel.

cephalopodium *n*. the head and arms constituting the head region in cephalopods.

cephalopsin *n*. photosensitive pigment resembling rhodopsin, in eyes of cephalopods and some other invertebrates.

cephalosporins β-lactam antibiotics produced by streptomycetes and strains of the fungus *Cephalosporium*.

cephalosporium *n*. a globular mucilaginous mass of spores.

cephalostegite *n*. anterior part of cephalothoracic shield in certain arthropods.

cephalothorax *n*. body region formed by fusion of head and thorax in crustaceans; in arachnids, the prosoma *q.v.*

cephalula *n*. free-swimming embryonic stage in certain brachiopods.

ceps boletes *q.v.*

ceraceous *a*. waxy, wax-like.

ceral *a. pert*. the cere in birds.

ceramide *n*. *N*-acyl sphingosine, widely distributed in plant and animal tissues, and a constituent of sphingolipids such as cerebrosides and gangliosides.

ceras *sing*. of cerata *q.v.*

cerata *n.plu*. lobes or leaf-like processes functioning as gills on the back of nudibranch molluscs (e.g. sea-slugs). *sing*. ceras.

ceratium *n*. a type of siliqua.

ceratobranchial *n*. a ventral skeletal element of the gill arch.

ceratohyal *n*. a ventral skeletal element of the hyoid arch in fishes, next below the epihyal.

ceratomorphs *n.plu*. suborder of mammals that contains the rhinoceroses and tapirs.

ceratotrichia *n.plu*. thin rods of collagen (elastoidin) which form sheets between rays of fins of elasmobranch fishes (e.g. sharks, dogfishes) and stiffen them.

cercal *a. pert*. the tail; *pert*. a cercus, *appl*. hairs, nerve.

cercaria *n*. a heart-shaped, tailed, larval stage of a trematode (fluke) produced in the snail host, and which is released from the snail, sometimes then encysting, and subsequently infects a vertebrate host. *plu*. cercariae.

cerci *plu*. of cercus *q.v.*

Cercidiphyllales *n*. order of dicot trees comprising the family Cercidiphyllaceae, with a single genus *Cercidiphyllum*.

cercoid *n*. one of paired appendages on the 9th or 10th abdominal segment of certain insect larvae.

cercopithecoid *a. appl*. monkeys of the superfamily Cercopithecoidea, the Old World monkeys (e.g. baboons) which together with apes and man are the only primates with a fully opposable thumb.

cercopod cercus *q.v.*

cercus *n*. jointed appendage at end of abdomen in many arthropods; appendage bearing acoustic hairs in some insects. *plu.* cerci.

cere *n.* swollen fleshy patch at base of bill in birds.

cereal *n.* plant of the family Gramineae, the grasses, whose seeds are used as food.

cerebellospinal tract tract of motor nerve fibres running from cerebellum in brain to anterior horn cells of the spinal cord, concerned with automatic regulation of muscle tone and posture.

cerebellum *n.* part of hindbrain, pair of finely convoluted rounded hemispherical masses of tissue situated behind the midbrain, concerned with regulation of muscle tone and posture and coordination of movement in relation to sensory signals received in other parts of the brain. *a.* cerebellar.

cerebral *a. pert.* the brain, more particularly *pert.* hemispheres of forebrain; *appl.* arteries supplying the frontal and middle parts of cerebral hemispheres.

cerebral aqueduct aqueduct of Sylvius *q.v.*

cerebral cortex the superficial layer of grey matter of the cerebral hemispheres, some 2 mm deep and consisting of several layers of nerve cell bodies and their complex interconnections. In humans and anthropoid apes the highly convoluted surface of the brain provides a much enlarged area of cortex. The cortex is the site of analysis and interpretation of sensory information, the generation of all voluntary motor action, and of higher cognitive functions such as learning and memory and conscious perception. It receives input from cranial nerves and peripheral sensory receptors via various nuclei in other parts of the brain. Different functions, such as the analysis of visual information, auditory information, generation of motor signals, etc., are localized to different areas of the cortex.

cerebral ganglia the supraoesophageal ganglia, or "brain" of invertebrates.

cerebral hemispheres pair of symmetrical rounded masses of convoluted tissue forming the bulk of the human brain, separated by a deep fissure but connected at the base by a bridge of fibres, the corpus callosum. *alt.* cerebrum.

cerebral peduncles twin short pillars of tissue in brain, supporting and carrying fibres to and from the cerebral hemispheres, and which together with the corpora quadrigemina and the aqueduct of Sylvius form the midbrain.

cerebrifugal *a. appl.* nerve fibres which pass from brain to spinal cord.

cerebropedal *a. appl.* nerve fibres connecting cerebral and pedal ganglia in molluscs.

cerebrose *a.* resembling the convolutions of the brain, *appl.* surface of spores, pilus, etc.

cerebroside *n.* any of a group of glycolipids containing the amino alcohol sphingosine and only one sugar residue, glucose or galactose, found chiefly in nerve cell membranes.

cerebrospinal *a. pert.* brain and spinal cord.

cerebrospinal fluid (CSF) in vertebrates, a fluid filling the cavity in the brain and spinal cord and the subarachnoid space surounding the brain and spinal cord, secreted by the choroid plexuses and reabsorbed by veins on the brain surface. *alt.* neurolymph.

cerebrovisceral *a. appl.* nerve fibres connecting cerebral and visceral ganglia in molluscs.

cerebrum cerebral hemispheres *q.v.*

cereous *a.* waxy.

ceriferous *a.* wax-producing.

cernuous *a.* drooping; pendulous.

ceroma cere *q.v.*

cerous *a. appl.* a structure resembling a cere (fleshy patch at base of bird's bill).

ceruloplasmin *n.* blue copper-containing protein present in blood plasma.

cerumen *n.* ear wax, secreted by the ceruminous glands of the external auditory meatus of the ear; wax secreted by scale insects; wax of the nests of certain bees.

cervical *a. appl.* or *pert.* structures connected with the neck, as nerves, bones blood vessels; *appl.* cervix or neck of an organ; *appl.* groove across carapace of certain crustaceans that appears to delimit a "head".

cervicum *n.* neck region of arthropods; the neck of vertebrates.

cervids *n.plu.* members of the mammalian family Cervidae: the deer.

cervix *n.* the neck or narrow mouth of an organ, as the uterine cervix, the neck of the uterus just above the vagina.

cespitose caespitose *q.v.*

Cestoda, cestodes *n.*, *n.plu.* tapeworms, a class of platyhelminths (flatworms) that are internal parasites of humans and animals. They have a long flattened ribbon-like body lacking a gut or mouth, usually divided into many identical segments (proglottids), and attach themselves to the wall of the gut through an attachment organ at the anterior end. Reproduction is via mature proglottids which detach to form a new reproductive unit. The complex life cycle involves two or more hosts.

Cetacea, cetaceans *n.*, *n.plu.* order of wholly aquatic placental mammals including the whales and dolphins, having a body highly adapted for swimming, with the forelimbs modified as flippers and the hind limbs often hardly developed and invisible externally.

cetology *n.* study of whales and dolphins.

C gene gene coding for the constant regions of immunoglobulin molecules, several different forms of which, specifying different classes of immunoglobulin, are present in the mammalian genome, and which is joined to the variable region genes (V, D and J) by RNA splicing when the mRNA for the individual polypeptide chains of an antibody molecule is being produced. *see also* V gene.

chaeta *n.* retractable bristle partly made of chitin projecting from the body wall in oligochaete worms and on parapodia of polychaete worms; chitinous sensory bristle on body and appendages of insects. *plu.* chaetae.

chaetiferous, chaetigerous *a.* bristle-bearing.

Chaetognatha *n.* arrow-worms, a phylum of marine coelomate animals found in swarms in plankton, having an elongated transparent body with head, trunk and tail.

chaetophorous *a.* bristle-bearing.

chaetopods *n.plu.* the annelid worms that bear chaetae (bristles): the Polychaeta and Oligochaeta.

chaetotaxy *n.* the arrangement of bristles or chaetae on an insect.

chain behaviour a series of actions, each being induced by the antecedent action and being an integral part of a unified performance.

chalaza *n.* (*zool.*) one of 2 spiral bands attaching yolk to membrane of a bird's egg; (*bot.*) base of nucellus in ovule, from which integuments arise.

chalazogamy *n.* fertilization in which the pollen tube pierces chalaza of the ovule.

chalcone *n.* any of a group of yellow flavonoid plant pigments.

chalcone synthetase a key enzyme in flavonoid biosynthesis in plants. EC 2.3.1.74, *r.n.* naringenin-chalcone synthase.

chalice *n.* arms and disc of a crinoid; simple gland cell or goblet cell.

chalice cell goblet cell *q.v.*

chalicotheres *n.plu.* extinct family of ungulates which had clawed feet.

chamaephyte, chamaeophyte *n.* perennial woody plant having overwintering buds at or just above ground level.

chambered organ in sea lilies, an aboral cavity with five compartments, which occupies the body region enclosed by the thecal plates and sends branches to the cirri.

channel *n.* pore in biological membrane formed by some membrane proteins which allows the simple diffusion of certain solutes across the membrane.

chaparral *n.* type of vegetation found in areas with a mediterranean climate, dominated by evergreen shrubs with broad hard leaves.

chaperonin *n.* member of one class of molecular chaperone proteins found in bacteria and mammalian cells.

characins *n.plu.* group of tropical freshwater bony fish (the Characinoidei) that includes the tetras and piranhas, characterized by complex teeth bearing 5-7 cusps, strong jaws and a scaly body.

character *n.* any feature or trait transmitted from parent to offspring.

character convergence the condition in which two newly evolved species interact in such a way that one or both converges in one or more traits towards the other.

character displacement the condition in which two newly evolved species interact in such a way that both diverge evolutionarily further from each other.

character state any of the range of values or expressions of a particular character.

characteristic species plant species that are almost always found within a particular association.

Charadriiformes *n.* a large diverse group of shore and wading birds including the gulls and terns, auks, oystercatchers, plovers, curlews, snipes and waders.

Charophyceae *n.* a class of algae of the Chlorophyta that are encrusted with calcium carbonate so are commonly called stoneworts. They have a filamentous or thalloid body bearing lateral branches in whorls. Sometimes considered as a division, the Charophyta.

chartaceous *a.* like paper.

chasmocleistogamy *n.* the condition of having both chasmogamous (opening) and cleistogamous (never opening) flowers.

chasmogamy *n.* opening of a mature flower in the normal way to ensure fertilization. *cf.* cleistogamy.

chasmophyte *n.* a plant which grows in rock crevices.

cheating *n.* in animal behaviour, any behaviour (e.g. exaggeration of body size) intended to mislead a rival or potential mate into an incorrect estimate of the animal's strength or genetic fitness.

cheiropterygium *n.* the pentadactyl limb, as of all vertebrates except fish.

chela *n.* the large claw borne on certain limbs of arthropods, as the pincers of a crab. *plu.* chelae.

chelate *a.* claw-like or pincer-like; *v.* to combine with a metal iron to form a stable compound, a chelate.

chelating agent compound that can react with a metal ion and form a stable compound, a chelate, with it. *v.* to chelate.

chelation *n.* structural combination of organic ions and metal ions, as in chlorophyll, cytochromes, haemoglobin.

chelicera *n.* one of the pair of prehensile appendages at the extreme anterior end of arachnids and horseshoe crabs, in arachnids often modified into fangs used to inject poison into prey. *plu.* chelicerae. *see also* Chelicerata.

Chelicerata, chelicerates *n., n.plu.* a class or subphylum of arthropods (*q.v.*) with a body generally in two parts, a prosoma bearing the paired chelicerae (poison jaws) and sensory pedipalps, and a posterior opisthosoma bearing usually four pairs of walking legs. The Chelicerata include the arachnids (e.g. spiders, ticks, mites, scorpions), pycnogonids (sea spiders), horseshoe crabs and the extinct eurypterids. *see also* Arachnida, Merostomata, Pycnogonida.

cheliferous *a.* having chelae or claws.

cheliform *see* chelate.

cheliped *n.* claw-bearing appendage in crustaceans.

Chelonia, chelonians *n., n.plu.* turtles and tortoises, an order of reptiles having a short broad trunk protected by a dorsal shield (carapace) and ventral shield (plastron) composed of bony plates overlain by epidermal plates of tortoiseshell.

chemical defences unpalatable or toxic chemicals, such as astringent tannins and toxic alkaloids, produced by plants in their tissues to deter herbivores.

chemical ecology the study of the secondary chemical compounds produced by plants and animals (e.g. antibiotics, alkaloids, unpalatable and/or toxic compounds) and their effect on the interaction of the organism with other animals and plants in the ecosystem, especially in respect of plants' defences against herbivores and animals' defences against predators.

chemical fossils supposed chemical traces of life, such as alkanes and porphyrins, found in rocks older than the earliest true fossil-bearing rocks.

chemical oxygen demand (COD) a chemical test for the degree of organic pollution of water which measures the amount of oxygen taken up from a sample of water by the organic matter in the sample, expressed as parts per million of oxygen taken up from a solution of boiling potassium dichromate in 2 hours.

chemical synapse *see* synapse.

chemiluminescence *n.* light production during a chemical reaction at ordinary temperature, as bioluminescence, *q.v.*

chemiosmotic theory generally accepted mechanism for the transduction of energy derived from breakdown of substrates during cellular aerobic respiration, or from sunlight in photosynthesis, to power the linked reaction of ATP synthesis. Passage of electrons along respiratory or photosynthetic electron transport chains in the membranes of mitochondria and chloroplasts leads to ejection of protons from the membranes and the formation of a proton gradient (the proton motive force) across them. The energy stored in these gradients drives ATP synthesis as protons flow back down the gradient through ATP synthetases located in the membranes, which are otherwise impermeable to protons and other ions. A similar process involving the cell membrane drives various reactions in some bacterial cells.

chemoattractant *n.* any chemical that attracts cells or organisms to move towards it.

chemoautotroph *n.* any organism using inorganic sources of carbon, nitrogen etc. as starting materials for biosynthesis, and an inorganic chemical energy source. *a.* chemoautotrophic.

chemoceptor chemoreceptor *q.v.*

chemoheterotroph *see* chemotroph, heterotroph.

chemokinesis *n.* a movement in response to the intensity of a chemical stimulus, including that of scent.

chemolithotroph chemoautotroph *q.v.*

chemonasty *n.* a nastic movement in response to diffuse or indirect chemical stimuli.

chemoreceptor *n.* a terminal sense organ or cell receiving chemical stimuli, as taste bud, olfactory sense cells etc.; protein or protein complex located on cell membrane which binds specific effector molecules, e.g. protein(s) that binds neurotransmitter on postsynaptic membrane of a chemical synapse in the nervous system.

chemoreflex *n.* reflex caused by chemical stimulus.

chemorepellant *n.* any chemical that causes cells or organisms to move away from it.

chemosensory *a.* sensitive to chemical stimuli; *pert.* sensing of chemical stimuli.

chemostat *n.* any organ concerned in maintaining constancy of chemical conditions, such as carotid body regulating pH of blood.

chemosynthesis *n.* biosynthesis of organic compounds; use of oxidation of inorganic compounds as energy source for biosynthesis in some bacteria. *a.* chemosynthetic.

chemotaxis *n.* reaction of motile cells or microorganisms to chemical stimuli by moving towards or away from source of chemical. *a.* chemotactic.

chemotroph *n.* any organism obtaining energy by taking in and oxidizing chemical compounds, either inorganic chemical sources of energy as chemoautotrophs, or by the breakdown of complex organic compounds (food) as chemoheterotrophs. *a.* chemotrophic. *cf.* phototroph.

chemotropism *n.* curvature of plant or plant organ in response to chemical stimuli.

Chenopodiales Caryophylales *q.v.*

chernozem *n.* black soil, formed under continental climatic conditions and characteristic of subhumid to temperate grasslands. *alt.* black earth.

chersophilous *a.* thriving on dry waste land.

chersophyte *n.* a plant which grows on wasteland or on shallow soil.

chestnut soils dark-brown soils of semiarid steppe-lands, fertile under adequate rainfall or when irrigated.

cheta *see* chaeta.

chevron *n.* V-shaped bones articulating with ventral surface of spinal column in caudal region of many vertebrates.

chiasma *n.* a partial crossing over of nerve fibres from either side of the body, as at optic chiasma in brain, *alt.* chiasm; a visible X-shaped structure formed by homologous chromatids in prophase of meiosis and which represents the site of crossing-over and exchange of segments of DNA between homologous chromatids (recombination) by the mechanism of breakage and reunion. *plu.* chiasmata.

chilaria *n.plu.* pair of processes between 6th pair of appendages in the king crab, *Limulus*.

chilidium *n.* a shelly plate covering the opening between hinge and beak in brachiopods.

Chilopoda, chilopods *n., n.plu.* in some classifications, a class of arthropods comprising the centipedes, which have numerous and similar body segments each with one pair of walking legs, except the first segment which bears a pair of poison claws. Considered as a subclass or order of class Myriapoda in some classifications.

chimaera, chimera *n.* single organism developing from embryo composed of cells from 2 different individuals and therefore composed of cells of 2 different genotypes; organism composed of cells of 2 different genotypes as a result of grafting, e.g. of bone marrow.

chimaeric *a. appl.* animals or plants consisting of some cells with one genetic constitution and some with another, *n.* chimaera; *appl.* DNA or proteins consisting of a mixture of sequences or subunits from different sources, *alt.* hybrid.

chimeric, chimera *see* chimaeric *q.v.*

chionophyte *n.* a snow-loving plant.

chirality *n.* property of molecules with handedness in their chemical structure (i.e. their mirror image cannot be exactly superimposed on the 'real' image in any orientation) and possessing optical activity. *a.* chiral. *cf.* prochirality.

chiro- cheiro-.

chironomid *a., n.* midge of the genus *Chironomus*, whose aquatic larvae are used as indicator species in assessing biotic indices for freshwater habitats. Abundance of the larvae of *C. riparius* (the bloodworm) on their own indicates heavy organic pollution.

Chiroptera, chiropterans *n., n.plu.* order of placental mammals including the small mainly insectivorous bats (microchiroptera) and the fruit-eating flying foxes (megachiroptera), having the forelimbs modified for flight supporting a wing membrane stretched between limbs, and body and hindlimbs reduced.

chiropterophilous *a.* pollinated by the agency of bats.

chiropterygium cheiropterygium *q.v.*

chisel teeth chisel-shaped incisors of rodents.

chi sequence short DNA sequence repeated many times in the bacterial (*E. coli*) chromosome at which RecA-mediated recombination is stimulated.

chi (χ) structure X-shaped structures derived from figure-of-eight DNA molecules isolated from bacterial cells undergoing genetic recombination and which are presumed to represent recombination intermediates of circular molecules.

chitin *n.* long chain polymer of *N*-acetylglucosamine units, the chief polysaccharide in fungal cell walls, and also found in the exoskeletons of arthropods.

chitinase *n.* enzyme that hydrolyses chitin. EC 3.2.1.14.

chitinous *a.* composed of, or containing chitin.

chitons *n.plu.* common name for the Polyplacophora, a class of marine molluscs with an elongated body bearing a shell of calcareous plates and a muscular foot on which they crawl about.

chlamydate *a.* having a mantle *q.v.*; ensheathed, enclosed in a cyst.

chlamydeous *a.* having a perianth.

chlamydiae *n.plu.* a group of prokaryotic microorganisms, obligate intracellular parasites, responsible for a variety of human and animal diseases.

chlamydocyst *n.* an encysted zoosporangium, as in certain saprobic fungi.

chlamydospore *n.* thick-walled resting spore of certain fungi and protozoans.

chloragen chloragogen *q.v.*

chloragogen *a.* cells in the roof of annelid intestine containing stores of oil and glycogen.

chloramphenicol *n.* antibiotic produced by the actinomycete *Streptomyces venezuelae*, which blocks translation in bacteria and mitochondria by attaching to the 50S ribosomal subunit and preventing the addition of an amino acid to the polypeptide chain by inhibiting the peptidyltransferase reaction.

chloramphenicol acetyltransferase (CAT) bacterial enzyme (EC 2.3.1.28) that confers resistance to the antibiotic chloramphenicol. The bacterial *cat* gene encoding the enzyme is widely used as a marker gene in gene transfer experiments, as cells carrying it can be selected by their survival when cultured in the presence of chloramphenicol.

chloranthy *n.* reversion of petals, sepals, etc. to ordinary green leaves.

chlorenchyma *n.* plant tissue containing chlorophyll.

chloride cell columnar cell of gill epithelium, specialized for excretion of chloride, in certain fishes.

chloride shift movement of chloride ions

into red blood cells and bicarbonate ions out.

chlorinated hydrocarbons compounds such as DDT and PCBs, which are persistent, entering food chains and often accumulating in other organisms.

chlorocruorin *n.* a green, haem-containing, oxygen-carrying protein found in the blood of certain polychaete worms.

chlorolabe *n.* green-sensitive pigment of the human eye.

chloronema *n.* in mosses, a type of protonemal branch which grows along the surface of the substrate or into the air for a short distance and contains many conspicuous chloroplasts.

chlorophore *n.* a chlorophyll granule in protists.

Chlorophyceae *n.* Chlorophyta *q.v.*

chlorophyll *n.* principal photosynthetic pigment of green plants and algae, consisting of a porphyrin (tetrapyrrole) ring with magnesium at the centre and esterified to a long chain aliphatic alcohol (phytol), different chlorophylls having different side chains. It is located in the thylakoid membranes of chloroplasts where it traps light energy, absorbing mainly in red and violet-blue regions of the spectrum, chemically distinct forms having different absorption maxima. Chlorophylls *a* and *b* are found in higher plants, chlorophylls *c* and *d* in algae. *see* Appendix 1 (60) for chemical structure. *see also* bacteriochlorophyll.

chlorophyllose cells elongated, very narrow living cells containing chloroplasts, separated from each other by large empty cells, in *Sphagnum* moss leaves.

Chlorophyta *n.* the green algae, in some modern classifications a phylum of the kingdom Protista, in more traditional classifications regarded as a division of the kingdom Plantae. Mostly freshwater, but some marine, they include unicellular and multicellular groups, have chlorophylls and carotenoids similar to those of vascular plants and appear green, store food as starch and have cellulose cell walls.

chloroplast *n.* semi-autonomous organelle found in the cytoplasm of cells of all green plants, and in which the reactions of photosynthesis take place, contains the green pigment chlorophyll and other pigments involved with photosynthesis and also contains DNA which specifies rRNAs, tRNAs and some chloroplast proteins. *alt.* plastid. *see also* photosynthesis, stroma, thylakoid. *see* Fig. 1.

chloroplast factor (CF) enzyme complex in thylakoid membranes of chloroplasts which is responsible for coupling of electron/proton transport to ATP synthesis in photosynthesis, and is composed of 2 protein components, CF_1 and CF_0, CF_1 catalysing formation of ATP from ADP, CF_0 forming a proton channel through membrane.

chloroplast pigments chlorophylls, carotene, and xanthophyll.

chloroquine *n.* antimalarial drug that raises the pH of cellular organelles such as lysosomes, endosomes and Golgi apparatus.

chlorosis *n.* abnormal condition characterized by lack of green pigment in plants, owing to lack of light, or to magnesium or ion deficiency, or to genetic deficiencies in chlorophyll synthesis. *a.* chlorotic.

chlorosome *n.* photosynthetic structure containing chlorophyll in green algae.

Chloroxybacteria prochlorophytes *q.v.*

choana *n.* a funnel-shaped opening; opening of the nostrils into the pharynx or throat.

Choanichthyes Sarcopterygii *q.v.*

choanocyte *n.* flagellated cell, with protoplasmic collar around base of the flagellum, lining body cavity of sponges. Involved in uptake of food particles from water drawn into cavity by the beating of the flagella. *alt.* collar cell.

choanoflagellates *n.plu.* a group of flagellates which have a protoplasmic collar around the base of the flagella.

cholecyanin bilicyanin *q.v.*

cholecystokinin (CCK) *n.* a peptide hormone produced by duodenal cells and acting on pancreatic acinar cells to induce secretion of digestive enzymes, also induces contraction of gall bladder and relaxation of sphincter around the duodenal end of the common bile duct. Also present in the central nervous system where its role is not yet known.

choleic *a. appl.* a bile acid.

cholesterol *n.* a neutral lipid, a sterol found in animal cell membranes, where it influences membrane fluidity, but absent from higher plants and most bacteria. Synthesized starting from acetyl CoA, it is the precursor of many biologically active steroids such as the steroid hormones and vitamin D. In humans the main site of synthesis is the liver, and cholesterol is transported in the blood mainly in the form of lipoprotein particles. In certain conditions it is deposited from plasma onto blood vessel walls forming atherosclerotic plaques. *see* Appendix 1 (45) for chemical structure. *see* plasma lipoproteins.

cholic *a. pert.*, present in, or derived from bile.

cholic acid *n.* an acid, $C_{24}H_{40}O_5$, present in bile, helps in fat digestion.

choline *n.* $(CH_3)_3NCH_2CH_2OH$, which is acetylated to form the neurotransmitter acetylcholine, and which is also a moiety of the membrane phospholipid phosphatidylcholine. Choline is sometimes considered as a vitamin, one of the components of the B vitamin complex. Glandular tissue, nervous tissue, egg yolk and some vegetable oils are particularly rich in choline.

choline acetyltransferase enzyme which synthesizes the neurotransmitter acetylcholine.

cholinergic *a. appl.* nerve fibres that liberate acetylcholine from their terminals.

cholinesterase *n.* enzyme found in the synaptic cleft at cholinergic synapses, which hydrolyses acetylcholine to choline and acetic acid and also other acylcholines to choline and carboxylic acids, EC 3.1.1.8. *alt.* acetylcholinesterase.

chondral *a. pert.* cartilage.

chondric *a.* gristly, cartilaginous.

Chondrichthyes *n.* class of fishes known from the Devonian to the present day, commonly known as the cartilaginous fishes, having a cartilaginous skeleton, spiral valve in the gut and no lungs or air bladder, and including the rays, skates and sharks. *cf.* Osteichthyes.

chondrin *n.* a gelatinous bluish-white substance forming the ground substance of cartilage, having a firm elastic consistency.

chondrioclast chondroclast *q.v.*

chondroblast, chondrocyte *n.* cartilage cell, which secretes chondrin, the extracellular matrix comprising cartilage.

chondroclast *n.* large multinuclear cell that breaks down cartilage matrix. *alt.* chondrioclast.

chondrocranium *n.* the skull when in a cartilaginous condition, either temporarily as in embryos, or permanently as in some fishes.

chondroid *a.* cartilage-like, *appl.* tissue: undeveloped cartilage or pseudo-cartilage, serving as support in certain invertebrates and lower vertebrates.

chondroitin *n.* glycosaminoglycan portion of a proteoglycan from cartilage and other connective tissues, composed largely of repeating sulphated disaccharide units D-glucuronic acid, *N*-acetylgalactosamine.

chondrophore *n.* structure which supports the inner hinge cartilage in a bivalve shell.

chondrophores *n.plu.* colonial hydrozoans of the order Chondrophora, showing a degree of division of labour and cooperation between zooids approaching that of siphonophores.

chondroseptum *n.* the cartilaginous part of the septum of the nose.

chondroskeleton *n.* a cartilaginous skeleton.

Chondrostei, chondrosteans *n., n.plu.* group of primitive actinopterygian bony fishes including the extant bichirs of the Nile, which have lungs, and paddlefishes and sturgeons, as well as many extinct groups. The bony skeleton has largely been substituted by cartilage. They have usually a spiral valve in the gut, and retain the spiracle, and paddlefishes and sturgeons have a heterocercal tail. *see also* palaeoniscids.

chondrosteosis *n.* the conversion of cartilage to bone.

chondrosteous *a.* having a cartilaginous skeleton.

chondrosternal *a. pert.* rib cartilages and sternum.

chorda *n.* any cord-like structure. *plu.* chordae.

chordacentra *n.plu.* vertebral centra formed by conversion of notochordal sheath into a number of rings.

chordae tendineae tendons connecting

papillary muscles with valves of heart.

chordal *a. pert.* a chorda or chordae; *pert.* the notochord.

Chordata, chordates *n., n.plu.* a phylum of coelomate animals having a notochord and gill clefts in the pharynx at some point in their life history, and a hollow nerve cord running dorsally with the anterior end usually dilated to form a brain. The chordates include the vertebrates, the cephalochordates (e.g. amphioxus) and the urochordates (e.g. sea squirts).

chordotonal *a. appl.* sensilla: rod-like or bristle-like receptor for mechanical or sound vibrations in various parts of body of insects.

chore *n.* an area showing a unity of geographical or environmental conditions.

chorioallantoic *a. appl.* placenta when chorion is lined by allantois, allantoic vessels conveying blood to embryo, as in some marsupials and all eutherian mammals.

choriocapillaris *n.* the innermost vascular layer of the chorion.

chorion *n.* embryonic membrane external to and enclosing the amnion and yolk sac; allantochorion *q.v.*; tough shell covering the eggs of insects; (*bot.*) outer membrane of a seed.

chorion frondosum villous placental part of chorion.

chorionic *a. pert.* the chorion; *pert.* hormone, *see* gonadotropin.

chorionic villus sampling method of antenatal diagnosis of various genetic diseases by analysing cells taken from chorionic villi. Can be carried out earlier than amniocentesis.

chorioretinal *a. pert.* choroid and retina.

choriovitelline *a. appl.* placenta when part of chorion is lined with yolk sac, vitelline blood vessels being connected with uterine wall, as in some marsupials.

chorisis *n.* increase in parts of floral whorl due to division of its primary members.

chorismate (chorismic acid) *n.* aromatic carboxylic acid, intermediate in synthesis of aromatic amino acids in microorganisms.

C-horizon lowest layer of a soil profile above the bedrock.

choroid *a.* infiltrated with many blood vessels, *appl.* delicate and highly vascular membranes; *n.* intermediate pigmented layer of eye, underlying the retina, and forming the diaphragm of the iris.

choroid plexus highly vascularized network of interlaced capillaries and nerves between retina and the tough opaque fibrous layer of the eyeball; vascular tissue lining the ventricles of the brain, which secretes cerebrospinal fluid.

chorology *n.* study of the geographical distribution of plants and animals.

choronomic *a.* external, *appl.* influences of geographical or regional environment.

chorotypes *n.plu.* local types.

Christmas disease *see* haemophilia.

Christmas factor Factor IX *q.v.*

chroma *n.* the hue and saturation of a colour.

chromaffin cell cell of neural origin that secretes catecholamines, present in carotid body, medulla of adrenal gland in vertebrates and in some invertebrates such as annelids.

chromatic *a.* of, produced by, or full of bright colour.

chromatic threshold minimal stimulus, varying with the wavelength of the light, which induces a sensation of colour.

chromatid *n.* one of two copies of a replicated chromosome, visible at prophase and metaphase of mitosis and meiosis as half of a double structure held together at the centromere and becoming a separate daughter chromosome at anaphase of mitosis and in the 2nd meiotic division.

chromatin *n.* nucleoprotein material staining with basic dyes, found in the nucleus of eukaryotic cells and becoming organized into visible chromosomes at cell division, and being composed of DNA associated with histones (*q.v.*) to form nucleosomes (*q.v.*). *see also* heterochromatin, euchromatin.

chromatocyte *n.* any cell containing a pigment.

chromatography *n.* separation of compounds from a mixture on the basis of their affinity for and migration with a nonpolar solvent such as water, on a polar support such as paper or a calcium carbonate column. *see also* affinity chromatography, partition chromatography, gel-filtration chromatography.

chromatophil(ous) chromophilous *q.v.*

chromatophore *n.* organelle containing pigment; a pigment cell or group of cells which, under the control of the nervous system or of hormones, can be altered in shape or colour.

chromatophyll *n.* the colouring matter of plant-like flagellates.

chromatotropism *n.* orientation in response to stimulus consisting of a particular colour.

chromidial substance material in cytoplasm of some cells staining with basophilic dyes and which is now known to be rough endoplasmic reticulum.

chromo-argentaffin *a.* staining with bichromates and silver nitrate, *appl.* flask-shaped cells in epithelium of crypts of Lieberkuhn.

chromoblast *n.* embryonic cell giving rise to a pigment cell.

chromocentre *n.* granule of heterochromatin, part of interphase chromosomes, many of which show up on staining of interphase nuclei.

chromocyte *n.* pigment cell, any cell containing pigment.

chromogen *n.* substance that is converted into a pigment.

chromogenesis *n.* production of colour or pigment.

chromogenic *a.* colour-producing.

chromomeres *n.plu.* structures in early meiotic chromosomes which are visible as densely staining granules which give a beaded appearance to the chromosomes.

chromonema *n.* threads of chromatin (chromosomes) as detected during interphase when they are extended and dispersed in the nucleus. *plu.* chromonemata.

chromophane *n.* any of a type of pigmented oil globule (red, yellow or green) found in retina of birds, reptiles, fishes, marsupials; any retinal pigments.

chromophil(ic) chromaffin *q.v.*; chromophilous *q.v.*

chromophilous *a.* staining readily.

chromophore *n.* group of atoms in a molecule responsible for absorbing light energy and to whose presence colour in a compound is due.

chromoplast *n.* a coloured plastid containing no chlorophyll but usually containing some red or yellow pigment.

chromoprotein *n.* any protein that is combined with a pigment, as haemoglobin, chlorocruorin, cytochromes.

chromosomal *a.* of or *pert.* a chromosome or chromosomes.

chromosomal incompatibility failure to interbreed due to differences in chromosome composition.

chromosomal mosaic an individual who has somatic cells of more than one type of chromosome constitution usually arising from an error in mitosis during development.

chromosome *n.* one of the small, rod-shaped, deeply staining bodies that become visible under the light microscope in the eukaryotic cell nucleus at mitosis and meiosis, and which carry the genes in linear order. Each chromosome consists of a single very long molecule of DNA which is packaged into a more compact structure by association with histones and other proteins to form chromatin. Chromosomes are generally extended throughout the nucleus, but at mitosis or meiosis become more highly compacted and visible. *see also* chromatin, DNA, gene. The single DNA molecule in prokaryotic cells is also often called a chromosome.

chromosome aberration any departure from the normal in gross chromosomal structure or number, such as trisomy, deletions, duplications, inversions and translocations.

chromosome complement karyotype *q.v.*

chromosome map a plan showing the position of genes on a chromosome.

chromosome mutation chromosomal aberration *q.v.*

chromosome number the usual constant number of chromosomes in a somatic cell that is characteristic of a particular species.

chromosome painting the delineation of a particular chromosome or set of chromosomes in a karyotype by staining with fluorescent dyes attached to a collection of DNA probes specific for that chromosome or chromosomes.

chromosome pair the pair of homologous chromosomes, one derived from the maternal, one from the paternal parent, which become associated at meiosis.

chromosome races races of a species differing in number of chromosomes, or in number of sets of chromosomes.

chromosome walking technique for mapping chromosomes from a collection of overlapping restriction fragments. Starting from a known DNA sequence, the overlapping sequences can be detected in other restriction fragments and a map of a particular area gradually built up.

chromosome 21 the chromosome found in a trisomic state in the cells of persons with Down syndrome.

chromotropic *a. appl.* any agent or factor controlling pigmentation.

chronaxia, chronaxie, chronaxy *n.* latent time between electrical stimulation and muscle contraction; minimal time required between successive firings of a nerve cell under a prolonged stimulus.

chronotropic *a.* affecting the rate of action.

chrysalis *n.* the pupa of insects with complete metamorphosis, enclosed in a protective case, which is sometimes itself called the chrysalis.

chryso- prefix derived from Gk. *chrysos,* gold.

chrysocarpous *a.* with golden-yellow fruit.

chrysolaminarin *n.* storage polysaccharide of golden-brown algae.

chrysophanic *a.* having a golden or brown colour, *appl.* an acid formed in certain lichens and in leaves.

Chrysophyceae *n.* the golden-brown algae, a class of freshwater and marine algae, unicellular or colonial, an important constituent of the marine nannoplankton. *see also* Chrysophyta.

chrysophyll *n.* a yellow colouring matter in plants, a decomposition product of chlorophyll.

Chrysophyta *n.* the diatoms (class Bacillariophyceae), the golden-brown algae (class Chrysophyceae) and the yellow-green algae (class Xanthophyceae), in some modern classifications a division of the kingdom Protista, in more traditional classifications regarded as a division of the kingdom Plantae. Almost entirely unicellular, they contain large amounts of carotenoid pigments, reserves of oil and the polysaccharide chrysolaminarin. Diatoms

have no cellulose cell wall but the membrane tends to become silicified, giving individual species their characteristic shapes. The Chrysophyceae often possess siliceous or organic skeletons or scales.

chyle *n.* lymph containing globules of emulsifed fat, found in the lymphatic vessels of small intestine during digestion.

chylifaction *n.* formation of chyle.

chyliferous *a. appl.* tubes and vessels conveying chyle.

chylific *a.* chyle-producing, *appl.* ventricle or true stomach of insects.

chylocaulous *a.* with fleshy stems.

chylomicrons *n.plu.* small lipoprotein particles in plasma and other body fluids, which transport cholesterol, triacylglycerols and other lipids from intestine to adipose tissue.

chylophyllous *a.* with fleshy leaves, *appl.* succulent plants adapted to dry conditions, such as stonecrops and sedums.

chymase rennin *q.v.*

chyme *n.* partially digested food leaving the stomach.

chymosin *n.* a proteinase in gastric juice and some digestive secretions of some insectivorous plants, which catalyses the conversion of caseinogen in milk to insoluble casein (paracasein). EC 3.4.23.4. *alt.* rennin.

chymotrypsin *n.* proteinase digestive enzyme found in pancreatic juice, which cleaves peptides and proteins at hydrophobic amino acids and acts on the products of pepsin and trypsin action during digestion, formed in pancreas by enzymatic cleavage of an inactive precursor, chymotrypsinogen, by trypsin. EC 3.4.21.1.

Chytridiomycota, chytrids *n., n.plu.* phylum of unicellular saprobic or parasitic protists living in soil or fresh water and formerly classified as fungi. They are unicellular or coenocytic, with cell walls containing chitin. Sexual reproduction is oogamous and they also reproduce asexually by motile zoospores. Sperm and zoospores all bear a single posterior whiplash flagellum, distinguishing them from the hypochytrids. *alt.* Chytridiomycetes.

chytridium *n.* structure containing spores in certain fungi.

cibarium *n.* part of buccal cavity anterior to

pharynx in insects.

cicatrice, cicatrix *n.* scar; small scar in place of previous attachment of organ.

cicatricial tissue newly-formed fine fibrous connective tissue that closes and draws together wounds.

Ciconiiformes *n.* an order of wading birds with long necks, legs and bill, including herons, bitterns, storks and flamingos.

cilia *n.plu.* motile hair-like outgrowths from the surface of many types of eukaryotic cell, capable of whip-like beating movement and which in free-living unicells (e.g. protozoans) propel the cell, and in stationary cells (e.g. of nasal epithelium) produce a flow of material over the cell surface. Cilia are composed of a central core of microtubules (the axoneme) anchored to the cell by a basal body, the whole enclosed in plasma membrane. The bending beating movement is caused by the microtubules sliding against each other. In multicellular organisms they chiefly occur on cells making up a ciliated epithelium lining various internal passages; various other hair-like structures, esp. eyelash. *sing.* cilium.

ciliaris *n.* muscle forming a ring outside anterior part of choroid in the eye, and which is responsible for regulating the convexity of the lens.

ciliary *a. pert.* cilia; *pert.* eyelashes, *appl.* sweat glands; *appl.* certain structures in the eyeball as arteries, body, processes, muscle; *appl.* branches of the nasociliary nerve and ganglion.

ciliary body muscular ring behind iris in vertebrate eye, which regulates size of pupil.

Ciliata, Ciliatea, ciliates *n., n., n.plu.* class of free-living and sessile protozoans of complex cellular structure, bearing cilia, often in rows on the surface or grouped into compound structures. Common in marine and fresh water, and in rumen of cattle and other ruminants where they digest cellulose. Almost all ciliates possess two nuclei, a possibly polyploid macronucleus that directs vegetative growth, and a diploid micronucleus involved in sexual reproduction.

ciliated *a.* bearing cilia.

ciliated epithelium epithelium lining various internal passages, as nasal passage, composed of cells with cilia projecting from the free surface, and whose beating action draws fluid etc. over the surface.

ciliograde *a. appl.* movement due to cilia.

Ciliophora *n.* phylum of protists comprising the unicellular ciliates (*q.v.*), which are characterized by possession of cilia, at least when immature, are usually binucleate, and are never amoeboid; in older classifications a subphylum of the protozoa. *see* Appendix 5.

ciliospore *n.* a ciliated protozoan swarm spore.

cilium *sing.* of cilia *q.v.*

cinchonine *n.* alkaloid found in various dicots of the family Rubiaceae.

cinereous *a.* ash-grey.

cingula *n.* ring formed on stalk of some mushroooms and toadstools at which the edge of the cap was attached to the stalk, *plu.* cingulae; *plu.* of cingulum *q.v.*

cingulate *a.* shaped like a girdle, *appl.* a gyrus (ridge) and sulcus (fissure) above the corpus callosum on the median surface of cerebral hemisphere.

cingulum *n.* any structure that is like a girdle; part of plant between root and stem; a ridge round base of crown of a tooth; tract of fibres connecting callosal and hippocampal convolutions of brain. *plu.* cingula.

circadian rhythm metabolic or behavioural rhythm with a cycle of about 24 hours.

circannual rhythm a rhythm or cycle of behaviour of approximately one year.

circaseptan *a. appl.* biological rhythm of around 7 to 8 days.

circinate, circinnate *a.* rolled on the axis so that apex is the centre, as young fern fronds.

circulation *n.* the regular movement of any fluid within definite channels in the body; the blood circulatory system.

circulus *n.* a ring-like arrangement.

circum- prefix derived from L. *circum*, around, surrounding.

circumduction *n.* form of movement exhibited by a bone describing a conical space with the joint cavity as the apex.

circumesophageal circumoesophageal *q.v.*

circumferential *a.* around the circumference; *appl.* primary lamellae parallel to circumference of bone.

circumfila *n.plu.* looped or wreathed filaments on antennae of some insects.

circumflex *a.* bending round.

circumgenital *a.* surrounding the genital pore; *appl.* glands secreting a waxy powder in oviparous species of coccid bugs.

circumnutation *n.* the irregular elliptical or spiral movement exhibited by a growing stem, shoot or tendril.

circumoesophageal *a.* structures or organs surrounding the gullet.

circumoral *a.* surrounding the mouth, *appl.* cilia, tentacles, nerve ring, etc.

circumorbital *a.* surrounding the orbit of the eye, *appl.* bones of the skull.

circumpolar *a. appl.* flora and fauna of Polar regions; *appl.* to distributions of plants and animals in northerly parts of Northern Hemisphere that extend through Asia, Europe and North America.

circumscissile *a.* splitting along a circular line, *appl.* dehiscence of a pyxidium.

circumvallate *a.* surrounded by a wall, as of tissue.

cirral *a. pert.* cirri or a cirrus.

cirrate *a.* having cirri.

cirrhus *n.* ribbon-like string of spores held together by mucus, issuing from a sporocarp.

cirri *plu.* of cirrus *q.v.*

Cirripedia, cirriped(e)s *n., n.plu.* a subclass of aquatic crustaceans, commonly called barnacles, which as adults are stalked or sessile sedentary animals with the head and abdomen reduced and the body enclosed in a shell of calcareous plates.

cirrose, cirrous *a.* with cirri or tendrils; *appl.* leaf with prolongation of midrib forming a tendril.

cirrus *n.* tendril or tendril-like structure; feathery feeding appendage of barnacle; respiratory and copulatory appendage of annelids; projection from the stalk of sea lilies, which anchors them to the substratum; organ having the function of penis in some molluscs and flatworms; hair-like structure on appendages of insects. *plu.* cirri.

cis in genetics two different mutations at the same locus in a diploid organism are said to be in *cis* if they are both on the same chromosome, *cf. trans*; *appl.* molecular configuration, one of two configurations

of a molecule caused by the limitation of rotation around a double bond, the alternative configuration being the *trans*-configuration.

***cis*-acting** *a. appl.* genes or their protein products produced by one chromosome and acting on or cooperating with neighbouring genes on the same chromosome.

***cis*-dominance** the property of a genetic site to control adjacent genes irrespective of the presence of other alleles of the site in the cell, but not to control corresponding genes on other DNA molecules or chromosomes. *alt. cis*-acting.

***cis-trans* complementation test** complementation test *q.v.*

Cistales Violales *q.v.*

cistern epiphyte an epiphyte lacking roots and gathering water between leaf bases.

cisterna *n.* closed space containing fluid, as any of the subarachnoid spaces; expanded flattened sacs of membrane in Golgi apparatus or as part of endoplasmic reticulum; any of various other flattened membrane-bounded fluid-filled vesicles. *plu.* cisternae.

cistron *n.* DNA sequence coding for a single polypeptide.

citrate (citric acid) *n.* 6-carbon tricarboxylic acid, a component of the tricarboxylic acid cycle, where it is converted to isocitrate by the enzyme aconitase, and which is also responsible for the sour taste of oranges and grapefruit. *see* Appendix 1 (53) for chemical structure.

citric acid cycle tricarboxylic acid cycle *q.v.*

citrin *n.* mixture of flavonoid pigments first isolated from lemon peel, of which the active constituent is hesperidin.

citrulline *n.* amino acid, first isolated from water melon, important as an intermediate in the urea cycle in vertebrates.

cladanthous *a.* having terminal archegonia on short lateral branches.

clade *n.* a branch of a phylogenetic tree containing the set of all organisms descended from a particular common ancestor which is not an ancestor of any non-member of the group.

cladistics *n.* method of classification of living organisms that makes use of lines of descent only, rather than phenotypic similarities, to deduce evolutionary relationships, and which groups organisms strictly on the relative recency of common ancestry. Cladistic methods of classification only permit taxa in which all the members share a common ancestor who is also a member of the taxon and which include all the descendants of that common ancestor, regardless of their different degrees of divergence.

cladocarpous cladanthous *q.v.*

cladode *n.* green flattened lateral shoot, arising from the axil of a leaf and resembling a foliage leaf.

cladodont *a.* having or *appl.* teeth with prominent central and small lateral cusps.

cladogenesis *n.* branching of evolutionary lineages so as to produce new types; evolutionary change as a result of multiplication of species at any one time and their subsequent evolution along different lines. *cf.* anagenesis.

cladogenous *a.* borne on stem, *appl.* certain roots; borne on branches.

cladogram *n.* tree-like diagram showing the evolutionary descent of any group of organisms.

cladophyll(um) cladode *q.v.*

cladoptosis *n.* annual or other shedding of twigs.

cladose *a.* branched.

cladosiphonic *a.* with insertion of leaf trace on periphery of the axial stele.

cladus *n.* a branch, as of a branched spicule.

clamp connections characteristic swellings on hyphae of certain basidiomycete fungi, through which a daughter nucleus has passed and at which a septum forms to make two binucleate cells.

clan phratry *q.v.*

clandestine *a. appl.* evolutionary change which is not apparent in adult forms; or of adult characters from ancestral embryonic characters.

Clara cell lung cell of unknown function.

claspers *n.plu.* modified organs or parts of various types enabling the two sexes to clasp one another during mating, e.g. the rod-like processes on pelvic fins of some fishes.

claspettes harpagones *q.v.*

class *n.* taxonomic group into which a phylum or a division is divided, and which is itself divided into orders.

class I MHC antigen(s) protein(s) encoded within the major histocompatibility complex and which occur on the surface of all body cells. Involved in immune responses against virus-infected cells and rejection of transplanted tissue. In humans represented by HLA-A, -B, and -C antigens.

class II MHC antigen(s) cell surface protein(s) encoded within the major histocompatibility complex and involved in antigen recognition by the immune system. Normally produced only by T cells, B cells, macrophages and other antigen-presenting cells. In humans represented by HLA-D antigens. *alt.* Ia antigen *q.v.*

classical conditioning experimental behavioural technique whereby a response (the unconditional response, e.g. salivation) elicited by a natural stimulus such as food (the unconditional stimulus) becomes a response (the conditional response) to an unrelated stimulus (the conditional stimulus, e.g. a bell) by repeated association of the conditional and unconditional stimuli. Positive conditioning increases an animal's response to the stimulus, negative conditioning results in an increased avoidance of the stimulus. *alt.* Pavlovian conditioning.

classification *n.* the arrangement of living organisms into groups on the basis of observed similarities and differences, modern classifications of plants and animals attempting wherever possible to reflect degrees of evolutionary relatedness. The smallest group in classification is usually the species although subspecies, races and varieties (in cultivated plants) below the level of the species are often recognized. Species are grouped into genera, genera into families, families into orders, orders into classes, classes into phyla (for animals) or divisions (for plants), and phyla and divisions into kingdoms. There are also intermediate categories such as superfamilies and infraclasses. *see* Appendices 2–7.

class switching intrachromosomal recombination in B lymphocytes in which a rearranged immunoglobulin heavy chain variable region coding sequence can be successively linked to different heavy chain constant region genes. This results

clathrate

in successive expression of antibody molecules with the same antigen specificity but different effector functions during an immune response.

clathrate, clathroid *a.* lattice-like.

clathrin *n.* fibrous protein that forms a characteristic polyhedral coat on the surface of coated pits and coated vesicles.

claustrum *n.* thin strip of grey matter in hemispheres of brain, located just under the cortex at each side.

clava *n.* club-shaped structure, e.g. club-shaped spore-bearing branch in certain fungi, the knob-like end of antenna in some insects. *plu.* clavae.

clavate *a.* club-shaped, thickened at one end.

clavicle *n.* collar bone, forming anterior or ventral portion of the shoulder girdle. *a.* clavicular.

clavicularium *n.* in turtles and tortoises, one of a pair of anterior bony plates in shell.

claviform *a.* club-shaped.

clavola *n.* the terminal joints of an insect antenna.

clavula *n.* a club-shaped fruiting body of certain fungi.

clavus *n.* the part of hemielytron lying next to the scutellum in Hemiptera; ergot disease in grasses.

claw *n.* (*bot.*) the stalk of a petal.

clay *n.* a soil with most particles below 0.002 mm in diameter, made of hydrated aluminosilicates, and poorly drained and aerated.

cleavage *n.* the series of mitotic divisions, usually occurring with no increase in cytoplasmic mass, that first transform the single-celled zygote into a multicellular morula or blastula.

cleavage nucleus nucleus of fertilized egg of zygote produced by fusion of male and female pronuclei; the egg nucleus of parthenogenetic eggs.

cleidoic *a.* having or *pert.* eggs enclosed within a shell or membrane.

cleisto- prefix derived from the Gk. *kleistos*, closed.

cleistocarp cleistothecium *q.v.*

cleistocarpous, cleistocarpic *a.* having closed fruiting bodies, in certain fungi and mosses.

cleistogamy *n.* the condition of having flowers that never open and are self-pollinated, and are often small and inconspicuous. *a.* cleistogamic, cleistogamous.

cleistothecium *n.* a closed fruiting body (ascocarp) characteristic of certain ascomycete fungi, in which spores are produced internally and are released by breakdown of the wall.

cleithrum *n.* clavicular element in some fishes.

cleptobiosis *n.* the robbing of food stores or scavenging in the refuse piles of one species by another that does not live in close association with it. *alt.* kleptobiosis.

cleptoparasitism kleptoparasitism *q.v.*

climacteric *n.* a critical phase, or period of changes, in living organisms; *appl.* change associated with menopause or with recession of male function; (*bot.*) appl. phase of increased respiratory activity at ripening of fruit.

climactic *a. pert.* a climax.

climatype *n.* a biotype resulting from selection in particular climatic conditions.

climax *n.* the mature or stabilized stage in a successional series of communities, when dominant species are completely adapted to environmental conditions.

clinandrium *n.* cavity in the column between anthers in orchids.

clinanthium *n.* a dilated floral receptacle, as in the flower-head of Compositae.

cline *n.* a graded series of different forms of the same species, usually distributed along a spatial dimension.

clinidium *n.* a filament that produces spores, in a pycnidium.

clisere *n.* succession of communities which results from a changing climate.

clitellum *n.* the swollen glandular portion of skin of certain annelids, such as earthworm, which secretes the cocoon in which an embryonic worm develops.

clitoris *n.* erectile organ, homologous with penis, at upper part of vulva.

clivus *n.* a shallow depression in sphenoid, behind dorsum sellae; the posterior sloped part of the monticulus of cerebellum.

cloaca *n.* the common chamber into which intestinal, genital and urinary canals open in vertebrates, except most mammals; posterior end of intestinal tract in some invertebrates. *a.* cloacal.

clonal *a. pert.* a clone.

clonal selection in an immune response, the proliferation, in response to stimulation by an antigen, of clones of lymphocytes of the corresponding specificity.

clone *n.* a group of genetically identical individuals or cells derived from a single cell by repeated asexual divisions, *see also* DNA clone; (*bot.*) an apomict strain; *v.* to produce a set of identical individual cells, or DNA molecules from a single starting cell or molecule.

cloning *n. see* clone; DNA cloning *q.v.*

cloning vector specially modified plasmid or phage into which "foreign" genes can be inserted for introduction into bacterial or other cells for multiplication, studies of gene expression, etc.

clonotype *n.* a specimen of an asexually propagated part of a type specimen or holotype.

clonus *n.* a series of muscular contractions in which individual contractions are discernible.

closed community ecological community into which further colonization is prevented as all niches are occupied.

closed forest forest where the crowns of the trees touch and produce a closed canopy for all or part of the year.

closterovirus group plant virus group containing very long thread-like single-stranded RNA viruses, with two subgroups: (1) aphid-transmitted viruses, type member beet yellows virus; (2) no known vectors, type member apple chlorotic leafspot virus.

clostridia *n.plu.* bacteria of the genus *Clostridium*, strictly anaerobic, spore-forming rods widely distributed in soil, some species of which produce powerful toxins, and which include *Cl. tetani*, the causal agent of tetanus, *Cl. botulinum*, whose toxin formed in contaminated food causes botulism, and *Cl. histolyticum*, the cause of gas gangrene.

clotting blood clotting *q.v.*; coagulation of milk by the action of enzymes such as rennin, which converts soluble proteins to insoluble forms.

clotting factor coagulation factor *q.v.*, *see* blood clotting.

clover-leaf structure description of secondary structure common to all transfer RNAs in which base pairing within the RNA chain forms 3 loops resembling a clover-leaf.

club moss common name for a member of the division Lycophyta *(q.v.)*, a group of seedless vascular plants.

clunes *n.plu.* buttocks.

clupeine *n.* a protamine obtained from herring sperm.

cluster-crystals globular aggregates of calcium oxalate crystals in plant cells.

cluster-cup aecidium *q.v.*

cluster of differentiation (CD) a set of monoclonal antibodies that distinguishes a particular cell-surface antigen (CD antigen) on white blood cells. Many cell-surface antigens have been characterized in this way and given a CD number, e.g. CD3 is a protein complex associated with the T-cell receptor, CD4 and CD8 are glycoprotein adhesion molecules on T cells, CD25 is the receptor for interleukin-2, CD45 is a tyrosine phosphatase involved in signal transduction in white blood cells.

clutch *n.* number of eggs laid by a female at one time.

clypeal *a. pert.* clypeus *q.v.*

clypeate *a.* round or shield-like.

clypeus *n.* shield-shaped plate of exoskeleton in the centre front of insect head; strip of cephalothorax between eyes and bases of chelicerae in spiders.

cnemial *a. pert.* tibia; *appl.* ridge along dorsal margin of tibia.

cnemidium *n.* lower part of bird's leg devoid of feathers and usually scaly.

cnemis *n.* shin or tibia.

Cnidaria *n.* phylum of simple, aquatic, mostly marine, invertebrate animals containing corals, sea fans and sea anemones (class Anthozoa), the hydroids and milleporine corals (Hydrozoa), the jellyfishes (Scyphozoa) and the box-jellies (Cubozoa). They include both colonial and solitary forms. Individuals are generally radially symmetrical, with only one opening (mouth) to the gut and a simple two-layered body with a primitive nerve net between the two layers. Cnidaria have hydroid (polyp) and/or medusa forms, and bear stinging cells (cnidoblasts) on the tentacles fringing the mouth. With the phylum Ctenophora, the Cnidaria form the large grouping known as the coelenterates.

cnidoblast *n.* stinging cell of sea anemone, jellyfish and other coelenterates, containing a coiled thread which is discharged on contact with prey. *alt.* nematoblast, nematocyst.

cnidocil *n.* minute process projecting from a cnidoblast (stinging cell), whose stimulation causes discharge of a nematocyst.

cnidophore *n.* a modified zooid in a hydrozoan colony that bears cnidoblasts.

Cnidospora *n.* name for the Cnidosporidia (*q.v.*) in older classifications.

Cnidosporidia *n.* in protist classification, a phylum of parasitic protozoans, e.g. *Myxobolus*, which have a multinucleate sporing stage, and which cause disease in fish and some other animals.

coacervate inorganic colloidal particle, e.g. clay, on which organic molecules have been adsorbed. Such particles may have been important in prebiotic evolution to bring together organic molecules in sufficient concentration for life to evolve.

coaction *n.* the reciprocal activity of organisms within a community.

coadaptation *n.* the correlated variation and adaptation displayed in two mutually dependent organs, organisms, etc.

coadapted *a. appl.* a set of genes all involved in effecting a complex process.

coagulation *n.* curdling or clotting; the change from a liquid to a viscous or solid state by chemical reaction.

coagulation factor any of a group of blood proteins which mediate blood clotting, such as Factors VIII (antihaemophilia factor), XII (Hageman factor), IX (Christmas factor), kallikrein, fibrinogen and prothrombin. *see individual entries*; vitamin K *q.v.*

coagulocyte n. a granular blood cell in insects.

coal ball a more-or-less spherical aggregate of petrified plant structures found in certain coal measures.

coagulum *n.* a coagulated mass or substance; a clot.

coancestry coefficient of kinship *q.v.*

coarctate *a.* compressed; closely connected; with abdomen separated from thorax by a constriction.

coarctate larva or pupa larval stage in certain Diptera in which the larval skin is retained as a protective puparium.

coastal zone the zone that extends from the high-tide mark on land to the edge of the continental shelf, comprising the flora and fauna of the beach and rocks and of the relatively warm, nutrient-rich shallow waters over the continental shelf.

coated pit *see* coated vesicle.

coated vesicle type of vesicle found in most eukaryotic cells which has a bristle-like coat of the protein clathrin on the cytoplasmic surface. It is generated from a "coated" portion of the plasma membrane (coated pit) which is involved in the endocytosis of molecules bound to cell-surface receptors.

cobalamin(e) *n.* cobalt-containing vitamin, vitamin B_{12}, synthesized by microorganisms, a prosthetic group of certain mammalian enzymes.

cobamide coenzyme coenzyme form of cobalamin(e) (vitamin B_{12})

coca *n.* the dried leaves of the coca plant *Erythroxylon coca*, which contains cocaine.

cocaine *n.* an alkaloid obtained from coca leaves, which has been used as a local anaesthetic, is taken as a stimulant and can result in addiction.

cocarboxylase *n.* a coenzyme needed for a carboxylase to work, e.g. thiamine pyrophosphate.

cocci *plu.* of coccus.

Coccidia *see* coccidiosis.

coccidioidomycosis *n.* fungal disease of humans and other animals caused by *Coccidioides immitis*, marked by fever and granulomas in lung.

coccidiosis *n.* disease of animals caused by sporozoan protozoan parasites of the order Coccidia, which infect the lining of the gut. *plu.* coccidioses.

coccids *n.plu.* the scale insects, minute bugs with winged males, scalelike females that are attached to the infested plant, and young that suck the sap.

coccogone *n.* reproductive cell in certain algae.

coccoid *a.* spherical or globose, like or *pert.* a coccus *q.v.*

coccolithophoroids *n.plu.* resting form of

some species of small motile golden algae (Chrysophyta or Haptophyta), having calcareous plates (coccoliths) covering the cells. *see also* Appendix 5.

coccoliths *n.plu.* thin calcareous plates formed on an organic base, which form a cellular covering in algae of the Haptophyta *q.v.*

coccospheres *n.plu.* remains of hard parts of certain algae and radiolarians.

coccus *n.* roughly spherical bacterial cell. *plu.* cocci.

coccygeal *a. pert.* or in region of coccyx.

coccygeomesenteric *a. appl.* a branch of the caudal vein, as in birds.

coccyges *plu.* of coccyx *q.v.*

coccyx *n.* the terminal part of the vertebral column beyond the sacrum. *a.* coccygeal.

cochlea *n.* the part of the inner ear concerned with hearing, a hollow structure spirally coiled like a snail's shell and containing sensory cells that respond to vibrations of fluid inside the cochlea caused by the transmission of sound from the outer ear. The cochlea is maximally stimulated at different points along its length by sound of different frequency. *a.* cochlear.

cochlear *a.* (*bot.*) appl. a mode of disposition of parts of flower in the bud where a wholly internal leaf is next but one to a wholly external leaf; (*zool.*) *pert.* cochlea.

cochleariform *a.* screw- or spoon-shaped.

cochleate *a.* screw-like; spiral; like a snail's shell.

cocoon *n.* the protective case of many larval forms before they become pupae; silky or other covering formed by many animals for their eggs.

cockroaches *n.plu.* common name for many members of the Dictyoptera *q.v.*

codeine *n.* an alkaloid found with morphine in opium and having effects similar to but weaker than morphine.

coding sequence nucleotide sequence in DNA or RNA that specifies a polypeptide.

coding strand the DNA strand in a duplex carrying the same base sequence as the corresponding RNA transcript. *alt.* antisense strand.

codominance *n.* property shown by some alleles which when present in a heterozygous state produce a different phenotype from that produced by either allele in the homozygous state.

codominant *a.* (*genet.*) *appl.* alleles showing codominance; (*bot.*) *appl.* two species being equally dominant in climax vegetation.

codon *n.* a group of 3 consecutive bases in DNA (or RNA) which specifies an amino acid or a signal for termination of translation, the RNA form of the codon being conventionally given, e.g. GAA (specifying glutamine), GUA (valine), UAA, UAG and UGA (termination codons). *alt.* triplet. *cf.* anticodon. *see* Fig. 5.

codon bias preferential use of a particular codon for an amino acid, which may vary between different genes or between different species.

codon family set of codons differing only in the third base and which all specify the same amino acid.

coefficient of consanguinity coefficient of kinship *q.v.*

coefficient of genetic relatedness coefficient of relationship *q.v.*

coefficient of inbreeding *see* inbreeding coefficient.

coefficient of kinship (f) of two individuals, the probability that two gametes taken at random, one from each individual, carry alleles at a given locus that are identical by virtue of common descent.

coefficient of relationship (r) probability that a gene in one individual will be identical by virtue of common descent to a gene in a particular relative, e.g. for monozygotic twins $r = 1$, for parents and offspring $r = 0.5$, for full siblings $r = 0.5$, for grandparents and grandchildren $r = 0.25$, etc. r also gives the fraction of genes identical by common descent between two individuals.

coefficient of selection (f) a measure of the strength of natural selection, calculated as the proportional reduction in contribution of one genotype to the gametes compared with a standard genotype, and which may have any value from zero to one.

coelacanth *see* Crossopterygii.

coelenterates *n.plu.* the animal phyla Cnidaria (corals, sea anemones, jellyfish and hydroids) and Ctenophora (sea combs or sea gooseberries) collectively. They have radial symmetry, a single body cavity (coelenteron) opening in a mouth, and a simple body wall of endoderm (gastrodermis) and ectoderm (epidermis) separated by usually non-cellular gelatinous mesogloea, and include both solitary and colonial (e.g. corals) forms. Contractile musculo-epithelial cells and a simple nerve net are present.

coelenteron *n.* the single body cavity of coelenterates.

coeliac *a. pert.* the abdominal cavity, *appl.* arteries, veins, nerves; *appl.* plexus: solar plexus *q.v.*

coelo- *alt.* spelling of celo-.

coelom *n.* body cavity in animals lined with epithelium and usually housing gonads and excretory organs and opening to the exterior, animals possessing a coelom are known as coelomates *q.v.*

coelomates *n.plu.* animals possessing a true coelom, comprising the phyla Mollusca, Annelida, Arthropoda, Phoronida, Bryozoa, Brachiopoda, Echinodermata, Chaetognatha, Hemichordata and Chordata.

coelomic *a. pert.* a coelom.

coelomocyte *n.* any of various types of cell found in the coelom in annelid worms, in the body cavity of nematodes, and in the coelomic, water and blood circulation systems in echinoderms.

coelomoduct *n.* channel leading from the body cavity to the exterior.

coelomostome *n.* the external opening of a coelomoduct.

coelomyarian *a.* having a longitudinal row of muscle cells bulging into the pseudocoel, *appl.* some nematodes.

coelozoic *a. appl.* a trophozoite when parasitizing some cavity of the body.

coen- *alt.* spelling of cen-.

coenangium *n.* a coenocytic sporangium.

coenanthium *n.* inflorescence with a nearly flat receptacle having upcurved margins.

coenenchyme *n.* the common tissue that connects individual polyps in a coral.

coenobium *n.* colony of unicellular organisms having a definite form and organization, which behaves as an individual and reproduces to give daughter coenobia.

coenocyte *n.* fungal or algal tissue in which constituent protoplasts are not separated by cell walls. *a.* coenocytic. *alt.* aseptate.

coenoecium *n.* the common ground work of a bryozoan colony.

coenogametangium *n.* a coenocytic gametangium, as in zygomycete fungi.

coenogamete *n.* a multinuclear gamete.

coenogamodeme *n.* a unit made up of all interbreeding subunits that can, under specified conditions, exchange genes.

coenogamy *n.* union of coenogametangia, as in some zygomycete fungi.

coenopopulation *n.* an aggregate of individuals in an assemblage of plants living in a particular locality.

coenosarc *n.* thin outer layer of tissue that unites all the members of a coral colony.

coenosis *n.* random assemblage of organisms with similar ecological preferences. *cf.* community.

coenosite *n.* an organism habitually sharing food with another.

coenospecies *n.* a group of taxonomic units such as species, ecospecies or varieties, which can intercross to form hybrids that are sometimes fertile, the group being equivalent to a subgenus or superspecies.

coenosteum *n.* the common colonial skeleton in corals and bryozoans.

coenozygote *n.* zygote formed by coenocytic gametes, as in some fungi.

coenurus *n.* a metacestode with large bladder, from whose walls many daughter cysts arise, each with one scolex.

coenzyme *see* cofactor.

coenzyme A (CoA) important coenzyme which acts as a carrier of activated acyl groups in many metabolic reactions, made up of phosphoadenosine diphosphate linked to a pantothenate (*q.v.*) unit linked to a β-mercaptoethylamine unit, to which an acyl group can be linked to form an acyl CoA (often acetyl CoA). *see* Appendix 1 (28) for structure.

coenzyme Q ubiquinone *q.v.*

coevolution *n.* evolution of two species in relation to each other, such as a predator and its prey; the evolution of two identical genes together so that they are both maintained in the original functional form and do not markedly diverge.

cofactor *n.* any non-protein substance required by a protein for biological activity, such as prosthetic groups and, especially in enzyme-catalysed reactions, other compounds (coenzymes) such as NAD, NADP, flavin nucleotides, coenzyme A, which are not consumed in the process and are found unchanged at the end of the reaction.

cognate *a. appl.* a tRNA recognized by a particular aminoacyl-tRNA synthetase, an antigen recognized by its corresponding antibody, etc.

cognition *n.* those higher mental processes in humans and animals, such as the formation of associations, concept formation and insight, whose existence can only be inferred and not directly observed.

coherent *a.* with similar parts united but capable of separating with a little tearing.

cohesion *n.* (*bot.*) condition of union of separate parts of floral whorl.

cohesive ends the complementary single-stranded ends of e.g. phage lambda DNA, or DNA treated with certain restriction enzymes, and which can pair to form a circular molecule or join to other similar cohesive ends. *alt.* sticky ends.

cohort *n.* a group of individuals of the same age in a population; an indefinite taxonomic group used in different ways.

cointegrate *n.* the product of the integration of one DNA replicon into another, often mediated by a transposon residing on one of the replicons.

coition, coitus *n.* sexual intercourse, copulation.

colitose *n.* a 3,6-dideoxyhexose sugar found in the lipopolysaccharide outer membrane of some enteric bacteria.

colonial *a. appl.* organisms that live together in large numbers, esp. where the individual organisms form part of a larger structure, as in some algae and in soft corals, reef-building corals, gorgonians and other anthozoans, and in the siphonophores.

colour phase an unusual but regularly occurring colour variety of a plant or animal, e.g. the white colour forms of many plants.

colour vision the ability to distinguish colours within the visible light spectrum, which is mediated by colour-sensitive cone cells in the retina. Many vertebrates, including humans, possess three types of cones with distinct spectral sensitivities, and are said to have trichromatic vision. Some possess only two types of cones and have dichromatic vision, in which some

wavelengths will be confused and be perceived as the same colour.

Col plasmid colicinogenic plasmid, any of a group of plasmids found in certain bacteria, especially strains of *Escherichia coli*, which carry genes directing the production of colicins.

colchicine *n.* toxic drug isolated from the autumn crocus *Colchicum* which arrests cells in metaphase by disrupting microtubule organization, less toxic derivatives are widely used in cell biology.

cold-blooded poikilothermal *q.v.*

cold-sensitive mutation a mutation resulting in a gene product that is functional at a normal or higher temperature but not at a lower temperature.

coleogen *n.* meristematic tissue in plants that gives rise to the endodermis.

Coleoptera *n.plu.* a very large order of insects, commonly called beetles, having complete metamorphosis, with the forewings modified as hard wing-cases (elytra) and covering membranous hindwings which may be reduced or absent.

coleoptile *n.* protective sheath surrounding the plumule (germinating shoot) of some monocotyledonous plants such as grasses.

coleorhiza *n.* protective sheath surrounding the radicle (germinating root) of some monocotyledons such as grasses.

colic *a. pert.* the colon.

colicin *n.* any of a group of proteins produced by various bacteria and which kill or inhibit the growth of other bacteria.

colicinogenic *a.* producing colicins; *appl.* plasmids: Col plasmid *q.v.*

coliform *a.* sieve-like; *n.* any of a group of colon bacteria typified by *Escherichia coli*. Their presence is used as a standard indicator of faecal pollution of water.

Coliiformes *n.* an order of birds including the colies or mousebirds.

coliphage *n.* bacteriophage that attacks *Escherichia coli*.

collagen *n.* one or all of a family of fibrous proteins found throughout vertebrates, most abundant protein in mammals, major element of skin, bone, cartilage, teeth, blood vessels etc., forming insoluble fibres of high tensile strength, and which contains the unusual amino acids hydroxyproline and hydroxylysine. It is rich in glycine but lacks cysteine and tryptophan, and has an unusually regular amino acid sequence. *see also* tropocollagen, procollagen.

collagenase *n*. any of a group of enzymes specifically attacking collagen, found in certain bacteria e.g. *Clostridium histolyticum*, the bacterium responsible for gas gangrene, and in developing and metamorphosing animal tissue.

collagenous *a*. containing collagen.

collaplankton *n*. plankton organisms rendered buoyant by a mucilaginous or gelatinous envelope.

collar *n*. any structure comparable to or resembling a collar, such as the fleshy rim projecting beyond the edge of a snail's shell, the junction between root and stem in a plant, the junction between blade and leaf sheath in grasses.

collar cell choanocyte *q.v.*

collarette *n*. line of junction between pupillary and ciliary zones on the anterior surface of the iris in the vertebrate eye.

collateral *a*. side by side; *(bot.) appl*. ovules; *appl*. vascular bundles with xylem and phloem on the same radius, the phloem to the outside of the xylem; *appl*. bud at side of axillary bud; *(zool.) appl*. fine lateral branches from the axon of a nerve cell; *appl*. circulation established through anastomosis with other parts when the chief vein is obstructed; *(genet.) appl*. inheritance of character from a common ancestor in individuals not lineally related.

collateral ganglia ganglia of the autonomic nervous system, lying in mammals e.g. at the roots of the coeliac, anterior mesenteric and posterior mesenteric arteries, and as the solar plexus, and which are not joined together by nerve fibres. *cf*. lateral ganglia.

collecting ducts or tubules more-or-less straight ducts that convey urine from the cortical to the pelvic region of the kidney.

collecting hair, collector one of the pollen-retaining hairs on stigma or style of certain flowers.

collective fruit anthocarp *q.v.*

Collembola *n*. order of insects containing the springtails, small wingless insects with only six segments in the abdomen and with two long projections from the abdomen that spring out from a folded position to make the animal jump. Common in soil and leaf litter.

collenchyma *n*. in plants, peripheral parenchymal supporting tissue with cells more-or-less elongated and thickened; the middle layer of sponges.

collet *n*. root zone of hypocotyl, where cuticle is absent.

colleterium *n*. a mucus-secreting gland in the female reproductive system of insects, for production of egg case or for cementing eggs to substrate. *alt*. colleterial glands.

colleters *n.plu*. in plants, the hairs, usually secreting a gluey substance, which cover many resting buds.

collicular *a. pert*. a colliculus.

collicular commissure tract of fibres connecting the two cerebral hemispheres at base of occipital lobes.

colliculate, colliculose *a*. having small elevations.

colliculus *n*. any of various structures forming a rounded eminence; in the brain, a part of the corpora quadrigemina *q.v.*; in the eye, the slight elevation formed by optic nerve at entrance to retina.

colligation *n*. the combination of persistently discrete units.

colloblast *n*. cell on tentacles and pinnae of ctenophores, which carries small globules of adhesive material. *alt*. lasso cell.

colloid *n*. substance of high molecular weight which does not readily diffuse through a semipermeable membrane; a substance composed of two homogeneous parts or phases, one of which is dispersed in the other.

collum *n*. neck, collar, or any similar structure.

colon *n*. in insects the second portion of the intestine; in vertebrates, a portion of the large intestine preceding the rectum and whose main function is reabsorption of water.

colonization *n*. the invasion of a new habitat by a species; the occupation of bare ground by seedlings; the establishment of a bacterial flora in intestine etc.

colony *n*. a group of individuals of the same species living together in close proximity, sometimes forming an aggregate structure with organic connections between the individual members (e.g. sponges, corals), or, as in the social insects (e.g. bees, ants, termites), large numbers of free-living individuals which habitually live together in a strictly-organized social group; a group of animals or plants living together and somewhat isolated, or recently established in a new area; a coenobium *q.v.*; (*microbiol.*) a discrete aggregate of certain microorganisms (esp. bacteria and fungi) formed when growing on solid media. Bacterial colonies comprise the progeny of a single cell, those of fungi and actinomycetes the mycelium arising from a single spore or fragment of hypha. Their colour, shape, texture, etc. on various media are important features in identification.

colony fission the production of new colonies by departure of some members while leaving the parent colony intact.

colony-forming unit (CFU) cell which can multiply to form a colony, e.g. of haematopoietic cells in spleen.

colostrum *n*. clear fluid secreted from mammary glands at the end of pregnancy and differing in composition from the milk secreted later.

colour blindness genetically determined inability in humans to distinguish all or certain colours, the most common red-green colour blindness being determined by various X-linked recessive genes.

colpate *a*. furrowed, *appl.* pollen grains.

Columbiformes *n*. an order of birds including the doves and pigeons.

columella *n*. (*bot.*) column of sterile cells in centre of capsule in mosses; central core in root cap; (*zool.*) central pillar in skeleton of some corals; the central pillar in some gastropod shells; small bone in skull of some reptiles; partly bony and partly cartilaginous rod connecting tympanum with inner ear in birds, reptiles and amphibians; the axis of the cochlea; lower part of nasal septum. *a*. columellar.

column(a) *n*. any structure like a column, such as spinal column, cylindrical body of hydroids up to the tentacles, stalk of a crinoid; (*bot.*) stamens in mallows; united stamens and style in orchids.

columnar *a*. *pert.*, or like, a column, *appl.* cells longer than broad; *appl.* epithelium composed of such cells.

Columniferae Malvales *q.v.*

coma *n*. terminal cluster of bracts, as in pineapple; hair tufts on certain seeds; compact head of clustered leaves or branches in certain mosses such as *Sphagnum*.

comb jellies common name for the Ctenophora, a phylum of marine coelenterates, also called sea gooseberries, which are free-swimming and biradially symmetrical with eight meridional rows of ciliated ribs (swimming plates) by which the organism propels itself.

combination colours colours produced by structural features of a surface in conjunction with pigment.

combining site site on antibody to which specific antigen binds.

comb-rib swimming plate *q.v.*

comes *n*. a blood vessel that runs alongside a nerve. *plu*. comites.

comitalia *n.plu*. small two- or three-rayed spicules in sponges.

comma bacillus former name for the bacterium *Vibrio cholerae*, the cause of cholera.

Commelinales *n*. order of herbaceous monocots including the family Commelinaceae (tradescantia) and others.

commensal *n*. a partner, usually the one that benefits, in a commensalism. *a*. *pert.* commensalism.

commensalism *n*. the association between two organisms of different species that live together and share food resources, one species benefiting from the association and the other not being harmed.

comminator *a*. *appl.* muscles which connect adjacent jaws of Aristotle's lantern in echinoderms.

commissure *n*. the joining line between two parts; a connecting band of nerve tissue.

community *n*. a well-defined assemblage of plants and/or animals, clearly distinguishable from other such assemblages.

community biomass total weight per unit area of the organisms in a community.

community production (PP) primary production, the quantity of dry matter formed by the vegetation covering a given area.

comose *a*. hairy; having a tuft of hairs.

comovirus group plant virus group containing small isometric single-stranded RNA viruses, type member cowpea mosaic virus. They are multicomponent viruses in which two genomic RNAs are encapsidated in three different virus particles, one of which lacks nucleic acid.

compact bone osteon bone *q.v.*

companion cell *see* phloem.

compartments *n.plu.* in insects, invariant, precisely delimited areas into which the epidermis of each segment is divided, whose boundaries do not follow any particular morphological feature and which arise from the progeny of a small group of founder cells. The delimitation of anterior and posterior compartments of prospective segments occurs early in embryonic development; metabolic compartments *q.v.*

compartmentation *n.* in eukaryotic cells, the segregation of different metabolic processes into the various membrane-bounded areas or organelles inside the cell.

compass *n.* a curved forked ossicle, part of Aristotle's lantern in echinoderms.

compass plants plants with a permanent north and south direction of their leaf edges.

compatible *a.* able to cross-fertilize; having the capacity for self-fertilization; able to coexist; in plant pathology, *appl.* an interaction between host and pathogen that results in disease.

compatibility group group of plasmids whose members are unable to coexist within the same bacterial cell.

compensation point the point at which respiration and photosynthesis are balanced at a given temperature, as determined by the intensity of light or the concentration of carbon dioxide. *see* carbon dioxide compensation point; limit of lake or sea depth at which green plants and algae lose more by respiration than they gain by photosynthesis. *alt.* compensation depth or level.

competence *n.* state in which a cell or organism is able to respond to a stimulus, refers esp. to the ability of embryonic cells to respond to a specific developmental stimulus.

competition *n.* active demand by 2 or more organisms for a material or condition, so that both are inhibited by the demand, e.g. plants competing for light, water etc., *cf.* amensalism; active demand by 2 or more substances for the same binding site on enzymes, receptor molecules, cells etc.

competitive exclusion principle the principle that two different species cannot indefinitely occupy the same ecological niche, one eventually being eliminated. *alt.* Gause's principle.

competitive inhibitor *see* inhibitor.

complement *n.* the proteins of the complement system (*q.v.*) collectively. *see also* C1, C2, C3, C4 etc.

complement fixation the binding of complement by antigen-antibody complexes.

complement system a group of blood proteins of the globulin class involved in the lysis of foreign cells after they have been coated with antibody, and which also promote the removal of antibody-coated foreign particles by phagocytic cells. Complement activation proceeds by a cascade reaction of successive binding and proteolytic cleavage of complement components, either by the "classical pathway" triggered by antigen-antibody complexes on cell surfaces, or by the "alternative pathway", triggered by certain initiating surfaces. Complement activation culminates in attack and lysis of foreign cells, and release of complement components which cause local inflammatory reactions and other complement components which bind to phagocytic cells facilitating the uptake of antigen-antibody complexes.

complemental air volume of air which can be taken in in addition to that drawn in during normal breathing.

complemental male pygmy male *q.v.*

complementary *a.* *appl.* non-suberized cells loosely arranged in cork tissue and forming air passages in lenticels; *appl.* the two strands of duplex DNA; *n.* the coronoid bone in mandible of reptiles.

complementary DNA (cDNA) single-stranded DNA synthesized *in vitro* on an RNA template by reverse transcriptase, used widely in genetic engineering to make DNA for cloning using mRNAs isolated from cells as the template RNA.

complementary male pygmy male *q.v.*

complementation group group of mutations which fail to complement each other in *trans* in a complementation test, formally defining a cistron (a DNA sequence specifying a single polypeptide chain).

complementation test test used to determine whether two recessive mutations giving the same phenotype lie in the same or different genes, by making a cross between homozygotes for each mutation and determining whether the heterozygote progeny have recovered the wild-type function, as they will if the mutations lie in different genes. Mutations in the same gene will only show wild-type function if the mutations are in the *cis* configuration (i.e. on the same chromosome) in the heterozygote. *alt. cis-trans* complementation test.

complete *a. appl.* flowers containing sepals, petals, stamens and carpels.

complete metamorphosis insect metamorphosis in which the young are usually different from the adult and are called larvae, and go through a resting stage, the pupa, before reaching the adult stage (the imago), the wings being developed inside the body during the pupal stage. *cf.* incomplete metamorphosis.

complex *n.* two or more molecules held together, usually by non-covalent bonding, so that they are quite easily separable; (*ecol.*) the meeting of several distinct communities related to each other by certain shared species.

complex locus genetic locus in eukaryotes which encodes an apparent single function, but which is divisible by various genetic tests into several subloci, each acting in some respects as a separate locus. Molecular analysis indicates that some complex loci at least represent a cluster of individual protein-coding genes and control regions involved in directing different steps in a particular process.

complex mRNA scarce mRNA *q.v.*

complex tissues tissues composed of more than one type of cell.

complexity *n.* the amount of sequence information in a DNA, measured as the total length of different sequences present.

complexus *n.* an aggregate formed by a complicated interweaving of parts; *appl.* a muscle, the semispinalis capitis.

complicant *a.* folding over one another, *appl.* wings of certain insects.

complicate *a.* folded, *appl.* leaves or insect wings folded longitudinally so that right and left halves are in contact; in fungi, *appl.*

fruit bodies of some hymenomycetes composed of several caps with stalks fusing to form a central stalk.

Compositae, composites *n., n.plu.* a very large family of dicotyledonous flowering plants, typified by dandelions, daisies, thistles, etc., in which the inflorescence is a capitulum made up of many tiny florets.

composite *a.* closely packed, as the florets in a capitulum; *appl.* fruits, *see* sorosis, syconus, strobilus; *appl.* a member of the plant family Compositae.

composite transposon a transposon with a central region (often carrying antibiotic-resistance genes etc.) flanked by insertion sequences.

compound *a.* made up of several elements, *appl.* e.g. to flowerheads made up of several or many individual flowers, eyes (of insects) made up of many identical elements, leaves made up of several leaflets.

compound chromosome isochromosome *q.v.*

compound eye eye characteristic of insects and most crustaceans, made up of many identical units, *see* ommatidium.

compound nest nest containing colonies of two or more species of social insects, up to the point where the galleries of the nest run together. Adults sometimes intermingle but the broods are kept separate.

compressed *a.* flattened transversely.

compression wood in conifers, reaction wood formed on the lower sides of crooked stems or branches, and having a dense structure and much lignification.

compressor *n.* muscle that serves to compress.

conalbumin *n.* glycoprotein of egg white, also produced in liver where it is known as transferrin (*q.v.*).

conarium *n.* transparent deep-sea larva of certain coelenterates.

concanavalin A (con A) a lectin (*q.v.*) binding specifically to α-mannosyl and glucose residues, used widely in experimental cell biology as a marker for glycoproteins on cell membranes, etc., also has mitogenic activity.

concatenate *a.* forming a chain, as spores; linked at their bases, as cellular processes; *n.* a chain of linked elements.

concentration factor the factor by which

the level of a toxic pollutant, e.g. a heavy metal or pesticide, accumulates in the tissue of a living organism compared with its concentration in the environment, or with its concentration in the previous organism in the food chain.

concentric *a*. (*bot*.), having a common centre, *appl*. vascular bundles with one kind of tissue surrounding another.

conceptacle *n*. a depression in the thallus of certain algae in which gametangia are borne.

conceptus *n*. the fertilized ovum in mammals.

concerted evolution coevolution *q.v*.

concha *n*. cavity of the external ear, opening into the external auditory meatus; a conch shell.

conchiform, conchoid *a*. shell-shaped (like a conch shell).

conchiolin *n*. the protein component of ligament and external layer of mollusc shells.

conchology *n*. that branch of zoology dealing with molluscs and their shells.

concolorate, concolorous *a*. similarly coloured on both sides or throughout; of the same colour as a specified structure.

Concorde effect in animal behaviour, the case where future behaviour is determined by past investment rather than future prospects.

concrescence *n*. the growing together of parts.

concrete *a*. grown together to form a single structure.

condensation *n*. *appl*. chromosomes, the compaction that takes place as chromosomes enter mitosis or meiosis; (*biochem*.) the formation of a larger molecule from the polymerization of smaller units accompanied by the elimination of the elements of a single water molecule during formation of each bond. It occurs e.g. during the formation of proteins from amino acids and nucleic acids from nucleotides.

condensed *a*. (*bot*.) *appl*. inflorescence with short-stalked or sessile flowers closely crowded.

condensing vacuoles large immature secretory vesicles associated with the Golgi apparatus.

conditional *a*. *appl*. dominance owing to influence of modifying genes.

conditional lethal a mutation or allele that causes the death of the embryo only under certain conditions, e.g. of temperature, nutrition, etc.

conditional mutations mutations that only produce an effect under certain environmental conditions, as e.g. of temperature, nutritional status.

conditional response (CR) *see* classical conditioning. *alt*. conditioned response.

conditional stimulus (CS) in classical conditioning, the stimulus used to provoke a conditional response or reflex. *alt*. conditioned stimulus.

conditioned reflex conditional response, *see* classical conditioning.

conditioning *see* classical conditioning, operant conditioning, pseudoconditioning, second-order conditioning.

conductance *n*. a measure of the permeability of a biological membrane or of a single ion channel to ions. It is the inverse of electrical resistance or impedance and is measured in siemens (S).

conducting *a*. conveying, *appl*. tissues, structural elements that convey material or a signal from one place to another.

conduction *n*. (*bot*.) the transfer of soluble material from one part of a plant to another; (*zool*.) the conveyance of an electrical impulse along a nerve fibre.

conduplicate *a*. *appl*. cotyledons folded to embrace the radicle; *appl*. leaves in which one half is folded longitudinally upon the other.

condyle *n*. a process on a bone at which it articulates at a joint; a rounded structure adapted to fit into a socket; (*bot*.) the antheridium of stoneworts. *a*. condylar.

condyloid *a*. shaped like, or situated near, a condyle.

cone *n*. cone- or flask-shaped light-sensitive sensory cell in the retina, responsible for colour vision and vision in good light, individual cones being sensitive to blue, green or red wavelengths; (*bot*.) cone-shaped reproductive structure in certain groups of plants, *see* strobilus.

conferted *a*. closely packed, crowded.

conflict *n*. situation in which two motivations compete for dominance in the control of behaviour, as when an animal is deciding which of two objects to approach (or avoid) or whether to approach or run away from an object.

confluence *n*. angle of union of superior sagittal and transverse sinuses at occipital bone; of cell cultures, the point at which cells have formed a continuous sheet over the dish, at which point they usually stop dividing.

conformation *n*. 3-dimensional arrangement of atoms in a structure.

conformer *n*. an animal that allows its internal environment to be influenced by external factors. *cf*. regulator.

congeneric *a*. belonging to the same genus.

congenetic *a*. having the same origin.

congenic *a*. *appl*. specially bred strains of mice etc., in which a particular allele or block of alleles from one strain is superimposed on the genetic background of another.

congenital *a*. present at birth, *appl*. physiological or morphological defects, not necessarily inherited.

congestin *n*. toxin of sea anemone tentacles.

conglobate *a*. ball-shaped.

conglomerate *a*. bunched or crowded together.

coni *plu*. of conus, *appl*. various cone-shaped structures; coni vasculosi: lobules forming head of epididymis.

conidia *plu*. of conidium *q.v*.

conidial *a*. *pert*. a conidium.

conidiophore *n*. a hypha with sterigmata that bear conidia.

conidiosporangium *n*. conidium, esp. in Phycomycetes, which may produce zoospores or germinate directly.

conidium, conidiospore *n*. an asexual fungal spore produced by the constriction of a tip of a hypha or sterigma and not enclosed in a sporangium.

Coniferales, Coniferae *n*. an order of gymnosperm trees, with reproductive organs as separate male and female cones, and usually needle-shaped leaves (e.g. pines, larches, cypresses).

conifer *n*. a cone-bearing tree, sometimes used to include all temperate gymnosperm trees (as opposed to broadleaved angiosperm trees) even when (as yews, junipers, etc.) they do not bear cones; sometimes refers just to the needle-leaved,

cone-bearing gymnosperms such as pines, cypresses, spruces, larches, etc. with reproductive organs as separate male and female cones.

Coniferophyta *n*. one of the five main divisions of extant seed-bearing plants, commonly called conifers. They have simple, often needle-like leaves, and bear their megasporangia usually in compound strobili (cones). In some classifications called Coniferopsida.

coniferous *a*. cone-bearing; *pert*. conifers or other cone-bearing plants.

conjoined adnate *q.v*.

conjugate *v*. to unite, as protozoan cells acting as gametes; to undergo conjugation; *a*. united in pairs; *appl*. pores united by a groove; coordinated, *appl*. movements of the two eyes.

conjugated *a*. united; *appl*. protein when molecule is united with a non-protein.

conjugation *n*. the temporary union or complete fusion of two gametes; the pairing of chromosomes; in unicellular organisms, the joining together and exchange of genetic material, as in the ciliate *Paramecium,* or one-way transfer of genes as in bacteria or the green alga *Spirogyra*.

conjugation canal in certain green algae, the tube formed in fused outgrowths from opposite cells of parallel filaments, for passage of male gametes to the other filament as in scalariform conjugation in the green alga *Spirogyra*.

conjunctiva *n*. delicate mucous membrane lining eyelids and reflected over cornea and sclera and constituting the corneal epithelium.

conjunctive *a*. *appl*. symbiosis in which the partners are organically connected.

Connarales *n*. order of dicots comprising the family Connaraceae.

connate *a*. firmly joined together from birth; *appl*. organs growing together and becoming joined, but separate at birth.

connate-perfoliate *a*. joined together at base so as to surround stem, *appl*. opposite sessile leaves.

connective *n*. connecting band of nerve fibres between two ganglia; tissue separating two lobes of anther.

connective tissue the supporting tissues of the animal body including bone, cartilage, adipose tissue and the fibrous tissues supporting and connecting internal organs. It is derived from the embryonic mesoderm and often contains large amounts of non-living extracellular matrix.

connexon *n*. unit composed of two protein particles from opposing membranes at a gap junction.

connivent *a*. converging; arching over so as to meet.

conodonts *n.plu*. abundant tooth-like fossils from the Paleozoic and early Triassic, thought to be the jaw elements of invertebrates of uncertain affinities and generally placed in a separate phylum, Conodonta.

conoid *a*. cone-like, but not quite conical.

conoid tubercle a small rough eminence on posterior edge of clavicle, at which the conoid ligament is attached.

conopodium *n*. conical receptacle or thalamus of a flower.

consanguineous *a*. related, in human genetics *appl*. matings between relatives.

consciousness *n*. an awareness of one's actions or intentions, and having a purpose and intention in one's actions; sometimes defined as the presence of mental images and their use by an animal to regulate its behaviour, although the question of whether animals possess consciousness is controversial.

conscutum *n*. dorsal shield formed by united scutum and alloscutum in certain ticks.

consensual *a*. *appl*. involuntary action correlated with voluntary action; relating to excitation of a corresponding organ; *appl*. contraction of both pupils when only one retina is directly stimulated.

consensus sequence the "ideal" form of a DNA sequence found in slightly different forms and positions in different organisms or in different genes, but which is believed to have the same function (e.g. recognition sites in promoters), giving for each position the nucleotide most often found. *alt*. canonical sequence.

conservative *a*. *appl*. characters that change little during evolution; *appl*. taxa retaining many ancestral characters.

conservative recombination breakage and reunion of pre-existing strands of DNA without any synthesis of new stretches of DNA, *see also* site-specific recombination.

conserved *a*. *appl*. structures, proteins, genes, DNA sequences, etc. that are identical or very similar in different organisms. Also *appl*. stretches of DNA sequences that are very similar in different genes.

consimilar *a*. similar in all respects; with both sides alike, as some diatoms.

consociation *n*. a unit of plant association, a climax community characterized by a single dominant species.

consocies *n*. a consociation representing a stage in the process of succession.

consortes *n.plu*. associate organisms other than symbionts, commensals, or hosts and parasites. *sing*. consors.

consortium *n*. a kind of symbiosis in which both gain benefit, as of the alga and fungus in a lichen.

conspecific *a*. belonging to the same species.

consperse *a*. densely scattered, *appl*. dot-like markings, pores, etc.

constancy *n*. in ecology, the frequency with which a particular species occurs in different samples of the same association.

constant *n*. in ecology, a species that occurs in at least 95% of samples taken from random within a community.

constant region *see* C region.

constitutive *a*. *appl*. enzymes synthesized by the cell in the absence of any specific stimulus; *appl*. genes which are expressed all the time and do not require a specific external stimulus for initiation of transcription; *appl*. heterochromatin, chromosomal regions that form heterochromatin in all cells. *cf*. inducible.

constricted *a*. narrowed, compressed at regular intervals.

constrictor *n*. a muscle that compresses or constricts.

consumer *n*. heterotrophic organism, i.e. one that must consume resources provided by autotrophic organisms. *see also* primary consumer, secondary consumer.

consummatory behaviour in classical ethology, a piece of behaviour, such as drinking, which is considered to be the end result of a physiological need (in this case thirst) which drives a search for a suitable stimulus (in this case water).

consute *a*. with stitch-like markings.

contact inhibition the cessation of cell growth and division in normal cultured cells when they come into contact with each other.

contact receptor a sensory receptor in dermis or epidermis.

contact sensitivity type of allergic response caused by direct contact with certain substances and producing severe inflammation at site of contact.

contest competition type of competition in which the successful competitor gains sufficient resources for survival and reproduction whereas the unsuccessful competitors gain nothing or insufficient for survival.

context *n.* (*mycol.*) fibrous layers developed between hymenium and true mycelium in cap of some basidiomycete fungi.

contiguous *a.* touching each other at the edges but not actually united.

continental drift the gradual movement of the continents over the surface of the Earth.

continuous culture methods of growing bacterial cultures so that nutrients and space do not become exhausted and the culture is always in the rapidly multiplying phase of growth.

contiguous gene syndrome clinical syndrome with a complex pathology associated with the deletion of a set of adjacent genes.

continental *a. appl.* climate characterized by some or all of the features of weather associated with continental interiors, namely hot summers and cold winters with a wide temperature range between extremes, short spring and autumn seasons, marked rainy and dry seasons.

continuous variation variation between individuals of a population in which differences are slight and grade into each other, continuously variable phenotypic characters being those that are determined by a large number of genes and/or considerable environmental influence.

continuum *n.* (*bot.*) a form of vegetation in which one type passes almost imperceptibly into another and no two types are repeated exactly.

contorted *a. appl.* arrangement of floral parts in which one petal overlaps the next with one margin and is overlapped by the previous on the other.

contortuplicate *a. appl.* bud with contorted and folded leaves.

contour *n.* outline of a figure or body, *appl.* outermost feathers that cover the body of a bird.

contractile *a.* capable of contracting.

contractile cell any cell in a sporangium or anther which by hygroscopic contraction helps to open the organ.

contractile fibre cells elongated spindle-shaped more-or-less polyhedral muscle cells, containing a central bundle of myofibrils.

contractile ring bundles of actin filaments immediately beneath the plasma membrane which are involved in forming the constriction that separates the two new cells in animal cell division.

contractile vacuole small spherical vesicle found in the cytoplasm of many freshwater protozoans, which expels surplus water.

contractility *n.* power by which muscle fibres are able to contract; the capacity to change shape.

contracture *n.* contraction of muscles persisting after stimulus has been removed.

contra-deciduate *a. appl.* foetal placenta and distal part of allantois that are absorbed by maternal tissues at birth.

contralateral *a. pert.* or situated on the opposite side. *cf.* ipsilateral.

contranatant *a.* swimming or migrating against the current.

contrasuppressor cell type of T lymphocyte that suppresses the activity of suppressor T lymphocytes.

control *n.* an experiment or test carried out to provide a standard against which experimental results can be evaluated.

controlling element any of several types of genetic element (DNA sequences) in maize which can become inserted at various sites on the chromosomes during somatic cell division and which influence the expression of nearby genes. The first "mobile" genetic element to be discovered as a result of effects such as patchy or speckled coloration in the maize grains due to activation or inactivation of certain genes in clones of somatic cells during development as a result of insertion of a controlling element near them. *alt.* transposable element.

conuli *n.plu.* tent-like projections on surface of some sponges, caused by principal skeletal elements.

conus *n.* any cone-shaped structure; diverticulum of right ventricle from which pulmonary artery arises.

conus arteriosus a funnel-shaped structure between ventricle and aorta in fishes and amphibians.

conus medullaris the tapering end of the spinal cord.

conventional behaviour any behaviour by which members of a population reveal their presence and allow another organism to assess their numbers.

convergence, convergent evolution *n.* similarity between two organs or organisms (or protein or DNA molecules) due to independent evolution along similar lines, rather than a common ancestor; coordinated movement of eyes when focusing a near point.

convolute *a.* rolled together, *appl.* leaves and cotyledons; *appl.* shells in which outer whorls overlap inner; coiled, *appl.* parts of renal tubule.

convolution *n.* a coiling or twisting as of brain, intestine.

Coombs test modification of antibody-mediated cell agglutination reaction using heterologous anti-immunoglobulin to enhance agglutination.

cooperation *n.* in animal behaviour, the sharing of a task between different animals, as in hunting by a wild dog pack, the sharing of child care by relatives of the mother in chimpanzees, feeding of nestlings by other members of the community in some communal birds. *see also* symbiosis.

cooperative binding phenomenon where binding of one substrate to one active site on a protein raises the affinity of the other active site(s) for substrate. *see also* allostery.

cooperative breeding a breeding system in which parents are assisted in the care of their offspring by other adults.

copal *n.* resin exuding from various tropical trees and hardening to a colourless, yellow, red or brown mass. Used in varnishes.

Copepoda, copepods *n., n.plu.* subclass

of free-living or parasitic small crustaceans that form a large part of the marine zooplankton and are also found in fresh water. They have no carapace and have one median eye in the adult.

copia a family of transposon-like DNA sequences found in the fruitfly *Drosophila melanogaster.*

copia-like elements transposon-like DNA sequences found in *D. melanogaster* resembling copia in general structure and including 412, gypsy, mgd1 etc.

coppicing *n.* woodland management practice in which trees such as willow, hazel and sweet chestnut are regularly cut back almost to the ground every few years so that they develop a low "stool" from which many long straight shoots arise. *cf.* pollarding.

coprodaeum *n.* the division of the cloaca that receives rectum.

coprolite, coprolith *n.* fossilized faeces.

coprophage *n.* animal that feeds on dung. *a.* coprophagous.

coprophagy *n.* habitual feeding on dung; reingestion of faeces.

coprophil, -ic, -ous *a.* growing in or on dung, *appl.* certain fungi, bacteria, etc.

coprophyte *n.* plant that grows on dung.

coprosterol *n.* sterol produced by bacterial reduction of cholesterol, found in faeces.

coprozoic *a.* living in faeces, as some protozoans.

coprozoite *n.* any animal that lives in or feeds on dung.

copula *n.* a ridge in development of tongue, formed by union of ventral ends of 2nd and 3rd arches; basihyal in certain reptiles; fused basibranchial and basihyal in birds; any bridging or connecting structure.

copulant *n.* a unit in conjugation with another, as nuclei, cells, hyphae, thalli, etc.

copularium *n.* a cyst formed around two associated gametocytes in gregarines.

copulation *n.* sexual union; in protozoans, complete fusion of two individuals.

copy number the number of plasmids of a particular type that can accumulate in a bacterial cell.

coracidium *n.* ciliated embryo of certain cestodes (tapeworms), developing into a procercoid within first intermediate host.

Coraciformes *n.* an order of birds including the kingfishers, rollers and hornbills.

coracoid *a. appl.* or *pert.* bone or part of the pectoral girdle between shoulder blade and breast-bone; *appl.* ligament which stretches over the suprascapular notch.

coracoid process the rudimentary coracoid bone fused to the shoulder-blade in most mammals.

coral *n.* member of a group of colonial coelenterates composed of individual polyps connected by living tissue. Some forms secrete a stony matrix binding the colony together and some build extensive reefs. *see* soft corals, stony corals.

coralliferous *a.* coral-forming; containing coral.

coralliform *a.* resembling or branching like a coral.

coralligenous *a.* coral-forming.

coralline *a.* resembling a coral, *appl.* to some lime-encrusted red agae; composed of or containing coral; *appl.* zone of coastal waters at about 30–100 m.

corallite *n.* cup of an individual coral polyp.

coralloid *a.* resembling or branching like a coral; *appl.* negatively geotropic roots of cycads (gymnosperms of the family Cycadaceae) which arise from the swollen hypocotyl and taproot, often deep in the soil, and are infected with nitrogen-fixing cyanobacteria.

corallum *n.* skeleton of a compound coral.

corbiculum *n.* the pollen basket on the hind legs of many bees, formed by stout hairs on the borders of the tibiae; fringe of hair on insect tibia.

Cordaitales *n.* order of fossil conifers, being mostly tall trees with slender trunks and a crown of branches, with spirally arranged simple grass-like or paddle-like leaves and with mega- and microsporangia in compound strobili.

cordate, cordiform *a.* heart-shaped.

cordycepin *n.* 3′deoxyadenosine, a nucleotide analogue that specifically inhibits polyadenylation of eukaryotic RNA.

core area that part of the home range of an animal in which it spends most of its time.

core enzyme enzyme containing a functional catalytic site but lacking one or more of its associated polypeptide subunits necessary for all its normal regulated functions.

coregonid *n.* freshwater fish of the genus *Coregonus*, found only in unpolluted waters, e.g. houting, powan and vendace.

coremata *n.plu.* accessory copulatory organ in moths, paired sacs bearing hairs, on membrane between seventh and eighth abdominal segments. *sing.* corema.

coremiform *a.* shaped like a broom or a sheaf.

coremiospore *n.* one of the series of spores on top of a coremium.

coremium *n.* a sheaf-like aggregation of conidiophores or of hyphae.

corepressor *n.* in bacteria, a small molecule that prevents the expression of the operon specifying the enzymes able to synthesize it by binding to a specific gene regulatory protein.

core temperature temperature of an animal's body measured at or near the centre.

Cori cycle conversion in the liver of lactate formed by glycolysis in contracting muscle into glucose (by gluconeogenesis), which is then transported to muscle.

coriaceous, corious *a.* leathery, *appl.* leaves.

corium *n.* dermis *q.v.*; central division of an insect elytron (wing-case); the main part of the front wing of a hemipteran bug.

cork *n.* external layer of plant tissue composed of dead cells filled with suberin, forming a seal impermeable to water. Present on woody stems and derived from the cork cambium. *alt.* phellem.

cork cambium a secondary meristematic layer in woody stems that gives rise to the corky layer of the bark.

corm *n.* enlarged, solid underground stem, rounded in shape, composed of two or more internodes, and covered externally with a few thin scales or leaves.

cormel *n.* secondary corm produced by an old corm.

cormidium *n.* a group of individuals of a siphonophore colony that can separate from the colony and live a separate existence.

cormoid *a.* like a corm.

cormophytes *n.plu.* plants differentiated into roots, shoots and leaves, and well adapted for life on land, comprising pteridophytes and the Spermatophyta.

cormous *a.* producing corms.

cormus *n*. body of a plant that is developed into root and shoot systems; body or colony of a compound animal.

Cornales *n*. order of dicot trees, shrubs and herbs, with leaves often much divided, and including the families Umbelliferaceae (carrot, etc.), Araliaceae (ginseng), Cornaceae (dogwood), Davidiaceae (dove tree) and others.

cornea *n*. the transparent covering of the front of the eye; outer transparent part of each element of a compound eye.

corneosclerotic *a. pert.* cornea and sclera.

corneoscute *n*. an epidermal scale.

corneous *a*. horny, *appl.* sheath covering bill of birds.

cornicle *n*. wax-secreting organ of aphids; any small horn or horn-like process, *alt.* corniculum, cornule.

corniculate *a*. having little horns.

corniculate cartilages two small, conical, elastic cartilages articulating with apices of arytaenoids in larynx.

cornification *a*. formation of outer horny layer of epidermis. *alt.* keratinization.

cornified *a*. keratinized, *appl.* epithelium; transformed into horn.

cornified layer of epidermis, stratum corneum *q.v.*

cornifying keratinizing *q.v.*

cornua *n.plu.* horns; horn-like prolongations, as of bones, nerve tissues, cavities, etc.; the dorsal, lateral and ventral columns of grey matter in the spinal cord. *sing.* cornu.

cornule *see* cornicle.

cornute *a*. with horn-like processes.

corolla *n*. the petals of a flower collectively. *a*. corollaceous.

corolliferous *a*. having a corolla.

corona *n*. (*bot.*) frill at mouth of corolla tube formed by union of scales on petals, as the trumpet of a daffodil; (*zool.*) the theca and arms of a crinoid; ciliated disc or circular band of certain animals such as rotifers; the head or upper portion of any structure.

corona radiata layer of cells surrounding mammalian egg.

coronal *a. pert.* a corona; *appl.* suture between frontal and parietal bones of skull; *appl.* plane of section through brain, being a vertical section at right-angles to the long axis.

coronary *a*. crown-shaped or crown-like; encircling.

coronary arteries/veins arteries/veins supplying tissue of heart.

coronary bone small conical bone in mandible of reptiles; small pastern bone in horses.

coronary sinus channel receiving most cardiac veins and opening into the right auricle.

coronary vessels coronary arteries/veins *q.v.*

coronate *a*. having a corona; having a row of tubercles encircling a structure, or mounted on whorls of spiral shells.

Coronaviridae, coronaviruses *n*. family of medium-sized, enveloped, single-stranded RNA viruses covered with protein petal-like projections, and which include human coronavirus and avian infectious bronchitis virus.

coronet *n*. small terminal ring of hairs, spines, etc.; corona of certain flowers; burr or knob at base of antler.

coronoid *a*. shaped like a beak; *n*. coronary bone of reptiles.

coronula *n*. a group of cells forming a crown on oosphere, as in green algae of the Charophyceae; a circle of pointed processes around frustule of certain diatoms.

corpora *plu*. of corpus *q.v.*

corpora albicantia white bodies or scars formed in ovarian follicle after disintegration of luteal cells.

corpora allata endocrine glands in insects situated just behind the brain, which secrete juvenile hormone, and which may be paired ovoid whitish structures, or may fuse during development to form a single median structure – the corpus allatum.

corpora amylacea spherical bodies of nucleic acid and protein, in alveoli of prostate gland becoming more numerous with age.

corpora bigemina the optic lobes of vertebrate brain, corresponding to the superior colliculi of corpora quadrigemina in mammals.

corpora cardiaca neurosecretory bodies between cerebral ganglia and corpora allata, in some insects.

corpora cavernosa erectile mass of tissue forming anterior part of penis; erectile tissue of clitoris.

corpora mamillaria *see* mamillary bodies.

corpora quadrigemina four rounded eminences or colliculi which form the dorsal part of the midbrain in mammals.

corpus *n.* body; any fairly homogeneous structure which forms part of an organ. *plu.* corpora.

corpus allatum *see* corpora allata.

corpus callosum wide tract of fibres connecting the two cerebral hemispheres and involved in the transfer of information and learning from one hemisphere to the other.

corpus luteum glandular mass of yellow tissue that develops from a Graafian follicle in the mammalian ovary after the ovum is extruded, and which secretes progesterone.

corpus spongiosum mass of erectile tissue forming posterior wall of penis.

corpus striatum in cerebral hemispheres of brain, mass of grey matter containing white nerve fibres and consisting of the caudate nucleus and lenticular nucleus, and situated on the outer side of each lateral ventricle.

corpuscle *n.* cell *q.v.*; any of various small multicellular structures, e.g. Malpighian corpuscle.

corpuscles of Ruffini Ruffini's organs *q.v.*

corpuscular *a.* like or *pert.* a corpuscle; compact or globular.

correlation centres regions in the brain where information from various sense organs is integrated and the resultant response determined.

corrin *n.* cobalt atom surrounded by 4 pyrrole units (corrin ring).

corrugator *n.* muscle that causes wrinkling.

cortex *n.* cerebral cortex *q.v.*; the outer layer of a structure or organ; in vascular plants, the tissue in stem and root surrounding and not part of the stele (vascular bundles); in ova, the outermost layer of cytoplasm, just under the cell membrane; in bacterial spores, the envelope between spore wall and spore coat. *plu.* cortices.

cortical *a. pert.* the cortex.

cortical granules in ovum, granules lying just beneath the plasma membrane which release their contents when a sperm enters the egg changing the composition of the outer coat and preventing entry of other sperm.

corticate *a.* having a special outer covering.

cortices *plu.* of cortex *q.v.*

corticiferous *a.* forming or having a barklike cortex.

corticoid corticosteroid *q.v.*

corticolous *a.* inhabiting, or growing on, bark.

corticospinal *a. pert.* or connecting cerebral cortex and spinal cord.

corticospinal tract pyramidal tract *q.v.*

corticosteroid *n.* any of a group of steroids secreted by the cortex of the adrenal glands, some of which are hormones, and including the glucocorticoids (*q.v.*) and the mineralocorticoids (*q.v.*). *see also* corticosterone, cortisone, cortisol, aldosterone. *a.* corticosteroid.

corticosterone *n.* steroid hormone produced by the cortex of the adrenal glands in response to adrenocorticotropic hormone, and which has effects on lipid, carbohydrate and protein metabolism and can act as an immunosuppressant. See Appendix 1(46) for chemical structure.

corticostriate *a. appl.* nerve fibres that join corpus striatum to cerebral cortex.

corticotrop(h)ic adrenocorticotrop(h)ic *q.v.*

corticotropin adrenocorticotropic hormone (ACTH) *q.v.*

corticotropin-releasing factor (CRF) small peptide secreted by the hypothalamus which stimulates the release of adrenocorticotropic hormone (ACTH) from the anterior pituitary. *alt.* corticotropin-releasing hormone (CRH).

cortina *n.* the veil in some agaric fungi.

cortinate *a.* of a cobwebby texture; having a veil, of mushrooms and toadstools.

cortisol hydrocortisone *q.v.*

cortisone *n.* steroid hormone secreted in small amounts by the human adrenal cortex, and which has many complex effects, including suppression of inflammation and the promotion of carbohydrate formation.

Corti's organ, Corti's membrane, Corti's rods *see* organ of Corti.

coruscation *n.* twinkle, rapid fluctuation in a flash or oscillation in light emission, as of fireflies.

corvine *a. appl.* birds of the Corvidae or crow family.

corymb *n.* an inflorescence in the form of a raceme with lower flower stalks elongated so that the top is nearly flat. *a.* corymbose.

corymbose cyme a flat-topped cyme which therefore resembles a corymb in appearance but not in mode of development.

corynebacteria *n.plu.* bacteria of the family Corynebacteriaceae, which are characterized by irregularly shaped cells dividing by "snapping fission", and which include the human pathogen *Corynebacterium diphtheriae*, the cause of diphtheria. *sing.* corynebacterium.

coryneform *a. appl.* bacteria characterized by irregular, often club-shaped, Gram-positive cells.

coscinoid *a.* sieve-like.

cosere *n.* a series of plant successions on the same site.

cosmid *n.* a type of cloning vector consisting of a bacterial plasmid into which have been inserted the *cos* sequences of phage lambda thus allowing growth as plasmids *in vivo* and subsequent purification of the DNA by packaging into phage particles *in vitro*.

cosmine *n.* type of dentine with tiny branched canals, found in the scales of some fishes.

cosmoid scale type of scale found in typical crossopterygian fishes. It has an outer layer of enamel, a layer of cosmine (a type of dentine), and then bone, growth in thickness being by addition of inner layers only. *cf.* ganoid scale.

cosmopolitan *a.* world-wide in distribution.

cosmopolite *n.* a species with a world-wide distribution.

cost *n.* the decrement in an animal's inclusive fitness that results from a particular behaviour. *cf.* benefit.

costa *n.* rib; anything rib-like in shape, as a ridge on a shell, coral, etc.; anterior vein or margin of insect wing; swimming plate of sea gooseberries; structure at base of undulating membrane in certain protozoans. *plu.* costae.

costaeform *a.* rib-like; *appl.* unbranched parallel leaf veins.

costal *a. pert.* ribs or rib-like structures; *appl.* bony shields of turtles and tortoises; *pert.* costa of insect wing.

costalia *n.plu.* the supporting plates in test of echinoderms.

costate *a.* with one or more longitudinal ribs; with ridges or costae.

cost function the combination of the various costs of an animal's behaviour, which is used to evaluate all aspects of the animal's state and behaviour.

coterie *n.* a social group which defends a common territory against other coteries.

coterminous *a.* of similar distribution; bordering on; having a common boundary.

cotransduction *n.* the phage-mediated transduction of two genes in a single event.

cotransfection *n.* the simultaneous transfection of cells with two different genes.

cotransformation *n.* the neoplastic transformation of cells by the simultaneous introduction of two different proteins, genes or transforming viruses, neither of which can completely transform the cells on their own.

cotranslational transfer or translocation process in eukaryotic cells in which proteins destined for cell membranes or for secretion start their passage across the membrane of the endoplasmic reticulum before translation is complete.

cotyle acetabulum *q.v.*

cotyledon *n.* (*bot.*) the first leaf or leaves of a seed plant, found in the embryo, and which may form the first photosynthetic leaves or may remain below ground; (*zool.*) a patch of villi on mammalian placenta. *a.* cotyledonary.

cotyloid, cotyliform *a.* cup-shaped.

cotylophorous *a.* with a cotyledonary placenta.

cotylosaurs *n.plu.* a group of primitive anapsid reptiles of the Upper Carboniferous to Triassic, the possible ancestral reptiles from which many later forms developed.

cotylosaurs Captorhinida *q.v.*

cotype syntype *q.v.*

coumarin *n.* substance found in many plants, esp. clover, having an odour of new-mown hay, and used in perfumery and to make the anticoagulant dicoumarin (dicumarol).

counteracting selection the operation of selection pressures on two or more levels of organization, e.g. individual, family or population, in such a way that certain genes are favoured at one level but disfavoured at another. *cf*. reinforcing selection.

counterevolution *n*. the evolution of traits in one population in response to adverse interactions with another population, as between prey and predator.

countershading *n*. condition of an animal being dark dorsally and pale ventrally, so that when lighting is from above the ventral shadow is obscured and the animal appears evenly coloured and inconspicuous.

countertranscript *n*. RNA that exerts some function because of its partial complementarity to and presumed base pairing with another RNA coded in the same region of DNA.

coupling *n*. of cells, chemical and electrical communication between adjacent animal cells via gap junctions.

coupling factor (F₁, F₀) multisubunit protein component of the mitochondrial inner membrane, composed of a spherical head, F_1, on the matrix side which contains the active site for ATP synthesis, joined by a short "stalk" to a transmembrane portion, F_0, which contains the proton channel. It catalyses the synthesis of ATP from ADP in the presence of a proton gradient across the membrane.

court *n*. a small area of a lek used by an individual male for display.

courtship *n*. behaviour pattern preceding mating in animals, often elaborate and ritualized.

coverts *n.plu*. feathers covering bases of quills.

Cowper's glands bulbo-urethral glands *q.v.*

coxa *n*. joint of leg nearest body in insects, arachnids and some other arthropods. *a*. coxal.

coxite *n*. one of paired lateral plates of exoskeleton next to insect sternum.

coxocerite *n*. the basal joint of an insect antenna.

coxopleurite catapleurite *q.v.*

coxopodite *n*. the joint nearest the body in crustacean limbs.

Coxsackie virus *a*. picornavirus *(q.v.)* which multiplies chiefly in the intestinal tract, but which can cause aseptic menin-

gitis and other conditions.

C3 pathway carbon dioxide fixation in plants via the Calvin cycle.

C3 plant any plant in which carbon dioxide fixation is solely via the Calvin cycle as in most temperate plants. *cf*. C4 plant.

C4 pathway alternative pathway of carbon dioxide fixation present in many tropical plants in which CO_2 is incorporated into 4-carbon compounds, first oxaloacetate, then malate and aspartate which are transported to chloroplasts of bundle sheath cells where the CO_2 is released and enters the Calvin cycle, ensuring an adequate and continuous supply of CO_2 for photosynthesis under tropical conditions. *alt*. Hatch-Slack pathway.

C4 plant any of a diverse group of chiefly tropical plants adapted to high temperatures and low humidity which possess the alternative C4 pathway (*q.v.*) of carbon dioxide fixation in which various metabolic pathways concerned with photosynthesis are compartmented between mesophyll cells and bundle sheath cells in the leaf. *cf*. C3 plant.

crampon *n*. an aerial root, as in ivy.

craniad *a*. towards the head.

cranial *a. pert*. skull, or that part which encloses the brain; *appl*. nerves arising from the brain.

Craniata, craniates *n., n.plu*. in some classifications, the Vertebrata *q.v.*

cranihaemal *a. appl*. anterior lower portion of a sclerotome.

cranineural *a. appl*. anterior upper portion of a sclerotome.

craniology *n*. the study of the skull.

craniometry *n*. the science of the measurement of skulls.

craniosacral *a. appl*. nerves, the parasympathetic system.

cranium *n*. the skull, more particularly that part enclosing the brain.

craspedote *a*. having a veil or cortina, *appl*. to some fungi.

Crassulacean acid metabolism (CAM) metabolic pathway in some succulent plants (Crassulaceae and Cactaceae) in which carbon dioxide is first fixed nonphotosynthetically into carboxylic acids (chiefly malate) which are then mobilized in the light and used as CO_2 sources for the Calvin cycle.

crateriform *a.* bowl-shaped.

craticular *a.* bowl-shaped, *appl.* stage in life history of diatom where new valves are formed before old ones are lost.

C-reactive protein inducible factor in blood serum involved in general resistance to bacterial infection.

creatine *n.* amino acid present in vertebrate muscle, which as creatine phosphate is a high energy compound donating phosphoryl groups to ADP during muscle contraction to recover ATP. *see* Appendix 1 (52) for chemical structure.

creatinine *n.* compound formed from creatine by dehydration, present in muscles, blood and urine.

Creationism *n.* the doctrine that the different types of living organisms have all arisen independently by divine creation. It has now been superseded as a serious doctrine in biology by the overwhelming evidence for evolution, but is still held by many non-scientists on religious grounds, and by a tiny minority of practising biologists.

C region the constant region of an immunoglobulin molecule, which has the same amino acid sequence in antibodies of the same class or subclass, and which comprises most of the N-terminal portion of the heavy chain with a shorter homologous constant region at the N-terminal end of the light chain, and which is the part of the molecule that determines antibody class, e.g. IgG, IgD, etc.

cremaster *n.* the cluster of hooks on the hind end of butterfly pupae, from which it is suspended; abdominal spine in subterranean insect pupae; thin muscle along the spermatic cord.

cremocarp *n.* a fruit composed of two bilocular carpels, as in umbellifers, in which the two carpels separate into one-seeded indehiscent capsules which remain attached to the supporting axis before dispersal.

crena *n.* notch in a crenate leaf, etc.; cleft, as anal cleft; deep groove as the longitudinal groove of heart.

crenate *a.* with scalloped margin.

crenation *n.* a scalloped margin or rounded tooth, as of leaf; notched or wrinkled appearance.

crenulate *a.* with minutely scalloped margin.

creodonts *n.plu.* a group of extinct placental mammals of the Cretaceous and Pliocene, which were probably archaic carnivores.

crepitaculum *n.* a stridulating organ as in some Orthoptera; rattle in rattlesnake's tail.

crepitation *n.* in insects, the discharge of fluid with an explosive sound for defence.

crepuscular *a. pert.* dusk, *appl.* animals flying before sunrise or at twilight. *alt.* vespertine.

crescentic, crescentiform *a.* crescent-shaped.

Cretaceous *n.* last period of the Mesozoic era, from approx. 140 to 70 million years ago, during which chalk was being laid down, occurring after the Jurassic and before the Tertiary.

cribellum cribrellum *q.v.*

cribi- cribri-.

cribrellate *a.* having many pores, as certain spores.

cribrellum *n.* a plate perforated by openings of silk ducts in certain spiders; a perforated chitinous plate in some insects.

cribriform *a.* sieve-like.

cribriform plate portion of ethmoid or mesethmoid bone perforated by many foramina for exit of olfactory nerves.

cribrose *a.* having sieve-like pitted markings.

crickets common name for many members of the Orthoptera *q.v.*

crico-thyroid *a. pert.* cricoid and thyroid cartilages, *appl.* tensor muscle of vocal chord.

cricoid *a.* ring-like; *appl.* cartilage in larynx, articulating with thyroid and arytaenoid cartilages; *appl.* placenta lacking villi on central part of the disc, as in certain Edentata.

crinite *a.* with hairy or hair-like structures or tufts; *n.* a fossil crinoid.

Crinoidea, crinoids *n., n.plu.* a class of echinoderms, commonly called sea lilies and feather stars, present since the Cambrian, having a cup-shaped body with feathery arms, attached to the substratum by a stalk in the case of sea lilies.

crinose *a.* with long hairs.

crisped, crispate *a*. curled, frizzled.

crissal *a. pert*. the crissum.

criss-cross inheritance pattern of inheritance of X-linked genes.

crissum *n. pert*. the region around the cloaca in birds; vent feathers or lower tail coverts. *a*. crissal.

crista *n*. a crest or ridge; a single fold of the inner membrane in mitochondria. *plu*. cristae.

cristae *n.plu*. the folds of the inner mitochondrial membrane. *sing*. crista.

cristate *a*. crested; shaped like a crest.

crithidial *a. appl*. long slender form of trypanosome with a partial undulating membrane and kinetoplast anterior to the nucleus.

critical frequency maximum frequency of successive stimuli at which they can produce separate sensations; minimum frequency for a continuous sensation.

critical group a taxonomic group containing organisms that cannot be divided into smaller groups, as in apomictic species.

critical links organisms in a food chain that are responsible for primary energy capture and nutrient assimilation, and which are critical in the transformation of nutrients into forms that can be used by organisms at higher trophic levels in the chain.

crochet *n*. a larval hook used in locomotion in insects; the terminal claw of chelicerae in arachnids.

Crocodylia, crocodilians *n., n.plu*. order of reptiles found first in the Triassic, typified by the present-day crocodiles and alligators, which are armoured, have front limbs shorter than hind and a body elongated for swimming.

Cro-Magnon man early type of modern man *Homo sapiens*, whose fossils were first found at Cro-Magnon in the Dordogne in France.

crop *n*. sac-like dilation of gullet of bird in which food is stored; similar structure in alimentary canal in insects.

crop-milk secretion of epithelium of crop in pigeons, stimulated by prolactin, for nourishment of nestlings.

crosier crozier *q.v.*

Crosopterygii, crossopterygians *n., n.plu*. a group of mainly extinct bony fishes, of which the coelacanth (*Latimeria*) is the only living member, also called tassel-finned or lobe-finned fishes.

cross *n*. an organism produced by mating parents of different strains or breeds; *v*. to hybridize.

cross-fertilization the fusion of male and female gametes from different individuals, especially of different genotypes. *alt*. allogamy, allomixis.

crossing-over the process of exchange of genetic material (recombination) between homologous chromosomes during meiosis, consisting of the breakage of homologous chromatids at corresponding sites and their reunion with each other, the physical structure formed being called a chiasma or crossover and being visible under the light microscope. *see also* unequal crossing-over.

crossover chiasma *q.v*. see also crossing-over.

crossover fixation model proposed mechanism for maintenance of fidelity of multiple repeated DNA sequences (such as the rRNA genes, and some satellite DNAs in arthropods) in which the entire cluster is continually rearranged by the mechanism of unequal crossing-over during recombination.

crossover value the percentage of gametes resulting from crossing-over of a particular gene.

cross-pollination transfer of pollen from the anther of one flower to the stigma of another, especially on a different plant.

cross-reacting *appl*. material that interacts with antibodies specific for another substance.

cross-reflex the reaction of an effector on one side of the body to stimulation of a receptor on the other side.

crotchet *n*. a curved bristle, notched at the end.

crown *n*. crest; head; the exposed part of a tooth, especially the grinding surface; head, cup and arms of a crinoid; leafy upper part of a tree; short rootstock with leaves.

crown gall disease a plant tumour caused by the bacterium *Agrobacterium tumefaciens*.

crozier *n*. (*bot*.) the coiled young frond of a fern; (*mycol*.) hook formed by the terminal cells of ascogenous hyphae.

crucial, cruciate *a*. in the form of a cross.

crucifer *n*. any member of large family of dicotyledons, the Cruciferae, whose flowers have four petals in the form of a cross, and which includes cabbage, turnip, wallflower, mustard, cress, etc. *a*. cruciferous.

cruciform *a*. arranged like the points of a cross; in the shape of a cross.

crumena *n*. a sheath for retracted stylets in bugs (hemiptera).

cruor *n*. the clots in coagulated blood.

crura *n.plu.* a variety of columnar structures in different organs and organisms. *sing.* crus.

crura cerebri the cerebral peduncles, two cylindrical masses forming the sides and floor of the midbrain.

crural *a. pert*. the thigh. *alt.* femoral.

crureus *n*. parts of the quadriceps muscle in thigh.

crus *sing*. of crura *q.v.*; the shank of hindlimb of vertebrates.

Crustacea, crustaceans *n., n.plu.* subphylum of arthropods (*q.v.*), considered as a class in older classifications. They are mainly aquatic, gill-breathing animals, such as crabs, lobsters and shrimps, having a body divided into a head bearing five pairs of appendages (two pairs of pre-oral sensory feelers and three pairs of post-oral feeding appendages) and a trunk and abdomen bearing a variable number of often biramous appendages which serve as walking legs and gills. Crustacea often have a hard carapace or shell.

crustaceous *a*. with characteristics of a crustacean; thin or brittle; forming a thin crust.

crustose *a*. forming a thin crust on the substratum, as certain lichens and sponges.

cryobiology *n*. study of the freezing of cells and tissues.

cryo-electron microscopy electron microscopy carried out on unstained thin frozen films of material, used to investigate the structure of large molecules and molecular complexes.

cryoenzymology *n*. study of properties of enzymes at very low temperatures (e.g. –50°C) at which the catalytic process is slowed down.

cryoglobulin *n*. any of a group of globulins that occur in very small amounts in normal human blood serum and have the property of precipitating, gelling or crystallizing spontaneously from solution when cooled.

cryophil(ic) *a*. thriving at low temperature.

cryophylactic *a*. resistant to low temperature, *appl*. bacteria.

cryophyte *n*. plant, alga, bacterium or fungus that lives in snow and ice.

cryoplankton *n*. plankton found around glaciers and in the polar regions.

crypsis *n*. the phenomenon of being camouflaged to resemble part of the environment.

crypt *n*. a simple glandular tube or cavity, e.g. the narrow epithelium-lined cavities in the intestinal wall, depression in uterine mucous membrane; (*bot*.) pit of stoma.

cryptic *n. appl*. protective colouring making concealment easier; *appl*. genetic variation due to the presence of recessive genes; *appl*. species extremely similar in external appearance but which do not normally interbreed.

crypto- prefix derived from the Gk. *kryptos*, hidden.

cryptocarp cystocarp *q.v.*

cryptococcosis *n*. a human disease affecting either lungs or central nervous system, caused by the yeast-like fungus *Cryptococcus neoformans* (*Lipomyces neoformans*).

cryptofauna *n*. organisms living in concealment in protected situations, such as crevices in coral reefs.

cryptogram *n*. a method of expressing in standard form a collection of data used in classification.

cryptogams *n.plu*. plants reproducing by spores, such as the mosses and ferns. The term has also been used for plants without flowers, or without true stems, roots or leaves.

cryptohaplomitosis *n*. type of unusual mitosis in some flagellate protozoans, where the total chromatin divides into two masses which pass to opposite poles without apparent condensation into chromosomes.

cryptomitosis *n*. type of nuclear division in some protozoans in which the mass of chromatin assembles in the equatorial region without apparent formation of chromosomes.

cryptonema *n*. a filamentous outgrowth in

cryptostomata of brown algae.

Cryptophyta, cryptophytes *n.*, *n.plu.* phylum of mostly free-living but some parasitic unicellular protists containing both photosynthetic and non-photosynthetic forms, with an ovoid flattened cell and a gullet (crypt) from which arise two unequal flagella. They are distinguished from other protists by their mode of cell division. Found in a wide range of moist habitats including the intestinal tracts of some animals. *alt.* cryptomonads.

cryptophyte *n.* a perennial plant persisting by means of rhizomes, corms, or bulbs underground, or by underwater buds; a member of the protist phylum Cryptophyta *q.v.*

cryptoptile *n.* a feather filament, developed from a papilla.

cryptorchid *a.* having testes abdominal in position.

cryptosphere *n.* the habitat of the cryptozoa.

cryptostomata *n.plu.* non-sexual apparent conceptacles in some large brown algae, bearing only sterile hairs. *sing.* cryptostoma.

cryptovirus group plant virus group containing isometric double-stranded RNA viruses that produce no symptoms on infected plants. There are two subgroups: A, with smooth isometric particles, type member white clover cryptic virus 1; B, with isometric particles with prominent subunits, type member white clover cryptic virus 2.

cryptozoa *n.* ecological term for the small terrestrial animals that live on the ground but above the soil in leaf litter and twigs, among and under pieces of bark, stones, etc.; animals that live in crevices. *a.* cryptozoic.

cryptozoite *n.* stage of the sporozoite of parasitic protozoa when living in tissues before entering blood.

crypts of Lieberkühn tubular exocrine glands in the intestines that secrete digestive enzymes.

crystal cells cells found in the ceoelom of echinoderms, containing rhomboid crystals.

crystal-containing body microbody *q.v.*

crystalline *a.* transparent or translucent.

crystalline cone cone-shaped extracellular jelly-like structure in ommatidium of a compound eye.

crystalline style a translucent proteinaceous rod containing carbohydrases and involved in carbohydrate digestion in the alimentary canal of some molluscs.

crystallins *n.plu.* family of small globular proteins which are the principal components of the lens in the eye.

crystalloid *n.* substance which in solution readily diffuses through a semi-permeable membrane, *cf.* colloid; protein crystals found in some plant cells.

ctene *n.* swimming plate of sea gooseberries (Ctenophora).

ctenidium *n.* feathery or comb-like structure, esp. respiratory apparatus in molluscs; rows of fused cilia forming the swimming plates of sea gooseberries (Ctenophora); row of spines forming a comb in some insects. *plu.* ctenidia.

cteniform, ctenoid, ctenose *a.* comb-shaped, with comb-like margin.

Ctenophora, ctenophores *n.*, *n.plu.* phylum of coelenterates containing the sea-gooseberries or comb jellies. They are free-living and biradially symmetrical with eight meridional rows of ciliated ribs (ctenes) by which they propel themselves. They have no nematoblasts (stinging cells).

ctenocyst *n.* sense organ of Ctenophora (sea gooseberries) borne on the aboral surface.

C-terminal denotes the end of a polypeptide chain containing a free COO^- group, the first part of the chain synthesized.

C-type viruses a class of retroviruses infecting many species of birds and mammals giving rise to C-type particles, whose virions contain 2 copies of a single-stranded RNA genome and the enzyme reverse transcriptase which synthesizes a DNA reproductive intermediate from the viral RNA in the infected cell. The DNA becomes integrated into the host genome as proviral DNA where it is transcribed to produce new viral genomes and viral mRNA. Replication-defective variants of C-type viruses in which part of the viral genome has been replaced by an oncogene and which can induce transformation and tumour formation are thought to have arisen by the acquisition of host cell sequences during infection. *see also* RNA tumour viruses, oncogene.

cubical *a. appl.* cells as long as broad.

cubital *a. pert.* elbow; *pert.* ulna or cubitus; *n.* in birds, a secondary wing quill connected with the ulna.

cubitus *n.* the ulna or the forearm generally; the main vein in an insect wing.

cuboid *n.* a bone in the ankle.

Cubozoa *n.* class of Cnidaria comprising the box-jellies *q.v.*

Cuculiformes *n.* an order of birds including the cuckoos.

cucullate *a.* hooded; (*bot.*) with hood-like petals or sepals; (*zool.*) with prothorax hood-shaped, in insects.

cucullus *n.* a hood-shaped structure; upper part of harpe in Lepidoptera.

cucumovirus group plant virus group containing small isometric single-stranded RNA viruses, type member cucumber mosaic virus. They are multicomponent viruses in which four genomic RNAs are encapsidated in three different virus particles.

cucurbit *n.* any member of the family of dicotyledonous plants, the Cucurbitaceae, the marrows, cucumbers and gourds.

Cucurbitales *n.* order of dicots, herbs and small trees, often climbing by tendrils and comprising the family Cucurbitaceae (gourds, melon, cucumber, etc.).

cuiller *n.* spoon-like terminal portion of male insect clasper.

cuirass *n.* bony plates or scales arranged like a cuirass.

culm *n.* the flowering stem of grasses and sedges.

culmen *n.* median longitudinal ridge of a bird's beak; part of superior vermis of cerebellum.

culmicole, culmicolous *a.* living on grass stems.

cultellus *n.* a sharp knife-like organ, one of the mouthparts of certain blood-sucking flies.

cultivar *n.* plant variety found only in cultivation, conventionally denoted by the species name followed by the abbreviation cv. followed by the cultivar name, e.g. *Rosa foetida* cv. Persian Yellow.

cultural eutrophication eutrophication as a result of human activity such as agricul-

ture, urbanization and sewage discharge.

cultural inheritance the transmission of particular traits and behaviours from generation to generation by learning rather than genetic inheritance.

culture *n.* microorganisms, cells or tissues growing in nutrient medium in the laboratory; *v.* to isolate and grow microorganisms, cells or tissues as above.

culture collection a reference collection of different species and strains of microorganisms or cultured cells.

cumarin coumarin *q.v.*

cumulose *a. appl.* deposits consisting mainly of plant remains, e.g. peat.

cumulus *n.* the mass of epithelial cells bulging into the cavity of an ovarian follicle and in which the ovum is embedded.

cuneate *a.* wedge-shaped; (*bot.*) *appl.* leaves with broad abruptly pointed apex and tapering to the base; (*neurobiol.*) *appl.* nucleus of grey matter situated in anterior medulla oblongata at which spinal neurones carrying sensory information (e.g. on touch, pressure, vibration) from skin and other tissues first make connections with other neurones that transmit the information onwards to other regions of the brain.

cuneiform *a.* wedge-shaped.

cuneiform nucleus area in midbrain, implicated in control of locomotion.

cuneus *n.* division of elytron in certain insects; (*neurobiol.*) a wedge-shaped area of the occipital lobe of cerebral cortex between calcarine fissure and medial part of parieto-occipital fissure.

cup fungi common name for the Discomycetes *q.v.*

cupula *n.* the bony apex of the cochlea; the jelly-like cup over a group of hair cells (a neuromast) in acoustico-lateralis system; cupule *q.v.*

cupulate *a.* cup-shaped; having a cupule.

cupule *n.* (*bot.*) the green involucre of bracts round the fruit of some trees, e.g. acorns; (*zool.*) a small sucker in various animals.

curare *n.* poison extracted from cinchona bark, blocks neuromuscular transmission causing paralysis.

curie (Ci) unit of radiation corresponding to an amount of radioactive material producing 3.7×10^{10} disintegrations per second, which is the activity of radium. It is being replaced by the derived SI unit, the becquerel (Bq). 1 Ci = 3.7×10^{10} Bq.

currant gall type of small spherical gall found on oak leaves or catkins caused by larvae of the gall wasp *Neuoterus quercus baccarum.*

cursorial *a.* having limbs adapted for running, *appl.* birds, bipedal dinosaurs, etc.

cusp *n.* a prominence, as on molar teeth; a sharp point. *alt.* tubercle.

cuspidate *a.* terminating in a sharp point.

cutaneous *a. pert.* the skin.

cuticle *n.* an outer skin or pellicle, sometimes referring to the epidermis as whole, esp. when impermeable to water; (*zool.*) an outer protective layer of material, of various composition, produced by the epidermal cells, that covers the body of many invertebrates; (*bot.*) layer of waxy material, cutin, on the outer wall of epidermal cells in many plants, making them fairly impermeable to water. *a.* cuticular.

cuticularization *n.* the formation of a cuticle; the laying down of cutin in the outer layers of epidermis.

cuticulin *n.* lipoprotein secreted by epidermal cells and forming the epicuticle of insects.

cutin *n.* a wax-like mixture of fatty substances impregnating epidermal walls of plant cells, and also forming a separate layer, the cuticle, on the outer wall of epidermis in plants, making the surface impermeable to water.

cutis *n.* the dermis of skin; the outer layer of cap and stalk of mushrooms and toadstools.

cuttlebone *n.* the shell of the cuttlefish, *Sepia*, which acts as a buoyancy organ.

cuttlefish *n.* common name for a group of cephalopod molluscs, e.g. *Sepia*, characterized by a shell of unusual structure (cuttlebone), and having eight short arms around the mouth and long tentacles.

Cuvierian ducts short paired veins opening into the sinus venosus, and formed by the union of anterior and posterior cardinal veins, in fish and in tetrapod embryos.

Cuvierian organs tubular organs in sea cucumbers (Holothuroidea) which secrete collagen and polysaccharide which is released as slime via the cloaca when the animal is harassed.

C value the total amount of DNA in the haploid genome of a species, either measured directly in picograms or expressed in base pairs or daltons.

C-value paradox the fact that C values for closely related species can differ widely and also appear not to be closely related to the relative organizational complexity of different organisms, and that the total amounts of DNA in higher multicellular organisms seem to be much more than is needed to encode the required number of genes.

cyanelle *n.* DNA-containing photosynthetic organelle within algae such as *Cyanophora paradoxa*, and which has many similarities with photosynthetic bacteria and which may represent a relic of a bacterial endosymbiont.

cyanidin *n.* violet flavonoid pigment present in many flowers.

cyanobacteria *n.plu.* a group of prokaryotic, photosynthetic, non-flagellate, unicellular, filamentous or colonial microorganisms, found in aquatic and terrestrial environments either free-living or in symbiotic association with fungi as lichens. They have an oxygen-evolving type of photosynthesis resembling that of algae and green plants. Some species (e.g. *Anabaena, Nostoc*) can fix atmospheric nitrogen. Commonly (but misleadingly) known as the blue-green algae and formerly classified in the plant kingdom as the Cyanophyta or Cyanophyceae. They contain the pigments chlorophyll (not contained in chloroplasts), α- and β-carotenes, phycoerythrin and phycocyanin and often appear blue. Some cyanobacteria produce toxins which can become a health hazard in conditions where cyanobacterial "algal blooms" appear. *sing.* cyanobacterium.

cyanocobalamin *n.* common commercial form of cobalamin (vitamin B_{12}) with CN substituted at the 6th position on the Co atom.

cyanogenesis *n.* the elaboration of hydrocyanic acid by some plants. *a.* cyanogenic.

cyanogenic glycosides plant glycosides that, when fully hydrolysed, liberate hydrogen cyanide, which is toxic to most cells. Found in species of *Sorghum, Prunus* and *Linum.*

cyanolabe *n.* the blue-sensitive pigment in

the human eye.

cyanophil *a.* with special affinity for blue or green stains.

cyanophoric *a. appl.* glycosides which on hydrolysis yield glucose, hydrocyanic acid, and benzaldehyde, such as amygdalin.

Cyanophyceae, Cyanophyta *see* cyanobacteria.

cyathiform *a.* cup-shaped.

cyathium *n.* the peculiar inflorescence of the spurges (*Euphorbia*), consisting of a cup-shaped involucre of bracts surrounding staminate flowers each with a single stamen, with a central pistillate flower.

cyathozooid *n.* the primary zooid in certain tunicates.

cyathus *n.* small cup-shaped organ or structure.

cybernetics *n.* science of communication and control, as by nervous system and brain.

Cycadeoidophyta, cycadeoids *n. n.plu.* division of fossil cycad-like plants that had massive stems, pinnate leaves and were usually monoecious, with sporophylls arranged in a flower-like structure. They have been considered as possible ancestors of angiosperms.

Cycadophyta, cycads *n.* one of the five main divisions of extant seed-bearing plants, commonly called cycads or sago palms. They are palm-like in appearance with massive stems which may be short or tree-like. Microsporangia and megasporangia are borne on sporophylls arranged in cones, male and female cones being borne on separate plants. In some classifications called the Cycadopsida or Cycadales. *see* Appendix 2.

Cyclanthales *n.* order of often palm-like monocots, also climbers and large herbs, comprising the family Cyclantheraceae.

cyclic *a.* having parts of flowers arranged in whorls. *alt.* cyclical.

cyclic AMP (cAMP) adenosine 3′,5′-cyclic monophosphate, a ubiquitous regulatory molecule, in which an oxygen of the phosphate is bonded to C of the ribose ring. Formed from ATP by adenylate cyclase, its formation is controlled by hormonal and other stimuli acting at the plasma membrane. It acts as a universal "second messenger", triggering reactions that produce a cellular response to the particular stimulus. In cellular slime moulds it is a chemoattractant causing the aggregation of myxamoebae into a slug and fruiting body. *see* Appendix 1 (24) for chemical structure.

cyclic-AMP-responsive element *see* CRE.

cyclic GMP molecule analogous to cyclic AMP, formed from guanosine monophosphate by the enzyme guanylate cyclase, and which functions as a second messenger in some cellular reactions.

cyclic phosphorylation type of photophosphorylation involving photosystem I only, in which ATP is generated without concomitant NADPH generation.

cyclins *n.plu.* a group of related proteins which accumulate in interphase of the eukaryotic cell cycle and are destroyed at the conclusion of mitosis, and which are believed to be involved in inducing mitosis.

cyclocoelic *a.* with the intestines coiled in one or more distinct spirals.

cyclogenous *a. appl.* a stem growing in concentric circles.

cyclogeny *n.* the production of a succession of different morphological types in a life cycle.

cycloheximide *n.* an antibiotic produced by the actinomycete *Streptomyces griseus*, which inhibits protein synthesis in eukaryotic cells only, by inhibiting translation.

cycloid *a. appl.* scales with an evenly curved free border.

cyclomorial *a. appl.* scales growing by apposition at margin only.

cyclomorphosis *n.* a cycle of changes in form, usually seasonal, esp. in marine zooplankton, possibly in response to changes in salinity.

cyclopean *a. appl.* to a single median eye developed under certain artificial conditions or as a mutation, instead of the normal pair.

cyclopoid naupliiform *q.v.*

cyclops *n.* a larval stage in some copepods, having many characteristics of the adult.

cyclosis *n.* the circulation of protoplasm within a cell.

cyclospermous *a.* with embryo coiled in a circle or spiral.

cyclospondylic, cyclospondylous *a.* *appl.* vertebral centra in which the calcified material forms a single cylinder surrounding the notochord.

cyclosporin A immunosuppressive drug isolated from the fungus *Tolypocladium,* used clinically to suppress unwanted immune responses leading to organ rejection after transplantation.

cyclostomes *n.plu.* lampreys and hagfish, primitive jawless fishes.

Cyclostomata *n.,* cyclostomes, *see* Agnatha; the Cyclostomata, an order of Ectoprocta *q.v.* in which the zooids are completely fused, the case enclosing them is completely calcified and the pore has no lid.

cydippid *n.* a ctenophore larva.

cygneous *a.* shaped like a swan's neck.

cylindrical *a.* *appl.* leaves rolled on themselves, or to solid cylinder-like leaves.

cymba *n.* upper part of cavity of external ear.

cymbiform *a.* boat-shaped.

cymbium *n.* boat-shaped tarsus of pedipalp in some spiders.

cyme *n.* repeatedly branching determinate inflorescence in which each growing point ends in a flower, with the oldest flowers at the end of the branch.

cymose *a.* *appl.* an inflorescence formed by successive growths of axillary shoots after growth of main shoot in each branch has stopped.

cymotrichous *a.* having wavy hair.

cynarrhodium, cynarrhodon *n.* fruit of the rose-hip type.

Cynodontia, cynodonts *n., n.plu.* group of extinct mammal-like reptiles, found from the late Permian to the middle of the Jurassic, possessing a secondary bony palate, complex crowns on the cheek teeth, and other mammalian features, and which are believed to be the direct ancestors of mammals.

cynopodous *a.* with non-retractile claws, as a dog.

Cyperales *n.* order of herbaceous monocots with rhizomes and solid stems triangular in cross-section and comprising the family Cyperaceae (sedge).

cyprinids, cyprinoids *n.plu.* group of freshwater fish (the Cyprinoidei) widespread in Europe, Asia, Africa and North America, and including the carps and minnows.

cyprinine *n.* a protamine present in carp sperm.

cyprinodonts *n.plu.* an order of small, mainly tropical fishes, the Cyprinidontiformes, including the toothed carps.

cypsela *n.* an inferior achene composed of two carpels, as in Compositae.

cyst *n.* protective coat surrounding resting cells of e.g. soil amoebae and some bacteria which is formed in dry conditions; a bladder or air vesicle in certain seaweeds; an abnormal fluid-filled sac developing in tissues; any bladder- or sac-like structure.

cysteine (Cys, C) *n.* a sulphur-containing amino acid, constituent of many proteins where it forms disulphide bonds cross-linking the protein chain.

cystic *a.* *pert.* gall bladder or urinary bladder.

cystic fibrosis (CF) inherited condition in humans caused by a recessive defect in a gene carried on chromosome 7. Symptoms appear only in individuals homozygous for the defective gene. It is characterized by malfunction of pancreas and abnormal secretion from the lungs, leading to serious secondary effects such as infection, and is fatal in early childhood if untreated. The primary defect is believed to be in the transport of chloride ions into and out of cells.

cysticercoid, cysticercus *n.* larval stage in tapeworm, consisting of a fluid-filled sac containing a scolex, which in the appropriate host attaches to the gut wall and develops into an adult. *a.* cysticercoid.

cysticolous *a.* living in a cyst.

cystidium *n.* inflated hair-like cell in the hymenium of some fungi. *plu.* cystidia.

cystine (Cys-Cys) *n.* amino acid residue formed in proteins by oxidation of the sulphydryl groups of 2 cystine residues to form a disulphide bridge cross-linking the protein chain.

crystoarian *a.* *appl.* gonads when enclosed in coelomic sacs, as in most teleost fishes.

cystochroic *a.* having pigment in cell vacuoles.

cystogenous *a.* cyst-forming; *appl.* large nucleated cells that secrete the cyst in cercaria.

cystolith *n.* a mass of calcium carbonate, occasionally of silica, formed on ingrowths of epidermal cell walls in some plants.

cystozooid *n.* body portion of a metacestode.

cytidine *n.* a nucleoside made up of the pyrimidine base cytosine linked to ribose.

cytidine diphosphate (CDP) cytosine nucleotide containing a diphosphate group.

cytidine monophosphate (CMP) nucleotide composed of cytosine, ribose and a phosphate group, product of the partial hydrolysis of RNA and also found in the activated intermediate CMP-*N*-acetylneuraminate in ganglioside synthesis. *alt.* cytidylate, cytidylic acid, cytidine 5′-phosphate. *see* Appendix 1 (22) for structure.

cytidine triphosphate (CTP) cytosine nucleotide containing a triphosphate group, one of the ribonucleotides required for RNA synthesis, also acts as an energy donor in metabolic reactions in a manner analogous to ATP, esp. in triacylglycerol synthesis where it forms activated CDP-acylglycerols.

cytidylate (cytidylic acid) cytidine monophosphate *q.v.*

cytidyl transferrase *see* nucleotidyl-transferase.

cytocentrum centrosome *q.v.*

cytochalasin *n.* any of a group of compounds, produced by various moulds, which inhibit actin polymerization into microfilaments.

cytochemistry *n.* the study of the chemical composition and chemical processes in cells by specific staining and microscopical examination. *alt.* histochemistry.

cytochimaera *n.* tissue or organism in which cells have different chromosome numbers.

cytochroic *a.* having pigmented cytoplasm.

cytochrome *n.* any of a group of electron-transporting proteins containing a haem prosthetic group, components of the respiratory and photosynthetic electron transport chains, in which the haem iron exists in oxidized or reduced state.

cytochrome *a* and *a₃* components of cytochrome oxidase (cytochrome *c* oxidase), the terminal electron carriers in the respiratory chain, electrons being trans-ferred from the copper prosthetic group of cytochrome a_3 to molecular oxygen to form water.

cytochrome *b* and *b₃* components of the respiratory electron transport chain, constituents of the QH_2-cytochrome *c* reductase complex.

cytochrome *c₅₅₂* formerly cytochrome *f*.

cytochrome *c* component of the respiratory electron transport chain, transfers electrons from the QH_2-cytochrome *c* reductase complex to the cytochrome c oxidase complex (cytochromes *a* and *a₃*).

cytochrome *c* oxidase cytochrome oxidase *q.v.*

cytochrome *f* now cytochrome c_{552}.

cytochrome oxidase enzyme complex catalysing the terminal transfer of electrons to oxygen (producing water) in the respiratory chain, contains cytochromes *a* and *a₃*. EC 1.9.3.1, *r.n.* cytochrome *c* oxidase. Originally known as Warburg's factor, Warburg's respiratory enzyme.

cytochrome P₄₅₀ specialized cytochrome, terminal component of an electron transport chain in adrenal mitochondria and liver microsomes, involved in hydroxylation reactions.

cytocidal *a.* cell-destroying.

cytocuprein superoxide dismutase *q.v.*

cytocyst *n.* envelope formed by remains of host cell within which a protozoan parasite multiplies.

cytode *n.* a non-nucleated protoplasmic mass.

cytodeme *n.* a local interbreeding unit in a taxon differing cytologically, usually in chromosome number, from the rest of the taxon.

cytofluorimetric *a. appl.* the identification and isolation of specific cell types by means of immunological or other specific staining with fluorescent antibodies or dyes.

cytogamy *n.* cell conjugation.

cytogenesis *n.* development or formation of cells.

cytogenetic *a. pert.* cytogenetics; *pert.* cytogenesis.

cytogenetics *n.* the study of the microscopic structure of chromsomes, esp. the mapping of genes by correlation of visible chromosomal defects with a particular mutation.

cytogenic *a. appl.* reproduction by cell division, as in a clone.

cytogenous *a.* producing cells.

cytokine *n.* any protein factor such as a growth factor, which is a product of a cell and which affects the growth and division, or other functions, of other cells.

cytokinesis *n.* cytoplasmic division into two daughter cells after mitosis or meiosis, in animal cells it is usually synonymous with cell division.

cytokinins *n.plu.* plant growth hormones, derivatives of adenine, and including naturally-occurring zeatin and i⁶Ade, and kinetin, which probably does not occur in nature. They act in concert with IAA (an auxin) to promote rapid cell division, and are widely used to induce the formation of plantlets from callus tissue in culture. *alt.* kinin, phytokinin.

cytology *n.* the study of the structure, functions and life history of cells.

cytolysin *n.* a substance causing cell lysis.

cytolysis *n.* cell lysis or disintegration.

cytolysosome *n.* a large lysosome apparently involved in autolysis.

cytomegaloviruses *n.plu.* a group of DNA viruses of the herpesvirus family, infecting vertebrates, the infection characterized by the formation of large inclusion bodies in infected glandular cells (often of salivary gland in humans), and which can be fatal in very young children.

cytomeres *n.* cells formed by schizont and giving rise to merozoites.

cytometry *n.* counting cells.

cytomorphosis *n.* a series of structural modifications of a cell or successive generations of cells; cellular change, as in senescence.

cytopathic effect (CPE) the destruction of cells infected by a virus.

cytopempsis transcytosis *q.v.*

cytopharynx *n.* a tube-like structure leading from "mouth" into cell interior in certain protozoans. *alt.* gullet.

cytophilic *a.* having an affinity for cells; binding to cells; staining cells.

cytophilic antibody antibody that becomes adsorbed to cells by its constant region, leaving the antigen-binding site free to subsequently bind antigen, as IgE.

cytoplasm *n.* all the living part of a cell inside the cell membrane and excluding the nucleus. *a.* cytoplasmic, *pert.* or found in the cytoplasm.

cytoplasmic determinant in developmental biology, substances laid down in cytoplasm of egg during its formation that direct the early development of the zygote.

cytoplasmic inheritance inheritance of genes carried in organelles such as chloroplasts and mitochondria, or in other cytoplasmic particles, which behave in a non-Mendelian fashion as they are usually inherited only from the maternal parent.

cytoplasmic male sterility a form of male sterility in plants determined by cytoplasmic factors, usually mitochondrial DNA.

cytoplasmic streaming the rapid movement of cytoplasm within eukaryotic cells, seen most clearly in plant and algal cells, some protozoa and in the plasmodia of slime moulds. It is thought to involve the action of actin filaments and myosin at the boundaries of the streams of cytoplasm.

cytoproct, cytopyge *n.* the "anus" of a unicellular organism, as certain protozoans.

cytosine (C) *n.* a pyrimidine base, constituent of DNA and RNA, and which is the base in the nucleoside cytidine. Pairs with guanine. Cytosine in DNA is sometimes methylated to form methylcytosine, *see* DNA methylation. *see* Appendix 1 (17) for structure.

cytosine arabinoside arabinosyl cytosine, an antiviral drug.

cytoskeleton *n.* internal proteinaceous framework of a eukaryotic cell. It is composed of actin microfilaments, intermediate filaments and microtubules and gives shape to a cell, provides support for cell extensions such as villi and axons of nerve cells, is involved in cell movement, in interactions with the substratum on which the cell is lying, and in intracellular transport. The cytoskeleton is not usually a permanent structure but is continually being disassembled, reassembled and rearranged. *see also* intermediate filament, microfilament, microtubule.

cytosol *n.* the cytoplasm other than the various membrane-bounded organelles.

cytosome microbody *q.v.*

cytostatic *a. appl.* any substance suppressing cell growth and multiplication.

cytostome *n*. the specialized region acting as a "mouth" of a unicellular organism, as in some protozoans.

cytotaxis *n*. movement of cells to or away from a stimulus.

cytotaxonomy *n*. classification based on characteristics of chromosome structure and number.

cytotoxic *a*. attacking or destroying cells.

cytotoxic T lymphocyte, cytotoxic T cell (T_C, CTL) type of T lymphocyte that interacts directly in an antigen-specific manner with virus-infected, parasite-infected, or otherwise antigenically altered or abnormal cells, and destroys them. It is the effector cell of cell-mediated immune responses.

cytotoxin *a*. any substance that can poison or destroy cells.

cytotrophoblast *n*. inner layer of trophoblast.

cytotropism *n*. the mutual attraction of two or more cells.

cytotype *n*. genetically determined characteristic carried by cytoplasmic part of cell, different cytotypes usually being revealed by their effects on the expression of the nuclear genotype.

cytozoic *a*. living inside a cell, *appl.* sporozoan trophozoite.

D

δ heavy chain class corresponding to IgD.

δ- *for all headwords with prefix δ- or delta-refer to headword itself.*

D aspartic acid *q.v.*

2,4-D dichlorophenoxyacetic acid, a synthetic auxin, has been used as a herbicide.

D-, L- prefixes denoting particular molecular configurations, defined according to convention, of certain optically active compounds esp. monosaccharides and amino acids, the L configuration being a mirror image of the D. In living cells such molecules usually occur in one or other of these configurations but not both (e.g. glucose as D-glucose, amino acids always in the L form in proteins).

DAF delay-accelerating factor *q.v.*

dal, Da dalton *q.v.*

DAP diaminopimelic acid *q.v.*

DBA a lectin isolated from *Dolichos biflorus*.

DBH diameter at breast height (1.4 m), standard measurement of a trunk of a tree.

D-DNA form of DNA taken up by synthetic molecules lacking guanine, with 8 base pairs per turn.

DDT dichlorodiphenyltrichloroethane, a persistent organochlorine insecticide.

Df in genetic terminology, a deficiency, the absence or inactivation of a given gene.

D gene *see* D segment.

DHFR dihydrofolate reductase *q.v.*

DIPF diisopropylphosphofluoridate, an enzyme inhibitor, reacting with serine residue of proteolytic enzymes and acetylcholinesterase, which is the basis for its use in insecticides and nerve gas.

DM dry matter *q.v.*

DMD Duchenne muscular dystrophy *q.v.*

DMS dimethylsulphate.

DNA deoxyribonucleic acid *see* DNA, and entries pertaining to it in body of dictionary.

DNase, DNAse deoxyribonuclease *q.v.*

DNase I an endonuclease that makes single-stranded nicks in duplex DNA, widely used to study the structure of chromatin as e. g. chromatin that is being actively transcribed seems to be more susceptible to DNase I than non-transcribing chromatin.

DNP dinitrophenol *q.v.*

DOC dissolved organic matter (carbon).

DOM dissolved organic material, in waters.

Dp in genetic terminology, a duplication of a given gene or chromosomal segment.

DPD diffusion pressure deficit *see* suction pressure.

DPG diphosphoglycerate *q.v.*

dsDNA double-stranded DNA.

DW dry weight *q.v.*

dacryocyst *n.* pouch in the angle of the eye between upper and lower eyelids, which receives tears from the lacrimal ducts.

dacryoid *a.* tear-shaped.

dactyl *n.* digit; finger or toe.

dactylognathite *n.* the terminal segment of a maxilliped, a maxillary appendage of arthropods.

dactyloid *a.* like a finger or fingers.

dactylopatagium *n.* the part of the flying membrane of bats that is carried on the metacarpals and phalanges.

dactylopodite *n.* joint furthest away from the body in limbs of certain crustaceans; tarsus and metatarsus of spiders.

dactylozooid palpon *q.v.*

dalton (dal, Da) unit of mass very nearly equal to that of a hydrogen atom, and equal to 1.00 on the atomic mass scale. Used as a unit to express molecular mass.

dahlia starch inulin *q.v.*

dammar *n.* a resin obtained from several

Malaysian trees.

damselflies *see* Odonata.

dance, of bees a series of movements performed by honeybees on their return to the hive after finding a food source, which informs other bees in the hive of the location of the food, *see also* waggle dance.

daphnid *n*. any of various small water fleas, esp. of the genus *Daphnia*.

dark adaptation the visual adaptations that occur in the eye for vision in dim light compared with bright light. The threshold of just-visible light is lowered as a result of e.g. the switch from the use of the retinal cones to the more sensitive rod cells, dilation of the pupil and the migration of choroid pigment away from the outer segments of the photoreceptor cells (in many fish, amphibians, reptiles and birds).

dark-field microscopy type of optical microscopy used for studying living cells, which produces an illuminated object on a dark background.

dark reaction in photosynthesis, reactions occurring in the stroma of chloroplasts, for which light is not required, in which carbon dioxide is reduced to carbohydrate. *alt*. Calvin cycle, carbon dioxide fixation, photosynthetic carbon reduction cycle (PCR cycle).

Darlington's rule the fertility of an allopolyploid is inversely proportional to the fertility of the original hybrid.

dart *n*. a crystalline structure in molluscs, used in copulation; in nematodes, a sharp point used to penetrate the host.

dart sac a small sac, containing a calcareous dart, attached to vagina near its opening in some gastropods.

Darwinian evolution *see* Darwinism.

Darwinian fitness the fitness of a genotype measured by its proportional contribution to the gene pool of the next generation.

Darwinian tubercle the slight prominence on the helix of external ear, near the point where it bends downwards.

Darwinism *n*. the theory of evolution by means of natural selection, put forward by Charles Darwin in his book *Origin of Species* published in 1859 but formulated some years earlier. The theory was based on his observation of the genetic variability that

exists within a species and the fact that organisms produce more offspring than can survive. Under any particular set of environmental pressures, those heritable characteristics favouring survival and successful reproduction would therefore be preferentially passed on to the next generation (natural selection). Selection for particular aspects of life-style or in different environmental conditions could therefore eventually lead to two populations differing in many ways from the original and to the development of complex adaptations to a particular mode of life or environment. A very similar theory was proposed independently by Alfred Russel Wallace. In the 1930s and 1940s the theory of Mendelian genetics was incorporated into Darwin's original theory to produce a modern version, the neo-Darwinian synthesis. *see also* neo-Darwinism, natural selection.

Darwin's finches the 14 species of finches found on the Galapagos Islands, which all possess features adapting them to a different mode of life, which were studied by Charles Darwin in the course of his voyage on the *Beagle*, and which are said to have helped to stimulate his ideas on evolution and the origin of species.

dasypaedes *n.plu*. birds whose young are downy at hatching.

dasypaedic *a. appl.* birds whose young are downy at hatching.

dasyphyllous *a*. with thickly haired leaves.

Datiscales Begoniales *q.v.*

daughter *n*. progeny of cell, nucleus, etc. arising from division, with no reference to sex, as daughter cell, daughter nucleus.

Davson-Danielli model one of the first models proposing a lipid bilayer structure for biological membranes – now superseded by the fluid mosaic model.

day-neutral *a. appl.* plants in which flowering can be induced by either a long or a short photoperiod or by neither.

de- prefix from L. *de*, away from, denoting removal of.

dealation *n*. the removal of wings, as by female ants after fertilization, or by termites.

deamination *n*. the removal of an amino group ($-NH_2$) from a molecule.

death point temperature, or other environmental variable, above or below which organisms cannot exist.

death rate the crude death rate is calculated for human populations as the annual number of deaths per 1000 population in a given geographical area, with the population number usually taken at the midpoint of the year in question.

Débove's membrane layer between tunica propria and epithelium of tracheal, bronchial, and intestinal membranes.

debranching enzyme enzyme which hydrolyses the α-1,6-glycosidic bonds at the branch points in glycogen molecules. EC 3.2.1.10, *r.n.* oligo-1,6-glucosidase.

deca- prefix derived from Gk. *deka*, ten, denoting having ten of, divided into ten, etc.

decagynous *a.* having ten pistils.

decalcify *n.* to treat with acid to remove calcareous parts.

decamerous *a.* with the various parts arranged in tens, of flowers.

decandrous *a.* having ten stamens.

decaploid *a.* having ten times the normal chromosome set.

Decapoda, decapods *n., n.plu.* order of freshwater, marine and terrestrial crustaceans having five pairs of legs on the thorax and a carapace completely covering the throat, and including the prawns, shrimps, crabs and lobsters; an order of cephalopods having two retractile arms as well as eight normal arms, including the squids and cuttlefish.

decem- prefix derived from the L. *decem*, ten, denoting having ten of, divided into ten, etc.

decemfid *a.* cut into ten segments.

decemfoliate *a.* ten-leaved.

decemjugate *a.* with ten pairs of leaflets.

decempartite *a.* having ten lobes.

deception *n.* any behaviour or feature that deceives a predator or other animal, and which ranges from physical mimicry to apparently cognitive deceptions. *cf.* honest behaviour.

decidua *n.* the mucous membrane lining the pregnant uterus, cast off after giving birth. *a.* decidual.

deciduate *a.* characterized by having a decidua; partly formed by the decidua.

deciduous *a.* falling at the end of growth period or at maturity; *appl.* teeth: milk teeth; *appl.* trees: those having leaves that all fall at a certain time of the year.

declinate *a.* bending aside in a curve.

declivis *n.* part of superior vermis of cerebellum, continuous laterally with lobulus simplex of cerebellar hemispheres.

decollated *a.* with apex of spire wanting.

decomposed *a.* not in contact; not adhering, *appl.* barbs of feather when separated; decayed; rather shapeless and gelatinous, *appl.* cortical hyphae in lichens.

decomposers *n.plu.* organisms that feed on dead plant and animal matter, breaking it down physically and chemically and recycling elements and organic and inorganic compounds to the environment, and which include chiefly microorganisms and small animals.

decompound *a. appl.* compound leaf whose leaflets are also compound.

deconjugation *n.* separation of paired chromosomes, as before end of meiotic prophase.

decorticate *a.* with bark (of trees) or cortex (of brain, other organs) removed.

decumbent *a.* lying on ground but rising at tip, *appl.* stems.

decurrent *a.* having leaf base prolonged down stem as a winged expansion of rib; prolonged down stipe, as gills of agarics.

decurved *a.* curved downwards.

decussate *a.* crossed; having paired leaves, succeeding pairs crossing at right angles.

decussation *n.* crossing of nerves (from either side of body) with interchange of fibres, as in optic and pyramidal tracts.

dedifferentiation *n.* loss of characteristics of a specialized cell and its regression to an undifferentiated state.

dediploidization *n.* in basidiomycete and ascomyete fungi, the production of haploid cells or hyphae from a dikaryotic diploid cell or mycelium.

defaunation *n.* removal of animal life, esp. of symbiotic protozoans from an insect.

defective viruses viruses that can infect cells but not reproduce within them, since they lack genes for essential viral components. They can reproduce if a related nondefective helper virus is also present to direct the synthesis of the appropriate viral components. *alt.* replication-defective viruses.

deferent *a.* carrying away from.

deferred *a. appl.* shoots arising from dormant buds.

deficiency *n.* in genetics, the inactivation or absence of a gene or segment of chromosome. *alt.* deletion.

deficiency, deficiency diseases pathological conditions in plants and animals due to lack of some vitamin, trace element or other minor nutrient, e.g. crown rot in sugar beet due to boron deficiency, vitamin deficiency diseases in mammals such as scurvy due to lack of vitamin C, beriberi due to lack of vitamin B_1, rickets due to lack of vitamin D.

definite *a.* fixed; constant; (*bot.*) *appl.* inflorescences with primary axis terminating early in a flower; *appl.* stamens limited to 20 or less in number.

definitive *a.* defining or limiting; complete; fully developed; final, *appl.* host of adult parasite.

deflorate *a.* after the flowering stage.

defoliate *a.* bared at the annual leaf fall. *v.* to strip of leaves.

deforestation *n.* complete and permanent removal of forest or woodland and its associated undergrowth.

degeneracy *n.* in the genetic code, the fact that the third base in many codons does not affect the amino acid specified.

degenerate *a. appl.* the genetic code, in which each of the twenty amino acids of which proteins are composed is encoded by several different triplets.

degeneration *n.* breakdown in structure; change to a less specialized or functionally less active form; evolutionary change resulting in change from a complex to a simpler form.

degenerative disease disease caused by deterioration of organs or tissues, rather than by infection, e.g. osteoarthritis, Alzheimer's disease, cardiovascular disease.

deglutition *n.* the process of swallowing.

degranulation *n.* release of granules, from e.g. mast cells, basophil leukocytes during local inflammatory reactions.

degree of relatedness coefficient of relationship *q.v.*

dehiscence *n.* the spontaneous opening of an organ or structure along certain lines or in a definite direction. *a.* dehiscent.

dehydration *n.* the removal of water from a material, either by heating, to obtain the dry weight (*q.v.*) of soil or biomass, or by chemical means e.g. soaking a tissue sample in ethanol before staining for microscopy.

dehydrogenase *n.* any enzyme catalysing the transfer of hydrogen from a donor to an acceptor compound, classified amongst the oxidoreductases in EC class 1.

deimatic display frightening behaviour consisting of adoption of a posture by one animal to intimidate another.

Deiter's cells non-neural supporting cells between rows of outer hair cells in organ of Corti in cochlea.

Deiters' nucleus lateral vestibular nucleus, area in medulla involved in relaying sensory information relating to balance to motor pathways.

delamination *n.* the splitting off of cells to form a new layer; splitting of a layer.

delay-accelerating factor (DAF) membrane glycoprotein that binds activated complement components C3b and C4b and inhibits further action of complement.

delayed hypersensitivity inflammatory immune reaction in skin, induced by activated T cells and arising 24–48 hours after T-cell induced damage.

deletion *n.* mutation involving loss of part of a chromosome, or a base or bases in a DNA sequence.

deliquescent *a.* becoming liquid; (*bot.*) having lateral buds the more vigorously developed, so that the main stem seems to divide into a number of irregular branches.

delitescence *n.* the latent period of a poison; the incubation period of a pathogenic organism.

delomorphic *a.* with a definite form.

delphinidin *n.* blue flavonoid pigment present in many flowers.

delphinology *n.* the study of dolphins.

delta- *for all headwords with prefix delta- refer to headword itself.*

delthyrium *n.* the opening between hinge and beak, where the stalk emerges, in many brachiopods.

deltidium *n.* a plate covering the opening

between hinge and beak, where the stalk emerges in many brachiopods.

deltoid *a*. more-or-less triangular in shape.

demanian *a. appl.* a complex system of paired efferent tubes connecting the intestine and uteri in nematodes, and associated with secretion of gelatinous material to protect the eggs.

deme *n*. (*bot*.) assemblage of individuals of a given taxon, usually qualified by a prefix, as ecodeme, gamodeme, topodeme, all *q.v*.; (*zool*.) a gamodeme, a local population unit of a species within which breeding is completely random.

demersal *a*. living on or near the bottom of sea or lake.

demersed *a*. growing under water, *appl.* parts of plants.

demibranch hemibranch *q.v*.

demifacet *n*. part of facet of parapophysis when divided between centra of two adjacent vertebrae.

demilunes *n.plu*. crescentic cells; crescentic bodies of mucous infoldings of salivary glands.

demisheath *n*. one of paired protecting covers of insect ovipositor.

demographic society a society that is relatively stable throughout time, being relatively closed to newcomers and whose composition is therefore the result largely of the demographic processes of birth and death.

demography *n*. the study of numbers of organisms in a population and their variation over time.

Demospongia *n*. a class of sponges (Porifera *q.v*.) which may have silica spicules, in the form of simple needles or a four-armed spicule whose points describe a tetrahedron, or may have no spicules, and which often have the body wall strengthened by a tangled mass of spongin fibres (e.g. in the bath sponge *Spongia*). They are found on shores, and down to depths of more than 5000 m.

denatant *a*. swimming, drifting, or migrating with the current.

denaturation *n*. alteration in the structural properties of a macromolecule such as a protein or a nucleic acid, leading to loss of function, as a result of heating, change in pH, irradiation, etc. In most cases denaturation refers to the disruption of noncovalent bonding leading to loss of secondary structure (e.g. unfolding of a protein chain or separation of the two strands of a DNA double helix).

dendriform *a*. branched like a tree.

dendrite *n*. fine cytoplasmic processes on neurones that receive signals from other neurones.

dendritic *a*. much branched; resembling, *pert*., or having dendrites or dendrons.

dendritic cell type of non-lymphocyte cell in some lymphoid tissues, which acts as an antigen-presenting cell.

dendrochronology *n*. the study of the age of trees and timber, generally by counting tree-rings, and the study and analysis of tree-rings in relation to changes in climate over time.

Dendrogaea *n*. a biogeographical region including all the neotropical region except temperate South America.

dendrogram *n*. any branching tree-like diagram illustrating the relationship between organisms or objects.

dendroid *a*. tree-like, much branched.

dendrology *n*. the study of trees.

dendrometer *a*. a device for measuring small changes in the diameter of a tree trunk, such as the minute amounts of shrinkage and swelling that accompany the daily fluctuations in transpiration.

dendron *n*. cytoplasmic process, usually much branched (into dendrites) that conducts impulses towards the nerve cell body.

denervation *n*. removal of the nerve supply to an organ, muscle, etc.

denitrification *n*. the conversion of nitrate to nitrite and nitrite to molecular nitrogen leading to the loss of nitrogen from the biosphere, carried out by a few genera of anaerobic bacteria; the reduction of nitrates to nitrites and ammonia, as in plant tissues.

denitrifier *n*. any of a group of diverse anaerobic bacteria capable of converting nitrate to nitrite and nitrite to molecular nitrogen (denitrification), e.g. *Pseudomonas, Achromobacter, Thiobacillus* and *Micrococcus* spp.

dens *n*. tooth or tooth-like process. *plu.* dentes.

density gradient centrifugation procedure for separating cell components or macromolecules by centrifugation in a sucrose or caesium chloride solution of graded density, in which a component of a particular density will collect at the band of identical density within the solution.

density-dependent *a. appl.* factors limiting the growth of a population which are dependent on the existing population density; *appl.* selection that either favours or disfavours the rarer forms of individual within a population.

density-independent *a. appl.* population variables that are independent of the existing density of the population.

dental *a. pert.* teeth, *appl.* pulp, the inner soft tissue of tooth, innervated and supplied with blood vessels, *appl.* papilla, the small mass of undifferentiated tissue from which a tooth forms.

dental formula method of representing the number of each type of tooth in a mammal, consisting of a series of fractions, the numerators representing the number of each type of tooth in one half of the upper jaw, and the denominators the number in the corresponding lower jaw.

dentaries, dentary *n.* bone(s) in lower jaw of many vertebrates.

dentate *a.* toothed; with large saw-like teeth on the margin.

dentate-ciliate with teeth and hairs on the margin, *appl.* leaves.

dentate-crenate with rounded teeth on the margin, *appl.* leaves.

dentes *plu.* of dens *q.v.*

denticidal *a. appl.* dehiscent fruit with tooth-like formation around top of capsule, as in the dicot family Caryophyllaceae.

denticle *n.* small tooth-like process; a type of scale present in many fish and covering the whole of the body surface in elasmobranchs, having a shape and structure similar to a small tooth.

denticulate *a.* (*zool.*) having denticles; (*bot.*) with tiny teeth on the margin.

dentin dentine *q.v.*

dentinal *a. pert.* dentine; *appl.* tubules: the minute canals in dentine of teeth.

dentine *n.* hard elastic substance, also called ivory, with same constituents as bone (collagen and calcium salts), constituting the interior hard part of vertebrate teeth and outer layer of denticles and dermal bone.

dentition *n.* the type, number and arrangement of teeth.

deoxyadenosine, deoxycytidine, deoxyguanosine nucleosides consisting of the relevant purine or pyrimidine base linked to the sugar deoxyribose. *see also* thymidine.

deoxyadenosine triphosphate (dATP) deoxyribonucleotide of adenine, one of the four deoxyribonucleotides needed for DNA synthesis.

deoxyadenylate *n.* deoxyadenosine monophosphate (dAMP), a product of partial hydrolysis of DNA.

deoxycytidine triphosphate (dCTP) deoxyribonucleotide of cytosine, one of the four deoxyribonucleotides needed for DNA synthesis.

deoxycytidylate deoxycytidine monophosphate (dCMP), a product of partial hydrolysis of DNA.

deoxyguanosine triphosphate (dGTP) deoxyribonucleotide of guanine, one of the four deoxyribonucleotides needed for DNA synthesis.

deoxyguanylate *n.* deoxyguanosine monophosphate (dGMP), a product of partial hydrolysis of DNA.

deoxyribonuclease (DNase) *n.* any of various enzymes that cleave DNA into shorter oligonucleotides or degrade it completely into its constituent deoxyribonucleotides. *alt.* nuclease. *see* endonuclease, exonuclease.

deoxyribonucleic acid *see* DNA and associated entries in body of dictionary.

deoxyribonucleoside *see* nucleoside.

deoxyribonucleotide *n.* a nucleotide containing the sugar deoxyribose.

deoxyribose *n.* a pentose sugar similar to ribose but lacking an oxygen atom, present in DNA. *see* Appendix 1 (5) for structure.

deoxythymidine thymidine *q.v.*

deoxythymidylate (deoxythymidylic acid) thymidine monophosphate *q.v.*

deperulation *n.* the pushing apart or throwing off of bud scales.

dephosphorylation *n.* the removal of a phosphate group. Often refers to the dephosphorylation of proteins by protein phosphatases, an important mechanism of control of cellular activities.

depigmentation *n.* the loss of pigment by

a cell.

depilation *n*. loss of hairy covering, as of plants when maturing.

deplanate *a*. levelled; flattened.

deplasmolysis *n*. re-entry of water into a plant cell after plasmolysis and reversal of shrinkage of protoplasm.

deplumation *n*. moulting, in birds.

depolarization *n*. reduction in electrical potential difference across a membrane.

depressant *a*. anything that lowers activity.

depressed *a*. flattened dorsoventrally.

depressomotor *a*. *appl*. any nerve that lowers muscular activity.

depressor *n*. any muscle that lowers or depresses any structure; *appl*. a nerve that lowers the activity of an organ; *appl*. compounds that slow down metabolic rate.

derepressed *a*. *appl*. genes which have been switched on from a previously repressed state. *alt*. induced.

derived *a*. *appl*. character or character state not present in the ancestral stock.

dermal *a*. *pert*. dermis or more generally, to skin.

dermal bone bone derived from the dermis of skin.

dermal tissue system of higher plants, the epidermis and associated tissues.

dermalia *n.plu*. small hard plates in the surface layer of certain sponges.

Dermaptera *n*. an order of insects, commonly called earwigs, having cerci modified as forceps, small leathery forewings and membranous hindwings. Undergo a slight metamorphosis.

dermatan sulphate glycosaminoglycan containing *N*-acetylgalactosamine, constituent of extracellular matrix in skin, blood vessels and other organs.

dermatic dermal *q.v.*

dermatitis *n*. inflammation of the surface of the skin.

dermato- *see* dermo-.

dermatogen *n*. the young epidermis in plants; tissue giving rise to the epidermis in plants.

dermatoglyphics *n*. the pattern of whorls or prints of skin, palm, finger, toe, or sole of foot.

dermatoid *a*. resembling a skin; functioning as a skin.

dermatomes *n.plu*. lateral parts of mesoderm in early vertebrate embryo that develop in connective tissue of dermis; areas of skin supplied by individual spinal nerves.

dermatomycosis *n*. a fungal infection of human or animal skin.

dermatophyte *n*. any fungal parasite of skin.

dermatopsy *n*. having a skin sensitive to light.

dermatosis *n*. any disease of the skin.

dermatosome *n*. a unit of cellulose in plant cell wall.

dermatozoon *n*. any animal parasite of the skin.

dermethmoid supraethmoid *q.v.*

dermic dermal *q.v.*

dermis *n*. the deeper layers of vertebrate skin underlying the epidermis, from which it is separated by a basal membrane. Composed of connective tissues and derived from the mesoderm. *see* Fig. 10.

dermo-, dermato-, -derm, -dermis, -dermal word elements from Gk. *dermis*, skin.

dermomyotome *n*. the dorsilateral part of mesodermal somites in embryo.

dermo-ossification *n*. a bone formed in the skin.

dermopharyngeal *n*. a superior or inferior plate of membrane bone supporting pharyngeal teeth in some fishes.

dermophyte dermatophyte *q.v.*

dermosclerites *n.plu*. masses of spicules found in tissues of some alcyonarians.

dermosphenotic *n*. a bone located around the orbit of eye, between supraorbitals and suborbitals, in teleost fishes.

dermotrichia *n.plu*. fin-rays made of dermal bone.

descending *a*. directed downwards, or towards caudal region, *appl*. blood vessels or nerves; (*bot*.) growing or hanging downwards; *appl*. arrangement of parts of a flower where each petal overlaps the one in front of it.

desegmentation *n*. fusion of segments previously separate.

desensitization *n*. loss of regulatory sites by an enzyme while retaining full catalytic activity.

desert *n*. biome where the average amount

of precipitation is erratic and less than 25 cm per annum, and evaporation exceeds precipitation. Such areas have sparse highly adapted vegetation, e.g. cacti, succulents and spiny shrubs. Hot deserts such as the Sahara have very high daytime temperatures. Cold deserts such as the Gobi and the northern Californian desert have very low winter temperatures.

deserticolous *a*. living in the desert.

desertification *n*. the conversion of pastureland and crop land into desert, or the gradual enlargement and encroachment of deserts into formerly marginal arid lands, caused by climatic factors such as prolonged drought and by overgrazing and overcultivation.

desmergate *n*. type of ant intermediate between a worker and a soldier.

desmids *n.plu*. a group of unicellular or colonial green freshwater algae whose cells are typically almost divided in two by a narrow constriction of the cell wall.

desmin *n*. a component protein of intermediate filaments in muscle cells.

desmoid *a*. band-like; forming a chain or ribbon; resembling desmids.

desmology *n*. the anatomy of ligaments.

desmoneme *n*. a nematocyst in which the distal end of the thread or closed tube coils around prey when discharged.

desmosome *n*. point at which an animal cell is strongly anchored to its neighbours, belt desomosomes being continuous bands of contact extending around the cell, spot desomsomes being roughly circular points of contact and hemidesmosomes resembling spot desmosomes but joining the cell to the basal lamina rather than to another cell. All desmosomes are characterized by thickening of apposed cell membranes separated by an enlarged intercellular space filled with various types of filamentous material connecting the two cells.

despotism *n*. social system in animals in which one individual dominates the rest of the flock which are all equally subservient to him and of equal rank with each other. *see also* dominance systems.

desquamation *n*. shedding of cuticle or epidermis in flakes.

desynapsis, desyndesis *n*. failure of synapsis, caused by disjunction of homologous chromosomes.

determinant *see* antigenic determinant, cytoplasmic determinant.

determinate *a*. with certain limits, *appl*. growth that stops when an organism or part of an organism reaches a certain size and shape; with a well-marked edge; *appl*. inflorescence with primary axis terminated early with a flower bud; *appl*. development in which the fate of each cell is already determined by the time it is formed, *cf*. regulative.

determination *n*. the irreversible commitment of a cell to a particular developmental pathway, which occurs in many cases long before overt differentiation.

determined *a. appl*. embryonic cells once their fate has been irrevocably established.

detoxication *n*. removal of toxic substances from the body by converting them to relatively harmless compounds that are then eliminated.

detritiphage detritivore *q.v.*

detritivore *n*. organism that feeds on detritus. *alt*. detritus feeder.

detritus *n*. small pieces of dead and decomposing plants and animals; detached and broken down fragments of a structure.

detrusor *n*. the outer of three layers of the muscular coat of urinary bladder, or may refer to all three layers.

detumescence *n*. subsidence of swelling.

deuterium *n*. isotope of the element hydrogen with mass number 2 (^{2}H), which is the hydrogen isotope in heavy water, deuterium oxide, D_2O.

deuterocerebrum *n*. that portion of crustacean brain from which antennular nerves arise.

deuterocoel coelom *q.v.*

deuterocone *n*. cusp on mammalian premolar teeth corresponding to the protocone of molar teeth.

deuteroconidium *n*. one of the conidia produced by division of protoconidium in dermatophytes (fungal parasites of skin).

deuterogamy *n*. secondary fertilization; pairing substituting for the union of gametes, as in fungi.

deuterogenesis *n*. second phase of embryonic development, involving growth in length and consequent bilateral symmetry.

Deuterolichenes *n*. class of lichens in

which a deuteromycete is the fungal partner.

Deuteromycota, Deuteromycotina, deuteromycetes *n.*, *n.*, *n.plu.* large group of fungi, known only in the asexual, conidia-bearing form, but which display strong affinities to the ascomycetes. They may be ascomycetes that have lost their ascus stage or in which the sexual phase has not yet been discovered. *alt.* Fungi Imperfecti. *see* Appendix 3.

deuterostoma *n.* a mouth formed secondarily in embryonic development, as distinct from the gastrula mouth.

deuterostomes *n.plu.* animals with a true coelom, radial cleavage of the egg, and in which the blastopore becomes the anus (pogonophorans, hemichordates, echinoderms, urochordates and chordates).

deuterotoky *n.* parthenogenesis where both sexes are produced.

deuterotype *n.* the specimen chosen to replace the original type specimen for designation of a species.

deutocerebrum *n.* portion of insect brain derived from fused ganglia of antennary segment of head.

deutomerite *n.* the posterior division of certain gregarines. *cf.* primite.

deutonymph *n.* the second nymphal stage or instar, either chrysalis-like or motile, in development of mites and ticks.

deutoscolex *n.* a secondary scolex produced by budding, in bladderworm stage of certain tapeworms.

deutosporophyte *n.* second sporophyte phase in life cycle of red algae.

deutosternum *n.* sternite of segment bearing pedipalps in mites and ticks.

development *n.* in biology, the changes that occur as a multicellular organism develops from a single-celled zygote, from the first cleavage of the fertilized ovum until maturity. *alt.* ontogeny.

developmental *a. pert.* or involved in development, *appl.* genes, hormones, etc. specifically active during development.

Devonian *a. pert.* or *appl.* geological period lasting from about 400 to 360 million years ago.

dexiotropic *a.* turning from right to left, as whorls, *appl.* shells.

dextral *a.* on or *pert.* the right.

dextran *n.* any of a variety of storage polysaccharides (usually branched) made of glucose residues joined by α-1,6 linkages, found in yeast and bacteria.

dextrin *n.* any of a group of small soluble polysaccharides, partial hydrolysis products of starch.

dextrorse *a.* growing in a spiral which twines from left to right; clockwise.

dextrose glucose *q.v.*

diabetes *n.* condition characterized by abnormally high levels of blood glucose, eventually resulting in coma and death if not controlled. May be due to various causes, e.g. non-production of the hormone insulin which stimulates glucose uptake, inability of cells to respond to insulin, or excessive production of hormones with antagonistic action to insulin.

diabetogenic *a.* causing diabetes.

diachaenium *n.* each part of a cremocarp.

diachronous *a.* dating from different periods, *appl.* fossils occurring in the same geological formation but of different ages.

diacoel(e) *n.* the third ventricle of the brain.

diacranteric *a.* with a distinct space between front and back teeth, as in snakes.

diactinal *a.* having two rays pointed at each end.

diacylglycerol *n.* an intracellular second messenger produced by the action of phospholipase C on membrane phosphatidylinositol phosphate in response to stimulation of cell-surface receptors by specific growth factors, hormones and neurotransmitters. Its main activity appears to be the activation of protein kinase C. *see* Appendix 1 (36) for chemical structure.

diacytic *a. appl.* stomata of a type with one pair of subsidiary cells, with their common axis at right angles to the long axis of the guard cells, surrounding the stoma.

diadelphous *a.* having stamens in two bundles due to fusion of filaments.

diadematoid *a.* of sea urchins, having three primary pore plates with occasionally a secondary between aboral and middle primary.

diadromous *a.* having veins radiating in a fan-like manner, *appl.* leaves; migrating between fresh and sea water.

diaene *n.* a two-pronged spicule.

diageotropism *n.* a growth movement in a

plant organ so that it assumes a position at right angles to the direction of gravity.

diagnosis *n*. a concise description of an organism with full distinctive characters; identification of a physiological or pathological condition by its distinctive signs.

diagnostic *a*. distinguishing; differentiating the species, genus, etc. from others similar.

diaheliotropism *see* diaphototropism.

diakinesis *n*. the last stage of meiotic prophase in which the nuclear membrane breaks down.

dialect *n*. local variant of bird songs, mating calls, bee waggle dances, etc.

diallelic *a. pert*. or involving two alleles, *appl*. polyploid with two different alleles at a locus.

dialysate *n*. any substance which passes through a semipermeable membrane during dialysis. *alt*. diffusate.

dialysis *n*. separation of colloids (such as proteins) from small molecules and ions by the inability of the larger molecules to pass through a semipermeable membrane.

dialystelic *a*. having the steles in the stem remaining more or less separate. *n*. dialystely.

diamine *n*. compound containing two amino groups, such as cadaverine or putrescine.

diaminopimelic acid (DAP) amino acid structurally similar to lysine, found in the peptidoglycans of bacterial cell walls.

diancistron *n*. a spicule in the shape of the letter sigma, but the inner margin of both hook and shaft thins to a knife-edge and is notched.

diandrous *a*. having two free stamens; in a moss, having two antheridia surrounded by each bract.

dianthovirus group plant virus group containing isometric single-stranded RNA viruses, type member carnation ringspot virus.

diapause *n*. a spontaneous state of dormancy occurring in the lives of many insects, esp. larval stages.

diapedesis *n*. migration of white blood cells from capillaries into surrounding tissue.

Diapensales *n*. order of dicot trees and herbs including the families Ebenaceae (ebony), Sapotaceae (sapote), Styracaceae (styrax) and others.

diaphorase *n*. general name for an enzyme that catalyses the oxidation of NADH or NADPH.

diaphototropism *n*. a growth movement in plant organs to assume a position at right angles to rays of light; when the light is sunlight known as diaheliotropism.

diaphragm(a) *n*. a sheet of muscle and tendon separating chest cavity from abdominal cavity in mammals and whose movement aids breathing; various other partitions in other organisms, such as the fibromuscular abdominal septum enclosing perineural sinus in some insects, transverse septum separating cephalothorax from abdomen in some arachnids.

diaphysis *n*. shaft of limb bone; abnormal growth of an axis or shoot.

diaplexus *n*. choroid plexus of 3rd ventricle of brain.

diapophysis *n*. lateral or transverse process of neural arch of a vertebra.

diapsid *a*. having skull with both dorsal and ventral temporal fenestrae on each side.

Diapsida *n*. subclass of reptiles with diapsid skulls, known from the late Carboniferous, to which most modern reptiles belong, and also including extinct forms such as the dinosaurs.

diarch *a*. with two xylem and two phloem bundles; *appl*. root in which protoxylem bundles meet and form a plate of tissue across cylinder with phloem bundle on each side.

diarthric biarticulate *q.v.*

diarthrosis *n*. an articulation allowing considerable movement, a moveable joint.

diaschistic *a. appl*. type of tetrads produced by one transverse and one longitudinal division during meiosis.

diaspore *n*. any spore, seed, fruit, or other part of plant when being dispersed and able to produce a new plant.

diastase *n*. original name for a mixture of α- and β-amylase, then for a time in 19th century used for any enzyme, later used for α- or β-amylase.

diastasis *n*. rest period preceding systole in heart contraction; abnormal separation of parts that are usually joined together.

diastema *n*. a toothless space usually between two types of teeth.

diaster *n*. the stage in mitosis where daughter chromosomes are grouped near spindle poles ready to form new nuclei.

diastole *n.* rhythmical relaxation of the heart; rhythmical expansion of a contractile vacuole. *cf.* systole.

diastomatic *a.* through stomata or pores; giving off gases from spongy parenchyma through stomata.

diathesis *n.* a constitutional predisposition to a type of reaction, disease or development.

diatom *n.* common name for a member of the class Bacillariophyceae, a group of algae characterized by delicately marked thin double shells of silica.

diatomaceous *a.* containing the shells of diatoms, *appl.* earth.

diatomin phycoxanthin *q.v.*

diatropism *n.* tendency of organs or organisms to place themselves at right angles to line of action of stimulus.

diauxic growth growth occurring in two phases separated by a period of inactivity.

diauxy *n.* adaptation of a microorganism to the use of two different sugars in a culture medium, having the constitutive enzymes for digestion of one sugar and inducing synthesis of the enzymes for the other.

diaxon *a.* with two axes.

diaxonic *a. appl.* neurones with two axons.

diazotroph *n.* organism able to fix elemental nitrogen to ammonia.

dibranchiate *a.* with two gills.

dicaryo- dikaryo-.

dicemyid *n.* mesozoan parasite of the kidneys of cephalopod molluscs. *see* Mesozoa.

dicentral *a. appl.* central canal in fish vertebral column.

dicentric *a. appl.* chromosomes or chromatids possessing two centromeres, which often leads to chromosome breakage at mitosis as the two centromeres are pulled towards opposite poles.

dicerous *a.* with two horns; with two antennae.

dichasium *n.* a cymose inflorescence in which two lateral branches occur at about same level. *alt.* dichasial cyme.

dichlamydeous *a.* having both calyx and corolla.

dichocarpous *a.* with two forms of fructification, *appl.* certain fungi.

dichogamy *n.* maturing of sexual elements at different times, ensuring cross-fertilization. *a.* dichogamous.

dichophysis *n.* a rigid, dichotomous hypha, as in hymenium and trama of some fungi.

dichoptic *a.* with eyes quite separate. *cf.* holoptic.

dichotomy *n.* branching that results from division of growing point into two equal parts; repeated forking. *a.* dichotomous.

dichroism *n.* property displayed by some substances of showing two colours, as one by transmitted and one by reflected light. *a.* dichroic.

dichromatic *a.* showing dichromatism; seeing only two colours.

dichromatism *n.* condition in which members of a species show one of only two distinct colour patterns. *a.* dichromatic.

diclesium *n.* a multiple fruit or anthocarp formed from an enlarged and hardened perianth.

diclinous *a.* with stamens and pistils on separate flowers; with staminate and pistillate flowers on same plant; with antheridia and oogonia on separate hyphae.

dicoccus *a.* having two joined capsules, each containing one seed.

dicots dicotyledons *q.v.*

Dicotyledones, dicotyledons *n., n.plu.* a class of flowering plants having an embryo with two cotyledons (seed leaves), parts of the flower usually in 2s or 5s or their multiples, leaves with net veins, and vascular bundles in the stem in a ring surrounding a central pith. In some classifications called the Magnoliopsida.

dicotyledonous *a. pert.* dicotyledons; *pert.* an embryo with two cotyledons (seed leaves).

dicratic *a.* with two spores of a meiotic tetrad being of one sex and the other two of the opposite sex.

dictyodromous *a.* net-veined, when the smaller veins branch and anastomose freely.

Dictyoptera *n.* order of insects including the cockroaches and praying mantises, winged but often non-flying, with long antennae, biting mouthparts, tough narrow forewings and broad membranous hindwings.

dictyosome Golgi body *q.v.*

dictyospore *n.* a spore with both vertical

and horizontal septa.

dictyostele *n.* a network formed by leaf traces; stele having large overlapping leaf gaps which dissect the vascular system into strands, each having phloem surounding xylem.

Dictyostelium discoideum a cellular slime mould, widely used as a model system for developmental and genetic studies.

dictyotene *n. appl.* stage of meiosis in which diplotene is prolonged, as in oocytes during yolk formation.

dicyclic *a.* with two whorls; biennial, *appl.* herbaceous plants.

Dicyemida(e) *n.* class of Mesozoa having a body that is not annulated.

dicystic *a.* with two encysted stages.

didactyl *a.* having two fingers, toes or claws.

didelphic *a.* having a paired uterus, as in certain nematodes; having two uteri, as in marsupials.

diductor *n.* muscle running posterior to axis of hinge and joining the two valves of the shell in articulate brachiopods, and which opens shell as it contracts.

Didymelales *n.* order of dicot trees comprising the family Didymelaceae with a single genus *Didymeles*.

didymous *a.* growing in pairs.

didynamous *a.* with four stamens, two long and two short.

dieback *n.* population crash *q.v.*; death of stems of woody plants from the tip backwards.

diecdysis *n.* in the moulting cycle of arthropods, a short period between pro- and metecdysis.

diecious *alt.* spelling of dioecious *q.v.*

diel *a.* during or *pert.* 24 hours; occurring at 24-hour intervals.

diencephalon *n.* part of forebrain underlying the cerebral hemispheres and containing the thalamus and hypothalamus.

diestrus *alt.* spelling of dioestrus *q.v.*

differentiated *a. appl.* cells that have developed their final specialized structure and function, as muscle, nerve, red blood cell, epidermal cell, etc.

differentiation *n.* in general sense, the increasing specialization of organization of the different parts of an embryo as a multicellular organism develops from the undifferentiated fertilized egg; of cells, the development of cells with specialized structure and function from unspecialized precursor cells, which occurs in embryonic development and in the subsequent replacement of certain types of cell from persisting unspecialized stem cells.

differentiation antigens antigens found on the surface of cells and which appear to be specific to different tissues.

diffluence *n.* disintegration by vacuolation.

diffraction colours colours produced not by pigment but by unevenness on the surface of an organism resulting in the diffraction of light reflected from it.

diffusate dialysate *q.v.*

diffuse *a.* widely spread; not localized; not sharply defined at the margin.

diffuse-porous *appl.* wood in which vessels of approximately the same diameter tend to be evenly distributed in a growth ring. *cf.* ring-porous.

diffusion *n.* the free passage of molecules, ions, etc. from a region of high concentration to a region of low concentration.

diffusion pressure deficit suction pressure *q.v.*

digalactosyl diacyglycerol a glycolipid found in plant cell membranes.

digametic heterogametic *q.v.*

digastric *a. appl.* muscles fleshy at ends with a central portion of tendon; *appl.* one of the suprahyoid muscles; *appl.* a branch of the facial nerve; *appl.* a lobule of cerebellum.

digenean *a. appl.* parasitic flatworms of the order Digenea, which include liver, blood and gut flukes such as *Schistosoma*, the cause of schistosomiasis in humans. As adults they are endoparasites of many vertebrates. They have complex life cycles with larval stages in molluscs and sometimes also in several other different hosts. *see also* cercaria, metacercaria, miracidium.

digenesis *n.* alternation of sexual and asexual generations.

digenetic *a. pert.* digenesis; requiring an alternation of hosts, *appl.* parasites.

digenic *a. pert.* or controlled by two genes.

digeny *n.* sexual reproduction.

digestion *n.* the process whereby nutrients are rendered soluble and capable of being absorbed by the organism or by a cell, by the action of various hydrolytic enzymes that break down proteins, large carbohydrates, fats, etc.

digestive *a. pert.* digestion, or having power of aiding in digestion.

digestive gland sac-like portion of intestine in molluscs and other invertebrates, which produces digestive enzymes and in which food is digested.

digestive vacuole type of lysosome resulting from phagocytosis of large particles in animal cells.

digger wasps solitary insects of the superfamilies Pompiloidea and Sphecoidea of the order Hymenoptera, somewhat resembling the true wasps in appearance, and including species that nest in burrows they dig in the ground.

digit *n.* terminal division of limb as finger or toe, in vertebrates other than fishes.

digital *n.* the terminal joint of pedipalp of spider.

digitaliform *a.* finger-shaped, *appl.* flowers that are shaped like the fingers of a glove, e.g. foxglove.

digitalin *n.* a glycoside from leaves of foxglove *Digitalis purpurea*, stimulates the heart and is used in the treatment of heart disease.

digitate *a.* having parts arranged like fingers of a hand; having fingers.

digitiform *a.* finger-shaped.

digitigrade *a.* walking with only digits touching the ground.

digitinervate *a.* having veins radiating out from base like the fingers of a hand, with usually five or seven veins, *appl.* leaves.

digitonin *n.* a glycoside from the leaves of foxglove.

digitoxin *n.* a glycoside from the leaves of foxglove.

digitule *n.* any small finger-like process.

digitus digit *q.v.*

diglyphic *a.* having two siphonoglyphs.

digoneutic *a.* breeding twice a year.

digynous *a.* with two carpels.

dihaploid *n.* an organism arising from a tetraploid but only containing half a normal tetraploid chromosome complement.

dihybrid *n.* the progeny of a cross in which the parents differ in two distinct characters; an organism heterozygous at two distinct loci.

dihydroxyacetone phosphate (DHAP) 3-carbon ketose monosaccharide phosphate intermediate in photosynthetic carbon dioxide fixation, glycolysis etc., important in cellular respiratory metabolism as part of the glycerol phosphate shuttle.

dihydrofolate reductase (DHFR) enzyme catalysing the regeneration of tetrahydrofolate from dihydrofolate, and which is inhibited by the folate antagonists methotrexate (amethopterin) and aminopterin.

dihydrotachysterol *n.* vitamin D_4, an irradiation product of the dihydro derivative of ergosterol, which counteracts impaired parathyroid function.

dihydrouridine *n.* unusual nucleotide found in tRNA, formed by addition of hydrogen to positions 5 and 6 of uracil, saturating the double bond.

dikaryon *n.* a pair of nuclei situated close to one another and dividing at the same time, as in some hyphae; the hypha containing such a pair of nuclei. *a.* dikaryotic.

dikont biflagellate *q.v.*

dilambodont *a. appl.* insectivores having molar teeth with W-shaped ridges.

dilatation *n.* dilation.

dilatator dilator *q.v.*

dilated *a.* expanded or flattened; *appl.* parts of insects, plants, etc. with a wide margin.

dilator *n.* any muscle that expands or dilates an organ.

Dilleniales *n.* order of woody, often climbing dicots, comprising the families Crossosomataceae and Dilleniaceae.

diluvial *a.* produced by a flood, *appl.* soil deposits.

dimastigote biflagellate *q.v.*

dimegaly *n.* condition of having two sizes, *appl.* sperm and ova.

dimer *n.* protein made up of two subunits.

dimeric *a.* of a protein, having two subunits; having two parts, bilaterally symmetrical.

dimerous *a.* in two parts; having each whorl of two parts, of flowers; with a two-jointed tarsus.

dimixis *n.* fusion of two kinds of nuclei, in heterothallism.

dimorphic *a.* having or *pert.* two different forms.

dimorphism *n.* condition of having two distinct forms within a species, of having two distinct sizes of stamens, two different kinds of leaves, etc.

dinergate *n.* soldier ant.

dineuronic *a.* with double innervation, *appl.* chromatophores with two sets of nerve fibres, one directing dispersion and the other concentration of pigment.

dinitrogen reductase the iron-containing protein of the bacterial nitrogenase complex, which is thought to transfer electrons from ferredoxin to the Mo-Fe protein component of nitrogenase.

dinitrophenol (DNP) small lipid-soluble organic compound with various uses in experimental biology: as an uncoupler of oxidative phosphorylation, as a marker for certain amino acids in protein sequencing, as a hapten in experimental immunology.

Dinoflagellata, dinoflagellates *n., n.plu.* phylum of unicellular protists (class Phytomastigophorea in animal classification or Pyrrophyta (Dinophyta) in plant classification), having two flagella, one pointing forwards, the other forming a girdle around the body. A major component of marine and freshwater plankton. Some are autotrophic and photosynthetic, some are heterotrophic.

dinokaryotic *a.* term sometimes applied to the cellular organization of dinoflagellates which especially in their nuclear and chromosome structure differ from a typical eukaryotic or prokaryotic type of organization.

Dinophyta Dinoflagellata *q.v.*

Dinornithiformes *n.* an order of flightless birds from New Zealand, including the moas.

dinosaur *n.* a member of either of two orders of reptiles that flourished during the Mesozoic, the Saurischia, the lizard-hipped dinosaurs, or the Ornithischia, the bird-hipped dinosaurs. The Saurischia included both bipedal carnivores and very large quadrupedal herbivores. The Ornithischia were mostly quadrupedal and all herbivorous.

dinucleotide *n.* two nucleotides linked together by a 3′,5′-phosphodiester bond.

diocoel *n.* cavity of the diencephalon, particularly in embryo.

dioecious *a.* having the sexes separate; having male and female flowers on different individuals. *n.* dioecism.

dioestrus *n.* the quiescent period between periods of heat in animals with more than one period of fertility each year.

dionychous *a.* having two claws, as tarsi of certain spiders.

dioptrate *a.* having eyes or ocelli separated by a narrow line.

dioptric *a. pert.* transmission and refraction of light, *appl.* structures in eye, as cornea, lens, aqueous and vitreous humor.

diorchic *a.* having two testes.

diosgenin *n.* a complex steroid obtained from certain species of yam and which can be converted into 16-dehydropregnenolone, one of the main active ingredients in oral contraceptives.

dipeptide *n.* two amino acids linked by a peptide bond. *see* Appendix 1 (30).

dipetalous *a.* having two petals.

diphasic *a.* having two distinct states, *appl.* life cycle, etc.

diphosphatidyl glycerol any phosphoglyceride with glycerol as the alcohol group, found chiefly in plants, cardiolipin being one of the few examples known from animals.

2,3-diphosphoglycerate (DPG) compound present in red blood cells in equimolar amounts with haemoglobin, binds to deoxyhaemoglobin reducing its affinity for oxygen and allowing oxygen unloading in tissues.

diphtheria toxin protein toxin from *Corynebacterium diphtheriae* which inhibits the action of elongation factor eEF2 in eukaryotic cells by catalysing ADP-ribosylation of the factor, and thus blocks protein synthesis in eukaryotic cells.

diphycercal *a.* with a caudal fin in which the vertebral column runs straight to tip, dividing the fin symmetrically.

diphygenetic *a.* producing embryos of two different types.

diphyletic *a. pert.* or having origin in two separate lines of descent.

diphyllous *a.* having two leaves.

diphyodont *a.* with deciduous and permanent sets of teeth, i.e. two sets of teeth.

diplanetic *a.* with two distinct kinds of zoospores.

diplanetism *n.* condition of having two periods of motility in one life history, as of zoospores in some fungi.

dipleurula *n.* a bilaterally symmetrical larva of echinoderms.

diplobiont *n.* organism characterized by at least two kinds of individual in its life-cycle, such as sexual and asexual. *cf.* haplobiont.

diplobivalent *n.* a bivalent containing two anomalous doubly duplicated chromosomes and hence eight chromatids.

diploblastic *a.* having only two germ layers, e.g. endoderm and ectoderm, as coelenterates and sponges. *cf.* triploblastic.

diplocardiac *a.* with the two sides of the heart quite distinct.

diplocaryon diplokaryon *q.v.*

diplocaulescent *a.* with secondary stems and branches.

diplochlamydeous dichlamydeous *q.v.*

diplochromosome *n.* anomalous chromosome, having four chromatids instead of two, attached to centromere.

diplococcus *n.* any of a genus *(Diplococcus)* of parasitic bacteria that occur usually in pairs and include some serious disease-causing agents.

diplocyte *n.* a cell having conjugate nuclei.

diplogangliate *a.* with ganglia arranged in pairs.

diplogenesis *n.* development of two parts instead of a single part.

diplohaplont *n.* an organism with alternation of diploid and haploid generations.

diploic *a.* occupying channels in cancellous tissue of bone.

diploid *a. appl.* organisms whose cells (apart from the gametes) have 2 sets of chromosomes, and therefore 2 copies of the basic genetic complement of the species, designated 2*n*; *n.* a diploid organism or cell.

diploid arrhenotoky the production of diploid males from unfertilized eggs, which is known only in the scale insect *Lecanium putnami*.

diploidization *n.* the doubling of chromosome number in haploid cells; the restoration of the diploid state.

diploidy *n.* the diploid state.

diplokaryon *n.* a nucleus with two diploid sets of chromosomes.

diplomycelium *n.* a diploid or dikaryotic mycelium.

diploneural *a.* supplied with two nerves.

diplont *n.* organism having diploid somatic nuclei. *alt.* diploid.

diploperistomous *a.* having a double projection or peristome.

diplophase *n.* stage in life history of an organism in which nuclei are diploid, *alt.* sporophyte phase; diplotene phase in meiosis.

diplophyll *n.* a leaf having palisade tissue on upper and lower side with intermedial spongy parenchyma tissue.

Diplopoda, diplopods *n., n.plu.* in some classifications, a class of arthropods commonly called millipedes, having numerous similar apparent segments each in fact made up of two segments and therefore bearing two pairs of legs. In some classifications considered a subclass or order of class Myriapoda.

diploptile *n.* double down feather, without a shaft, formed by precocious development of barbs of the adult feather.

diplosis *n.* doubling of chromosome number in syngamy.

diplosome *n.* a double centrosome lying outside the nuclear membrane; a paired heterochromosome.

diplosomite *n.* a body segment consisting of two annular parts, prozonite and metazonite, in diplopods.

diplospondyly *n.* the condition of having two centra to each myotome, or with one centrum and well-developed intercentrum. *a.* diplospondylic.

diplospory *n.* type of apomixis in which a diploid megaspore mother cell gives rise directly to the embryo.

diplostemonous *a.* with two whorls of stamens in regular alternation with perianth leaves; with stamens double the number of petals.

diplostichous *a.* arranged in two rows or series.

diplotegia *a.* an inferior fruit with dry dehiscent pericarp.

diplotene *n.* stage in meiosis at which bivalent chromosomes split longitudinally.

Diplura *n.* order of wingless insects with a pair of cerci and two "tails" on last segment, sometimes called two-pronged bristletails. Minute white insects with no eyes, found in soil and under stones.

dipnoan

dipnoan *a.* breathing by lungs and gills.

Dipnoi, dipnoans *n., n.plu.* group of bony fishes, commonly called lungfish, known from the Devonian, possessing lungs and broad, crushing toothplates. The three genera of modern lungfish (found in Australia, South America and Africa) are air-breathing and live in tropical areas with a dry season, and have a reduced skeleton.

diprotodont *a.* having two anterior incisors large and prominent, the rest of incisors and canines being smaller or absent.

Dipsacales *n.* order of dicot herbs and shrubs, rarely small trees, comprising the families Adoxaceae (moschatel), Caprifoliaceae (honeysuckle), Dipsacaceae (teasel) and Valerianaceae (valerian).

Diptera, dipterans *n., n.plu.* large order of insects including the housefly and other two-winged (true) flies, mosquitoes, fruitflies (*Drosophila*), having one pair of wings only, the second pair being reduced to small halteres. The eruciform larvae (grubs) undergo a complete metamorphosis.

dipterocecidium *n.* a gall caused by a dipterous insect.

dipterous *a.* with two wings, or wing-like expansions; *pert.* the Diptera (true flies).

directional selection selection that acts on one extreme of the range of variation in a particular character, and therefore tends to shift the entire population to the opposite end.

dirhinic *a.* having two nostrils; *pert.* both nostrils.

disaccharide *n.* any of a group of carbohydrates which are the condensation products of two monosaccharide units with the elimination of a molecule of water, and which include sucrose, lactose, maltose.

disarticulate *a.* separated at a joint or joints.

disassortative mating mating between organisms of unlike phenotype.

disc *n.* (*bot.*) middle portion of capitulum in Compositae; adhesive end of tendril; base of seaweed thallus; (*zool.*) area around mouth in many animals, as in starfish.

disc florets inner florets borne on much reduced stalks in many flower-heads of the Compositae type, as the yellow florets in a daisy. *cf.* ray florets.

discal *n.* a large cell at base of wing of Lepidoptera completely enclosed by veins, also in some Diptera.

disciflorous *a.* with flowers in which the receptacle is large and disc-like.

disciform discoid *q.v.*

disclimax *n.* a subclimax stage in plant succession replacing or modifying true climax, usually due to animal and human agency, e.g. cultivated crops.

discoblastic *a. pert.* blastula formed from egg with a disc-like blastoderm.

discocarp *n.* special enlargement of thalamus below calyx in certain flowers; (*mycol.*) disc-shaped apothecium *q.v.*

discocellular vein cross vein between 3rd and 4th longitudinal veins of insect wing.

discoctasters *n.plu.* sponge spicules with eight rays terminating in discs, each disc corresponding in position to the corners of a cube.

discodactylous *a.* with a sucker at end of digit.

discohexactine *n.* a sponge spicule with six equal rays meeting at right angles.

discohexaster *n.* a hexactine sponge spicule with rays ending in discs.

discoid *a.* flat and circular; disc-shaped.

discoidal *a.* disc-like; *appl.* segmentation in which blastoderm initially forms a one-layered disc or cap on top of yolk, as in eggs of birds and reptiles.

discoidins *n.* family of lectins produced by the slime mould *Dictyostelium discoideum.*

Discomycetes *n.* a group of ascomycete fungi, commonly called cup fungi, and including also the earth tongues, truffles and morels, in which the fruiting body is in the form of an apothecium, usually black or brightly coloured, which may be open and cup-shaped or disc-like, or closed and subterranean (truffles).

discontinuity *n.* occurrence in two or more separate areas of geographical regions; *appl.* layer: thermocline.

discontinuous distribution pattern of geographical distribution where the same or similar species are found in widely separated parts of the world, which is taken to indicate that the species was formerly distributed over the whole area but has become extinct in the intervening regions.

discontinuous variation variation between individuals of a population in which differences are marked and do not grade into each other, brought about by the effects of different alleles at a few major genes. *alt.* qualitative inheritance. *cf.* continuous variation.

disconula *n.* eight-rayed stage in larval development of certain coelenterates.

discoplacenta *n.* a placenta with villi on a circular cake-like disc.

discoplasm *n.* colourless stroma or framework of a red blood cell.

discous discoid *q.v.*

discrimination learning situation in which although an animal is capable of discriminating between two stimuli, it only does so after it has learned to tell them apart.

discus proligerus granular zone in a Graafian follicle, the mass of cells in which the ovum is embedded.

disease gene the mutant allele responsible for an inherited disease.

dishabituation *n.* the abolition of habituation to a particular stimulus, seen for example after administration of a strong generalized stimulus of a different type.

disinhibit *v.* to remove inhibition. *n.* disinhibition.

disjunct *a.* with body regions separated by deep constrictions; *appl.* distribution in which potentially interbreeding populations are separated by sufficient distance to preclude gene flow.

disjunction *n.* separation of paired chromosomes in anaphase of mitosis or meiosis; geographical distribution in discontinuous areas, *alt.* discontinuity.

disjunction mutants mutants in which chromosomes are divided unequally between daughter cells at meiosis.

disjunctive symbiosis a mutually helpful condition of symbiosis although there is no direct connection between the partners.

disjunctor *n.* a weak connective structure, or an intercalary cell, and zone of separation between successive conidia.

disk *alt.* spelling of disc *q.v.*

disomic *a. pert.* or having two homologous chromosomes or genetic loci. *n.* disomy.

disoperation *n.* co-actions resulting in disadvantage to individual or group; indirectly harmful influence of organisms upon each other.

dispermous *a.* having two seeds.

dispermy *n.* the penetration of an ovum by two sperm.

displacement activity the performance of a piece of behaviour, usually in moments of frustration or indecision, that is not directly relevant to the situation at hand.

display *n.* series of stereotyped movements, sounds, etc. which cause a specific response in another animal, usually of the same species, often used in courtship or territorial defence.

disporocystid *n. appl.* oocyst of sporozoans when two sporocysts are present.

disporous *a.* with two spores.

disruptive coloration colour patterns that obscure the outline of an animal and so act as camouflage and protection against predators.

disruptive selection selection that operates against the middle range of variation in a particular character, tending to split populations into two showing the extreme at either end of the range.

dissected *a.* having leaf blade cut into lobes, with incisions nearly reaching the midrib; with parts displayed.

disseminule diaspore *q.v.*

dissepiment *n.* a partition in compound plant ovary; a calcareous partition in corals; trama *q.v.*

dissilient *a.* springing open; *appl.* capsules of various plants which dehisce explosively.

dissimilation *n.* the breakdown of nutrients to provide energy and simple compounds for intermediary metabolism. *a.* dissimilatory.

dissociation constant the inverse of the equilibrium constant (k) (*q.v.*), a measure of the affinity of two molecules for each other.

dissogeny, dissogony *n.* the condition of having two sexually mature periods in the same animal, one in larva and the other in adult.

dissolved oxygen level (DO) amount of oxygen gas (O_2) dissolved in a given volume of water at a given temperature and pressure, and usually expressed as a concentration of oxygen in parts per million of water. *see also* biological oxygen demand.

distad *adv.* towards or at a position away

from centre or from point of attachment; in a distal direction.

distal *a.* far apart, distant, *appl.* bristles, etc.; *pert.* the end of any structure furthest away from middle line of organism or point of attachment; *appl.* region of a gene furthest away from the promoter. *cf.* proximal.

distalia *n.plu.* the distal or 3rd row of carpal or of tarsal bones.

distance receptor a sense organ that responds to stimuli emanating from distant objects, an olfactory, visual or auditory receptor.

distemonous *a.* having two stamens.

distichous *a.* arranged in two rows; *appl.* alternate leaves arranged so that 1st is directly below 3rd and so on.

distractile *a.* widely separated, *appl.* long-stalked anthers.

distraction display behaviour in female birds which distracts an enemy from the eggs or chicks, and often takes the form of feigning injury to entice the predator away.

distribution *n.* geographical range of a species or group of species.

disturbance climax disclimax *q.v.*

disulphide bond S–S bond formed between two cysteine residues in proteins, contributing to tertiary and quaternary structure.

dithecal *a.* two-celled.

ditokous *a.* producing two eggs or two young at one time.

ditrematous *a.* with genital and anal openings separate.

ditrochous *a.* with a divided trochanter.

ditypism *n.* occurrence or possession of two types ; sex differentiation, represented by + and –, of two apparently similar haplonts.

diuresis *n.* increased or excessive secretion of urine.

diuretic *a.* increasing the secretion of urine; *n.* any agent causing the above.

diurnal *a.* occurring every day; active in the daytime; opening in the daytime.

diurnal rhythm metabolic or behavioural rhythm with a cycle of about 24 hours.

divaricate *a.* widely divergent; forked; bifid.

divergence *n.* of protein or nucleotide sequences, the percentage of amino acid or nucleotide residues that are different in corresponding proteins or genes from different species or related proteins or genes from the same species, etc.

divergency *n.* the fraction of stem circumference, usually constant for a species, which separates two consecutive leaves in a spiral.

divergent *a.* separated from another, having tips further apart than the bases; *appl.* evolutionary change tending to produce differences between two organisms, genes, etc.

diversion behaviour distraction display *q.v.*; behaviour likely to confuse an enemy, e.g. squids ejecting a cloud of black "ink".

diversity (*imm.*) the variety of specificities displayed within the total repertoire of antigen receptors, i.e. immunoglobulins and T-cell receptors; (*ecol.*) *see* biodiversity.

diversity index of a community, the ratio between number of species and number of individuals; a measure of the biological diversity within an environment which can be used to detect stress on an environment, e.g. pollution, and which is calculated in various ways from the number of species present, sometimes in combination with their relative abundances. *cf.* biotic index.

diverticulate *a.* having a diverticulum; having short offshoots approximately at right angles to axis.

diverticulum *n.* a blind-ended tube or sac opening off a canal or cavity.

divided *a.* with leaf blade cut by incisions reaching to midrib.

division *n.* a major taxonomic grouping in plants, corresponding to a phylum in animals. Examples: Bryophyta (mosses and liverworts), Pterophyta (ferns, etc.), Spermatophyta (seed-bearing plants, the gymnosperms and angiosperms).

dixenous *a.* parasitizing or able to parasitize two host species.

dizygotic *a.* originating from two fertilized ova, as nonidentical or fraternal twins (DZ twins).

D loop single-stranded "displacement loop" seen in replication of mitochondrial and chloroplast DNA, where replication proceeds for a short length along one strand of parental DNA only, displacing the other parental strand; single-stranded loop seen in duplex DNA when invaded by a homologous single strand of DNA or RNA which pairs with one strand of the duplex displacing the other to form a loop, formed during genetic recombination and also during certain types of *in vitro* DNA and RNA hybridization reactions.

DNA deoxyribonucleic acid, very large linear molecule composed of carbon, oxygen, nitrogen, hydrogen and phosphorus, found in all living cells, the physical carrier of genetic information. A DNA molecule is made up of two complementary chains of deoxyribonucleotide subunits, and may be up to millions of nucleotides in length. Each chain is composed of nucleotides covalently linked through regularly repeating sugar (deoxyribose) phosphate ester bonds between the hydroxyl group on C5 of one deoxyribose and the phosphate group on C3 of another, and contains the nucleotide bases adenine (A), thymine (T), cytosine (C) and guanine (G). The two antiparallel chains are wound round each other to form the Watson–Crick right-handed "double helix" held together by weak specific pairing between A and T and between C and G. Genetic information is encoded in the sequence of the bases along the polynucleotide chains which form a genetic code directing the synthesis of RNAs and proteins. In eukaryotic cells DNA is present in a complex with histones, forming chromatin, which is packaged with other proteins into discrete chromosomes. DNA replicates by semiconservative replication, in which the two strands of the helix separate and two new complemen-

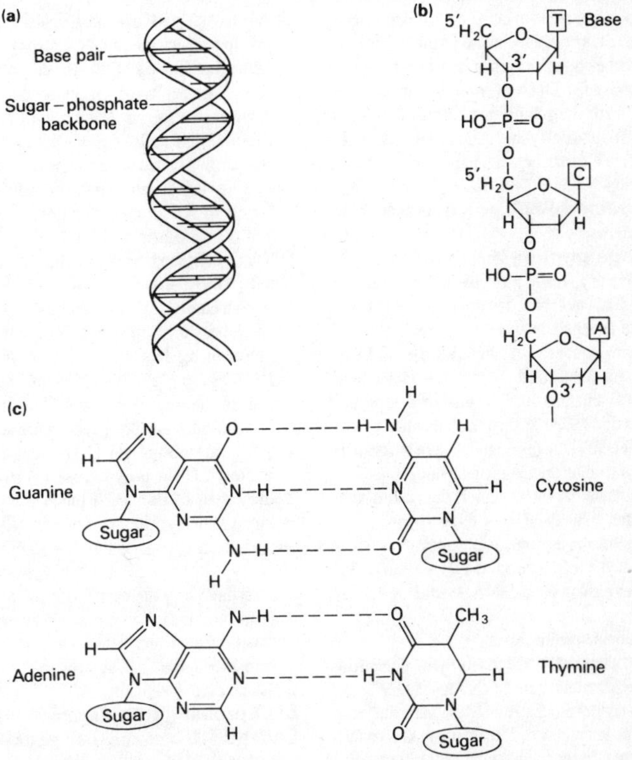

Fig. 2 (a) The double-helical structure of DNA, (b) polydeoxyribo-nucleotide chain showing the sugar–phosphate backbone, (c) hydrogen-bonded pairing between the bases.

tary strands are synthesized using the old ones as templates. DNA can occur in several configurations: B-DNA is the classical right-handed double helix with 10–10.4 nucleotide residues per complete turn and is the form generally found *in vivo*; A-DNA consists of a more tightly wound right-handed helix with around 11 residues per turn and is formed by dehydration of B-DNA; Z-DNA is a left-handed double helix proposed on the basis of the crystal structure of the duplex trinucleotide $d(CG)_3$, and containing 12 residues per turn, not yet found *in vivo*. *see also* chromatin, chromosome, DNA polymerase, gene, genetic code, transcription, translation. *see Fig.* 2.

DNA cloning the isolation and multiplication of a particular gene by incorporating it into a specially modified phage or plasmid and introducing it into a bacterial cell where the DNA of interest is replicated along with the phage or plasmid DNA and can subsequently be recovered from the bacterial culture in (relatively) large amounts.

DNA-directed RNA polymerase RNA polymerase *q.v.*

DNA fingerprinting method of ascertaining identity, family relationships, etc. by means of a DNA fingerprint which is unique to each individual. The DNA fingerprint consists of the pattern of DNA fragments obtained on restriction analysis *(q.v.)* of certain highly variable repeated DNA sequences within the genome whose number and arrangement are virtually unique to each person or animal.

DNA gyrase bacterial type II topoisomerase enzyme involved in DNA replication, which removes positive supercoils from circular double-stranded DNA by nicking and resealing the DNA backbone. *alt.* gyrase.

DNA helicase helicase *q.v.*

DNA hybridization technique for determining the similarity of 2 DNAs (or DNA and RNA) by reassociating single strands from each molecule and determining the extent of double-helix formation (indicating similar base sequences); a general method involving reassociation of complementary DNA or RNA strands used widely to iden-

tify and isolate particular DNA or RNA molecules from a mixture. *see also in situ* hybridization.

DNA library a collection of cloned DNA fragments, usually representing an entire genome or copies of the mRNAs present in a single type of cell.

DNA ligase enzyme acting on double-stranded DNA only, which joins 2 DNAs end to end or closes breaks in the sugar-phosphate backbone, acting in normal DNA synthesis and repair of damaged DNA, and which is used to splice DNAs in genetic engineering work. EC 6.5.1.1 and 6.5.1.2, *r.n.* polydeoxyribonucleotide synthetase.

DNA-mediated transfection transfection *q.v.*

DNA methylation addition of a methyl group to cytosine in DNA, usually at CG's. Modification of DNA in this way is believed to be involved in long-term gene control in differentiating and differentiated mammalian cells, genes that are available for transcription sometimes being less heavily methylated than the same genes in a cell in which they are never expressed. *see also* modification.

DNA modification modification *q.v.*

DNA polymerase any of several enzymes which catalyse DNA synthesis by addition of deoxyribonucleotide units to a DNA chain using DNA or (in the retroviruses) RNA as template. In eukaryotic cells, DNA polymerase α is reponsible for nuclear DNA synthesis, β for repair functions, and γ for mitochondrial DNA replication. In bacteria, DNA polymerase I is involved in the repair of damaged DNA, DNA III catalyses the synthesis of new DNA strands, and the function of DNA polymerase II is not yet known. Some DNA polymerases also have an editing/proofreading activity with which they can recognize and excise mispaired nucleotides. *see also* reverse transcriptase.

DNA probe *see* probe.

DNA profiling DNA fingerprinting *q.v.*

DNA puff chromosome puff generated in the salivary glands of sciarid insects which contains locally amplified sequences of DNA as well as RNA.

DNA rearrangement *see* immunoglobulin

genes, antigenic variation, mating type switch.

DNA recombination *see* recombination.

DNA repair various biochemical processes by which DNA damaged by the action of chemicals or irradiation can be restored. Altered and incorrectly matched bases are recognized by enzymes that excise them and new DNA is then synthesized by reference to the undamaged strand. Breaks in DNA are repaired by DNA ligases. The repair of certain types of damage may result in alteration of the base sequence of the DNA and thus in mutation.

DNA replication the process by which a new copy of a DNA molecule is made. The two strands of the double helix are separated and each acts as a template for the enzymatic synthesis of a new complementary strand, resulting in two new identical double-stranded DNA molecules. This type of replication is termed semiconservative. It is catalysed by DNA polymerase. *see also* Okazaki fragments.

DNA sequence order of nucleotides or base pairs in a DNA molecule, and which in protein-coding DNA determines the order of amino acids in the proteins specified. *alt.* base sequence, nucleotide sequence.

DNA sequencing determination of a DNA sequence, by either of two methods: the chemical cleavage method (developed by Maxam and Gilbert) or controlled interruption of enzymatic replication (developed by Sanger and associates) or by various automated methods developed from these. *alt.* nucleotide sequencing, base sequencing.

DNA splicing the rearrangement of DNA sequences into different combinations which occurs naturally, e.g. in the generation of antibody genes in somatic cells, and artificially in DNA cloning and genetic engineering procedures.

DNA topoisomerases enzymes that change the degree of supercoiling of DNA by introducing a transient break into the molecule, allowing limited unwinding, and then resealing it. Type I topoisomerases make a break in a single strand only; type II topoisomerases (e.g. bacterial DNA gyrase) make a double-stranded break.

DNA transcription *see* transcription.

DNA tumour viruses a group of unrelated DNA viruses that can cause cancers by various means. They include adenovirus, SV40 (simian virus 40) and polyoma, which are only tumorigenic in certain highly susceptible newborn animals, but which are widely studied in cultured cells as models of tumorigenesis, and also viruses such as Epstein-Barr virus, hepatitis B and certain human papilloma viruses, which are implicated in the development of some types of cancers in humans.

DNA-uracil glycosidase uracil-DNA glycosidase *q.v.*

DNA viruses viruses containing DNA as the genetic material and including the Adenoviridae, Herpesviridae, Poxviridae, Papovaviridae, Parvoviridae and Iridoviridae amongst vertebrate viruses, and the caulimoviruses and geminiviruses amongst the plant viruses.

dodecagynous *a.* having 12 pistils.

dodecamerous *a.* having each whorl composed of 12 parts.

dodecandrous *a.* having at least 12 stamens.

dolabriform, dolabrate *a.* axe-shaped.

dolichocephalic *a.* long-headed; with a cephalic index of under 75.

dolichofacial *a.* long-faced.

dolichohieric *a.* having a sacral index below 100.

dolichol phosphate very long-chain lipid, carrier of activated oligosaccharides for attachment to glycoproteins in lumen of endoplasmic reticulum.

dolichostylous *a. pert.* long-styled anthers in dimorphic flowers.

dolioform *a.* barrel-shaped.

doliolaria *n.* in some sea cucumbers, a barrel-shaped larva which develops from an auricularia.

dolipore *n.* the pore in the septum (cross wall) of the hyphae of Basidiomycetes, which is a barrel-shaped structure.

dolipore septum a septum which flares out in the middle of the hypha, forming a barrel-shaped structure with open sides.

Dollo's rule or law that evolution is irreversible and that structures and functions once lost are not regained.

dolphin *n.* member of the marine family Delphinidae or of the Platanistidae (river dolphins) of the suborder Odontoceti (toothed whales) of the mammalian order Cetacea *q.v.* They are slim and fast-moving with a prominent elongated snout or 'beak'.

domain *n.* structurally defined compact globular section of a protein molecule, different domains often joined by a flexible portion of polypeptide chain; length of looped chromatin in a chromosome; region of cell membrane of particular lipid and protein composition.

domatium *n.* a crevice or hollow in some plants, serving as a lodging for insects or mites.

dome cell the penultimate cell of a hyphal crozier, containing two nuclei which fuse, being the first stage in ascus formation.

dominance *n.* (*genet.*) property possessed by some alleles of solely determining the phenotype for any particular gene when present as one member of a heterozygous pair, when they mask the effects of the other allele (the recessive allele) to give a phenotype identical to that when the dominant allele is present as two copies. This phenomenon is known as complete dominance. Incomplete dominance is exhibited when the effects of the other allele are not completely masked. *see also* codominance; (*ecol.*) the extent to which a particular species predominates in a community and affects other species; (*behav.*) *see* dominance systems.

dominance frequency proportion of samples in which a particular species is predominant.

dominance systems social systems in which certain individuals aggressively dominate others. In the case where one individual dominates all the others with no intermediate ranks it is known as a despotism. In the more common dominance hierachies or social hierarchies there are distinct ranks, with individuals of any rank dominating those below them and submitting to those above them.

dominant *a. appl.* plants which by their numbers and extent determine the biotic conditions in an area; *appl.* species most prevalent in a particular community, or at a given period; *appl.* an individual which is high ranking in the social hierarchy or

peck order; *appl.* a phenotypic character state or an allele that masks an alternative character state or allele when both are present in a hybrid, *see also* heterozygote. *cf.* recessive.

Domin scale scale (1–10) used to indicate the approximate percentage cover of individual plant species in a given area, with 1 corresponding to insignificant cover, 3 corresponding to 1–5% cover, 8 to 50–75% cover, and 10 equal to 100% cover.

Donnan free space the fraction of a tissue available for ion-exchange reactions, i.e. in plant tissue the cell wall.

DOPA, dopa (L-dopa) 3,4-dihydroxy-phenylalanine, formed from tyrosine in the adrenal medulla, brain, and sympathetic nerve terminals by the enzyme tyrosine hydroxylase (*r.n.* tyrosine-3-monooxygenase, EC 1.14.16.2) and which is a biosynthetic precursor of noradrenaline, adrenaline and dopamine. Also oxidized by dopa-oxidase to a melanin precursor, as in the basal layers of skin, etc. L-dopa is used in the treatment of Parkinson's disease.

dopamine *n.* a biogenic amine, a catecholamine, a neurotransmitter in the central nervous system. Also produced in small amounts by adrenal medulla. A deficiency of dopamine as a result of destruction of dopaminergic neurones in forebrain is an underlying cause of Parkinsonism. *see* Appendix 1 (31) for structure.

dopaminergic *a. appl.* neurones that release dopamine from their terminals.

dormancy *n.* resting or quiescent condition with reduced metabolism, as in seeds, spores, etc. under unfavourable conditions for germination. *a.* dormant.

dormin abscisic acid *q.v.*

dorsad *a.* towards back or dorsal surface.

dorsal *a. pert.* or nearer back (not hind end) of an animal; upper surface of leaf, wing, etc.; *appl.* placentation of ovule, ovules attached to midrib of carpels. *cf.* ventral.

dorsal aorta major artery carrying oxygenated blood to the rest of the body in vertebrates and cephalochordates. In mammals the dorsal aorta is formed from the left branch of the fourth (systemic) aortic arch.

dorsal fin the large fin on the back of most fish and cetaceans.

dorsalis *n.* an artery which supplies the dorsal surface of any organ.

dorsal lip of blastopore *see* organizer.

dorsiferous *a.* bearing sori on back of leaf.

dorsifixed *n.* having filament attached to back of anther.

dorsigerous *a.* carrying young on back.

dorsigrade *a.* having back of digit on the ground when walking.

dorsilateral *a.* of or *pert.* back and sides.

dorsispinal *a. pert.* or referring to back and spine.

dorsiventral *a.* flattened and having upper and lower surface of distinctly different structure, *appl.* leaves.

dorsobronchus *n.* one of a set of tubes in lungs of birds which branch off the bronchi and are connected with the posterior air sacs.

dorsocentral *a. pert.* mid-dorsal surface; *pert.* aboral surface in echinoderms.

dorsolumbar *a. pert.* lumbar region of back.

dorsoventral *a. appl.* axis, from back to belly of the animal body, from upper to lower surface of a limb, leaf, etc.; *pert.* structures, axis, gradients, etc. stretching from dorsal to ventral surface.

dorsum *n.* tergum and notum of insects and crustaceans; inner margin of insect wing; the back of higher animals; upper surface, as of tongue.

dorylaner *n.* exceptionally large male ant of driver-ant group.

dosage compensation regulation of the dosage of genes carried on the sex chromosomes in the sex carrying two (or more) copies of the same chromosome. In mammals for example, one copy of the X chromosome is permanently inactivated in the somatic cells of the female.

dose equivalent a standardized measure of the effect of ionizing radiation on tissue, which is the measured absorbed dose multiplied by weighting factors for particular tissues and types of radiation, and which is expressed in sieverts (SI unit), and which represents the risk to health from that amount of radiation if it had been absorbed uniformly throughout the body.

dose response curve for any assay, the relation between the concentration of active agent (virus, hormone, enzyme, etc.) in the sample, and the quantitative response in that particular assay.

double fertilization a characteristic feature of flowering plants, in which one male haploid nucleus fuses with the polar nuclei to form the triploid primary endosperm nucleus, and the other fuses with the egg cell nucleus to produce the diploid zygote.

double helix the typical conformation of double-stranded DNA in which the two strands are wound around each other with base pairing between the strands. Also found in double-stranded RNA. *see* Fig. 2.

double recessive cell or organism homozygous for a recessive allele and thus showing the recessive phenotype.

double-minute chromosomes self-replicating extrachromosomal DNA elements lacking centromeres, seen in eukaryotic cells after certain treatments resulting in selective gene amplification.

doubling time the time it takes for the quantity of something growing exponentially (e.g. a population of living organisms) to double.

down feather, down the first fluffy feathers of young birds, with a short quill and with barbules not interlocking to form a flat vane. Some birds retain a down layer under the adult plumage.

downland *n. appl.* grassland vegetation typical of the chalk downs of southern England, which is generated and maintained by continuous grazing (by sheep and rabbits), and which is short turf rich in small flowering plants.

down mutation mutation in which transcription of a particular gene(s) is much reduced but the gene product is unaltered, usually due to a mutation in DNA regions involved in regulation of gene expression.

downregulation *n.* decrease in number, as of receptors on a cell surface, or in rate of production.

Down syndrome condition due to the presence of three copies of chromosome 21 (trisomy 21), characterized by some degree of mental retardation, short stature and poor muscle tone.

downstream *a. appl.* DNA sequences, control sites, etc. on the distal side of any given point in relation to the direction of transcription. Generally refers to points after the startpoint of transcription. *cf.* upstream.

dragonflies *see* Odonata.

drepanium *n.* a helicoid cyme with secondary axes developed in a plane parallel with that of main peduncle and its first branch, so that the whole inflorescence is sickle-shaped.

drepanoid *a.* sickle-shaped.

drift *see* genetic drift, continental drift.

drive *n.* the motivation of an animal resulting in its achieving a goal or satisfying a need.

dromaeognathous *a.* having a palate in which palatines and pterygoids do not articulate, owing to intervention of vomer.

dromotropic *a.* bent in a spiral.

drone *n.* male social insect, esp. honeybee.

drop mechanism the mechanism by which a pollen grain, trapped in a drop of liquid in gymnosperm ovules, is drawn into the micropyle.

dropper *n.* downward outgrowth of a bulb which may form a new bulb.

Drosophila melanogaster a fruitfly, a member of the Diptera, a favoured subject for experimental genetics and developmental biology.

drosopterin *n.* red pteridine pigment in eyes and other organs of some insects, including *Drosophila*.

drought *n.* situation in which a region does not receive sufficient water because of decreased rainfall, increased evaporation due to higher than normal temperatures or a combination of both.

drug-resistance factors or plasmids (R factors) plasmids in enterobacteria and other medically important bacteria, which carry genes for resistance to various commonly used antibiotics.

drupaceous *a.* bearing drupes.

drupe *n.* a more-or-less fleshy fruit with one compartment and one or more seeds, having the pericarp differentiated into a thin epicarp, a fleshy mesocarp, and a hard stony endocarp, e.g. plum, cherry, etc. *alt.* stone fruit.

drupel *n.* one of a collection of small drupes forming an aggregate fruit, e.g. raspberry, blackberry. *alt.* drupelet.

druse *n.* a globular compound crystal whose component crystals project from the surface.

dry matter (DM) the material left after removal of water from organic matter such as plant biomass or soil, obtained by heating to constant weight in an oven at 90–95 °C.

dryopithecids, dryopithecines *n.plu.* a group of Miocene ape-like fossils from India and Africa, including the genera *Dryopithecus* and *Proconsul*, dating from around 20 to 8 million years ago, and which are thought to include the ancestors of modern apes.

dry weight (DW), dry mass the weight or mass of organic matter or soil after removal of water by heating to constant weight. *see* dry matter.

dry weight rank method technique for estimating the contribution each plant species makes to the total yield of a pasture.

D segment any of a set of DNA sequences specifying the "diversity" region of the variable region in the heavy chains of immunoglobulin molecules, one being joined to a V gene and subsequently to a J gene during the DNA rearrangements that produce a DNA sequence coding for the variable region of the heavy chain. *alt.* D gene.

Duchenne muscular dystrophy (DMD) an X-linked heritable disease leading to muscular atrophy and eventual death, affecting around 1 in 4000 newborn males. It is due to a defect in the gene encoding dystrophin, a large protein associated with the sarcolemma of muscle fibres.

duct *n.* any tube that conveys fluid or other material.

ductless glands glands such as endocrine glands that do not release their secretions into a duct.

ductule *n.* minute duct.

ductulus efferens one of the ductules leading from the testis to the vas deferens known collectively as the vasa efferentia.

ductus *n.* a duct.

ductus arteriosus the connection between the pulmonary arch and dorsal aorta in mammalian foetus.

duetting *n.* rapid antiphonal calling back and forth between mated pairs in some birds, presumed to be a recognition and bonding device.

Dufour's gland gland that leads into the poison sac at the base of the sting in certain Hymenoptera. *alt.* alkaline gland.

dulosis *n.* slavery among ants, in which those of one species are captured by another species and work for them, an extreme example of social parasitism.

dumose *a.* shrub-like in appearance.

duodenum *n.* that part of intestine next to the pyloric end of the stomach.

duplex *a.* double; compound, *appl.* flowers; consisting of two distinct structures; having two distinct parts.

duplex DNA double-stranded DNA, *see* DNA.

duplication *n.* the case in which a chromosome, segment of a chromosome, or gene is present in more than the normal number of copies.

duplicident *a.* with two pairs of incisors in upper jaw, one behind the other.

duplicity *n.* condition of being two-fold.

duplicocrenate *a.* with scalloped margin, and each rounded tooth again notched, *appl.* leaf.

duplicodentate *a.* with marginal teeth on leaf bearing smaller teeth-like indentations.

duplicoserrate *a.* with marginal saw-like teeth and smaller teeth directed towards leaf tip.

dura mater the tough membrane forming the outermost covering of brain and spinal cord.

dura spinalis the tough membrane lining the spinal canal.

dural *a. pert.* dura mater.

duramen *n.* the hard darker central wood of a tree trunk. *alt.* heartwood.

dust lice *see* Psocoptera.

dwarf male a small, usually simply formed male individual in many classes of animal, either free-living or carried by the female. *alt.* pygmy male.

dyad *n.* one member of a pair of homologous replicated chromosomes synapsed at meiosis; *appl.* symmetry about a twofold axis.

dynein *n.* protein with ATPase activity, attached to microtubules in eukaryotic cilia and flagella, and which powers their movement.

dynorphin *n.* any of several endorphin-like peptides found in brain and gut.

dysgenesis *n.* infertility of hybrids in matings between themselves, although fertile with individuals of either parental stock.

dysgenic *a. appl.* traits inimical to the propagation of the organism such as sterility, chromosomal aberrations, mutations, abnormal segregation at meiosis etc.

dysmerogenesis *n.* segmentation resulting in unlike parts.

dysphotic *a.* dim; *appl.* zone: waters at depths between 80 and 600 m, between euphotic and aphotic zones.

dyspnoea *n.* difficulty in breathing.

dystrophic *a.* wrongly or inadequately nourished; inhibiting adequate nutrition; *pert.* faulty nutrition; *appl.* lakes rich in undecomposed organic matter so that nutrients are scarce.

dystrophin *n.* protein associated with the muscle cell membrane, and whose absence or abnormality is thought to be the cause of Duchenne muscular dystrophy (*q.v.*).

DZ twins dizygotic twins *q.v.*

Dzierzon theory belief that males of the honey bee are always produced from unfertilized eggs, first put forward by the Polish apiculturist J. Dzierzon.

E

ε- heavy chain class corresponding to IgE.

ε- *for all headwords with prefix* ε- *or epsilon- refer to headword itself.*

E glutamic acid *q.v.*

EBV Epstein–Barr virus, a herpes virus, the cause of glandular fever, and also involved in Burkitt's lymphoma and nasopharyngeal carcinoma.

EC effective concentration *q.v.*

E-DNA form of DNA taken up by synthetic molecules lacking guanine, with 7½ base pairs per turn.

EDTA ethylenediaminetetraacetate, a chelating agent that binds magnesium and calcium ions. Used widely in experimental biochemistry to study, e.g., the role of metal ions in the structure of proteins and protein assemblies.

eEF general designation for eukaryotic elongation factors *q.v.*

EEG electroencephalogram.

EF general designation for bacterial elongation factors *q.v.*

EFA essential fatty acids *q.v.*

EGF epidermal growth factor *q.v.*

EGTA a chelating agent.

EI energy index, *see* phosphorylation potential.

eIF followed by a numeral, a eukaryotic initiation factor *q.v.*

ELISA enzyme-linked immunosorbent assay *q.v.*

EMP Embden–Meyerhoff–Parnas pathway, *see* Embden–Meyerhoff pathway.

env gene of C-type viruses specifying envelope components of the virus particle.

epp end-plate potential *q.v.*

epsp excitatory postsynaptic potential *q.v.*

ER endoplasmic reticulum *q.v.*

erb-A gene for the thyroid hormone receptor, a potential oncogene. *v-erb*, oncogene carried by the RNA tumour virus avian erythroblastosis virus. *c-erb*, corresponding proto-oncogene found in normal cells.

erb-B gene for the receptor for epidermal growth factor, a potential oncogene.

eRF eukaryotic release factor *q.v.*

ERG electroretinogram.

ESS evolutionarily stable strategy *q.v.*

EWL evaporative water loss *q.v.*

e- prefix derived from L. *ex*, out of or *ex* without, often denoting a lack of, e.g. ebracteolate, lacking bracts, ecaudate, without a tail.

ear *n.* sense organ in vertebrates concerned with hearing and gravity detection, consisting in mammals of an external ear (pinna) surrounding the auditory canal or auditory meatus leading to the membranous eardrum. A row of small bones transmits vibrations of the eardrum caused by sound waves across the air-filled space of the middle ear to the fluid-filled inner ear, which contains the cochlea and the semicircular canals. The cochlea contains sensory cells (hair cells) that detect vibrations of the surrounding fluid and transmit signals encoding tone and pitch of the original sound to the brain via the auditory nerve. The semicircular canals are concerned largely with gravity detection and maintenance of balance. *see also* acoustico-lateralis system, and individual entries for cochlea, semicircular canals, etc.; (*bot.*) the spike of grasses, usually *appl.* cereals.

ear sand otoconia *q.v.*

eardrum tympanic membrane *q.v.*

early *a. appl.* the first phase of the lytic cycle of phage or virus infection, from the entry of virus DNA (or RNA) to the start

165

of its replication; *appl*. phage or viral genes expressed early in infection, before replication.

earth ball common name for basidiomycete fungi of the genus *Scleroderma* and their relatives, with hard tuberous unstalked fruiting bodies that crack open to release a mass of spores.

earthworm *n*. common name for a number of terrestrial oligochaete worms (*Lumbricus* spp. and others) inhabiting the soil, and which contribute to soil aeration through their tunnels and to soil fertility by bringing humus-containing soil to the surface.

earwigs common name for insects of the order Dermaptera *q.v*.

East African Floral Region part of the Palaeotropical Realm comprising Africa from north of Lake Victoria south to southern Mozambique and westward to include southern Angola.

eavesdropping *n*. behavioural strategy in which rival males are attracted to a female by another male's courtship display.

Ebner's gland von Ebner's gland *q.v*.

ebracteate, ebracteolate *a*. without bracts, without bracteoles.

ecad *n*. a plant or animal form modified by the environment; habitat form.

ecalcarate *a*. having no spur or spur-like process, *appl*. petals.

ecardinal, ecardinate, *a*. having no hinge, *appl*. shells.

ecarinate *a*. having no keel or keel-like ridge.

ecaudate *a*. without a tail.

eccentric excentric *q.v*.

eccentric cell nerve cell in ommatidium of compound eye of *Limulus* which receives signals from retinula cells and transmits them to the central nervous system.

eccrine *a. appl*. glands that secrete without disintegration of secretory cells. *cf*. apocrine.

eccritic *a*. causing or *pert*. excretion; preferred, *appl*. temperature or other environmental state.

ecdemic *a*. not native.

ecdysial *a. pert*. ecdysis or moulting.

ecdysis *n*. moulting, the periodic shedding of cuticular exoskeleton in insects and some other arthropods, to allow for growth.

ecdysiotropic hormone insect protein hormone produced in "brain" and which acts on the resting ovary in concert with juvenile hormone, resulting in the secretion of the steroid hormone ecdysone.

ecdysone, ecdysterone *n*. steroid hormone produced by the prothoracic gland of insects and the Y-organs of crustaceans, which stimulates growth and moulting.

ecesis *n*. the invasion of organisms into a new habitat.

echidna *n*. spiny anteater, a monotreme (*q.v*.).

echinate *a*. bearing spines or bristles.

echinidium *n*. marginal hair with small pointed or branched outgrowths on cap of fungi.

echinococcus *n*. a vesicular metacestode developing a number of daughter cysts, each with many heads. *see* polycercoid.

Echinodermata, echinoderms *n., n.plu*. a phylum of marine coelomate animals that are bilaterally symmetrical as larvae but show five-rayed symmetry as adults and have a calcareous endoskeleton and a water vascular system. It includes the classes Crinoidea (sea lilies and feather stars), Asteroidea (starfish), Ophiuroidea (brittle stars), Echinoidea (sea urchins) and Holothuroidea (sea cucumbers).

echinoid *a. pert*. or like a sea urchin.

Echinoidea *n*. class of echinoderms, commonly called sea urchins, having a typically globular body with skeletal plates fitting together to form a rigid test.

echinopluteus *n*. the pluteus larva of sea urchins.

echinulate *a*. having small spines; *appl*. bacterial colonies, having pointed outgrowths.

Echiura, echiurans *n., n.plu*. phylum of unsegmented coelomate marine worms, with soft plump bodies, which live in U-shaped tubes or in rock crevices down to abyssal depths. They have an extensible but not eversible proboscis with a ciliated groove for collecting food. *alt*. spoon worms.

echolocation *n*. locating objects by sensing the echoes returned by very high frequency sounds emitted by the animal, as used, e.g., by bats.

echoviruses, ECHO viruses group of

picornaviruses *(q.v.)* infecting the intestinal tract and which may also cause respiratory illnesses and meningitis.

eclipse *n.* plumage assumed after spring moult, as in drake; the period of multiplication of a virus when it is not easily detectable in the host cell.

eclosion *n.* hatching from egg or pupa case.

ecobiotic *a.* adaptation to a particular mode of life within a habitat.

ecoclimatic *a. appl.* adaptation to the physical and climatic conditions in a particular region.

ecocline *n.* a continuous gradient of variation of ecotypes in relation to variation in ecological conditions.

ecodeme *n.* a deme occupying a particular ecological habitat.

E. coli the bacterium *Escherichia coli* q.v.

ecological *a. pert.* or concerned with ecology.

ecological diversity the diversity of ecosystems (e.g. forest, desert, grassland, oceans) within a given region.

ecological efficiency the efficiency of use of energy by an ecosystem, which is described in terms of several energy coefficients: the energy coefficient of the first order (production, P, divided by consumption, C), the energy coefficient of the second order (P divided by assimilated energy, A) and the assimilation efficiency (A/C).

ecological niche the role of an organism in a community in terms of the habitat it occupies, its interactions with other organisms and its effect on the environment. A given niche, e.g. the small herbivore niche, may be occupied by different species in different ecosystems and different parts of the world. *alt.* niche. *see also* realized niche.

ecological pyramid diagram showing the biomass, numbers or energy levels of individuals of each trophic level in an ecosystem, starting with the primary producers (e.g. green plants) at the base. *see* Fig. 3.

ecological succession the natural process whereby communities of plant and animal species are replaced by others, usually more complex, over time as a mature ecosystem develops. *see* climax, primary succession, secondary succession, succession.

ecology *n.* the interrelationships between organisms and their environment and each other; the study of these interrelationships.

economic density of a population, the number of individuals per unit of inhabited area.

ecoparasite *a.* a parasite that can infect a healthy and uninjured host; parasite restricted to a specific host or small group of host species.

ecophysiology *n.* the study of animal physiology in relation to life-style and adaptation to environment.

ecorticate *a.* without a cortex, as certain lichens.

ecospecies *n.* a group of individuals associated with a particular ecological niche and behaving like a species, but capable of interbreeding with neighbouring species.

ecosphere *n.* the planetary ecosystem, consisting of the living organisms of the world and the components of the environment with which they interact.

ecostate *a.* without ribs or costae.

ecsoma *n.* the retractile posterior part of

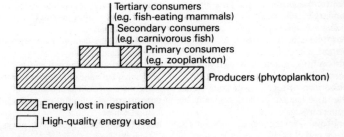

Fig. 3 Typical pyramid of energy flow through a river ecosystem.

body in certain trematodes.

ecosystem *n.* a community of different species interdependent on each other together with their non-living environment, which is relatively self-contained in terms of energy flow, and is distinct from neighbouring communities. Different types of ecosystem are defined by the collection of organisms found within them, e.g. forest, soil, grassland. Continuous ecosystems covering very large areas, such as the northern coniferous forest or the steppe grassland are known as biomes *q.v.*

ecotone *n.* zone where two ecosystems overlap, and which supports species from both ecosystems as well as species found only in this zone.

ecotope *n.* a particular kind of habitat within a region; the total relationship of an organism with its environment, being the interaction of niche, habitat and population factors.

ecotype *n.* a subspecific form within a true species, resulting from selection within a particular habitat and therefore adapted genetically to that habitat, but which can interbreed with other members of the species.

ect-, ecto- word elements derived from Gk. *ektos*, outside.

ectad *adv.* towards the exterior; outwards.

ectadenia *n.plu.* ectodermal accessory genital glands in insects.

ectal *a.* outer; external.

ectamnion *n.* ectodermal thickening in proamnion, beginning of head-fold.

ectangial *a.* outside a vessel; produced outside a primary sporangium.

ectendotrophic *a.* partly ectotrophic and partly endotrophic, *appl.* mycorrhizas.

ectethmoid *n.* lateral ethmoid bone of skull.

ectoascus *n.* outer membrane of ascus wall in certain ascomycete fungi. *cf.* endoascus.

ectobatic *a.* efferent; centrifugal.

ectoblast epiblast *q.v.*

ectobronchus *n.* lateral branch of main bronchus in birds.

ectochondrostosis *n.* ossification beginning in perichondrium and gradually invading cartilage.

ectochroic *a.* having pigmentation on surface of cell or hypha.

ectocommensal *n.* commensal living on the surface of another organism.

ectocrine *a. appl.* and *pert.* substances or decomposition products in the external medium that inhibit or stimulate plant growth.

ectoderm *n.* outer layer of embryonic epithelium in multicellular animals, which gives rise to epidermis and nervous system; in coelenterates the outer layer of epithelium. *a.* ectodermal.

ectoenzyme exoenzyme *q.v.*

ectogenesis *a.* embryonic development outside the maternal organism; development in an artificial environment. *a.* ectogenetic.

ectogenic *a.* of external origin, not produced by organisms themselves.

ectogenous *a.* able to live an independent life; originating outside the organism.

ectoglia *n.* an outer layer of glia in the nervous system.

ectolecithal *a. appl.* eggs having yolk around the periphery.

ectoloph *n.* the ridge stretching from paracone to metacone in a lophodont molar tooth.

ectomeninx *n.* outer membrane covering embryonic brain and giving rise to the dura mater.

ectomere *n.* a blastomere that gives rise to ectoderm.

-ectomy suffix derived from Gk. *ek*, out, and *temnein*, to cut, signifying an excision, e.g. thyroidectomy, gonadectomy, etc.

ectomycorrhiza *n.* type of mycorrhiza in which the fungal hyphae form a superficial covering and do not extensively penetrate the root, found on both coniferous and broadleaved forest trees, the infecting fungi being chiefly higher basidiomycetes.

ectoneural *a. appl.* system of oral ring, radial and subepidermal nerves in echinoderms.

ectoparasite *n.* parasite that lives on the surface of an organism.

ectopatagium *n.* the part of the wing-like flying membrane of bats that is carried on the metacarpals and phalanges.

ectophagous *a.* feeding on the outside of a food source.

ectophloeodeic *a.* growing on plants, *appl.* lichens.

ectophloic *a.* with phloem outside xylem.

ectophyte *n.* any external plant parasite of plants and animals. *a.* ectophytic.

ectopic *a.* not in normal position, *appl.* organs, pregnancy, etc.

ectoplasm *n.* the external layer of cytoplasm in a cell, next to the cell membrane and usually clear and non-granular; ectosarc *q.v.* cf. endoplasm.

ectoplast *n.* the plasma membrane nearest to the cell wall in plant cells.

Ectoprocta, ectoprocts *n., n.plu.* phylum of small marine and freshwater colonial animals, which superficially resemble mosses, hence the common name of moss animals, composed of zooids each bearing a crown of ciliated tentacles (a lophophore), and which live in horny, calcareous or gelatinous cases. *alt.* (formerly) Bryozoa, bryozoans, Polyzoa.

ectopterygoid *n.* a ventral membrane bone behind palatine and extending to quadrate.

ectosarc *n.* external layer of cytoplasm, esp. in protozoans. *alt.* ectoplasm.

ectospore *n.* the spore formed at end of each sterigma in basidiomycete fungi.

ectosporous *a.* with spores borne exteriorly.

ectostosis *n.* formation of bone in which ossification begins under the perichondrium and either surrounds or replaces the cartilage.

ectostracum *n.* the outer primary layer of exocuticle of exoskeleton in ticks and mites.

ectostroma *n.* fungal tissue penetrating cortical tissue of host and bearing conidia.

ectotherm *n.* animal whose source of body heat is primarily external, as fish, amphibians and reptiles, formerly known as "cold-blooded" animals. *a.* ectothermic.

ectotrophic *a.* finding nourishment from outside; *appl.* mycorrhizas in which the fungal hyphae form a superficial covering on the host roots and do not extensively penetrate the root itself.

ectotropic *a.* tending to curve or curving outwards.

edaphic *a. pert.* or influenced by conditions of soil or substratum.

edema oedema *q.v.*

Edentata, edentates *n., n.plu.* an order of placental mammals known from the Paleocene, extant members of which include the armadillo, anteater and two-toed sloth, having reduced teeth and often an armoured body.

edentate *a.* without teeth or tooth-like projections.

edge effect tendency to have greater variety and density of organisms in the boundary zone between communities.

edge species species living primarily or most frequently or numerously at junctions of communities.

Ediacaran fauna a fauna of soft-bodied animals of problematic affinities present in late Precambrian strata, and bearing little resemblance to any later organisms.

editing in DNA synthesis the exonuclease functions of DNA polymerase which correct mistakes (mismatched bases). *see also* proofreading, RNA editing.

Edman degradation method of determining the amino acid sequence of peptides by sequentially removing amino acids for identification from the amino terminal end.

edriophthalmic *a.* having sessile eyes, *appl.* to certain crustaceans.

eelworms *n.plu.* group of soil nematode worms, some of which cause serious damage to crop plants.

E face in freeze-fractured membranes, the face representing the interior side of the outer half of the lipid bilayer.

effective concentration (EC) the concentration of a toxic substance that is sufficient to cause adverse symptoms (in cases where effects other than death are being studied) within a given period, and which is expressed as e.g. 48-hour EC_{50}, the concentration required to cause symptoms in 50% of the animals tested within 48 hours.

effective population number the number of individuals in an ideal, randomly breeding population with a 1/1 sex ratio that would have the same rate of heterozygosity decrease as the actual population under consideration.

effector *n.* any organ, cell, etc. that reacts to a stimulus by producing something, carrying out a specific set of functions, or doing mechanical work, as muscle, electric and luminous organs, chromatophores, glands, plasma cells and mature T cells of immune system, etc.

efferent *a.* conveying from, *appl.* motor

nerves carrying impulses outwards from central nervous system, *appl.* vessels conveying blood or lymph outwards from an organ or lymph node respectively; *appl.* ductules from rete testis opening into epididymis. *cf.* afferent.

effigurate *a.* having a definite shape or outline.

efflorescence *n.* flowering; time of flowering; bloom as on surface of grapes and other fruits.

effodient *a.* having the habit of digging.

effoliation *n.* shedding of leaves.

effuse *a.* spreading loosely, *appl.* flowerheads; spreading thinly, *appl.* bacterial colonies.

eft *n.* juvenile phase in life cycle of a newt.

egesta *n.plu.* the sum total of material and fluid discharged from the body; material passed out of the body in egestion.

egestion *n.* the process of ridding the body of any waste material as by defaecation and excretion; specifically, the excretion of material that has never been taken out of the gut, as defaecation.

egg *n.* ovum *q.v.*; in certain animals, e.g. reptiles, birds, amphibians and insects, the fertilized ovum and nutritive and protective tissues, and from which a young individual emerges.

egg apparatus egg cell and two cellular synergids that develop at the micropylar end of megagametophyte in flowering plants.

egg cell ovum *q.v.*

egg coat outer glycoprotein coat of ovum, the inner layer of which is the zona pellucida or vitelline layer.

egg membrane vitelline layer *q.v.*; chorion *q.v.*; layer of tough tissue lining an egg shell.

egg nucleus the female pronucleus *q.v.*

egg tooth small structure on tip of upper jaw or beak with which hatchling breaks out of shell.

eglandular *a.* without glands.

egocentric *a. appl.* behaviour that benefits the survival of the individual that exhibits it.

Eichler's rule groups of hosts with more variation are parasitized by more species than taxonomically uniform groups.

eicosanoid *n.* member of a class of compounds derived from the fatty acid arachidonic acid, which includes prostaglandins, prostacyclins, thromboxanes, leukotrienes and lipoxins. They can act as intracellular messengers and be secreted as local chemical signals in e.g. inflammatory and hypersensitivity reactions and have many different effects on a variety of cells.

Eimer's organs organs in the snout of moles, probably tactile organs.

einkorn *n.* a primitive cultivated diploid wheat (*Triticum monococcum*) first cultivated in the Near East and south-west Asia around 11,000 years ago, and which is derived from the wild *T. boeoticum*.

ejaculate *a.* the emitted seminal fluid.

ejaculatory *a.* throwing out, *appl.* certain ducts that discharge their secretions with some force.

ejaculatory sac organ pumping ejaculate from vas deferens through ejaculatory duct to penis, in certain insects.

elaborate *v.* to form complex organic substances from simple materials.

Elaeagnales *n.* an order of shrubs, often with leathery leaves and thorns, comprising the family Elaeagnaceae (oleaster).

elaeoblast *n.* a mass of nutrient material at posterior end of body in some tunicates.

elaeocyte *n.* cell containing lipid droplets, in coelomic fluid of annelid worms.

elaeodochon *n.* oil gland in birds.

elaio- elaeo-.

elaioplankton *n.* planktonic organisms rendered buoyant by oil globules in their cells.

elaioplast *n.* colourless organelle in plant cells in which oils or fats are formed and stored.

elaiosome *n.* fleshy oil-containing appendages present on seeds that are to be dispersed by ants, such as those of castor oil plant.

elaiosphere *n.* oil globule in a plant cell.

elasmobranchs *n.plu.* a group of cartilaginous fishes including the sharks, dogfishes, skates and rays, which have an outer covering of bony tooth-like scales, a spiracle and no covering over the gill openings.

elastase *n.* proteolytic enzyme secreted in the pancreas and by some bacteria and acting *inter alia* on elastin. EC 3.4.21.11.

elastic fibres yellow fibres *q.v.*

elastin *n*. connective tissue protein, major component of elastic fibres as in blood vessels, ligaments etc., formed from a soluble precursor, proelastin.

elater *n*. (*bot*.) cell in spore capsule of liverworts with spiral thickening in its wall which is sensitive to changes in humidity, expanding or twisting and aiding the dispersal of spores; flattened appendage of spore of horsetails, which unfolds in dry conditions and aids in dispersal by wind; (*zool*.) springing organ in Collembola (springtails).

electric organ modifications of muscle or epithelium which discharge electric energy, found chiefly in certain fishes, e.g. electric eel.

electric receptors receptors present in some fish, sensitive to potential differences across the skin, used to detect distortions of the electric field that the fish sets up around itself.

electrical coupling the passive flow of electric current from one animal cell to its neighbour through gap junctions, as in cells of heart muscle where it mediates the synchronous contraction of individual fibres.

electrical synapse junction between cells, including those of the nervous system, at which electric current flows directly from one cell to another, identified with gap junction in many instances. *cf*. chemical synapse (*see* synapse).

electrically active cells muscle and nerve cells.

electroblast *n*. modified muscle cell that gives rise to an electroplax.

electrochemical gradient the gradient across a membrane with respect to an ion or other solute, which comprises both the concentration gradient and the gradient of electrical charge across the membrane.

electrocyte *n*. modified muscle or nerve cell in electric organ of gymnotoid fish (American knife-fish) that produces an electric discharge.

electrogenic *a*. generating an electrical potential across a membrane, *appl*. certain ion pumps in cell membrane.

electrolemma *n*. membrane surrounding an electroplax.

electromagnetic senses senses that de-

tect electric fields and which are used by some animals, e.g. some fish, to detect distortions in the Earth's electric field caused by objects in the locality. *see also* magnetotaxis.

electromyography *n*. measurement of electrical activity of muscle.

electron microscope *see* scanning electron microscope, transmission electron microscope, tunnelling electron microscope.

electron transfer chain, electron transport chain general term for a series of electron carriers as found in mitochondria and chloroplasts along which electrons are transferred in a series of redox reactions, and which consists of cytochromes, quinones, ferredoxin (in chloroplasts), flavoproteins (in mitochondria) and other components. *see also* respiratory chain.

electroporation *n*. the entry of DNA molecules into animal cells under the influence of a strong electric field.

electrophoresis *n*. technique for separating molecules such as proteins or nucleic acid fragments on the basis of their net charge and mass, by their differential migration through paper, or through a polyacrylamide or agarose gel (gel electrophoresis) in an electric field.

electrophysiology *n*. the study of physiological processes in relation to electrical phenomena.

electroplax *n*. one of the constituent plates of electric organ of e.g. electric eel.

electroreceptor *n*. sensory receptor sensitive to an electric field. *see* electrosensory.

electrosensory *a. appl*. sensory systems in animals, such as the ampullae Lorenzini in dogfish (*Scyliorhinus*), that detect changes in electric fields.

electrostatic bond bond formed by electrostatic interaction between oppositely charged groups, important in biological molecules. *alt*. ionic bond, ion pair, salt bond, salt linkage.

electrotaxis *n*. orientation of movement of a cell or organism within an electric field.

electrotonic *a. appl*. localized potentials produced by subthreshold ionic currents in nerve cell membranes, determined by the passive electrical properties of cell; *appl*. passive propagation of an electrical impulse in nerve cell membrane which gradually diminishes and dies away over a distance of a few millimetres. *cf.* action potential.

electrotonus *n*. the modified condition of a nerve when subjected to a constant current of electricity.

electrotropism *n*. plant curvature in an electric field.

elephant-tooth shells common name for the Scaphopoda *q.v.*

elicitor *n*. in plant pathology, a compound that induces a plant defence response, and which may either be derived from a plant pathogenic microorganism (a biotic elicitor) or may be an inorganic material such as mercury and other heavy metals.

elittoral *a. appl.* zone out from the coast where light ceases to penetrate to the sea bottom.

elliptical *a. appl.* leaves of about the same breadth at equal distances from apex and base, which are themselves slightly acute.

elongation factor any of several accessory proteins (e.g. EF-Tu, EF-Ts, EF-G in *E. coli*) required for translation to proceed, which are involved in correct positioning of aminoacyl-tRNA on the mRNA-ribosome complex.

eluvial *a. appl.* layer in soils which is impoverished and leached, above the illuvial layer, *alt.* A-horizon; *appl.* gravels formed by breakdown of rocks *in situ*.

elytra *plu.* of elytron *q.v.*

elytroid *a.* resembling an elytron.

elytron *n*. the forewing of beetles, which is hard and stiff and is not flapped in flight, but serves at rest as a protective covering (wing case or wing sheath) for the membranous hindwing. *plu.* elytra.

elytrum elytron *q.v.*

emarginate *a.* with a distinct notch or indentation.

emasculation *n*. removal of anthers to prevent self-pollination; removal of testes.

Embden–Meyerhof(–Parnas) pathway glycolysis *q.v.*

embedding *n*. method used in preparing permanent microscope slides or specimens for electron microscopy. For optical microscopy, the prepared material is impregnated with molten paraffin wax before sectioning. For electron microscopy, the specimen is embedded in a colourless resin before sectioning.

embiids common name for the Embioptera *q.v.*

Embioptera *n*. order of insects with soft flattened bodies, and incomplete metamorphosis, commonly called embiids or foot spinners, which live in groups in silken tunnels and have wingless females and winged males.

embolism *n*. a blood clot blocking a blood vessel.

embolium *n*. outer or costal part of forewing in certain insects.

embolomerous *a. appl.* type of vertebra having two vertebral rings in each segment, due to union of hypocentra with neural arch and union of two pleurocentra below notochord.

embolus *n*. a projection closing foramen of an ovule; apical division of palp in some spiders; the core of horn of ruminants; embolism *q.v.*

emboly *n*. invagination.

embryo *n*. animal or plant in the earliest stages of development, when it is entirely dependent directly or indirectly on resources provided by the parent (e.g. in seed or egg), from the fertilized egg (zygote) up to the stage of a free-living miniature adult or larva in animals, or a photosynthesizing seedling in plants. In human development, the conceptus is technically known as an embryo either from the moment of conception, or sometimes from the blastocyst stage, until the main parts of the body and main internal organs have started to take shape at around the 7th week of gestation. After this it is called a foetus. *see also* pro-embryo.

embryo cell in some plants, the one of the two cells formed from the 1st division of the fertilized egg that becomes the embryo, the other developing into the suspensor.

embryo sac in flowering plants, the mature female gametophyte (megagametophyte) which develops in the ovule, comprising an egg cell and accessory cells.

embryo transfer reproductive technology in which very early embryos produced by *in vitro* fertilization or artificial insemination are transferred into a surrogate mother for further development, used in cattle and sheep breeding to produce many more offspring from a prize female than she could produce naturally.

embryogenesis *n.* the development of an embryo from a fertilized ovum.

embryogeny *n.* formation of the embryo, used esp. in botany.

embryology *n.* the study of the formation and development of embryos.

embryonated *a. appl.* eggs in which embryo has developed.

embryonic *a. pert.* an embryo.

embryonic membranes the various membranes surrounding the developing embryo, including the amnion.

embryophore *n.* ciliated mantle enclosing embryo in many tapeworms, and formed from superficial blastomeres of embryo.

embryotega *n.* small hardened portion of testa which marks micropyle in some seeds and separates like a little lid at germination.

emergence *n.* an outgrowth from subepidermal tissue; an epidermal appendage.

emersed *a.* rising above surface of water, *appl.* leaves.

eminence *n.* ridge or projection on surface of bones.

emissary *a.* coming out; *appl.* veins passing through apertures in cranial wall and establishing connection between sinuses inside and veins outside.

emmenophyte, emmophyte *n.* a water plant without any floating parts.

emmer *n.* a primitive cultivated wheat (*Triticum dicoccum*) first domesticated in the Near East and south-west Asia around 11,000 years ago, and which is thought to be derived from wild emmer (*T. dicoccoides*), which is a hybrid between *T. monococcum* and goat-grass (*Aegilops*).

empodium *n.* outgrowth between the claws on the last tarsal segment in flies.

enamel *n.* hard material containing over 90% calcium and magnesium salts which forms cap over dentine or may form complete tooth or scale.

enamel cell ameloblast *q.v.*

enamel organ complex structure of tall columnar epithelium (ameloblasts or enamel cells) forming the surface of the dental papilla, from which tooth enamel is developed.

enameloid *a. appl.* enamel-like material in fish.

enantiomer *n.* one of two isomers of a molecule that are mirror images of each other.

enantiomers optical isomers *q.v.*

enantiomorphic *a.* similar but contraposed, as mirror image; deviating from normal symmetry.

enantiostylous *a.* having flowers whose styles protrude right or left of the axis, with the stamens on the other side.

enarthrosis ball-and-socket joint *q.v.*

enation *n.* outgrowth from a previously smooth surface.

encephalic *a. pert.* the brain.

encephalization *n.* evolutionary tendency towards formation of an anterior brain.

encephalomyelic *a. pert.* brain and spinal cord.

encephalon *n.* the brain.

enchytracheids *n.plu.* group of small oligochaete worms, living in soil.

encyst *v.* of a cell or small organism, to surround itself with an outer tough coat or capsule. *n.* encystment.

endangered *a.* IUCN definition *appl.* species or larger taxa whose numbers have become so low, or whose habitats have been so drastically reduced, that they are thought to be in immediate danger of extinction in the wild in the foreseeable future if there is no change in circumstances. *see also* rare, rarity, vulnerable.

endangium *n.* innermost lining or tunica intima of blood vessels.

endarch *a.* with central protoxylem, or several protoxylem groups surrounding pith, produced when xylem matures centrifugally so the oldest protoxylem is closest to the centre of the axis.

end-bulbs minute oval or cylindrical bodies, representing the terminals of sensory neurones in mucous and serous membranes, in skin of genitalia, and in synovial layer of certain joints.

endemic *a.* restricted to a certain region or part of region; *appl.* disease, present at relatively low levels in the population all the time. *n.* endemism.

endergonic *a.* absorbing or requiring energy, *appl.* metabolic reactions.

endexine *n.* the inner membranous layer of extine (exine) of a spore.

endites *n.plu.* offshoots from the border of

certain appendages of arthropods.

endo- prefix derived from Gk. *endon*, within, signifying within, inside, acting inside, opening to the inside, etc.

endoascus *a*. inner coat of ascus wall, protruding after rupture of the ectoascus, typical of certain ascomycete fungi.

endobasal *a*. *appl*. body in nucleus which acts as a centrosome in certain protozoans.

endobiotic *a*. living within a substratum or within another living organism.

endoblast hypoblast *q.v.*

endobronchus entobronchus *q.v.*

endocardiac, endocardial *a*. situated within the heart.

endocardium *n*. membrane lining internal cavities of heart.

endocarp *n*. the innermost layer of pericarp of fruit, usually fibrous, hard or stony, as the "stone" enclosing the seed in plums, cherries, etc.

endocarpy angiocarpy *q.v.*

endochondral *a*. *appl*. ossification in cartilage beginning inside and working outwards.

endochorion *n*. inner lamina of chorion of insect eggs.

endochroic *a*. having pigment within a cell or hypha.

endochrome *n*. any pigment within a cell, esp. other than chlorophyll.

endocoel *n*. coelom *q.v.*; the cavities in proboscis, collar and trunk of certain hemichordates.

endocoelar *a*. *pert*. inner wall of coelom.

endoconidium *n*. a conidium formed within a conidiophore.

endocranium neurocranium *q.v.*

endocrine *a*. *appl*. ductless glands secreting hormones directly into blood; *pert*. such glands. *cf*. exocrine.

endocrine system the system of endocrine glands secreting a variety of hormones, which are controlled by peptide hormones released from the pituitary and by direct neural input. *see* adrenal glands, neuroendocrine system, ovary, steroid hormones, testis, thyroid gland.

endocrinology *n*. study of hormones, produced in endocrine glands and their effects.

endocuticle *n*. innermost layer of the cuticle (exoskeleton) of arthropods.

endocycle *n*. layer of tissue separating internal phloem from endodermis.

endocyclic *a*. with the mouth remaining in axis of coil of gut, *appl*. crinoids; having an apical system with double circle of plates surrounding anus, *appl*. echinoids; *pert*. endocycle.

endocyst *n*. the soft body wall of a polyzoan zooid; the membranous inner lining of a protozoan cyst.

endocytosis *n*. process by which eukaryotic cells take up material from the outside by invagination of the plasma membrane to form vesicles enclosing the external material. *see also* pinocytosis, phagocytosis. *a*. endocytotic.

endodeme *n*. a gamodeme composed of predominantly inbreeding dioecious plants or bisexual animals.

endoderm *n*. inner layer of embryonic epithelium in multicellular animals, which gives rise to digestive tract, respiratory tract and associated glandular epithelium; in coelenterates the inner epithelial layer surrounding the coelenteron. *a*. endo-dermal.

endodermis *n*. innermost layer of cortical cells in stems and roots of vascular plants, the layer surrounding the pericycle.

endoenzyme *n*. any intracellular enzyme.

endogamy *n*. zygote formation within a cyst by the fusion of two of the products of preceding division; self-pollination; inbreeding.

endogastric *a*. having curvature of body with enclosing shell towards ventral side; within the stomach.

endogenote *n*. in bacterial sexual processes, the part of the recipient cell's chromosome that is homologous to the incoming DNA.

endogenous *a*. originating within the organism, cell, or system being studied; originating from a deep-seated layer, *appl*. lateral roots; *appl*. metabolism, biosynthetic and degradative processes in tissues. *cf*. exogenous.

endogenous rhythm a metabolic or behavioural rhythm that originates within the organism and persists although external conditions are kept constant. There may be some slight change in the periodicity, when it is said to be free running, such as a circadian rhythm changing to 23 or 25 hours. Endogenous circadian rhythms are maintained at the regular 24-hour cycle by an external stimulus such as light or temperature, the *zeitgeber*.

endogenous virus viruses (e.g. C-type viruses) carried permanently in an inactive proviral state in the genome and inherited from generation to generation, and which may become activated on infection with another virus.

endognath, endognathite *n.* inner branch of oral appendages of crustaceans.

endogonidium *n.* the colony-forming cell in colonial protists such as *Volvox*.

endolaryngeal *a.* within the larynx.

endolithic *a.* burrowing or existing in a stony substratum, as algal filaments.

endolymph *n.* the fluid filling the semicircular canals and cochlea of the ear. *a.* endolymphatic.

endolymphangial *a.* situated in a lymphatic vessel.

endolymphatic *a. pert.* lymphatic system, or labyrinth of ear.

endolysin *n.* formerly used for intracellular enzymes of macrophages, etc. that destroy engulfed bacteria.

endolysis *n.* destruction of a cell or other material within another cell that has engulfed it.

endomembrane system the intracellular membrane system of a eukaryotic cell, comprising the endoplasmic reticulum, Golgi apparatus, lysosomes and plasma membrane. These are all connected by a flow of membrane from one to the other by means of small membrane vesicles. Mitochondria and chloroplasts, with their double membranes, are not part of this endomembrane system.

endomeninx *n.* single inner membrane covering embryonic brain, giving rise to pia mater and arachnoid.

endomere *n.* a blastomere that gives rise to endoderm.

endometrium *n.* mucous epithelium lining the uterus.

endomitosis *n.* multiplication and separation of interphase chromosomes without subsequent nuclear division, leading to endopolyploidy (*q.v.*).

endomixis *n.* self-fertilization in which a male and female nucleus from the same individual fuse.

endomycorrhiza *see* mycorrhiza.

endomysium *n.* connective tissue binding muscle fibres.

endoneurium *n.* delicate connective tissue holding nerve fibres together in a bundle within a nerve.

endonuclease *n.* nuclease that splits nucleic acid chain at internal sites.

endoparasite *n.* parasite that lives inside another cell or organism.

endopeptidase *n.* a proteinase that splits a protein molecule into smaller units by hydrolysing internal bonds EC 3.4.21–24.

endoperidium *n.* inner layer of peridium.

endophagous *a.* feeding inside a food source.

endophragm *n.* an internal septum formed from projections from the cephalic and thoracic regions in crustaceans.

endophyllous *a.* sheathed by a leaf; living within a leaf.

endophyte *n.* bacterium, fungus, alga or other plant living inside the body or cells of another organism. *alt.* endosymbiont.

endoplasm *n.* the inner part of the cytoplasm of a cell, which is often granular; endosarc *q.v.* cf. ectoplasm.

endoplasmic reticulum (ER) extensive, convoluted internal membrane in eukaryotic cells, continuous with the outer nuclear membrane and enclosing a continuous internal space (lumen). Involved in the synthesis and transport of membrane proteins and lipids and of proteins destined for secretion from the cell. The cytoplasmic face of the membrane may be studded with protein-synthesizing ribosomes, when it is known as rough endoplasmic reticulum (RER), or lack ribosomes, when it is called smooth endoplasmic reticulum (SER). *see* Fig. 1.

endopleura *n.* the inner seed coat.

endopodite *n.* the inner or mesial branch of a two-branched crustacean limb, or if not branched, the only part of the limb remaining.

endopolyploidy *n.* polyploidy resulting from repeated doubling of chromosome number without normal mitosis.

Endoprocta Entoprocta *q.v.*

Endopterygota, endopterygotes *n.,* *n.plu.* a division of insects having complete metamorphosis with wings developing internally and larvae different from adults. Includes the Coleoptera, Neuroptera, Mecoptera, Trichoptera, Lepidoptera, Diptoptera, Siphonaptera, Hymenoptera and other orders. *cf.* Exopterygota.

endoral *a. pert.* structures situated within the "mouth" of certain protozoans.

endorachis *n.* layer of connective tissue lining canal of vertebral column and skull. *alt.* endorhachis.

endoreplication *n.* continued replication of DNA without separation of new chromosomes or cell division, which produces giant polytene chromosomes.

end organ a structure at the end of a nerve, such as a sensory receptor or motor end plate.

endorhizal *a.* with the radicle enclosed, as in seed of monocotyledons.

endorhizosphere *n.* the epidermis and cortex of normal healthy roots which is invaded by soil microorganisms.

endorphins *n.*, *n.plu.* family of peptides produced in gut, brain and pituitary, which can mimic the narcotic effects of morphine by binding to opiate receptors, and which together with the enkephalins and the dynorphins are known as the endogenous opioids. β-endorphin is a peptide of 31 amino acids, generated from proopiomelanocorticotropin, which produces effects such as analgesia, rigidity and changes in mood and behaviour.

endosarc *n.* internal layer of cytoplasm, esp. in protozoans.

endosclerite *n.* any sclerite of the endoskeleton of arthropods.

endosiphuncle *n.* the tube leading from the protoconch to siphuncle in certain cephalopods.

endoskeleton *n.* any internal skeleton or supporting structure.

endosmosis *n.* osmosis in an inward direction, such as into a cell.

endosome *n.* vesicle formed in animal cells by fusion of endocytotic vesicles.

endosperm *n.* the nutritive tissue surrounding embryo in most seeds. *alt.* albumen.

endospore *n.* any spore formed within a sex organ or sporangium; (*bact.*) thick-walled heat- and drought-resistant asexual spore formed within the cells of certain bacteria

e.g. *Bacillus* and *Clostridium* spp. in unfavourable growing conditions, and which can remain dormant for many years.

endosporium *n.* inner coat of spore wall.

endosporous *a.* having spores borne inside an organ such as a sporangium, *appl.* fungi and plants; producing endospores, *appl.* bacteria.

endosteal *a. pert.* endosteum.

endosternite *a.* internal skeletal plate for muscle attachment.

endosteum *n.* membrane lining cavities in bones.

endostosis *n.* ossification that begins in cartilage.

endostracum *n.* the inner layer of mollusc shell.

endostyle *n.* longitudinal groove in ventral wall of pharynx in urochordates and some primitive chordates, involved in mucus secretion.

endosymbiont *see* endosymbiosis.

endosymbiosis *n.* symbiosis in which one partner (the endosymbiont) lives inside the cells of the other, e.g. photosynthetic cyanobacteria living in the cells of nonphotosynthetic dinoflagellates. *a.* endosymbiotic.

endotergite *n.* an infolding from tergite of insects, for muscle attachment.

endotheca *n.* the system of membranes lining and joining polyp cavities in a coral.

endothecium *n.* in bryophytes, inner tissue of sporogonium, formed from central region of embryo; inner lining of an anther. *a.* endothecial.

endothelial *a. pert.* or part of endothelium.

endothelin *n.* polypeptide produced by endothelial cells and which stimulates contraction of the underlying smooth muscle of blood vessel walls.

endotheliochorial *a. appl.* placenta with chorionic epithelium in contact with endothelium of uterine capillaries, as in carnivores and some other mammals.

endothelium *n.* a single layer of flattened cells of mesodermal origin that lines internal body surfaces such as blood and lymphatic vessels, heart, and other fluid-filled cavities. *cf.* epithelium.

endotherm *n.* animal that generates its own body heat, as birds and mammals, formerly known as "warm-blooded" animals.

endothermic *a. appl.* animals able to generate their own body heat.

endotoxin *n.* bacterial toxin that is an integral component of the cell and remains within or attached to the cell, as bacterial lipopolysaccharide. *cf.* exotoxin.

endotrophic *a. appl.* mycorrhizas in which the fungal hyphae extensively penetrate the cells of the host roots.

endozoic *a.* living within an animal or involving passage through an animal as in the distribution of some seeds.

endozoochore *n.* any spore, seed or organism dispersed by being carried within an animal.

endozoochory *n.* dispersal of seeds by ingestion by animals and passage through them.

end-piece filament forming the tail of a sperm.

end plate the contact of a nerve fibre terminal with a muscle fibre.

end-plate potential (epp) change in potential in muscle cell membrane seen on stimulation of its associated nerve fibre.

end-product inhibition feedback control *q.v.*

energy budget the balance of energy input and use in a biological system, expressed as consumption (C) = production (P) + respiration (R) + rejecta (faeces and urine) (F + U). *see also* assimilated energy.

energy charge (EC) index of the energy status of a cell in terms of proportions of ATP, ADP and AMP: EC = ([ATP] + [ADP])/([ATP] + [ADP] + [AMP]). *see also* phosphorylation potential.

energy flow the transfer of energy from one organism to another through the trophic levels in an ecosystem. About 90% of the chemical energy is lost at each transfer.

energy of activation *see* free energy of activation.

enervose *a.* having no veins, *appl.* certain leaves.

engram *n.* a memory trace, a supposed permanent change in the brain accounting for the existence of memory.

engraved *a.* with irregular linear grooves on the surface.

enhalid *a.* containing salt water, *appl.* soils; growing in saltings or on loose soil in salt water, *appl.* plants.

enhancer *n.* type of gene-regulatory site in DNA, present in the control regions of many eukaryotic genes, and also found in prokaryotes, whose activation by the binding of specific regulatory proteins dramatically increases the level of transcription. Enhancers differ from other control sites in that they are still effective if inverted or placed in a different position, often some distance away from the gene they affect; *a. appl.* mutations that intensify the phenotypic effect of other mutations.

enkephalin *n.* pentapeptide found in brain, gut, and adrenal gland. Two main types, Leu-enkephalin and Met-enkephalin. Binds to opiate receptors in brain, mimicking the effects of morphine, producing effects such as analgesia, rigidity and changes in mood and behaviour. Together with endorphins known as the endogenous opiates.

enneagynous *a.* having nine pistils.

enneandrous *a.* having nine stamens.

enolase *n.* enzyme catalysing the formation of phosphoenolpyruvate from 2-phosphoglycerate in glycolysis and other metabolic pathways. EC 4.2.1.11.

Enopla *n.* class of proboscis worms having the mouth anterior to the brain, and usually an armed proboscis.

enphytotic *a.* afflicting plants, *appl.* diseases restricted to a locality.

enrichment culture a technique for isolating a particular microorganism or microorganisms from a natural mixed population by culture in conditions particularly favourable for their growth, so that they become the predominant form in the resulting culture.

ensiform *a.* sword-shaped.

entad *adv.* towards the interior.

ental *a.* inner; internal.

entangial *a.* within a vessel; produced inside a sporangium.

entelechy *n.* in vitalist theories, a vital principle or influence, distinct from physicochemical forces, inherent in living organisms and directing their vital processes.

enteral *a.* within intestine; *appl.* the parasympathetic portion of the autonomic nervous system.

enteric *a. pert.* alimentary canal; *pert.* intestines or gut; *appl.* parasympathetic ganglia.

entero-, -enteron word elements derived from Gk. *enteron*, gut.

enterobacteria *n.plu.* bacteria belonging to the family Enterobacteriaceae, Gram-negative facultatively anaerobic rods, which include some spp. living in soil, some normal inhabitants of the mammalian intestine (e.g. *Escherichia coli*), and some which cause disease (e.g. *Salmonella* spp.).

enterococcus *n.* streptococcus present in the intestinal tract, esp. *Streptococcus faecalis*.

enterocoel *n.* a coelom arising as a pouchlike outgrowth of the archenteron, or as a series of such outgrowths.

enterogastrone *n.* a duodenal hormone which inhibits secretion and motility of stomach.

enterokinase enteropeptidase *q.v.*

enteron *n.* alimentary tract; coelenteron *q.v.*; any structure corresponding to the archenteron of gastrula.

enteronephric *a.* with nephridia opening into gut, *appl.* oligochaetes.

enteropathogenic *a.* producing disease in the intestinal tract.

enteropeptidase enzyme produced by duodenal cells that activates trypsinogen by cleavage of a peptide bond to produce trypsin. EC 3.4.21.9. *alt.* (formerly) enterokinase.

Enteropneusta *n.* class of solitary, worm-like burrowing hemichordates, having many gill slits and no lophophore, commonly called acorn worms.

enteroproct *n.* opening from the endodermal gut into the posterior (ectodermally derived) gut.

enterostome *n.* in coelenterates, the aboral opening of the actinopharynx, leading to the coelenteron; the posterior opening of the stomodaeum into endodermal gut.

enterosympathetic *a. appl.* that part of the sympathetic nervous system supplying the gut.

enterotoxin *n.* toxin produced by certain bacteria infecting the intestine.

enteroviruses *n.plu.* picornaviruses (*q.v.*) that typically occur in the intestine and stomach, and which include, e.g. Coxsackie viruses, echoviruses, poliovirus.

enterozoon *n.* any animal parasite inhabiting the intestines.

enthalpy *n.* a thermodynamic term describing the energy lost as heat to the environment in an exothermic chemical reaction or in an open thermodynamic system such as a living organism.

enthetic *a.* introduced; implanted.

entire *a.* unimpaired; with continuous margin, *appl.* leaves, bacterial colonies, etc.

Entner–Doudoroff pathway metabolic pathway in some bacteria in which glucose is converted to 2-ketogluconic acid.

ento- endo- *q.v.*

entobranchiate *a.* having internal gills.

entobronchus *n.* the dorsal secondary branch of bronchus in birds.

entochondrite *n.* carapace or endosternum of the king crab *Limulus*.

entocoel *n.* space enclosed by a pair of mesenteries in Anthozoa.

entoconid *n.* the postero-internal cusp of a lower molar.

entocuneiform *n.* the most internal of distal tarsal bones.

entoectad *a.* from within outwards.

entoglossal *a.* lying within tissue of tongue.

entoglossum *n.* extension of basihyal into tongue in some fishes.

entomochory *n.* dispersal of seeds or spores through the agency of insects. *a.* entomochoric.

entomofauna *n.* the insects of a particular environment or region.

entomogenous *a.* growing in or on insects, as certain fungi.

entomology *n.* the study of insects.

entomophagous insectivorous *q.v.*

entomophilous *a.* pollinated by insects. *n.* entomophily.

Entomophthorales *n.* order (in some classifications considered as a class) of zygomycete fungi chiefly parasitic on insects and also known as fly fungi, which reproduce by forcibly discharged conidia, and have a mycelium with septa and a tendency to fragment.

entomophyte *n.* any fungus growing on or in an insect.

entomostracans *n.plu.* large group of small crustaceans including branchiopods, Branchiura (fish lice), copepods, barnacles and ostracods.

entoneural *a. appl.* system of aboral ring and genital nerves in enchinoderms.

entopic *a.* in the normal position. *cf.* ectopic.

entoplastron *n.* the anterior median plate in the shell of turtles or tortoises, often called episternum.

Entoprocta, entoprocts *n., n.plu.* phylum of solitary or colonial small marine pseudocoelomate animals, in some classifications included in the Bryozoa, having a U-shaped gut with anus opening to the exterior within a circle of ciliated tentacles.

entopterygoid *n.* a dorsal membrane bone behind palatine in some fishes.

entoptic *a.* within the eye; *appl.* visual sensations caused by eye structures and processes not by light.

entosternum *n.* entoplastron *q.v.*; an internal process of sternum of numerous arthropods.

entoturbinals *n.plu.* a division of ethmoturbinals.

entotympanic *n.* a separate tympanic element in some mammals.

entovarial *a.* inside the ovary.

entozoic *a.* living inside the body of another animal or plant.

entrainment *n.* process by which a free-running endogenous circadian rhythm is synchronized to an exact 24-hour cycle.

entropy *n.* a thermodynamic term describing the total disorder or randomness of a system.

enucleate *n.* lacking a nucleus; *v.* to remove nucleus from a cell.

envelope *n.* of certain viruses, layer of lipid and protein surrounding capsid, derived largely from host cell plasma membrane as virus is discharged from the cell. Such viruses are known as enveloped viruses. *see also* cell envelope, floral envelope, nuclear envelope.

environment *n.* the sum total of external influences acting on an organism.

environmental audit a survey of all aspects of an environment, including e.g. the geology, hydrology, and habitats and species present, and human influences, often undertaken as part of an environmental impact assessment.

environmental impact assessment (EIA) an evaluation of the likely environmental consequences of a proposed development and of the measures to be taken to minimize adverse effects, which is now a legal requirement in some countries.

environmental resistance the factors limiting the population growth of a species in a given environment.

environmental science the study of how humans and other species interact with their non-living and living environment, which includes the study of ecology, resource management and conservation, population dynamics, economics, politics and ethics.

enzootic *a.* afflicting animals; *appl.* disease in animals restricted to a locality.

enzymatic *a. pert.* enzymes, enzyme action or reactions catalysed by enzymes. *alt.* enzymic.

enzyme *n.* any of a large and diverse group of (mainly) proteins that function as biological catalysts in virtually all biochemical reactions, essential in all cells, different enzymes being highly specific for a particular chemical reaction and reactants. Enzymes are classified according to the type of chemical reaction catalysed and substrate acted on, e.g. hydrolases, carboxylases, oxidoreductases, nucleases, proteinases. Certain RNAs can also act as enzymes, *see* ribozyme. *see also* immobilized enzyme.

enzyme kinetics the progress of an enzymatic reaction over time, the exact course of which is dependent on substrate concentration, temperature and the molecular properties of the enzyme (e.g. allostery).

enzyme-linked immunosorbent assay (ELISA) a type of serological assay in which the antibodies used to detect a particular substance are labelled by linkage to an enzyme. The test substance is immobilized on a plastic surface and a positive reaction, i.e. antibody binding to the surface, is detected by the action of the enzyme on a colourless substrate to produce a coloured product.

enzymic enzymatic *q.v.*

enzymology *n.* study of enzymes and their action.

eobiogenesis *n.* the formation of living matter from prebiotic macromolecular systems.

Eocene *n.* early epoch of the Tertiary period, between Paleocene and Oligocene, lasting from around 55 million years to 40 million years ago.

Eogaea *n.* a zoogeographical division including Africa, South America and Australasia.

eolian *alt.* spelling of aeolian *q.v.*

eosin *n.* red/brown acidic dye, sodium or potassium salt of eosin ($C_{20}H_8Br_4O_5$).

eosinophil *n.* type of white blood cell classed as a granulocyte and characterized by staining with the red acidic dye eosin, involved in destruction of internal parasites and modulation of allergic inflammatory reactions. *alt.* acidophil.

eosinophilia *n.* abnormal increase in number of eosinophils, seen in certain allergic states and parasitic infections.

eosinophilic *a.* staining readily with eosin.

epacme *n.* the stage in the phylogeny of a group just before its highest point of development; stage in development of an individual just before adulthood.

epactal *a.* supernumerary; intercalary; *appl.* bones situated in sutures between two other bones.

epalpate *a.* lacking palps.

epanthous *a.* living on flowers, *appl.* certain fungi.

epapillate *a.* lacking papillae.

epapophysis *n.* a median process arising from centre of vertebral neural arch.

eparterial *a.* situated above an artery.

epaulettes *n.plu.* branched or knobbed processes projecting from oral arms of some jellyfish; tegula (*q.v.*) in Hymenoptera.

epaxial *a.* above the axis, dorsal, usually *appl.* axis formed by the vertebral column.

ependyma *n.* layer of cells lining cavities of brain and spinal cord. *a.* ependymal.

ephedrine *n.* a sympathomimetic alkaloid obtained from *Ephedra* spp., having the same effects as adrenaline and used as a nasal decongestant, and to treat hayfever and asthma.

ephemeral *a.* short-lived; taking place once only, *appl.* plant movements as expanding buds; completing life cycle within a brief period; *n.* a short-lived plant or animal species.

Ephemeroptera, ephemerids *n., n.plu.* order of insects including the mayflies, whose adult life is very short, sometimes less than a day. They have an aquatic nymph (larval) stage. They have three appendages, "tails", projecting from the end of the abdomen, and the hindwing is smaller than the forewing.

ephippial *a. appl.* winter eggs, as of rotifers and daphnids.

ephippium *n.* the pituitary fossa, the cavity in which the pituitary body lies in the sphenoid bone; saddle-shaped modification of cuticle detached from carapace and enclosing winter eggs in daphnids.

ephyra *n.* immature medusa in some jellyfish, formed by strobilization from a polyp. *plu.* ephyrae.

epi- prefix derived from Gk. *epi*, upon, signifying above or upon.

epibasal *n.* upper segment of plant zygote or embryo, ultimately giving rise to the shoot.

epibasidium *n.* upper part of cell of septate basidium from which sterigmata and basidiospores are produced.

epibenthos *n.* fauna and flora of sea bottom between low-water mark and 200-metre line.

epibiotic *a. appl.* endemic species that are relics of a former flora or fauna; growing on the exterior of living organisms; living on a surface, as of sea bottom.

epiblast *n.* outer layer of the gastrula, giving rise to ectoderm. *a.* epiblastic.

epiblem(a) *n.* the outermost layer of tissue in roots, which may be the piliferous layer, or the exodermis in an older root where the piliferous layer has worn away.

epiboly *n.* the growth of one part down over another, as the growth of ectoderm down over mesoderm in amphibian gastrulation.

epibranchial *n.* a dorsal skeletal element of gill arch.

epicalyx *n.* calyx-like structure formed outside but close to the true calyx, formed from stipules fused in pairs, or by aggregation of bracts, bracteoles or small sepal-like structures.

epicanthus *n.* a prolongation of upper eyelid over inner angle of eye. *alt.* epicanthic fold.

epicardia *n.* that part of oesophagus running from diaphragm to stomach; *plu.* of epicardium *q.v.*

epicardium *n.* that part of the pericardium that is reflexed over the viscera (internal organs) including the heart.

epicarp exocarp *q.v.*

epicentral *a.* attached to or arising from vertebral centra, *appl.* intermuscular bones.

epicerebral *a.* situated above the brain.

epichilium *n.* in orchid flowers, the outer part of the lip where there are two distinct parts.

epichondrosis *n.* formation of cartilage on the fibrous membrane surrounding a bone, as in formation of antlers.

epichordal *a.* upon the notochord; *appl.* vertebrae in which the ventral cartilaginous portions are almost completely suppressed; *appl.* upper lobe of caudal fin in fishes.

epichroic *a.* discolouring, as after injury.

epiclinal *a.* situated on the receptacle or torus of a flower.

epicoel *n.* cavity of mid-brain in lower vertebrates; cerebellar cavity; a perivisceral cavity formed by invagination.

epicondyle *n.* a medial and a lateral protuberance at distal end of humerus and femur. *a.* epicondylar, epicondylic.

epicone *n.* the part anterior to the girdle in dinoflagellates.

epicoracoid *a. pert.* an element, usually cartilaginous, at sternal end of cora- coid in amphibians, reptiles, and monotremes.

epicormic *a.* growing from a dormant bud.

epicortex *n.* an outer layer, as of filaments covering cortex of some fungi.

epicotyl *n.* the stem-like axis of the young plant embryo above the cotyledons, terminating in an apical meristem and sometimes bearing one or more young leaves.

epicranial *a. pert.* the cranium; *pert.* the epicranium of insects.

epicranium *n.* region between and behind eyes in insect head; scalp; the structures covering the cranium in vertebrates.

epicritic *a. appl.* stimuli and nerves concerned with delicate touch and other special sensations in skin.

epicuticle *n.* outer waxy layer of the exoskeleton of arthropods.

epicutis *n.* outer layer of cutis in agarics.

epicyst *n.* the external resistant cyst of an encysted protozoan.

epicytic *a.* on the surface of a cell.

epideictic display a suggested display by which members of a population reveal their presence and allow others to assess the density of the population.

epidemic *a.* affecting a large number of individuals at the same time, *appl.* disease; *n.* an outbreak of epidemic disease.

epidemiology *n.* the study of the occurrence of infectious diseases, their origins and pattern of spread through the population; the cause and pattern of spread of a disease; the study of the incidence of non-infectious disease (e.g. cancer) with a view to finding causes, e.g. the causal link between smoking and lung cancer was found by epidemiological studies.

epidermal *a. pert.* epidermis.

epidermal growth factor (EGF) a polypeptide that stimulates the division of cells of ectodermal and mesodermal origin, and promotes precocious eyelid opening in mice. Role *in vivo* is not yet known. *alt.* urogastrone.

epidermis *n.* outer layer or layers of the skin, derived from embryonic ectoderm. In vertebrates a non-vascular stratified tissue, often keratinized; outer epithelial covering of roots, stems and leaves in plants. *a.* epidermal. *see* Fig. 10.

epidermoid *a.* resembling epidermis.

epididymis *n.* mass at back of testicle composed mainly of ductules leading from testis to vas deferens. *a.* epididymal. *plu.* epididymides.

epidural *a. pert.* dura mater; *appl.* space between dura mater and wall of vertebral canal around spinal cord.

epifauna *n.* animals living on the surface of the ocean floor; any encrusting fauna.

epigaeous epigeal *q.v.*

epigamic *a. appl.* any trait related to courtship and sex other than the essential sexual organs and copulatory behaviour; tending to attract the opposite sex, e.g. *appl.* colour displayed in courtship.

epigaster *n.* that part of embryonic intestine that later develops into the colon.

epigastric *a. pert.* anterior wall of abdomen; middle region of upper zone of artificial divisions of abdomen.

epigastrium *n.* the middle region of upper part of abdomen above the navel.

epigeal, epigean, epigeous *a.* borne

above ground; *appl.* type of germination in which cotyledons are carried above ground as shoot grows; *appl.* insects, living near the ground.

epigenesis n. the accepted central concept of embryonic development, that an embryo is formed by the gradual differentiation and organization of its parts from an undifferentiated single fertilized egg cell. This was originally opposed to the earlier theory of preformation, which held that the sperm (or the egg) contained a fully formed miniature individual, and that development consisted only in an increase in size.

epigenetic *a. appl.* the chain of processes linking genotype and phenotype, other than the initial gene action.

epigenous *a.* developing or growing on a surface.

epigeous epigeal *q.v.*

epiglottis n. moveable flap of fibrocartilage which bends over the opening of the windpipe in throat when food is being swallowed.

epignathous *a.* having upper jaw longer than lower.

epigonial *a. appl.* sterile posterior portion of genital ridge.

epigonium n. the cover over the young sporogonium of a liverwort; any calyptra.

epigyne, epigynium, epigynum n. external female genitalia in arachnids.

epigynous *a. appl.* flowers in which sepals, petals and stamens are attached on top of the ovary. n. epigyny.

epilimnion n. upper water layer, above thermocline in lakes, rich in oxygen.

epilithic *a.* attached on rocks, *appl.* algae, lichens.

epimandibular *a. pert.* a bone in the lower jaw of vertebrates.

epimeletic *a. appl.* animal behaviour relating to the care of others.

epimer n. either of two molecules, esp. monosaccharides, that differ only in the arrangement of atoms around a single carbon atom other than the carbonyl carbon atom.

epimerase n. any enzyme that can convert one epimer into the other, included in EC 5.1.

epimere n. the upper part of a somite in vertebrate embryos, giving rise to muscle.

epimerite n. prolongation of protomerite of certain gregarines for attachment to host.

epimerization n. the conversion of one epimer into another.

epimeron n. the posterior part of the side wall of any segment in insects.

epimorphic *a.* maintaining the same form in successive stages of growth.

epimorphosis n. type of regeneration in which proliferation of new material precedes development of new part.

epimysium n. sheath of areolar tissue which invests the entire muscle.

epinasty n. the more rapid growth of upper surface of a dorsoventral organ, e.g. a leaf, thus causing unrolling or downward curvature.

epinekton n. nekton that are incapable of actively swimming themselves but are attached to actively swimming organisms.

epinephrin(e) preferred name for adrenaline (*q.v.*) in North America.

epineural *a.* arising from vertebral neural arch; *pert.* canal external to radial nerve in some echinoderms; *appl.* sinus between embryo and yolk, beginning of body cavity in insects.

epineurium n. the fibrous connective tissue sheath around a nerve.

epineuston n. those animals living at the surface of water, in the air.

epinotum propodeon *q.v.*

epiostracum n. thin cuticle or epicuticle covering exocuticle or exostracum in mites and ticks.

epiotic *a. pert.* upper element of bony capsule of ear.

epiparasite ectoparasite *q.v.*

epipelagic *a. pert.* deep-sea water between surface and bathypelagic zone; or, inhabiting oceanic water at depths not exceeding around 200 m, i.e. above mesopelagic zone.

epipelon n. algal community living in or on the surface of sediments in shallow waters where light penetrates.

epipetalous *a.* having stamens inserted upon petals.

epipharynx n. the upper lip of dipteran insects (e.g. flies), adapted as a piercing organ in some cases.

epiphenomenon n. something produced as a side-effect of a process.

epiphloem outer bark, periderm *q.v.*

epiphragm *n.* membrane or plate that closes an opening, e.g. the shell of certain molluscs, or capsule of certain mosses.

epiphyll *a.* a plant that grows upon leaves, e.g. various lichens.

epiphyllous *a.* growing upon leaves; united to perianth, *appl.* stamens.

epiphysis *n.* any part or process of a bone which is formed from a separate centre of ossification and later fuses with the bone, becoming its terminal portion; pineal body *q.v.*

epiphyte *n.* a plant that lives on the surface of other plants but does not derive water or nourishment from them.

epiphyton *n.* community of plants living attached to other plants, as many algae in aquatic environments.

epiphytotic *a. pert.* disease epidemic in plants.

epiplankton *n.* that portion of plankton from surface to about 200 m.

epiplastron *n.* in turtles and tortoises, one of a pair of anterior bony plates in shell.

epipleura *n.* a rib-like structure in teleost fishes which is not preformed in cartilage; hooked process on rib in birds; turned down outer margin of elytra in some beetles.

epiploic *a. pert.* omentum *q.v.*

epiploic foramen opening between bursa omentalis and large sac of peritoneum.

epiploon *n.* greater omentum *q.v.*; fat-body (*q.v.*) of insects.

epipodite *n.* process arising from basal joint of crustacean limb, modified for various functions.

epiproct *n.* the central "tail" of silverfish and other members of the insect order Thysanura.

epipteric *a.* winged at tip, *appl.* certain seeds; *pert.* or shaped like, or placed above wing.

epipterygoid *n.* small bone extending nearly vertically downwards from pro-otic to pterygoid in skull of some reptiles.

epipubic *a. pert.* or borne upon the pubis.

epipubis *n.* unpaired cartilage or bone borne anteriorly on pubis.

epirhizous *a.* growing upon a root.

episclera *n.* connective tissue between sclera and conjunctiva.

episematic *a.* aiding in recognition, *appl.* coloration, markings. *cf.* aposematic, parasematic, sematic.

episeme *n.* a marking or colour aiding in recognition.

episepalous *a.* adnate to sepals.

episkeletal *a.* outside the skeleton.

episome *n.* autonomous self-replicating DNA in bacteria which can integrate into the bacterial chromosome semi-permanently, e.g. F factor, some plasmids; often used interchangeably with plasmid. *a.* episomal.

episperm *n.* the outer coat of seed.

epistasis, epistasy *n.* the suppression or masking of the effect of a gene by another non-allelic gene. *a.* epistatic.

epistatic *a. appl.* gene that masks or suppresses the expression of another non-allelic gene.

episternalia *n.plu.* two small elements preformed in cartilage frequently intervening in development between clavicles and sternum, and ultimately fusing with sternum.

episternum *n.* a bone between the clavicles; the anterior part of the side wall of any of the thoracic segments in insects.

epistome *n.* small lobe overhanging mouth in Polyzoa and containing part of body cavity; region between antenna and mouth in Crustacea; portion of insect head immediately behind labrum; portion of rostrum in some dipterans.

epistrophe *n.* the position assumed by chloroplasts along outer and inner wall of cell when exposed to diffuse light.

epistropheus *n.* the second cervical or axis vertebra.

epithalamus *n.* part of brain comprising habenula, pineal body, and posterior commissure.

epithalline *a.* growing upon the thallus.

epithallus *n.* cortical layer of hyphae in lichens.

epitheca *n.* an external layer surrounding theca in many corals; older half of frustule in diatoms.

epithelia *plu.* of epithelium.

epitheliochorial *a. appl.* placenta with apposed chorionic and uterine epithelia, and villi pitting the uterine wall, as in marsupials and ungulates.

epithelioid *a.* resembling epithelium.

epithelium *n.* sheet of cells tightly bound together, lining any external or internal surface in multicellular organisms, e.g. the epidermis, surfaces of mucous membranes, the lining of the gut, and the linings of ducts and glands. Epithelia variously serve protective, secretory or absorptive functions. *plu.* epithelia. *a.* epithelial.

epitope antigenic determinant *q.v.*

epitreptic *a. appl.* animal behaviour causing another animal of the same species to approach.

epitrichium *n.* an outer layer of foetal epidermis of many mammals, usually shed before birth.

epitrochlea *n.* inner condyle at distal end of humerus.

epizoic *a.* living on or attached to the body of an animal; having seeds or fruits dispersed by being attached to the surface of an animal.

epizoite *n.* organism that lives on the shell or surface of another animal but is not parasitic on it.

epizoochory *n.* dispersal of seeds by being carried on the body of an animal.

epizoon *n.* animal that lives on the body of another animal.

epizootic *n.* epidemic disease amongst animals.

eplicate *a.* not folded; not plaited.

epoch *n.* in geological time, the subdivision of a period.

eponychium *n.* the thin fold of cuticle which overlaps the lunula (half-moon) of nail.

eponym *n.* name of a person used in designation of an entity, as species, organ, law, disease, etc.

Epstein-Barr virus *see* EBV.

equatorial furrow division around equator of egg undergoing the first cleavage divisions.

equatorial plate imaginary plane halfway between opposite poles of the spindle and at right angles to the spindle axis and on which the centromeres of the chromosomes lie at metaphase of mitosis, *alt.* metaphase plate; site of new cell wall formation after cell division.

equiaxial *a.* having axes of equal length.

equibiradiate *a.* with two equal rays.

equicellular *a.* composed of equal number of cells, or composed of cells of equal size.

equifacial *a.* having equivalent surfaces or sides, as vertical leaves.

equifinality *n.* the arrival at a common end point in behavioural development by different routes.

equilateral *a.* having the sides equal; *appl.* shells symmetrical about a transverse line drawn through umbo.

equilibrium constant *see* affinity constant.

equilibrium potential the membrane potential (*q.v.*) at which a particular ion can pass easily in either direction across the membrane.

equimolecular *see* isotonic.

equinoctial *a. appl.* flowers that open and close at definite times.

equipotent *a.* totipotent *q.v.*; able to perform the function of another cell, part or organ.

equipotential *a.* of equal developmental potential.

equitant *a.* overlapping like a saddle, as leaves in leaf bud, leaves on stem of e.g. iris.

Equisetales horsetails. *see* Sphenophyta.

equivalence group a group of embryonic cells with the same developmental potential.

equivalve *a.* having two halves of shell alike in form and size.

era *n.* a main division of geological time, such as Paleozoic, Mesozoic, Cenozoic, and divided into periods.

eradication *n.* the extinction of a species in a particular area.

erect *a.* directed towards summit of ovary, *appl.* ovule; not decumbent, *appl.* plants.

erectile *a.* capable of being raised or erected, as of a penis or crest.

erector *n.* a muscle which raises an organ or part.

Eremian *a. appl.* or *pert.* part of the Palaearctic region including deserts of North Africa and Asia.

eremic *a. pert.*, or living in, deserts.

eremobic *a.* growing or living in isolation; having a solitary existence.

eremophyte *n.* a desert plant.

ergaloid *a.* having adults sexually capable though wingless, *appl.* insects.

ergatandrous *a.* having worker-like males.

ergataner *n.* male ant resembling worker.

ergate(s) *n.* a worker ant.

ergatogyne *n.* a female ant resembling a worker, intermediate between queen and worker.

ergatoid *a.* resembling a worker, *appl.* male ant.

ergocalciferol *see* calciferol.

ergonomics *n.* the anatomical, physiological and psychological study of humans in their working environment.

ergonomy *n.* the differentiation of functions; physiological differentiations associated with morphological specialization.

ergosterol *n.* a sterol present in yeasts, moulds and certain algae, and in animal tissues, and which is converted to vitamin D_2 (ergocalciferol) on irradiation with ultraviolet light.

ergot *n.* the hardened mycelial mass (sclerotium) of the fungus *Claviceps purpurea* which replaces the grain of infected rye and some other grasses, and which contains poisonous alkaloids that cause ergotism – abortion, hallucinations and sometimes death – in animals and humans who eat the infected grain.

ericaceous *a.* of or *pert.* the Ericaceae, the heather family, which includes the heaths, heathers and rhododendrons.

Ericales *n.* order of dicot shrubs, rarely trees or herbs, and including the families Actinidiaceae, Empetraceae (crowberry), Ericaceae (heaths and rhododendrons), Monotropaceae (Indian pipe) and Pyrolaceae (wintergreen).

ericoid *a. appl.* endomycorrhizas formed on members of the Ericaceae, lacking a fungal sheath and with extensive penetration of the cell of the root cortex, with formation of intracellular hyphal coils.

erineum *n.* an outgrowth of abnormal hairs on leaves produced by certain gall mites.

Eriocaules *n.* order of moncot herbs comprising the family Eriocaulaceae (pipewort).

eriocomous *a.* having woolly hair; fleecy.

eriophyid *n.* any of a large family (Eriophyidae) of minute plant-eating mites that have two pairs of legs at front and no respiratory system.

eriophyllous *a.* having leaves with a cottony appearance.

erose *a.* having margin irregularly notched, *appl.* leaf, bacterial colony.

erostrate *a.* having no beak, *appl.* anthers.

errantia *n.* mobile organisms *cf.* sedentaria.

eruca *n.* a caterpillar; an insect larva having the shape of a caterpillar.

erucic acid unusual unsaturated fatty acid $(C_{22}H_{42}O_2)$ found in large quantities in rapeseed (*Brassica campestris*) and some other related plants.

eruciform *a. appl.* insect larvae with a more-or-less cylindrical body and stumpy legs on the abdomen as well as the true thoracic legs, as caterpillars and grubs; having the shape of a caterpillar or grub.

erumpent *a.* breaking through suddenly, *appl.* fungal hyphae.

erythrism *n.* abnormal prescence, or excessive amount, of red colouring matter, as in petals, feathers, hair, eggs.

erythroaphins *n.plu.* red pigments formed by the postmortem enzymatic transformation of yellow plant pigments in aphids.

erythroblast *n.* nucleated cell of bone marrow which gives rise to erythrocytes.

erythrocruorin *n.* formerly used for the haemoglobin respiratory pigments of annelids and molluscs.

erythrocuprein superoxide dismutase *q.v.*

erythrocyte *n.* the predominant type of cell in vertebrate blood. It is small, disc-shaped, lacks most internal organelles (including the nucleus in mammals) and contains the oxygen-binding protein haemoglobin, by means of which it transports oxygen. *alt.* red blood cell, red blood corpuscle.

erythrocyte colony-forming unit (CFU-E) erythrocyte precursor cell which forms colonies of erythrocytes in culture if erythropoietin is present.

erythrocytic *a. appl.* phase of the malarial plasmodium life cycle in humans in which merozoites invade red blood cells.

erythrocytic burst-forming unit (BFU-E) bone marrow cell which in culture forms large colonies of erythrocytes if stimulated with erythropoietin, presumed to be a precursor of the erythrocyte colony-forming unit *q.v.*

erythrogenic *a.* producing reddening, as in inflammation.

erythrolabe *n.* the red-sensitive pigment of the human eye.

erythromycin *n.* an antibiotic synthesized

by the actinomycete *Streptomyces erythreus*, which inhibits bacterial protein synthesis.

erythron *n.* the red cells of blood and bone marrow collectively.

erythrophilous *a.* having an affinity for red stains.

erythrophyll *n.* a red anthocyanin as found in some leaves and red algae.

erythropoiesis *n.* the production of red blood cells.

erythropoietin *n.* a glycoprotein hormone produced chiefly by the kidney and which stimulates the final differentiation of red blood cells from precursor cells.

erythrose *n.* 4-carbon sugar, as the phosphate derivative involved esp. in carbon fixation in green plants.

escape *n.* plant or animal originally domesticated and now established in the wild; *a.* appl. behaviour in which an animal moves away from an unpleasant stimulus.

Escherichia coli a generally harmless bacterial inhabitant of the colon of humans and some other mammals, widely used as an experimental subject in bacterial genetics and as a host bacterium in recombinant DNA work.

escutcheon *n.* area on rump of many quadrupeds which is either variously coloured or has the hair specially arranged.

eseptate *a.* without septa.

eserine *n.* plant alkaloid, specific inhibitor of cholinesterase. *alt.* physostigmine.

esophageal, esophagus *alt.* spelling of oesophageal, oesophagus *q.v.*

esoteric *a.* arising within the organism.

espathate *a.* having no spathe.

esquamate *a.* having no scale.

essential amino acids amino acids which cannot be synthesized by the body, or only in insufficient amounts, and must be supplied in the diet: for humans these are Arg, His, Ile, Leu, Lys, Met, Phe, Thr, Trp, Val.

essential elements trace elements *q.v.*

essential fatty acids (EFA) fatty acids that cannot be synthesized by mammals and must be present in the diet, e.g. linoleate and linolenate.

essential oils mixtures of various volatile oils derived from benzenes and terpenes found in plants and producing characteristic odours, and having various functions

such as attracting insects or warding off fungal attacks.

esterase *n.* any hydrolytic enzyme that attacks an ester, splitting off the acid, EC 3.1.

esterification *n.* formation of an ester.

esters *n.plu.* compounds formed by condensation of an acid with an alcohol, and including the fats and oils, which are esters of fatty acids and glycerol.

esthesia *alt.* spelling of aesthesia *q.v.*

estipulate *a.* having no stipules.

estival *alt.* spelling of aestival *q.v.*

estivation *alt.* spelling of aestivation *q.v.*

estradiol, estrin, estriol, estrogen, estrone, estrous, estrus *alt.* spellings of oestradiol, oestrin, oestriol, oestrogen, oestrous, oestrus, *all q.v.*

estriate *a.* not marked by narrow parallel grooves or lines; not streaked *cf.* striate.

estuarine *a.* living in the lower part of a river or estuary where freshwater and seawater meet.

etaerio *n.* an aggregate fruit composed of achenes, berries, drupels, follicles or samaras.

ethanol *n.* ethyl alcohol, CH_3CH_2OH, which is produced from sugars by fermentation by yeasts and other microorganisms, in which the metabolic intermediate pyruvate is converted into acetaldehyde and then into ethanol. Large-scale industrial biotechnological processes for producing ethanol include brewing and wine-making as well as the production of fuel alcohol from organic wastes.

ethanolamine *n.* small hydrophilic molecule, forms the headgroup in phosphatidylethanolamine.

ethereal *a. appl.* a class of odours including those of ethers and fruits; *appl.* fragrant oils in many seed plants.

ethidium bromide reagent showing orange fluorescence on binding to double-stranded DNA.

Ethiopian *a. pert.* a zoogeographical region including Africa south of the Sahara and the south-western part of the Arabian peninsula, and divisible into African and Malagasy subregions; *pert.* a floral region that is part of the Palaeotropical floral realm, comprising Ethiopia and the south-western tip of the Arabian peninsula.

ethmohyostylic *a.* with mandibular suspension from ethmoid region and hyoid bar.

ethmoid *a. pert.* bones that form a considerable part of the nasal cavity.

ethmoidal *a. pert.* ethmoid bones or region.

ethmoidal notch a quadrilateral space separating the two orbital parts of the frontal bone.

ethmopalatine *a. pert.* ethmoid and palatine bones or their region.

ethmoturbinals *n.plu.* cartilages or bones in the nasal cavity which are folded so as to increase area of olfactory epithelium.

ethmovomerine *a. pert.* ethmoid and vomer regions; *appl.* the cartilage that forms the nasal septum in early embryo.

ethnobotany *n.* the study of the use of plants by humans.

ethnography *n.* the description and study of human races.

ethnology *n.* science dealing with the different human races, their distribution, relationship and activities.

ethnozoology *n.* the study of the use of animals by humans.

ethogram *n.* a catalogue of the natural behaviours of an animal and the contexts in which they occur.

ethological isolation the prevention of interbreeding between species as the result of behavioural differences.

ethology *n.* the study of the behaviour of animals in their natural habitats. *a.* ethological.

ethylene *n.* C_2H_4, a gas produced by plants in minute amounts and which has various developmental effects as a hormone, including regulation of fruit ripening.

ethylenediaminetetraacetate EDTA *q.v.*

etiolation *n.* the appearance of plants grown in the dark, having no chlorophyll, chloroplasts not developing, internodes being greatly elongated so the plant is tall and spindly, and having small, rudimentary leaves. *a.* etiolated.

etiolin protochlorophyll *q.v.*

etiological *alt.* spelling of aetiological *q.v.*

etiology *alt.* spelling of aetiology *q.v.*

etioplast *n.* chloroplast formed in the absence of light, found in etiolated leaves, lacking thylakoid membranes, chlorophyll, and which will develop into a functional chloroplast on illumination; chloroplast precursor.

-etum in ecology, a suffix used to indicate a plant commmunity dominated by a particular species, e.g. a callunetum, a community dominated by heather or ling (*Calluna vulgaris*).

euapogamy *n.* development of a diploid sporophyte from one or more cells of the gametophyte without fusion of gametes. For haploid apogamy, *see* meiotic apogamy.

Euascomycetae *n.* large class of ascomycete fungi including the black moulds, blue moulds, powdery mildews, discomycetes, tar spots, morels and truffles, in which the asci are enclosed in an ascocarp *q.v.*

euaster *n.* aster in which rays meet at a common centre.

eubacteria *n.plu.* the "true" bacteria, unicellular prokaryotic microorganisms possessing cell walls, with cells in the form of rods, cocci or spirilla, many species motile with cells bearing one or more flagella. They are distinguished from the archaebacteria (*q.v.*) by the posession of peptidoglycan cell walls and ester-linked lipids. They include the Gracilicutes, and the Firmicutes. *see* Appendix 6.

eucarpic *a.* having a thallus differentiated into a soma and fruiting body, *appl.* Chytridiomycetes and related fungi having rhizoids or haustoria.

eucaryote, eucaryotic eukaryote, eukaryotic *q.v.*

eucentric pericentric *q.v.*

eucephalous *a.* with well-developed head, *appl.* certain insect larvae.

euchroic *a.* having normal pigmentation.

euchromatin *n.* that portion of the chromatin other than the heterochromatin, found in an uncoiled state in the interphase nucleus and containing DNA sequences that can be transcribed (the active genes). *a.* euchromatic.

Eucommiales *n.* an order of dicot trees comprising a single family Eucommiaceae with one genus *Eucommia*.

eucone *a.* having crystalline cones fully developed in each ommatidium, *appl.* compound eyes.

eudominant *n.* a dominant species that is more or less restricted to a particular climax vegetation.

eudoxome *n.* free-swimming stage of a siphonophore lacking the nectocalyx.

eugamic *a.* appl. mature period rather than youthful or senescent.

eugenic *a. pert.* or able to increase the fitness of a race or breed.

eugenics *n.* a pseudoscientific philosophy at its height in Europe and the United States in the early 20th century which aimed to "improve" the genetic quality of the human population, and which eventually led to abuses such as compulsory sterilization of those deemed "unfit" and persecution of racial minorities.

euglenoid *a. pert.* or like the protistan *Euglena*; *appl.* movement: movement resulting from a change in shape, as in *Euglena*.

Euglenophyta, euglenoids *n., n.plu.* phylum of unicellular flagellate protists typified by *Euglena*, which have no rigid cell wall, have chlorophyll and carotenoid pigments (although pigment may be absent) and store food as fat or the polysaccharide paramylon. In older botanical classifications they are treated as a division of the algae. In zoological classifications they are included in the Mastigophora. *see* Appendices 2 and 5.

eugonic *a.* prolific; growing profusely, *appl.* bacterial colonies.

euhaline *a. appl.* seawater or water of comparable salinity, i.e. *ca.* 35 parts per thousand of sodium chloride (salt); living only in saline waters.

euhyponeuston *n.* organisms living in the top 5 cm of water for the whole of their lives.

Eukaryota *n.* one of two proposed "superkingdoms", the other being the Prokaryota, into which all living organisms would be divided, the Eukaryota to include all organisms with eukaryotic cells.

eukaryotes *n.plu.* organisms with cells possessing a membrane-bounded nucleus in which the DNA is complexed with histones and organized into chromosomes. Eukaryotic cells also have an extensive cytoskeleton of protein filaments and tubules, and many cellular functions are sequestered in membrane-bounded organelles such as mitochondria, chloroplasts, endoplasmic reticulum and Golgi apparatus. The eukaryotes comprise protozoans, algae, fungi, slime moulds, plants and animals. *a.* eukaryotic. *cf.* prokaryotes.

eulamellibranch *a. appl.* gills of bivalve molluscs whose filaments are attached to adjacent ones by bridges of tissue.

eumelanin *n.* black melanin.

eumerism *n.* an aggregation of like parts.

eumeristem *n.* meristem composed of isodiametric thin-walled cells with dense cytoplasm and large nuclei.

eumerogenesis *n.* segmentation in which the units are similar for at least some time.

eumetazoa *n.* the multicellular animals excluding the sponges.

eumitosis *n.* typical mitosis, as occurs in the cells of most multicellular plants and animals.

Eumycota *n.* the "true" fungi, comprising the classes Ascomycotina, Basidiomycotina, Deuteromycotina, Mastigomycotina and Zygomycotina, and excluding the slime moulds (Myxomycota). *see* Appendix 3.

euphausiid *n.* a member of the order Euphasiacea, small, usually luminescent shrimp-like crustaceans forming an important part of the marine plankton.

Euphorbiales *n.* order of dicot trees, shrubs and occasionally herbs, including the families Buxaceae (boxwood), Euphorbiaceae (spurge), Simmondsiaceae (jojoba) and others.

euphotic *a.* well-illuminated, *appl.* zone of surface waters to depth of around 80 100 m; upper layer of photic zone.

euphotometric *a. appl.* leaves oriented to receive maximum diffuse light.

euplankton *n.* the plankton of open water.

euploid *a.* having an exact multiple of the haploid number of chromosomes, e.g. being diploid, triploid, tetraploid, etc.; also *appl.* chromosomal abnormalities that do not disrupt relative gene dosage, *alt.* balanced. *n.* euploidy.

eupotamic *a.* thriving both in streams and in their backwaters, *appl.* plankton.

Euptales *n.* order of dicot trees and shrubs comprising the family Eupteleaceae with a single genus *Euptelea*.

Euro-Siberian Floral Region part of the

Holarctic Realm comprising the whole of Europe from southern Scandinavia to northern Spain, and Asia north of the Caspian Sea to northern Japan.

eurybaric *a. appl.* animals adaptable to great differences in altitude or pressure.

eurybathic *a.* having a wide range of vertical distribution.

eurybenthic *a. pert.* or living within a wide range of depths in the sea, of organisms that live on the ocean floor.

eurychoric *a.* widely distributed.

euryhaline *a. appl.* marine organisms adaptable to a wide range of salinity.

euryhygric *a. appl.* organisms adaptable to a wide range of atmospheric humidity.

euryoecious *a.* having a wide range of habitat selection.

euryphotic *a.* adaptable to a wide range of illumination.

Eurypterida, eurypterids *n., n.plu.* subclass (or order) of giant (2 m long) predatory fossil aquatic arthropods of the class Merostomata, present in the Ordovician, having a short non-segmented prosoma and a long segmented opisthosoma, and resembling scorpions.

eurythermic *a. appl.* organisms adaptable to a wide range of temperature. *alt.* eurythermous.

eurytopic *a.* having a wide range of geographical distribution.

euryxerophilous *a. appl.* plants adaptable to a wide range of dry conditions within a temperate climate.

eusocial *a. appl.* social insects which display cooperative care of the young, reproductive division of labour with more-or-less sterile individuals working on behalf of those engaged in reproduction, and an overlap of at least two generations able to contribute to colony labour. They include all ants, termites and some bees and wasps.

eusporangiate *a. appl.* ferns in which a sporangium develops from a series of superficial initial cells which form the sporangium wall. *cf.* leptosporangiate.

Eustachian tube canal connecting middle ear and pharynx (throat).

Eustachian valve valve guarding orifice of inferior vena cava in atrium of heart.

eustele *n.* type of stele in which strands of vascular tissue (vascular bundles) surround a central pith and are separated from each other by parenchymatous ground tissue. Present in stems of horsetails and of dicotyledons and some gymnosperms.

Eustigmatophyta *n.* phylum of unicellular yellow-green photosynthetic protists, possessing a distinctive eyespot composed of drops of carotenoid pigments, and usually a single flagellum.

eustomatous *a.* having a distinct mouth or mouth-like opening.

eustroma *n.* in lichens, stroma formed of fungal cells only.

Eutheria, eutherians *n., n.plu.* an infraclass of mammals, including all mammals except the monotremes and marsupials, which are viviparous with an allantoic placenta, and have a long period of gestation, after which the young are born as immature adults. *alt.* placental mammals.

euthycomous *a.* straight-haired.

eutrophic *a.* of water bodies, rich in plant nutrients and therefore usually highly productive, with very large numbers of plankton, often dominated by cyanobacteria, and often with turbid water in summer. Eutrophic waters suffer frequent algal blooms. Coarse fish (e.g. perch, roach and carp) are dominant. Larger aquatic plants may be absent as the water can become depleted of dissolved oxygen through the decay of large amounts of organic matter. *n.* eutrophy. *see* eutrophication.

eutrophication *n.* the enrichment of bodies of fresh water by inorganic plant nutrients (e.g. nitrate, phosphate). It may occur naturally but can also be the result of human activity (cultural eutrophication from fertilizer runoff and sewage discharge) and is particularly evident in slow-moving rivers and shallow lakes. The biomass of phytoplankton and herbivorous zooplankton increases, and species diversity decreases. The water becomes turbid in summer, the growth of the larger aquatic plants may eventually become suppressed and algal blooms are frequent. The water may become anoxic through the decay of large amounts of organic matter. Increased sediment deposition can eventually raise the level of the lake or river bed, allowing land plants to colonize the edges, and eventually converting the area to dry land.

eutropic *a*. turning sunward; dextrorse *q.v*.

eutropous *a. appl.* insects adapted to visiting special kinds of flowers; *appl.* flowers whose nectar is available to only a restricted group of insects.

euxerophyte *n*. plant that shows adaptations to and thrives in very dry conditions.

evaginate *v*. to evert from a sheathing structure; to protrude by eversion.

evagination *n*. the process of unsheathing, or the product of this process; the process of turning inside out.

evanescent *a*. disappearing early; *appl.* flowers that fade quickly.

evaporative water loss (EWL) *(bot.) see* evapotranspiration; (*zool.*) in mammals and birds, the loss of heat from the body through the evaporation of water, which may occur from the body surface (sweating) in some mammals and/or from the respiratory tract (thermal panting) in some mammals and birds. Although EWL is actively employed by some birds and animals for cooling, in small mammals it could lead to dehydration and is minimized.

evapotranspiration *n*. loss of water from the soil by evaporation from the surface and by transpiration from the plants growing thereon.

evelate *a*. lacking a veil, *appl.* certain agaric fungi.

even-toed ungulates artiodactyls *q.v*.

evergreen *a. appl.* vascular plants that do not shed all their leaves at the same time and therefore appear green all the year round.

eviscerate *a*. to disembowel; to eject the internal organs, as do holothurians (sea cucumbers) on capture.

evocation *n*. in developmental biology, the ability of an inducer to elicit a particular pathway of differentiation in the induced tissue.

evolute *a*. turned back; unfolded.

evolution *n*. the development of new types of living organisms from pre-existing types by the accumulation of genetic differences over long periods of time. It is studied by reference to the fossil record and to the anatomical, physiological and genetical differences between extant organisms. Present-day views on the process of evo-

lution are based largely on the theory of evolution by natural selection formulated by Charles Darwin and Alfred Russel Wallace in the 19th century. Darwin's theory has undergone certain modifications to incorporate the principles of Mendelian genetics, unknown in his day, and the more recent discoveries of molecular biology, but still remains a basic framework of modern biology. *see also* Creationism, Darwinism, natural selection, neo-Darwinism, macroevolution, microevolution, molecular evolution.

evolutionarily stable strategy (ESS) in evolutionary theory, a behaviour pattern or strategy which, if most of the population adopt it, cannot be bettered by any other strategy and will therefore tend to become established by natural selection. Using games theory the results of various different strategies (e.g. in contests between males) can be worked out and a theoretical ESS determined and compared with actual behaviour.

evolutionary clock a measure of time elapsed since the divergence of different present-day lineages from their common ancestor, which can, in principle, be estimated by comparing suitable corresponding DNA sequences or protein sequences from two extant species and counting the differences that have accumulated between them. Over long periods of time the rate of certain types of unselected nucleotide change appears to be directly proportional to time elapsed, if measurements are restricted to appropriate sequences and closely related lineages. Such clocks may be calibrated in real time by comparison of sequences from species whose point of divergence is well established from the fossil record. *alt.* molecular clock.

evolutionary grade the level of development of a structure, physiological process or behaviour occupied by a species or group of species, not necessarily related.

evolutionary taxonomy taxonomic philosophy and method that utilizes both phenotypic characters and lines of descent in the classification of organisms. One difference from the cladistic method of classification is the acceptance of taxa which do not contain all the descendants of the common ancestor, e.g. the class Reptilia, which does not contain the birds, who are descendants of a reptile ancestor. *cf.* cladistics.

evolutionary transformation series pair of homologous characters, one of which is derived from the other.

evolvate *a.* lacking a volva, *appl.* certain agaric fungi.

evolve *v.* to undergo evolution.

ex- prefix derived from Gk. *ex*, without.

exafferent *a. appl.* stimulation that results solely from factors outside the body.

exalate *a.* wingless.

exalbuminous *a.* lacking albumen; *appl.* seeds without endosperm or perisperm.

exannulate *a.* having sporangia not furnished with an annulus, *appl.* certain ferns.

exanthema *n.* a skin rash, or a disease in which such a rash appears, e.g. measles.

exarate *a. appl.* insect pupae in which all the appendages are free.

exarch *n.* stele with protoxylem strands to the outside of the metaxylem, produced when xylem matures centripetally so that the oldest protoxylem is farthest from the centre of the axis.

exarillate *a.* lacking an aril.

exasperate *a.* furnished with hard, stiff points.

excavate *a.* hollowed out.

excentric *a.* one-sided; having the two portions of a lamina or pileus unequally developed. *alt.* eccentric.

exchange diffusion *see* antiport *q.v.*

exciple, excipulum *n.* the outer covering of apothecium in some lichens.

excision repair process whereby abnormal or mismatched nucleotides are enzymatically cut out of a strand of a DNA molecule and the correct nucleotides replaced by enzymatic synthesis using the remaining intact strand as template.

excitability *n.* capability of a living cell or tissue to respond to an environmental change or stimulus.

excitation *n.* act of producing or increasing stimulation; immediate response of a cell, tissue or organism to a stimulus.

excitation contraction coupling the process by which the contractile fibrils of a muscle are stimulated to contract by excitation by a neurone.

excitatory *a.* tending to excite, *appl.* stimuli, cells, etc.; *appl.* neurones whose activity causes adjacent neurones to fire; *appl.* neurotransmitters whose action tends to cause a neurone to fire; *appl.* synapses at which transmission of a signal tends to cause the postsynaptic neurone to fire. *cf.* inhibitory.

excitatory cells motor neurones in the sympathetic nervous system.

excitatory postsynaptic potential (epsp) electrical potential generated in a neurone by the action of neurotransmitter liberated at a synapse, and which tends to produce an action potential.

exclusive species a species that is confined to one community.

exconjugant *n.* an organism which is leading an independent life after conjugation with another.

excorticate decorticate *q.v.*

excreta *n.plu.* waste material eliminated from body or any tissue thereof; harmful substances formed within a plant.

excretion *n.* the elimination of waste material from the body of a plant or animal, specifically the elimination of waste materials produced by metabolism.

excretophores *n.plu.* cells of coelomic epithelium in which waste substances from blood accumulate, for discharge into coelomic fluid in invertebrates.

excretory *a. pert.* or functioning in excretion, *appl.* organs, ducts, etc.

excurrent *a. pert.* ducts, channels, or canals in which there is an outgoing flow; with undivided main stem; having midrib projecting beyond apex, *appl.* leaves.

excurvate *a.* curved outwards from centre.

excystation *n.* emergence from a cyst.

exendospermous *a.* without endosperm.

exergonic *a.* releasing energy, *appl.* metabolic reactions.

exflagellation *n.* process of microgamete formation by microgametocyte in protozoan blood parasites.

exfoliation *n.* the shedding of leaves or scales from a bud; shedding in flakes, as of bark.

exhalant, exhalent *a.* carrying from the interior outwards.

exine *n.* tough and durable outer layer of wall of pollen grain, often intricately sculptured, composed mainly of sporopollenin.

exinguinal *a.* occurring outside the groin; *pert.* 2nd joint of arachnid leg.

exites *n.plu.* offshoots on outer lateral border of axis of certain arthropod limbs.

exo- prefix derived from Gk. *exo*, without, signifying outside, acting outside, opening to the outside, etc.

exobiology *n.* the study of life originating outside the Earth.

exobiotic *a.* living on the exterior of a substrate or the outside of an organism.

exocardiac *a.* situated outside the heart.

exocarp *n.* the outermost layer of pericarp of fruit, the skin. *alt.* epicarp.

exoccipital *a. pert.* a skull bone on each side of the foramen magnum.

exochorion *n.* the outer layer of membrane secreted by follicular cells surrounding the egg in ovary of insects.

exocoel *n.* space between adjacent mesenteries in sea anemones and their relatives; exocoelom *q.v.*

exocoelar *a. pert.* parietal wall of coelom.

exocoelom *n.* extraembryonic cavity of embryo.

exocone *a. appl.* insect eye with cones of cuticular origin.

exocrine *a. appl.* glands whose secretion is drained by ducts; *pert.* such glands. *cf.* endocrine.

exocuticle *n.* the main layer of the cuticle (exoskeleton) of arthropods, which in crustaceans often contains calcium salts.

exocytosis *n.* the process by which molecules are secreted from eukaryotic cells. They are packaged in membrane-bounded vesicles which then fuse with the plasma membrane releasing their contents to the outside of the cell. *a.* exocytotic.

exodermis *n.* specialized cell layer in root immediately underneath the epidermal layer that produces root hairs.

exoenzyme *n.* any enzyme secreted by a cell and which acts outside the cell, i.e. an extracellular enzyme.

exo-erythrocytic *a.* outside red blood cells, *appl.* phase of the malarial plasmodium life cycle in humans in which merozoites produced from schizonts reinvade tissue cells.

exogamy *n.* outbreeding; cross-pollination; disassortative mating.

exogastric *a.* having the shell coiled towards dorsal surface of body.

exogenote *n.* in bacterial conjugation, the chromosome fragment that passes from donor to recipient to form part of the merozygote.

exogenous *a.* originating outside the organism, cell or system being studied; developed from superficial tissue, the superficial meristem; growing from parts that were previously ossified; *appl.* metabolism concerned with motor and sensory activities, hormone production and action, temperature control, etc. *cf.* endogenous.

exogenous rhythm a metabolic or behavioural rhythm which is synchronized by some external factor and which ceases to occur when this factor is absent.

exognath, exognathite *n.* the outer branch of oral appendages of crustaceans.

exogynous *a. appl.* flower with style longer than corolla and projecting above it.

exo-intine *n.* middle layer of a spore-covering, between exine and intine.

exomixis *n.* union of gametes derived from different sources.

exon *n.* a block of DNA encoding part of a polypeptide chain (or tRNA or rRNA), which forms part of the coding sequence of a eukaryotic gene, and which is separated from the next exon by a non-coding region of DNA (an intron).

exonephric *a.* with nephridia opening to exterior, *appl.* oligochaetes.

exonuclease *n.* any of various enzymes that degrade DNA or RNA by progressively splitting off single nucleotides from one end of the chain. *alt.* nuclease, deoxyribonuclease, ribonuclease.

exopeptidase *n.* any of a class of enzymes that degrade peptides and proteins by successively splitting off terminal amino acids. EC 3.4.11–17. *cf.* proteinase.

exoperidium *n.* the outer layer of spore covering (peridium) in certain fungi.

exophytic *a.* on, or *pert.* exterior of plants.

exopodite *n.* the outer branch of a typical two-branched (biramous) crustacean limb.

Exopterygota *n.* major division of the insects including those with only slight metamorphosis and no pupal stage. Includes the Dictyoptera, Isoptera, Phasmida, Orthoptera, Embioptera, Ephemeroptera, Odonata, Plectoptera, Dermaptera,

Psocoptera, Mallophaga, Anoplura, Thysanoptera, Hemiptera.

exopterygote, exopterygotous *a.* appl. insects in which the wings develop gradually on the outside of the body and there is no pupal stage, e.g. dragonflies, and whose young are called nymphs. *see* Exopterygota.

exoskeleton *n.* hard supporting structure secreted by and external to the epidermis, as the calcareous exoskeletons of some sponges and the chitinous exoskeleton of arthropods.

exosmosis *n.* osmosis in an outward direction, as out of a cell.

exostosis *n.* formation of knots on surface of wood; formation of knob-like outgrowths of bone at a damaged portion, or of dental tissue in a similar way.

exoteric *a.* produced or developed outside the organism.

exothecium *n.* the outer specialized dehiscing cell layer of anther.

exothermic *a. appl.* chemical reactions that release energy as heat.

exothermic *a. appl.* animals that gain heat primarily from external sources such as sunlight, *n.* exotherm; *appl.* chemical reactions that release energy mainly as heat.

exotic *n.* a foreign plant or animal which has not acclimatized or naturalized.

exotoxin *n.* toxin secreted by bacteria. *cf.* endotoxin.

experimental extinction blotting out an acquired behavioural response.

expiration *n.* the act of emitting air or water from the respiratory organs; emission of carbon dioxide by plants and animals.

explanate *a.* having a flat extension.

explant *n.* small fragment of tissue taken from a living organism and grown in culture.

explosive *a. appl.* flowers in which pollen is suddenly discharged on decompression of stamens by alighting insects, as of broom and gorse; *appl.* fruits with sudden dehiscence, seeds being discharged to some distance; *appl.* evolution, rapid formation of numerous new types; *appl.* speciation, rapid formation of species from a single species in one locality.

exponential growth type of growth in which numbers increase by a fixed percentage of the total population in a given time period, and which gives a J-shaped curve when numbers are plotted over time. *alt.* logarithmic growth. *cf.* linear growth.

exponential growth phase phase of maximum population growth.

expression vector DNA vector which has been constructed in such a way that the "foreign" gene it contains will be expressed.

expressivity *n.* the degree to which a gene produces a phenotypic effect.

exsculptate *a.* having the surface marked with more-or-less regularly arranged raised lines with grooves between.

exscutellate *a.* having no scutellum, *appl.* certain insects.

exserted *a.* protruding beyond; *appl.* stamens that protrude beyond corolla.

exsertile *a.* capable of extrusion.

exsiccata *n.plu.* dried specimens, as in an herbarium.

exstipulate *a.* without stipules.

exsuccate, exsuccous *a.* sapless, without juice.

exsufflation *n.* forced expiration from lungs.

extended phenotype the concept of the phenotype of an organism extended to include its behaviour, and its relations with its family group, who share some of its genes, and other members of its own species.

extensor *n.* any muscle which extends or straightens a limb or part. *cf.* flexor.

exterior *a.* situated on side away from axis or definitive plane.

external *a.* outside or near the outside; away from the mesial plane.

external fertilization fertilization that takes place outside the body.

external respiration respiration considered in terms of the gaseous exchange between organism and environment and the transport of gases to and from cells.

exteroceptor *n.* sense organ or receptor that detects stimuli originating outside the body. *cf.* interoceptor.

extinction *n.* the complete disappearance of a species from the Earth; (*behav.*) the process by which learned behaviour patterns cease to be performed when they are no longer appropriate; (*phys.*) absorbance *q.v.*

extinction point the minimum level of illumination below which a plant is unable to survive in natural conditions.

extine exine *q.v.*

extra- prefix derived from L. *extra*, outside, signifying located outside, etc.

extrabranchial *a.* arising outside the branchial arches.

extracapsular *a.* arising or situated outside a capsule; *appl.* ligaments at a joint; *appl.* protoplasm lying outside the central capsule in some protozoans.

extracellular *a.* occurring outside the cell; secreted by or diffused out of the cell.

extracellular matrix macromolecular ground substance of connective tissue, secreted by fibroblasts and other connective tissue cells, and which generally consists of proteins, polysaccharides and proteoglycans.

extrachorion *n.* in certain insect eggs, an outermost layer external to the exochorion.

extrachromosomal *a. appl.* DNA molecules such as plasmids and episomes in bacteria and the rRNA genes in certain animals which are independent of the chromosomes.

extracortical *a.* not within the cortex, *appl.* part of brain.

extraembryonic *a.* situated outside the embryo proper, as the various foetal membranes (amnion, chorion, allantois, yolk sac).

extraenteric *a.* situated outside the alimentary tract.

extrafloral *a.* outside the flower, *appl.* nectaries.

extrafoveal *a. pert.* yellow spot surrounding the fovea centralis.

extrafusal *a. appl.* muscle fibres outside muscle spindle.

extrahepatic *a. appl.* cystic duct and common bile duct.

extranuclear *a. pert.* structures or processes occurring outside the nucleus; situated outside the nucleus; *appl.* e.g. mitochondrial, chloroplast and other cytoplasmic genes.

extraocular *a.* exterior to the eye, *appl.* antennae of insects.

extraperitoneal subperitoneal *q.v.*

extrapulmonary *a.* external to the lungs, *appl.* bronchial system.

extravaginal *a.* forcing a way through the sheath, as shoots of many plants.

extravasate *v.* to force its way from proper channel into the surrounding tissue, said of blood, etc.

extraxylary *a.* on the outside of the xylem, *appl.* fibres.

extrinsic *a.* acting from the outside; *appl.* muscles not entirely within the part or organ on which they act; *appl.* membrane proteins which are embedded in the outer layer of a biological membrane.

extrinsic isolating mechanism an environmental barrier which isolates potentially interbreeding populations.

extrinsic pathway series of reactions in blood leading to clot formation, triggered by trauma to tissue or addition of tissue extracts to plasma *in vitro*, the final stages being the same as in the intrinsic pathway (*q.v.*).

exudate *n.* any substance released from a cell, organ or organism by exudation, e.g. sweat, gums, resins.

exudation *n.* the discharge of material from a cell, organ or organism through a membrane, incision, pore or gland.

exumbral *a. pert.* the rounded upper surface of a jellyfish.

exuviae *n. plu.* the cast-off skin, shells, etc. of animals.

exuvial *a. appl.* insect glands whose secretions facilitate moulting.

eye *n.* light-sensitive organ, the organ of sight or vision, taking various forms in different groups of animals. Insects and most crustaceans have compound eyes as well as simple single eyes in some cases, made up of many separate units or ommatidia (*q.v.*). The vertebrate eye consists of a jelly-filled ball, the back of which is lined with a photosensitive layer, the retina (*q.v.*). Light is focused onto the retina through a single transparent lens. The amount of light entering the eye is regulated by varying the size of the pupil (*q.v.*). Cephalopods also have a very similar type of eye; (*bot.*) the bud of a tuber. *see also* aqueous humor, cornea, eye-spot, iris, lens, sclera, vitreous humor.

eye-spot *n.* small cup-shaped pigmented spot of sensory tisssue in invertebrates, and also in some vertebrates, which have a light-detecting or visual function; orange carotenoid-containing structure in some flagellates; eye-like marking on wings of some butterflies and moths, or on the bodies of other animals, and which are exposed to distract predators when the animal is disturbed. *alt*. stigma.

eye teeth upper canine teeth.

F

F phenylalanine *q.v.*

f coefficient of kinship *q.v.*

F_1, F1 denotes 1st filial generation, hybrids arising from a first cross, successive generations arising from this one being denoted by F_2, F_3, etc. P_1 denotes the parents of the F_1 generation, P_2 the grandparents, etc.

F_1, F_0 coupling factor *q.v.*

Fab the portion of an antibody molecule containing an antigen-binding site, comprising one light chain and its paired portion of heavy chain, obtained on digestion of the molecule with papain.

FACS fluorescence-activated cell sorter. A piece of equipment for counting and separating cells in a mixed population after staining for distinctive cell-surface molecules with antibodies linked to different coloured fluorescent dyes. A stream of labelled cells is run through a fluorescence detector, which counts the cells of different types and which can also deflect appropriately labelled cells from the main stream, thus separating them from the mixture, *see also* flow cytometry.

FAD familial Alzheimer's disease, *see* Alzheimer's disease; flavin adenine dinucleotide *q.v.*

$FADH_2$ reduced form of flavin adenine dinucleotide *q.v.*

Fc the portion of an antibody molecule comprising the constant regions of the two heavy chains up to the hinge region, obtained on digestion of the molecule with papain.

FDNB fluorodinitrobenzene.

FeLV feline leukaemia virus, an RNA tumour virus.

FeSV feline sarcoma virus, an RNA tumour virus, various strains carry oncogenes *v-*

yes, *v-fms*, *v-fgr*.

FFA free fatty acid *q.v.*

FGF fibroblast growth factor *q.v.*

fMet formylmethionine.

fMet-$tRNA_f$ formylmethionyl-tRNA, the first aminoacyl-tRNA to be bound to the initiation complex in bacterial protein synthesis.

FMN flavin mononucleotide *q.v.*

$FMNH_2$ reduced form of flavin mononucleotide *q.v.*

fms *v-fms*, oncogene carried by a feline sarcoma virus, probably corresponding to an altered form of the receptor for macrophage colony-stimulating factor (CSF-1). *c-fms*, corresponding proto-oncogene in normal cells, which encodes the CSF-1 receptor.

α-FP α-foetoprotein *q.v.*

FSH follicle-stimulating hormone *q.v.*

FU fluorouracil *q.v.*

Fabales *n.* an order of dicots, also known as legumes, whose fruit is a pod, and whose roots contain nitrogen-fixing bacteria of the genus *Rhizobium*. They include the families Mimosaceae, Cesalpinaceae, and Papilionaceae (Fabaceae), and are also known as the Leguminosae.

fabavirus group group of isometric single-stranded RNA plant viruses similar in structure to comoviruses, but producing different symptoms. The type member is broad bean wilt virus. They are multicomponent viruses in which two genomic RNAs are encapsidated in three different virus particles, one of which lacks nucleic acid.

fabiform *a.* bean-shaped.

facet *n.* a smooth, flat or rounded surface for articulation; surface of an ommatidium, (one of the units making up a compound eye), or the ommatidium itself.

facial *a. pert.* face, *appl.* arteries, bones etc.; *appl.* nerve: 7th cranial nerve, which supplies facial muscles, activates some salivary glands and conveys taste sensation from front of tongue.

faciation *n.* formation or character of a facies; grouping of dominant species within an association; geographical differences in abundance or proportion of dominant species in a community.

facies *n.* a surface, in anatomy; aspect, as superior or inferior; a grouping of dominant plants in the course of a successional series; one of different types of deposit in a geological series or system, and the palaeontological and lithological character of a deposit.

facilitated diffusion transport of molecules or ions across a membrane along a concentration gradient by a carrier system without the expenditure of energy.

facilitation *n.* in neurophysiology, the process whereby the amount of neurotransmitter liberated at an axon terminal, and therefore the postsynaptic potential, increases with the frequency of stimulation of the presynaptic nerve cell; in animal behaviour, an improvement in a pre-existing capability in response to a particular stimulus; social facilitation: the initiation or increase in an ordinary behaviour pattern by the presence or actions of another animal.

faciolingual *a. pert.* or affecting face and tongue.

F actin *see* actin.

factor *n.* any agent (biological, climatic, nutritional, etc.) contributing to a result or effect; in physiology and cell biology, used for any ill-defined endogenous substance that appears to have a physiological effect. Many such "factors" have subsequently been isolated and characterized as proteins and polypeptides, as e.g., nerve growth factor, epidermal growth factor, atrial natriuretic factor.

Factor V a blood clotting factor, a modifier protein which stimulates the conversion of prothrombin to thrombin by the enzyme, Factor X. *alt.* accelerin.

Factor VII a blood clotting factor, an enzyme

which together with tissue factor activates Factor X. EC 3.4.21.21, *r.n.* coagulation factor VIIa. *alt.* proconvertin.

Factor VIII non-enzyme protein involved in blood clotting, accelerating the conversion of factor X to its active form by the proteolytic enzyme factor IX_a, and whose deficiency causes haemophilia A. *alt.* antihaemophilia factor.

Factor IX blood clotting factor, a precursor to a proteinase (factor IXa), and whose deficiency is the cause of haemophilia B (Christmas disease). *alt.* Christmas factor.

Factor X a proteolytic blood clotting factor which converts prothrombin to thrombin. EC 3.4.21.6, *r.n.* coagulation factor Xa. *alt.* Stuart factor (formerly known as thrombokinase or thromboplastin).

Factor XI a blood clotting factor, enzymatically activates Factor IX. EC 3.4.21.27, *r.n.* coagulation factor XIa. *alt.* plasma thromboplastin antecedent.

Factor XII enzyme precursor found in blood which on contact with an abnormal surface such as wounded tissue is converted to a proteolytic enzyme, Factor XIIa, which then activates Factor XI. *alt.* Hageman factor.

Factor XIII plasma protein involved in blood clotting, cross-linking fibrin and stabilizing clots.

facultative *a.* having the capacity to live under different conditions, e.g. *appl.* aerobes, anaerobes, symbionts, parasites, etc., organisms that can live in this way but are not obliged to and may under certain conditions adopt another mode of life. *cf.* obligate.

facultative heterochromatin *see* heterochromatin.

faeces *n.* excrement from alimentary canal.

Fagales *n.* an order of dicots including many deciduous forest trees such as beech, birch, oak, sweet chestnut, hazel and hornbeam. *see also* Betulales.

falcate *a.* sickle-shaped; hooked.

falces *n.plu.* chelicerae (poison claws) of arachnids; *plu.* of falx *q.v.*

falcial *a. pert.* falx, esp. of the falx cerebri.

falciform *a.* sickle- or scythe-shaped.

Falconiformes *n.* the birds of prey, an order of birds including eagles, hawks, buzzards, kestrels, falcons, and vultures.

falcula *n.* a curved scythe-like claw.

fallopian tube one of a pair of narrow ducts in mammals each leading from an ovary to the uterus, into which ova are released on ovulation, and in which fertilization normally takes place.

false foot pseudopodium *q.v.*

false fruits fruits formed from the receptacle or other part of the flower, in addition to the ovary, or from complete inflorescences.

false ribs those ribs whose cartilaginous ventral ends do not join the breastbone directly. *alt.* floating ribs.

false scorpions pseudoscorpions *q.v.*

false yeasts yeast-like fungi which have no known ascus stage and are therefore classified in the Deuteromycetes. They include *Torulopsis*, which has been utilized as a source of protein for food, and several human pathogens. *Candida albicans*, an opportunistic pathogen, causes common thrush and also more severe forms of systemic candidiasis in immunosuppressed patients. *Cryptococcus neoformans* is the cause of cryptococcosis, affecting either lungs or central nervous system.

falx *n.* a sickle-shaped structure; falx cerebri: a sickle-shaped fold in the dura matter. *plu.* falces. *a.* falcate, falcular, falciform.

familial *a. pert.* family; *appl.* traits that tend to occur in several members and subsequent generations of a family. They may not necessarily be genetically based but be due to the family environment; *appl.* disease, usually signifies a genetically based heritable condition, as opposed to the spontaneous and sporadic occurrence of the same condition, *see* genetic disease.

familial Alzheimer's disease *see* Alzheimer's disease.

familial cancers cancers in which there is an underlying inheritable predisposition to develop the disease, usually much earlier than is normal. *see also* tumour suppressor genes.

familial emphysema α_1-antitrypsin deficiency *q.v.*

familial hypercholesterolaemia *see* cholesterolaemia.

family *n.* taxonomic group of related genera, related families being grouped into orders. Familial names usually end in -aceae in plants and -idae in animals.

far red light light of wavelength between 700–800 nm, which reverses the effect of red light on phytochrome.

farina *n.* flour or meal; a fine mealy powder present on the surface of some plants, as certain primulas, and insects.

farinaceous *a.* containing flour or starch; mealy; covered with a fine powder or dust; covered with fine white hairs that can be detached like dust. *alt.* farinose.

fascia *n.* an ensheathing band of connective tissue; a transverse band of different colour, as in some plants; any band-like structure. *a.* fascial.

fasciated *a.* banded; arranged in bundles or tufts; *appl.* stems or branches malformed and flattened.

fasciation *n.* the formation of bundles; the coalescent development of branches of a shoot system, as in cauliflower; abnormal development of flattened, malformed fused stems or branches.

fascicle *n.* a bundle, as of pine needles; a small bundle, as of muscle fibres, nerve fibres; tuft, as of leaves. *alt.* fasciculus.

fascicular *a. pert.* a fascicle; arranged in bundles or tufts; (*bot.*) *appl.* cambium or tissue within vascular bundle, *alt.* intrafascicular.

fasciola *a.* a narrow colour band.

fasciole *n.* ciliated band on certain sea urchins for sweeping water over surrounding parts.

fastigiate *a.* with branches close to stem and erect.

fast-twitch *appl.* muscle fibres capable of rapid contraction, *see* FG fibres, FOG fibres.

fat *n.* a compound of fatty acids and glycerol, having a large proportion of saturated fatty acids and being solid at 20°C. Fats are hydrolysed by lipases to fatty acids and glycerol and are food stores in animals and plants. Adipose tissue in animals consists of cells filled with globules of fats. *cf.* oils.

fat body diffuse gland dorsal to gut in insects, with function analogous to that of liver in vertebrates, in which fats, glycogen and protein are stored and which is a major site of intermediary metabolism; structure filled with fat globules and associated with gonads in amphibians; other fat storage organs in animals and plants.

fat cell adipocyte *q.v.*

fat index ratio of dry weight of total body fat to that of non-fat.

fat-soluble *appl.* vitamins A, D and K.

fate map a "map" of the surface of fertilized egg or early embryo predicting which regions will form various tissues or parts of the body.

fatigue *n.* effect produced by unduly prolonged stimulation on cells, tissues or other structures so that they are less responsive to further stimulation.

fatty acid *n.* long chain organic acid of the general formula $CH_3(C_nH_x)COOH$, where the hydrocarbon chain is either saturated (x=2n) (e.g. palmitate, $C_{15}H_{31}COO^-$) or unsaturated (e.g. oleate, $C_{17}H_{33}COO^-$), unbranched (in animals), a constituent of lipids, and a fuel molecule in cells.

fatty acid synthetase a multifunctional protein complex of the 7 enzyme activities involved in fatty acid synthesis from acetyl CoA, found in eukaryotes.

fatty acid thiokinase acyl CoA synthetase *q.v.*

fauces *n.plu.* upper or anterior portion of throat between palate and pharynx; mouth of a spirally coiled shell.

fauna *n.* the animals peculiar to a country, area, specified environment or period, microscopic animals usually being called the microfauna.

faunal collapse local extinction of an animal or a number of animal species.

faunal region area characterized by a special group or groups of animals; zoogeographical region *q.v.*

favella *n.* conceptacle of certain red algae.

faveolate *a.* honeycombed.

faveolus *n.* a small depression or pit.

favoid *a.* resembling a honeycomb.

Fc receptor receptor on macrophages and some other cells to which the constant region of antibody heavy chains (the Fc region) attaches.

feather *n.* keratinous epidermal structure forming the body cover (plumage) of birds. Each feather consists of a midrib (rachis) from which project on either side many delicate thread-like barbs. In the fluffy down feathers of nestlings, the barbs do not interlock, whereas in the outer feathers and flight feathers of older birds, the barbs interlock to form a flat flexible wind-resistant surface (the vane).

feather epithelium epithelium on inner surface of nictitating membrane in birds and reptiles, whose cells each have a process with numerous lateral filaments, and which acts to clean the eye surface.

feather follicle the epithelium surrounding the base of a feather, and from which the feather has developed.

feature detection columns columns or slabs of nerve cells in visual cortex, perpendicular to the surface, which all respond to a particular type of visual stimulus such as a line in a particular orientation, neighbouring columns responding to a line in a slightly different orientation.

feces faeces *q.v.*

fecundate *a.* to impregnate; to fertilize; to pollinate.

fecundity *n.* the capacity of an individual or a species to multiply rapidly; in a stricter sense the number of eggs produced by an individual.

feedback control, inhibition, regulation type of metabolic regulation in which the first enzyme in a metabolic pathway (usually a biosynthetic pathway) is inhibited by reversible binding of the final product of the pathway. *alt.* end-product inhibition.

feedback mechanism general mechanism operative in many biological and biochemical processes, in which once a product or result of the process reaches a certain level it inhibits or promotes further reaction.

felid *n.* member of the mammalian family Felidae, the cats.

feloids *n.plu.* mammals of the cat (Felidae) or mongoose (Viverridae) families.

female *n.* individual whose gonads contain only female gametes, symbol ♀.

female pronucleus the haploid nucleus of the egg.

femoral *a. pert.* femur.

femur *n.* the thigh bone, the large bone in the upper part of the hindlimb of vertebrates; 3rd segment of insect, spider and myriapod leg, counting from the body. *plu.* femora. *a.* femoral.

fen *n.* plant community on alkaline, neutral or slightly acid wet peat, characterized by tall herbaceous plants, e.g. reeds, reed canary grass. *cf.* bog.

fenestra *n.* an opening in a bone or between two bones, e.g. the openings in the wall of bony labyrinth of ear between the tympanic cavity and vestibule of inner ear; a transparent spot on wings of insects.

fenestra ovalis the upper of two membrane-covered openings in the bony wall between the tympanic cavity (middle ear) and vestibule of inner ear.

fenestra pseudorotunda opening covered by the endotympanic membrane in birds, the fenestra rotunda in mammals having a different embryonic origin.

fenestra rotunda the lower of two membrane-covered openings in the bony wall between the tympanic cavity (middle ear) and vestibule of inner ear.

fenestra tympani fenestra rotunda *q.v.*

fenestra vestibuli fenestra ovalis *q.v.*

fenestrate(d) *a.* having small perforations or transparent spots, *appl.* insect wings; having numerous perforations, *appl.* leaves, partitions in corals etc.

fenestrated membrane a close network of elastic connective tissue resembling a membrane with perforations, as in inner tunic of arterial wall; basal membrane of compound eye, perforated by ommatidial nerve fibres.

feral *a.* wild, or escaped from domestication and reverted to wild state.

fermentation *n.* glycolysis *q.v.*; anaerobic breakdown of carbohydrates by living cells, esp. microorganisms, often with the production of heat and waste gases (as in alcoholic fermentation in yeasts) and a wide variety of end-products (e.g. ethanol, lactic acid).

fern *n.* common name for a member of the Pterophyta *q.v.*

ferralitic soils deep red soils, acid in reaction, found on freely drained sites in humid tropical regions.

ferredoxin *n.* widely distributed iron-sulphur protein, acting as an electron carrier and as a biological reducing agent in its reduced form, a component of the photosynthetic electron transport chain and an electron donor in nitrogen fixation in microorganisms.

ferrihaemoglobin methaemoglobin *q.v.*

ferritin *n.* protein containing a large amount of iron as ferric hydroxide phosphate, found in spleen, liver (where much iron is stored as ferritin) and bone marrow, the protein part of the molecule (apoferritin) being synthesized by intestinal mucosa cells and involved in the uptake of dietary iron.

ferrocyte *n.* iron-containing cell in ascidians, apparently concerned with the production of the cellulose-like polysaccharide tunicin.

ferruginous *a.* having the appearance of iron rust; rust-coloured.

fertile *a.* producing viable gametes; capable of producing living offspring; of eggs or seeds, capable of developing; *appl.* a soil containing the necessary nutrients for plant growth.

fertilis- fertiliz-.

fertility *n.* the reproductive performance of an individual or population, measured as the number of viable offspring produced per unit time.

fertility factor F factor *q.v.*

fertility schedule demographic data giving the average number of female offspring that will be produced by a female at each particular age.

fertilization *n.* the union of male and female gametes (e.g. sperm and egg) to form a zygote.

fertilization cone protuberance on ovum at point of contact and entry of spermatozoon before fertilization.

fertilization membrane a membrane formed by the ovum in response to penetration by a sperm, which grows rapidly from the point of penetration and covers the ovum, excluding other sperm.

fertilization tube process of an antheridium, penetrating oogonial wall, for passage of male gamete in certain fungi.

fetal, fetus *alt.* spelling of foetal, foetus *q.v.*

Feulgen stain histological stain that shows up DNA as purple.

F factor, F plasmid (F′) transmissible plasmid in the bacterium *Escherichia coli* that acts as a sex factor, directing synthesis of the sex pilus (F pilus), conjugation, and chromosomal gene transfer from an F^+ to an F^- bacterium. Can exist as a free element or is integrated into the bacterial chromosome, in which state it mediates the transfer of chromosomal genes at greater frequency. *alt.* fertility factor, sex factor. *see also* Hfr strain.

FG fibres fast glycolytic muscle fibres in muscle of mammalian limbs, white muscle fibres adapted for mainly anaerobic metabolism, and which are used only when the animal is running fast. *cf.* FOG fibres, SO fibres.

F₁ generation *see* F₁.

F₁ hybrid in horticulture and experimental genetics, the first cross between two pure-breeding lines.

fiber *see* fibre.

Fibonacci series the unending sequence 1, 1, 2, 3, 5, 8, 13, 21, 34..., where each term is defined as the sum of its two predecessors.

fibre *n.* an elongated cell or aggregation of cells forming a strand of, e.g. muscle, nerve or connective tissue; protein filament as of keratin in wool and hair; (*bot.*) a delicate root; a tapering elongated sclerenchyma cell providing mechanical strength in stem.

fibril *n.* small thread-like structure or fibre; component part of a fibre; root hair.

fibrilla *n.* fine root hair or branch of root; minute muscle-like thread found in some ciliates. *plu.* fibrillae.

fibrillar *a. pert.*, or like, fibrils or fibrillae.

fibrillar flight muscles wing muscles of dipteran (flies) and hymenopteran (wasps and bees) insects, which, unlike ordinary striated muscle, do not need an action potential to initiate every new contraction.

fibrillate *a.* possessing fibrillae or hair-like structures.

fibrillate *v.* undergo quivering movement of muscular fibrils (as in heart muscle).

fibrillation *n.* spontaneous, asynchronous contractions of muscle fibres seen after nerve supply has been cut.

fibrillose *a.* furnished with fibrils, *appl.* mycelia of certain fungi.

fibrin *n.* an insoluble protein produced from fibrinogen in the blood by the action of thrombin, and forming a mesh of fibres (a clot) in which platelets and blood cells are caught.

fibrinogen *n.* soluble blood plasma protein, precursor to fibrin.

fibrinolysin plasmin *q.v.*

fibrinolysis *n.* the dissolving of blood clots

as a result of fibrin degradation.

fibrinopeptide *n.* peptide that is cleaved from the protein fibrinogen when it is activated to form fibrin during blood clotting.

fibrin-stabilizing factor Factor XIII *q.v.*

fibroblast *n.* a flattened, irregular shaped connective tissue cell, ubiquitous in fibrous connective tissue and which secretes components of the extracellular matrix, including type I collagen and hyaluronic acid.

fibroblast growth factor (FGF) protein produced by many cells and which amongst other effects, stimulates the division of fibroblasts. A similar protein in amphibian eggs is thought to act as a morphogen.

fibrocartilage *n.* type of cartilage whose matrix is mainly composed of fibres similar to connective tissue fibres, found at articulations, cavity margins and grooves in bones.

fibrocyte *n.* fibroblast, *q.v.*; inactive cell produced from a fibroblast.

fibrohyaline chondroid *q.v.*

fibroin *n.* the protein of silk fibres, produced by proteolysis from a precursor protein, fibroinogen.

fibroinogen *n.* a protein secreted by the silk glands of certain insects, and which is cleaved to form fibroin.

fibronectin *n.* glycoprotein located on external surface of plasma membrane of most animal cells, involved in cell-substratum interactions.

fibrosarcoma *n.* a tumour of fibrous connective tissue.

fibrous *a.* composed of or resembling fibres, *appl.* tissue, mycelium, etc; forming fibres, *appl.* proteins such as collagen, elastin, keratin, fibrin, fibroin, etc.

fibrous astrocytes astrocytes found mainly in white matter, having thick processes which branch, some having foot-like processes that abut on blood vessels.

fibrous root system a root system in which the roots form a fibrous mass without a tap root.

fibula *n.* in tetrapod vertebrates, the bone posterior to the tibia in the shank of the hindlimb; in humans the outer and smaller shin bone; in some insects, a structure holding fore- and hindwings together. *a.* fibulate.

fibulare *n.* outer bone of proximal row of tarsus.

fibularis peroneus *q.v.*

ficin *n.* an endopeptidase enzyme found in fig trees. EC 3.4.22.3.

fidelity *n.* the degree of limitation of a species to a particular habitat; of DNA replication, transcription and translation, the probability of an error being made during the copying of DNA into DNA or RNA, or during the translation of RNA into protein.

field *n.* a dynamic system in which all the parts are interrelated, so that a change in any part affects the whole.

field layer herb layer *q.v.*

field metabolic rate the metabolic rate as measured in a freely ranging animal, most commonly by the doubly labelled water method, in which water labelled with either deuterium (^{2}H) or tritium (^{3}H) and the oxygen isotope ^{18}O is injected at the start of the experiment and the decline of ^{18}O and labelled hydrogen in the blood after a period of days or weeks is measured and the concomitant rate of CO_2 production calculated.

Fijian Floral Region part of the Palaeotropical Realm comprising the islands of Fiji.

fijivirus group group of isometric double-stranded RNA plant viruses of the family Reoviridae, type member Fiji disease virus of sugar-cane, which are transmitted by plant-hoppers. The RNA genome is composed of 10 different RNAs.

filament *n.* used generally for any slender threadlike structure such as a fungal or actinomycete hypha, a chain of bacterial or algal cells, very fine fibres in cells, etc., *a.* filamentous; specifically: the stalk of an angiosperm anther; the rachis of a down feather; slender apical end of egg tube of insect ovary. *see also* thick filament, thin filament.

filamin *n.* cytoplasmic protein which when added to a solution of actin filaments changes it from a viscous fluid to a solid gel, and which may be involved in similar changes *in vivo*.

filaria *n.* parasitic nematode worm. *plu.* filariae.

filator *n.* part of the spinning organ of silk-worms which regulates the size of the silk fibre; the spinnerets of other caterpillars.

filial generation F_1, F_2 *q.v.*

filial imprinting imprinting (*q.v.*) resulting in attachment of an offspring to parents or foster parents.

filibranch *a. appl.* gills of bivalve molluscs whose filaments are attached to adjacent ones by cilia.

Filicales *n.* group of ferns (Pterophyta) including most extant species, mainly terrestrial, with typically large compound leaves and rhizomatous roots, and which produce spores on the undersides of the leaves.

filicauline *a.* with a thread-like stem.

filiciform *a.* shaped like the frond of a fern; fern-like.

Filicinophyta *n.* name sometimes given to the Pterophyta (*q.v.*)

filiform *a.* thread-like.

filiform papillae papillae on tongue ending in numerous minute slender processes.

filigerous *a.* with thread-like outgrowths or flagella.

filipendulous *a.* thread-like, with tuberous swelling at middle or end, *appl.* roots.

filoplume *n.* delicate hair-like feather with long axis and a few free barbs at apex.

filopodia *n.plu.* thread-like pseudopodia of some protozoans.

Filoviridae *n.* family of enveloped single-stranded RNA animal viruses, with long thread-like particles, comprising Marburg fever and Ebola fever viruses, which cause highly contagious haemorrhagic fevers.

filose *n.* slender and thread-like.

filter feeders organisms that feed on small organisms in water or air, straining them out of the surrounding medium by various means.

fimbria *n.* any fringe-like structure; (*bact.*) one of numerous filaments, smaller than flagella, fringing certain bacteria; (*neurobiol.*) a posterior prolongation of fornix to hippocampus. *plu.* fimbriae.

fimbriate(d) *a.* bordered with fine hairs.

fimbrin *n.* protein associated with actin filaments in the supporting cytoskeleton of intestinal microvilli.

fin *n.* fold of skin supported by bony or cartilaginous rays in fishes and used for locomotion, balancing, steering, display, etc. Most fishes have an upright dorsal fin on back, a caudal fin at end of tail, an anal fin on the underside just anterior to the anus, a pair of pelvic fins on the underside and a pair of pectoral fins just behind the gills. The pectoral and pelvic fins represent the fore- and hindlimbs of other vertebrates; any similarly shaped structure in other aquatic animals.

fingerprinting *n.* in biochemistry, a technique for detecting small differences in amino acid composition/sequence between different proteins by selective cleavage into small peptides which are then separated by electrophoresis in 1 dimension and chromatography in the 2nd dimension resulting in a pattern of peptide spots characteristic for each protein. *see also* DNA fingerprinting.

fin-rays stiff rods of connective tissues, generally cartilage or bone, which support the fins.

fire climax plant community maintained as climax vegetation by natural or man-made fires which destroy the plants that would otherwise become dominant.

Firmibacteria *n.* a class of prokaryotes including all Gram-positive "true" bacteria, e.g. staphylococci, bacilli, streptococci.

Firmicutes *n.* major division of prokaryotes including the Gram-positive bacteria and the Actinomycetes and their relatives.

fishes *n.plu.* group of aquatic limbless vertebrates, breathing mainly by means of gills, with streamlined bodies and fins and with the body covered in scales (in bony fishes), and comprising the Chondrichthyes (*q.v.*) (cartilaginous fishes) and the Osteichthyes (*q.v.*) (bony fishes). *see* Appendix 4.

fish lice *see* Branchiura.

fissile *a.* tending to split; cleavable.

fissilingual *a.* with a forked tongue.

fission *n.* cleavage of cells; division of a unicellular organism into two or more parts.

fission fungi old and obsolete name for the bacteria.

fission yeast *Schizosaccharomyces* and related species, whose cells multiply by division into two equal daughter cells. *cf.* budding yeast.

fissiparous *a.* reproducing by fission.

fissiped *a.* with digits of feet separated, as

the Fissipedia, the name for the carnivores such as cats, dogs, bears, etc. in some zoological classifications.

Fissipedia *n.* in some classifications the name given to the order of terrestrial carnivorous mammals containing the cats, dogs, bears, hyenas, etc., the marine carnivores, the seals, etc., being placed in the order Pinnipedia.

fissirostral *a.* with deeply cleft beak.

fissure *n.* deep groove or furrow dividing an organ into lobes, or subdividing and separating certain areas of the lobes; sulcus *q.v.*

fistula *n.* a pathological or artificial pipe-like opening; a water-conducting vessel.

fistular *a.* hollow and cylindrical, *appl.* stems of umbellifers, *appl.* leaves surrounding the stem in some monocotyledons.

fitness *n.* the fitness of an individual is defined as the relative contribution of its genotype to the next generation relative to the contributions of other genotypes, i.e. it is determined by the number of offspring it manages to produce and rear successfully. *see also* inclusive fitness.

fix **genes** group of genes required for symbiotic nitrogen fixation, found in *Rhizobium* spp.

fixation *n.* of carbon and nitrogen *see* carbon dioxide fixation, nitrogen fixation; (*genet.*) of an allele, its spread throughout a population until it is the only allele found at that locus; (*behav.*) in experimental psychology, a stereotyped response shown by an animal regardless of whether the stimulus is accompanied by positive or negative reinforcement, and often shown in an insoluble problem situation; (*cytol.*) treatment of specimens to preserve structure, for microscopy.

fixation index in population genetics a measure of genetic differentiation between subpopulations, being the proportionate reduction in average heterozygosity compared with the theoretical heterozygosity if the different subpopulations were a single randomly mating population.

fixative *n.* a chemical such as ethanol or formaldehyde which is used to preserve cells and cellular structure.

fixed-action pattern a stereotyped and

fixed response found in animal behaviour where learning has not occurred.

flabellate *a*. with projecting flaps on one side, *appl*. to certain insect antennae.

flabellum *n*. any fan-shaped organ or structure.

flaccid *a*. limp, *appl*. leaves that do not have enough water and are wilting or about to wilt.

flagellate(d) *a*. bearing flagella; like a flagellum.

flagellated chambers in sponges the central cavities lined with choanocytes, flagellated cells specialized for uptake of food particles from water.

flagellates *n.plu*. a highly diverse group of unicellular eukaryotic microorganisms, including photosynthetic and non-photosynthetic, heterotrophic species, and classified in various schemes as protozoans, protists or algae. They are motile in the adult stage, swimming by means of flagella, and include both free-living marine and freshwater species and some important human parasites such as trypanosomes. In zoological classifications they are often placed in a superclass, Mastigophora, and are divided into the non-photosynthetic Zoomastigophorea and the photosynthetic Phytomastigophorea. In botanical classifications the photosynthetic flagellates are variously classed within the divisions Chrysophyta, Euglenophyta, Eustigmatophyta, Pyrrophyta (Pyrrophyceae), Prymnesiophyta and Xanthophyta.

flagellum *n*. a long whip-like or feathery structure borne either singly or in groups by the motile cells of many bacteria and unicellular eukaryotes and by the motile male gametes of many eukaryotic organisms, which propels the cell through a liquid medium. Bacterial and eukaryotic flagella differ in internal structure and mechanism of action; in insects, the distal part of the antenna, beyond the 2nd segment. *plu*. flagella.

flame cell *see* protonephridial system.

flash colours the sudden flash of colour displayed by some species during an attempt to escape from a predator, which may startle and distract the predator or deceive it into thinking the prey has gone.

flask fungi a common name for the Pyrenomycetes *q.v.*

flatworms Platyhelminthes *q.v.*

flavedo *n*. the outer layer, or rind, of pericarp in citrus fruits.

flavescent *a*. growing or turning yellow.

flavin *n*. any of a group of yellowish pigments showing greenish yellow fluorescence, containing a nitrogenous base, usually isoalloxazine, and occurring free in higher animals and plants usually as riboflavin, and bound to enzymes as the flavin nucleotide prosthetic groups FAD and FMN.

flavin adenine dinucleotide (FAD) important coenzyme, a derivative of riboflavin, tightly bound as a prosthetic group to certain oxidative enzymes, known as flavoproteins. Exists in an oxidized form (FAD) and a reduced form (FADH$_2$) and acts as an electron carrier in many metabolic reactions. *see also* flavin mononucleotide.

flavin adenine mononucleotide flavin mononucleotide *q.v.*

flavin mononucleotide (FMN) riboflavin 5′ phosphate, formed by the phosphorylation of riboflavin by ATP, a prosthetic group of several enzymes including NADH dehydrogenas (NADH-Q reductase) in the mitochon- drial respiratory chain, where it is reduced to FMNH$_2$ during electron transfer. *see* Appendix 1 (27) for chemical structure.

Flaviviridae *n*. family of enveloped single-stranded RNA animal viruses with icosahedral particles, including the viruses of yellow fever and dengue haemorrhagic fever.

flavone *n*. any of a group of pale yellow flavonoid plant pigments, with the C$_3$ part of the molecule forming an oxygen-containing ring.

flavonoid *n*. any of various compounds containing a C$_6$–C$_3$–C$_6$ skeleton, the C$_6$ parts being benzene rings and the C$_3$ part varying in different compounds, often forming an oxygen-containing ring, and which include many water-soluble plant pigments.

flavonol *n*. any of a group of pale yellow flavonoid plant pigments.

flavoprotein *n*. any protein which contains a flavin prosthetic group (FAD or FMN), important as enzymes in electron transfers such as oxidation reactions in respiration where the flavin group is alternately oxidized and reduced, and which are yellow when oxidized but colourless when reduced.

flavoxanthin *n.* yellow carotenoid pigment in petals as is in the buttercup family, Ranunculaceae.

fleas *n.pl.* common name for the Siphonaptera (Aphaniptera) *q.v.*

flexor *n.* any muscle which bends a limb or its part by its contraction. *cf.* extensor.

flexuose, flexuous *a.* curving in a zig-zag manner.

flexure *n.* a curve or bend.

flimmer *n.* minute hairs borne on some types of flagella giving them the appearance of tinsel.

floating ribs false ribs *q.v.*

floccose *a.* covered with wool-like tufts.

flocculation *n.* clumping of small particles in the disperse phase of a colloidal system, such as the clumping of clay particles which can be brought about by lime.

flocculence *n.* adhesion in small flakes, as of a precipitate.

flocculent *a.* covered in a soft waxy substance, giving appearance of wool.

flocculus *n.* a small lobe on each lateral lobe of the cerebellum; a posterior hairy tuft in some Hymenoptera.

floccus *n.* the tuft of hair terminating a tail; downy plumage of young birds; any tuft-like structure; a mass of hyphal filaments in fungi or algae.

flor *n.* covering of yeasts and bacteria and other microorganisms which forms on the surface of some wines during fermentation.

flora *n.* the plants peculiar to a country, area, specified environment or period; a book giving descriptions of these plants; the microorganisms that naturally live in and on animals and plants, e.g. gut flora, skin flora.

floral *a. pert.* flora of a country or area; *pert.* flowers.

floral axis receptacle *q.v.*

floral diagram a conventional way of representing a flower, indicating the position of the parts relative to each other.

floral envelope the perianth or calyx and corolla considered together.

floral formula an expression summarizing the number and position of parts of each whorl of a flower.

floral kingdom *see* floral realm.

floral leaf petal or sepal.

floral realm large geographical area of the world distinguished by a particular flora; the highest level recognized in the geographical grouping of plants, a realm being divided into regions, which are further divided into provinces or domains. Four floral realms are generally recognized: the Holarctic, the Neotropical, the Palaeotropical and the Austral realms. *see also* phytogeographical kingdoms.

floret *n.* one of the small individual flowers of a crowded inflorescence such as a capitulum of Compositae; individual flower of grasses.

floridean starch a type of starch found in red algae which gives a brown reaction with iodine instead of blue and is diagnostic for that group.

florigen *n.* hypothetical developmental hormone in plants that causes a bud to develop into a flower.

floristic *a. pert.* the species composition of a plant community; *pert.* or *appl.* flora.

floristics *n.* the study of an area of vegetation in terms of the species of plants in it.

flow cytometry technique for counting cells and distinguishing different types of cells in a mixed cell population. The cells are usually stained with different fluorescent-labelled antibodies to distinctive cell-surface molecules and a stream of labelled cells is then run through a fluorescence detector, which counts the numbers of cells of each type. *see also* FACS.

flower *n.* the reproductive structure of Anthophyta (*q.v.*), being derived evolutionarily from a leafy shoot in which leaves have become modified into petals, sepals and calyx, and into the carpels and stamens in which the gametes are formed. Although flowers can take many different forms, they can all be represented by concentric whorls of different parts inserted on a base (the receptacle). The outermost whorl of sepals (often green) forms the calyx, inside that is a whorl of often brightly coloured petals, next is a ring of stamens (the male reproductive organs), and in the centre are the carpels (the female reproductive organs). *see Fig. 4.*

flowering plant *see* Anthophyta.

fluid mosaic model most recent and widely accepted model for the structure of biological membranes, proposing that membranes are composed of a phospholipid/glycolipid bilayer in which proteins are embedded and are generally free to diffuse laterally but cannot rotate from one side of the membrane to the other. *see* Fig. 7.

flukes *n.plu.* a group of parasitic flatworms (platyhelminths), including the Monogenea, which are ectoparasitic on skin and gills of fishes, and the Trematoda, which include endoparasitic blood, liver and gut flukes, such as *Fasciola*, the common liver fluke of sheep, and *Schistosoma*, the cause of schistosomiasis in humans. *see also* digenean, Platyhelminthes, Trematoda.

fluorescamine *n.* a dye that reacts with amino acids to give a fluorescent product, used in analysis of amino acid composition of proteins.

fluorescence-activated cell sorter *see* FACS.

fluorescence microscopy technique for locating particular molecules in cells by labelling with specific antibodies tagged with fluorescent dyes to make them visible in the microscope.

fluorescyanine *n.* a mixture of pterins with a yellow or blue fluorescence, found in the eyes, eggs and luminous organs of some insects.

fluoroacetate *n.* a poison isolated first from leaves of *Dichapetalum cymosum*, converted *in vivo* to fluorocitrate which inhibits aconitase and thus the tricarboxylic acid cycle.

fluorochrome *n.* any fluorescent compound.

fluorouracil (FU) *n.* fluorine-substituted analogue of uracil which as the deoxynucleotide blocks dTMP synthesis and is used as an anticancer drug.

flush *n.* a patch of ground where water lies but does not run into a channel; a period of growth, esp. in a woody plant; a rapid increase in the size of a population.

flushing *n.* the washing of dissolved substances upwards in the soil so that they are deposited near the surface. *cf.* leaching.

fluviatile *a.* growing in or near streams; inhabiting and developing in streams, *appl.* certain insect larvae; caused by rivers, *appl.* deposits.

fluvioterrestrial *a.* found in streams and in the land beside them.

F-mediated transduction sexduction *q.v.*

fodrin *n.* protein identical to spectrin, present in neurones.

foet- *alt.* fet-.

foetal *a. pert.* a foetus.

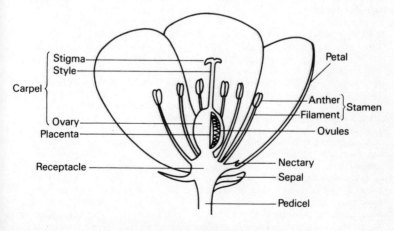

Fig. 4 Cross-section of a generalized angiosperm flower.

foetal haemoglobin (HbF) type of haemoglobin present in foetal blood consisting of 2 α and 2 γ subunits and possessing a higher oxygen affinity under physiological conditions than adult haemoglobin.

foetal membranes membranes that protect and nourish the foetus, such as the chorion, amnion, allantois and yolk sac in mammals. *alt.* extraembryonic membranes.

foetation *n.* development of the foetus within the uterus.

α-foetoprotein (α-FP, AFP) a protein secreted by yolk sac and embryonic liver epithelial cells. In adults, produced by proliferating liver cells and can be used as an indicator of a liver cancer. In pregnant women raised blood levels of α-foetoprotein may indicate a foetus with spina bifida.

foetus *n.* a mammalian embryo after the stage at which it becomes recognizable, technically, the human embryo from 7 weeks after fertilization.

fog-basking behaviour of some small insects in desert environments in which nocturnal fogs regularly occur, by which they can collect drinking water from the condensation of the fog on the body surface.

FOG fibres fast oxidative glycolytic muscle fibres in muscle of mammalian limbs, red muscle fibres adapted for aerobic metabolism and which are used when the animal is walking or running. *cf.* FG fibres.

folate (folic acid) *n.* pteroylglutamic acid, a water-soluble vitamin, sometimes considered as part of the vitamin B complex, found esp. in yeast, liver and green vegetables, deficiency causes megaloblastic anaemia, important in nucleic acid metabolism as tetrahydrofolate is a coenzyme for purine and pyrimidine biosynthesis. *alt.* PGA.

foldback DNA DNA consisting of inverted repeats that have base-paired on reassociation of denatured DNA.

foliaceous *a.* having the texture of a foliage leaf; thin and leaf-like; bearing leaves.

Folian process anterior process of malleus, one of the auditory ossicles of mammalian middle ear.

foliar *a. pert.* or consisting of leaves; bearing leaves (as opposed to flowers), *appl.* side spurs.

folic acid *see* folate.

folicole, folicolous *a.* growing on leaves, *appl.* certain fungi.

foliobranchiate *a.* possessing leaf-like gills.

foliolate *a. pert.,* having, or like, leaflets.

foliole *n.* small leaf-like organ or appendage; leaflet, as of a compound leaf.

foliose *a.* with many leaves; having leaf-like lobes, *appl.* the thallus of some lichens and liverworts.

folium *n.* a flattened structure in the cerebellum, expanding laterally into superior semilunar lobes; one of the folds on sides of tongue.

folivorous *a.* leaf-eating.

follicle *n.* small sac-like structure; (*zool.*) in the ovary, a group of cells surrounding the oocyte and probably concerned with its nutrition. *see also* Graafian follicle; hair follicle, sheath of epithelium surrounding hair root, *see* Fig. 10; (*bot.*) dry dehiscent fruit consisting of a single carpel and opening along one side only.

follicle-stimulating hormone (FSH) a glycoprotein hormone secreted by the anterior pituitary, acting on the gonads to stimulate growth of Graafian follicles and oestrogen secretion in the ovaries, and spermatogenesis in the testes.

follicular *a. pert.,* like or consisting of follicles; *appl.* an ovarian hormone: oestrone *q.v.*

folliculate *a.* containing, consisting of or enclosed in follicles.

folliculose *a.* having follicles.

following response the innate response shown by the young of many species (e.g. chicks, ducklings), which will indiscriminately follow moving objects.

fontanel(le) *n.* a gap or space between bones of cranium, closed only by a membrane.

food bodies Beltian bodies *q.v.*

food chain a sequence of organisms within an ecosystem in which each is the food of the next member in the chain. A chain starts with the primary producers, which are photosynthetic organisms (e.g. algae, plants, bacteria) or chemoautotrophic bacteria. These are eaten by herbivores (primary consumers) which are in turn eaten by carnivories (secondary consumers). Small carnivores may be eaten by larger carnivores. *see also* food web.

food pollen pollen present in flowers to provide food for visiting insects, instead of or as well as nectar, and which may be sterile and produced in special anthers.

food vacuole small vacuole enclosing fluid and ingested food particles, in many unicellular protists.

food web the interconnected food chains in an ecosystem.

foot-jaws poison claws or 1st pair of legs in centipedes *alt.* maxillipeds.

foot plates terminal enlargements of processes of astrocytes in contact with minute blood vessels.

foot spinners common name for the Embioptera *q.v.*

footprinting technique for determining the area of DNA to which a protein binds, based on digestion of the DNA-protein complex by an endonuclease so that only the area protected by the protein remains intact.

forage *n.* the vegetation eaten by grazing and browsing animals.

foraging *n.* the collection of food by animals.

foramen *n.* any small perforation; aperture through a shell, bone or membrane. *plu.* foramina.

foramen magnum the opening in back of skull for passage of the spinal cord.

foramen of Monro passage between third and lateral ventricles of brain.

foramen ovale aperture in great wing of sphenoid bone, passage for mandibular nerve; opening between atria in foetal heart.

foramen Panizzae an opening at point of contact between left and right systemic arteries in crocodilians.

foramen rotundum aperture in great wing of sphenoid bone, passage for maxillary nerve.

foramina *plu.* of foramen *q.v.*

foraminate *a.* pitted, having foramina or perforations.

Foraminifera, foraminiferans *n., n.plu.* phylum of mainly marine unicellular protists (classified as protozoans of class Sarcodina in animal classifications), having a calcareous, siliceous or composite shell through which project fine pseudopodia, and a highly vacuolated outer

layer of cytoplasm. Chalk is largely composed of foraminiferan shells and they are also major components of many deep-sea oozes. *see* globigerina ooze.

foraminiferous *a.* having foramina; containing shells of foraminiferans.

forb *n.* herbaceous plant, esp. a pasture plant other than grasses.

forceps *n.* the clasper-shaped anal cercus of some insects; the large fighting or seizing claws of crabs and lobsters; (*neurobiol.*) fibres of corpus callosum curving into frontal and occipital lobes.

forcipate *a.* forked like forceps.

forebrain *n.* the telencephalon (cerebral hemispheres) and diencephalon (thalamus, hypothalamus). *alt.* prosencephalon.

fore-gut stomodaeum *q.v.*

fore-kidney pronephros *q.v.*

fore-milk colostrum *q.v.*

forest *n.* biome consisting of continuous or semi-continuous tree cover, which may develop in areas where the average annual precipitation is sufficient (> 75 cm) to support the growth of trees and shrubs. Broadleaved forest, coniferous forest, pine forest, etc. describe forest in which the named types of tree comprise at least 80% of the canopy. *see also* ancient forest, mixed forest, monsoon rain forest, old-growth forest, rain forest, secondary forest, taiga, temperate rain forest, tropical rain forest, etc. *cf.* desert, grassland.

forficiform, forficulate *a.* scissor-shaped.

form *n.* a taxonomic unit consisting of individuals that differ from those of a larger unit by a single character, therefore being the smallest category of classification, *alt.* forma; one of the kinds of a polymorphic species; a taxonomic group whose status is not clear but may be species or subspecies; the concealed resting place of a hare.

form genus a genus whose species may not be related by common ancestry.

form species the members of a form genus *q.v.*

formation *n.* the vegetation proper to a definite type of habitat over a large area, as tundra, coniferous forest, prairie, tropical rain forest, etc.

formic acid HCOOH, an organic acid present in some ants and other insects, and in some plants.

formicarian *a. pert.* ants; *appl.* plants which attract ants by means of sweet secretions.

formicarium *n.* an ants' nest, particularly an artificial arrangement for purposes of study.

formylmethionine (fMet) *n.* a modified amino acid (methionine) which is always the first amino acid of polypeptide chains in bacteria.

fornical *a. pert.* fornix.

fornicate(d) *a.* concave within, convex without, arched.

fornices *plu.* of fornix.

fornix *n.* an arched recess, as between eyelid and eyeball, or between vagina and cervix; arched sheet of white fibres beneath corpus callosum in brain. *plu.* fornices.

forward mutation a mutation which inactivates a gene, a mutation from normal wild type.

fos a potential oncogene, encoding a transcription factor.

fossa *n.* a pit or trench-like depression.

fossette *n.* small pit or depression.

fossil *n.* the remains or traces of animal and plant life, of the past, found embedded in rock either as petrified hard parts or as moulds, casts or tracks.

fossiliferous *a.* containing fossils.

fossorial *a.* adapted for digging.

fossulate *a.* with slight grooves and hollows.

fossulet *n.* a long narrow depression.

founder effect genetic differences between an original population and an isolated offshoot due to alleles in the founder members of the new population being unrepresentative of the alleles in the original population as a whole.

fourchette *n.* the united clavicles of a bird, the wishbone; the frog of a horse's hoof; the junction of labia minora.

fourth ventricle the cavity of the hindbrain in vertebrates.

fovea *n.* a small pit, fossa or depression; shallow pit in centre of retina, point of greatest acuity of vision, present in diurnal birds, lizards, and primates, formed only of cones, the thinnest part of retina, *alt.* fovea centralis; (*bot.*) small hollow at leaf base in quillworts (Isoetales), containing a sporangium; pollinium base in orchids. *plu.* foveae.

foveal *a. pert.* fovea; *pert.* fovea centralis, *appl.* cone vision.

foveate *a.* pitted.

foveola *n.* a small pit; a shallow cavity in bone.

foveolate *a.* having regular small depressions.

fovilla *n.* the contents of a pollen grain.

F pilus a filamentous appendage produced by strains of bacteria carrying F factors and through which the F factor may be transmitted to another bacterium. *alt.* sex pilus.

fragile X syndrome a form of heritable mental retardation which is associated with an abnormally fragile section of the X-chromosome.

fragmentation mapping of chromosomes, the determination of relative positions of genetic loci, esp. on human chromosomes, in cultured somatic cells by the loss of genetic markers from the cell on radiation-induced fragmentation of chromosomes, neighbouring loci being more likely to be lost together.

frame silk one type of silk produced by spiders, which forms the supporting frame and radii of a typical orb web. *see also* viscid silk.

frameshift *n.* mutation which causes a change in reading frame as a result of the insertion or deletion of non-multiples of three consecutive nucleotides in a DNA sequence.

frameshift suppression the case where the effect of a previous frameshift mutation is overcome, and which may be due to insertions or deletions in the same gene which restore the original reading frame, or to the action of certain tRNAs (extragenic frameshift suppressors) which can recognize 4-base "codons".

frankincense *n.* a balsam obtained from plants of the genus *Boswellia*.

Fraser Darling effect the stimulation of reproductive activity by the presence and activity of other members of the species in addition to the mating pair.

fraternal *a. appl.* twins produced from two separately fertilized eggs.

free *a.* motile; unattached; distinct; separate; of pupa, exarate.

free central *appl.* placentation of plant ovary, ovules borne on a free-standing central column of tissue.

free energy (G) (Gibbs free energy) a thermodynamic function used to describe chemical reactions: the change in free energy, ΔG, being given by the equation $\Delta G = \Delta E - T\Delta S$ for a system undergoing change at constant temperature (T) and pressure, where ΔE is the change in internal energy of the system and ΔS is the change in entropy of the system, any reaction only being able to occur spontaneously if ΔG is negative. Denoted by F in the older literature.

free energy of activation ($\Delta G^{\ddagger}$) in a chemical reaction A↔B, the difference in free energy between A and the intermediate transition state of higher energy, on which the rate of reaction depends. Enzymes act as catalysts by lowering the free energy of activation.

free nuclear division division of nucleus not followed by formation of new cell walls, as in the endosperm of angiosperm seeds.

free running *appl.* an endogenous rhythm unaffected by any external influence.

freemartin *n.* sterile female or intersex twin born with a male, the abnormality being due to sharing of blood circulation in the uterus and consequent masculinization of the female by male hormones.

freeze-etching an extension of the freeze-fracture method of preparing specimens for electron microscopy. Before platinum shadowing, freeze-fractured material is subjected to freeze-drying which exposes the cytoplasmic and exterior faces of membranes and structures in the interior of the cell.

freeze-fracture method of preparing specimens for electron microscopy which enables the interior of cell membranes to be visualized. A frozen block of cells is split, often fracturing along the middle of the lipid bilayer of cell membranes, a platinum replica of the exposed surface is prepared by shadowing and the organic material dissolved away.

frenulum *n.* a fold of membrane, as of tongue, clitoris, etc.; a process on hindwing of Lepidoptera for attachment to forewing.

frenum *n.* fold of integument at junction of mantle and body of barnacles, carrying

eggs in some species.

frequency-dependent selection selection occurring when the fitness of particular genotypes is related to their frequency in the population. When rare, the particular genotype is at an advantage compared with the other possible genotypes, but when it is common it is at a disadvantage.

friable *a.* crumbly, easily powdered.

frilled organ organ at anterior end of some cestodes, for attachment to host.

frog *n.* common name for a member of the amphibian order Anura. *see* Amphibia; a triangular mass of horny tissue in the middle of sole of the foot of a horse and related animals.

frond *n.* a leaf, esp. of fern or palm; flattened thallus of certain seaweeds or liverworts; a leaf-like outgrowth from thallus, as of lichen; any leaf-like structure.

frondescence *n.* development of leaves.

frons *n.* forehead or comparable part of head in other animals.

frontal *n.* frontal scale in reptiles; frontal bone in vertebrates; *a.* in region of forehead; *appl.* plane at right angles to median longitudinal or sagittal plane.

frontal lobe of brain, the front part of cerebral hemisphere.

fronto-ethmoidal *a. pert.* frontal and ethmoidal bones, *appl.* suture.

frontocerebellar fibres nerve fibres passing from frontal lobes of cortex to cerebellum.

frontoclypeus *n.* frons and clypeus, fused, in insects.

frontoparietal *a. pert.* frontal and parietal bones, *pert.* suture: the coronal suture.

frontosphenoidal *a. pert.* frontal and sphenoidal bones, *appl.* process of zygomatic bone articulating with frontal.

fructicole *a.* inhabiting fruits, *appl.* fungi.

fructification *n.* a fruit or fruiting body or any spore-producing structure.

fructokinase phosphofructokinase *q.v.*

fructosan *n.* polysaccharide made of condensed fructose units, as inulin.

fructose *n.* hexose sugar found in many plants, as fructose phosphates an intermediate in gluconeogenesis and photosynthetic carbon fixation. *see* Appendix 1 (3) for chemical structure.

fructose bisphosphatase EC recommended name for fructose-1,6-diphosphatase which catalyses conversion of fructose 1,6-bisphosphate to fructose 6-phosphate in e.g. gluconeogenesis and photosynthetic carbon fixation. EC 3.1.3.11.

fructose diphosphate fructose 1,6-bisphosphate.

fructose 1-phosphate pathway metabolic pathway in liver for the conversion of dietary fructose into glyceraldehyde phosphate for glycolysis.

frugivorous *a.* fruit-eating.

fruit *n.* the developed ovary of a flower containing the ripe seeds, and any associated structures. In mosses the spore-containing capsule is often called the fruit.

fruit body, fruiting body in fungi, slime moulds, algae, etc. any specialized structure that produces spores or gametes.

fruit wall outer part of a fruit, either the pericarp derived from the ovary wall, or a structure derived from the ovary wall together with receptacle or other parts of the flower.

fruitfly any member of the dipteran genus *Drosophila*, especially *Drosophila melanogaster*, a favoured organism for experimental genetics and developmental biology.

frustose *a.* cleft into polygonal pieces; covered with markings resembling cracks.

frustration *n.* situation where an animal cannot make an appropriate response to a stimulus, and resulting in displacement activities, etc.

frustule *n.* the silicaceous two-part wall of a diatom.

frutex *n.* shrub. *plu.* frutices.

fruticose *a.* shrubby, *appl.* lichens that grow in the form of tiny shrubs.

fucoid *a. pert.* or resembling a seaweed.

fucosan *n.* polysaccharide composed of fucose units, found in vesicles (fucosan vesicles) in cells of brown algae where it may be a storage polysaccharide or a waste metabolic product.

fucose *n.* 5-carbon (pentose) sugar, constituent of some plant polysaccharides.

fucoxanthin *n.* brown xanthophyll carotenoid pigment found in brown algae, diatoms and Chrysophyceae.

fugacious *a.* withering or falling off very rapidly.

fugitive species a species of newly disturbed habitats, which has a high ability to disperse, and which is usually eliminated from established habitats by interspecific competition.

fulcral *a. pert.* or acting as a fulcrum; *appl.* triangular plates aiding movement of stylets in Hymenoptera.

fulcrate *a.* having a fulcrum.

fulcrum *n.* a supporting structure such as a tendril or stipule; the pivot of a lever, *appl.* points of articulation of some bones; the hinge-line in brachiopod shells.

fulvic acid fraction remaining in a solution of humus in weak alkali after removal of humin and humic acid.

fulvous *a.* deep yellow, tawny.

fumarate (fumaric acid) *n.* 4-carbon dicarboxylic acid of the tricarboxylic acid cycle, hydrated by fumarase to malate, also involved in the urea cycle in vertebrates.

fundament primordium *q.v.*

fundamental niche the largest niche an organism could occupy in the absence of competition or other interacting species.

fundamental tissue ground tissue *q.v.*

fundatrix *n.* a female founding a new colony by oviposition, *appl.* aphids.

fundic *a. pert.* a fundus, *appl.* cells of the stomach.

fundiform *a.* looped.

fundus *n.* the base of an organ, as stomach or bladder, etc. *a.* fundic.

fungal *a.* of or *pert.* fungi.

Fungi, fungi *n., n.plu.* kingdom of heterotrophic, non-motile, non-photosynthetic and chiefly multicellular organisms that absorb nutrients from dead or living organisms. Although traditionally grouped with plants they are now considered an independent evolutionary line. Multicellular terrestrial fungi comprise four main groups, the Zygomycota (zygomycetes) (e.g. bread moulds), the Ascomycota (ascomycetes) (the unicellular yeasts and multicellular sac fungi), the Basidiomycota (basidiomycetes) (e.g. mushroooms and toadstools), and the form group Deuteromycota (Fungi Imperfecti), which lack a sexual stage. The lower fungi include the slime moulds and water moulds, which are sometimes alternatively classified as protists. Many fungi are serious plant pathogens. Multicellular fungi grow vegetatively as a mycelium, a mat of thread-like hyphae from which characteristic fruiting bodies arise (e.g. the blue or black spore-heads of many common moulds, the mushrooms and toadstools of agarics, and the brackets of bracket fungi). In some fungi the hyphae are divided into uninucleate or binucleate cells by transverse partitions (septa). Hyphae possess rigid cell walls which differ in composition from those of plants, chitin rather than cellulose being a main constituent in most fungi. *see also* Appendix 3.

Fungi Imperfecti Deuteromycota *q.v.*

fungicolous *a.* living in or on fungi.

fungiform *a.* mushroom-shaped, *appl.* certain rounded papillae scattered irregularly on the tongue and having a few taste buds in the epithelium of their walls.

fungistasis mycostasis *q.v.*

fungistatic *a.* inhibiting the growth of, but not killing, fungi.

fungivore *n.* an organism feeding on fungi.

fungivorous *a.* feeding on fungi.

fungoid, fungous *a. pert.* fungi; with the character or consistency of a fungus.

fungus *sing.* of fungi *q.v.*

funicle *n.* small stalk of an ovule or seed; a small cord or band as of nerve fibres.

funicular *a. pert.* a funicle; consisting of a small cord or band.

funiculose *a.* rope-like.

funiculus *n.* funicle *q.v.*; one of the ventral lateral or dorsal columns of white matter in the spinal cord.

funnel *n.* siphon of cephalopods.

furanose *n.* a monosaccharide in the form of a 5-membered ring with 4 carbon and 1 oxygen atoms. *cf.* pyranose.

furca *n.* any forked structure.

furcal *a.* forked.

furcate *a.* branching like the prongs of a fork.

furcula *n.* the forked "spring" of springtails (Collembola); the united clavicles of birds, the wishbone.

furcular furcate *q.v.*

furfuraceous *a.* scurfy, covered with scurf-like or bran-like particles.

furovirus group group of fungus-transmitted rod-shaped single-stranded RNA plant viruses, type member soil-borne wheat mosaic virus. They are multicomponent viruses in which two genomic RNAs are encapsidated in two different-sized virus particles.

fuscin *n.* brown pigment in retinal epithelium.

fuscous *a.* of a dark, almost black, colour.

fuseau *n.* a spindle-shaped structure; a spindle-shaped thick-walled spore divided by septa, in some fungi.

fusi *n.plu.* in spiders, organs composed of two retractile processes, which form the threads of silk.

fusiform *a.* spindle-shaped, tapering gradually at both ends, *appl.* innermost layer of cerebral cortex, *appl.* a gyrus of temporal lobe.

fusiform initial in vascular cambium, a spindle-shaped cell that gives rise to secondary xylem or phloem.

fusimotor *a. appl.* motor nerve fibres which cause contraction of muscle fibres within muscle spindles.

fusion gene gene composed of the 3′ part of one gene and the 5′ part of another, and which may be generated spontaneously by mutation or constructed by recombinant DNA techniques.

fusion nucleus central nucleus of embryo sac of ovule, formed by fusion of odd nuclei from each end.

fusocellular *a.* having, or *pert.* spindle-shaped cells.

fusoid *a.* somewhat fusiform.

fusulae *n.plu.* minute tubes of spinnerets of spiders.

futile cycle substrate cycle *q.v.*

G

γ heavy chain class corresponding to IgG.

γ- *for all headwords with prefix γ- or gamma-refer to headword itself.*

G Gibbs free energy, *see* free energy; glycine *q.v.*; guanine *q.v.*; the invertible segment of phage Mu which controls strain specificity of infection.

ΔG‡ Gibbs free energy of activation. *see* free energy of activation.

G₀, G0 phase of eukaryotic cell cycle in which the cycle is arrested.

G₁, G1 phase of eukaryotic cell cycle following mitosis and from which cells may either enter a resting phase, G_0, or continue on to DNA replication (S phase) and mitosis.

G₂, G2 phase of eukaryotic cell cycle between DNA replication and mitosis.

G4 single-stranded DNA phage of *E. coli*.

GA gibberellic acid or gibberellin *q.v.*; Golgi apparatus *q.v.*

GABA γ-aminobutyric acid *q.v.*

gag gene of C-type viruses specifying proteins of the viral nucleoprotein core.

GAG glycosaminoglycan *q.v.*

GalNAc *N*-acetylgalactosamine *q.v.*

Gal galactose *q.v.*

GAP glyceraldehyde 3-phosphate *q.v.*; GTPase activating protein *q.v.*

G-CSF granulocyte colony-stimulating factor *q.v.*

GDP guanosine diphosphate *q.v.*

GFAP glial fibrillary acidic protein *q.v.*

GH growth hormone *q.v.*

GHRF growth-hormone releasing factor, somatoliberin *q.v.*

Gᵢ G proteins (*q.v.*) that inhibit adenylate cyclase, leading to a suppression of cyclic AMP production.

Glc glucose *q.v.*

GlcN glucosamine *q.v.*

GlcNAc *N*-acetylglucosamine *q.v.*

Gln glutamine *q.v.*

Glu glucose *q.v.*; glutamic acid *q.v.*

Glx glutamic acid or glutamine.

Gly glycine *q.v.*

GM₁, GM₂, etc. ganglioside *q.v.*

GM-CSF granulocyte/macrophage colony-stimulating factor *q.v.*

GMP guanosine monophosphate *q.v.*

GnRF gonadotropin-releasing factor *q.v.*

GOGAT the enzyme glutamate synthase *q.v.*

G_olf G proteins (*q.v.*) found in olfactory neurons, which couple olfactory receptors to the cyclic AMP second messenger pathway.

gp glycoprotein.

G_p G proteins (*q.v.*) involved in coupling receptors to the phosphoinositide second messenger pathway.

GPP gross primary production *q.v.*

G3P glyceraldehyde 3-phosphate *q.v.*

G6P glucose 6-phosphate, *see* glucose.

GRE glucocorticoid response element, control site in DNA at which glucocorticoid hormone/receptor complexes bind to activate gene expression.

G_s G proteins (*q.v.*) that couple receptors to the cyclic AMP second messenger pathway. They stimulate adenylate cyclase, leading to a rise in cyclic AMP production.

GS glutamine synthetase *q.v.*

G_t transducin *q.v.*

GTP guanosine triphosphate *q.v.*

GTPase guanosine triphosphatase *q.v. see also* guanine-nucleotide binding proteins.

Gy the SI unit of ionizing radiation absorbed by tissue, the gray *q.v.*

GABA receptors receptors on neurones and other cells for the neurotransmitter γ-aminobutyric acid (GABA). They are ion channels.

G actin *see* actin.

gadfly *see* Brachycera.

Gadidae, gadids *n.*, *n.plu.* large and economically important family of marine bony fishes including the cod, whiting and haddock.

gain-of-function *appl.* mutant alleles directing overexpression of normal gene product.

galactan *n.* polygalactose, galactose residues linked together in a chain, a hydrolysis product of, e.g., pectins and plant gums.

galactolipid *n.* any glycolipid in which the sugar is galactose, such as cerebrosides.

galactosaemia *n.* inborn error of metabolism leading to inability to metabolize dietary galactose because of a deficiency of the enzyme galactose 1-phosphate uridyl transferase, leading to failure to thrive in affected infants due to accumulation of toxic compounds.

galactosamine *n.* amino derivative of the sugar galactose, substituted at carbon 2, as *N*-acetylgalactosamine a common constituent of the sugar side chains of glycoproteins.

galactose *n.* 6-carbon aldose sugar, a constituent with glucose of the disaccharide lactose. Also found in various complex carbohydrates such as the pectins of plant cell walls and in some glycolipids and glycoproteins. *see* Appendix 1 (2) for chemical structure.

galactosemia *alt.* spelling of galactosaemia *q.v.*

galactosidase *n.* an enzyme that hydrolyses lactose to galactose and glucose. EC 3.2.1.22–23.

galactoside *n.* any compound in which galactose is linked via a glycosidic bond to another sugar or a non-sugar alcohol, and including the disaccharide lactose and the galactolipids such as cerebrosides.

galactosyl transferases enzymes that add monosaccharide units to the sugar side chains of glycoproteins.

galactotropic *a.* stimulating milk secretion.

galacturonate (galacturonic acid) a sugar acid derived from galactose by oxidation, a component of plant cell walls.

galbulus *n.* a closed globular female cone with shield-shaped scales which are fleshy or thickened, as in the false-cypresses.

galea *n.* (*bot.*) helmet-shaped petal; (*zool.*) outer branch of maxilla in insects.

Galen, veins of internal cerebral veins and great cerebral vein formed by their union.

galericulate *a.* bearing or covered by a small cap.

galeriform *a.* shaped like a cap.

gall *n.* an abnormal outgrowth from plant stem or leaf caused by the presence of young insects (e.g. gall wasps or gall mites) in the tissues (as e.g. oak-apple) or by infection by certain fungi or bacteria (e.g. crown gall caused by *Agrobacterium*); bile *q.v.*

gall bladder pear-shaped or spherical sac that stores bile.

gall flower in fig trees, an infertile female flower in which fig wasp lays its eggs.

gall wasps minute hymenopteran insects belonging to the superfamily Cynipoidea, which lay their eggs in the leaf and stem tissue of oak and a few other plants (e.g. roses), inducing the formation of a gall, inside which the grub develops, feeding on the gall tissue.

gallic acid organic acid obtained from hydrolysis of tannin, and present in tea, galls and some plants.

gallicolous *a.* living in plant galls.

Galliformes *n.* an order of heavy-bodied, chicken-like land birds, including grouse, partridges, pheasants, quails and domestic fowl.

gallinaceous *a.* resembling the domestic fowl, *appl.* birds of the same family.

gallotannin *n.* a tannin found in many types of galls, esp. on oak, and which is a glucoside of glucose and digallic acid.

Galton's law of filial regression the tendency of offspring of outstanding parents to revert to the average for species.

galvanotaxis, galvanotropism *n.* movement in response to an electrical stimulus.

game theory mathematical theory concerned with determining the optimal strategy in situations of competition or conflict. This theory can be applied to the relationships within a community, and the computer simulation of such relationships to determine winning strategies can help to throw light on ecological and social relationships and their evolution.

gametal *a. pert.* a gamete; reproductive.

gametangiogamy *n.* union of gametangia.

gametangium *n.* any structure producing gametes. *plu.* gametangia.

gamete *n.* haploid reproductive cell produced by sexually reproducing organisms which fuses with another gamete of opposite sex or mating type to produce a zygote, the male gamete being variously called a spermatozoon, spermatozoid or antherozoid and the female gamete an ovum or egg. *alt.* germ cell, sex cell, sexual cell.

gametic *a. pert.* a gamete; *appl.* a mutation that occurs before the gamete matures.

gametic number the haploid, n, number of chromosomes present in gamete.

gametocyst *n.* cyst surrounding two associated individuals in which sexual reproduction takes place, as in some gregarine protozoans.

gametocyte *n.* cell from which gamete is produced.

gametogamy *n.* the union of gametes.

gametogenesis *n.* gamete formation.

gametogenetic *a.* stimulating gamete formation.

gametogenic variation variation arising from mutations in gametes.

gametoid *n.* a structure behaving like a gamete, as the multinucleate masses of protoplasm that fuse to form a zygotoid in some fungi.

gametophore *n.* a special part of the gametophyte on which gametangia are borne, esp. the upright leafy shoot of a moss; in fungi, a hyphal outgrowth that fuses with a similar neighbouring outgrowth to form a zygospore.

gametophyll *n.* modified leaf bearing sexual organs; a microsporophyll or megasporophyll.

gametophyte *n.* the haploid gamete-forming phase in the alternation of generations in plants. *cf.* sporophyte.

gametothallus *a.* a thallus that produces gametes.

gametropic *a.* movement of plant organs before or after fertilization.

gamic *a.* fertilized.

gamma *for all headwords with prefix gamma- refer also to headword itself, e.g.*

for γ-aminobutyric acid look under A.

gamma globulins group of blood serum proteins with a particular range of electrophoretic mobility, and including the immunoglobulins, as well as some non-immunoglobulins.

gammopetalous *a.* having petals joined into a tube at least at the base, *alt.* sympetalous. *cf.* monopetalous, polypetalous.

gammosepalous *a.* having sepals joined into a tube at least at the base, *alt.* monosepalous. *cf.* polysepalous.

gamobium *n.* the sexual generation in alternation of generations, i.e. the gametophyte.

gamodeme *n.* a deme forming a relatively isolated interbreeding community.

gamodesmic *a.* having the vascular bundles fused together instead of separated by parenchyma tissue.

gamogenesis *n.* sexual reproduction.

gamogenetic, gamogenic *a.* sexual, produced from union of gametes.

gamogony *n.* formation of gametes or gametocytes from a gamont in certain protozoans.

gamont *n.* in some Protozoa, a generation or individual which produces gametes which then unite in pairs to form the zygote or sporont; sporont *q.v.*

Gamophyta *n.* phylum of green pigmented photosynthetic protists comprising (1) those multicellular filamentous green algae that lack flagella at all stages of the life cycle and engage in sexual reproduction by conjugation between haploid vegetative cells (e.g. *Spirogyra*), and (2) the desmids (*q.v.*).

gamostely *n.* the condition in stems with several steles when the separate steles are fused together and surrounded by pericycle and endodermis.

ganglia *plu.* of ganglion.

gangliar *a. pert.* a ganglion or ganglia.

gangliate *a.* having ganglia.

gangliform *a.* in the shape of a ganglion.

ganglioblast *n.* the precursor to a ganglionic neurone (gangliocyte) of the peripheral nervous system.

ganglioid *a.* like a ganglion.

ganglion *n.* structure within the nervous system formed of a mass of nerve cell bodies. *plu*. ganglia.

ganglion cells in mammalian retina, nerve cells forming the outermost layer of nervous tissue, and whose fibres feed into the optic nerve.

ganglionated gangliate *q.v.*

ganglioneural *a. appl.* a system of nerves, consisting of a series of ganglia connected by nerve fibres.

ganglionic *a. pert.*, consisting of, or near a ganglion, *appl.* layer of retina containing ganglion cells.

ganglioplexus *n.* a diffuse ganglion.

ganglioside *n.* any of a group of complex glycolipids, found chiefly in nerve cell membranes, and containing sphingosine, fatty acids, and an oligosaccharide chain containing at least 1 acid sugar such as *N*-acetylglucosamine, *N*-acetylgalactosamine or *N*-acetylneuraminic acid. *see* Appendix 1 (42) for structure.

ganoid scales rhomboidal scales, found in primitive fish, with many outer layers of enamel, below which is dentine (cosmine) then layers of bone, growth in thickness being by layers above and below.

gap genes developmental genes acting early in *Drosophila* embryogenesis, which are involved in specifying broad regions of the embryo. If mutant they result in an embryo lacking chunks of contiguous segments.

gap junction small area where plasma membranes of two neighbouring cells are separated by a very narrow gap, and where the plasma membranes contain multisubunit channel proteins – gap junction proteins – that are aligned opposite each other to form a continuous aqueous channel connecting the cytoplasm of the two cells. Gap junctions allow the passage of ions and small molecules up to 1000–1500 molecular weight.

gape *n.* the distance between the open jaws of birds, fishes, etc.

Garryales *n.* in some classifications, an order of woody dicots with opposite evergreen leaves and flowers in hanging catkin-like panicles.

gas bladder swimbladder *q.v.*

gas gland glandular portion of air bladder of certain fishes which secretes gas into the bladder.

gas transport transport of gases between respiratory surface and tissues.

gas vacuoles vacuoles in cyanobacteria which appear as black bodies.

gaseous exchange the exchange of gases between an organism and its surroundings, including uptake of oxygen and release of carbon dioxide in respiration in animals and plants, and the uptake of carbon dioxide and release of oxygen in photosynthesis in plants.

gaster *n.* an abdomen, esp. a swollen one; the swollen portion of abdomen of hymenopterans, which lies behind the waist.

Gasteromycetes *n.* a group of basidiomycete fungi in which the hymenium is completely enclosed in a basidiocarp and never exposed. It comprises the puffballs, earth stars, stinkhorns and bird's nest fungi.

Gasteropoda, gasteropods Gastropoda, gastropods *q.v.*

gastraeum *n.* the ventral side of the body.

gastral *a. pert.* stomach.

gastral layer in sponges the layer of flagellated cells (choanocytes) lining the internal cavities which take up food particles from the water.

gastralia *n.plu.* abdominal ribs *q.v.*, as in some reptiles.

gastric *a. pert.* or in region of stomach.

gastric filaments in some jellyfish, endodermal filaments lined with nematocysts which kill any live prey entering the stomach.

gastric gland simple or compound tubular gland at the base of the stomach in the wall, which secretes gastric juice containing pepsin, rennin, hydrochloric acid and mucus.

gastric intrinsic factor intrinsic factor *q.v.*

gastric mill in decapod crustaceans, the hard lining of the gizzard and its associated muscles, which grinds and strains food.

gastric shield in bivalve molluscs, a hard structure in the stomach against which the crystalline style rubs and is worn away, releasing amylase.

gastrin *n.* peptide hormone produced by stomach which stimulates secretion of gastric juice (digestive enzymes). Also found in central nervous system.

gastrocnemius *n.* the large calf muscle of leg.

gastrocolic *a. pert.* stomach and colon.

gastrodermal *a. pert.* gastrodermis *q.v.*

gastrodermis *n.* single layer of epithelium lining the gut cavity in simple animals such as coelenterates, flatworms, nematodes, etc.

gastroepiploic *a. pert.* stomach and the main peritoneal fold, *appl.* arteries, veins.

gastrohepatic *a. pert.* stomach and liver, *appl.* a portion of lesser omentum, a mesentery connecting liver and stomach in reptiles.

gastrointestinal *a. pert.* stomach and intestines.

gastrolienal *a. pert.* stomach and speen.

gastrolith *n.* mass of calcareous matter found on each side of gizzard of crustaceans before a moult.

Gastromycetes Gasteromycetes *q.v.*

gastroparietal *a. pert.* stomach and body wall.

gastrophrenic *a. pert.* stomach and diaphragm, *appl.* ligament.

Gastropoda, gastropods *n., n.plu.* class of molluscs including the winkles and whelks, sea slugs, water snails, and land snails and slugs. They are characterized by a large flat muscular foot on which they crawl about, and, where a shell is present, it is a rounded or conical spirally coiled shell in one piece.

gastropores *n.plu.* in milleporine corals, the larger pores in the surface of the colony, through which protrude polyps with four knobbed tentacles.

gastropulmonary *a. pert.* lungs and stomach.

gastrosplenic *a. pert.* stomach and spleen.

Gastrotricha *n.* a phylum of marine and freshwater microscopic pseudocoelomate animals which have an elongated body and move by ventral cilia.

gastrozooid *n.* individual specialized for feeding in siphonophore colony.

gastrula *n.* the cup- or basin-shaped early embryo that develops by invagination of the blastula, and in which the three primary germ layers (ectoderm, meso- derm and endoderm) begin to be distinguished.

gastrulation *n.* stage in early embryo-genesis involving extensive cell movements, and in which the gut cavity is formed and the three primary layers of the animal body (ectoderm, mesoderm and ectoderm) are placed in position for further development.

gate *n.* mechanism by which certain ion conductance channels in membranes may be opened and closed, such channels being known as gated channels.

gated *a. appl.* ion channels in cellular membranes whose activation is dependent on a specific stimulus, either binding by a specific molecule (ligand), or a change in membrane potential to some threshold level; *appl.* any activity that requires a stimulus at a certain level to allow it to occur.

gating current ionic current associated with the opening or closing of gated ion channels in membranes.

Gause's principle an ecological principle stating that usually only one species may occupy a particular niche in a habitat. *alt.* competitive exclusion principle.

Gaussian curve the symmetrical curve representing the frequency distribution of a normally distributed population.

Gaussian distribution a symmetrical distribution about the mean.

Gaviiformes *n.* an order of birds, including divers and loons.

G-banding technique in which mitotic chromosomes are subjected to various treatments and stained with Giemsa producing a pattern of bands specific and invariant for each chromosome, which enables different chromosomes to be distinguished.

GC box short DNA sequence rich in guanine and cytosine, typically found in the promoters of many eukaryotic genes transcribed by RNA polymerase II, and which may be the recognition site for the general transcription factor Sp1.

gel electrophoresis *see* electrophoresis; polyacrylamide gel electrophoresis.

gel-filtration chromatography technique for separating molecules on the basis of size by passage through a column of beads of an insoluble, highly hydrated polymer (e.g. Sephadex), larger molecules being unable to enter the spaces within the beads and thus being more rapidly eluted from the column.

gelatin *n.* jelly-like substance obtained from animal tissue on heating, and which is denatured collagen.

gelatinous *a.* jelly-like in consistency.

gelsolin *n.* a cytoplasmic protein that can fragment actin filaments by inserting between actin subunits.

gemellus *n.* either of two muscles, superior and inferior, running from ischium to greater trochanter and to trochanteric foassa of femur respectively.

geminate *a.* growing in pairs; *appl.* species or subspecies: corresponding forms in similar but separate regions, as reindeer and caribou.

geminiflorous *a. appl.* a plant whose flowers are arranged in pairs.

geminivirus group the only group of plant viruses with single-stranded circular DNA genomes, and which have "twinned" particles composed of two isometric virus particles attached to each other. There are three subgroups: A, leafhopper-transmitted viruses with genomes composed of a single type of DNA, infecting monocotyledonous plants, type member maize streak virus; B, whitefly-transmitted viruses with two types of DNA, infecting dicotyledonous plants, type member African cassava mosaic virus; C, leafhopper-transmitted viruses with genomes composed of a single type of DNA, infecting dicots, type member beet curly top virus.

geminous *a.* in pairs, paired.

gemma *n.* a bud or outgrowth from a plant or animal that develops into a new organism. *plu.* gemmae.

gemma cup cup-shaped or crescent-shaped structure in which gemmae are produced in some liverworts.

gemmaceous *a. pert.* gemmae or buds.

gemmate *a.* having or reproducing by buds or gemmae.

gemmation *n.* budding; the development of new independent individuals by budding off from the parent; the arrangement of buds on a twig or branch.

gemmiform *a.* bud-shaped.

gemmiparous *a.* reproducing by gemmae or buds.

gemmule *n.* any small asexual propagative unit.

gena *n.* cheek or side part of head. *a.* genal.

gene *n.* the basic unit of inheritance, by which hereditary characteristics are transmitted from parent to offspring. At the molecular level a single gene consists of a length of DNA (or in some viruses, RNA) which exerts its influence on the organism's form and function by encoding and directing the synthesis of a protein, or in some cases, a tRNA, rRNA or other structural RNA. Each living cell carries a full complement of the genes typical of the species, borne in linear order on the chromosomes.

gene amplification the repeated duplication of a gene to produce a number of identical copies in the nucleus.

gene bank gene library, *see* DNA library.

gene centres geographical regions in which certain species of cultivated plant are represented in their greatest number of different varieties or forms, and which may in some cases correspond to their centre of origin and initial domestication.

gene cloning *see* DNA cloning.

gene cluster two or more contiguous genes of related or identical nucleotide sequence.

gene complex a group of genes whose combined effects determine a phenotypic character.

gene conversion phenomenon which can be followed in some ascospore-producing fungi, where abnormal ratios of a pair of parental alleles are seen after meiosis in some asci as a result of "correction" of heteroduplex DNA to one genotype or the other before segregation and replication of the products of crossing-over.

gene diminution the routine reduction or elimination of certain genes during development which occurs in some animals such as the nematode *Parascaris equorum* and ciliates.

gene disruption *see* gene knock-out, homologous recombination.

gene dosage the number of genes, or in some instances the numbers of copies of a particular gene, in any given nucleus, cell etc.

gene duplication the, very rare, generation of additional copies of a gene during normal cellular processes such as recombination, which is thought to be the origin of families of related genes such as the globin genes.

gene evolution *see* molecular evolution.

gene expression the realization of genetic information encoded in the genes to produce a functional protein or RNA. In its broadest sense it encompasses both transcription and translation, but often refers to transcription only.

gene family set of genes with similarities in their nucleotide sequences, and which are thought to be descended by duplication and subsequent variation from the same ancestral gene.

gene flow the spread of particular alleles in a population and between populations resulting from outbreeding and subsequent intercrossing.

gene-for-gene resistance in plant pathology, a type of resistance in which a so-called avirulence gene in the pathogen is matched by a resistance gene in the plant. The outcome of an infection depends on which alleles of each gene are present.

gene frequency the frequency of a particular variant (allele) of a gene in a population.

gene gun colloquial term for the equipment used to introduce DNA into a cell using biolistics *(q.v.)*.

gene knock-out technique by which specific genes can be disrupted and rendered non-functional in unicellular microorganisms and in cultured cells. When the technique is applied to mouse embryonic stem cells, the mutant cells can then be used to generate transgenic mice mutant for a specific gene. *see* homologous recombination. *alt.* gene disruption.

gene library *see* DNA library.

gene locus the site on a chromosome occupied by a gene.

gene manipulation alteration of genes by mutation *in vitro*; genetic engineering *q.v.*

gene mapping determination of the location of a gene on a chromosome, which may be carried out by genetic mapping and/or physical mapping.

gene pool all the genes, and their different alleles, present in an interbreeding population.

gene product the protein, tRNA, rRNA or other structural RNA encoded by a gene.

gene rearrangement *see* antigenic variation, immunoglobulin genes, mating-type switch, T-cell receptor.

gene regulatory proteins proteins that regulate the expression of a gene by interacting with a control site in DNA, and including transcriptional activators, repressors, transcription factors. *see individual entries.*

gene sequencing *see* DNA sequencing.

gene superfamily a set of genes that are thought originally to have derived from a single ancestral gene, but which have diverged to such an extent that they now encode proteins with many different functions and roles, as the immunoglobulin superfamily and the serine proteinase superfamily. The former, e.g., contains the immunoglobulins, the T-cell antigen receptor, various other proteins carried on the surface of lymphocytes, and various cell adhesion molecules.

gene targeting the replacement or mutation of a particular gene, using recombinant DNA techniques.

gene therapy *n.* the amelioration of a genetic disease by the replacement or supplementation of affected cells with genetically corrected cells, or by introduction of correct copies of the gene directly into affected cells. Trials of somatic gene therapy are in process for a small number of human genetic diseases.

gene transfer the introduction of genes from one species into another, using recombinant DNA techniques, *see* genetic engineering, recombinant DNA, transgenic, transgene.

genecology *n.* ecology studied in relation to the population genetics of the organisms concerned.

genera *plu.* of genus *q.v.*

general recombination genetic recombination between homologous DNA sequences, the typical form of reciprocal recombination occurring at crossing-over during meiosis in the eukaryotic cell.

generalist, generalist species organism or species with a very broad ecological niche, such as humans, rats and house-sparrows, which can tolerate a wide range of environmental conditions and eat a variety of foods. *cf.* specialist species.

generalized *a.* combining characteristics of two groups, as in many fossils, not specialized.

generation of diversity in immunology, the various processes that combine in the development of lymphocytes to generate the almost infinitely variable repertoire of antibodies and T-cell receptors.

generation time the time between formation of a somatic cell by mitosis and its own division; the average span of time between the birth of parents and the birth of their offspring.

generative *a.* concerned in reproduction; *appl.* the smaller of two cells into which a pollen grain primarily divides.

generative apogamy the condition where the sporophyte plant is developed from the ovum or another haploid cell of the gametophyte, with no fertilization. *alt.* haploid apogamy, meiotic apogamy, reduced apogamy.

generative cell cell in male gametophyte of conifers that gives rise to sperm (male gamete).

generator potential graded electrical potential difference across the cell membrane produced in a sensory receptor in response to a stimulus, which is graded in strength proportionally to the strength of the stimulus, and which triggers an action potential when it reaches a certain threshold.

generic *a.* common to all species of a genus; *pert.* a genus.

genesis *n.* formation, production or development of a cell, organ, individual, or species.

genet *n.* a unit or group of individuals deriving by asexual reproduction from a single zygote.

genetic *a. appl.* anything involving, caused by or *pert.* genes; *pert.* genetics; *pert.* genesis. *see also* gene.

genetic adaptation adaptation to a new habitat or changed environmental conditions as a result of genetic change.

genetic code rules by which the amino acid sequence of a polypeptide chain is specified by the order of bases in DNA, in which groups of 3 bases (triplets or codons) specify one of the 20 amino acids of which proteins are composed, the code being redundant in that each amino acid is specified by more than one triplet. *see* Fig. 5.

genetic disease or defect heritable disease or condition, such as haemophilia or sickle-cell anaemia, which is caused by a specific defect in a single gene.

genetic diversity variability within a species due to genetic differences between individuals.

genetic drift random changes in gene frequency in small isolated populations owing to factors other than natural selection such as sampling of gametes in each generation, *alt.* Sewall Wright effect; random nucleotide changes in a gene not subject to natural selection.

genetic engineering any change in the genetic constitution of a living organism that has been brought about by artificial means (i.e. not by conventional breeding) and which usually would not occur in nature, such as the introduction of a gene from another species. This was first accomplished in bacteria and other microorganisms and strains of bacteria and yeast altered in this way are used to produce large quantities of valuable proteins such as hormones, viral proteins for vaccines, etc. which are difficult to obtain by other means. Genes can also be introduced into mammalian and other animal cells in culture and into plant cells. In mammals a permanently altered strain of animal can be produced by introducing the required gene into the fertilized egg or very early embryo *in vitro*, and then replacing it in the uterus. The genetic constitution of certain other animals can also be altered in various ways. Genetically engineered plants are produced by introducing new genes into a single protoplast or piece of cultured plant tissue, from which a complete new plant can be regenerated. *see also* recombinant DNA.

genetic equilibrium condition in a population where gene frequencies stay constant from generation to generation.

genetic fingerprinting *see* DNA fingerprinting.

genetic fitness *see* fitness.

genetic information the information for synthesizing RNAs and proteins, which is contained in DNA or RNA.

genetic isolation lack of interbreeding between groups of a population or between different populations due to geographic isolation or in humans, to cultural preferences also.

genetic load the average number of mutant alleles that reduce fitness per individual in a population, considered as an accumulated decrease in fitness from a theoretical optimum.

genetic locus *see* gene locus.

genetic manipulation any deliberate alteration made to the genome of an organism by genetic engineering techniques.

genetic map a map of the positions of gene loci on a chromosome.

genetic mapping determining the position of a gene on the chromosomes by means of recombination frequencies, and other purely genetic means. *cf.* physical mapping.

genetic marker term used generally to describe a piece of DNA or a gene whose properties, and sometimes position on the chromosome, are known and which may

be used to identify particular cells or organisms, or as a point of reference in a genetic mapping experiment.

genetic modification *see* genetic engineering, recombinant DNA.

genetic polymorphism the stable, long term existence of multiple alleles at a gene locus. Technically a locus is said to be polymorphic if the most common homozygote occurs at a frequency of less than 90% in the population.

genetic recombination *see* recombination.

genetic spiral in a spiral arrangement of leaves around an axis, the imaginary spiral line following points of insertion of successive leaves.

genetic transformation *see* transformation.

genetic variation heritable variation in a population as a result of the presence of

		Second base			
		U	C	A	G
First base	U	UUU⎫ Phe UUC⎭ UUA⎫ Leu UUG⎭	UCU⎫ UCC⎪ Ser UCA⎪ UCG⎭	UAU⎫ Tyr UAC⎭ UAA Stop UAG Stop	UGU⎫ Cys UGC⎭ UGA Stop UGG Trp
	C	CUU⎫ CUC⎪ Leu CUA⎪ CUG⎭	CCU⎫ CCC⎪ Pro CCA⎪ CCG⎭	CAU⎫ His CAC⎭ CAA⎫ Gln CAG⎭	CGU⎫ CGC⎪ Arg CGA⎪ CGG⎭
	A	AUU⎫ AUC⎬ Ile AUA⎭ AUG Met	ACU⎫ ACC⎪ Thr ACA⎪ ACG⎭	AAU⎫ Asn AAC⎪ AAA⎫ Lys AAG⎭	AGU⎫ Ser AGC⎭ AGA⎫ Arg AGG⎭
	G	GUU⎫ GUC⎪ Val GUA⎪ GUG⎭	GCU⎫ GCC⎪ Ala GCA⎪ GCG⎭	GAU⎫ Asp GAC⎭ GAA⎫ Glu GAG⎭	GGU⎫ GGC⎪ Gly GGA⎪ GGG⎭

Fig. 5 The genetic code.

different variants (alleles) of any gene and, in eukaryotes, their shuffling into new combinations by sexual reproduction and recombination.

genetics *n.* that part of biology dealing with heredity and variation and their physical basis in DNA, the genetic material; of an organism, the physical basis of its inherited characteristics, i.e. the sequence and arrangement of its genes, and those processes involved in their transmission from one generation to the next.

genial *a. pert.* chin.

genic *a. pert.* genes.

genicular *a. pert.* region of the knee; *pert.* geniculum, *appl.* a ganglion of the facial nerve; *appl.* bodies, the lateral and medial corpora geniculata, comprising part of the thalamus, *see* lateral geniculate nucleus.

geniculate *a.* bent like a knee.

geniculum *n.* sharp bend in a nerve; part of facial nerve in temporal bone where it turns abruptly towards the stylomastoid foramen.

genioglossal, geniohyoglossal *a.* connecting chin and tongue, *pert.* muscle that moves tongue in vertebrates.

geniohyoid *a. pert.* chin and hyoid, *appl.* muscles running from hyoid to lower jaw.

genital *a. pert.* region of the reproductive organs.

genital bursae in brittle stars, sacs into which the gonads open, and which open on each side of the base of each arm, also concerned with respiration and sometimes with the brooding of larvae.

genital canals in crinoids, the canals in the arms carrying genital cords that enlarge into gonads when they reach the pinnule.

genital coelom in cephalopods, a coelom at the apex of the visceral hump.

genital cord the cord formed by the posterior ends of the Müllerian and Wolffian ducts in the mammalian embryo.

genital disc in *Drosophila* and other flies, a disc of cells in larva from which the external genitalia and reproductive system of adult will develop at morphogenesis.

genital duct duct leading from gonads to the exterior.

genital operculum in some arachnids, a soft rounded median lobe divided by a cleft on the sternum of the 1st preabdominal segment, with the opening of the genital

duct at its base.

genital plates the plates in sea urchins bearing the opening of the gonads.

genital pleurae in some hemichordates, a pair of lateral ridges or folds in the region of the gills in which the gonads lie.

genital rachis in many echinoderms, a ring of genital cells on which the gonads are borne.

genital ridge germinal ridge *q.v.*

genital sinus fused male and female genital ducts in some trematodes; in cartilaginous fish, a paired sinus opening into the posterior cardinal sinus.

genital tubercle tissue in mammalian embryo that develops into the clitoris in females and the penis in males.

genitalia, genitals *n.plu.* the organs of reproduction, the gonads and accessory organs, esp. the external organs.

genito-anal *a.* in the region of the genitalia and anus.

genitocrural *a.* in the region of the genitalia and thigh, *appl.* a nerve originating from the 1st and 2nd lumbar nerves.

genitofemoral genitocrural *q.v.*

genito-urinary urinogenital *q.v.*

Gennari's band a layer of white nerve fibres in middle layer of cerebral cortex in occipital lobe.

genocline *n.* a gradual reduction in the frequency of various genotypes within a population in a particular spatial direction.

genocopy *n.* production of the same phenotype by different mimetic non-allelic genes.

genodeme *n.* a deme differing from others genotypically, but not necessarily phenotypically.

genoholotype *n.* a species defined as typical of its genus.

genome *n.* the genetic complement of a living organism or a single cell, more specifically the total haploid genetic complement of a diploid organism or the total number of genes carried by a prokaryotic microorganism or a virus; sometimes used for the total DNA content of any nucleus.

genome library DNA library (*q.v.*) consisting of fragments of chromosomal DNA (as opposed to cDNA).

genomic *a. pert.* genome; *appl.* chromosomal DNA (as opposed to cDNA synthesized on an RNA template).

genomic imprinting situation where a particular gene is expressed only if it derives from the male parent (paternal imprinting) or from the female parent (maternal imprinting). *alt.* parental genomic imprinting.

genospecies *n.* a group of bacterial strains capable of gene exchange.

genosyntype *n.* a series of species together defined as typical of their genus.

genotype *n.* the genetic constitution of an organism, which acting together with environmental factors determines phenotype. *a.* genotypic.

genotypic *a. pert.* genotype. *cf.* phenotypic.

genotypic variance a statistical measure used in determining the heritability of a particular trait, being that part of the phenotypic variance that can be attributed to differences in genotype between individuals.

gens *n.* a taxonomic group used in different ways by different writers and never precisely defined. *plu.* gentes.

gentes *plu.* of gens *q.v.*

Gentianales *n.* an order of dicot trees, shrubs or herbs and including the families Asclepiadaceae (milkweed), Gentianaceae (gentian), Rubiaceae (madder) and others.

genu *n.* the knee; a joint between femur and tibia in some ticks and mites; a knee-like bend in an organ or part; anterior end of corpus callosum.

genus *n.* taxonomic group of closely related species, similar and related genera being grouped into families. Generic names are italicized in the scientific literature, e.g. *Homo* (man), *Quercus* (oaks), *Canis* (wolves and dogs), *Salmo* (salmon and relatives). *plu.* genera. *a.* generic.

geobiotic terrestrial *q.v.*

geobotany phytogeography *q.v.*

geocarpic *a.* having fruits maturing underground due to young fruits being pushed underground by curvature of stalk after fertilization.

geochronology *n.* science dealing with the measurement of time in relation to the Earth's evolution.

geocline *n.* a gradual and continuous change in a phenotypic character over a considerable area as a result of adaptation to changing geographical conditions.

geocoles *n.plu.* organisms that spend part of their life in the soil and affect it by aeration, drainage, etc.

geocryptophyte *n.* a plant with dormant parts hidden underground.

geographical isolation the separation of two populations originally of the same species from each other by a physical barrier such as mountains, oceans, rivers, etc. Eventually this may lead to such differences evolving in one or both of the isolated populations that they are unable to interbreed and a new species is formed.

geographical race a population separated from other populations of the same species by geographical barriers such as mountain ranges or oceans, and showing little or no differences from the rest of the population, and which can usually interbreed to produce fertile offspring if brought into contact with the rest of the species.

geology *n.* science dealing with the physical structure, activity and history of the Earth.

geonasty *n.* a curvature, usually a growth curvature, in response to gravity. *a.* geonastic.

geophilous *a.* living in or on the soil; having leaves borne at soil level on short stout stems.

geophyte *n.* a land plant; a perennial herbaceous plant with dormant parts (tubers, bulbs, rhizomes) underground.

geoplagiotropic *n.* growing at right angles to the surface of the ground, in response to the stimulus of gravity.

geosere *n.* series of climax formations developed through geological time; the total plant succession of the geological past.

geosmin *n.* substance produced by certain cyanobacteria and actinomycetes which contributes to the earthy smell of damp soil.

geotaxis *n.* movement in response to gravity. *a.* geotactic.

geotonus *n.* normal position in relation to gravity.

geotropism *n.* movement or growth in relation to gravity, either in the direction of gravity (positive geotropism, as plant roots) or away from the ground (negative geotropism, as plant shoots).

Geraniales *n.* order of dicot trees, shrubs or

herbs including the families Balsaminaceae (balsam), Eryththroxylaceae (coca), Geraniaceae (geranium), Linaceae (flax), Tropaeolaceae (nasturtium) and others.

germ *n.* vernacular for a microorganism, esp. one that causes disease; the embryo of a seed.

germ band primitive streak *q.v.*; in a developing insect blastoderm, the cells that give rise to the embryo.

germ cell a reproductive cell; gamete, or cell giving rise to gamete, often set aside early in embryonic life.

germ layer early differentiated layer of cells in embryo, *see* primary germ layer.

germ line line of cells from which gametes (e.g. ova and sperm) are produced and through which hereditary characteristics (genes) are passed on from generation to generation in animals.

germ nucleus egg or sperm nucleus.

germ plasm originally, a term coined by the 19th-century biologist, A. Weismann, to denote the idea of protoplasm that was transmitted unchanged from generation to generation in the germ cells (as opposed to the inheritance of acquired characteristics). Nowadays usually denotes cells from which a new plant or animal can be regenerated, as in collections of plant seeds in seed banks. *alt.* germplasm.

germ pore thin region in pollen grain wall through which the pollen tube emerges; similar area in a spore wall for exit of a germ tube.

germ tube short filament put forth by a germinating spore.

germiduct *n.* oviduct of trematodes.

germinal *a. pert.* a seed, a germ cell or reproduction.

germinal centre a focus of dividing lymphocytes in a lymphoid organ, formed after exposure to a foreign antigen.

germinal crescent a region of blastoderm forming a crescent of primordial germ cells partially surrounding anterior end of primitive streak.

germinal epithelium epithelial cells covering stroma of an ovary; layer of epithelial cells lining the testis which give rise to the spermatogonia and Sertoli cells.

germinal pore germ pore *q.v.*

germinal ridge mesodermal ridge in vertebrate embryo, into which migrate primordial germ cells, and giving rise ultimately to interstitial cells of testis or to follicle cells of ovary. *alt.* genital ridge.

germinal spot the nucleolus of the germinal vesicle or ovum.

germinal streak primitive streak *q.v.*

germinal vesicle the nucleus of an oocyte before the formation of polar bodies.

germination *n.* resumption of growth of plant embryo in favourable conditions after maturation and dispersal of seed, and emergence of young shoot and root from the seed. Germination is taken to be complete when photosynthesis commences and the plant is no longer dependent on the food stored in the seed. Germination requires a suitable temperature and the presence of sufficient water and oxygen; emergence of thallus, hypha, etc. from spore.

germline configuration of immunoglobulin and T-cell receptor genes, the state of such genes before DNA rearrangement has occurred.

gerontal, gerontic *a. pert.* old age.

gerontogaeous *a. pert.* or originating in the Old World.

gerontology *n.* the study of senescence and aging.

gestalt *n.* an organized or fixed response to an arrangement of stimuli; coordinated movements or configuration of motor reactions; a mental process considered as an organized pattern, involving explanation of parts in terms of the whole; a pattern considered in relation to background or environment, *appl.* morphology irrespective of taxonomic or phylogenetic relationships.

gestation *n.* the period between conception and birth in animals that give birth to live young.

ghosts red cell ghosts *q.v.*

giant cells large nerve fibres in annelids.

giant chromosomes *see* polytene chromosome.

giant fibres nerve fibres of very large diameter running longitudinally through ventral nerve cord of many invertebrates.

giant-grass community grassland with grasses up to 4 m high.

gibberellins (GA) *n.plu.* a group of substances (*see* Appendix 1 (44) for chemical structure) affecting plant growth which were first discovered in the fungus *Gibberella fujikuroi*, and later in many flowering plants where they are believed to act as natural growth hormones. When applied to plants they increase both cell division and cell elongation in stems and leaves, producing long spindly plants, or in dwarf mutant plants increasing stem growth so that it resembles a normal plant. Gibberellins are able to break dormancy in some seeds and are used to induce uniform germination in malt barley. They also induce stem elongation and flowering in some plants and are used to induce early seed production in, e.g., lettuce. In germinating seeds, gibberellin stimulates growth by inducing the synthesis of enzymes that break down the starch stored in the endosperm into sugars.

gibbose, gibbous *a*. inflated, pouched.

Gibbs free energy *see* free energy.

Gibbs free energy of activation *see* free energy of activation.

Gibbs standard free energy the free energy (*q.v.*) of a chemical reaction under a specified set of standard conditions.

Giemsa banding *see* G-banding.

gigantism *n*. growth of an organ or organism to an abnormally large size.

gilgai *n*. a black clay soil found in Australia, which cracks widely. The soil moves down into the cracks and inverts itself.

gill *n*. respiratory organ of many aquatic animals (e.g. crustaceans, fishes, amphibians), a plate-like or filamentous outgrowth well-supplied with blood vessels at which gas exchange between water and blood occurs; (*mycol.*) one of the radial spore-bearing lamellae on the underside of cap of toadstools and mushrooms (agarics).

gill arch the bony or cartilaginous skeleton forming an arch supporting the tissues, blood vessels and nerves of an individual gill.

gill bar in chordates, the tissue separating the gill slits from each other and containing blood vessels, nerves and skeletal support.

gill basket branchial basket *q.v.*

gill book type of gill in horseshoe crabs, consisting of a large number of leaf-like structures between which the water circu-

lates.

gill cavity in agaric fungi, the cavity in the developing fruit body in which the gills are formed.

gill clefts gill slits *q.v.*

gill cover operculum covering the gills in bony fish.

gill filaments fine lateral processes of gills, increasing the area for gas exchange.

gill fungi members of the basidiomycete fungi whose hymenium is borne on gills, including most mushrooms and toadstools.

gill heart in cephalopods, the auricle attached to each gill and which connects to a common ventricle.

gill plume the ctenidium in most gastropod molluscs.

gill pouches oval pouches containing gills and communicating with the exterior, as in lampreys and hagfish; outpushings of side wall of pharynx in all chordate embryos which develop into gill slits in fish and some amphibians.

gill rakers bristle-like processes on gill arches.

gill rays slender skeletal structures on outer margin of gill arches, which stiffen gills.

gill remnants epithelial, postbranchial or suprapericardial bodies arising in pharynx of vertebrates other than fish.

gill rods gelatinous bars supporting the pharynx in cephalocordates.

gill slits series of perforations on each side of pharynx to exterior, persistent in fish and some amphibia, embryonic only in reptiles, birds and mammals. When functional they allow the passage of water over the gill tissue.

gingivae *n.plu.* the gums. *a*. gingival.

ginglymus *n*. a hinge joint. *a*. ginglymoid.

Ginkgophyta *n*. one of the five main divisions of extant seed-bearing plants, containing only one living genus, *Ginkgo*, characterized by its distinctive fan-shaped leaves. Ovules are borne in pairs on long stalks and microsporophylls in cones on separate plants.

girdle bundles leaf trace vascular bundles which girdle the stem and converge at the leaf insertion, as in cycads.

girdle scar a series of scale scars on axis of twig where bud scales have fallen off.

girdling *n*. the removal of a complete ring

of bark from around the trunk of a tree, thus preventing the flow of solutes from leaves to roots through the phloem, which is removed with the bark.

gizzard *n*. muscular chamber in alimentary canal in which food is ground up in some animals, e.g. birds and insects, posterior to the crop.

glabrate *a*. with a nearly smooth surface.

glabrescent *a*. becoming hairless.

glabrous *a*. with a smooth even surface; without hairs.

glacial *a. pert.* or *appl.* the Pleistocene epoch of the Quaternary period, characterized by periodic glaciation.

gladiate *a*. sword-like.

gladiolus mesosternum *q.v.*

gladius *n*. the pen or chitinous shell in colonial hydroids of the Chondrophora.

gland *n*. single cell or mass of cells specialized to secrete substances such as hormones, mucus, etc., either for use inside the body or for release to the exterior; a small vesicle containing oil, resin, or other liquid on any part of a plant.

glandiform *a*. acorn-shaped.

glands of Brunner glands in duodenum which secrete an alkaline fluid containing glycoproteins but no enzyme activity.

glands of Nuhn anterior lingual glands.

glandula *n*. gland *q.v.*

glandular *a. pert.* a gland or glands; with a secretory function.

glandular epithelium the epithelium lining a gland, composed of polyhedral, columnar or cubical cells, which contain or manufacture the material to be secreted.

glandular tissue in plants, parenchymatous tissue adapted for secretion of aromatic or other substances; the tissue making up a gland.

glandulose-serrate *a*. having the serrations tipped with glands, of leaves.

glans *n*. nut; gland.

glans clitoridis the small rounded mass of erectile tissue at the end of the clitoris.

glans penis the bulbous tip of the penis in mammals.

glareal *a. pert.*, or growing on, dry gravelly ground.

Glaserian fissure a fissure in temporal bone of mammals which holds the Folian process of malleus of middle ear.

glass sponges hexactinellid sponges *q.v.*

Glaucophyta *n*. in some classifications a division of the algae comprising unicellular eukaryotes lacking chlorophyll but carrying modified cyanobacteria (blue-green algae) as endosymbionts.

glaucous *a*. bluish green; covered with a pale green bloom.

gleba *n*. in Gasteromycete fungi, the spore-bearing tissue enclosed in the peridium of sporophore.

glebula *n*. small prominence on lichen thallus.

glenoid *a*. like a socket.

glenoid cavity or fossa cavity in pelvic girdle into which the head of femur fits in tetrapods; cavity on the squamosal for articulation of the lower jaw in mammals, *alt.* mandibular fossa.

gley *n*. a soil formed under conditions of poor drainage and water-logged all or part of the time.

glia neuroglia *q.v.*

gliadin *n*. seed storage protein found in wheat.

glial cell any of the various cells of the neuroglia (*q.v.*) of the central nervous system and similar cells associated with the peripheral nervous system, e.g. Schwann cells.

glial fibrillary acidic protein (GFAP) a component of intermediate filaments in glial cells.

glioblast *n*. immature glial cell.

gliogenesis *a*. the generation of glial cells.

glioma *n*. a tumour of glial cells in central nervous system.

globigerina ooze mud formed largely from the shells of foraminiferans, esp. the calcareous shells of *Globigerina*.

globin *n*. the protein constituent of haemoglobin, each haemoglobin molecule consisting of four globin subunits of various types, e.g. $\alpha\alpha\beta\beta$ in adult human haemoglobin; generally, any member of the large globin superfamily of related proteins which also includes myoglobin and leghaemoglobin.

α-globin one of several types of globins found in animals, and one of the constituents of foetal and adult human haemoglobins.

α-globin genes the gene cluster encoding

α-globin and several related globins, i.e. in humans, ζ-globin, foetal and adult α-globins, and θ1-globin.

β-globin one of several types of globins found in animals, and a constituent of adult human haemoglobin ($\alpha\alpha\ \beta\beta$).

β-globin genes the gene cluster encoding β-globins and several related globins, i.e. in humans ε-globin, γ-globins, δ-globin and β-globin.

γ-globin a type of globin, related to β-globin, a constituent of human embryonic and foetal haemoglobins ($\zeta\zeta\gamma\gamma$ and $\alpha\alpha\gamma\gamma$).

δ-globin type of globin, related to β-globin, a constituent of a minor type of human adult haemoglobin ($\alpha\alpha\delta\delta$).

ε-globin type of globin, related to β-globin, a constituent of human embryonic haemoglobins ($\alpha\alpha\varepsilon\varepsilon$ and $\zeta\zeta\varepsilon\varepsilon$).

θ1-globin possible globin encoded by the θ1 gene in the α-globin gene cluster in primates.

ζ-globin type of globin, related to α-globin, a constituent of human embryonic haemoglobin ($\zeta\zeta\varepsilon\varepsilon$).

globin genes the genes specifying the various types of globin subunits, in mammals comprising two clusters of genes on different chromosomes, the α-globin cluster containing genes for ζ-globin, α-globin and δ-globin, the β-globin cluster containing those for γ-globin, θ-globin and β-globin.

globoid *n.* spherical body in aleurone grains, made up of double phosphate of magnesium and calcium and protein.

globose *a.* spherical or globular.

globule *n.* any small spherical structure.

globulin *n.* any of a large group of compact proteins which are water-insoluble but soluble in dilute salt solution, from which they can be salted out; a major class of proteins found in blood serum and other secretions, and as storage proteins in plant seeds.

β-globulin fraction of blood protein that contains some immunoglobulins.

globulose *a.* spherical; consisting of, or containing globules.

globus *n.* a globe-shaped structure.

globus pallidus pallidum *q.v.*

glochidiate *a.* covered with barbed hairs.

glochidium *n.* hairs bearing barbed proc-

esses; the larva of freshwater mussels such as *Unio* and *Anodon*.

gloea *n.* adhesive secretion of some protozoans.

gloeocystidium *n.* a cystidium containing a slimy or sticky substance.

glomerular *a. pert.* or like a glomerulus.

glomerulate *a.* arranged in clusters; bearing glomerulus.

glomerule *n.* a condensed head of almost sessile flowers; a compact cluster, as of spores.

glomerulus *n.* network of blood capillaries; in the kidney the knot of capillaries surrounded by the dilated end (Bowman's capsule) of a kidney tubule; (*bot.*) a compact mass of almost sessile flowers; a compact mass of spores. *plu.* glomeruli.

glomus *n.* a number of glomeruli run together.

glossa *n.* the tongue in vertebrates; in insects, one of a pair of small lobes at tip of labium (lower lip) which are long in honey- and bumblebees and used to suck nectar; any tongue-like structure. *plu.* glossae. *a.* glossal.

glossarium *n.* the slender-pointed glossa of some Diptera.

glossate *a.* having a tongue or tongue-like structure.

glosso-epiglottic *a. pert.* tongue and epiglottis, *appl.* folds of mucous membrane.

glossohyal *n.* extension of the basihyal bone into the mouth of some fishes.

glossopalatine *a.* connecting tongue and soft palate.

glossopalatine nerve branch of facial nerve that suplies the tongue and the palate.

glossopalatinus, glossopalatine muscle a thin muscle which arises on each side of the soft palate and is inserted into the tongue.

glossophagine *a.* securing food by means of the tongue.

glossopharyngeal *a. pert.* tongue and pharynx; *appl.* nerve: 9th cranial nerve, chiefly conveying sensation including taste from pharynx, tonsil and back of tongue.

glottis *n.* the opening from the trachea (windpipe) into the throat.

glucagon *n.* polypeptide hormone synthesized by α-cells of pancreas when blood glucose level is low, and which stimulates breakdown of glycogen to glucose in liver.

glucan *n.* any of a variety of polysaccharides made of glucose units linked together and including cellulose, starch, glycogen.

glucocerebroside *n.* a lipid found in cell membranes, a cerebroside containing glucose rather than galactose.

glucocorticoid *n.* any of the steroid hormones produced by adrenal cortex and which influence carbohydrate metabolism, as the formation of carbohydrate from fat and protein, and which include corticosterone, cortisone etc., but not aldosterone, which is a mineralocorticoid.

glucocorticoid response element *see* GRE.

glucogenic *a. appl.* amino acids that are degraded to tricarboxylic acid cycle intermediates which can be converted to glucose and thus enter the normal carbohydrate metabolism of the body, e.g. glycine, alanine, aspartic acid, glutamic acid, arginine, ornithine; stimulating glucose formation.

glucokinase *n.* enzyme catalysing the formation of glucose 6-phosphate from glucose in the liver as a precursor to glycogen synthesis, also found in microorganisms. EC 2.7.1.2.

glucokinin *n.* an insulin-like protein found in plants, which can reduce blood sugar.

glucomannan *n.* any of a group of hemicellulose polysaccharides composed of mannose and glucose residues linked together, usually in random order, found esp. in cell walls of conifers.

gluconeogenesis *n.* metabolic pathway by which glucose is synthesized from non-carbohydrates such as lactate, some amino acids and glycerol, chiefly in liver and kidney, and in plants, especially in seeds.

glucosamine *n.* amino derivative of the sugar glucose, substituted at carbon 2, as *N*-acetylglucosamine a common constituent of the sugar side chains of glycoproteins and the subunit of the polysaccharide chitin. *see* Appendix 1 (9).

glucosan glucan *q.v.*

glucose *n.* a hexose sugar found in all living cells, especially plant sap and in the blood and tissue fluids of animals, being the chief end-product of carbohydrate di-

gestion, and as glucose 6-phosphate being the chief substrate for cellular respiration and taking part in many other metabolic pathways. *alt.* dextrose, grape sugar, starch sugar, blood sugar. *see* Appendix 1 (1).

glucose-6-phosphatase enzyme from liver and kidney that converts glucose 6-phosphate into glucose, enabling it to be released into the blood. EC 3.1.3.9.

glucose-6-phosphate dehydrogenase (G6P dehydrogenase) enzyme catalysing the formation of phosphogluconate from glucose 6-phosphate. An inherited deficiency in the enzyme causes a haemolytic anaemia induced by the antimalarial drug primaquine but appears to confer some resistance to falciparum malaria. EC 1.1.1.49.

glucose-6-phosphate dehydrogenase deficiency an X-linked genetically determined lack of the enzyme glucose-6-phosphate dehydrogenase, which is generally symptomless, but which can lead to severe anaemia if the individual is exposed to certain drugs (e.g. the antimalarial drug chloroquine) or a lack of oxygen.

glucosephosphate isomerase a widely distributed enzyme catalysing the conversion of glucose 6-phosphate to fructose 6-phosphate. EC 5.3.1.9. *alt.* phosphoglucose isomerase, phosphohexose isomerase.

glucosidase *n.* any of a group of enzymes that split off terminal glucose units from oligosaccharides, and which are divided into 2 types, α-D-glucosidase (EC 3.2.1.20) which hydrolyses 1,4-α-D-glucosidic bonds and includes maltase, and β-D-glucosidase (EC 3.2.1.21) which hydrolyses β-D-glucosides, and which includes gentiobiase and cellobiase.

glucoside *n.* any glycoside yielding a sugar, usually glucose, on hydrolysis.

glucosinolate *n.* sulphur-containing compound found in some crucifers (e.g. brassicas, horseradish, mustard, oilseed rape) and responsible for the pungent flavour of these plants.

glucuronate (glucuronic acid) *n.* a sugar acid derived from glucose by oxidation, a component of polysaccharides in, e.g., plant cell walls, bacterial capsules.

glumaceous *a.* dry and scaly like glumes,

formed of glumes.

glume *n*. a dry chaffy bract, present in a pair at the base of spikelet in grasses.

glumella palea *q.v.*

glumiferous *a*. bearing or producing glumes.

Glumiflorae Poales *q.v.*, the grasses.

glumiflorous *a*. having flowers with glumes or bracts at their base.

glutaeal, glutaeus gluteal, gluteus *q.v.*

glutamate (glutamic acid) (Glu, E) *n*. amino acid, negatively charged form of α-aminoglutaric acid, constituent of protein and also acts as a neurotransmitter in the central nervous system.

glutamate dehydrogenase widely distributed enzyme involved in amino acid metabolism which catalyses the conversion of glutamate into a keto acid and ammonium ion, and which is also involved in the incorporation of ammonium into glutamate in nitrogen assimilation. EC 1.4.1.2–4.

glutamate receptors cell-surface receptors for the neurotransmitter glutamate, found on neurones and other cells. There are two main structural types: ion channels (e.g. the NMDA receptor) and G protein-coupled receptors (the metabotropic glutamate receptors).

glutamate synthase (GOGAT) enzyme involved in nitrogen assimilation in bacteria, algae and plants, which catalyses the reductive transfer of an amino group from glutamine (formed by the action of glutamine synthase on glutamate + ammonia) to 2-oxoglutarate to form glutamate. There are different forms of the enzyme using different electron donors (i.e. ferredoxin, NADH and NADPH). EC 1.4.7.1, 1.4.1.13, 1.4.1.14.

glutaminase *n*. enzyme catalysing the conversion of glutamine to glutamate and ammonium ion. EC 3.5.1.2.

glutamine (Gln, Q) *n*. amino acid, uncharged monoamide derivative of glutamate, constituent of protein, key component in control of nitrogen metabolism.

glutamine synthetase (GS) key enzyme in regulation of nitrogen metabolism and nitrogen assimilation, catalysing the incorporation of ammonium ion into glutamine. EC 6.3.1.2.

glutaredoxin *n*. protein cofactor acting with

glutathione in the synthesis of deoxyribonucleotides from ribonucleotides.

glutathione *n*. a tripeptide, Glu-Cys-Gly, found in many tissues esp. red blood cells where it acts as a buffer for haemoglobin, exists in reduced or oxidized form and acts as a coenzyme in certain redox reactions.

gluteal *a*. *pert*. or in region of buttocks or hindquarters.

glutelins *n.plu*. storage proteins in plant seeds which are insoluble in water but soluble in dilute acid and alkali and which include glutenin and oryzenin.

glutinous *a*. having a sticky, slimy surface.

glycan *n*. a carbohydrate polymer, *see* glycosaminoglycan, polysaccharide; the oligosaccharide or polysaccharide portions of any macromolecule containing a considerable amount of carbohydrate, as proteoglycans and glycoproteins.

glycation *n*. direct non-enzymatic addition of a free sugar to a compound.

glyceraldehyde 3-phosphate (GAP, G3P) triose phosphate intermediate in the Calvin cycle of photosynthesis, glycolysis and gluconeogenesis. *see* Appendix 1 (49) for structure.

glyceride *n*. any ester of glycerol, including some fats and oils in which three molecules of fatty acids combine with one molecule of glycerol.

glycerine *n*. pure glycerol, a sweet viscous liquid.

glycerol *n*. a 3-carbon sugar alcohol, important precursor of many lipids, as a breakdown product of dietary lipids forms a substrate for glycolysis. *see* Appendix 1 (35) for structure.

glycerol phosphate shuttle process by which electrons from NADH generated by glycolysis in the cytoplasm are carried into mitochondria to enter the respiratory chain by glycerol 3-phosphate which is converted in mitochondria to dihydroxy-acetone phosphate, which diffuses back to the cytoplasm, esp. prominent in insect flight muscle.

glycerolipid *n*. a lipid based on glycerol. Membrane glycerolipids (e.g. phosphatidylcholine, phosphatidylinositol) have two long-chain fatty acids attached to the glycerol along with another group (e.g. phosphocholine, phosphoinositol).

Triacylglycerols, in which three long-chain fatty acids are attached to glycerol, are not found in membranes but are carbon storage compounds in many plant seeds (as oils) and in animal fatty tissue.

glycine (Gly, G) *n.* the simplest amino acid, having a hydrogen atom as its side chain. It is a constituent of protein and also acts as an inhibitory neurotransmitter in the central nervous system. *see* Appendix 1, table.

glycinin *n.* protein from soya beans.

glycobiology *n.* the study of carbohydrates.

glycocalyx cell coat *q.v.*

glycocholate *n.* the major bile salt, formed by the breakdown of cholesterol in the liver.

glycoconjugate *n.* any molecule composed of a glycan portion or portions attached to a non-carbohydrate.

glycogen *n.* a branched chain polysaccharide made up of glucose units, and which acts as a storage substance in vertebrate liver and muscle, is mobilized by phosphorolysis (*q.v.*) to glucose 1-phosphate, and can also be hydrolysed via dextrins and maltose to glucose, and which is also found in invertebrates and some plants, formerly called animal starch or amylum. *see* Appendix 1 (15) for structure.

glycogen granules granules in cytoplasm of liver and muscle cells which contain glycogen and also contain the enzymes involved in its synthesis and degradation.

glycogen phosphorylase enzyme that catalyses the cleavage of glycogen by inorganic phosphate (phosphorolysis) to glucose 1-phosphate in liver and muscle. EC 2. 4.1.1.

glycogen synthase enzyme catalysing the transfer of glucose from UDP-glucose to a growing glycogen chain. EC 2.4.1.1, *r.n.* glycogen(starch)synthase. *alt.* glycogen synthetase.

glycogenase *n.* general name for two enzymes that catalyse synthesis of glycogen in liver - glycogen (starch) synthetase, EC 2.4.1.11, and glucose-1-phosphate uridylyltransferase, EC 2.7.7.9.

glycogenesis *n.* the synthesis of glycogen from glucose, as in liver and muscle.

glycogenolysis *n.* the breakdown of glycogen to produce glucose, stimulated in mammalian cells by adrenaline and glucagon, and inhibited by insulin.

glycogenosis *n.* disease caused by disturbance of glycogen metabolism.

glycolate (glycolic acid) *n.* organic acid, formerly known as glycollate (glycollic acid), found chiefly in plants where it is a substrate for photorespiration, and also gives sour taste to some unripe fruit.

glycolate cycle complex metabolic cycle underlying photorespiration in plants, in which glycolate is used as a substrate, producing serine and glycine via glyoxylate (in peroxisomes).

glycolipid *n.* any complex lipid containing one or more carbohydrate residues, and including cerebrosides, gangliosides, sulphatides (from animal brain) and sulpholipids (from plants).

glycollate (glycollic acid) glycolate *q.v.*

glycolysis *n.* anaerobic breakdown of glucose to pyruvate in living cells, with the production of ATP, the pyruvate formed either being converted to acetyl CoA and entering the tricarboxylic acid cycle, or being anaerobically converted to lactic acid (e.g. in muscle) or ethanol, carbon dioxide and other organic products (e.g. in yeasts and other microorganisms); sometimes used for the complete pathway of anaerobic respiration (*q.v.*) or fermentation (*q.v.*). *alt.* Embden–Meyerhof(–Parnas) pathway.

glyconeogenesis *n.* synthesis of glycogen starting from non-carbohydrate compounds.

glycophorins *n.plu.* proteins rich in sialic acid spanning the red blood cell membrane, which carry the Gerbich blood group antigens.

glycophyte *n.* a plant unable to tolerate saline conditions.

glycoprotein *n.* any of a large class of proteins that contain carbohydrate in the form of chains of monosaccharide units attached to specific amino acid residues. Ubiquitous components of cell membranes and cellular secretions.

glycosaminoglycan (GAG) *n.* any of a group of polysaccharides made up of repeating disaccharide units of amino sugar derivatives, and including hyaluronate, chondroitin sulphate, keratan sulphate and heparin, and which are found in the proteoglycans of connective tissue. *alt.* (formerly) mucopolysaccharide.

glycoside *n.* any of a class of compounds which on hydrolysis give a sugar and an non-sugar (aglutone) residue, e.g. glucosides give glucose, galactosides give galactose, widely distributed in plants and including anthrocyanins and anthoxanthins.

glycosidic bond type of chemical linkage between sugar residues in oligo- and polysaccharides, and often between sugar residues and non-carbohydrates, such as purines and pyrimidine bases in nucleosides.

glycosome *n.* an organelle in some protozoans, including trypanosomes, containing the enzymes concerned with glycolysis.

glycosphingolipid *n.* membrane glycolipid containing the long-chain amino alcohol sphingosine and no glycerol.

glycosylation *n.* addition of oligosaccharide side chains to proteins to form glycoproteins in eukaryotic cells, taking place in the lumen of the endoplasmic reticulum and in Golgi apparatus.

glyoxalase system enzyme system converting methylglyoxal to lactic acid (the glyoxalate reaction) in animal tissues, and comprising 2 enzymes, lactoyl-glutathione lyase (EC 4.4.1.5) and hydroxyacyl-glutathione hydrolase (EC 3.1.2.6) formerly known as glyoxalase I and glyoxalase II, and utilizing glutathione as a cofactor.

glyoxylate (glyoxylic acid) *n.* compound important in glycine and hydroxyproline metabolism, an end-product of purine nucleotide catabolism, and used in the glyoxylate cycle.

glyoxylate (glyoxylic acid) cycle a modified version of the tricarboxylic acid cycle found in higher plants (especially germinating seeds) in which isocitrate is converted to glyoxylate and succinate, glyoxylate being converted to malate, occurs in glyoxysomes. *alt.* Kornberg cycle.

glyoxysome *n.* plant cell organelle, esp. found in germinating seeds, concerned with the breakdown and conversion of fatty acids to acetyl CoA for the glyoxylate

cycle. *alt.* microbody.

gnathal, gnathic *a. pert.* the teeth and jaws.

gnathism *n.* formation of jaw with reference to the degree of projection.

gnathites *n.plu.* mouth appendages of arthropods.

gnathopod *n.* any arthropod limb in the mouth region which is modified to assist in feeding.

gnathopodite maxilliped *q.v.*

gnathosoma *n.* the mouth region, including oral appendages, of some archnids.

Gnathostomata, gnathostomes *n.*, *n.plu.* the jawed vertebrates, a subphylum of Chordata comprising the fishes, amphibians, reptiles, birds and mammals; group of irregularly shaped sea urchins.

gnathostomatous *a.* with jaws at the mouth.

Gnathostomulida *n.* phylum of tiny acoelomate marine worms living in sediments and on plant and algal surfaces in shallow waters.

gnathothorax *n.* that part of the cephalothorax posterior to protocephalon in crabs and lobsters, etc.

Gnetophyta *n.* one of the five divisions of extant seed-bearing plants, comprising only three genera, *Gnetum*, *Ephedra* and *Welwitschia*, which differ considerably in form and reproduction. They are considered as gymnosperms although having some angiosperm features.

gnotobiosis *n.* the rearing of laboratory animals in a germ-free state or containing only a known prespecified flora of microorganisms.

gnotobiotics *n.* the study of organisms or species when other organisms or species are absent; germ-free culture; study of germ-free animals.

goblet cell vase-shaped epithelial cell that secretes mucus.

golden-brown algae common name for the Chrysophyceae *q.v.*

Golgi apparatus (GA) stacks of flattened membrane sacs present in eukaryotic cells, and which are involved in directing membrane lipids and proteins and secretory proteins to their correct destination in the cell, and also in trimming and adding to the sugar side chains of glycoproteins. *alt.* Golgi bodies, dictyosomes. *see* Fig. 1.

Golgi body one of the stacks of the Golgi apparatus. *alt.* (plants) dictyosome.

Golgi cell type of interneurone found in cerebellum with cell body in granular layer, acts on granule cells.

Golgi staining staining of brain tissue by a silver impregnation method, devised by the neuroanatomist Camillo Golgi in the 19th century, which allows visualization of a whole neurone and its axon and dendrites, and which is still in use.

Golgi tendon organ tendon organ *q.v.*

Golgi vesicles small membrane vesicles derived from the cisternae of the Golgi stacks. Some contain material being exported to the exterior of the cell or to the plasma membrane, others will become lysosomes.

gomphosis *n.* articulation by insertion of a conical process into a socket, as of roots of teeth into their sockets.

gonad *n.* organ in which reproductive cells are produced, as the ovary and testis, or in lower animals an ovotestis.

gonadal *n.* of, *pert.* or *appl.* gonads.

gonadectomy *n.* excision of gonad, castration in male, spaying in female.

gonadotropic *a.* affecting the gonads, *appl.* hormones: the gonadotropins *q.v.*

gonadotropin-releasing factor (GnRF) hypothalamic peptide that stimulates the release of gonadotropins (follicle-stimulating hormone and luteinizing hormone) from the pituitary. *alt.* gonadotropin-releasing hormone.

gonadotropins *n.plu.* hormones that stimulate gonadal function: follicle-stimulating hormone (*q.v.*), and luteinizing hormone (*q.v.*) which are produced by the anterior pituitary, and chorionic gonadotropin, produced by the placenta, which is similar but not identical to luteinizing hormone. Prolactin (*q.v.*) also has gonadotropic activity.

gonal *a.* giving rise to a gonad, *appl.* middle section of germinal ridge which alone forms a functional gonad.

gonapophyses *n.plu.* chitinous outgrowths aiding copulation in insects; the component parts of a sting; any genital appendages.

Gondwanaland *n.* a southern land mass composed of South America, Africa, India, Australia and Antarctica, before continental drift moved these continents to their present positions.

gongylidia *n.plu.* hyphal swellings or modifications in fungi cultivated by ants.

gongylus *n.* a round reproductive body in certain algae and lichens.

gonia, gonial(e) *n.plu.* primitive sex cells, spermatogonia or oogonia.

gonidangium *n.* a structure producing or containing gonidia.

gonidia *n.plu.* asexual non-motile reproductive cells such as gemmae of mosses, formed on the gametophyte in some plants; spores in some cyanobacteria (blue-green algae).

gonidial *a. pert.* gonidia.

gonidiferous *a.* bearing or producing gonidia.

gonidioid *a.* like a gonidium, *appl.* certain algae.

gonidiophore *n.* an aerial hypha supporting a gonidangium.

gonidiophyll *n.* a gametophyte leaf bearing gonidia.

gonimium *n.* a bluish-green spore in the cyanobacterial cells of some lichens.

gonimoblasts *n.plu.* filamentous outgrowths of a fertilized carpogonium of red algae.

gonimolobe *n.* a group of carposporangia borne on a gonimoblast, in some red algae.

goniocyst *n.* in lichens, a cluster of gonidia.

gonion *n.* the angle point on the lower jaw.

gonocalyx *n.* the bell of a reproductive medusoid individual in a siphonophore colony.

gonocoel *n.* the body cavity containing the gonads.

gonocytes *n.plu.* in sponges, the mother cells of ova and spermatozoa or the gametes themselves.

gonoduct *n.* a duct leading from gonad to the exterior.

gonoecium *n.* a reproductive individual in a polyzoan colony.

gonomery *n.* separate grouping of maternal and paternal chromosomes during cleavage stages in development of some organisms.

gononephrotome *n.* embryonic segment containing primordia of the urinogenital system.

gonophore *n*. a reproductive zooid in a hydrozoan colony.

gonopod(ium) *n*. modified anal fin serving as copulatory organ in some male fishes; clasper of male insects, centipedes and millipedes.

gonopore *n*. a reproductive aperture.

gonosome *n*. the reproductive zooids of a coelenterate colony collectively.

gonostyle *n*. in hydrozoans, a columnar zooid with or without mouth and tentacles, bearing gonophores; sexual palpon or siphon of siphonophores; a bristle-like process on base of clasper of some male insects; part of sting in some hymenopterans.

gonotheca *n*. transparent protective expansion of perisarc around blastostyle in some colonial hydrozoans.

gonotome *n*. the embryonic segment containing primordia of the gonad.

gonotrema, gonotreme *n*. genital aperture of arachnids.

gonozooid *n*. a reproductive individual of a hydrozoan colony.

gordian worms Nematomorpha *q.v.*

gorgonians *n.plu.* the sea fans, horny corals with large, delicate, fan-shaped, branching colonies anchored by a "stalk" to the substrate.

gorgonin *n*. fibrous protein in the mesoglea of sea fans (gorgonians) which forms the stiff skeleton of the colony.

G phases of cell cycle, *see* G_0, G_1, G_2.

G proteins class of guanine-nucleotide binding proteins (*q.v.*) associated with the cytoplasmic face of the plasma membrane of mammalian cells, which are involved in transmitting signals from many hormone and neurotransmitter receptors to intracellular pathways. They are composed of three subunits, α, β and γ, of which the α subunit binds guanine nucleotides and is usually the effector subunit. *see also* G_i, G_{olf}, G_p, G_s, transducin. *alt.* heterotrimeric GTP-binding proteins.

GPI anchor glycosylphosphatidylinositol anchor, a linkage that attaches some membrane proteins to the lipid bilayer.

Graafian follicle in the mammalian ovary, a spherical vesicle containing a developing ovum and a liquid, the liquor folliculi, surrounded by numerous follicle cells, and from which the ovum is released on ovulation.

gracile *a*. lightly-built, small and slender, *appl.* australopithecines: *Australopithecus africanus*.

gracile nucleus area in medulla of brain involved in relaying sensory information (touch, pressure and vibration) from skin, deep tissues and joints.

Gracilicutes *n*. division of the kingdom Monera or Prokaryotae, including all Gram-negative "true" bacteria (e.g. enterobacteria, pseudomonads, Gram-negative cocci, spirilla, Gram-negative rods, etc.) and photosynthetic bacteria (green and purple photosynthetic sulphur bacteria and the cyanobacteria).

gracilis *n*. a slender muscle on the inside of the thigh; fasciculus gracilis: a bundle of fibres in the medulla oblongata; a nucleus of grey matter ventral to clava in medulla oblongata.

grade *n*. a taxonomic category representing a level of morphological organization in which a group of organisms shares a number of characteristics but may not owe them to a common ancestor, e.g. the protozoa.

gradient *see* positional information.

graduate sorus in ferns, a sorus in which sporangia develop from the tip towards the base.

graduated *a*. becoming longer or shorter in discrete steps.

graft *n*. tissue, organ or part of an organism inserted into and uniting with a larger part of the same or another organism; *v*. to insert scion into stock, or animal tissue from donor to recipient in a transplant.

graft chimaera, graft hybrid an individual into which tissue from another individual of different genotype has been grafted or transplanted, and which shows characteristics of both individuals.

graft-versus-host reaction the reaction seen when immunocompetent cells in a graft attack and destroy host tissue in a cell-mediated immune response.

Gram stain a staining procedure for bacteria devised by the Danish physician H.C.J. Gram, which correlates well with certain morphological features of bacteria and is widely used in identification, consisting essentially of staining with crystal violet and subsequent treatment with alcohol, those bacteria which are decolorized by alcohol being termed Gram-negative and those retaining the stain Gram-positive, a property related to the structure of the cell wall.

gramicidin S cyclic peptide antibiotic produced by a *Bacillus* sp., and which is not synthesized by the mRNA/tRNA/ribosome system.

Gramineae *n*. the grasses, a large and ubiquitous family of monocotyledonous plants, mostly herbaceous but with some woody species (e.g. bamboos). The linear leaves arise from the nodes of jointed stems. Flowering stems bear clusters of wind-pollinated or cleistogamous flowers. The Gramineae contain many important crop plants including wheat (*Triticum* spp.), barley (*Hordeum vulgare*), rye (*Secale cereale)*, oats (*Avena sativa*), maize (*Zea mays)*, rice (*Oryza sativa*) and sugar-cane (*Saccharum officinale*). Pasture grasses include cocksfoot (*Dactylis*), fescues (*Festuca* spp.) and rye-grasses (*Lolium* spp.). In many grasses, the growing shoot remains near or under the ground with the leaves arising mainly from ground level. Leaves can therefore be continually cropped to near the ground by grazing without damaging the plant.

graminaceous *a. pert.* grasses, *appl.* members of the grass family (Gramineae) such as cereals; grass coloured, *appl.* insects.

graminicolous *a*. living on grasses.

graminifolious *a*. with grass-like leaves.

graminivorous *a*. grass-eating.

graminoid *a. appl.* grasses and grass-like forms.

graminology *n*. the study of grasses.

grammate *a*. striped; marked with lines or slender ridges.

grana *n.plu.* in chloroplasts, groups of stacked disc-shaped structures (thylakoids) on whose membranes the photosynthetic pigments and energy-transducing components are borne. *sing.* granum.

grand postsynaptic potential the sum of all the synaptic potentials received by a neurone, and whose magnitude determines whether the neurone will generate an impulse.

grandifoliate, grandifolious *a*. large-leaved, particularly when the leaves are the dominant organ, as in water-lilies.

Grandry's corpuscle a touch receptor in skin of beak and tongue of birds.

granellae *n.plu.* refractile granules consisting chiefly of barium sulphate in the tubes of certain protozoans (Sarcodina).

granite mosses a small group of mosses, class Andreaeidae, sometimes called rock mosses, small blackish-green or olive-brown tufted plants, in which the gametophyte arises from a plate-like protonema and spores are shed from slits in the capsule wall.

granivorous *a*. feeding on grain.

granose *a*. in appearance like a chain of grains, as some insect antennae.

granular *a. appl.* soil crumbs which are rounded and rather small.

granular, granulate, granulose *a*. appearing as if made up of, or covered with, granules.

granule cell type of interneurone found in cerebellum with cell body in granular layer and axon projecting into molecular layer where it forms parallel fibres synapsing with dendrites of Purkinje cells.

granules of Nissl Nissl granules *q.v.*

granulocyte *n*. class of white blood cell characteristically containing numerous secretory vesicles or granules, and which includes polymorphonuclear leukocytes, eosinophils and basophils.

granulocyte colony-stimulating factor (G-CSF) protein that stimulates formation of granulocytes from bipotential macrophage-granulocyte precursor cells.

granulocyte/macrophage colony-stimulating factor (GM-CSF) glycoprotein required for the survival and differentiation of the precursor granulocyte/macrophage cell, produced by many cells of the body, and which increases in response to infection.

granuloma *n*. inflammatory nodule caused by activated phagocytes in certain conditions.

granulopoiesis *n*. the formation of granulocytes.

granulosa cells cells in mammalian ovarian follicles that produce oestrogens and eventually form the corpus luteum.

granum *sing.* of grana *q.v.*

grape sugar glucose *q.v.*

Graptolita, graptolites *n., n.plu.* a group of fossil invertebrates from the Paleozoic, of doubtful affinity but thought to be allied with the hemichordates.

grasshoppers common name for many members of the Orthoptera *q.v.*

grassland *n.* biome found in regions where the average annual precipitation (*ca.* 25–76 cm) is sufficient to support the growth of grasses and other herbaceous plants but generally insufficient to support continuous tree cover, which may also be suppressed as a result of grazing by herbivores. *see* alpine grassland, prairie, savanna, short-grass community, steppe, tall-grass community, veld.

graveolent *a.* having a strong or offensive smell.

gravid *a. appl.* female with eggs, or pregnant uterus.

graviperception *n.* the sensation or perception of gravity.

graviportal *a.* with the legs adapted to supporting great weights, as in elephants.

gravitational *a. appl.* water in excess of soil requirements, which sinks under action of gravity and drains away.

gravitaxis *n.* geotaxis *q.v.*

gravitropism geotropism *q.v.*

gray *see* grey.

gray (Gy) the SI unit of the dose of ionizing radiation absorbed by living tissue, being equal to 1 J of energy imparted to 1 kg of mass, and which replaces the non-SI unit, the rad. 1 Gy = 100 rad.

greater omentum a fold of peritoneum attached to the colon and stomach and hanging over the small intestine.

green algae common name for the Chlorophyta *q.v.*

green deserts areas of little ecological interest despite their "green" appearance, e.g. extensive areas of crop monocultures, heavily fertilized pastureland replanted with alien grass species, and large areas of closely mown grass in urban parks.

green glands paired excretory organs at base of antennae in some crustaceans.

Green Revolution the introduction of scientifically bred high-yielding varieties of staple crop plants such as wheat, rice and maize from the 1960s onwards into areas of the world where lower-yielding traditional varieties adapted to local conditions were formerly grown, and which has considerably increased food production, particularly in the countries of South-east Asia and India.

green sulphur bacteria photoautotrophic bacteria, mainly aquatic, which oxidize sulphide to sulphur.

greenhouse effect the trapping of a proportion of the heat radiated from the Earth's surface by water vapour, carbon dioxide and other compounds in the lower atmosphere, and its re-radiation back to the surface. If the levels of e.g. carbon dioxide in the atmosphere progressively increase as a result of human activity it is thought that this natural effect will result in a rise in the temperature of the lower atmosphere eventually leading to widespread climatic change.

gregaloid *a. appl.* protozoan colony of indefinite shape, usually with gelatinous base, formed by incomplete division of individuals or partial union of adults.

gregarines *n.plu.* a group of parasitic protozoans of invertebrates, in which only the adults live outside the host cell, and in which the male and female gametes are smaller than the normal cell.

gregariniform *a. pert.* gregarine protozoans; *appl.* spores moving with the gliding motion characteristic of gregarines.

gregarious *a.* tending to herd together; growing in clusters; colonial.

gressorial *a.* adapted for walking, *appl.* certain insects and birds.

grey crescent lightly-pigmented crescent-shaped band visible on side of fertilized eggs of some amphibia, which forms opposite site of sperm entry and through which the first cleavage plane cuts, delimiting right and left sides of the body.

grey matter regions of brain and spinal cord that appear grey, consisting chiefly of cell bodies of neurones.

grinding teeth molars *q.v.*

griseofulvin *n.* antibiotic produced by some species of *Penicillium*, esp. *P. griseofulvum*, which is toxic to some fungi by inhibiting mitotic metaphase.

grooming *n.* the cleaning of fur or feathers

by an animal (generally called preening in birds), performed either by itself (self grooming) or to another member of the same species (social grooming), functioning not only to keep the animal clean, but as a displacement activity or to improve the social cohesion of the group.

gross efficiency a measure of the efficiency of an animal in converting food consumed into body substance.

gross primary production (GPP) the total assimilation of inorganic nutrients in a plant community per unit time, *cf*. net primary production.

ground layer moss layer *q.v.*

ground meristem *see* ground tissue, primary meristem.

ground substance non-living matrix of connective tissue.

ground tissue of plants, the parenchyma, collenchyma and sclerenchyma, i.e. all tissues other than the epidermis, reproductive tissues and vascular tissues. It arises from the ground meristem, a primary meristem derived from the apical meristem in the developing embryo.

groundwater water that sinks down through soil and rock and collects in underground aquifers; underground water in the zone of saturation, below the water table, *cf*. soil water.

group selection selection that operates on two or more members of a lineage group as a unit such that characters that benefit the group rather than the individual may be selected for.

group translocation type of active transport of molecules into the cell carried out by some bacteria, in which molecules entering the cell by passive transport are chemically modified during transport to prevent their outward leakage.

growing point a part of plant body at which cell division is localized, generally terminal and composed of meristematic cells.

growth *n*. increase in mass and size by cell division and/or cell enlargement.

growth cone organelle in developing neurones which forms the leading end of developing axons and dendrites.

growth curvature the curved shape imparted to a plant organ by the difference in the rates of growth of its sides.

growth curve generally a plot of log numbers of a population of living organisms against time, which for a culture of microorganisms with limiting nutrients, etc. is typically bell-shaped. A short lag phase of little or no increase in numbers is followed by a steep rise in numbers (exponential growth or log phase) to a plateau, followed by a rapid fall in growth rate as nutrients become exhausted, waste products accumulate and the available "habitat" becomes overcrowded.

growth factor receptors cell-surface receptors for protein and peptide growth factors. They fall into several different structural classes and include the tyrosine protein kinase receptors for growth factors such as insulin, epidermal growth factor and platelet-derived growth factor.

growth factors organic compounds other than those required as carbon and energy sources which are needed by many organisms for proper growth and development, and which may include vitamins, amino acids, purines, pyrimidines, etc.; general term for specific peptides or proteins required by particular cells for division and/or differentiation, e.g. nerve growth factor, epidermal growth factor. *see under individual names.*

growth hormone (GH) in mammals, the growth-promoting protein hormone produced by the anterior pituitary which stimulates the production of somatomedins by liver resulting in stimulation of bone and muscle growth, *alt*. somatotropin; in animals generally, any of various growth-promoting hormones; in plants, any of the growth-promoting phytohormones, e.g. giberellic acid, indole acetic acid (IAA).

growth rate (r) increase in the size of a population per unit of time.

growth rings the rings seen on the cut trunk of some trees, each representing the growth of wood in one year. The width of each ring reflects climatic conditions (e.g. temperature and rainfall) in that year, wider rings being produced in favourable years, *see also* dendrochronology; (*zool.*) the layer of shell laid down in each growth period in various animals such as bivalve molluscs; a layer of a scale in fishes.

growth substance generally refers to any

compound, natural or artificial, that when present in small amounts has a marked effect on the growth and/or development of a plant. *alt*. plant growth substance.

grub *n*. legless larva of insects of the Diptera, Coleoptera and Hymenoptera.

Gruiformes *n*. an order of birds including cranes, bustards, rails, gallinules and coots and the hemipodes.

Gruinales Geraniales *q.v.*

grumose, grumous *a*. knotted: clotted.

GTPase-activating protein (GAP) protein that interacts with the GTP-bound forms of guanine-nucleotide binding proteins to stimulate the intrinsic GTPase activity.

GTP-binding protein guanine-nucleotide binding protein *q.v.*

guaiacyl coniferyl alcohol, *see* lignin.

guanidine *n*. a base produced by oxidation of guanine, whose metabolism is regulated by the parathyroids.

guanine (G) *n*. a purine base, constituent of DNA and RNA, and which is the base in the nucleoside guanosine, and which is also found as iridescent granules or crystals in certain chromatophores, *see* guanophore. *see* Appendix 1 (20) for chemical structure.

guanine-nucleotide binding proteins a diverse family of proteins, including the G proteins, transducin, the *ras* proteins, and the protein synthesis elongation factors, that occur complexed with guanine nucleotides, usually GDP. The replacement of GDP with GTP activates the protein, which remains active until the GTP is hydrolysed to GDP by the protein's intrinsic GTPase activity. The G proteins and transducin are involved in relaying signals from cell-surface receptors to enzymes that generate or regulate second messenger production, such as adenylate cyclase and cyclic GMP phosphodiesterase (for transducin).

guano *n*. deposits of bird droppings rich in phosphates, used as fertilizer, formerly collected from islands off the west coast of South America.

guanophore *n*. a chromatophore containing guanine, either as pale granules usually giving a yellow colour, or as iridescent crystals as in the skin of some reptiles and fishes.

guanosine *n*. a nucleoside made up of the purine base guanine linked to ribose.

guanosine diphosphate (GDP) guanine nucleotide containing a diphosphate group.

guanosine monophosphate (GMP) nucleotide composed of guanine, ribose and a phosphate group, a product of the partial hydrolysis of RNA, and synthesized *in vivo* from xanthylate by amination. *alt*. guanylate, guanylic acid, guanosine 5′-phosphate.

guanosine pentaphosphate (pppGpp) unusual nucleotide with a 5′ triphosphate and a 3′ diphosphate, accumulated by bacterial cells in conditions of amino acid starvation.

guanosine tetraphosphate (ppGpp) an unusual nucleotide with diphosphate at the 3′ and 5′ positions, accumulated by bacterial cells in conditions of amino acid starvation.

guanosine triphosphatase (GTPase) enzyme catalysing the hydrolysis of GTP to GDP with the release of free energy. GTPase activity is part of several groups of proteins, notably the guanine-nucleotide binding proteins, in which an intrinsic GTPase activity hydrolyses bound GTP to GDP, thus converting the protein from an activated to an inactive state.

guanylate (guanylic acid) guanosine monophosphate *q.v.*

guanylate cyclase the enzyme catalysing the formation of cyclic GMP from guanosine monophosphate (GMP). *alt*. guanylyl cyclase.

guanyltransferase *see* nucleotidyltransferase.

guard cells two specialized epidermal cells that contain chloroplasts and surround the central pore of a stoma. Changes in the turgor of these cells open and close the stomatal opening.

guard polyp nematocalyx *q.v.*

gubernaculum *n*. a cord stretching from epididymis to scrotal sac and supporting testis; tissue between gum and dental sac of permanent teeth.

Guérin's glands racemose mucous glands of the female urethra. *alt*. para-urethral glands, Skene's glands.

guest *n*. an animal living and breeding in the nest of another, esp. an insect.

guild *n.* a group of species having similar requirements and foraging habits and so having similar roles in the community.

gula *n.* the upper part of the throat. *a.* gular, *appl.* the pouch of skin below the beak in pelicans and their relatives.

gular *n.* an unpaired anterior horny shield in shell of tortoises and turtles.

gullet *n.* oesophagus *q.v.*, in protozoans, the cytopharynx *q.v.*; any cavity by which food may be taken into the body.

gummiferous *a.* gum-producing or exuding.

gummosis *n.* condition of plant tissue when cell walls become gummy, caused by certain bacteria.

gums *n.plu.* (*bot.*) various materials, composed largely of polysaccharides, resulting from breakdown of plant cell walls and exuding from wounds; trees of the genus *Eucalyptus*; (*zool.*) fibrous tissue covering the jaws around the base of the teeth.

gustatory *a.* pert. sense of taste.

gut *n.* the alimentary canal or part of it.

gutta *n.* small spot of colour on e.g. insect wing; oil drop as in fungal hypha; latex of various Malaysian trees, including gutta percha and balata.

guttation *n.* formation of drops of water on leaves, forced out of leaves through special pores (hydathodes) by root pressure, seen, e.g., on tips of leaves of many grasses in early morning; formation of drops of water on surface of plant from moisture in air; exudation of aqueous solutions, as by sporangiophores, nectaries, etc.

guttiferous *a.* exuding a resin or gum.

guttiform *a.* drop-like; in the form of a drop.

guttula *n.* a small drop-like spot.

guttulate *a. pert.* or containing guttae; in the form of a small drop, as markings; containing oily droplets, *appl.* spores.

guttulose *a.* covered with, or containing, droplets.

gymnetrous *a.* without an anal fin.

gymno- prefix derived from Gk. *gymnos*, naked.

gymnoarian *a. appl.* gonads when naked and not enclosed in coelomic sacs.

gymnoblastic *a.* without hydrothecae and gonothecae, as certain coelenterates.

gymnocarpic, gymnocarpous *a.* having the fruit naked, not covered by some kind of masking structure; having hymenium exposed during maturation of spores.

gymnogynous *a.* with exposed ovary.

Gymnolaemata *n.* a class of marine Ectoprocta (Bryozoa) in which individual zooids have a circular crown of tentacles (lophophore) and are enclosed in a box-like calcareous exoskeleton, found encrusting seaweeds, stones and shells.

Gymnomycota *n.* in modern classifications, a division of the Fungi comprising the cellular and acellular slime moulds, when considered as fungi.

Gymnophiona Apoda *q.v.*

gymnopterous *a.* having bare wings, without scales.

gymnosomatous *a.* having no shell or mantle, as certain molluscs.

gymnosperms *n.* large group of seed-bearing woody plants, having seeds not enclosed in an ovary but borne on the surface of the sporophylls, either treated as a division (Gymnophyta or Pinophyta) of the seed plants, or as a grouping of four divisions, the cycads (Cycadophyta), *Ginkgo* (Ginkgophyta), the conifers (Coniferophyta) and the gnetophytes (Gnetophyta), taxa which have been given various ranks in other classifications. Extinct taxa of gymnosperms include the seed ferns, cycadeoids (Bennettitales) and cordaites (Cordaitales). *alt.* (formerly) Gymnospermae.

gymnospore *n.* a naked spore not enclosed in a protective coat.

gymnostomatous, gymnostomous *a.* having no peristome, *appl.* mosses.

gymnotoids *n.plu.* small group of slender freshwater bony fish from Central and South America (the Gymnotodei) with very long anal fins, and an electric organ that sets up an electric field around the fish, commonly called American knife-fishes, and including the electric eel *Electrophorus*.

-gyn- word element derived from Gk. *gynē*, woman, indicating female.

gynaecaner *n.* a male ant resembling a female.

gynaecium gynoecium *q.v.*

gynaecoid *n.* an egg-laying worker ant.

gynaecophore *n.* a canal or groove of certain trematodes, formed by inrolling of the sides, in which the female is carried.

gynander gynandromorph *q.v.*

gynandrism hermaphroditism *q.v.*

gynandromorph *n.* an individual exhibiting a spatial mosaic of male and female characters. *alt.* gynander, sex mosaic.

gynandromorphism *n.* condition of being a gynandromorph or manifesting a mosaic of male and female sexual characteristics, such as having one side characteristically male, the other female.

gynandrosporous *a.* with androspores adjoining the oogonium, as in some algae.

gynandrous *a.* having stamens fused with pistils, as in some orchids.

gynantherous *a.* having stamens converted into pistils.

gyne *n.* female ant, esp. a queen.

gynecium gynoecium *q.v.*

gynic *a.* female.

gynobasic *a. appl.* a style arising from the base of the ovary.

gynodioecious *a.* having female and hermaphrodite flowers on different plants.

gynoecious *a.* having female flowers only.

gynoecium *n.* the female reproductive organ of a flower, consisting of one or more carpels forming one or more ovaries, together with their stigmas and styles. *alt.* pistil, gynaecium.

gynogenesis *n.* development from eggs penetrated by sperm but not fertilized.

gynomerogony *n.* the development of an egg fragment obtained before fusion with male nucleus, and containing maternal chromosomes only.

gynomonoecious *a.* having female and hermaphrodite flowers on the same plant.

gynomorphic *a.* having a morphological resemblance to females.

gynosporangium *n.* female sporangium; megasporangium *q.v.*

gynostemium *n.* the column composed of united pistil and stamens in orchids.

gypsophil(ous) *a.* thriving in soils containing chalk or gypsum.

gyral, gyrate *a. pert.* a gyrus; *pert.* spiral or circular movement.

gyrase DNA gyrase *q.v.*

gyration *n.* rotation, as of a cell; a whorl of a spiral shell.

gyre *n.* circular movement.

gyrencephalic, gyrencephalous *a.* having a convoluted surface to the cerebral hemispheres.

gyri *plu.* of gyrus.

gyrodactyloid *n.* ciliated larva of certain ectoparasitic flukes which swims to find new host.

gyrose *a.* with undulating lines; sinuous; curving.

gyrus *n.* a convolution of the surface of the brain; a ridge winding between two grooves. *plu.* gyri.

H

H histidine *q.v.*

H-2 the major histocompatibility complex (*q.v.*) in the mouse. *see also* MHC molecules.

HA hyaluronic acid *q.v.*; haemagglutinin *q.v.*

Ha-MSV Harvey mouse sarcoma virus, the source of the oncogene *v-H-ras*.

Hb haemoglobin *q.v.*

HbA adult human haemoglobin.

HbCO carboxyhaemoglobin *q.v.*

HbF foetal haemoglobin *q.v.*

HbO₂ oxyhaemoglobin *q.v.*

HCG human chorionic gonadotropin *q.v.*

HDL high-density lipoprotein *q.v.*

Hfr *see* Hfr strain.

HGH human growth hormone.

HGPRT hypoxanthine guanine phosphoribosyltransferase, enzyme catalysing the formation of inosine monophosphate or guanine monophosphate in the minor pathway of nucleic acid biosynthesis, widely used as a genetic marker in somatic cell genetics.

HI haemagglutination inhibition *q.v.*

HIV human immunodeficiency virus *q.v.*

His histidine *q.v.*

HLA human leukocyte antigens, *see* histocompatibility antigens, major histocompatibility complex. *alt.* transplantation antigens. *see also* MHC molecules.

HLA-A, -B, -C the class I human major histocompatibility antigens.

HLA-D the class II human major histocompatibility antigens.

HMM heavy meromyosin, *see* meromyosin.

hnRNA heterogeneous nuclear RNA *q.v.*

HPFH hereditary persistence of foetal haemoglobin *q.v.*

HPLC high-performance liquid chromatography.

5-HT 5-hydroxytryptamine (serotonin) *q.v.*

HTLV-I human T-cell leukaemia virus, the cause of adult T-cell leukaemia, a rare cancer.

HTLV-II human T lymphotropic virus II, a retrovirus isolated from humans but causing no known disease.

HTLV-III human lymphotropic virus III, now known as human immunodeficiency virus (HIV) *q.v.*

hv hypervariable region (*q.v.*) of antibody molecule.

Hyp hydroxyproline *q.v.*

habenula *n.* a name *appl.* various band-like structures. *a.* habenular.

habilines *n.plu.* fossil hominids from East Africa, dated at around 2 million years ago, assigned by some palaeoanthropologists to the species *Homo habilis*, thus making them possibly the earliest human fossils yet known.

habit *n.* the external appearance or way of growth of a plant, e.g. climbing, bushy, erect, shrubby, etc.; the normal or regular behaviour of an animal.

habit formation trial-and-error learning *q.v.*

habitat *n.* the locality or environment in which a plant or animal lives.

habitat diversification beta diversity *q.v.*

habitat form the way a plant grows resulting from conditions in that particular habitat.

habitat space the habitable part of a space or area available for establishing a population.

habitat type a group of plant communities that produce similar habitats.

habituation *n.* adjustment, by a cell or an organism, by which subsequent contacts with the same stimulus produce diminishing effects; a form of learning in which reflex behaviour is extinguished when the animal finds it has no adaptive value.

habitus *n*. the general appearance or conformation characteristic of a plant or an animal.

hackle *n*. long erectile feather on the necks of some birds.

hadal *a. appl.* or *pert.* abyssal deeps below 6000 m.

haem (heme) *n*. a porphyrin (protoporphyrin IX) with an iron atom in its centre, found as an oxygen-carrying group in various proteins (e.g. haemoglobin), the Fe atom being in either the ferrous or ferric state, only the ferrous state being able to bind oxygen. *see* Appendix 1 (59) for chemical structure.

haem-, haema-, haemo- *alt.* hem-, hema-, hemo-.

haemacyte *n*. a blood cell in insects and other invertebrates.

haemadsorption *n*. the binding of red blood cells by some virus-infected cells, used in diagnosis and infectivity assays.

haemagglutination *n*. the clumping together of red blood cells, either spontaneously or in response to treatment with a specific antibody or other agent, and which is used as a diagnostic assay for some viruses that contain haemagglutinating proteins.

haemagglutination inhibition (HI) an immunological assay for certain viruses, in which antibody combining with the virus prevents it agglutinating red blood cells.

haemagglutinin *n*. any substance that causes the agglutination of red blood cells.

haemal *a. pert.* blood or blood vessels; situated on the same side of the vertebral column as the heart.

haemal arches in many vertebrates, lateral processes from the vertebrae of the tail which fuse ventrally to form the haemal canal enclosing an artery and a vein.

haemal ridge haemapophysis of vertebra.

haemal strands strands of spongy tissue in echinoderms which are part of the blood vascular system and may be concerned with the phagocytosis and destruction of invading microorganisms.

haemangioblast blood island *q.v.*

haemapophysis *n*. one of the plate-like or

spine-like processes projecting out to the side from the lower edge of the centrum of a vertebra.

haematal *a. pert.* blood or blood vessels.

haematin *n*. hydroxide of haemin (*q.v.*), decomposition product of haemoglobin.

haemato- *see* haem, haema-, haemo-.

haematobium *n*. an organism living in blood. *a.* haematobic.

haematoblast, haemoblast *n*. precursor cell to an erythroblast.

haematochrome *n*. carotenoid pigment in some red algae.

haematogenous *a.* formed in blood; derived from blood.

haematoidin bilirubin *q.v.*

haematology *n*. the study of the blood and its formation.

haematolymphoid *a. pert.* the lymphatic system and the blood.

haematolysis haemolysis *q.v.*

haematophagous *a.* feeding on or obtaining nourishment from blood.

haematopoiesis *n*. formation of blood, development of blood cells from stem cells.

haematopoietic *a.* blood-forming, *pert.* haematopoiesis; *appl.* stem cells.

haematosis haematopoiesis *q.v.*

haematoxylin *n*. pink dye that preferentially stains DNA and RNA in a cell.

haematozoon *n*. any animal parasitic in blood.

haemerythrin haemoerythrin *q.v.*

haemic haemal *q.v.*

haemin *n*. haem (*q.v.*) with the iron atom in the oxidized (ferric) state, usually found as the chloride.

haemobilirubin *n*. breakdown product of haemoglobin which is converted to bilirubin and biliberdin in the liver.

haemoblast haematoblast *q.v.*

haemochorial *a. appl.* placenta with branched chorionic villi penetrating blood sinuses after breaking down uterine tissues, as in insectivores, rodents and primates.

haemoclastic *a.* breaking down blood cells.

haemocoel *n*. blood-filled cavity consisting of spaces between organs, which is the main body cavity in molluscs.

haemoconia *n.plu.* minute particles of red blood cells taken up by phagocytes of the reticuloendothelial system.

haemocuprein superoxide dismutase *q.v.*

haemocyanin *n.* a copper-containing protein, blue in the oxidized form, that acts as a respiratory pigment in the blood of many annelids and arthropods.

haemocyte *n.* a blood cell in insects and other invertebrates.

haemocytoblast myeloid stem cell *q.v.*

haemodynamics *n.* study of the principles of blood flow.

haemoerythrin *n.* a red, iron-containing, non-haem protein that acts as a respiratory pigment in the blood of sipunculids, various molluscs and crustaceans. *alt.* hemerythrin.

haemogenesis haematopoiesis *q.v.*

haemoglobin (hemoglobin, Hb) *n.* oxygen-carrying haem protein, a compact globular molecule consisting of 4 globin subunits comprising 2 pairs of polypeptide chains ($\alpha\alpha\beta\beta$ in adult human haemoglobin), each chain associated with 1 haem group. The principal oxygen carrier in vertebrate blood, located in red blood cells and giving blood its colour, existing as oxyhaemoglobin in arterial blood, oxygen binding to the haem prosthetic group. *see also* carboxyhaemoglobin, foetal haemoglobin, haemoglobinopathies, leghaemoglobin, methaemoglobin, sickle cell haemoglobin, thalassaemias.

haemoglobinopathies *n.plu.* any of various inherited diseases, such as sickle-cell anaemia, in which the genetic defect affects the structure, function or production of haemoglobin.

haemoid *a.* resembling blood.

haemolymph *n.* a fluid in the coelom of some invertebrates, regarded as equivalent to blood and lymph of more complex animals.

haemolysin *n.* activity present in blood serum, capable of destroying red blood cells.

haemolysis *n.* the lysis of red blood cells. *alt.* erythrocytolysis, haematolysis, haemocytolysis, laking.

haemolytic *pert.* or causing haemolysis.

haemoparasite *n.* blood parasite, a parasitic organism that lives in the blood of its host.

haemopathic *a.* affecting the blood circulatory system.

haemophilia *n.* genetically determined condition characterized by excessive bleeding due to inability of blood to clot normally, haemophilia A caused by a deficiency of a blood protein Factor VIII essential to the clotting reaction, the rarer haemophilia B or Christmas disease by a deficiency of Factor IX.

haemopoetins *n.plu.* growth factors involved in the proliferation and differentiation of blood cells.

haemopoiesis haematopoiesis *q.v.*

haemoproteins *n.plu.* proteins having an iron-porphyrin (haem) prosthetic group.

haemorrhoidal *a.* rectal, *appl.* blood vessels, nerve.

haemosiderin *n.* a complex of protein and ferric hydroxide, a yellow pigment of blood and found in most tissues, stored as large granules esp. in liver, spleen and bone marrow.

haemostatic *a. appl.* an agent that stops bleeding.

haemotoxin *n.* toxin that produces lysis of red blood cells.

haemotropic *a.* affecting or acting upon blood.

haerangium *n.* in some ascomycete fungi, an adhesive droplet containing spores.

Hageman factor Factor XII *q.v.*

hagfishes *n.plu.* common name for the Myxiniformes, a small order of bottom-living jawless marine fish, eel-like in shape, lacking pectoral or pelvic fins and with no scales.

hair (*bot.*) trichome *q.v.*; root hair *q.v.*; (*zool.*) in mammals, a thread-like epidermal structure consisting of cornified epithelial cells which grows by cell division from a hair follicle at its base.

hair cell sensory cell in the organ of Corti in the ear, involved in hearing; sensory cell in the vestibular apparatus of the ear, involved in sense of balance; similar cell in the acoustico-lateralis system of fishes and amphibia.

hair follicle *see* follicle.

hairpin region of duplex structure in RNA and DNA, formed by base-pairing between adjacent or nearby complementary sequences on the same strand, the unpaired bases between the sequences forming a single-stranded loop (hairpin loop) at the end of the hairpin.

hairy root disease abnormal proliferation

of root hairs caused by the bacterium *Agrobacterium rhizogenes*.

halarch succession halosere *q.v.*

Haldane's rule a rule stating that when offspring of one sex produced from a cross are inviable or infertile, it is always the heterogametic sex.

half-inferior *a.* having ovary only partly adherent to calyx.

half-life time required for the disappearance of half the original quantity of a given substance, e.g. from the circulation, assuming that the substance disappears at a regular rate; of radioactive elements, time required for the radioactive decay of half the original amount of material.

half-sibs individuals having only one parent in common.

half-spindle one half of a mitotic or meiotic spindle, comprising the microtubules arising from one pole as far as the equator.

half-terete *a.* rounded on one side, flat on the other.

hallucinogen *n.* any drug capable of evoking hallucinations, e.g. LSD, cannabis, psilocybins.

hallux *n.* first digit of hindlimb in vertebrates, the big toe.

halobacteria *n.plu.* group of bacteria included in the archaebacteria, requiring high salt concentrations and living in salt lakes etc.

halobenthos *n.* marine benthos.

halobios *n.* sum total of organisms living in sea; animals living in sea or any salt water. *a.* halobiotic.

halodrymium *n.* a mangrove association.

halolimnic *a.* marine organisms adapted to live in fresh water.

halomorphic *a. appl.* soils containing an excess of salt or an alkali.

halophilic *a.* salt-loving; thriving in the presence of salt.

halophilic bacteria group of archaebacteria (e.g. *Halobacterium, Halococcus*) adapted to life in saturated salt solutions, e.g. brine pools. *alt.* halobacters. *see also* Appendix 6.

halophobe *n.* a plant intolerant of salt.

halophyte *n.* a sea-shore plant; a plant that can thrive on soils impregnated with salt. *a.* halophytic.

haloplankton *n.* the organisms drifting in the sea.

Haloragales Hippuridales *q.v.*

halosere *n.* a plant succession originating in a saline area, as in salt marshes.

haloxene *a.* tolerating salt water.

halteres *n.plu.* a pair of small rounded bodies borne on metathorax of dipteran flies and representing rudimentary posterior wings.

Hamamelidales *n.* order of woody dicots including the families Hamamelidaceae (witch hazel), Myrothamnaceae, and Platanaceae (plane).

hamate *a.* hooked, or hook-shaped at the tip.

hamatum *n.* the hooked bone in the carpus (wrist).

hamiform *a.* hook-shaped.

hamirostrate *a.* having a hooked beak.

hammer malleus *q.v.*

hamose hamate *q.v.*

hamstrings tendons of insertion of the posterior femoral muscles of fore- and hindlimbs.

hamular *a.* hooked; hook-like.

hamulate *a.* having small hook-like processes.

hamuli *n.plu.* minute hooks on front edge of hindwing of Hymenoptera which link fore- and hindwings together.

hamulose hamate *q.v.*

hamulus *n.* hooklet or hook-like process on bone, feathers and other structures. *plu.* hamuli.

handling time the time it takes a predator to catch and eat its prey.

hanger *n.* a wood situated on a hillside.

hapanthous *a.* reproducing only once, towards the end of a plant's life.

hapaxanthic, hapaxanthous *a.* with only a single flowering period.

haplobiont *n.* organism characterized by only one type of individual in its lifecycle. *cf.* diplobiont.

haplocaulescent *a. appl.* plants with a simple axis, i.e. capable of producing seed on the main axis.

haplocheilic *a. appl.* type of stomata in gymnosperms in which the subsidiary cells are not related developmentally to the guard cells.

haplochlamydeous *a.* having only one whorl of perianth segments.

haplodioecious heterothallic *q.v.*

haplodiploid *a. appl.* species in which sex is determined by the male being haploid, the female diploid, as in bees and wasps. *n.* haplodiploidy.

haplodiplont *n.* an organism exhibiting the haplodiploid condition; a plant with haploid and diploid vegetative phases.

haploid *a. appl.* cells having one set of chromosomes representing the basic genetic complement of the species, usually designated *n; n.* a haploid organism or cell.

haploid apogamy/apogamety generative apogamy *q.v.*

haploidization *n.* in certain fungi, an event occurring in the parasexual cycle during which a diploid cell becomes haploid by loss of one chromosome after another by non-disjunction.

haploidy *n.* the state of being haploid.

haplo-insufficient *a. appl.* genetic loci in diploid organisms where the presence of only one copy of the wild-type allele is not sufficient to produce a normal phenotype.

haplomonoecious homothallic *q.v.*

haplomycelium *n.* a haploid mycelium.

haploneme *a.* having threads of uniform diameter.

haplont *n.* any organism having haploid somatic nuclei or cells.

haploperistomic, haploperistomatous *a.* having a single peristome; having a peristome with a single row of teeth, *appl.* mosses.

haplopetalous *a.* with a single row of petals.

haplophase *n.* stage in life history of an organism when nuclei are haploid.

haplophyte *n.* a haploid plant or gametophyte.

haploptile *n.* a down feather without a shaft, formed by precocious development of the barbs of the adult feather.

haplosis *n.* halving of the chromosome number during meiosis; reduction and disjunction.

haplostele *n.* a simple stele having a cylindrical core of xylem surrounded by phloem.

haplostemonous *a.* having one whorl of stamens.

haplo-sufficient *a. appl.* genetic loci in diploid organisms where the presence of only one copy of the wild-type allele is sufficient to produce a normal phenotype.

haplotype *n.* the set of alleles borne on one of a pair of homologous chromosomes, esp. in relation to complex loci such as the major histocompatibility complex (MHC).

haploxylic *a.* possessing only one vascular bundle.

hapten *n.* small molecule capable of eliciting an immune response only when attached to a larger macromolecule. *a.* haptenic.

hapteron *n.* holdfast, a disc-like outgrowth from the stalk of some algae that attaches them to substrate. *plu.* haptera.

haptic *a. pert.* touch, *appl.* stimuli and reactions.

haptoglobin *n.* serum protein that combines with haemoglobin, having the function of ridding serum of free haemoglobin.

haptomonad *n.* an attached form of certain parasitic flagellates.

haptonasty *n.* a plant movement elicited by touching, as the drooping of the leaves of mimosa.

haptonema *n.* a distinctive very long flagellum in certain algae, having no locomotory function and consisting of three concentric membranes surrounding a central space.

haptophore *n.* the part of a molecule such as a toxin, agglutinin, antigen, antibody, etc. that carries the site at which it binds to other free molecules or cells.

Haptophyta, haptomonads *n., n.plu.* group of small motile golden algae, mainly marine, bearing a coiled thread-like haptoneme between the two flagella, which acts as a holdfast, and a cell surface covered with thin scales. Many species have a resting stage, when the surface becomes covered with elaborate calcareous scales (coccoliths), when the organisms are known as coccolithophoroids.

haptor *n.* attachment organ in skin flukes (flatworms).

haptospore *n.* an adhesive spore.

haptotropism *n.* response by curvature to a contact stimulus, as in tendrils or stems that twine around a support. *alt.* thigmotropism.

hard pan hard layer developed in the B-horizon of the soil, consisting of deposited salts, which restricts drainage and root growth.

hard-wired *appl.* neuronal circuits whose connections and properties are fixed during development, as opposed to those whose properties can be modified by experience.

Harderian gland an accessory lacrimal gland of 3rd eyelid or nictitating membrane.

hardwood *n.* wood of broad-leaved trees, although not all hardwoods are in fact harder than all softwoods (wood from coniferous trees).

Hardy-Weinberg Law in a large randomly mating population, in the absence of migration, mutation and selection, allele frequencies stay the same from generation to generation according to the following rule (the Hardy-Weinberg rule) which states that if the allele frequency of one of the alleles at a locus is p and that of the other is q then the frequencies of the two homozygotes and the heterozygote are given by p^2, q^2 and $2pq$ respectively.

harem *n.* a group of breeding females which a dominant male mates with and guards from rival males.

harmonic suture an articulation formed by apposition of edges or surfaces as between palatine bones.

harp *n.* region on wings of grasshoppers, crickets, etc. whose vibration helps produce and amplify the characteristic sounds produced by these insects.

harpagones *n.plu.* claspers of certain male insects.

harpes *n.plu.* claspers of male lepidoptera; chitinous processes between claspers of mosquito. *sing.* harpe.

Hartig net network of fungal hyphae within the epidermis and outer cortex of roots of plants with mycorrhizas, the hyphae not penetrating into the endodermis, and only rarely entering the root cells.

harvestmen common name for the Opiliones, an order of arachnids having very long legs and with the prosoma and opisthosoma forming a single structure.

Hassall's corpuscles structures in medulla of thymus containing epithelial cells, macrophages and cell debris.

hastate *a.* spear-shaped, more or less triangular with the two basal lobes divergent, *appl.* leaves.

HAT selection technique for identifying cell hybrids in somatic cell genetics, based on selection of HGPRT$^+$;TK$^+$ cells in a special medium (HAT) containing hypoxanthine, thymidine and aminopterin, when the major pathway of DNA synthesis is blocked by aminopterin.

Hatch-Slack pathway C4 pathway *q.v.*

haulm *n.* stem of peas, and of grasses.

haustellate *a.* having a haustellum, a proboscis adapted for sucking liquids.

haustorium *n.* an outgrowth of stem, root or hyphae of certain parasitic plants or fungi, through which they obtain food from the host plant. *a.* haustorial. *plu.* haustoria.

Haversian canals small canals in bone in which run blood capillaries, nerves and lymph.

Haversian system the unit of concentric rings of bone (Haversian lamellae) surrounding a central canal (Haversian canal) with the bone cells and canaliculi, which forms the basic structural unit of compact bone. *alt.* osteon.

Hawaiian Floral Region part of the Palaeotropical Realm comprising the Hawaiian islands.

H band, H disc, H line a singly refracting light band in the centre of the sarcomere of striated muscle under the microscope, representing thick filaments only.

Hb Kenya abnormal haemoglobin produced in a type of thalassaemia in which a deletion has occurred between one of the γ-globin genes and the β-globin gene, producing a hybrid protein consisting of the N-terminal portion of a γ-globin and the C-terminal of β-globin.

Hb Lepore abnormal haemoglobin produced in a type of β-thalassaemia in which the δ- and β-globin genes have become fused and produce a hybrid protein consisting of the N-terminal sequence of δ and the C-terminal sequence of β.

HbH disease type of α-thalassaemia in which only one α-globin gene is functional in the diploid cell, which causes the formation of an abnormal β-globin tetramer from the excess β-globin chains.

H chain heavy chain (*q.v.*) of immunoglobulin.

HD protein helix-destabilizing protein, *see* ss-binding protein *q.v.*

head-cap acrosome (*q.v.*) of sperm.

head-case the hard outer covering of insect head.

head-kidney the pronephric portion of kidney, in vertebrates represented only in embryo; a nephridium usually developed in the head segment of invertebrates.

heart *n.* hollow muscular organ which by rhythmic contractions pumps blood around the body. The mammalian heart consists of four chambers, an upper thin-walled atrium and lower thick-walled ventricle on either side. Venous blood from the body enters the right atrium and leaves for the lungs from the right ventricle, oxygenated blood re-enters the heart through the left atrium and leaves the heart from the left ventricle. The muscular contraction of the heart is self-sustaining and is synchronized by pacemaker cells located in the sinuatrial node. *see also* muscle.

heartwood *n.* the darker, harder, central wood of trees, containing no living cells.

heat spot in skin an area in which sensory nerve endings sensitive to heat are found.

heat shock in microbiology and cell biology, a short period of exposure of a cell to temperatures above its normal range, which results in changes of gene expression and protein synthesis aimed at protecting the cell against the effects of the heat stress.

heath *n.* vegetation developing on poor, usually acid, sandy or gravelly soils in the lowlands and dominated by heathers (*Calluna* and *Erica*).

heavy chain (H chain) one of the 2 larger identical polypeptide chains in an immunoglobulin molecule, each containing a variable and a constant region. The variable region contributes to an antigen-binding site, the constant region determines antibody class (*q.v.*); *see* myosin.

heavy chain switching DNA rearrangement in B lymphocytes resulting in cells switching from making IgM to either IgG, IgA or IgE, as the pre-existing heavy chain variable region gene segment is rejoined to an alternative C region gene.

hebbian synapse a synapse whose transmitting properties can be modulated by the input it receives.

hebetate *a.* blunt-ended.

hebetic *a.* pert. adolescence.

hectocotylus *n.* arm of male cephalopod specialized for transfer of spermatophore to female.

hederiform *a.* shaped like an ivy-leaf, *appl.* nerve endings, as pain receptors in skin.

hedonic *a. appl.* skin glands of certain reptiles, which secrete a musk-like substance and are specially active at mating season.

Heidelberg man type of primitive man known from fossils found near Heidelberg in Germany, which is now considered as a subspecies of *Homo erectus*.

hekistotherm *n.* a plant that grows well in generally cold conditions (e.g. arctic and alpine plants).

HeLa cells an aneuploid line of human epithelial cells originating from a cervical carcinoma and which have been propagated in tissue culture since 1952.

heleoplankton *n.* plankton of marshy ponds and lakes.

helical *a. appl.* long rod-shaped virions in which the protein subunits of the coat and associated nucleic acid are arranged in helical spirals, as tobacco mosaic virus; *appl.* arrangement of myofibrils in some smooth muscle; *appl.* a type of cell wall thickening in xylem.

helicase *n.* any enzyme which can unwind the DNA double helix, e.g. rep protein *q.v.* *alt.* unwindase.

helicine *a.* spiral, convoluted; *pert.* outer rim of external ear.

helicoid *a.* spiral; shaped like a snail's shell; *pert.* type of sympodial branching in which sympodium consists of fork branches off same side.

helicoid cyme a cymose inflorescence produced by suppression of successive axes on the same side, thus causing the sympodium to be spirally twisted so that the blooms are on only one side of the axis.

helicoid dichotomy a type of branching in which there is repeated forking but with the branches on one side uniformly more vigorous than on the other.

helicone *n.* in gastropods, a shell coiled in a helical spiral.

helicorubin *n.* a red pigment of gut of pulmonates (e.g. snails and slugs) and of liver of certain crustaceans.

helicospore *n.* a convoluted or spirally twisted spore.

heliophil, heliophilic, heliophilous *a.* adapted to a relatively high intensity of light.

heliophobe *n.* plant that thrives in shade.

heliophyll *n.* plants with leaves of similar structure on both sides and arranged vertically.

heliophyte *n.* a plant requiring full sunlight to thrive.

heliosis *n.* production of discoloured spots on leaves through sunlight.

heliotaxis *n.* movement in response to the stimulus of sunlight.

heliotropism *n.* a plant growth movement in response to the stimulus of sunlight.

helioxerophil *n.* plant that thrives in full sunlight and in arid conditions.

heliozoan *n.* member of the Heliozoa, an order of mostly freshwater protozoans of the Sarcodina, having a radially symmetrical body, stiff slender pseudopodia and often a skeleton of spicules.

helix *n.* the outer rim of external ear in humans; double helix *see* DNA.

α-helix regular periodic secondary structure common in proteins in which the polypeptide backbone is twisted in a right-handed spiral to form a rigid rod-like structure.

helix-destabilizing protein ss-binding protein *q.v.*

helix-turn-helix motif a structural motif in proteins associated with the capacity to bind to specific recognition sites in DNA.

helmet *n.* the process of bill of hornbills; (*bot.*) the helmet-shaped petal or galea of certain flowers.

helminth *n.* parasitic flatworm (fluke and tapeworm) or roundworm.

helminthoid *a.* shaped like a worm.

helminthology *n.* the study of worms; the study of parasitic flatworms and roundworms.

helobious *a.* living in marshes.

helophyte *n.* a marsh plant, esp. a perennial herbaceous plant of marshes with the perennating parts lying in the mud.

helotism *n.* symbiosis in which one organism enslaves another and forces it to labour on its behalf, e.g. in some species of ants.

helper T lymphocytes, helper T cells a class of T lymphocytes which interact with B cells in the presence of their cognate antigen to stimulate B-cell differentiation and proliferation into antibody-secreting plasma cells.

helper virus a replication competent virus needed for the multiplication of related viruses which have lost the capacity to replicate as a result of e.g. the replacement of part of their genome with host cell genes (as in some RNA tumour viruses). Helper viruses assist the replication of defective viruses by providing the necessary viral components.

hem-, hema-, hemo- haem-, haema-, haemo-.

hemelytron *n.* forewing of heteropteran bug, with a distal membranous section.

hemeranthic, hemeranthous *a.* flowering by day.

hemi- prefix derived from Gk. *hēmi,* half.

Hemiascomycetae *n.* class of unicellular and simple mycelial ascomycete fungi including the budding yeasts and leaf-curl fungi, which bear naked asci not enclosed in an ascocarp. *alt.* Hemiascomycetidae.

hemiascus *n.* an atypical multinucleate ascus produced by some ascomycete fungi, and in which hemiascospores are formed.

hemi-autophyte *n.* parasitic plant that produces its own chlorophyll.

hemibasidium *n.* a septate basidium as found in some basidiomycete fungi. *alt.* heterobasidium.

hemibiotroph *n.* fungus that is only partly dependent on a living host. *cf.* biotroph.

hemibranch *n.* gill with gill filaments on only one side.

hemicellulase *n.* any of a group of enzymes that hydrolyse hemicelluloses.

hemicellulose *n.* any of a diverse group of polysaccharides found in plant cell walls and as storage carbohydrates in some seeds, and which contain a mixture of sugar residues including xylose, arabinose, mannose, glucose, glucuronic acid and galactose, and which include xylans, glucomannans, arabinoxylans, xyloglucans.

hemicephalous *a. appl.* insect larvae with a reduced head.

Hemichordata, hemichordates *n., n.plu.* a phylum of marine, worm-like, coelomate invertebrate animals, which have pharyngeal gill slits and a body divided into three regions.

hemichordate *a.* possessing a rudimentary notochord.

hemicryptophyte *n.* herbaceous perennial plant in which the perennating parts are at soil level, often protected by the dead leaves.

hemicyclic *a.* with some floral whorls cyclic, some spiral; (*mycol.*) lacking summer stages, in life cycle of rust fungi.

hemidesmosome *see* desmosome.

hemielytron hemelytron *q.v.*

hemiepiphyte *n.* a plant that does not spend its whole life cycle as a complete epiphyte: either a plant whose seeds germinate on another plant, but which later sends roots to the ground, or a plant that begins life rooted but later becomes an epiphyte.

hemigamy *n.* activation of ovum by male nucleus without nuclear fusion.

hemignathous *a.* having one jaw shorter than the other, as some fishes and birds.

hemikaryon *n.* a nucleus with the gametic or haploid number of chromosomes.

hemikaryotic *a. pert.* a hemikaryon; haploid *q.v.*

hemimetabolous *a. appl.* the orders of insects having an incomplete metamorphosis, with no pupal stage in the life history, comprising the Orthoptera (crickets, locusts), Dictyoptera (cockroaches), Plecoptera (stoneflies), Dermaptera (earwigs), Ephemeroptera (mayflies), Odonata (dragonflies), Embioptera (foot spinners), Isoptera (termites), Psocoptera (book-lice), Anoplura (biting and sucking lice), Thysanoptera (thrips) and Hemiptera (bugs).

hemimethylated *a. appl.* to DNA in which a CG doublet is methylated (at the C residue) on only one strand.

hemiparasite *n.* an individual which is partly parasitic but which can survive in the absence of its host; a parasitic plant that develops from seeds germinating in the soil rather than in host body; a parasite that can exist as a saprophyte.

hemipenis *n.* one of paired grooved copulatory structures present in males of some reptiles. *plu.* hemipenes.

hemipneustic *a.* with one or more pairs of spiracles closed.

hemipodes *n.plu.* an order of small quail-like birds of the order Gruiformes, related to cranes and rails.

Hemiptera, hemipterans *n., n.plu.* order of sucking insects commonly known as bugs, and including water boatmen and pond skaters as well as blood-sucking bugs parasitic on mammals and the sap-sucking aphids, scale insects, leaf-hoppers, mealy bugs and cicadas. Some blood-sucking bugs transmit disease, and the aphids are important vectors of plant viral diseases.

hemisaprophyte *n.* a plant living partly by photosynthesis, partly by obtaining food from humus; a saprophyte that can also survive as a parasite.

hemisphere cerebral hemisphere *q.v.*

hemisystole *n.* contraction of one ventricle of heart.

hemitropous *a.* turned half round, having an ovule with hilum on one side and micropyle opposite in plane parallel to placenta; *appl.* flowers restricted to medium-length tongued insects for pollination; *appl.* insects with medium-length tongues visiting such flowers.

hemixis *n.* fragmentation and reorganization of macronucleus without involving micronucleus, in the ciliate *Paramecium*.

hemizygous *a. appl.* genetic locus present in only one copy, either as a gene in a haploid organism, or a sex-linked gene in the heterogametic sex (e.g. X-linked genes in human males), or a gene in a segment of chromosome whose partner has been deleted.

Henle's layer outermost layer of nucleated cubical cells in inner root sheath of hair follicle.

Henle's loop loop of Henle *q.v.*

Henle's sheath perineurium, or its prolongation surrounding branches of nerve.

Hensen's cells columnar supporting cells on basal membrane in ear.

Hepadnaviridae *n.* family of single-stranded DNA viruses, which includes the hepatitis B virus, with an unusual mode of replication, in which viral DNA is synthesized from an RNA template. Chronic infection with hepatitis B virus has been implicated in the increased incidence of liver cancer in areas where the virus is widespread.

hepar *n*. liver, or an organ having a similar function.

heparan *n*. glycosaminoglycan, composed of repeating disaccharides of *N*-acetylglucosamine and glucuronic acid or iduronic acid, as the sulphate a component of extracellular matrix.

heparin *n*. a polysaccharide (a glycosaminoglycan) found in mast cells and which has anticoagulant activity, inhibiting blood clotting by increasing the rate of inactivation of thrombin by antithrombin III.

heparinocyte mast cell *q.v.*

hepatectomy *n*. removal of the liver.

hepatic *a. pert.*, like, or associated with the liver; (*bot.*) *pert.* liverworts; *n*. a liverwort.

Hepaticae, Hepaticopsida Hepatophyta *q.v.*

hepatic portal system in vertebrates, the part of the vascular system carrying blood to the liver and consisting of the hepatic portal vein, which carries blood from gut to liver, and the hepatic artery which carries blood away from liver.

hepatocystic *a*. liver and gall bladder.

hepatocyte *n*. liver cell, specialized for the synthesis, degradation and storage of a large number of substances, and which secretes bile.

hepatoduodenal *a. pert.* liver and duodenum.

hepatoenteric *a*. of or *pert.* liver and intestine.

hepatogastric *a. pert.* liver and stomach.

hepatopancreas *n*. gland in many invertebrates secreting digestive enzymes, and which is presumed also to perform a function similar to the liver.

Hepatophyta *n*. division of non-vascular spore-bearing green plants commonly called liverworts, which with the hornworts and mosses are known as the bryophytes (*q.v.*). Liverworts are small, generally inconspicuous plants, growing in low clumps on the ground or on rocks, tree-bark etc. The photosynthetic gametophyte is thalloid or leafy, most liverworts having stems with three rows of leaves. The sporophyte is a stalked capsule growing

from the gametophyte. The plant is anchored to the ground by fine unicellular rhizoids. Liverworts lack specialized conducting tissue (with a few possible exceptions), a cuticle and stomata, and are the simplest of multicellular plants. *alt.* hepatics. *see also* Appendix 2.

hepatoportal system hepatic portal system *q.v.*

hepatorenal *a. pert.* liver and kidney.

hepatoumbilical *a*. joining liver and umbilicus.

hepta- prefix derived from Gk. *hepta*, seven, and denoting having seven of, or arranged in sevens, etc., as heptaploid, having seven times the haploid number of chromosomes; heptamerous, having seven of each part, of flowers.

heptagynous *a*. with seven pistils.

heptarch *a. appl.* stele having seven initial groups of xylem.

heptastichous *a*. arranged in seven rows, *appl.* leaves.

heptose *n*. any sugar having the formula $(CH_2O)_7$, e.g. sedoheptulose.

herb *n*. any seed plant with non-woody green stems.

herb layer a horizontal ecological stratum of a plant community comprising the herbaceous plants. *alt.* field layer.

herbaceous *a. appl.* seed plants with non-woody green stems; soft, green, with little woody tissue, *appl.* plant organs.

herbarium *n*. a collection of dried or preserved plants, or of their parts, and the place where they are kept.

herbicide *n*. a chemical that kills plants.

herbivore *n*. an animal that feeds exclusively on plants. *a*. herbivorous.

herbivory *n*. feeding on plants.

herbosa *n*. vegetation composed of herbaceous plants.

hercogamy *n*. the condition in which self-fertilization is impossible.

hereditary *a. appl.* characteristics that can be transmitted from parent to offspring, i.e. characters that are genetically determined.

hereditary persistence of foetal haemoglobin (HPFH) type of β-thalassaemia in which δ- and β-globins are absent but there are no clinical symptoms because of the continued synthesis of foetal haemoglobin (ααγγ) into adulthood.

heredity *n*. the genetic constitution of an individual; the transmission of genetically based characteristics from parents to offspring.

heritability *n*. capacity for being transmitted from one generation to the next; in quantitative genetics can be used in two senses: (1) broad-sense heritability, which is the proportion of phenotypic variation in a particular trait that is attributable to differences in genotype among individuals, (2) narrow-sense heritability, which is used, e.g., to determine the amount of improvement possible in a particular trait by selective breeding.

hermaphrodite *n*. an organism with both male and female reproductive organs, as occurs normally in some groups of animals and many plants, *alt*. bisexual; in mammals and some other groups of animals, an individual with a mixture of male and female organs arising as a result of a developmental abnormality, more properly called a pseudohermaphrodite. *a*. hermaphrodite, *alt*. androgynous, bisexual.

hermaphroditism *n*. condition of being a hermaphrodite.

heroin *n*. an addictive alkaloid obtained from morphine by acetylation, and which acts as a narcotic.

Herpesviridae, herpesviruses *n*., *n.plu*. family of DNA viruses including the various herpesviruses that cause cold sores, genital herpes, etc., and the Epstein–Barr virus which causes glandular fever, and is also involved in Burkitt's lymphoma in children in Africa and nasopharyngeal carcinoma in China and South-east Asia.

herpetology *n*. that part of zoology dealing with the study of reptiles.

hesmosis *n*. the splitting of some ant and termite colonies by the departure of reproductive individuals with an attendant group of sterile workers.

hesperidin *n*. a flavone derivative, the active constituent of citrin, and which affects the permeability and fragility of blood capillaries.

hesperidium *n*. type of indehiscent fruit exemplified by oranges and lemons, formed from a superior, multilocular ovary, having epicarp and mesocarp joined together, with endocarp projecting into the interior as membranous partitions which divide the pulp into segments.

hesthogenous *a*. covered with down at hatching.

heteracanthous *a*. having the spines in dorsal fin asymmetrically turning alternately to one side then the other.

heteractinal *a. pert*. sponge spicules having a disc of six to eight rays in one plane and a stout ray at right angles to these.

heterandrous *a*. with stamens of different length or shape.

heterauxesis *n*. irregular or asymmetric growth of organs; relative growth rate of parts of an organism.

heteraxial *a*. with three unequal axes.

heterecious, heterecism *alt*. spelling of heteroecious, heteroecism *q.v*.

hetero- prefix from Gk. *heteros*, other: indicating difference in structure, from different sources, of different origins, containing different components, etc.

heteroagglutinin *n*. agglutinin of ova which reacts with sperm of a different species.

heteroallelic *a. appl*. mutant alleles which have mutations at different sites so that intragenic recombination can yield a functional gene.

heteroantibody *n*. an antibody from one species that reacts with antigen from another species.

heteroantigen *n*. an antigen that is antigenic in a species other than that from which it was obtained.

Heterobasidiomycetae, heterobasidiomycetes *n*., *n.plu*. basidiomycete fungi producing their basidiospores from septate basidia (heterobasidia). They include the rusts, smuts and jelly fungi. *cf*. Homobasidiomycetae.

heterobasidium *n*. a septate basidium of lower basidiomycete fungi, composed of a hypobasidium, and an epibasidium from which sterigmata and basidiospores arise.

heteroblastic *a*. arising from dissimilar cells.

heterocarpous *a*. bearing more than one distinct kind of fruit.

heterocaryo- heterokaryo-.

heterocellular *a*. composed of cells of more than one sort.

heterocephalous *a*. having pistillate flowers on separate heads from staminate.

heterocercal *a.* having vertebral column terminating in upper lobe of tail fin which is usually larger than lower lobe, as in dogfish and other sharks.

heterochlamydeous *a.* having a calyx differing from corolla in colour, texture, etc.

heterochromatin *n.* regions of densely staining chromatin existing in a condensed form in interphase nuclei and which are never transcribed, and which can be divided into constitutive heterochromatin which consists of regions never transcribed in any cell (such as satellite DNA), and facultative heterochromatin which comprises whole chromosomes that are not expressed in one cell lineage but are in another (e.g. one or other of the mammalian X chromosomes). *a.* heterochromatic.

heterochromosome *n.* chromosome composed mainly of heterochromatin.

heterochromous *a.* differently coloured, *appl.* disc and marginal florets of some composite flowers.

heterochronic *a. appl.* mutations that affect the timing of a developmental pro - cess.

heterochrony *n.* a departure from typical sequence in time of formation of organs.

heterochrosis *n.* abnormal coloration.

heterocoelous *a. pert.* vertebrae with saddle-shaped articulatory centra.

heteroclitic *a. appl.* antibody raised against one antigen but having a higher affinity for another antigen not used in the original immunization.

heterocotyledonous *a.* having cotyledons unequally developed.

heterocyst *n.* rounded, thick-walled, seemingly empty cell found at intervals in filaments of some cyanobacteria, and which lacks the photosynthetic apparatus and in which nitrogen fixation is thought to take place.

heterodactylous *a.* with the 1st and 2nd toes turned backwards

heterodimer *n.* protein composed of 2 different subunits.

heterodont *a.* having teeth differentiated for various purposes.

Heterodontiformes *n.* order of selachians, including the hornsharks, having the notochord only partially replaced in extant species.

heterodromous *a.* having genetic spiral of stem leaves turning in different direction to that of branch leaves.

heteroduplex DNA DNA duplex comprising 2 strands of different origin, formed for example by crossing-over during meiosis. *alt.* hybrid DNA.

heteroecious *a.* passing different stages of life history in different hosts; requiring two hosts to complete its life cycle, *appl.* some rust fungi.

heterofacial *a.* showing regional differentiation.

heterogameon *n.* a species consisting of races which, when selfed, produce a morphologically stable population, but when crossed may produce several types of viable and fertile progeny.

heterogametangic *a.* having more than one kind of gametangium.

heterogamete anisogamete *q.v.*

heterogametic sex the sex possessing a pair of non-homologous sex chromosomes (e.g. the male, which is XY, in mammals, and the female, which is WZ, in birds), and therefore producing two different types of gametes: one possessing one type of sex chromosome and one possessing the other.

heterogamous *a.* having unlike gametes; having two or more types of flower, such as male, female, hermaphrodite or neuter.

heterogamy *n.* alternation of generations, *q.v.*; alternation of two sexual generations, one being true sexual, the other parthenogenetic.

heterogangliate *a.* with widely spaced and asymmetrically placed ganglia.

heterogeneity *n.* heterogenous state; heterogenetic or genotypic dissimilarity.

heterogeneous *a.* consisting of dissimilar parts.

heterogeneous nuclear RNA (hnRNA) the unstable RNA of very broad size distribution found in the nucleoplasm of eukaryotic cells as ribonucleoprotein particles (hnRNPs), comprising the primary transcripts of nuclear genes (other than the rRNA genes), most of which never leaves the nucleus, a small proportion only being processed to yield mRNA.

heterogeneous summation, law of rule that the different independent features of an environmental stimulus (e.g. the shape, size and coloration of eggs) are additive in their effect on an animal's behaviour.

heterogenetic *a*. descended from different ancestral stock; *appl*. induction or stimulation by a complex of stimuli of different sorts.

heterogenous *a*. having a different origin, not originating in the body; *pert*. heterogeny.

heterogeny *n*. having several different distinct generations succeeding each other in a regular series.

heterograft *n*. tissue graft originating in a donor of a different species from the recipient. *alt*. xenograft.

heterogynous *a*. with two types of females.

heteroimmune *a*. displaying immunity to an antigen from another species; *appl*. sera, containing antibodies raised in one species to an antigen from another species.

heterokaryon *n*. a cell containing two (or more) genetically different nuclei, formed naturally in various organisms, such as many fungi, or artificially in culture by the fusion of two animal cells or plant protoplasts. *a*. heterokaryote, heterokaryotic.

heterokaryosis *n*. the presence of genetically dissimilar nuclei within individual cells.

heterokont *a*. bearing different kinds of flagella, *appl*. zoospores, etc.

heterolecithal *a. appl*. eggs with the yolk distributed unevenly.

heterologous *a*. of different origin; derived from a different species; differing morphologically, *appl*. alternation of generations; *appl*. various substances such as agglutinins that affect cells from species other than their own; (*imm*.) *appl*. antibody that reacts with an antigen other than that against which it was raised and vice versa.

heterologous anti-immunoglobulin antibody against immunoglobulins raised in a different species, used in many types of immunoassay. *alt*. secondary antibody.

heterology *n*. non-correspondence of parts owing to different origin or different elements. *alt*. non-homology.

heterolysis *n*. the dissolution of cells or tissue by the action of exogenous enzymes or other agents. *a*. heterolytic.

heteromallous *a*. spreading in different directions.

heteromastigote *a*. having two different kinds of flagella.

heteromeric *a. pert*. another part; *appl*. neurone with axon extending to other side of spinal cord.

heteromerous *a*. having, or consisting of, an unequal number of parts; in insects, having unequal numbers of tarsal segments on the three pairs of legs.

heterometabolous *a. appl*. insects having incomplete metamorphosis.

heteromixis *n*. the union of genetically different nuclei, as in heterothallism.

heteromorphic *a*. having different forms at different times; *appl*. chromosomes of different size and shape, or chromosome pairs different in size; *appl*. species in which haploid and diploid generations are morphologically dissimilar.

heteromorphosis *n*. development of a part in an abnormal position; regeneration when the new part is different from that which was removed.

heteromorphous *a. pert*. an irregular structure, or departure from the normal.

heteromultimer *n*. a protein consisting of different subunits.

heteronomous *a*. subject to different laws of growth; specialized along different lines; *appl*. segmentation into dissimilar segments.

hetero-oligomer *n*. a protein composed of different types of subunits.

heteropetalous *a*. with dissimilar petals.

heterophagous *a*. having very immature young.

heterophil *a. appl*. nonspecific antigens and antibodies present in an organism, affording natural immunity.

heterophilic *a*. binding like-to-unlike, *appl*. certain cell adhesion molecules that bind to different receptor molecules on other cells. *cf*. homophilic.

heterophyllous *a*. bearing foliage leaves of different shape on different parts of the plant.

heterophylly *n*. the bearing of foliage leaves of different shape on different parts of the plant.

heterophyte *n*. a saprophytic or parasitic plant.

heterophytic *a.* with two kinds of spores, borne by different sporophytes.

heteroplanogametes *n.plu.* motile gametes that are unlike one another.

heteroplasia *n.* the development of a tissue from another of a different kind.

heteroplasm *n.* tissue formed in abnormal places.

heteroplastic *a. appl.* grafts of unrelated material; *appl.* grafts between individuals of different species or genera.

heteroploid *a.* having an extra chromosome through non-disjunction of a pair in meiosis; not having a multiple of the basic haploid number of chromosomes; *n.* an organism having heteroploid nuclei.

heteropolymer *n.* a macromolecule (protein or carbohydrate) composed of different sorts of subunits.

heteropolysaccharide *n.* any polysaccharide made up of several different monosaccharide units, e.g. gums, chitin, pectins.

Heteroptera *n.* in some classifications, an order of insects including the water boatmen, capsids and bed bugs.

heteropycnotic *a. appl.* regions of chromosomes that remain compact and densely staining, even in interphase nuclei, when the remainder of the chromatin is more dispersed.

heterorhizal *a.* with roots coming from no determinate point.

heterosexual *a.* of, or *pert.* the opposite sex, *appl.* hormones, etc.

heterosis *n.* cross-fertilization; hybrid vigour; result of heterozygosis *q.v.*

heterosomal *a.* occurring in, or *pert.,* different bodies; *appl.* rearrangements in two or more chromosomes.

heterosporangic *a.* bearing two kinds of spores in separate sporangia.

heterosporous *a. appl.* plants producing two kinds of spores by meiosis, megaspores and microspores, i.e. all seed plants, some ferns and club mosses. Megaspores give rise to the female gametophyte, microspores to the male gametophyte, both much reduced in these plants. *n.* heterospory.

heterostemonous *a.* with unlike stamens.

heterostrophic *a.* coiled in a direction opposite to normal.

heterostyly *n.* condition in which individuals within a species differ in the length of style in their flowers, as in primroses with their pin-eyed (long-styled) and thrumeyed (short-styled) flowers. Anthers in one type of flower are at the same level as stigmas in the other, thus ensuring cross-pollination. *a.* heterostylic, heterostylous.

heterosynapsis *n.* the pairing of two non-homologous chromsomes.

heterosynaptic facilitation *see* presynaptic faciliation.

heterotaxis *n.* abnormal or unusual arrangement of organs or parts.

heterothallic *a. appl.* cells, thalli or mycelia of algae or fungi which can only undergo sexual reproduction with members of a physiologically different strain. *see also* mating type.

heterotic *a. pert.* heterosis, *appl.* vigour: hybrid vigour.

heterotopic *a.* in a different or unusual place, *appl.* transplantation of tissue or organ.

heterotopy *n.* displacement; abnormal habitat.

heterotrichous *a.* having two types of cilia; having a thallus consisting of prostrate and erect filaments, as certain algae.

heterotroph *n.* any organism only able to utilize organic sources of carbon, nitrogen etc. as starting materials for biosynthesis, using either sunlight as a primary energy source (photoheterotrophs – a few bacteria and algal flagellates) or energy from chemical processes (chemoheterotrophs – all animals, fungi, most bacteria, some parasitic plants). *a.* heterotrophic. *alt.* organotroph, *a.* organotrophic.

heterotropous *a. pert.* ovule with micropyle and hilum at opposite ends in a plane parallel with that of placenta.

heterotypic *a. pert.* mitotic division in which daughter chromosomes remain united and form rings; heterophilic *q.v.*

heterotypical *a. appl.* genus comprising species that are not truly related.

heteroxylous *a. appl.* wood containing vessels and fibres as well as tracheids.

heterozygosity *n.* proportion of heterozygotes for a given locus in a population.

heterozygote *n.* a heterozygous organism or cell. *alt.* hybrid.

heterozygote advantage the case where the heterozygote for a given pair of alleles is of superior fitness than either of the two homozygotes.

heterozygous *a. appl.* diploid organism that has inherited different alleles (of any particular gene) from each parent, i.e. carries different alleles at the corresponding sites on homologous chromosomes, *appl.* also to cells or nuclei from such an organism. *alt.* hybrid.

hexa- prefix derived from Gk. *hex*, six, signifying having six of, arranged in sixes, etc.

hexacanth *a.* having six hooks.

hexactinal *a.* with six rays.

hexactine *n.* a sponge spicule with six equal and similar rays meeting at right angles.

Hexactinellida *n.* a class of Porifera, the glass sponges or hexactinellid sponges, typically radially symmetrical with a skeleton of large six-rayed spicules of silica, often fused to form a three-dimensional network.

hexactinian *a.* with tentacles or mesenteries in multiples of six, *appl.* certain coelenterates.

hexacyclic *a.* having floral whorls consisting of six parts.

hexaene *n.* sponge spicule like a trident but with six branches.

hexagynous *a.* having six pistils or styles; with six carpels to a gynaecium.

hexamerous *a.* occurring in sixes, or arranged in sixes.

hexandrous *a.* having six stamens.

hexapetaloid *a.* with petaloid perianth of six parts.

hexapetalous *a.* having six petals.

hexaphyllous *a.* having six leaves.

hexaploid *a.* having six sets of chromosomes; *n.* an organism having six times the haploid chromosome number.

hexapod *a.* having six legs; *n.* an insect.

Hexapoda older name for Insecta *q.v.*

hexapterous *a.* having six wings or wing-like expansions.

hexarch *a. appl.* stele having six alternating xylem and phloem groups; having six vascular bundles.

hexasepalous *a.* having six sepals.

hexaspermous *a.* having six seeds.

hexasporous *a.* having six spores.

hexastemonous *a.* having six stamens.

hexaster *n.* a hexactine spicule in which the rays branch and produce star-shaped figures.

hexastichous *a.* having the parts arranged in six rows.

hexokinase *n.* an enzyme that catalyses the phosphorylation of glucose and some other hexose sugars. EC 2.7.1.1.

hexosamine *n.* an amino sugar in which the sugar is a hexose, e.g. galactosamine, glucosamine.

hexosaminidase *n.* enzyme which catalyses the cleavage of a terminal hexose amino sugar from gangliosides, etc., hexosaminidase A being involved in the breakdown of ganglioside GM_2 in the brain by removing the terminal N-acetylgalactosamine residue.

hexosan *n.* any polysaccharide made of linked hexose units, such as starch, glycogen, inulin, cellulose.

hexose *n.* a monosaccharide containing six carbon atoms (formula $C_6H_{12}O_6$), e.g. glucose, fructose, galactose, mannose.

hexose monophosphate shunt pentose phosphate pathway *q.v.*

Hfr strain bacterial strain in which the F factor is integrated into the chromosome, leading to an increased frequency of transfer of chromosomal genes (i.e. high frequency of recombination).

hiatus *n.* any large gap or opening.

hibernaculum *n.* a winter bud.

hibernal *a.* of the winter.

hibernating glands former term for brown adipose tissue *q.v.*

hibernation *n.* the condition of passing the winter in a resting state of deep sleep, when metabolic rate and body temperature drop considerably. Only a few small mammals, e.g. some rodents, hedgehogs, bats, and other small insectivores, undergo a "true" hibernation. Obligate hibernators enter hibernation spontaneously as the result of a circannual behavioural rhythm. Facultative hibernators enter hibernation when food becomes scarce and temperatures drop below a certain level. Related conditions include winter torpor in reptiles and winter lethargy in larger mammals, e.g. bears, badgers, skunks and racoons. *see also* aestivation.

hidrosis *n.* sweating, perspiration.

hiemal *a. pert.* winter, *appl.* aspect of a community.

hiemilignosa *n.* monsoon forest composed of small-leaved trees and shrubs which shed their leaves in the dry season.

hierarchy *n. see* dominance systems; a natural classification system in which organisms are grouped according to the number of characteristics they have in common and ranked one above another.

high density lipoprotein (HDL) any of a group of lipoproteins found in blood plasma, synthesized in the liver, rich in phospholipids and cholesterol and which amongst other things transport cholesterol from liver to peripheral tissues.

high-energy bond a somewhat misleading term denoting any chemical linkage whose breakage releases a large amount of free energy, as the bond between the 2 last phosphate groups in ATP.

hilar *a.* of or *pert.* a hilum.

hiliferous *a.* having a hilum.

Hill coefficient (n$_H$) a number obtained from equilibrium binding experiments that gives information on the number of binding sites for a ligand present on a protein and about whether they show cooperativity.

Hill reaction the reaction showing that isolated chloroplasts could, on illumination, cause the reduction of suitable electron acceptors such as ferricyanide to ferrocyanide and generate oxygen, first demonstrated by Robert Hill in 1939.

hilum *n.* (*bot.*) scar on ovule and on seed where it was attached to ovary; nucleus of a starch grain; (*zool.*) small notch, opening or depression, usually where vessels, nerves, etc. enter, of kidney, lung, spleen, etc.

hilus hilum *q.v.*

hindbrain *n.* cerebellum, pons and medulla oblongata, concerned with basic body activities independent of conscious control such as regulation of muscle tone, posture, heartbeat, respiration and blood pressure. *alt.* rhombencephalon.

hind-gut an outgrowth of the yolk sac extending into tail-fold in human embryo; posterior portion of alimentary tract; proctodaeum *q.v.*

hind-kidney metanephros *q.v.*

hinge cells large epidermal cells which, by changes in turgor, control rolling and unrolling of a leaf.

hinge ligament tough elastic substance that joins the two parts of a bivalve shell.

hinge tooth one of the projections found on the hinge line, or line of articulation, of a bivalve shell.

hinoid *a.* with parallel veins at right angles to midrib, *appl.* leaves.

hip *n.* common name for the type of pome fruit produced by some members of the Rosaceae (e.g. roses); region of articulation of vertebrate hindleg with trunk, *see also* coxa.

hip girdle pelvic girdle *q.v.*

hippocampal commissure tract of fibres connecting the two cerebral hemispheres in the region of the hippocampal areas.

hippocampus *n.* area in centre of brain lying around the thalamus and just above the corpus callosum, showing well-organized neuronal structure, function still uncertain but possibly involved in spatial and cognitive mapping of the physical environment and/or response to novel stimuli. Damage to the hippocampus is associated with amnesia.

hippomorphs *n.plu.* group of the Perissodactyla including the extinct brontotheres and the horses (family Equidae).

hippuric acid benzoyl glycine, a constituent of the urine of herbivorous animals.

Hippuridales *n.* order of dicots, land, marsh or water plants, comprising the families Gunneraceae (gunnera), Haloragaceae and Hippuridaceae (mare's tail).

hirsute *a. appl.* birds, covered with hair-like feathers; having stiff, hairy bristles or covering.

hirsutidin *n.* a blue anthocyanin pigment.

hirudin *n.* protein obtained from buccal secretions of leech, which inhibits action of thrombin on fibrinogen, preventing clotting of blood.

Hirudinea *n.* class of carnivorous or ectoparasitic annelids, commonly called leeches, which have 33 segments, cirumoral and posterior suckers and usually no chaetae.

His' bundle band of muscle fibres, with nerve fibres, connecting auricles and ventricles of heart. *alt.* atrioventricular bundle.

hispid *a.* having stiff hairs, spines or bristles.

histamine *n.* a powerful vasodilator, causing blood vessels to dilate and become leaky, synthesized from histidine by decarboxylation, produced by mast cells and others, and responsible for many of the symptoms of allergies such as hayfever. Has also been identified as a neurotransmitter in the central nervous system.

histidine (His, H) *n.* amino acid, constituent of protein, probably non-essential in the human diet, precursor of histamine.

histioblast *n.* formerly used for a precursor of monocytes or macrophages; an embryonic cell of sponges.

histiocyte *n.* a reticuloendothelial cell; a fixed macrophage in loose connective tissue; formerly used for a precursor of monocytes and macrophages.

histiogenic histogenic *q.v.*

histioid *a.* like a web.

histiotypic *a. appl.* uncontrolled growth of cells in tissue culture. *cf.* organotypic.

histoblast nests small groups of cells in abdomen of dipteran larvae which develop into the adult epidermal structures of the abdomen at metamorphosis.

histochemistry *n.* the chemistry of living cells and tissues, esp. in relation to their staining properties.

histocompatibility antigens highly variable glycoproteins present on the surfaces of almost all body cells and which are involved in antigen recognition generally. The particular set of histocompatibility antigens displayed by any individual is known as their tissue type. The major histocompatibility antigens (*q.v.*) are encoded by the major histocompatibility complex (MHC). Reactions against different histocompatibility antigens on tissues and organs grafted from another individual are responsible for transplant rejection. *alt.* transplantation antigens, MHC antigens, HLA antigens (in humans), H-2 antigens (in mice).

histocyte *n.* formerly used term to distinguish a differentiated tissue cell as opposed to a germ cell.

histogen *n.* zone of tissue in apical meristems in plants from which new tissue develops.

histogenesis *n.* development of tissues.

histogenic *a.* producing tissues.

histoid histioid *q.v.*

histology *n.* the study of the detailed structure of living tissue.

histolysis *n.* the dissolution of tissues.

histomorphology histology *q.v.*

histone *n.* any one of a set of simple basic proteins (H1, H2A, H2B, H3, H4), rich in arginine and lysine, bound to DNA in eukaryote chromosomes to form nucleosomes *q.v.*

histotrophic *a. pert.* or connected with tissue formation or repair.

histozoic *a.* living within tissue, *appl.* the trophozoite stage of certain sporozoan parasites.

hives urticaria *q.v.*

HMG proteins the high-mobility (on electrophoresis) group of nonhistone proteins in chromatin.

hoary *n.* greyish white, having a frosted appearance.

hock *n.* in horses and other hoofed mammals, the joint on hindleg corresponding to the tarsal joint.

Hogness box TATA box *q.v.*

holandric *a.* transmitted from male to male through Y chromosomes, *appl.* sex-linked characters.

holandry *n.* having the full number of testes, as two pairs in oligochaete worms.

Holarctic *a.* (*zool.*) *appl.* or *pert.* Holarctica *q.v.*; (*bot.*) *appl.* or *pert.* the floral realm consisting of the Northern Hemisphere south to the Tropic of Cancer and consisting of eight floral regions: the Arctic, the Atlantic North American, the Euro-Siberian, the Hudsonian, the Irano-Turanian, the Mediterranean, the Pacific North American, and the Sino-Japanese.

Holarctica *n.* zoogeographical region comprising the Nearctic and Palaearctic Regions.

holcodont *a.* having the teeth in a long continuous groove.

holdfast appressorium *q.v.*

holistic *a. appl.* explanations that attempt to explain complex phenomena in terms of the properties of the system as a whole. *n.* holism. *cf.* reductionist.

Holliday junction the structure formed between two double-stranded DNAs at the point of crossing-over during recombination.

Holobasidiomycetidae Homobasidiomycetae *q.v.*

holobenthic *a.* living on sea bottom or in depths of sea throughout life.

holoblastic *a. pert.* eggs with total cleavage.

holobranch *n.* a gill in which gill filaments are borne on both sides.

holocarpic *a.* having fruit body formed by entire thallus, *appl.* certain algae; *appl.* parasitic fungi without rhizoids or haustoria, living in host cell.

Holocene *n.* recent geological epoch following Pleistocene, began *ca.* 10,000 years ago. *alt.* Recent.

holocentric *a. appl.* chromosomes having a "diffuse centromere" so that when fragmented each part of the chromosome behaves at mitosis as though it possesses a centromere.

holocephalian *a. pert.* cartilaginous fishes of the subclass Holocephali, the rabbit fishes, with crushing teeth, a whip-like tail and an operculum covering the gills. *alt.* chimaeras, rat-fish.

holocephalous *a. appl.* a rib with a single head.

holochroal *a.* having eyes with globular or biconvex lenses closely crowded together, so that cornea is continuous over whole eye.

holocrine *a. appl.* glands whose secretion is accompanied by complete breakdown of the secretory cells, as sebaceous glands.

holocyclic *a. pert.* or completing alternation of sexual or parthenogenetic generations.

holoenzyme *n.* complete, fully functional enzyme molecule, consisting of a protein portion (apoenzyme), a non-protein prosthetic group(s) or any other regulatory or accessory protein subunit if appropriate.

hologamodeme *n.* group of freely interbreeding individuals of the same taxon in a local area.

hologamy *n.* condition of having gametes similar to somatic cells; fusion between mature individuals as in some protozoans.

holognathous *a.* having jaw in a single piece.

hologynic *a.* transmitted directly from female to female, *appl.* sex-linked characters.

holomastigote *a.* having one type of flagellum scattered evenly over the body.

holometabolous *a. appl.* the orders of insects that undergo a full metamorphosis, with a four-stage life history (egg, larva, pupa, adult), and comprising the Neuroptera (alderflies, lacewings), Mecoptera (scorpion flies), Trichoptera (caddis flies), Lepidoptera (butterflies and moths), Coleoptera (beetles), Strepsiptera, Hymenoptera (ants, bees and wasps), Diptera (two-winged flies), and Siphonaptera (fleas).

holomorphosis *n.* regeneration in which the entire part is replaced. *a.* holomorphic.

holoparasite *n.* parasite that cannot exist independently of its host, or on a dead host.

holophyte *n.* any green, phototrophic, independent plant.

holophytic *a.* autotrophic *q.v.*; sometimes only used for phototrophic *q.v.*

holoplankton *n.* organisms that complete their life cycle in the plankton.

holopneustic *a.* with all spiracles open for respiration.

holoptic *a.* with eyes touching or almost touching on top of head.

holoschisis *n.* division of the nucleus by constriction without the formation of chromosomes or a spindle and without the breakdown of the nuclear membrane.

holosericeous *a.* completely covered with silky hairs; having a silky lustre or sheen.

holostean, holosteous *a.* having a bony skeleton, *appl.* fishes.

Holostei, holosteans *n., n.plu.* group of bony fishes present from the Mesozoic but now represented only by the garpike and bowfin.

holostomatous *a.* with mouth of aperture entire.

holostylic *a. appl.* type of jaw suspension in which the palatoquadrate is fused with the cranium without involving the hyoid arch, typical of rabbit-fishes.

holosystolic *a. pert.* a complete systole.

Holothuroidea, holothurians *n., n.plu.* class of sausage-shaped echinoderms commonly called sea cucumbers having minute skeletal plates embedded in the fleshy body wall.

Holotrichia, holotrichans *n., n.plu.* a group of ciliate protozoans having no obvious zone of composite cilia around the mouth, and swimming by cilia distributed all over the body.

holotype type specimen *q.v.*

holozoic *a.* obtaining food in the manner of animals, by ingesting food material and then digesting it.

holozygote *n.* zygote containing the entire genomes of both uniting cells.

homaxial *a.* built up around equal axes.

home range teritory *q.v.*

homeobox *n.* a nucleotide sequence first identified in homeotic genes in *Drosophila* and also found in other developmental genes in fruitflies and in a wide range of other organisms, including plants and humans. It encodes a DNA-binding sequence, the homeodomain.

homeodomain *n.* DNA-binding protein domain which is encoded by the homeobox sequence, and which is found in many transcriptional regulator proteins involved in development.

homeosis *n.* the transformation of one part into another, as in the modification of antenna into a minute leg in the drosophila mutant *Antennapedia*, or of petal into stamen. *alt.* homoeosis, metamorphy, metamorphosis.

homeostasis *n.* maintenance of the constancy of internal environment of the body or part of body; maintenance of equilibrium between organism and environment; the balance of nature. *a.* homeostatic. *alt.* homoeostasis.

homeotely *n.* evolution from homologous parts, but with less close resemblance.

homeothermic homoiothermic *q.v.*

homeotic homoeotic *q.v.*

homeotic *a. appl.* mutations that transform part of the body into another part, e.g. the mutation *bithorax* in *Drosophila*, in which thoracic segment 3, which normally does not bear wings, is transformed into segment 2, which does, thus producing a four-winged fly; *appl.* genes identified by these mutations, which in insects are now known to specify the identity of individual segments of the body. *alt.* homoeotic.

hominid *n.* a member of a human (*Homo*

spp.) or human-like (*Australopithecus*) species characterized by upright posture and other features distinguishing it from the ape lineage (the pongids).

Hominidae, hominids *n., n.plu.* the family of primates that comprises true humans (*Homo* spp.) and human-like hominids (*Australopithecus* spp.).

hominoid *a.* having similarities to humans, *appl.* African apes and various ape-like fossils, as well as early hominids.

Hominoidea, hominoids *n., n.plu.* the superfamily of primates which includes the families Hominidae (humans and human-like hominids), Pongidae (great apes) and Hylobatidae (gibbons).

homiothermic homoiothermic *q.v.*

Homo the genus of true men, including several extinct forms (*H. habilis, H. erectus, H. neanderthalensis*) and modern man, *H. sapiens*, who are or were primates characterized by completely erect stature, bipedal locomotion, reduced dentition, and above all by an enlarged brain size.

homo- prefix from Gk. *homo*, the same: indicating similarity of structure, from the same source, of similar origins, containing similar components, etc.

homoallelic *a. appl.* allelic mutant genes which have mutations at the same site, so that intragenic recombination cannot yield a functional cistron.

Homobasidiomycetae, homobasidiomycetes *n., n.plu.* basidiomycete fungi producing their basidiospores on typically club-shaped, non-septate basidia (homobasidia or holobasidia). They comprise the mushrooms and toadstools, bracket fungi, coral fungi, puffballs, earthstars, stinkhorns and bird's nest fungi. *alt.* Holobasidiomycetidae. *cf.* Heterobasidiomycetae.

homoblastic *a.* arising from similar cells.

homocarnosine *n.* a dipeptide, Ala-γ-aminobutyric acid, found chiefly in brain.

homocarpous *a.* bearing only one kind of fruit.

homocellular *a.* composed of cells of one type only.

homocercal *a. appl.* to type of tail fin in which vertebral column ends before it and the upper and lower lobes are more-or-less equal.

homochlamydeous *a.* having the outer and inner perianth whorls alike, not distinguishable as calyx and corolla.

homochromous *a.* of one colour, *appl.* florets of a composite flowerhead.

homochronous *a.* occurring at the same age or period, in successive generations.

homocysteine *n.* an amino acid not found in protein, an intermediate in the biosynthesis of methionine.

homocytotropic antibody IgE *q.v.*

homodimer *n.* protein composed of two identical subunits.

homodont *a.* having teeth all alike, not differentiated.

homodromous *a.* having the genetic spiral alike in direction in stem and branches; moving or acting in the same direction.

homoecious *a.* occupying the same host or shelter throughout the life cycle.

homoeo- homeo- *q.v.*

homoeo box homeobox *q.v.*

homoeologous *a.* partly homologous, *appl.* sets of chromosomes in a polyploid species that are similar to but not identical to, and therefore not of the same origin, as the other chromosome sets in the genome.

homoeostasis homeostasis *q.v.*

homoeotic homeotic *q.v.*

homogametic sex the sex possessing a pair of homologous sex chromosomes and therefore producing gametes all of one sex. In mammals it is the female, which is XX. In birds, reptiles and lepidopterans the homogametic sex (ZZ) is the male.

homogamy *n.* inbreeding due to some type of isolation; condition of having flowers all alike; having stamens and pistils mature at same time.

homogangliate *a.* having ganglia symmetrically arranged.

homogenetic *a.* having the same origin.

homogentisate *n.* intermediate compound in the degradation of the amino acids phenylalanine and tyrosine, in the inherited enzyme deficiency alcaptonuria accumulates in urine which turns dark on exposure to air.

homogeny *n.* correspondence between parts due to common descent.

homoiosmotic *a. appl.* organisms with constant internal osmotic pressure.

homoiotherm *n.* a warm-blooded animal.

homoiothermic *a.* warm-blooded; *appl.* animals (birds and mammals) that maintain a more-or-less constant body temperature regardless of external temperature variations. *n.* homoiothermy.

homokaryon *n.* a hypha or mycelium having more than one haploid nucleus of identical genetic constitution.

homokaryotic *a.* having genetically identical nuclei in a multinucleate cell, or in different cells of a hypha.

homolateral *a.* on, or *pert.* the same side.

homolecithal *a. appl.* eggs having little, evenly distributed yolk.

homologous *a.* resembling in structure and origin; *appl.* chromosomes in a diploid cell which contain the same sequence of genes but are derived from different parents; *appl.* genes determining the same character; *appl.* DNAs of identical or very similar base sequence; *appl.* structures having the same phylogenetic origin but not necessarily the same final structure or function, e.g. wings and legs in insects, which are examples of the phenomenon of homology; (*imm.*) allogeneic *q.v.*

homologous recombination recombination between similar DNA sequences; often refers to a technique for targeted gene disruption in which a chromosomal gene is disrupted by the introduction into the cell of a mutant copy of the gene, which then undergoes recombination with the chromosomal gene, replacing it with the mutant copy.

homologue *n.* any structure of similar evolutionary and developmental origin to another structure, but serving different functions.

homology *n.* resemblance by virtue of common descent; refers to structures, DNA sequences and behaviours, even though they may now have different functions in the different taxa in which they occur.

homomallous *a.* curving uniformly to one side, *appl.* leaves.

homomixis *n.* the union of nuclei from the same thallus, as in homothallism.

homomorphic *a.* of similar size and structure.

homomorphism *n.* condition of having perfect flowers of only one type; similarity of larva and adult. *a.* homomorphic.

homomorphosis *n*. having a newly regenerated part like the one removed.

homomultimer *n*. a protein consisting of two or more identical subunits.

homonomous *a. appl.* segmentation into similar segments; following the same stages or processes, as of development or growth.

homonym *n*. a name which has been given to two different species. When a case is discovered the second named species must be renamed.

homo-oligomeric *a. appl.* proteins composed of several identical subunits.

homopetalous *a*. having all the petals alike.

homophilic *a*. binding like-to-like, *appl.* certain cell adhesion molecules that bind to the same molecules on other cells. *cf.* heterophilic.

homophyllous *a*. bearing leaves all of one kind.

homoplast *n*. organism or organ formed from similar cells, as a coenobium.

homoplastic *a*. similar in shape and structure but not origin; *appl.* graft made into another individual of the same species.

homoplasy *n*. resemblance in form or structure between different organs or organisms due to evolution along similar lines rather than common descent. *alt.* homoplasty, convergent evolution.

homopolysaccharide *n*. any polysaccharide made up of only one type of monosaccharide unit.

Homoptera *n*. group of insects, including the plant bugs, aphids, cicadas, and scale insects.

homopterous *a*. having wings alike.

homosequential *a. appl.* species of Diptera with polytene chromosomes that have exactly the same banding pattern.

homoserine *n*. amino acid not found in protein, involved in the biosynthesis of methionine and threonine.

homosporous *a. appl.* plants producing only one type of spore by meiosis, e.g. most mosses and ferns. *n*. homospory.

homostyly *n*. the condition that all flowers of the same species have styles of the same length. *cf.* heterostyly.

homotaxis, homotaxy *n*. similar assemblage or succession of species or types in different regions or strata, not necessarily contemporaneous. *a*. homotaxial.

homothallic *a. appl.* cells, thalli or mycelia of algae or fungi which can undergo sexual reproduction with a similar strain, or a branch of its own mycelium or thallus; *appl.* strains of the yeast *Saccharomyces cerevisiae* in which switches of mating type take place in some individuals (and in which, therefore, conjugation can take place between members of the same strain). *see also* mating type.

homotherm *n*. any homoiothermic animal *q.v.*

homotropous *a*. turned in the same direction; *appl.* ovules having micropyle and chalaza at opposite ends.

homotypic homophilic *q.v.*

homotypy *n*. the equality of structures on both sides of the main axis of body; serial homology, as of successive segments of some animals; reversed symmetry.

homoxylous *a. appl.* wood without xylem vessels and consisting of tracheids.

homozygote *n*. a homozygous organism or cell.

homozygous *a. appl.* diploid organism that has inherited the same allele (of any particular gene) from both parents, i.e. carries identical alleles at the corresponding sites on homologous chromosomes, also *appl.* cells or nuclei from such an organism. *n*. homozygosity.

homunculus *n*. the miniature human foetus supposed to be present in sperm, according to proponents of 18th century preformation theory.

honest behaviour behaviour that conveys the individual's real intentions to another individual.

honey bee generally refers to *Apis mellifera*, the hive bee. *see also* bees.

honey guides nectar guides *q.v.*

honeydew *n*. sugary exudate on leaves of many plants; sweet liquid secreted by aphids.

honey-stomach in some insects, an expansion of the oesophagus in the anterior portion of the abdomen, serving to store ingested liquid which is regurgitated as required.

hookworms *n.plu.* parasitic nematode worms that cause severe disease in humans, and including *Ancylostoma* *duodenale* and *Necator americanus*. Common and widespread in tropical areas, the larvae enter the body through the skin, and the adult worm lives in the intestine, abrading the intestinal walls and eventually causing severe anaemia and general debilitation.

hordaceous *a. pert.* or resembling barley.

hordein *n.* a simple storage protein present in barley grains.

hordeivirus group plant virus group containing rigid rod-shaped single-stranded RNA viruses, type member barley stripe mosaic virus. They are multicomponent viruses in which three genomic RNAs are encapsidated in different virus particles.

horizon *n.* soil layer of more-or-less well-defined character; a layer of deposit characterized by definite fossil species and formed at a definite time.

horizontal cell type of nerve cell in retina, forming a layer with bipolar and amacrine cells and making lateral connections.

hormogonium *n.* cyanobacterial filament between two heterocysts, which propagates a new organism when it breaks away.

hormone *n.* a substance that is produced by one tissue and transported to another tissue where it induces a specific physiological response; animal hormones are produced by ductless glands of the endocrine system and pass into and travel in the blood; plant hormones include auxins, gibberellins and cytokinins.

horn *n.* the hollow processes on the head of many ruminants, consisting of layers of keratinized epidermis laid down on a bony base; any projection resembling a horn; anterior part of each uterus when posterior parts are united; a tuft of ear feathers in owls; a spine in fishes; a tentacle in snails; cornu *q.v.*; keratin *q.v.*

horn core the bony core of a hollow horn of ruminants, fusing with frontal bone, and over which the keratinized tissue is laid down.

hornworts *n.plu.* common name for members of the plant division Anthocerophyta *q.v.*

horny corals another name for the gorgonians *q.v.*

horological *a. appl.* flowers opening and closing at a particular time of day and night.

horotelic *a.* evolving at a standard rate.

horsehair worms Nematomorpha *q.v.*

horseshoe crabs common name for the Xiphosura, also called king crabs, a group of aquatic arthropods with affinities with the arachnids rather than the crustaceans, and often placed in the separate class Merostomata. They have a heavily chitinized body with the cephalothorax covered by a horseshoe-shaped carapace.

horsetails common name for the Sphenophyta *q.v.*

host *n.* any organism in which another spends part or all of its life, and from which it derives nourishment or gets protection; the recipient of grafted or transplanted tissue.

host-induced modification of enveloped viruses, the incorporation of host cell membrane material into the envelope, causing differences in the physical properties of virions propagated in different types of cell.

hotspot *n.* region of a gene at which mutations preferentially occur; site in the chromosome at which mutation or recombination is markedly increased.

housekeeping genes genes that are expressed in most cell types and which are concerned with basic metabolic activities common to all cells.

H strand heavy strand of DNA (esp. *pert.* mammalian mitochondrial DNA which can be separated into H and L (light) strands on the basis of their density).

H substance complex carbohydrate antigen on red blood cells, the unmodified form of the basic ABO blood group antigen, found in persons of blood group O.

Hudsonian Floral Region part of the Holarctic Realm comprising North America from southern Alaska and the northern shores of the Great Lakes north to northern Alaska and Labrador.

human chorionic gonadotropin (HCG) protein hormone produced by the developing conceptus and placenta after implantation and which is involved in maintenance of pregnancy.

Human Genome Project international project with the aim of mapping and eventually sequencing the complete human genome.

human immunodeficiency virus (HIV) a retrovirus transmitted maternally or by transfer of body fluids (sexually or by transfusion of infected blood), which infects a subtype of T lymphocyte, leading to their eventual depletion and resulting in the severe and eventually fatal immunodeficiency, acquired immune deficiency syndrome (AIDS).

humanized antibodies antibodies constructed by genetic engineering in which a desired antigen-binding site from a mouse antibody is inserted into a human antibody framework.

humeral *a. pert.* shoulder region.

humerus *n.* the bone of the upper arm, or upper part of vertebrate forelimb.

humic *a. pert.* or derived from humus.

humic acid fraction that precipitates from a solution of humus in weak alkali on addition of acid.

humicolous *a.* living in the soil. *n.* humicole.

humification *n.* the production of humus in the soil by the action of microorganisms on plant and animal residues.

humin *n.* black insoluble residue left when humus is dissolved in dilute alkali.

humor *n.* any body fluid, nowadays chiefly used in connection with the fluids of the eye, *see* aqueous humor, vitreous humor.

humoral *a. appl.* immunity mediated by antibodies in the circulation; *appl.* antibodies circulating in blood and lymph.

humulone *n.* a bitter compound obtained from hops.

humus *n.* black organic material of complex composition which is the end-product of the microbial breakdown of plant and animal residues in the soil.

Huntington's chorea autosomal dominant genetic disease characterized by the onset of mental and physical deterioration in middle age.

Huxley's layer the middle layer of polyhedral cells in the inner root sheath of hair follicle.

hyaline *a.* clear; transparent; free from inclusions; *appl.* cartilage of smooth glassy appearance, lacking obvious fibres.

hyalocyte *n.* cell secreting the vitreous humor of eye.

hyaloid *a.* glassy; transparent.

hyaloid artery central artery of retina running through hyaloid canal to back of lens, in foetal eye.

hyaloid canal canal running through vitreous body of eye, from optic nerve to back of lens.

hyaloid fossa anterior concavity of vitreous body in the eye, receptacle of lens.

hyaloid membrane delicate membrane enveloping vitreous body of the eye.

hyaloplasm *n.* old term for the ground substance of cytoplasm in which organelles are found, now more usually known as the cytosol; (*bot.*) outer cytoplasm in plant cells.

hyalopterous *a.* having transparent wings.

hyalospore *n.* transparent unicellular spore in some fungi.

hyaluronate (hyaluronic acid) any of a group of viscous high molecular weight polymers of *N*-acetylglucosamine and glucuronic acid, abundant in connective and other tissues, which act as lubricating agents in synovial fluid and form the cementing substance between animal cells.

hyaluronidase enzyme that increases tissue permeability by reducing the viscosity of hyaluronate. EC 3.2.1.36, *r.n.* hyaluronoglucuronidase.

H-Y antigen antigen specified by a Y-linked gene, appearing early in embryonic development in human males and thought to be important in sex determination.

hybrid *n.* any cross-bred animal or plant; heterozygote *q.v.*; any macromolecule (esp. DNA) composed of 2 or more portions of different origins. *v.* hybridize. *a.* hybrid.

hybrid-arrested translation technique that identifies the cDNA corresponding to an mRNA by its ability to pair with the mRNA *in vitro* to inhibit translation.

hybrid cell cell formed by fusion of cells from two different species in which the chromosomes are contained in a single large nucleus. *cf.* heterokaryon.

hybrid cline the serial arrangement of characters or forms produced by crossing species.

hybrid DNA heteroduplex DNA *q.v.*; DNA molecule composed of segments of different origin, as in recombinant DNA. *alt.*

chimeric DNA.

hybrid dysgenesis the production of sterile progeny showing chromosomal abnormalities, mutations etc., on crossing of certain strains of the fruitfly *Drosophila melanogaster*, and which may involve either the I-R system or the P-M system. Dysgenesis is seen in crosses of I males with R females and in crosses of P males with M females but not vice versa. *see* P factors, M cytotype.

hybrid sterility sterility in an individual arising from the fact that it is a hybrid.

hybrid swarms populations consisting of descendants of species hybrids, as at borders between geographical areas populated by these species.

hybrid vigour the phenomenon often seen in crosses between two pure-bred lines of plants, that the hybrid is more vigorous than either of its parents, presumably owing to increased heterozygosity. *see also* overdominance.

hybrid zone a geographic area in which two populations once separated by a geographical barrier hybridize after the barrier has broken down and in the absence of reproductive isolation.

hybridization *n.* formation of a hybrid (*q.v.*); state of being hybridized; cross-fertilization. *see also* DNA hybridization.

hybridoma *n.* an artificially produced hybrid cell line producing monoclonal antibodies, formed by fusion of a single antibody-producing B cell from the spleen with a myeloma cell. The resulting cell can both multiply indefinitely in culture and produce antibodies and is used to produce monoclonal antibodies for research and medical diagnostic procedures.

hydathode *n.* an epidermal structure in plants specialized for secretion, or for exudation, of water.

hydatid *n.* any vesicle or sac filled with clear watery fluid; sac containing encysted stages of larval tapeworms; vestige of Mullerian duct constituting appendix of testis.

hydatiform *a.* resembling a hydatid.

hydatiform mole cyst-like growth arising in uterus from implantation of abnormal embryo.

hydranth *n.* an individual specialized for feeding in a hydrozoan colony.

hydrarch succession hydrosere *q.v.*

hydratase *n.* enzyme catalysing the hydration of a compound, i.e. the acceptance by it of a molecule of water without splitting it, and the reverse reaction. EC sub-subgroup 4.2.1. *r.n.* hydro-lyase.

hydric *a.* having an abundant supply of moisture.

hydrobiology *n.* the study of aquatic plants and animals and their environment.

hydrobiont *n.* an organism living mainly in water.

hydrocarbon *n.* a chemical compound composed of hydrogen and carbon only.

hydrocarpic *a. appl.* aquatic plants having flowers that are fertilized out of the water but submerged for development of fruit.

hydrocaulus *n.* "stem" and "branches" of a colonial hydroid.

Hydrocharitales *n.* order of aquatic herbaceous monocots comprising the family Hydrocharitaceae (frog's-bit).

hydrochoric *a.* dispersed by water; dependent on water for dissemination. *n.* hydrochory.

hydrocladia *n.plu.* the branches of certain hydrozoan colonies.

hydrocoel *n.* the water vascular system in echinoderms.

hydrocoles *n.plu.* animals living in water or a wet environment.

hydrocortisone *n.* glucocorticosteroid hormone produced by the cortex of the adrenal gland and very similar to cortisone in structure and properties. It has marked effects on carbohydrate metabolism and is an immunosuppressant. *alt.* cortisol.

hydrofuge *a.* water-repelling.

hydrogenase *n.* EC recommended name for ferredoxin:H^+ oxidoreductase (EC 1.18.3.1), an enzyme present in many bacteria, and which can use molecular hydrogen for the reduction of a variety of substances. Other enzymes often termed hydrogenases are hydrogen dehydrogenase (EC 1.12.1.2) and cytochrome C_3 hydrogenase (EC 1.12.2.1).

hydrogen bond weak bond in which a hydrogen atom is shared by 2 other atoms, an important bond in many large biological molecules in stabilizing secondary and tertiary structure and in the binding of substrate to enzyme.

hydroid *n.* (*bot.*) empty water-conducting cell, joined with others to form a strand of water-conducting tissue in the stems of many mosses; (*zool.*) one of the forms of individuals in the Hydrozoa, a class of solitary and colonial coelenterates, having a hollow cylindrical body closed at one end and with a mouth at the other surrounded by tentacles, *alt.* polyp.

hydrolase *n.* any enzyme that catalyses a hydrolysis. EC group 3.

hydrological cycle water cycle *q.v.*

hydro-lyase hydratase *q.v.*

hydrolysis *n.* the addition of the hydrogen and hydroxyl ions of water to a molecule, with its consequent splitting into two or more simpler molecules.

hydrolytic *a.* *pert.* or causing hydrolysis.

hydrome *n.* any tissue that conducts water.

hydromesophyte *n.* aquatic plant of temperate climates.

hydromorphic *a.* *appl.* soils containing excess water.

hydronasty *n.* plant movement induced by changes in atmospheric humidity.

hydropathy plot analysis of a protein sequence to determine regions of hydrophobic amino acids, which, in a membrane protein, may indicate transmembrane regions.

hydrophilic *a.* water-attracting or attracted to water, as polar groups on compounds such as lipids, proteins etc.

hydrophilous *a.* pollinated by the agency of water.

hydrophily *n.* pollination by water.

hydrophobic *a.* water-repelling or repelled by water, as non-polar groups on lipids, proteins etc. which tend to aggregate, excluding water from between them.

hydrophoric *a.* carrying water, *appl.* canal: the stone canal in echinoderms.

hydrophyllium *n.* one of leaf-like transparent bodies arising above and partly covering the sporosacs in a siphonophore.

hydrophyte *n.* an aquatic plant living on or in the water; acquatic perennial herbaceous plant in which the perennating parts lie in water.

hydrophyton *n.* a complete hydrozoan colony.

hydroplanula *n.* stages between planula and

actinula in larval stages of coelenterates.

hydropolyp *n.* a polyp of a hydrozoan colony.

hydroponics *n.* cultivation of plants without soil in nutrient-rich water, which is usually irrigated over some inert medium such as sand.

hydropote *n.* a cell or cell group, in some submerged leaves, easily permeable by water and salts.

hydropyle *n.* a specialized area in cuticular membrane of some insect embryos, for passage of water.

hydrorhiza *n.* branching root-like foot of a hydroid colony, which attaches it to the substratum.

hydrosere *n.* a plant succession originating in a wet environment.

hydrosinus *n.* an extension of the mouth cavity in some cyclostomes.

hydrosoma, hydrosome *n.* the conspicuously hydra-like stage in a coelenterate life history.

hydrosphere *n.* the portion of the planet which is water, i.e. the oceans, rivers, lakes, streams, etc., and including soil water.

hydrospire *n.* long pouches running at the side of the ambulacral grooves and acting as respiratory structures in certain echinoderms.

hydrospore *n.* a zoospore when moving in water.

hydrostatic *a.* *appl.* organs of flotation, as air sacs in aquatic larvae of insects.

hydrostome *n.* the mouth of a hydroid polyp.

hydrotaxis *n.* a movement or locomotion in response to the stimulus of water.

hydrotheca *n.* cup-like extension of perisarc around individual polyps in some colonial hydrozoans, into which the polyp may withdraw.

hydrothermal vent community community of organisms living around volcanic vents in the sea floor at great depths. Primary producers include chemoautotrophic sulphide-oxidizing bacteria, which use the energy of sulphide oxidation to fix CO_2. These include free-living species (e.g. *Beggiatoa*) and intracellular symbionts living in giant vestimentiferan tube worms such as *Riftia pachyptila*.

hydrotropic *a.* *appl.* curvature of a plant

organ towards a greater degree of moisture.

hydroxy(l)apatite *n.* hydrated calcium phosphate $(Ca_{10}(PO_4)_6(OH)_2)$, major constituent of inorganic phase of bone.

3-hydroxybutyrate a ketone body, formed by reduction of acetoacetate, can act as a substrate for cellular respiration.

hydroxycobalamin(e) vitamin B_{12b}. *see* cobalamine.

hydroxylysine *n.* hydroxylated derivative of lysine, modified after lysine incorporation into a polypeptide chain, found in collagen.

hydroxyproline (Hyp) *n.* hydroxylated derivative of the amino acid proline, modified after incorporation into the protein chain, found in collagen.

5-hydroxytryptamine (5-HT) amine identified as a neurotransmitter in mammalian and other central nervous systems, the cell bodies of neurones containing 5-HT (5-hydroxytryptaminergic neurones) being concentrated in the raphe nuclei of the midbrain in mammals and 5-HT pathways being implicated inter alia in regulation of wakefulness and pain sensation. Also produced by platelets, mast cells and others and causes constriction of blood vessels by stimulating contraction of smooth muscle. *alt.* serotonin. *see* Appendix 1 (33) for chemical structure.

5-hydroxytryptaminergic serotoninergic *q.v.*

Hydrozoa, hydrozoans *n., n.plu.* class of coelenterates with two body forms, hydroid (polyp) and medusa, generally occurring as different stages of the life cycle. They include solitary forms such as *Hydra*, branching colonial forms, and the siphonophores such as the Portuguese Man o' War which are colonies of several different types of modified polyps and medusae.

hygric *a.* humid; tolerating, or adapted to humid conditions.

hygrochasy *n.* dehiscence of fruits when induced by moisture.

hygrokinesis *n.* movement induced by a change in humidity.

hygromesophyte *n.* plant of temperate climates that lives in water but is not aquatic.

hygromorphic *a.* structurally adapted to a moist habitat.

hygropetric *a. appl.* fauna of submerged rocks.

hygrophanous *a.* as if impregnated with water.

hygrophilic, hygrophilous *a.* inhabiting moist or marshy places.

hygrophyte *n.* plant that thrives in plentiful moisture, but is not aquatic.

hygroreceptor *n.* a specialized cell or structure sensitive to humidity.

hygroscopic *a.* sensitive to moisture; absorbing water.

hygrotaxis *n.* movement in response to moisture or humidity.

hygrotropism *n.* a plant growth movement in response to moisture or humidity.

hylea *n.* the primaeval forest, esp. tropical.

hylophagous *a.* wood-eating, *appl.* certain insects.

hylophyte *n.* a fungus growing in wood.

hylotomous *a.* wood-cutting, *appl.* certain insects.

hymen *n.* thin fold of mucous membrane at mouth of vagina.

hymeniferous *a.* having a hymenium.

hymenium *n.* in ascomycete and basidiomycete fungi, the distinct layer of spore-bearing structures, asci and basidia respectively, often interspersed with barren cells (paraphyses). *a.* hymenial. *plu.* hymenia.

Hymenomycetes *n.* a group of basidiomycete fungi bearing their basidia in a well-defined layer (hymenium) which becomes exposed while the basidia are still immature. It comprises, in modern classifications, mushrooms and toadstools and bracket fungi.

Hymenoptera, hymenopterans *n., n.plu.* order of insects, including solitary and social species, comprising the ants, bees and wasps. They have two pairs of wings, and many have a pronounced waist between second and third abdominal segments. Males are haploid and females diploid, males developing from unfertilized ova. In colonial forms, a colony usually contains one reproductive female (the queen), sterile female workers, a few reproductive males, and (in ants) sterile soldiers.

hyobranchial *a. pert.* to the hyoid and branchial arches.

hyoepiglottic *a*. connecting hyoid and epiglottis.

hyoglossus *n*. an extrinsic muscle of the tongue, arising from greater cornu of hyoid bone.

hyoid *a. pert.* or designating a bone or series of bones lying at base of tongue and developed from hyoid arch of vertebrate embryos.

hyoid arch the most anterior of the gill arches in fishes.

hyoideus *n*. a nerve supplying mucosa of mouth and muscles of hyoid region.

hyomandibular cartilage the dorsal skeletal element of hyoid arch in fishes.

hyomental *a. pert.* hyoid and chin.

hyoplastron *n*. the 2nd lateral plate in shell of tortoises and turtles.

hyostyly *n*. condition of having jaw articulated with skull by hyomandibular cartilage or corresponding part, as in elasmobranch fishes. *a*. hyostylic.

hyothyroid *a. pert.* hyoid bone and thryoid cartilage of larynx.

hypanthium *n*. in some flowers, a cupshaped extension of the margin of the receptacle to which sepals, petals and stamens are attached.

hypanthodium *n*. an inflorescence with a concave capitulum on whose walls the flowers are arranged.

hypantrum *n*. notch on vertebra of certain reptiles for articulation with a wedge-shaped process (the hyposphene) on neural arch of neighbouring vertebra.

hypapophysis *n*. a ventral process on vertebra.

hyparterial *a*. situated below an artery, *appl.* branches of bronchi below pulmonary artery.

hypaxial *a*. ventral or below vertebral column, *appl.* muscles.

hypercholesterolaemia *n*. raised levels of cholesterol in the blood; the inherited condition familial hypercholesterolaemia, which in homozygotes results in a deficiency of LDL receptors leading to increased blood cholesterol and deposition of cholesterol in nodules in tendons, premature atherosclerosis and childhood coronary artery disease.

hyperchromism *n*. increased absorbance (*q.v.*) as of DNA on melting into separate strands.

hypercoracoid *a. pert.* or designating upper bone at the base of pectoral fin in fishes.

hyperdactyly polydactyly *q.v.*

hyperdiploid *n*. cell or organism which, as a result of a translocation, has more than two copies of a particular chromosome segment.

hyperfeminization *n*. condition of a feminized male with female characteristics exaggerated, as in small size and weight.

hyperglycaemia *n*. excess glucose in blood.

hyperhaploid *n*. cell or organism containing supernumerary chromosomes.

hyperkinetic *a*. over-active.

hypermasculinization *n*. condition of a masculinized female with male characteristics exaggerated, as in large proportions, appearance of male secondary sexual characters.

hypermetamorphosis *n*. kind of insect life history which includes two or more different kinds of larva.

hypermorph *n*. a mutant allele which produces a more exaggerated version of the effect of the wild-type gene.

hypermutation *n*. mutation occurring at a higher rate than the normal for that particular gene or species.

hyperosmotic *a. appl.* a solution of higher osmotic concentration than a given reference solution.

hyperparasite *n*. an organism which is a parasite of, or in, another parasite.

hyperpharyngeal *a*. dorsal to the pharynx.

hyperpituitarism *n*. overactivity of the pituitary gland, resulting in gigantism.

hyperplasia *n*. excessive development due to an increase in the number of cells; an abnormal increase in cell proliferation.

hyperploid *a*. having extra chromosomes; having too many copies of the gene in question.

hyperpnoea *n*. rapid breathing due to insufficient supply of oxygen.

hyperpolarization *n*. an increase in the electrical potential difference across a membrane.

hyperpolyploid *n*. a polyploid cell or organism containing more than the normal number of chromosomes in each of its haploid sets.

hypersensitive *n*. showing an exaggerated

or otherwise unduly sensitive response to a stimulus; in immunology, showing an inappropriate or uncontrolled response of some part of the immune system, as in allergic reactions, anaphylactic shock etc. *n.* hypersensitivity.

hypersensitive response rapid death of plant cells in response to infection with a fungal, bacterial or viral pathogen, a common defence mechanism in plants. It is generally also associated with other defence responses e.g. accumulation of phytoalexins and lignification in neighbouring living cells. *alt.* hypersensitive reaction.

hypersensitivity response in plant pathology, a response made by a plant on infection by a pathogen or on wounding, which involves tissue death and other biochemical changes with the aim of restricting the spread of the pathogen.

hyperstriatum ventralis pars caudalis in brain of birds, an integration centre for auditory and motor information in song control.

hypertely, hypertelia *n.* excessive imitation of colour or pattern, being of problematical utility; overdevelopment of canines of babirusa, an East Indian pig, the male of which has four large tusks.

hyperthermia *n.* a rise in body temperature above normal, which is used adaptively by some animals living in hot climates as a water-conserving mechanism.

hyperthyroidism *n.* overactivity of the thyroid gland, with excess production of thyroid hormone, resulting in increased metabolic rate, high blood pressure, protrusion of the eyeballs, rapid heart rate, thinness and emotional disturbances.

hypertonia *n.* excessive muscle tone.

hypertonic *a.* having a higher osmotic pressure than that of another fluid, that is, if the two solutions were separated by a semipermeable membrane water would flow into the hypertonic solution from the other.

hypertriploid *a. appl.* cells with more than three sets of chromosomes, as many cell lines.

hypertrophic *a. appl.* waters grossly enriched with plant nutrients.

hypertrophy *n.* excessive growth due to increase in the size of the cells. *cf.*

hyperplasia.

hypervariable (hv) regions portions of the variable regions of antibody molecules that are particularly variable in amino acid sequence, 3 such regions being present in each variable region, and which make up the antigen-binding site.

hypha *n.* a tubular filament which is the basic structural unit of a mycelium, the vegetative growth phase of fungi which is produced by hyphal extension and branching. Hyphae may comprise continuous tubes of multinucleate protoplasm or may be partially or completely subdivided along their length by transverse partitions (septa) into uninucleate or binucleate compartments. Fungal hyphae are covered by a rigid cell wall, which in most cases contains chitin as well as or instead of cellulose. The acellular filaments produced by the prokaryotic actinomycetes are also known as hyphae, as sometimes are similar filamentous vegetative structures in some algae. *plu.* hyphae. *a.* hyphal.

hyphopodium *n.* a hyphal branch with enlarged terminal cell or haustorium for attaching the hypha, as in some ascomycetes.

Hyphochytridiomycota, hyphochytrids *n., n.plu.* phylum of freshwater protists commonly known as water moulds and formerly classified as fungi. They are parasitic on algae or fungi or saprobic on plant and insect debris, growing as fine threads and producing motile zoospores with one anterior tinsel flagellum. *alt.* Hyphochytridiomycetes.

hypnobasidium sclerobasidium *q.v.*

hypnody *n.* the long resting period of certain larvae.

hypnogenic *a.* sleep-inducing.

hypo- prefix from the Gk. *hypo*, under. In anatomical terms often denoting situated under, in physiological and biochemical terms denoting a decrease in.

hypoachene *n.* an achene developed from an inferior ovary.

hypobasal *n.* the lower segment of developing ovule, which ultimately gives rise to the root.

hypobasidium *n.* basal part of cell of septate basidium, in which nuclei unite and which gives rise to the epibasidium from which the basidiospores are budded off.

hypobenthos *n.* the fauna of the sea bottom below 1000 m.

hypoblast *n.* the inner germ layer of a gastrula.

hypobranchial *a. pert.* lower or 4th segment of gill arch; *appl.* space under gills in decapod crustaceans.

hypocalcaemic, hypocalcemic *n.* reducing the level of calcium in the blood.

hypocarp *n.* fleshy modified stalk of some fruits, as cashew-apple.

hypocarpogenous *a.* having both flowers and fruit borne underground.

hypocentrum *n.* a transverse cartilage that develops below nerve cord and becomes part of vertebral centrum. *plu.* hypocentra.

hypocercal *a.* having notochord terminating in lower lobe of tail fin.

hypocerebral *a. appl.* ganglion of stomatogastric system in insects, linked to frontal and ventral ganglia, also to corpora cardiaca.

hypochile, hypochilium *n.* in orchid flowers, the inner or basal part of lip when in two distinct parts.

hypochondrium *n.* abdominal region lateral to epigastric and above lumbar. *a.* hypochondriac.

hypochordal *a.* below the notochord.

hypochromicity *n.* the decrease in optical density of a duplex DNA in comparison to the value expected from the optical density of a mixture of its constituent nucleotides in free form, caused by interactions between the stacked bases in the duplex. *alt.* hypochromism.

hypocone *n.* posterior internal cusp of upper molar teeth; the part posterior to girdle in dinoflagellates.

hypoconid *n.* posterior cusp of lower molar on the cheek side.

hypoconule *n.* fifth or distal cusp of upper molar.

hypoconulid *n.* posterior middle cusp of lower molar teeth.

hypocoracoid *a. pert.* lower bone at base of pectoral fin in fishes.

hypocotyl *n.* that portion of stem below cotyledons in plant embryo, which eventually bears the roots.

hypodermis *n.* in leaves, a layer of cells immediately underlying the epidermis; layer of cells, often a syncytium, underlying the cuticle in nematode worms.

hypodiploid *a. appl.* cells with less than a complete diploid set of chromosomes, as many cell lines.

hypogastric *a. pert.* lower abdomen.

hypogastrium *n.* lower central region of abdomen.

hypogeal, hypogean *a.* living or growing underground; *appl.* germination when cotyledons remain underground.

hypogenesis *n.* development without occurrence of alternation of generations.

hypogenous *a.* growing on the undersurface of anything.

hypoglossal nerve 12th cranial nerve, controls muscles of tongue and floor of mouth, in anamniotes it is a spinal nerve.

hypoglottis *n.* the under part of the tongue.

hypoglycaemia *n.* abnormally low levels of glucose in the blood.

hypoglycaemic, hypoglycemic *a. appl.* agents tending to lower blood glucose level, as insulin; *pert.* hypoglycaemia.

hypognathous *a.* having the lower jaw slightly longer than the upper, with mouthparts on the underside, *appl.* insects.

hypogynium *n.* structure supporting ovary in flowers of sedges.

hypogynous *a. appl.* flowers having petals, sepals and stamens attached to the receptacle below the ovary. *n.* hypogyny.

hypohaploid *n.* cell or organism with one or several chromosomes missing from their haploid complement.

hypohyal *n.* a ventral skeletal element of the hyoid arch in fishes.

hypolimnion *n.* the water between the thermocline and bottom of lakes.

hypolithic *a.* found or living under stones.

hypomorph *n.* a mutant allele which behaves in a similar way to the normal gene but has a weaker effect.

hyponasty *n.* the state of growth in a flattened structure when the under side grows more vigorously than the upper.

hyponeural *a. appl.* system of transverse and radial nerves in echinoderms.

hyponeuston *n.* organisms swimming or floating immediately under the water surface.

hyponychium *n.* epidermal layer on which nail rests. *a.* hyponychial.

hyponym *n.* generic name not founded on

a type species; a provisional name for a specimen.

hypo-osmotic *a. appl.* a solution of lower osmotic concentration than a given reference solution.

hypopetalous *a.* having corolla inserted below, and not adjacent to, gynaecium.

hypopharyngeal *a. pert.* or situated below or on lower surface of pharynx.

hypopharynx *n.* a projection from floor of mouth in dipteran insects (e.g. flies) forming part of the food canal.

hypophragm *n.* lid closing opening of shell in some gastropod molluscs.

hypophyllous *a.* located or growing under a leaf.

hypophyseal, hypophysial *a. pert.* the hypophysis.

hypophysectomy *n.* excision or removal of the pituitary gland.

hypophysis pituitary body *q.v.*

hypopituitarism *n.* deficiency of pituitary hormones, resulting in a type of infantilism.

hypoplasia *n.* developmental deficiency; deficient growth.

hypoplastron *n.* the 3rd lateral bony plate in shell of turtles and tortoises.

hypoploid *a.* having too few copies of the gene in question; lacking one or more of the chromosomes of the normal haploid set.

hypopneustic *a.* having a reduced number of spiracles.

hypopolyploid *n.* polyploid cell or organism lacking one or more chromosomes.

hypoptilum *n.* a small tuft of down near base of a feather.

hypopyge, hypopygium *n.* clasping organ of male dipterans.

hyporachis *n.* the stem of aftershaft (hypoptilum) of a feather.

hyporadiolus *n.* a barbule of aftershaft (hypoptilum) of a feather.

hyporadius *n.* a barb of aftershaft (hypoptilum) of a feather.

hyporheic zone zone around a river, esp. those with gravel beds, in which river water and its microflora and fauna extends as ground water throughout the surrounding land.

hyposkeletal *a.* lying beneath or internal to the endoskeleton.

hyposomite *n.* ventral part of body segment, as in certain cephalochordates such as amphioxus.

hyposperm *n.* the lower region of ovule or seed, below the level at which the integument or testa is free from the nucellus.

hyposphene *n.* wedge-shaped process on neural arch of vertebra of certain reptiles, which fits into a notch on next vertebra, in certain reptiles.

hypostasis *n.* case where expression of one gene is suppressed by another non-allelic gene.

hypostomatous *a.* having stomata on under side (of leaf); having mouth placed on lower or ventral side.

hypostatic *a. appl.* a gene whose expression is prevented by another non-allelic gene.

hypostome *n.* conical projection containing mouth in hydrozoans; the fold bounding posterior margin of oral aperture in crustaceans; anterior and ventral part of insect head; lower mouthpart in ticks, used to anchor the animal to the skin while it feeds.

hypostracum *n.* inner primary layer or endocuticle of exoskeleton in ticks and mites.

hypothalamus *n.* region of brain located below thalamus and forming greater part of floor of 3rd ventricle, secreting various peptide hormones, including releasing factors for pituitary hormones. Involved in the control of motivated behaviour such as eating, drinking and sex.

hypothallus *n.* a thin layer under sporangia in slime moulds; undifferentiated hyphal growth or marginal outgrowth in lichens.

hypotheca *n.* younger or inner half of frustule in diatoms.

hypothecium *n.* the dense layer of hyphal threads below the hymenium in certain fungi.

hypothenar *a. pert.* the prominent part of palm of hand below base of little finger.

hypothermia *n.* a drop in body temperature below normal limits, which leads to death if an external heat source is not applied, and which is usually distinguished from the decreased body temperature that occurs in hibernating animals.

hypothyroidism *n.* condition due to underactivity of thyroid gland and deficiency of thyroid hormones, resulting in mental defectiveness and other conditions.

hypotonic *a*. having a lower osmotic pressure than that of another fluid, that is, if the two solutions were separated by a semipermeable membrane water would flow from the hypotonic solution to the other.

hypotrematic *a*. *appl*. the lower lateral bar of branchial basket of cyclostomes.

hypotrichous *a*. having cilia mainly restricted to under surface.

hypotrochanteric *a*. running below the trochanter.

hypotrophy *n*. condition where wood or cortex grows more thickly on underside of a horizontal branch or other organ; the condition where stipules or buds form on the underside.

hypotympanic *a*. situated below the tympanum, *pert*. quadrate bone.

hypoxanthine *n*. 6-oxypurine, the purine base in the ribonucleoside inosine, chiefly in tRNA, similar to adenine, but with the amino group replaced by a hydroxyl group, also a breakdown product of purines during uric acid formation.

hypoxia *n*. low levels of oxygen. *a*. hypoxic.

hypselodont hypsodont *q.v.*

hypsilophodont *a*. having high-crowned teeth with transverse ridges on the cheek-teeth grinding surfaces.

hypsodont *a*. *appl*. molar and premolar teeth with high crowns and short roots, as those of grazing mammals such as horses, cows, etc.

hypural *a*. *pert*. a bony structure formed by fused haemal spines of last few vertebrae, which supports the tail in certain fishes.

hyracoids *n.plu*. a group of placental mammals including the hyrax, which have a rodent-like body and skull, but digits over a pad and bearing nails like elephants.

hysteranthous *a*. coming into leaf after flowering.

hysteresis *n*. a lag in one of two associated processes or phenomena; lag in adjustment of external form to internal stresses.

hysterochroic *a*. gradually discolouring from base to tip, *appl*. ageing fruit bodies.

hysterogenic *a*. of later development or growth.

hysterotely *n*. the retention or manifestation of larval characteristics in pupa or imago, or of pupal characters in imago.

hysterothecium *n*. an apothecium with slits opening in moist conditions and closing in drought, as in certain fungi and lichens.

I

I inosine *q.v.*; isoleucine *q.v.*

Ia immune response associated antigen *see* class II MHC antigen *q.v.*

i⁶Ade ⁶N-isopentenyladenine, a naturally occurring cytokinin in plants, and a constituent of some transfer RNAs.

IAMS International Association of Microbiological Societies.

IAN indole acetonitrile, a naturally occurring auxin.

IAP islet-activating protein *see* pertussis toxin.

IAPT The International Association of Plant Taxonomy.

ICBN International Code of Botanical Nomenclature.

ICM inner cell mass *q.v.*

ICNB International Committee on Nomenclature of Bacteria.

ICNV International Committee on Nomenclature of Viruses.

ICSB International Committee on Systematic Bacteriology.

ICSH interstitial-cell stimulating hormone *see* luteinizing hormone *q.v.*

ICTV International Committee on Taxonomy of Viruses.

ICZN The International Commission on Zoological Nomenclature.

ID$_{50}$ the dose of virus, etc. at which 50% of the test units (e.g. animals, tissue cultures) become infected.

IEP isoelectric point *q.v.*

IF1 initiation factor 1 in bacterial protein synthesis.

IF2 initiation factor 2 in bacterial protein synthesis, involved in initiator tRNA binding.

IF3 initiation factor 3 in bacterial protein synthesis, involved in mRNA binding to 30S ribosomal subunit, and subunit dissociation.

IFN interferon *q.v.* as IFN-α, IFN-β and IFN-γ.

Ig immunoglobin *q.v.*

IgA immunoglobulin A, the type of immunoglobulin produced by certain lymphoid tissues and secreted locally in gut, saliva, tears, milk and colostrum. It occurs as either a monomer or a dimer.

IgD immunoglobulin D, appears as surface bound immunoglobulin on B cells subsequent to and in conjunction with IgM. Only very small amounts are secreted and its function is unknown.

IgE immunoglobin E, class of antibody involved in local inflammatory reactions and reactions to intestinal parasites, chiefly found bound to receptors on mast cells and basophil leukocytes, where binding by specific antigen triggers release of cell contents - histamine, heparin and leukotrienes (SRS-A). Formerly known as reaginic antibody or homocytotropic antibody.

IGF-I somatomedin C *q.v.*

IGF-II insulin-like growth factor II, role *in vivo* uncertain.

IgG imunoglobulin G, the main type of immunoglobulin produced towards the end of a primary immune response and in a secondary response.

IgM immunoglobulin M, the first class of immunoglobulin synthesized in a primary immune response. The IgM secreted by plasma cells is a pentamer of five IgM molecules joined together by a joining polypeptide (J chain).

IL-1, IL-2 etc. interleukins *q.v.*

IL-4 interleukin-4, *see* B-cell stimulatory factor.

Ile isoleucine *q.v.*

IMP inosine monophosphate *q.v.*

InsP$_3$ inositol trisphosphate *q.v.*

ipsp inhibitory postsynaptic potential *q.v.*

IPTG isopropylthiogalactoside, a non-metabolized gratuitous inducer of the *lac* operon in *E. coli.*

IR infrared radiation *q.v.*

IRM innate releasing mechanism *q.v.*

IS1, 2, 3, 4 etc. insertion sequence *q.v.*

IU International Unit *q.v.*

IVF *in vitro* fertilization *q.v.*

IVS intervening sequence *see* intron.

I-band, I-disc the light band at either end of a sarcomere of striated muscle, representing a region of actin filaments only. The I-band is bounded at its outer edge by a dark Z-line, the membrane to which the actin filaments are anchored at one end.

I-cell disease disease involving lysosomes, in which hydrolytic enzymes are secreted into the extracellular fluid instead of being packaged in lysosomes.

I-cells interstitial cells, as in coelenterates.

Ice Age Pleistocene *q.v.*

ichneumon flies insects of the Ichneumonidae, a family of hymenopterans that are parasitoids, laying their eggs in the larvae of other insects, especially butterflies and moths.

ichthyic, ichthyoid *a. pert.*, characteristic of, or resembling fishes.

ichthyodont *n.* a fossil tooth of fish.

ichthyofauna *n.* the fishes of a particular region, area or habitat.

ichthyoid *a. pert.*, characteristic of, or resembling a fish.

ichthyolite *n.* a fossil fish or part of one.

ichthyology *n.* the study of fishes.

ichthyopterygium *n.* the vertebrate limb when it is in the form of a fin.

Ichthyosauria, ichthyosaurs *n., n.plu.* group of Mesozoic aquatic reptiles with spindle-shaped body with fins and fin-like limbs.

ichthyostegalians, ichthyostegids *n.plu.* group of primitive extinct amphibians from the Devonian and Carboniferous, having many fish-like characteristics and sometimes considered to be primitive labyrinthodonts.

iconotype *n.* a representation, drawing or photograph of a type specimen.

icosandrous *a.* of flowers, having 20 or more stamens.

ideal angle in phyllotaxis, the angle between successive leaf insertions on a stem when no leaf would be exactly above any lower leaf, $137°30'28''$.

ideomotor *a. pert.* involuntary movement in response to a mental image.

idiobiology *n.* the study of individual organisms.

idiopathic *a. appl.* diseases or conditions not preceded or occasioned by another, or by a known cause.

idiosoma *n.* the body, prosoma and opisthosoma, of ticks and mites.

idiotope *n.* any one of the antigenic determinants formed by the variable regions of antibodies, the set of such idiotopes carried by a single antibody molecule being its idiotype.

idiotrophic *a.* capable of selecting food.

idiotype *n.* collectively, the unique antigenic determinants of the variable region of an antibody molecule; the total hereditary determinants of an individual.

ilarvirus group plant virus group containing quasi-isometric single-stranded RNA viruses, type member tobacco ringspot virus. They are multicomponent viruses in which four genomic RNAs are encapsidated in three different virus particles.

ileac, ileal *a. pert.* the ileum.

ileocaecal *a. pert.* ileum and caecum.

ileocolic *a. pert.* ileum and colon.

ileum *n.* the lower part of small intestine; anterior end of hind-gut in insects.

iliac *a. pert.* or in region of ilium.

iliacus *n.* muscle stretching from upper part of iliac fossa to side of tendon of posoas major.

ilicium *n.* dorsal spine with modified tip for luring prey of angler fish (Lophiidae).

iliocaudal *a.* connecting ilium and tail, *appl.* muscle.

iliocostal *a.* in region of ilium and ribs, *appl.* muscles.

iliofemoral *a. pert.* ilium and femur, *appl.* ligament.

iliohypogastric *a. pert.* ilium and lower anterior part of abdomen, *appl.* a nerve.

ilio-inguinal *a.* in the region of ilium and groin, *appl.* a nerve.

ilio-ischadic *a. pert.* opening between ilium and ischium when these are fused at both ends.

iliolumbar *a.* in region of ilium and loins.

iliopectineal *a. appl.* an eminence marking the point of union of ilium and pubis.

iliopsoas *n.* the iliacus and psoas major considered as one muscle.

iliotibial *a. appl.* tract of muscle at lower end of thigh.

iliotrochanteric *a.* uniting ilium and trochanter of femus, *appl.* a ligament.

ilium *n.* the dorsal bone in each half of pelvic girdle.

Illicidales *n.* an order of woody dicots comprising shrubs, climbers and small trees, and including the two families Iliaceae (star anise) and Schisandraceae.

illuvial *a. appl.* layer of deposition and accumulation below the alluvial layer in soils. *alt.* B-horizon.

imaginal *a. pert.* an imago.

imaginal discs small sacs of undifferentiated epithelium in body of many insect larvae, which on metamorphosis, produce the adult epidermal structures appropriate to each segment, each disc specifying a single structure, as, e.g., leg disc, antennal disc, genital disc.

imago *n.* the last or adult stage of insect metamorphosis, the perfect insect.

imbibition *n.* the passive uptake of water, esp. by substances such as cellulose and starch, as in uptake of water by seeds before germination.

imbricate *a.* overlapping, of scales, etc.

imbrication lines parallel growth lines of dentine.

imbricational *a.* overlapping, *appl.* layers of enamel deposited on sides of teeth during growth.

imitative *a. appl.* form, habit, colouring, etc. assumed for protection or aggression when it imitates that of another organism.

immaculate *a.* without spots or marks of a different colour.

immarginate *a.* without a distinct margin.

immediate early genes in the response of eukaryotic cells to agents that induce mitosis and cell division, a group of genes that are rapidly and transiently induced and which include, in mammalian cells, the genes c-*myc*, c-*fos*, and c-*jun*, all of which produce proteins involved in transcriptional regulation and which may activate genes involved in stimulating cell division.

immediate hypersensitivity inflammatory immune reaction in skin, induced by antibody and mediated by complement activation, arising shortly after initial antibody-mediated damage.

immigrant species species that migrate into an ecosystem or are introduced accidentally or deliberately by humans.

immobilization *n.* the incorporation of inorganic elements into organic compounds in living tissue, where they are not available for circulation in the environment.

immobilized enzyme in biotechnology, a purified enzyme, or whole cells containing the required enzyme, that is immobilized by attachment to an inert solid matrix to increase the efficiency of enzyme use in industrial processes.

immortalization *n.* production of a permanent cell line from cells isolated directly from tissues.

immortalized *a. appl.* mammalian cells that have become able to continue cell division indefinitely, as the cells of a permanent cell line which has been derived on prolonged culture from a primary tissue culture.

immune *a. appl.* animal that has produced protective antibodies and/or activated T lymphocytes against an invading microorganism and is therefore resistant to infection; in its widest sense *appl.* any animal that has produced antibodies or activated T lymphocytes against a specific antigen.

immune complex a complex of antibody, antigen, and complement. If deposited in blood vessels, etc. activation of the complement pathway results in a hypersensitivity reaction.

immune response a selective response mounted by the immune system of vertebrates in which specific antibodies and/or cytotoxic cells are produced against invading microorganisms, parasites, transplanted tissue and many other substances which are recognized as foreign by the body (antigens). The production of antibodies circulating in the blood is known as a humoral immune response, the production of cytotoxic cells as a cell-mediated or cellular immune response. *see also* allergy, antibody, antigen, B lymphocyte, clonal selection, immune system, immunity, plasma cell, primary immune response, secondary immune response, T lymphocyte.

immune response genes Ir genes *q.v.*

immune system collectively the various cells and tissues in vertebrates which enable them to mount a specific protective response to invading microorganisms, parasites, etc., protecting the body from infection and in some cases setting up a long-lasting specific immunity to reinfection. It is also involved in recognition and rejection of foreign cells and tissues as in organ transplantation and blood transfusion. *see also* antibody, antigen, histocompatibility, immune response, immunity, lymphocyte, lymphoid, primary lymphoid organs, secondary lymphoid tissues.

immune tolerance *see* tolerance.

immunity *n.* resistance to the onset of disease after infection by harmful microorganisms (bacteria, viruses, fungi) or internal parasites. Innate nonspecific immunity is due to mechanical barriers, nonspecific enzymes such as lysozyme, which kills bacteria, etc. Long-lived specific immunity is a result of the action of the immune system and may be acquired naturally by previous infection by the particular microorganism, and can also be induced by vaccination with suitably treated microorganisms or their products. It may be of two types: humoral immunity, which is mediated by specific antibodies in the circulation and which is effective chiefly against bacteria and viruses, and cellular immunity, which is mediated by certain types of cells in the immune system, and is chiefly effective against fungi, parasites, intracellular viral infections, cancer cells and foreign transplanted tissue, *see also* immune response; in bacteria, the inability of a bacterium carrying a prophage or a plasmid to be infected with another phage or plasmid of the same type.

immunization *n.* the administration of an antigen (e.g. by injection) that results in a specific immune response against the antigen. *alt.* inoculation or vaccination, when the antigen is derived from a pathogen and confers immunity against a disease.

immunize *v.* to render resistant to harmful microorganisms, or their toxins, by the induction of specific antibodies and T lymphocytes, usually by administration of killed or otherwise harmless forms of the particular organism (vaccination), or by repeated injections of small amounts of toxin; in the widest sense, to induce the production of specific antibodies and T lymphocytes by injection of a foreign antigen.

immunoassay *n.* any quantitative assay of a substance using its reaction with specific antibodies.

immunocompetence *n.* the capacity to respond to antigen stimulation, the capacity to mount an immune response. *a.* immunocompetent.

immunodeficiency *n.* any deficiency in the ability to mount an effective immune response, and which may be due to various causes, such as the destruction of a class of helper T lymphocytes (in AIDS), the non-production of immunoglobulin due to defects in the immunoglobulin genes, or the non-development of lymphocytes due to various genetic defects.

immunoelectron microscopy electron microscopy where specimens have been stained with electron-dense material (e.g. gold particles) attached to a specific antibody in order to highlight specific structures.

immunoelectrophoresis *n.* technique for analysing antigens in a complex mixture by prior separation of the antigen components by electrophoresis before testing against antibodies by variants of e.g. the Ouchterlony double diffusion technique.

immunofluorescence microscopy optical microscopic technique in which cells are stained with antibodies labelled with immmunofluorescent dyes in order to highlight specific structures within the cell. The specimen is viewed in a special microscope under ultraviolet illumination that makes the dye fluoresce.

immunogen *a.* any substance that causes an immune response. *see also* antigen.

immunogenic *a.* causing an immune response. *see also* antigenic.

immunogenicity *n.* the ability to provoke an immune response, *appl.* to antigens.

immunoglobulin A *see* IgA.

immunoglobulin D *see* IgD.

immunoglobulin E *see* IgE.

immunoglobulin G *see* IgG.

immunoglobulin M *see* IgM.

immunoglobulin genes genes encoding the polypeptide chains of immunoglobulins. They comprise many similar but not identical "gene segments" for parts of the immunoglobulin molecule. DNA rearrangements during lymphocyte development bring together one of each type of gene segment to form a complete heavy or light chain gene. *see* C gene, D gene, J gene, V gene.

immunoglobulin receptor usually refers to the antigen receptor carried on the surface of the B cells of the immune system, which is a transmembrane form of IgM; may occasionally be used to describe a receptor for immunoglobulin, *see* Fc receptor, polyIg receptor.

immunoglobulins (Ig) *n.plu.* highly variable proteins synthesized by the B lymphocytes of the immune system and the plasma cells derived from them. They occur in two forms: as membrane-bound immunoglobulin on the surface of B cells, where they act as the antigen receptors, and as antibodies secreted by plasma cells after the differentiation of B cells into plasma cells during an immune response. Antibodies are found in blood and body secretions. An immunoglobulin molecule consists of two identical light chains (*q.v.*) and two identical larger heavy chains (*q.v.*). Light and heavy chains have a variable N-terminal region which varies from one antibody to another, and a constant C-terminal region which is characteristic of the particular class to which the antibody belongs. Each antigen-binding site is composed of the variable regions of one heavy and one light chain. Some classes of antibodies (IgA and IgM) are composed of more than one immunoglobulin molecule. *see also* antibody, antibody class, B-cell receptor, C region, V region, IgA, IgD, IgE, IgG, IgM, immunoglobulin genes.

immunological memory the ability of the immune system to mount a larger and more rapid response (the secondary immune response) to an antigen already encountered, and which is mediated by long-lived T and B cells activated in the initial immune response (the primary immune response).

immunology *n.* the study of the immune system.

immunosuppression *n.* the (usually temporary) suppression of the ability to mount an immune response, as by radiation and certain drugs which preferentially kill lymphocytes.

immunotoxin *n.* an antitoxin that confers immunity against a disease; an antibody conjugated with a toxin or drug that is specifically targeted to destroy or react with certain cells.

imparidigitate *a.* having an unequal number of digits.

imparipinnate *a.* pinnate with an odd terminal leaflet.

impedicellate *a.* having a very short or no pedicel (stalk).

imperfect *a. appl.* flowers lacking either stamens or carpels.

imperfect fungi Deuteromycetes *q.v.*

imperfect stage in ascomycete fungi, the asexual reproductive phase in their life history in which they bear conidia.

imperforate *a.* not pierced, *appl.* shells of foraminiferans that lack fine pores.

implantation *n.* the embedding of the fertilized ovum in lining of uterus.

impregnation *n.* transfer of sperm from male to body of female.

imprinting *n.* a process usually occurring shortly after birth in animals in which a particular stimulus normally provided by a parent becomes permanently associated with a particular response, as when young birds follow any large moving object, usually the parent bird, or when young mammals follow their mother in response to the smell of her milk. *see also* genomic imprinting, sexual imprinting.

impulse *see* action potential.

in situ in the original place.

***in situ* hybridization** technique for locating the site of a specific DNA sequence on a chromosome, by treating mitotic cells with a radioactive nucleic acid probe exactly complementary for that sequence and which binds to it. Its position can then be detected by autoradiography. The technique is also applied to detect and locate the synthesis of specific mRNAs in tissues.

in vitro *appl.* biological processes and reactions occurring in (i) cells or tissues grown in culture or (ii) in cell extracts or synthetic mixtures of cell components.

in vitro fertilization (IVF) the fertilization of an ovum outside the mother's body, generally followed by replacement of the fertilized egg into the mother or into a pseudopregnant "foster mother" where it develops normally.

in vitro translation or transcription translation or transcription carried out in isolated extracts of cells, or defined systems consisting of the appropriate enzymes and accessory proteins.

in vivo *appl.* biological processes occurring in a living organism.

inactivation *n.* in neurophysiology, decline of sodium current component of action potential.

inantherate *a.* lacking anthers.

inappendiculate *a.* lacking appendages.

Inarticulata *n.* class of brachiopods in which the shells are joined by muscles only and not by a hinged joint. *cf.* Articulata.

inarticulate *a.* not segmented; not jointed.

inborn error of metabolism any of a diverse group of inherited conditions caused by a genetically determined deficiency of a particular enzyme or production of an impaired protein. *alt.* enzyme deficiency disease.

inbreeding *n.* matings between related individuals, *alt.* consanguineous matings; successive crossing between very closely related individuals, in laboratory animals, plants, etc., leading to the establishment of pure-breeding strains or varieties in which individuals are homozygous at a large proportion of their loci, or at selected loci.

inbreeding coefficient (*F*) of an individual, the probability that the pair of alleles carried by the male and female gametes that produced it are identical by descent from a common ancestor, as a result of inbreeding; a measure of the reduction of heterozygosity as a result of inbreeding, given by $F_S = (H_I - H_S)/H_S$, where H_I and H_S are heterozygosity among an inbred and outbred group of individuals of the same population respectively.

inbreeding depression loss of vigour following inbreeding, due to the expression of numbers of deleterious genes in the homozygous state.

incipient lethal level concentration of a toxin at which 50% of the population of test organisms can live for an indefinite time.

incipient species populations that are diverging towards the point of becoming separate species, but which can still interbreed although they are prevented from doing so by some geographical barrier.

incised *a.* with deeply notched margin.

incisiform *a.* shaped like incisors.

incisive *a. pert.* or in region of incisors.

incisors *n.plu.* the cutting front (premaxillary) teeth of mammals.

included *a.* having stamens and pistils not protruding beyond the corolla.

inclusion bodies intracellular particles such as crystals formed of viruses, etc.

inclusive fitness the sum of an individual's own fitness plus all its influence on fitness in its relatives other than direct descendants.

incompatible *a. see* incompatibility; in plant pathology, *appl.* an interaction between host and pathogen that does not result in disease.

incompatibility *n.* genetically determined inability to mate successfully; the rejection of transplanted tissue of a different tissue type; of plasmids, *see* plasmid incompatibility.

incomplete *a. appl.* flowers lacking sepals or petals or stamens or carpels.

incomplete dominance codominance *q.v.*

incomplete metamorphosis insect metamorphosis in which young are hatched in general adult form (but without wings and without mature sexual organs), and develop without a quiescent (pupal) stage. *cf.* complete metamorphosis.

incomplete penetrance the lack of expression of a genetic trait in some individuals possessing the genotype associated with the character.

incongruent *a.* not suitable or fitting, *appl.* surfaces of joints that do not fit properly.

incoordination *n.* want of coordination; irregularity of movement due to loss of muscle control.

incrassate *a.* thickened, becoming thicker.

incubation *n.* the hatching of eggs by means

of heat, natural or artificial; the growth of a bacterial culture, etc., by keeping it for some time at an optimum temperature.

incubation period period between infection and the appearance of symptoms induced by pathogenic bacteria, viruses or other parasites, during which the bacteria, etc. are multiplying.

incubatorium *n.* temporary pouch surrounding mammary area, in which the egg of the echidna, the spiny anteater, is hatched.

incudal *a. pert.* the incus, one of the small bones of the middle ear.

incumbent *a.* lying upon; bent downwards to lie along a base.

incurrent *a.* leading into; afferent; *appl.* ectoderm-lined canals which admit water in sponges; *appl.* inhalant siphons of molluscs; *appl.* ostia in insect heart which admit blood.

incurvate *a.* curved upwards or bent back.

incus *n.* the middle ossicle (small bone) of middle ear in mammals, shaped like an anvil.

indeciduate *a.* not withering away, of sepals on fruit; with the maternal part of the placenta not coming away at birth.

indeciduous *n.* persistent; not falling off at maturity; everlasting; evergreen.

indefinite *n.* not limited in size, number, etc.

indehiscent *a. appl.* fruits which do not open to release the seeds, the whole fruit being shed from the plant; *appl.* spore capsules, sporangia, etc., which do not open to release the spores.

independent assortment the second of Mendel's laws, which describes the fact that each gene will be inherited independently of another, i.e. the gametes may contain any combination of the parental alleles. This law has had to be modified by the subsequent discovery of linkage, i.e. that alleles carried on the same chromosome tend to be inherited together.

indeterminate *a.* indefinite; undefined; not classified.

indeterminate growth *appl.* growth from an apical meristem which forms an unrestricted number of lateral organs such as stems, branches or shoots indefinitely, not stopped by the development of a terminal bud; indefinite prolongation and subdivision of an axis.

indeterminate inflorescence an inflorescence that grows by indeterminate branching because unlimited by the development of a terminal bud.

index *n.* the forefinger or digit next to the thumb; a number or formula expressing ratio of one quantity to another.

index fossil a fossil which typically occurs in a particular zone of a rock stratum and after which the zone is known.

index species an organism that lives only within a narrow range of environmental conditions and whose presence therefore indicates places where those conditions exist.

Indian Floral Region part of the Palaeotropical Realm comprising most of the Indian subcontinent except for the extreme north-west.

indicator species species characteristic of climate, soil and other conditions in a particular region or habitat; the dominant species in a biotype; species whose disappearance or disturbance gives early warning of the degradation of an ecosystem.

indifference curves iso-utility curves *q.v.*

indifferent species species that is not found in any particular community.

indigenous *a.* belonging to the locality; not imported; native.

individual distance the distance around a bird in a flock, which it defends while feeding, etc.

individualism *n.* symbiosis in which the two parties together form what appears to be a single organism.

individuation *n.* the formation of interdependent functional units, as in a colonial organism; process of developing into an individual.

indole acetonitrile (IAN) a naturally occurring auxin *q.v.*

indoleacetic acid (IAA) a naturally occurring auxin *q.v.*

Indo-Malaysian subkingdom a subdivision of the Palaeotropical kingdom comprising the Indian subcontinent and the Himalayas, China, Japan, South-East Asia, the Malaysian archipelago, Indonesia and New Guinea.

induced mutations those occurring as a result of deliberate treatment with a mutagen. *cf.* spontaneous mutations.

induced-fit model model proposed in which substrates induce conformational changes in the active site of an enzyme or other protein on binding to provide a good fit for substrate and enzyme. *cf.* lock–and–key model.

inducer *n.* any compound which specifically causes the synthesis of an enzyme, *cf.* repressor; in bacteria, any small molecule that specifically causes the expression of genes specifying the enzymes required to metabolize it; any chemical or physical stimulus that causes the expression of a specific gene; in embryology, any substance produced by cells which influences neighbouring cells or tissues.

inducible *a. appl.* proteins whose synthesis is stimulated in the presence of a specific agent (the inducer) which may be a chemical compound or a physical stimulus such as heat or light, *cf.* constitutive; *appl.* genes whose transcription is similarly specifically stimulated.

induction *n.* act or process of causing to occur; (*dev.*) process in development whereby a cell or tissue directs neighbouring cells or tissues to develop in a particular way; (*genet.*) the specific synthesis of proteins (used esp. of enzymes) in response to some stimulus, involving the specific activation and transcription of the genes encoding the required proteins; (*neurobiol.*) lowering by one reflex of the threshold of another, spinal induction; (*virol.*) of viruses, the production of infectious particles from a cell carrying a provirus or prophage.

inductive stimulus an external stimulus that influences the growth or behaviour of an organism; an internal stimulus causing the phenomenon of induction in development.

inductor organizer *q.v.*

indumentum *n.* plumage of birds; a hairy covering in plants or animals.

indurated *a.* becoming firmer or harder.

indusial *a.* containing insect larval cases, as certain limestones; *pert.* an indusium.

indusiate *a.* having an enveloping case, *appl.* insect larvae; having an indusium.

indusium *n.* (*bot.*) an outgrowth of epidermis in plants, covering and protecting a sorus, as in ferns; outgrowth hanging from top of stipe in some fungi; (*zool.*) case of a larval insect. *plu.* indusia.

industrial melanics dark-coloured forms of otherwise light-coloured moths and other insects that have increased in industrial areas since the Industrial Revolution, selected by the need for camouflage against soot-blackened walls, trees, etc.

industrial melanism the takeover of the population by a dark or melanistic form of an insect or other animal, compared to the previously preponderant light coloured form, that has occurred in industrial areas. This is due to the darker forms being better camouflaged against sooty walls and trees and therefore being selected for.

induviae *n.plu.* scale leaves.

inequilateral *a.* having two sides unequal; having unequal portions on either side of a line drawn from umbo to gape of a bivalve shell.

inequivalve *a.* having the valves of shell unequal, *appl.* molluscs.

inerm(ous) *a.* without means of defence and offence; without spines.

inert *a.* physiologically inactive; *appl.* regions of heterochromatin, in which genes are not active.

inertia *n.* the ability of a living system to resist being disturbed.

infection *n.* invasion by endoparasites, i.e. bacteria, viruses, fungi, protozoans, etc.

infection thread structure formed by invagination of root hair cell via which the nitrogen-fixing rhizobia enter host tissue.

infectious *a.* capable of causing an invasion; capable of being transmitted from one organism to another.

infectivity *n.* a measure of the strength of infectious virus, bacteria or other parasite in a sample, measured by various assays.

inferior *a.* below; growing or arising below another organ or structure; *appl.* plant ovary having perianth inserted around the top; *appl.* sepals, petals or stamens attached to the top of the ovary.

inferobranchiate *a.* with gills under margin of mantle, as certain molluscs.

inferolateral *a.* below and at or towards the side.

inferomedian *a.* below and about the middle.

inferoposterior *a.* below and behind.

infertile *a.* not fertile; non-reproductive.

infestation *n.* invasion by ectoparasites.

inflammation *n.* painful swelling and redness caused as a response to infection and tissue damage by some microorganisms, or as a result of certain types of immune reactions.

inflected, inflexed *a.* curved or abruptly bent inwards or towards the axis.

inflorescence *n.* flower-head in flowering plants; in mosses and liverworts the area bearing the antheridia and archegonia.

influents *n.plu.* the animals present in a plant community, or those primarily dependent and acting upon the dominant plant species.

infra- prefix derived from Gk. *infra*, below; in classification, denotes a group just below the status of a subgroup of the taxon following it, as in infraclass, the group below the subclass.

infra-axillary *a.* branching off just below the axil.

infrabranchial *a.* below the gills.

infrabuccal *a.* below the cheeks; below the buccal mass in molluscs.

infracentral *a.* below a vertebral centrum.

infraclass *n.* taxonomic grouping between subclass and order, e.g. Eutheria (mammals lacking pouches) and Metatheria (marsupials which bring forth live young) in the subclass Theria.

infraclavicle *n.* membrane bone occurring in pectoral girdle of some fishes.

infraclavicular *a.* below the clavicle.

infracortical *a.* below the cortex.

infracostal *a.* beneath the ribs, *appl.* muscles.

infradentary *a.* below the dentary bone.

infradian *a. appl.* a biological rhythm with a period of less than 24 hours.

infrahyoid *a.* beneath the hyoid, *appl.* muscles.

infralabial *a.* beneath the lower lip.

infralittoral *a. appl.* depth zone of lake which is permanently covered with rooted or floating macroscopic vegetation; upper subdivision of the marine sublittoral zone.

inframarginal *a.* under the margin, or marginal structure.

inframaxillary *a.* beneath maxilla, *appl.* nerves.

infraneuston *n.* the animals living on the underside of the surface film of water, as some mosquito larvae.

infraorbital *a.* beneath the orbit of the eye.

infrapatellar *a. appl.* pad of fat beneath patella.

infrared radiation (IR) electromagnetic radiation with wavelength longer than 780 nm.

infrascapular *a.* beneath the shoulder blade.

infraspecific *a.* occurring within a species, *appl.* variation, competition, etc.; *pert.* a subdivision of a species, as subspecies and varieties.

infrastapedial *a.* beneath the stapes of ear.

infrasternal *a.* below the breast bone.

infratemporal *a.* beneath the temporal bone.

infructescence anthocarp *q.v.*

infundibula *n.plu.* passages surrounded by air cells in the lung; *plu.* of infundibulum.

infundibulum *n.* funnel-shaped organ or structure; the siphon of a cephalopod; outpushing of floor of brain that develops into the hypophysis; the conus arteriosus *q.v..*

infusoria, infusorians *n., n.plu.* term originating in the 19th century for microscopic animals such as protozoans and rotifers. Until quite recently used to denote the ciliate protozoans, but now no longer in general scientific use.

ingesta *n.* sum total of material taken in by ingestion.

ingestion *n.* the swallowing or taking in of food into the gut or food cavity. *v.* ingest.

ingression *n.* the penetration of superficial cells individually into the interior of the embryo.

inguinal *a. pert.* or in region of the groin.

inhalant, inhalent *a.* adapted for breathing in or drawing in water, as terminal pores of sponges and siphons of molluscs.

inheritance *see* heredity, Mendelian inheritance, non-Mendelian inheritance, polygenic inheritance, simple Mendelian traits.

inhibin A, inhibin B polypeptide hormones produced by gonads which inhibit the secretion of follicle-stimulating hormone.

inhibition *n.* prevention or checking of an action, process or biochemical reaction; action of one neurone on another tending to prevent it from generating an impulse.

inhibitor *n*. any agent which checks or prevents an action or process; a substance which reversibly or irreversibly prevents the normal action of an enzyme without destroying the enzyme, competitive inhibitors acting by binding to the active site and preventing binding of substrate, non-competitive inhibitors acting by binding to other parts of the enzyme.

inhibitory *a*. tending to inhibit, *appl.* stimuli, cells, compounds etc.; *appl.* neurones whose activity prevents adjacent neurones from firing; *appl.* neurotransmitters whose action tends to prevent firing; *appl.* synapses at which transmission of a signal tends to prevent firing in the postsynaptic neurone, *cf.* excitatory.

inhibitory postsynaptic potential (ipsp) potential produced in a postsynaptic neurone by the action of neurotransmitter at a synapse and which tends to inhibit the generation of an action potential.

inion *n*. the exernal protuberance of occipital bone.

initial *n*. self-perpetuating cell in a plant meristem, which remains undifferentiated and in the meristem and continues to divide, producing both further initials and sister cells that go on to differentiate.

initiation *n*. in animal behaviour, the response to a stimulus with behaviour that was not present before.

initiation codon the first codon of the coding region in mRNA and the point at which translation starts, usually specifying methionine (in eukaryotes) or formyl-methionine (in bacteria and mitochondria).

initiation factor any of several accessory proteins needed for the initiation of translation and which are involved in mRNA binding to the ribosome, initiator tRNA binding and ribosome subunit association and dissociation. Designated IF1, IF2, etc. in bacteria, eIF1, eIF2, etc. in eukaryotic cells.

ink sac in certain cephalopods such as *Sepia*, a pear-shaped body in wall of the mantle cavity which contains the ink gland, secreting a black substance, ink or sepia, which is ejected as a means of defence.

innate *a*. inborn, *appl.* behaviour which does not need to be learned; *appl.* immunity: non-specific defence mechanisms against disease-causing microorganisms.

innate immunity protection against infection as a result of pre-existing non-specific defence mechanisms. *cf.* acquired immunity.

innate releasing mechanism (IRM) an internal and instinctive mechanism in an animal which is activated to produce a response by some external stimulus.

inner cell mass (ICM) thickening of the inner wall of the mammalian blastocyst which is composed of a mass of cells destined to become the embryo.

inner ear bony cavity in skull behind middle ear, contains a fluid-filled membranous structure, the membranous labyrinth, consisting of 3 semicircular canals (*q.v.*) at right angles to each other which are concerned with balance and sensing position, separated from a spirally-coiled canal, the cochlea (*q.v.*), which is concerned with sensing pitch, by an intermediate portion, the utricle and the saccule.

innervation *n*. the nerve supply to an organ or part.

innominate *a*. nameless.

innominate artery artery that gives rise to arteries in head and forelimb.

innominate bone the hip bone or lateral half of the pelvic girdle.

innominate vein vein that joins up veins from head and forelimb.

inoculation *n*. the administration of a vaccine in order to induce protective immunity; the introduction of bacteria or other microorganisms, or plant and animal cells, into nutrient medium to start a new culture; introduction of pathogen into a host.

inoculum *n*. the cells, bacteria, spores, etc. used for inoculation.

inoperculate *a*. without an operculum or lid, *appl.* spore capsules, etc., *appl.* fish lacking a gill cover.

inordinate *a*. not in any regular arrangement, *appl.* spores in an ascus.

inorganic *a. appl.* material or molecules which do not contain carbon.

inorganic chemistry the chemistry of substances other than those containing carbon.

inosculate *v*. to intercommunicate or unite, as vessels, ducts, etc.

inosine (I) *n*. nucleoside found in tRNA,

formed from adenosine by replacement of the amino group with an oxygen, often present in the first anticodon position where it can pair with U, C or A. The base is hypoxanthine.

inosine monophosphate (inosinic acid, inosinate) (IMP) *n.* a nucleotide, biosynthetic precursor to adenine monophosphate and guanosine monophosphate.

inositol *n.* cyclic hexahydric alcohol, occurring in various forms, of which *myo*-inositol is the most important, a constituent of phospholipids (e.g. phosphatidyl inositol) and also of phytic acid and phytin in plants. *see* Appendix 1 (8) for chemical structure (as inositol trisphosphate).

inositol trisphosphate (InsP₃, IP₃) inositol, 1,4,5-trisphosphate, an intracellular second messenger produced by the action of phospholipase C on membrane phosphatidylinositol phosphate in response to stimulation of cell-surface receptors by specific growth factors, hormones and neurotransmitters. Its chief effect is to stimulate the release of Ca^{2+} from intracellular stores such as the endoplasmic reticulum, thus raising the level of free Ca^{2+} in the cytosol, and indirectly stimulating the appropriate cellular response. *see* Appendix 1 (8) for chemical structure.

inquilinism *n.* type of symbiosis in which an animal lives in the nest of another, is tolerated by the host and shares its food.

Insecta, insects *n.*, *n.plu.* very large class of arthropods (*q.v.*), found as fossils from the Devonian onwards, containing some three-quarters of all known extant species of animal. The insects include flies, bees and wasps, ants, butterflies and moths, beetles, dragonflies, grasshoppers and crickets, and many other orders. The segmented body is divided into distinct head, thorax and abdomen. The head bears one pair of antennae and paired mouthparts, and the thorax bears three pairs of walking legs and usually one or two pairs of wings. Other types of appendage may be present on the abdomen. The life history usually includes metamorphosis. *alt.* Hexapoda. *see* Appendix 4 for orders.

insecticide *n.* chemical that kills insects.

Insectivora, insectivores *n.*, *n.plu.* large order of primitive insect-eating and omnivorous placental mammals known from Cretaceous times to the present, including hedgehogs, moles, shrews.

insectivorous *a.* insect-eating, *appl.* certain animals and carnivorous plants. *alt.* entomophagous. *n.* insectivore.

insemination *n.* the introduction of semen or sperm into female genital tract; transfer of a fertilized ovum from one female to another.

inserted *a.* attached; united by natural growth, as petals to ovary, etc.

insertion *n.* point of attachment of organs, as of muscles, leaves; point on which force of a muscle is applied; mutation in which a segment of chromosome or a short sequence of bases is inserted in a gene.

insertional mutagenesis the production of a mutation as a result of the insertion of a piece of DNA into a gene by *in vitro* gene manipulation.

insertion sequence (IS) a simple type of transposon found in bacteria, consisting of around 800–1500 bp for different insertion sequences and carrying only the genetic functions for its own transposition. Insertion sequences or sequences derived from them are also found at the ends of composite transposons containing genes such as those for antibiotic resistance (e.g. Tn5, Tn10) and are responsible for their transposition.

insessorial *a.* adapted for perching.

insight learning learning involving reasoning, as in humans and to a lesser extent in some animals.

insistent *a. appl.* hind-toe of certain birds, whose tip only reaches the ground.

insolation *n.* exposure to the sun's rays.

inspiration *n.* the act of drawing air or water into the respiratory organs.

instaminate *a.* lacking stamens.

instar *n.* insect or other arthropod at a particular stage between moults.

instinct *n.* behaviour that occurs as an inevitable stereotyped response to an appropriate stimulus, sometimes equivalent to species-specific behaviour. *a.* instinctive.

instipulate *a.* lacking stipules.

instructive *a. appl.* developmental signal that changes the course of development of a tissue, *cf.* permissive.

instrumental learning type of learning in which the animal initially succeeds as a result of trial and error, is rewarded, and eventually learns to perform the task correctly at the first attempt, the correct response being "instrumental" in eliciting the reward.

insula *n.* a triangular eminence lying deep in lateral fissure of temporal lobe of brain; islet of Langerhans *q.v.*; blood island *q.v.*

insulin *n.* polypeptide hormone produced by β-cells of the islets of Langerhans in the pancreas, which decreases the amount of glucose in the blood by promoting glucose uptake by cells and increasing the capacity of the liver to synthesize glycogen. Its action is antagonistic to glucagon, adrenal glucocorticoids and adrenaline, and its deficiency or reduced activity produces diabetes with a raised blood sugar level. It consists of 2 polypeptide chains linked by 2 disulphide bridges. The first protein whose amino acid sequence was determined and one of the first proteins to be produced from genetically engineered bacteria.

insulin-like growth factor I (IGF-I) somatomedin C *q.v.*

insulin-like growth factor II (IGF-II) polypeptide with insulin-like activity but whose role *in vivo* is not yet known.

intectate *a.* without a tectum.

integral *a. appl.* membrane proteins that are firmly embedded in the membrane and difficult to remove.

integrate *n.* a piece of DNA integrated into another DNA.

integrated pest management combined use of biological, chemical and cultivation methods to keep pests at an acceptable level.

integration *n.* in nervous system, the coordination of inputs from many neurones in a single neurone which combines them to produce a new signal for onward transmission; of DNA, the insertion of a viral or other duplex DNA into a chromosome or other replicon to form a covalently linked DNA continuous with the host DNA, which is subsequently replicated as part of the replicon, and which may express some or all of its genes.

integrifolious *a.* with entire leaves.

integrin *n.* any of a large structurally related family of cell-surface protein complexes found in vertebrates and invertebrates which cause adhesion of cells to other cells and to extracellular matrix by binding to receptor proteins on these surfaces.

integument *n.* a covering, investing or coating structure or layer; coat of ovule. *a.* integumentary.

intentional behaviour behaviour that involves a mental representation of a goal that guides the behaviour.

intention movement preparatory motions an animal goes through before a complete behavioural response, e.g. the snarl before the bite.

inter- prefix derived from L. *inter*, between.

interalveolar *a.* among alveoli, *appl.* cell islets.

interambulacral *a. appl.* area of echinoderm test between two ambulacral areas.

interatrial *a. appl.* groove and partition separating the two atria of the heart.

interatrial septum thin transparent sheet of nervous tissue from heart, containing neurones receiving input from the vagus nerve, and innervating the muscles of the atrium.

interauricular *a.* between auricles of heart.

interaxillary *a.* placed between the axils.

interband *n.* in polytene chromosomes, a lightly staining region of variable size alternating with densely staining regions (bands) along the length of the chromosome, each chromosome having a characteristic and invariant banding pattern.

interbrachial *a.* between arms or rays (e.g. of echinoderms).

interbranchial *a. appl.* partitions between successive gill slits.

interbreed *v.* to cross different varieties, species, or genera of plants and animals.

interbreeding *a. appl.* a population whose members can breed successfully with each other.

intercalary *a.* inserted between others; *appl.* growth elsewhere than at growing point; (*bot.*) *appl.* meristematic layers between masses of permanent tissues; (*zool.*) *appl.* veins between main veins of insect wings; *appl.* discs: transverse wavy bands formed by boundaries of sarcomeres in heart muscle. *alt.* intercalated.

intercalation *n*. insertion between two other structures, as of certain mutagens such as acridine dyes, which insert between the base pairs in DNA.

intercarpal *a*. among or between carpal bones, *appl*. joints.

intercarpellary *a*. between the carpels.

intercellular *a*. among or between cells, as spaces, material etc.

intercentral *a*. between two centra.

intercentrum *n*. a 2nd central ring in some types of vertebrae.

interchondral *a*. *appl*. articulations and ligaments between costal cartilages.

interchromosomal *a*. between chromosomes; *appl*. fibrils playing a part in the beginning of cell wall formation in plants.

interclavicle *n*. bone between clavicles of shoulder girdle in most reptiles.

interclavicular *a*. between the clavicles.

intercostal *a*. between the ribs, as arteries, veins, glands, muscles; between veins of a leaf.

intercostobranchial *a*. *appl*. lateral branch of 2nd intercostal nerve which supplies upper arm.

intercropping *n*. growing two types of crop on the same land, usually where one benefits the other, as legumes and cereals.

intercross *n*. the crossing of heterozygotes of the F_1 generation amongst themselves.

interdeferential *a*. between the vasa deferentia.

interdemic selection selection of entire breeding groups (demes) as the basic unit.

interdigitating cells non-lymphocyte cells in lymph nodes that trap foreign antigens and act as antigen-presenting cells in the immune system.

interfascicular *a*. *appl*. parenchyma separating vascular bundles from each other in stems of some conifers and dicotyledons; situated between the vascular bundles, *appl*. cambium.

interfemoral *a*. between the thighs.

interference *n*. in virology, the case where the presence of one type of virus in a cell prevents concurrent infection with, or multiplication of, another virus; in genetic recombination, the effect that the presence of one cross-over reduces the probability of another occurring in its vicinity.

interference colours colours produced by optical interference between reflections from different layers of the surface.

interferons *n.plu*. family of small proteins produced by vertebrate cells in response to viral infection and as cytokines during an immune response. Interferons α and β (fibroblast and leukocyte interferons) are induced by viral infection and have antiviral effects as a result of their inhibition of translation and stimulation of the breakdown of viral nucleic acid within the infected cell. $β_2$-interferon (now called interleukin-6) is structurally unrelated to the other interferons. It is synthesized by fibroblasts and some tumour cells, has some anti-viral activity, and also stimulates immunoglobulin synthesis and secretion from B lymphocytes. Interferon γ (immune interferon) is secreted by activated T lymphocytes during an immune response and is an important growth factor for lymphocytes and other immune system cells.

interfertile *a*. able to interbreed.

interganglionic *a*. connecting two ganglia, as nerve cords.

intergemmal *a*. between taste buds, *appl*. nerve fibres.

intergeneric *a*. between genera, *appl*. hybridization.

intergenic *a*. situated between genes.

interglacial *a*. *appl*. or *pert*. the ages between glacial ages, particularly of the Pleistocene.

interlamellar *a*. *appl*. vertical bars of tissue joining gill lamellae of molluscs; *appl*. compartments of lung book in scorpions and spiders; *appl*. spaces between lamellae or gills of agarics.

interleukins diverse group of proteins produced by activated macrophages and lymphocytes during an immune response and which act on lymphocytes and other leukocytes to stimulate proliferation of activated antigen-specific lymphocytes and thus expand the population of activated antigen-specific B and T lymphocytes. They also have many other effects on lymphocytes and other white blood cells. Abbreviated IL-1, IL-2, IL-3, etc., with more than 13 different interleukins having been identified to date.

interlittoral *a*. shallow marine zone to a

depth of around 20 m.

interlobar *a*. between lobes; *appl*. sulci and fissures dividing cerebral hemispheres into lobes.

intermalar *a*. situated between the cheek bones.

intermaxilla *n*. bone between maxillae; premaxilla *q.v.*

intermediary *a. appl*. nerve cells receiving impulses from afferent cells and transmitting them to efferent cells.

intermediary metabolism metabolic pathways by which the basic molecular building blocks in a cell (e.g. monosaccharides, amino acids, nucleotides, etc.) are interconverted and incorporated into larger molecules.

intermediate filaments protein filaments which form a tough and durable network in eukaryotic cells. Exact role still uncertain.

intermediate host a host in which a parasite lives for part of its life cycle but in which it does not become sexually mature.

intermediate junction type of adhesive junction between adjacent epithelial cells which runs in a band around the circumference of the upper part of the cell, just below the tight junction. *alt*. belt desmosome, adhaerens junction, zonula adhaerens.

intermedin melanocyte-stimulating hormone *q.v.*

intermedium *n*. a small bone of carpus and tarsus.

intermembrane space space between the double membrane of organelles such as mitochondria and chloroplasts and within the double membranes of bacterial envelopes.

intermuscular *a*. between or among muscle fibres.

internal *a*. located on inner side; nearer middle axis; located or produced within.

internal phloem primary phloem found internal to primary xylem.

internal respiration the biochemical, intracellular reactions of respiration.

internarial *a*. between the nostrils, *appl*. septum.

International Unit unit used to measure vitamin content of foods.

interneurone *n*. small neurone in grey matter of central nervous system, interposed between afferent and efferent neurones in spinal cord, as between sensory and motor neurones in a reflex arc, and also forming cross-connections between neural pathways generally.

internode *n*. the part between two nodes (between two leaves on a plant stem) or two joints.

internuncial *a*. intercommunicating, as paths of transmission or nerve fibres.

interoceptor *n*. a sensory receptor which receives stimuli from within the body.

interoceptor *n*. sense organ or receptor that detects stimuli originating inside the body. *cf*. exteroceptor.

interocular *a*. between the eyes.

interoptic *a*. between the optic lobes of brain.

interorbital *a*. between the orbits of the eye, *appl*. sinus.

interosseous *a*. between bones, *appl*. arteries, ligaments, muscles, membranes, nerves.

interparietal *n*. in many vertebrates a bone arising between parietals and supraoccipital.

interpeduncular *a. appl*. fossa between cerebral peduncles, and to ganglion.

interphase *n*. the period between one mitosis and the next, the period during the cell cycle during which no visible changes take place, the chromosomes are in the uncondensed state, transcription occurs, and during which the DNA is replicated; period sometimes occurring between the 1st and 2nd meiotic division.

interpositional growth of cells, by interposition between neighbouring cells without loss of contact.

interradius *n*. a radius between the four primary radii of a radially symmetrical animal.

interrenal *a*. between the kidneys, *appl*. veins.

interrenal body a gland, situated between kidneys of elasmobranch fishes, representing the adrenal cortex of higher vertebrates.

interrupted *a*. with continuity broken; irregular; asymmetrical.

interrupted gene *see* split gene.

interruptedly pinnate pinnate with pairs

of small leaflets occurring between larger ones.

interscapular *a.* between the shoulder blades; *appl.* feathers; *appl.* brown fat, the so-called hibernating gland of some rodents.

interscutal *a.* between scuta or scutes.

intersegmental *a.* between segments; between spinal segments, *appl.* axons, septa.

interseminal *a.* between seeds or ovules, *appl.* scales in certain gymnosperms.

interseptal *a. appl.* spaces between septa or partitions.

intersex *n.* a organism with characteristics intermediate between a typical male and a typical female of its species; an organism first developing as a male or female, then as an individual of the opposite sex.

interspecific *a.* between distinct species, *appl.* crosses, as mule, hinny, cattalo, tigon.

interspecific competition competition between the members of different species for the same resource. *cf.* intraspecific competition.

interspersed repeated DNA repeated DNA sequences occurring as individual copies interspersed throughout the genome, rather than as blocks of tandemly repeated sequences.

interspinal *a.* occurring between spinal processes or between spines, *appl.* bones, ligaments, muscles.

intersterility *n.* incapacity to interbreed.

intersternal *a.* between the sterna; *appl.* ligaments connecting manubrium and body of sternum.

interstitial *a.* occurring in interstices or spaces; *appl.* flora and fauna living between sand grains or soil particles; *appl.* cells, *see* interstitial cells; *appl.* fluid, *see* interstitial fluid.

interstitial cell stimulating hormone luteinizing hormone *q.v.*

interstitial cells connective tissue cells producing testosterone, lying between the tubules in vertebrate testis; small cells lying in the interstices between other cells in coelenterates and which give rise to various cell types, *alt.* I cells.

interstitial fluid the fluid filling intercellular spaces. *alt.* tissue fluid.

intertemporal *n.* paired membrane bone, part of sphenoid complex.

intertentacular organ structure in certain bryozoans through which eggs are released.

intertidal *a. appl.* shore organisms living between high- and low-water marks.

intertrochanteric *a.* between trochanters, *appl.* crest, line.

intertrochlear *a. appl.* an ulnar ridge fitting into a groove of the humerus.

intertubular *a.* between tubules, between kidney tubules, *appl.* capillaries; between seminiferous tubules.

intervarietal *a. appl.* crosses between two distinct varieties of a species.

intervening sequence intron *q.v.*

interventricular *a.* between the ventricles (of heart or brain); *appl.* foramen: foramen of Monro, the passage between the third and lateral ventricles in brain.

intervertebral *a.* occurring between the vertebrae, *appl.* discs, fibrocartilages, veins, etc.

intervillous *a.* occurring between villi; *appl.* spaces in placenta filled with maternal blood.

interxylary *a.* between xylem strands, *appl.* phloem.

intestine *n.* that part of the alimentary canal from the pyloric end of the stomach to the anus, comprising, in humans, the duodenum, jejunum, ileum, caecum, colon and rectum in that order; in other animals the part of the gut corresponding to this, i.e. from end of stomach to rectum. *a.* intestinal.

intima *n.* the layer of tissue forming the innermost lining of a part or organ. *alt.* tunica intima.

intine *n.* the cellulosic inner layer of wall of pollen grain.

intra- prefix derived from L. *intra*, within, and signifying within, inside. *cf.* inter-.

intrabulbar *a.* within a taste bud.

intracapsular *a.* contained within a capsule.

intracardiac endocardiac *q.v.*

intracellular *a.* within a cell.

intracellular transport the directed transport of material and movement of organelles within the cell. Material is transported between certain organelles, and to the plasma membrane for secretion, in the form of small membrane-bounded vesicles. In some cases directed transport is known to involve movement of vesicles along underlying tracks of microtubules, as in axonal transport.

intrachromosomal *a. appl.* duplication or rearrangement occurring within a chromosome.

intracistronic complementation the ability of two mutations in one gene (cistron) to produce a normal phenotype (to complement each other) when present in the same cell.

intraclonal *a.* within a clone, *appl.* differentiation.

intracortical *a.* within the cortex, *appl.* nerve tracts linking different parts of the cerebral cortex.

intrademic selection selection within a local breeding group.

intradermal *a.* within the dermis of skin.

intraepithelial *a.* occurring within epithelium *appl.* glands.

intrafascicular *a.* within a vascular bundle.

intrafoliaceous *a. appl.* stipules encircling stem and forming a sheath.

intrafusal *a. appl.* muscle fibres in muscle spindle; *appl.* nerve endings in tendons.

intragemmal *a.* within a taste bud.

intrageneric *a.* among members of the same genus.

intragenic *a.* within a gene, *appl.* recombination, mutation, etc.

intragenic suppression *see* back mutation.

intrajugular *n.* process in middle of jugular notch of occipital bone.

intralamellar *a.* within a lamella, *appl.* trama of gill-bearing fungi.

intramembrane particles particles seen studding the interior faces of freeze-fractured membranes, thought to be protein.

intramembranous *a.* within a membrane, *appl.* bone development.

intramolecular *a.* occurring or existing within a molecule.

intramolecular respiration formation of carbon dioxide and organic acid by normally aerobic organisms if deprived of oxygen.

intranarial *a.* inside the nostrils.

intranuclear *n.* within the nucleus.

intraparietal *a.* enclosed within an organ; in the parietal lobe of the brain.

intrapleural *a.* within the thoracic cavity.

intrasexual *a. appl.* selection amongst competing individuals of the same sex.

intraspecific *a.* within a species, as variation; *appl.* competition between the members of the same species for the same resource, e.g. mates, food, territory.

intrastelar *a.* within the stele (central vascular tissue) of a stem or root.

intrathecal *a.* within the meninges of the spinal cord.

intrauterine *a.* within the uterus.

intravaginal *a.* within vagina; contained within a sheath, as grass leaves or branches.

intravascular *a.* within blood vessels.

intraventricular *a.* within a ventricle, *appl.* caudate nucleus of corpus striatum, seen within a ventricle of brain.

intravesical *a.* within the bladder.

intravitelline *a.* within the yolk of an egg or ovum.

intrazonal *a.* within a zone; *appl.* locally limited soils, differing from prevalent or normal soils of the region or zone.

intrinsic *a.* inherent; inward; *appl.* inner muscles of a part or organ; *appl.* rate of natural increase in a population having a balanced age distribution; *appl.* sensation of brightness due to the differential response of retinal cells to different wavelengths; *appl.* membrane proteins, refers to proteins which span the whole membrane. *cf.* extrinsic.

intrinsic factor glycoprotein secreted by stomach, binding cobalamin (vitamin B_{12}) in the intestine and carrying it into the blood, and whose deficiency causes pernicious anaemia due to impaired cobalamin absorption. *alt.* gastric intrinsic factor.

intrinsic isolating mechanism any genetic mechanism preventing interbreeding.

intrinsic pathway series of reactions in blood leading to clot formation, triggered by contact with an abnormal surface such as the walls of a glass vessel, the final stages being the same as in the extrinsic pathway (*q.v.*).

intrinsic rate of increase the fraction by which a population is growing at each instant of time, symbolized by *r*.

introduced *a. appl.* plants and animals not native to the country and thought to have been brought in by man.

introgression *n.* the gradual diffusion of genes from the gene pool of one species into another when there is some hybridization between them as a result of incomplete genetic isolation.

intromittent *a.* adapted for insertion, *appl.* male copulatory organs.

intron *n.* a non-coding nucleotide sequence, one or more of which interrupt the coding sequences in many eukaryotic genes, and which are transcribed into RNA and subsequently removed by RNA splicing to leave a functional mRNA or other RNA. *alt.* intervening sequence.

introrse *a.* turned inwards or towards the axis; of anthers, opening towards the centre of the flower.

introvert *a.* that which can be drawn inwards, as anterior region of some polyps, of certain annelid worms, etc.

intumescence *n.* the process of swelling up; a swollen or tumid condition.

intussusception *n.* growth in surface extent or volume by intercalation of new material among that already present. *cf.* accretion, apposition.

inulase inulinase *q.v.*

inulin *n.* linear polysaccharide made up of fructose units, a storage carbohydrate in the roots, rhizomes and tubers of many Compositae.

inulinase *n.* enzyme hydrolysing inulin to fructose. EC 3.2.1.7. *alt.* inulase.

invagination *n.* the insinking of a wall of vessel, blastula, etc. which draws an exterior layer into the interior, forming a cavity; involution or turning inside out (of a tube); drawing into a sheath. *v.* invaginate.

inversion *n.* a turning inward, or inside out, or upside down of a part; (*genet.*) chromosomal rearrangement in which a sequence of genes appears in reverse of its normal order on the chromosome, due to the repositioning of a chunk of chromosomal material; (*biochem.*) hydrolysis of sucrose to glucose and fructose.

invert sugar sucrose *q.v.*

invertase *n.* enzyme hydrolysing terminal non-reducing β-D-fructofuranoside residues in β-D-fructofuranosides, e.g. sucrose to glucose and fructose, sometimes wrongly called sucrase, found in plants, fungi and bacteria. EC 3.2.1.26, *r.n.* β-D-fructofuranosidase.

invertebrates *n.plu.* a general term for all animals without backbones, i.e. all groups except the vertebrates.

inverted repeats in nucleic acids, two adjacent or nearby copies of an identical sequence, one being in reverse orientation.

investing bone membrane bone *q.v.*

investment *n.* the outer covering of a part, organ, animal or plant.

involucrate *a.* bearing involucres.

involucre *n.* a circle of bracts at the base of a compact flowerhead; group of leaves surrounding the groups of antheridia and archegonia in mossess and liverworts. *a.* involucral.

involucrum *n.* the notum of metathorax in Orthoptera; layer of bone formed around dead bone in some diseased conditions.

involuntary *a.* not under the control of the will, *appl.* movements, etc.

involuntary muscle smooth muscle *q.v.*

involute *a.* having edges rolled inwards at each side, *appl.* leaves; of shells, tightly coiled.

involution *n.* reduction to normal of enlarged, modified or deformed conditions; decrease in size, or structural or functional changes, as in old age; a rolling inwards, as of leaves; movement of exterior cells to interior, as in gastrulation; resting, *appl.* spores, stage, etc.

iodophilic *a.* staining darkly with iodine solution; *appl.* bacteria that stain blue with iodine.

iodopsin *n.* photosensitive protein pigment with rhodopsin-type protein component, in the cones of the retina. *alt.* visual violet.

iodothyroglobulin *n.* compound of iodine and thyroglobulin, extractable from the thyroid gland.

ion channel membrane protein that forms an aqueous pore in the membrane, allowing the passive flow of ions across the membrane, and which may be gated, i.e. only open in response to a specific simulus. Different ion channels convey different ions, being divided broadly into those conveying anions and those that carry cations.

ion exchange the adsorption of ions of one type onto a resin, or clay particles in soil, in exchange for others which are lost into solution.

ion-exchange chromatography separation of molecules such as proteins on the basis of their net charge by differential binding to a column of carboxylated polymer, positively charged molecules binding to the column.

ion pair electrostatic bond *q.v.*

ion pump protein that actively transports an ion across a biological membrane against a concentration gradient.

ionic bond electrostatic bond *q.v.*

ionic coupling the direct passive flow of ions from the cytoplasm of one animal cell to an adjacent cell through gap junctions.

ionizing radiation short wavelength high-energy radiation such as gamma-rays and fast-moving particles such as alpha- and beta-particles emitted by radioisotopes, which cause the formation of ions in tissues, thus contributing to DNA and tissue damage.

ionophore *n.* small hydrophobic molecules that dissolve in lipid bilayers and increase the permeability of the membrane to ions.

ionophoresis *n.* movement of ions under the influence of an electric current. *a.* ionophoretic.

ionotropic receptor cell-surface receptor for neurotransmitter, which is an ion channel.

ipsilateral *a. pert.* or situated on the same side. *cf.* contralateral.

Ir genes immune response genes, genes of the major histocompatibility complex which control the ability of an animal to respond to a particular antigen, now identified as genes for class II MHC proteins.

Irano-Turanian Floral Region part of the Holarctic Realm comprising central Asia north of the Himalayas and from the western edge of the Black Sea to central China.

iridal, iridial *a. pert.* the iris.

Iridales *n.* an order of herbaceous monocots and including the families Corsiaceae, Iridaceae (iris) and others.

iridocytes, iridophores *n.plu.* guanine-containing granules, bodies or plates of which the reflecting, silvery or iridescent tissue of skin of fish and reptiles is composed; iridescent cells in integument of some cephalopods.

Iridoviridae *n.* family of enveloped, double-stranded DNA viruses including that causing African Swine Fever. Most are insect viruses.

iris *n.* thin circular contractile disc with a central aperture, the pupil, lying in front of the lens in the vertebrate eye.

iris cells pigment cells surrounding cone and retinula of an ommatidium of a compound eye.

iron-sulphur (FeS) protein any of a group of proteins containing non-haem iron complexed either with sulphur atoms derived from cysteine (FeS centres containing 1 Fe complexed with 4 S atoms) or a mixture of inorganic and cysteine-derived sulphur atoms (Fe_2S_2 and Fe_4S_4 centres). A component of photosynthetic electron transport chains. *alt.* non-haem iron protein.

irreciprocal *a.* not reversible; one-way.

irregular *a. appl.* flowers showing bilateral rather than radial symmetry. *alt.* zygomorphic.

irritability *n.* capacity to receive external stimuli and respond to them.

irrorate *a.* covered as if by minute droplets; with minute colour markings, as the wings of some butterflies.

isandrous *a.* having similar stamens, their number equalling the number of sections of the corolla.

isanthous *a.* having uniform or regular flowers.

ischaemia *n.* tissue damage and localized death due to lack of oxygen. *a.* ischaemic.

ischiadic *a. pert.* or in region of hip.

ischiatic sciatic *q.v.*

ischiopodite *n.* proximal joint of walking legs of certain crustaceans, or of maxillipeds.

ischiopubic *a. appl.* gap between ischium and pubis.

ischiopubis *n.* a fused ischium and pubis.

ischiorectal *a. pert.* ischium and rectum, *appl.* fossa and muscles.

ischium *n.* the ventral and posterior bone of each half of pelvic girdle of vertebrates except fishes, often fused to pubis.

iscom vaccine vaccine composed of immune stimulatory complexes (iscoms) of the required protective antigens complexed with some inert material that prevents its degradation in the body and helps to stimulate an immune response.

isidia *n.plu.* coral-like soredia on surface of some lichens.

island biogeography the study of the flora and fauna of islands with a view to understanding the nature and evolution of biodiversity in an isolated environent.

islets of Langerhans spherical or oval groups of cells scattered throughout the pancreas and which secrete glucagon (from α or A cells) and insulin (from β or B cells).

isoaccepting *a. appl.* tRNAs which pick up the same amino acid and are charged by the same aminoacyl-tRNA synthetase.

isoagglutination *n.* agglutination of erythrocytes by blood from another member of the same species or of the same blood group; agglutination of sperm by agglutinins of ova of the same species.

isoagglutinin *n.* agglutinin of ova which reacts with sperm of the same species.

isoalleles *n.plu.* alleles that are identical in their gross phenotypic effects but can be distinguished by biochemical means at protein or DNA level.

isoantigen alloantigen *q.v.*

isobilateral *a. appl.* a form of bilateral symmetry where a structure is divisible in two planes at right angles.

isocarpous *a.* having carpels and perianth divisions equal in number.

isocercal *a.* with vertebral column ending in median line of caudal fin.

isochromatic, isochromous *a.* uniformly coloured; equally coloured.

isochromosome *n.* abnormal chromosome with two genetically identical arms.

isochronous *a.* having equal duration; occurring at the same rate.

isocitrate (isocitric acid) 6-carbon intermediate of the tricarboxylic acid cycle, formed from citrate via *cis*-aconitate in reactions catalysed by the enzyme aconitase.

isocitrate dehydrogenase either of two enzymes, one (EC 1.1.1.41, NAD requiring) located in mitochondria and catalysing formation of α-ketoglutarate from isocitrate, a regulatory step in the tricarboxylic acid cycle, the other (EC 1.1.1.42, NADP requiring) located in the cytoplasm.

isocortex *n.* the six outermost cell layers of the cerebral cortex.

isocytic *a.* with all cells equal.

isodactylous *a.* with all digits of equal size.

isodemic *a.* with, or *pert.,* populations composed of an equal number of individuals; *appl.* lines on a map which pass through points representing equal population density.

isodiametric *a.* having equal diameters, *appl.* cells or other structures; *appl.* rounded or polyhedral cells.

isodont homodont *q.v.*

isodynamic *a.* of equal strength; providing the same amount of energy, *appl.* foods.

isoelectric focusing the separation of proteins on the basis of charge, by their differential migration on electrophoresis in a pH gradient, proteins of different charge each ceasing their migration at their isoelectric point.

isoelectric point (IEP) the pH at which an amphoteric molecule, such as a protein, carries no net charge, being a definite value for each protein.

isoenzyme *n.* any one of several different forms in which some enzymes may be found, each having a similar enzyme specificity but differing in properties such as optimum pH or isoelectric point. *alt.* isozyme.

Isoetales *n.* order of Lycopsida having linear leaves, and a "corm" with complex secondary thickening, and including the quillworts.

isoforms *n.plu.* forms of a protein, such as a receptor, with slightly different amino acid sequences, and often differences in activity, function, distribution, etc.; two or more proteins or RNAs that are produced from the same gene by differential transcription and/or differential RNA splicing.

isogametangiogamy *n.* the union of similar gametangia.

isogamete *n.* one of a pair of gametes that are morphologically similar.

isogamy *n.* the fusion of gametes that are morphologically similar, i.e. of equal size and similar structure. *a.* isogamous.

isogeneic syngeneic *q.v.*

isogenes *n.plu.* lines on a map which connect points where same gene frequency is found.

isogenetic *a.* arising from the same or a similar origin; of the same genotype.

isogenic homozygous *q.v.*

isogenomic *a.* containing similar sets of chromosomes, *appl.* nuclei.

isogenous *a.* of the same origin.

isognathous *a.* having both jaws alike.

isogonal *a.* forming equal angles, *appl.* branching.

isograft *n.* tissue graft taken from another individual of the same genotype as the recipient.

isohaline *a. appl.* water having the same level of salinity.

isokont isomastigote *q.v.*

isolate *n.* a breeding group limited by isolation; the first pure culture of a microorganism derived from soil, tissues, etc.

isolateral *a.* having equal sides, *appl.* leaves with palisade tissue on both sides.

isolating mechanisms mechanisms that prevent breeding between two populations and eventually lead to speciation, chiefly including geographical isolation, and also the development of genetic, anatomical and behavioural barriers to successful interbreeding.

isolation *n.* prevention of mating between breeding groups owing to spatial, topographical, ecological, morphological, physiological, genetic, behavioural, or other factors.

isolecithal *a. appl.* eggs with yolk distributed nearly equally throughout.

isoleucine (Ile, I) *n.* an amino acid, stereoisomer of leucine, constituent of proteins, and essential in diet of man and other animals. *see* table in Appendix 1 for chemical formula.

isomastigote *a.* having flagella or cilia of the same length.

isomer *n.* one of two or more chemical compounds each having the same kind and number of atoms but differing in arrangement of the atoms and in physical and (sometimes) chemical properties. *see also* anomer, enantiomer, epimer, optical isomer, tautomerism, stereoisomer.

isomerase *n.* any enzyme that catalyses atomic rearrangements within molecules and including epimerases, racemases, tautomerases, mutases. EC class 5.

isomere *n.* a homologous structure or part.

isomerous *a.* having equal numbers of different parts; *appl.* flowers with equal numbers of parts in each whorl.

isometric *a.* of equal measure or growth rate; *appl.* contraction of muscle under tension without change in length.

isometry *n.* growth of a part at the same rate as the standard or as the whole.

isomorphic, isomorphous *a.* superficially alike; *appl.* alternation of haploid and diploid phases in morphologically similar generations.

isomorphism *n.* apparent similarity of individuals of different race or species.

isonym *n.* a new name, of species, etc., based upon oldest name or basinym.

iso-osmotic *a. appl.* two solutions of the same osmotic concentration. *see* isotonic.

isopetalous *a.* having similar petals.

isophagous *a.* feeding on one or allied species, *appl.* fungi.

isophane *n.* a line connecting all places within a region at which a biological phenomenon, e.g. flowering of a plant, occurs at the same time.

isophene *n.* a contour line delimiting an area corresponding to a given frequency of a variant form.

isophyllous *a.* having uniform foliage leaves, on the same plant.

isophytoid *n.* an "individual" of a compound plant not differentiated from the rest.

isoplankt *n.* a line representing on a map the distribution of equal amounts of plankton, or of particular species.

isoploid *a.* with an even number of chromosome sets in somatic cells.

Isopoda, isopods *n., n.plu.* group of marine, freshwater and terrestrial malacostracan crustaceans, including the woodlice and water slaters, having a dorsoventrally flattened body and no carapace.

isopodous *a.* having the legs alike and equal.

isopogonous *a.* having the two sides of feather equal and similar.

isopolyploid *a., n.* polyploid with an even number of chromosome sets, as tetraploid, hexaploid, etc.

isoprenoid *n.* any of a large and varied group of organic compounds built up of 5-carbon isoprene units and including carotenoids, terpenes, natural rubber and the side chains of e.g. chlorophyll and vitamin K.

Isoptera *n*. order of social insects comprising the termites and their relatives, which live in large organized colonies containing reproductive forms (the queen and the king) and non-reproductive wingless soldiers and workers, all offspring of the king and queen. Termites have a gut flora that enables them to digest wood, and they can be serious pests, devouring wooden buildings, trees and paper.

isosporous homosporous *q.v.*

isostemonous *a*. having stamens equal in number to that of sepals or petals.

isotelic *a*. exhibiting, or tending to produce, the same effect; homoplastic *q.v.*

isotomy *n*. forking repeatedly in a regular manner.

isotonic *a*. of equal tension; of equal osmotic pressure, *appl*. solutions; *appl*. contraction of muscle with change in length, *cf*. isometric contraction.

isotonicity *n*. normal tension under pressure or stimulus.

isotope *n*. a form of a chemical element having the same atomic number (number of protons) and identical chemical properties as another but differing in atomic mass as a result of a different number of neutrons in the atomic nucleus. *a*. isotopic.

isotropic, isotropous *a*. singly refracting in polarized light, *appl*. the light stripes of striated muscle under the microscope; symmetrical around longitudinal axis; not influenced in any one direction more than another, *appl*. growth rate; without predetermined axes, as some ova. *n*. isotropy.

isotype *n*. antigenic determinant on immunoglobulin molecules which distinguishes the constant regions of the different heavy chain classes and light chain types, all isotypes characteristic of the species being found in serum of every individual.

iso-utility curves in the study of animal behaviour and ecology, curves joining all points of equal utility or benefit, and which show how different behaviours (e.g. feeding behaviours) can result in the same quantitative benefit. *alt*. indifference curves.

isoxanthopterin *n*. a colourless pterin in the wings of cabbage butterflies and in eyes and bodies of other insects. *alt*. leucopterin B.

isozoic *a*. inhabited by similar animals.

isozyme isoenzyme *q.v.*

isthmus *n*. a narrow structure connecting two larger parts.

iter *n*. a passage or canal, as those of middle ear, brain, etc.

iteration *n*. repetition, as of similar trends in successive branches of a taxonomic group.

iteroparity *n*. production of offspring by an organism in successive groups. *a*. iteroparous. *cf*. semelparity.

ivory *n*. dentine of teeth, usually that of tusks, as of elephant and narwhal.

J

J joule *q.v.*

jacket cell a cell in an antheridium that gives rise to the wall not to the androcytes.

Jacobson's cartilage vomeronasal cartilage supporting Jacobson's organ.

Jacobson's organ a diverticulum of olfactory organ in many vertebrates, often developing into an epithelium-lined sac opening into the mouth.

jactitation *n*. scattering of seeds by a censer mechanism.

jaculatory *a*. darting out; capable of being emitted.

jaculatory duct portion of vas deferencs which can be protruded, in many animals.

jaculiferous *a*. bearing dart-like spines.

jarovization vernalization *q.v.*

Java man fossil hominid found in Java and originally called *Pithecanthropus erectus*, now called *Homo erectus*, and dating from mid-Pleistocene.

jaw-foot maxilliped *q.v.*

jaws *n.plu.* skeletal structure present in vertebrates other than agnathans. The jaws comprise the upper and lower jaws, forming part of the mouth, bearing the teeth or horny tooth-plates, the upper jaw articulating with the braincase, the lower jaw movable, articulating with the upper jaw to open and shut the mouth; structures of similar function and mechanism in invertebrates.

J chain short peptide chain joining individual immunoglobulin molecules in IgM pentamers and in oligomers of IgA.

jecoral *a*. of or *pert*. the liver.

jejunum *n*. part of small intestine between duodenum and ileum in mammals. It has large villi and is the main absorptive region.

jelly fungi common name for a group of basidiomycete fungi whose fruiting bodies are typically of a jelly-like consistency, e.g. the funnel-shaped *Tremella* and the ear-shaped *Auricularia*.

jelly of Wharton gelatinous connective tissue surrounding the vessels of the umbilical cord.

jellyfish common name for the Scyphozoa *q.v.*

Jerne plaque assay assay for lymphoid cells that produce antibodies specific for a cell-surface antigen, based on complement-mediated lysis.

J gene *see* J segment.

Johnston's organ a sensory organ concerned with balance or with sensing sound or mechanical vibrations, in 2nd segment of insect antennae.

joining segment *see* J segment.

joint *n*. strictly speaking the articulation between two neighbouring parts, such as the tibia and femur of a leg. Also used as a synonym of segment, of, e.g., insect appendages.

jordanon *n*. a true breeding unit below the species level, with little variability, such as a race, subspecies, or variety. *alt.* microspecies.

J segment any of the "joining" gene segments, short DNA sequences one of which is spliced to a V gene (or V + D gene for heavy chain genes) to produce a DNA sequence coding for a complete variable region of a heavy or light chain of an immunoglobulin molecule. *alt.* J gene.

joule (J) the derived SI unit of energy, which may be used in place of the calorie, with 1 joule = 0.24 calorie.

jubate *a*. with a mane-like growth.

jugal *n*. a component of the cheekbone in

vertebrate skull.

jugate *a.* having pairs of leaflets.

Juglandales *n.* an order of dicot trees, often aromatic, with pinnate leaves and comprising two families Juglandaceae (walnut) and Rhoiptelaceae.

jugular *a. pert.* neck or throat; *appl.* veins, the main paired veins carrying blood from the head in vertebrates; *appl.* ventral fins beneath and in front of pectoral fins.

jugum *n.* (*bot.*) a pair of opposite leaflets or leaves; (*zool.*) small lobe on posterior of forewing of some moths; ridge or depression connecting two structures.

jumping gene colloquial (but misleading) term for a variety of "mobile" DNA sequences (transposable elements) found in prokaryotes and eukaryotes which can "move" to different positions on the chromosomes or other DNAs in the cell, sometimes causing mutations and influencing the action of nearby genes with disruption of normal gene action during cellular development. *see* transposon, transposition.

jun potential oncogene encoding a protein (Jun) that can form a complex with the protein Fos to form the transcription factor AP-1.

Juncaceae *n.* the rushes, a family of monocotyledons that live mainly in wet or cold habitats, with small flowers having a perianth of six brown or green segments.

Juncales *n.* order of herbaceous monocots with long narrow channelled or grass-like leaves and comprising the families Juncaceae (rush) and Thurniaceae.

junctional complex in epithelial cells, the region of attachment between neighbouring cells. *see* desmosome, hemidesmosome, tight junction.

junctional diversity *see* N-regions.

junk DNA a vernacular term for the large amounts of DNA in the genomes of plants and animals that is composed of simple repetitive non-coding sequences and appears to have no effect on phenotype.

Jurassic *a. pert.* or *appl.* geological period lasting from about 213 to 144 million years ago, after the Triassic and before the Cretaceous.

juvenal, juvenile *a.* youthful, *appl.* plumage replacing nestling down of 1st plumage.

juvenile *n.* a young bird or other animal, before it has acquired full adult plumage or form.

juvenile hormone lipid hormone that prevents metamorphosis to the adult form in insects, secreted by the corpus allatum.

juxta- prefix derived from L. *juxta*, close to.

juxta-articular *a.* near a joint or articulation.

juxtaglomerular cells cells surrounding the arteriole feeding the network of blood capillaries surrounding the dilated end of a kidney tubule, and which secrete renin.

juxtamedullary *a.* near medulla, *appl.* inner portion of zona reticularis of adrenal glands.

juxtanuclear *a.* beside the nucleus, *appl.* bodies: basophil deposits in cytoplasm of vitamin D deficient parathyroid cells.

K

κ one of the two classes (κ and λ) of antibody light chains.

K lysine *q.v.*; symbol for the carrying capacity (*q.v.*) of the environment.

K$_M$ Michaelis constant. *see* Michaelis-Menten kinetics.

kb, kbp kilobase = 1000 bases or base pairs of DNA.

kD, kdal kilodalton *q.v.*

Ki-MSV Kirsten mouse sarcoma virus, the source of the oncogene *v-K-ras*.

kainate *n.* structural analogue of glutamate, used to define a class of glutamate receptors in central nervous system.

kairomone *n.* a chemical messenger or pheromone emitted by one species which has an effect on a member of another species, sometimes to the detriment of the transmitter, such as a chemical that attracts a male to a female also attracting a predator.

kallikrein *n.* proteolytic enzyme found in animal tissues, converting kininogen to kinin in the blood clotting pathway. EC 3.4.21.8.

kanamycins *n.plu.* antibiotics produced by the actinomycete *Streptomyces kanamyceticus* that interfere with bacterial protein synthesis. A gene for kanamycin resistance is widely used as a selectable marker gene in the construction of recombinant DNA.

kappa *n.* a self-replicating DNA particle in cytoplasm of some strains of *Paramecium*, which confers the ability on these strains to kill other paramecia; one of the two types of light chain found in mammalian antibodies.

karyaster *n.* a star-shaped group of chromosomes.

karyo- prefix derived from Gk. *karyon*, nut (= nucleus), signifying *pert.* nucleus or chromosomes.

karyoclasis *n.* breaking down of cell nucleus. *a.* karyoclastic.

karyogamy *n.* fusion of the nuclei of two gametes after cytoplasmic fusion.

karyology *n.* the study of the nucleus and the chromosomes.

karyolysis *n.* disintegration of cell nucleus. *a.* karyolytic.

karyomixis *n.* mingling or union of nuclear material of gametes.

karyon *n.* the cell nucleus. *plu.* karya.

karyoplasm nucleoplasm *q.v.*

karyoplast *n.* an isolated nucleus, with a small amount of cytoplasm adhering.

karyorhexis *n.* fragmentation of cell nucleus.

karyoschisis karyorhexis *q.v.*

karyota *n.plu.* nucleated cells.

karyotype *n.* representation of the chromosome complement of a cell, with individual mitotic chromosomes arranged in pairs in order of size.

Kaspar Hauser experiment experiment in which an animal is reared completely isolated from members of its own species.

kasugamycin *n.* antibiotic that blocks initiation of translation in bacteria.

katagenesis *n.* retrogressive evolution.

katakinetic *a. appl.* processes leading to discharge of energy.

kataphoric *a. appl.* passive action.

kataplexy cataplexis *q.v.*

katharobic *a.* living in clean waters, *appl.* protists.

K cell *see* killer cells.

keel *n.* a narrow ridge; ridge on sternum of flighted birds and bats to which wing muscles are attached; boat-shaped structure formed by two anterior petals in flowers of the pea family.

kelp *n*. common name for seaweeds of the Laminariales, marine multicellular brown algae with a large broad-bladed thallus attached to the substratum by a tough stalk and holdfast.

kenenchyma *n*. tissue devoid of its living contents, as cork.

kenosis *n*. process of voiding, or condition of having voided; exhaustion; inanition.

kentragon *n*. a larval stage following the cypris stage of parasitic cirripeds (barnacles), which penetrates into the body of host.

keraphyllous *a*. *appl*. layer of a hoof between horny and sensitive parts.

kerasin *n*. cerebroside from brain yielding on hydrolysis a fatty acid (lignoceric acid), galactose and sphingosine.

keratan *n*. a polysaccharide (galactose and *N*-acetylglucosamine residues), part of a proteoglycan from cartilage and other connective tissues, where it occurs as keratan sulphate.

keratin *n*. fibrous protein rich in cystine, and which is the chief constituent of horn, hair, nails and the upper flaky layer of skin.

keratinization *n*. intracellular deposition of keratin to form an inert horny material, e.g. nails, claws, horns, outer layers of skin. *alt*. cornification.

keratinizing *a*. becoming horny due to intracellular deposition of keratin, *appl*. cells of epidermis (skin cells, hair shaft cells, cells of fingernails and toenails, epidermal cells at base of horns). *alt*. cornifying.

keratinocyte *n*. epidermal (skin) cell that synthesizes keratin.

keratinolytic *a*. breaking down keratin, *appl*. enzymes.

keratinophilic *a*. growing on horn or other keratinized substrate, *appl*. certain fungi.

keratinous *a*. *pert*., containing, or formed by, keratin.

keratogenous *a*. horn-producing.

keratohyalin *n*. substance formed in the middle layer of vertebrate epidermis (skin), preceding full keratinization.

keratoid *a*. resembling horns.

keratose *a*. having horny fibres in skeleton, as certain sponges.

kernel *n*. the inner part of a seed containing the embryo.

keroid keratoid *q.v.*

ketogenesis *n*. the production of ketone bodies, which occurs especially during starvation or fasting.

ketogenic *a*. *appl*. amino acids that give rise to ketone bodies during their oxidation to carbon dioxide and water, e.g. leucine, phenylalanine and tyrosine.

ketogenic hormone lipotropin *q.v.*

α-ketoglutarate (ketoglutaric acid) a 5-carbon carboxylic acid, intermediate of the tricarboxylic acid cycle, which is decarboxylated to yield succinyl CoA, a reaction catalysed by the α-ketoglutarate dehydrogenase complex, and which can also be aminated to form glutamate.

ketone body any of a group of compounds such as acetoacetate, D-3-hydroxybutyrate, and acetone, formed in the liver from acetyl CoA produced during fatty acid oxidation, especially during fasting or in conditions such as diabetes, can be used by the brain as an alternative fuel to glucose.

ketose *n*. any monosaccharide containing a keto (C=O) group. *cf.* aldose.

ketosis *n*. change in energy source for brain from glucose to ketone bodies which occurs during, e.g., starvation.

key *n*. a means of identifying objects or organisms by a series of questions with alternative answers, each answer leading on to another question or a positive identification; winged nutlet hanging in clusters, as in ash, *alt*. samara.

keystone species species that have a key role in an ecosystem, affecting many other species, and whose removal leads to a series of extinctions within the system.

kidney *n*. one of pair of organs in vertebrates concerned with excretion of nitrogenous waste as urine and regulation of water balance. It is made up of a mass of convoluted tubules in intimate contact with small blood vessels.

kidney duct in fishes and amphibians the duct through which urine is excreted to the exterior.

killed vaccine vaccine made of killed or inactivated viruses, bacteria or other parasites.

killer cells non-phagocytic cells related to

monocytes and macrophages which can lyse foreign cells in the presence of antibody. *see also* natural killer cells.

kilobase (kb) unit of length used for nucleic acids and polynucleotides, corresponding to 1000 base pairs or bases.

kilodalton (kD, kdal) unit of mass equal to 1000 daltons, or 1000 units of molecular mass, also sometimes abbreviated to K (e.g. a 30K protein), used chiefly for proteins.

kinaesthesis *n.* perception of movement due to stimulation of muscles, tendons and joints.

kinaesthetic *a. pert.* sense of movement or muscular effort.

kinase *n.* any enzyme that transfers a phosphoryl (phosphate) group from ATP or other high-energy phosphates to an acceptor. EC 2.7.1–4. *alt.* phosphokinase; kinase was used in the older literature for "substances" that transformed zymogens to enzymes, now known to be proteases.

kindling long-term electrophysiological and morphological changes produced in neuronal tissue following repeated electrical stimulation, and which can e.g. result in a seizure after a mild stimulus many months afterwards.

kindred *n.* in human genetics, a group of people related by marriage or ancestry.

kinesin *n.* protein with ATPase activity that can move along microtubules and thus act as a molecular motor for various types of movement within cells. *see* axonal transport.

kinesis *n.* random movement; an orientation movement in which the organism swims at random until it reaches a better environment, the movement depending on the intensity, not the direction, of the stimulus.

kinesth- kinaesth-.

kinethmoid *n.* small bone in fish skull intermediate between premaxilla and cranium, involved in protrusion of premaxillae.

kinetic *a. pert.* movement; active; energy employed in producing or changing motion.

kinetin *n.* an adenine derivative with cytokinin activity, probably not occurring naturally.

kinetium *n.* a row of kinetosomes within a kinetodesmata. *plu.* kinetia. *alt.* kinety.

kinetoblast *n.* ciliated membrane of aquatic larvae, used for locomotion.

kinetochore *n.* densely-staining fibrillar region within the centromere of a chromosome, to which spindle microtubules attach during meiosis or mitosis.

kinetochore fibres or microtubules microtubules of spindle which are attached to kinetochores of chromosomes at one end and to pole of spindle at the other.

kinetodesma *n.* a fibril alongside a row of kinetosomes in ciliate protozoans.

kinetonucleus kinetoplast (*q.v.*) of parasitic flagellate protozoans.

kinetoplast *n.* DNA-rich region at flagellar end of mitochondrion in some flagellate protozoans.

kinetosome *n.* one of a group of granules occupying the polar plate region in sporogenesis in mosses; a self-duplicating granule, basal body, at base of cilium in ciliate protozoans.

kinetospore *n.* a motile spore.

king *n.* in social hymenopterans or termites, a male reproductive individual.

king crabs horseshoe crabs *q.v.*

kingdom *n.* in taxonomy, a primary division of the living world, five kingdoms being now generally recognized: Prokaryotae or Monera (bacteria and other prokaryotes), Protoctista (or Protista) (simple eukaryotic organisms such as the protozoans and algae), Fungi, Plantae (multicellular green plants other than algae) and Animalia (multicellular animals). Older classifications placed the fungi, algae and bacteria in the Plantae and protozoans in the Animalia. Kingdoms are divided into phyla, for animals, and divisions for plants and other organisms; biogeographical kingdom *q.v.*

kinins *n.plu.* bradykinins, *q.v.*, cytokinins *q.v.*

kinoplasmasomes *n.plu.* phragmoplast fibres seen at periphery of cell plate in plant cell division.

Kinorhyncha *n.* a phylum of marine microscopic pseudocoelomate invertebrate animals having a body of jointed spiny segments and a spiny head.

kinship *n.* possession of a common ancestor in the not too distant past.

kin selection the selection of genes due to

individuals favouring or disfavouring the survival of relatives, other than offspring, who possess the same genes by common descent.

kinship, coefficient of coefficient of kinship *q.v.*

kirromycin *n.* antibiotic that inhibits elongation of peptide chain during translation in bacteria by inhibiting elongation factor EF-Tu.

Kjeldahl analysis a technique widely used to determine the total N content of tissue.

K⁺ leak channel protein channel in plasma membrane of animal cells that allows K^+ ions to leak out of the cell down their concentration gradient. Together with the Na^+-K^+ pump it is involved in generating and maintaining plasma membrane electrical potential.

Klenow fragment fragment of the DNA polymerase I from *Escherichia coli* that contains both the polymerase and $3' \rightarrow 5'$ exonuclease but lacks the $5' \rightarrow 3'$ exonuclease activity, and which is used in various *in vitro* genetic manipulation techniques.

kleptobiosis *see* cleptobiosis.

kleptoparasitism *n.* type of parasitism in which the female searches out the prey or stored food of another female, usually of a different species, and takes it for her own offspring. *alt.* cleptoparasitism.

Klinefelter's syndrome syndrome occurring in men having genetic constitution XXY, resulting in underdevelopment of male sexual organs and sterility.

klinokinesis *n.* movement in which an organism continues to move in a straight line until it meets an unfavourable environment, when it turns, resulting in its remaining in a favourable environment, the frequency of turning depending on the intensity of the environmental stimulus; change in rate of change of direction or angular veolcity due to intensity of stimulation.

klinotaxis *n.* a taxis in which an organism orients itself in relation to a stimulus by moving its head or whole body from side to side symmetrically in moving towards the stimulus, and so compares the intensity of the stimulus on either side.

knee *n.* joint between tibia and femur; root process that emerges above water or ground in certain trees living in swampy land; joint in stem of some grasses.

knob *n.* in cytogenetics, a darkly staining chromomere that identifies a particular chromosome; in parasitology, cells parasitized with some strains of the malaria parasite *Plasmodium falciparum*, which produce knobs that attach to endothelial cells and block cerebral vessels, giving rise to cerebral malaria.

knot *n.* in wood, the base of a branch surrounded by concentric layers of new wood and hardened under pressure.

Kornberg cycle glyoxylate cycle *q.v.*

Kornberg enzyme a DNA polymerase involved in DNA repair.

Kranz anatomy type of leaf anatomy in plants with C4 photosynthesis, in which photosynthetic outer mesophyll cells are arranged in a "wreath" around inner bundle sheath cells. This enables the intercellular transport of the C4 acids, which are the first products of photosynthesis, from the mesophyll cells to the bundle sheath cells in which they are decarboxylated to provide CO_2 for the Calvin cycle. C3 metabolites are transported the other way to act as a substrate for the initial CO_2 fixation.

krasnozems deep friable red loamy soils found in the subtropics and developed from base-rich parent materials.

Krebs cycle tricarboxylic acid cycle *q.v.*

Krebs-Henseleit cycle urea cycle *q.v.*

krill *n.* planktonic crustaceans, which are abundant in the oceans and form the principal food of the whalebone (filter-feeding) whales.

kringle *n.* a structural motif shared by various proteins.

***K*-selected species** species selected for its superiority in a stable environment, typically with slow development, relatively large size, and producing only a small number of offspring at a time.

***K* selection** selection favouring superiority in stable, predictable environments in which rapid population growth is unimportant. *cf.* r selection.

Kupffer cells star-shaped phagocytic cells (macrophages) found in liver sinusoids and which ingest defunct red blood cells.

kwashiorkor *n.* deficiency disease caused by an insufficiency of protein.

kynurenine *n.* a metabolic product of tryptophan, and a precursor of some ommatochromes and other pigments in insects.

L

λ one of the two classes (κ and λ) of antibody light chains; wavelength, of light; DNA phage of *E coli*.

L leucine *q.v.*

L-, D- prefixes denoting particular molecular configurations, defined according to convention, of certain optically active compounds esp. monosaccharides and amino acids, the L configuration being a mirror image of the D. In living cells such molecules usually occur in one or other of these configurations but not both (e.g. glucose as D-glucose, amino acids always in the L form in proteins).

LAI leaf area index *q.v.*

LAK lymphokine-activated killer cells *q.v.*

LAR leaf area ratio *q.v.*

LAV lymphadenopathy virus, now called HIV *q.v.*

LC lethal concentration *q.v.*

LC₅₀ concentration of any toxic chemical that kills 50% of the organisms in a test population per unit time.

LCA family of lectins isolated from lentils, *Lens culinaris*.

LCM lymphocytic choriomeningitis virus *q.v.*

LD₅₀ a measure of infectivity for viruses, toxicity of chemicals, etc., the dose at which 50% of test animals die.

LDL low-density lipoprotein *q.v.*

Leu leucine *q.v.*

LH luteinizing hormone *q.v.*

LHC light-harvesting complex *q.v.*

LHCP light-harvesting chlorophyll-protein *see* photosynthetic unit.

LHRH luteinizing hormone releasing hormone *q.v.*

LMM light meromyosin, *see* meromyosin.

LP protein with lectin-like activity isolated from *Limulus polyphemus*.

LPH lipotropin *q.v.*

LPS lipopolysaccharide *q.v.*

LSD lysergic acid diethylamide, a powerful hallucinogenic drug whose effects sometimes mimic the symptoms of some psychotic conditions, esp. schizophrenia.

LTH prolactin *q.v.*

LTM long-term memory *q.v.*

LTR long terminal repeat *q.v.*

Lys lysine *q.v.*

labella *n.* pair of grooved lobes at end of labium in some dipteran insects, for mopping up liquid food.

labellate *a.* furnished with labella or small lips.

labelled *a. appl.* molecule made detectable and traceable by incorporation of a radioactive element, or other detectable chemical tag.

labia *n.plu.* lips; lip-like structures; *plu.* of *labium q.v.*

labia majora outer lips of vulva.

labia minora inner lips of vulva.

labial palp lobe-like structure near mouth of certain insects.

labiate *a.* lip-like; possessing lips or thickened margins; *n.* a member of the dicot flower family Labiatae, which includes the mints and balsams, characterized by typically square stems, opposite decussate leaves and flowers with a corolla divided into two lips.

labidophorous *a.* possessing pincer-like organs.

labiella *n.* mouthpart of millipedes.

labile *a.* readily undergoing change; unstable; *appl.* genes that have a tendency to mutate.

lability *n.* in evolutionary theory, the ease and speed with which particular categories of traits evolve.

labiodental *a. pert.* lip and teeth, *appl.* the surface of tooth nearest to lip.

labium *n.* a lip or lip-shaped structure; in insects the fused 2nd maxillae, forming the lower lip; inner margin of mouth of gastropod shell. *plu.* labia. *a.* labial.

Laboulbeniomycetes *n.* a group of highly specialized ascomycete fungi, exoparasitic on insects and arachnids, which have an ascogonium with a trichogyne and fertilization by spermatia.

labrum *n.* anterior lip of some arthropods; outer margin of gastropod shell. *a.* labral.

labyrinth *n.* the complex convoluted membranous and bony structures of the inner ear (*see* ear); much folded accessory respiratory organ above gills in some fishes; any of various other convoluted structures.

labyrinthodont *a.* having teeth with a complicated arrangement of dentine.

Labyrinthodontia, labyrinthodonts *n. plu.* subclass of early amphibians, of the late Paleozoic and Mesozoic, with labyrinthodont teeth, now extinct. They included temnospondyls, anthracosaurs and ichthyostegalians.

Labyrinthulomycota, labyrinthulids *n., n.plu.* the slime nets, colonies of cells that move and grow within a slime track that they secrete. Formerly classified as fungi, they are now considered a phylum of protists.

lac *n.* a resinous secretion of lac glands of certain insects, some types used to make shellac.

lac operon bacterial operon specifying the synthesis of three proteins involved in the uptake and metabolism of lactose. Consists of three structural genes whose transcription is controlled by a repressor protein encoded by the adjacent *I* gene. When lactose or another inducer binds to the repressor it dissociates from DNA allowing transcription to be initiated.

laccate *a.* appearing as if varnished.

lacertiform *a.* shaped like a lizard.

Lacertilia, lacertids *n., n.plu.* the suborder of reptiles containing the lizards. *see* Squamata.

lacewings *see* Neuroptera.

lacinia *n.* (*bot.*) segment of finely cut leaf or petal; slender projection from a thallus; (*zool.*) extension of posterior portion of proglottis; the inner branch of maxilla in insects.

laciniate *a.* irregularly cut, as of petals; fringed.

laciniform *a.* fringe-like.

laciniolate *a.* minutely incised or fringed.

lacinula *n.* a small lacinia; the inflexed sharp point of a petal. *plu.* lacinulae.

lacinulate *a.* having lacinulae.

lacrimal *a.* secreting or *pert.* tears; *pert.* or situated near the tear gland, *appl.* artery, duct, nerve.

lacrimal bone a bone in skull near the tear gland.

lacrimal gland gland in the corner of the eye which secretes tears. *alt.* tear gland.

lacrimiform *a.* tear-shaped, *appl.* spores.

lacrimonasal *a. pert.* lacrimal and nasal bones and duct.

lacrimose *a.* bearing tear-shaped appendages, as the gills of certain fungi.

lacrioid *a.* tear-shaped.

lactalbumin *n.* an albumin present in milk.

β-lactam antibiotic any of a large group of antibiotics, including the penicillins and cephalosporins, that contain a β-lactam group.

lactase *n.* the enzyme which hydrolyses terminal non-reducing β-D-galactose residues in β-D-galactosides, e.g. lactose, to glucose and galactose. EC 3.2.1.23. *r.n.* b-D-galactosidase.

lactate (lactic acid) 3-carbon organic acid formed in animal cells, esp. muscle when insufficient oxygen is supplied, and which is also produced by fermentation of sugars by certain bacteria esp. lactobacilli.

lactate dehydrogenase enzyme catalysing reduction of pyruvate by NADH to lactate in animal tissues and some bacteria. EC 1.1.1.27.

lactation *n.* secretion of milk in mammary glands; period during which milk is secreted.

lacteals *n.plu.* lymphatic vessels of the small intestine; (*bot.*) ducts that carry latex.

lacteous *a.* milky in appearance or texture.

lactescent *a.* producing milk or latex.

lactic *a. pert.* milk.

lactic acid *see* lactate.

lactic acid bacteria lactobacilli *q.v.*

lactifer *n.* any latex-containing cell, series of cells or duct.

lactiferous *a.* forming or carrying milk; carrying latex.

lactific *a.* milk-producing.

lactobacillus *n.* any of a group of rod-shaped, gram-positive bacteria (the genus *Lactobacillus*) characteristically producing lactic acid as an end product of anaerobic respiration, responsible for the souring of milk. *plu.* lactobacilli.

lactobiose lactose *q.v.*

lactogenesis *n.* the initiation of milk secretion.

lactogenic *a. pert.* or stimulating secretion of milk; *appl.* hormone: prolactin *q.v.*; *appl.* interval between parturition and ovulation, or between parturition and menstruation.

lactoglobulin *n.* milk protein soluble in ammonium sulphate but insoluble in water alone.

lactoprotein *n.* any of the proteins in milk.

lactose *n.* a disaccharide, esp. abundant in milk, composed of glucose and galactose residues.

lactose intolerance genetically determined inability of some adult humans to digest lactose due to a deficiency of the enzyme lactase, leading to abdominal symptoms. It is common in peoples of south-east and eastern Asia, but rare in peoples of northern Europe.

lactosis lactation *q.v.*

lacuna *n.* a space or cavity. *plu.* lacunae.

lacunar *a.* having, resembling, or *pert.* lacunae.

lacunate *a.* possessing or forming lacunae.

lacunose *a.* having many cavities.

lacunosorugose *a.* having deep furrows or pits, as some seeds and fruits.

lacunula *n.* a minute cavity or lacuna; a minute air space.

lacustrine *a. pert.*, or living in or beside, lakes.

laeotropic *a.* inclined, turned or coiled to the left.

laetrile *n.* amygdalin, extracted from apricot stones.

laevigate levigate *q.v.*

laevulose fructose *q.v.*

lag phase the first phase of growth of a bacterial culture, in which there is no appreciable increase in cell numbers.

lageniform *a.* shaped like a flask.

Lagomorpha, lagomorphs *n., n.plu.* the rabbits, hares and pikas, an order of herbivorous mammals known from the Eocene, with skulls and dentition similar to rodents, but with a second pair of incisors, and with hindlimbs modified for leaping.

lagopodous *a.* having hairy or feathered feet.

laking of blood, haemolysis *q.v.*

Lamarckism *n.* the theory of evolution chiefly formulated by the French scientist J.B. de Lamarck in the 18th century, which embodied the principle, now known to be mistaken, that characteristics acquired by an organism during its lifetime can be inherited.

lambda (λ) *n.* a bacterial DNA virus with an icosahedral head and a cylindrical "tail", which infects the bacterium *Escherichia coli*, and whose genetic structure and function have been minutely dissected. Lambda is a temperate bacteriophage that may either persist as a prophage integrated into the bacterial DNA or multiply within the bacterial cell, eventually destroying it. It is widely used as a vector in recombinant DNA work; (*anat.*) the junction of the lambdoid and sagittal sutures of skull; (*imm.*) one of the two types of light chain found in antibodies.

lambda particles cytoplasmic inclusions seen in the ciliate *Paramecium*; particles of phage lambda.

lambdoid *a. appl.* phages resembling phage lambda, e.g. λ80, λ21; lambda-shaped, *appl.* cranial suture joining occipital and parietal bones.

lamella *n.* any thin or plate-like structure; a gill of a mushroom or toadstool; a layer of cells. *plu.* lamellae.

lamellar, lamellate *a.* composed of thin plates.

lamellasome *n.* layered membranous structures in cyanobacteria which bear the biochemical apparatus of photosynthesis.

lamellated corpuscles Pacinian bodies *q.v.*

lamellibranch(iate) *a.* having plate-like gills on each side; with bilaterally compressed symmetrical body, like a bivalve.

lamellibranchs *n. plu.* the Lamellibranchia, a large subclass of bivalve molluscs, including clams, cockles, mussels, etc.

lamellicorn *a.* having segments of antennae expanded into flattened plates.

lamelliferous *a.* having small plates or scales.

lamelliform, lamelloid *a.* plate-like.

lamellipodium thin sheet-like pseudopodial extension temporarily put forward by some animal cells (e.g. fibroblasts) when moving over a surface. *plu.* lamellipodia.

lamellirostral *a.* having inner edges of bill bearing lamella-like ridges.

lamellose *a.* containing lamellae; having a lamellar structure.

Lamiales *n.* an order of dicot herbs, shrubs and trees including Verbenaceae (verbena), Lamiaceae (mint, etc.) and others.

lamin *n.* any of a small family of proteins that are constituents of the nuclear lamina.

lamina *n.* a thin layer, plate or scale; a layer of cell bodies in cerebral cortex and other brain structures; the blade of leaf or petal; the flattened part of a thallus.

lamina basalis Bruch's membrane *q.v.*

lamina cribrosa region of sclera at site of attachment of optic nerve and with perforations for axons of retinal ganglion cells.

lamina fusca inner layer of sclera, adjoining lamina suprachoroidea.

lamina propria layer of loose connective tissue in the mucosa of gut wall which houses the bases of glands and contains blood and lymph vessels.

lamina suprachoroidea delicate tissues layer between choroid and sclera of eye.

lamina terminalis thin layer of grey matter forming anterior boundary of 3rd ventricle of brain.

lamina vasculosa outer layer of choroid beneath suprachoroid membrane.

laminar *a.* consisting of plates or thin layers; (*bot.*) *appl.* placentation of ovule, attachment over the surface of carpel.

laminarian *a. appl.* zone between low tide line to about 30 m, i.e. the zone typically inhabited by *Laminaria* seaweeds.

laminarin *n.* any of various carbohydrates which are the main food reserves in brown algae and are stored in solution, consisting mainly of glucose units but some containing mannitol.

laminated *a.* composed of thin plates, *appl.* plant cuticle.

lamination *n.* the formation of thin plates or layers; arrangement in layers, as nerve cell bodies of cerebral cortex.

laminiform *a.* like a thin layer or layers; like a leaf blade; laminar *q.v.*

laminin *n.* a glycoprotein of the extracellular matrix.

laminiplantar *a.* having scales of metatarsus meeting behind in a smooth ridge.

lamp shells common name for the Brachiopoda *q.v.*

lampbrush chromosome type of bivalent chromosome formed in many vertebrate oocyte nuclei during extended meiosis, esp. prominent in certain amphibians, in which the chromosomes are in an extended state, with loops of chromatin folded out from each chromatid at the chromomeres to give an appearance rather like a bottle-brush, the chromatin loops being actively transcribed as can be seen in the electron microscope.

lampreys *n.plu.* common name for primitive fish-like freshwater and marine chordates of the order Petromyzoniformes, in the class Agnatha, which as adults have sucking and rasping mouthparts.

lanate *a.* covered with short woolly hairs.

lance-oval *a.* having a shape intermediate between lanceolate and ovate, *appl.* leaves.

lancelets common name for the Cephalocordata *q.v.*

lanceolate *a.* slightly broad or tapering at base, and tapering to a point at tip; lance-shaped. *appl.* leaves.

lancet *n.* one of the paired parts ventral to stylet of sting in bees and wasps.

Langerhans cells cells found in the skin and parts of gastrointestinal tract, involved in presentation of antigen to initiate an immune response.

Langerhans, follicles of *see* follicles of Langerhans.

Langerhans, islets of *see* islets of Langerhans.

laniary *a.* adapted for tearing, *appl.* canine tooth.

laniferous, lanigerous *a.* wool-bearing; fleecy.

lantern *n.* Aristotle's lantern *q.v.*; a light-emitting organ, as of lantern fishes.

lanuginose, lanuginous *a.* covered in down.

lanugo *n.* the downy covering on a foetus

which begins to be shed before birth.

lapidicolous *a. appl.* animals that live under stones.

lappaceous *a.* like a burr; prickly.

lappet *n.* any of various hanging, lobe-like structures; wattle of a bird.

large intestine the caecum, colon and appendix in some vertebrates, sometimes used for the colon only.

large T protein specified by polyoma virus and produced in infected cells (large T, middle T and small T all being produced from overlapping genes by differential transcription and splicing). It is involved in the neoplastic transformation of cells by polyoma virus.

larmier dacryocyst *q.v.*

larva *n.* independently living, post-embryonic stage of an animal that is markedly different in form from the adult and which undergoes metamorphosis into the adult form, e.g. caterpillar, grub, tadpole. *plu.* larvae. *a.* larval.

Larvacea *n.* class of tunicates (urochordates) which retain the larval "tadpole" form throughout their lives.

larviform *a.* shaped like a larva.

larviparous *a.* giving birth to offspring at the larval stage.

larvivorous *a.* larva-eating.

larvule *n.* a young larva.

laryngeal *a. pert.* or near the larynx, *appl.* artery, vein, nerve, etc.

laryngeal prominence in primates, a subcutaneous projection of the thyroid cartilage in front of the throat, causing a ridge on the ventral surface of the neck, and more pronounced in males. *alt.* Adam's apple.

larynges *plu.* of larynx *q.v.*

laryngopharynx *n.* part of pharynx between soft palate and oesophagus.

laryngotracheal *a. pert.* larynx and trachea; *appl.* chamber into which lungs open in amphibians.

larynx *n.* in mammals, the organ in throat that produces sound, the voice box. It is the upper end of the windpipe stiffened by cartilage and with two membranes (vocal chords) each extending half-way across the windpipe leaving a narrow slit between them. Sound is produced when air is driven through the slit setting the vocal chords vibrating.

lash flagellum flagellum in which the main filament ends in a thinner portion, the lash.

lasso cell colloblast *q.v.*

lasso *n.* a contractile filamentous noose used in trapping nematodes by certain soil fungi.

late *a. appl.* the period from the start of virus nucleic acid replication to release of infectious phage or virus in a lytic infection; *appl.* phage or viral genes expressed at later stages of infection, i.e. during virus nucleic acid replication and particle assembly.

late wood wood formed in the later part of an annual ring, having denser and smaller cells than early wood. *alt.* summer wood.

latebra *n.* the bulb or flask-shaped mass of white yolk in eggs.

latebricole *a.* living in holes.

latent *a.* lying dormant but capable of development under certain circumstances, *appl.* buds, resting stages; *appl.* characteristics that will become apparent under certain conditions.

latent bodies the resting stage of certain flagellate blood parasites.

latent period reaction time *q.v.*

laterad *a.* towards the side; away from the axis.

lateral *a. pert.* or situated at a side, or at a side of an axis.

lateral ganglia ganglia of the autonomic nervous system, lying in mammals in two chains alongside the aorta and which are linked to each other. *cf.* collateral ganglia.

lateral geniculate nucleus termination of optic tract (*see* optic nerves) in brain from which impulses are relayed via the optic radiation to visual area of cerebral cortex.

lateral inhibition a mechanism in compound eyes which enhances contrast at boundaries between lighter and darker parts of the visual field. Illumination of one part of the eye inhibits the response of neighbouring ommatidia even when their level of illumination remains unchanged; *(dev.)* a mechanism in developmental biology which ensures that repeating structures such as bristles are evenly spaced, in which the developing structure inhibits the differentiation of similar elements nearby.

lateral line longitudinal line on each side of the body in fishes which marks the position of cutaneous sensory cells of the acoustico-lateralis system concerned with the perception of movement and sound waves in water, the cells on the lateral line being known collectively as the lateral line system.

lateral line organ neuromast *q.v.*

lateral meristems dividing tissues in plants that are concerned with the production of secondary tissues rather than with the primary apical growth of the plant body, and which comprise the vascular cambium (producing the vascular bundles) and the cork cambium (producing the outer layer of cork or bark).

lateral plate mesoderm mesoderm that gives rise to the splanchnic and somatic mesoderm in vertebrates.

lateral roots roots which branch off the primary root, and which themselves may give rise to further lateral roots.

lateral ventricle large fluid-filled cavity in centre of cerebral hemisphere of brain.

lateralia *n.plu.* the lateral plates of barnacles.

lateralis organ neuromast *q.v.*

laterigrade *a.* walking sideways, like a crab.

laterinerved *a.* with lateral veins, *appl.* leaves.

laterite *a. appl.* tropical red soils containing alumina and iron oxides and little silica owing to leaching under hot moist conditions.

laterobronchi *n.plu.* secondary bronchi arising from the mesobronchus in birds.

laterocranium *n.* skull of insect head comprising genae and postgenae.

laterosensory *a. appl.* lateral-line system in fishes.

laterosphenoid *n.* bone on the mid-line of the reptilian skull, behind the orbit and above the palate.

latex *n.* thick milky or clear juice or emulsion of diverse composition present in plants such as rubber trees, spurges, and in certain agaric fungi.

lathyrism disease of animals, characterized by fragile collagen, caused by eating seeds of *Lathyrus odoratus* (sweet pea) which contain β-aminopropionitrile, which inhibits essential post-translational modification of collagen.

laticifer lactifer *q.v.*

laticiferous *a.* conveying latex, *appl.* cells,

tissues.

latifoliate *a.* with broad leaves.

latiplantar *a.* having hinder tarsal surface rounded.

latirostral *a.* broad-beaked.

latiseptate *a.* having a broad septum in the siliqua.

latitudinal furrow one running round the segmenting egg above and parallel to the equatorial furrow.

latosol *n.* leached red or yellow tropical soils.

Laurales *n.* an order of dicot trees, shrubs and climbers, with ethereal oils in cells, and including the families Calycanthaceae (calycanthus), Lauraceae (laurels), and others.

Laurasia *n.* the former northern land mass comprising present-day North America, Europe and Northern Asia, before they were separated by continental drift. *cf.* Gondwanaland.

laurilignosa *n.* a type of subtropical forest and bush composed of laurel.

laurinoxylon *n.* fossil wood.

law of independent assortment *see* independent assortment.

law of segregation *see* segregation of alleles.

law of the minimum *see* minimum, law of.

lax *a.* loose, *appl.* panicle of flowers.

layer *n.* horizontal stratum in a plant community, i.e. the tree layer comprising the canopy, the shrub layer comprising the shrubby understorey, the herb layer comprising grass and herbaceous plants, and the ground (moss) layer comprising the ground surface and lichens and mosses.

L chain light chain *q.v.*

LDL receptor receptor on non-hepatic cells which is specific for low density lipoprotein (LDL), important in uptake and clearance of cholesterol from the circulation.

leaching *n.* washing ions and nutrients out of soil downwards during drainage.

leaching *n.* the process by which chemicals in the upper layers of the soil are dissolved and carried down into lower layers.

leader *n., a.* region in mRNA (and DNA) preceding the coding region, which is transcribed, contains the ribosome-binding site, but is not translated; a translated sequence at the N-terminal end of secretory proteins and certain membrane proteins, which is essential for their passage across or into a membrane, and which is subsequently cleaved, *see also* signal sequence; topmost growing shoot or main branch of tree.

leader peptide a short peptide translated from a leader sequence in mRNA.

leaf *n.* an expanded outgrowth from plant stem, usually green and the main photosynthetic organ of most plants.

leaf area index (LAI) of a given area of vegetation, the total area of photosynthetic leaf surface divided by the area of soil covered.

leaf area ratio (LAR) the ratio of the photosynthetic surface area of a leaf to its dry weight.

leaf buttress lateral prominence on shoot axis due to underlying leaf primordium, representing leaf base.

leaf cushions prominent persistent leaf bases in palms etc., and furnishing diagnostic characters in some fossil plants.

leaf divergence the fixed proportion of the circumference of the stem by which each leaf is separated from the next.

leaf gap region of ground tissue interrupting the pattern of the stele, resulting from the divergence of vascular tissue (the leaf trace) away from stele to leaf.

leaf insects common name for some members of the Phasmida *q.v.* whose bodies mimic leaves in form.

leaf mosaic the arrangement of leaves on a plant which results in minimum overlap and maximum exposure to sunlight.

leaf scar the trace, usually covered with a corky layer, left on stem after leaf has fallen.

leaf sheath extension of leaf base sheathing the stem, as in grasses.

leaf stalk petiole *q.v.*

leaf trace vascular tissue extending from stele of stem into base of leaf.

leaflet *n.* a small leaf; individual unit of a compound leaf; one layer of a lipid bilayer.

leaky *a. appl.* mutations with some residual function. *see also* hypomorph.

learning *n.* any process in an animal in which its behaviour becomes consistently modified as a result of experience, and including conditioning, habituation or imprinting. From work in invertebrates on simple learning phenomena such as habituation, the ability to learn is believed to be due at least in part to the ability to permanently or semipermanently modify the transmitting properties of synapses between neurones, the modified neural circuit then producing a different response to a given stimulus. *see also* long-term potentiation, memory.

learning set in animal behaviour, the apparent learning of a general rule for solving a set of problems.

leberidocytes *n. plu.* cells containing glycogen, present in blood of arachnids at moulting.

lechriodont *a.* with vomerine and pterygoid teeth in a row nearly transverse.

lecithin phosphatidylcholine *q.v.*

lecithinases *see* phospholipase.

lecithoprotein *n.* lipoprotein in which lecithin (phosphatidylcholine) is the lipid component.

lecithotrophic *a.* feeding on stored yolk, as in some sea urchin larvae.

lecithovitellin *n.* lipoprotein composed of lecithin (phosphatidylcholine) and the yolk protein vitellin.

lectin *n.* any of a group of plant proteins that agglutinate animal cells *in vitro* by binding to specific sugar residues in membrane glycoproteins. Some lectins also have mitogenic activity. Used widely as probes of membrane structure. Natural role in plants uncertain.

lectotype *n.* a specimen chosen from syntypes to designate type species.

leeches common name for the Hirudinea *q.v.*

leghaemoglobin *n.* red oxygen-binding protein pigment, resembling haemoglobin, found in the nitrogen-synthesizing root nodules of leguminous plants.

legume *n.* type of fruit derived from a single carpel that splits down both sides at maturity, characteristic of the pea family; a member of the Leguminosae, e.g. peas, beans, clovers, vetches, gorse, broom.

legumin *n.* a protein present in seeds of leguminous plants.

Leguminosae *n.* large family of dicotyledonous plants, commonly called legumes or leguminous plants, including trees, shrubs, herbs and climbers, with typical sweet-pea shaped flowers and fruit as pods, and including peas, beans, clovers, vetches, etc. *see also* Fabales.

leguminous *a. pert.* Leguminosae; *pert.* legumes; *pert.* or consisting of, peas, beans or other legumes.

leimocolous *a.* inhabiting damp meadows.

leiosporous *a.* with smooth spores.

leiotrichous *a.* having straight hair.

Leishmania genus of parasitic protozoa, infecting humans and other mammals, with sandflies as the intermediate host and vector. *L. donovani* causes the chronic and often fatal tropical disease visceral leishmaniasis, *L. tropica* cutaneous leishmaniasis or tropical sore.

leishmanial *a. appl.* short stout forms of trypanosome lacking a free flagellum.

Leitneriales *n.* an order of resinous dicot shrubs comprising the family Leitneriaceae with the single genus *Leitneria*.

lek *n.* a special arena removed from nesting and feeding grounds, used for communal courtship display (lekking) preceding mating in some birds (e.g. ruffs, many grouse species). The term is now also sometimes applied to similar areas used by other animals for communal displays.

lekking *n.* a highly ritualized sexual display by birds such as grouse, which takes place on a particular display ground, the lek, and which precedes mating.

lemma *n.* the lower of the two bracts enclosing a floret (individual flower) in grasses.

lemniscus *n.* band of white matter in midbrain and medulla oblongata.

lens *n.* transparent structure in the eye through which light is focused onto the retina, the crystalline lens of the vertebrate eye being formed from prism-shaped, refractile dead cells filled with the protein crystallin; modified portion of the cornea in front of each element of a compound eye; modified cells of luminescent organ in certain fishes.

lens fibres elongated, lifeless, crystallin-filled cells making up the lens of the eye.

lens placode local thickening of the ectoderm opposite the optic vesicle, which invaginates to form the lens pit, which then closes to become the lens vesicle, developing into the lens. *alt.* lens rudiment.

lentic *a. appl.* standing water; *appl.* organisms living in swamp, pond, lake or any other standing water.

lenticel *n.* pore in periderm of trees and shrubs, allowing the passage of air to internal tissues.

lenticula *n.* a spore case in certain fungi; lenticel *q.v.*; lentigo or freckle.

lenticular *a.* shaped like a double convex lens; *pert.* lenticels.

lenticulate *a.* meeting in a sharp point; depressed, circular and often ribbed.

lentiform *a.* lentil-shaped.

lentiform glands lymphoid glands situated between pyloric glands.

lentiform nucleus *see* corpus striatum.

lentigerous *a.* having a lens.

lentiginose, lentiginous *a.* freckled; speckled; bearing many small dots.

lentiviruses *n.plu.* a subfamily of non-oncogenic slow-acting retroviruses, which cause chronic infections that only become manifest years after infection, including HIV (human immunodeficiency virus).

lepidic *a.* consisting of scales; *pert.* scales.

lepidodendroid *a.* having scale-like leaf scars.

Lepidodendron a fossil tree-fern with small leaves producing scale-like leaf scars.

lepidoid *a.* resembling a scale or scales.

lepidomorium *n.* a small scale or unit of composite scale, with bony base and conical crown of dentine, containing a pulp cavity and sometimes covered with enamel. *a.* lepidomorial.

lepidophyte *n.* fossil fern.

Lepidoptera *n.* order of insects commonly known as moths and butterflies. Their bodies and wings are covered by small scales, often brightly and variously coloured, forming characteristic patterns. They undergo complete metamorphosis, the larval (caterpillar) stage, giving rise to a pupa in which metamorphosis with development of adult structures such as the two pairs of membranous wings, the legs and compound eyes occurs. Adult Lepidoptera feed largely on nectar, through a hollow proboscis. *a.* lepidopterous.

Lepidosauria, lepidosaurs *n., n.plu.* subclass of reptiles comprising the lizards, snakes and amphisbaenians, and the tuatara, with a diapsid skull, and with limbs and limb girdles unspecialized, reduced or absent.

lepidosis *n.* character and arrangement of scales on an animal.

lepidote *a.* covered with minute scales.

lepidotrichia *n.plu.* bony unjointed rays at edge of fins in teleost fishes.

lepospondylous *a.* having hour-glass shaped vertebrae.

lepto- prefix derived from Gk. *leptos*, slender.

leptocaul *a.* having a slender primary stem.

leptocentric *a. appl.* concentric vascular bundle with phloem at centre.

leptocephaloid *a.* resembling or having the shape of eel larvae.

leptocephalus *n.* translucent larva of certain eels, before the elver stage.

leptocercal *a.* with a long slender tapering tail, *appl.* some fishes.

leptodactylous *a.* having slender fingers.

leptodermatous *a.* thin-skinned.

leptoid *n.* living food-conducting cell joined with others to form a simple conducting tissue in stems of some mosses.

leptoma *n.* thin area in the wall of a gymnosperm pollen grain, through which the pollen tube emerges.

leptome *n.* sieve elements and parenchyma of phloem; similar conducting elements in bryophytes.

leptomonad *a. appl.* long slender form of trypanosome with a free flagellum.

leptonema *n.* fine thread of chromosome that appears at leptotene.

leptophyllous *a.* with slender leaves; having a small leaf area, under 25 mm^3.

leptosome *a.* tall and slender.

leptosporangiate *a. appl.* ferns in which the sporangia develop from a single initial cell, which first produces a stalk and then a capsule. *cf.* eusporangiate.

leptotene *n.* early stage in the prophase of meiosis in which the chromatin is beginning to become compacted, showing as fine threads.

leptotrombicula *n.* the larval form of a trombicula, a mite transmitting scrub typhus (tsutsugamushi disease), a rickettsial disease.

leptoxylem *n.* rudimentary wood tissue.

leptus *n.* the six-legged larva of mites.

Lesch-Nyhan syndrome inborn error of metabolism characterized by an almost complete deficiency of the enzyme hypoxanthine-guanine ribosyltransferase and by symptoms of self-mutilation, mental deficiency and spasticity.

lesser omentum a fold of peritoneum which connects the stomach and liver and supports the hepatic vessels.

lestobiosis *n.* the relation in which colonies of small species of insect nest in the walls of the nests of larger species and enter their chambers to prey on the brood or rob food stores.

lethal *a.* causing death; of a parasite, fatal or deadly in relation to a particular host; *appl.* mutations or alleles which when present cause the death of the embryo at an early stage.

lethal concentration (LC) where death is the criterion of toxicity, the results of toxicity tests are expressed as a number (LC$_{50}$, LC$_{70}$) which indicates the percentage of test organisms killed at a particular concentration over a given exposure time, e.g. the 48-hour LC$_{70}$ is the concentration of a toxic material that kills 70% of the test organisms in 48 hours.

lethal dose (LD) dose of a toxic chemical or of a pathogen that kills all the animals in a test sample within a certain time. *cf.* median lethal dose.

lethal synthesis the synthesis *in vivo* of a metabolic poison from a substance that is not itself toxic.

lethality *n.* the ratio of fatal cases to the total number of cases affected by disease or other harmful agent.

leucine (Leu, L) *n.* α-amino isocaproic acid, an amino acid with a hydrocarbon side chain, constituent of protein, essential in human and animal diet. *see* table in Appendix 1 for chemical formula.

leucine zipper structural motif found in many proteins involved in gene regulation and which is thought to be involved in protein–protein interactions.

leucism *n.* the presence of white plumage or fur in animals with pigmented eyes and skin. *cf.* albinism.

leuco- prefix derived from Gk. *leukos*, white.

leucoanthocyanidins *n.plu.* a group of colourless flavonoids.

leucocarpous *a*. with white fruit.

leucocidin *n*. a toxin synthesized by some staphylococci, and which destroys leukocytes.

leucocyan *n*. a pigment found in some algae.

leucocyte, and its derivative terms *alt*. spelling of leukocyte (*q.v.*) and some of its corresponding derivative terms.

leucoplast(id) *n*. colourless plastid that develops into chloroplasts, etc.

leucopterin *n*. a pterin constituting the white pigment of cabbage white butterflies and other Lepidoptera and wasps, and which can be reduced to xanthopterin. *see also* isoxanthopterin.

leucosin *n*. a storage polysaccharide forming whitish granules in some yellow-brown algae.

leukaemia *n*. a malignant disorder of white blood cells in which precursors proliferate and fail to differentiate.

leukaemogenesis *n*. the generation of a leukaemia.

leukemia *alt*. spelling of leukaemia *q.v.*

leukin *n*. basic polypeptide extracted from leukocytes and active against Gram-positive bacteria.

leuko- leuco-. *q.v.*

leukoblast myeloblast *q.v.*

leukocyte *n*. any of a group of colourless cells of blood, commonly known as white blood cells, and which include monocytes, granulocytes, and lymphocytes. All derive from a common progenitor in bone marrow and are involved in the protection of the body from infection.

leukocytolysis *n*. the breakdown or disintegration of white blood cells.

leukocytosis leukosis *q.v.*

leukopenia *n*. reduction in the number of leukocytes in the blood, characteristic of many diseases.

leukopoiesis *n*. the generation of white blood cells.

leukosis *n*. an increase in the numbers of circulating white blood cells. *alt*. leukocytosis.

leukotrienes *n.plu.* class of compounds released from mast cells in local inflammatory reactions, formerly known as slow-reacting substance of anaphylaxis (SRS-A), and which variously cause smooth muscle contraction (leukotriene E4) and attract polymorphonuclear leukocytes and eosinophils to sites of injury or infection (leukotriene B4).

levan *n*. any polysaccharide made up of fructose units.

levator *n*. a muscle serving to raise an organ or part.

levigate *a*. made smooth, polished.

levulose fructose *q.v.*

ley *n*. temporary agricultural grassland, which is sown and used as a crop.

Leydig cells cells in interstitial tissue of testis which secrete testosterone.

Leydig's organs minute organs on antennae of arthropods, possibly chemoreceptors.

L-form *n*. stage of certain mycoplasmas in which they will pass through the normal bacterial filter and which consists of specialized reproductive bodies, produced in extreme conditions.

liana *n*. any woody climbing plant of tropical and semitropical forests.

Lias *n*. marine and estuarine deposits of the Jurassic period, containing remains of fossil cycads, insects, ammonites, and saurians. *a*. Liassic.

libriform *a. appl*. woody fibres with thick walls and simple pits.

Librium trade name for one of the commonly used benzodiazepines *q.v.*

lice *n.plu.* common name for various small insects of the orders Psocoptera (book lice) and Anoplura (Siphunculata, the sucking lice, and Mallophaga, the biting lice), and for crustaceans of the orders Isopoda (woodlice) and Branchiura.

lichen *n*. a composite organism formed from the symbiotic association of certain fungi and a green alga or cyanobacterium, forming a simple thallus, found encrusting rocks, tree trunks, etc., often in extreme environmental conditions.

lichen starch lichenin *q.v.*

lichenase *n*. enzyme that breaks down lichenin to glucose, and so digests lichens, as found in the gut of reindeer and caribou and some gastropods. EC 3.2.1.6. *r.n.* endo-1,3(4)-β-D-glucanase.

lichenicole, lichenicolous *a*. living or growing on lichens.

lichenin *n*. a polysaccharide present in the walls of lichen fungi, and also in some seeds, esp. oat seeds, and which is hydrolysed by the enzyme lichenase to glucose.

lichenization *n*. production of a lichen by alga and fungus; spreading or coating of lichens over a substrate; effect of lichens on their substrates.

lichenoid *a*. resembling a lichen.

lichenology *n*. the study of lichens.

Lieberkühn's crypts crypts of Lieberkühn *q.v.*

Liebig's law the food element least plentiful in proportion to the requirements of plants limits their growth; law of the minimum *q.v.*

lienal *a. pert.* spleen.

lienculus *n*. an accessory spleen.

lienogastric *a. pert.* spleen and stomach, *appl.* artery supplying parts of spleen and parts of stomach and pancreas.

life living organisms can be distinguished from other complex physico-chemical systems by their storage and transmission of molecular information in the form of nucleic acids, their possession of enzyme catalysts, their energy relations with the environment and their internal energy conversion processes (e.g. photosynthesis, respiration and other enzyme-catalysed metabolic activities), their ability to grow and reproduce, and their ability to respond to stimuli (irritability). Entities such as viruses, which satisfy only some of these criteria, are also generally considered as part of the living world.

life cycle the various phases an individual passes through from birth to maturity and reproduction.

life expectancy *see* average life expectancy.

life form the typical adult form of a species.

life tables demographic data required to calculate, e.g. the intrinsic rate of increase of a population. They comprise the survivorship schedule, which gives the number of individuals surviving to each particular age, and the fertility schedule, which gives the average number of female offspring that will be produced by a single female at each particular age. From these the net reproductive rate, R_0, the average number of female offspring produced by each female during her lifetime, can be calculated. The intrinsic rate of increase, r, of the population can be computed from survivorship and fertility schedules using the Euler-Lotka equation.

ligament *n*. a strong fibrous band of tissue connecting two or more moveable bones or cartilages; band of elastic tissue forming the hinge of a bivalve shell.

ligand *n*. any molecule that binds to another, usually used for signalling molecules binding to receptor proteins, regulatory molecules binding to enzymes, etc.

ligand-gated *appl.* receptors and ion channels in cellular membranes which are stimulated to open or close by the binding of a specific molecule (ligand).

ligase *n*. any of a class of enzymes that catalyse the joining together of 2 molecules coupled with the breakdown of a pyrophosphate bond in ATP or a similar triphosphate, and which include the synthetases, carboxylases, and DNA and RNA ligases. EC 6.

ligation *n*. the joining of two pieces of DNA end-to-end by DNA ligase to form a continuous DNA molecule.

light chain (L chain) in an immunoglobulin molecule, one of 2 identical polypeptide chains each containing a V region (*q.v.*) and a short C region (*q.v.*); in myosin, any of 4 polypeptide chains attached to the globular heads of the molecule.

light-harvesting complex (LHC) complex of chlorophyll (and other pigments) and protein that collects light energy and passes it on to the photosynthetic reaction centre. In green plants there are two light-harvesting complexes, one associated with photosystem I and one with photosystem II.

light reaction in photosynthesis, reactions occurring in the thylakoid membranes of chloroplasts, in which light energy drives the synthesis of NADP and ATP. *cf.* dark reaction.

ligneous *a*. woody or resembling wood in structure.

lignescent *a*. developing the character of woody tissue.

lignicole, lignicolous *a*. growing or living on or in wood.

lignification *n*. wood formation; the thickening of plant cell walls by deposition of lignin, occurs in both primary and secondary walls.

lignin *n.* a hard material found in walls of cells of xylem and sclerenchyma fibres in plants, a very variable cross-linked polymer of phenylpropane units such as coniferyl alcohol (guaiacyl), sinapyl alcohol (syringyl) or hydroxycinnamyl alcohol, and which stiffens the cell wall.

lignivorous *a.* eating wood, *appl.* various insects.

lignocellulose *n.* lignin and cellulose combined, a constituent of woody tissue.

lignolytic *a.* lignin-degrading.

lignosa *n.* vegetation made up of woody plants.

ligula *a.* band of white nerve fibres in dorsal wall of 4th ventricle of brain.

ligula, ligule *n.* a membranous outgrowth at junction of blade and leaf sheath or petiole; small scale on upper surface of leaf base in some club mosses and quillworts; a tongue-shaped corolla, as of some florets.

ligular *a.* tongue-shaped.

ligulate *a.* having or *pert.* ligules; strap-shaped, as ray florets of Compositae; *appl.* capitulum of strap-shaped florets.

liguliferous *a.* having ligulate flowers only.

Ligustrales Oleales *q.v.*

Liliales *n.* an order of monocot plants, growing from rhizomes or bulbs, mostly herbaceous, and including the families Liliaceae (lily), Agavaceae (agave), Alliaceae (onion), Amaryllidaceae (daffodil), Dioscoreaceae (yam) and others. *alt.* Liliiflorae.

Liliopsida *n.* in some plant classifications the name for the class containing the monocotyledons.

limacel(le) *n.* concealed vestigial shell of slugs.

limaciform *a.* slug-shaped; like a slug.

limacine *a. pert.* slugs.

limb *n.* branch; arm; leg; wing; (*bot.*) expanded part of calyx or corolla, the base of which is tubular.

limb bud small protuberance on vertebrate embryo from which a limb develops.

limbate *a.* with a border; bordered and having a differently coloured edge.

limbic *a.* bordering.

limbic system region in the anterior part of cerebral hemispheres, including the amygdala and septum, and believed to be involved in the generation and control of emotions like rage, fear and joy.

limbus *n.* any border if distinctly marked off by colour or structure; transitional zone between cornea and sclera.

lime *n.* calcium oxide; calcium hydroxide; any calcium salt.

limicolous *a.* living in mud.

liminal *a. pert.* a threshold, *appl.* minimal stimulus or quantitative difference in stimulus that is perceptible. *cf.* subliminal.

limit dextrins branched oligosaccharide fragments formed as one of the final products of starch hydrolysis by α-amylase.

limiting factor any single factor that limits e.g. a biochemical process, the growth of an organism, its abundance or distribution.

limiting membrane basal connective tissue membrane underneath retina.

limivorous *a.* mud-eating, *appl.* certain aquatic animals.

limnetic *a.* living in, or *pert.* marshes or lakes; living in open water; *appl.* zone of deep water between surface and compensation depth (depth at which photosynthesis cannot be supported owing to insufficient light).

limnium *n.* a lake community.

limnobiology *n.* the study of life in standing waters, i.e. ponds, marshes, lakes.

limnobios *n.* freshwater plants and animals collectively.

limnobiotic *a.* living in freshwater marshes.

limnology *n.* the study of the biological and other aspects of standing waters.

limnophilous *a.* living in freshwater marshes.

limnophyte *n.* a pond plant.

limnoplankton *n.* the floating microscopic life in freshwater lakes, ponds and marshes.

Limulus a genus of horseshoe crab *q.v.*

linea *n.* a line-like structure or mark.

lineage *n.* a line of common descent. *see also* cell lineage.

lineage group group of species allied by common descent.

linear *a. appl.* leaves, the long narrow leaves which are characteristic of monocotyledons.

linear-ensate *a.* between linear and sword-shaped, *appl.* leaves.

linear growth type of growth in which the amount increases by the same amount over each set period (e.g. a year). *cf.* exponential growth.

linear-lanceolate *a.* between linear and lanceolate in shape, *appl.* leaves.

linear-oblong *a.* between linear and oblong in shape, *appl.* leaves.

lineolate *a.* marked by fine lines or striae.

LINEs long interspersed DNA sequence elements, a class of interspersed repetitive retrotransposon-like sequences found in thousands of copies in mammalian genomes. *see also* SINEs.

line transect the recording of types and numbers of plants along a measured line.

lingua *n.* a tongue or tongue-like structure.

lingual *a. pert.* tongue, *appl.* artery, nerve, vein, etc.

linguiform *a.* tongue-shaped.

lingula *n.* a small tongue-like process of bone or other tissue.

lingulate *a.* shaped like a short broad tongue. *cf.* ligulate.

linin *n.* protein of flax seed; a bitter purgative substance obtained from purging flax.

linkage *n.* the tendency for some parental alleles to be inherited together, in opposition to Mendel's law of independent assortment, and which is due to their presence close together on the same chromosome.

linkage disequilibrium condition in which certain alleles at two linked loci are nonrandomly associated with each other, either because of very close physical proximity which virtually precludes recombination between them, or because the combination is under some form of selective pressure.

linkage group the genes carried on any one chromosome.

linkage map genetic map *q.v.*

linker (linker sequence) a short stretch of synthetic DNA, most usually containing a restriction enzyme site, which is often used to connect two different DNA molecules to form a recombinant DNA, thus allowing the DNA of interest to be easily recovered after e.g. cloning in a bacterial or other cell.

linking number number of turns of one strand of a closed circular DNA about the other, $L = T$(degree of twisting of the dou-

ble helix) + W(extent of supertwisting*)*.

Linnean *a. pert.* or designating the system of binomial nomenclature and classification established by the 18th century Swedish biologist Carl von Linné or Linnaeus.

linneon *n.* a taxonomic species distinguished on purely morphological grounds, esp. one of the large species described by Linnaeus or other early naturalists.

linoleate (linoleic acid) common unsaturated fatty acid, $CH_3(CH_2)_4$ $(CH=CHCH_2)_2(CH_2)_6COO^-$, essential for growth in mammals and necessary in the diet.

linolenate (linolenic acid) common unsaturated fatty acid, CH_3CH_2 $(CH=CHCH_2)_3(CH_2)_6COO^-$, necessary for growth in mammals but not essential in the diet as it can be synthesized from linoleate. *see* Appendix 1 (39) for chemical structure.

lipase *n.* any of a group of widely distributed enzymes (produced esp. by pancreas) hydrolysing triacylglycerols to diacylglycerols plus a fatty acid anion. EC 3.1.1.3, *r.n.* triacylglycerol lipase, and EC 3.1.1.34, *r.n.* lipoprotein lipase; sometimes wrongly used for enzymes that break down fats during digestion.

lipid *n.* any of a diverse class of compounds found in all living cells, insoluble in water but soluble in organic solvents such as ether, acetone and chloroform, and which include fats, oils, triacylglycerols, fatty acids, glycolipids, phospholipids and steroids, some lipids being essential components of biological membranes, others acting as energy stores and fuel molecules for cells.

lipid bilayer *see* bilayer.

lipid bodies lipid storage structures found in oil-rich plant seeds, composed of a large droplet of triacylglycerol surrounded by a single-layered membrane. *alt.* oil bodies.

lipid droplet large droplet of (usually) triacylglycerols found in cells specialized for fat storage.

lipoamide *n.* enzyme cofactor derived from lipoic acid, and which is an acyl group carrier esp. important in carbohydrate metabolism.

lipoate (lipoic acid) 1,2-dithiolane-3-valeric acid, composed of a fatty acid (valeric acid) and disulphide. Required for carbohydrate metabolism as a precursor for the enzyme cofactor lipoamide.

lipochroic *a.* with pigment in oil droplets.

lipochromes *n.plu.* fat-soluble pigments.

lipocyte *n.* cell specialized for lipid production and storage. In animals, fat cells or adipocytes.

lipofection the transfer of material into a cell by enclosing it in liposomes, which fuse with the cell membrane.

lipofuscin granules residual orange fluorescent bodies seen in ageing cells, derived from lysosomes. *alt.* age pigments.

lipogenesis *n.* synthesis of fatty acids and lipids.

lipogenous *a.* fat-producing.

lipoic acid *see* lipoate.

lipoid *a.* resembling a fatty substance.

lipoid *n.plu.* substances which are not true lipids but resemble them in certain properties and are extracted in fat solvents, including sterols and steroids.

lipolysis *n.* enzymatic breakdown of fats, as during digestion; breakdown of triacylglycerols in adipose cells during mobilization of food reserves.

lipolytic *a. pert.* lipolysis; *appl.* enzymes, capable of digesting or dissolving fat; *appl.* hormone: lipotropin (LPH) *q.v.*

lipopalingenesis *n.* the loss of some developmental stage or stages during evolution.

lipopexia *n.* deposition and storage of fats in tissues.

lipopolysaccharide (LPS) *n.* any molecule consisting of a lipid joined to a polysaccharide, one of the main constituents of the outer cell envelope of Gram-negative bacteria.

lipoprotein *n.* complex of lipid and protein.

liposome *n.* artificially constructed spheres of lipid bilayer enclosing an aqueous compartment, used in experimental biology to study properties of biological membranes and as a possible means of delivering drugs to cells more efficiently, *alt.* lipid vesicle; fatty droplet in cytoplasm, esp. of an egg.

lipotropic *a.* concerned with the mobilization of storage lipids and their breakdown into fatty acids and triacylglycerols.

lipotropin (LPH) *n.* in mammals, either of two peptide hormones produced by the anterior pituitary gland which stimulate

lipolysis.

lipovitellin *n.* a lipoprotein which is part of amphibian egg yolk.

lipoxenous *a.* leaving the host before development is complete, *appl.* parasites.

lipoxidase lipoxygenase *q.v.*

lipoxin *n.* member of a class of lipid-derived chemical signal molecules produced in cells from arachidonic acid in response to injury, during inflammation, etc. They can cause contraction of smooth muscle, vasodilation, chemotaxis, hyperfiltration in the kidney, and inhibition of natural killer cell activity.

lipoxygenases *n.plu.* enzymes that catalyse the addition of a molecule of oxygen to the double bonds of certain unsaturated fatty acids and their derivatives. They are involved *inter alia* in the synthesis of some eicosanoid chemical mediators (e.g. leukotrienes and lipoxins).

liquor folliculi fluid surrounding the ovum in a Graafian follicle.

lirella *n.* type of long apothecium in some lichens.

Lissamphibia *n.* in some classifications a subclass of amphibians containing all extant species and divided into three orders Salientia (Anura), Urodela and Apoda.

lissencephalous *a.* having few or no convolutions on the surface of the brain.

lithite *n.* a calcareous secretion in ear, or, with otocysts, lithocysts, and tentaculocysts, sensory organs of many invertebrates. *alt.* statolith.

lithocarp *n.* a fossil fruit.

lithocysts *n.plu.* minute sacs or grooves containing lithites, found in many invertebrates; enlarged cells of plant epidermis, in which cystoliths are formed.

lithocyte *n.* large cell in hydrozoan statocyst (gravity detector) containing a concretion of calcium salts, whose movement under gravity is detected by an adjacent sensory cell.

lithodomous *a.* living in holes or clefts in rock.

lithogenous *a.* rock-forming or rock-building, as the reef-building corals.

lithophagous *a.* stone-eating, as some birds; rock-burrowing, as some molluscs and sea urchins.

lithophyll *n.* fossil leaf or leaf impression.

lithophyte *n.* plant growing on rocky ground.

lithosere *n.* a plant succession originating on rock surfaces.

lithosols *n.plu.* soils which develop at high altitudes on resistant parent materials which withstand weathering and result in a humus-rich, shallow, stony soil.

lithosphere *n.* the Earth's crust; the non-living, non-organic part of the environment, such as rocks, the mineral fraction of soil, etc.

lithotomous *a.* stone-boring, as certain molluscs.

lithotroph *n.* an autotroph, esp. a chemo-autotroph *q.v. a.* lithotrophic.

litter *n.* (*bot.*) partly decomposed plant residues on the surface of soil, mainly in woodlands; (*zool.*) offspring produced at a single multiple birth.

littoral *a.* growing or living near the sea-shore; *appl.* zone between high- and low-water marks, *see* Fig. 9; *appl.* zone of shallow water and bottom above compensation depth (depth at which photosynthesis cannot be supported) in lakes.

Littré's glands mucus-secreting glands of the urethra.

lituate *a.* forked, with prongs curving outwards.

live vaccine vaccine made from active but non-pathogenic viruses of the same or similar type as that which it is intended to protect against. Such viruses can multiply to a limited extent in the vaccinated host and stimulate protective immunity, but do not cause any severe clinical symptoms.

liver *n.* in vertebrates a glandular organ closely associated with gut, developing mainly from embryonic gut epithelium. It secretes bile and is a key organ in the metabolism and storage of foodstuffs. The specialized cells of the liver are known as hepatocytes; in some invertebrates, a glandular digestive organ.

liverworts *n.plu.* common name for members of the plant division Hepatophyta (*q.v.*).

living fossil extant species of ancient lineage that has remained morphologically unchanged for a very long time, and whose only close relatives are fossils and which in some cases was itself thought to be ex-tinct, such as the coelacanth, the ginkgo and the metasequoia.

loam *n.* a rich friable soil consisting of a fairly equal mixture of sand and silt and a smaller proportion of clay.

lobar *a. pert.* a lobe.

lobate *a.* divided into lobes.

lobe-finned fishes common name for lungfishes and crossopterygians.

lobed *a. appl.* leaves having margin cut up into rounded divisions that reach less than halfway to midrib.

lobopodia *n.plu.* blunt-ended pseudopodia of some protozoans.

lobose lobate *q.v.*

lobular *a.* like or *pert.* small lobes.

lobulate *a.* divided into small lobes.

lobule *n.* a small lobe.

local-circuit neurones neurones in the brain that form part of a localized network or circuit within a particular region. *cf.* projection neurones.

local faciation, lociation local differences in abundance or proportion of dominant species.

localization of function in brain, the situation that different areas of the brain are concerned with different tasks, e.g. the visual area in the occipital lobe with processing incoming signals from the eye, the motor areas with generating outgoing signals to muscles, and so on.

localization of sensation identification on surface of body of exact spot affected.

locellate loculate *q.v.*

locellus *n.* a small compartment of plant ovary.

loci *plu.* of locus *q.v.*

lock-and-key model theory first proposed at end of 19th century, that the active site on an enzyme is an exact 3-dimensional fit for the substrate, also applied to other specific interactions such as antigen-antibody binding, and now accepted with modifications as correct, but *see also* induced-fit model.

locular, loculate *a. pert.* locules; containing or composed of locules.

locule *n.* a small chamber or cavity, e.g. chamber containing ovules in ovary of flower, cavity in fungal stroma containing asci. *plu.* loculi.

loculi *plu.* of loculus *q.v.*

Loculoascomycetes *n*. a group of ascomycete fungi parasitic on plants and insects, and which bear their asci in cavities in a loose hyphal stroma. They include the sooty moulds and agents of various scab (e.g. *Venturia*, apple scab), leaf spot (e.g. *Mycosphaerella* on strawberries) and anthracnose diseases of plants.

loculose *a*. having several locules; partitioned into small cavities.

loculus locule *q.v.*

locus *n*. the position on a chromosome occupied by a particular gene; location of a stimulus. *plu*. loci.

locus coeruleus pair of minute bodies of grey matter lying just beneath the floor of the midbrain. Contains the cell bodies of noradrenergic neurones.

locusta *n*. a spikelet of grasses; a locust.

locusts common name for many of the Orthoptera *q.v.*

lod score a statistical measure of the likelihood that two genes are linked.

lodicule *n*. small body at the base of each carpel in grass florets.

loess *n*. a clay soil formed from wind-blown particles which is very fertile when mixed with humus.

logarithmic phase the rapid stage in growth of a bacterial culture when increase follows a geometric progression.

logistic curve an S-shaped curve initially rising slowly, then steeply, and finally flattening out, and which is characteristic of the growth and stabilization of a population.

loma *n*. thin membranous flap forming a fringe round an opening.

lomasomes *n.plu*. invaginations of cell membranes found in certain algae, vascular plants and fungi.

loment, lomentum *n*. a pod constricted between seeds.

lomenta *plu*. of lomentum, loment *q.v.*

lomentaceous *a. pert*. resembling or having lomenta (pods).

long-day plants plants that will only flower if the daily period of light is longer than some critical length: they usually flower in summer. The critical factor is in fact the period of continuous darkness they are exposed to. *cf*. short-day plants, day-neutral plants.

long period interspersion type of sequence arrangement within some eukaryotic genomes in which rather long moderately repetitive DNA sequences alternate with long sequences of non-repetitive DNA.

long-term memory (LTM) the process whereby information is stored for weeks, months or years in the brain and the behaviour of recalling that information. *cf*. short-term memory.

long-term potentiation (LTP) phenomenon demonstrated in hippocampus, and some other parts of brain, in which a particular type of short-term stimulus results in an altered pattern of activity on subsequent normal stimulation, this effect lasting for many hours.

long terminal repeat (LTR) a sequence repeated at each end of integrated retroviral proviruses, generated from the ends of the virus by its mode of insertion, and which contains control sites for initiation of transcription.

longicorn *a*. having long antennae, *appl*. certain beetles.

longipennate *a*. having long wings or long feathers.

longirostral, longirostrate *a*. having a long beak.

longisection *n*. section along or parallel to a longitudinal axis.

loop of Henle a loop of mammalian kidney tubule that passes from cortex into medulla and back into cortex, having thin walls and in which a complicated series of movements of water, sodium and urea through the walls of the tubule lead eventually to a reabsorption of water in the collecting ducts, and production of concentrated urine.

loph *n*. crest which may connect cones in teeth and so form a ridge.

lophium *n*. a community living on a ridge.

lophobranchiate *a*. with tufted gills.

lophocercal *a*. having a ray-less caudal fin in the form of a ridge around the end of the vertebral column.

lophodont *a*. having transverse ridges on the cheek-teeth grinding surface.

lophophorate *a*. having a lophophore.

lophophore *n*. horseshoe-shaped crown of tentacles surrounding the mouth, characteristic of animals of the phyla Brachiopoda, Ectoprocta (Bryozoa) and Phoronida.

lophoselenodont *a.* having cheek teeth ridged with crescentic cuspid ridges on grinding surface.

lophotrichous *a.* having long whip-like flagella; with a tuft of flagella at one pole.

loral *a. pert.* or situated at the lore.

lorate *a.* strap-shaped.

lore *n.* space between bill and eyes in birds.

Lorenzini's ampulla ampulla of Lorenzini *q.v.*

lorica *n.* a protective external case, as in diatoms, some protozoans and rotifers.

loricate *a.* covered with a protective coat of scales.

Loricifera *n.* small phylum of tiny marine multicellular pseudocoelomate animals, with spiny heads and abdomens covered with spiny plates called lorica, living in sediments.

lorulum *n.* the small strap-shaped and branched thallus of some lichens.

loss-of-function *appl.* mutant alleles which produce no functional gene product.

lotic *a. appl.* or *pert.* running water; living in a brook or river.

low-density lipoprotein (LDL) a group of plasma lipoproteins rich in cholesterol esters, synthesized from very low-density lipoprotein (VLDL), and which transport cholesterol to peripheral tissues and regulate *de novo* cholesterol synthesis. *see also* LDL receptor.

lower shore zone of seashore that extends from the lowest low-water level to the average low-water level and which is therefore only uncovered occasionally and for short periods.

lowland *n.* land up to *ca.* 700 m, depending on geographical area.

loxodont *a.* having molar teeth with shallow grooves between the ridges.

L strand light strand of DNA (esp. *pert.* mammalian mitochondrial DNA which can be separated into H (heavy) and L strands on the basis of their density).

luciferase *n.* an enzyme present in all luminescent organisms which reacts with the activated substrate luciferin, the reaction resulting in light emission as luciferin returns to the ground state as a result of enzymatic oxidation.

luciferin *n.* light-emitting compound whose enzymatic oxidation is responsible for bioluminescence in organisms such as many deep-sea fishes, coelenterates, fireflies and glow-worms.

lumbar *a. pert.* or near the region of the loins.

lumbarization *n.* fusion of lumbar and sacral vertebrae.

lumbocostal *a. pert.* loins and ribs, *appl.* arch, ligament.

lumbosacral *a. pert.* region of loins and termination of vertebral column (sacrum).

lumbrical *a.* like a worm in appearance; *appl.* four small muscles in palm of hand and sole of foot.

lumbricid *a. appl.* worms of the genus *Lumbricus* and close relatives, i.e. earthworms.

lumbriciform *a.* like a worm in appearance.

lumbricoid *a.* resembling an earthworm.

lumen *n.* internal space of any tubular or sac-like organ or sub-cellular organelle.

luminal *a.* within or *pert.* a lumen.

luminescent organs light-emitting organs, present in various animals.

lumirhodopsin *n.* transient orange-red product of the bleaching of rhodopsin by light, which is converted into metarhodopsin.

lunar, lunate *a.* somewhat crescent-shaped.

lunar bone, lunatum *n.* a carpal bone, the middle of the three proximal carpals.

lunar rhythms physiological or behavioural patterns influenced by the lunar cycle, which occur in both marine and terrestrial organisms. *see also* tidal rhythms.

lunette *n.* transparent lower eyelid in snakes.

lung *n.* organ specialized for the respiratory uptake of oxygen directly from air and release of carbon dioxide to the air. In vertebrates, lungs are present in air-breathing fishes (the lungfishes, Dipnoi), and tetrapods. In mammals the lungs are a paired mass of spongy tissue made up of finely divided airways lined with moist epithelium extending from the bronchi and ending in small sacs, the alveoli, providing a large surface area for gas exchange between air and bloodstream. The lung of terrestrial molluscs is a cavity under the mantle lined with vascular tissue.

lung book one type of respiratory organ in spiders and some other arachnids, formed of thin hollow "leaves" filled with blood, at which gas exchange takes place.

lungfishes *n.plu*. lobe-finned bony air-breathing fish of the subclass Dipnoi, represented by only four extant genera.

lunula, lunule *n*. small crescent-shaped marking; white opaque crescent of nail near its root.

lunulet *n*. a small bundle.

lupulin *n*. a yellow resinous powder on the flower scales of hops, containing humulone and lupulone.

lupuline, lupulinous *a*. resembling a group of hop flowers.

lupulon(e) *n*. a bitter antibiotic compound obtained from lupulin, effective against fungi and various bacteria.

Lusitanian *a. appl.* certain plants and animals that occur both in the Iberian Peninsula and in coastal regions of the far west of the British Isles, e.g. in western Ireland and Cornwall, and which are thought to be relicts of an interglacial period.

luteal *a. pert.* or like the cells of the corpus luteum; *appl.* hormones: progesterone, relaxin.

lutein *n*. a yellow carotenoid pigment present in the leaves and flowers of many plants, in green and brown algae, in egg yolk and in the corpus luteum.

lutein cells yellow cells found in the corpus luteum of the ovary and formed either from the follicle cells or the cells of the theca interna.

luteinization *n*. formation of the corpus luteum.

luteinizing hormone (LH) glycoprotein hormone secreted by the anterior pituitary, acting on the gonads to stimulate testosterone production and interstitial cell formation in the testes, and oocyte maturation, ovulation and formation of corpus luteum in the ovary.

luteinizing hormone releasing hormone (factor) (LHRH) decapeptide hormone produced by the hypothalamus, causes secretion of luteinizing hormone by cells of anterior pituitary, a similar peptide also known to act as a neurotransmitter in frog.

luteolysis *n*. the breakdown of the corpus luteum which occurs during the cycle of ovulation in mammals.

luteotropin prolactin *q.v.*

luteovirus group plant virus group containing isometric single-stranded RNA viruses that typically cause yellowing of the leaves, type member barley yellow dwarf virus.

luxury genes genes that are expressed only in a particular cell type and specify proteins produced only by that cell type (usually in large amounts), e.g. ovalbumin in the chick oviduct, globin in red blood cell precursors.

lyase *n*. any of a large class of enzymes catalysing the cleavage of C–C, C–O, C–S and C–N bonds by means other than hydrolysis or oxidation, having 2 substrates in one reaction direction, and 1 in the other, in this direction eliminating a molecule (of carbon dioxide, water etc.) creating a double bond, and including the decarboxylases, aldolases, hydratases, dehydratases, and synthases. EC group 4.

lychnidiate *a*. luminous.

lycopene *n*. the red carotenoid pigment of tomatoes and other fruits.

Lycophyta, lycopods, lycophytes *n*., *n.plu.*, *n.plu.* one of the four major divisions of extant seedless vascular plants, with 10–15 living genera comprising club mosses, Selaginella and the aquatic quillwort. They are characterized by a sporophyte with roots, stems and small leaves arranged spirally on the stem, with sporangia solitary and borne on or associated with a sporophyll. Fossil lycophytes are found from the Devonian onwards and include extinct forms such as the woody tree-like lepidodendroids, which are the dominant plants of the Carboniferous coal measures.

Lycopodophyta, Lycopodiales, Lycopsida *n*. Lycophyta *q.v.*

lycopods *see* Lycophyta.

lygophil *a*. preferring shade or darkness.

lymph *n*. fluid bathing all tissue spaces and which drains into the lymphatic vessels and which is in communication with the blood in lymphoid tissues.

lymph gland lymph node *q.v.*

lymph heart contractile expansion of lymph vessel where it opens into a vein, in many vertebrates.

lymph node small structure of lymphoid tissue present at intervals along lymphatics, in which antigens entering blood and lymphatic system are entrapped and stimulate the proliferation of antigen-specific T lymphocytes and antibody-producing plasma cells.

lymphatic a. pert. or conveying lymph; n. vessel carrying lymph.

lymphatic system network of fine capillaries extending throughout the body in vertebrates, connected at points to the blood circulatory system. The lymphatic vessels contain fluid (lymph) draining from the intercellular spaces and which also includes particles of debris, bacteria etc. Lymph contains lymphocytes and macrophages.

lymphoblast n. precursor cell in lymphoid tissue which divides to form mature lymphocytes.

lymphocyte n. small mononuclear white blood cell present in large numbers in lymphoid tissues and circulating in blood and lymph. There are two main functional types, B lymphocytes (q.v.) and T lymphocytes (q.v.), which take part in antigen-specific immune reactions. Natural killer cells (q.v.) are a type of non-specific lymphocyte.

lymphocytic choriomeningitis virus (LCM) virus of the Arenaviridae, infects the membranes of the brain provoking a cellular immmune response.

lymphogenic a. produced in lymphoid tissue.

lymphogenous a. lymph-forming.

lymphoid a. appl. tissue in which lymphocytes are produced, consisting of aggregations of lymphocytes either as discrete organs such as tonsils and lymph nodes, or as diffuse aggregations in connective tissue. Primary lymphoid organs are the bone marrow and thymus in which lymphocytes are produced which then migrate to the secondary lymphoid tissues (e.g. lymph nodes, spleen) where they mature, encounter foreign antigens, and participate in immune reactions.

lymphokine-activated killer cells (LAK) cytotoxic cells produced by the activation of peripheral blood lymphocytes or T cells by the interleukin IL-2.

lymphokines n.plu. polypeptides released by activated T cells in immune reactions and which have a variety of effects, and which include macrophage chemotactic factor, macrophage migration inhibition factor, lymphocyte growth factors (e.g. IL-2, IL-3, B cell growth factor), lymphotoxin, γ interferon, and transfer factor. see individual entries.

lymphoma n. a cancer of lymphoid tissues.

lymphopoiesis n. the production and differentiation of lymphoid cells.

lymphotoxin n. protein toxin produced by T lymphocytes which can kill tumour cells and stimulate the phagocytic activity of neutrophils. alt. tumour necrosis factor-β (TNF-β).

Lyon hypothesis the physical inactivation of one copy (chosen at random) of the X chromosome in somatic cells of early female embryos in humans and other mammals. alt. single active X principle.

Lyonnet's glands paired accessory silk glands in lepidopterous larvae.

lyophil a. appl. solutes which after evaporation to dryness go readily into solution again after addition of liquid.

lyosphere n. a thin film of water surrounding a colloidal particle.

lyra n. triangular lamina or psalterium joining lateral part of fornix of brain, marked with fibres as a lyre; a lyrate pattern as on some bones; a series of chitinous rods forming the stridulating organ in certain spiders.

lyrate a. lyre-shaped, appl. leaves.

lyriform a. lyre-shaped.

lysate a. a suspension of cells that have been broken up (lysed).

Lysenkoism n. a doctrine promoted by the Soviet agriculturalist T. Lysenko in the 1930s and 1940s, which was based on the idea that acquired characteristics could be inherited. This mistaken theory became for some time the only officially permitted theory of genetics in the USSR and led to the suppression of the teaching of Mendelian genetics and work based on modern genetic concepts, and the persecution of geneticists who opposed Lysenkoism.

lysergic acid diethylamide see LSD.

lysigenic, lysigenous a. appl. formation

of tissue cavities caused by degeneration of cell walls in centre of mass.

lysin *n.* any substance that can promote the lysis or dissolution of cells.

lysine (Lys, K) *n.* diaminocaproic acid, a basic amino acid, constituent of protein, essential in the human diet. *see* table in Appendix 1 for chemical formula.

lysis *n.* the breaking down or dissolution of cells.

lysogen *n.* bacterium in which lysogeny has been established.

lysogenesis *n.* the action of lysins.

lysogenic *a. pert.* or involved in lysogeny; *appl.* bacterium carrying a prophage integrated into its DNA; *appl.* a bacteriophage capable of causing lysogeny, *alt.* temperate.

lysogenic immunity ability of a prophage to prevent another phage of the same type from becoming established in the same bacterium.

lysogenize *v.* to produce lysogeny.

lysogeny *n.* in bacteria, the condition where a bacterium infected with a phage carries the phage DNA integrated into the bacterial chromosome as a prophage in which most phage functions are repressed; (*bot.*) formation of cavities within tissues by breakdown of cell walls in centre of mass.

lysosome *n.* membrane-bounded organelle in eukaryotic cells, esp. animal cells, which has a relatively acid internal environment and contains digestive enzymes that process material taken in by phagocytosis and damaged or redundant cellular components. Lysosomes are part of the cell's endomembrane system. In plant cells, the function of the lysosome is taken by the vacuole. *a.* lysosomal. *see* Fig. 1.

lysozyme *n.* widespread enzyme found esp. in animal secretions such as tears, in white of egg, and in some microorganisms, and which splits the glycosidic bond between certain residues in mucopolysaccharides and mucopeptides of bacterial cell walls, resulting in bacteriolysis. EC 3.2.1.17.

lytic *a. pert.* lysis; *pert.* a lysin.

lytic complex the complex of complement components $(C5, 6, 7, 8, 9)_2$ which is formed on the surface of a cell and forms a pore through the plasma membrane, leading to cell lysis.

lytic infection type of infection of a cell by a virus (or bacterium by a phage) in which the virus multiplies inside the cell eventually causing lysis of the cell with release of progeny virus.

lytta *n.* a worm-like structure of muscle, fatty and connective tissue or cartilage, under the tongue of carnivores such as dog; cantharis, a blister beetle.

M

μ heavy chain class corresponding to IgM.

M methionine *q.v.*

M$_r$ relative molecular mass *q.v.*

M13 single-stranded DNA phage of *E. coli.*

MAP microtubule-associated protein, any of various proteins found in close association with the microtubules of the cytoskeleton; mitogen-associated protein, *see* MAP kinase.

MAT mating type locus *q.v.*, of the yeast *S. cerevisiae.*

mC methylcytosine, *see* also 5-methy-lcytosine, DNA methylation.

M-CSF macrophage colony-stimulating factor *q.v.*

MCP Methyl-accepting chemotaxis proteins *q.v.*

mDNA DNA corresponding to the cellular mRNA sequences.

MeNA α-napththylacetic acid methyl ester, a volatile compound used as an artifical auxin.

mepp miniature end plate potential *q.v.*

Met methionine *q.v.*

MetHb methaemoglobin *q.v.*

MHC major histocompatibility complex *q.v.*

MIF macrophage migration inhibition factor *q.v.*

mitDNA mitochondrial DNA.

MLD minimum lethal dose *q.v.*

MLV murine leukaemia virus, an oncogenic retrovirus.

MLR mixed lymphocyte response *q.v.*

MMTV mouse mammary tumour virus, an oncogenic retrovirus.

Mn symbol for the element manganese.

Mo symbol for the element molybdenum.

MPF maturation-promoting factor *q.v.*

mRNA messenger RNA *q.v.*

MSH melanocyte-stimulating hormone *q.v.*

MSV murine sarcoma virus, an RNA tumour virus, various strains, carry oncogenes *v-fos, v-K-ras, v-H-ras.*

mtDNA mitochondrial DNA.

MTOC microtubule-organizing centre *q.v.*

MW molecular weight, *see* relative molecular mass.

myc potential oncogene encoding a protein active in nucleus and believed to be involved in regulating cell division. N-*myc* and c-*myc* are the cellular genes, v-*myc* is an altered counterpart present in the avian RNA tumour virus, avian myelocytomatosis virus.

Macaronesian Floral Region part of the Palaeotropical Realm comprising the islands off the west coast of Africa, e.g. Madeira, the Canary Islands and the Cape Verde Islands.

macerate *v.* to soften or wear away by digestion or other means; *n.* softened tissue in which cells have been separated.

machair *n.* herb-rich calcareous grassland on shell-sand on the western Scottish coast.

machairodont *a.* sabre-toothed.

machlovirus group plant virus group containing isometric single-stranded RNA viruses that typically cause chlorosis of the plant, type member maize chlorotic dwarf virus.

macrander *n.* a large male plant.

macrandrous *a.* having large male plants or male elements.

macraner *n.* male ant of unusually large size.

macrergate *n.* worker ant of unusually large size.

macro- prefix derived from Gk. *makros*, large.

macroalga *n.* multicellular alga, e.g. a seaweed. *cf.* microalga.

macroarthropod *n*. medium or large-sized arthropod whose size is measured in millimetres or centimetres rather than microscopic units.

macrobiota *n*. the population of organisms of a size larger than a few centimetres in any habitat or ecosystem (esp. applied to soil). *cf*. microbiota, mesobiota.

macrobiotic *a*. long-lived.

macroblast *a*. a large cell.

macrocarpous *a*. having large fruit.

macrocephalous *a*. with a large head.

macrochaeta *n*. a large bristle, as on body of some insects. *plu*. macrochaetae.

macrochromosomes *n.plu*. the relatively large chromosomes in a nucleus.

macroclimate *n*. the climate over a relatively large area, generally synonymous with climate in the usual sense.

macroconidium *n*. a large asexual spore or conidium.

macroconjugant *n*. the larger individual of a conjugating pair.

macrocyclic *a. appl*. rust fungi that produce basidiospores, teleutospores and at least one other type of spore (aeciospores and/or uredospores) during their life cycle.

macrocyst *n*. a large reproductive cell of certain fungi; a large cyst or case for spores.

macrocystidium *n*. a long bladder-like sterile cell in the hymenium of some gasteromycete fungi.

macrodactylous *a*. with long digits.

macrodont *a*. with large teeth.

macroelement *n*. element required in large quantities in the nutrition of living organisms. *alt*. macronutrient.

macroevolution *n*. evolutionary processes extending through geological time, leading to the evolution of markedly new genera and higher taxa. *cf*. microevolution.

macrofauna *n*. animals whose size is measured in centimetres rather than microscopic units.

macrofibrils *n.plu*. cellulose fibrils made of bundles of microfibrils, in plant cell walls. They are arranged scattered in primary wall and more ordered in secondary wall.

macrogamete *n*. the larger of two gametes in a heterogametic organism, usually considered equivalent to the ovum or egg.

macrogametocyte *n*. the mother cell of a macrogamete, esp. in protists. *alt*. megagametocyte.

macroglia *n*. general term for certain neuroglial cells which may apply to astrocytes and oligodendroglia, or also to ependyma.

macroglossate *a*. having a large tongue.

macrognathic *a*. having especially well-developed jaws.

macrogyne *n*. female ant of unusually large size.

macroinvertebrate *n*. any invertebrate or invertebrate larva whose size is measured in millimetres or centimetres rather than microscopic units. Such species are one of the main groups of organisms sampled in surveys of water quality.

macrolecithal megalolecithal *q.v.*

macrolymphocyte *n*. any large lymphocyte.

macromere *n*. in cleavage of eggs with unequally distributed yolk, a larger cell of the lower half of morula.

macromerozoite *n*. one of many divisions produced by the macroschizont stage of sporozoan parasites.

macromesentery *n*. one of the larger complete mesenteries of Anthozoa (sea anemones and their relatives).

macromolecule *n*. very large organic molecule such as a protein, nucleic acid or polysaccharide.

macromutation *n*. simultaneous mutation of several characters; a hypothetical step involving a single large mutational change in an organism in some evolutionary theories.

macronotal *a*. with a large notum, as queen ant.

macront *n*. the larger of two sets of cells formed after schizogony in some sporozoans, the macront giving rise to the macrogametes.

macronucleocyte *n*. a white cell of blood in insects, having a large nucleus.

macronucleus *n*. the larger of the two types of nucleus found in cells of ciliate protozoans, corresponding in function to the nucleus of eukaryotic somatic cells. Formed from the micronucleus and disappears and is renewed after sexual reproduction. It contains much of the genome in multiple and fragmented form.

macronutrients *n.plu*. elements required in

relatively large amounts by living organisms for proper growth and development, being major constituents of living matter, e.g. carbon, nitrogen, oxygen, phosphorus, calcium, etc.

macrophage *n.* type of large phagocytic mononuclear white blood cell also widely distributed in tissue, and which ingests invading microorganisms and also scavenges damaged cells and cellular debris.

macrophage colony-stimulating factor (M-CSF) protein growth factor that stimulates the growth of monocytes and macrophages. It is produced by e.g. bone marrow stromal cells and macrophages. The M-CSF receptor is the proto-oncogene *fms*.

macrophage-like cells cells of the reticuloendothelial system in lymphoid tissue, lungs, liver and kidneys and the Langerhans cells of the dermis, which can take up foreign particles, antigens etc.

macrophage migration inhibition factor (MIF) protein produced by activated T lymphocytes which prevents the movement of macrophages.

macrophagous *a.* feeding on relatively large masses of food. *cf.* microphagous.

macrophanerophytes *n.plu.* trees.

macrophyll, macrophyllous megaphyll, megaphyllous *q.v.*

macrophyte *n.* large aquatic plant (e.g. water lily, water crowfoot) as opposed to the phytoplankton and small plants like duckweed.

macroplankton *n.* the larger organisms drifting with the surrounding water, as jellyfish, sargassum weed, etc.

macropodous *a.* having a long stalk, as a leaf or leaflet; having hypocotyl large in relation to rest of embryo; long-footed.

macropterous *a.* with unusually large fins or wings.

macroschizogony *n.* multiplication of macroschizonts; schizogony giving rise to macromerozoites.

macroschizont *n.* stage in life cycle of some sporozoan blood parasites, developing from sporozoite and giving rise to macromerozoites.

macrosclere megasclere *q.v.*

macrosclereids *n.plu.* relatively large columnar sclereids, as in coat of some leguminous seeds.

macroscopic *a.* visible to the naked eye.

macrosepalous *a.* with especially large sepals.

macroseptum *n.* a primary or perfect septum of Anthozoa (sea anemones and their relatives).

macrosiphon *n.* large internal siphon of certain cephalopods.

macrosmatic *a.* with a well-developed sense of smell.

macrospecies *n.* a large polymorphic species, usually with several to many subdivisions.

macrosplanchnic *a.* large-bodied and short-legged.

macrosporangium megasporangium *q.v.*

macrospore *n.* large anisospore or gamete of some protozoans; megaspore *q.v.*

macrosporogenesis *n.* production of macrospores.

macrosporophore *n.* a leafy megasporophyll, i.e. a structure bearing the female sex organs in plants.

macrosporophyll megasporophyll *q.v.*

macrosporozoite *n.* a large sporozoite in which fusion takes place between the products of cell division.

macrostomatous *a.* with a very large mouth.

macrostylous *a.* with long styles.

macrosymbiont *n.* the larger of two symbiotic organisms.

macrotous *a.* with large ears.

macrotrichia *n.plu.* the larger bristles on the wings or body of insects.

macrozoospore *n.* a large motile spore.

macruric *a.* long-tailed.

macula *n.* a spot or patch of colour; a small pit or depression. *plu.* maculae.

macula adhaerens spot desmosome. *see* desmosome.

macula cribrosa area on wall of vestibule of ear, perforated for passage of auditory nerve fibres.

macula lutea oval yellow area at fovea of retina in eyes of some mammals, also called the yellow spot.

maculae *plu.* of macula.

macular *a. pert.* a macula; *pert.* macula lutea.

maculate *a.* spotted.

maculation *n.* the arrangement of spots on

an animal.

Madagascan Floral Region part of the Palaeotropical Realm comprising the island of Madagascar and its neighbouring islands.

madescent *a.* becoming moist; slightly moist.

madrepore *n.* a branching, stony, reef-building coral of the order Scleractinia.

madreporite *n.* in echinoderms, a perforated plate at the end of the stone canal of the water vascular system.

Magendie's foramen central aperture in roof of 4th ventricle of brain, connecting with the subarachnoid space.

maggot *n.* a worm-like insect larva, without appendages or distinct head, as that of some dipterans such as the blowfly.

magnetotaxis *n.* directed movement within a magnetic field, shown by some bacteria which contain small particles of magnetite within their cells. *a.* magnetotactic.

magnetotropism *n.* a tropism in response to lines of magnetic force.

magnocellular layer layer of lateral geniculate nucleus of primate, comprising layers 1 and 2 and composed of large cells.

Magnoliales *n.* an order of dicot trees and shrubs including the families Annonaceae (custard apple), Magnoliaceae (magnolia), Myristicaceae (nutmeg), and others.

Magnoliophyta *n.* in some classifications the name for the angiosperms.

Magnoliopsida *n.* in some classifications an alternative name for the class comprising the dicotyledons.

maintenance behaviour animal behaviour involved in carrying out day-to-day activities such as search for food, mating, reproduction or avoidance of extreme environments.

maintenance ration food required to maintain an animal when the production term (P) in the energy budget is zero.

maize *n. Zea mays*, a member of the Gramineae, originating and first domesticated in the Americas and now widely grown as a crop plant for human and animal feed.

major element macronutrient *q.v.*

major gene a gene having a pronounced phenotypic effect, as distinguished from a modifying gene.

major histocompatibility antigens (MHC antigens) cell-surface glycoproteins carried on the surface of all body cells in mammals, which are involved in antigen recognition in immune responses and are the antigens reacted against in rejection of transplanted organs. Known as HLA antigens in humans and H-2 antigens in mice. MHC antigens are highly variable between individuals, many different variants of each type being present in the population. *see also* antigen recognition, class I MHC antigen, class II MHC antigen, MHC restriction.

major histocompatibility complex (MHC) a cluster of genes encoding the major histocompatibility antigens, some complement proteins, and other surface proteins of immune system cells. In humans it is known as the HLA complex, in mice the H-2 complex. *see also* minor histocompatibility locus.

major locus genetic locus that plays the major role in determining a phenotypic trait.

major worker member of the largest worker subcaste. In ants this is equivalent to a soldier.

mala *n.* cheek, in vertebrates, and similar regions in invertebrates.

malacoid *a.* soft-textured.

malacology *n.* the study of molluscs.

malacophilous *a.* pollinated by the agency of gastropod molluscs, generally snails and slugs.

malacophyllous *a.* with soft and fleshy leaves.

malacopterous *a.* with soft fins.

malacospermous *a.* having seeds covered by a soft coat.

Malacostraca, malacostracans *n., n.plu.* a subclass of crustaceans containing the crabs, lobsters, crayfish, shrimps, woodlice, etc., having a carapace developed to a variable extent.

malacostracous *a.* soft-shelled.

Malagasy *a. appl.* or *pert.* the zoogeographical subregion including Madagascar and adjacent islands.

malar *a. pert.* or in region of cheek; *appl.* bone: the jugal bone or cheekbone.

malaria *n.* sometimes fatal disease characterized by recurrent fevers, caused by parasitic protozoa of the genus *Plasmodium*, of which *P. vivax* causes the most severe form, falciparum malaria. The intermediate hosts and vectors of *Plasmodium* spp. are anopheline mosquitoes.

malate (malic acid) 4-carbon acid of the tricarboxylic acid cycle, where it is oxidized to oxaloacetate by malate dehydrogenase (EC.1.1.1.37), and of other metabolic pathways where it may be oxidized to pyruvate by malate dehydrogenases EC.1.1.1.38–40, and also gives tartness to some fruits such as apples. *see* Appendix 1 (56) for chemical structure.

malate dehydrogenase any of several enzymes that catalyse the reversible conversion of pyruvate or oxaloacetate to malate using either NAD (EC 1.1.1.37–39) - in the tricarboxylic acid cycle – or NADP (EC 1.1.1.40 and 1.1.1.82) – in β-oxidation reactions and in the C4 pathway of carbon dioxide fixation in tropical plants, the NADP-dependent enzymes being also known as malic enzyme.

Malayan *a. appl.* and *pert.* the zoogeographical subregion including Malaysia, Indonesia west of Wallace's line and the Philippines.

Malaysian-Papuan Floral Region part of the Palaeotropical Realm comprising the southern part of the Malaysian peninsula, the islands of Indonesia and New Guinea.

male *n.* individual whose sex organs contain only male gametes, symbol .

male haploidy arrhenotoky *q.v.*

male pronucleus the nucleus the sperm contributes to the zygote.

male sterile *appl.* mutants of normally hermaphrodite plants which do not produce viable pollen, and which are of importance in plant-breeding as they can be used to block self-fertilization and force cross-fertilization, thus avoiding the need for laborious emasculation of the plants used as the female parent. *see* cytoplasmic male sterility.

malic enzyme *see* malate dehydrogenase.

malignant *a. appl.* neoplasms that have become invasive and spread to other sites in the body. *alt.* cancerous.

malleate *a.* hammer-shaped.

mallee scrub vegetation consisting of low bushes of *Eucalyptus* spp. typical of dry subtropical regions of south-east and south-west Australia.

malleoincudal *a. pert.* malleus and incus of middle ear.

malleolar *n.* the vestigial fibula of ruminants. *a.* malleolar.

malleolus *n.* prolongation of lower end of tibia or fibula; a club-shaped or hammer-shaped projection.

malleus *n.* the ear ossicle connecting the eardrum and incus in middle ear of mammals.

mallochorion *n.* the primitive mammalian chorion.

Mallophaga *n.* in some classifications an order of insects known as bird or biting lice, which are ectoparasitic on birds, and have biting mouthparts, secondarily no wings and slight metamorphosis.

malloplacenta *n.* non-deciduate placenta with villi evenly distributed, as in cetaceans and some ungulates.

malonate (malonic acid) *n.* metabolic poison, blocks cellular respiration by reversibly inhibiting succinate dehydrogenase. Occurs as an end-product of metabolism in some plants.

Malpighian body in spleen, a nodular mass of lymphoid tissue ensheathing the smaller arteries; in vertebrate kidney, the glomerulus of convoluted capillaries and the Bowman's capsule that encloses it. *see* Fig. 10.

Malpighian corpuscle *see* Malpighian body.

Malpighian layer in vertebrate skin, the innermost layer of the epidermis, containing dividing cells and cells containing melanin. *alt.* rete mucosum, basal layer.

Malpighian tubule fine, thin-walled excretory tubule, present in large numbers, leading into posterior part of gut in insects.

malt sugar maltose *q.v.*

maltase *n.* enzyme which splits maltose into 2 glucose residues, *see* glucosidase.

maltobiose maltose *q.v.*

maltodextrins *n.plu.* polysaccharides formed during incomplete hydrolysis of starch.

maltose *n.* a disaccharide of glucose, produced by hydrolysis of starch with amylase. It does not occur widely in the free state, but is produced by germinating barley. Hydrolysed by maltase to glucose.

Malvales *n.* order of dicot trees, shrubs and

herbs, often mucilaginous, and including the families Malvaceae (mallow, cotton), Sterculiaceae (cocoa), Tiliaceae (linden), Bombacaceae (baobab, silk cotton) and others.

malvidin *n.* a mauvish anthocyanin pigment.

mamilla *n.* nipple; nipple-shaped structure. *plu.* mamillae.

mamillary bodies two white bodies enclosing grey matter in hypothalamus in the brain, situated beneath floor of third ventricle. *alt.* corpora mamillaria.

mamillary process or tubercle metapophysis *q.v.*

mamillate *a.* covered with small protuberances.

mamillothalamic tract a bundle of nerve fibres running from the corpora mamillaria to the thalamus. *alt.* bundles of Vicq d-Azyr.

mamma *n.* mammary gland. *plu.* mammae.

Mammalia, mammals *n., n.plu.* a class of homoiothermic vertebrates, known from the late Triassic to present, having the body covered with hair, a four-chambered heart, and possessing mammary glands producing milk with which the female suckles her young. Except in the egg-laying monotremes, the young develop inside the mother in the uterus and are born at a more or less mature stage.

mammal-like reptiles a common name for extinct reptiles of the subclass Synapsida, living from the Carboniferous to the Triassic, with synapsid skulls, including the orders Pelycosauria and Therapsida, of which the therapsids were direct ancestors of the mammals.

mammalology *n.* the study of mammals.

mammary *a. pert.* the breast, *appl.* arteries, veins, tubules etc.

mammary gland gland secreting milk in female mammals. *alt.* mamma.

mammiferous *a.* developing mammae; milk secreting.

mammiform *a.* shaped like a breast, *appl.* cap of some fungi.

mammilla mamilla *q.v.*

manchette armilla *q.v.*

mandible *n.* in vertebrates, the lower jaw, either a single bone or comprised of several; in arthropods, paired mouthpart, usually used for biting.

mandibular *a. pert.* the lower jaw in vertebrates, or arthropod mandible, *appl.* arch, canal, foramen, nerve, notch. *alt.* submaxillary.

mandibulate *a.* having a lower jaw; having functional jaws; having mandibles.

mandibuliform *a.* resembling or used as a mandible, *appl.* certain insect maxillae.

mandibulohyoid *a.* in region of the mandible and hyoid.

manducation *n.* chewing or mastication.

manganese (Mn) an essential micro-nutrient for plants.

manicate *a.* covered with entangled hairs or matted scales.

manna *n.* the hardened exudate of bark of certain trees such as the European Ash, *Fraxinus ornus*, and similar substances in other plants such as tamarisk, where its production is caused by infestation with scale insects.

mannan *n.* any of a group of polysaccharides composed of (predominantly) mannose residues linked together, found esp. in cell walls of conifers. *see also* glucomannan.

mannitol *n.* a sweet-tasting polyhydroxy-alcohol derivative of mannose or fructose found in many plants and some algae. Used commercially as a sweetening agent.

mannoglycerate *n.* food storage product in red algae.

mannose *n.* a 6-carbon aldose sugar found in glycoproteins and in many polysaccharides, esp. those of plant cell walls.

manoxylic *a.* having soft wood containing much parenchyma, as cycads.

mantids common name for many members of the Dictyoptera *q.v.*

mantle *n.* (*bot.*) external sheath of fungal mycelium covering the plant rootlets in a mycorrhiza; (*zool.*) fold of soft tissue underlying shell in molluscs, barnacles and brachiopods and which usually encloses a space, the mantle cavity, between it and the body proper; body wall of ascidians; feathers of bird between neck and back.

mantle cell a cell of coat or outer covering of a sporangium.

mantle layer layer of embryonic tissue in spinal cord that represents the future columns of grey matter.

mantle lobes dorsal and ventral flaps of mantle in bivalve molluscs.

manual *n.* quill feather borne on the "hand" region of wing. *alt.* primary feather.

manubrium *n.* tube bearing the mouth hanging down from the undersurface of a medusa.

manus *n.* hand, or part of forelimb corresponding to it, as present in amphibians, reptiles and mammals.

manyplies omasum *q.v.*

map distance the relative distance apart of two gene loci on the same chromosome measured by the extent of recombination between them, expressed as the percentage of recombinants in the total progeny, 1 map unit = 1% recombination.

MAP kinase serine-threonine protein kinase whose activity is stimulated by the action of many growth factors and mitogens. It lies at the end of a multicomponent intracellular signal transduction pathway and can phosphorylate transcription factors, which are then able to stimulate gene expression.

map unit *see* map distance. *alt.* centimorgan.

maquis *n.* vegetation composed of low-growing xerophilous shrubs, found in the Mediterranean area.

marafivirus group plant virus group containing isometric single-stranded linear RNA viruses, type member maize rayado fino virus.

marasmus *n.* a deficiency disease caused by general undernutrition, generally found in infants.

marble gall type of gall found on oak and caused by larvae of a species of gall wasp.

marcescent *a.* withering but not falling off, *appl.* calyx or corolla persisting after fertilization.

marcid *a.* withered, shrivelled.

marginal *a. pert.*, at, or near, the margin, edge or border; *appl.* venation of leaf or insect wing; *appl.* a convolution of frontal lobe of brain; *appl.* plates around edge of carapace of turtles and tortoises; (*bot.*) *appl.* placentation of ovules, the attachment of ovules to margin of carpel.

marginal meristem meristematic tissue that arises on either side of axis of developing leaf and from which growth occurs.

marginal veil a secondary growth around

edge of cap in boletes and agaric fungi.

marginalia *n.plu.* sponge spicules which project beyond the body surface.

marginate *a.* having a distinct margin in structure or colouring.

marginella *n.* ring formed by part of cutis proliferating beyond margin of gills, in certain fungi with an exposed hymenium.

marginiform *a.* like a margin or border in appearance.

marginirostral *a.* forming the edges of a bird's bill.

marijuana, marihuana *n.* the drug cannabis *q.v.*; sometimes used for the whole plant.

marita *n.* sexually mature stage in trematode life history. *a.* marital.

marker *n.* an identifying factor; a gene or other DNA of known location and effect which is used to track the inheritance etc. of other genes whose exact location is not yet known.

marmorate *a.* of marbled appearance.

marrow bone marrow *q.v.*

marsh *n.* plant community developing on wet but not peaty soil.

Marsupialia, marsupials *n., n.plu.* the only order in the Metatheria, a group of mammals found only in Australia and South America, in which the placenta does not develop or is not as efficient as that of therian mammals, so that the young are borne in a very immature state. They then migrate to a pouch (marsupium) where they are suckled until relatively mature. Marsupials include kangaroos, wallabies and opossums, and, in Australia, many other species adapted to fill ecological niches filled by therian mammals elsewhere. *alt.* metatherians.

marsupium *n.* the pouch in which marsupials suckle their young.

mask *n.* a hinged prehensile structure, corresponding to adult labium, peculiar to dragonfly nymph.

masked *a. appl.* virus whose presence in cell is difficult to detect because its multiplication is prevented by superinfection with another virus; personate, *appl.* corolla.

mass extinction any of the various episodes in evolutionary history in which numerous large groups of organisms disappear from the fossil record over a relatively short time (as the extinction of the dinosaurs and other groups at the end of the Cretaceous). Mass extinctions are generally explained by sudden changes in climate, of various causes.

mass flow a theory of the way in which materials are translocated through phloem, which proposes that the cause of movement is the difference in the hydrostatic pressure at each end of a sieve tube, resulting in flow of contents along the tube.

mass provisioning the act of storing all of the food required for the development of a larva at the time the egg is laid.

masseter *n.* muscle that raises lower jaw and assists in chewing.

massula *n.* a mass of microspores in a sporangium of some pteridophytes; a mass of pollen grains in orchids.

mast *n.* the fruit of beech and some related trees.

mast cells cells of the haematopoietic lineage, often amoeboid, with large nucleus and very granular cytoplasm, found in connective and fatty tissue where they secrete heparin and histamine. Involved in some hypersensitivity reactions.

mastax *n.* pharynx of rotifers, containing structures used to grind small food.

mastication *n.* process of chewing food until reduced to small pieces or a pulp. *v.* masticate.

masticatory stomach gastric mill *q.v.*

mastigium *n.* defensive posterior lash of some larvae.

mastigobranchia *n.* an outgrowth from basal joint of crustacean limb and extending upwards in a thin sheet between gills.

Mastigomycota *n.* in modern classifications, a major division of the Fungi including the classes Chytridiomycetes, Hyphochytridiomycetes, Plasmodiophoromycetes and Oomycetes, i.e. simple, generally aquatic, often microscopic fungi with motile flagellate zoospores and/or gametes.

mastigoneme *n.* filaments projecting laterally from the flagellar sheath in the "tinsel" flagella of some flagellates.

mastoid *a.* nipple-shaped, *appl.* a process of temporal bone, and to foramen, fossa and notch associated with it.

mastoideosquamous *a. pert.* mastoid and

squamous parts of temporal bone.

masto-occipital *a. pert.* occipital bone and mastoid process of temporal bone.

mastoparietal *a. pert.* parietal bone and mastoid process of temporal.

maternal developmental determinants products laid down in the ovum in the mother and which direct the development of the zygote.

maternal effect *appl.* genes that must be functional in the mother to produce a normally developing embryo, regardless of the genetic constitution of the embryo itself. In general they are thought to encode products that are laid down in the egg as it is being formed in the mother and which direct the earliest stages of development, which are completed before the embryo's own genes are activated. The embryo's genes (the zygotic genes) cannot therefore compensate for any deficiencies in the action of maternal effect genes. Maternal effect mutations are mutations that render such genes inactive in the mother, resulting in an inviable or abnormal embryo, even when the embryo is heterozygous for the mutant gene.

maternal inheritance the inheritance of genes carried by mitochondria, chloroplasts, and any other cytoplasmic genes through the maternal line only, as only the egg contributes cytoplasm to the zygote; preferential survival in a cross of genetic markers provided by one parent. *cf.* maternal effect.

mating factors protein pheromones secreted by cells of different mating types in yeasts and other unicellular organisms, which attract cells of an appropriate mating type and induce conjugation.

mating type genetically determined property of e.g. bacteria, ciliates, fungi and algae, determining their ability to conjugate and undergo sexual reproduction with other individuals in the population. In the case of yeast only cells of opposite mating type can conjugate, in e.g. ciliates, individuals which can conjugate are said to belong to the same mating type.

mating type locus genetic locus in fission and budding yeasts which determines mating type and which in some strains undergoes frequent rearrangements of DNA sequence resulting in a switch in mating type. *see also* cassette model.

matriclinal *a*. with inherited characteristics more maternal than paternal.

matrifocal *a. pert.* a society in which most of the activities and personal relationships are centred on the mothers.

matrilineal *a*. passed from a mother to her offspring.

matrilinear inheritance the inheritance of cytoplasmic genes, such as those carried in mitochondria, through the mother only, as only the mother contributes cytoplasm to the zygote.

matrix *n*. medium in which a substance is embedded; ground substance of connective tissue; part beneath body and root of nail; body upon which a lichen or fungus grows; substance in which a fossil is embedded; in mitochondria, the inner region enclosed by the inner mitochondrial membrane.

matromorphic *a*. resembling the female parent in morphological characters.

mattula *n*. fibrous network covering petiole bases of palms.

maturase RNA maturase *q.v.*

maturation *a. appl.* maturing processes.

maturation *n*. ripening; the process of becoming mature, fully-differentiated and fully functional; *appl.* divisions by which gametes are produced from primary gametocytes, during which meiosis occurs; the automatic development of a behaviour pattern which becomes progressively more complex as the animal matures and which does not involve learning.

maturation-promoting factor (MPF) cytoplasmic protein factor first identified in mature *Xenopus* eggs which can induce the onset of meiosis in immature eggs. It is also present in the cytoplasm of mammalian and other eukaryotic cells, where it induces mitosis in the normal cell cycle. Now identified as a complex of a cyclin and a protein kinase of the cdc2 family.

Mauthner cell one of a pair of large neurones in medulla of brain of teleost fish, controls tail movement.

Mauthner's cells cells forming a layer between myelin sheath and axon membrane.

maxilla *n*. the upper jaw; a paired appendage on head or cephalothorax of most arthropods, posterior to the mandible, present in one or two pairs and modified in various ways in different groups.

maxillary *a. pert.* or in region of upper jaw in vertebrates, or of arthropod maxilla.

maxillary glands paired excretory organs opening at base of maxilla in Crustacea.

maxilliferous *a*. bearing maxillae.

maxilliform *a*. like a maxilla.

maxilliped *n*. appendage on cephalothorax of crustaceans and some other arthropods, present in one, two or three pairs, posterior to the maxillae, and generally used to hold and manipulate food.

maxillodental *a. pert.* jaws and teeth.

maxillojugal *a. pert.* jaws and cheek bone.

maxillolabial *a. pert.* maxilla and labium, *appl.* dart in ticks.

maxillomandibular *a. appl.* arch forming jaws of primitive fishes.

maxillopalatine *a. pert.* jaw and palatal bones.

maxillopharyngeal *a.* pert. lower jaw and pharynx.

maxillopremaxillary *a. pert.* whole of upper jaw; *appl.* jaw when maxilla and premaxilla are fused.

maxilloturbinal *n*. bone arising from the lateral wall of nasal cavity, which supports sensory epithelium.

maxillule *n*. 1st maxilla in crustaceans, where there is more than one pair. *alt.* maxillula.

maxim *n*. an ant of the large worker or soldier caste. *cf.* minim.

maximum allowable concentration of pollutants, the concentration deemed in regulations to be safe to healthy adults in the work-place, assuming they are not exposed to the pollutant outside working hours.

maximum permissible body burden the concentration of a radioisotope that will not deliver more than the maximum permissible dose to any organ, if inhaled or ingested at a normal rate.

maximum permissible dose the dose of ionizing radiation, accumulated over a given time, that is considered not to result in any harmful effects to the individual over their lifetime or to cause genetic damage that might affect their descendants.

maximum sustainable yield the maximum crop or yield that can be harvested

from a plant or animal population each year
without harming it.

maxithermy *n.* the maintenance of body
temperature at a maximum for as long as
possible. *a.* maxithermic, *appl.* animals that
can do this.

mayflies common name for the Ephemeroptera *q.v.*

M cytotype a maternal cytotype of *D.
melanogaster* which when crossed with
males carrying P factors produce dysgenic
progeny presumably as the result of activation of P element transposition when
exposed to the M cytotype. *see* hybrid
dysgenesis.

meadow *n.* permanent grassland, esp. one
that is mown for hay and not grazed in summer. *cf.* ley, pasture, water meadow.

mealworm *n.* larva of the beetle *Tenebrio*,
which lives in grain stores.

mealy *a.* covered with a powder resembling
flour or coarse ground cereal.

mealy bug scale insect *q.v.*

meatus *n.* passage or channel, as auditory
meatus, nasal meatus, etc.

mechanical tissue supporting tissue *q.v.*

mechanoreceptor *n.* specialized sensory
structure sensitive to mechanical stimuli
such as extension, contact, pressure, or
gravity.

Meckel's cartilage in elasmobranch fish,
the skeletal element forming lower jaw, and
which in other vertebrates forms the axis
around which the membrane bones of
lower jaw are arranged.

meconidium *n.* the sessile or stalked
medusa lying outside the capsule of the reproductive structure of certain hydrozoans.

meconium *n.* waste products of pupa or
other embryonic form; contents of intestine of newborn mammal.

Mecoptera *n.* an order of slender carnivorous insects with complete metamorphosis, commonly called scorpion flies, having
biting mouthparts, long slender legs, and
membranous wings lying along the body
in repose.

media *n.* the longitudinal vein running
through the central region of most insect
wings; *plu.* of medium.

media worker in ants with three or more
worker subcastes, an individual belonging
to the medium-sized subcaste(s).

mediad *a.* towards but not quite on the
midline or axis.

medial *a.* situated in the middle; *n.* the main
middle vein of insect wing.

median *a.* lying or running in axial plane;
intermediate; middle; *n.* the middle variate
when a set of variates are arranged in order of magnitude.

median eminence portion of hypothalamus situated immediately above the
pituitary gland and which synthesizes and
secretes peptide-releasing hormones that
are transported to the anterior pituitary via
the hypophyseal portal blood   system.

median lethal dose (LD$_{50}$) dose of a toxic
chemical or of a pathogen that kills 50%
of the animals in a test sample within a
certain time.

median nerve nerve arising from median
and lateral cord of brachial plexus, with
branches in forearm.

mediastinal *a. pert.* or in region of
mediastinum.

mediastinum *n.* the part of thoracic cavity
on the midline between the right and left
pleura.

mediator *n.* a nerve cell connecting or intermediate between a receptor and effector; any enzyme, hormone, etc. which is
involved in and influences a metabolic or
other biochemical pathway.

mediocentric *a. appl.* chromosomes having a centrosome in the centre, dividing
them into two more-or-less equal arms.

Medio-Columbian Sonoran *q.v.*

mediocubital *n.* a cross vein between posterior media and cubitus of insect wing.

mediodorsal *a.* in the dorsal middle line.

mediolateral *a. appl.* axis, from the median
plane outward to both sides.

mediolecithal *a. appl.* eggs having a moderate amount of yolk.

mediopalatine *n.* between palatal bones,
appl. a cranial bone of some birds.

mediopectoral *a. appl.* middle part of sternum (breast bone).

mediostapedial *a.* that part of columella
of the ear external to stapes.

mediotarsal *a.* between tarsal bones.

medioventral *a.* in the middle ventral line.

mediterranean *appl.* European climate
characterized by hot dry summers and mild
wet winters.

Mediterranean Floral Region that part of the Palaearctic Realm comprising southern Europe and North Africa, around the Mediterranean Sea.

medithorax *n.* middle part of the thorax.

medium *n.* nutritive material in which microorganisms, cells and tissues are grown in the laboratory. *plu.* media.

medulla *n.* central part of an organ or tissue, interior to the cortex; marrow of bones; medulla oblongata *q.v.*; pith or central region of plant stem; loose hyphae in a tangled fungal structure such as rhizomorphs or fruit bodies.

medulla oblongata bulbous upward prolongation of the spinal cord, the lowest part of the brain, controlling heartbeat, respiration and blood pressure.

medullary *a. pert.* or in region of medulla.

medullary canal central cylindrical hollow of a long bone which contains the marrow.

medullary canal, folds, groove, plate neural canal, folds, groove, plate *q.v.*

medullary phloem internal phloem in a vascular bundle with two layers of phloem, as in cucurbits.

medullary ray (*bot.*) rays of parenchyma tissue extending from pith to outer edge of vascular tissue in plant stem; (*zool.*) bundles of straight urine-carrying tubules in medulla of kidney.

medullary sheath the fatty (myelin) sheath surrounding the axis of some nerve fibres. *alt.* myelin sheath; (*bot.*) ring of protoxylem around the pith in certain stems.

medullated *a.* of nerve fibres, possessing a myelin sheath; (*bot.*) *appl.* plant stems having pith.

medulliblasts *n.plu.* cells of embryonic nervous tissue, arising in the medulla or central portion, and which produce neuroblasts (embryonic neurones) and spongioblasts (embryonic glial cells).

medullispinal *a. pert.* the medulla or central portion of the spinal cord.

medusa *n.* one of the forms of individuals of coelenterates of the classes Hydrozoa (hydroids) and Scyphozoa (jellyfish). It is bell-shaped, with a tube hanging down in the centre ending in a mouth, and tentacles around the edge of the bell. It forms the free-swimming sexual reproductive stage of most hydrozoans, and is large and conspicuous in jellyfish. *plu.* medusae.

medusoid *a.* resembling or developing into a medusa.

mega- prefix derived from Gk. *megas*, large.

megacephalic *a.* with abnormally large head; having a cranial capacity, in humans, of over 1450 cm³.

megachiroptera *n.* fruit-eating bats (flying foxes), of the mammalian order Chiroptera. *cf.* microchiroptera.

megachromosomes *n.plu.* large chromosomes forming an outer set in certain sessile ciliate protozoans. *cf.* microchromosomes.

megacins *n.plu.* a group of bacteriocins produced by strains of *Bacillus megatherium*.

megagamete *n.* a cell regarded as an ovum or its equivalent, that develops from a megagametocyte in sporozoans.

megagametocyte *n.* a cell developing from a merozoite and giving rise to a megagamete in sporozoans.

megagametophyte *n.* in heterosporous plants, the female gametophyte, which develops from a megaspore.

megakaryocyte *n.* giant amoeboid cell in bone marrow with a single lobed nucleus, and which gives rise to blood platelets.

megalaesthetes *n.plu.* sensory organs, sometimes in the form of eyes, in chitons (Amphineura).

megalecithal megalolecithal *q.v.*

megaloblast *n.* a large erythroblast precursor cell.

megalolecithal *a.* containing large amounts of yolk.

megalopic *a.* belonging to the megalops stage.

megalops *n.* a larval stage of certain crustaceans such as crabs, having large stalked eyes and a crab-like cephalothorax.

megalospheric *a. appl.* to many-chambered shells of foraminifera which have a large initial chamber.

megamere macromere *q.v.*

megameric *a.* with relatively large parts.

meganephridia *n.plu.* large nephridia, occurring as one pair per segment.

meganucleus macronucleus *q.v.*

megaphanerophyte *n.* a tree exceeding 30 m in height.

megaphyll *n.* type of leaf present in most vascular plants, being relatively large and having a network of veins (vascular tissue), and which is associated with the presence of leaf gaps in the stele of the stem; a large leaf, esp. as produced by ferns.

megaphyllous *a*. having relatively large leaves.

megaplankton macroplankton *q.v.*

megasclere *n*. skeletal spicule of general supporting framework of a sponge.

megasorus *n*. a sorus containing megasporangia.

megasporangiate *a*. composed of or producing megasporangia, *appl.* cones.

megasporangium *n*. a megaspore-producing sporangium; nucellus of ovule; sometimes used incorrectly for whole ovule.

megaspore *n*. in heterosporous plants, the spore that gives rise to the female gametophyte, and which is formed in a megasporangium; in any organism that produces two types of spore, the larger spore.

megaspore mother cell the diploid cell in the megasporangium of heterosporous plant, in which meiosis will occur to produce one or more megaspores.

megasporocyte *n*. the embryo sac mother cell, a diploid cell in ovary of plant that undergoes meiosis, producing four haploid megaspores.

megasporophyll *n*. leaf or leaf-like structure on which a megasporangium develops, in flowering plants known as a carpel.

megatherm *n*. a tropical plant; a plant requiring moist heat and thriving in temperatures between 20 and 35°C.

megazooid *n*. the larger zooid resulting from binary or other fission.

megazoospore *n*. a large zoospore, as in reproduction of certain radiolarian protozoans; a zoogonidium of certain algae.

megistotherm *n*. a plant that thrives at more-or-less uniformly high temperature.

meio- prefix derived from Gk. *meion*, less.

meiocyte *n*. a reproductive cell prior to meiosis.

meiofauna mesofauna *q.v.*

meiogenic *a*. promoting nuclear division, esp. meiosis.

meiolecithal *a*. having little yolk.

meiomery *n*. condition of having fewer than the normal number of parts.

meiophase *n*. in life cycle the stage at which meiosis occurs with reduction of chromosome number from diploid to haploid.

meiophylly *n*. the suppression of one or more leaves in a whorl.

meiosis *n*. a type of nuclear division which results in daughter nuclei each containing half the number of chromosomes of the parent, i.e. chromosome number is reduced from diploid to haploid, and comprises two distinct nuclear divisions, the 1st and 2nd meiotic divisions, which may be separated by cell division, the actual reduction in chromosome number taking place during the first division. *see* Fig. 6. *cf*. mitosis. *see also* reduction division.

meiosporangium *n*. thick-walled diploid sporangium producing haploid zoospores by meiosis.

meiospore *n*. spore produced by meiosis; uninucleate haploid spore formed in a meiosporangium.

meiosporocyte megaspore mother cell *q.v.*

meiostemonous *a*. having fewer stamens tham sepals and petals.

meiotaxy *n*. suppression of a whole whorl of leaves or floral parts.

meiotherm *n*. a plant that thrives in a cool temperate environment.

meiotic *a*. *pert*. or produced by meiosis.

meiotic apogamy generative apogamy *q.v.*

meiotic drive any mechanism that operates during meiosis in heterozygotes to produce a disproportionate representation of one member of a chromosome pair in the gametes. Usually due to a mutation, e.g. the *SD* allele at the segregation distorter locus in *Drosophila*.

meiotic spindle *see* spindle.

Meissner's corpuscles sensory nerve endings associated with sense of pain in skin of fingers, lips, etc. *see* Fig. 10.

Meissner's plexus a plexus of nerve fibres in submucous coat of small intestine wall.

Melanconia *n*. form class of deuteromycete fungi that reproduce by conidia borne in acervuli, and which includes many plant pathogenic fungi causing anthracnose diseases (e.g. *Colletotrichum*, *Marssonina*). *alt*. Melanconiales.

melanin *n*. any of a range of black or brown pigments produced from tyrosine by the enzyme tyrosinase and giving colour to animal skin etc., and also found in some plants.

melanism *n.* excessive development of black pigment; industrial melanism *q.v.*

melanoblast, melanocyte *n.* cell in which melanin is formed in the Malpighian layer of the skin.

melanocyte-stimulating hormone (MSH) peptide hormone produced by the pars intermedia of the adenohypophysis (pituitary gland) from the precursor proopiocorticotropin, and which causes dispersal of melanin granules in melanophores of some animals, resulting in a generalized darkening of the skin. *alt.* intermedin.

melanogen *n.* a colourless compound formed by reduction of the red oxidation product of tyrosine, and which is oxidized to melanin.

melanogenesis *n.* formation of melanin.

melanoids *n.plu.* dark-brown or black pigments related to melanin.

melanoma *n.* a dark pigmented mole on skin, some forms of which, malignant melanoma, are highly invasive.

melanophore *n.* cell containing melanin; chromatophore *q.v.*

melanosome *n.* pigment granule synthesizing melanin and rich in the enzyme tyrosinase, found in animal tissue; dark pigment mass associated with ocellus, as in certain dinoflagellates.

melanospermous *a. appl.* seaweeds with dark-coloured spores.

melanosporous *a.* with dark-coloured spores.

melanotic *a. appl.* animals and plants which are much darker than the usual colour, as a

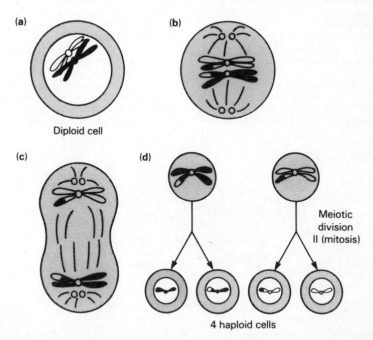

Fig 6 Meiosis. (a) Diploid nucleus, prophase I, (b) metaphase I, (c) nuclear division I, (d) meiotic division II (equivalent to a mitotic division), producing 4 haploid nuclei from the original diploid nucleus. For clarity, only one pair of homologous chromosomes is shown throughout.

result of unusual or excessive production of melanin.

melatonin *n*. *N*-acetyl-5-methoxytrypt-amine, a compound isolated from pineal gland, hypophysis and peripheral nerve tissue, and which causes the aggregation of melanin granules in melanocytes causing lightening of skin colour.

meliphagous *a*. honey-eating.

melittophile *n*. an organism that must spend at least part of its life cycle with bee colonies.

melliferous *a*. honey-producing.

mellisugent *a*. honey-sucking.

mellivorous *a*. feeding on honey.

melting separation of the strands of the DNA double helix by heating, or in acid or alkaline conditions.

melting temperature (T_m) temperature at which half the macromolecules in a solution are denatured (i.e. protein tertiary structure lost or DNA dissociated into single strands).

member *n*. a limb or organ; a well-defined part or organ of a plant.

membrana propria basement membrane *q.v.*

membranaceous *a*. having the consistency or structure of a membrane.

membranal *a*. pert., or within membranes.

membrane *n*. a thin film, skin or layer of tissue covering a part of an animal or plant, or separating different layers of tissue; of cells, an organized layer a few molecules thick forming the boundary of the cell (the cell or plasma membrane) and of intracellular organelles, and composed of 2 oriented lipid layers in which proteins are embedded, and which acts as a selective permeability barrier. *see also* fluid mosaic model.

membrane anchor type of linkage involving a lipid molecule that attaches some proteins to one face of a membrane.

membrane bone dermal bone *q.v.*

membrane domain an area of biological membrane distinguished by a particular protein and/or lipid composition, and which e.g. in epithelial cells is isolated from other membrane domains by morphological features such as tight junctions.

membrane potential electrical potential difference present across the plasma membranes of all living cells, so that the cytoplasmic side of the membrane is negative with respect to the exterior. It is generated by the movements of ions (largely Na^+ and K^+) in and out of the cell, and in most eukaryotic cells is maintained by the activity of membrane Na^+-K^+ ATPases.

membrane protein any protein that forms part of a biological membrane, and which

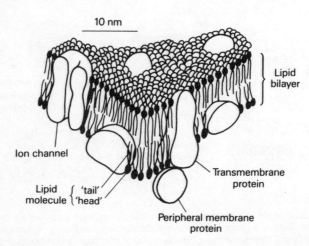

Fig 7 The structure of a cell membrane.

may be attached to either the external or cytoplasmic face, or be firmly anchored in the membrane, spanning it completely.

membrane skeleton highly regular cytoskeleton underlying the red blood cell plasma membrane and giving shape to the cell.

membrane traffic the flow of membrane between different endomembrane compartments within eukaryotic cells (e.g. between endoplasmic reticulum, Golgi apparatus and plasma membrane) in the form of small vesicles of membrane which pinch off one set of membranes and fuse with another.

membrane transport proteins proteins that carry specific molecules across biological membranes which are otherwise impermeable to them.

membrane vesicle small intracellular sac of membrane, in eukaryotic cells usually contains molecules being transported around the cell.

membranelle *n.* undulating structure formed by fusion of clumps of cilia in some ciliate protozoans and some rotifers. *alt.* membranella.

membraniferous *a.* enveloped in or bearing a membrane.

membranoid *a.* resembling a membrane.

membranous *a.* formed of membrane; thin, dry and delicate, often transparent.

membranous cranium during development a covering derived from mesoderm, enclosing the brain.

membranous labyrinth the cochlea, semicircular canals, etc. of the inner ear, separated from the surrounding bone by perilymph.

membranula membranelle *q.v.*

membranule *n.* a small opaque area in anal area of wing of some dragonflies.

memory cells long-lived B and T cells which undergo only limited differentiation after a first exposure to an antigen and which are the basis of immunological memory. They are stimulated into activity on subsequent encounters with the same antigen, when they rapidly differentiate into effector cells. *cf.* plasma cells. *see* secondary immune response.

menacme *n.* interval between first and last menstruation, i.e. life between menarche and menopause.

menadione *n.* a vitamin K_2 analogue (2-methyl-1,4-naphthoquinone) sometimes known as vitamin K_3, *see* vitamin K.

menarche *n.* first menstruation; age at first menstruation.

Mendelian inheritance inheritance of genes or characters which are inherited according to Mendel's laws, i.e. nuclear genes.

Mendel's laws laws of inheritance first proposed by Gregor Mendel in the 19th century from breeding experiments in plants, which describe some basic principles of inheritance in sexually-reproducing organisms, the first law being that of the segregation of alleles, the second being the law of independent assortment of genes. *see* segregation of alleles, independent assortment.

Mendosicutes *n.* a division of the kingdom Monera or Prokaryotae, comprising a single class, the Archaebacteria.

meningeal *a. pert.* or in region of meninges.

meninges *n.plu.* the three membranes enclosing the brain and spinal cord, from ouside to inside: the dura mater, arachnoid and pia mater. *sing.* meninx.

meningitis *n.* inflammation of the meninges.

meningocyte *n.* a phagocytic cell of the subarachnoid space (the space between arachnoid and pia mater).

meningosis *n.* attachment by means of membranes.

meningospinal *a. pert.* membranes surrounding the spinal cord.

meningovascular *a. pert.* meningeal blood vessels.

meninx *sing.* of meninges *q.v.*

meniscus *n.* a fibrocartilage sandwiched between articulating faces of joints subject to violent concussion; intervertebral disc; the end of a touch receptor, being the terminal extension of axon and tactile corpuscle.

menognathous *a.* with biting jaws, *appl.* insects.

menopause *n.* cessation of ovulation and menstruation in women.

menorhynchous *a.* with persistent sucking mouthparts, *appl.* insects.

menotaxis *n.* compensatory movements to

maintain a given direction of body axis in relation to sensory stimuli, esp. light, but not necessarily moving towards or away from it; maintenance of visual axis during locomotion.

mensa *n.* grinding surface of a tooth.

menses *n.plu.* menstrual discharge.

menstrual *a. pert.* menstruation, *see* menstrual cycle; monthly; lasting for a month, as of flower.

menstrual cycle monthly cycle of ovulation and menstruation in human females and some other primates, in which the lining of the uterus thickens in preparation to receive the fertilized ovum, and if fertilization and conception does not occur is shed along with the unfertilized ovum in a short period of menstrual bleeding.

menstruation *n.* discharge of unfertilized ovum plus layer of uterine wall that occurs periodically in humans and some other higher mammals.

mental *a. pert.* the mind; *pert.* or in the region of the chin, *appl.* nerve, spines, tubercle, muscle.

mentigerous *a.* supporting or bearing the chin.

mentum *n.* the chin, or a similar region in some invertebrates.

mericarp *n.* a one-seeded unit that breaks off a composite fruit at maturity.

mericlinal *a. appl.* structures in which inner tissue of one sort is only partly surrounded by tissue of another.

meridional *a.* running from pole to pole of a structure, as along a meridian.

meridional furrow a longitudinal furrow extending from pole to pole of a fertilized egg undergoing cleavage.

meridiungulates *n.plu.* extinct group of South American hoofed mammals, present during the Tertiary, which included the camel-like litopterns and the notoungulates.

merisis *n.* increase in size owing to cell division.

merismatic *a.* dividing or separating into cells or segments.

merispore *n.* a segment or spore of a multicellular spore body.

meristele *n.* the branch of a stele supplying a leaf. *alt.* leaf trace.

meristem *n.* plant tissue capable of undergoing mitosis and so giving rise to new cells and tissues. It is located at growing tip of shoot and root (apical meristems), in the cambium and cork cambium encircling some plant stems (lateral meristems) and in leaves, fruits, etc.

meristematic, meristemic *a. pert.* meristem.

meristematic ring in developing plant axis, tube of meristematic tissue between cortex and pith, subtending the apical meristem and giving rise to the vascular tissues.

meristic *a.* segmented; divided off into parts.

meristic variation changes in numbers of parts or segments, and in the geometrical relations of parts.

meristo- word element from Gk. *meristos*, divided.

meristogenetic, meristogenous *a.* developing from a meristem; developing from a single hyphal cell or group of contiguous cells.

Merkel cell sensory cell sensitive to touch, in skin and in submucosa of mouth.

mermaid's purse the horny egg case of elasmobranch fishes.

mero- word element from Gk. *meros*, part, or from Gk. *mēros*, thigh.

meroandry *n.* the condition of having a reduced number of testes.

meroblast *n.* intermediate stage between schizont and merozoite in some sporozoan parasites.

meroblastic *a. appl.* eggs that undergo only partial cleavage during development, as a result of the presence of large amounts of yolk, the embryo developing from a small cap of cells on top of the yolk, as eggs of birds, reptiles.

merocerite *n.* fourth segment of crustacean antenna.

merocrine *a. appl.* glands whose secretion accumulates below the free surface of the cells through which it is released without destroying the cells, e.g. goblet cells and sweat glands.

merocytes *n.plu.* nuclei formed by repeated division of supernumerary sperm nuclei, as in egg of selachians, reptiles and birds.

merodiploid *n.* cell that is partially diploid, e.g. a bacterium that is partially diploid as a result of the introduction (e.g. by transduction) of part of a genome from another bacterium of the same species.

merogamete *n.* protozoan individual smaller than normal cells and functioning as a gamete.

merogenesis *n.* formation of parts; segmentation.

merognathite *n.* fourth segment of crustacean mouthpart.

merogony *n.* development of normal young of small size from part of an egg in which there was no female pronucleus.

meroistic *a. appl.* ovariole containing nutritive or nurse cells.

meromixis *n.* a type of genetic exchange in bacteria where transfer of genetic material is in one direction only.

meromorphosis *n.* regeneration of a part with the new part less than that lost.

meromyarian *a.* with only a few longitudinal rows of muscle cells, as in some nematodes.

meromyosin *n.* either of 2 fragments formed by trypsin cleavage of myosin, light meromyosin (LMM) containing the "tail" of the molecule, and heavy meromyosin (HMM) the globular "heads" of the molecule and ATPase activity.

meron *n.* posterior portion of coxa of insects; sclerite between coxa of middle and hind leg in dipterans.

meront *n.* any unicell formed by cleavage or schizogony; a uninucleate schizont stage in some sporozoan parasites, succeeding the planont stage.

meroplankton *n.* temporary plankton, consisting of eggs and larvae; seasonal plankton.

meropodite *n.* femur in spiders; fourth segment of thoracic appendage in crustaceans.

merosome *n.* body segment or somite.

merospermy *n.* condition where the nucleus of a sperm does not fuse with that of the egg, and development of a new individual is parthenogenetic.

merosporangium *n.* outgrowth from the apex of a sporangiophore, producing a row of spores.

merosthenic *a.* with unusually developed hindlimbs.

Merostomata *n.* a class of aquatic arthropods, the horseshoe crabs, which breathe by gills and have chelicerae (claws) and

walking legs on the prosoma, and an opisthosoma with some segments lacking appendages.

merotomy *n.* segmentation or division into parts.

merozoite schizozoite *q.v.*

merozygote *n.* a zygote containing only part of the genome of one of the two cells or gametes from which it is formed.

mesadenia *n.plu.* mesodermal accessory genital glands in insects.

mesal *a.* medial (*q.v.*) or mesial (*q.v.*).

mesarch *a. appl.* xylem having metaxylem developing in all directions from the protoxylem, characteristic of ferns; having the protoxylem surrounded by metaxylem; beginning in a mesic environment, *appl.* seres.

mesaxonic *a.* with the line dividing the foot passing up the middle digit, as in perissodactyls.

mescaline *n.* a hallucinogenic alkaloid obtained from the mescal cactus, *Lophophora williamsii*.

mesectoderm *n.* upper layer of germinal disc before ectoderm and mesoderm become differentiated.

mesencephalic *a. pert.* or involving mesencephalon.

mesencephalon midbrain *q.v.*

mesenchyme *n.* undifferentiated cells of embryo, derived from mesoderm, and which differentiate into muscle and connective tissue.

mesendoderm *n.* cells in embryo that will give rise to both endoderm and mesoderm.

mesenterial *a. pert.* a mesentery, *appl.* filaments of anthozoans.

mesenteric *a. pert.* a mesentery, *appl.* arteries, glands, nerves, veins etc.; *appl.* lymph nodes associated with the small intestine.

mesenteriole *n.* a fold of peritoneum derived from mesentery and retaining vermiform process or appendix in position.

mesenterium mesentery *q.v.*

mesenteron *n.* the mid-gut, portion of alimentary canal lined with endoderm and derived from the archenteron; in corals and sea anemones, the main digestive cavity as distinct from spaces between mesenteries.

mesentery *n.* a fold of the peritoneum serving to hold viscera in position; a muscular partition extending inwards from body wall in coelenterates.

mesepimeron *n.* portion of lateral exoskeletal plate of insect mesothorax.

mesethmoid *n.* bone of skull separating the nasal cavities.

mesiad *a.* towards or near the middle plane.

mesial, mesian *a.* in the middle vertical or longitudinal plane.

mesic *a.* conditioned by a temperate moist climate.

meso-, mesi-, mes- word elements from Gk. *mesos*, middle, signifying situated in the middle, intermediate, neither to one end or the other of a range of conditions.

mesobenthos *n.* animal and plant life of the sea bottom at depths between 200 and 1000 m.

mesobiota *n.* the population of organisms in an ecosystem, habitat, etc. (esp. soil) which range in size from approx. 200 μm to 1 cm, i.e. larger than bacteria and unicellular algae, and smaller than the large soil organisms such as earthworms, and including, e.g., mites, nematode worms.

mesoblastic *a. pert.* or developing from the mesoderm of an embryo.

mesobranchial *a. pert.* middle gill region.

mesobronchus *n.* in birds, the main trunk of a bronchus giving rise to secondary bronchi.

mesocaecum *n.* the mesentery connected with the caecum.

mesocardium *n.* an embryonic mesentery binding heart to pericardial walls; part of pericardium enclosing veins or aorta.

mesocarp *n.* the middle layer of the pericarp, e.g. comprising the flesh of fruits such as plums and cherries.

mesocentrous *a.* ossifying from a median centre.

mesocephalic *a.* having a cranial capacity (in humans) of between 1350 and 1450 cm^3.

mesocercaria *n.* trematode larval stage between cercaria and metacercaria.

mesochilium *n.* the middle portion of labellum of orchid flower.

mesocoel *n.* middle portion of coelomic cavity; the second of three main parts of coelom in molluscs; the cavity of the mesencephalon, *alt.* aqueduct of Sylvius.

mesocole *n.* animal living in conditions not very wet or very dry.

mesocolic *a. pert.* mesocolon, *appl.* lymph glands.

mesocolon *n.* mesentery or fold of the peritoneum attaching colon to dorsal wall of abdomen.

mesocoracoid *a.* situated between hypo- and hypercoracoid, *appl.* middle part of coracoid arch of certain fishes.

mesocotyl *n.* internode between scutellum and coleoptile in grass seeds.

mesocycle *n.* layer of tissue between phloem and xylem in stems with a single stele.

mesodaeum *n.* endodermal part of embryonic digestive tract, between stomodaeum and proctodaeum.

mesoderm *n.* the layer of embryonic cells lying between ectoderm and endoderm in gastrula, and from which muscle, blood, connective tissues, vascular system and heart, much of the kidney, and the dermis of skin develop in vertebrates.

mesodermal *a. pert.*, derived, or developing from mesoderm.

mesodont *a. appl.* stag beetles having a medium development of mandible projections.

mesofauna *n.* animals of size from 200 μm to 1 cm. *alt.* meiofauna.

mesogamy *n.* entry of the pollen tube into an ovule through the funicle or the integument.

mesogaster *n.* the mesentery or fold of peritoneum supporting the stomach.

mesogastric *a. pert.* mesogaster, mesogastrium, or to middle gastric region.

mesogastrium *n.* mesentery connecting stomach with dorsal abdominal wall in embryo; middle abdominal region.

mesogenous *a.* produced at or from the middle.

mesoglia oligodendroglia *q.v.*

mesogloea *n.* a gelatinous, non-cellular layer between the inner and outer body wall in sponges and coelenterates.

mesognathion *n.* the lateral segment of premaxilla, bearing lateral incisor.

mesohaline *a. appl.* brackish water of salinity between 5–15 parts per thousand.

mesohepar *n.* mesentery supporting the liver.

mesohydrophytic *a*. growing in temperate regions but requiring much moisture.

mesokaryote *a*. term sometimes used to describe the nucleus of dinoflagellates in which the chromosomes are permanently condensed and which does not undergo a conventional mitosis.

mesolecithal *a*. having a moderate yolk content.

mesomere *n*. the middle zone of coelomic pouches in embryo; a somite *q.v.*

mesome *n*. the axis of a plant regarded as a morphological unit.

mesometrium *n*. the mesentery of uterus and connecting tubes.

mesomicrotherm *n*. plant living in a temperate climate which can resist low winter temperatures.

mesomitosis *n*. mitosis within the nuclear envelope.

mesomorph *n*. a mesomorphic animal or plant.

mesomorphic *a*. of normal or average structure, form or size, or intermediate between extremes. *n*. mesomorph (usually *appl*. animals).

meson *n*. the central plane or a region of it.

mesonephric *a*. *pert*. mesonephros, *appl*. tubules, *appl*. duct.

mesonephridium *n*. nephridium or excretory organ of some invertebrates, derived from mesoderm.

mesonephros *n*. one of the middle of the three pairs of renal organs of vertebrate embryos, which persist as the adult kidney of anamniotes.

mesonotum *n*. dorsal part of insect metathorax.

mesopelagic *a*. *pert*. or inhabiting the ocean at depths between 200 and 1000 m.

mesoperidium *n*. middle layer, between endoperidium and exoperidium, of the peridium of some fungi.

mesopetalum *n*. labellum of orchid.

mesophanerophyte *n*. tree from 8 to 30 m in height.

mesophil(ic) *a*. thriving at moderate temperatures, between 20 and 45 °C when *appl*. bacteria. *n*. mesophile.

mesophilous *a*. *appl*. plants associated with neutral soils.

mesophloem *n*. middle or green bark.

mesophragma *n*. a chitinous piece of exoskeleton descending into interior of insect body with postscutellum for base.

mesophyll *n*. the internal parenchyma of a leaf, which is usually photosynthetic; a leaf of moderate size.

mesophyllous *a*. having leaves of moderate size, between microphyllous and macrophyllous.

mesophyte *n*. a plant thriving in temperate climate with normal amount of moisture.

mesoplankton *n*. plankton at depths of 200 m downwards; drifting organisms of medium size.

mesoplastron *n*. bony plate of shell of some turtles, between second and third lateral plates.

mesopleurite, mesopleuron *n*. lateral sclerite of mesothorax as in dipterans.

mesopodial *a*. having a supporting structure, such as a stipe, in a central position.

mesopodium *n*. leaf stalk or petiole region of leaf; middle part of molluscan foot; the metacarpus or metatarsus.

mesopostscutellum *n*. postscutellum of mesothorax in insects.

mesoprescutum *n*. prescutum of mesothorax in insects.

mesopterygium *n*. the middle of three basal pectoral fin cartilages in recent elasmobranchs.

mesopterygoid *n*. the middle of three pterygoid bone elements in teleost fishes.

mesoptile *n*. feather following protoptile.

mesorchium *n*. mesentery supporting testis.

mesorectum *n*. mesentery supporting rectum.

mesorhinal *a*. between nostrils.

mesorhinium *n*. the internarial surface region of a bird's bill.

mesosalpinx *n*. the portion of broad ligament enclosing uterine tube.

mesosaprobic *a*. *appl*. aquatic habitats having a decreased quantity of oxygen and substantial organic decomposition.

α-mesosaprobic category in the saprobic classification of river organisms comprising those that can live in polluted water in which decomposition is partly aerobic and partly anaerobic, e.g. the water-louse (*Asellus*). *cf*. β-mesosaprobic, mesosaprobic, oligosaprobic, polysaprobic.

β-mesosaprobic category in the saprobic

classification of river organisms comprising those that can live in water mildly polluted with organic pollutants, in which organic decomposition is mainly aerobic, e.g. the three-spined stickleback (*Gasterosteus aculeata*) and Canadian pondweed (*Elodea canadensis*). *cf.* α-mesosaprobic, oligosaprobic, polysaprobic.

mesoscapula *n.* the spine on the scapula when considered a distinct unit.

mesosaur *n.* member of the order Mesosauria, an order of lower Permian anapsid reptiles which were fish-eating and slender with long jaws.

mesoscutellum *n.* scutellum of insect mesothorax.

mesoscutum *n.* scutum of insect mesothorax.

mesosoma *n.* the anterior, broader part of abdomen in scorpions; the middle portion of body of certain invertebrates, esp. when their original segmentation is obscured.

mesosome *n.* an invagination of the bacterial plasma membrane.

mesosperm *n.* the integument surrounding nucellus of ovule.

mesosporium *n.* the middle of three layers in coat of some spores.

mesosternum *n.* middle part of sternum (breast bone) of vertebrates; the sternum of mesothorax in insects.

mesostylous *a.* having styles of intermediate length, *appl.* heterostylous flowers.

mesotarsus *n.* a tarsus of middle limb in insects. *a.* mesotarsal.

mesothecium *n.* the middle layer of coat of anther sac.

mesotheic *a.* neither highly susceptible nor completely resistant to parasites or infection.

mesothelioma *n.* rare type of tumour, derived from mesothelial cells of peritoneium, pleura or pericardium. Pleural mesotheliomas have been found at a greater than normal frequency in workers exposed to asbestos.

mesothelium *n.* epithelium-like layer of mesoderm-derived flattened squamous cells lining serous cavities (e.g. pericardial, pleural cavities); mesoderm bounding embryonic coelom and giving rise to muscular and connective tissue.

mesotherm *n.* plant thriving in moderate heat, within the range 12–19 °C, as in a warm temperate climate.

mesothoracic *a. pert.* or in region of mesothorax.

mesothorax *n.* the middle segment of thoracic region of insects. *cf.* prothorax, metathorax.

mesotrochal *a. appl.* larva of annulates with a circlet of cilia round middle of body.

mesotrophic *a.* having partly autotrophic and partly saprobic nutrition; obtaining nourishment partly from an outside source; partly parasitic; providing a moderate amount of nutrition, *appl.* environment.

mesotropic *a.* turning or directed toward the middle or the median plane.

mesovarium *n.* mesentery supporting ovary.

mesoventral *a.* in middle ventral region.

mesoxerophilous *a. appl.* plant of temperate climates showing adaptation to dry conditions and thriving in such conditions.

mesoxerophyte *n.* plant of a temperate climate that thrives in dry conditions.

Mesozoa, mesozoans *n.* phylum of extremely simple small multicellular marine invertebrates which live as parasites in the kidneys of other marine invertebrates. The adult has a body composed of two layers of cells, no muscular or nervous system, and lacking a body cavity and all organs except a gonad. Infusoriform and vermiform larvae are produced. The Mesozoa include the dicyemids, heterocyemids and orthonectids.

Mesozoic *a. pert.* or *appl.* geological era lasting from about 248 to 65 million years ago and comprising the Triassic, Jurassic and Cretaceous periods.

message in molecular biology, messenger RNA *q.v.*

messenger RNA (mRNA) type of RNA found in all cells which acts as a template for protein synthesis, each different mRNA being a copy of a single protein-coding gene (or, in bacteria, often a set of adjacent genes), and which in eukaryotes is the product of extensive processing of the primary RNA transcript in the nucleus. *alt.* message. *see also* transcription, RNA processing, splicing.

meta- prefix derived from Gk. *meta*, after,

signifying posterior, as in metathorax, the third and last thoracic segment in insects, or, derived from Gk. *meta*, change of, as in metamorphosis, a change in form.

metabasidium *n.* cell in which meiosis occurs in basidium of basidiomycete fungi.

metabiosis *n.* the beneficial exchange of factors (e.g. nutrients, vitamins) between species.

Metabola pterygotes *q.v.*

metabolic *a. pert.* metabolism.

metabolic activation conversion of foreign compounds into chemically reactive (often toxic or carcinogenic) forms in the body, esp. in the liver by microsomal enzymes.

metabolic compartments in eukaryotic cells, the membrane-bounded organelles in which particular metabolic processes are segregated.

metabolic pathway chain of enzyme-catalysed biochemical reactions in living cells which, e.g., convert one compound into another, or build up large macromolecules from smaller units, or break down compounds to release usable energy.

metabolic rate a measure of the rate of metabolic activity in a living organism, the rate at which an organism uses energy to sustain essential life processes such as respiration, growth, reproduction and, in animals, processes such as blood circulation, muscle tone and activity. It can be determined in numerous ways: as the total heat produced over a given period; as oxygen consumption (and sometimes also carbon dioxide production) over a given period, which although easier to measure, only gives the contribution of aerobic metabolism; as the energy content of the food eaten over a given period; and by the fate of isotopically labelled water. *see* average daily metabolic rate, basal metabolic rate, energy budget, field metabolic rate, respiratory quotient, resting metabolic rate, standard metabolic rate.

metabolic scope the range of metabolic rate shown by an animal, which is the difference between the resting metabolic rate and the maximum rate of energy expenditure of which the animal is capable at maximum activity.

metabolic water water produced by

oxidative processes (e.g. respiration) within the body.

metabolism *n.* integrated network of biochemical reactions in living organisms, often referring to the biochemical changes occurring in the living organism or cell as a whole. *see also* anabolism, catabolism, metabolic pathway.

metabolite *n.* any substance involved in or a product of metabolism.

metaboly *n.* change of shape resulting in movement, as in euglenoids and other flagellates.

metabotropic *a. appl.* non-ion channel glutamate receptors.

metabranchial *a. pert.* or in region of posterior gills.

metacarpal *n.* one of the bones joining wrist and fingers.

metacarpus *n.* the skeletal part of hand between wrist and fingers, typically consisting of five cylindrical bones, the metacarpals. *a.* metacarpal.

metacentric *a. appl.* chromosomes with the centromere halfway along and which appear V-shaped when segregating during mitosis or meiosis; *n.* a metacentric chromosome.

metacercaria *n.* the stage in the life cycle of endoparasitic flukes that develops from a cercaria, in which the cercaria loses its tail. In schistosomes this occurs subsequent to infection of the vertebrate host by cercariae, in other species metacercariae infect the vertebrate host. The adult develops directly from a metacercaria.

metacestode *n.* larval tapeworm found in the intermediate host.

metachroic *a.* changing colour, as older tissue in fungi.

metachromatin *see* volutin granules.

metachromy *n.* change in colour, as of flowers.

metachronal *a.* one acting after another, *appl.* rhythm of movement of cilia, the legs of centipedes and millipedes, etc.

metachrosis *n.* ability to change colour by expansion or contraction of pigment cells.

metacoel *n.* the posterior part of the coelom as of molluscs; the anterior extension of fourth ventricle in brain.

metacommunication *n.* communication about the meaning of other acts of communication, such as the posture adopted by a dominant male, which signals to other males its status and likely behaviour if attacked.

metacone *n*. posterior external cusp of upper molar tooth.

metaconid *n*. posterior internal cusp of lower molar tooth.

metaconule *n*. posterior secondary cusp of upper molar tooth.

metacoracoid *n*. posterior part of coracoid.

metacromion *n*. posterior branch process of acromion process of scapular spine.

metacyclic *a. appl.* infective short broad forms of trypanosome that develop in the insect vector salivary gland and which pass on to the next host.

metadiscoidal *a. appl.* placenta in which villi are at first scattered and later restricted to a disc, as in primates.

metadromous *a.* with primary veins of segment of leaf arising from upper side of midrib.

metafemale *n*. a female *Drosophila* with a normal diploid set of autosomes but having three X chromosomes. *alt.* superfemale.

metagastric *a. pert.* posterior gastric region.

metagenesis *n*. alternation of generations, esp. of asexual and sexual generations, esp. in animals.

metagnathous *a.* having mouthparts for biting in the larval stage and sucking in the adult, as certain insects; having the points of the beak crossed, as in crossbills.

metagyny protandry *q.v.*

metallic *a.* iridescent, *appl.* colours due to interference by fine striations or thin plates, as in insects.

metalloenzyme *n*. any enzyme containing metal ions.

metalloprotein *n*. any protein containing metal ions.

metallothionein *n*. protein involved in cell detoxification mechanisms for various heavy metals.

metaloph *n*. the posterior crest of a molar tooth, uniting metacone, metaconule and hypocone.

metamale *n*. a male *Drosophila* with one X chromosome and three sets of autosomes. *alt.* supermale.

metamere *n*. a body segment or somite.

metameric *a.* having the body divided into a number of segments more-or-less alike. *n.* metamerism.

metamerized *a.* segmented.

metamitosis *n*. mitosis in which nuclear membrane breaks down.

metamorphosis *n*. change in form and structure undergone by animal from embryo to adult stage, as in insects and amphibians. In insects, incomplete metamorphosis occurs in, e.g. locusts and grasshoppers, where the larval form is relatively similar to the adult and changes gradually towards the adult form at each moult, and in which there is no non-feeding pupal stage. Complete metamorphosis occurs in, e.g. butterflies and flies, in which the larval caterpillar or maggot stage is quite unlike the adult in form and internal structure, and undergoes a radical remodelling during a non-feeding pupal stage; transformation of one structure into another, as of stamens into petals, *alt.* homoeosis; interference with normal symmetry in flowers.

metamorphy homoeosis *q.v.*

metanauplius *n*. larval stage of crustaceans, succeeding the nauplius stage.

metandry *n*. meroandry with retention of posterior pair of testes only.

metanephric *a. pert.* or in region of kidney.

metanephridium *n*. nephridial tubule with opening into the coelom.

metanephros *n*. organ arising behind mesonephros, and replacing it as functional kidney of fully-developed amniotes.

metanotum *n*. notum of insect metathorax.

metanucleus *n*. egg nucleolus after extrusion from germinal vesicle.

metaphase *n*. stage in mitosis or meiosis when the chromosomes have become aligned on the equator of the cell with all the centromeres lying along the spindle equator (mitosis or 2nd meiotic division) or equidistant from it (1st meiotic division).

metaphase plate equatorial plate *q.v.*

metaphloem *n*. primary phloem formed after the protophloem.

metaphysis *n*. vascular part of diaphysis of bone adjoining epiphyseal cartilage.

Metaphyta, metaphytes *n., n.plu.* multicellular plants.

metaplasia transdifferentiation *q.v.*

metaplastic *a. pert.* metaplasia.

metapleural *a.* posteriorly and laterally situated.

metapneustic *a. appl.* insect larvae with only the terminal pair of spiracles open.

metapodeon *n.* that part of insect abdomen behind petiole or podeon.

metapodium *n.* portion of foot between tarsus and digits; in four-footed animals, metacarpus and metatarsus.

metapophysis *n.* a prolongation of a vertebral articular process esp. in lumbar region, developed in some vertebrates.

metapostscutellum *n.* postscutellum of insect metathorax.

metaprescutellum *n.* prescutellum of insect metathorax.

metapterygium *n.* the posterior basal fin cartilage, pectoral or pelvic, of recent elasmobranchs.

metapterygoid *n.* posterior of three pterygoid elements in some lower vertebrates.

metarhodopsin *n.* transient orange product of lumirhodopsin, dissociating into *trans*-vitamin A aledehyde and opsin.

metarteriole *n.* a small branch of an arteriole between arteriole and arterial capillaries.

metascolex *n.* a massive organ formed by enlargement of the neck area directly behind scolex in some cestodes.

metascutellum *n.* scutellum of insect metathorax.

metascutum *n.* scutum of insect metathorax.

metasoma *n.* the posterior region of opisthosoma of arachnids and some crustaceans.

metasomatic *a. pert.* or situated in metasoma.

metastases *n.plu.* foci of proliferating cancer cells in organs and tissues other than the site of origin. *sing.* metastasis.

metastasis *n.* a change in state, position, form or function; the migration of cancer cells to colonize and grow in other tissues and organs. *v.* metastasize.

metasternum *n.* the sternum of insect metathorax; posterior part of sternum of higher vertebrates.

metasthenic *a.* with well-developed posterior part of body.

metastoma *n.* the two-lobed lip of crustaceans; the hypopharynx of myriapods.

metastomial *a.* behind the mouth region, *appl.* segment posterior to peristomium or buccal segment in annelids.

metastructure *n.* ultramicrosopic organization.

metatarsal *n.* one of the bones joining ankle and toes.

metatarsus *n.* skeleton of vertebrate foot between tarsus and toes, comprising the metatarsals; 1st segment of insect tarsus. *a.* metatarsal.

Metatheria, metatherians marsupials *q.v.*

metathorax *n.* posterior (third) segment of insect thorax.

metatracheal *a.* appl. wood in which xylem parenchyma is located independently of the vessels and scattered throughout the annual ring.

metatroch *n.* in a trochophore, a circular band of cilia behind the mouth.

metatype *n.* a topotype of the same species as the holotype or lectotype.

metaxenia *n.* the case where there is a difference in the influence of pollen from different parents on the development of a fruit.

metaxylem *n.* primary xylem developing after the protoxylem and before the secondary xylem, if present, and which is distinguished by wider vessels and tracheids.

Metazoa, metazoans *n., n.plu.* multicellular animals, sometimes more strictly applied only to those multicellular animals with cells organized into tissues and possessing nervous tissue.

metecdysis *n.* in arthropods, period after moult when the new cuticle hardens.

metencephalon *n.* part of hindbrain, consisting of cerebellum, pons and intermediate part of 4th ventricle; the whole hindbrain.

metepimeron *n.* epimeron of insect metathorax.

metepisternum *n.* episternum of insect metathorax.

metestrus *alt.* spelling of metoestrus *q.v.*

methaemoglobin (MetHb) haemoglobin with the haem iron in the ferric state and unable to bind oxygen, found in small amounts in the blood and produced by the action of oxidizing agents such as nitrite and chlorate poisons. *alt.* ferrihaemoglobin.

methane *n.* CH_4, a gas derived from the anaerobic breakdown of organic matter. *alt.* marsh gas.

methanogenesis *n.* the generation of the gas methane (CH_4) by living organisms, mainly bacteria.

methanogenic *a.* generating methane, as do some bacteria and rumen ciliates. *see* methanogens.

methanogens *n.plu.* archaebacteria that produce the gas methane (CH_4, marsh gas) by the reduction of CO_2 or carbonate coupled with the oxidation of hydrogen (H_2). Some can also use formate, methanol or acetate as a source of electrons for reducing CO_2. They are found in anoxic environments, such as marine and freshwater muds, the intestinal tracts of animals and sewage treatment plants. *see also* Appendix 6.

methionine (Met, M) *n.* a sulphur-containing amino acid, constituent of proteins, essential in human diet, provides sulphur and methyl groups for metabolic reactions. It is the first amino acid to be inserted in all eukaryotic polypeptide chains, formylmethionine being used in bacteria and mitochondria, and which is often removed after synthesis. *see* table in Appendix 1 for chemical formula.

methotrexate amethopterin *q.v.*

methyl-accepting chemotaxis proteins (MCPs) receptor proteins that detect external stimuli, involved in bacterial chemotaxis.

methylase *n.* an enzyme that attaches a methyl group to a molecule, e.g. DNA methylase which methylates C residues in DNA. *alt.* methyltransferase.

methylation *n.* the addition of a methyl group (–CH_3) to a chemical compound or to a macromolecule such as a protein or DNA. *see also* DNA methylation.

5-methylcytosine *n.* modified base found in DNA, formed by enzymatic methylation of cytosines at specific sites in DNA, and which can cause mutations by spontaneous deamination to thymine resulting in base substitution at subsequent replication.

methyltransferase *n.* any of a group of enzymes that transfer methyl groups from one compound to another. EC 2.1.1. *alt.* transmethylase.

metochy *n.* relationship between a neutral guest insect and its host.

metoecious *a. appl.* parasites that are not host-specific.

metoestrus *n.* the luteal phase, period when activity subsides after oestrus.

metope *n.* the middle frontal portion of a crustacean.

metopic *a. pert.* forehead; *appl.* frontal suture.

metoxenous metoecious *q.v.*

metra uterus *q.v.*

mevalonate (mevalonic acid) 6-carbon organic acid, an intermediate in cholesterol biosynthesis.

MHC antigens major histocompatibility antigens. *see* MHC molecules.

MHC complex *see* major histocompatibility complex.

MHC molecules cell-surface glycoproteins encoded by the major histocompatibility complex of genes. They consist of MHC class I molecules, which are found on most cells of the body, and MHC class II molecules, which are usually restricted to certain cells of the immune system. They are involved in the presentation of peptide antigens to the immune system, and also determine the tissue type of an individual. Transplantation of organs not matched for MHC type with the recipient leads to transplant rejection unless normal immune function is suppressed by drugs. *alt.* MHC antigens, H-2 antigens (in mouse), histocompatibility antigens, HLA antigens (in humans), transplantation antigens. *see* major histocompatibility complex.

MHC restriction the phenomenon that T cells will only recognize foreign antigen if presented on the surface of a cell in conjunction with MHC antigens of the genetically-determined type it has encountered in its passage through the thymus. T cells with different functions recognize foreign antigen in conjunction with different sorts of MHC antigens, and are thus also restricted to interacting with certain types of cell only. The class I antigens (HLA-A, B and C in humans) occur on almost all the cells of the body, the class II MHC antigens (HLA-D) occur normally only on T cells, B cells and antigen-presenting cells such as macrophages. Helper T cells recognize antigen only in conjunction with class II MHC antigens, cytotoxic T cells recognize antigen in conjunction with class I MHC antigens.

micelle *n*. orderly arrangement of molecules, as in cellulose microfibrils in plant cell walls, or phospholipids in aqueous solution; *n*. spherical structure formed by aggregates of amphipathic molecules such as phospholipids and glycolipids in water, with hydrophilic groups on the outside and polar groups inside.

Michaelis-Menten kinetics set of equations describing the properties of many enzyme-catalysed reactions, from which two characteristic parameters of such reactions are derived, K_m (the Michaelis constant) and V_{max}. K_m = substrate concentration at which reaction rate is half its maximal value, V_{max} = maximal rate of an enzyme-catalysed reaction under steady-state conditions.

micraesthetes *n.plu.* the smaller sensory organs of chitons (Amphineura).

micrander *n*. a dwarf male, as of certain green algae.

micraner *n*. a dwarf male ant.

micrergate *n*. a dwarf worker ant.

micro- prefix derived from Gk. *mikros*, small.

microaerophile *n*. any aerobic organism that thrives at a reduced oxygen tension.

microaerophilic *a*. tolerating only a small amount of oxygen, *appl*. certain bacteria. *n*. microaerophile.

microalga *n*. unicellular alga visible only under a microscope. *cf.* macroalga.

microarthropod *n*. arthropod of microscopic size. *cf.* macroarthropod.

microballistics *n*. biolistics *q.v.*

microbe *n*. any microorganism (*q.v.*), esp. a bacterium. *alt.* microorganism.

microbial *a. pert.* or caused by microorganisms.

microbiology *n*. the science dealing with the study of microorganisms.

microbiota *n*. the population of organisms of microscopic size in any ecosystem, habitat, etc. (esp. applied to soil) and which includes bacteria, unicellular algae, fungi and protozoa. *cf.* macrobiota, mesobiota.

microbivore *n*. animal that feeds on microorganisms. *a*. microbivorous.

microbody *n*. any of a diverse class of small spherical bodies bounded by a single membrane, found in plant and animal cells (esp. liver and kidney), and including glyoxysomes (*q.v.*), peroxisomes (*q.v.*). *alt.* cytosome, phragmosome, crystal-containing body.

microcephalic *a*. with an abnormally small head, having cranial capacity (in humans) of less than 1350 cm^3.

microchaeta *n*. small bristle, as on body of some insects.

microchiroptera *n*. small, mainly insectivorous, bats of the mammalian order Chiroptera. *cf.* megachiroptera.

microchromosomes *n.plu.* chromosomes considerably smaller than the other chromosomes of the same type in nucleus, and usually centrally placed in the equatorial plate during metaphase. *alt.* M chromosomes.

microclimate *n*. the climate within a very small area or in a particular habitat.

micrococcus *n*. bacterium of the family Micrococcaceae, aerobic or facultatively anaerobic Gram-positive cocci found in soil and water. *plu.* micrococci.

microconidium *n*. minute conidium, produced by some ascomycete fungi which can behave as a male sex cell or germinate to give rise to a mycelium. *plu.* microconidia.

microconjugant *n*. a motile free-swimming conjugant or gamete that attaches itself to a macroconjugant and fertilizes it.

microcosm *n*. a world in miniature, a community that is a miniature version of a larger whole.

microcyclic *a. appl.* rust fungi that produce only teleutospores and basidiospores during the life cycle; *appl.* short and simple life cycles; having a haplophase or gametophyte stage only.

microcyst *n*. a resting spore of slime moulds.

microcytes *n.plu.* red blood cells about half the size of normal, numerous in certain diseases.

microdissection *n*. technique for manipulating and operating upon live microorganisms and individual cells using a microscope to view the specimen being manipulated.

microdont *a.* with comparatively small teeth.

microelectrode *n.* very fine electrode used for recording from single cells.

microelements trace elements *q.v.*

microendemic *a.* restricted to a very small area.

microenvironment *n.* microhabitat *q.v.*

microevolution *n.* evolutionary change consisting in alterations in gene frequencies, chromosome structure or chromosome numbers due to mutation and recombination and which can be noticed over a relatively short time, e.g. the acquisition of resistance to a pesticide amongst insects; the relatively small-scale evolutionary change that differentiates the members of geographical races, subspecies, sibling species, etc. *cf.* macroevolution.

microfauna *n.* animals less than 200 μm long, such as protozoa, only visible under the microscope.

microfibril *n.* microscopic fibre composed of protein (as keratin fibres of hair) or polysaccharide (e.g. cellulose microfibrils in plant cell walls).

microfilament *n.* a protein filament composed of actin subunits, an element of the cytoskeleton in eukaryotic cells and a constituent of muscle. *alt.* actin filament.

microfilaria *n.* the embryo of certain parasitic threadworms.

microflora *n.* the microorganisms (bacteria, unicellular fungi and algae) living in or on an organism, or in a particular habitat or ecosystem; (*bot.*) the dwarf flora of high mountains.

microfossil *n.* any microscopic fossil, as of prokaryotic and eukaryotic microorganisms, spores, pollen, microscopic animals and plants, etc.

microfungi *n.plu.* microscopic forms of fungi, the yeasts and moulds, as opposed to mushrooms and toadstools.

microgamete *n.* the smaller of two conjugant gametes, regarded as the male.

microgametoblast *n.* intermediate stage between microgametocyte and microgamete in certain sporozoans.

microgametocyte *n.* cell developed from merozoite in certain protozoans, giving rise

to microgametes.

microgametogenesis *n.* development of microgametes or spermatozoa.

microgametophyte *n.* in heterosporous plants, the male gametophyte, which develops from a microspore.

microglia *n.plu.* type of neuroglia more common in grey than white matter of brain, composed of small elongated cell bodies which are phagocytic. *a.* microglial.

β₂-microglobulin small polypeptide resembling portion of immunoglobulin heavy chain constant region and which forms a subunit of class I MHC antigen molecules.

micrograph *n.* photograph of an image obtained by microscopy.

microgyne *n.* dwarf female ant.

microhabitat *n.* the immediate environment of an organism, esp. a small organism; a small place in the general habitat distinguished by its own set of environmental conditions.

microheterogeneity *n.* small variations in nucleotide sequence seen, e.g., in individual repeating units of a tandemly repeated gene cluster.

microinjection the introduction of substances into a single cell by injection with special instruments (e.g. micropipette).

microlecithal *a. appl.* ova containing little yolk.

micromere *n.* small blastomere of upper or animal hemisphere in cleaving eggs containing unequally distributed yolk. *cf.* macromere.

micromerozoite *n.* cell derived from microschizont and developing into gametocyte in certain sporozoan blood parasites.

micromesentery *n.* a secondary incomplete mesentery in sea anemones.

micrometre (μm) *n.* unit of microscopic measurement, being 10^{-6} m, one-thousandth of a millimetre. Formerly called a micron.

micromutation point mutation *q.v.*

micron micrometre *q.v.*

micronemic, micronemous *a. pert.* or having small hyphae.

micronephridia *n.* small nephridia.

micront *n.* a small cell formed by schizogony, itself giving rise to microgametes.

micronucleus *n.* the smaller of the two types of nuclei found in cells of ciliate protozoans, a diploid nucleus involved in meiosis and sexual reproduction but which does not produce RNA. *cf.* macronucleus.

micronutrients *n.plu.* elements and compounds required in small amounts by living organisms for proper growth and development, e.g. trace elements (such as zinc, iron, copper) and vitamins.

microorganism *n.* any microscopic organism, including bacteria, viruses, unicellular algae and protozoans, and microscopic fungi (yeasts and moulds). *alt.* microbe.

microparasite *n.* any parasite of microscopic size.

microphage *n.* small phagocytic blood cell, chiefly polymorphonuclear leukocyte.

microphagic, microphagous *a.* feeding on minute organisms or particles; feeding on small prey. *n.* microphagy. *alt.* filter feeding.

microphanerophyte *n.* small tree or shrub from 2 to 8 m in height.

microphil(ic) *a.* tolerating only a narrow range of temperature, *appl.* certain bacteria.

microphyll *n.* simple leaf containing only a single strand of vascular tissue, leaf type present in horsetails, club mosses and the Psilophyta; a small leaf, esp. as produced by ferns and their relatives.

microphyllous *a.* having small leaves.

micropipette *n.* very fine glass pipette used to impale a single cell for electrical recording, introduction of DNA, etc.

microplankton *n.* small organisms drifting with the surrounding water, somewhat larger than those of nannoplankton.

Micropodiformes Apodiformes *q.v.*

micropodous *a.* with rudimentary or small foot or feet.

micropterous *a.* having small hindwings invisible until tegmina are expanded, as some insects; having small fins.

micropyle *n.* (*bot.*) small pore or channel in coat at apex of ovule through which the pollen tube enters; corresponding aperture in testa of seed between hilum and point of radicle; (*zool.*) aperture in the membrane of some animal eggs through which sperm enter; pore in the coat of sponges through which gemmules escape.

micropyle apparatus raised processes or porches, sometimes of elaborate structure, developed round micropyle of certain insect eggs.

microsatellite *n.* region of DNA composed of a very short nucleotide sequence repeated in tandem many times, which is very variable within a population and can be used for DNA fingerprinting.

microsaurs *n.plu.* extinct order of lepospondyl amphibians having an elongated or reptilian shape and developed limbs each with four or fewer fingers.

microschizogony *n.* schizogony (fission into several new cells) resulting in small merozoites, as of some protozoans.

microschizont *n.* a cell functioning as a male gamete, produced by schizogony, in some protozoans.

microsclere *n.* one of small spicules scattered in tissues of sponges.

microsclerotium *n.* a microscopic sclerotium.

microscopy *see* dark-field microscopy, light-field microscopy, phase-contrast microscopy, scanning electron microscope, transmission electron microscope, tunnelling electron microscope.

microseptum *n.* an incomplete mesentery of sea anemones.

microsere *n.* a successional series of plant communities in a microhabitat.

microsmatic *a.* with a feebly developed sense of smell.

microsomal *a. appl.* enzymes associated with the endoplasmic reticulum and which spin down in the microsomal faction in the ultracentrifuge; *appl.* enzymes associated with peroxisomes (*q.v.*). *see* microsomes.

microsomes *n.plu.* the smallest size particles spun down from cell homogenates in the ultracentrifuge and including broken parts of other fractions esp. endoplasmic reticulum with ribosomes attached; formerly used by cytologists for any small granules in the cytoplasm, now usually restricted to peroxisomes (*q.v.*); aggregations of ribosomes.

microsomia *n.* dwarfism.

microsorus *n.* a sorus containing microsporangia.

microspecies *n.* a unit below the species level, such as a race, subspecies or variety.

microsphere *n.* structure of organic material formed by heating polypeptides, which can absorb various organic molecules from aqueous solution.

microspheric *a. appl.* foraminiferans when initial chamber of shell is small.

microspike *n.* one of a number of hair-like extensions put out by animal cells in tissue culture as they settle on to a surface or migrate.

microsplanchnic *a.* small-bodied and long-legged.

microsporangiate *a.* composed of or producing microsporangia, *appl.* cones.

microsporangium *n.* a sporangium containing a number of microspores; pollen sac in spermatophytes.

microspore *n.* in heterosporous plants, the spore that gives rise to the male gametophyte (microgametophyte), and which is formed in a microsporangium; in any organism that produces two types of spore, the smaller spore.

Microsporidea *n.* class of parasitic protozoa of the Cnidospora. Classified in the protist phylum Cnidosporidia (*q.v.*).

microsporocyte *n.* microspore mother cell, which undergoes meiosis to produce haploid microspores.

microsporophore microsporangium *q.v.*

microsporophyll *n.* small leaf or leaf-like structure on which a microsporangium develops.

microsporozooite *n.* a smaller endogenous sporozoite of sporozoan parasites.

microstome *n.* a small opening or orifice.

microstrobilus *n.* a male cone com- posed of microsporophylls, as in gymnosperms.

microstylous *a.* having short styles, *appl.* heterostylous flowers.

microsymbiont *n.* the smaller of two symbiotic organisms.

microteliospore *n.* a spore produced in a microtelium.

microtelium *n.* sorus of microcyclic rust fungi.

microtherm *n.* a plant of the cold temperate zone, which can grow below 12 °C.

microthorax *n.* in insects a term applied to the cervix or neck when thought to represent a reduced prothorax.

microtome *n.* machine with a sharp metal blade for slicing tissue into sections for microscopy.

microtomy *n.* the cutting of thin sections of tissues or other material, for examination by microscopy.

microtrabecular network a three-dimensional network of very fine filaments, seen in electron micrographs of eukaryotic cells after certain treatments.

microtrichia *n.plu.* small unjointed hairs in insect wings.

microtubule *n.* hollow tube of protein made up of the globular polypeptide tubulin, a component of the cytoskeleton in eukaryotic cells, forming esp. the spindle and which is the chief constituent of flagella and cilia. Particularly abundant in vertebrate neurones.

microtubule-associated proteins *see* MAP.

microtubule-organizing centre (MTOC) structures within eukaryotic cells which initiate microtubule formation, e.g. the centrosome of mammalian cells.

microvilli *n.plu.* thin fingerlike infoldings of the cell surface, which on intestinal and renal epithelium form brush borders. *sing.* microvillus.

microzoid *n.* male gamete, as in algae.

microzooid *n.* a free-swimming ciliated cell budded off certain protozoans.

microzoospore *n.* a small motile spore.

mictic *a.* capable of reproducting by apomixis *q.v.*

micton *n.* a species resulting from interspecific hybridization and whose individuals are interfertile.

micturition *n.* urination.

mid-blastula transition the point in amphibian embryonic development at which the zygote's own genes become active.

mid-body a cell plate or group of granules in equatorial region of spindle in anaphase of mitosis.

midbrain *n.* that portion of the brain comprising the corpora quadrigemina, cerebral peduncles and the aqueduct of Sylvius.

middle commissure grey matter connecting thalami across the 3rd ventricle.

middle ear cavity bounded by ear drum on one side, bones of the skull, and the bony wall of the inner ear on the other, spanned by 3 small bones, the auditory ossicles, through which sound vibrations are conducted from eardrum to inner ear.

middle lamella the layer composed largely of pectin between two adjoining plant cell walls. It is derived from the cell plate formed as a plant cell divides. Cellulose is laid down on both sides of it to form the new walls of the daughter cells.

middle shore zone of seashore between the average low-tide level and average high-tide level, which is usually the most extensive zone and is covered by the sea twice a day.

middle T protein specified by polyoma virus and produced in infected cells, and which is primarily responsible for inducing transformation in susceptible cells, and which has been identified as a protein kinase. *see also* large T, small t.

mid-grass community grassland containing grasses of medium height, over 60 cm but under 2 m.

mid-gut mesenteron *q.v.*

midpiece *n.* the middle portion of a sperm, consisting of a portion of the flagellum sheathed by mitochondria.

midrib *n.* the large central vein of a leaf.

migration *n.* change of habitat according to season, climate, food supply, etc. that many animals undergo, often travelling very long distances along predetermined routes; the movement of plants into new areas.

migratory cells certain leukocytes, macrophages, etc. that routinely enter tissues from the bloodstream.

mildew *n.* plant fungal disease manifested by a downy or powdery fungal coating on leaf or other affected parts, caused by fungi from several different groups. Powdery mildews are caused by ascomycetes, downy mildews by phycomycetes.

miliary *a.* of granular appearance; consisting of small and numerous grain-like parts.

milk *n.* secretion from mammary glands with which mammals suckle their young. It is rich in fat, protein and sugar (mainly lactose).

milk sugar lactose *q.v.*

milk teeth first dentition of mammals, shed after or before birth. *alt.* deciduous teeth.

milkweed *n.* plants of the genus *Asclepias*, which contain high concentrations of cardiac glycosides.

mill chamber part of alimentary canal in some crustaceans, in which food is broken down by movement of chitinous plates.

milleporine *n. appl.* stony corals of the order Milleporina, which have colonies of two kinds of polyp living in pits on the surface of a massive calcareous skeleton (a corallium) and a brief medusoid stage.

millet *n.* refers to any or all of several plants of the monocot family Gramineae cultivated for their grain and including *Sorghum vulgare*, *Setaria italica*, *Pennisetum typhoideum* and *Panicum* spp.

millimicron *n.* former term for nanometre, being one-thousandth of a micron (micrometre).

millipedes *n.plu.* common name for the diplopods (Diplopoda).

milt *n.* testis or sperm of fishes.

mimesis *n.* mimicry *q.v.*; the effect of the actions of one animal of a group on the activity of the others.

mimetic *a. pert.* or exhibiting mimicry.

mimicry *n.* the resemblance of one animal to an animal of a different species so that a third animal is deceived into confusing them; the resemblance of an animal or plant to an inanimate object, or of an animal or part of an animal to a plant or part of a plant, usually for the purposes of camouflage. *a.* mimetic. *see also* Batesian mimicry, Müllerian mimicry.

mineralization *n.* the breakdown of organic matter into its constituent inorganic components, carried out chiefly by decomposer microorganisms, and, for carbon, during respiration when carbon dioxide is returned to the environment.

mineralocorticoid *n.* any of a group of corticosteroid hormones, secreted by the adrenal cortex, including aldosterone, and which helps in regulation of water and electrolyte balance in the body.

mineralocorticoid receptor (MR) intracellular receptor protein for mineralocorticoid steroid hormones, which when complexed with the hormone acts as a transcription factor, regulating the expression of particular genes.

miniature end plate potential (mepp) potential produced in muscle cell membrane by release of very small amounts of acetylcholine from unstimulated nerve fibre.

minima minor worker *q.v.*

minimal medium culture medium containing a basic set of nutrients only, on which normal wild-type organisms can grow, but which cannot support the growth of metabolic mutants.

minimum, law of the that factor for which an organism or species has the narrowest range of tolerance or adaptability limits its existence.

minimum lethal dose (MLD) minimum dose of any agent sufficient to cause 100% mortality in the test population.

minimus *n.* fifth digit of hand or foot.

minisatellite *n.* region of DNA composed of repeats of a short DNA sequence, which is very variable within the population, and can be used in DNA fingerprinting.

minor elements trace elements *q.v.*

minor gene a gene which has a small effect individually but contributes to a multifactorial phenotypic trait.

minor histocompatibility antigen any of the many antigens encoded outside the major histocompatibility complex (*q.v.*) that also seem to be involved in rejection of transplanted tissues. They represent various proteins that can differ antigenically between genetically different members of the same species.

minor worker a member of the smallest worker subcaste, esp. in ants.

Miocene *n.* a geological epoch of the Tertiary, between Oligocene and Pliocene, lasting from about 25 to 5 million years ago.

miracidium *n.* ciliated larval stage of gut, liver and blood flukes, which hatches out of the egg and infects the snail host.

mire *n.* bog or fen, usually referring to a peatland but also used to describe a fen developing on mineral soils. *see* blanket mire, raised mire, topogenous mire.

mismatch repair type of DNA repair in which an incorrect mismatched nucleotide residue, or sequence of mismatched residues, is excised from DNA and a new correct base inserted or sequence synthesized, using the intact DNA strand as template.

misogamy *n.* antagonism to mating; reproductive isolation.

mispairing *n.* the condition of having a base in one strand of DNA which does not pair correctly with its opposite number in the other chain of the double helix.

missense mutation mutation in which one base pair is altered causing an amino acid change in the protein product of the gene.

missense suppressor mutant gene specifying a tRNA with an altered anticodon that overcomes the effects of a missense mutation in another gene by inserting an acceptable amino acid at the site.

Mississippian *a.* Lower Carboniferous in North America.

miter mitra *q.v.*

mites *n.plu.* common name for many of the Acarina *q.v.*

mitochondria *n.plu.* organelles in the cytoplasm of eukaryotic cells, having a double membrane, the inner one invaginated, and which are the site of the tricarboxylic cycle and oxidative phosphorylation of oxidative respiration, generating ATP. They contain a small circular DNA which specifies tRNAs, rRNAs and some mitochondrial proteins. *sing.* mitochondrion. *see* Fig. 1.

mitochondrial ATPase *see* coupling factor.

mitochondrial sheath an envelope containing mitochondria sheathing the beginning of the flagellum of a sperm.

mitochondrion *sing.* of mitochondria *q.v.*

mitochondriopathy *n.* disease caused by defects in mitochondrial function.

mitogen *n.* any substance that causes mitosis and (usually) cell division.

mitogenic *a.* inducing mitosis and cell division.

mitomycin C antibiotic produced by *Streptomyces caespitosus* which inhibits nuclear division, DNA and protein synthesis in mammalian cells, and is used clinically as an anti-tumour agent.

mitosis *n.* the typical process of nuclear division in eukaryotic cells, in which each member of a duplicated chromosome segregates into a daughter nucleus, resulting in daughter nuclei containing identical sets of chromosomes, identical to that of the parent nucleus, and which is followed by cell division. *see* Fig. 8. *cf.* meiosis.

mitospore *n.* a uninucleate diploid zoospore produced by mitosis.

mitotic *a. pert.* or produced by mitosis.

mitotic index the number of dividing cells per 1000, at any time.

mitotic non-disjunction *see* non-disjunction.

mitotic recombination genetic recombination occurring between homologous chromosomes during mitosis in a somatic cell. If the cell is heterozygous for the genetic loci exchanged, a daughter cell of a different phenotype from that of the rest of the tissue may be produced.

mitotic spindle *see* spindle.

mitra *n.* a helmet-shaped piece of calyx or corolla.

mitral cell pyramidal neurone with thick basal dentrites, found in molecular layer of olfactory bulb of brain.

mitral valve valve of the left auriculoventral orifice of the heart. *alt.* bicuspid valve.

mitriform *a.* shaped like a mitre.

mixed forest, mixed woodland forest in which at least 20% of the trees are of species other than the dominant species.

mixed-function oxygenase, mixed

function oxidase monooxygenase *q.v.*

mixed lymphocyte response (MLR) the reaction seen when lymphocytes from two genetically distinct individuals are cultured together, the T cells from each individual responding to the histocompatibility antigens of the other by differentiating and proliferating.

mixed spinal nerves spinal nerves after union of the dorsal (afferent) and ventral (efferent) roots.

mixed tissue tissue containing different cell types, all originating from the same group of founder cells.

mixipterygium *n.* clasper of male elasmobranch fishes, the medial lobe of the pelvic fin. *alt.* mixopterygium.

mixis *n.* sexual reproduction, esp. the fusion of gametes. *a.* mictic.

mixoploidy *n.* condition of having cells or tissues with different chromosome numbers in the same individual, as in a chimaera or a mosaic.

mixotrophic mesotrophic *q.v.*

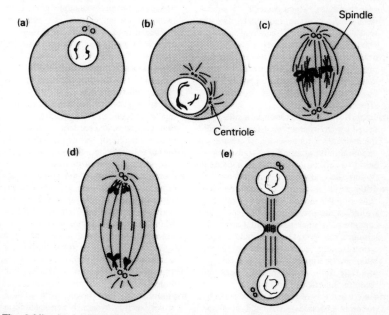

Fig. 8 Mitosis. (a) Diploid nucleus; (b) prophase; (c) anaphase; (d) metaphase; (e) telophase and cytokinesis. For clarity only two chromosomes are shown.

mnemonic *a. pert.* memory.

mnemotaxis *n.* movement directed by memory, as returning to feeding place or homing.

mobile genetic elements *see* transposable elements.

modal number the most frequently occurring chromosome number in a taxonomic group.

modality *n.* the qualitative nature of a sense, stimulus, etc., e.g. taste, smell, hearing, sight are different modalities of sensory experience.

mode n. in a distribution, the most frequently occurring value.

moderate *a. appl.* animal viruses that can form a stable association with an infected cell, not destroying the cell, producing few if any virus particles, and being passed on from a cell to its progeny. *cf.* virulent.

moderator *n.* band of muscle checking excessive distention of right ventricle, as in heart of some mammals.

modification *n.* in bacteria, the selective methylation of DNA (at cytosine or adenine residues) which protects a bacterium's DNA against degradation by its own restriction enzymes; a phenotypic change due to environment or use.

modifier *a. appl.* mutations that modify the phenotypic effect of other mutations; gene that modifies the effect of a gene at a different locus; any factor that modifies the effect of another factor.

modulation *n.* an alteration in a cell, produced by environmental stimuli, without impairment of its essential character; a variation in the strength of a signal delivered to a cell, or its response to it, caused by some second stimulus a modulator.

modulator *n.* a band of the spectrum, localized in the red-yellow, green, and blue regions, which evokes colour sensation; the component in processes essential for maintaining a steady state which controls specific reactions, such as a catalyst, gene, brain, etc; the agency that selects the appropriate route of transmission between receptor and effector.

molality *n.* a way of expressing the concentration of a chemical solution: a molal solution contains 1 mole of solute per kg of solvent.

molar *a.* adapted for grinding, as *appl.* teeth; containing one gram molecule or mole per litre, *appl.* solutions.

molarity *n.* a way of expressing the concentration of a chemical solution: a molar solution contains 1 mole of solute per litre of solution.

mold *alt.* spelling of mould *q.v.*

mole, mol the SI unit for the amount of a substance which contains as many elementary units as there are atoms in 0.012 kilograms of ^{12}C. The elementary units must be specified. One mole of a substance has a mass equal to its molecular weight in grams.

molecular biology study of biological phenomena at the molecular level.

molecular chaperone protein involved in facilitating the correct folding of newly synthesized proteins but which is not itself changed in the process and does not itself provide folding information.

molecular clock evolutionary clock *q.v.*

molecular evolution the changes that occur in DNA and in proteins as a result of mutation, chromosomal duplications, chromosomal rearrangements, etc. over long periods of time, which may alter function, eventually giving rise to novel genes and proteins.

molecular evolution changes that occur at the level of DNA, gene arrangement and proteins during evolution.

molecular genetics the study of the molecular structure of DNA and the information it encodes, and the biochemical basis of gene expression and its regulation.

molecular layer external layer of cortex and cerebellum. *alt.* plexiform layer.

molecular mass sum of the atomic masses of all the atoms in a molecule. It is generally expressed in daltons or kilodaltons (1kDa = 1000 daltons), where 1 dalton = 1.000 on the atomic mass scale, a unit of mass very nearly equal to the mass of a hydrogen atom. The term molecular weight, although not strictly equivalent, is often used as a synonym. *see also* relative molecular mass.

molecular motors proteins with ATPase activity such as myosin, dynein and kinesin that can power the movement of microfilaments or microtubules.

molecular phylogeny the tracing of evolutionary relationships by the comparison of DNA and protein sequences from different organisms.

molecular weight *see* molecular mass, relative molecular mass *q.v.*

Mollicutes *n.* a class of prokaryotes including diverse wall-less microorganisms of doubtful origins, e.g. rickettsiae, chlamydiae, and mycoplasmas.

Moll's glands modified sweat glands between follicles of eyelashes. *alt.* ciliary glands.

Mollusca, molluscs *n., n.plu.* a large and diverse phylum of soft-bodied, usually unsegmented, coelomate animals, many of which live enclosed in a hard shell. They include the classes Gastropoda (winkles, whelks, slugs, snails, sea slugs, etc.), Bivalvia (clams, cockles, etc.), and other smaller classes of shells, and the Cephalopoda (nautilus, squids and octopuses). The coelom is small and the main body cavity is a blood-filled haemocoel. Molluscs have well-developed sense organs and nervous system, esp. in the Cephalopoda, a heart and blood system.

molluscicide *n.* a chemical that kills molluscs, e.g. snails.

molluscoid *a. pert.* or resembling a mollusc.

mollusk *alt.* spelling of mollusc *q.v.*

molt *alt.* spelling of moult *q.v.*

moltinism *n.* the condition in which different strains of a species undergo a different number of larval moults.

molybdenum (Mo) an essential micronutrient for plants. It is also an essential component of bacterial nitrogenases.

monacanthid *a.* with one row of ambulacral spines, as some starfishes.

monactinal *a.* with a single ray, *appl.* spicules of sponges, etc.

monactinellid *a. appl.* certain sponges that bear spicules with only a single ray. *cf.* hexactinellid.

monad *n.* a single-celled organism or flagellate cell; a single cell, instead of a tetrad, arising from meiosis.

monadelphous *a.* having stamens united into one bundle by union of filaments.

monamniotic *a.* having one amnion, *appl.* uniovular (identical) twins.

monandrous *a.* having only one stamen; having only one antheridium; having only one male mate.

monanthous *a.* having only one flower.

monarch *a.* with only one protoxylem strand, or only one vascular bundle, of plant axis.

monaxial *a.* having one line of axis; having inflorescence developed on the primary axis.

monaxonic *a.* with one axon, *appl.* nerve cell.

monecious *alt.* spelling of monoecious *q.v.*

monembryonic *a.* producing one embryo at a time.

Monera Prokaryotae *q.v.*

monestrous *alt.* spelling of monoestrus *q.v.*

Monilia *n.* large form class of deuteromycete fungi that reproduce by oidia or by budding, or by conidia not borne in pycnidia or acervuli. It includes *Penicillium* and *Aspergillum*, the false yeasts (e.g. *Cryptococcus*), and fungi that cause skin diseases in humans and animals (e.g. ringworm, *Microsporum*) and the serious human fungal pathogens *Blastomyces* and *Histoplasma*.

monilicorn *a.* having antennae with appearance like a chain of beads.

moniliform *a.* arranged like a chain of beads, *appl.* spores, hyphae, antennae; constricted at regular intervals, so appearing like a chain of beads.

monimostylic *a.* having quadrate united to squamosal, and sometimes to other bones, as in some reptiles.

monkeys *see* Primates.

mono- prefix derived from Gk. *monos*, single, signifying one, having one of, borne singly, etc.

monoallelic *a. appl.* a polyploid in which all the alleles at a locus are the same.

monoamine neurotransmitters dopamine, adrenaline, noradrenaline, *all q.v.*

monoamine oxidase enzyme that inactivates the catecholamine neurotransmitters dopamine, adrenaline and noradrenaline by oxidative removal of the amino group. *alt.* monoaminooxidase.

monoblast *n.* cell that develops into a monocyte.

monoblastic *a. appl.* embryos with a single undifferentiated germ layer.

monocardian *a.* having one auricle and ventricle.

monocarpellary *a.* containing or consisting of a single carpel.

monocarpic *a. appl.* plants that die after bearing fruit once. *n.* monocarp.

monocarpous *a.* having only one carpel to a gynaecium.

monocaryon monokaryon *q.v.*

monocentric *a.* having a single centromere.

monocephalous *a.* of flowers, with one capitulum only.

monocercous uniflagellate *q.v.*

monocerous *a.* having one horn.

monochasium *n.* a branched flowerhead with main axes producing one branch each.

monochlamydeous *a.* having a calyx but no corolla, or having only one whorl of perianth segments.

monochorionic *a.* having only one chorion, *appl.* uniovular twins.

monochromatic *a.* having only one colour; colour blind, seeing variations in brightness but no colours.

monochronic *a.* happening or originating only once.

monocistronic *a. appl.* mRNA coding for only one polypeptide chain, i.e. some bacterial and all eukaryotic mRNAs.

monoclinous *a.* hermaphrodite; having stamens and pistil in each flower; having antheridium and oogonium originating from the same hypha.

monoclonal antibody antibody produced by a single clone of B cells and thus consisting of a population of identical antibody molecules all specific for a single antigenic determinant. Produced from cultured hybridoma cell lines for research and commercial purposes. Widely used in research and medical diagnostics as highly specific reagents in the assay and identification of macromolecules, viruses, etc. or for accurately pinpointing structural features of cells in such techniques as immunofluorescence microscopy.

monocolpate *a. appl.* pollen grains with one groove, through which the pollen tube emerges.

monocondylar *a.* having a single occipital condyle (point of articulation of skull with spinal column), as birds and reptiles.

monocots monocotyledons *q.v.*

Monocotyledon(e)ae, Monocotyledones, monocotyledons *n., n.plu.* a class of angiosperms having an embryo with only one cotyledon, parts of the flower usually in threes, leaves with parallel veins, and vascular bundles scattered throughout the stem. They include familiar bulbs such as daffodils, snowdrops, lilies, etc., and the cereals and grasses such as maize, wheat, rice, etc. *a.* monocotyledonous.

monocotyledonous *a. pert.* monocotyledons; *pert.* embryo with only one cotyledon.

monocratic *a.* with the four spores of a meiotic tetrad being of the same sex.

monocule *n.* a one-eyed animal, as certain insects and crustaceans.

monoculture *n.* a large area covered by a single species (or for crops, a single variety) of plant, esp. if grown year after year.

monocyclic *a.* having a single cycle; (*bot.*) *appl.* annual herbaceous plants; *appl.* flowers with a single whorl.

monocystic *a.* with one stage of encystation.

monocyte *n.* large phagocytic white blood cell with a single oval or horseshoe-shaped nucleus, wanders into tissue where it becomes a macrophage. *alt.* mononuclear leukocyte.

monodactylous *a.* with a single digit, or a single claw.

Monodelphia Eutheria, *see* eutherians.

monodelphic *a.* having uteri more or less united, as in placental mammals; having a single uterus, as *appl.* certain nematodes.

monodelphous monadelphous *q.v.*

monodesmic *a.* (*zool.*) *appl.* scales formed of fused small bony scales with a continuous covering of dentine; (*bot.*) having a single vascular bundle.

monodont *a.* having a single persistent tooth, as male narwhal.

monoecious *a.* having male and female flowers on the same plant; with male and female sex organs on same gametophyte; having microsporangia and megasporangia on same sporophyte.

monoestrous *a.* having only one period of oestrus in a sexual season. *cf.* polyoestrus.

monofactorial unifactorial *q.v.*

monogalactosyl diacylglycerol glycolipid found in plant cell membranes.

monogamous *a.* consorting with one mate only, usually for the whole of the animal's lifetime. *n.* monogamy.

monoganglionic *a.* having a single ganglion.

monogastric *a.* with only one gastric cavity.

Monogenea, monogeneans *n., n.plu.* class of parasitic flatworms comprising the skin and gill flukes, ectoparasites mainly of fish and amphibians. They have a flattened leaf-shaped body and a simple life cycle on one host.

monogenetic *a. appl.* parasites completing their life cycle in a single host; *appl.* origin of a new form at a single place or period. *n.* monogenesis.

monogenic *a.* controlled by a single gene; producing offspring all of the same sex.

monogenomic *a.* having a single set of chromosomes.

monogenous *a.* asexual, as *appl.* reproduction.

monogeny *n.* production of offspring consisting of one sex, either male or female. *a.* monogenic.

monogoneutic *a.* breding once a year.

monogony *n.* asexual reproduction, including schizogony and budding.

monogynaecial *a.* developing from one pistil.

monogynous *a.* having one pistil; having one carpel to a gynaecium; consorting with only one female.

monogyny *n.* in animals generally, the tendency of each male to mate with only one female; in social insects, the existence of a single functional queen in the colony.

monohybrid *n.* hybrid offspring of parents differing in one character; *a.* heterozygous for a single pair of alleles.

monokaryon *n.* cells of a hypha containing one nucleus.

monokont *a.* having a single flagellum.

monolayer *n.* a single homogeneous layer of units, as of molecules, cells, etc.

monolepsis *n.* transmission of characteristics from only one parent to progeny. *a.* monoleptic.

monomastigote *a.* having one flagellum, as certain protists.

monomeniscous *a.* having an eye with one lens.

monomer *n.* molecule consisting of a single unit, as e.g. a protein consisting of one polypeptide chain.

monomeric *a. pert.* one segment; derived from one part; *appl.* protein molecules composed of a single polypeptide chain.

monomerosomatous *a.* having body segments all fused together, as certain arthropods.

monomerous *a.* consisting of one part only, *appl.* flowers.

monometrosis *n.* colony foundation by one female, as by queen in some social hymenopterans.

monomolecular layer monolayer *q.v.*

monomorphic *a. appl.* species in which all individuals look alike; developing with no or very slight change from stage to stage, as certain protozoans and insects; producing spores of one kind only. *cf.* dimorphic, polymorphic.

monomorphic loci genetic loci at which the most common homozygote has a frequency of more than 90% in a given population.

monomorphism *n.* in entomology, the existence of only a single worker subcaste within an insect species or colony.

mononemic *a.* consisting of one strand.

mononeuronic *a.* with one nerve; *appl.* chromatophores with single type of innervation.

mononuclear, mononucleate *a.* with a single nucleus.

mononychous *a.* having a single or uncleft claw.

mononym *n.* a designation consisting of one term only; name of a monotypic genus.

monooxygenase *n.* any of a class of oxidoreductase enzymes in which 1 atom of oxygen is incorporated into the product and 1 is reduced to form water. *alt.* mixed-function oxygenase, mixed-function oxidase.

monopectinate *a.* having one margin furnished with teeth like a comb.

monopetalous *a.* having one petal only; having petals united all round.

monophagous *a.* subsisting on one kind of food, *appl.* insects feeding on plants of one genus only, or insects restricted to one species or variety of food plant.

monophasic *a. appl.* the shortened life cycle of some trypanosomes, lacking the active stage.

monophenol monooxygenase polyphenol oxidase *q.v.*

monophyletic *a.* derived from a common ancestor, *appl.* taxa derived from and including a single founder species.

monophyllous *a.* having one leaf only; having calyx in one piece.

monophyodont *a.* having only one set of teeth, the milk teeth being absorbed in the foetus or being absent altogether.

monoplacid *a.* with one plate only.

Monoplacophora *n.* a mainly extinct class of molluscs with a shell like a limpet, the living forms (e.g. *Neopilina*) being known only from the deep-sea bed.

monoplanetic *a.* with one stage of motility in life history, *appl.* certain fungi.

monoplastic *a.* persisting in one form; having one chloroplast, *appl.* cells.

monoploid *a.* having one set of chromosomes, true haploid; in a polyploid series, having the basic haploid chromosome number.

monopodal *a.* having one supporting structure; with one pseudopodium.

monopodial *a.* branching from one primary axis with the youngest branches arising at the apex.

monopodium *n.* a single main or primary axis from which all main lateral branches develop.

monopyrenic, monopyrenous *a.* singlestoned, as of fruit.

monorchic *a.* having one testis.

monorefringent isotropic *q.v.*

monorhinal *a.* having only one nostril; *pert.* one nostril.

monosaccharide *n.* any of a class of simple carbohydrates, all being reducing sugars, with the general formula $(CH_2O)_n$, n greater than 3, such as glyceraldehyde (a triose, n = 3), ribose (a pentose, n = 4), glucose (a hexose, n = 6).

monosaccate *a. appl.* pollen grains with one air bladder.

monosepalous *a.* having one sepal.

monose monosaccharide *q.v.*

monosiphonic, monosiphonous *a. appl.* algae having a single central tube in filament.

monosomic *a. appl.* a chromosome lacking its homologous partner or to an unpaired X chromosome; *appl.* diploid cells or organisms in which one partner of a pair of homologous or sex chromosomes has been lost, e.g. XO cells are monosomic for X.

monosomy *n.* the absence of a single chromosome from the diploid set.

monospecific *a. appl.* antibody reacting with only one antigen.

monospermic, monospermous *a.* oneseeded; fertilized by entrance of only one sperm into ovum.

monospermy *n.* normal fertilization by penetration of one sperm into ovum.

monospondylic *a. appl.* vertebrae with only one vertebral ring or central hole.

monospore *n.* a simple or undivided spore.

monosporic *a.* originating from a single spore.

monosporous *a.* having only a single spore.

monostachyous *a.* having only one spike.

monostele protostele *q.v.*

monostichous *a.* arranged in a single row; along one side of an axis.

monostigmatous *a.* with one stigma only.

monostromatic *a.* having a single-layered thallus, *appl.* algae.

monostylous *a.* having one style only.

monosulcate *a. appl.* pollen grains with a single groove on the surface away from that through which pollen tube emerges.

monosy *n.* separation of parts normally fused.

monosymmetrical zygomorphic *q.v.*

monotaxic *a.* belonging to the same taxonomic group.

monothalamous *a.* with a single chamber or locule; *appl.* fruits formed from single flowers.

monothecal *a.* single-chambered; having one loculus.

monothetic *a. appl.* classification based on only one or a few characteristics, such as of plants based on number of stamens. *cf.* polythetic.

monotokous *a.* uniparous, having one offspring at birth. *alt.* monotocous.

Monotremata, monotremes *n., n.plu.* order of primitive mammals that lay eggs, have mammary glands without nipples, and no external ears, of which the only extant species are the duck-billed platypus of Australia (*Ornithorhynchus*) and the spiny anteaters (*Tachyglossus* and *Zaglossus*), which are found in Australia and New Guinea.

monotrichous *a.* having a single polar flagellum.

monotrochous *a.* having a trochanter in a single piece, as most stinging Hymenoptera.

monotrophic *a.* subsisting on one kind of food.

monotropic *a.* turning in one direction only; visiting only one kind of flower, *appl.* insects.

monotropoid *a. appl.* mycorrhizas formed on members of the Monotropaceae, plants lacking chlorophyll which are dependent on the mycorrhiza for their carbon and energy source, in which an extensive rootball of fungal and root tissue is formed, which also forms connections with the ectomycorrhizal roots of nearby green plants.

monotype *n.* single type which constitutes species or genus.

monotypic *a. appl.* genera having only one species; *appl.* species having no subspecies.

monovalent univalent *q.v.*

monovoltine univoltine *q.v.*

monovular uniovular *q.v.*

monoxenic, monoxenous *a.* inhabiting one host only, *appl.* parasites.

monoxylic *a.* having wood formed as a continuous ring.

monozoic *a.* producing one sporozoite only.

monozygotic *a.* originating from a single fertilized ovum (zygote), as identical twins (MZ twins).

Monro, foramen of foramen of Monro *q.v.*

monsoon rain forest type of rain forest that develops in tropical and subtropical regions with a high annual rainfall but marked dry and rainy seasons (monsoon rainfall), consisting of deciduous trees and shrubs that lose their leaves in the dry season. *alt.* monsoon forest.

mons pubis, mons Veneris prominence of fatty subcutaneous tissue in front of pubic bone.

montane *a. pert.* mountains.

monticolous *a.* inhabiting mountainous regions.

monticulus *n.* largest part of the superior vermis of cerebellum.

moor *n.* open area of upland acid peat, with vegetation of heathers, sedges and certain grasses (e.g. *Molinia, Caerulea*).

mor *n.* acid humus of cold wet soils which inhibits action of soil organisms and may form peat. *cf.* mull.

morbilliviruses *n.* a group of RNA viruses of the paramyxovirus family, related to canine distemper virus.

mores *n.plu.* groups of organisms preferring the same habitat, having the same reproductive season, and agreeing in their general reactions to the physical environment.

morgan *n.* a unit of distance on genetic map, in which the mean number of recombinations is unity.

moriform *a.* shaped like a mulberry; formed in a cluster resembling an aggregate fruit.

morph *n.* one of the forms present in a polymorphic population.

morphactins *n.plu.* a group of substances derived from fluorine-9-carboxylic acids, which affect plant growth and development.

morphallaxis *n.* transformation of one part into another when parts are regenerating; gradual growth or development into a particular form.

morphine *n.* the chief alkaloid of opium, used clinically to relieve pain, but produces dependency on long-term use.

morphogen *n.* any substance that influences morphogenesis or embryonic development generally.

morphogenesis *n.* the development of shape and structure; origin and development of organs or parts of organisms. *alt.* morphogeny.

morphogenetic *n. pert.* morphogenesis; *appl.* hormones: e.g. thyroxine, ecdysone, juvenile hormone, etc., hormones that influence growth, development and/or metamorphosis of organisms.

morphogenetic furrow a furrow that moves over the eye imaginal disc in certain larval insects. Cells through which the furrow has passed begin to differentiate into prospective ommatidia of the compound eye.

morphologic index ratio expressing relation of trunk to limbs.

morphological species *see* morpho- species.

morphology *n*. the form and structure of an organism as distinct from its physiology, etc.; the study of form and structure. *a*. morphologic, morphological.

morphoplankton *n*. plankton organisms rendered buoyant by small size, or body shape, or structures containing oily globules, mucilage, gas, etc.

morphosis *n*. the manner of development of part of an organism; the formation of tissues. *a*. morphotic.

morphospecies *n*. a group of individuals which are considered to belong to the same species on grounds of morphology alone.

morphotype *n*. type specimen of one of the forms of a polymorphic species.

morula *n*. a solid globular mass of cells, the product of the first rounds of cell division of the fertilized egg, the stage preceding the gastrula.

mosaic *n*. disease of plants characterized by mottling of leaves, caused by various viruses, e.g. tobacco mosaic, cucumber mosaic; organism whose body cells are a mixture of two or more different genotypes, e.g., human and other mammalian females, who have one of their X chromosomes inactivated at random early in development so that adult tissues usually contain a mixture of cells containing different active X chromosomes.

mosaic eggs eggs whose development appears to be directed by cytoplasmic determinants localized to different parts of the egg, the blastomeres formed developing to a large extent independently of each other, and being unable to substitute for each other if one is removed.

moschate *a*. having or resembling the odour of musk; musky.

moss animals common name for the Bryozoa *q.v.*

mosses *n.plu.* common name for members of the Bryophyta (*q.v.*), a division of non-vascular, spore-bearing green plants. The mosses are divided into three classes: the "true" mosses, the sphagnum mosses and the granite mosses (*see individual entries*). Several plants commonly called mosses belong to other groups: reindeer moss is a lichen, club mosses and Spanish moss are vascular plants, sea moss and Irish moss are algae.

moss layer the lowest horizontal ecological stratum of a plant community comprising the ground surface and plant cover such as mosses and lichens. *alt.* ground layer.

mossy fibres nerve fibres branching profusely around cells of the internal layer of the cerebellar cortex.

mother cell cell which gives rise to other cells by division.

moths *n.plu.* the common name for many members of the Lepidoptera, having antennae tapering to a point and not clubbed.

motile *a*. capable of spontaneous movement.

motivation *n*. internal factors controlling behaviour in an animal, that lead to its achieving a goal or satisfying a need.

motivational state the combined effect of the physiological state of an animal and its perception of stimuli from the environment, which determines behaviour.

motoneurone motor neurone *q.v.*

motor *a. pert.* or connected with movement, *appl.* nerves etc.

motor areas areas of the brain where muscular activity is initiated and coordinated.

motor cortex area of cerebral cortex in brain concerned with initiation of voluntary movement.

motor end plate or end organ the structure in which the axon of a motor neurone terminates in a skeletal muscle fibre.

motor neurone nerve cell that carries impulses away from the central nervous system (brain and spinal cord) to an effector muscle. *alt.* motoneuron(e).

motor unit a motor neurone and associated muscle fibres.

mould *n*. common name for many members of the Fungi that grow as a fluffy mycelium over a substrate.

moult *n*. the periodic shedding of outer covering, whether of feathers, hair, skin or cuticle. In crustaceans and arthropods, it is necessary during larval growth as the exoskeleton, once hardened, does not allow further internal growth.

moulting glands ecdysial glands *q.v.*

moulting hormone ecdysone *q.v.*

moultinism moltinism *q.v.*

mouthpart *a*. head or mouth appendage of

arthropods.

M phase the period of mitosis during the cell cycle.

MR mutator system in *Drosophila*, increase in frequency of crossing-over and elevation of recessive mutation rate, caused by introduction into laboratory strains of second chromosomes from some wild flies. *see* hybrid dysgenesis.

Mu DNA bacteriophage of *E. coli* which can insert into the bacterial genome at random within a specific region in a manner analogous to transposons.

mu particles particles borne by some *Paramecium aurelia* strains, which kill or injure susceptible partners with which they conjugate.

mucid *a.* mouldy; slimy.

muciferous body protrusible organelle found in euglenoids and dinoflagellates.

mucific *a.* mucus-secreting.

mucigel *n.* gelatinous material on the surface of roots in soil, comprising a mixture of plant mucilages, bacterial capsules and slime layers and colloidal soil particles.

mucilage *n.* general term for complex substances composed of various types of polysaccharides, becoming viscous and slimy when wet, widely occurring in plants, and secreted by plant roots and by bacteria (the capsule or slime layer).

mucilaginous *a. pert.*, containing or composed of mucilage.

mucin *n.* general term for various glycoproteins found in secretions such as saliva, mucus, etc.

mucinogen *n.* substance occuring in granules or globules in mucin-producing cells, giving rise to mucin.

muciparous *a.* mucus-secreting.

mucivorous *a.* feeding on plant juices, *appl.* insects.

mucocellulose *n.* cellulose mixed with various mucins as in some seeds and fruit.

mucocutaneous *a. appl.* skin and mucous membranes.

mucoid *a. appl.* method of feeding employed by some molluscs, which extrude mucus from the mouth and then ingest it with the small particles attached to it.

mucoids *n.plu.* glycoproteins (*q.v.*) of bone, tendon and other connective tissues.

mucolytic *a.* breaking down mucus or mucilage.

mucopeptide *n.* peptidoglycan (*q.v.*), formerly used for the degradation product of the peptidoglycans of bacterial cell walls.

mucopolysaccharide glycosaminoglycan *q.v.*

mucopolysaccharidosis *n.* disease characterized by massive accumulation of mucopolysaccharides (glycosaminoglycans) in lysosomes.

mucoprotein *n.* glycoprotein *q.v.*, esp. those found in mucous secretions; proteoglycan *q.v.*

Mucorales *n.* order (in some classifications considered as a class) of zygomycete fungi with a well-developed mycelium and non-motile spores contained in a stalked sporangium, most living as saprophytes on dung or decaying plant and animal matter. It includes the bread moulds (e.g. *Mucor*, *Rhizopus*) and the dung fungus *Pilobolus*.

mucosa *n.* mucous membrane *q.v.*; the lining of the gut which consists of three layers, the inner epithelium containing glands, the lamina propria *q.v.*, and the mucosa muscularis *q.v.*

mucoserous *a.* secreting mucous and fluid.

mucous *a.* secreting, containing or *pert.* mucus.

mucous membrane any epithelial layer secreting mucus, e.g. the linings of the nasal pasages, reproductive tract, gut, etc. *alt.* mucosa.

mucro *n.* a sharp point at termination of an organ or other structure; a small awn; a pointed keel or sterile third carpel, as in pine.

mucronate, mucroniferous *a.* abruptly terminated by a sharp spine.

mucronulate *a.* tipped with a small mucro.

mucus *n.* slimy material rich in glycoproteins, secreted by goblet cells of mucous membranes or by mucous cells of a gland; similar slimy secretion produced on the external body surface of many animals.

mull *n.* humus of well-aerated moist soils, formed by action of soil organisms on plant debris and favouring plant growth.

Müller cells glial cells of retina.

Müller's fibres neuroglia forming a framework supporting the nervous tissue of the retina.

Müllerian ducts a pair of ducts developing in the early vertebrate embryo that will give rise to the oviduct in females.

Müllerian mimicry the resemblance of two animals to their mutual advantage, for example the yellow and black stripes of wasps and of the unpleasant-tasting cinnabar moth caterpillars, which leads to a predator that has encountered one subsequently also avoiding the other. *cf.* Batesian mimicry.

multi- prefix derived from L. *multus*, many.

multiarticulate *a.* many-jointed.

multiaxial *a.* having or *pert.* several axes; allowing movement in many planes.

multicamerate *a.* with many chambers; multiloculate.

multicarinate *a.* with many ridges or keels.

multicarpellary *a.* with compound gynaecium consisting of many carpels.

multicauline *a.* with many stems.

multicellular *a.* many-celled, *appl.* eukaryotic organisms consisting of many cells specialized for different functions and organized into a cooperative structure; consisting of more than one cell. *n.* multicellularity.

multicentral *a.* with more than one centre of growth or development.

multiciliate *a.* with some or many cilia.

multicipital *a.* with many heads or branches arising from one point.

multicomponent viruses viruses in which different parts of the viral genome are packaged into separate virus particles.

multicopy plasmid plasmid present in a bacterial cell in several copies, the number being characteristic for different plasmids.

multicostate *a.* with many ribs or veins (*appl.* leaves); with many ridges.

multi-CSF interleukin 3, a glycoprotein stimulating the growth *in vitro* of white blood cell precursors of all types.

multicuspid(ate) *a.* with several cusps, *appl.* molar teeth.

multidentate *a.* with many teeth or indentations.

multideterminant *a. appl.* antigen carrying more than one antigenic determinant.

multidigitate *a.* many-fingered.

multidrug resistance simultaneous resistance to a variety of unrelated anti-cancer drugs shown by many tumour cells.

multienzyme complex a complex of different enzyme molecules, usually all catalysing different steps in a particular pathway, as the cellulose-synthesizing enzyme complex that is formed in the Golgi apparatus and delivered to the cell wall in plant cells, or pyruvate dehydrogenase, which is a complex of 44 molecules with three distinct enzymatic activities.

multifactorial *a. appl.* phenotypic traits, *see* polygenic.

multifactorial inheritance inheritance of phenotypic characters determined by the action of several independent genes.

multifascicular *a.* containing or *pert.* many small bundles or fascicles.

multifarious polystichous *q.v.*

multifid *a.* having many clefts or divisions.

multiflagellate *a.* having several or many flagella.

multiflorous *a.* having many flowers.

multifoliate, -foliolate *a.* having many leaves or many small leaves respectively.

multiform *a.* occurring in, or containing, different forms; *appl.* layer: the inner cell layer of cerebral cortex.

multifunctional proteins proteins containing various enzymic or other activities within a single molecule.

multigene family a set of similar but not identical genes that encode the different members of a family of related proteins such as the interferons, the actins, the globins etc. Multigene families are presumed to have arisen by duplication and divergence of an ancestral gene.

multigyrate *a.* intricately folded.

multijugate *a.* having many pairs of leaflets.

multilacunar *a.* having many lacunae; having a number of leaf gaps, *appl.* nodes.

multilaminate *a.* composed of several layers.

multilobate, -lobulate *a.* with many lobes or lobules respectively.

multilocular, multiloculate *a.* many-chambered, *appl.* ovary (of flower), shells.

multimer *n.* protein molecule made up of more than one polypeptide chain (protein subunit); protein complex made up of several different proteins. *a.* multimeric.

multinervate *a.* with many nervures (of wing or leaf) or nerves.

multinodal, multinodate *a.* with many nodes.

multinomial *a. appl.* a name or designation composed of several terms.

multinucleate *a.* with several or many nuclei.

multinucleolate *a.* with several or many nucleoli.

multiovulate *a.* with several or many ovules.

multiparous *a.* bearing several, or more than one, offspring at a birth; (*bot.*) developing several or many lateral axes.

multiperforate *a.* having more than one perforation.

multipennate *a. appl.* muscle containing a number of extensions of its tendon of insertion.

multipinnate *a.* having many leaflets; with each leaflet or division divided pinnately, and so on, *appl.* leaves.

multiple alleles a series of more than two alleles for a given gene locus, loci with multiple alleles being known as polymorphic. *alt.* genetic polymorphism.

multiple corolla a corolla with two or more whorls of petals.

multiple drug resistant bacteria bacteria carrying resistance genes to a number of commonly used antibiotics, often together on a plasmid.

multiple fission repeated division; division into a large number of parts or spores.

multiple fruit fruit developing from the gynaecia (carpels) of more than one flower, e.g. pineapple.

multiplicate *a.* consisting of many; having many folds.

multipolar *a. appl.* nerve cells with more than two main cellular processes (i. e. more than one main dendron and axon); *appl.* mitosis in which more than two poles are formed, normal in certain sporozoan protozoans, but otherwise pathological.

multiporous *a.* having many pores.

multipotent *a.* capable of giving rise to several different kinds of structure or types of cell.

multiradiate *a.* many-rayed, *appl.* spicules of certain sponges.

multiradicate *a.* with many roots or rootlets.

multiramose *a.* much branched.

multiseptate *a.* having numerous partitions.

multiserial, multiseriate *a.* arranged in many rows; *appl.* xylem rows more than one cell wide; *appl.* spores in rows in ascus.

multispecificity *n.* ability of a single type of antibody molecule to combine with a range of different antigens.

multistaminate *a.* having many stamens.

multisulcate *a.* much furrowed.

multitentaculate *a.* having numerous tentacles.

multituberculate *a.* having several or many small humps or protuberances.

multituberculates *n.plu.* a class of extinct herbivorous mammals existing in the Jurassic to Eocene, with affinities to present-day monotremes.

multivalent *a. appl.* antibodies with more than one antigen-binding site; *n.* structure formed by the association of more than two chromosomes during meiosis in polyploids.

multivalve *a. appl.* a shell composed of more than two pieces or parts.

multivoltine *a.* having more than one brood in a year, *appl.* some birds.

multocular *a.* many-eyed.

multungulate *a.* with the hoof in more than two parts.

mune mores *q.v.*

mural *a.* constituting or *pert.* a wall; growing on a wall.

muralium *n.* a structure formed by layers one cell thick, as internal structure of liver.

muramic acid monosaccharide found in the peptidoglycans of bacterial cell walls, an *N*-acetylamino acid comprising *N*-acetylglucosamine condensed with lactic acid at carbon 3 of the sugar. *alt.* *N*-acetylmuramic acid.

murein *n.* any of the peptidoglycans found in the cell walls of bacteria and cyanobacteria (blue-green algae).

muricate *a.* formed with sharp points; covered with short sharp outgrowths; studded with oxalic acid crystals.

muriform *a.* (*bot.*) like a brick wall, *appl.* a parenchyma tissue so arranged with cells in overlapping rows, in medullary ray of dicotyledons and in cork; (*zool.*) shaped like a morula, *appl.* coelomocytes.

muscarine *n.* a ptomaine base, found in the

fly agaric toadstool, *Amanita muscaria*, and other plants.

muscarinic *a. appl.* ACh receptors blocked by muscarine and similar drugs (as opposed to the nicotinic ACh receptors).

Musci mosses *q.v.*

muscicoline *a.* living or growing among or on mosses.

muscimol *n.* hallucinogenic plant alkaloid, binds to GABA receptors in brain.

muscle *n.* contractile animal tissue involved in movement of the organism and which also forms part of many internal organs. Muscle cells contain contractile protein microfibrils which contract simultaneously, usually in response to a nervous or chemical stimulus. There are three main types of muscle in vertebrates: striated or striped muscle which forms the muscles attached to the skeleton, smooth muscle associated with many organs and which forms the contractile layer of arteries, and the cardiac muscle of the heart. *see* cardiac muscle, smooth muscle, striated muscle.

muscle cell any of various types of specialized contractile cell making up the muscular tissues of the body, the four main types in mammals being skeletal muscle cells, cardiac muscle cells, smooth muscle cells, and myoepithelial cells. All contract by means of a contractile apparatus formed of the proteins actin and myosin. Skeletal muscle cells, also called muscle fibres, are often very large, roughly spindle shaped, multinucleate, and develop from the fusion of several immature cells. They contain orderly arrays of the proteins actin and myosin giving the cells a striated appearance. Heart muscle cells are smaller and uninucleate and also striated. Smooth muscle cells (found e.g. in the wall of the gut and blood vessels) are small, unstriated and uninucleate. Myoepithelial cells are contractile epithelial cells, forming e.g. the dilator muscle of the iris, and the contractile tissue of some glands.

muscle fibre *see* muscle cell.

muscle segment myomere *q.v.*

muscle spindle sensory receptor in muscle which monitors degree of stretch, consisting of a spindle-shaped connective tissue sheath containing small modified fibres, with nerve fibres entering each spin-

dle and forming spirals or arborizations around individual muscle fibres.

muscoid *a.* moss-like; mossy.

muscology *n.* study of mosses.

Muscopsida mosses *q.v.*

muscular *a. pert.* or consisting of muscle.

muscular dystrophy *see* Becker muscular dystrophy, Duchenne muscular dystrophy.

muscularis externa layer of the gut wall between the submucosa and serosa consisting of a sheet of longitudinal and a sheet of circular muscles.

muscularis mucosa the outermost layer of the mucosa of the gut wall, made of smooth muscle fibres.

musculature *n.* the system or arrangement of muscles as a whole.

musculo-epithelial cells cells in the gastrodermis of coelenterates, with a cell body and long contractile processes.

musculocutaneous *a. pert.* muscles and skin, *appl.* veins and nerves of limbs which supply muscles and skin.

musculophrenic *a.* supplying diaphragm and body wall muscles; *appl.* artery: a branch of the internal mammary artery.

musculospiral *a. appl.* radial nerve which passes spirally down humerus; *appl.* spiral arrangement of muscle fibres.

musculotendinous *a. pert.* muscle and tendon.

mushroom *n.* common name for edible basidiomycete fungi, esp. of the genus *Agaricus.*

mushroom bodies corpora pedunculata *q.v.*

mushroom gland large seminal vesicles of certain insects.

mustelids *n.plu.* members of the family Mustelidae: weasels, stoats, badgers, otters, polecats, martens.

mutagen *n.* any agent that can cause a mutation. *a.* mutagenic.

mutagenesis *n.* the production of mutations by agents such as X-rays or chemicals.

mutagenic *a.* capable of causing a mutation, as radiation, chemicals or other extracellular agents.

mutagenize *v.* to treat with a mutagen.

mutant *n.* organism or cell carrying altered genetic material owing to which it usually differs from its parent or immediate precursor cell in some physical or biochemical characteristic(s). *a.* mutant.

mutase *n.* any of a group of enzymes that catalyse the intramolecular shift of a chemical group and which are classified amongst the isomerases in EC class 5.

mutate *v.* to undergo mutation.

mutation *n.* a change in the amount or chemical structure of DNA resulting in a change in the characteristics of an organism or an individual cell as a result of alterations in, or non-production of, proteins (or RNAs) specified by the mutated DNA. Mutations occurring in body cells of multicellular organisms are called somatic mutations and are only passed on to the immediate descendants of those cells, mutations occurring in germline cells can be inherited by the offspring. Mutations can occur spontaneously as a result of errors in normal cell processes e.g. DNA replication, or can be induced by certain chemicals, types of radiation etc. Alterations in DNA that do not cause any phenotypic change are also sometimes called mutations (silent mutations). *see also* point mutation, base substitution, translocation, transversion, transposition, insertion, deletion, revertant, wild type, back mutation, silent mutation.

mutation pressure changes in gene frequencies brought about by mutational change alone.

mutation rate the rate at which mutations arise in a population. The spontaneous rate of base-pair changes (due to errors in DNA replication and environmental influences) is *ca.* 1 in 10^9 rounds of replication on experimental evidence, from which it has been calculated that a mutation will occur in a protein at a rate of *ca.* 1 per 10^6 cell generations. The spontaneous mutation rate as detected by the appearance of detectably mutant organisms appears to vary between different organisms and between different genes.

mutational load the reduction in population fitnes due to the accumulation of deleterious mutations.

mutator *a. appl.* genes which increase the general mutation rate.

muticate, muticous *a.* without a point; without defensive structures.

mutilous *a.* without defensive structures, as clawless, harmless, toothless, blunt.

mutualism *n.* a special case of symbiosis in which both partners benefit from the association. *a.* mutualistic.

mutuality *n.* evolutionary strategy in regard to animal communication where both signaller and receiver benefit from the interaction.

myarian, myaric *a. appl.* classification according to musculature; *pert.* musculature.

myasthenia gravis autoimmune disease in which antibodies against skeletal muscle acetylcholine receptors are produced.

Mycelia Sterilia a diverse group of fungi without any presently known conidial (asexual) or sexual reproductive stages.

mycelioid *a.* resembling a mycelium; growing in the form of a mycelium.

mycelium *n.* a network of hyphae forming the characteristic vegetative phase of many fungi, often visible as a fluffy mass or mat of threads (hyphae).

Mycetae *n.* in some classifications an alternative name for the Fungi *q.v.*

mycetocyte *n.* one of follicle cells at posterior pole of oocyte through which the egg of aphids and other bugs may be infected by symbionts.

mycetogenetic *a.* produced by a fungus.

mycetoid *a.* like a fungus.

mycetophage *n.* an organism that eats fungi.

mycina *n.* a spherical stalked apothecium of certain lichens.

myco-, myce-, mycet- prefixes derived from Gk. *mykēs*, fungus.

mycobacterium *n.* bacterium of the family Mycobacteriaceae, Gram-positive non-motile rods, some spp. found in soil, others pathogenic for man and animals, e.g. *Mycobacterium tuberculosis*, and *M. leprae*, the causal agent of leprosy.

mycobiont *n.* the fungal component of a lichen.

mycobiota *n.* the fungi of an area or region.

mycoderm *n.* a superficial film of bacteria or yeast that develops during alcoholic fermentation.

mycoecotype *n.* the habitat type of mycorrhizal or parasitic fungi.

mycoflora *n.* all fungi growing in a specified area or region, or within an organism.

mycogenetics *n.* the genetics of fungi.

mycoid *a*. like a fungus.

mycology *n*. the study of fungi.

mycolysis *n*. the lysis or disintegration of fungi, as by bacteria.

mycophage *n*. a virus infecting fungi.

mycophagy *n*. feeding on fungi.

mycophthorous *a*. fungus-destroying, *appl*. or *pert*. fungi parasitizing other fungi.

Mycophycophyta lichens *q.v.*

mycoplasma *n*. any of a group of almost sub-microscopic prokaryotic microorganisms of very simple internal structure, generally classified with the bacteria, but which lack the typical rigid bacterial cell wall and differ from bacteria in their complex life cycle. They occur in a variety of morphological forms, are obligate intracellular parasites and are responsible for several animal diseases. Formerly called the pleuropneumonia-like organisms (PPLOs). *alt*. mycoplasm. Mycoplasmas infecting plants are also known as mycoplasma-like organisms. *see also* spiroplasmas.

mycoplasma-like organism (MLO) almost sub-microscopic plant-pathogenic, motile prokaryotes lacking cell walls, which are the cause of some "yellows" diseases of plants. They are similar to spiroplasmas but are not helical.

mycorhiz- mycorrhiz-

mycorrhiza *n*. a symbiotic association between plant roots and certain fungi, in which a sheath of fungal tissue (the mantle) encloses the smallest rootlets, with fungal hyphae penetrating between the cells of the epidermis and cortex (ectomycorrhizas), or invading the cells themselves (endomycorrhizas, in which the external fungal sheath is often lacking), and which are essential for optimum growth and development in many trees, shrubs and herbaceous plants. *a*. mycorrhizal.

mycorrhizoma *n*. association of fungi and a rhizome.

mycosis *n*. animal disease caused by a fungus. *plu*. mycoses.

mycostasis *n*. the inhibition of germination of fungal spores in the soil in the absence of nutrients; inhibition of fungal growth generally. *alt*. fungistasis.

mycosterols *n.plu*. sterols originally found in fungi but now known to be much more widely distributed, as ergosterol, fucosterol, etc.

mycotic *a*. caused by fungi.

mycotoxin *n*. any toxin produced by a fungus.

mycotrophic *a*. *appl*. plants living symbiotically with fungi.

mycteric *a*. *pert*. nasal cavities.

mydriasis *n*. dilation of pupil of the eye.

myelencephalon *n*. the posterior part of hindbrain, comprising medulla oblongata and lower part of fourth ventricle.

myelic *a*. *pert*. medulla of spinal cord.

myelin *n*. a highly refractory fatty material forming a multilayered insulating sheath around certain nerve fibres.

myelin sheath fatty insulating sheath formed of Schwann cell body wrapped round and round axon of nerve fibre to form concentric layers of cell membrane.

myelination *n*. acquisition of a myelin sheath by a nerve cell.

myelo- prefix derived from Gk. *myelos*, marrow.

myeloblast *n*. an undifferentiated non-granular lymphoid cell of bone marrow.

myelocoel *n*. the canal of the spinal cord.

myelocyte *n*. a bone marrow cell.

myeloid *a*. *pert*. bone marrow; *appl*. tissue in which haemopoiesis occurs in vertebrates, as bone marrow, and liver and spleen in embryos; *appl*. cells: monocytes, macrophages, mast cells, and leukocytes other than lymphocytes; (*neurobiol*.) resembling myelin.

myeloid lineage bone-marrow derived blood cells with the exception of lymphocytes.

myeloid stem cell stem cell in bone marrow that gives rise to all blood cells except lymphocytes.

myeloma *n*. a cancer of bone marrow or plasma cells (B cells) which sometimes produces very large quantities of a single antibody species as a result of the proliferation of a single B cell clone.

myelomere *n*. a segment of the spinal cord.

myelopoiesis *a*. formation and development of white blood cells; formation of bone marrow.

myenteric plexus layer of ganglia between the circular and longitudinal layers of muscular coat of small intestine. *alt*.

Auerbach's plexus.

myenteron *n*. the muscular coat of the intestine.

myiasis *n*. invasion of living tissue by the larvae of certain flies.

mylohyoid *a*. in region of hyoid and posterior part of mandible.

myo- prefix derived from Gk. *mys*, muscle.

myoblast *n*. undifferentiated mesenchymal cell present in embryo and adult (as satellite cell), several of which fuse together to form a multinucleate skeletal muscle cell.

myocardium *n*. the muscular wall of the heart. *a*. myocardial.

myocoel *n*. part of the coleom enclosed in a myotome.

myocomma *n*. a ligamentous connection between successive myomeres. *alt*. myoseptum.

myocyte muscle cell *q.v.*

myodynamic *a. pert*. muscular force or contraction.

myoelastic *a. appl*. tissue composed of unstriated muscle fibres (smooth muscle) and elastic connective tissue fibres.

myoepicardial *a. appl*. a covering consisting of the mesocardium walls, destined to form the muscular and epicardial walls of the heart.

myoepithelial cell contractile epithelial cell; musculo-epithelial cell *q.v.*

myofibril *n*. contractile protein fibril of muscle cells, composed of many myofilaments.

myofilament *n*. protein thread making up the myofibrils of muscle cells, *see* thin filament, thick filament.

myogenesis *n*. the differentiation and development of muscle.

myogenic *a*. having origin in muscle cells; *appl*. contractions arising in muscle cells spontaneously and independently of nervous stimulation, as heartbeat.

myoglobin *n*. oxygen-carrying globular haem protein consisting of a single polypeptide chain, involved in oxygen storage and transport in vertebrate muscle.

myohaemoglobin myoglobin *q.v.*

myoid *a*. resembling or composed of muscle fibres, *appl*. striated cells of thymus; *n*. contractile proximal part of filament of rods and cones of retina.

myokinase adenylate kinase *q.v.*

myolemma sarcolemma *q.v.*

myology *n*. the study of muscles.

myomere *n*. segment of muscle separated from the next by a thin sheet of connective tissue.

myometrial *a. pert*. myometrium; *appl*. glandular tissue of uterus supposed to produce a hormone affecting growth of mammary glands.

myometrium *n*. the muscular uterine wall.

myonema, myoneme *n*. minute contractile thread in protists.

myoneural neuromuscular *q.v.*

myopathy *n*. any disease or condition leading to degeneration and loss of function of muscles.

myoplasm *n*. contractile portion of a muscle cell or fibre.

myoseptum myocomma *q.v.*

myosin *n*. ubiquitous protein of eukaryotic cells, with ATPase activity, which interacts with actin to form a contractile complex. There are various types. The myosin I that comprises the thick filaments of striated muscle has a long α-helical "tail" and two globular "heads" containing ATPase activity, which form cross-bridges with the actin thin filaments. A cycle of ATPase hydrolysis powers the movement of the myosin head and contraction of the muscle. Other types of myosin (e.g. myosin II) are found in non-muscle cells. *see also* meromyosin.

myosis *n*. contraction of pupil of eye.

myotendinal musculotendinous *q.v.*

myotic *a*. causing or *pert*. myosis or pupillary contraction.

myotome *n*. one of a series of hollow cubes of mesenchyme that develops into muscle tissue in early vertebrate embryo; a muscular segment of primitive vertebrates and segmented invertebrates.

myotonia *n*. muscular tension or tonicity.

myotube *n*. a stage in development of a skeletal muscle cell, formed from fusion of individual immature myoblasts resulting in an elongated syncytium. It differentiates into a muscle fibre.

Myriapoda, myriapods *n., n.plu.* centipedes and millipedes and their relatives, terrestrial arthropods characterized by possession of a distinct head with a pair of antennae followed by numerous similar segments each bearing legs.

Myricales *n*. an order of dicot trees and

shrubs comprising the family Myriaceae (sweet gale).

myriophylloid *a.* having a much-divided thallus, *appl.* certain algae.

myriosporous *a.* having many spores.

myristic acid a 14-carbon saturated fatty acid.

myristoylation *n.* the covalent addition of myristic acid to a protein, thought to be important in attaching some proteins to cell membranes.

myrme- prefix derived from Gk. *myrmex*, an ant.

myrmecochore *n.* an oily seed modified to attract and be spread by ants.

myrmecioid complex one of the two major taxonomic subgroups of ants, exemplified by the subfamily Myrmeciinae.

myrmecole *n.* an organism occupying ants' nests.

myrmecology *n.* the study of ants.

myrmecophobic *a.* repelling ants, *appl.* plants with special glands, hairs etc., that check ants.

myrmecophagous *a.* ant-eating.

myrmecophil(e) *n.* a guest insect in an ants' nest; organism that must spend some part of its life cycle with ant colonies.

myrmecophilous *a. appl.* flowers, pollinated by the agency of ants; *appl.* fungi, serving as food for ants; *appl.* spiders, living with, preying on or mimicking ants.

myrmecophyte *n.* a plant pollinated by ants, or one that benefits from ant inhabitants and has special adaptations for housing them.

myrosin sinigrin *q.v.*

myrrh *n.* fragrant resin obtained from plants of the genus *Commiophora*.

Myrtales *n.* order of dicots, mostly shrubs and trees, including Lythraceae (loosestrife), Myrtaceae (myrtle), Onagraceae (evening primrose), Punicaceae (pomegranate), Rhizo-phoraceae (mangrove) and others.

Myrtiflorae Myrtales *q.v.*

mystacial *a. appl.* a pad of thickened skin on either side of snout of some mammals, and to tactile hairs and vibrissae (whiskers) borne on it.

Mystacocarida *n.* class of small crustaceans, similar to copepods, living in marine sands and having no carapace, no clear divisions of the body, and simple appendages.

mystax *n.* group of hairs above mouth of certain insects; the tactile hairs or whiskers on an animal's snout.

Mysticeti *n.* an order of placental mammals, the baleen or whalebone whales, including the blue whale, the right whales, and rorquals. *alt.* Balaemoidea.

myxamoeba *n.* in slime moulds, an amoeboid cell produced from a germinating spore.

myxinoids *n.plu.* an order of cyclostomes comprising the hagfish as distinct from lampreys.

myxo- prefix derived from Gk. *myxa*, slime.

Myxobacteria *n.* bacterial grouping including the myxobacteria (*q.v.*) and the gliding bacteria that do not form fruiting bodies.

myxobacteria *n.plu.* group of flexible rod-shaped bacteria with a gliding movement, which aggregate into multicellular "fruiting bodies" containing resting spores called myxospores.

Myxomycetes acellular slime moulds *q.v.*

Myxomycota *n.* in classifications of the fungi, a major division containing all slime moulds; in protist classification, a phylum containing the plasmodial (acellular) slime moulds.

Myxophyta former name for the cyanobacteria (*q.v.*) in some classifications.

myxopodium *n.* a slimy pseudopodium.

myxosporangium *n.* fruit body of slime moulds.

myxospore *n.* a spore of a slime mould; a spore separated off by slimy disintegration of the hypha.

Myxosporidea *n.* class of parasitic protozoa of the Cnidospora, with a multinucleate spore-producing stage, which infect fish, e.g. *Myxobolus*, the cause of a skin disease in cyprinid fish. Classified in the protist phylum Cnidosporidia (*q.v.*).

Myzostomaria *n.* a class of annelids, ectoparasitic on echinoderms, and almost circular in shape.

MZ twins monozygotic twins.

N

N asparagine *q.v.*; denotes a position in a consensus DNA sequence at which there is no apparent preference for any particular nucleotide.

NA neuraminidase *q.v.*; noradrenaline *q.v.*

NAA α-naphthaleneacetic acid, a synthetic auxin, used to induce roots in cuttings and prevent premature fruit drop in commercial crops.

nAChR nicotinic acetylcholine receptor *q.v.*

NAD nicotinamide adenine dinucleotide *q.v.*

NADH, NADH$_2$ reduced form of nicotinamide adenine dinucleotide.

NADP nicotinamide adenine dinucleotide phosphate *q.v.*

NADPH, NADPH$_2$ reduced form of nicotine adenine dinucleotide phosphate.

NAM *N*-acetylmuramic acid *q.v.*

NANA *N*-acetylneuraminic acid *q.v.*

NAR nett assimilation rate *q.v.*

NDV Newcastle disease virus, a paramyxovirus.

NE norepinephrine, *see* noradrenaline.

NEFA non-essential fatty acids, those fatty acids that can be synthesized *de novo* and therefore do not need to be supplied in the diet.

NGF nerve growth factor *q.v.*

NMDA *N*-methyl-D-aspartate, structural analogue of glutamate, used to define a class of glutamate receptors in central nervous system.

NOR nucleolar organizer (region) *q.v.*

NPT neomycin phosphotransferase *q.v.*

NST non-shivering thermogenesis *q.v.*

NVC National Vegetation Classification *q.v.*

nacre *n.* mother of pearl, iridescent inner layer of many mollusc shells, and the substance of pearls.

nacreous *a.* yielding or resembling nacre.

nacrine *a.* mother-of-pearl colour; *pert.* nacre.

NADH dehydrogenase respiratory chain enzyme, an iron-sulphur flavoprotein with a FMN prosthetic group, which transfers electrons from NADH to ubiquinone. EC 1.6.99.3. *alt.* NADH-Q reductase.

NADH-Q reductase NADH dehydrogenase *q.v.*

naiad *n.* the aquatic nymph stage of certain insects such as dragonflies and mayflies.

naidid *a. appl.* freshwater worms of the genus *Nais*, which are often found in increased numbers in water subject to organic pollution.

Najadales *n.* order of aquatic and semiaquatic monocots including Potamogetonaceae (pondweed), Zosteraceae (eel-grass) and others.

Na⁺-K⁺ ATPase plasma membrane protein with ATPase activity which actively transports Na⁺ out of and K⁺ into animal cells using energy derived from ATP hydrolysis. The ion gradients set up across the membrane by its action are largely responsible for the membrane potential, control of cell volume, and active transport of some sugars and amino acids into some cells. *alt.* sodium pump.

Na⁺-K⁺ pump Na⁺-K⁺ ATPase *q.v.*

naloxone *n.* compound that blocks opiate receptors.

nanander *n.* a dwarf male, in plants. *a.* nanandrous.

nanism *n.* dwarfism.

nanno- nano-.

nano- prefix derived from Gk. *nanos*, dwarf, signifying small, or smallest.

nanoid *a.* dwarfish.

nanometre (nm) *n.* a unit of microscopic measurement and of wavelength of some electromagnetic radiation, being 10^{-9} m, a thousandth of a micrometre, 10 Ångström units, formerly called millimicron.

nanophanerophyte *n.* shrub under 2 m tall.

nanophyllous *a.* having minute leaves.

nanoplankton *n.* microscopic floating plant and animal organisms.

nanous *a.* dwarfed; dwarfish.

naphthaquinone *n.* a derivative of quinone from which vitamin K is synthesized.

napiform *a.* turnip-shaped, *appl.* roots.

narcosis *n.* state of unconsciousness or stupor produced by a drug.

narcotic *a. appl.* drugs that can produce a state of unconsciousness, sleep or numbness.

nares *n.plu.* nostrils *q.v.* sing. naris.

nares, anterior openings of olfactory organ to exterior. *alt.* nostrils.

nares, posterior openings of olfactory organ into pharynx or throat.

narial *a. pert.* the nostrils; *appl.* septum, the partition between nostrils.

naricorn *n.* the terminating horny part of nostril of certain birds such as albatross.

nariform *a.* shaped like nostrils.

naris *sing.* of nares *q.v.*

nasal *a. pert.* the nose; *n.* nasal scale, plate or bone.

nascent DNA, or RNA newly synthesized DNA or RNA.

nasion *n.* middle point of nasofrontal suture.

nasoantral *a. pert.* nose and cavity of upper jaw.

nasobuccal *a. pert.* nose and cheek; *appl.* nose and mouth cavity.

nasociliary *a. appl.* branch of ophthalmic nerve, with internal and external nasal branches, and giving off the long ciliary and other nerves.

nasofrontal *a. appl.* part of superior ophthalmic vein which communicates with the angular vein; *appl.* suture between nasal and frontal bones.

nasolabial *n. pert.* nose and lip.

nasolacrimal *a. appl.* canal from lacrimal sac to inferior meatus of nose through which the tear duct passes.

nasomaxillary *a. pert.* nose and upper jaw.

naso-optic *a. appl.* embryonic groove between nasal and maxillary process.

nasopalatine *a. pert.* nose and palate; *pert.* canal communicating with vomeronasal

organs.

nasopharyngeal *a. pert.* nose and pharynx, or nasopharynx.

nasopharynx *n.* that part of pharynx or throat continuous with the posterior openings of the nasal passsages.

nasoturbinal *a. appl.* outgrowths from lateral wall of nasal cavity increasing the area of sensory surface.

nastic movement a plant movement caused by a diffuse non-directional stimulus and which is usually a growth movement but which may be a change in turgidity as in the sensitive plant that droops on contact.

nasty nastic movement *q.v.*

nasus *n.* snout-like organ of soldiers of some species of termite, used to eject poisonous or sticky fluid at invaders.

nasute *a.* possessing a nasus.

nasute, nasutus *n.* a type of soldier termite.

natal *a. pert.* birth.

natality *n.* birth rate.

natant *a.* floating on surface of water.

natatorial *a.* formed or adapted for swimming.

natatory *a.* swimming habitually; *pert.* swimming.

nates *n.plu.* buttocks.

National Vegetation Classification (NVC) United Kingdom national survey of plant communities and vegetation types, which started in 1975, and which has devised a standard nomenclature for the types of plant community found in Britain.

native *a. appl.* animals and plants which originate in district or area in which they live.

native species indigenous species that is normally found as part of a particular ecosystem.

natriferic *a.* transporting sodium.

natriuresis *n.* excessive loss of sodium in urine, disturbing electrolyte balance in body.

natural classification a classification that groups organisms or objects together on the basis of the sum total of all their characteristics, and tries to indicate evolutionary relationships. *cf.* artificial classification.

natural increase the rate of growth of a population, calculated by subtracting the numer of deaths from the number of births in a given period, or of deaths from births if the population is decreasing.

natural killer cells (NK cells) large granular lymphocytes, which do not possess specific antigen receptors, but that recognize and kill certain types of tumour cells and cells infected with some viruses.

natural selection the process by which evolutionary change is chiefly driven according to Darwin's theory of evolution. Environmental factors such as climate, disease, competition from other organisms, availability of certain types of food, etc., will lead to the preferential survival and reproduction of those members of a population genetically best fitted to deal with them. Continued selection will therefore lead to certain genes becoming more common in subsequent generations. Such selection, operating over very long periods of time is believed to be able to give rise to the considerable differences now seen between different organisms.

naturalized *a. appl.* alien species that have become successfully established.

naupliiform *a.* superficially resembling a nauplius, *appl.* larvae of certain hymenopterans.

nauplius *n.* earliest larval stage of many crustaceans, with three pairs of appendages. *plu.* nauplii.

nautiliform *a.* shaped like a nautilus shell.

nautiloid *n.* member of the subclass Nautiloidea of the cephalopod molluscs, typified by the pearly nautilus (*Nautilus*), bearing a spiral, many-chambered shell from which the head and tentacles emerge.

navicular *a.* boat-shaped.

naviculare *n.* a boat-shaped bone of the mammalian carpus; tarsal bone between talus and cuneiform bones.

NC cells natural cytotoxic lymphocytes *q.v.*

neala *n.* fan-like posterior lobe of hindwing of some insects.

neallotype *n.* a type specimen of the opposite sex to that of the specimen previously chosen for designation of a new species.

nealogy *n.* the study of young animals.

Neanderthal man a subspecies of *Homo sapiens*, *H. sapiens neanderthalensis*, living in the Old World during the Pleistocene.

Nearctic *a. appl.* or *pert.* zoogeographical region, or subregion of the Holarctic region, comprising Greenland and North America, and including northern Mexico.

necrocytosis *n.* cell death.

necrogenic *a.* promoting necrosis in the host, *appl.* certain parasitic fungi.

necrogenous *a.* living or developing in dead bodies.

necrophagous, necrophilous *a.* feeding on dead bodies.

necrophoresis *n.* transport of dead members of the colony away from the nest.

necrophoric *a.* carrying away dead bodies, *appl.* certain insects.

necrosis *n.* the death of cells or tissues as a result of external damage. *a.* necrotic. *cf.* apoptosis.

necrotoxin *n.* a toxin whose effects cause necrosis of tissue.

necrotroph *n.* fungus living off dead host plant tissue.

necrovirus group plant virus group containing isometric single-stranded RNA viruses that typically cause necrosis of infected tissue, type member tobacco necrosis virus.

nectar *n.* sweet liquid secreted by the nectaries of flowers and certain leaves to attract insects, and some birds, for pollination; in some fungi, a liquid exuding from fruiting body and containing spores.

nectar gland, nectary *n.* a group of cells secreting nectar in flowers, and in some leaves; the gland secreting the sweet honeydew in aphids.

nectar guides markings on petals of flowers that guide insects to the nectar, thus making cross-fertilization more likely.

nectariferous *a.* producing or carrying nectar.

nectarivorous, nectivorous *a.* nectar-eating.

nectocalyx nectophore *q.v.*

necton nekton *q.v.*

nectophore *n.* in a siphonophore, a medusoid individual modified for swimming, clusters of which propel the colony through the water. *alt.* nectocalyx.

nectopod *n.* an appendage modified for swimming.

negative control type of control of gene expression (in bacteria) in which genes are expressed unless they are switched off by a repressor protein.

negative feedback type of control mechanism common in physiological homeostasis, in which a departure from the norm in a particular variable is sensed by the cell or organism and a response is made that restores the variable to the norm and thus shuts off the signal.

negative reinforcement a stimulus or series of stimuli which are unpleasant to an animal and so diminish its response to the stimulus or cause avoidance reactions.

negative staining staining technique used in electron microscopy, esp. of virus particles, in which the specimen is surrounded by a heavy metal stain which outlines its shape and also penetrates into surface clefts, providing a "negative impression" of the viral outline and surface features. Analogous techniques in which the background is stained to show up an unstained specimen are used in light microscopy.

negative-strand RNA viruses viruses with a single-stranded RNA genome that is complementary to the messenger RNA, and from which a mRNA is made by a viral transcriptase.

negative tropism tendency to move or grow away from the source of the stimulus, e.g. plant shoots show negative geotropism.

nekton *n*. the organisms swimming actively in water.

Nelumbonales *n*. order of large aquatic herbaceous dicots and including the single family Nelumbonaceae (Indian lotus).

nematoblast *n*. the cell from which a nematocyst develops.

nematocalyx *n*. in some colonial hydrozoans, a small polyp that has no mouth but engulfs organisms by pseudopodia.

nematoceran, nematocerous *a*. possessing thread-like antennae.

nematocide *n*. chemical that kills nematode worms.

nematocyst *n*. stinging cell of sea anemones, jellyfishes and other coelenterates, containing a long coiled thread which is discharged on contact and pierces prey. Sometimes refers only to the contents of the cell, the cell itself being termed a nematoblast or cnidoblast.

Nematoda, nematodes *n., n.plu.* roundworms. Slender, pseudocoelomate, unsegmented worms circular in cross-section. Some (eelworms) are serious parasites of plants, others are parasitic in animals and some are free-living in soil and marine muds. Parasitic nematodes causing severe diseases in humans include the hookworms *Ancyclostoma* and *Necator*, *Trichinella* (causing trichinellosis) and *Wucheria* (the cause of elephantiasis). The soil nematode *Caenorhabditis elegans* is an important experimental organism in genetic and developmental research.

nematoid *a*. thread-like; filamentous.

nematology *n*. the study of nematodes.

Nematomorpha *n*. phylum of pseudocoelomate worms, sometimes known as horsehair worms, that are free-living in soil or fresh water as adults, and parasitic in arthropods when young. *alt*. threadworms.

nematophore nematocalyx *q.v.*

nematosphere *n*. the enlarged end of a tentacle in some sea anemones.

nematozooid *n*. a defensive zooid in hydrozoans.

Nemertea, nemerteans *n., n.plu.* phylum of long, slender, acoelomate, marine worms, flattened dorsoventrally, e.g. the bootlace worm (*Lineus*). Most live on shores around the low tide line. They have a mouth and anus, a simple blood system, and a typical muscular proboscis, which is extended to catch prey. *alt*. proboscis worms, ribbon worms.

Nemertina, nemertines Nemertea, nemerteans *q.v.*

nemoral *a*. living at the edges of woodlands, or in open woodland.

nemorose, nemoricole *a*. inhabiting open woodland places.

neo- prefix derived from Gk. *neos*, young, signifying young or new.

neoblast *n*. one of the undifferentiated cells forming primordium of regeneration tissue after wounding.

neocarpy *n*. production of fruit by an otherwise immature plant.

neocerebellum *n*. region of cerebellum that receives nerve fibres predominantly from the pons.

neocortex *n*. in mammalian brain, the cerebral cortex, excluding the hippocampus and pyriform lobe.

Neo-Darwinism the modern version of the Darwinian theory of evolution by natural selection, incorporating the principles of genetics and still placing emphasis on natural selection as a main driving force of evolution.

neoencephalon *n.* the telencephalon, or latest evolved anterior portion of brain.

Neogaea *n.* zoogeographical region comprising southern Mexico, Central and South America, and the West Indies. *alt.* Neotropical region, Neogea.

neogamous *a. appl.* forms of protozoans showing precocious association of gametes.

Neogea Neogaea *q.v.*

neogenesis *n.* new tissue formation; regeneration.

Neolaurentian *a. pert.* or *appl.* early Proterozoic era.

Neolithic *a. appl.* or *pert.* the New or polished Stone Age, characterized by the use of polished stone tools and weapons and the appearance of settled cultivation.

neomorph *n.* a structural variation from the type; a mutant allele that produces changes in developmental processes, resulting in the appearance of a new character.

neomorphosis *n.* regeneration when the new part is unlike anything in body.

neomycin *n.* an antibiotic synthesized by *Streptomyces fradiae.*

neomycin phosphotransferase (NPT) enzyme (EC 2.7.1.95) conferring resistance to the antibiotics neomycin, kanamycin and related compounds, and whose gene is often used as a selectable marker for recombinant DNAs.

neonate *n.* newly born animal. *a.* neonatal.

neontology *n.* the study of existing organic life. In the study of evolutionary biology neontologists are those who study evolution by comparisons between living animals and plants, palaeontologists study evolution through the fossil record.

neonychium *n.* a soft pad enclosing each claw of embryos of clawed vertebrates, to prevent tearing of foetal membranes; a horny claw pad in birds before hatching.

neopallium neocortex *q.v.*

neoplasia *n.* cell proliferation, often uncontrolled, producing additional tissue, often

used with reference to the malignant proliferation of cancer cells.

neoplasm *n.* an uncontrolled growth of cells, sometimes self-limiting as in benign tumours, sometimes malignant.

neoplastic *a. appl.* cells or tissue arising as a result of uncontrolled growth. *alt.* (in some cases) malignant, tumorous.

neoplastic transformation the changes that take place in a cell as it becomes cancerous.

neoptile *n.* down feather.

Neornithes *n.* subclass of birds (Aves) including all extant modern birds, the other subclass including only the extinct *Archaeopteryx.*

neossoptile neoptile *q.v.*

neostigmine *n.* a plant alkaloid, an inhibitor of the enzyme acetylcholinesterase. *alt.* prostigmine.

neoteinic *n.* a supplementary reproductive termite.

neotenous *n. appl.* animals in which larval characters are retained beyond the normal period, e.g. some species of amphibian which retain the tadpole form as sexually mature adults. *n.* neoteny.

neoteny *n.* retention of larval characters beyond normal period and into sexually mature adult, as in some amphibia. *a.* neotenous.

neothalamus *n.* that part of the thalamus consisting of nuclei with connections to association areas in the cerebral cortex.

Neotropical *a. appl.* or *pert.* a zoogeographical region consisting of southern Mexico, Central and South America, and the West Indies; *appl.* a floral realm which is divided into three floral regions, *see* Central American Floral Region, Pacific South American Floral Region, Parano-Amazonia Floral Region. *alt.* Austro-Columbian.

neotype *n.* a new type; a new type specimen from the original locality.

neoxanthin *n.* xanthophyll carotenoid pigment found in algae, esp. green algae and euglenaphyceae.

Nepenthales *n.* order of herbaceous carnivorous dicots with leaves adapted for trapping small animals, and comprising the families Droseraceae (sundew) and Nepenthaceae (pitcher plants).

nephric *a. pert.* kidney.

nephridial *a. pert.* kidney, usually *appl.* the small excretory tubules; *pert.* excretory organ or nephridium of invertebrates.

nephridioblast *n.* ectodermal cell that gives rise to a nephridium.

nephridiopore *n.* external opening of excretory organs (nephridia) in invertebrates.

nephridiostome *n.* ciliated opening of a nephridium into the coelom.

nephridium *n.* excretory organ having function of kidney in invertebrates; embryonic kidney tubule in vertebrates.

nephroblast *n.* embryonic cell that gives rise ultimately to nephridia.

nephrocoel *n.* the cavity of a nephrotome.

nephrocyte *n.* any of various cells in sponges, ascidians and insects, that store and discharge waste products.

nephrodinic *a.* having one duct serving for both excretory and genital purposes.

nephrogenic *a. pert.* development of kidney; *appl.* cord or column of fused mesodermal cells giving rise to tubules of mesonephros.

nephrogenous *a.* produced by the kidney.

nephroid reniform *q.v.*

nephrolytic *a. pert.* or designating enzymatic action destructive to kidneys.

nephromere nephrotome *q.v.*

nephromixium *n.* a compound excretory organ comprising flame cells and coelomic funnel and acting both as an excretory organ and genital duct.

nephron *n.* the individual structural and functional unit of a vertebrate kidney, comprising glomerulus, Bowman's capsule and a convoluted tubule.

nephros *n.* a kidney, or usually the functional portion of a kidney. *plu.* nephroi.

nephrostome *n.* opening of nephridial tubule into body cavity.

nephrotome *n.* that part of a somite developing into an embryonic excretory organ. *alt.* nephromere.

nepovirus group plant virus group containing nematode-transmitted isometric single-stranded RNA viruses, type member tobacco black ring virus. They are multicomponent viruses in which two genomic RNAs are encapsidated in three different virus particles, one of which lacks nucleic acid.

neritic *a. pert.* or living only in coastal waters, as distinct from oceanic.

neritopelagic *a. pert.* or inhabiting the sea above the continental shelf.

nervate *a.* having nerves or veins, *appl.* leaves, insect wings.

nervation, nervature *n.* venation of leaves or insect wings.

nerve *n.* bundle of many nerve fibres (axons) of individual neurones, connecting the central nervous system with other parts of the body. Discrete bundles of nerve fibres within the nerve are each enclosed in a sheath of connective tissue (perineurium), and the whole nerve is covered in an epineurium of fibrous connective tissue; (*bot.* & *zool.*) vein of leaf or insect wing, leaf veins being strands of vascular tissue, the veins of insect wings being extensions of the tracheal system.

nerve canal a canal for passage of nerve to pulp of a tooth.

nerve cell neurone *q.v.*

nerve centre group of nerve cells associated with a particular function.

nerve cord in invertebrates, a bundle of nerve fibres, or chain of ganglia and interconnecting nerve fibres, running the length of the body.

nerve ending the terminal portion of axon of a neurone, or receptor portion of a sensory neurone, modified in various ways.

nerve fibre axon (*q.v.*) of nerve cell, many individual nerve fibres making up a nerve.

nerve growth factor (NGF) protein homodimer produced by some nerve cell target tissues, such as smooth muscle, which is needed for maintenance of some types of neurone, esp. sympathetic and sensory neurones. It also stimulates the outgrowth of neurites from developing neurones of these types.

nerve impulse *see* action potential.

nerve net a simple network of nerve cells in body wall of coelenterates, and some other invertebrates, connecting sensory cells and muscular elements.

nervicolous *a.* living or growing in the veins of a leaf.

nerviduct *n.* passage for nerves in cartilage or bone.

nervous *a. pert.* nerves; *appl.* tissue composed of neurones.

nervous system highly organized system of electrically active cells (nerve cells or neurones), which generate and convey signals in the form of electrical impulses. A nervous system is present in all multicellular animals except sponges, and is most highly developed in vertebrates. It receives and coordinates input from the environment and from the body through sensory receptors and conveys executive commands to muscles and glands, enabling the animal to sense and respond rapidly to external and internal stimuli. In all but the most primitive nervous systems, the nerve cells are organized into nerves and aggregates of nerve cell bodies (ganglia). The vertebrate nervous system consists of a brain and spinal cord, which constitute the central nervous system, and a peripheral nervous system, consisting of sensory cells and the peripheral nerves and their branches which convey signals to and from the central nervous system. *see also* brain, central nervous system, nerve, nerve net, neuromuscular junction, neurone, peripheral nervous system, synapse.

nervule *n.* tiny branch of nervure of insect wing.

nervure *n.* (*zool.*) one of the rib-like structures which support the membranous wings of insects, branches of the tracheal system; (*bot.*) a leaf vein.

nervus lateralis a branch of the vagus nerve in fishes, connecting sensory lateral line with brain.

nervus terminalis preoptic nerve *q.v.*

nessoptile neoptile *q.v.*

nest epiphyte an epiphyte which builds up a store of humus around itself for growth.

nest provisioning returning regularly to nests to bring food to developing offspring, as in some solitary wasps.

net assimilation rate (NAR) the increase in dry weight of a single plant per unit time, with reference to the total area involved in assimilation.

net efficiency a measure of the efficiency of an organism in converting its assimilated food to protoplasm.

net photosynthesis photosynthesis measured as the net uptake of carbon dioxide into the leaf, equal to gross photosynthesis less respiration.

net plasmodium the kind of plasmodium found in some slime moulds, where the cells are connected by cytoplasmic strands, forming a net.

net production the amount of food in an ecosystem available for the primary consumers, being the gross primary production minus the amount of biomass used in respiration by primary producers.

net reproductive rate average number of offspring a female produces during her lifetime, symbolized by R_0.

netted-veined with veins in the form of a fine network, *appl.* leaves, insect wings.

network theory in immunology, the theory that the antigen-specific cells of the immune system are linked by interactions other than those directly mounted against the incoming antigen. These additional interactions involve recognition of idiotypic antigenic determinants on antigen receptors and antibodies by lymphocytes not directly concerned with mounting a response against the original antigen.

neurad *adv.* dorsally, being on the same side of the backbone as the spinal cord.

neural *a. pert.* or closely connected with nerves, or nervous system, or nervous tissue. *alt.* neuric.

neural arc the afferent and efferent neuronal connections running between sensory receptor and the effector.

neural arch arch on dorsal surface of vertebra for passage of the spinal cord.

neural canal canal through backbone formed by neural arches of individual vertebrae, through which runs the spinal cord.

neural crest ridge of ectoderm that forms above the neural tube during early embryogenesis in vertebrates, the cells of which migrate to develop into melanocytes, the dorsal root ganglia of the sensory nervous system, the autonomic nervous system, the adrenal medulla and some skeletal elements in the face.

neural folds the edges of the neural plate which rear up and join to form the neural tube.

neural induction in vertebrates, the induction of the neural tube from ectoderm by the underlying mesoderm.

neural lobe of pituitary body, *see* neurohypophysis.

neural network a system of interconnected neurones, proposed as the basis of higher brain functions such as perception, memory and learning, in which a perception, etc. results from the pattern of activity within the network, which can be varied by varying the strength and number of the connections between the elements.

neural plate band of thickened ectoderm down midline of back of early chordate embryo from which the neural tube develops; a lateral member of neural arch; one of a median row, usually of eight bony plates, in carapace of turtle.

neural retina the photoreceptor and nerve cell layers of the retina.

neural shields horny shields above neural plates of turtles.

neural spine dorsal process on a vertebra.

neural stalk infundibulum of neurohypophysis.

neural tube tube of ectoderm formed down the back of early vertebrate embryo from which will develop the brain and spinal cord.

neuraminic acid *see* N-acetylneuraminic acid.

neuraminidase *n.* enzyme that removes terminal N-acetylneuraminic acid from carbohydrate side chains of glycoproteins. Sometimes known as receptor-destroying enzyme because by splitting neuraminic acid from the cell surface glycoproteins that act as receptors for some viruses it destroys their receptor properties. Some viruses, e.g. myxoviruses and paramyxoviruses, carry neuraminidases on their surface.

neurapophysis *n.* one of the two plates growing from centrum of vertebra and meeting over the spinal cord to form the neural spine.

neuraxis *n.* the cerebrospinal axis; axon of neurone, *alt.* neuraxon.

neurectoderm *n.* the ectodermal cells forming the earliest rudiment of the nervous system, as distinct from epidermal ectoderm.

neurenteric *a. pert.* cavity of neural tube and enteric cavity, *appl.* canal temporarily connecting posterior end of neural tube with posterior end of archenteron.

neuric neural *q.v.*

neurilemma, neurolemma *n.* thin sheath investing the myelin sheath of a nerve fibre. *a.* neurilemmal, neurolemmal.

neurine *n.* a ptomaine with a fishy smell obtained mainly from brain, bile and egg yolk, and which is formed from choline in putrefying meat.

neurite *n.* general term for a nerve cell process, especially in developing neurones in culture when axons and dendrites cannot be distinguished.

neurobiology *n.* the study of the morphology, physiology, biochemistry and development of the brain and nervous system, and the biochemical and cell biological basis of brain function, not generally including psychology and cognitive psychology.

neuroblast *n.* cell from which a neurone is formed.

neuroblastoma *n.* tumour originating from immature cells of the nervous system.

neurocele neurocoel *q.v.*

neurocentral *a. appl.* two vertebral synchondroses (*q.v.*) persisting during the first few years of human life.

neurochord *n.* a giant nerve fibre; a primitive tubular nerve cord.

neurocoel *n.* the central cavity of the central nervous system.

neurocranium *n.* the cartilaginous or bony case containing the brain and capsules of special sense organs (e.g. eyes and ears).

neurocrine *a. pert.* secretory function of nervous tissue or cells. *alt.* neurosecretory.

neurodegenerative *a. appl.* diseases and conditions of the brain or other parts of the nervous system which result from the progressive death of neurones and loss of function, e.g. Parkinson's disease and Alzheimer's disease in humans, scrapie in sheep and bovine spongiform encephalopathy.

neuroectoderm *n.* that portion of the ectoderm giving rise to the nervous system.

neuroendocrine neurosecretory *q.v.*

neuroendocrine system the hypothalamus and pituitary, secretion of hormones from the pituitary being regulated by homones secreted by the neurones of the hypothalamus.

neuroepithelium *n.* a superficial layer of

cells where it is specialized for sensory reception.

neurofibrillary tangles abnormal aggregation of paired helical protein filaments within neurones, destroying their function, which occurs in brains of patients with Alzheimer's disease.

neurofibrils *n.plu.* very fine protein fibres (neurofilaments, microtubules and actin filaments) running longitudinally in axons and dendrites of nerve cells and forming a complex meshwork in the cell body.

neurofilaments *n.plu.* longitudinal intermediate filaments supporting axon of a neurone.

neurogenesis *n.* formation of the nervous system during development; development of nerves.

neurogenic *a.* induced by nervous stimulation, as muscular contraction and secretion from some glands; giving rise to nervous tissue or nervous system.

neuroglia *n.plu.* cells other than neurones found in the central nervous system, including astrocytes, oligodendrocytes, microglia, and ependymal cells. *alt.* glia. *a.* neuroglial.

neurohaemal *a. appl.* nerve endings in close relationship with blood vessels and discharging neurosecretory material into blood; *appl.* organ: organ such as corpora cardiaca in insects in which secretion from numbers of neurosecretory cells is stored and released into the blood.

neurohormone *n.* any hormone produced by neurosecretory cells, usually in brain. Neurohormonal activity is distinguished from that of classical neurotransmitters by its effects on cells distant from the source of the hormone and not in a one-to-one relationship to the secretory neurone.

neurohumoral *a. pert.* hormones released into the general circulation from the brain.

neurohypophysis *n.* the posterior part of the pituitary gland, containing secretory nerve endings from the hypothalamus and producing vasopressin and oxytocin amongst other hormones. It comprises the pars nervosa or neural lobe, and the neural stalk (infundibulum).

neurolemma neurilemma *q.v.*

neuroleukin *n.* glucose-6-phosphate isomerase, promotes growth and survival

of cultured spinal and sensory neurones.

neurology *n.* the study of the morphology, physiology and pathology of the nervous system, in present-day usage sometimes implying a more clinical orientation than e.g. neurobiology or neuroscience.

neurolymph cerebrospinal fluid *q.v.*

neurolysis *n.* the lysis or disintegration of nerve tissue.

neuromast *n.* a group of hair cells (sensory cells) comprising a sensory unit of the acoustico-lateralis system in fishes and some amphibians.

neuromere *n.* a segment of vertebrate spine between the attachment points of successive pairs of spinal nerves; segmental ganglion of annelids and arthropods.

neuromodulation *n.* proposed mode of action of some chemical transmitters in brain, which affect the activity of a neuronal pathway by influencing the efficiency of synaptic transmission.

neuromuscular *a. pert.* or involving both nerves and muscles.

neuromuscular junction the site at which an axon terminal contacts a muscle cell.

neuronal *a. pert.* neurones or nerve cells.

neuronal transplantation the transplantation of foetal neurones into the brains of adults to try and restore functions lost by disease or damage. Neuronal transplantation in humans has so far been limited to attempts to provide a replacement source of dopamine in patients suffering from Parkinson's disease.

neurone *n.* nerve cell, basic unit of the nervous system, specialized for the conveyance and transmission of electrical impulses. Typically consists of a cell body, which contains the nucleus and other organelles, from which cytoplasmic processes project – the dendrites, which receive signals from other neurones, and the axon, which conducts impulses outward from the cell body. *alt.* neuron.

neuronephroblast *n.* in some annelids, a cell that gives rise to part of germinal bands from which nerve cord and nephridia develop.

neuropeptide *n.* any of many small peptides produced by the nervous system, some of which act as neurotransmitters, others as neuromodulatory hormones.

neuropil n. network of axons, dendrites and synapses.

neuroplasm n. cytoplasm of neurone not including the fibrillar components.

neuropodium n. ventral lobe of polychaete parapodium.

neuropore n. the opening of neural tube or neurocoel to the exterior.

Neuroptera n. order of insects with complete metamorphosis, including alder flies, lacewings and ant lions, having long antennae, biting mouthparts and two pairs of membranous wings held roof-like over the abdomen in repose.

neurose a. having numerous veins, *appl.* leaves and insect wings.

neurosecretion n. the release of neurotransmitters and neuromodulatory compounds from nerve cells.

neurosecretory a. *appl.* nerve cells that secrete substances that travel via the blood to their targets.

neurosensory a. *appl.* the epithelial sensory cells of coelenterates.

neuroskeleton n. the cytoskeleton of a neurone.

neurosynapse synapse q.v.

neurotendinous a. *pert.* nerves and tendons; *appl.* nerve endings in tendons.

neurotome neuromere q.v.

neurotoxic a. *appl.* any toxin affecting nervous system function.

neurotoxin n. poison acting on the nervous system.

neurotransmitter n. chemical liberated by the axon terminal of a neurone in response to an electrical impulse in the neurone and which transmits the neuronal signal across a synapse to another neurone or muscle fibre. Neurotransmitters are released by the pre-synaptic neurone and interact with receptors on the dendrites (or other parts) of the post-synaptic neurone. Some common neurotransmitters are acetylcholine, noradrenaline (norepinephrine), dopamine, and 5-hydroxytryptamine (serotonin).

neurotrophic a. nourishing nervous tissue.

neurotropic a. *appl.* viruses and bacteria that infect nervous tissue, toxins that act on nerve cells, etc.

neurotubule n. microtubule of neurone, numbers of which run longitudinally along axon.

neurovascular a. *appl.* nerves and blood vessels.

neurula n. very early embryo (in chordates) at the stage of formation of the neural tube, and which develops from gastrula.

neurulation n. formation of the neural tube, the precursor of the nervous system, which occurs early in chordate embryogenesis.

neuston n. organisms floating or swimming in surface water, or inhabiting surface film.

neuter a. sexless, neither male nor female; having neither functional stamens nor pistils; n. a non-fertile female of social insects; a castrated animal.

neutral a. neuter q.v.; neither acid nor alkaline, pH 7.0; achromatic, as white, grey, and black; day-neutral q.v.; *appl.* changes in nucleotide and, sometimes, amino acid sequence, that have no effect on function of the gene product and therefore no effect on phenotype.

neutral allele neutral mutation q.v.

neutral fat triacylglycerol q.v.

neutral mutation a mutation that confers no selective advantage or disadvantage on the individual.

neutral polymorphism a genetic polymorphism within a population in which the relative frequencies of the different forms are the result of chance and the action of intrinsic genetic mechanisms and are not being maintained by selection.

neutralization n. inactivation of virus by complexing with specific antibody.

neutrophil a. *appl.* white blood cells whose granules stain only with neutral stains; n. polymorphonuclear leukocyte q.v.

neutrophilic a. staining only with neutral stains

New Caledonian Floral Region part of the Palaeotropical Realm comprising the islands of Vanuatu (formerly New Caledonia).

New Zealand Floral Region part of the Austral Realm comprising New Zealand and its offshore islands.

newt n. common name for the genera *Triturus*, *Taricha* and *Notophthalamus* (in the family Salamandridae) of tailed amphibians (urodeles). They return to the water to breed and lay their jelly-coated eggs singly.

newts *see* urodeles.

nexin *n*. protein linking adjacent microtubules in cilia and flagella. *see also* protease nexin.

nexus *n*. region of fusion of plasma membrane between two excitable cells.

niacin *n*. nicotinic acid, a member of the vitamin B complex, vitamin B₇, found in all living cells as the nicotinamide moiety of the enzyme cofactors NAD and NADP, blood, liver, legumes and yeast being particularly rich sources, and which can be used to treat pellagra in humans. *alt.* pellagra-preventive factor.

niche ecological niche *q.v.*

niche diversification alpha diversity *q.v.*

niche overlap the situation where two or more species use the same resources or the same habitat within a community and thus share the same ecological niche, leading to competition between them. Two species with identical niche requirements cannot co-exist in the same community, but niche overlap may be seen where there are small and difficult to determine ecological differences between species.

nick translation limited DNA synthesis initiated at a single-strand break (nick) and which displaces the homologous DNA strand from the template, carried out by bacterial DNA polymerase I and having a repair function *in vivo*, widely used *in vitro* to introduce radioactively labelled nucleotides into DNA.

nicotinamide *n*. an amide derived from nicotinic acid (niacin), constituent of the enzyme cofactors NAD and NADP, may be used like niacin to treat pellagra in humans. *alt.* vitamin B₇, pellagra-preventive factor.

nicotinamide adenine dinucleotide (NAD, NAD⁺) important coenzyme, composed of nicotinamide, adenine, 2 riboses and 2 phosphate groups, found in all living cells where it acts as a hydrogen (electron) acceptor and is reduced to NADH (+ H⁺), in which form it is an important source of reducing power in the cell, formerly called DPN (dipyridine nucleotide) or coenzyme I. *see* Appendix 1 (26) for structural formula.

nicotinamide adenine dinucleotide phosphate (NADP) important coenzyme, composed of NAD with an extra phosphate group attached, found in all living cells where it acts as a hydrogen (electron) acceptor, esp. in biosynthetic pathways, being reduced to NADPH (+ H⁺), in which form it is an important source of reducing power in the cell, formerly called TPN (tripyridine nucleotide) or coenzyme II.

nicotine *n*. an alkaloid obtained from the tobacco plant *Nicotiana tabacum*, toxic to many animals because it blocks the normal action of the neurotransmitter acetylcholine at neuromuscular junctions.

nicotinic *a. pert.* nicotine, *appl.* acetylcholine receptors stimulated by nicotine; resembling nicotine in its effects, *appl.* various substances that also act at the class of acetylcholine receptors sensitive to nicotine.

nicotinic acid (nicotinate) niacin *q.v.*

nictitating membrane the third eyelid, a membrane that can be passed over eye and helps to keep it clean in reptiles, birds and some mammals.

nidation implantation *q.v.*

nidicolous *a*. living in the nest for a time after hatching.

nidification *n*. nest building and the behaviour associated with it.

nidifugous *a*. leaving the nest soon after hatching.

nidulant *a*. partially surrounded or lying free in a hollow or cup-like structure.

nidus *n*. nest; a nest-like hollow; a cavity for development of spores.

nif **genes** the genes in nitrogen-fixing bacteria that direct the synthesis of nitrogenase and some other proteins required for nitrogen fixation.

nigrescent *a*. nearly black; blackish; turning black.

nigropunctate *a*. black-spotted.

nimbospore *n*. a spore having a gelatinous coat, as of some fungi.

ninhydrin *n*. reagent that gives an intense blue colour with amino acids (yellow with proline).

Nissl granules former name for granules in nerve cells subsequently identified as ribosomes.

nitid, nitidous *a*. glossy.

nitrate bacteria bacteria in the soil that convert nitrite to nitrate. *see* nitrifier.

nitrate reductase general name for enzymes that catalyse the conversion of nitrate to nitrite. EC 1.6.6.1, 1.6.6.2, 1.6.6.3, 1.9.6.1., and 1.7.99.4.

nitrification *n.* the oxidation of ammonium ion to nitrite, and the oxidation of nitrite to nitrate, carried out chiefly by a few groups of soil bacteria (nitrifiers), mainly genera *Nitrosomonas* and *Nitrobacter* and also by a few species of fungi.

nitrifier *n.* any of a group of autotrophic aerobic soil bacteria that can either oxidize ammonia to nitrite (e.g. *Nitrosomonas*) or nitrite to nitrate (e.g. *Nitrobacter*). *alt.* nitrifying bacteria.

nitrite bacteria bacteria in the soil that convert ammonium to nitrite. *see* nitrifier.

nitrite reductase enzyme that catalyses the conversion of nitrite to ammonium hydroxide. EC 1.6.6.4.

nitrocobalamine *n.* vitamin $B_{12}c$, *see* cobalamine.

nitrogen-15 (^{15}N) naturally occurring, stable isotope of nitrogen, accounting for about 0.4 per cent of atmospheric N_2.

nitrogen assimilation in plants, the uptake of nitrogen from the soil in the form of ammonia, nitrites and nitrates.

nitrogen balance equilibrium state of body in which nitrogen intake and excretion are equal. *alt.* nitrogen equilibrium.

nitrogen cycle the sum total of processes by which nitrogen circulates between the atmosphere and the biosphere or any subsidiary cycles within this overall process. Atmospheric elemental nitrogen is converted by a few groups of soil and aquatic microorganisms into inorganic nitrogenous compounds (nitrogen fixation) which are utilized by other living organisms and the nitrogen incorporated into complex organic molecules in their tissues. These are subsequently broken down by bacteria and fungi (ammonification and nitrification) which generate inorganic nitrogen compounds, ammonia, nitrites and nitrates, which may be utilized by plants as nutrients, or may be converted to elemental nitrogen or nitrous oxide by certain bacteria (denitrification) thus releasing nitrogen to the atmosphere. The cycle also incorporates non-biological exchanges of nitrogen between atmosphere and biosphere as in the precipitation of inorganic nitrogen compounds in rainwater, and the fixation of atmospheric nitrogen by lightning.

nitrogen equilibrium nitrogen balance *q.v.*

nitrogen fixation the process whereby atmospheric elemental nitrogen is reduced to ammonia, and which is carried out in the living world only by some free-living bacteria and cyanobacteria (blue-green algae) and by a few groups of bacteria in symbiotic association with plants (the *Rhizobium*–legume association and the actinomycete–non-legume associations). The reaction is catalysed by the enzyme nitrogenase. Biological nitrogen fixation is the chief process by which atmospheric nitrogen enters the biosphere and becomes available as a nutrient to other organisms. A smaller amount of atmospheric nitrogen is also fixed by conversion into nitrogen oxides by the action of lightning.

nitrogenase *n.* enzyme complex in nitrogen-fixing microorganisms which catalyses the reduction of elemental nitrogen (N_2) to ammonium ions. The most widespread form contains molybdenum; other nitrogenases containing vanadium and iron have been found.

nitrogenous *a. pert.* or containing nitrogen.

nitrophilous *a.* nitogen-loving, *appl.* plants.

NK cells natural killer cells *q.v.*

NMDA receptor one type of receptor for the transmitter glutamate in the brain, containing an intrinsic ion channel, so-called from its specific interaction with *N*-methyl-D-aspartate, which distinguishes it from other glutamate receptors.

nociception *n.* sensing of painful or injurious stimuli.

nociceptive *a. appl.* stimuli which tend to injure tissue or induce pain; *appl.* reflexes which protect from injury.

nociceptor *n.* receptor sensitive to injurious or painful stimuli.

noctilucent *a.* phosphorescent; luminescent.

nocturnal *a.* seeking food and moving about only at night; occurring only at night.

nodal *a. pert.* a node or nodes.

Nodaviridae *n.* family of insect RNA viruses.

node *n.* (*bot.*) knob or joint of a stem at which leaves arise; (*zool.*) aggregation of specialized cardiac cells, as atrioventricular and sinuatrial nodes; one of the constrictions in the myelin sheath of peripheral nerve fibres: node of Ranvier.

nodes of Ranvier constrictions of the myelin sheath occuring at intervals along myelinated nerve fibres, and at which the axon membrane is exposed.

nod **genes** genes carried by nitrogen-fixing rhizobia that are required for infection of a legume root and formation of root nodules.

nodose *a.* having knots or swellings.

nodular *a. pert.,* or like a nodule or knot.

nodulated *a.* bearing nodules, in plants *appl.* esp. to roots bearing nodules containing nitrogen-fixing bacteria.

nodulation *n.* formation of nitrogen-fixing root nodules on plant roots.

nodule *n. see* root nodule.

noduliferous *a.* bearing nodules.

nodulins *n.plu.* nodule-specific proteins of host origin produced by legumes infected with nitrogen-fixing rhizobia, and which are involved in establishing symbiotic nitrogen fixation.

Nomarski differential-interference-contrast microscopy a type of optical microscopy which produces a high-contrast image of unstained living cells and tissue.

nomen nudum a name not valid because when it was originally published the organism to which it referred was not adequately described, defined or sketched.

nomenspecies *n.* a group of individuals bearing a binomial name, whatever its status in other respects.

non-adaptive *a. appl.* traits that tend to decrease an organism's genetic fitness.

non-allelic *a. appl.* mutations that produce the same or very similar phenotypes when either is homozygous but a normal phenotype when both are heterozygous, usually showing that they are affecting separate loci or cistrons, but *see* intracistronic complementation; *appl.* similar genes at two or more different loci.

non-competitive inhibitor *see* inhibitor.

non-conjugative *a. appl.* plasmids that cannot direct their own transfer by conjugation between two bacteria.

non-conjunction *n.* the failure of homolo-gous chromosomes to pair at meiosis.

non-cyclic photophosphorylation a type of photophosphorylation in which both ATP and $NADPH_2$ are generated using light energy.

non-degradable *a. appl.* material which cannot be broken down by the natural processes of decomposition by microorganisms, and which therefore persists in the environment.

non-disjunction *n.* failure of a pair of chromatids to separate and go to opposite poles at mitosis or meiosis, which results in aneuploidy in the daughter cells.

non-essential amino acids amino acids which can be synthesized in the body and are not required in the diet: for humans these are Ala, Asn, Asp, Cys, Glu, Gln, Gly, Pro, Ser, Tyr.

non-haem iron protein iron-sulphur protein *q.v.*

non-histone protein any of several types of protein, other than histones, associated with eukaryotic chromosomes where they are thought to be involved in stabilizing higher-order packing structure.

non-medullated fibres grey or yellowish-grey nerve fibres, lacking myelin sheaths, comprising most of the fibres of the sympathetic system and some of the central nervous system.

non-Mendelian *a. appl.* genes or characters which are not inherited according to Mendel's laws, e.g. mitochondrial or chloroplast genes.

non-permissive *a.* (*genet.*) *appl.* conditions under which an organism or cell carrying a conditional lethal mutation displays the mutant phenotype and dies or becomes severely defective; (*virol.*) *appl.* conditions in which virus infection and multiplication in a cell does not occur, and which may be environmental or genetic (i.e. species differences in susceptibility).

non-porous wood secondary xylem with no vessels.

non-reciprocal recombination a genetic recombination event such as the integration of one DNA molecule into another, in which there is not an exact exchange of parts of the two DNAs undergoing recombination. *cf.* reciprocal recombination.

non-repetitive DNA single-copy DNA *q.v.*

non-shivering thermogenesis (NST) the generation of large amounts of heat by metabolism, without shivering, of which some mammals are capable. *see* brown fat.

non-spiking *a. appl.* neurones that do not produce impulses.

non-storied *a. appl.* cambium in which initials are not arranged in horizontal series on tangential surfaces.

non-striated muscle smooth muscle *q.v.*

non-sulphur purple bacteria group of photosynthetic heterotrophic bacteria containing purple pigments.

non-transmissible *a. appl.* diseases not caused by infection with microorganisms, e.g. cancer, diabetes, cardiovascular disease, and which therefore cannot be transmitted from one individual to another.

non-viable incapable of surviving or developing.

nonsense codon termination codon *q.v.*

nonsense mutation mutation which generates one of the nonsense (termination) codons UAA, UAG or UGA, resulting in premature termination of polypeptide synthesis during translation.

nonsense suppressor mutation which generates a tRNA that inserts an amino acid at a nonsense codon.

nopaline *n.* an unusual amino acid produced in plant cells infected with the Ti plasmid of *Agrobacterium tumefaciens*, the crown gall bacterium, which carries the gene for its synthesis.

noradrenalin(e) (NA) *n.* catecholamine closely related to adrenaline and with similar effects, secreted with it in small amounts by the adrenal medulla. Also acts as a neurotransmitter in sympathetic and enteric nervous systems, being secreted by sympathetic nerve endings in internal organs such as the gut, heart and spleen, and in various pathways in the central nervous system. *alt.* norepinephrine. *see* Appendix 1 (32) for chemical structure.

noradrenergic *a. appl.* nerve fibres, of the sympathetic system, that liberate noradrenaline from their terminals.

norepinephrine (NE) preferred usage for noradrenaline *q.v.* in North America.

norma *n.* view of the skull as a whole from certain points: basal, vertical, frontal, occipital, and lateral.

normalizing selection stabilizing selection *q.v.*

normoblast *n.* immature nucleated mammalian red blood cell, derived from an erythroblast and which develops into an erythrocyte.

Northern blotting technique in which RNAs separated by gel electrophoresis are transferred to a suitable medium for subsequent hybridization with radioactive probes for identification and isolation of RNAs of interest, named by analogy with the Southern blotting technique.

nosogenic pathogenic *q.v.*

nosology *n.* branch of medicine dealing with the classification of diseases; pathology *q.v.*

nostrils *n.plu.* exterior openings of the nose. *alt.* nares.

notal *a.* dorsal; *pert.* the back; *pert.* notum.

notate *a.* marked with lines or spots.

nothocline hybrid cline *q.v.*

Nothosauria, nothosaurs *n., n.plu.* an extinct order of streamlined fish-eating marine reptiles.

notocephalon *n.* dorsal shield of leg-bearing segments in some ticks and mites.

notochord *n.* slender rod of cells of mesodermal origin running along the back in the early chordate embryo and which directs formation of the neural tube. It persists in primitive chordates but in vertebrates is replaced by the spinal column.

notochordal *a. pert.* or enveloping notochord, *appl.* sheath, tissue, etc.

Notogaea *n.* zoogeographical region comprising Australia, New Zealand, New Guinea and the islands of the Pacific Ocean south and east of Wallace's line (which runs between the Indonesian islands of Bali and Lombok, between Borneo and Sulawesi and east of the Philippines. *alt.* Notogea.

notogaster *n.* posterior dorsal shield of certain mites and ticks.

Notogea Notogaea *q.v.*

notogenesis *n.* the development of the notochord.

notonectal *a.* swimming with back downwards.

notopleural suture in dipterans, a lateral suture separating the mesonotum from the pleuron.

notopodium *n.* dorsal lobe of polychaete

parapodium.

nototribe *a. appl.* flowers whose anthers and stigma touch back of insect as it enters corolla, a device for securing cross-fertilization.

notum *n.* dorsal portion of an insect segment.

novobiocin *n.* antibiotic synthesized by the actinomycete *Streptomyces niveus*.

N-regions short sequences of nucleotides, not encoded in the genome, which are added at the junctions between V, J and D gene segments during the rearrangement of immunoglobulin heavy chain genes and T-cell α and β receptor genes and which contribute to the generation of antibody and T-cell receptor diversity.

nucellus *n.* in seed plants, the megasporangial tissue which persists around the megaspore (embryo sac) and the megagametophyte that develops within it, and eventually forms the inner layer of the ovule wall.

nuchal *a. pert.* nape of neck.

nuciferous *a.* nut-bearing.

nucivorous *a.* nut-eating.

nuclear *a. pert.* a nucleus.

nuclear disc star-like structure formed by chromosomes at equator of spindle during mitosis.

nuclear envelope the double membrane surrounding the eukaryotic nucleus.

nuclear gene, nuclear genome those genes present on the chromosomes in the nucleus as opposed to mitochondrial or chloroplast genes.

nuclear lamina fibrous protein layer underlying the inner nuclear membrane.

nuclear layer internal layer of cerebral cortex; inner nuclear layer of retina, between inner and outer plexiform layers, and outer nuclear layer, between outer plexiform layer and basement membrane.

nuclear localization squences, nuclear location sequences *see* nuclear recognition sequences.

nuclear matrix dense fibrillar network lying on the inner side of the nuclear membrane, with which ribonucleoprotein particles are associated.

nuclear membrane the double membrane surrounding the eukaryotic nucleus, or one of these membranes.

nuclear plate equatorial plate *q.v.*

nuclear pore structure formed where inner and outer nuclear membranes join and which connects cytoplasm and nucleoplasm. They are present in large numbers in the nuclear envelope and form channels through which macromolecules pass from nucleus to cytoplasm or vice versa, larger proteins being selectively transported by some means as yet unknown.

nuclear pore complex disc of protein granules surrounding a nuclear pore.

nuclear receptors receptors for steroid and thyroid hormones, and some other compounds, which are intracellular, and which enter the nucleus after binding to their ligand, the receptor–ligand complex then acting as a gene regulatory protein.

nuclear recognition sequences sequences of amino acids that occur in proteins destined to enter the nucleus, and which apparently aid their transport through nuclear pores and their maintenance in the nucleus.

nuclear sap nucleoplasm *q.v.*

nuclear spindle an organized system of microtubules which forms a guiding framework for chromosome separation and movement during mitosis and meiosis. It develops from material surrounding the centrioles in animal cells and a similar region in plant cells as the chromosomes begin to become more compact preceding nuclear division. In mitosis, sister chromatids become connected to opposite poles of the spindle by microtubules attached to their centrosomes and are drawn towards the poles by shortening of the microtubules, thus ensuring an equal distribution of chromosomes to the two daughter nuclei.

nuclear transplantation the transfer of an intact nucleus from one cell to another which has usually had its nucleus removed, a technique used to study the expression of genes in different cytoplasmic environments.

nuclease *n.* any of a class of enzymes that degrade nucleic acids into shorter oligonucleotides or single nucleotide subunits by hydrolysing sugar-phosphate bonds in the nucleic acid backbone. *see* deoxyribonuclease, endonuclease, exonuclease, ribonuclease.

nucleate *a.* containing a nucleus.

nucleation *n.* formation of a nucleus.

nuclei *plu.* of nucleus *q.v.*

nucleic acids *n.plu.* deoxyribonucleic acid (DNA) and ribonucleic acid (RNA). Very large linear molecules containing C, H, O, N and P, which on partial hydrolysis yield nucleotides and nucleosides, and which are composed of one (RNA) or two (DNA) polynucleotide chains. They are essential components of all living cells, where they are the carriers of genetic information (DNA and mRNA), components of ribosomes (rRNA), and involved in deciphering the genetic code (tRNA). *see also* DNA (deoxyribonucleic acid), genetic code, messenger RNA, nucleotide, polynucleotide, ribosomal RNA, RNA (ribonucleic acid), transfer RNA.

nucleic acid hybridization *see* DNA hybridization, *in situ* hybridization.

nucleiform *a.* shaped like a nucleus.

nucleocapsid *n.* the nucleic acid plus protein coat of enveloped viruses.

nucleohistone *n.* a complex of DNA and histone, found in the nucleus of the eukaryotic cell and comprising the chromatin.

nucleoid *a.* resembling a nucleus; *n.* in prokaryotes, the region of the cell containing the "chromosome"; a nucleus-like body occurring in some red blood cells; the dense region seen in some viruses which represents the nucleic acid.

nucleolar *a. pert.* a nucleolus.

nucleolar organizer (NOR) region of a chromosome associated with the nucleolus, corresponding to a cluster of rRNA genes.

nucleolus *n.* region of the nucleus where rRNA is synthesized and consisting of a fibrillar core surrounded by a granular region in which ribonucleoprotein particles, possibly ribosomes, are being assembled. *plu.* nucleoli.

nucleolysis *n.* disintegration of a cell nucleus.

nucleoplasm *n.* the ground substance of the nucleus (excluding the nucleolus) internal to the nuclear membrane and excluding the chromatin. *alt.* nuclear sap.

nucleoplasmin *n.* protein first found in the nucleoplasm of *Xenopus laevis* oocytes, which is a molecular chaperone involved in the assembly of nucleosomes in eukaryotic cells.

nucleoprotein *n.* any complex of protein and nucleic acid.

nucleosidase *n.* any of a class of enzymes that split nucleosides (and nucleotides) into a nitrogenous base and a pentose (or pentose phosphate), EC 3.2.2, *alt.* nucleotide phosphorylase; *r.n.* for enzyme splitting *N*-ribosyl-purine into purine and ribose, EC 3.2.2.1.

nucleoside *n.* any of a group of compounds consisting of a purine or pyrimidine base (commonly adenine, guanine, cytosine, thymine) linked to the sugar ribose or deoxyribose, and including adenosine, cytidine, uridine, thymidine, guanidine.

nucleoside phosphorylase nucleosidase *q.v.*

nucleosome *n.* repeating structural unit in eukaryotic chromatin in which DNA is wound round a protein core composed of 2 each of the 4 histones H2A, H2B, H3 and H4, nucleosomes being linked by a stretch of DNA associated with histone H1.

nucleotidase *n.* any of several enzymes that hydrolyse nucleotides into nucleosides and orthophosphate, EC 3.1.3.5–7.

nucleotide *n.* phosphate ester of a nucleoside, consisting of a purine or pyrimidine base linked to ribose or deoxyribose phosphates, the purine nucleotides having chiefly adenine or guanine as the base, the pyrimidine nucleotides cytosine, thymine or uracil, and which are the basic repeating units in DNA and RNA. The nucleotide adenosine triphosphate (ATP) is a universal energy currency in all living cells. Other nucleotide triphosphates are also sources of stored chemical energy and nucleotides are also constituents of coenzymes such as NAD and FAD. *see* Appendix 1 (22, 23) for chemical structural formulae.

nucleotide sequence order of nucleotide residues in a nucleic acid, and which in mRNA and protein-coding DNA determines the amino acid sequence of the proteins specified. *alt.* base sequence.

nucleotide sequencing *see* DNA sequencing.

nucleotidyltransferase *n.* any of a class of enzymes that transfer a nucleotide (in this case a nucleoside monophosphate) from one compound to another, and which include adenylyltransferases, uridylyltransferases etc. and the DNA and RNA polymerases. EC 2.7.7.

nucleus *n.* a large dense organelle bounded by a double membrane (nuclear envelope), present in eukaryotic but not prokaryotic cells, and which contains the chromatin and in which DNA replication and transcription take place. Its function is essential to the survival of most cells. In most eukaryotes it disappears during mitosis or meiosis when the chromosomes divide, and is reformed after cytoplasmic division preceding cell division. *see* Fig 1; the centre of any structure, around which it grows; (*neurobiol.*) any of various masses of grey matter (nerve cell bodies) in the central nervous system. *plu.* nuclei.

nucleus ambiguus cells in medulla oblongata from which originate the motor fibres of glossopharyngeal and vagus, and of cerebral part of spinal accessory nerves.

nucleus pulposus the soft core of an intervertebral disc, remnant of notochord.

nucleus robustus archistrialis in brain of birds, a main centre for motor organization in the forebrain.

nuculanium *n.* berry formed from a superior ovary.

nucule *n.* nutlet.

nudibranch *a.* lacking a protective cover over the gills, *appl.* a group of shell-less molluscs (Nudibranchia), the sea slugs.

nudibranchiate *a.* having gills not covered by a protective shell or membrane, in a branchial chamber, *appl.* certain molluscs.

nudicaudate *a.* having a tail not covered in hair or fur.

nudicaulous *a. appl.* or having a stem without leaves.

nudiflorous *a.* having flowers without glands or hairs.

Nuhn, glands of *see* glands of Nuhn.

null allele mutant allele which results in an absence of functional gene product, the mutation itself being termed a null or amorphic mutation. Null alleles are usually recessive.

null hypothesis in planning a scientific experiment, the hypothesis that would give a certain set of experimental results in the conditions of the experiment. If the observed results depart significantly from these expected results the null hypothesis is unlikely to be true.

nullipennate *a.* without flight feathers.

nullisomic *a. appl.* an organism or cell which has both members of a pair of chromosomes missing. *n.* nullisomic.

numerical *a. appl.* hybrid of parents with different chromosome numbers.

numerical abundance the number of individuals of a species present in a given area.

numerical chromosome mutation a mutation that involves a change in the number of chromosomes in a cell.

numerical taxonomy classification of organisms by a quantitative assessment of their phenotypic similarities and differences, not necessarily leading to a phylogenetically based classification.

nummulation *n.* the tendency of red blood cells to adhere together like piles of coins (rouleaux).

nummulitic *a.* containing nummulites, a type of fossil foraminiferan.

nunatak *a.* an area, on a mountain or plateau, which has escaped past environmental changes, such as glaciation, and in which plants and animals of earlier floras and faunas have survived.

nuptial flight flight taken by queen bee when fertilization takes place.

nurse cells single cells or layers of cells attached to or surrounding an oocyte and which provide nutrients to it and elaborate specific proteins and mRNAs which pass into the oocyte and are stored until after the egg is fertilized and development begins; Sertoli cells in testis; certain other cells that provide nourishment and sometimes storage material to other cells.

nurture *n.* the sum total of environmental influences on a developing individual.

nut *n.* a dry, indehiscent, one- or two-seeded, one-chambered fruit with a hard woody pericarp (the shell), such as an acorn.

nutant *a.* bent downwards; drooping.

nutation *n.* rotational curvature of the growing tip of a plant; slow rotating movement by pseudopodia.

nutlet *n*. a small nut; an individual achene of a fruit such as beechmast; the stone of a stone fruit.

nutrient *n*. any substance used or required by an organism as food. *see also* macronutrients, micronutrients.

nutrient cycles the exchanges of elements between the living and non-living components of an ecosystem.

nutrient person gastrozooid *q.v.*

nutrition *n*. the process by which an organism obtains from its environment the energy and the chemical elements and compounds it needs for its survival and growth. *see* autotrophic, chemotrophic, heterotrophic, phototrophic.

nutritive *a*. concerned with nutrition.

nyctanthous *a*. flowering by night.

nyctinasty *n*. "sleep" movements in plants, involving a change in the position of leaves, petals, etc. as they close at night or in dull weather in response to a change in the level of light and/or temperature. *a.* nyctinastic. *alt.* nyctitropism, nyctitropic.

nyctipelagic *a*. rising to the surface of sea only at night.

nyctitropism nyctinasty *q.v.*

nyctoperiod *n*. daily period of exposure to darkness.

nymph *n*. a juvenile form without wings or with incomplete wings in insects with incomplete metamorphosis, i.e. when the change at each moult is small and the larvae are relatively similar to the adult form. *a.* nymphal.

Nymphales *n*. order of herbaceous aquatic dicots including families Ceratophyllaceae (hornwort) and Nymphaceae (water lily).

nymphochrysalis *n*. pupa-like resting stage between larval and nymphal form in certain mites.

nymphosis *n*. the process of changing into a nymph or pupa.

nystatin *n*. an antibiotic with antifungal activity.

O

OD optical density, *see* absorbance.
OP osmotic pressure *q.v.*
ori origin of replication.
OTU operational taxonomic unit *q.v.*

oak-apple *n.* type of hard spherical gall found on stems and leaves of oak and caused by larvae of a species of gall wasp.
oar feathers the wing feathers used in flight.
oat *n. Avena sativa,* a cereal plant of the Gramineae grown for its grain in cool temperate regions.
ob- prefix derived from L. *ob,* against, signifying the other way round, obversely, esp. when prefixed to names of leaf shapes.
obcompressed *a.* flattened in a vertical direction.
obconic *a.* shaped like a cone, but attached at its apex.
obcordate, obcordiform *a.* inversely heart-shaped; *appl.* leaves which have stalk attached to apex of heart.
obcurrent *a.* converging, and attaching at point of contact.
obdeltoid *a.* more or less triangular with point of attachment at apex of triangle.
obdiplostemonous *a.* with stamens in two whorls, the inner opposite the sepals and the outer opposite the petals.
obelion *n.* the point on skull between parietal foramina, on sagittal suture.
obimbricate *a.* with regularly overlapping scales, with the overlapped ends downwards.
oblanceolate *a.* inversely lanceolate, of leaves.
obligate *a.* obligatory; limited to one mode of life or action.
obligate parasite an organism that can only live as a parasite.
oblique *a. appl.* leaves, asymmetrical.

obliquely striated muscle type of muscle found in nematodes, molluscs and annelid worms, in which the arrangement of the myofibrils leads to a diagonal striation, and which is capable of greater extension and contraction than vertebrate striated muscle.
obliquus *n.* an oblique muscle, as of ear, eye, head, abdomen.
obliterate *a.* indistinct or profuse, *appl.* markings on insects; suppressed.
obliterative coloration type in which parts of an organism exposed to the brightest light are shaded more darkly, ensuring that it blends with its background more effectively.
obovate *a.* inversely egg shaped, *appl.* leaf with narrow end attached to stalk.
obpyriform *a.* inversely pear-shaped.
obsolescence *n.* the gradual reduction and eventual disappearance of a species; gradual cessation of a physiological process, or of a structure becoming disused, over evolutionary time; a blurred portion of a marking on an animal.
obsolete *a.* wearing out or disappearing; *appl.* any character that is becoming less and less distinct in succeeding generations; (*bot.*) *appl.* calyx united with ovary or reduced to a rim.
obsubulate *a.* reversely awl-shaped; narrow and tapering from tip to base.
obtect *a. appl.* pupa with wings and legs held close to body.
obturator *n.* any of various structures which close off a cavity; *a. pert.* any structure in the neighbourhood of the obturator foramen, an oval foramen between ischium and pubis.
obtuse *a.* with blunt or rounded end, *appl.* leaves, etc.

obtusilingual *a*. with a blunt short tongue.

obumbrate *a*. with some structure over-hanging the parts so as partially to conceal them.

obverse *a*. with base narrower than apex.

obvolute *a*. overlapping.

obvolvent *a*. bent downwards and inwards, *appl*. wings and elytra of some insects, etc.

Occam's or Ockham's razor the principle that where several hypotheses are possible, the simplest is chosen, first proposed by William of Ockham, a medieval scholastic philosopher.

occasional species one which is found from time to time in a community but is not a regular member of it.

occipital *a. pert*. to back of the head or skull.

occipital condyles two knob-like protrusions on back of skull of amphibians and mammals which articulate with the atlas vertebra of backbone.

occipital foramen posterior opening of head in insects; the foramen magnum of skull in vertebrates.

occipital lobe of brain, the hind part of cerebral hemisphere.

occipitalia *n.plu*. the group of parts of cartilaginous brain case forming back of head. *alt*. occipital bones.

occipito-atlantal *a. appl*. membrane closing gap between skull and neural arch of atlas in amphibians; *appl*. dorsal (posterior) and ventral (anterior) membranes between margin of foramen magnum and atlas in mammals.

occipito-axial *a. appl*. ligament or tectorial membrane connecting occipital bone with axis.

occipitofrontal *a. appl*. longitudinal arc of skull; *appl*. tract of fibres running from frontal to occipital lobes of cerebral hemispheres.

occiput *n*. back of head or skull; dorsolateral region of insect head.

occlusal *a*. contacting the opposing surface; *appl*. teeth which touch those of the other jaw when jaws are closed.

occlusion *n*. closure or blocking of a duct, tubule, etc.; overlapping activation of several motor neurones by simultaneous stimulation of several afferent nerves.

occlusor *n*. a closing muscle; *a. appl*. muscles of operculum or moveable lid.

oceanic *a*. inhabiting the open sea, where the sea is deeper than 200 m.

oceanodromous *a*. migrating only within the ocean, *appl*. fishes. *cf*. potamodromous.

ocellate *a*. like an eye or eyes, *appl*. markings.

ocellated *a*. having ocelli, or eye-like spots or markings.

ocelli *plu*. of ocellus *q.v.*

ocellus *n*. (*zool*.) a simple eye or eye-spot found in many invertebrates; a dorsal eye in insects; an eye-like marking as in many insects, fishes, on feathers, etc.; (*bot*.) large cell in leaf epidermis specialized for reception of light. *plu*. ocelli. *a*. ocellar.

ochraceous, ochreous *a*. ochre-coloured.

ochre the termination codon UAA.

ochre suppressor mutant gene specifying a tRNA with an altered anticodon that inserts an amino acid at the ochre termination codon UAA thus overcoming the effect of an ochre mutation in another gene.

ochrea ocrea *q.v.*

ochroleucous *a*. yellowish-white, buff-coloured.

ochrophore *n*. a cell bearing yellow pigment.

ochrosporous *a*. having ochre-coloured spores.

Ockham's razor *see* Occam's razor.

ocrea *n*. a tubular sheath-like expansion at base of petiole; a sheath; partial covering of stipe of some toadstools, formed by fragments of the disintegrated veil.

ocreaceous *a*. like an ocrea, *appl*. various structures in plants and animals.

ocreate *a*. having an ocrea; booted; sheathed; intrafoliaceous *q.v.*

octa- prefix derived from Gk. *octa*, eight, signifying having eight of, arranged in eights, etc.

octactine *n*. a sponge spicule with eight rays.

octad *n*. group of eight cells originating by division of a single cell.

octagynous *a*. having eight pistils or styles; having eight carpels to a gynaecium.

octamerous *a. appl*. organs or parts of organs when arranged in eights.

octandrous *a*. having eight stamens.

octant *n*. one or all of the eight cells formed by division of the fertilized ovum in plants and animals.

Octapoda *n.* an order of Cephalopoda whose members have eight tentacles and no shell (e.g. octopus).

octarch *a. appl.* stems with steles having eight alternating groups of phloem and xylem; with eight vascular bundles.

octopamine *n.* a neurotransmitter, related to the catecholamines.

octopetalous *a.* having eight petals.

octophore *n.* modified ascus with eight spores arranged radially, as in some fungi.

octopine *n.* an unusual amino acid produced in plant cells infected with the Ti plasmid of *Agrobacterium tumefaciens*, the crown gall bacterium, which carries the gene for its synthesis.

octoploid *a.* having eight haploid chromosome sets in somatic cells.

octopod *a.* having eight tentacles, arms or feet.

octoradiate *a.* having eight rays or arms.

octosepalous *a.* having eight sepals.

octospore *n.* one of eight spores, as formed at end of carpogonial filaments, or in an octophore.

octosporous *a.* having eight spores.

octostichous *a.* arranged in eight rows; having leaves in eight's.

octozoic *a. appl.* a spore, of gregarines, containing eight sporozoites.

ocular *a. pert.,* or perceived by, the eye.

ocular dominance columns slabs of cells in visual cortex, perpendicular to the surface, which respond to stimuli from one or the other eye.

oculate *a.* having eyes or eye-like spots.

oculiferous, oculigerous *a.* having eyes.

oculofrontal *a. pert.* region of forehead and eye.

oculomotor *a.* causing movements of the eyeball, *appl.* 3rd cranial nerve, which controls four of the six small muscles moving the eyeball.

oculonasal *a. pert.* eye and nose.

oculus *n.* the eye; a leaf bud in a tuber.

Oddi's sphincter muscle fibres surrounding duodenal end of common bile duct.

odd-pinnate pinnate with one terminal leaflet.

odd-toed ungulates perissodactyls *q.v.*

odogen *n.* substance that stimulates the sense of smell.

Odonata *n.* order of insects including the dragonflies and damselflies. Winged carnivorous insects with brilliant metallic colouring, whose eggs are laid in water and develop through an aquatic nymph (larval) stage which has gills.

odontoblast, odontocyte *n.* one of the columnar cells on outside of dental pulp that secretes dentine of tooth; one of the cells giving rise to teeth of molluscan radula.

odontobothrion *n.* tooth socket.

Odontoceti *n.* the toothed whales, a suborder of the Cetacea, which are all predatory, feeding on fish and other marine animals, and which include the sperm whale, killer whale, narwhal, porpoises and dolphins.

odontoclast *n.* one of the large multi-nucleate cells that absorb roots of milk teeth and destroy dentine.

odontogeny *n.* the origin and development of teeth.

odontoid *a.* tooth-like; *pert.* the odontoid process.

odontoid process tooth-like peg around which the atlas or first cervical vertebra rotates.

odontology *n.* dental anatomy, histology, physiology and pathology.

odontophore *n.* the tooth-bearing organ in molluscs.

odontorhynchous *a.* having inner edge of bill bearing plate-like ridges.

odontosis *n.* dentition *q.v.*; tooth development.

odontostomatous *a.* having tooth-bearing jaws.

odorimetry *n.* measurement of the strength of the sense of smell, using substances of known ability to stimulate olfaction.

odoriphore osmophore *q.v.*

oecium *n.* the calcareous or chitinous covering of a hydrozoan colony.

oedema *n.* swelling of a tissue through increase in tissue fluid. *alt.* edema.

oenocyte *n.* one of the large cells from clusters which surround trachea and fat-body of insects and undergo changes related to the moulting cycle.

oesophageal *a. pert.* or near oesophagus.

oesophageal bulbs two swellings on the oesophagus in nematodes, the posterior of which, the pharynx, exhibits rhythmical pumping movements.

oesophageal glands or pouches in

earthworms, outgrowths of the oesophagus which secrete calcium carbonate.

oesophagus *n.* that part of alimentary canal between pharynx and stomach or part equivalent thereto. *alt.* esophagus, gullet.

oestr- *alt.* estr-.

oestradiol *n.* a principal oestrogen in ovarian follicular fluid and also produced by the placenta. It is responsible for the development and maintenance of secondary female sexual characteristics, the maturation and cyclic function of accessory sexual organs and the development of the duct system in mammary glands. *see* Appendix 1 (47) for chemical structure

oestrin oestrone *q.v.*

oestriol *n.* an oestrogen present in urine of pregnant women.

oestrogens *n.plu.* a group of vertebrate steroid hormones, the principal female sex hormones, synthesized chiefly by the ovary and placenta in females and responsible for the development and maintenance of secondary female sexual characteristics and the growth and function of female reproductive organs. Similar compounds have been found in plants.

oestrogenic *a.* inducing oestrus, *appl.* hormones; *pert.* oestrogen.

oestrone *n.* a derivative of oestradiol with similar activity, excreted in urine of some pregnant mammals.

oestrous *a. pert.* oestrus.

oestrus *n.* the period of sexual heat and fertility in a female mammal when she is receptive to the male.

oestrus cycle reproductive cycle in female mammals in the absence of pregnancy, comprising oestrus, when ovarian follicles mature and ovulation takes place, metoestrus (luteal phase) *q.v.*, and pro-oestrus *q.v.*

officinal *a.* used medicinally, *appl.* plants.

oidia *plu.* of oidium, *see* oidiospore.

oidiophore *n.* specialized hypha that cuts off oidia from its tip.

oidiospore, oidium *n.* fungal spore formed by transverse segmentation of a hypha. *plu.* oidia.

oil bodies, oil storage bodies lipid bodies *q.v.*

oil gland any gland secreting oils; in birds, a gland in the skin that secretes oil used in preening the feathers.

oils *n.plu.* glycerides and esters of fatty acids which are liquid at 20°C, the fatty acids being in general less saturated than in fats. *cf.* fats.

Okazaki fragments short lengths of DNA of around 1000 nucleotides formed during DNA replication by discontinuous DNA synthesis on the lagging ($3'\rightarrow5'$) DNA strand at the replication fork, due to the fact that DNA polymerase can only synthesize DNA in the $5'\rightarrow3'$ direction, and which are rapidly joined by DNA ligase to form a continuous strand.

old-growth forest uncut virgin forest containing trees that are hundreds and sometimes thousands of years old, as in the forests of Douglas fir, western hemlock, giant sequoia and redwoods in the western United States.

oleaginous *a.* containing or producing oil.

Oleales *n.* order of dicot trees, shrubs and climbers comprising the family Oleaceae (olive, privet).

oleate (oleic acid) common widely distributed unsaturated fatty acid, $CH_3(CH_2)_7CH=CH(CH_2)_7COO^-$. *see* Appendix 1 (37) for chemical structure.

olecranon *n.* large bony process at upper end of ulna.

oleiferous *a.* producing oils.

olein *n.* a fat, containing oleic acid, liquid at ordinary temperatures, found in animal and plant tissues.

oleoplast, oleosome elaioplast, elaiosome *q.v.*

olfaction *n.* the sense of smell; the process of smelling.

olfactory *a. pert.* sense of smell, *appl.* stimuli, organs, tract of nerve fibres etc.

olfactory bulb or lobe lobe projecting from anterior lower margin of cerebral hemispheres, concerned with sense of smell, contains the terminations of the olfactory nerves.

olfactory epithelium epithelial lining of nasal cavity containing olfactory neurones.

olfactory lobe olfactory bulb *q.v.*

olfactory nerves 1st cranial nerves in vertebrates, sensory nerves running from olfactory organs to olfactory bulb in forebrain.

olfactory neurones ciliated chemoreceptor

cells in nasal epithelium that sense odours and transmit signals to the olfactory nerve.

olfactory pit olfactory organ in the form of a small pit or hollow, in certain invertebrates; embryonic ectodermal sac which gives rise to nasal cavity, olfactory epithelium and vomeronasal organ in vertebrates.

olfactory spindle sensory cell structure associated with olfactory nerve in antennule of decapod crustaceans. *alt.* lobus osphradicus.

olig- prefix derived from Gk. *oligos*, few, signifying having few, having little of, etc.

oligacanthous *a.* bearing few spines.

oligandrous *a.* having few stamens.

oligarch *a.* having few vascular bundles or elements.

oligocarpous *a.* having few carpels.

Oligocene *n.* a geological epoch in the Tertiary, between Eocene and Miocene, lasting from about 38 to 25 million years ago.

Oligochaeta, oligochaetes *n., n.plu.* class of annelid worms characterized by possession of a few bristles (setae or chaetae) on each segment and no parapodia, and which includes the earthworms.

oligodendrocyte *n.* type of small neuroglial cell predominant in white matter of the central nervous system, where they form myelin sheaths around axons, collectively known as oligodendroglia.

oligodendroglia *see* oligodendrocyte.

oligogenic *a. appl.* characters controlled by a few genes responsible for major heritable changes.

oligoglia oligodendroglia *q.v.*

oligogyny *n.* the occurrence in a single colony of social insects of from two to several functional queens.

oligohaline *a. appl.* brackish water with salinity of 0.5–5 parts per thousand.

oligolecithal *a. appl.* eggs containing little yolk.

oligolectic *a.* selecting only a few, *appl.* insects visiting only a few different food plants or flowers.

oligolobate *a.* divided into only a small number of lobes.

oligomer *n.* a molecule composed of only a few monomer units.

oligomerous *a.* having one or more whorls with fewer members than the rest.

oligomycin *n.* antibiotic which inhibits ATP synthesis in mitochondria by interacting with one of the proteins in mitochondrial coupling factor.

oligonucleotide *n.* short chain of nucleotides (*q.v.*). *see also* polynucleotide.

oligopeptide *n.* a small polypeptide.

oligophagous *a.* restricted to a single order, family or genus of food plants, *appl.* insects.

oligophyletic *a.* derived from a few different lines of descent.

oligopneustic *a. appl.* insect respiratory system in which only a few spiracles are functional.

oligopod *a.* having few legs or feet; having thoracic legs fully developed.

oligorhizous *a.* having few roots, *appl.* certain marsh plants.

oligosaccharide *n.* a molecule composed of only a few (4–20) monosaccharide units.

oligosaprobic *a. appl.* category in the saprobic classification of river organisms comprising those that can only live in water unpolluted by organic pollutants, e.g. brown trout (*Salmo trutta*) and stonefly nymphs (plecopterans). *cf.* α-mesosaprobic, β-mesosaprobic, polysaprobic; *appl.* aquatic environment with a high dissolved oxygen content and little organic decomposition.

oligospermous *a.* having few seeds.

oligosporous *a.* producing or having few spores.

oligostemonous *a.* having few stamens.

oligotaxy *n.* reduction in the number of whorls, of floral leaves, etc.

oligothermic *a.* tolerating relatively low temperatures.

oligotokous *a.* bearing few young. *alt.* oligotocous.

oligotrophophyte *n.* plant that will grow on poor soil.

oligotrophic *a.* providing or *pert.* inadequate nutrition; *appl.* waters relatively low in nutrients, as the open oceans compared with the continental shelves, and as some lakes whose waters are low in dissolved minerals and which cannot support much plant life; *appl.* microorganism that thrives and predominates in a nutrient-poor environment. *n.* oligotroph.

oligotrophy *n.* the ability to live in a nutrient-poor environment, as e.g. many soil actinomycetes.

oligotropic *a.* visiting only a few allied species of flowers, *appl.* insects.

oligoxenous *a. appl.* parasites adapted for life in only a few species of hosts.

olive, olivary nucleus one of several nuclei in brain situated in the medulla just below the pons.

omasum *n.* in ruminants, the third chamber of the stomach, through which food must pass from rumen to abomasum, and whose structure prevents mixing of rumen and abomasum contents.

ombrogenous *a. appl.* wet habitats arising from precipitation rather than from water in the ground.

ombrophile *a.* adapted to living in a rainy place, *appl.* plants, leaves.

ombrophobe *n.* a plant that does not thrive under conditions of heavy rainfall.

ombrophyte *n.* a plant adapted to rainy conditions.

omentum *n.* fold of peritoneum either free or acting as connecting link between viscera. *a.* omental.

ommateum *n.* a compound eye.

ommatidium *n.* individual facet of insect or crustacean compound eye, comprising a hexagonal tube with a lens of transparent cuticle at the external face, from which a cone of transparent jelly (the crystalline cone) projects backwards. The sides of the ommatidium are composed of pigmented cells. At the base is a cup-shaped retinula of sensory photoreceptor cells containing light-sensitive pigment. *plu.* ommatidia. *a.* ommatidial.

ommatin(e)s *n.plu.* group of ommatochrome pigments which are small dialysable molecules, and including xanthommatin, the brown pigment of some insect eyes.

ommatochromes *n.plu.* group of pigments of arthropods, found in eye and in rest of body, being yellow, red and brown.

ommatophore *n.* a moveable process bearing an eye.

ommochromes ommatochromes *q.v.*

omni- prefix derived from L. *omnis*, all.

Omnibacteria *n.* name sometimes given to a large grouping of aerobic or facultatively anaerobic heterotrophic Gram-negative bacteria that includes the enterobacteria,

vibrios, aeromonads, and the stalked, budding and sheathed bacteria.

omnicolous *a.* capable of growing on different substrates, *appl.* lichens.

omnivore *n.* animal that eats both plant and animal food. *a.* omnivorous.

omohyoid *a. pert.* shoulder and hyoid, *appl.* muscle.

omphalic umbilical *q.v.*

omphalodisc *n.* an apothecium with a small central protuberance, as in certain lichens.

omphalogenesis *n.* development of the umbilical vesicle and cord.

omphaloid *a.* like a navel; umbilicate *q.v.*

omphalomesenteric *a. pert.* umbilicus and mesentery, *appl.* arteries, veins, ducts.

onchocerciasis *n.* tropical disease that causes impaired vision and eventual blindness, caused by infection with the parasitic roundworm *Onchocerca volvulus*, which is transmitted to humans by blackflies. *alt.* river blindness.

onchosphere oncosphere *q.v.*

oncofoetal antigens antigens found on the surface of cancer cells and also on embryonic cells but not on normal adult cells.

oncogene *n.* a gene carried by a tumour virus or a cancer cell, which is solely or partly responsible for tumorigenesis. Many oncogenes have been identified as altered versions of normal cellular genes involved in the control of cell division or differentiation. The oncogenes in RNA tumour viruses derive from cellular genes which have been picked up during virus infection of a cell. The normal counterpart to an oncogene is sometimes known as a proto-oncogene, but the term oncogene is often used to denote any gene, altered or not, which is a potential oncogene. Viral oncogenes are symbolized as v-*onc*, where *onc* may be a wide range of genes. The normal counterpart is often symbolized as c-*onc*. *see also* tumour suppressor genes.

oncogenesis *n.* the generation and development of a tumour.

oncogenic *a.* capable of causing a tumour.

oncolytic *a.* capable of destroying cancer cells.

oncomiracidium *n.* the free-swimming ciliated larval form of skin and gill flukes (monogenean flatworms). *plu.* oncomiracidia.

oncomouse *n*. strain of transgenic mice that carries an activated oncogene, and which is used experimentally to study onco-genesis.

oncoprotein *n*. protein encoded by an oncogene, and which can cause transfor-mation if introduced into a cell.

oncornavirus acronym for oncogenic RNA virus, *see* oncoviruses, RNA tumour virus.

oncosphere *n*. spherical, hooked larva that hatches from tapeworm egg and develops into a cysticercoid.

oncoviruses *n.plu.* a subfamily of retro-viruses, comprising the RNA tumour vi-ruses. Divided into types C, B and D. For-merly known as oncornaviruses.

one gene–one enzyme hypothesis the original form of the idea that each gene specifies one polypeptide chain, developed in the 1930s and 1940s from the study of biochemical mutants.

ontogenesis ontogeny *q.v.*

ontogeny *n*. the history of development and growth of an individual. *alt.* ontogenesis. *a.* ontogenetic. *cf.* phylogeny.

onychium *n*. the layer below the nail; pulvillus *q.v.*; a special false articulation to bear claws at end of tarsus in some spi-ders.

onychogenic *a*. capable of producing a nail or nail-like substance; *appl.* material in nail matrix, and cells forming fibrous substance and cuticula of hairs.

Onychophora, onychophorans *n., n.plu.* class of primitive worm-like terrestrial ar-thropods with a soft flexible cuticle, some-times treated as a separate phylum, which live in damp habitats in warm climates. They possess a pair of antennae and a pair of stiff jaw appendages, followed by a number of pairs of unjointed walking legs. There is only one extant genus, *Peripatus*. *alt.* velvet worms.

ooangium archegonium *q.v.*

ooapogamy *n*. the parthenogenetic devel-opment of an unfertilized ovum.

ooblast *n*. in some red algae, a tubular out-growth from the carpogonium through which the fertilized nucleus or its deriva-tives pass into the auxiliary cell.

oocyst *n*. cyst formed around two conjugat-ing gametes in sporozoan protozoans.

oocyte *n*. female ovarian cell in which meiosis occurs to form the egg. Cells undergoing the 1st meiotic division are often termed primary oocytes, after which they become secondary oocytes which un-dergo the 2nd meiotic division to become mature eggs.

oocyte injection technique for studying gene expression by injecting purified DNA or mRNAs into the nucleus or cytoplasm of (usually) *Xenopus* oocytes where they are transcribed and/or translated by the oocyte's own enzymes.

ooecium ovicell *q.v.*

oogamete *n*. a female gamete, esp. a large non-motile one containing food material for the zygote.

oogamous *n*. reproducing by oogamy.

oogamy *n*. the union of unlike gametes, usu-ally a large non-motile female gamete and a small motile male gamete.

oogenesis *n*. formation, development and maturation of the female gamete or ovum.

oogloea *n*. egg cement.

oogonial *a. pert.* an oogonium.

oogonium *n*. diploid precursor to female germ cells. In animals, it becomes an oocyte which undergoes meiosis to pro-duce the mature ovum; the female repro-ductive organ in fungi and algae. *plu.* oogonia.

ooid *a*. egg-shaped; oval.

ookinesis *n*. the mitotic stages of nuclear division in maturation and fertilization of eggs.

oology *n*. the study of birds' eggs.

Oomycota, oomycetes *n., n.plu.* phylum of simple non-photosynthetic, saprobic or parasitic, unicellular or filamentous protists formerly classifed as fungi. Unlike most fungi their cell walls contain cellulose. Sexual reproduction is oogamous and they reproduce asexually by motile zoospores. They include the water moulds (e.g. *Saprolegnia*), and the causative organisms of several important plant diseases, e.g. downy mildew of grapes (*Plasmopora*) and potato blight (*Phytophthora infestans*).

oophoridium *n*. the megasporangium in certain plants.

oophyte *n*. the gametophyte in certain lower plants.

ooplasm *n*. the cytoplasm of an egg.

oopod *n*. component part of sting or

ovipositor.

oosphere *n.* a female gamete or egg, esp. as produced in an oogonium by algae and oomycete fungi.

oosporangium oogonium *q.v.*

oospore *n.* a thick-walled zygote that arises from a fertilized oosphere in oomycete fungi, algae and protozoa.

oostegite *n.* a plate-like structure on basal portion of thoracic limb in crustaceans, which helps to form receptacle for the egg and acts as a brood pouch.

oostegopod *n.* a thoracic appendage bearing an oostegite (egg receptacle) in crustaceans.

ootheca *n.* egg case in certain insects.

ootocoid *a.* giving birth to young at a very early stage and then carrying them in a pouch, as marsupials.

ootokous *a.* egg-laying. *alt.* ootocous.

ootype *n.* structure in the female reproductive system of flatworms in which the fertilized zygote and yolk cells are enclosed in a capsule to form the egg.

ooze *n.* a deposit containing skeletal parts of minute organisms and covering large areas of the ocean floor; soft mud.

oozoid *n.* any individual developed from an egg.

oozoite *n.* the asexual parent, in ascidians.

opal the termination codon UAG.

Opalinata, opalinids *n.*, *n.plu.* group of multiflagellate protists of the phylum Zoomastigina, which are parasitic in the guts of amphibians, reptiles and fish. They are characterized by the falx, a structure made up rows of kinetosomes, each supporting a flagellum.

open *a. appl.* arrangement of floral parts where perianth segments do not meet at the edges, as in Cruciferae; *appl.* plant community that does not completely cover the ground but leaves bare areas that can be colonized.

open forest area covered with trees sufficiently widely spaced not to form a closed canopy. *cf.* closed forest.

open reading frame a stretch of DNA that contains a signal for the start of translation followed in the correct register by a sufficient length of amino acid-encoding triplets to form a protein, followed by a signal for termination of translation, and which may therefore indicate the presence of a protein-coding gene.

operant behaviour spontaneous animal behaviour that occurs without any apparent stimulus.

operant conditioning type of procedure for studying animal behaviour in which rewards and punishments are used to select, strengthen or weaken behaviour patterns.

operational taxonomic unit (OTU) any group such as genus, species, etc., evaluated on taxonometric methods.

operator *n.* region of DNA found in many bacterial operons to which the repressor (or apoinducer) binds, thus preventing (or allowing) transcription of the operon.

opercula *plu.* of operculum.

opercular *n.* posterior bone of operculum in fishes; *a. pert.* operculum; *appl.* fold of skin covering gills in tadpoles.

operculate *a.* having a lid, *appl.* spore capsules, etc.; having a covering (operculum) over gills, as most fishes.

operculiform *a.* lid-like.

operculigenous *a.* producing or forming a lid.

operculum *n.* lid, or covering flap, as on spore capsules, eggs of some invertebrates; gill cover in fishes; flap covering the nostrils and ears in some birds; moveable plates in shell of barnacle; lid-like structure closing mouth of shell in some gastropod molluscs; small bone in middle ear of amphibians, lying on the oval window. *plu.* opercula.

operon *n.* a type of genetic unit in bacteria, in which several genes coding for the enzymes of a metabolic pathway are clustered and transcribed together into a polycistronic mRNA. The transcription of each operon is initiated at a promoter region and controlled by a neighbouring regulatory gene. This specifies a regulatory protein (a repressor or apoinducer) which binds to an operator sequence in the operon to respectively prevent or allow its transcription.

ophidians *n.plu.* snakes.

ophiocephalous *a.* snake-headed; *appl.* small pedicellariae of some echinoderms having broad jaws with toothed edges.

ophiopluteus *n.* the pluteus larva of brittle stars.

ophiurans Ophiuroidea *q.v.*

ophiuroid *a. pert.* or resembling a brittle star; *appl.* cells: multiradiate or spiculate sclereids.

Ophiuroidea *n.* a class of echinoderms commonly known as brittle stars, having a star-shaped body with the arms clearly marked off from the disc.

ophryon *n.* point of junction of median line with a line across narrowest part of forehead.

ophthalmic *a. pert.* eye; *appl.* nerve: division of 5th cranial nerve conveying sensation from eye orbit, forehead and front of scalp; *appl.* an artery arising from internal carotid; *appl.* inferior and superior veins of eye orbit.

ophthalmic rete network of arteries and veins in the head of birds, which acts as a heat exchanger to allow the brain to be kept at a temperature below that of the arterial blood leaving the heart.

ophthalmogyric *a.* bringing about movement of the eye.

ophthalmophore ommatophore *q.v.*

ophthalmopod *n.* eye-stalk, as of decapod crustaceans.

opiate *n.* any of a class of compounds mimicking the effects of opium in the brain.

opiate receptors class of chemoreceptors in brain and gut that bind morphine and other opiates, and whose natural substrates are the enkephalins and endorphin.

Opiliones *see* harvest men.

opioid *a.* having opiate-like activity, *appl.* peptides such as the enkephalins and endorphin.

opisth- prefix derived from Gk. *opisthe*, behind.

opisthial *a.* posterior, *appl.* pore or stomatal margin.

opisthion *n.* median point of posterior margin of foramen magnum.

opisthobranch *n.* member of the mollusc subclass Opisthobranchia (e.g. sea slugs, sea hares), which are marine, and in which the shell is much reduced or absent.

opisthocoelous *a.* having the centrum concave behind, *appl.* vertebrae.

opisthodetic *a.* lying posterior to beak or umbo, *appl.* ligaments in some bivalve shells.

opisthogenesis *n.* development of seg-

ments or markings proceeding from the posterior end of the body.

opisthoglossal *a.* having tongue fixed in front, free behind.

opisthognathous *a.* having retreating jaws; with mouthparts directed backwards, *appl.* head of insects.

opisthogoneate *a.* having the genital aperture at hind end of body, *appl.* arthropods: insects and centipedes.

opisthohaptor *n.* posterior sucker or disc in trematodes.

opisthokont *a.* with flagellum or flagella at posterior end.

opisthomere *n.* terminal plate on abdomen of female earwig.

opisthonephros *n.* renal organ of embryo, consisting of meso- and metanephric series of tubules.

opisthosoma *n.* posterior section of the body in arachnids; also used for the posterior section of body in some other invertebrates.

opisthotic *a. pert.* inferior posterior bony element of otic capsule.

opisthure *n.* the projecting tip of vertebral column.

opium *n.* addictive drug obtained from the opium poppy *Papaver somniferum*, consisting of the dried milky juice from the slit poppy capsules, acts as a stimulant, narcotic and hallucinogen, formerly widely used to ease pain but now replaced by its derivative alkaloids such as morphine.

opophilous *a.* feeding on sap.

opponens *a. appl.* muscles which cause digits to approach one another.

opportunistic *a. appl.* microorganisms that are normally non-pathogenic but can cause disease in certain conditions, e.g. in immunosuppressed or otherwise debilitated individuals; *appl.* species specialized to exploit newly opened habitats.

opposable *a.* of digits, as of the thumb of primates, which may be brought together with fingers in a grasping action enabling objects to be held.

opposite *a. appl.* leaves or other organs that form a pair opposite each other on the stem.

opsiblastic *a.* with delayed cleavage, *appl.* eggs having a dormant period before hatching. *cf.* tachyblastic.

opsigenes *n.plu.* structures formed or becoming functional long after birth.

opsin *n.* protein component of the various vertebrate visual pigments, differing slightly from pigment to pigment, these differences determining the spectral sensitivity.

opsonic *a. pert.* or affected by opsonins.

opsonin *n.* type of antibody (e.g. IgM and IgG) whose binding to antigens on virus or bacterium facilitates their subsequent ingestion by phagocytic cells.

opsonization *n.* process whereby foreign particles become coated with specific antibody making them more readily ingested by phagocytic cells.

optic *a. pert.* vision or the eye.

optic axis line between central points of anterior and posterior curvature or poles of eyeball.

optic bulb expansion of the embryonic optic vesicle, later invaginated to form the optic cup from which retina develops.

optic chiasm(a) X-shaped structure below frontal lobes of brain in which the optic nerves from right and left eyes meet, nerve fibres from the inner half of each retina crossing over to form two optic tracts composed of fibres from the right and left halves respectively of both retinae, the right tract carrying all sensation from the left visual field and the left tract from the right visual field.

optic lobes part of brain concerned with processing of visual signals, consisting of large part of occipital lobes. *alt.* visual areas, visual cortex.

optic nerves 2nd cranial nerves in vertebrates, concerned with vision, each optic nerve running from cells of retina of one eye and partially crossing over at the optic chiasm below the frontal lobes of the brain, before terminating in the lateral geniculate nuclei, from where connections then run to the visual cortex.

optic radiation pathways from lateral geniculate nucleus to visual cortex.

optical *a. pert.* vision; of or using light, as optical microscope.

optical density (OD) absorbance *q.v.*

optical isomer either of 2 optically active isomeric forms of a compound, one of which rotates a beam of plane-polarized light to the right (dextrorotatory, *d*) and the other to the left (laevorotatory, *l*).

opticociliary *a. pert.* optic and ciliary nerves.

opticon *n.* inner zone of optic lobe in insects.

opticopupillary *a. pert.* optic nerve and pupil.

optimal *a.* the most efficient, the most cost-effective; *appl.* various animal behaviours as optimal foraging, optimal reproductive strategy, etc. It may be used in a long-term sense to indicate a behaviour that results in an animal leaving the largest possible number of viable offspring, or in a short-term sense to indicate behaviour that optimizes the energy collected in a certain time, as in optimal foraging, or minimizes the metabolic energy expended in achieving a goal.

optimal yield the highest rate of increase that a population can sustain in a given environment.

optimum *n.* the most suitable degree of any environmental factor or set of factors for full development of organism concerned; point at which best response is obtained in any system.

optocoel *n.* the cavity in optic lobes of brain.

optokinetic *a. pert.* movement of the eyes.

optomotor *a. appl.* reflex of turning head or body in response to stimulus of moving stripes.

Opuntiales Cactales *q.v.*

ora *n.* a margin; ora serrata: the wavy margin of retina, where nervous tissue ends; *n.plu.* mouths (*plu.* of os).

orad *a.* towards the mouth or mouth region.

oral *a. pert.* or belonging to the mouth; on the same side as the mouth (of radially symmetrical animals such as echinoderms).

oral disc in anthozoan polyps, circular flattened area surrounded by tentacles with the mouth at the centre.

oral siphon in urochordates, a narrowing of the body near the mouth.

oral valve in crinoids, one of five low triangular flaps separating the ambulacral grooves.

orbicular *a.* (*bot.*) of leaves, round or shield-shaped with petiole attached in the centre; (*zool.*) *appl.* eye muscles.

orbicularis *n.* muscle whose fibres surround an opening.

orbiculate *a.* nearly circular in outline.

orbit *n.* bony cavity in which eye is situated; skin around eye in bird; conspicuous zone around compound eye in insects; hollow in arthropod cephalothorax from which eye-stalk arises.

orbital *a. pert.* the orbit.

orbitomalar *a. pert.* orbit of eye and malar bone.

orbitonasal *a. pert.* orbit of eye and nasal portions of adjoining bones.

orbitosphenoid *a. pert.* paired cranial elements lying between presphenoid and frontal; *appl.* bone with foramen for optic nerve.

orb web the most familiar type of spider's web, with a spiral centre supported on radiating threads anchored to a roughly triangular frame.

orchid mycorrhiza mycorrhiza formed by basidiomycete fungi on the embryos of orchids (family Orchidaceae), which are necessary for the embryo's successful development.

Orchidales *n.* an order of monocot herbs, with tubers or stems swollen into pseudobulbs, often epiphytic or sometimes saprophytic, often with showy flowers, and comprising the family Orchidaceae (orchids).

orchitic *a.* testicular *q.v.*

orculaeform *a.* cask-shaped, *appl.* spores of certain lichens.

order *n.* taxonomic group of related organisms ranking between family and class.

ordinate *a.* having markings in rows.

ordinatopunctate *a.* indicating presence of serial rows of dots etc.

Ordovician *a. pert.* or *appl.* geological period lasting from about 500 to 440 million years ago.

organ *n.* any part or structure of an organism adapted for a special function or functions, e.g. heart, stomach, kidney, etc.

organ of Corti a structure running the length of the mammalian cochlea, located on the basilar membrane and concerned with sound perception. It consists of rows of sensory hair cells supported on a double row of arching rods (Corti's rods) and which are sensitive to vibrations in the surrounding fluid set up by sound waves transmitted from the outer ear. It is covered by a membrane, the tectorial or Corti's membrane.

organ of Golgi cylindrical sensory receptor at junction of tendons and muscles.

organelle *n.* a structure within a eukaryotic cell in which certain functions and processes are localized.

organic *a. pert.*, derived from, or showing the properties of, a living organism; *appl.* molecules, containing carbon.

organic chemistry the chemistry of carbon-containing compounds.

organic farming farming without the use of artificial fertilizers and pesticides.

organicism *n.* the integration of an organism as a unit.

organific *a.* making an organized structure.

organism *n.* any living thing.

organismal, organismic *a. pert.* an organism as a whole; *appl.* or *pert.* factors or processes involved in maintaining the integrity and life of an individual.

organized *a.* exhibiting characteristics of, or behaving like, an organism; *appl.* growth of cells in tissue culture resembling their normal growth.

organizer *n.* a part of an embryo which directs the development of other parts, often refers esp. to the mesodermal cells of the dorsal lip of the blastopore in amphibians and similar regions in other vertebrates, which trigger gastrulation and much subsequent development, and are known as the primary organizer.

organogen *n.* any of the elements C, H, O, N, also S, P, Cl.

organogenesis *n.* the formation and development of organs.

organogenic *a.* due to the activity of an organ; *pert.* organogenesis.

organoid *a.* having a definite organized structure, *appl.* certain plant galls.

organoleptic *a. appl.* a stimulus capable of affecting the sensory organs.

organology *n.* the study of organs of plants and animals.

organophyly *n.* the phylogeny of organs.

organoplastic *a.* capable of forming, or producing, an organ; *pert.* formation of organs.

organotroph heterotroph *q.v.*

organotrophic *a. pert.* formation and nourishment of organs; heterotrophic *q.v.*

organotypic *a. appl.* organized growth of

cultured cells in a form resembling a tissue.

orgasm *n.* immoderate excitement, esp. sexual; turgescence of an organ.

ori origin of replication in DNA.

Oriental Region zoogeographical region, which is part of Arctogaea, and which comprises the Indian subcontinent south of the Himalayas, South-east Asia and Indonesia apart from Sulawesi and New Guinea. It is separated from the Australian Region by Wallace's line *q.v.*

orientation *n.* alteration in position shown by organs or organisms under stiumlus.

orientation columns feature detection columns *q.v.*

orientation response the physiological response of an organism to a sudden change (e.g. a sound or a novel sight) in its environment.

orifice *n.* mouth or aperture.

origin (ori) in DNA, a sequence at which replication is initiated.

origin of life modern scientific theories on the origin of life almost all propose an origin on the early Earth, some 4000 million years ago, from simple carbon-containing compounds, beginning with the evolution of some form of self-replicating molecule, possibly of RNA, whose considerable catalytic potential as well as its role in information storage and transmission is now recognized, *see* ribozyme.

ornis avifauna *q.v.*

ornithic *a. pert.* birds.

ornithine *n.* an amino acid, diaminovaleric acid, involved in the urea cycle (*q.v.*), and in birds excreted with one of its derivatives, ornithuric acid.

ornithine cycle urea cycle *q.v.*

Ornithischia, ornithischians *n., n.plu.* order of Mesozoic dinosaurs, commonly called the bird-hipped dinosaurs, having a pelvis resembling that of a modern bird. They were all herbivorous and included both bipedal (ornithopods) and quadrupedal (ceratopians, stegosaurs and ankylosaurs) members.

Ornithogaea *n.* the zoogeographical region which includes New Zealand and Polynesia.

ornithology *n.* the study of birds.

ornithophilous *a. appl.* flowers pollinated

through the agency of birds. *n.* ornithophily.

ornithuric acid dibenzoylornithine, a constituent of bird droppings.

oroanal *a.* serving as mouth and anus; connecting mouth and anus.

orobranchial *a. pert.* mouth and gills, *appl.* epithelium.

oronasal *a. pert.* or designating groove connecting mouth and nose.

oropharynx *n.* the cavity of mouth and pharynx.

orotate *n.* a free pyrimidine, precursor of the pyrimidine nucleotide orotidylate.

orotidylate (orotidylic acid) *n.* a pyrimidine nucleotide which is decarboxylated to form uridine monophosphate (UMP).

orphan virus a virus not known to cause a disease.

orphon *n.* isolated gene that is related in sequence to members of a gene cluster.

ortet *n.* original single ancestor of a clone.

orthal *a.* straight up and down, *appl.* jaw movement.

orthaxial *a.* with a straight axis, or vertebral axis, *appl.* tail fin.

ortho- prefix derived from Gk. *orthos*, straight.

orthoblastic *a.* with a straight germ band.

orthochromatic *a.* of the same colour as the stain; staining positively.

orthocladous *a.* straight-branched.

orthodentine *n.* dentine pierced by numerous more-or-less parallel dentinal tubules.

orthodromic *a.* moving in the normal direction, *appl.* conduction of nerve impulse.

orthoenteric *a.* having alimentary canal along ventral body surface, *appl.* certain ascidians.

orthogenesis *n.* evolution along some apparently predetermined line independent of natural selection or other external forces.

orthognathous *a.* having straight jaws; having axis of head at right angles to body, as some insects.

orthograde *a.* walking with the body in a vertical position.

orthokinesis *n.* movement in which the organism changes its speed when it meets an unfavourable environment, the speed depepnding on the intensity of the stimulus, resulting in the dispersal or aggregation of organisms; variation in linear velocity with intensity of stimulation.

orthologous *a. appl.* genes in different species that are homologous because they are derived from a common ancestral gene (e.g. α-globin genes from humans and horses).

Orthomyxoviridae, orthomyxoviruses *n., n.plu.* family of large, enveloped RNA viruses with single-stranded segmented genomes, and including influenza.

Orthonectidea *n.* class of the invertebrate phylum Mesozoa (*q.v.*) that includes the orthonectids, parasites of the kidneys of flatworms, nemertines, polychaetes, bivalve molluscs and echinoderms. They have a free-swimming sexual generation.

orthophosphate (P$_i$) HPO$_4^{2-}$ (+ H$_2$PO$_4^-$).

orthoploid *a.* with even chromosome number; polyploid with complete and balanced genomes.

Orthoptera, orthopterans *n., n.plu.* order of insects that includes crickets, locusts and grasshoppers (in some classifications also cockroaches and mantises), with long antennae, biting mouthparts, long narrow tough forewings, and broad membranous hindwings, usually also having enlarged hindlegs for jumping, and stridulating organs, which produce the characteristic sounds of these insects.

orthoradial *a. appl.* cleavage where divisions are disposed symmetrically around egg axis.

orthoselection *n.* natural selection acting continuously in the same direction over a long period.

orthosomatic *a.* having a straight body, *appl.* certain larval insects.

orthospermous *a.* with straight seeds.

orthospiral *a. appl.* coiling of parallel sister chromatids interlocked at each twist. *alt.* plectonemic.

orthostichous *a.* arranged in a vertical row, *appl.* leaves.

orthostichy *n.* the vertical row of leaves formed by the leaves immediately above each other in a spiral arrangement.

orthotopic *a.* in the proper place, *appl.* transplantation. *cf.* heterotopic.

orthotopy *n.* natural placement; existence in natural habitat.

orthotropism *n.* growth in a straight line,

as of tap roots; condition of tending to be oriented in the line of action of a stimulus. *a.* orthotropic.

orthotropous *a. appl.* ovules having chalaza, hilum and micropyle in a straight line so that ovule is not inverted.

orthotype *n.* genetype originally designated.

oryzenin *n.* a protein found in rice.

os *n.*, **ora** *plu.* mouth.

os *n.*, **ossa** *plu.* bone.

oscheal *a.* scrotal; *pert.* scrotum.

oscitate *v.* to yawn, to gape.

oscula *plu.* of osculum *q.v.*

osculate *a.* to have characteristics intermediate between two groups.

osculum *n.* large pore in body wall of sponge through which water flows out from the body cavity. *plu.* oscula. *a.* oscular.

osmatic *a.* having a sense of smell.

osmeterium *n.* organ borne on the 1st thoracic segment of larva of some butterflies, emitting a smell.

osmiophilic *a.* staining readily with osmium stains.

osmoconformer *n.* organism which does not regulate the osmotic concentration of its internal fluids, which therefore vary with the osmotic concentration of the external environment, as some estuarine invertebrates.

osmolality *n.* the osmotic concentration of a solution, usually expressed in osmoles (1 osmole = 1 molal, *see* molality).

osmomorphosis *n.* change in shape or structure due to changes in osmotic pressure, such as due to changes in salinity.

osmophore *n.* the group of atoms responsible for the odour of a compound, and which combines with the chemoreceptors on olfactory neurones in nasal epithelium or other sensory organ.

osmoreceptor *n.* cell stimulated by changes in osmotic pressure, e.g. the cells reacting to osmotic changes in the blood.

osmoregulation *n.* in animals, regulation of the osmotic presure of body fluids by controlling the amount of water and/or salts in the body.

osmoregulator *n.* organism which actively regulates the osmotic concentration of its internal fluids.

osmosis *n.* diffusion of a solvent, usually

water, through a semipermeable membrane from a dilute to a concentrated solution, or from the pure solvent to a solution.

osmotaxis *n.* a taxis in reponse to changes in osmotic pressure.

osmotic *a. pert.* osmosis.

osmotic pressure (OP) a measure of the osmotic activity of a solution, defined as the minimum pressure that must be exerted to prevent the passage of pure solvent into the solution when the two are separated by a semipermeable membrane. Osmotic pressure is proportional to solute concentration.

osmotroph *n.* any heterotrophic organism (as fungi and bacteria) that absorbs organic substances in solution. *cf.* phagotroph.

osphradium *n.* chemical sense organ associated with visceral ganglia in many molluscs.

osphresis *n.* the sense of smell.

ossa *n.plu.* bones, *plu.* of os.

ossa sutura or triquetra sutural bones *q.v.*

osseous *a.* composed of or resembling bone.

osseous labyrinth the vestibule, semicircular canals and cochlea of inner ear embedded in the temporal bone.

ossicle *n.* any small bone or other calcified hard structure such as a plate of exoskeleton in echinoderms; in particular any of the three small bones of the middle ear.

ossicone *n.* the bony core of horn of ruminants (sheep, cows, etc.).

ossicular *a. pert.* ossicles.

ossiculate *a.* having ossicles.

ossification *n.* formation of bone, replacement of cartilage by bone.

ossify *v.* to change to bone.

Ostariophysi *n.* a superorder of teleost fishes that includes the carps, minnows, catfishes and loaches, characterized by the presence of Weberian ossicles.

Osteichthyes *n.* the bony fishes, a class comprising all fish except for the Agnatha and the Selachii, and which have a bony skeleton, usually an air bladder (swimbladder) or lung, and a cover (operculum) over gill openings.

osteoblast *n.* bone-forming cell that secretes the bone matrix. *alt.* bone cell.

osteochondral *a. pert.* bone and cartilage.

osteochondrous *a.* consisting of both bone and cartilage.

osteoclasis *n.* destruction of bone by osteoclasts.

osteoclast *n.* large multinucleate cell, derived from macrophage, that destroys bone or any matrix, whether calcified or cartilaginous, during bone formation and remodelling. *alt.* giant cell.

osteocranium *n.* the bony cranium as distinguished from cartilaginous or chondrocranium.

osteocyte *n.* non-secreting, non-dividing cell derived from an osteoblast, found in calcified bone. *alt.* bone cell.

osteodentine *n.* a variety of dentine which closely approaches bone in structure.

osteodermis *n.* a dermis that is more-or-less ossified; a bony dermal plate.

osteogen *n.* tissue which forms and alters bone.

osteogenesis *n.* formation and growth of bones. *a.* osteogenic, osteogenetic, *pert.* or causing formation of bone.

osteoid *n.* collagenous material secreted by bone cells (osteoblasts) and which forms the basis of bone, the uncalcified bone matrix; *a.* bone-like.

osteolepid *a.* having a skin armoured with bony scales, as the Crossopterygii.

osteology *n.* the study of the structure, nature and development of bones.

osteolysis *n.* breakdown and dissolution of bone.

osteomere *n.* a segment of vertebrate skeleton.

osteon bone type of bone composed of overlapping cylinders (osteons) each comprised of concentric layers (lamellae), the collagen fibres of each layer running at right angles to the next, surrounding a central canal (Haversian canal).

osteonectin *n.* protein specific to bone and which is involved in the growth of hydroxyapatite crystals during calcification.

osteoplastic *a.* producing bone; *appl.* cells: osteoblasts.

osteosarcoma *n.* tumour of bone.

osteoscute *n.* a bony external shield or plate, as in armadillos.

ostia *plu.* of ostium *q.v.*

ostiate *a.* having ostia.

ostiolar *a. pert.* an ostiole.

ostiolate *a.* having ostioles.

ostiole *n.* a small aperture or opening, as in

the wall of sponges. *alt.* ostium.

ostium *n.* any mouth-like opening; opening in arthropod heart through which blood enters the pericardial cavity; opening from exterior into body cavity of sponges through lateral wall, through which water is drawn in. *plu.* ostia. *a.* ostial.

Ostracoda, ostracods *n., n.plu.* group of small aquatic crustaceans having a bivalved carapace enclosing head and body, and reduced trunk and abdominal limbs.

ostracoderms *n.plu.* extinct Palaeozoic jawless fishes (Agnatha) which were armoured with an exoskeleton of dermal bone.

otic *a. pert.* ear and region of ear.

otoconia, otoconites *n.plu.* minute grains of calcium carbonate found in membranous labyrinth of inner ear.

otocyst *statocyst q.v.*

otolith *n.* calcareous particle found in fluid of semicircular canals, utricle and saccule of inner ear and whose movement under gravity in response to changes in position of the head stimulates sensory cells, giving the animal its position with respect to gravity and allowing it to balance. *alt.* statolith.

otolith organ structure in ear of fish comprising an otolith and associated hair cells (sensory cells).

oto-occipital *n.* bone formed by fusion of opisthotic with exoccipital.

otostapes *n.* portion of columella primordium which in adult may give rise to stapes and part of columella.

ouabain *n.* G-strophanthidin, a plant glycoside which is a specific inhibitor of the Na^+-K^+ ATPase in eukaryotic cell membranes, and whose digitalis-like effects on the heart are mediated by inhibition of Na^+-K^+ ATPase.

Ouchterlony double diffusion technique for measuring the amount of antigen in a sample and/or the degree of identity of two antigens, based on diffusion of antigen and antibody from wells in an agar slab, a line of precipitation being formed in the gel where they meet if the antigen reacts with the antibody.

outbreeding *n.* the mating of individuals who are not closely related. *a.* outbred. *cf.*

inbreeding, cross-fertilization.

outcrossing outbreeding *q.v.*

outer phalangeal cells Deiter's cells *q.v.*

ova *plu.* of ovum.

ovalbumin *n.* a glycoprotein, chief protein constituent of egg white.

oval window fenestra ovalis *q.v.*

ovarian *a. pert.* an ovary.

ovarian follicle *see* Graafian follicle.

ovariole *n.* egg tube of insect ovary.

ovariotestis *n.* reproductive organ when both male and female elements are formed, as in case of sex reversal.

ovarium ovary *q.v.*

ovary *n.* in plants and animals the reproductive organ in which female gametes or egg cells are produced. In flowering plants it comprises the enlarged portion of the carpel(s) containing the ovules and after fertilization develops into the fruit containing the seeds.

ovate *a.* egg-shaped and attached by the broader end, *appl.* leaves.

ovate-acuminate *a.* having an ovate blade with a very sharp point.

ovate-ellipsoidal *a.* ovate, approaching ellipsoidal, *appl.* leaves.

ovate-lanceolate *a.* having leaf-blade intermediate betwen ovate and lanceolate.

ovate-oblong *a.* having leaf-blade oblong with one end narrower.

overdominance *n.* the condition where a heterozygote has a more extreme phenotype than either of the corresponding homozygotes.

overflow *a. appl.* behaviour in which an inappropriate response occurs to a certain stimulus in order to satisfy certain drives, such as a dog displaying maternal care to a bone.

overlapping genes an uncommon genetic arrangement where part of the protein-coding DNA sequence of one gene forms part or all of the protein-coding sequence of another, usually not in the same reading frame.

overshoot *n.* situation in which the population size of a species temporarily exceeds the carrying capacity of its habitat, leading to a sharp reduction in the population.

ovicapsule *n.* egg case or ootheca.

ovicell *n.* a specialized chamber in which embryos develop in certain bryozoans.

oviducal *a. pert.* oviduct.

oviduct *n.* a tube that carries eggs from ovary to exterior. *a.* oviducal.

oviferous, ovigerous *a.* serving to carry eggs.

oviform *a.* oval.

oviger *n.* egg-carrying leg of certain arachnids.

oviparous *a.* egg-laying. *n.* oviparity.

oviposition *n.* the deposition of eggs on a surface, esp. in insects and fish.

ovipositor *n.* a specialized structure in insects for depositing eggs; a tubular extension of genital orifice in fishes. ˙

ovisac *n.* an egg case or receptacle; a brood pouch *q.v.*

ovocyst, ovocyte, ovogenesis oocyst, oocyte, oogenesis *q.v.*

ovoid *a.* egg-shaped.

ovomucoid *n.* glycoprotein in white of egg.

ovo-testis *n.* the hermaphrodite reproductive organ of certain gastropod molluscs, which produces both eggs and sperm; organ composed of ovarian and testicular tissue found in some cases of human pseudohermaphroditism.

ovovitellin *see* vitellin.

ovovitelline duct duct lined with yolk glands that produce yolk cells.

ovoviviparous *a. pert.* organisms that produce an egg with a persistent membrane, but which hatches within the maternal body. *n.* ovoviviparity.

ovular *a.* like or *pert.* an ovule.

ovulary *n.* the ovary in plants.

ovulate *a.* producing ovules, *appl.* cones.

ovulation *n.* the shedding of an egg or eggs from the ovary. *v.* to ovulate.

ovulatory *a. pert.* ovulation.

ovule *n.* in seed plants, the structure consisting of the megagametophyte and megaspore, surrounded by the nucellus and enclosed in an integument, and which develops into a seed after fertilization.

ovuliferous *a.* bearing or containing ovules; *appl.* the scales bearing one or more ovules that develop on the bract scales of cones.

ovum *n.* a female gamete *q.v. alt.* egg, egg cell. *plu.* ova. *see also* oocyte.

oxalic acid (oxalate) simple organic carboxylic acid $(COOH)_2$, excreted as calcium oxalate by many fungi, forming crystals, oxalates also occurring as by-products in various plant tissues and in urine and also found in the mantles of certain bivalves.

oxaloacetate (oxaloacetic acid) 4-carbon carboxylic acid, component of the tricarboxylic acid cycle (*q.v.*), accepting acetyl groups from acetyl CoA to form citrate. *see* Appendix 1 (54) for chemical structure.

oxalosuccinate (oxalosuccinic acid) 6-carbon carboxylic acid, component of the tricarboxylic acid cycle (*q.v.*), decarboxylated to α-ketoglutarate.

oxea *n.* a sponge spicule, rod-shaped and pointed at both ends.

oxidase *n.* any enzyme which catalyses oxidation/reduction reactions using molecular oxygen as acceptor. *cf.* reductase, dhydrogenase.

oxidation *n.* the addition of oxygen, loss of hydrogen or loss of electrons from a compound, atom or ion. *cf.* reduction.

β-oxidation pathway metabolic pathway by which fatty acids are degraded in the mitochondria to yield acetyl CoA, 2 carbon units being removed at each round of reactions.

oxidation reduction enzymes oxidoreductases *q.v.*

oxidation reduction potential redox potential *q.v.*

oxidative phosphorylation process in aerobic organisms in which ATP is formed from ADP and orthophosphate, the process being driven by electron flow along an electron transport chain with O_2 as the final acceptor, the electrons being derived originally from the oxidation of fuel molecules during aerobic respiration. *cf.* photophosphorylation. *see also* chemiosmotic hypothesis.

oxidoreductase *n.* any of a class of enzymes catalysing the oxidation of one compound with the reduction of another, and including the dehydrogenases, catalases, oxidases, peroxidases and reductases.

oxoglutarate *n.* present chemically correct name for ketoglutarate, which is however, still widely used.

oxy- when prefixing the name of an oxygen-carrying blood pigment such as haemoglobin denotes the oxygenated form.

oxyaster *n.* stellate sponge spicule with

pointed rays.

oxybiotic *a.* living in the presence of oxygen.

oxychlorocruorin *n.* chlorocruorin combined with oxygen, as in the aerated blood of certain polychaete worms.

oxydactylous *a.* having slender tapering digits.

oxyerythrocruorin *n.* erythroruorin combined with oxygen, as in the aerated blood of many annelids and molluscs.

oxygen debt a deficit in stored chemical energy which builds up when a normally aerobic tissue such as muscle is working with an inadequate oxygen supply. It then consumes oxygen above the normal rate for some time until energy supplies are restored by respiration.

oxygen dissociation curve a graph of percentage saturation of haemoglobin with oxygen against concentration of oxygen, which gives information about the dissociation of oxyhaemoglobin under different environmental conditions or in different animals; amy graph showing the dissociation of oxygen from a substance.

oxygen quotient (Q_{O_2}) the volume of oxygen in microlitres gas at normal temperature and pressure taken in per hour per milligram dry weight.

oxygen sag curve curve that is obtained when the dissolved oxygen in a watercourse receiving a source of organic pollution is plotted against distance downstream of the discharge, which shows a sharp decrease in dissolved oxygen immediately downstream of the discharge followed by a gradual increase as one goes downstream.

oxygenic photosynthesis type of photosynthesis in which oxygen is produced, and which is the type of photosynthesis carried out by green plants, algae and cyanobacteria.

oxygenotaxis oxytaxis *q.v. a.* oxygenotactic.

oxygenotropism oxytropism *q.v.*

oxyhaemocyanin *n.* haemocyanin combined with oxygen, as in the aerated blood of many molluscs and arthropods.

oxyhaemerythrin *n.* haemerythrin combined with oxygen as in the blood of some annelids.

oxyhaemoglobin (HbO$_2$) *n.* haemoglobin combined with oxygen, formed when the concentration of oxygen is high, as in lungs, and dissociating into its component parts and releasing oxygen in tissues, where oxygen concentration is low.

oxygnathous *a.* with more or less sharp jaws.

oxymyoglobin *n.* myoglobin combined with oxygen, formed when oxygen concentration is high and releasing oxygen when oxygen concentration is low.

oxyntic *a.* acid-secreting, *appl.* cells in the gastric gland of the stomach which secrete hydrochloric acid (HCl).

oxyphil(ic) *a.* having a strong affinity for acidic stains. *alt.* acidophil, acidophilic.

oxyphilous *a.* tolerating only acid soils and substrates.

Oxyphotobacteria *n.* class of photosynthetic bacteria comprising the cyanobacteria, i.e. those bacteria that produce oxygen as a by-product of photosynthesis.

oxyphobe *a.* unable to tolerate acid soils.

oxyphyte *a.* a plant thriving on acid soils. *alt.* calcifuge.

oxytaxis *n.* a taxis in response to the stimulus of oxygen. *a.* oxytactic.

oxytocic *a.* accelerating parturition.

oxytocin *n.* peptide hormone secreted by the neurohypophysis (the posterior lobe of the pituitary), and which in mammals induces contraction of smooth muscle, especially uterine muscle.

oxytropism *n.* tendency of organisms or organs to be attracted by oxygen.

ozone *n.* O$_3$, gas formed from oxygen (O$_2$) under the action of short wavelength ultraviolet radiation in the stratosphere, where it forms the ozone layer, which absorbs considerable solar ultraviolet radiation and shields the Earth's surface from its harmful effects. Ozone is also formed as a pollutant in the lower atmosphere from e.g. nitrogen oxides. It is damaging to herbaceous plants at levels greater than 100 parts per billion.

P

P proline *q.v.*

P$_f$, P$_{fr}$ the plant pigment phytochrome in its activated form, produced by the action of red light on P$_r$.

P$_i$ inorganic phosphate or orthophosphate, *q.v.*

P$_r$ the inactive form of phytochrome *q.v.*

P$_1$ parents, P$_2$, grandparents, etc. in genetic crosses.

Δp proton motive force *see* chemiosmotic theory.

P430 bound ferredoxin *q.v.*, a component of photosystem I *q.v.*

P680 the reaction centre of photosystem II *q.v.*

P700 reaction centre of photosystem I *q.v.*

P730 the plant pigment phytochrome in its activated form.

PABA *p*-aminobenzoic acid *q.v.*

PAGE polyacrylamide gel electrophoresis *q.v.*

PAH polycyclic aromatic hydrocarbons *q.v.*

PAR photosynthetically active radiation *q.v.*

PAS periodic acid-Schiff reagent, a dye used for staining proteins rich in carbohydrate side chains.

PBL peripheral blood lymphocytes *q.v.*

PCB polychlorinated biphenyl *q.v.*

PD$_{50}$ a measure of activity for certain viruses, the dose at which 50% of test animals show paralysis.

PDGF platelet-derived growth factor *q.v.*

PDI protein disulphide isomerase *q.v.*

PEP phosphoenolpyruvate *q.v.*

PFU plaque-forming unit *q.v.*

PG prostaglandin *q.v.*

PGA phosphoglycerate *q.v.*; pteroylglutamic acid. *see* folic acid; prostaglandin A.

PGB, PGE, PGF classes of prostaglandins *q.v.*

pH a measure of the acidity of a solution, the negative log$_{10}$ of the hydrogen ion concen-

tration. The pH of a neutral solution is 7, that of acid solutions less than 7 and of alkaline solutions greater than 7.

PHA phytohaemagglutinin *q.v.*

Phe phenylalanine *q.v.*

PI phosphoinositide. *see* phosphoinositide pathway.

PIP phosphatidylinositol phosphate *q.v.*

PLP pyridoxal phosphate *q.v.*

PMN polymorphonuclear leukocyte, *q.v.*

PMS pregnant mare's serum.

POMC pro-opiomelanocortin *q.v.*

PP primary production *q.v.*

p.p.b. parts per billion (10^9).

ppGpp guanosine tetraphosphate *q.v.*

PP$_i$ pyrophosphate *q.v.*

PPI peptidyl prolyl *cis-trans* isomerase *q.v.*

PPLO pleuropneumonia-like organisms *q.v.*

p.p.m. parts per million.

PPP oxidative pentose phosphate pathway, *see* pentose phosphate pathway.

pppGpp guanosine pentaphosphate *q.v.*

Pro proline *q.v.*

PRPP 5-phosphoribosyl-1-pyrophosphate, an activated donor of a ribose group.

PSI photosystem I *q.v.*

PSII photosystem II *q.v.*

PSTV potato spindle tuber viroid.

PT pertussis toxin *q.v.*

PTA phosphotungstic acid, a staining agent for electron microscopy; plasma thromboplastin antecedent *q.v.*

PtdInsP phosphatidylinositol phosphate *q.v.*

PTH parathyroid hormone *q.v.*

PWM pokeweed mitogen *q.v.*

Pacchionian bodies eminences of subarachnoid tissue covered by arachnoid membrane and pressing into dura mater.

pacemaker *n.* cell or region of organ determining rate of activity in other cells or organs; in heart, the sino-auricular or sinuatrial node, which initiates and maintains the normal heartbeat; *appl*. neurones that set up and maintain rhythmic activity without further external stimuli.

pachy- prefix derived from Gk. *pachys,* thick.

pachycarpous *a*. with a thick pericarp, *appl*. fruit.

pachycaulous *a*. with a thick or massive primary stem and root.

pachycladous *a*. with thick shoots.

pachydermatous *a*. with thick skin or covering.

pachyderms *n.plu*. any of various non-ruminant hoofed mammals, e.g. an elephant, or rhinoceros, which are generally large, and have thick tough skins.

pachynema *n*. a thick thread, formerly used to describe chromosomes at the pachytene stage of meiosis.

pachynosis *n*. growth in thickness, as of plants.

pachyphyllous *a*. thick-leaved.

pachytene *n. appl*. stage in prophase of meiosis in which homologous chromosomes are associated as bivalents.

Pacific North American Floral Region part of the Holarctic Realm comprising North America west of the Rocky Mountains from southern Alaska south to the Mexican border.

Pacific South American Floral Region part of the Neotropical Realm comprising the Andes and the coastal strip to their west from Ecuador south to central Chile.

Pacinian bodies or corpuscles sensory receptors in joints and skin consisting of a connective tissue capsule with core of cells innervated by sensory nerve endings, and sensitive to pressure. *see* Fig. 10.

packing ratio of DNA, the ratio of the length of the DNA molecule to the unit length of the fibre containing it.

paedogamy pedogamy *q.v.*

paedogenesis pedogenesis *q.v.*

paedomorphosis pedomorphosis *q.v.*

Paeoniales *n*. order of herbaceous dicots, some shrubs, with large deeply cut leaves and large showy flowers. Comprises the family Paeoniaceae (peony).

paired bodies small bodies lying close to sympathetic nerve chain in elasmobranch fishes, representing the adrenal medulla.

paired fins pelvic and pectoral fins of fishes.

pairing *n*. the synapsis of homologous chromosomes during the zygotene stage of meiosis, when they come to lie side by side.

pair-rule genes developmental genes in the fruitfly *Drosophila* involved in delimiting segments in early embryo, so-called because a given pair-rule gene is expressed in either odd-numbered or even-numbered segments only.

Palaearctic *a. appl*. or *pert*. zoogeographical region, or subregion of the Holarctic Realm, including Europe, North Africa, western Asia, Siberia, northern China and Japan.

palaeo- prefix derived from Gk. *palaios* ancient. *alt*. paleo-.

palaeobiology *n*. the study of the biology of extinct plants, animals and microorganisms.

palaeobotany *n*. study of fossil plants and plant impressions.

Palaeocene Paleocene *q.v.*

palaeocerebellum *n*. the phylogenetically older region of cerebellum, receiving spinal and vestibular afferent nerve fibres. *cf.* neocerebellum.

palaeocranium *n*. type of skull or stage in development extending back no farther than the vagus nerve.

palaeodendrology *n*. the study of fossil trees and tree impressions.

palaeoecology *n*. the study of the relationship between past organisms and the environment in which they lived.

palaeoencephalon *n*. the primitive vertebrate brain.

Palaeogaea *n*. the area comprising the Palaearctic, Ethiopian, Indian, and Australian zoogeographical regions.

palaeogenetic *a. appl*. atavistic features which are fully developed in adult although normally only characteristic of embryonic stage.

palaeogenetics *n*. the application of the principles of genetics to interpretation of the fossil record and the evolution of now extinct species.

Palaeognathae *n*. the ratites, flightless birds of the subclass Neornithes such as the kiwis, cassowaries and ostrich, which are secondarily flightless.

Palaeolaurentian *a. pert.* or *appl.* Archaeozoic era.

Palaeolithic *a. appl.* or *pert.* the Old Stone Age, characterized by a hunter-gatherer economy and chipped stone tools.

Palaeoniscoidei, Palaeonisciformes, palaeoniscids *n., n.plu., n.plu.* group of actinopterygian fishes, most now extinct, existing from the Devonian to the present day and including the bichars of the Nile. They are carnivorous with large sharp teeth and usually a very heterocercal tail.

palaeontology *n.* the study of past life on Earth from fossils and fossil impressions.

palaeosere *n.* the development of vegetation throughout the Palaeozoic.

palaeospecies *n.* a group of extinct organisms that are assumed to have been capable of interbreeding and so are placed in the same species.

Palaeotropical Realm floristic area comprising Africa except for the northern part around the Mediterranean Sea and the tip of southern Africa (but including Madagascar), Arabia and Asia, including the Indian subcontinent and the islands of the Indian Ocean south from the Himalayas to New Guinea. It is made up of 14 floral regions: the East African, the Ethiopian, the Fijian, the Hawaian, the Indian, the Macaronesian, the Madagascan, the Malaysian-Papuan, the New Caledonian, the Polynesian, the Saharo-Arabican, the South-East Asian, the Sudanian-Sindian, and the West African.

Palaeozoic Paleozoic *q.v.*

palaeozoology *n.* the study of the biology of extinct animals from fossils and fossil impressions.

palama *n.* the foot-webbing of aquatic birds.

palatal, palatine *a. pert.* palate.

palate *n.* roof of mouth in vertebrates; roof of pharynx in insects; (*bot.*) projection of lower lip of personate corolla.

palatine *n.* a bone or cartilage of the hard palate in vertebrates.

palatoglossal *a. pert.* palate and tongue; *appl.* a muscle: glossopalatinus *q.v.*

palatonasal *a. appl.* palate and nose.

palatopharyngeal *a. pert.* palate and pharynx.

palatopterygoid *a. pert.* palate and pterygoid.

palatoquadrate *a.* connecting palatine and

quadrate bones or cartilages, *appl.* dorsal cartilage of mandibular arch.

palatoquadrate cartilage paired skeletal element forming upper jaw in elasmobranch fishes.

pale *n.* palea *q.v.*; upper or inner pale: palea *q.v.*; lower or outer pale: lemma *q.v.*

palea *n.* the upper of the two bracts enclosing a floret (individual flower) in grasses. *plu.* paleae.

paleaceous *a.* chaffy, *appl.* a capitulum furnished with small scaly bracts or paleae.

paleo- palaeo-.

Paleocene *n.* the earliest epoch of the Tertiary period, before the Eocene and lasting from around 65 to 55 million years ago.

paleola lodicule *q.v.*

Paleozoic *a. pert.* or *appl.* geological era lasting from about 590 to 250 million years and comprising the Cambrian, Ordovician, Silurian, Devonian, Carboniferous and Permian periods.

palet palea (*q.v.*) of grasses.

palette *n.* the modified cupule-bearing tarsus of anterior leg in male beetles.

pali *n.plu.* series of small pillars projecting upwards from the base of theca towards stomatodaeum of madrepore corals. *sing.* palus.

paliform *a.* like an upright stake.

palinal *a.* from behind forwards, *appl.* jaw movements as in elephants.

palindrome *n.* in DNA, a base sequence that reads exactly the same from left to right as from right to left.

palingenesis *n.* abrupt metamorphosis; rebirth of ancestral characters.

palingenetic *a.* of remote or ancient origin; *pert.* palingenesis.

palisade tissue the layer or layers of photosynthetic cells beneath the epidermis of many foliage leaves.

pallaesthesia *n.* vibratory sensation, as felt, e.g., in bones.

pallet *n.* a shelly plate on a bivalve siphon.

palliate *n.* having a mantle or similar structure.

pallidium *n.* the central portion of the lentiform nucleus in the midbrain.

palliopedal *a. pert.* molluscan mantle and foot.

pallioperitoneal *a. appl.* complex of organs in some molluscs, including heart, renal organs, gonads and ctenidia.

pallium *n*. the mantle (*q.v.*) of molluscs and brachiopods; the cerebral cortex *q.v.*

Palmaceae, palms *n.*, *n.plu.* tropical and subtropical family of monocots typically with large leathery fan-like leaves. They include climbing and tree species. The fruit is a berry or drupe. Cultivated species include the date palm (*Phoenix dactylifera*), oil palms (*Elaeis* spp.) and the coconut palm (*Cocos nucifera*).

palmaesthesia *n*. sensing sounds by the vibrations they make in the bones.

palmar *a. pert.* palm of hand.

palmate *a.* (*bot.*) *appl.* leaves divided into lobes arising from a common centre; *appl.* tuber shaped like a hand, as in some orchids; (*zool.*) having anterior toes webbed, as in most aquatic birds.

palmatifid *a. appl.* leaves divided into lobes to about the middle, at acute angles to each other.

palmatilobate *a.* palmate with rounded lobes and divisions halfway to base, of leaves.

palmatipartite *a.* palmate with divisions more than halfway to base, of leaves.

palmatisect *a.* palmate with divisions nearly to base, of leaves.

palmella *n.* a sedentary stage of certain algae, the cells dividing within a jelly-like mass and producing motile gametes. *a.* palmelloid.

palmigrade *a.* walking on the sole of the foot.

palmiped *a.* web-footed; *n.* a web-footed bird.

palmitate (palmitic acid) common widely distributed saturated fatty acid, $CH_3(CH_2)_{14}COO^-$.

palmitin *n.* a fat present in adipose tissue, milk and palm oil.

palmitylation *n.* the covalent addition of the fatty acid palmitic acid to a protein, by which some proteins are attached to cell membranes.

palmoid *a. appl.* palms and palm-like forms.

palmula *n.* terminal lobe or process between paired claws of insect feet.

palp *n.* labial feeler of an insect; other similarly situated appendages in other invertebrates. *plu.* palps. *a.* palpal.

palpacle *n.* the tentacle of a dactylozooid

individual of a siphonophore.

palpal *a. pert.* a palp.

palpate *a.* having palps; *v.* to examine by touch.

palpebral *a. pert.* eyelids.

palpi *plu.* of palpus (palp).

palpifer, palpiger *n.* in insects, a lobe of maxilla or other mouthpart which bears a palp.

palpiform *n.* resembling a palp.

Palpigrada *n.* order of very small arachnids having a jointed flagellum on the last segment of the opisthosoma.

palpimacula *n.* sensory area on labial palps of certain insects.

palpocil *n.* stiff sensory filament of tactile sensory cells of some coelenterates.

palpon *n.* in colonial hydrozoans, an individual hydroid modified for catching prey and for defence, being long and slender, usually with tentacles and without a mouth.

palpule, palpulus *n.* a small palp or feeler.

palpus palp *q.v.*

paludal *a.* marshy; *pert.*, or growing in, marshes or swamps.

paludicole *a.* living in marshes.

palus *n.* a small stake-like structure. *plu.* pali.

palustral, palustrine *a.* growing in marshes or swamps.

palynology *n.* the study of pollen and its distribution, *alt.* pollen analysis; the study of spores.

***p*-aminobenzoic acid** part of folic acid, necessary as a vitamin in rats, but does not seem to be required in humans.

pampiniform *a.* like a tendril; *appl.* convoluted vein plexus of the spermatic cord which acts as a countercurrent heat exchanger to cool the testes; *appl.* body: a small collection of tubules anterior to ovary, the remnant in adult of embryonic mesonephros.

pamprodactylous *a.* with all the toes pointing forward.

pancolpate *a.* of pollen grains, having many furrows.

pancreas *n.* a compound glandular organ associated with the gut in most vertebrates, and secreting the hormones insulin and glucagon from endocrine glands and digestive enzymes from exocrine glands.

pancreastatin *n.* peptide from porcine pancreas which inhibits glucose-induced release of insulin from pancreas.

pancreatic *a. pert.* pancreas.

pancreatic juice a secretion containing digestive enzymes and enzyme precursors secreted into the gut from the pancreas, including trypsinogen, chymotrypsinogen, procarboxypeptidases, lipase, α-amylase, maltase and ribonuclease.

pancreatic ribonuclease an endoribonuclease (originally isolated from mammalian pancreas) that specifically cleaves RNA on the 3′ side of C or U.

pancreatic trypsin inhibitor protein produced in the pancreas which binds specifically to the enzyme trypsin, inhibiting it. *alt.* antitrypsin.

pancreaticoduodenal *a. pert.* pancreas and duodenum, *appl.* arteries, veins.

pancreozymin cholecystokinin *q.v.*

Pandanales *n.* order of monocots, mainly sea coast or marsh plants with tall stems supported by aerial roots, and leaves running in spirals, comprising the family Pandanaceae (screw pine).

pandemic *a.* very widely distributed; *n.* epidemic disease with a worldwide distribution.

panduriform *a.* fiddle-shaped, *appl.* leaves.

Paneth cells secretory cells at base of crypts of Lieberkühn in small intestine.

Pangaea *n.* the supercontinent made up of all the present continents fitted together before their separation by continental drift.

pangamic *a. appl.* indiscriminate or random mating. *n.* pangamy.

pangenesis *n.* a now discarded theory that hereditary characteristics were carried and transmitted by gemmules from individual body cells.

panicle *n.* a branched flower-head, strictly one in which the branches are alternate and side-branches also branch alternately.

panicoid *a. appl.* millets of the genus *Panicum.*

paniculate *a.* having flowers arranged in panicles.

panmictic *a.* characterized by, or resulting from, random matings.

panmixia *n.* indiscriminate interbreeding. *alt.* panmixis.

panniculus *n.* a layer of tissue.

panniculus carnosus a thin layer of muscle fibres in dermis and which is involved in moving or twitching the skin.

pannose *a.* resembling cloth in texture.

panoistic *a. appl.* ovariole in which nutritive cells are absent, egg yolk being formed by epithelium of follicle.

panphotometric *a. appl.* leaves oriented to avoid maximum direct sunlight.

panphytotic *n.* a pandemic affecting plants.

panspermia *n.* a theory popular in the 19th century and which enjoys periodic revivals, that life did not originate on Earth but arrived in the form of bacterial spores or viruses from an extraterrestrial source.

pansporoblast *n.* a cell complex of certain sporozoan protozoans, the Neosporidia, producing sporoblasts and spores.

panthalassic *a.* living in both coastal and oceanic waters.

panting centre region of hypothalamus whose stimulation causes the rate of respiration to quicken.

pantodonts *n.plu.* group of extinct North American and Asian herbivorous placental mammals from the Paleocene to Oligocene.

pantonematic *a. appl.* flagella with longitudinal rows of fine hairs along their axis.

pantostomatic *a.* capable of ingesting food at any part of the surface, as amoebae and similar organisms.

pantothenate (pantothenic acid) organic acid, vitamin B_3 or B_5, a precursor of the important enzyme cofactor coenzyme A, and is necessary for growth in various animals, and is the rat anti-grey hair and chick antidermatitis factor.

pantotheres *n.plu.* an order of trituberculate mammals of the Jurassic, possible ancestors of living therians, having molar teeth showing the basic pattern found in living forms.

pantropic *a.* invading many different tissues, *appl.* viruses; turning in any direction.

pantropical *a.* distributed throughout the tropics, *appl.* species.

papain *n.* an endopeptidase found in the fruit juice and leaves of pawpaw (papaya), *Carica papaya*, and used commercially as a meat tenderizer. EC 3.4.22.2.

Papaverales *n.* order of mainly herbaceous dicots, some shrubs and small trees, and including the families Fumariaceae (fumitory), Hypecoaceae and Papaveraceae (poppy).

paper chromatography chromatography (*q.v.*) in which paper is used as the support medium on which the substances to be separated migrate differentially in the solvent.

papilionaceous *a*. like a butterfly, *appl.* flowers like sweet peas, gorse, broom, etc. with a corolla of five petals, one enlarged upright standard or vexillum, two anterior united to form a keel or carina, and two lateral, the wings or alae.

papilla *n*. a small projection or protuberance, as clusters of taste buds on tongue; a conical structure, as nipple. *plu.* papillae.

papillary *a. pert.* or with a papilla; *appl.* a process of caudal lobe of liver; *appl.* muscle between walls of ventricles of heart and chordae tendineae; *appl.* a layer of the dermis.

papillate *a*. covered with papillae; like a papilla.

papilliform *a*. like a papilla in shape.

papilloma *n*. a wart or similar small, benign tumour.

papillomaviruses *see* Papovaviridae.

Papovaviridae, papovaviruses *n., n.plu.* family of small, non-enveloped, doublestranded DNA viruses that includes polyoma and SV40 (simian virus 40), and the papillomaviruses that cause common warts. Polyoma and SV40 are oncogenic in the cells of certain species, and some papillomaviruses may be involved in human cervical cancer.

pappiferous *a*. bearing a pappus.

pappose, pappous *a*. having limb of calyx developed as a tuft of hairs or bristles (pappus); downy, or covered with feathery hairs.

pappus *n*. a circle or tuft of bristles, hairs or feathery processes in place of a calyx on florets of flowers of the Compositae, which persists on the seed aiding dispersal by wind.

Papuan *a. appl.* subregion of Australian zoogeographical region: New Guinea and islands westwards to Wallace's line.

papulae *n.plu.* hollow contractile pustules on epidermis of some echinoderms, such as starfish, having a respiratory function.

papyraceous, papyritic *a*. like paper in texture.

para- prefix derived from Gk. *para*, beside, signifying situated near, or surrounding.

para-aminobenzoic acid *see p-*aminobenzoic acid.

para-aortic *a. appl.* chromaffin bodies situated alongside the abdominal aorta.

parabasal body kinetosome *q.v.*

parabiosis *n*. the condition of being conjoined either from birth, as Siamese twins, or experimentally as laboratory animals; use of the same nest by different species of ant, which, however, keep their broods separate.

parabiotic *a. appl.* ants of different species living together amicably in a compound nest; conjoined to a greater or lesser extent, as Siamese twins.

parablastic *a. appl.* large nuclei of eggs laden with yolk granules.

parabranchia *n*. a feathery osphradium (chemoreceptor organ) in molluscs, socalled from its superficial resemblance to gills.

parabranchial cavity space between gill opening and gill.

parabronchus *n*. one of numerous small tubules in lungs of birds, connecting dorsobronchi and ventrobronchi and in which respiratory gas exchange takes place.

Paracanthopterygii *n*. an advanced group of teleost fishes, existing from the Eocene to the present day and including the cod family.

paracardial *a*. surrounding the neck of the stomach.

paracentral *a*. situated near the centre, *appl.* lobule, gyrus, fissure; *appl.* retinal area surounding fovea centralis.

paracentric *a*. on same side of centromere of a chromosome; *appl.* chromosomal rearrangements involving only one chromosome arm; *appl.* inversions which do not include the centromere.

parachordal *a*. on either side of the notochord.

parachrosis *n*. process or condition of changing colour; discoloration; fading.

parachute *n*. special structure of seeds such as aril, caruncle, pappus, or wing, which aids dispersal by wind; a fold of skin used for gliding as in flying squirrels.

paracme *n.* the evolutionary decline of a taxon after reaching the highest point of development; the declining or senescent period in the life history of an individual.

paracoel *n.* lateral ventricle or cavity of cerebral hemisphere.

paracondyloid *a. appl.* process of occipital bones occurring beside condyles in some mammals.

paracone *n.* the front outside cusp of upper molar tooth.

paraconid *n.* the front inside cusp of lower molar tooth.

paracorolla *n.* a corolla appendage such as a corona.

paracrine *a. appl.* actions of substances produced by cells and acting at short-range on neighbouring cells.

paracymbium *n.* accessory part of the cymbium, between tarsus and tibia in some spiders.

paracyte *n.* modified cell extruded from embryonic tissue into yolk as in some insects.

paracytic *a. appl.* stomata of a type in which subsidiary cells lie alongside the stoma parallel to the long axis of guard cells, formerly called rubiaceous.

paracytoids *n.plu.* minute pieces of chromatin extruded from nuclei of embryonic cells and shed into the blood, as in certain insects.

parademe *n.* an apodeme arising from the edge of a sclerite.

paraderm *n.* the delicate limiting membrane of a pronymph.

paradermal *a. appl.* section cut parallel to the surface of a flat organ such as a leaf. *cf.* tangential.

paradesmus *n.* secondary connection between centrioles outside nucleus in mitosis of some flagellate protozoans.

paradidymis *n.* a body of convoluted tubules anterior to the lower part of spermatic cord, representing posterior part of embryonic mesonephros.

paraesophageal paraoesophageal *q.v.*

parafacialia *n.plu.* narrow parts of head capsule between frontal suture and eyes, as in certain Diptera.

parafibula *n.* an accessory element outside fibula at proximal end, seen in some lizards and young marsupials.

paraflagellum *a.* a subsidiary flagellum.

paraflocculus *n.* cerebellar lobule lateral to flocculus.

parafrons *n.* area between eyes and frontal suture in certain insects.

parafrontals *n.plu.* the continuation of genae between eyes and frontal suture in insects.

paraganglia *n.plu.* scattered clusters of cells secreting adrenaline alongside aorta.

paragaster *n.* a central cavity lined with choanocytes, into which ostia open, in sponges.

paragastric *a. appl.* passages or cavities in branches of sponges; *appl.* paired blind canals from infundibulum to oral cone of ctenophores.

paragenesis *n.* condition in which an interspecific hybrid is fertile with the parental species but not with other similar hybrids; a subsidiary mode of reproduction.

paragenetic *a. appl.* mutation affecting the expression rather than the structure of a gene.

paraglenal hypercoracoid *q.v.*

paraglobulin *n.* one of the globulins present in blood serum.

paraglossa *n.* an appendage on either side of labium of insects, *alt.* labella; a paired cartilage of the chondrocranium.

paraglossum *n.* median cartilaginous or bony prolongation of copula supporting the tongue, as in birds.

paraglycogen *n.* carbohydrate food reserve in protozoans, resembling glycogen.

paragnatha *n.plu.* paired, delicate, unjointed processes of maxilla of certain arthropods; the buccal denticles of certain polychaete worms.

paragnathous *a.* with mandibles of equal length, *appl.* birds.

paragula *n.* a region beside gula on insect head.

paraheliode *n.* special parasol-like arrangement of spines in certain cacti.

paraheliotropism paraphototropism *q.v.*

parahormone *n.* substance that acts like a hormone but is a product of the ordinary metabolism of cells.

paralectotype *n.* specimen of a series used to designate a species, which is later designated as a paratype.

paralimnic *a. pert.* or inhabiting the lake

shore.

parallel descent or evolution evolution in similar direction in different groups; the independent acquisition of similar traits in two related species.

parallel distributed processing one model of brain function in which various aspects of information from, e.g., the visual field (such as colour, movement, form) are encoded in many separate neuronal channels at source and throughout much of their analysis in the cortex, a final perception being due to activation of connections between channels and the simultaneous activity of neurones in several different areas.

parallel flow the flow of two fluids in the same direction.

parallel venation leaf venation in which veins run parallel longitudinally along the leaf, as in monocotyledons.

parallelinervate, parallelodrome *a. appl.* leaves with veins parallel.

parallelotropism orthotropism *q.v.*

paralogous *a. appl.* similarities in anatomy that are not related to common descent or similar function; *appl.* two genes in a genome that are similar because they derive from a gene duplication (e.g. α- and β-globin). *n.* paralogy. *cf.* orthologous.

paralogy *n.* similarities in anatomy that are not related to common descent or similar function. *a.* paralogous.

paramastigote *a.* having one long principal flagellum and a short accessory one, as certain flagellate protozoans.

paramastoid *a.* beside the mastoid; *appl.* two paraoccipital processes of exoccipitals; *appl.* a process projecting from the jugular process.

paramecin *n.* a toxin produced by strains of the ciliate *Paramecium* containing kappa particles, and which is lethal to other sensitive strains of *Paramecium*.

paramere *n.* half of a bilaterally symmetrical structure; one of paired lobes exterior to penis in some insects.

paramesonephric ducts Müllerian ducts *q.v.*

parametrium *n.* fibrous tissue partly surrounding uterus.

paramitosis *n.* nuclear division, as in some protozoans, in which the chromosomes are not regularly arranged on equator of spindle and tend to cohere at one end when separating.

paramorph *n.* any variant form or variety; a form induced by environmental factors without underlying genetic change.

paramutation *n.* the condition when one allele influences the expression of another at the same locus when they are combined in a heterozygote.

paramutualism facultative symbiosis *q.v.*

paramylon *n.* substance allied to starch, present in certain algae and flagellates.

paramylum paramylon *q.v.*

paramyosin *n.* protein present in filaments of unstriated muscle, as of molluscs.

Paramyxoviridae, paramyxoviruses *n., n.plu.* family of large, enveloped, single-stranded RNA viruses including measles and Newcastle disease virus of swine.

paranasal *a. appl.* air sinuses in bones of upper jaw and face.

paranema *n.* sterile filament in reproductive organs of algae, mosses and ferns.

paranemic *a. appl.* DNA structures in which the two strands are not intertwined in a double helix, as may occur in limited regions of heteroduplex DNA during recombination; *appl.* a double spiral structure in which the two strands do not interlock at each turn but lie side by side. *alt.* anorthospiral.

paranephric *a.* beside the kidney, *appl.* a fatty body behind the connective tissue ensheathing the kidney.

paranephrocyte arthrocyte *q.v.*

Parano-Amazonian Floral Region part of the Neotropical Realm comprising Brazil and Bolivia.

paranotum *n.* one of the lateral expansions of the arthropod notum or tergum, thought to be the structures that have evolved into wings in some insects. *a.* paranotal.

Paranthropus genus of fossil hominids from southern Africa, subsequently renamed *Australopithecus robustus*.

paranuchal *a. appl.* bone on each side of nuchal bone (at nape of neck) in placoderms.

paranucleus *n.* micronucleus (*q.v.*) of ciliates; an aggregation of mitochondria in spermatid, destined to form the mitochondrial sheath.

paraoesophageal *a. appl.* nerve fibres connecting "brain" or cerebral ganglion with suboesophageal ganglion in some inverte-brates.

parapatric *a. appl.* distribution of species or other taxa that meet in a very narrow zone of overlap. *n.* parapatry.

parapet *n.* circular fold in body wall below margin of disc in sea anemones.

paraphototropism *n.* tendency of plants to turn edges of leaves towards intense illumination, thus protecting the surfaces.

paraphyletic *a. appl.* groups such as the reptiles which have evolved from and include a single ancestral species (known or hypothetical) but which do not contain all the descendants of that ancestor. *n.* paraphyly.

paraphyll(ium) *n.* one of the branching chlorophyll-containing outgrowths arising between leaves or from the bases, in mosses.

paraphysis *n.* in fungi, sterile elongated cell interspersed amongst asci or basidia in hymenium of some ascomycetes and basidiomycetes. *plu.* paraphyses.

parapineal *a. appl.* the eye-like parietal organ of brain of cyclostomes and some reptiles, the pineal body in other vertebrates.

parapleuron *n.* anterior lateral plate of exoskeleton at base of meta- and mesothoracic segments, in insects.

parapodium *n.* a paired lateral flap of tissue on segment of polychaete worms, bearing numerous chaetae and used for locomotion; lateral undulating extension of foot in some molluscs, used for propulsion. *plu.* parapodia. *a.* parapodial.

parapolar *a.* beside the pole.

parapophysis *n.* a transverse process arising from centrum of vertebra.

parapostgenal *a. appl.* thickened portion of occiput in insects.

paraprostate *a.* anterior bulbo-urethral glands.

parapsid *a.* having skull with one dorsal temporal fenestra on each side.

parapsidal *a. pert.* a parapsis; *appl.* furrows between dorsal portion of mesonotum and the parapsides in hymenopterans.

parapsides *plu.* of parapsis *q.v.*

parapsis *n.* lateral portion of mesonotum, as in ants. *plu.* parapsides.

parapteron *n.* small lateral sclerite of exoskeleton like a shoulder lappet on the

side of mesothorax of some insects, esp. the tegula of hymenopterans and lepidopterans.

paraquadrate squamosal *q.v.*

pararectal *a.* beside rectum.

parasegment *n.* a developmental unit in *Drosophila* and other insect embryos, delimited before visible segmentation, and which consists of the posterior compartment of one prospective segment and the anterior compartment of the next.

parasematic *a. appl.* markings, structures or behaviour tending to mislead or deflect attack by an enemy. *cf.* aposematic, episematic.

paraseme *n.* misleading appearance of markings, as an ocellus near the tail of fishes.

paraseptal *a. appl.* cartilage almost enclosing vomeronasal organ.

parasexual *a. appl.* or *pert.* the operation of genetic recombination other than by means of the alternation of karyogamy and meiosis characteristic of sexual reproduction.

parasexual cycle a cycle of plasmogamy, karyogamy and haploidization in some fungi that superficially resembles true sexual reproduction but which may take place at any time in the life cycle.

parasite *n.* an organism that for all or some part of its life derives its food from a living organism of another species (the host), usually living in or on the body or cells of the host, which is usually harmed to some extent by the association. *a.* parasitic.

parasite chain a food chain passing from large to small organisms.

parasitic castration castration caused by the presence of a parasite, as in male crabs infested by the barnacle, *Sacculina*; sterility in various other plants or animals caused by a parasite attacking sex organs.

parasitic male a dwarf male which is parasitic on its female and has a reduced body in all but the sex organs, as in some deep-sea fish.

parasitism *n.* a special case of symbiosis in which one partner (the parasite) receives advantage to the detriment of the other (the host).

parasitocoenosis *n.* the whole complex of parasites living in any one host.

parasitoid *n.* organism alternately parasitic and free living and whose parasitism ultimately kills its host, such as certain insects (e.g. ichneumon flies) in which the adults are free living but which lay their eggs in the bodies of other insect larvae, in which the larvae develop, consuming host tissues and killing the host on hatching and emergence.

parasitology *n.* the study of parasites, esp. animal parasites.

parasocial *a. appl.* social group in which some features of complete eusociality are absent.

parasol paraheliode *q.v.*

parasphenoid *n.* skull bone forming floor of cranium in certain vertebrates.

parasporal *a. appl.* protein bodies formed within the cell during sporulation in some bacteria.

parasporangium *n.* a sporangium containing paraspores.

paraspore *n.* a spore formed from a cortical somatic cell, as in some algae.

parastamen, parastemon staminode *q.v.*

parasternum *n.* the sum total of abdominal ribs in some reptiles.

parastichy *n.* a descending curved row formed by leaf primordia at the growing apex of a shoot.

parasymbiosis *n.* the living together of organisms without mutual harm or benefit.

parasympathetic *a. appl.* components of the parasympathetic nervous system, as parasympathetic ganglion.

parasympathetic nervous system part of the autonomic nervous system controlling involuntary muscular movement of blood vessels and gut and glandular secretions from the eye, salivary glands, bladder, rectum and genital organs. Also includes the vagus nerve supplying the viscera as a whole. Parasympathetic nerve fibres are contained within the last five cranial nerves and the last three spinal nerves and terminate at parasympathetic ganglia near or in the organ they supply. The actions of the parasympathetic system are broadly antagonistic to those of the sympathetic system, lowering blood pressure, slowing heartbeat, stimulating the process of digestion etc. The chief neurotransmitter in the parasympathetic system is acetylcholine.

parately *n.* evolution from material unrelated to that of type, but resulting in superficial resemblance.

parateminal *a. appl.* bodies constituting part of the anterior median wall of lateral ventricles of brain in amphibians and reptiles.

paratestis *n.* small reddish-yellow fatty body in some male newts and salamanders which produces hormones regulating the appearance of mating coloration, etc.

parathecium *n.* peripheral layer of apothecium, as in cup fungi; peripheral layer of hyphae in lichens.

parathormone parathyroid hormone *q.v.*

parathyroid glands four small brownish-red endocrine glands near or within the thyroid in mammals, which secrete parathyroid hormone.

parathyroid hormone (PTH) protein hormone secreted by parathyroid glands which stimulates increase in bone resorption, resulting in increase of calcium and phosphate in blood, also increase in resorption of calcium and magnesium and decrease in resorption of phosphate in kidney tubules. It is essential for normal skeletal development.

paratoid *a. appl.* double row of poison glands extending along back of certain amphibians, as of salamanders.

paratomium *n.* side face of a bird's beak, between the cutting edge and the median longitudinal ridge.

paratomy *n.* reproduction by fission with regeneration, in certain annelids.

paratonic *a.* stimulating or retarding; *appl.* movements induced by external stimuli, as tropisms or nastic movements.

paratose *n.* a 3,6-dideoxyhexose sugar found in the lipopolysaccharide outer membrane of some enteric bacteria.

paratracheal *a.* with xylem parenchyma cells around or close to vascular tissue.

paratrophic *a. appl.* method of nutrition of obligate parasites.

paratympanic *a.* medial and dorsal to the tympanic cavity; *appl.* a small organ with sensory epithelium innervated from geniculate ganglion in many birds.

paratype *n.* specimen described at same time as the one regarded as type specimen of a new genus or species.

para-urethral *a. appl.* mucus-secreting glands associated with urethra, Littrée's glands.

paravertebral *a.* alongside the spinal column, *appl.* trunk of sympathetic nerve.

paravesical *a.* beside the bladder.

paraxial *a.* alongside the axis.

paraxon *n.* a lateral branch of the axon of a neurone.

paraxonic *a. pert.* or having an axis outside the usual axis; with axis of foot between 3rd and 4th digits, as in artiodactyls.

Parazoa *n.* a term sometimes used for those multicellular animals, such as sponges, having a loose organization of cells and not forming distinct tissues or organs.

parencephalon cerebral hemisphere *q.v.*

parenchyma *n.* (*bot.*) soft plant tissue composed of thin-walled, relatively undifferentiated cells, which may vary in structure and function; (*zool.*) solid layer of tissue between muscle layer and gut in platyhelminths, composed of many different types of cell; ground tissue of an organ.

parenchymatous *a. pert.* or found in parenchyma.

parenchymula *n.* a flagellate sponge larva with cavity filled with gelatinous connective tissue.

parental generation the parent individuals in a genetic cross, designated P_1.

parental genomic imprinting *see* genomic imprinting.

parental investment any behaviour towards offspring that increases the chances of the offspring's survival at the cost of the parent's ability to invest in other offspring.

parenthosome *n.* a structure formed from endoplasmic reticulum and covering the pore in the dolipore septa of certain fungi.

parethmoid ectethmoid *q.v.*

parhomology *n.* apparent similarity of structure.

paries *n.* the central division of a compartment of barnacles; the wall of a hollow structure, as of tympanum, or of honeycomb. *plu.* parietes.

parietal *a. pert.*, next to, or forming part of the wall of a structure; *appl.* placentation of plant ovary: ovules borne in rows on carpel wall or extensions of it.

parietal bone a paired bone of roof of skull.

parietal cells oxyntic cells *q.v.*

parietal lobe of brain, dorsal lobe of cerebral cortex, lying behind central sulcus, part lying between frontal and occipital lobes and above temporal lobe.

parietal organ the pineal body in some lower vertebrates, where it has a photoreceptor function.

parietal vesicle dilated distal part of pineal stalk.

parietes *plu.* of paries *q.v.*

parietobasilar *a. appl.* muscles between pedal disc and lower part of body wall in sea anemones.

parietofrontal *a. appl.* a skull bone in place of parietals and frontals, as in lungfishes.

parietomastoid *a.* connecting mastoid and parietal bone, *appl.* a suture.

parieto-occipital sulcus fissure between parietal and occipital lobes of cerebral hemisphere.

parietotemporal *a. pert.* parietal and temporal regions; *appl.* a branch of the middle cerebral artery.

parietovaginal *a. appl.* a paired muscle for retracting tentacles in bryozoans.

paripinnate *a.* pinnate without a terminal leaflet.

parity *n.* the number of times a female has given birth, regardless of the number of offspring produced at any one birth.

parivincular *a.* bivalve hinge ligament attached along whole edge of shell.

parkinsonism *n.* the symptoms of muscular rigidity, weakness and tremor at rest characteristic of Parkinson's disease, apparently caused by disorders in the dopaminergic pathways in brain.

parocciput *n.* in insects, a thickening of the occiput for articulation of neck sclerites.

paroecious *a.* with antheridium and archegonium close to one another.

parolfactory *a. appl.* an area and sulcus adjoining the small triangular space in olfactory lobe.

paronychia *n.* bristles on pulvillus of insect foot.

parosteosis *n.* bone formation in tracts normally composed of fibrous tissue.

parotic *n.* a process formed by fusion of exoccipital and opisthotic elements of skull in some reptiles; a similar process in some fishes.

parotid glands paired salivary glands opening into the mouth in some mammals.

parotoid glands in some amphibians, large swellings on side of head, formed of aggregated cutaneous glands, sometimes poisonous.

parovarium *n*. a small collection of tubules anterior to ovary, the remnants in adult of embryonic mesonephros.

pars *n*. a part of an organ.

pars distalis a part of the adenohypophysis *q.v.*

pars intercerebralis in an insect's forebrain, the region containing neurosecretory cells.

pars intermedia a part of the glandular tissue of the pituitary, producing lipotropin and melanocyte-stimulating hormone.

pars nervosa the neural lobe of the pituitary, containing neurosecretory neurones projecting from the hypothalamus and secreting e.g. oxytocin and vasopressin.

pars tuberalis a part of the adenohypophysis *q.v.*

parsimony principle in molecular taxonomy, the principle that organisms that are closely related (i.e. that diverged more recently) will have fewer differences in their DNA than those that diverged longer ago.

parthenapogamy *n*. somatic or diploid parthenogenesis.

parthenita *n*. unisexual stage of trematodes in intermediate host.

parthenocarpic *a. appl.* fruit lacking seeds, which has developed without fertilization.

parthenogamy parthenomixis *q.v.*

parthenogenesis *n*. reproduction from a female gamete without fertilization by a male gamete.

parthenogenetic *a. appl.* organisms produced by parthenogenesis; *appl.* agents that can activate an unfertilized ovum.

parthenogonidia *n.plu.* zooids of a protozoan colony, witht he function of asexual reproduction.

parthenokaryogamy *n*. the fusion of two female haploid nuclei.

parthenomixis *n*. the fusion of two nuclei produced within one gamete or gametangium.

parthenosperm *n*. a seed or spore produced

without fertilization, but resembling a zygote.

parthenote *n*. a parthenogenetically produced haploid organism.

partial diploid a bacterium or other cell carrying two copies of some but not all its genes (in bacteria the extra genes are usually introduced via plasmids).

partial dominance codominance *q.v.*

partial parasite semiparasite *q.v.*

partial veil inner veil of certain basidiomycete fungi, growing from stipe towards edge of cap and becoming separated to constitute the cortina or superior ring.

particulate inheritance refers to the fact that hereditary characteristics are transmitted by discrete entities (genes) which themselves remain unchanged from generation to generation. *cf.* blending inheritance.

partite *a*. divided nearly to the base.

partition chromatography separation technique in which the materials to be separated are selectively partitioned between two solvents.

parturition *n*. the act of giving birth.

parumbilical *a. appl.* small veins from anterior abdominal wall to portal and iliac veins.

parvicellular *a*. consisting of small cells.

parvocellular layer layer of lateral geniculate nucleus of primate, comprising layers 3, 4, 5 and 6 and composed of small cells.

Parvoviridae, parvoviruses *n., n.plu.* family of small, non-enveloped, single-stranded DNA viruses that includes Aleutian disease of mink, adeno-associated viruses and human parvoviruses.

pascual *a. pert.* pastures, *appl.* flora.

pasculomorphosis *n*. changes in the structural features of plants as a result of grazing.

passage cells thin-walled exodermal or endodermal cells of root which permit passage of solutes, and are usually associated with cells with thick walls, as companion cells of phloem.

Passeriformes, passerines *n., n.plu.* large order of birds, which includes small and medium-sized perching birds and songbirds such as crows, tits, warblers, thrushes and finches.

Passiflorales *n*. order of dicot shrubs, herbs,

climbers and small trees and including the families Caricaceae (pawpaw) and Passifloraceae (passion flower).

passive immunity short-lived immunity acquired by transfer of preformed antibodies, as from mother to foetus across placenta and in mother's milk to infant.

passive transport simple diffusion of small uncharged molecules or protein-mediated transport (facilitated diffusion) of ions and other charged molecules across a biological membrane in the direction of concentration gradient and electrochemical gradient and which does not require energy.

Pasteur effect the observation of Louis Pasteur (originally in yeast) that glycolysis is inhibited by aerobic conditions (i.e. by respiration); sometimes incorrectly said to be the ability of a normally anaerobic organism to oxidize sugar completely to carbon dioxide and water in the presence of oxygen, which is a later interpretation of the effect.

pasteurization *n.* method of partial sterilization, used for milk, wine and other beverages, by heating at 62 °C for 30 minutes or at 72 °C for 15 seconds, followed by rapid cooling.

pasture *n.* grassland used for grazing.

patabiont *n.* animal that spends all its life in the litter on the forest floor.

patacole *n.* animal that lives temporarily in the litter on the forest floor.

patagial *a.* of or *pert.* a patagium, or flying membrane.

patagiate *a.* having a patagium.

patagium *n.* the membranous expansion between fore- and hindlimbs of bats; the extension of skin between fore- and hindlimbs of flying lemurs and flying squirrels.

patch clamp technique in neurophysiology which isolates a tiny portion of nerve cell membrane allowing recording from single ion channels.

patching *n.* clustering of membrane proteins seen on eukaryotic cells when treated with lectins or specific antibody, caused by lateral movement of membrane protein in the fluid lipid bilayer as a result of crosslinking by lectins or antibodies.

patella *n.* the knee cap or elbow cap; segment between femur and tibia in certain

arachnids; 4th segment of spider's leg; a genus of limpet; a rounded apothecium of lichens.

patellar *a. pert.* a patella.

patellaroid, patelliform *a.* shaped like a patella; pan-shaped; like a bordered disc.

patent *a.* open; spreading widely.

pateriform *a.* saucer-shaped.

pathetic *a. appl.* nerve: 6th cranial nerve, controls a muscle of eye.

pathogen *n.* any disease-causing microorganism.

pathogenesis *n.* the origin or cause of the pathological symptoms of a disease.

pathogenic *n.* causing disease, *appl.* a parasite (esp. a microorganism) in relation to a particular host.

pathology *n.* science dealing with disease or dysfunction; the characteristic symptoms and signs of a disease.

patina *n.* circle of plates around calyx of crinoids.

patoxene *n.* animal that occurs accidentally in the litter on the forest floor.

patriclinal, patriclinous *a.* with hereditary characteristics more paternal than maternal. *alt.* patroclinal, patroclinous.

patrilineal *a. pert.* paternal line; passed from male parent.

patristic *a.* in plants, *pert.* similarity due to common ancestry.

pattern formation the generation of spatial patterns of differentiated cells to form tissues, and eventually organs and morphological structures, within a multicellular embryo.

patulent, patulose, patulous *a.* spreading open; expanding.

paturon *n.* basal segment of spider's chelicera, used for crushing prey.

paucilocular *a.* containing, or composed of, few small cavities or loculi.

paucispiral *a.* with few coils or whorls.

paulospore *n.* a resting stage in development, as a cyst.

paunch rumen *q.v.*

Pauropoda *n.* class of arthropods (*q.v.*) allied to the myriapods, which have 12 body segments, 9 of which carry appendages.

pavement epithelium epithelium of cuboid cells, or a simple squamous epithelium.

Pavlovian conditioning classical condi-

paxilla

tioning *q.v.*

paxilla *n.* a thick plate, supporting calcareous pillars, the summit of each covered by a group of small spines, in certain starfish.

paxillate *a.* having paxillae.

paxilliform *a.* shaped like a paxilla.

paxillus paxilla *q.v.*

PCR cycle photosynthetic carbon reduction cycle. *see* Calvin cycle.

pea enation mosaic virus group plant virus group with a single member, pea enation mosaic virus, an isometric single-stranded RNA virus. It is a multicomponent virus in which two genomic species of RNA are encapsidated in different particles.

pearl *n.* the abnormal growth formed around a minute grain of sand or other foreign matter which gets inside the shell of certain bivalve molluscs, and which consists of many layers of nacre.

peat *n.* type of soil formed by partly decomposed plant material in anaerobic waterlogged conditions.

peat mosses *see* sphagnum mosses.

peck order a social hierarchy, esp. in birds, ranging from the most dominant and aggressive animal down to the most submissive.

pecorans *n.plu.* giraffes, deer and cattle.

pectate (pectic acid) polygalacturonan (*q.v.*), formerly thought to be a discrete component of pectin, now considered to be part of a more complex polysaccharide.

pecten *n.* any comb-like structure; a process of inner retinal surface in reptiles, expanded into a folded quadrangular plate in birds; a ridge of the superior branch of the pubic bone; part of anal canal between internal sphincter and anal valves; a part of the stridulating organ of certain spiders; a sensory abdominal appendage of scorpions; a genus of scallops; a comb-like assemblage of sterigmata.

pectic *a. pert.* pectin.

pectic acids, pectate the galacturonans occurring in pectin once their methyl groups have been removed, which form a colloidal solution.

pectic enzymes enzymes that hydrolyse the components of pectin, and including polygalacturonase (*q.v.*), pectinesterase (*q.v.*), pectin lyase and pectate lyase.

pectin *n.* a group of highly variable complex polysaccharides found in middle lamella of plant primary cell walls, rich in galacturonic acid, and forming a gel when isolated, and also containing arabinose, galactose and rhamnose residues.

pectinal *a. pert.* a pecten.

pectinase *n.* commercially usually a mixture of glycosidases that hydrolyse the various components of pectin; sometimes refers to polygalacturonase (EC 3.2.1.15). (*see* polygalacturonan).

pectinate *a.* shaped like a comb.

pectineal *a.* comb-like; *appl.* a ridge-line on pubis of birds; *appl.* a ridge on femur and its attached muscle, the pectineus muscle.

pectinellae *n.plu.* transverse comb-like membranelles constituting adoral ciliary spiral of some ciliate protozoans.

pectines *plu.* of pecten *q.v.*

pectinesterase *n.* an enzyme catalysing the removal of methyl groups from the methylated galacturonans occurring in pectin, producing methanol and "pectate". EC 3.1.1.11.

pectineus *n.* a flat muscle between pecten of pubis and the upper medial part of femur.

pectinibranch *a.* having one margin only furnished with teeth like a comb, *appl.* gills in certain molluscs.

pectiniform pectinate *q.v.*

pectocellulose *n.* pectin mixed with cellulose as in fleshy roots and fruits.

pectolytic *a.* breaking down pectin, *appl.* enzymes.

pectoral *a. pert.* chest, in region of chest.

pectoral fin the fin on the side of body of fish.

pectoral girdle in vertebrates, a skeletal support in the shoulder region for attachment of fore-fins or forelimbs, made up of a hoop of cartilages or bones, usually the scapula, clavicle and coracoid.

pectoralis *n.* breast muscle connecting breastbone with humerus, much enlarged in birds where it is a main flight muscle.

pectoralis major and minor outer and inner chest muscles connecting ventral chest wall with shoulder and humerus.

pectose protopectin *q.v.*

pectosinase protopectinase *q.v.*

ped *n*. soil particles cemented together by humus, inorganic salts and mucilage to form clumps of various sizes and shapes.

pedal *a. pert.* foot or feet; *appl.* disc: base of a sea anemone.

pedalfer *n*. any of a group of soils, in humid regions, usually characterized by the presence of aluminium and iron compounds and the absence of carbonates.

pedate *a*. with toe-like parts.

pedatipartite *a. appl.* a type of palmate leaf having three main divisions and the two outer divisions subdivided one or more times.

pedatisect *a*. in a pedate arrangement and with divisions nearly to midrib.

pedicel *n*. small, short stalk of leaf, fruit or sporangium (*see* Fig. 4), or foot-stalk of a fixed organism or of an organ.

pedicellariae *n.plu.* minute pincer-like structures studding the surface of some echinoderms (e.g. starfish) which grab, kill and discard small animals that touch them. *sing.* pedicellaria.

pedicellate *a*. supported by a small stalk; *appl.* Hymenoptera whose thorax and abdomen are joined by a short stalk.

pedicellus pedicel *q.v.*

pedicle *n*. short stem or stalk.

pediculates *n.plu.* an order of very specialized marine bony fishes, including anglerfish and batfishes, which have anterior portion of dorsal fin modified as a lure.

pediferous *a*. having feet; having a foot-stalk.

pedigree *n*. in genetics, a diagram showing the ancestral history of a group of related individuals.

pedipalp *n*. in spiders and other arachnids, a small paired leg-like appendage, immediately anterior to the walking legs, in male spiders used in mating to transfer sperm into female, sometimes modified as a pincer or claw.

pedo- paedo-.

pedocal *n*. any of a group of soils, of arid and semiarid regions, characterized by the presence of carbonate of lime.

pedogamy *a*. type of autogamy in protozoans where gametes are formed after multiple division of nucleus.

pedogenesis *n*. reproduction in young or larval stages, as axolotl; spore production in immature fungi; formation of soil.

pedogenic *a. pert.* the formation of soil.

pedology *n*. soil science.

pedomorphosis *n*. retention of juvenile traits in adults. *a*. pedomorphic.

pedonic *a. appl.* organisms of freshwater lake bottoms.

peduncle *n*. (*bot.*) the stalk of a flowerhead; (*zool.*) the stalk of sedentary protozoans, crinoids, brachiopods and barnacles; link between thorax and abdomen in arthropods; (*neurobiol.*) band of nerve fibres joining different parts of brain.

pedunculate *a*. growing on or having a peduncle.

Pekin(g) man an extinct fossil hominid found near Peking and at first called *Sinanthropus pekinensis*, then *Pithecanthropus pekinensis*, and now classified as *Homo erectus*.

pelage *n*. the hairy, furry or woolly coat of mammals.

pelagic *a*. living in the sea or ocean at middle or surface levels.

pelargonidin *n*. red flavonoid pigment present in many flowers, such as geraniums.

pelasgic *a*. moving from place to place.

Pelecaniformes *n*. an order of large aquatic, fish-eating birds with all four toes webbed, including the pelicans, cormorants, boobies and gannet.

Pelecypoda Bivalvia *q.v.*

P elements family of transposable elements found in *D. melanogaster*, involved in hybrid dysgenesis and also used experimentally to introduce genes into the germline to produce transgenic insects.

pellagra *n*. deficiency disease caused by insufficient tryptophan and nicotinate (niacin) in the diet, can be treated with niacin or nicotinamide.

pellagra-preventive factor nicotinamide or niacin *q.v.*

pellicle *n*. a thin flexible outer layer, as formed e.g. by the plasma membrane and its underlying protein strips in *Euglena*; any delicate surface or skin-like growth; the skin-like aggregation of bacteria or yeasts on the surface of liquid media.

pelliculate *a*. having a pellicle on outer surface.

pellions *n.plu.* ring of plates supporting

suckers of echinoids.

pelma *n.* the sole of the foot.

pelophilous *a.* growing on clay.

peloria, pelory *n.* condition of abnormal regularity; a modification of structure from irregularity to regularity.

peloric *a. appl.* a flower which, normally irregular, becomes regular.

pelotons *n.plu.* coils of hyphae of the mycorrhizal fungus in meristematic tissue of orchid embryos.

pelta *n.* a shield-shaped apothecium of certain lichens.

peltate *a.* shield-shaped; having the stalk inserted at or near the middle of the under surface, not near the edge.

peltinervate *a.* having veins radiating from near the centre, as of a peltate leaf.

pelvic *a. pert.* or situated at or near pelvis.

pelvic fins paired fins on underside of body in fish, representing the hindlimbs of land vertebrates.

pelvic girdle in vertebrates, a skeletal support in the hip region for attachment of hind-fins or hindlimbs, made of a hoop of cartilages or bones, in tetrapods usually ilium, ischiopubis or ischium, and pubis.

pelvis *n.* the cavity in vertebrates formed by pelvic girdle along with coccyx and sacrum; of kidney, the expansion of ureter at its junction with kidney. *a.* pelvic.

pelvisternum *n.* the epipubis when separate from pubis.

pelycosaurs *n.plu.* an order of aberrant and primitive mammal-like reptiles of the Carboniferous to Permian, having a primitive sprawling gait and including the sail-back lizards.

pen *n.* a midrib of leaf; the horny beak of certain cephalopods; a primary wing feather; a female swan.

pendent *a.* hanging down, as certain lichens, flowers, leaves, etc.

pendulous *a.* bending downwards from point of origin; overhanging; *appl.* ovules, flowers, branches, etc.

penetrance *n.* the percentage of individuals possessing a particular genotype who show the associated phenotype, i.e. in whom the gene is expressed, complete penetrance being shown by a trait if it is expressed in all persons who carry it, incomplete penetrance being shown if it is

not expressed at all in some individuals who carry it.

penetration path path of sperm in ooplasm to the female pronucleus.

penicillate *a.* resembling a small paintbrush, *appl.* heads of conidia of fungi of the genus *Penicillium*.

penicillia *n.plu.* ascomycete fungi of the large and ubiquitous genus *Penicillium*, characterized by heads of chains of spores (conidia) resembling an artist's brush (penicillum), which include spp. producing antibiotics (e.g. penicillin produced by *P. notatum*), important in cheese-making (e.g. *P. roquefortii*) and the commercial production of certain organic compounds.

penicillin *n.* any of various antibiotics, based on a β-lactam ring structure, produced by the mould *Penicillium notatum* and related species, and which inhibit bacterial cell wall synthesis leading to osmotic lysis.

penicillinase *n.* enzyme secreted by penicillin-resistant bacteria, which destroys penicillin's antibiotic activity, hydrolysing the β-lactam ring. EC 3.5.2.6.

penis *n.* the male copulatory organ.

penna *n.* a contour feather in birds, as distinguished from a plume or down feather.

pennate *a.* divided in a feathery manner; feathered; having a wing; in the shape of a wing; *appl.* diatoms that are bilaterally symmetrical in valve view.

Pennsylvanian *a. appl.* and *pert.* an epoch of the Carboniferous era, lasting from 320 to 286 million years ago; *appl.* coal measures in North America.

pensile *a.* hanging.

penta- prefix derived from Gk. *pente*, five, signifying having five of, arranged in fives, etc.

pentactinal *a.* five-rayed; five-branched.

pentadactyl *a.* having all four limbs normally terminating in five digits.

pentadactyl limb the limb with five digits characteristic of tetrapods.

pentadelphous *a.* applied to flowers having anthers consisting of five clusters of more-or-less united filaments.

pentafid *a.* in five divisions or lobes.

pentagynous *a.* having five pistils or styles; having five carpels to the gynaecium.

pentamerous *a.* composed of five parts; *appl.* flowers with parts arranged in whorls of five or a multiple of five.

pentandrous *a.* having five stamens.

pentaploid *a.* having five times the normal haploid number of chromosomes.

pentapterous *a.* with five wings, as some fruits.

pentaradiate *a.* with a body built on a five-rayed plan.

pentarch *a.* with five alternating groups of xylem and phloem; with five vascular bundles.

pentasepalous *a.* with five sepals.

pentaspermous *a.* with five seeds.

pentastemonous *a.* with five stamens.

pentasternum *n.* sternite of 5th segment of prosoma or 3rd segment of podosoma in ticks and mites.

pentastichous *a.* arranged in five vertical rows, *appl.* leaves.

Pentastomida, pentastomids *n., n.plu.* phylum of flat, soft-bodied, worm-like parasites, commonly known as tongue-worms, that live in the nasal passages of vertebrates, mainly tropical and sub-tropical. They have three larval stages, the first taking place in the egg in the nasal passages of herbivorous mammals. *alt.* Pentastoma, pentastomes.

pentosan *n.* any polysaccharide made of linked pentose units, such as xylans and arabinans.

pentose *n.* any monosaccharide having the formula $(CH_2O)_5$, e.g. arabinose, xylose and ribose.

pentose phosphate pathway the oxidative pentose phosphate pathway (PPP) comprises a group of reactions that generates biochemical reducing power in the form of NADPH from the partial oxidation of glucose-6-phosphate, producing carbon dioxide and ribose-5-phosphate which can then undergo conversion to triose phosphates, sometimes in a cycle, regenerating glucose-6-phosphate. It is the only source of NADPH in red blood cells, and is active in adipose cells rather than muscle. *alt.* hexose monophosphate shunt, phosphogluconate oxidative pathway; the reductive pentose phosphate pathway (RPPP) is also present in plants and some bacteria and is involved in carbon dioxide fixation into hexoses in photosynthesis. *see* Calvin cycle.

peonidin *n.* red anthocyanin pigment found in paeonies and other plants.

pepo *n.* an inferior many-seeded pulpy fruit, as melons, cucumbers, squashes, etc.

peppered moth *Biston betularia. see* industrial melanism.

pepsin *n.* digestive enzyme, a proteinase in gastric juice which hydrolyses proteins to peptides, formed from pepsinogen by cleavage of one peptide bond by hydrochloric acid in the stomach. EC 3.4.23.1, *r.n.* pepsin A (pepsin B, EC 3.4.23.1 and pepsin C, EC 3.4.23.1. have more restricted substrate specificity).

pepsinogen *n.* inactive precursor of pepsin, secreted into the gastric lumen by cells of the gastric mucosa and activated by hydrochloric acid from oxyntic cells.

pepstatin *n.* hexapeptide inhibitor of the enzyme pepsin and other acid proteases.

peptic *a.* relating to or promoting digestion.

peptic cells cells of the gastric gland that secrete prorennin and pepsinogen, precursors to the digestive enzymes rennin and pepsin.

peptidase *n.* formerly used for any enzyme that attacked peptide bonds in proteins; now only used for several groups of exopeptidases which split off 1 or 2 units at the end of a polypeptide chain. EC 3.4.11–17.

peptide *n.* a chain of a small number (up to around 20) of amino acids linked by peptide bonds, dipeptides consisting of 2 amino acids, tripeptides of 3, hexapeptides of 6, etc., and which may be synthesized on the mRNA/tRNA/ribosome system (e.g. the various peptide hormones), or may be the product of partial hydrolysis of a protein, or may be made by chemical synthesis.

peptide bond covalent bond joining the α-amino group of one amino acid to the carboxyl group of another with the loss of a water molecule, and which is the bond linking amino acids together in a protein chain. *see* Appendix 1 (30) for chemical structure.

peptide hydrolase general term for enzymes splitting peptide bonds by hydrolysis, including the proteinases and peptidases. EC 3.4.

peptide map *see* fingerprinting.

peptide transporters membrane transport proteins (the TAP proteins) that carry peptides produced in the cytosol by breakdown of foreign protein antigens into the endoplasmic reticulum for eventual binding to MHC molecules and transport to the cell surface.

peptidoglycan *n.* any of a class of macromolecules in which linear polysaccharide chains are extensively cross-linked by short peptides, components of bacterial cell walls.

peptidyl prolyl *cis-trans* isomerase (PPI) enzyme involved in the folding of newly synthesized proteins, and which is identical with cyclophilin.

peptidyl transferase enzyme catalysing the formation of peptide bonds during protein synthesis on the ribosomes, an integral part of the large ribosomal subunit. EC 2.3.2.12.

peptidyl-tRNA a tRNA attached to the whole polypeptide chain synthesized so far, formed on ribosome at each cycle of amino acid addition during protein synthesis.

peptone *n.* polypeptide product of hydrolysis of proteins by enzymes such as pepsin.

peptonephridia *n.plu.* the anterior nephridia, which function as digestive glands, in some oligochaete worms.

per os by mouth.

percentage successive mortality apparent mortality *q.v.*

percept *n.* a conscious mental image of a perceived object.

perception *n.* the mental interpretation of physical sensations produced by stimuli from the outside world.

percurrent *a.* extending throughout length, or from base to apex.

perdominant *n.* a species present in almost all the associations of a given type.

pereion *n.* the thorax of crustaceans.

pereiopod *n.* walking limb of crustaceans such as crayfish, crabs and lobsters, present as four pairs on cephalothorax.

perennating *a.* overwintering, *appl.* roots, buds, etc.

perennation *n.* of a plant, survival from year to year; survival for a number of years. *a.* perennating.

perennial *n.* plant which persists for several years.

perennibranchiate *a.* having gills persisting throughout life, as certain amphibians.

perfect *a. appl.* flowers containing both stamens and carpels.

perfect stage *a.* in ascomycete fungi, the sexual reproductive phase in their life history in which they produce asci.

perfoliate *a. appl.* a leaf with basal lobes united so that it entirely surrounds the stem.

perforate *a.* having pores, as corals, foraminiferans; *appl.* leaves having small translucent spots when held up against light, as Perforate St. John's Wort.

perforation plate perforate septum or area of contact between cells or elements of xylem vessels.

perforator *n.* a barbed spear-like head and process of some spermatozoa, as of salamander.

perforatorium acrosome *q.v.*

perforin *n.* protein produced by cytotoxic T cells and natural killer cells which forms pores in target cell membrane leading to cell lysis.

peri- prefix derived from Gk. *peri*, around, and signifying surrounding, or situated around.

perianth *n.* the outer whorl of floral leaves of a flower when not clearly divided into calyx or corolla; collectively, the calyx and corolla; the cover or sheath surrounding the archegonia in some mosses; tubular sheath surrounding developing sporophyte in leafy liverworts.

periaxial *a.* surrounding an axis or an axon; *appl.* space between the axon membrane and the myelin sheath in a medullated nerve fibre.

periblast *n.* the outside layer, epiblast, or blastoderm of an insect embryo; syncytium formed by fusion of small marginal blastomeres and not forming part of mammalian embryo.

periblastesis *n.* envelopment by surrounding tissue, as of lichen gonidia.

periblastic *a. pert.* periblast; superficial, as *appl.* segmentation of fertilized ovum.

periblem *n.* the meristem that produces the cortex.

peribranchial *a.* around the gills; *appl.* type of budding in ascidians; *appl.* atrial cavity in ascidians and lancelet; *appl.* circular spaces surrounding basal parts of papulae of starfish.

peribulbar *a*. around the eyeball; surrounding a taste bud, *appl*. nerve fibres.

pericambium *n*. layer of dividing cells around the stele of plant axis, also called the pericycle.

pericapillary *a. appl*. cells in contact with outer surface of walls of capillaries.

pericardiac, pericardial *a. pert*. the pericardium; surrounding the heart, *appl*. cavity, septum; *appl*. paired excretory organs in lamellibranchs; *appl*. cells: cords of nephrocytes in certain insects.

pericardium *n*. membrane surrounding the heart, delimiting a pericardial cavity which is sometimes itself called the pericardium.

pericarp *n*. the tissues of a fruit that develop from the ovary wall, comprising an outer skin, sometimes a fleshy mesocarp, and an inner endocarp.

pericaryon *alt*. spelling of perikaryon *q.v.*

pericellular *a*. surrounding a cell.

pericentral *a*. around or near centre.

pericentric *a. pert*. or involving the centromere of a chromosome; *appl*. breaks in arms of a chromosome on either side of the centromere; *appl*. inversions including the centromere.

perichaetial *a. pert*. a perichaetium.

perichaetine *a*. having a ring of chaetae encircling the body.

perichaetium *n*. one of the membranes or leaves enveloping archegonia or antheridia of bryophytes. *a*. perichaetial.

perichondral *a. appl*. ossification in cartilage beginning on outside and working inwards.

perichondrium *n*. fibrous envelope of connective tissue surrounding cartilage. *a*. perichondrial.

perichondrostosis *n*. ossification in cartilage beginning outside and working inwards.

perichordal *a*. enveloping or near the notochord.

perichoroidal *a*. surrounding the choroid, *appl*. lymph space.

perichylous *a*. with water-storage cells outside chlorenchyma, *appl*. leaves.

pericladium *n*. the lowermost clasping portion of a sheathing petiole.

periclinal *a. appl*. division of cells parallel to the surface of the structure in which they occur; *appl*. system of cells parallel to surface of apex of growing point; *appl*. two tissues in which one completely surrounds the other.

periclinium *a*. the involucre of a composite flower.

pericranium *n*. fibrous membrane investing skull.

pericycle *n*. the external layer of stele (primary vascular tissue in stem), the layer between endodermis and conducting tissue.

pericyclic *a. appl*. fibre situated on the outer edge of the vascular region in plant axis and usually arising in the primary phloem.

pericyte *n*. macrophage in outgrowths of small blood vessels.

pericytial pericellular *q.v.*

peridental periodontal *q.v.*

periderm *n*. tissue that replaces the epidermis in most stems and roots having secondary growth. It is made up of an outer layer of cork tissue containing suberin which is non-living at maturity, a cork cambium which produces cork on its outer side and the phelloderm, a living parenchyma tissue on its inner side. Commonly known as the bark.

peridesm *n*. tissue surrounding a vascular bundle.

peridial *a. pert*. a peridium *q.v.*

perididymis *n*. the dense white connective tissue surrounding testis.

peridinin *n*. a xanthophyll pigment found in algae of the Pyrrophyta.

peridiole *n*. one of the small "eggs" of bird's nest fungi, having a hard waxy wall and containing basidiospores, and which acts as a propagative unit, being splashed out of the peridium (the "nest") by rain.

peridium *n*. the outer wall of a fruiting body, esp. in fungi.

peridural *a. appl*. perimeningeal space at a later stage in development.

perienteric *a*. surrounding the enteron, or gut.

periesophageal perioesophageal *q.v.*

perifibrillar *a*. surrounding a fibril, *appl*. the cytoplasm surrounding the neurofibrils in axon.

perifoliary *a*. surrounding a leaf margin.

periganglionic *a*. surrounding a ganglion.

perigastric *a*. surrounding the viscera, *appl*.

abdominal cavity.

perigastrium *n*. the body cavity.

perigemmal *a*. surrounding a taste bud, *appl*. nerve fibres, spaces.

perigenous *a*. borne or growing on both sides of a leaf or other structure, or borne on all sides of a structure.

perigonium *n*. a floral envelope or perianth; involucre around antheridium in mosses.

perigynium *n*. membranous envelope or pouch of archegonium in liverworts; involucre in mosses; fruit-investing utricle of the sedge, *Carex*.

perigynous *a*. *appl*. flowers in which petals and stamens are attached to the extended margin of the receptacle. *n*. perigyny.

perihaemal *a*. *appl*. the canals of the blood vascular system in echinoderms, which enclose the haemal strands.

perikaryon *n*. the cell body containing the nucleus in nerve cells, and in other structures such as the syncytial tegument of flatworms in which many cell bodies in the deep layer of the skin are connected to an outer continuous layer of cytoplasm.

perikymata *n.plu*. ridges in enamel of teeth caused by incremental growth.

perilymph *n*. fluid surrounding membranous labyrinth of the ear separating it from the bony labyrinth, contained in the perilymphatic space.

perimeningeal *a*. *appl*. a space between the layer of connective tissue lining canal of vertebral column and skull and the meninges.

perimysium *n*. connective tissue binding numbers of fibres into bundles and muscles, and continuing into tendons.

perinaeum, perinaeal perineum, perineal *q.v.*

perineal *a*. *pert*. perineum, *appl*. artery, nerve, gland, etc.

perinephrium *n*. the enveloping adipose and connective tissue around kidney.

perineum *n*. part of body surface delimited by scrotum or vulva in front, anus behind and side of thigh to the side.

perineural *a*. surrounding a nerve or nerve cord.

perineurium *n*. the fibrous connective tissue sheath around a bundle of nerve fibres within a nerve.

perineuronal *a*. surrounding a nerve cell or cells.

perinium *n*. outer coat of microspores of certain pteridophytes.

perinuclear space the space between the outer and inner nuclear membranes.

periocular *a*. surrounding the eyeball within the orbital cavity.

period *n*. in geological time, a subdivision of an era, e.g. the Jurassic is a period of the Mesozoic era.

periodicity *n*. the fulfilment of functions at regular intervals or periods; in a cyclic or rhythmic reaction or process, the interval between two peaks of activity; of DNA, the number of base pairs per turn of the double helix.

periodontal *a*. covering or surrounding a tooth, *appl*. membrane, tissue of gums, etc.

perioesophageal *a*. surrounding oesophagus, *appl*. a nerve ring.

periople *n*. thin outer layer of hoof of equines.

periopticon *n*. in insects, the region of the optic lobes nearest to the eye.

periorbital *a*. surrounding the eye socket.

periosteum *n*. fibrous membrane investing the surface of bones.

periostracum *n*. the external layer of most mollusc and brachiopod shells.

periotic *n*. bone enclosing parts of membranous labyrinth of inner ear.

peripetalous *a*. surrounding petals or a petaloid structure.

peripharyngeal *a*. surrounding or encircling the pharynx, *appl*. cilia of ascidians and cephalochordates.

peripheral *a*. distant from the centre; near circumference; *appl*. membrane proteins that are weakly attached to membrane and easily removed.

peripheral blood lymphocytes (PBL) lymphocytes circulating in the blood.

peripheral lymphocytes lymphocytes present in the blood (as opposed to lymphoid tissues).

peripheral nervous system the nervous system of vertebrates other than the brain and spinal chord, comprising sensory receptors of trunk, limbs and internal organs and nerves other than the cranial nerves.

peripherical *a*. *appl*. a plant embryo more-

or-less completely surounding the endosperm in seed.

periphloem *n.* phloem sheath or pericambium *q.v.*

periphloic *a.* having phloem outside a centric xylem in vascular bundle.

periphoranthium periclinium *q.v.*

periphorium *n.* fleshy structure supporting ovary and to which stamens and corolla are attached.

periphyllum *n.* scale at base of ovary in grasses, supposed to represent part of perianth.

periphysis *n.* in some fungi, short sterile hair fringing the inside of a pore or aperture in fruiting body.

periphyton *n.* plants and animals adhering to parts of rooted aquatic plants.

peripileic *a.* arising from around the margin of cap in agaric fungi.

periplasm *n.* cytoplasm surrounding yolk of ova with central yolk; the layer of protoplasm surrounding the oosphere in some Oomycetes.

periplasmic space space between the plasma membrane and the cell wall in bacteria.

periplasmodium *n.* a mass of multinucleate protoplasm derived from tapetal cells and enclosing developing spore.

periplast *n.* outer covering of cells of some algae of the Cryptophyta; intercellular substance or stroma of tissues.

peripneustic *a.* having spiracles arranged all along sides of body, the usual condition in most insect larvae.

peripodial *a.* surrounding an appendage; *appl.* membrane covering wing bud in insects.

periportal *a. pert.* transverse fissure of the liver; *appl.* connective tissue partially separating lobules and forming part of the hepatobiliary capsule.

peripyle *n.* one of the apertures, additional to the astropyle, of the central capsule of some radiolarians.

perisarc *n.* chitinous outer casing of common tissue connecting individuals in some colonial hydrozoans.

periscleral *a. appl.* lymph space external to sclera of the eye.

perisome *n.* a body wall; integument of echinoderms.

perisperm *n.* in some seeds, a storage tissue formed by proliferation of the nucellus rather than the endosperm.

perispiracular *a.* surrounding a spiracle; *appl.* glands with oily secretion in certain aquatic insect larvae.

perisporangium *n.* membrane covering a sorus; indusium of ferns.

perispore *n.* spore-covering; transient outer membrane enveloping a spore; mother cell in algal spores.

perissodactyl *a.* with an uneven number of digits.

Perissodactyla, perissodactyls *n., n.plu.* order comprising the odd-toed ungulate mammals, in which the weight is borne mainly on the 3rd toe, and which includes the horse, tapir and rhinoceros. *cf.* Artiodactyla.

peristalsis *n.* successive contractions of muscular walls of gut that moves gut contents along. *a.* peristaltic.

peristasis *n.* the environment, including internal physiological action, that produces a certain phenotype from a genotype. *a.* peristatic.

peristigmatic perispiracular *q.v.*

peristome, peristomium *n.* (*bot.*) ring of teeth around mouth of spore capsule of mosses; (*zool.*) region surrounding the mouth, in ciliate protozoans, starfish, annelid worms, insects, echinoderms, etc.

perisystole *n.* interval elapsing between diastole and systole of heart.

perithecium *n.* a flask-shaped structure opening in a terminal hole (ostiole) and which contains the asci in some groups of ascomycete fungi, the flask fungi or Pyrenomycetes.

perithelium *n.* connective tissue associated with capillaries.

peritoneal *a. pert.* peritoneum, *appl.* cavity, membrane, fossa, etc.

peritoneum *n.* membrane partly applied to abdominal walls, partly extending over the organs contained in abdominal cavity, delimiting the peritoneal cavity.

peritreme, peritrema *n.* margin of a shell opening; small plate perforated by spiracle opening in ticks and mites.

Peritrichia, peritrichans *n., n.plu.* a group of ciliate protozoans which are usually fixed permanently to the substratum and have few cilia, e.g. *Vorticella*.

peritrichous *a. appl.* flagella, distributed all over the cell surface.

peritrochium *n.* a band of cilia; a circularly ciliated larva.

peritrophic *a.* (*zool.*) *appl.* a fold of membrane in mid-gut of insects and to space between it and gut lining; (*bot.*) *appl.* mycorrhiza with special fungal populations on root surfaces.

perittogamy *n.* random plasmogamy of undifferentiated cells in gametophytes.

periurethral *a.* surrounding the urethra, *appl.* glands, homologues of prostate glands.

perivascular *a.* surrounding a blood vessel, *appl.* lymph channels, fibres, spaces; (*bot.*) surrounding the vascular cylinder, *appl.* fibres.

perivascular feet terminal enlargements of processes of astrocytes in contact with minute blood vessels.

perivisceral *a.* surrounding the viscera, *appl.* body cavity.

perivitelline *a. appl.* space between ovum and zona pellucida.

perixylic *a.* having xylem outside centric phloem, *appl.* vascular bundles.

perizonium *n.* the membrane or siliceous wall enveloping the autospore or zygote in diatoms.

permafrost *n.* a layer, usually sub-surface, of permanently frozen soil which occurs where temperatures are low enough, and which is a feature of the tundra in polar regions. It is generally covered in warmer months with a thin partially melted layer, which in the tundra supports a vegetation of mosses, lichens, grasses and small herbaceous plants. The permafrost layer may be up to hundreds of metres thick.

permanent cartilage cartilage that remains unossified throughout life as opposed to temporary cartilage which is ossified into bone.

permanent hybrid a heterozygote which breeds true because of the elimination of certain homozygous genotypes by lethal factors in the genotype.

permanent teeth or dentition a set of teeth developed after milk or deciduous teeth, the second set of most, third set of some mammals. Some mammals do not develop a second set of teeth.

permanent tissue tissue consisting of cells which have completed their period of growth, are fully differentiated, and subsequently change little until they die.

permeants *n.plu.* animals which move freely from one community or habitat to another.

permease *n.* general term for any protein that specifically transports a solute across the plasma membrane; any of various proteins in bacteria responsible for carrying solutes into the cell.

Permian *a. pert.* or *appl.* geological period lasting from about 286 to 250 million years ago.

permissive *a.* (*genet.*) *appl.* conditions under which an organism or cell carrying a conditional lethal mutation does not display the mutant phenotype; (*virol.*) *appl.* conditions allowing normal infection and multiplication of a virus in a cell, and which may be environmental or genetic (i.e. species differences in susceptibility); (*dev.*) *appl.* a developmental signal that merely allows a certain process to proceed, *cf.* instructive.

peronaeus peroneus *q.v.*

peronate *a.* covered with woolly hairs; surrounded by volva, *appl.* stalk; powdery or mealy externally.

peroneal *a. pert.,* or lying near, the fibula, *appl.* artery, nerve, tubercle.

peroneotibial *a.* in region of fibula and tibia, *appl.* certain muscles.

peroneus *n.* any of several muscles arising from the fibula in the lower leg, as two lateral muscles, longus and brevis, and an anterior muscle, tertius.

peronium *n.* in some jellyfish, one of the cartilaginous processes ascending from the margin of the disc towards the centre.

peropod *a.* with rudimentary limbs.

peroral *a. appl.* membrane formed by the concrescence of rows of cilia around the cytopharynx of ciliate protozoans.

peroxidases *n.plu.* enzymes present in plants and animals, esp. mammalian spleen and lung, which act on hydrogen peroxide and organic peroxides only in the presence of an oxygen acceptor, EC 1.11.1; specifically, an enzyme using several oxidizing donors with hydrogen peroxide, EC

1.11.1.7.

peroxisome *n*. a small organelle bounded by a single membrane and containing catalase and peroxidases. In liver cells is believed to be important in detoxification reactions (e.g. of ethanol). *see also* glyoxysome. *alt.* microbody.

perradius *n*. one of the four main radii of a radially symmetrical animal.

perseveration *n*. the persistence of a response after the original stimulus has ceased.

persistent *a*. remaining attached until maturity, as corolla or perianth on a developing fruit; *appl.* teeth with continuous growth.

person *n*. an individual or zooid in a colonial coral, hydrozoan, siphonophore, etc.

Personatae Scrophulariaceae *q.v.*

personate *a. appl.* corolla of two lips, touching and with a projection of the lower closing the throat of the corolla, as in snapdragons.

perthophyte *n*. a parasitic fungus that obtains nourishment from host tissues after having killed them by a poisonous secretion.

pertusate *a*. pierced at apex.

pertussis toxin (PT) protein toxin from *Bordetella pertussis*, the bacterium causing whooping cough. Used experimentally to study G protein activity, as it induces permanent activation of some G proteins.

perula *n*. a leaf bud scale.

pervalvar *a*. dividing a valve longitudinally.

pervious *a. appl.* nostrils with no septum between nasal cavities.

pes *n*. a foot, base or foot-like structure, as in certain parts of brain, branches of facial nerve. *plu.* pedes.

pessulus *n*. an internal dorsoventral rod at lower end of trachea in syrinx of birds.

pesticide *n*. general term for any chemical agent that kills unwanted weeds, fungi or animal pests.

petal *n*. modified sterile leaf, often brightly coloured and with others forming the corolla, or inner series of perianth segments, of a flower. *see* Fig. 4.

petaliferous *a*. bearing petals.

petaliform *a*. petal-shaped, petal-like.

petalody *n*. the conversion of other parts of a flower into petals.

petaloid *a*. resembling a petal.

petaloideous *a. appl.* monocotyledons with a coloured perianth.

petalomania *n*. an unusual multiplication of petals.

petasma *n*. a complex membranous plate on inner side of peduncle with interlocking coupling hooks, in certain crustaceans.

petiolar *a. pert.* a petiole.

petiolate *a*. having a petiole.

petiole *n*. (*bot*.) the stalk of a leaf; (*zool*.) slender stalk connecting thorax and abdomen in certain insects such as wasps; any short, slender stalk-like structure.

petiolule *n*. the stalk of a leaflet of a compound leaf.

petites *n.plu.* yeast strains isolated as slow-growing small colonies on solid medium, which lack mitochondrial function due to mutations in nuclear or mitochondrial genes.

Petri dish shallow circular glass or plastic dish used for growing microorganisms or cultured cells, etc. in a suitable medium. *alt.* plate.

petrification *n*. fossilization through saturation with mineral matter in solution, subsequently turned to solid.

petrohyoid *a. pert.* hyoid and petrous part of temporal bone of skull.

petromastoid *a. pert.* mastoid process and petrous portion of temporal bone of skull.

petro-occipital *a. pert.* occipital and petrous part of temporal bone of skull, *appl.* a fissure.

petrophyte *n*. a rock plant.

petrosal *a*. made of compact bone; *appl.* otic bones of fishes; *appl.* a sphenoidal process, to a ganglion of glossopharyngeal nerve, and to nerves and sinus in region of petrous portion of temporal bone; *appl.* bone: the periotic *q.v.. alt.* petrous.

petrosphenoidal *a. pert.* sphenoid and petrous portion of temporal bone, *appl.* fissure.

petrosquamosal *a. pert.* squamosal and petrous portion of temporal bone, *appl.* sinus and suture.

petrotympanic *a. pert.* tympanum and petrous portion of temporal bone; *appl.* fissure: the fissure in the temporal bone of mammals which holds the Folian process of the malleus of middle ear.

petrous *a.* very hard or stony; *appl.* a pyramidal portion of the temporal bone behind sphenoid and occipital; *appl.* a ganglion on its lower border. *alt.* petrosal.

petunidin *n.* a purple anthocyanin pigment found in petunias and other plants.

Peyer's patches lymph glands scattered along the inner lining of the walls of the intestine.

P face in freeze-fractured membranes, the face representing the hydrophobic interior of the cytoplasmic half of the lipid bilayer.

P factors transposable DNA sequences (now known as P elements) present on the chromosomes of certain strains of *D. melanogaster* which are responsible for hybrid dysgenesis in progeny of matings of males carrying these factors with females of certain other strains (M cytotype).

Pflüger's cords columns of cells growing from germinal epithelium into stromatic tissue of embryo and which give rise to the gonads.

phacoid *a.* lentil or lens shaped.

phaeic *a.* of a dusky colour.

phaeism *n.* duskiness; *appl.* colouring of butterflies due to incomplete melanism.

phaenantherous *a.* with anthers extending beyond rim of flower; with stamens extending beyond rim of flower.

phaeo- pheo-.

phaeochrome chromaffin *q.v.*

phaeochrous *a.* of a dusky colour.

phaeodium *n.* in some radiolarians, an aggregation of food and excreta forming a mass around the central aperture of capsule.

phaeomelanin *n.* a brownish melanin.

phaeophyll *n.* the colouring matter of brown algae, a mixture of fucoxanthin, xanthophyll, chlorophyll and carotene.

Phaeophyta *n.* the brown algae or brown seaweeds, in some modern classifications a division of the kingdom Protoctista (Protista), in more traditional classifications regarded as a division of the kingdom Plantae. Mainly multicellular and marine, they can reach large sizes and some groups possess complex internal structure. Their green chlorophyll is masked by the brown pigment fucoxanthin so that they appear brownish. Carbohydrate reserves are in the form of laminarin.

phaeophytin pheophytin *q.v.*

phaeoplast *n.* pigmented plastid of brown algae.

phaeosome *n.* an optic organelle in some epidermal cells of annelid worms.

phaeospore *n.* a spore containing phaeoplasts.

phage bacteriophage *q.v.*

phage conversion phenomenon in which genes carried by a temperate phage change the phenotype of the host bacterium, e.g. alteration of O-chain polysaccharide in salmonellae, and the production of phage-specified toxin by *Corynebacterium diphtheriae.*

phagocyte *n.* a cell specialized to carry out phagocytosis, in mammals chiefly macrophages and polymorphonuclear leukocytes; in plants, a root cell with a lobed nucleus, capable of digesting endotrophic fungal filaments.

phagocytic *a. pert.* or effecting phagocytosis.

phagocytic vacuole a membrane-bounded vesicle in the cytoplasm, formed by budding off of an invagination of the cell membrane and containing material taken up by phagocytosis. *alt.* phagosome.

phagocytose *v.* to carry out phagocytosis.

phagocytosis *n.* uptake of large solid particles (including other cells) into a cell by the process of endocytosis, seen, e.g., in amoeboid protozoans engulfing their prey and in specialized cells of the vertebrate immune system which ingest and destroy invading microorganisms and scavenge damaged and senescent cells.

phagolysis *n.* the lysis or disintegration of phagocytes.

phagolysosome *n.* a cytoplasmic vesicle formed by the fusion of a phagosome and lysosome.

phagosome phagocytic vacuole *q.v.*

phagotroph *n.* any heterotrophic organism (as many animals and protozoans) that ingests nutrients as solid particles. *cf.* osmotroph.

phagozoite *n.* an animal which feeds on disintegrating or dead tissue.

phalange phalanx *q.v.*

phalangeal *a. pert.* or resembling phalanges, like segmented fingers.

phalanges *n.plu.* the bones of the fingers

and toes of vertebrates; various finger-like processes in other organs.

phalanx *n*. one of several bones (phalanges) in fingers or toes; a bundle of stamens united by filaments; a taxonomic group, never precisely defined, but usually used for a group resembling a subfamily.

phallic *a. pert.* phallus; *appl.* gland secreting substance for spermatophores, as in certain insects.

phalloidin *n*. alkaloid obtained from *Amanita phalloides* which binds to actin filaments and prevents cell movement.

phallomere *n*. valve of penis in insects.

phallosome *n*. a structure of tissue from inner surface of base of clasper and penis valves in male mosquitoes.

phallus *n*. the embryonic structure which becomes penis or clitoris; the penis; external genitalia of a male insect; the fruiting body of the stinkhorn fungi.

phanerogams *n.plu.* all seed-bearing plants; formerly used for plants with conspicuous flowers. *a.* phanerogamic.

phanerophyte *n*. tree or shrub with aerial dormant buds; plant whose size is not appreciably less during cold or dry season.

phaneroplasmodium *n*. in slime moulds, a thick opaque plasmodium with veins having clearly defined endoplasm and ectoplasm.

Phanerozoic *n*. eon comprising the Paleozoic, Mesozoic and Cenozoic eras.

phaoplankton *n*. surface plankton living at depths at which light penetrates.

pharate *a. appl.* instar (larval stage of some insects) within previous cuticle before moulting.

pharmacodynamics *n*. the study of the action of drugs, including all aspects of their behaviour in the body, i.e. transport to t issues, persistence in blood stream and tissues, as well as their immediate biochemical activity.

pharmacogenetics *n*. the study of genetically-determined responses to drugs.

pharmacology *n*. the study of the action of medicinal drugs and other biologically active chemicals.

pharmacophore *n*. the part of a molecule causing the specific physiological effects of a drug.

pharotaxis *n*. the movement of an animal towards a definite place, the stimulus for which is acquired by conditioning or learning.

pharyngeal *a. pert.* pharynx.

pharyngobranchial *n*. a dorsal skeletal element of gill arch.

pharyngopalatine *a. pert.* pharynx and palate, *appl.* arch and muscle.

pharyngotympanic *a. pert.* tube connecting pharynx and tympanic cavity, the auditory or Eustachian tube.

pharynx *n*. in humans and other vertebrates the throat, in other animals the gullet or anterior part of the alimentary canal or enteron following the mouth.

phase variation in *Salmonella* spp. the change from expression of one gene for flagellin (a flagellar protein) to another non-allelic flagellin gene which occurs at a regular frequency within a population (around once every 1000 cell divisions), originally detected by a change in antigenic specificity within the population.

phase-contrast microscopy type of optical microscopy which enables living unstained cells and tissue to be studied by using the way different parts of the cell diffract light to giving a high-contrast image.

phaseolin *n*. a globulin protein obtained from the seeds of the bean *Phaseolus*.

phasmid *n*. one of a pair of posterior sense organs in nematodes, possibly detecting chemical stimuli; stick or leaf insect, a member of the insect order Phasmida.

Phasmida *n*. order of insects including the stick insects, which are long slender insects with long legs, and the leaf insects, which have flattened bodies with leaf-like flaps on their limbs, both being excellently camouflaged in the bushes and trees in which they live.

phellem(a) *n*. cork; cork and the non-suberized layers forming an external zone of periderm produced by the cork cambium.

phelloderm *n*. the secondary cortex of parenchyma cells filled with suberin that is formed by and on the inner side of the cork cambium. *alt.* secondary cortex.

phellogen cork cambium *q.v.*

phelloid *a.* cork-like; *n.* non-suberized cell layer in outer periderm.

phencyclidine *n.* a hallucinogenic drug.

phene *n.* a phenotypic character which is genetically determined.

phenetic *a. appl.* classification purely based on similarities in phenotypic characters, not necessarily reflecting relationships by evolutionary descent.

phenocontour *n.* a contour line on a map showing the distribution of a certain phenotype; a line connecting all places within a region at which a biological phenomenon, e.g. flowering of a plant, occurs at the same time; a contour line delimiting an area corresponding to a given frequency of a variant form.

phenocopy *n.* a modification produced by environmental factors which simulates a genetically determined change.

phenocritical period for a particular gene, the time during development when its expression is required.

phenogram *n.* a tree-like diagram showing the conclusions of numerical taxonomy.

phenological *a. pert.* phenology; *appl.* isolation of species owing to differences in flowering or breeding season.

phenology *n.* recording and study of periodic biological events, such as flowering, breeding, migration, etc., in relation to climate and other environmental factors; the study of cyclical events in constructive metabolism (anabolism).

phenome *n.* all the phenotypic characteristics of an organism.

phenomenology phenology *q.v.*

phenon *n.* group of organisms placed together by numerical taxonomy.

phenotype *n.* the visible or otherwise measurable physical and biochemical characteristics of an organism, a result of the interaction of genotype and environment; a group of individuals exhibiting the same phenotypic characters. *a.* phenotypic.

phenotypic *a. pert.* phenotype. *cf.* genotypic.

phenotypic plasticity the range of variability shown by the phenotype in response to environmental fluctuations.

phenylalanine (Phe, F) *n.* amino acid with an aromatic side chain, constituent of protein, essential in human diet. *see* table in Appendix 1 for chemical formula.

phenylketonuria *n.* inborn error of metabolism due to absence or deficiency of phenylalanine hydroxylase, leading to accumulation of phenylalanine in all body fluids, and to mental retardation and early death if untreated.

pheophytin *n.* either of 2 blue-black pigments, components of the photosynthetic electron transport chain, derived from chlorophylls *a* and *b* by removal of the magnesium atom. *alt.* phaeophytin.

pheromone *n.* a chemical released, usually in minute amounts, by one organism which is detected and acts as a signal to another member of the same species, such as the volatile sexual attractants released by some female insects, which can attract males from a distance. Some pheromones act as alarm signals.

phialid(e) *n.* a small bottle-shaped outgrowth of hypha in some fungi, from which spores (phialospores) are produced.

phialiform *a.* cup-shaped; saucer-shaped.

phialophore *n.* a hypha which bears a phialide.

phialopore *n.* the opening in a hollow daughter colony or gonidium of the protist *Volvox*.

phialospore *n.* a spore or conidium borne at tip of a phialide.

Philadelphia chromosome abnormal chromosome 22 (22q-) present in leukaemic cells of many patients with chronic myelogenous leukaemia, in which material from the long arm of 22 has been exchanged with material from the end of the long arm of chromosome 9, the translocation producing a novel gene on chromosome 22, *abl-bcr*, which may be involved in oncogenesis.

-phil, -philous suffixes derived from Gk. *philein*, to love, denoting loving, or thriving in.

philopatry *n.* tendency of an organism to stay in or return to its home area.

philoprogenitive *a.* having many offspring.

philtrum *n.* the depression on upper lip beneath septum of nose.

phlebenterism *n.* the condition of having branches of the intestine extending into other organs, as arms and legs.

phlobaphaenes *n.plu.* phenolic compounds, derivatives of tannins, producing yellow, red or brown colours in fern ramenta, roots and sections of wood.

phloem *n.* the principal food-conducting tissue of vascular plants, extending throughout the plant body. It is composed of elongated conducting vessels, sieve tubes (in angiosperms) or sieve cells (in ferns and gymnosperms), both containing clusters of pores (sieve areas) in the walls, through which the protoplasts of adjacent cells communicate. Sugars and amino acids are the main nutrients transported via the phloem. Parenchymatous companion cells (or albuminous cells in gymnosperms) closely associated with the conducting elements are involved in the delivery to and uptake of material from the phloem. Phloem also contains supporting fibres (bast). *see also* vascular bundle, xylem.

phloem loading the active transport of sugars into the phloem at their site of synthesis.

phloem-mobile *appl.* ions that can be transported via the phloem, e.g. K^+, Cl^-, $H_2PO_4^{2-}$, but not Ca^{2+}, which is termed phloem-immobile.

phloem parenchyma thin-walled parenchyma associated with the sieve tubes of phloem.

phloem sheath pericycle *q.v.*

phloeodic *a.* having the appearance of bark.

phloic *a. pert.* phloem.

phlorizin *n.* plant glucoside from roots, used experimentally as an uncoupler of electron transport and ATP synthesis in chloroplasts.

phobotaxis *n.* avoiding reaction in some protozoans.

phocids *n.plu.* members of the Phocidae: the seals.

Phoenicopteriformes *n.* an order of birds in some classifications, including the flamingoes.

pholadophyte *n.* a plant living in hollows, shunning bright light.

pholidosis *n.* the arrangement of scales, as on scaled animals.

Pholidota *n.* an order of placental mammals known from the Pleistocene or possibly Oligocene, the only living member being the pangolin (scaly anteater), having no teeth, and the body covered with imbricated scales.

phonation *n.* production of sounds, e.g. by insects.

phonoreceptor *n.* a receptor for sound waves.

phoranth(ium) *n.* the receptacle of flowerheads of Compositae.

phorbol esters polycyclic alcohols derived from croton oil (e.g. 12-O-tetradecanol phorbol-13 acetate) which activate protein kinase C as a result of their resemblance to diacylglycerol. Although not in themselves carcinogenic, they can act as tumour promoters through persistent activation of protein kinase C and thus of signalling pathways involved in cell proliferation.

phoresia, phoresy *n.* the carrying of one organism by another, without parasitism, as in certain insects.

Phoronida *n.* a small phylum (only 15 species) of marine worm-like coelomate animals that secrete chitinous tubes in which they live. The mouth is surrounded by a horseshoe-shaped crown of tentacles (a lophophore) which projects from the tube.

phoront *n.* an encysted stage leading to formation of trophont in life cycle of some ciliate protozoans.

phosphagen creatine phosphate *q.v.*

phosphatase *n.* any of a large group of widely distributed enzymes catalysing the hydrolysis and synthesis of organic phosphate esters, including the hydrolytic removal of phosphate (phosphoryl) groups from proteins, and also the transfer of phosphate groups from one compound to another. EC 3.1.3. *see also* phosphorylation, protein phosphatase.

phosphatidalcholine, phosphatidalethanolamine, phosphatidalserine *see* plasmalogen *q.v.*

phosphatide phospholipid *q.v.*

phosphatidylcholine *n.* a phosphoglyceride (*q.v.*) with a choline alcohol group, the principal phospholipid in most membranes of higher organisms, occurring in different forms depending on the particular fatty acid substituents. *alt.* lecithin. *see* Appendix 1 (40) for chemical structure.

phosphatidylethanolamine *n.* a phosphoglyceride (*q.v.*) with ethanolamine as the alcohol group, a common phospholipid of cell membranes. *alt.* (formerly) cephalin.

phosphatidylinositide (PI) pathway a biochemical pathway for signal transduction from receptor proteins on cell surface to the intracellular biochemical response machinery. Stimulation of certain receptors results in the breakdown of phosphatidylinositol phosphates in the membrane into inositol trisphosphate and diacylglycerol. This is catalysed by phospholipase C. Both inositol trisphosphate and diacylglycerol have second messenger activity, stimulating the release of Ca^{2+} from intracellular stores and the activation of protein kinase C respectively.

phosphatidylinositol *n.* a phosphoglyceride (*q.v.*) with inositol as the alcohol group, a common phospholipid of cell membranes.

phosphatidylinositol phosphate (PIP) phosphate derivative of inositol, a membrane phospholipid that is broken down to form the second messengers diacylglycerol and inositol trisphosphate. *see* Appendix 1 (41) for chemical structure.

phosphatidylserine *n.* a phosphoglyceride (*q.v.*) with serine as the alcohol group, a common phospholipid of cell membranes.

phosphene *n.* a light impression on retina due to stimulus other than rays of light.

phosphodiester bond the covalent bond -O-P- that e.g. forms the linkage between nucleotide residues in a polynucleotide chain.

phosphodiesterase *n.* any of a group of enzymes that hydrolyse phosphodiester bonds, and including cyclic AMP phosphodiesterase which converts cyclic AMP to adenosine monophosphate.

phosphoenolpyruvate (PEP) *n.* 3-carbon intermediate in the conversion of phosphoglycerate to pyruvate in glycolysis, and whose conversion to pyruvate generates ATP, also an important intermediary metabolite in biosynthesis of glycogen, neuraminic acid and phenylalanine. *see* Appendix 1 (51).

phosphofructokinase enzyme that phosphorylates fructose 6-phosphate to fructose 1,6-bisphosphate in glycolysis and other metabolic pathways (*r.n.* 6-phosphofructokinase, EC 2.7.1.11) or the enzyme using fructose 1-phosphate in the same reaction (*r.n* 1-phosphofructokinase, EC 2.7.1.56). *alt.* fructokinase.

phosphoglucomutase *n.* a widely distributed phosphotransferase which catalyses the conversion of glucose 1-phosphate to glucose 6-phosphate using glucose 1,6-bisphosphate (diphosphate) as donor. EC 2.7.5.1.

phosphogluconate oxidative pathway pentose phosphate pathway *q.v.*

phosphoglucose isomerase glucose-phosphate isomerase *q.v.*

phosphoglycerate (phosphoglyceric acid) (PGA) 3-carbon monosaccharide, occurring in all cells as 2- or 3-phosphoglycerate, an important intermediate in photosynthesis, respiration and carbohydrate metabolism. *see* Appendix 1 (50).

phosphoglyceride *n.* any of a group of phospholipids based on the 3-carbon alcohol glycerol phosphate, and consisting of a glycerol backbone, (usually) 2 fatty acid chains and a phosphorylated alcohol (choline, ethanolamine etc.), and which are constituents of cell membranes.

phosphoglyceromutase a widely distributed phosphotransferase which catalyses the conversion of 2-phosphoglycerate to 3-phosphoglycerate using 2,3-bisphosphoglycerate as donor. EC 2.7.5.3.

phosphohexose isomerase glucose-phosphate isomerase *q.v.*

phosphoinositide *n.* any sphingolipid containing inositol.

phosphokinase *n.* enzyme that catalyses the addition of phosphate groups to a molecule. *alt.* kinase.

phospholipase *n.* any of a group of enzymes which catalyse the hydrolysis of membrane phospholipids to give diacylglycerol and a phosphate of the headgroup.

phospholipid *n.* any of a group of amphipathic lipids with either a glycerol or a sphingosine backbone, fatty acid side chains and a phosphorylated alcohol headgroup, which form the lipid bilayer in all biological membranes. They include the glycerolipids phosphatidylcholine, phosphatidylethanolamine, phosphatidylinositol and phosphatidylserine, and the sphingolipid sphingomyelin.

phosphoprotein *n.* a protein carrying phosphoryl groups, which are added by protein kinases after the protein has been synthesized.

phosphorescence *n.* the luminescence of marine protozoans, copepods and the majority of deep-sea animals, which is produced without accompanying heat.

phosphoribosylpyrophosphate (PRPP) an activated form of ribose phosphate which is an important intermediate in biosynthesis of aromatic amino acids and purine and pyrimidine nucleotides.

phosphorolysis *n.* cleavage of a chemical bond by orthophosphate (P_i), in biochemical reactions catalysed by phosphorylases. *cf.* hydrolysis.

phosphorylase *n.* recommended name for a group of enzymes that catalyse the progressive breakdown of glucose polysaccharides such as starch or glycogen by phosphorolysis, giving glucose 1-phosphate, EC 2.4.1.1; any of a group of enzymes that catalyse the transfer of glucose residues from a glucose oligo- or polysaccharide to orthophosphate giving glucose 1-phosphate as a product, e.g. maltose phosphorylase (EC 2.4.1.8).

phosphorylase kinase protein kinase that phosphorylates glycogen phosphorylase and activates it.

phosphorylation *n.* the addition of a phosphate (phosphoryl) group to a molecule. The enzymatic phosphorylation of proteins at specific amino acids by protein kinases is a widespread means of altering a protein's activity.

phosphorylation potential index of the energy status of a cell in terms of potential transferable phosphate groups, being calculated as the ratio of the concentration of ATP to the product of the concentrations of ADP and inorganic phosphate.

phosphoserine *n.* phosphorylated derivative of the amino acid serine in proteins, modified after incorporation into the protein chain, enzymatic phosphorylation and dephosphorylation of serine being an important mechanism for regulating the activity of certain enzymes and other proteins.

phosphothreonine *n.* phosphorylated derivative of the amino acid threonine in proteins, modified by protein kinases after incorporation into the protein chain, enzymatic phosphorylation and dephos-

phorylation of threonine being an important mechanism for regulating the activity of certain enzymes and other proteins.

phosphotyrosine *n.* phosphorylated derivative of the amino acid tyrosine in proteins, modified after incorporation into the protein chain, enzymatic phosphorylation and dephosphorylation of tyrosine being an important mechanism for regulating the activity of some hormone receptors and other proteins.

phosphovitin, phosvitin *n.* a phosphoprotein of amphibian egg yolk.

photic *a. appl.* zone of surface waters penetrated by sunlight.

photo- prefix derived from Gk. *phôs*, light, indicating response to, sensitivity to or causation by light.

photoassimilate *n.* the carbon-containing compounds produced as a result of photosynthesis.

photoautotroph *n.* organism using light as an energy source and carbon dioxide as the main source of carbon, as green plants and some bacteria.

photobleaching *n.* loss of colour by photosensitive pigments such as rhodopsin on exposure to light.

photoceptor photoreceptor *q.v.*

photochemical *a. appl.* and *pert.* chemical changes brought about by light.

photochromatic *a. appl.* interval between achromatic and chromatic thresholds.

photochromic effect change of colour brought about by light.

photodinesis *n.* protoplasmic streaming induced by light.

photodynamics *n.* the study of the effects of light on plants.

photogen *n.* light-producing organ or substance.

photogene *n.* a gene whose expression is controlled by light.

photogenic *a.* light-producing; luminescent.

photogenin luciferase *q.v.*

photoheterotroph *n.* an organism that uses light as a source of energy, but derives much of its carbon from organic compounds, such as the photosynthetic nonsulphur purple bacteria.

photoinhibition *n.* inhibition by light, e.g. of germination.

photokinesis *n.* a kinesis (random movement) in response to certain regions of visible light spectrum.

photolabile *a. appl.* substances such as retinal pigments which undergo a chemical change on exposure to light.

photolithotroph photoautotroph *q.v.*

photolyase *n.* enzyme that catalyses the repair of pyrimidine dimers in UV-irradiated DNA by light. EC 4.1.99.3. *r.n.* deoxyribopyrimidine photo-lyase. *alt.* photoreactivating enzyme.

photolysis *n.* splitting a compound or molecule by the action of light, as the splitting of water into hydrogen and oxygen.

photomorphogenesis *n.* any effect on plant growth produced by light.

photomotor reflex change in size of pupil of eye with sudden change in light intensity.

photon *n.* the particulate unit of light, carrying a quantum of energy. 1 mol photons (or 1 mol quanta) is the number of photons corresponding to Avogadro's number of particles (6.023×10^{23}) and is the number of photons required to convert 1 mol of a substance to another form with 100% efficiency if captured in a single step. Photon number incident on a surface normal to the beam in a given time is the photon flux (often called photon flux density, not recommended) and is measured in mol $m^{-2} s^{-1}$.

photonasty *n.* response of plants to diffuse light stimuli, or to variations in illumination.

photoorganotroph photoheterotroph *q.v.*

photopathy *n.* a pronounced movement in relation to light, usually away from it as in negative phototaxis or phototropism.

photoperiod *n.* duration of daily exposure to light; the length of day favouring optimum functioning of an organism.

photoperiodicity, photoperiodism *n.* the response of an organism to the relative duration of day and night, such as the flowering of many plants and mating of many animals, which is triggered by the lengthening or shortening of the days as the seasons change.

photophase *n.* the developmental stage of a plant when it shows definite requirements as to duration and intensity of light and temperature.

photophilous *a.* seeking, and thriving in, strong light.

photophobic *a.* not tolerating light, shunning light.

photophore *n.* a light-emitting organ which directs light ventrally in some deep-sea fish, crustaceans and cephalopods, and so camouflages the silhouette from ventral view; any light-emitting organ.

photophosphorylation *n.* the formation of ATP using energy from light during photosynthesis. *cf.* oxidative phosphorylation.

photophygous *a.* avoiding strong light.

photopia *n.* adaptation of the eye to light.

photopigment *n.* any light-sensitive pigment.

photoplagiotropy *n.* tendency to take up a position transverse to the incident light.

photopsin *n.* protein component of the violet retinal cone pigment iodopsin.

photoreactivation *n.* the reactivation or repair of some inactivated or damaged protein, DNA, etc. by the stimulus of light, which initiates a light-dependent enzymatic reaction.

photoreceptor *n.* sense organ responding to light (e.g. eye); cell or part of a cell sensitive to light (e.g. rod or cone cells in the retina); molecule sensitive and responding to light, such as rhodopsin in retina, chlorophyll in plants.

photoregulation *n.* the regulation of genes, or a physiological or developmental process, by light.

photorespiration *n.* type of "wasteful" respiration occurring in green plants in the light, different from normal mitochondrial respiration, consuming oxygen and evolving carbon dioxide using chiefly glycolate derived from the primary photosynthate as substrate, occurring to a much greater extent in C3 plants (*q.v.*) than in C4 plants (*q.v.*).

photospheres *n.plu.* luminous organs of crustaceans.

photosynthate *n.* product(s) of photosynthesis.

photosynthesis *n.* in green plants the synthesis of carbohydrate in chloroplasts from carbon dioxide as a carbon source and water as a hydrogen donor with the release of oxygen as a waste product, using light energy trapped by the green pigment, chlorophyll, to drive the synthesis of ATP (the light reaction). This is subsequently used as an energy source in carbohydrate synthesis (the dark reaction); in certain bacteria, a similar process but sometimes using hydrogen donors other than water and producing waste products other than oxygen. *a.* photosynthetic. *see also* photosystem I and II, reaction centre, chlorophyll, photophosphorylation, Calvin cycle.

photosynthetic carbon reduction cycle (PCR cycle) Calvin cycle *q.v.*

photosynthetic cycle Calvin cycle *q.v.*

photosynthetic efficiency the conversion factor of the energy falling per unit area on a photosynthetic tissue and the energy value of the biochemical compounds produced.

photosynthetic quotient ratio between the volume of oxygen produced and the volume of carbon dioxide used in photosynthesis.

photosynthetic reaction centre *see* reaction centre.

photosynthetic unit proposed functional unit composed of several hundred chlorophyll molecules, a reaction centre and accessory pigments, which is required to generate one oxygen molecule in photosynthesis.

photosynthetic zone of sea or lakes, the vertical zone in which photosynthesis can take place, between surface and compensation point.

photosynthetically active radiation (PAR) radiation capable of driving the light reactions of photosynthesis, wavelength 380 710 nm.

photosystem I (PSI) and photosystem II (PSII) the 2 multimolecular complexes found in thylakoid membranes of chloroplasts and involved in the light reactions of photosynthesis, PSI consisting of chlorophyll *a* bound to protein (the reaction centre P700) and generating NADPH, and being linked to PSII by an electron transport chain, and PSII consisting of chlorophylls *a* and *b* bound to protein (reaction centre P680), generating a strong oxidant, and splitting water to produce oxygen.

phototaxis *n.* movement in response to

light. Positive phototaxis: movement towards a light source; negative phototaxis: movement away from a light source. *a.* phototactic.

phototransduction *n.* the reception and interpretation by a cell of a signal in the form of light.

phototroph *n.* organism using sunlight as a source of energy. *a.* phototrophic.

phototropism *n.* growth movement of plants in response to stimulus of light. Where the stimulus is sunlight, sometimes called heliotropism.

phototropy *n.* a reversible change in the colour of a substance while it is illuminated.

phragma *n.* a septum or partition. *plu.* phragmata.

Phragmobasidiomycetes, Phragmobasidiomycetidae *n.* basidiomycete fungi which form basidiospores on a septate basidium, and including the rust and smut fungi.

phragmobasidium *n.* a septate basidium forming four cells.

phragmocone *n.* in certain molluscs, the cone of a shell divided internally by a series of partitions perforated by a siphuncle.

phragmocyttarous *a.* building, or *pert.,* to combs attached to a supporting surface, as of certain wasps.

phragmoplast *n.* system of vesicles laid down across middle of plant cell undergoing cell division which determine the formation of the cell plate and new cell wall.

phragmosis *n.* the use of part of the body to close a burrow (in reptiles or amphibians).

phragmosome microbody *q.v.*

phragmospore *n.* a septate spore, in fungi.

phratry *n.* a loose term in classification, never generally adopted or precisely defined, but often used to mean a subtribe.

phreaticolous *a. appl.* organisms living in underground fresh water.

phreatophyte *n.* plant with very long roots reaching water table.

phrenic *a.* in region of diaphragm, *appl.* artery, nerve, etc.

phrenicocostal *a. appl.* a narrow slit or sinus between costal and diaphragmatic pleurae.

phrenicolienal *a. appl.* ligament forming part of peritoneum reflected over spleen and extending to diaphragm.

phrenicopericardiac *a. appl.* ligament extending from diaphragm to pericardium.

phrenosin *n.* a glycolipid (cerebroside) obtained from the brain, which on hydrolysis yields a fatty acid (phrenosinic acid), galactose and sphingosine.

phthinoid *a.* withered; weak; underdeveloped.

phthisaner *n.* pupal male ant parasitized by an *Orasema* larva.

phthisergate *n.* a pupal worker ant parasitized by an *Orasema* larva.

phthisogyne *n.* a pupal female ant parasitized by an *Orasema* larva.

phyad *n.* a plant or animal form due to inheritance rather than environment.

phyco- prefix derived from Gk. *phykos,* seaweed, signifying to do with algae.

phycobiliproteins *n.plu.* protein pigments with phycobilin chromophores (e.g. phycoerythrin and phycocyanin) found in some algae and blue-green algae, where they act as accessory photosynthetic pigments.

phycobilisome *n.* one of a number of small particles present on photosynthetic lamellae of some red algae and cyanobacteria, and which contain phycobilin.

phycobiont *n.* the algal partner in a symbiosis, e.g. in lichens and in certain marine invertebrates.

phycochrome *n.* general term for a pigment found in algae.

phycochrysin *n.* a golden-yellow pigment present in chromophores of the golden-brown algae (Chrysophyta).

phycocoenology *n.* the study of algal communities.

phycocyanin *n.* a chromoprotein, giving cyanobacteria their colour and also present in red algae and Cryptophyceae, acts as an accessory photosynthetic pigment.

phycoerythrin *n.* a chromoprotein giving red algae their colour and also present in some cyanobacteria and Cryptophyceae, acts as an accessory photosynthetic pigment.

phycology *n.* the study of the algae.

Phycomycetes *n.plu.* a general name for the simple, mainly aquatic fungi of the modern classes Chytridiomycetes, Hyphochytridiomycetes, Plasmodiophoromycetes and Oomycetes.

phycophaein *n.* a brown pigment in brown algae, now thought to be an oxidation product of fucosan.

Phycophyta *n.* in some classifications the name for the algae.

phycoplast *n.* a system of microtubules that develops parallel to the plane of nuclear division in dividing cells of green algae of the class Chlorophyceae.

phycoxanthin *n.* a yellow or brownish-yellow pigment present in some algae and diatoms.

phyla *plu.* of phylum *q.v.*

phylacobiosis *n.* mutual or unilateral protective behaviour, as of certain ants. *a.* phylacobiotic.

phylactocarp *n.* a modification of hydrocladium in some hydrozoans, for protection of the gonophore.

Phylactolaemata *n.* in some classifications a class of freshwater Bryozoa.

phylembryo *n.* a developmental stage in brachiopods at completion of the embryonic shell.

phyletic *a. pert.* a phylum or a major branch of an evolutionary lineage; *appl.* a group of species related to each other by common descent; *pert.* a line of direct descent.

phyletic evolution sequence of evolutionary changes leading to a sequence of species or forms arising through time in a single line of descent.

phyletic gradualism the idea that evolutionary change is built up in small steps, the change in any single generation being extremely small. Each stage must have a selective advantage, however marginal, eventually giving rise to new forms, organs and functions by a cumulative effect. The term phyletic gradualism is sometimes also used to refer to the view that evolution proceeds at a steady rate by such imperceptible small changes, and in this usage is often contrasted with the idea of punctuated equilibria and a non-uniform rate of evolution.

phyllade *n.* a reduced scale-like leaf.

phyllary *n.* a bract of the involucre in Compositae.

phyllidium bothridium *q.v.*

phyllo- prefix derived from Gk. *phyllon,* leaf.

phyllobranchia *n.* a gill consisting of numbers of lamellae or thin plates.

phylloclade *n.* a green flattened photosynthetic stem; in cacti, a green, rounded stem functioning as a leaf; an assimilative branch of a "shrubby" thallus of lichens.

phyllocyst *n.* the rudimentary cavity of a hydrophyllium or protective medusoid.

phyllode *n.* winged leaf-stalk with flattened surfaces placed laterally to stem, functioning as a leaf.

phyllody *n.* metamorphosis of an organ into a foliage leaf.

phylloerythrin *n.* red pigment derived from chlorophyll and occurring in bile of herbivorous mammals.

phyllogen *n.* meristematic cells that give rise to primordial leaf.

phyllogenetic *a.* producing or developing leaves.

phylloid *a.* leaf-like; *n.* the leaf regarded as a flattened branch or telome.

phyllomania *n.* abnormal leaf production.

phyllome *n.* the leaf structures of a plant as a whole.

phyllomorphosis *n.* the metamorphosis of a plant organ into a foliage leaf; variation of leaves at different seasons.

phyllophagous *a.* feeding on leaves.

phyllophore *n.* terminal bud or growing point of palms.

phyllophorous *a.* bearing or producing leaves.

phylloplane *n.* the leaf surface.

phyllopode *n.* a sheathing leaf-base in quillworts (Isoetales).

phyllopodium *n.* the axis of a leaf; the stem when regarded as a pseudoaxis formed of fused leaf bases (as in some palms); a leaf-like swimming appendage in some crustaceans such as the branchiopods.

phylloquinone *see* vitamin K.

phyllorhiza *n.* a young leaf with a root.

phyllosiphonic *a.* with insertion of leaf trace disturbing axial stele tissue. *cf.* cladosiphonic.

phyllosoma *n.* larval stage of crawfish, being a broad thin schizopod larva.

phyllosperm *n.* a plant having seeds or spores borne on the leaves.

phyllosphere *n.* the leaf surfaces.

phyllospondylous *a. appl.* vertebrae consisting of a hypocentrum and neural arch, both contributing to hollow transverse

processes.

phyllosporous *a.* having sporophylls like foliage leaves.

phyllotactic *a. pert.* phyllotaxis, *appl.* the fraction of circumference of stem between successive leaves, representing the angle of their divergence.

phyllotaxis, phyllotaxy *n.* the arrangement of leaves on an axis or stem, which for spirally arranged leaves may be expressed as the number of circuits of the stem that have to be made and the number of leaves that have to be passed to progress from the point of attachment of one leaf to that immediately above it (e.g. 1/2 for leaves positioned 180° apart). *see also* orthostichy, parastichy.

phyllozooid *n.* a shield-shaped medusa with a protective function in a siphonophore colony.

phylogenesis phylogeny *q.v.*

phylogenetic *a. pert.* the evolutionary history and line of descent of a species or higher taxonomic group; *appl.* tree: diagram setting out the genealogy of a species or other taxon.

phylogenetics *n.* the line of descent of a species or higher taxon; approach to classification that attempts to reconstruct evolutionary genealogies and the historical course of speciation.

phylogeny *n.* the evolutionary history and line of descent of a species or higher taxonomic group. *cf.* ontogeny.

phylum *n.* in classification, a primary grouping consisting of animals constructed on a similar general plan, and thought to be evolutionarily related. In plants the similar category is called a division. Examples: Cnidaria (sea anemones, jellyfish, corals, etc.), Porifera (sponges), Platyhelminthes (flatworms, flukes and tapeworms), Mollusca (molluscs), Arthropoda (spiders, insects, crustaceans) and Chordata (includes the vertebrates). *plu.* phyla.

physa *n.* the rounded base of burrowing sea anemones.

physical containment in genetic engineering and in microbiology generally, the level of physical security and safety required in laboratory procedures, different levels being recommended for work involving microorganisms of differing degrees of pathogenicity.

physical mapping methods of gene mapping that rely on finding the positions of genes and other DNA sequences by means such as cytogenetics and restriction mapping which do not involve recombination-based genetic mapping methods.

physiogenic *a.* caused by the activity of an organ or part; caused by environmental factors.

physiographic succession plant succession influenced mainly by topography and local climate.

physiological races outwardly similar races within a species which differ in their physiology as a result of genetically determined factors, as some plant pathogenic fungi which differ in their virulence towards different varieties of the host species.

physiology *n.* that part of biology dealing with the functions and activities of organisms, as opposed to their structure. *a.* physiological.

physiological race a race characterized by different physiological features, not morphological.

physoclistous *a.* having no channel connecting swim bladder and digestive tract, as in most teleosts.

physodes *n.plu.* spindles of phlorglucin contained in plasmodium of certain Sarcodina; fucosan (*q.v.*) vesicles.

physostigmine *n.* an alkaloid derived from the Calabar bean, an inhibitor of the enzyme acetylcholinesterase. *alt.* eserine.

phytal zone shallow lake bottom and its rooted vegetation.

phytic acid hexaphosphoinositol, the phosphate derivative of the sugar alcohol *myo*-inositol, found chiefly in seeds.

phytin *n.* magnesium calcium phytate, a phosphate storage substance in seeds, a salt of phytic acid.

phyto- prefix derived from Gk. *phyton,* plant.

phytoactive *a.* stimulating plant growth.

phytoalexin *n.* any of a group of substances produced by plant cells in response to wounding or attack by parasitic fungi or bacteria, and which are involved in resistance to infection and in limiting the damage caused by accidental wounding.

phytobiotic *a.* living within plants, *appl.* certain protozoans.

phytochemistry *n.* the chemistry of plants.

phytochoria *n.plu.* phytogeographical realms and regions.

phytochory *n.* dissemination of pathogens through the agency of plants.

phytochrome *n.* a light-sensitive protein pigment in plants. It exists in two forms: P_r, which is sensitive to red light, which converts it into P_{fr} which is sensitive to far-red light. Far-red light converts P_{fr} back to P_r. P_{fr} is active in stimulating developmental processes such as flowering in short-day plants and germination in some seeds, whereas P_r is inactive.

phytocoenology *n.* the study of plant communities.

phytocoenosis *n.* the assemblage of plants living in a particular locality.

phytoedaphon *n.* microscopic soil flora.

phytogenesis *n.* the evolution, or development of plants.

phytogenetics *n.* plant genetics.

phytogenous *a.* of vegetable origin, produced by plants.

phytogeographical kingdoms major geographical divisions of the world according to their flora: Antarctic, Australian, Neotropical, Boreal, Palaeotropical and South African; floral realm (*q.v.*).

phytogeography *n.* the study of the geographical distribution of plants.

phytoglycogen *n.* large highly branched polymer of glucose, found in some plants, and which is abundant in "waxy" mutants of maize.

phytohaemagglutinin (PHA) *n.* a lectin, a protein isolated from the kidney bean *Phaseolus* sp. and which acts as a mitogen on certain animal cells.

phytohormones *n.plu.* plant hormones, including auxins, gibberellins, cytokinins, abscisic acid. *see individual entries*.

phytoid *a.* plant-like.

phytokinin cytokinin *q.v.*

phytol *n.* a product of hydrolysis of chlorophyll used in the synthesis of vitamins E and K, being a long-chain alcohol forming the tail of the chlorophyll molecule.

phytolith *n.* minute mineral particle, as hydrate of silica, in plant tissue, particularly of grasses.

phytomass *n*. biomass *(q.v.)* composed of plants.

phytome *n*. plants considered as an ecological unit.

phytomorphosis *n*. changes in structural features of plants as a result of fungal and bacterial infection.

phytoparasite *n*. any parasitic plant.

phytopathology *n*. the study of plant diseases.

phytophage *n*. animal that feeds on plants, usually used for the smaller sap-sucking and leaf eating insects etc. rather than the larger herbivores. *a*. phytophagous.

phytophysiology *n*. plant physiology.

phytoplankton *n*. all photosynthetic plankton, including e.g. unicellular algae and cyanobacteria.

phytoreovirus group one of the two genera of plant viruses of the family Reoviridae, the other being the fijiviruses, containing isometric double-stranded RNA viruses which are considered to be insect viruses becoming adapted to plants. The genome is composed of 12 separate RNAs.

phytosociology *n*. the study of all aspects of the ecology of plants and the influences on them.

phytosociology *n*. the study of all aspects of the ecology of plants and the influences on them.

phytosis *n*. a disease caused by fungi.

phytosphingosine *see* sphingosine.

phytosterols *n.plu*. sterols such as sitosterol, originally isolated from plant material.

phytosuccivorous *a*. living on plant juices.

phytotoxic *a*. toxic to plants.

phytotoxin *n*. any toxin originating in a plant.

phytotrophic autotrophic *q.v.*

phytotype *n*. representative type of plant.

pia mater delicate vascular membrane, the innermost of the three membranes surrounding the brain and spinal cord.

Piciformes *n*. an order of birds including the woodpeckers.

Picornaviridae, picornaviruses *n*., *n.plu*. family of small, non-enveloped, single-stranded RNA viruses including poliomyelitis, the rhinoviruses that cause common colds, and the aphthoviruses which include foot and mouth disease.

piezoelectric *a*. becoming electrically polarized when subjected to mechanical stress.

pigment *n*. colouring matter in plants and animals.

pigment cell chromatophore *q.v.*; chromatocyte *q.v.*

pigmentation *n*. disposition of colouring matter in an organ or organism.

pigmy *see* pygmy.

pileate *a*. having a pileus.

pileated *a*. crested, *appl*. birds.

pileocystidium *n*. sterile hair-like structure on the cap of certain basidiomycete fungi.

pileolus *n*. a small cap, as of many small toadstools.

pileorhiza *n*. a root covering; root cap *q.v.*

pileum *n*. the top of the head in birds. *plu*. pilea.

pileus *n*. the umbrella-shaped cap of mushrooms and toadstools. *plu*. pilei.

pili *plu*. of pilus *q.v.*

pilidium *n*. the characteristic helmet-shaped larva of nemertine worms; a hemispherical apothecium of certain lichens.

pilifer *n*. part of labrum of some lepidopterans.

piliferous *a*. bearing or producing hair, *appl*. outermost layer of root which gives rise to root hairs.

pilomotor *a*. causing hairs to move; *appl*. non-myelinated nerve fibres innervating muscles to hair follicles.

pilose *a*. downy; hairy.

pilus *n*. one of slender hair-like structures covering some plants; fine filamentous appendage of bacteria, shorter and straighter than flagella, *alt*. fimbria.

pin feather a young feather, esp. one just emerging through the skin and still enclosed in sheath.

Pinaceae, pines *n*., *n.plu*. family of coniferous trees and shrubs that bear their needles in bunches.

pinacocytes *n.plu*. the flattened plate-like cells of outer epithelium of sponges.

pinacoderm *n*. the external layer of body wall of sponges.

pincushion gall reddish thread-like growth found on roses and caused by larvae of the gall wasp *Diplolepis roasae*.

pineal gland or body in vertebrates, an outgrowth from the 1st cerebral vesicle, which may have endocrine functions. It secretes vasotocin in mammals and melatonin. In lower vertebrates it is visible as the pineal or median eye. In higher vertebrates it is embedded in nervous tissue. May be involved in the regulation of certain biological rhythms.

pineal region portion of brain giving rise to pineal and parapineal organs.

pineal sac end vesicle of epiphysis, as in *Sphenodon*, the tuatara.

pineal stalk the connection between pineal body and rest of brain.

pineal system the parietal organ *q.v.* and associated structures, as pineal sac, stalk, and nerves, parapineal organ and epiphysis.

pin-eyed having stigma at mouth of tubular corolla, with shorter stamens. *cf.* thrumeyed.

pinna *n.* outer ear, thin cartilaginous structure covered with skin; a bird's feather or wing; fin or flipper; (*bot.*) leaflet of pinnate leaf; branch of pinnate thallus. *plu.* pinnae.

pinnate *a.* divided in a feathery manner; having lateral processes; (*bot.*) of a compound leaf, having leaflets on each side of an axis or midrib.

pinnatifid *a. appl.* leaves lobed halfway to midrib.

pinnatilobate *a.* with leaves pinnately lobed.

pinnation *n.* the pinnate condition.

pinnatipartite *a.* with leaves lobed threequarters of the way to the base.

pinnatiped *a.* with lobed toes, as certain birds.

pinnatisect *a.* with leaves lobed almost to base or midrib.

pinnatodentate *a.* pinnate with toothed lobes, of leaves.

pinnatopectinate *a.* pinnate with comblike lobes.

pinniform *a.* feather-shaped or fin-shaped.

pinninervate *a.* with veins disposed like parts of a feather, *appl.* leaves.

Pinnipedia, pinnipeds *n., n.plu.* the seals, walruses and sealions.

pinnule *n.* (*bot.*) a secondary branch of a compound pinnate leaf; (*zool.*) projection from arms of sea lilies, two rows of which fringe each arm.

pinocytosis *n.* uptake of droplets of liquid

into a cell by the process of endocytosis.

Pinophyta *n.* one name for the gymnosperms (*q.v.*), when treated as a division of the kingdom Plantae.

Pinophyta the gymnosperms *q.v.*

pinosome *n.* a vesicle containing material taken up by pinocytosis.

pinulus *n.* a spicule resembling a fir tree owing to development of small spines from one ray.

pioneer community the organisms that establish themselves on bare ground at the start of a primary succession.

pioneer species first species, usually mosses, lichens and microorganisms, that colonize a bare site as the first stage in a primary succession.

PI pathway the phosphoinositide pathway, a biochemical pathway generating the second messengers inositol trisphosphate and diacylgylcerol from breakdown of membrane inositol phospholipids in response to stimulation of cell-surface receptors by certain hormones, growth factors, neurotransmitters, etc.

Piperales *n.* order of woody and herbaceous dicots comprising the families Piperaceae (pepper, the spice) and Saururaceae (lizard's tail).

piperidine *n.* an alkaloid obtained from pepper, *Piper nigrum*.

piriform *a.* pear-shaped, *appl.* muscle of the buttocks, the musculus piriformis.

piroplasm(a) *n.* parasitic protozoan of the class Piroplasmea, which infect red blood cells. *Babesia* is the causal agent of red water fever of cattle. Piroplasms are characterized by an intracellular stage in which the parasite is not contained in a vacuole, but has its cell membrane in direct contact with the cytoplasm of the host cell.

Pisces *n.* the fishes.

piscicolous *a.* living in fish.

pisciform *a.* like a fish.

piscine *a. pert.* fishes.

piscivorous *a.* fish-eating.

pisiform *a.* pea-shaped, *appl.* to a carpal bone.

pisohamate *a. appl.* a ligament connecting pisiform and hamate bones.

pisometacarpal *a. appl.* a ligament connecting pisiform bone with 5th metacarpal.

pistil *n.* the carpels collectively when fused

into a single structure; each carpel with its stigma and style in flowers with carpels separate.

pistillate *a.* bearing pistils; *appl.* a flower bearing pistils but no stamens. *cf.* staminate.

pistillidium archegonium *q.v.*

pistillode *n.* a rudimentary or non-functional pistil.

pistillody *n.* the conversion of any organ of a flower into carpels.

pit *n.* minute, wall-free area in cell wall of plant cell. *see also* coated pit.

pit fields areas of small pits or depressions in the plant cell primary wall.

pit-pair the two corresponding pits in cell walls of adjacent plant cells and the plasma membrane separating them.

pitcher *n.* a modification of a leaf or part of a leaf into a hollow organ to trap and digest insects, in the pitcher-plants *Nepenthes*, *Sarracenia*, etc.

pith *n.* the central parenchymatous tissue present in the stems of some dicotyledons.

pithecanthropines *n.plu.* fossil hominids formerly placed in the genus *Pithecanthropus* and now considered to be the species *Homo erectus* and including Java man.

pituicyte *n.* a glial cell in the neurohypophysis (the neural lobe of pituitary body).

pituitary, pituitary body or gland in vertebrates, an endocrine gland attached to the undersurface of the brain below the hypothalamus by a short stalk, secreting a number of important hormones such as adrenocorticotropin (ACTH), prolactin, the gonadotropins, thyroid-stimulating hormone, oxytocin and vasopressin, and consisting of two parts, the glandular adenohypophysis and the neuroendocrine neurohypophysis. *alt.* hypophysis.

piuitrin *n.* former term for an extract of the neural lobe of pituitary body containing oxytocin and vasopressin.

pivot joint joint in which movement is limited to rotation.

placenta *n.* (*bot.*) that part of the plant ovary where the ovules originate and remain attached until maturity; (*zool.*) in mammals, a double-layered spongy, vascular tissue, formed from maternal and foetal tissue in wall of uterus, and in which the blood vessels of mother and foetus are in close proximity, allowing exchange of nutrients, respiratory exchange etc. The placenta also produces various hormones believed to be involved in the maintenance of pregnancy, e.g. chorionic gonadotropin. Eutherian mammals (placental mammals) form various types of long-lived placenta involving both yolk sac and chorion. Marsupials form a short-lived placenta involving only the yolk sac.

placental *a. pert.* a placenta; *appl.* mammals which develop a persistent placenta, as eutherians; secreted by placenta.

placental mammals mammals which develop a persistent placenta, as eutherians. *cf.* marsupials.

placentate *a.* having a placenta developed.

placentation *n.* the manner in which ovules are attached to the plant ovary wall; the type of placenta in mammals.

placentiferous, placentigerous placentate *q.v.*

placochromatic *a.* with plate-like arrangement of chromatophores.

placode *n.* plate-like structure, e.g. lens placode, the structure from which the lens of the vertebrate eye develops.

Placodermi, placoderms *n., n.plu.* extinct class of early Devonian to early Carboniferous jawed primitive fish, with archaic jaw suspension, crushing dental plates and bony dermal plates on the head and thorax.

Placodontia, placodonts *n., n.plu.* extinct order of fully aquatic Triassic marine reptiles, having a short armoured body, some with a turtle-like carapace.

placoid *a.* plate-like.

placoid scale *see* denticle.

Placozoa *n.* phylum of extremely simple metazoans consisting of a single known species, *Trichoplax adhaerens*, a round flattened sac-like organism with a fluid-filled internal cavity, and cilia covering the body surface, and which lacks differentiated tissues, organs, or any discernible head or tail or bilateral symmetry.

placula *n.* an embryonic stage in urochordates, a flattened blastula with a small blastocoel; a stage in the colonial protistan *Volvox*.

plagioclimax *n.* stage in plant succession

preceding the natural climax but which persists because of e.g. human intervention.

plagiopatagium *n*. part of bat wing membrane posterior to arm; part of the patagium between fore- and hindfeet in flying lemurs.

plagiosaurs *n.plu*. an advanced order of labyrinthodont amphibians, existing from Permian to Triassic times and having a very wide flat body in advanced forms, and a body armour of interlocking plates.

plagiosere *n*. an ecological succession deviating from its natural course as a result of continuous human intervention.

plagiotropism *n*. growth tending to incline a structure from the vertical plane to the oblique or horizontal as in lateral roots and branches. *a*. plagiotropic.

plagiotropous *a*. obliquely inclined.

plagula *n*. ventral exoskeletal plate protecting the pedicle in spiders.

plakea *n*. plate-like early stage in formation of a coenobium.

planarian *n*. any member of the order Tricladida, free-living flatworms (platyhelminths) living in streams, ponds, lakes and the sea, having a broad, flattened body, well-developed sense organs at the anterior end, and an intestine with three main branches.

planation *n*. the flattening of branched structures, as has occurred, e.g., in the evolution of fronds of ferns, etc.

planetism *n*. the condition of having motile or swarm cell stages. *a*. planetic.

planidium *n*. active migratory larva of certain insects.

planiform *a*. with nearly flat surface, *appl.* certain articulating surfaces.

plankter *n*. an individual planktonic organism.

planktohyponeuston *n*. aquatic organisms which gather near the surface at night but spend their days in the main water mass.

plankton *n*. the usually small marine or freshwater plants (phytoplankton) and animals (zooplankton) drifting with the surrounding water.

Planktosphaeroidea *n*. class of hemichordates known only from a free-swimming pelagic larval form that resembles that of pterobranchs.

planktotrophic *a*. feeding on plankton.

planoblast *n*. the free-swimming medusa form of a hydrozoan.

planoconidium *n*. zoospore of certain fungi.

planocyte *n*. a wandering or migratory cell; planospore *q.v.*

planogamete *n*. a motile, usually flagellate gamete, esp. of algae and fungi.

planont *n*. any motile spore, gamete or zygote; the initial amoebula stage in some sporozoans; a swarm spore produced in thick-walled or resting sporangia of certain lower fungi.

planospore *n*. a motile spore.

planozygote *n*. a motile zygote.

plant defence response the response of a plant to infection or wounding, which aims to localize damage and prevent the spread of infection, and which may include the production of phytoalexins, tissue death, lignification and callus formation.

plant genetic resources the genetic diversity present within plants, usually *appl.* crop plants and their wild relatives, and which is of potential use in plant breeding.

plant hormone a substance produced by a plant which regulates its growth and development. Plant hormones include auxins, ethylene and gibberellins.

plant pathogen any agent (bacterium, fungus, virus or mycoplasma) causing disease in plants.

plant pathology the study of the diseases of plants.

plant rhabdovirus group single-stranded RNA plant viruses of the family Rhabdoviridae, e.g. sonchus yellow net virus, resembling the animal and insect members of the family, having bullet-shaped or rod-shaped particles with a lipid outer coat in which glycoproteins are embedded. They are considered to be insect viruses becoming adapted to plants.

plant viruses viruses that infect plants.

planta *n*. the sole of the foot; the 1st tarsal joint of insects.

Plantae *n.plu*. the plant kingdom, in most modern classifications comprising the algae, bryophytes (mosses and liverworts), seedless vascular plants (ferns, club mosses, horsetails) and the seed plants (gymnosperms and angiosperms).In older classifications, the fungi and even bacteria are also included, but these groups are now usually placed in separate kingdoms. The algae are sometimes placed in the Protoctista (Protista). *see* Appendix 2.

plantar *a. pert.* sole of the foot.

plantaris *n.* a muscle of the lower leg.

plantigrade *a.* walking with the whole sole of the foot touching the ground.

plantula *n.* a pulvillus-like adhesive pad on the tarsal joints of some insects.

planula *n.* the ovoid free-swimming ciliated larva of coelenterates.

planum *n.* a plane or area, *appl.* some cranial bone surfaces.

plaque *n.* a clear area in a continuous sheet of cultured cells or bacterial growth, which indicates destruction of infected cells by a virus or phage.

plaque assay an assay for the presence and concentration of infectious virus in a sample, by counting the number of clear plaques in a continuous sheet of cultured susceptible cells infected with the virus sample. Each plaque is due to a single virus and its progeny destroying a group of contiguous cells.

plaque-forming unit (PFU) a quantitative measure of the number of infectious virus particles in a given sample, as each infectious virus can give rise to a single clear plaque on infection of a continuous "lawn" of bacteria. *see* plaque assay.

plasma *n.* the liquid part of body fluids, such as blood, milk or lymph, as opposed to suspended material such as cells and fat globules.

plasma cell antibody-secreting cell of the immune system, differentiates from B lymphocyte after recognition of specific antigen.

plasma lipoproteins lipid transport proteins in blood, carrying cholesterol, triacylglycerols and phospholipids complexed with a protein (apolipoprotein). There are three main types: high-density lipoproteins (HDL), carrying free cholesterol and phospholipids, low-density lipoproteins (LDL), carrying mainly cholesteryl esters, and very low-density lipoproteins (VLDL) rich in triacylglycerols.

plasma membrane membrane (*q.v.*) bounding the surface of all living cells, formed of a fluid lipid bilayer in which proteins carrying out the functions of enzymes, ion pumps, transport proteins, receptors for hormones etc. are embedded. It regulates the entry and exit of most solutes and ions, few substances being able to diffuse through unaided. *alt.* cell membrane, plasmalemma. *see* Fig. 7.

plasma thromboplastin antecedent (PTA) factor XI *q.v.*

plasmacyte plasma cell *q.v.*

plasmagel *n.* the more solid part of the cytoplasm, which usually may be reversibly converted to the more fluid plasmasol.

plasmagene *n.* any gene other than those carried in the nucleus of a eukaryotic cell, such as the genes of mitochondria and chloroplasts.

plasmalemma *n.* plasma membrane (*q.v.*), esp. in plants.

plasmalemmasome membranous structure formed in plant cells, consisting of tubules, cisternae and vesicles, between the plasma membrane and the cell wall.

plasmalogen *n.* any of a class of phospholipids found chiefly in animal heart and brain and including phosphatidal ethanolamine, phosphatidal choline etc., and which differ from the corresponding phosphoglycerides in having one of the fatty acid chains replaced by an α,β-unsaturated ether.

plasmasol *n.* the more fluid part of the cytoplasm, which may be reversibly converted to the more solid plasmagel.

plasmaspore *n.* an adhesive spore.

plasmatic *a. pert.* blood plasma; protoplasmic.

plasmatoparous *a.* developing directly into a mycelium upon germination, instead of into zoospores, *appl.* spores of some fungi, as downy mildew of grapes (*Plasmopora viticola*).

plasmid *n.* small circular DNA replicating independently of the chromosome in bacteria and unicellular eukaryotes such as yeasts, which is maintained at a characteristic stable number from generation to generation. Plasmids typically carry genes for antibiotic resistance, colicin production, the breakdown of unusual compounds, etc. They are widely used in genetic engineering as vectors into which foreign genes are inserted for subsequent cloning or expression in bacterial cells.

plasmid immunity *see* immunity.

plasmid incompatibility the inability of a plasmid to be maintained in a bacterial cell containing another plasmid of the same type or same compatibility group.

plasmin *n*. an enzyme in blood plasma that degrades fibrin. EC 3.4.21.7. *alt.* fibrinolysin.

plasminogen *n*. the inactive precursor of the proteolytic enzyme plasmin.

plasminogen activator *see* tissue plasminogen activator.

plasmodesmata *n.plu.* cytoplasmic threads running transversely through plant cell walls and connecting the cytoplasm of adjacent cells. *sing.* plasmodesma.

plasmodial *n*. growing as or *pert.* a plasmodium, *appl.* the acellular slime moulds.

plasmodial slime moulds acellular slime moulds *q.v.*

plasmodiocarp *n*. fruiting body of some acellular slime moulds in which the whole plasmodium develops into a sporangium.

Plasmodiophoromycota, plasmodiophorans *n., n.plu.* phylum of protists, mainly plant pathogens (e.g. the causal organism of clubroot, *Plasmodiophora brassicae*), which have a multinucleate plasmodial vegetative stage and uninucleate motile zoospores. Formerly classified in the fungi.

plasmodium *n*. in the plasmodial slime moulds, a multinucleate, fan-shaped mass of streaming protoplasm without a cell wall which may cover several square metres, and which forms the nonreproductive stage of the organism; a genus of parasitic protozoans including the causal agent of malaria *Plasmodium falciparum*. *a.* plasmodial.

plasmogamy *n*. in protozoans, fusion of several individuals into a multinucleate mass; fusion of cytoplasm without nuclear fusion.

plasmolysis *n*. the withdrawal of water from a plant cell by osmosis if placed in a strong salt or sugar solution, resulting in contraction of cytoplasm away from the cell walls.

plasmonema *n*. thread of protoplasm in connection with plastids. *plu.* plasmo-

nemata.

plasmoptysis *n*. emission of cytoplasm from tips of hyphae in host cells, in certain endomycorrhizas; localized extrusion of cell contents through cell wall of bacteria.

plasmotomy *n*. division of a plasmodium by cleavage into multinucleate parts.

plastic *a*. formative; capable of change, *appl.* connections in brain, etc.

plasticity *n*. the capacity for change under the influence of stimuli, as of connections between neurones in brain.

plastid *n*. a cellular organelle containing pigment, esp. in plants, e.g. chloroplast.

plastidome *n*. the plastids of a cell collectively.

plastiquinone *n*. one of the electron carriers in the photosynthetic electron transport chain, closely resembling ubiquinone.

plastochron *n*. the time interval between the successive similar developmental events at the shoot apex, e.g. initiation of successive leaf primordia.

plastocyanin *n*. blue copper protein in chloroplasts, a component of the photosynthetic electron transport chain.

plastodeme *n*. a deme that differs from others phenotypically but not genotypically.

plastogamy *n*. union of distinct unicellular individuals with fusion of cytoplasm but not nuclei.

plastogenes *n.plu.* cytoplasmic factors, controlled by or interacting with the nucleus, which determine the differentiation of plastids, now known to be genes carried in plastids.

plastolysis *n*. dissolution of mitochondria.

plastome *n*. the genome of a plastid.

plastoquinone (Q) *n*. any of various quinones found in chloroplasts as components of the photosynthetic electron transport chain.

plastorhexis *n*. the breaking up of mitochondria into granules.

plastron *n*. bony plate on underside of body of turtles and tortoises; thin film of air trapped on the surface of the bodies of some aquatic insects. *a.* plastral.

plate meristem a ground meristem in plant parts with a flat form such as a leaf, consisting of parallel layers of cells dividing only anticlinally in relation to surface.

platelet *n.* non-nucleated disc-shaped cell fragments present in blood, produced by fragmentation of megakaryocytes, and which are involved in blood clotting, gathering at sites of damage and releasing clotting Factor X and other active products.

platelet-derived growth factor (PDGF) a glycoprotein produced by platelets and other cells, and which stimulates the proliferation of cells of mesenchymal origin and which has been implicated in repair of the vascular system *in vivo*.

platy *a. appl.* soil crumbs in which the vertical axis is shorter than the horizontal.

platybasic *a. appl.* the primitive chondrocranium with wide hypophysial fenestra.

platydactyl *a.* with flattened out fingers and toes, as certain amphibians.

Platyhelminthes, platyhelminths *n.,* *n.plu.* a phylum of multicellular, acoelomate animals, commonly called flatworms, which are flattened dorsoventrally and are bilaterally symmetrical, and have the epidermis (ectoderm) and gut separated by a solid mass of tissue. They include the free-living Turbellaria and the parasitic Monogea (skin and gill flukes), Trematoda (gut, liver and blood flukes), and Cestoda (tapeworms).

platyhieric *a.* having a sacral index above 100.

platymyarian *a.* having flat muscle cells, *appl.* some nematode worms.

platypus *n.* the duck-billed platypus of Australia, *see* Monotremata.

platyrrhines *n.plu.* the New World monkeys.

platysma *n.* broad sheet of muscle between superficial fascia of neck.

platyspermic *a.* having seed which is flattened in transverse section.

play *n.* behaviour exhibited esp. by young animals in which they explore the environment and learn by trial and error during the time when life is fairly easy for them.

β-pleated sheet regular periodic secondary structure common in proteins, in which fully extended polypeptide chains lying adjacent to each other are held together by hydrogen bonding to form a sheet structure – in a parallel β sheet the adjacent chains all run in the same direction, in an antiparallel β sheet adjacent chains run in opposite directions.

Plecoptera *n.* order of insects commonly called stoneflies, similar in many resects to mayflies (Ephemoptera), but with two "tails" and hindwings larger than forewings.

plectenchymatous *a. appl.* a tissue of interwoven cell filaments or hyphae in algae or fungi.

plectoderm *n.* outer tissue of fungal fruit body when composed of densely interwoven branched hyphae.

Plectomycetes *n.* a group of ascomycete fungi, commonly called the blue, green and black moulds from the colour of their conidia, and generally bearing asci in closed ascocarps (cleistothecia). The asexual (imperfect) stages of many plectomycetes are similar to those of *Aspergillus* and *Penicillium*.

plectonemic *a. appl.* double spiral having the 2 strands interlocked at each twist, as in the structure of DNA.

plectonephridia, plectonephria *n.plu.* nephridia formed of networks of fine excretory tubules lying on body wall and septa of certain oligochaete worms.

plectostele *n.* a modified type of actinostele found in some club mosses of the genus *Lycopodium*, being deeply fissured in cross-section.

plectron *n.* hammer-like form of certain bacilli during sporulation.

plectrum *n.* styloid process of temporal bone. *alt.* malleus.

pleioblastic *a.* having several buds germinating at several points, as spores of certain lichens.

pleiochasium *n.* axis of a cymose inflorescence bearing more than two lateral branches. *alt.* pleiochasial cyme.

pleiocotyl *a.* having more than two cotyledons. *n.* pleiocotyledony.

pleiocyclic *a.* living through more than one cycle of activity, as a perennial plant.

pleiomerous *a.* having more than the usual number of parts, as of sepals and petals in a whorl. *n.* pleiomery.

pleiomorphic pleomorphic *q.v.*

pleiopetalous *a.* having more than the normal number of petals; having double flowers.

pleiophyllous *a.* having more than the normal number of leaves or leaflets.

pleiosporous polysporous *q.v.*

pleiotaxy *n.* a multiplication in the number of whorls, as in double flowers.

pleiotropy, pleiotropism *n.* multiple effects of a single gene which affects more than one phenotypic character. *a.* pleiotropic, *appl.* genes.

pleioxenous *a.* living on more than one host during life cycle, *appl.* parasites.

Pleistocene *n.* the glacial and postglacial epoch following the Tertiary, lasting from around 2 million to 10,000 years ago.

pleo- pleio-.

pleochromatic *a.* exhibiting different colours under different environmental or physiological conditions.

pleogamy *n.* maturation, and therefore pollination, at different times, as of flowers of one plant.

pleometrosis *n.* colony foundation by more than one female as in some social hymenopterans. *a.* pleometrotic.

pleomorphic *a.* being able to change shape; existing in different shapes at different stages of the life-cycle. *see also* polymorphic.

pleon *n.* the abdominal region in crustaceans.

pleopod swimmeret *q.v.*

pleotropy pleiotropy *q.v.*

plerocercoid *a.* a solid elongated metacestode of certain tapeworms, esp. when found in fish muscle. *alt.* plerocestoid.

plerocestoid metacestoid *q.v.*; plerocercoid *q.v.*

plerome *n.* the core or central part of an apical meristem.

plerotic *a.* completely filling a space; *appl.* oospore filling oogonium.

plesiobiosis *n.* the close proximity of two or more nests of social insects, accompanied by little or no direct communication between the colonies.

plesiobiotic *a.* living in close proximity, *appl.* colonies of ants of different species; building contiguous nests, *appl.* ants and termites.

plesiometacarpal *a. appl.* condition of retaining the proximal elements of metacarpals, as in many cervids.

plesiomorphic *a.* in cladistics, *appl.* the original pre-existing member of a pair of homologous characters.

plesiomorphous *a.* having a similar form.

Plesiosauria, plesiosaurs *n., n.plu.* order of Mesozoic reptiles that were fully aquatic with a barrel-shaped body and paddleshaped limbs.

pleura *n.* membrane lining thoracic cavity (chest) and extending to cover surface of lungs, delimiting the pleural cavity.

pleural *a. pert.* a pleura or pleuron, as pleural ganglia; *appl.* recesses: spaces within pleural sac not occupied by lung; *appl.* costal plates of carapace of turtles and tortoises.

pleural cavity body cavity occupied by lung(s).

pleurapophysis *n.* a lateral vertebral process or true rib.

pleurethmoid *n.* the compound ethmoid and prefrontal of some fishes.

pleurilignosa *n.* rain forest.

pleurite *n.* a sclerite of the lateral piece of a body segment of arthropods.

pleuroblastic *a.* producing, having or *pert.* lateral buds or outgrowths.

pleurobranchiae *n.plu.* gills arising from lateral walls of thorax of some arthropods, esp. crustaceans.

pleurocarpic *a.* with lateral fructifications.

pleuroccipital exoccipital *q.v.*

pleurocentrum *n.* a lateral element of vertebral centrum in many fishes and fossil amphibians.

pleurocerebral *a. pert.* pleural and cerebral ganglia, in molluscs.

pleurocystidium *n.* a sterile hair in hymenium of surface of gills, in agaric fungi.

pleurodont *a. appl.* teeth that are attached to the inside surface of jaw, as opposed to the outer edge (acrodont) or in sockets (thecodont).

pleurogenous *a.* originating or growing from the side or sides.

pleuron *n.* one of the external pieces on side of body segments of arthropods.

pleuropedal *a. pert.* pleural and pedal ganglia, in molluscs.

pleuroperitoneum *n.* pleura and peritoneum combined, a membrane lining the body cavity in vertebrates without a diaphragm.

pleuropneumonia-like organism (PPLO) *see* mycoplasma.

pleuropodium *n.* glandular process on abdomen of some insect embryos.

pleuropophysis *n.* a lateral vertebral process or true rib.

pleurosphenoid sphenolateral *q.v.*

pleurospore *n.* spore formed on sides of basidium.

pleurosteon *n.* lateral process of sternum of young birds, later becoming costal process.

pleurosternal *a.* connecting or *pert.* pleuron and sternum, *appl.* thoracic muscles in insects.

pleurotribe *a. appl.* flowers whose anthers and stigma are so placed as to rub the sides of insects entering them, a device for cross-pollination.

pleurovisceral *a. pert.* pleural and visceral ganglia, of molluscs.

pleurum pleuron *q.v.*

pleuston *n.* free-floating organisms especially those possessing a gas-filled bladder or float.

plexiform *a.* entangled or complicated; like a network, *appl.* two of the layers of the vertebrate retina, the outer and inner plexiform layers, containing synaptic connections between photoreceptors, horizontal and bipolar cells and between bipolar, amacrine and ganglion cells respectively, also *appl.* outermost layer of grey matter of cerebral cortex.

plexus *n.* network of interlacing vessels, nerves or fibres.

plexus myentericus Auerbach's plexus *q.v.*

plica *n.* a fold of skin, membrane or lamella; a corrugation of brachiopod shell.

plicate *a.* pleated, folded like a fan, *appl.* leaf; folded or ridged.

pliciform *n.* resembling a fold; disposed in folds.

Pliocene *n.* the geological epoch that followed the Miocene and preceded the Pleistocene, lasting from around 5 million years ago to 2 million years ago.

ploidy *n.* number of chromosomes or DNA molecules in a cell or organelle, or the typical chromosome number of a multicellular organism.

plotophyte *n.* a plant adapted for floating.

pluma *n.* a contour feather of birds.

plumate *a.* like a plume.

Plumbaginales *n.* order of dicot herbs or small shrubs and comprising the family Plumbaginaceae (leadwort), which are xerophytes or halophytes of steppes, semideserts and sea coasts, and include sea lavender and thrift.

plume *n.* feather or feather-like structure; feathery structure at "head" end of vestimentiferan tubeworms through which carbon dioxide, oxygen and sulphide are absorbed from sea water.

plumicone *n.* a spicule with plume-like tufts.

plumicorn *n.* horn-like tuft of feathers on bird's head.

plumigerous *a.* feathered.

plumiped *n.* a bird with feathered feet.

plumose *a.* feathery; having feathers; feather-like; *appl.* feathers without tiny hooks on barbules, i.e. down feathers.

plumula plumule *q.v.*

plumulaceous plumulate *q.v.*

plumular *a. pert.* a plumule.

plumulate *a.* with a covering of down.

plumule *n.* (*bot.*) the developing shoot of a plant embryo, comprising the epicotyl and young leaves; (*zool.*) down feather of adult birds. *a.* plumular.

plumule sheath coleoptile *q.v.*

pluriascal *a. pert.* or containing several asci.

pluriaxial *a.* having flowers developed on secondary shoots.

plurilocular *a.* having several compartments.

pluriparous *a.* giving birth to, or having given birth to, a number of offspring.

pluripartite *a.* with many lobes or partitions.

pluripolar *a.* with several poles, *appl.* neurones in ganglia, etc.

pluripotent *a. appl.* cells capable of developing into several different cell types.

plurisegmental *a. pert.* or involving a number of segments, *appl.* nerve conduction, reflexes.

pluriseptate *a.* with multiple septa.

pluriserial *a.* arranged in two or more rows.

pluristratose *a.* arranged in a number of layers; much stratified.

plurivalent multivalent *q.v.*

plurivorous *a.* feeding on several substrates or hosts.

pluteus *n.* free-swimming larval stage of sea urchins and brittle stars, characterized by long processes stiffened by tiny spicules.

pneumathode *n.* an aerial or respiratory root.

pneumatic *a. appl.* bones penetrated by air-filled cavities connecting with lungs in birds.

pneumaticity *n.* the condition of having air cavities, as bones of flying birds.

pneumatized *a.* having air cavities.

pneumato- *see* pneumo-.

pneumatocyst *n.* the air bladder or swim float of fishes; air cavity used as a float; air bladder of bladderwrack.

pneumatophore *n.* (*bot.*) air bladder of marsh or shore plants; aerating outgrowth in some ferns; a root rising above level of water or soil and acting as a respiratory organ in some trees; (*zool.*) the air sac or float of siphonophores.

pneumatopyle *n.* a pore of a pneumato-phore, opening above to exterior in certain siphonophores.

pneumatotaxis pneumotaxis *q.v.*

pneumo- *see* pneumato-.

pneumococcus *n.* general name for a bacterium of the genus *Pneumococcus*, forms of which cause pneumonia in man and some mammals, and in which the phenomenon of bacterial transformation was first discovered. *plu.* pneumococci. *a.* pneumococcal.

pneumocyte *n.* epithelial cell lining the air spaces of lung.

pneumogastric *a. appl.* 10th cranial or vagus nerve, supplying pharynx, larynx, heart, lungs and viscera.

pneumostome *n.* aperture through which air passes in and out of the mantle cavity in land snails.

pneumotaxis *n.* reaction to stimulus consisting of a gas, esp. carbon dioxide in solution.

poad *n.* a meadow plant.

Poales *n.* order of herbaceous monocots, the grasses, comprising the family Poaceae (Graminae).

poculiform *a.* cup-shaped, goblet-shaped.

pod *n.* legume *q.v.*; a husk; the cocoon in which eggs are laid in locusts; a school of fish in which the bodies of the individuals actually touch; a group of whales.

podal *a.* pertaining to feet. *alt.* pedal.

podeon *n.* the slender middle part of abdomen of hymenopterans (e.g. ants, bees and wasps), uniting propodeon and meta-podeon. *alt.* petiole.

podetiiform *a.* resembling a podetium.

podetium *n.* a stalk-like elevation; out-growth of thallus in certain lichens bearing apothecium.

podeum podeon *q.v.*

podia *plu.* of podium *q.v.*

Podicipediformes *n.* an order of small water birds without webbed feet, including the grebes.

podite *n.* a crustacean walking leg.

podium *n.* a foot or foot-like structure, such as a tube foot of echinoderm; a stem axis of a plant. *plu.* podia. *a.* podial.

podobranchiae, podobranchs *n.plu.* gills arising from the basal segments of thoracic appendages of certain arthropods. *alt.* foot-gills.

podocephalous *a.* having a head of flowers on a long stalk.

podoconus *n.* conical mass of endoplasm connecting the central capsule with the disc of rhizopod protozoans.

podocyst *n.* a sinus in the foot or a small sac at tail end of some gastropod molluscs.

podocyte *n.* epithelial cell of Bowman's capsule of kidney, having numerous processes which rest on the basement membrane.

podocytosis transcytosis *q.v.*

pododerm *n.* dermal layer of hoof, within the horny layer.

podogynium *n.* a stalk supporting gynaecium.

podomere *n.* a limb segment in arthropods.

podophthalmite *n.* in crustaceans, eye-stalk segment farthest from the head.

podosoma *n.* the body region of mites and ticks, bearing the four pairs of walking legs.

Podostemales *n.* an order of dicots living in fast-flowing water and on rocks in rivers, with often filamentous or ribbon-like leaves and stems. Comprises the family Podostemaceae.

Podostemonales Podostemales *q.v.*

podotheca *n.* a foot covering as of reptiles or birds; a leg sheath of a pupa.

podsol, podzol *n.* grey forest soil, the soil type of cold temperate regions, and formed on heathlands and under coniferous forest.

pogonion *n.* most prominent point on chin as represented on mandible (lower jaw).

Pogonophora, pogonophorans *n., n.plu.* phylum of marine sessile worm-like invertebrates, with similarities to hemichordates, which live in chitin tubes. *alt.* beard worms. *see also* Vestimentifera.

poikilochlorophyllous *a.* completely losing and regaining their chlorophyll in response to changes in environmental conditions, *appl.* some angiosperms.

poikilocyte *n.* a distorted form of erythrocyte present in certain pathological conditions.

poikilohydrous *a.* becoming dormant in the dry season after losing most of their water, *appl.* some angiosperms.

poikilosmotic *a.* having internal osmotic pressure varying with that of the surrounding medium, as with salinity.

poikilotherm *n.* a "cold-blooded" animal, *see* poikilothermic.

poikilothermic *a. appl.* animals whose temperature varies with that of the surrounding medium. *n.* poikilothermy.

point mutation a mutation involving a change at a single base-pair in DNA.

poiser haltere *q.v.*

pokeweed mitogen (PWM) lectin isolated from *Phytolacca americana*.

pol **gene** gene of C-type retroviruses specifying the enzyme reverse transcriptase.

polar *a.* situated at one end of cell or structure; *appl.* flagella, one or a small group situated at one end of the cell.

polar body the smaller of the products of meiotic division in human oocytes, one being produced at 1st and 2nd meiotic divisions and lying just outside the membrane of the larger cell which will become the egg.

polar capsules capsules of spores containing coiled extrusible filaments, in certain sporozoans.

polar cartilage posterior portion of cartilages surrounding hypophysis in embryo, or independent cartilage in that region.

polar fibres or microtubules microtubules araising from poles of spindle and which are not attached to kinetochores of chromosomes.

polar granule centromere *q.v.*; ribonucleoprotein granule in polar plasm of insect egg.

polar lobe cytoplasmic protrusion that appears during the early cleavages of many gastropod embryos.

polar microtubules polar fibres *q.v.*

polar nuclei nuclei at each end of angiosperm embryo, which later form secondary nucleus.

polar organ the cluster of pole cells (*q.v.*) at tail of early embryo in insects.

polar plates two narrow ciliated areas produced in transverse plane, part of equilibrium apparatus of certain coelenterates; areas of cytoplasm without centrosomes beyond poles of spindle in dividing nucleus of certain protozoans.

polar plasm granular cytoplasm at posterior end of fertilized insect egg, from which the pole cells (*q.v.*) are formed.

polar rings two ring shaped masses of cytoplasm formed near poles of ovum after fertilization.

polar translocation or transport movement of materials through tissues or cells in one direction only.

polarilocular *a. appl.* a cask-shaped spore with two cells separated by a partition having a perforation, of certain lichens.

polarity *n.* of macromolecular structures, cells, embryos, organs or organisms, having one end morphologically and/or functionally distinct from the other; tendency to develop from opposite poles (as of plants); of membranes, having one face of a different composition and functionally distinct from the other; existence of opposite qualities; differential distribution or gradation along an axis; in some bacterial transcription units, the fact that a nonsense mutation in an early part of the unit can prevent the expression of subsequent genes in the unit.

polarization *n.* the development of polarity, in cells, organs, etc.; the setting up of an electrical potential difference across a membrane.

polarizing regions small parts of developing embryo which direct the development of a limb, etc. *alt.* organizing regions.

pole cell large cell which buds forth rows of smaller cells, as in annelid and mollusc embryos.

pole cells large cells distinguishable at tail

end of insect embryo very early, and from which the germ cells are produced.

pole plates end plates or masses of cytoplasm at spindle poles in some protozoan mitosis.

Polemoniales *n.* an order of dicot trees, shrubs, vines and herbs, including the families Boraginaceae (borage), Convolvulaceae (morning glory), Cuscutaceae (dodder) and others.

Polian vesicles interradial vesicles opening into ring vessel of ambulacral system of most starfish and sea cucumbers.

poliovirus *n.* small icosahedral non-enveloped single-stranded RNA virus of the genus *Enterovirus* of the family Picornaviridae, which causes poliomyelitis.

pollakanthic *a.* having several flowering periods.

pollarding *n.* a method of tree management in which trees are cut back at some distance above the ground, producing a crown of long straight shoots on a relatively short single trunk. *cf.* coppicing.

pollen *n.* fine powder produced by anthers and male cones of seed plants, composed of pollen grains which each enclose a developing male gamete.

pollen analysis quantitative and qualitative determination of pollen grains preserved in deposits such as peat, from which the former vegetation of the area can be reconstructed.

pollen basket the pollen-carrying hairs at back of tibia on worker bees.

pollen brush enlarged hairy tarsal joint of bee leg, which brushes pollen from anther.

pollen case *see* theca.

pollen chamber pit formed at apex of nucellus in ovule below micropyle, into which the germinating pollen tube grows.

pollen comb comb-like structure of bristles on leg of bee below pollen basket, serving to sweep pollen into basket.

pollen flower a flower without nectar, which attracts pollen-feeding insects.

pollen grain the haploid microspore of a seed plant, enclosing the partly developed microgametophyte from which the male gamete develops.

pollen profile the vertical distribution of preserved pollen grains in a deposit such as peat, giving an indication of the former vegetation at different periods.

pollen sac the cavity in anther in which pollen is produced, being the microsporangium of seed plants.

pollen spectrum the relative numerical distribution or percentage of pollen grains of different species preserved in a deposit.

pollen tube tube that develops from a pollen grain after attachment to stigma and which grows down towards ovule, entering it at the micropyle and delivering male gametes, or in some cases just male nuclei, to fuse with the female gamete.

pollex *n.* the thumb, or corresponding innermost digit of the normal five of a forelimb.

pollinarium *n.* in orchids, the mass of pollen (pollinium) and its stalk and adhesive disc.

pollination *n.* transfer of pollen from anther (in angiosperms) or male cone (gymnosperms) to stigma or female cone respectively.

pollination drop mucilaginous drop exuded from micropyle and which detains pollen grains as in gymnosperms.

pollinator *n.* any insect or other animal whose activities effect pollination.

polliniferous, pollinigerous *a.* pollen-bearing; adapted for transferring pollen.

pollinium *n.* an agglutinated mass of pollen in orchids and some other flowers.

pollution *n.* any harmful or undesirable change in the physical, chemical or biological quality of air, water or soil as a result of the release of e.g. chemicals, radioactivity, heat, large amounts of organic matter (as in sewage). Usually *appl.* changes arising from human activity although natural pollutants, e.g. volcanic dust, sea salt, are known.

polospore *n.* fossil pollen.

polster *n.* a low compact perennial or cushion plant.

poly- prefix derived from Gk. *polys*, many.

poly A (poly (A)) polyribonucleotide (poly dA being the corresponding polydeoxyribonucleotide) composed exclusively of adenylate residues, and which is added to the 3′ ends of many eukaryotic mRNAs after transcription, where it may be involved in stabilizing the mRNA.

poly (A)⁺ denotes a messenger RNA carrying a poly (A) tail.

poly (A)⁻ denotes a messenger rRNA lacking a poly (A) tail.

poly (A) tail stretch of polyadenylic acid residues found at the 3′ ends of many eukaryotic messenger RNAs, which is added in the nucleus by the enzyme poly (A) polymerase after transcription.

polyacrylamide gel electrophoresis (PAGE) widely used technique for separating proteins, nucleic acid fragments etc. for further analysis, separation being primarily on the basis of size. *alt.* gel electrophoresis.

polyadelphous *a.* having stamens united by filaments into more than two bundles.

polyadenylation *n.* the addition of a poly (A) tail to eukaryotic mRNA precursors in the nucleus.

polyamine *n.* any of a group of compounds composed of one or more basic units of two amino groups joined by a short hydrocarbon chain, formed from amino acids (in some cases during breakdown of protein by bacteria) and including putrescine, cadaverine and spermidine.

polyandrous *a.* mating with more than one male at a time; having twenty or more stamens. *n.* polyandry.

polyarch *a. appl.* roots, stems in which the primary xylem of the stele is made up of numerous alternating bundles of phloem and xylem.

polyarthric multiarticulate *q.v.*

polyaxon *n.* type of spicule laid down along numerous axes.

polyblastic *a.* having spores divided by a number of septa, *appl.* lichens.

poly (C) polyribonucleotide composed exclusively of cytidylate residues, poly (dC) being the corresponding polydeoxynucleotide.

polycarp *n.* gonad of some ascidians, on inner surface of mantle.

polycarpellary *a.* with compound gynaecium consisting of many carpels.

polycarpic, polycarpous *a.* with numerous, usually free carpels; producing seed season after season.

polycentric *a. appl.* chromosomes with several centromeres; with several growth centres.

polycercous, polycercoid *a. appl.* bladderworms developing several cysts, each with a head, as in an echinococcus.

Polychaeta, polychaetes *n., n.plu.* the bristle worms, a class of mainly marine annelid worms, the ragworms, lugworms, etc., characterized by possession of parapodia bearing numerous chaetae which are used for crawling, and a pronounced head bearing tentacles, palps and often eyes.

polychasium *n.* a cymose branch system when three or more branches arise about the same point.

polychlorinated biphenyls (PCBs) a large group of toxic synthetic lipid-soluble chlorinated hydrocarbons, which are used in various industrial processes and which have become persistent and ubiquitous environmental contaminants which can be concentrated in food chains.

polychromatic *a.* with several colours, as pigmented areas.

polychromatophil *a.* having a staining reaction characterized by various colours; *appl.* erythrocytes with small haemoglobin content.

polycistronic *a. appl.* mRNA containing more than one polypeptide-coding sequence.

polyclad *n.* member of the order Polycladida, marine turbellarian flatworms with a very broad flattened leaf-shaped body, and large numbers of eyes at the anterior end.

polyclimax *n.* a climax community consisting of several different climax associations none of which shows a tendency to give way to any other.

polyclonal *a. appl.* tissue or structure derived from a number of founder cells; derived from several cell clones, *appl.* specific antibodies that have been obtained by immunization of an animal and therefore represent the products of different clones of antibody-producing cells. *cf.* monoclonal.

polyclone *n.* a discrete area of tissue derived from several clones of cells.

polycotyledon *n.* plant with more than two cotyledons.

polycotyledonary *a.* having placenta in many divisions.

polycotyledony *n.* a great increase in the number of cotyledons.

polycrotism *n*. condition of having several secondary elevations in pulse curve.

polycyclic *a*. having many whorls (of flowers); or ring structures (of organic compounds); with vascular system forming several concentric cylinders (of plant stems).

polycyclic aromatic hydrocarbons (PAH) group of compounds found e.g. in soot, coal tar and cigarette smoke, some of which are known carcinogens, e.g. benzo[a]pyrene.

polycystid *a*. partitioned off.

polydactyly *n*. condition of having more than five fingers or toes.

polydelphic *a*. having more than one set of ovaries, oviducts and uteri, *appl*. nematodes.

polyderm *n*. a protective tissue in plants made up of alternating layers of endoderm and parenchyma cells.

polydesmic *a*. *appl*. a type of fish scale growing by apposition at margin, and made up of several units of fused small bony teeth covered with dentine.

Polydnaviridae *n*. family of enveloped insect DNA viruses. Each particle contains many double-stranded DNAs of variable molecular weight.

polydomous *a*. *pert*. single colonies (of social insects) that occupy more than one nest.

polyembryony *n*. the development of several embryos within a single ovule; the case where a zygote gives rise to more than one embryo, as in identical twins.

polyenergid *a*. *appl*. nucleus with more than one centriole.

polyethism *n*. division of labour amongst members of an animal society, as in the social insects.

polyethylene glycol an agent used to induce cell fusion in the formation of somatic cell hybrids.

poly (G) polyribonucleotide composed exclusively of guanidylate residues, poly (dG) being the corresponding polydeoxynucleotide.

polygalacturonan *n*. a polysaccharide made of galacturonic acid residues linked together, a component of plant cell walls, found as part of the complex polysaccharide of the pectin fraction, and which

is hydrolysed by the enzyme polygalacturonase. *alt*. galacturonan.

polygalacturonase *n*. enzyme that degrades polygalacturonans in plant cell walls during fruit ripening.

Polygales *n*. order of dicot herbs, shrubs and small trees, and including the families Polygalaceae (milkwort) and others.

polygamy *n*. having more than one mate at a time; (*bot*.) having male, female or hermaphrodite flowers on the same plant. *a*. polygamous.

polygenes *n.plu*. genes which each have a small effect and which collectively produce a multifactorial or polygenic phenotypic trait.

polygenesis *n*. derivation from more than one source; origin of a new type at more than one place or time.

polygenic *a*. *appl*. phenotypic characters (such as height in humans) that are determined by the collective effects of a number of different genes; *appl*. inheritance, the inheritance of such traits, *alt*. multifactorial inheritance. *see also* quantitative variation.

polygenic inheritance inheritance of phenotypic characters (such as height, eye colour, etc. in humans) that are determined by the collective effects of several or many different genes. *alt*. multifactorial inheritance.

polyglucosan *n*. a polymer of glucose units. *alt*. glucan.

Polygonales *n*. order of herbaceous dicots, rarely trees, and comprising the family Polygonaceae (buckwheat).

polygoneutic *a*. rearing more than one brood in a season.

polygynaecial *a*. having multiple fruits formed by united gynaecia.

polygynous *a*. consorting with more than one female at a time; *appl*. flowers with numerous styles. *n*. polygyny.

polygyny *n*. having more than one female mate at a time; (*bot*.) having numerous styles. *a*. polygynous.

polyhaline *a*. *appl*. brackish water of salinity 18–30 parts per thousand, approaching that of sea water.

polyhaploid *a*. *pert*. the gametic chromosome number of a polyploid organism; *n*. a haploid organism derived from a normally polyploid species.

polyhybrid *n.* a hybrid heterozygous for many genes.

polyhydric *a. appl.* alcohol or acid with three, four or more hydroxyl groups.

polyIg receptor receptor for immunoglobulins on the basal membrane of epithelial cells which binds multimeric immunoglobulins such as IgA and IgM, during their secretion. Part of the receptor becomes secretory component.

polyisomeres *n.plu.* parts all homologous with each other, as leaves of plants of the same species.

polyisoprenoid *n.* any of a variety of compounds containing long-chain polymers of isoprene.

polykaryocyte megakaryocyte *q.v.*; multinucleate cell, produced in some viral infections by virus-induced fusion of infected cells.

polykaryon *n.* a nucleus with more than one centriole.

polykont multiflagellate *q.v.*

polylecithal *a. appl.* ova containing relatively large amounts of yolk.

polylepidous *a.* having many scales.

polymastia polymastism *q.v.*

polymastigote multiflagellate *q.v.*

polymastism *n.* occurrence of more than the normal number of mammae.

polymegaly *n.* occurrence of more than two sizes of sperm in one animal.

polymeniscous *a.* having many lenses, as compound eye.

polymer *n.* large organic molecule made up of repeating identical, or similar, subunits.

polymerase *see* DNA polymerase, RNA polymerase.

polymerase chain reaction (PCR) technique for selectively replicating a particular stretch of DNA *in vitro* to produce a large amount of a particular gene, which can be carried out using uncloned genomic DNA as the starting material, thus obviating the need for DNA cloning in microorganisms.

polymeric *a. appl.* system of independently segregating genes, additive in affecting the same phenotypic character.

polymerization *n.* the formation of a polymer from smaller subunits.

polymerous *a.* consisting of many parts or members.

polymitosis *n.* excessive cell division.

polymorph polymorphonuclear leukocyte *q.v.*

polymorphic *a.* existing in two or more different forms within a species or population, *appl.* genes, characters, morphological forms, etc.; showing a marked degree of variation in body form during the life-cycle or within the species; *pert.* or containing variously shaped units (i.e. cells, or individuals in a colony).

polymorphic loci genetic loci with two or more alleles, conventionally defined as loci at which the most common homozygote has a frequency of less than 90% in a given population.

polymorphism *n.* the existence within a species or a population of different forms of individuals; occurrence of different forms of, or different forms of organs in, the same individual at different periods of life. *see also* enzyme polymorphism, genetic polymorphism.

polymorphonuclear leukocyte type of phagocytic white blood cell characterized by multipartite nucleus, and which ingests invading organisms. *alt.* polymorph, neutrophil, granulocyte.

polymyarian *n.* having more than five longitudinal rows of muscle cells, *appl.* some nematodes.

polynemic *a.* consisting of several strands.

Polynesian Floral Region part of the Palaeotropical Realm comprising the Pacific islands east of Indonesia, except for the Fijian and Hawaiian islands.

polynuclear, polynucleate *a.* with several or many nuclei. *alt.* multinucleate.

polynucleotide *n.* unbranched chain of nucleotides (*q.v.*) linked through an alternating sugar/phosphate backbone. *see* Appendix 1 (25) for structure.

polynucleotide kinase enzyme which adds a phosphoryl group to the 5′-end of a DNA strand, used to label ^{32}P DNA for sequencing by Maxam and Gilbert's method. *r.n.* polynucleotide 5′-hydroxylkinase, EC 2.7.1.78.

polynucleotide phosphorylase enzyme catalysing the synthesis of polyribonucleotides from ribonucleoside diphosphates without the need for a DNA or RNA template. EC 2.7.7.8, *r.n.*

polyribonucleotide nucleotidyltransferase.

polyoestrous *a.* having a succession of periods of oestrus in one sexual season. *cf.* monoestrous.

polyoma *n.* a small DNA virus, normally found in mice and which can cause tumours in newborn rats, hamsters and some strains of laboratory mice.

polyp *n.* sedentary individual or zooid of a colonial animal; in coelenterates, an individual having a tubular body, usually with a mouth and ring of tentacles on top, like a miniature sea anemone; a small stalked outgrowth of tissue, usually benign, but may become malignant, from a mucous surface such as the intestine.

polyparium, polypary *n.* the common base and connecting tissue of a colony of polyps.

polypeptide *n.* a chain of amino acids linked together by peptide bonds *(q.v.).* A polypeptide chain is the basic structural unit of a protein, some proteins consisting of one, some of several polypeptides. The polypeptide chains of proteins are synthesized on the ribosomes, using messenger RNA as a template. *see also* genetic code, protein.

polypetalous *a.* having separate, free or distinct petals.

polyphagous *a.* eating various kinds of food; of insects, using many different food plants.

polyphenism *n.* the occurrence in a population of several phenotypes which are not genetically controlled.

polyphenol oxidase any of several enzymes which catalyse the oxidation of tyrosine to dioxyphenylalanine using molecular oxygen. EC 1.14.18.1. *r.n.* monophenol monooxygenase.

polypheny pleiotropy *q.v.*

polyphyletic *a. appl.* a taxonomic group having origin in several different lines of descent.

polyphyllous *a.* many-leaved.

polyphyodont *a.* having many successive sets of teeth.

Polyplacophora *n.* a class of molluscs which includes the chitons *q.v.*

polyplanetic *a.* having several motile phases with intervening non-motile or resting stages.

polyplastic *a.* capable of assuming many forms.

polyploid *a.* having more than 2 chromosome sets, as triploid (3), tetraploid (4), etc.; *n.* an organism with more than 2 chromosome sets per somatic cell.

polyploidy *n.* the polyploid condition, which may be the normal state of the somatic tissues of the whole organism as in some plants, or a reduplication of chromosome number found in only some tissues or cells, and which can be induced artificially by chemicals such as colchicine or β-naphthol.

polypneustic *a. appl.* lateral lobes bearing multiple spiracle pores, in some insects.

polypoid *a.* resembling a polyp.

polypores *n.plu.* group of basidiomycete fungi including the bracket fungi, coral fungi and the cantharelles, in which the basidocarp is usually leathery, papery or woody and the hymenium may be smooth, ridged, warty or spiny or form the lining of tubes or gills on the underside of the basidiocarp. *cf.* agarics.

polyprotein *n.* long polypeptide chain produced on translation by, e.g., poliovirus, which is subsequently cleaved into separate proteins.

polyprotodont *a.* with four or five incisors on each side of upper jaw, and one or two fewer in lower.

polyrhizal multiradicate *q.v.*

polyribosome polysome *q.v.*

polysaccharide *n.* any of a diverse class of high-molecular weight carbohydrates formed by the linking together by condensation of monosaccharide, or monosaccharide derivative, units into linear or branched chains, and including homopolysaccharides (composed of one type of monosaccharide only) and heteropolysaccharides (composed of a mixture of different monosaccharides). Found as storage products (e.g. starch and glycogen) and structural components of cell walls (e.g. cellulose, xylans and arabinans), and as components of glycoconjugates, *see* proteoglycans. *alt.* glycan.

polysaprobic *a. appl.* category in the saprobic classification of river organisms comprising those that can live in water heavily polluted with organic pollutants, in which decomposition is mainly anaerobic, e.g. sewage fungus, bloodworms and the rat-tailed maggot (*Eristalis tenax*); *appl.* aquatic habitats with heavy pollution by organic matter with little or no dissolved oxygen, the formation of sulphides, abundant bacteria, but few animals feeding on them or on the decaying matter. *cf.* α-mesosaprobic, β-mesosaprobic, oligosaprobic.

polysepalous *a.* having free or distinct sepals.

polysiphoneous, polysiphonic *a.* consisting of several rows of cells, *appl.* thallus of red or brown algae; consisting of several tubes.

polysome *n.* aggregate of ribosomes on messenger RNA during protein synthesis. *alt.* polyribosome.

polysomic *a. appl.* cells or organisms carrying more than the normal number of any particular chromosome.

polysomitic *a.* having many body segments; formed from fusion of primitive body segments.

polysomy *n.* condition in which more than two copies of any particular chromosome are present in diploid cells.

polyspermous *a.* having many seeds.

polyspermy *n.* entry of several sperms into one ovum.

polyspondyly *n.* condition of having vertebral parts multiple where the myotome has been lost.

polysporocystid *a. appl.* oocyst of sporozoans when more than four sporocysts are present.

polysporous *a.* many-seeded; many-spored.

polystachyous *a.* with numerous spikes.

polystely *n.* arrangement of vascular tissue in several steles, each containing more than one vascular bundle. *a.* polystelic.

polystemonous *a.* having stamens more than double the number of petals or sepals.

polystichous *a.* arranged in numerous rows or series.

polystomatous *a.* having many pores, openings or mouths.

polystomium *n.* a suctorial mouth in certain jellyfish.

polystylar *a.* having many styles.

polysymmetrical *a.* divisible through several planes into bilaterally symmetrical portions.

poly (T) polynucleotide composed exclusively of thymidylate residues.

polytene chromosome type of giant chromosome formed in some tissues (e.g. salivary gland) of dipteran larvae by successive replications of a synapsed pair of homologous chromosomes without separation of the new material, and which bears a characteristic pattern of dark and light bands visible in the light microscope.

polyteny *n.* the duplication of haploid chromosome content in a polytene chromosome, the degree of polyteny being the number of haploid chromosomes contained in each giant chromosome.

polythalamous *a.* aggregate or collective, as *appl.* to fruits; *appl.* shells made up of many chambers formed successively.

polythelia *n.* occurrence of supernumerary nipples.

polythermic *a.* tolerating relatively high temperatures.

polythetic *a. appl.* a classification based on many characteristics, not all of which are necessarily shown by every member of the group.

polytocous, polytokous *a.* producing several young at a birth; fruiting repeatedly.

polytomous *a.* having more than two secondary branches; with a number of branches originating in one place.

polytopic *a.* occurring or originating in several places.

polytrichous *a.* having the body covered with an even coat of cilia, as some ciliate protozoans; having many hair-like outgrowths.

polytrochal *a.* having several circlets of cilia between mouth and posterior end.

polytrophic *a. appl.* ovariole in which nutritive cells are enclosed in oocyte follicles; nourished by more than one organism or substance; obtaining food from many sources; eutrophic *q.v.*

polytropic pantropic *q.v.*

polytypic *a.* having or *pert.* many types; *appl.* species having geographical subspecies; *appl.* genus having several species.

poly (U) polyribonucleotide composed exclusively of uridylate residues.

polyunsaturated *a. appl.* fatty acids with

more than one C=C double bond in their hydrocarbon chain.

polyvoltine *a.* producing several broods in one season.

polyxenous *a.* adapted to life in many different hosts, *appl.* parasites.

polyxylic *a.* having many strands of xylem and several concentric vascular rings.

Polyzoa Bryozoa *q.v.*

polyzoic *a. appl.* a colony of many zooids.

polyzooid *n.* an individual in a polyzoan colony.

pome *n.* fruit derived from a compound inferior ovary in which the fleshy portion is largely the enlarged base of the perianth or receptacle, e.g. apples, pears.

pompetta *n.* an organ forcing sperm into penis, as in flies of the genus *Phlebotomus*.

ponderal *a. pert.* weight; *appl.* growth by increase in mass.

poneroid complex one of two major taxonomic groups of ants, exemplified by the subfamily Ponerinae.

pongid *n.* any anthropoid ape other than the gibbons or siamang, i.e. chimpanzee, gorilla and orang-utan.

pons *n.* structure connecting two parts; broad band of nerve fibres, pons Varolii, in the mammalian brain, connecting the two sides of the cerebellum and medulla oblongata.

pontic, pontile *a. pert.* a pons.

pontine *a. pert.* pons Varolii.

pooid *a. appl.* grasses of the genus *Poa*, e.g. *Poa annua* (meadow-grass).

popliteal *a. pert.* region behind and above knee joint.

population *n.* a group of individuals of a species living in a certain area.

population crash sharp reduction in the population of a species when its numbers exceed the carrying capacity of the habitat. *alt.* dieback.

population density number of individuals, usually with reference to a given species, living in a specified area.

population dispersion the distribution of the members of a population throughout its habitat.

population distribution the variation in population density over a given area.

population dynamics the changes in the structure of a population over time, i.e. the changes in the relative numbers of individuals of particular ages, different sexes, or different forms.

population ecology the study of factors influencing the numbers and structure of a given population.

population genetics the study of how genetic principles apply to groups of interbreeding individuals (a population) as a whole.

porcellanous *a.* resembling porcelain, white and opaque, *appl.* calcareous shells, as of foraminiferans and some molluscs.

pore *n.* a minute opening or passage, as of the skin, sieve plates, stomata, etc.

pore canals minute spiral tubules passing through the cuticle but not the epicuticle of insects.

pore cell in sponges, cells of the outer layer that are perforated by a pore through which water enters.

pore chains in wood, extensive radial, tangential or oblique groups of xylem vessels, as seen in cross-sections.

pore complex complex structure present at pores in the nuclear membrane.

pore organ structure surrounding canal for excretion of mucilage through pores, in desmids.

pore space in soil, the spaces between particles of soil collectively.

poricidal *a.* dehiscing by valves or pores.

Porifera *n.* phylum of simple multicellular animals, commonly called sponges, with a simple body enclosing a single central cavity (in the simple sponges) or penetrated by numerous interconnected cavities. The body wall consists of an outer layer of epithelium separated from an inner layer of ciliated choanocytes (feeding cells) by a mesogloeal layer. There are no nerve or muscle cells. Water is drawn into the internal cavities through pores (ostia), food particles are taken up by the choanocytes, and the water flows out through a large pore (the osculum). There are three classes: the Calcarea, the calcareous sponges (e.g. *Leucosolenia*), which have spicules of calcium carbonate embedded in the mesogloea and projecting to the outside, the Hexactinellida, the glass sponges (e.g. *Euplectella*, Venus's flower basket), with silica spicules, and the Demospongia, which includes some species with silica spicules and some species without, and which often have the body wall strengthened by a tangled mass of fibres (e.g. the bath sponge *Spongia*).

poriferous *a*. furnished with numerous openings.

poriform *see* poroid.

porin *n*. protein in outer lipopolysaccharide membrane of Gram-negative bacteria, and which enables small polar molecules to cross the membrane.

porogamy *n*. entrance of a pollen tube into ovule by micropyle.

poroid *a*. like a pore or pores; having porelike depressions, *alt*. poriform; *n*. minute depression in theca of dinoflagellates and diatoms.

porophyllous *a*. having or *appl*. leaves with numerous transparent spots.

porose *a*. having or containing pores.

porphyrophore *n*. a reddish-purple pigment-bearing cell.

porphyropsin *n*. light-sensitive visual pigment (rhodopsin) found in freshwater fish and amphibian tadpoles.

porpoise *n*. marine mammal, a member of the family Phocoenidae of the suborder Odontoceti (toothed whales) of the Cetacea *q.v.* Porpoises are smaller and dumpier than dolphins and lack the typical dolphin "beak". They also cannot leap completely out of the water like a dolphin.

porrect *a*. extended outwards.

porta *n*. any gate-like structure as transverse fissure of liver.

portal *a. appl*. veins leading from alimentary canal, spleen and pancreas to liver; also a vascular system to kidney in lower vertebrates.

ports, porters membrane transport proteins *q.v.*

position effect the influence of its location on the chromosome on the activity of a gene, demonstrated by the changes in gene expression which occur when genes are translocated to different positions on the chromosome, and which can produce changes in the phenotype without any quantitative change in the genotype.

positional cloning the isolation of a gene starting from a knowledge of its position on the chromsome.

positional information in embryonic development, the developmental signals a cell receives by virtue of the position in the embryo it has arrived at and which direct its further development. In the developing vertebrate limb, for example, certain mesenchymal cells are thought to be directed to eventually become cartilage or fibrous connective tissue by virtue of their position at a certain point in the developmental process. Positional information may be given to the cell by means of gradients of a chemical signal emanating from a particular source, or by cell-surface molecules on neighbouring cells, or by the composition of the extracellular matrix, etc.

positive control type of control of gene expression (in bacteria) in which genes are expressed only when an active regulator protein is present.

positive interference the effect that the occurrence of one chiasma decreases the likelihood of another forming nearby.

positive reinforcement a stimulus, or series of stimuli, which is pleasant to an animal and increases its response.

positive-strand RNA viruses RNA viruses (e.g. poliovirus) in which the genome has the same base sequence as the messenger RNA, a complementary copy of the genome RNA first being made, which then acts as a template for mRNA synthesis.

positive taxis or tropism tendency to move or grow towards the source of the stimulus.

post- prefix derived from L. *post*, after, signifying situated behind, the hindmost part of an organ or structure, or occurring after.

postabdomen *n*. in scorpions, the metasoma or posterior narrower five segments of abdomen; anal tubercle in spiders.

postalar *a*. situated behind the wings.

postanal *a*. behind the anus.

postantennal *a*. situated behind antennae.

postarticular *a*. posterior process of surangular, behind articulation with quadrate.

postaxial *a*. on posterior side of the axis, as e.g. on fibular side of leg.

postbacillary *a*. having nuclei behind sensory part of retinal cells, *appl*. ocellus, inverted eye, as of spiders.

postbranchial *a*. behind the gill clefts; *appl*.

bodies: ultimobranchial bodies *q.v.*

postcardinal *a.* behind the region of the heart.

postcava, postcaval vein the vein bringing blood to heart from posterior part of body. *alt.* posterior vena cava.

postcentral *a.* behind central region; *appl.* region of cerebral cortex immediately posterior to central sulcus; *appl.* part of the intraparietal sulcus.

postcentrum *n.* the posterior part of centrum of vertebrae, in some vertebrates.

postcerebral *a.* posterior to the brain, *appl.* salivary glands in the head of hymenopterans.

postcingular *a.* posterior to cingulum.

postclavicle *n.* a membrane bone present in pectoral girdle of some teleost fishes.

postcleithrum postclavicle *q.v.*

postclimax *a. appl.* stable plant community whose composition reflects climatic conditions which are more favourable (e.g. moister, cooler) than usual for the region and which therefore differs from the usual climax vegetation for the region.

postclisere *n.* a series of vegetative formations (*q.v.*) that arise when the climate becomes wetter.

postclitellian *a.* situated behind clitellum.

postclival *a.* situated behind the clivus of cerebellum, *appl.* a fissure.

postclypeus *n.* the posterior part of clypeus of an insect.

postcolon *n.* part of gut between colon and rectum in certain mites.

postcornual *a. appl.* glands situated behind horns, as in chamois.

postcranial *a. appl.* posterior head region.

postdicrotic *a. appl.* secondary wave of a pulse, or that succeeding the dicrotic.

postembryonic *a. pert.* the age or states succeeding the embryonic.

posterior *a.* situated behind; nearer the tail end; dorsal in human anatomy; behind the axis; superior or next to the axis. *cf.* anterior.

posterior commissure tract of fibres connecting the two cerebral hemispheres at posterior end of corpus callosum.

posterior horn of spinal cord, that part of grey matter containing sensory cells. *cf.* anterior horn.

posterior lobe of pituitary gland the pars intermedia and the pars nervosa (neurohypophysis).

posterior vena cava postcava *q.v.*

posterolateral *a.* placed posteriorly and towards the side, *appl.* arteries.

posteromedial *a.* placed posteriorly and towards the middle, *appl.* arteries.

postesophageal postoesophageal *q.v.*

postestrum metoestrus *q.v.*

postflagellate *a. appl.* to forms of trypanosomes intermediate between flagellates and cyst.

postfrons *n.* portion of frons of insect head posterior to antennary base line.

postfrontal *a. appl.* a bone present behind the orbit of eye in some vertebrates.

postfurca *n.* forked process from sternum or an apodeme of metathorax in insects.

postganglionic *a. appl.* autonomic nerve fibres issuing from ganglia.

postgena *n.* the posterior part of gena (side) of insect head.

postgenital *a.* situated behind the genital segment, in arthropods and other segmented animals.

postglacial *appl.* Holocene *q.v.*

postglenoid *a.* situated behind the glenoid fossa, *appl.* a process or tubercle.

posthepatic *a. appl.* latter part of alimentary canal, that from liver to end.

posthypophysis postpituitary *q.v.*

postical, posticous *a.* an outer or posterior surface; *appl.* lower or back surface of a thallus, leaf or stem, esp. in liverworts.

postischium *n.* a lateral process on the hinder side of ischium in some reptiles.

postlabrum *n.* posterior portion of insect labrum, where differentiated.

postmeiotic segregation the segregation of the 2 strands of a heteroduplex DNA (formed during genetic recombination) in a subsequent round of DNA replication that succeeds meiosis.

postmentum *n.* the united sclerites constituting the base of labium of insects.

postminimus *n.* a rudimentary additional digit occasionally present in amphibians and reptiles.

postmitotic *a. appl.* a cell that once it has been formed by mitosis does not undergo another mitosis and cell division before its death.

postneural *a.* situated at the end of the back,

appl. plates of shell of tortoises and turtles.

postnodular *a. appl.* a cerebellar fissure between nodule and uvula.

postnotum *n.* posterior portion of insect notum.

postocular *a.* behind the eye, *appl.* scales.

postoesophageal *a. appl.* nerve fibres connecting ganglia serving antennae in crustaceans; *appl.* nerve fibres connecting cerebral ganglia in various invertebrates.

postoestrus metoestrus *q.v.*

postoral *a.* behind the mouth.

postorbital *a.* behind the eye socket, *appl.* bone forming posterior wall of eye socket; *appl.* luminescent organ in certain fishes.

postorbital *n.* bone of vertebrate skull lying immediately behind orbit of eye.

postotic *a.* behind the ear.

postparietal *a. appl.* paired skull bones sometimes occurring between parietal and interparietal.

postpatagium *n.* in birds, small fold of skin extending between upper arm and trunk.

postpermanent *a. appl.* traces of a dentition succeeding the permanent.

postpetiole *n.* in ants, the 2nd segment of abdominal stalk.

postphragma *n.* a phragma developed in relation with a postnotum in insects.

postpituitary *a. pert.* or secreted by the posterior lobe of the pituitary gland (the pars intermedia and the neurohypophysis).

postpubic *a.* at posterior end of pubis, *appl.* processes of pubis parallel to ischium.

postpyramidal *a.* behind the pyramid of cerebellum, *appl.* a fissure.

postreduction *n.* halving the chromosome number in the 2nd meiotic division instead of the 1st.

postretinal *a.* situated behind the retina; *appl.* nerve fibres connecting periopticon and inner ends of ommatidia in a compound eye.

postscutellum *n.* a projection under the scutellar lobe of mesothorax in insects; a sclerite behind the scutellum.

postsegmental *a.* posterior to body segments or somites.

postsphenoid *n.* the posterior part of sphenoid in skull.

poststernellum *n.* most posterior portion of an insect sternite.

poststernite *n.* posterior sternal sclerite of insects.

postsynaptic *a. appl.* cell or part of cell (e.g. dendrite membrane or muscle cell membrane) on the receiving side of a synapse in nervous system or at a neuromuscular junction.

postsynaptic potential membrane potential generated in a receiving neurone by the action of a neurotransmitter at a synapse.

post-temporal *a.* behind the temporal bone, *appl.* bone and fossa.

post-transcriptional *a. appl.* processes occurring after transcription, as capping and poly(A) addition to eukaryotic RNAs.

post-translational *a. appl.* processes occurring after translation, as modifications to proteins such as glycosylation, and cleavage of preproteins, signal sequences, etc.

post-trematic *a.* behind an opening such as a gill cleft, *appl.* nerves running in posterior wall of 1st gill cleft in pharynx.

postzygapophyses *n.plu.* the posterior zygapophyses of a vertebra, articulating with the prezygapophyses of the vertebra immediately succeeding it.

potamobenthos *n.* the bottom-living organisms in a river or other fresh water.

potamodromous *a.* migrating only in fresh water.

Potamogetonales Najadales *q.v.*

potamoplankton *n.* the plankton of rivers, streams and their backwaters.

potassium (K) element essential for the growth and survival of living organisms. At the cellular level it is involved in maintaining intracellular ion balance and generating the membrane potential in all cells and in producing electrical signals in neurones. One of the major elements required for plant growth.

potassium channel ion channel in a cell membrane that is permeable to potassium ions.

potato *n. Solanum tuberosum*, a crop plant originating in South America and now grown widely in the cooler, wetter parts of the world, whose tubers are eaten as food. *cf.* sweet potato.

potential *a.* latent, as *appl.* characteristics. *see also* action potential, membrane potential.

potexvirus group group of single-stranded

RNA plant viruses with flexuous rod-shaped particles, type member potato virus X.

potyvirus group group of single-stranded RNA plant viruses with flexuous rod-shaped particles, type member potato virus Y.

Poupart's ligament the ligament of the groin.

powder-down feathers those which do not develop beyond an early stage.

powdery mildews parasitic fungi of the order Erysiphales in the Pyrenomycetes (*q.v.*), the powdery appearance being due to the large numbers of conidia (spores) formed on the surface of the host tissue.

Poxviridae, poxviruses *n., n.plu.* family of large, double-stranded DNA viruses that includes vaccinia and smallpox, fowl pox, sheep pox and myxoma.

P-particle a kappa particle that has been liberated into the medium and releases paramecin which kills sensitive strains of *Paramecium*.

prae- pre- *q.v.*

praeabdomen *n.* the anterior, broader part of abdomen in scorpions.

praecoces *n.plu.* newly hatched birds able to take care of themselves. *a.* praecocial.

praeputium prepuce *q.v.*

prairie *n.* in North America, the natural grassland covering the middle of the continent in the mid-latitudes, and which consists of tall-grass prairie in the cooler moister areas, most of which has now been converted into agricultural land, and short-grass prairie.

pratal *a. pert.* meadows; *appl.* flora of rich humid grasslands.

Prausnitz-Kästner reaction specific allergy produced in non-allergic individual after injection of serum from allergic individual.

pre- prefix derived from L. *prae*, before, signifying situated before or occurring before.

preadaptation *n.* any previously existing anatomical structure, physiological process or behaviour pattern that makes new forms of evolutionary adaptation more likely.

preanal *a.* anterior to the anus.

preantenna *n.* one of the pair of feelers on the 1st segment in onychophorans.

preaxial *a.* in front of the axis; on anterior border or surface.

pre-auricular *a. appl.* a groove at anterior part of auricular surface of hip-bone.

prebacillary *a.* having nuclei distal to sensory portion of retinal cells, *appl.* occellus, converted or erect eye, as of spiders.

pre-B cells an early stage in the B lymphocyte lineage, in which the immunoglobulin heavy chain gene has been rearranged, expressing heavy chains of class μ in the cytoplasm.

prebiotic *a.* before life appeared on earth.

prebranchial pretrematic *q.v.*

Precambrian *a. pert.* or *appl.* time before the Cambrian, reckoned generally as the era lasting from the earliest formation of rocks until around 590 million years ago, and divided into two eons, the Proterozoic and the earlier Archaean. The Precambrian saw the origin of life, the evolution of living cells and the evolution of the eukaryotic cell. The first multicellular animals arose towards the end of the era. *alt.* Archaean.

precapillary *a. appl.* arterioles having an incomplete muscular layer. *n.* a small vessel conducting blood from arteriole to capillary.

precartilage *n.* type of cartilage preceding other kinds, or persisting as in fin-rays of certain fishes.

precava, precaval vein the vein bringing blood to the heart from the anterior part of the body. *alt.* anterior vena cava.

precentral *a.* situated anterior to the centre; *appl.* region of cerebral cortex immediately anterior to central sulcus, *appl.* a sulcus anterior to and parallel with the central sulcus; *appl.* a gyrus.

precentrum *n.* the anterior part of centrum of vertebrae of certain vertebrates.

precheliceral *a.* anterior to chelicerae, *appl.* segment of mouth region in arachnids.

prechordal *a.* anterior to notochord or spinal chord, *appl.* to part of base of skull.

precingular *a.* anterior to cingulum.

precipitins *n.plu.* specific antibodies that form a precipitate with their corresponding antigen.

precipitin reaction assay for measuring the amount of antigen or antibody in a sample, based on the precipitation of antigen-antibody complexes from solution.

preclavia *n*. an element of the pectoral girdle.

preclimacteric *a*. a period before the climateric or time of ripening.

preclimax *n*. the plant community immediately preceding the climax community.

preclival *a*. *appl*. fissure in front of clivus of cerebellum.

precocial *a*. *appl*. young that are able to move around and forage at a very early stage, esp. in birds. *cf*. altricial.

preconnubia *n*. gatherings of animals before the mating season.

precoracoid *a*. an anterior ventral bone of pectoral girdle in amphibians and reptiles.

precostal *a*. *appl*. short spurs on basal portion of hind wing of lepidopterans.

precoxa subcoxa *q.v.*

precoxal subcoxal *q.v.*

precrural *a*. on anterior side of leg or thigh.

precuneus *n*. the medial surface of parietal lobe of cerebral hemisphere.

precursor *n*. cell from which other cells will develop; protein from which an enzyme, hormone, etc. will be produced by further chemical modification, enzymatic cleavage, etc.; any substance that precedes and is involved in the formation of a compound.

precystic *a*. *appl*. small forms appearing before the encystment stage in some protozoans.

predaceous *alt*. spelling of predacious *q.v.*

predacious *a*. *appl*. fungi of the family Zoopagaceae, which trap and feed on protozoans and nematode worms.

predator *n*. any organism that catches and kills other organisms for food.

predator chain food chain that starts from plants and passes from herbivores to carnivores. *cf*. parasite chain, saprophyte chain.

predator–prey relationship interaction between two organisms of different species in which one (the predator) captures and feeds on the other (the prey).

predentine *n*. immature dentine which is not yet calcified but made mainly of fibrils.

pre-embryo *n*. in mammals name sometimes given to the fertilized ovum and its cleavage stages up to blastocyst formation and the specification of the cells of the inner cell mass that will develop into the embryo proper.

preen gland oil gland *q.v.*

pre-epistome *n*. a plate covering base of epistome in certain arachnids.

pre-erythrocytic *a*. *appl*. phase of the malarial plasmodium life cycle in which it lives in the tissues of the human host, developing large schizonts that produce merozoites that infect erythrocytes.

prefemur *n*. second trochanter, as in walking legs of pycnogonids.

preferential species species that are present in several different communities, but are more common or thriving in one particular community.

preflagellate *a*. *appl*. forms of trypanosomes intermediate between cyst and elongated flagellates.

prefloration *n*. the form and arrangement of floral leaves in the flower bud.

prefoliation *n*. the form and arrangement of foliage leaves in the bud.

preformation theory theory current up to the 18th century, according to which it was supposed that a sperm contained a miniature adult (homunculus).

prefrontal *a*. *appl*. a bone anterior to frontal bone in certain vertebrates.

preganglionic *a*. nerve fibres running from spinal cord and ending in synapses in sympathetic ganglia.

pregenital *a*. situated anterior to genital opening; *appl*. segment behind 4th pair of walking legs in arachnids.

pregnenolone *n*. cholesterol derivative, precursor of all steroid hormones. Its synthesis is stimulated by ACTH.

prehallux *n*. a rudimentary additional digit on hindlimb.

prehaustorium *n*. a rudimentary root-like sucker.

prehensile *a*. adapted for grasping and holding.

prehepatic *a*. *appl*. part of digestive tract anterior to liver.

prehyoid *a*. *pert*. mandible and hyoid; *appl*. cleft between mandible and ventral parts of hyoid arch.

prehypophysis prepituitary *q.v.*

preimaginal *a*. preceding the imaginal or adult stage.

preimaginal conditioning in insects a response learnt in a larva which is retained preimplantation

by the adult.

preimplantation *a. appl.* mammalian embryo before it has become implanted in the ovarian wall.

preinterparietal *a.* one of two small upper membranous centres of formation of supraoccipital bone.

prelacteal *a. pert.* a dentition that may occur previous to the milk teeth.

premandibular *a.* anterior to mandible; *appl.* somites of amphioxus; *appl.* a bone of certain reptiles.

premaxilla *n.* a paired bone or cartilage anterior to the maxilla of jaw in most vertebrates. In terrestrial vertebrates it bears the front teeth. In most teleost fish it can be protruded independently of the maxilla. *plu.* premaxillae.

premaxillary *a.* anterior to maxilla; *pert.* premaxilla *q.v.*

premedian *a.* anterior to middle of body or part; *appl.* vein in front of median vein in certain insect wings.

prementum *n.* the united stipites bearing ligula and labial palps of insects.

pre-messenger RNA pre-mRNA *q.v.*

premetaphase prometaphase *q.v.*

premolars *n.plu.* teeth located between canines and molars in mammalian dentition.

premorse *a.* with irregular and abrupt termination, as if end were bitten off.

pre-mRNA the primary messenger RNA transcript. esp. in eukaryotic cells, before removal of introns and other processing.

premyoblast *n.* cell that gives rise to a myoblast.

prenasal *a. appl.* a bone developed in septum in front of mesethmoid in skulls of certain vertebrates.

prenatal *a.* before birth, *appl.* tests for genetic defects performed on a foetus in the womb.

preoccipital *a. appl.* an indentation or notch in front of posterior end of cerebral hemispheres.

preocular *a.* in front of the eye, as antennae, scales, etc.

preopercular *a.* anterior to gill cover; *appl.* luminescent organ in certain fishes; *appl.* bone: preoperculum *q.v.*

preoperculum *n.* anterior membrane bone of gill cover of fishes.

preoptic nerve cranial nerve associated

with vomeronasal organ and ending in nasal mucosa.

preoral *a.* situated in front of the mouth.

preorbital *a.* anterior to orbit of eye, *appl.* a bone in teleost fishes; *appl.* glands in ruminants.

prepatagium *n.* fold of skin extending between upper arm and forearm of birds; part of the patagium between neck and fore-feet in flying lemurs.

prepatellar *a. appl.* a pouch between lower part of knee-cap (patella) and skin.

prepattern *n.* a system of spatially organized molecular developmental cues already laid down in the fertilized egg and which guide development.

prepenna *n.* a nestling down feather which is succeeded by adult contour feather.

prepharynx *n.* narrow thin-walled structure connecting oral sucker and pharynx, in trematodes.

prephragma *n.* a phragma developed in relation with the notum in insects.

prepituitary *n.* anterior lobe of pituitary gland, the adenohypophysis *q.v.*

preplacental *a.* occurring before placenta formation or development.

preplumula *n.* a nestling down feather which is succeeded by adult down feather.

prepollex *n.* a rudimentary additional digit occurring sometimes before thumb in certain amphibians and reptiles.

prepotency *n.* the fertilization of a flower by pollen from another flower in preference to pollen from its own stamens, when both are offered simultaneously; capacity of one parent to transmit more characteristics to offspring than other parent.

prepotent *a.* transmitting the majority of characteristics, *appl.* one of the parents in certain genetic crosses; *appl.* a flower exhibiting a preference for cross-pollination; having priority, as one reflex among other reflexes.

preprohormone *n.* a precursor to a polypeptide prohormone, representing the original polypeptide chain synthesized from mRNA, and which is later modified, usually by enzymatic cleavage at specific points, to produce first a prohormone and then an active hormone.

preprotein *n.* transient protein precursor bearing terminal regions which are rapidly cleaved to produce a stable proprotein (such as proinsulin from preproinsulin) or a functional protein (as in many membrane proteins).

prepuberal *a.* anterior to pubis; prepubertal *q.v.*

prepubertal *a. pert.* age or state before puberty or sexual maturity.

prepubic *a. pert.* prepubis; *appl.* processes of pelvic arch, in certain fishes; on anterior part of pubis; *appl.* elongated processes on pubis of certain vertebrates.

prepubis *n.* part of pelvic girdle of certain reptiles and fishes.

prepuce *n.* foreskin, part of integument of penis which leaves surface at neck and is folded upon itself.

prepupa *n.* a quiescent stage preceding the pupal in some insects.

preputial *a. pert.* the prepuce or foreskin.

prepygidial *a.* anterior to pygidium, *appl.* growth zone in polychaete worms.

prepyloric *a. appl.* ossicle hinged to pyloric ossicle in gastric mill of crustaceans.

prepyramidal *a.* in front of pyramid of cerebrellum, *appl.* a fissure; *appl.* tract: the rubrospinal tract *q.v.*

prereduction *n.* halving the chromosome number in the 1st meiotic division.

prescutum *n.* anterior sclerite of insect notum.

presegmental *a.* anterior to body segments or somites.

presentation time minimum duration of continuous stimulus necessary for production of a response.

presequence *n.* sequence of amino acids at the N-terminal end of newly-synthesized proteins destined to enter the mitochondria, and which is required for their transport across the mitochondrial membranes from their site of synthesis in the cytoplasm.

prespermatid *n.* secondary spermatocyte.

presphenoid *n.* in many vertebrates, a bone of skull anterior to the basisphenoid; the anterior part of the sphenoid bone.

prespiracular pretrematic *q.v.*

pressor *a.* causing a rise in arterial pressure.

pressure-flow hypothesis of transport of assimilates through phloem. A generally accepted theory that sugars and other solutes are transported via phloem along a gradient of hydrostatic pressure developed

osmotically. Assimilates are actively transported into the phloem at their source, lowering the water potential in the phloem. Water enters by osmosis, and passively carries the assimilates to a sink, such as a storage root, where they are actively transported out of the phloem, causing an increase in water potential within the phloem, and causing water to leave the sieve tube by osmosis.

presternal *a.* situated in front of sternum; *pert.* anterior part of sternum; *appl.* jugular notch, on superior border of sternum.

presternum *n.* the manubrium or anterior part of sternum; anterior sclerite of insect sternum.

prestomium *n.* in biting and sucking insects the aperture between tips of the mouthparts serving for the intake of food.

presumptive *a. appl.* embryonic cells or tissues that will, in the normal course of development, give rise to a particular tissue, as presumptive endoderm, presumptive mesoderm, etc., but which does not necessarily mean that they are irreversibly committed to this course of development.

presynaptic *a. appl.* neurone or part of neurone, e.g. membrane of axon terminal, on the transmitting side of a synapse; *appl.* vesicles liberating neurotransmitter at axon terminals.

presynaptic facilitation enhanced transmission between one neurone and another at a synapse as the result of the action of a third neurone synapsing on the axon terminal of the transmitting neurone.

presynaptic inhibition action of an inhibitory neurone exerted on the axon terminal of another.

pretarsus *n.* terminal part or outgrowth on leg or claw of insects and spiders.

pretrematic *a.* anterior to an opening such as a gill cleft or spiracle.

prevenule *n.* small vessel conducting blood from capillaries to venule.

prevernal *a. pert.*, or appearing in, early spring.

prevertebral *a. pert.* or situated in region in front of vertebral column; *appl.* portion of base of skull; *appl.* ganglia of sympathetic nervous system.

prey *n.* organism that is captured and used as food by another organism (the predator).

prevomer *n.* a bone anterior to pterygoid in skull of some vertebrates; the vomer of non-mammalian vertebrates; in monotremes, a membrane bone in floor of nasal cavities.

prezygapophyses *n.plu.* the anterior zygapophyses of a vertebra, articulating with the postzygapophyses of the vertebra immediately preceding.

Priapulida, priapulids, *n., n.plu.* a phylum of burrowing and marine worm-like pseudocoelomate animals with a warty and superficially ringed body, with spines around the mouth.

Pribnow box short sequence of bases in the promoter regions of prokaryotic genes which appears to be the key recognition signal for RNA polymerase, the consensus sequence being TATAATG centred approx. 10 base pairs before the start of transcription.

prickle cells cells of the deeper layers of stratified squamous epithelium, connected by bundles of keratin fibres, which in isolated cells under the microscope project in tufts ("prickles") from the surface.

primaquine *n.* anti-malarial drug, which can cause haemolysis in people with glucose-6-phosphate dehydrogenase deficiency.

primaries *n.plu.* one type of main flight feather in bird's wing, attached in the region of the posterior edge of the hand bones.

primary *a.* first; principal; original.

primary cambium procambium *q.v.*

primary cell culture culture of cells prepared directly from tissues.

primary cell wall in plant cells, the first cell wall, laid down before and while the cell is growing and formed of cellulose microfibrils in a matrix of hemicelluloses, glycoprotein and pectic substances. It is relatively plastic and is the only cell wall present in most actively dividing cells or those engaged in photosynthesis, secretion, etc. *cf.* secondary cell wall.

primary centre a part of central nervous system, especially of sensory cortex, which is the first to receive the input from a sensory organ.

primary consumer herbivore *q.v.*

primary ecological succession primary succession *q.v.*

primary endosperm nucleus generally triploid nucleus that derives from fusion of one generative nucleus with two polar nucleus in the embryo sac of flowering plants. It divides by mitosis to give endosperm.

primary feathers primaries *q.v.*

primary germ layers *see* germ layers.

primary growth growth of roots and shoots from the time of initiation at the apical meristem in plant embryo to when their expansion and differentiation is completed. *cf.* secondary growth.

primary host host in which a parasite lives for much of its life cycle and in which it becomes sexually mature.

primary immune response the immune response made on first contact with an antigen, taking several days to develop and in which the antibody is initially IgM.

primary induction neural induction *q.v.*

primary lymphoid organs in mammals, the bone marrow and thymus, which produce B and T lymphocytes respectively, but which do not participate directly in immune reactions.

primary meristems the meristematic tissue in developing plant embryo, consisting of outer protoderm (which gives rise to epidermis), surrounding the ground meristem (which gives rise to highly vacuolated loose ground tissue), which encloses the central procambium or procambial strand from which the vascular bundles arise.

primary mycelium haploid mycelium originating from a basidiospore.

primary oocyte oogonia which have begun the 1st meiotic division.

primary oocyte diploid female germ cell at the point of entering the first meiotic division of oogenesis. It is a large cell with a large nucleus (germinal vesicle).

primary organizer *see* organizer.

primary phloem collectively, the protophloem and metaphloem, the phloem derived from the primary cambium during primary growth.

primary pit fields areas of plant cell wall containing many plasmodesmata, and in which pits are clustered.

primary plant body the plant body formed

from growth at the apical meristems.

primary producer autotroph *q.v.*

primary production (PP) the assimilation and fixation of inorganic carbon and other inorganic nutrients into organic matter by autotrophs, which are therefore called primary producers.

primary productivity the amount of organic matter fixed by the autotrophic organisms in an ecosystem per unit time.

primary root root that develops as a continuation of the radicle in plant seedling.

primary sere plant succession on area previously without vegetation, from bare ground to climax community.

primary sexual characters differences between the sexes relating to the reproductive organs and gametes.

primary spermatocyte *see* spermatocyte.

primary structure in proteins, the amino acid sequence, in nucleic acids, the nucleotide sequence.

primary succession a plant succession that begins on bare ground.

primary transcript original RNA product of a transcription unit which has not yet been modified by splicing, capping, polyadenylation etc.

primase *n*. RNA polymerase which synthesizes the RNA primer for DNA synthesis during DNA replication.

Primates *n*. an order of mammals known from the Paleocene and including tree shrews, lemurs, monkeys, apes and man. They are largely arboreal with limbs modified for climbing, leaping or brachiating (swinging), large brains in relation to body size, a shortening of the snout and elaboration of the visual apparatus, often with stereoscopic vision.

prime movers the ultimate factors that determine the direction of evolutionary change. They are of two kinds: basic genetic mechanisms, preadaptations and constraints imposed by an organism's existing developmental programme on the one hand, and the set of all environmental influences that constitute the agents of natural selection on the other.

primer *n*. short RNA which must be synthesized on a DNA template before DNA polymerase can start elongation of a new DNA chain. It is subsequently removed and

the gap infilled with DNA; in the polymerase chain reaction, a pair of synthetic oligonucleotides complementary to flanking regions of the gene to be copied, which are bound to the DNA before the reaction commences to ensure that DNA replication is initiated at the required points.

primite *n*. the first of any pair of individuals in the chain-like colonies of gregarine protozoans, in which the front end of one (the satellite) becomes attached to the posterior end of another (the primite).

primitive *a*. of earliest origin; not differentiated or specialized; *appl*. traits that appeared first in evolution and which give rise to other, more advanced, traits. They are often, but not always, less complex than the advanced ones.

primitive node area of proliferating cells in which the primitive streak begins, thickened anterior wall of primitive pit.

primitive pit enclosure at anterior end of the confluent folds of the primitive streak.

primitive streak in the flat, disc-like early embryos of reptiles, birds and mammals, the two parallel longitudinal folds that develop on the epiblast and which represent the region at which cells are moving into the interior of the embryo to form the notochord, the anterior end of the primitive streak corresponding to the dorsal lip of the blastopore in amphibian gastrulas.

primordial *a*. primitive; original, first begun; first formed; *appl*. e.g. to embryonic cells which will develop into particular cell types or tissues, as in primordial germ cell.

primordial cell initial *q.v.*

primordial follicle immature human ovarian follicle in the earliest stages of development, in which the oocyte is arrested in prophase of the first meiotic division and is surrounded by a single layer of follicle cells.

primordial germ cell in the early sexually undifferentiated embryo, cells whose descendants will eventually give rise to eggs or sperm.

primordium *n*. original form; a developing structure at the stage at which it starts to assume a form, *alt*. anlage; (*bot*.) group of immature cells that will form a particular structure, as leaf or flower primordia. *plu*.

primordia.

primosome *n.* assembly of proteins concerned with the initiation of RNA primer formation in DNA replication.

Primulales *n.* order of dicot herbs, shrubs and trees comprising the woody tropical families Myrsinaceae and Theophrastaceae and the temperate family Primulaceae (primrose).

Principes Arecales *q.v.*, the palms.

principle of antithesis the principle, first formulated by Charles Darwin, that animals often convey signals with opposite meanings by expressions or postures that are opposites.

priodont *a.* saw-toothed; *appl.* stag beetles with smallest development of mandible projections.

prion *n.* proteinaceous infectious particle, a protein complex lacking nucleic acid, which has been implicated in the transmission of transmissible spongiform encephalopathies such as scrapie in sheep and Kreutzfeldt-Jakob disease in humans.

prisere *n.* natural plant succession starting on bare ground and ending in a climax.

prismatic *a.* like a prism, *appl.* cells, leaves; consisting of prisms, as the prismatic layer of shells; *appl.* soil crumbs in which the vertical axis is longer than the horizontal.

private *a. appl.* antigenic determinants unique to a particular haplotype in serological analysis of histocompatibility antigens. *cf.* public.

pro- prefix derived from Gk. *pro*, before, denoting previous to, in front of, the precursor of, *or* from L. *pro*, forward, for.

pro-acrosome *n.* structure in spermatids which develops into the acrosome.

proamnion *n.* an area of blastoderm in front of head of early embryos of reptiles, birds and mammals.

proandry *n.* meroandry with retention of anterior pair of testes only.

proangiosperm *n.* a fossil type of angiosperm.

proatlas *n.* a median bone intercalated between atlas and skull in certain reptiles.

proband *n.* an individual affected by a genetic disease who is crucial in enabling a pedigree for the disease to be deduced. *alt.* propositus.

probasidium *n.* a thick-walled resting spore in Heterobasidiomycetes, which germinates to form a basidium (promycelium); an immature basidium, before basidiospore formation.

probasidium *n.* binucleate hyphal cell from which the basidium develops in basidiomycete fungi, and which in some groups becomes encysted and acts as a resting spore.

probe *n.* a defined, labelled fragment of DNA or RNA used to detect and identify corresponding sequences in nucleic acids by selectively hybridizing with them. *see also* DNA hybridization; labelled antibody used to detect and identify proteins.

Proboscidea *n.* an order of herbivorous placental mammals, known from the Eocene to the present, including the elephants and the extinct mammoths and mastodons. They are of great size, having a massive skeleton, stout legs, trunks and incisors modified as tusks.

proboscidiform *a.* like a proboscis.

proboscis *n.* a trunk-like projection of head, as of insects, annelid and nemertean worms, used for feeding; in mammals an elongated nose, as the trunk of an elephant, elephant seal, or snout generally.

proboscis worms a common name for the Nemertea *q.v.*

probud *n.* a larval bud from the stolon of certain salps (tunicates of the order Doliolidae), which moves by pseudopodia to a specialized outgrowth and there divides to form definitive buds.

procambium *n.* the meristematic tissue from which vascular bundles are developed. *alt.* procambial strand.

procarp *n.* the female organ of red algae, consisting of the carpogonium, trichogyne and auxiliary cells.

procartilage *n.* the early stage of cartilage formation.

procary- prokary-.

Procellariiformes *n.* order of ocean birds with external tubular nostrils and hooked beaks, including the albatrosses, shearwaters and petrels. *alt.* tubenoses.

procercoid *n.* early larval form of certain cestodes in 1st intermediate host.

procerebrum *n.* the forebrain developed in the preantennary region of insects.

procerus *n.* pyramidal muscle of the nose.

process *n*. a reaction or procedure; an elongated portion of a cell, such as the axon and dendrites of nerve cells; an elongated projection from any structure.

processed pseudogene pseudogene (*q.v.*) that seems to be derived from a reverse transcript of an mRNA that has become inserted into the genome.

processing *see* RNA processing.

processive enzyme enzyme that does not dissociate from its substrate between repeated catalytic events.

prochirality *n*. property of molecules lacking handedness in their chemical structure (i.e. their mirror images can be superimposed on each other) and which are optically inactive. *a*. prochiral. *cf*. chirality.

Prochlorophyta, prochlorophytes *n*., *n.plu*. photosynthetic prokaryotes containing chlorophyll *a* and chlorophyll *b* but lacking phycobiliproteins, and therefore resembling plant chloroplasts rather than cyanobacteria. They include both ectosymbiotic and free-living species.

prochorion *n*. an enveloping structure of blastodermic vesicle preceding formation of chorion.

proclimax *n*. stage in a sere appearing instead of usual climatic climax and not determined by climate. *alt*. subclimax.

procoelous *a*. with concave anterior face, as vertebral centra.

procollagen *n*. soluble biosynthetic precursor to the fibrous protein collagen, forming tropocollagen after specific cleavage of terminal peptides.

proconvertin Factor VII *q.v.*

procoracoid *n*. an anteriorly directed process of the glenoid fossa of urodele amphibians.

procruscula *n.plu*. a pair of blunt locomotory outgrowths on posterior half of a redia.

procrypsis *n*. shape, pattern, colour or behaviour tending to make animals less conspicuous in their normal environment; camouflage. *a*. procryptic.

proctal *a*. anal, *appl*. fish fins.

proctodaeum *n*. most posterior portion of the alimentary canal, lined by ectodermally derived epithelium.

proctodone *n*. insect hormone thought to be secreted by anterior part of intestine and which ends diapause.

proctolin *n*. neuromodulatory neuropeptide hormone active in nervous system of certain crustaceans.

procumbent *a*. trailing on the ground; lying loosely along a surface; *appl*. ray cells elongated in radial direction.

procuticle *n*. the colourless cuticle of insects, composed of protein and chitin, before differentiation into endocuticle and exocuticle.

prodeltidium *n*. a plate which develops into a pseudodeltidium.

prodentine *n*. a layer of uncalcified matrix capping tooth cusps before formation of dentine.

prodrome *n*. a preliminary process, indication or symptom. *a*. prodromal.

producer *n*. an autotrophic organism, usually a photosynthetic green plant in an ecosystem, which synthesizes organic matter from inorganic materials and is an early stage in a food chain. *alt*. primary producer.

production *n*. in ecology, the assimilation of nutrients into biomass. *see* net primary production, primary production.

productive infection of a virus, an infection in which new infectious viral particles are produced.

productivity *n*. the amount of organic matter fixed by an ecosystem per unit time. *see* primary production.

proecdysis *n*. in arthropods, the period of preparing for moulting with the laying down of new cuticle and the detachment of the older one from it.

proelastin *see* elastin.

proembryo *n*. in plant embryos, stage before differentiation of embryo into embryo proper and the stalk-like suspensor; an embryonic structure preceding true embryo.

proenzyme zymogen *q.v.*

proepimeron *n*. a sclerite posterior to propleura; posterior pronotal lobe of dipterans.

proerythrocyte reticulocyte *q.v.*

proestrum pro-oestrus *q.v.*

proeusternum *n*. sclerite between propleura, forming ventral part of prothorax in dipterans.

profile transect a profile of vegetation, drawn to scale and intended to show the heights of plant shoots. *alt.* stratum transect.

profilin *n.* a ubiquitous cytoplasmic protein in mammalian cells, to which unpolymerized actin subunits are bound.

proflavin *n.* a mutagenic acridine dye.

profunda *a.* deep-seated, *appl.* a branch of brachial, femoral or costocervical artery, to the ranine artery, terminal part of lingual artery, and to a vein of femur; *n.* a deep artery or vein.

profundal *a. appl.* or *pert.* the zone of deep water and bottom below compensation depth (point at which photosynthesis ceases) in lakes.

profundal zone the zone of a lake lying below the compensation point, comprising the deep water and the lake bottom.

progamete *n.* a structure giving rise to gametes by abstriction, in certain fungi.

progenesis *n.* the maturation of gametes before completion of body growth.

progeotropism *n.* positive geotropism.

progestagens *n.plu.* a group of steroids that have effects like progesterone.

progestational *a. appl.* phase of the oestrous cycle during luteal and endometrial activity; *appl.* hormones controlling uterine cycle and preparing uterus for implantation of conceptus.

progesterone *n.* steroid hormone secreted by the female mammalian corpus luteum and placenta, which prepares the uterus to receive the fertilized egg, maintains the uterus during pregnancy, inhibits ovulation during pregnancy and is the precursor to several other hormones. It is a component of some types of contraceptive pill. *alt.* progestin, progestone. Formerly also known as luteal hormone or luteosterone.

progestin, progestone forms of progesterone *q.v.*

progestogen *n.* any compound with progesterone-like effects in a female mammal.

proglottid, proglottis *n.* individual segment of an adult tapeworm, containing a set of reproductive organs. Eggs are formed in the posterior proglottids, are shed in the faeces and then enter the alternative host where they hatch. *plu.* proglottids, proglottides.

prognathous *a.* having prominent or projecting jaws; with mouthparts projecting downwards, *appl.* insects.

progonal *a. appl.* sterile anterior portion of genital ridge.

progoneate *a.* having the genital aperture anteriorly, as in some arthropods.

programmed cell death the death of cells at a specific stage of development as a part of the normal development process. *alt.* apoptosis.

progressive provisioning the feeding of a larva in repeated meals. *cf.* mass provisioning.

progress zone In the developing vertebrate limb, a population of actively dividing mesodermal cells immediately underneath the apical ectodermal ridge of the limb bud, from which derive the cells of the limb as it grows.

Progymnophyta, progymnophytes, progymnosperms *n., n.plu., n.plu.* division of extinct spore-bearing woody plants and trees, with secondary xylem similar to that of gymnosperms, and which are believed to be possible ancestors of the gymnosperms.

prohaptor *n.* anterior adhesive organ in trematodes, as sucker, suctorial grooves or glands.

prohormone *n.* an inactive precursor of a hormone, esp. of a polypeptide or peptide hormone, many of which are produced by enzymatic cleavage of an active portion from a longer polypeptide prohormone.

prohydrotropism *n.* positive hydrotropism.

proiospory *n.* the premature development of spores.

projectile *a.* protrusible; *appl.* structures that can be thrust forwards.

projection *n.* the perception of external sensation as external despite the fact that the stimuli are interpreted in the brain.

projection neurones neurones in the brain that have their cell bodies in one region and their axon terminals in another. *cf.* local-circuit neurones.

projicient *a. appl.* sense organs reacting to distant stimuli, as light, sound.

prokaryon *n.* former term for the nucleoid of a prokaryotic cell.

Prokaryotae *n.* kingdom of living organisms comprising all prokaryotes (*q.v.*). *alt.* Monera. *see* Appendix 6.

prokaryotes *n.plu.* the bacteria (including

mycoplasmas and actinomycetes) and the cyanobacteria (formerly known as the blue-green algae), unicellular organisms whose small, simple cells lack a membrane-bounded nucleus, mitochondria, chloroplasts and other membrane-bounded organelles typical of plant, animal, fungal, protozoan or algal cells. Their DNA is in the form of a single circular molecule not complexed with histones. In modern classifications placed in a separate kingdom, Monera or Prokaryotae. *a.* prokaryotic. *cf.* eukaryotes.

prolabium *n.* middle part of upper lip.

prolactin (LTH) *n.* glycoprotein hormone secreted by the anterior pituitary which stimulates the production of milk in mammals and of crop milk in birds such as the pigeon. It assists in maintaining the corpus luteum in mammals and has a range of effects in lower vertebrates. *alt.* lactogenic hormone, (formerly) luteotropin, luteotropic hormone.

prolamellar body paracrystalline arrangement of membranes in etioplasts.

prolarva *n.* a newly hatched larva during the first few days when it feeds on its supply of embryonic yolk, as in some fish.

prolamines *n.plu.* simple proteins found in seeds of cereals, soluble in ethanol, and including gliadin from wheat, zein from maize, hordein from barley.

proleg *n.* an unjointed abdominal appendage of larvae of Lepidoptera and some other arthropods.

proleukocyte *n.* in insects, a small leukocyte with basophil cytoplasm and large nuclei, and developing into a macronucleocyte.

proliferation *n.* increase by frequent and repeated reproduction; increase by cell division.

proliferous *a.* multiplying quickly, *appl.* bud-bearing leaves; developing supernumerary parts abnormally.

prolification *n.* shoot development from a normally terminal structure.

proline (Pro, P) *n.* a cyclic amino acid (more properly an imino acid) with a hydrocarbon side chain, constituent of proteins. *see also* hydroxyproline.

proloculus *n.* the first chamber, microspheric when formed by conjugation of swarm spores, megalospheric when formed asexually by fission, in polythalamous foraminiferans.

prolymphocyte *n.* an immature lymphocyte.

promeristem *n.* the part of the apical meristem consisting of the actively dividing cells and their most recent derivatives. *alt.* protomeristem.

prometaphase *n.* stage between prophase and metaphase in mitosis and meiosis.

promitochondria *n.plu.* abnormal mitochondria which are found in yeast grown anaerobically and which lack some cristae and cytochromes.

promonocyte *n.* a cell developing into a monocyte.

promonostelic *a. appl.* stem or root with a protostele or central cylinder of vascular tissue.

promontory *n.* prominence or projection, as of cochlea and sacrum.

promoter *n.* DNA region involved in and necessary for initiation of transcription, and including the RNA polymerase binding site, the startpoint of transcription and various other sites at which gene regulatory proteins may bind; in carcinogenesis, any agent that hastens the process of carcinogenesis while not being a carcinogen on its own; a protractor muscle.

promuscis *n.* the proboscis of Hemiptera.

promycelium *n.* a short hypha that germinates from teleutospore in rust and smut fungi and on which basidiospores develop.

promyelocyte *n.* a bone marrow cell that develops either into a granulocyte or macrophage.

pronase *n.* microbial proteolytic enzyme. EC 3.4.24.4.

pronate *a.* prone; inclined.

pronation *n.* movement by which palm of hand is turned downwards by means of pronator muscles.

pronephros *n.* the kidney that develops first in embryonic or larval life, later replaced by mesonephros and metanephros. *a.* pronephric.

pronotum *n.* the dorsal part of the prothorax (1st segment of thorax) of insects.

pronucleus *n.* haploid nucleus of unfertilized egg or of sperm.

pronymph *n.* in insect metamorphosis, the

stage preceding the nymph.

pro-oestrus *n*. the phase before oestrus or heat; period of preparation for pregnancy.

proofreading *n*. in DNA synthesis, the ability of DNA polymerase to recognize mismatched bases, *see also* editing; mechanism for ensuring that the correct amino acid is inserted in a protein chain during translation.

pro-opiomelanocortin (POMC) polypeptide synthesized in the pituitary and which is differentially processed to produce β-lipotropin and ACTH in the anterior lobe, and α-MSH and β-endorphin in the intermediate lobe.

pro-otic *n*. the anterior bone of otic capsule of vertebrates; *a. pert*. a centre of ossification of petromastoid part of temporal bone.

prop roots adventitious aerial roots growing downwards from stem, as in mangrove and maize, and helping to support stem.

propagate *v*. to travel along a nerve fibre without losing strength, *appl*. electrical impulses; to multiply, as of plants.

propagative *a*. reproductive, *appl*. a cell, a phase in life cycle, an individual in a colonial organism.

propagule *n*. any spore, seed, fruit or other part of a plant or microorganism capable of producing a new plant and used as a means of dispersal. *alt*. diaspore.

propatagium *n*. part of bat wing membrane anterior to arm.

properdin *n*. protein factor in serum, involved in the alternative pathway of complement activation.

properithecium *n*. a young perithecium which contains a single zygote giving rise ultimately to ascospores.

propes proleg *q.v*.

prophage *n*. bacteriophage DNA integrated into and replicating with the bacterial chromosome.

prophase *n*. the first stage in mitosis and meiosis in which the replicated chromosomes condense and become visible as double structures (sister chromatids).

prophialide sporocladium *q.v*.

prophloem protophloem *q.v*.

prophototropism *n*. positive phototropism.

prophyll(um) *n*. a small bract; first foliage leaf.

proplastid *n*. an immature plastid, as found in meristem cells.

propleuron *n*. a lateral plate of exoskeleton of prothorax in insects.

propneustic *a*. with only prothoracic spiracles open for respiration.

propodeon, propodeum *n*. the first segment of abdomen fused to thorax in some hymenopterans.

propodite *n*. foot segment of some crustacean limbs; the tibia in arachnids.

propodium *n*. small anterior portion of a molluscan foot.

propodosoma *n*. region of body bearing 1st and 2nd legs in mites and ticks.

propolis *n*. resinous substance from buds of certain trees, used by worker bees to fasten comb portions and fill up crevices.

propons *n*. delicate bands of white matter crossing anterior end of pyramid of cerebellum below pons Varolii.

propositus proband *q.v*.

proprioception *n*. reception of stimuli originating within the organism.

proprioceptor *n*. an internal sensory receptor sensitive to stimuli originating within the body, such as stretch receptors in muscle, and by which an animal receives information on its position and movements. *alt*. proprioreceptor.

propriogenic *a. appl*. effectors other than muscle, or organs which are both receptors and effectors.

proprioreceptor proprioceptor *q.v*.

propriospinal *a. pert*. wholly to the spinal cord; *appl*. fibres, etc. confined to spinal cord.

propterygium *n*. the foremost of three basal cartilages supporting pectoral fin of elasmobranch fishes.

propupa prepupa *q.v*.

propus propodite *q.v*.

propygidium *n*. dorsal plate anterior to pygidium in Coleoptera and some other insects.

proral *a*. from front backwards, *appl*. jaw movement in rodents.

prorennin *n*. precursor of rennin, secreted by the peptic cells and converted to active rennin by the action of hydrochloric acid.

proscapula clavicle *q.v*.

proscolex *n*. cysticercus *q.v*.; the inverted scolex inside a cysticercus.

prosencephalization *n*. the progressive

shifting of controlling centres towards the forebrain and the increasing complexity of the cerebral cortex in the course of vertebrate evolution.

prosencephalon forebrain *q.v.*

prosenchyma *n.* elongated pointed cells, with thick or thin walls, as in mechanical and vascular tissues of plants; fungal tissue formed of loosely woven hyphae. *a.* prosenchymatous.

prosequence *n.* part of a protein that assists in its folding but which is removed from the mature protein molecule.

prosethmoid *n.* an anterior bone of cranium of teleost fishes.

prosimian *n.* any primate, such as lemurs and tarsiers, belonging to the primitive suborder Prosimii.

prosiphon endosiphuncle *q.v.*

prosocoel *n.* a narrow cavity in epistome of molluscs, the 1st main part of coelom; median cavity between 3rd and lateral ventricles of brain.

prosodetic *a.* anterior to beak, *appl.* certain bivalve ligaments.

prosodus *n.* a delicate canal between chamber and incurrent canal in some sponges.

prosoma *n.* the anterior part of the body in arachnids and some other invertebrates, corresponding to a cephalothorax.

prosome *n.* small ribonucleoprotein particle associated with mRNA, composed of >20 different proteins and one of several small prosomal RNAs.

prosoplectenchyma *n.* a false tissue of fungal hyphae where the cells are oriented parallel and the walls are indistinct, as in some lichens.

prosopyle *n.* the aperture of communication between adjacent incurrent and flagellate canals in some sponges.

prosorus *n.* thick-walled cell that develops into a sorus, in some lower fungi.

prospory *n.* seed production in plants which are not fully developed.

prostacyclins *n.plu.* compounds derived from fatty acids, produced in endothelial cells and others in response to damage and various other stimuli, have vasodilator activity as a result of their relaxation of smooth muscle.

prostaglandins *n.plu.* any of a group of compounds formed from C_2 fatty acids and containing a 5-carbon ring. There are four main classes, PGA, PGB, PGE and PGF. They are present in many mammalian tissues. They modify the effects of other hormones, stimulate contraction of smooth muscle of uterus, inducing labour or abortion, and are also involved in inflammatory reactions. The anti-inflammatory effect of aspirin is due to its inhibition of the enzyme prostaglandin synthetase, which catalyses the formation of prostaglandins from eicosatrienoate.

prostalia *n.plu.* sponge spicules which project beyond the body surface. *sing.* prostal.

prostanoid *n.* compound derived from prostanoic acid, e.g. prostacyclins (*q.v.*) and prostaglandins (*q.v.*)

prostate *n.* the prostate gland in male mammals, a muscular and glandular organ around the beginning of the urethra in the pelvic cavity; the spermiducal gland in annelids. *a.* prostatic.

prosternum *n.* ventral part of prothorax of insects; ventral part of cheliceral segment in some arachnids.

prostheca *n.* moveable inner lobe of mandibles in certain insect larvae.

prosthetic group non-protein chemical group (e.g. haem, flavin or a metal atom) bound to a protein, as in many enzymes, usually forming part of the active site, and essential for biological activity, and which in some cases may be dissociated from the protein leaving an inactive apoprotein. *alt.* cofactor.

prosthion *n.* the middle point of the upper alveolar arch.

prosthomere *n.* most anterior or preoral somite.

prostigmine neostigmine *q.v.*

prostomiate *a.* having a portion of head in front of the mouth.

prostomium *n.* in some annelids and molluscs, that part of the head anterior to the mouth. *a.* prostomial.

prostrate *a.* trailing on the ground; lying closely along a surface.

Protacanthopterygii *n.* group of generalized and fairly primitive teleost fishes existing from Cretaceous times to the present day, including the salmonids, pike and stickleback.

protamine *n.* any of a group of small highly basic arginine-rich proteins associated with DNA in fish sperm in a similar manner to histones in other eukaryotic DNAs.

protandrism protandry *q.v.*, in zoology.

protandry *n.* condition of hermaphrodite plants and animals where male gametes mature and are shed before female gametes mature. *a.* protandrous.

protaphin *n.* a natural yellow pigment which is water-soluble and deep magenta in alkaline solutions and is readily converted to the aphins.

pro-T cells early stage in T lymphocyte development, immature thymocytes in which the T-cell receptor genes have not yet been rearranged.

Proteales *n.* an order of xerophytic shrubs and trees with entire or much divided leaves with thick cuticle and hairs, and showy flowers, comprising the family Proteaceae (protea).

protease proteinase *q.v.*

protease nexin a proteinase inhibitor of the serpin family.

protegulum *n.* the semicircular or semi-elliptical embryonic shell of brachiopods.

protein *n.* one of the chief constituents of living matter, any one of a group of large organic molecules containing chiefly C, H, O, N and S and consisting of unbranched chains constructed from a set of 20 different amino acids, one or more such polypeptide chains comprising a protein molecule and sometimes associated with non-protein compounds (e.g. haem, flavin) termed prosthetic groups. Each protein has a unique genetically determined amino acid sequence. Essential in living organisms as enzymes, structural constituents of cells and tissues and in control of gene expression etc. Each polypeptide chain is synthesized by translation of an mRNA template at the ribosomes. After synthesis it folds up into a characteristic three-dimensional conformation.

protein bodies membrane-bounded granules composed of storage proteins, formed via the endoplasmic reticulum and Golgi apparatus in seeds and other plant tissues.

protein degradation the breakdown of proteins into their constituent amino acids by proteolysis; *pert.* esp. to the cellular

processes by which abnormal proteins are targeted for destruction by the cell and to the degradative part of the normal turnover of cellular proteins.

protein disulphide isomerase (PDI) enzyme involved in the rearrangement of disulphide bonds during protein folding.

protein engineering the alteration of the structure of a protein through the artificial modification of the DNA that encodes it.

protein family group of proteins of related sequence and function, and which arise from the duplication and divergence of their genes from a single ancestral gene.

protein fingerprint pattern of peptide fragments obtained on two-dimensional gel electrophoresis of a protein digested with trypsin or other proteinases, and which is characteristic for each protein.

protein folding the folding up of a newly synthesized polypeptide chain into its final native three-dimensional conformation. The final conformation is determined only by the amino-acid sequence of the protein, and is a spontaneous process, although folding into the correct conformation may be aided by other proteins (molecular chaperones).

protein kinase enzyme that phosphorylates specific amino acid residues in a protein. Protein kinases phosphorylate either serine and threonine residues, or tyrosine residues, and are an important part of intracellular pathways that regulate enzyme activity and relay extracellular signals to the intracellular response machinery. *see* serine-threonine protein kinase, tyrosine protein kinase.

protein kinase A a serine threonine protein kinase activated by cyclic AMP.

protein kinase C a ubiquitous serine threonine protein kinase of mammalian and other eukaryotic cells which occurs in a variety of different forms. It phosphorylates and alters the activity of a wide range of substrates *in vitro* and *in vivo* and is implicated in the modification of, e.g., enzyme activity and ion channel conductivity in response to external signals.

protein phosphatase enzyme that removes a phosphate group from a protein, often reversing the effect of a previous phosphorylation. *alt.* phosphoprotein phosphatase.

protein phosphorylation *see* phosphorylation.

protein pump membrane protein or protein complex that pumps ions (e.g. Na^+, K^+, Cl^-) or simple organic compounds in and out of the cell against their concentration gradients in an energy-requiring reaction.

protein quality the nutritional value of a protein, which is determined both by its digestibility and by whether it contains adequate amounts of the essential amino acids that animals cannot synthesize for themselves.

protein sequence the sequence of amino acid residues in a protein chain.

protein subunit a constituent polypeptide chain in a multimeric protein or a protein complex.

protein superfamily group of proteins descended from a common ancestral protein but which have subsequently diverged considerably and acquired different functions.

protein synthesis synthesis of a protein at the ribosomes using messenger RNA as a template. *see also* genetic code, translation, transfer RNA.

protein targeting the process of transporting proteins synthesized in the cytoplasm to their correct destinations in other parts of a cell.

protein translocation the movement of certain proteins from the cytosol where they are synthesized into organelles or out of the cell, which involves their transport (translocation) across a membrane.

proteinaceous *a. pert.* or composed of protein.

proteinase *n.* any enzyme that degrades proteins by splitting internal peptide bonds to produce peptides, and including proteolytic enzymes, endopeptidases and peptidyl-peptide hydrolases. EC 3.4.21–24. *alt.* protease.

proteinase inhibitors small proteins, such as antitrypsin, which inhibit various proteinase enzymes.

proteinoid *n.* molecule like a protein, produced when trying to mimic primeval conditions by heating or other treatment of an amino acid mixture.

proteinoplast *n.* a storage plastid containing protein.

proteinuria *n.* the presence of proteins in the urine.

proteism *n.* the ability to change shape, as of amoeba and other cells.

protembryo *n.* the fertilized ovum and its cleavage stages preceding formation of blastula.

proteoclastic *a.* breaking down proteins, proteolytic *q.v.*

proteoglycan *n.* any of a class of compounds consisting of polysaccharide (95%) and protein (5%) units, forming the ground substance of connective tissue, important in determining the viscoelastic properties of joints etc.

proteolipid *n.* a type of complex macromolecule composed of protein and lipid and which has the solubility properties of a lipid. *cf.* lipoprotein.

proteolysis *n.* breakdown of proteins and peptides into their constituent amino acids by enzymatic or chemical hydrolysis of peptide bonds. *a.* proteolytic.

proter *n.* the anterior individual produced when a protozoan divides transversely.

proterandry protandry *q.v.*

proteranthous *a.* flowering before foliage leaves appear.

proterogenesis *n.* the foreshadowing of adult or later forms by youthful or earlier forms.

proteroglyph *a.* with specialized canine teeth in upper jaw.

proterogyny protogyny *q.v.*

proterosoma *n.* body region comprising mouth and the region bearing first and second legs in mites and ticks; prosoma *q.v.*

Proterozoic *n.* a geological eon of the Precambrian, before the Cambrian, lasting from around 2500 million years to 590 million years ago, and whose rocks contain few fossils, mainly blue-green algae (cyanobacteria) and soft-bodied animals of problematical affinity.

prothallial *a. pert.* a prothallus; *appl.* cell in microspore of gymnosperms and some pteridophytes, considered as a vestige of a prothallus.

prothalloid *a.* like a prothallus.

prothallus *n.* the hyphae of lichens during the initial growth stages; a small haploid gametophyte as in algae, ferns and some gymnosperms, bearing antheridia or archegonia or both, and developing from a spore.

protheca *n*. basal part of coral calicle, which is formed first.

prothecium *n*. a primary perithecium of many fungi.

protherians *n.plu*. egg-laying mammals, including the extinct triconodonts and multituberculates, and the extant monotremes (duck-billed platypus).

prothetely *n*. the development or manifestation of pupal or of imaginal characters in insect larvae.

prothoracic *a. pert.* prothorax; *appl.* glands secreting ecdysone; *appl.* anterior lobe of pronotum.

prothorax *n*. the first segment of insect thorax.

prothrombin *n*. blood plasma protein, involved in blood clotting, formed in liver in the presence of vitamin K, and which is converted to the proteinase thrombin in the blood at site of trauma by a proteolytic enzyme, Factor Xa, in the presence of calcium ions. *alt*. thrombinogen.

proticity *n*. term coined to denote flow of protons (by analogy with electricity). *see* chemiosmotic hypothesis.

protist *n*. a unicellular organism, usually nowadays referring to a unicellular eukaryotic organism e.g. microalgae or protozoa; a member of the Protoctista *q.v.*

Protista *n.plu*. simple unicellular eukaryotic organisms, such as unicellular algae, water moulds, slime moulds and protozoa; Protoctista *q.v.*

proto- prefix derived from Gk. *protos,* first.

proto-aecidium a cell mass surrounded by layers of hyphae, containing cells eventually producing aecidiospores and disjunctor cells.

protoaphins *n.plu*. yellow pigments found in some aphids.

Protoavis a possible fossil bird from the late Triassic, some 75 million years earlier than *Archaeopteryx*, but whose identification as a bird is still controversial.

protoaxis *n*. the primordial filament or axis in evolution of plant stem.

protobasidium *n*. a basidium of four cells, from each of which a basidiospore is developed by abstriction.

protoblast *n*. a blastomere that develops

into a definite organ or part; an internal bud stage in life history of some sporozoans.

protobranch *a. appl.* gills of bivalve molluscs having flat non-reflected filaments.

protocephalic *a. appl.* or *pert.* primary head region of insect embryo.

protocephalon *n*. head part of cephalothorax in crabs, lobsters and similar crustaceans; the first six body segments, which are fused to form the head of an insect.

protocercal *a*. with a caudal fin in which the vertebral column runs straight to tip, thereby dividing the fin symmetrically.

protocerebrum *n*. anterior part of "brain" of insects and other arthropods.

protochlorophyll *n*. a yellowish pigment in chloroplasts of plants grown in darkness, which is converted into chlorophyll by the agency of light.

protochordates *n.plu*. group of animals comprising the hemichordates, urochordates and cephalochordates, having gill slits, a dorsal hollow central nervous system, a persistent notochord and a postanal tail.

protocnemes *n.plu*. the six primary pairs of mesenteries of sea anemones and their relatives.

protocoel *n*. the front portion of the coelomic cavity.

protoconch *n*. the larval shell of molluscs, indicated by a scar on adult shell.

protocone *n*. inner cusp of upper molar tooth.

protoconid *n*. external cusp of lower molar tooth.

protoconidium *n*. a rounded or club-shaped cell at the tip of a filament, giving rise to conidia, as in dermatophytes.

protoconule *n*. anterior intermediate cusp of upper molar tooth.

protocorm *n*. swelling of rhizophore, preceding root formation as in certain club mosses that have mycorrhizal fungi in the early stages; in orchids, structure produced usually underground from germinating seedling and heavily infected with mycorrhizal fungus; undifferentiated cell mass of archegonium in Ginkgoales; (*zool.*) the posterior portion of germ band, which gives rise to trunk segments in insects.

protocranium *n*. posterior part of insect

head.

Protoctista, protoctists *n.*, *n.plu.* in some modern classifications a kingdom comprised of eukaryotic unicellular, colonial and simple multicellular organisms that do not fall easily into either the plant or animal kingdoms. The Protoctista are usually held to comprise the algae, including the multicellular seaweeds and other macroalgae, diatoms, protozoa, the water moulds and the cellular and acellular slime moulds. *alt.* Protista, protists.

protoderm *n.* embryonic epidermis in plant embryos; the outer layer of cells of apical meristem.

protoepiphyte *n.* a plant which is an epiphyte all its life and does not start life rooted to the ground or come to root in the ground later.

protofilament *n.* row of tubulin dimers joined end-to-end which forms a longitudinal element in the wall of a microtubule.

protogenic *a.* persistent from beginning of development.

protogyny *n.* the condition of hermaphrodite plants and animals in which female gametes mature and are shed before maturation of male gametes. *a.* protogynous.

protologue *n.* the printed matter accompanying the first description of a name.

protoloph *n.* anterior transverse crest of molar tooth.

protomala *n.* a mandible of millipedes.

protomer *n.* inactive form of enzyme.

protomeristem promeristem *q.v.*

protomerite *n.* anterior part of medullary protoplasm of adult gregarines.

protomitosis *n.* a primitive form of mitosis as in some slime moulds.

protomorphic *a.* first-formed; primordial; primitive.

proton gradient gradient of [H⁺] set up across a membrane by active pumping of protons from one side to the other, used by cells to power many processes such as ATP synthesis, bacterial flagellar rotation. *see also* proton motive force.

proton motive force the electrochemical gradient of protons formed across mitochondrial and chloroplast membranes during the passage of electrons down the electron transport chains in membrane, and which powers ATP synthesis linked to aerobic respiration or photosynthesis. *see* chemiosmotic theory.

proton pump *n.* active transport of protons across a membrane, mediated by membrane proteins such as bacteriorhodopsin.

protonema *n.* early filamentous stage in development of a fern prothallus; filamentous stage in the development of some algae; filamentous structure from which a moss plant buds.

protonemata *plu.* of protonema *q.v.*

protonematoid *a.* like a protonema.

protonephridial system excretory system in some simple invertebrates such as flatworms, nematodes, annelids and rotifers, consisting of a system of branching ducts (protonephridia) each closed at its internal end by a flame cell, and opening into a central duct or to the surface through pores. Flame cells bear a large bundle of flagella which project into the duct, and whose motion gives a flickering appearance under the light microscope similar to a flame.

protonephridium *n.* branching excretor and of the protonephridial system *(q.v.)*. *plu.* protonephridia.

proto-oncogene the normal cellular gene from which an oncogene has been derived.

protopathic *a. appl.* stimuli and sensory systems involved in the perception of pain and of marked variations in temperature.

protoperithecium *n.* primary haploid perithecium, as in some Pyrenomycetes.

protophloem *n.* first-formed phloem cells.

protoplasm *n.* living matter, the total substance of a living cell, cytoplasm and nucleoplasm in the case of a eukaryotic cell.

protoplasmic *a. pert.* protoplasm; *appl.* astrocytes: astrocytes found in grey matter and having thick branched processes similar to pseudopodia.

protoplast *n.* plant cell with cell wall removed; the living component of a cell, i.e. the protoplasm not including any cell wall.

protopod *a.* with feet or legs on anterior segments, not on abdomen, *appl.* insect larvae.

protopodite *n.* basal segment of arthropod limb.

protoporphyrin *n.* a precursor of porphyrin without the metal ion.

protosoma, protosome prosoma *q.v.*

protospore *n.* a spore of the 1st generation; a fungal spore germinating to produce mycelium

protosporophyte *n.* the filament produced by the fertilized female cell, the first sporophyte stage in life cycle of red algae.

protostele *n.* type of stele in which the vascular tissue forms a solid central strand with the phloem either surrounding the xylem or interspersed within it, regarded as the most primitive type and present in most roots and in stems of club mosses and some other groups.

protosterigma *n.* the basal portion of a sterigma.

protosternum *n.* sternite of cheliceral segment of prosoma in mites and ticks.

protostigmata *n.plu.* two primary gill slits of embryo. *sing.* prostigma.

protostoma blastopore *q.v.*

protostomes *n.plu.* collectively all animals with a true coelom, spiral cleavage of the egg, and in which the blastopore becomes the mouth (molluscs, annelids, arthropods, phoronids, bryozoans and brachiopods). *cf.* deuterosomes.

protostylic *a.* having lower jaw connected with cranium by original dorsal end of arch. *n.* protostyly.

protothallus *n.* the first-formed structure which gives rise to a thallus.

prototheca *n.* a skeletal cup-shaped plate at aboral end of coral embryo, the 1st skeletal formation.

Prototheria, prototherians *n., n.plu.* subclass of primitive egg-laying mammals which includes the orders Triconodonta, Multituberculata and Monotremata, of which only the monotremes (duck-billed platypus and spiny anteater) are extant.

prototroch *n.* a preoral circlet of cilia of a trochosphere larva.

prototroph *n.* nutritionally independent, wild-type strain of bacterium or fungus that has no special nutritional requirements. *cf.* auxotroph.

prototrophic *a.* nourished from one supply or in one manner only; feeding on inorganic matter, *appl.* iron, sulphur and nitrifying bacteria and green plants.

prototype *n.* an original type species or example; an ancestral form.

protovertebrae *n.plu.* a series of primitive mesodermal segments in a vertebrate embryo.

protoxylem *n.* the first-formed primary xylem elements of a plant body or organ.

Protozoa, protozoans *n., n.plu.* in zoological classifications, a phylum of unicellular heterotrophic, generally non-photosynthetic, aquatic eukaryotes, lacking cell walls. Commonly called protozoans, they include the Mastigophora (the flagellates, including the photosynthetic "plant" flagellates such as *Chlamydomonas*), the Sarcodina (amoebas and foraminiferans, and radiolarians and heliozoans), the Ciliophora (the ciliates), the Sporozoa (parasitic protozoans such as *Eimeria*, which causes coccidiosis, *Plasmodium*, the malaria parasite, and the piroplasms, such as *Babesia*), and the Cnidospora, which cause disease in fish and other animals. Protozoans are often now classified along with algae and other simple unicellular eukaryotes in a separate kingdom, Protista. *a.* protozoan.

protozoology *n.* that branch of biology dealing with protozoans.

protozoon *n.* individual protozoan cell.

protractor *n.* muscle that extends a part or draws it out from the body.

protriaene *n.* a trident-shaped sponge spicule with prongs of trident directed anteriorly.

Protura *n.* order of insects including the bark-lice, minute insects with 12 segments in the abdomen, no antenna or compound eyes and very small legs. Found under the bark of trees, in turf and in soil.

proventriculus *n.* gizzard of insects and crustaceans; in annelid worms, the portion of alimentary canal anterior to gizzard; in birds, the glandular stomach anterior to the gizzard.

provinculum *n.* a primitive hinge on shells of young stages of certain lamellibranch molluscs.

provirus *n.* virus DNA that has become integrated into a host cell's chromosome and is carried from one cell generation to the next in the chromosome, not producing infective virus particles.

provisioning *n.* providing food for young, as in mass provisioning, nest provisioning, progressive provisioning, *all* q.v.

provitamin *n.* precursor to a vitamin.

proxi- prefix derived from L. *proximus,* next.

proximad *adv.* towards, or placed nearest the body or base of attachment.

proximal *a.* nearest to the body, or centre or place of attachment; *appl.* region of a gene close to the promoter. *cf.* distal.

proximate *a.* nearest to, next to; *appl.* cause, direct immediate cause.

proximate analysis of a food a rough estimate of the nutritive value of a food made by first determining the total nitrogen and multiplying by 6.25 to get a rough value for total protein, then determining the fat content by ether extraction, and finally determining the carbohydrate content by the difference between the above two values added together and the total dry weight of the sample.

proximoceptor *n.* a sensory receptor which reacts only to nearby stimuli, as a touch receptor.

proximo-distal axis axis running from thumb to little finger.

prozonite *n.* the anterior part of a body segment consisting of two distinct parts.

prozymogen *n.* a precursor to a zymogen.

pruinose *a.* covered with whitish particles or globules; covered with bloom.

Prymnesiophyta *n.* in some classifications a division of minute golden motile algae. *see* Haptophyta.

Prymnesiophyta, prymnesiophytes *n.*, *n.plu.* mainly marine division of algae containing the coccolithophoroids, unicellular flagellate microorganisms armoured in calcareous "scales" (coccoliths), and which are important in the marine phytoplankton.

psalterium *n.* omasum *q.v.*; (*neurobiol.*) a thin triangular plate joining lateral portions of the fornix.

psammo- prefix derived from Gk. *psammos,* sand.

psammon *n.* the organisms living between sand grains, as of freshwater and marine shores.

psammophilous arenicolous *q.v.*

psammophore *n.* one of rows of hairs under mandibles and sides of head in desert ants, used for removal of sand grains.

psammophyte *n.* plant growing in sandy or gravelly ground.

psammosere *n.* a plant succession originating in a sandy area, as on dunes.

pseud- prefix derived from Gk. *pseudes,* false.

pseudambulacrum *n.* the lancet plate with adhering side plates and covering plates of certain echinoderms.

pseudannual *n.* a plant that completes its growth in one year but provides a bulb or other means of surviving the winter.

pseudanthium *n.* a flowerhead condensed to such an extent that it looks like a single flower.

pseudapogamy *n.* fusion of a pair of vegetative nuclei, as in certain fungi and fern prothalli.

pseudaposematic *a.* imitating warning coloration or other protective features of harmful or distasteful animals, i.e. showing Batesian mimicry.

pseudapospory *n.* spore formation without meiosis, resulting in a diploid spore that gives rise to the gametophyte.

pseudaxis *n.* an apparent main axis that really consists of a number of lateral branches running parallel.

pseudepipodite *n.* a flattened outer region of the 2nd maxilla and trunk appendages in certain crustaceans.

pseudepisematic *a.* having false coloration or markings, as in protective mimicry or for allurement or aggressive purposes.

pseudergate *n.* juvenile worker form in some termites, which can mature to a winged reproductive adult.

pseudholoptic *a. appl.* insect eyes intermediate between holoptic (where eyes meet in a coadapted line of union on top of the head) and dichoptic (where eyes are completely separate).

pseudimago *n.* a stage between pupa and imago in life history of some insects.

pseudoacrorhagus *n.* a tubercle near the margin of certain sea anemones containing ordinary epidermal nematocysts (stinging cells). *cf.* acrorhagus.

pseudoaethalium *n.* a dense aggregation of distinct sporangia, as in some slime moulds. *cf.* aethalium.

pseudoalleles *n.plu.* subdivisions of a compound locus that can be distinguished by recombinational analysis.

pseudoallelic *a. appl.* two or more mutations that behave as alleles of the same locus in a complementation test but which can be separated by crossing-over, and which indicate the presence of a complex locus *q.v.*

pseudoalveolar *a. appl.* a cytoplasmic inclusion containing starch grains.

pseudoangiocarpic *a.* with an exposed hymenium temporarily enclosed by incurved edge of cap or by a secondary veil.

pseudoaposematic pseudaposematic *q.v.*

pseudoaquatic *a.* thriving in wet ground.

pseudoarticulation *n.* incomplete subdivision of a segment, or groove having the appearance of a joint, as in limbs of arthropods.

pseudoautosomal region small region in the mammalian X-chromosome which is homologous with a region at the tip of the short arm of the Y-chromosome, and which is involved in pairing of the sex chromosomes at meiosis.

pseudobrachium *n.* locomotory appendage formed from elongated ray of pectoral fin, used by anglerfish and batfish to "walk" on the bottom.

pseudobulb *n.* a thickened internode of orchids and some other plants, for storage of water and food reserves.

pseudobulbil *n.* an outgrowth of some ferns, that substitutes for sporangia.

pseudobulbous *a.* adapted to hot dry conditions through development of pseudobulbs.

pseudocarp false fruit *q.v.*

pseudocartilage *n.* a cartilage-like substance serving as skeletal support in some invertebrates.

pseudocellus *n.* one of the scattered sense organs of unknown function in insects.

pseudocentrous *a. appl.* vertebra composed of two pairs of small cartilages which meet and form a suture later.

pseudocilia *n.plu.* threads of cytoplasm projecting from cell through surrounding sheath of mucilage in certain unicellular green algae.

pseudocoel *a.* fluid-filled cavity between epidermis and internal organs in rotifers, nematodes and some other invertebrates lacking a true coelom. It is derived from the embryonic blastocoel.

pseudocoelomate *a. appl.* animals whose body cavity is a pseudocoel and not a true coelom.

pseudoconch *n.* a structure developed above and behind the true sphenoidal bone in crocodilians; in some gastropod molluscs, a non-spiral shell.

pseudoconditioning *n.* situation where an unconditional response comes to be elicited by stimuli other than the unconditional stimulus even though there is no contingent relationship between them.

pseudocone *a. appl.* insect compound eye having ommatidia filled with transparent gelatinous material.

pseudoconidium *n.* one of the spores formed on lateral projections of pseudomycelium of certain yeasts.

pseudoconjugation *n.* conjugation in certain sporozoans, in which two individuals, temporarily and without fusion, join end to end, or side to side.

pseudocopulation *n.* in orchids, the case where the resemblance of the orchid flower to a female insect (as in bee orchids) leads to an attempt by the male to copulate with it and so effect pollination.

pseudocortex *n.* a cortex composed of gelatinous hyphae, as in some lichens.

pseudocostate *a.* false-veined, having a marginal vein uniting all others, *appl.* leaves, insect wings.

pseudoculus *n.* an oval area on either side of head of certain millipedes, possibly a receptor for mechanical vibrations.

pseudocyphella *n.* a structure in lichens similar to a cyphella but smaller, also thought to be used in aeration of the thallus.

pseudocyst *n.* a residual mass of protoplasm which swells and ruptures, liberating spores of sporozoans.

pseudodeltidium *n.* a plate partly or entirely closing deltidial fissure in ventral valve of some brachiopods.

pseudoderm *n.* a skin-like covering of certain compact sponges.

pseudodominance *n.* expression of a recessive allele in the absence of the dominant allele.

pseudodont *n.* having horny pads or ridges instead of teeth, as monotremes.

pseudoelater *n.* one of the chains of cells in sporogonium of some liverworts, probably functioning as a true elater.

pseudofoliaceous *a.* with expansions resembling leaves.

pseudogamy *n.* union of hyphae from different thalli; activation of ovum by sperm which plays no part in further development; pseudomixis *q.v.*

pseudogaster *n.* an apparent gastral cavity of certain sponges, opening to exterior by a pseudo-osculum and having no true oscula opening into itself.

pseudogastrula *n.* the stage in the development of certain sponges in which the archaeocytes become completely enclosed by flagellate cells.

pseudogene *n.* a stretch of DNA related in sequence to a functional gene but which is itself inactive as a result of the changes it has accumulated.

pseudohaptor *n.* a large discoidal organ in some trematodes, with a ventral armature of spines arranged in radial rows.

pseudoheart *n.* the axial organ in echinoderms; one of the contractile vessels pumping blood from dorsal to ventral vessel in annelids.

pseudoidium *n.* a separate hyphal cell which may germinate.

pseudolamina *n.* expanded apical portion of a phyllode.

pseudometamerism *n.* apparent serial segmentation; an approximation to metamerism, as in certain cestodes.

pseudomitotic diaschistic *q.v.*

pseudomixis *n.* sexual reproduction by fusion of vegetative cells instead of gametes, leading to zygote formation.

pseudomonads *n.plu.* bacteria of the family Pseudomonadaceae, widely distributed in soil and water, typically aerobic or facultatively anaerobic heterotrophs, Gram-negative polarly flagellated rods, some spp. containing blue or green fluorescent pigments.

pseudomonocarpous *a.* with seeds retained in leaf bases until liberated, as in cycads.

pseudomonocotyledonous *a.* with two cotyledons coalescing to appear as one.

pseudomorph *n.* a structure having an indefinite form.

pseudomycelium *n.* an assemblage of chains of single cells, as in some yeasts.

pseudomycorrhiza *n.* mild pathological fungal infection of plant roots, superficially resembling mycorrhiza.

pseudonavicella *n.* a small boat-shaped spore containing sporozoites, in sporozoans.

pseudonotum postscutellum *q.v.*

pseudonychium *n.* a lobe or process between claws of insects.

pseudo-osculum *n.* the exterior opening of a pseudogaster in sponges.

pseudopallium *n.* in some gastropod molluscs parasitic on echinoderms, a ring-like fold of skin developing at the base of the proboscis and eventually extending like a sac over the whole parasite.

pseudoparaphysis basidiolum *q.v.*

pseudoparasitism *n.* accidental entry of a free-living organism into the body and its survival there.

pseudoparenchyma *n.* fungal tissue formed of a tightly woven mass of hyphae, in which the hyphae have lost their individuality and which superficially resembles parenchyma tissue of plants; similar tissue formed from algal filaments.

pseudopenis *n.* the protruded evaginated portion of male deferent duct, in some oligochaete worms; copulatory structure in Orthoptera.

pseudoperculum *n.* a structure resembling an operculum or closing membrane.

pseudoperianth *n.* an envelope investing the archegonium in certain liverworts.

pseudoperidium *n.* the aecidiospore envelope in certain fungi.

pseudoplasmodium *n.* an aggregation of myxamoebae of a cellular slime mould, in which cells do not fuse.

pseudopod(ium) *n.* a protrusion of cytoplasm put out by a cell, particularly amoeboid protozoa and phagocytic cells of animals and plants where it serves for locomotion and feeding, *plu.* pseudopodia, *alt.* pseudopod; (*bot.*) the stalk supporting the sporangium in some mosses; slender branch of the gametophyte that bears gemmae in some mosses.

pseudopregnancy *n.* condition of development of accessory reproductive organs simulating true pregnancy although fertilization has not taken place.

pseudopregnant *a. appl.* mice and other

mammals which have been treated with hormones so that the uterus is receptive to *in vitro* fertilized ova and blastocysts which implant and develop normally.

pseudopupa coarctate pupa *q.v.*

pseudoramose *a.* having false branches.

pseudoramulus *n.* a spurious branch of certain algae.

pseudoraphe *n.* a smooth axial area in some diatoms.

pseudorhiza *n.* a root-like structure connecting mycelium in the soil with the fruit body of a fungus.

pseudosacral *a. appl.* sacral vertebra attached to pelvis by transverse process and not by sacral rib.

pseudoscolex *n.* modified anterior proglottids of certain cestodes where the true scolex is absent.

Pseudoscorpiones, psedoscorpions *n., n.plu.* order of small arachnids, commonly called false scorpions, which resemble scorpions but whose opisthosoma is not divided into two regions.

pseudosematic pseudepisematic *q.v.*

pseudoseptate *a.* apparently, but not morphologically septate, having a perforate or incomplete septum.

pseudoseptum *n.* a septum with pores, as in certain fungi; septum-like structure deposited at intervals in hyphae of some chytrids.

pseudosessile *a. appl.* abdomen of petiolate insects when petiole is so short that abdomen is close to thorax.

pseudosperm *n.* a small indehiscent fruit resembling a seed.

pseudospore *n.* an encysted resting myxamoeba.

pseudostele *n.* an apparently stelar structure, as midrib of leaf.

pseudostigma *n.* a cup-like pit in integument, as the socket of a sensory seta in ticks and mites.

pseudostiole, pseudo-ostiole *n.* a small opening formed by breaking down of cell walls or tissues, in certain fungi without perithecia.

pseudostipe *n.* a stem-like structure formed by presumptive spore-forming tissue, as in gasteromycete fungi.

pseudostipula, pseudostipule *n.* part of lamina at the base of a leaf stalk, which resembles a stipule.

pseudostoma *n.* a temporary mouth or mouth-like opening.

pseudostroma *n.* a mass of mixed fungal and host cells.

pseudosuckers *n.plu.* powerful organs of attachment, with gland cells, on either side of the oral sucker of trematodes.

pseudothecium *n.* fruiting body resembling a perithecium.

pseudotrachea *n.* a trachea-like structure; one of the trachea-like food channels of labellum, as in dipterans.

pseudotroch *n.* inner ring of cilia around mouth of some rotifers.

pseudotrophic *a. appl.* mycorrhiza when the fungus is parasitic.

pseudouracil (ΨU) unusual pyrimidine base found in tRNA, formed by exchange of a carbon and nitrogen atom at positions 1 and 5 in the uracil ring.

pseudovacuole gas vacuole *q.v.*

pseudovarium *n.* an ovary producing pseudova, ova that can develop without fertilization.

pseudovelum *n.* in jellyfish, a velum without muscular or nerve cells; in agaric fungi, a structure formed by outgrowths from cap and stalk, protecting immature hymenium, *alt.* pseudoveil.

pseudovitellus *n.* a mass of fatty cells in abdomen of aphid.

pseudovum *n.* ovum that can develop parthenogenetically; the earlier condition of viviparously produced aphids.

Psilophyta *n.* one of the four major divisions of extant seedless vascular plants, represented by only two living genera. They are tropical plants of simple structure, having a rootless sporophyte, dichotomously branching rhizomes, and aerial branches with small scale-like appendages (*Psilotum*) or larger bract-like outgrowths (*Tmesipteris*). *alt.* Psilopsida.

P-site site on the ribosome occupied by the last mRNA codon to have been read and by the peptidyl-tRNA.

Psittaciformes *n.* an order of birds including the parrots.

psoas *n.* either of two muscles in the loin: psoas major and psoas minor.

Psocoptera, psocids *n., n.plu.* an order of small insects, commonly called book lice and bark lice, having incomplete metamorphosis, a globular abdomen and often no wings.

psychogenetic *a. pert.* mental development; caused by the mind.

psychogenic *a.* of mental origin, *appl.* physiological and somatic changes.

psychophysiology *n.* physiology in relation to mental processes.

psychosomatic *a. pert.* relationship between mind and body; *pert.* or having bodily reactions to mental stimuli.

psychrophil(ic) *a.* thriving at relatively low temperatures; for bacteria, below 20°C.

ptera-, ptero- prefixes derived from Gk. *pteryx*, wing.

pteralia *n.plu.* axillary sclerites forming articulation of wing with the process of the mesonotum in insects.

pterate *a.* winged.

pterergate *n.* worker or soldier ant with vestigial wings.

pteridine *n.* organic compound composed of 2 fused 6-membered rings of nitrogen and carbon with various substituents, which is a constituent of many natural compounds such as pterins (leucopterin, xanthopterin) and folic acid.

pteridology *n.* the branch of botany dealing with ferns.

pteridophytes *n.plu.* major group of spore-bearing vascular plants: the ferns, club mosses, horsetails and the Psilophyta, sometimes treated as a division, Pteridophyta. *see* Lycophyta, Psilophyta, Pterophyta, Sphenophyta.

Pteridospermophyta, pteridosperms *n., n.plu.* an extinct division of seed-bearing vascular plants, the seed ferns (*q.v.*).

pterin *n.* any of a group of pigments, derivatives of pteridine, widespread in insects as eye pigments and in wings, and which are also found in vertebrates and plants.

pterion *n.* the point of junction of parietal, frontal and great wing of sphenoid, *appl.* ossicle, a sutural bone.

Pterobranchia, pterobranchs *n., n.plu.* class of small hemichordates with a vase-shaped body, U-shaped digestive tract, and a collar surrounding the mouth extended into pairs of hollow arms bearing ciliated tentacles. They are mostly colonial and some species are enclosed in a secreted

tube.

pterocardiac *a. appl.* ossicles with curved ends in gastric mill of crustaceans.

pterocarpous *a.* with winged fruit.

pterodactyls *n.plu.* the common name for the pterosaurs *q.v.*

pterodium *n.* a winged fruit or samara.

pteroic acid *n.* a compound of a pterin and *p*-aminobenzoic acid, which combines with glutamine to form folic acid.

pteroid *a.* resembling a wing; like a fern.

pteropaedes *n.plu.* birds able to fly when newly hatched.

pteropegum *n.* an insect's wing socket.

Pterophyta *n.* one of the four major divisions of the spore-bearing vascular plants, commonly called the ferns. The sporophyte has roots, stems and large leaves (fronds) which are megaphylls and bear the sporangia.

pteropleura *n.* thoracic sclerite between wing insertion and mesopleura in dipterans.

pteropodium *n.* a winged foot as in certain bats.

pteropods *n.plu.* group of marine gastropod molluscs with wing-like extensions to the foot, commonly called sea butterflies.

Pteropsida *n.* plant classification which has been used in different ways, as an alternative to Filicophyta (ferns), or for a larger grouping containing the ferns and seed plants.

Pterosauria, pterosaurs *n., n.plu.* an order of Jurassic and Cretaceous archosaurs, flying reptiles commonly called pterodactyls, which have a membranous wing supported by a greatly elongated fourth finger.

pterospermous *a.* with winged seeds.

pterostigma *n.* an opaque cell on wing of insect.

pterote *a.* winged, having wing-like outgrowths.

pterotheca *n.* the wing-case of a pupa.

pterothorax *n.* fused mesothoracic and metathoracic segments, in dragonflies.

pterotic *n.* a cranial bone overlying horizontal semicircular canal of ear; *a. appl.* the bone between pro-otic and epiotic.

pterygia *plu.* of pterygium *q.v.*

pterygial *a.* a wing or fin; *appl.* a bone or cartilage supporting a fin-ray.

pterygiobranchiate *a.* having feathery or

spreading gills, as certain crustaceans.

pterygiophore *n.* cartilaginous rod forming ray of fins of, e.g., sharks and dogfishes; similar bony ray in fins of bony fishes.

pterygium *n.* a prothoracic process in weevils; a small lobe at base of underwings in Lepidoptera; a vertebrate limb. *plu.* pterygia.

pterygoda *n.plu.* the tegulae of an insect.

pterygoid *a.* wing-like; *appl.* a wing-like process of sphenoid bone; *n.* a cranial bone, forming part of roof of mouth.

pterygoideus *n.* muscles causing protrusion and raising of mandible (lower jaw), pterygoideus externus and internus.

pterygomandibular *a. pert.* pterygoid and mandible; *appl.* a band of tendon or raphe of buccopharyngeal muscle.

pterygomaxillary *a. appl.* a fissure between maxilla and pterygoid process of sphenoid.

pterygopalatine *a. pert.* a region of pterygoid and palatal cranial bones, *appl.* canal, fossa, groove; *appl.* ganglion: sphenopalatine ganglion *q.v.*

pterygoquadrate *n.* a cartilage constituting dorsal half of mandibular arch of some fishes.

pterygospinous *a. appl.* a ligament between lateral pterygoid plate and spinous process of sphenoid.

pterygote, pterygotous *a. appl.* a large group of insects, the subclass Pterygota, which contains all the winged insects (and some wingless ones such as fleas and lice, considered to be descended from winged forms), and whose members undergo some form of metamorphosis and have no organs of locomotion on the abdominal segments. Applies to all insect orders except Thysanura, Diplura, Protura and Collembola.

pterylae *n.plu.* feather tracts.

pterylosis *n.* arrangement of feather tracts in birds.

ptilinum *n.* a head vesicle or bladder-like expansion of head of a fly emerging from pupa.

ptilopaedic *a.* covered with down when hatched.

ptomaine *n.* any of a group of amino compounds, usually poisonous, produced during protein breakdown during putrefaction of dead animal matter and by some plants, and including choline, putrescine, cadaverine, and muscarine.

ptosis *n.* drooping of the eyelids, congenital ptosis in humans usually being due to inheritance of a simple Mendelian dominant allele.

ptyalin *n.* salivary amylase, found in man and some herbivores, which digests broken starch grains. EC 3.2.1.1, *r.n.* a-amylase.

ptyophagous *a. appl.* digestion, by host cells, of the cytoplasm emitted from tips of mycorrhizal hyphae invading the cells.

ptyxis *n.* the form in which young leaves are folded or rolled on themselves in the bud.

puberty *n.* the beginning of sexual maturation.

puberulent *a.* covered with down or fine hair.

pubes *n.* the lower portion of hypogastric region of abdomen, the pubic region.

pubescent *a.* covered with soft hair or down.

pubic *a.* in the region of pubes; *pert.* pubis.

pubis *n.* in each half of pelvic girdle of vertebrates except fishes, bone forming anteroventral part of girdle, fused with ischium to form hipbone.

public *a. appl.* antigenic determinants shared by other haplotypes in serological analysis of histocompatibility antigens; *appl.* any antigenic determinant shared by different antigens.

puboischium *n.* fused pubis and ischium, bearing acetabulum and ilium on each side.

pudendum *n.* vulva, or external female genitalia. *plu.* pudenda.

puffball *n.* common name for fungi of the order Lycoperdales in the Gasteromycetes *q.v.*

puffs *n.plu.* temporarily swollen regions visible on polytene chromosomes at which chromosomal material is extruded and DNA is being actively transcribed.

pullulation *n.* reproduction by vegetative budding.

pulmo- prefix from L. *pulmo*, lung, usually denoting lung-like, to do with the lungs or breathing, etc.

pulmobranch(ia) *n.* a gill-like organ adapted to air-breathing conditions; lung book, as of spiders.

pulmogastric *a. pert.* lungs and stomach.

pulmonary *a. pert.* lungs.

pulmonary cavity or sac the mantle cavity, modified as a lung, in pulmonate molluscs (slugs and snails).

pulmonates *n.plu.* molluscs of the subclass Pulmonata, the snails and slugs, characterized by lack of ctenidia and in which the mantle cavity is used as a lung.

pulmones *n.plu.* lungs. *sing.* pulmo.

pulp *n.* internal cavity of vertebrate tooth, containing connective tissue, nerves and blood vessels.

pulsating or pulsatile vacuole contractile vacuole *q.v.*

pulse *n.* the seed of a legume, e.g. peas, beans, lentils, etc.

pulse-chase experiments experiments in which cells are very briefly labelled with a radioactive precursor of a particular molecule or pathway and the fate of the label is followed during subsequent incubation with non-labelled precursor.

pulsed-field gel electrophoresis electrophoretic technique for separating large pieces of DNA, e.g. chromosomes, in which an electric field is applied first in one direction and then in a direction at an angle to the first.

pulse wave a wave of increased pressure over the arterial system, started by ventricular systole.

pulverulent *a.* powdery; powdered.

pulvillar *a. pert.* or at a pulvillus.

pulvilliform *a.* like a small cushion.

pulvillus *n.* pad, process or membrane on foot or between claws, sometimes serving as an adhesive organ, in insects; lobe beneath each claw on feet of some mammals. *a.* pulvillar.

pulvinar *a.* cushion-like; *pert.* a pulvinus; *n.* an angular prominence on thalamus in brain.

pulvinate *a.* cushion-like, *appl.* a defensive or offensive gland in ants; having a pulvinus.

pulvinoid *a.* resembling a pulvinus, *appl.* modified petiole.

pulvinus *n.* a cellular swelling at junction of axis and leaf stalk, which plays a part in leaf or leaflet movement.

pulviplume *n.* powder-down feather *q.v.*

pump *n.* active transport mechanism for ions or small molecules in cell membrane.

punctae *n.plu.* small pores, holes, or dots on a surface, esp. the markings on valves of diatoms.

punctate *a.* dotted; having surface covered with small holes, pores or dots; having a dot-like appearance.

punctiform *a.* having a dot-like appearance; *appl.* distribution, as of cold, warm and pain spots on skin.

punctuated equilibrium the view that the course of evolution has been marked by long periods of little or no evolutionary change (stasis) punctuated by short periods of rapid evolution. The view is based on an interpretation of the fossil record which in some cases appears to show such a pattern. *cf.* phyletic gradualism.

punctulate *a.* covered with very small dots.

punctum *n.* a minute, dot, point or orifice.

puncture *n.* a small round surface depression; a perforation.

pungent *a.* producing a prickling sensation, *appl.* stimuli affecting chemoreceptors; bearing a sharp point, *appl.* apex of leaf or leaflet.

punishment negative reinforcement *q.v.*

Punnett square a conventional representation used to calculate the proportions of different genotypes in progeny of a genetic cross, e.g. for parents *Aa* and *aa*:

	A	*a*
a	*Aa*	*aa*
a	*Aa*	*aa*

pupa *n.* in insects with complete metamorphosis, a resting stage in the life cycle where the larval insect is enclosed in a protective case, within which tissues are reorganized and metamorphosis into a new form, usually the adult, occurs.

puparium *n.* the casing of a coarctate pupa, formed from the last larval skin, esp. in Diptera; a larval instar.

pupate *v.* to pass into the pupal stage.

pupiform *a.* resembling a pupa in shape.

pupigerous *a.* containing a pupa.

pupil *n.* central aperture of iris in vertebrate eye through which light enters, and whose size is varied by contraction of the ciliary body around the iris; central spot of an eye-spot (ocellus).

pupillary *a. pert.* pupil of an eye; *appl.* reflex: variation in aperture of pupil due to change in illumination, closing in bright light, opening in dim light.

pupillate *a. appl.* eye-like markings with a differently coloured central spot.

pupiparous *a.* bringing forth young already developed to the pupa stage, as in certain parasitic insects.

pupoid pupiform *q.v.*

pure line series of generations of organisms originating from a single homozygous ancestor or identical homozygous ancestors, and which are therefore themselves homozygous for a given character or characters.

purine *n.* a type of nitrogenous organic base, of which adenine and guanine are most common in living cells, occurring in nucleic acids where they pair with pyrimidines. Forms a nucleotide when linked to ribose or deoxyribose phosphates. Purine nucleotides (esp. those of adenine) are important cofactors and enerygy-rich compounds in metabolism. *see* Appendix 1 (20, 21) for chemical structural formulae.

purinergic *a.* secreting purines, *appl.* neurones that secrete purine neurotransmitters (e.g. adenosine and ATP).

purinoceptor *n.* cell-surface receptor for purines (e.g. adenosine and ATP), found on some neurones and other cells.

Purkinje cell, Purkinje neurone type of neurone found in cerebellum, whose cell bodies constitute a distinct layer in cerebellar cortex between the molecular and granular (nuclear) layers, and whose axons carry the output from cerebellum.

Purkinje fibres muscle fibres in the band of muscle and nerve fibres connecting auricles and ventricles of heart (His' bundle), differing from typical cardiac fibres especially in a higher rate of conduction of contractile impulse.

puromycin *n.* antibiotic which becomes incorporated into a polypeptide chain as it is being synthesized causing the release of the incomplete chain from the ribosome.

purple bacteria group of photoautotrophic bacteria, e.g. *Rhodopseudomonas*, which contain bacteriochlorophyll and the purple protein bacteriorhodopsin.

purple membrane membrane in certain bacteria, e.g. the halophile *Halobacterium*, which contains the purple protein bacteriorhodopsin, a light-driven transmembrane proton pump, in an organized array.

purple sulphur bacteria photoautotrophic bacteria, mainly aquatic, which oxidize sulphide to sulphur.

purposive behaviour goal-related behaviour.

pusule *n.* non-contractile vacuole containing watery fluid, filling or emptying by a duct, present in many dinoflagellates; a contractile vacuole in some algae.

pustule *n.* a blister-like prominence.

putamen *n.* lateral part of lentiform nucleus in cerebral hemispheres.

putrefaction n. decomposition of organic material, esp. the usually anaerobic splitting of proteins by microorganisms, resulting in incompletely oxidized, ill-smelling compounds such as mercaptans, alkaloids and polyamines.

putrescine *n.* a foul-smelling polyamine, $H_2N(CH_2)_4NH_2$, formed by decarboxylation of ornithine and often produced during breakdown of protein by bacteria, e.g. in putrefying meat, also found in small amounts in various mammalian tissues such as liver, pancreas, semen.

pycnia *plu.* of pycnium, *see* pycnidium.

pycnic *a.* thick-set; *appl.* type of body build, short stocky, with broad face and head.

pycnidia *plu.* of pycnidium *q.v.*

pycnidial *a. pert.* pycnidium.

pycnidiophore *n.* a conidiophore producing pycnidia.

pycnidiospore *n.* a spore produced in a pycnidium.

pycnidium *n.* small hollow spherical or flask-shaped structure enclosing conidiophores in rust and smut fungi. *a.* pycnidial. *plu.* pycnidia.

pycnium pycnidium *q.v.*; spermogonium *q.v.*

Pycnogonida, pycnogonids *n. n.plu.* class of chelicerate marine arthropods (*q.v.*) commonly known as sea spiders, with a long slender body consisting of an anterior cephalon, a trunk with four pairs of long walking legs, and a short segmented abdomen. Some species bear chelicerae and feelers, others have neither.

pycnosis *n*. cell degeneration including condensation of nuclear contents and formation of an intensely staining clump of chromosomes.

pycnospore pycnidiospore *q.v.*

pycnotic *a*. characterized by or *pert.* pycnosis, *appl.* small irregular nucleus of degenerated cells.

pycnoxylic *a*. having compact wood.

pygal *a*. situated at or *pert.* posterior end of back; *appl.* certain plates of shell of tortoises and turtles.

pygidium *n*. an exoskeletal shield covering tail region of some arthropods, and various structures in the same region in other insects. *a*. pygidial.

pygmy male a purely male form, usually small, found living close to the ordinary hermaphrodite form in certain animals, as in some polychaete worms and barnacles.

pygofer *n*. the last abdominal segment in leaf hoppers.

pygostyle *n*. a compressed upturned bone at end of vertebral column in birds, composed of fused vertebrae.

pykn- pycn-.

pylangium *n*. proximal portion of amphibian or foetal heart, through which blood is driven to the ventricles.

pylocyte *n*. a pore cell at inner end of small funnel-shaped depression, the porocyte of certain sponges.

pylome *n*. in some radiolarians, an aperture for emission of pseudopodia and intake of food.

pyloric *a. pert.* or in the region of the pylorus; *appl.* sphincter between mid-gut and hind-gut of insects; *appl.* posterior region of gizzard in crustaceans.

pylorus *n*. the lower opening of stomach, into duodenum.

pyogenic *a*. pus-producing, *appl.* bacteria.

pyramid *n*. conical structure, protuberance, eminence, as of cerebellum, medulla oblongata, temporal bone, kidney.

pyramid of biomass a representation of the total biomass at each level of a food chain, which forms a pyramid, the biomass at lower levels (e.g. primary producers) being greater than that at higher levels (e.g. carnivores).

pyramid of energy a representation of the energy available per unit time at each trophic level in an ecosystem, usually expressed in kilocalories per square metre per year.

pyramid of numbers a representation of the numbers of organisms at different levels of a food chain, which forms a pyramid, greater numbers of organisms being present at the lower levels (e.g. primary producers) than at higher levels (e.g. carnivores).

pyramidal *a*. conical, like a pyramid, *appl.* leaves, a carpal bone, tract of nerve fibres in brain.

pyramidal cell a type of neurone found in the cerebral cortex, of characteristic shape.

pyramidal tract tract of motor nerve fibres running from cerebral motor cortex of brain to the anterior horn cells of spinal cord at all levels, concerned with voluntary motion. *alt.* corticospinal tract.

pyranose *n*. a monosaccharide in the form of a 6-membered ring of 5 carbons and one oxygen. *cf.* furanose.

pyrene *n*. a seed surrounded by a hard body forming a fruit stone or kernel, often with several in one fruit.

pyrenocarp perithecium *q.v.*; drupe *q.v.*

pyrenoid *n*. in algae and certain liverworts, a protein body in the chloroplast which is the centre of starch formation.

Pyrenomycetes *n*. group of ascomycete fungi, commonly called the flask fungi, in which the generally club-shaped asci are usually borne in a hymenial layer in flask-shaped or spherical ascocarps (perithecia) which open in a terminal pore. They include the plant parasitic powdery mildews, the saprophytic pink bread mould *Neurospora*, the agents of several plant cankers (including the coral-spot fungus), anthracnoses, and leaf spot diseases.

pyretic *a*. increasing heat production; causing rise in body temperature.

pyrexia *n*. an increase in body temperature above normal to a new set point, which causes the symptoms of fever often accompanying an infection.

pyridine nucleotides nicotinamide adenine dinucleotide *q.v.*, and nicotinamide adenine dinucleotide phosphate *q.v.*

pyridoxal *see* pyridoxal phosphate.

pyridoxal phosphate (PLP) a coenzyme

derived from pyridoxine (vitamin B₆), the prosthetic group of many enzymes including all transaminases.

pyridoxamine *see* pyridoxine.

pyridoxine *n.* a form of vitamin B₆, a phenolic alcohol derived from pyridine, can be converted to pyridoxal and pyridoxamine in the body and is a precursor to the coenzyme pyridoxal phosphate. *see also* vitamin B₆.

pyriform *a.* pear-shaped.

pyrimidine *n.* a type of nitrogenous organic base, of which cytosine, uracil and thymine are most common in living cells, occurring in nucleic acids where they pair with purines. Forms a nucleotide when linked to ribose or deoxyribose phosphates. Some pyrimidine nucleotides also act as phosphate donors and energy-rich compounds in metabolism. *see* Appendix 1 (17, 18, 19) for structural chemical formulae.

pyrimidine dimer structure produced in DNA by ultraviolet light in which adjacent pyrimidines on the same strand become covalently linked, blocking replication and transcription.

pyrogen *n.* a substance that can cause pyrexia (fever), such as the endogenous pyrogens released by white blood cells in response to an infection.

pyrophosphatase *n.* enzyme hydrolysing inorganic pyrophosphate to orthophosphate. EC 3.6.1.1.

pyrophosphate (PP$_i$) $HP_2O_7^{3-}$.

pyrophyte *n.* plant that likes to grow on burnt ground.

pyrotheres *n.* a group of South American placental mammals of Eocene to Oligocene with tusk-like teeth and tending to large size like elephants.

Pyrrophyta *n.* the dinoflagellates, a group of largely unicellular biflagellated organisms sometimes known as whirling whips, considered either as protists or as part of the plant kingdom. They include both photosynthetic and heterotrophic forms and are important members of both marine and freshwater plankton. A feature of many dinoflagellates is the plates of cellulose immediately under the plasma membrane which form a sculptured wall (theca) around the cell.

pyruvate (pyruvic acid) *n.* 3-carbon organic acid ($CH_3COCOOH$) produced during glycolysis, an important intermediate in many metabolic pathways, converted to acetyl CoA, in which form it is the starting point of the tricarboxylic acid cycle of aerobic respiration. *See* Appendix 1 (48).

pyruvate decarboxylase key enzyme in alcoholic fermentation, catalysing the irreversible decarboxylation of pyruvate to yield acetaldehyde and CO_2. EC 4.1.1.17.

pyruvate dehydrogenase complex an assembly of 3 enzymes catalysing the oxidative decarboxylation of pyruvate to form acetyl CoA for the tricarboxylic acid cycle, composed of pyruvate dehydrogenase (EC 1.2.4.1), dihydrolipoamide transacetylase (EC 2.3.1.12) and dihydrolipoamide dehydrogenase (EC 1.6.4.3).

pyruvate kinase enzyme catalysing the transfer of phosphate from phosphoenolpyruvate to ADP with formation of pyruvate and ATP (and vice versa) in glycolysis and elsewhere. EC 2.7.1.40.

pyxidiate *a.* opening like a box by transverse dehiscence; *pert.,* or like a pyxidium or pyxis.

pyxidium *n.* a capsular fruit which dehisces transversely, the top coming off as a lid.

pyxis *n.* swelling of a podetium in lichens; pyxidium *q.v.*

Q

Q glutamine *q.v.*; queuosine *q.v.*; ubiquinone *q.v.*

Q$_{CO_2}$ the volume of carbon dioxide in microlitres of gas at normal temperature and pressure given out per hour per milligram dry weight.

Q$_{O_2}$ oxygen quotient *q.v.*

Q$_n$ ubiquinone *q.v.*

Q$_{10}$ temperature coefficient *q.v.*; ubiquinone *q.v.*

QH$_2$ reduced form of ubiquinone *q.v.*

Q bands characteristic pattern of fluorescent bands on mitotic chromosomes stained with quinacrine, used in identification of individual chromosomes.

quadrangular *a. appl.* four-cornered stems; *appl.* lobes of cerebellar hemispheres, connected by the largest part of the superior vermis of cerebellum.

quadrant *n.* all the cells derived from one of the first four blastomeres.

quadrat *n.* a sample area enclosed within a frame, usually a square, within which a plant community, or sometimes an animal community, is analysed.

quadrate bone part of hyomandibular in vertebrate skull, with which lower jaw articulates in birds, reptiles, amphibians and fishes.

quadratojugal *n.* membrane bone connecting quadrate and jugal bones.

quadratomandibular *a. pert.* quadrate and mandibular bones.

quadratomaxillary quadratojugal *q.v.*

quadratus *n.* the name of several rectangular muscles, such as quadratus femoris.

quadri- prefix derived from L. *quattuor*, four, signifying having four of, divided into four, in four parts, etc.

quadricarpellary *a.* containing four carpels.

quadriceps *n.* muscle in front of thigh, extending lower leg and divided into four portions at the upper end.

quadrifarious *a.* in four rows, *appl.* leaves.

quadrifid *a.* cleft into four parts.

quadrifoliate *a.* four-leaved; *appl.* compound palmate leaf, with four leaflets arising at a common point.

quadrigeminal bodies corpora quadrigemina *q.v.*

quadrihybrid *n.* a cross whose parents differ in four distinct characteristics; *a.* heterozygous for four pairs of alleles.

quadrijugate *a. appl.* pinnate leaf having four pairs of leaflets.

quadrilateral *n.* the discal cell in wing of dragonflies, a large cell at base of wing completely enclosed by veins.

quadrilobate *a.* four-lobed.

quadrilocular *a.* having four loculi or chambers, as ovary, or anthers of certain flowers.

quadrimaculate *a.* having four spots.

quadrimanous quadrumanous *q.v.*

quadrinate quadrifoliate *q.v.*

quadripennate *a.* having four wings.

quadripinnate *a.* divided pinnately four times.

quadriradiate *a.* having four rays.

quadriseriate *a.* arranged in four rows or series. *alt.* tetrastichous.

quadritubercular *a. appl.* teeth with four cusps.

quadrivalent *n.* association of four chromosomes that forms during meiosis in an individual carrying a heterozygous reciprocal translocation. This type of quadrivalent is formed of four complete chromosomes, the two chromosomes carrying the translocation and their untranslocated homologues, and has four centromeres; the term is also sometimes used for the pair of duplicated homologous chromosomes, made up of four chromatids, that are linked by chiasmata during meiosis. This type of quadrivalent has only two centromeres.

quadrivoltine *a.* having four broods in a year.

quadrumanous *a.* having hind-feet as well as fore-feet constructed as hands, as most Primates except man.

quadrupedal *a.* walking on four legs.

quadruplex *a. appl.* polyploids having four dominant genes.

qualitative inheritance the inheritance of phenotypic characters that occur in two or more distinct states and do not grade into each other, the states representing combinations of different alleles at a single locus, sometimes called simple Mendelian inheritance. *cf.* quantitative or polygenic inheritance.

quantitative inheritance the inheritance of characters determined by many different genes acting independently, and which appear as continuously variable characters within a population. *cf.* qualitative inheritance.

quantitative trait in genetics, a phenotypic character determined by the effects of many genes, which shows a continuously graded spectrum of variation which can only be measured quantitatively, such as height, weight, etc.

quantitative variation continuous variation *q.v.*

quantosomes *n.* particles on the thylakoid membranes of chloroplasts, which contain the photosynthetic apparatus.

quantum *n.* the smallest amount in which a neurotransmitter is normally secreted; a unit of light. *plu.* quanta. *a.* quantal.

quartet *n.* a group of four nuclei or four cells resulting from meiosis; four cells derived from a segmenting ovum during cleavage.

quasidiploid *a. appl.* cells with the diploid number of chromosomes but with an abnormal genetic makeup, as e.g. three copies of one chromosome and only one of another, as many cell lines.

quasisocial insects those social insects in which there is cooperative care of the brood but each female still lays eggs at some time.

quaternary *a. appl.* flower symmetry when there are four parts in each whorl.

Quaternary *a. pert.* or *appl.* geological period lasting from about 2 million years ago to present, comprising Pleistocene and Holocene epochs.

quaternary structure the relationships of the various subunits to each other in a protein composed of several polypeptide chains (subunits).

quaternate *a.* in sets of four; *appl.* leaves growing in fours from one point.

queen *n.* a member of the reproductive caste in eusocial and semisocial insects, sometimes, but not always, morphologically different from the workers.

queen substance the set of pheromones by which a queen honey-bee attracts workers and controls their reproductive activities, generally denotes *trans*-9-keto-2-decenoic acid, the most powerful of the components.

queuosine (Q) *n.* unusual nucleoside found only in tRNA, a modified form of guanosine with an aromatic ring substituent at position 7 and which can pair with U as well as C.

quiescence *n.* temporary cessation of development or other activity, owing to an unfavourable environment.

quiescent centre a group of cells, with few mitoses, between root meristem and root cap.

quill *n.* the central shaft of a feather; a hollow horny spine, as of porcupine.

quill feathers feathers of wings (remiges) and tail (rectrices) of birds.

quill knobs tubercles or exostoses on ulna of birds, for attachment of fibrous ligaments connecting with follicle of feather.

quillworts *n.plu.* an order, the Isoetales, of vascular, non-seed bearing plants of the division Lycophyta, having linear leaves and a "corm" with complex secondary thickening.

quinary *a. appl.* flower symmetry when there are five parts in a whorl.

quinate *a. appl.* five leaflets growing from one point.

quincuncial *a.* arranged in a quincunx.

quincunx *n.* arrangement of five structures

of which four are at corners of a square with the fifth in centre.

quinine *n*. an alkaloid produced from bark of a species of *Cinchona* and used medicinally as an antimalarial drug and a febrifuge.

quinone *n*. any of various compounds derived from benzene, which function in biological oxidation-reduction systems.

quinone-tanning hardening of arthropod cuticle and other invertebrate skeletal material by cross-linking proteins with quinones.

quinque- prefix derived from L. *quinque*, five, signifying having five of, divided into five, etc.

quinquecostate *a*. having five ribs on the leaf.

quinquefarious *a*. in five directions, rows or parts.

quinquefid *a*. cleft in five parts.

quinquefoliate *a*. with five leaves.

quinquefoliolate quinate *q.v.*

quinquelobate *a*. with five lobes.

quinquelocular *a*. with five cavities or loculi.

quinquenerved *a*. having the midrib divided into five, giving five main veins.

quinquetubercular *a. appl.* molar teeth with five cusps.

quintuplinerved quinquenerved *q.v.*

quisqualate *n*. structural analogue of glutamate, used to define a class of glutamate receptors in the central nervous system.

R

r coefficient of relationship *q.v.*; intrinsic rate of increase *q.v.*

r, R roentgen *q.v.*

R arginine *q.v.*

R_0 net reproductive rate *q.v.*

ras v-*ras*^{H} and v-*ras*^{K}, oncogenes carried by Harvey and Kirsten sarcoma viruses of rodents. c-*ras* (several variants), corresponding proto-oncogenes found in normal cells, specifying guanine nucleotide binding proteins possibly involved in the transduction of signals from growth factors, etc. Altered and activated *ras* genes have been found in many human tumours.

RaSV rat sarcoma virus, an RNA tumour virus, various strains of which are the source of *ras* oncogenes.

RCA family of lectins isolated from the castor oil bean, *Ricinis comunis*.

rcp reciprocal translocation.

rDNA DNA specifying ribosomal RNA *q.v.*; recombinant DNA *q.v.*

RecA protein in *E. coli* involved in the exchange of single DNA strands in recombination and which also has a protease activity that is activated by treatments that block replication and which activates the genes responsible for DNA repair and retrieval systems in the SOS response by inactivating the LexA repressor protein.

REM *see* REM sleep.

RER rough endoplasmic reticulum, *see* endoplasmic reticulum.

RES reticuloendothelial system *q.v.*

REV reticuloendotheliosis virus, an avian RNA tumour virus, carries oncogene v-*rel*.

RF general designation for release factors *q.v.*; replicative form *q.v.*

RFLP restriction fragment length polymorphism *q.v.*

Rh rhesus factor *q.v.*

RIA radioimmunoassay q.v.

RLO rickettsia-like organism q.v.

RMR resting metabolic rate *q.v.*

RNA ribonucleic acid, *see* RNA and related entries in the body of the dictionary.

5S RNA minor rRNA species in the large ribosomal subunit.

5.8S RNA minor rRNA species in the large ribosomal subunit of eukaryotic cells.

16S RNA rRNA component of the small subunit of bacterial ribosomes.

18S RNA rRNA component of the small subunit of a typical eukaryotic ribosome.

23S RNA major rRNA component of the large subunit of bacterial ribosomes.

28S RNA major rRNA component of the large subunit of a typical eukaryotic ribosome.

RNase, RNAse ribonuclease *q.v.*

RNase H endonuclease with specificity for RNA in the form of an RNA-DNA hybrid.

RNase P endoribonuclease from *E. coli* involved in processing of tRNA precursors and which has a protein and an RNA component, neither of which is active alone.

RNase III endoribonuclease from *E. coli* involved in processing of rRNA precursors.

RNP ribonucleoprotein *q.v.*

ROS rod outer segment, the photoreceptor portion of rod cells of vertebrate retina.

RPPP reductive pentose phosphate pathway, *see* Calvin cycle.

RQ respiratory quotient *q.v.*

rRNA ribosomal RNA *q.v.*

RSS recombination signal sequence *q.v.*

RSV Rous sarcoma virus, an avian RNA tumour virus, carries oncogene *src*.

RT reaction time *q.v.*

RTF resistance transfer factor *q.v.*

RuBP ribulose 1,5-bisphosphate *q.v.*

rabbit-fishes a small group of marine fishes, class Holocephalii, with long slender tails and large pectoral fins.

race *n.* a group of individuals within a species which forms a permanent and distinguishable variety; a rhizome, as of ginger.

racemase *n.* any of a group of enzymes catalysing the conversion of L-isomers to D-isomers, esp. of amino acids, classified in the isomerases in EC 5.1.

racemate *n.* a mixture of two optical isomers, dextrorotatory (D or +) and laevorotatory (L or -), whose steric formulae are mirror images of each other and not superimposable.

racemation *n.* a cluster, as of grapes.

raceme *n.* flower-head having a common axis bearing stalked flowers arranged spirally around it, the bottom flowers opening first, as in hyacinth.

racemiferous *a.* bearing racemes.

racemiform *a.* in the form of a raceme.

racemose *a. appl.* a flower-head whose growing points continue to add to the head and in which there are no terminal flowers, individual flowers or side branches being arranged spirally or alternately along a single main axis, with the lowest flowers opening first. *alt.* indefinite; *appl.* any structure resembling a raceme.

racemule *n.* a small raceme.

racemulose *n.* in small clusters.

rachial *a. pert.* a rachis.

rachides *pl.* of rachis *q.v.*

rachidial rachial *q.v.*

rachidian *a.* placed at or near a rachis; *appl.* median tooth in row of teeth of radula.

rachiform *a.* in the form of a rachis.

rachiglossate *a.* having a radula with pointed teeth, as whelks.

rachilla *n.* axis bearing the florets in a grass spikelet; small or secondary rachis.

rachiodont *a. appl.* egg-eating snakes with well-developed hypophyses of anterior thoracic vertebrae, which function as teeth.

rachiostichous *a.* having a succession of somactids as axis of fin skeleton, as in lungfish.

rachis *n.* the shaft of a feather; a stalk or axis.

rachitomous temnospondylous *q.v.*

racket mycelium raquet mycelium *q.v.*

rad unit formerly used to measure the amount of ionizing radiation absorbed by living tissue, 1 rad being equal to 100 erg per gram tissue. It has been replaced by the gray (Gy), with 1 rad = 10^{-2} Gy.

radial *a. pert.* the radius; growing out like rays from a centre; *pert.* ray of an echinoderm; *appl.* leaves or flowers growing out like rays from a centre; *n.* crossvein of an insect wing; supporting skeleton of fin-ray.

radial apophysis a process on palp of male arachnids, inserted into groove of female epigynum during mating.

radial cleavage *appl.* type of cleavage in which the first divisions of the fertilized egg occur at right angles to each other resulting in four upper blastomeres sitting directly on top of four lower blastomeres.

radial fibres fibrous tissue supporting the retina.

radial glial cells glial cells in the developing neural tube which stretch from the lumen to the outer surface and which disappear from the brain and spinal cord towards the end of development, possibly differentiating into astrocytes.

radial notch lesser sigmoid cavity of coronoid process of ulna.

radial symmetry having a plane of symmetry about each radius or diameter, as many flowers and some animals such as sea anemones and starfish.

radiale *n.* a carpal bone in line with radius.

radiant *a.* emitting rays; radiating; *pert.* radiants; *pert.* radiation; *n.* an organism or group of organisms dispersed from an original geographical location.

radiate *a.* radially symmetrical; radiating; stellate; *v.* to diverge or spread from a point; to emit rays.

radiate-veined veined in a palmate manner.

radiatiform *a.* with radiating marginal florets.

radiation *n.* the emission of radiant energy in the form of waves or particles; energy radiated in the form of waves or particles, as e.g. electromagnetic radiation (radio waves, infrared, visible light, ultraviolet, X-rays and gamma rays) or emissions from radioactive sources (e.g. beta rays), esp. that radiation potentially harmful to living organisms, e.g. ionizing radiation (e.g. X-

and gamma rays and streams of α- and β-particles, emitted from radioactive elements); (*evol.*) the relatively rapid increase in numbers of new species of a particular type of animal or plant and their diversification and spread into many new habitats, e.g. the mammalian radiation that occurred after the end of the Cretaceous period when most present-day types of mammals arose.

radiation biology the study of the effects of potentially damaging radiation on living organisms.

radiation chimaera experimental animal, usually a mouse, whose immune system has been destroyed by irradiation and in which a new immune system has been reconsituted by bone marrow transplantation from a genetically different individual.

radiation dose equivalent dose equivalent *q.v.*

radiation ecology the study of radiation as affecting the relationship between living organisms and environment, and of the ecological effects and destination of radioactive elements.

radical *a.* arising from root close to ground, as basal leaves and flower stems; *n.* group of atoms that does not exist in the free state but as a unit in a compound, as OH, NH₄, C₆H₅, etc.

radicant *a.* with roots arising from the stem.

radicate *a.* rooted; possessing root-like structures; fixed to substrate as if rooted.

radication *n.* the rooting pattern of a plant.

radicel *n.* a small root; rootlet.

radicicolous *a.* having roots.

radiciferous *a.* bearing roots.

radiciflorous *a.* with flowers arising at extreme base of stem, so apparently arising from root.

radiciform, radicine *a.* resembling a root.

radicivorous *a.* root-eating.

radicle *n.* embryonic plant root, developing at the lower end of the hypocotyl.

radicle sheath coleorhiza *q.v.*

radicolous *a.* living in or on roots.

radicose *a.* with a large root.

radicular *a. pert.* a radicle.

radicule *n.* rootlet.

radiculose *a.* having many rootlets or rhizoids.

radii *plu.* of radius.

radioactivity *n.* the disposition of some el-

ements to undergo spontaneous disintegration of their nuclei associated with the emission of ionizing particles and electromagnetic radiation, as α-particles or β-particles and gamma radiation.

radioautography autoradiography *q.v.*

radiobiology *n.* study of the effects of radiation, esp. potentially harmful ionizing radiation such as X-rays, on living cells and organisms.

radiocarbon *n.* radioactive isotope of carbon, ¹⁴C, occurring naturally in small amounts, used in biochemical and physiological research and as an indicator for dating in archaeology.

radiocarbon dating the use of the differential uptake of the rare radioactive isotope of carbon, ¹⁴C, and the much more abundant isotope, ¹²C, during carbon fixation by plants to date the remains of organic material in archaeology. The difference between the proportion of ¹⁴C in the material that would be expected if the organic material were newly synthesized and the actual proportion of ¹⁴C reflects the time since the plant died and over which the ¹⁴C has decayed. The radiocarbon method can be used to date material between 3000 and 40,000 years old.

radiocarpal *a. pert.* radius and wrist.

radioecology radiation ecology *q.v.*

radioimmunoassay (RIA) *n.* very sensitive method for the detection and measurement of substances using radioactively labelled specific antibodies or antigens.

radioiodine *n.* radioactive isotope of iodine, ¹³¹I, used for studying the thyroid and in treatment of thyroid cancers.

radioisotope *n.* radioactive isotope of an element, such as tritium (³H), ³²P, radiocarbon (¹⁴C), and radioiodine (¹³¹I), widely used in experimental biology to label tracer compounds, biological molecules, *etc.*

radiolarians *n.plu.* group of marine planktonic protists of the phylum Actinopoda (*q.v.*) (formerly classified as protozoans of the class Sarcodina), characterized by a symmetrical skeleton of siliceous spicules.

radiole *n.* a spine of sea urchins.

radioligand *n.* radioactively labelled ligand (hormone, neurotransmitter, etc.) used for receptor binding experiments.

radiomedial *n*. a cross vein between radius and medius of insect wing.

radiomimetic *a*. resembling the effects of radiation, *appl*. chemicals causing mutations.

radionuclide *n*. an unstable atomic nucleus, which undergoes spontaneous radioactive decay, emitting radiation and changing from one element into another.

radiophosphorus *n*. radioactive isotope of phosphorus, ^{32}P, used widely in biochemical and physiological research, and therapeutically.

radioreceptor *n*. a sensory receptor for receiving light or temperature stimuli.

radioresistant *a*. offering a relatively high resistance to the effects of radiation, esp. ionizing radiation such as X-rays.

radiosensitive *a*. sensitive to the effects of radiation, esp. ionizing radiation such as X-rays.

radiospermic *a*. having seeds which are circular in transverse section; *appl*. plants, esp. fossils, having such seeds.

radiosymmetrical *a*. having similar parts similarly arranged around a central axis.

radiotherapy *n*. the treatment of disease, such as cancer, by means of X-rays or radioactive substances.

radioulna *n*. radius and ulna combined as a single bone.

radius *n*. bone of vertebrate forelimb between humerus and carpals; a main vein of insect wing.

radix *n*. a root; point of origin of a structure.

radula *n*. short, broad organ with rows of chitinous teeth in mouth of most gastropod molluscs. Used for feeding.

radulate, raduliferous *a*. having a radula.

raduliform *a*. resembling a radula or a flexible file.

raffinose *n*. a trisaccharide found in sugar beet, cereals and some fungi, giving glucose, fructose and galactose on hydrolysis. *see* Appendix 1 (13) for chemical structure.

Rafflesiales *n*. order of plant parasitic dicots in which the body is reduced to a simple thallus, and which comprises two families, Hydnoraceae and Rafflesiaceae.

Rainey's corpuscles spores of *Sarcocystis*, an elongated sporozoan parasite found in skeletal muscle.

Rainey's tubes elongated sacs found in voluntary muscle which are the adult stages of certain sporozoan parasites.

rain forest forest biomes that develop in areas with an annual rainfall of more than 254 cm. *see* monsoon rain forest, temperate rain forest, tropical rain forest.

raised bog, raised mire convex lens-shaped acid peatland developed in fen basins or river flood plains in wet climates.

ramal *a*. belonging to branches; originating on a branch.

ramapithecids, ramapithecines *n.plu*. a group of Miocene ape-like fossils from Asia and Africa, including the genus *Ramapithecus*, which show hominid-like features in the teeth.

ramate *a*. branched.

ramellose *a*. having small branches.

ramenta *n.plu*. brown scales present on stems, leaves and petioles of ferns; long epidermal hairs, of leaves. *sing*. rament, ramentum.

ramentaceous *a*. covered with ramenta.

rameous *a*. branched; *pert*. a branch.

ramet *n*. an individual member of a clone, as an offshoot of a plant reproducing by stolons etc.

rami *plu*. of ramus *q.v.*

rami communicantes nerve fibres connecting sympathetic ganglia and spinal nerves. *sing*. ramus communicans.

ramicorn *a*. having branched antennae, as some insects.

ramification *n*. branching; a branch of a tree, nerve, artery, etc.

ramiflory *n*. the state of flowering from the branches. *a*. ramiflorous.

ramiform *a*. branch-like.

ramigenous ramiparous *q.v.*

ramiparous *a*. producing branches.

ramose *a*. much branched.

ramule, ramulus *n*. a small branch, a twig.

ramuliferous *a*. bearing small branches.

ramulose *a*. with many small branches.

ramulus ramule *q.v.*

ramus *n*. any branch-like structure; a barb of a feather.

ramus communicans, *see* rami communicantes.

ramuscule ramule *q.v.*

random drift random accumulation of changes in nucleotide sequence occurring over long periods of time and which do not appear to be due to any selective forces; the random changes in gene frequency that can occur in a small population over time as a result of sampling of gametes in each generation. *alt.* genetic drift.

range *n.* the area within which an animal or group of animals seeks food; an area of unenclosed, unintensively managed grassland on which livestock are allowed to graze freely, *alt.* rangeland.

rangeland *n.* land that provides food for browsing and grazing animals and is not intensively managed.

ranine *a. pert.* undersurface of the tongue.

ranivorous *a.* feeding on frogs.

ranunculaceous *a. pert.* a member of the dicot flower family Ranunculaceae, the buttercups and their relatives.

Ranunculales *n.* order of herbaceous and woody plants, climbers or shrubs, and including the families Berberidaceae (barberry), Ranunculaceae (buttercup) and others.

Ranvier's nodes nodes of Ranvier *q.v.*

raphe *n.* a seam-like suture, as junction of some fruits; line of fusion of funicle and anatropous ovule; slit-like line in diatom valves; line, or ridge, of perineum, scrotum, hard palate, medulla oblongata, etc.

raphe nucleus any of a cluster of nuclei in midbrain in which lie cell bodies of 5-hydroxytryptaminergic neurones.

raphide *n.* a needle-like crystal of calcium oxalate, produced as a metabolic by-product in many plant cells.

raphidiferous *a.* containing raphides.

raptatory *a.* preying.

raptorial *a.* adapted for snatching or robbing, *appl.* birds of prey.

raptors *n.* birds of prey, e.g. hawks, eagles, owls etc.

raquet mycelium hyphae enlarged at one end of each segment, small and large ends alternating.

rare *a.* IUCN definition *appl.* species or larger taxa that have small populations, and, although not at present considered endangered or vulnerable, are at risk e.g. because of their highly restricted distribution within a habitat, or because they are thinly spread over a very large area, as some large carnivores. *see also* endan-

gered, rarity, vulnerable.

rarity *n.* categories of rarity of plant and animal species have been defined by the International Union for Conservation of Nature and Natural Resources (IUCN). *see* endangered, rare, vulnerable.

ras protein protein encoded by the *ras* genes. *see* ras.

rasorial *a.* adapted for scratching or scraping the ground, as fowls.

rastellus *n.* group of teeth on paturon of arachnid chelicera.

Rathke's pouch a diverticulum of ectoderm from the mouth cavity in vertebrate embryos that eventually forms the pituitary gland.

ratite *a.* having a sternum with no keel, *appl.* birds.

ratites *n.plu.* a group of flightless birds comprising the ostrich, emus, rheas, cassowaries and kiwis. They have rudimentary wings, a breast bone without a keel and fluffy feathers with no barbs.

rattle *n.* the series of horny joints at the end of a rattlesnake's tail, which produces the rattling sound.

Raunkiaer's life forms a classification of plants by the type of perennating organs they possess and the position of these organs in relation to soil or water level. *see* chamaephyte, cryptophyte, geophyte, helophyte, hemicryptophyte, hydrophyte, phanerophyte, therophyte.

ray *n.* (*bot.*) a band of parenchyma tissue penetrating from cortex towards centre of stem into secondary xylem and phloem, or from pith outwards to edge of vascular tissue (medullary ray); the stalk of a group of flowers in an umbel; (*zool.*) one of the bony or cartilaginous spines supporting fins in fishes; a division of a radially symmetrical animal, such as arm of starfish; one of the straight urine-carrying tubules passing from medulla through cortex of kidney.

ray-finned fishes the common name for the Actinopterygii *q.v.*

ray florets large outer florets of some inflorescences, esp. of some Compositae, which are often sterile or carpellate. *cf.* disc florets.

ray initial in vascular cambium of plant stem, a cell that gives rise to medullary rays.

re- prefix derived from L. *re*, again.

reaction centre the protein–chlorophyll complex in chloroplasts and other photosynthetic membranes in which the light-induced electron transfer across the membrane occurs. A pair of chlorophyll molecules near one side of the membrane acts as the electron donor and a quinone (in purple bacteria and photosystem II) or iron–sulphur centre (photosystem I and green sulphur bacteria) on the opposite side acts as acceptor.

reaction time (RT) the time interval between completion of presentation of a stimulus and the beginning of the response. *alt*. latent period.

reaction wood wood modified by bending of stem or branches, apparently trying to restore the original position, and including compression wood in conifers and tension wood in dicotyledons.

read abomasum *q.v.*

reading frame starting point on DNA or messenger RNA from which the base sequence is read off in triplet codons, the correct reading frame specifying the amino acid sequence of the corresponding protein.

readthrough the continuation of transcription or translation past a termination signal in DNA or mRNA respectively.

reafferent *a. appl.* stimulation that occurs as a result of an animal's bodily movements.

reaginic antibody IgE *q.v.*

realized niche the acutal place and role in an ecosystem an organism or species occupies, as opposed to its niche under ideal conditions.

reannealing reassociation *q.v.*

reassociation *n.* of DNA, the pairing of complementary single strands to form a double helix. *alt*. reannealing.

recalcitrant *a.* non-biodegradable, *appl*. organic, usually man-made compounds in the soil.

recapitulation theory the theory, due largely to the 19th century biologist E. Haeckel, that ontogeny tends to repeat phylogeny, and the similar theory due to von Bauer that the individual's life history reproduces certain stages in the evolution-ary history of the species. Now considered to be a misinterpretation of the similarities in early embryonic development of, e.g., vertebrate embryos due to their common evolutionary ancestry.

Recent *n.* geological epoch following Pleistocene and lasting until the present day. *alt*. Holocene.

receptacle *n.* of a flower, the point from which floral organs such as ovary, anthers, petals, etc. arise. *alt*. floral axis.

receptacular *a. pert.* a receptacle of any kind; largely composed of the receptacle, as certain fruits.

receptaculum *n.* a receptacle of any kind.

receptaculum ovarum an internal sac in which ova are collected in oligochaetes such as earthworms.

receptive field of a neurone, restricted area on sensory organ, such as eye, that when stimulated influences the signalling of that neurone.

receptive hyphae female sex organs of rust fungi.

receptor *n.* specialized tissue or cell sensitive to a specific stimulus; sensory organ; sensory nerve ending; any site on or in a cell to which a neurotransmitter, hormone, drug, metabolite, virus, etc. binds specifically, in some cases to activate a specific cellular response, in others to gain access to the cell. Such a site is composed of a specific protein, glycoprotein or polysaccharide.

α-receptor *see* adrenergic receptors.

β-receptor *see* adrenergic receptors.

αβ-receptor highly variable type of antigen receptor found on the majority of T lymphocytes. *see* T-cell receptor. *cf*. γδ receptor

γδ-receptor type of antigen receptor found on a subclass of T lymphocytes that appear early in an animal's development, and which is less variable than the αβ T-cell receptors that form the bulk of the mature T-cell repertoire.

receptor-mediated endocytosis specialized pathway in most animal cells for taking up certain substances (e.g. cholesterol) which first bind to specific receptors on the cell surface in the area of coated pits and are then internalized by endocytosis of receptor and ligand. Also used for internalizing used-up cell-surface receptors.

receptor potential graded local depolarization of sensory nerve terminal membrane as a result of stimulation.

receptor tyrosine kinase tyrosine kinase activity which is an integral part of many receptors, e.g. those for growth factors, and which is activated when ligand binds to the receptor.

recess *n.* a fossa, sinus, cleft, or hollow space.

recessive *a. appl.* alleles which are not reflected in the phenotype when present as one member of a heterozygous pair, only determining the phenotype when present in the homozygous state; *appl.* phenotypic characters expressed only in the homozygous state.

recessivity *n.* the property displayed by a recessive allele.

reciprocal *a. appl.* translocations involving an interchange of parts between nonhomologous chromosomes; *appl.* recombination involving an equal and balanced interchange of DNA between homologous chromosomes, *cf.* non-reciprocal recombination.

reciprocal altruism social behaviour in which altruistic acts by one individual towards another are reciprocated, rare in most animals, but seen e.g. in some monkeys and anthropoid apes which will band together to aid each other in disputes against other members of the troop.

reciprocal cross two crosses between the same pair of genotypes or phenotypes in which the sources of the gametes are reversed in one cross.

reciprocal feeding trophallaxis *q.v.*

reciprocal hyrids two hybrids, one descended from male of one species and female of the other, the other from female of first species and a male of the second, such as the mule and the hinny.

reciprocal recombination genetic recombination in which corresponding parts of the two DNAs undergoing recombination are exchanged. *cf.* integration, non-reciprocal recombination.

reclinate *a.* curved downwards from apex to base.

reclining *a.* leaning over; not perpendicular.

recognition sites in molecular genetics,

conserved regions within promoters and other DNA sequences presumed to be recognized by proteins that bind to DNA such as RNA polymerase and regulatory proteins.

recombinant *a. appl.* genotypes, phenotypes, gametes, cells or organisms produced as a result of natural genetic recombination; *appl.* DNA *see* recombinant DNA; *appl.* proteins, e.g. recombinant insulin, recombinant growth hormone: proteins produced from cells containing recombinant DNA directing their synthesis; *appl.* somatic mammalian cells, organisms, bacteria, yeasts, viruses, into which recombinant DNA has been introduced, or whose genomes have been modified *in vitro* by recombinant DNA techniques.

recombinant *n.* any chromosome, cell or organism which is the result of recombination, either natural or artificial.

recombinant DNA DNA produced by joining together *in vitro* genes from different sources or which has in some way been modified *in vitro* to introduce novel genetic information; DNA produced as a result of natural genetic recombination.

recombinant DNA techniques techniques of molecular genetics, including restriction enzyme analysis, DNA cloning, DNA hybridization, DNA sequencing, etc., that are used to produce and exploit recombinant DNAs *q.v.*

recombinant DNA technology the techniques used to produce recombinant DNAs and artificially genetically modified organisms, cells and microorganisms, and their applications in biotechnology.

recombinant fraction, frequency, or value proportion of recombinant gametes produced by an individual (with respect to two genetic loci), calculated as the number of recombinant gametes divided by the total number of gametes, used to calculate distance apart of two loci.

recombinant protein any protein produced from a recombinant DNA template.

recombinase *n.* enzyme activity that is responsible for the joining of gene segments during immunoglobulin and T-cell receptor gene rearrangement.

recombination *n.* the process in sexually

reproducing organisms by which DNA is exchanged between homologous chromosomes (reciprocal or balanced recombination) by chromosome pairing and crossing over at meiosis during gamete formation. This produces gametes containing genes from both parents on the same chromosome, *alt.* general or generalized recombination; any exchange between or integration of one DNA molecule into another, which may be reciprocal or non-reciprocal. *see also* site-specific recombination.

recombination nodules spherical or cylindrical structures seen lying across the synaptonemal complex in fungi and insects and which may be sites of recombination.

recombination-repair type of DNA repair and retrieval in which gaps left in one DNA strand opposite damaged sites after replication are filled in by recombination with a normal DNA strand, the consequent gap in this strand being filled in by DNA synthesis.

recombination signal sequence (RSS) sequence at which the recombination and joining of individual gene segments takes place during immunoglobulin and T-cell receptor gene rearrangement.

reconstitution *n.* the reassembly of isolated and artificially separated differentiated cells to form a new tissue or even individual, as can be done for sponges.

recrudescence *a.* state of breaking out into renewed growth; fresh growth from a ripe part.

recruitment *n.* activation of additional motor neurones, causing an increased reflex when stimulus of the same intensity is continued; entry of new individuals into a population by reproduction or immigration.

rectal *a. pert.* rectum; *appl.* gland: small vascular sac of unknown function near end of gut in some fishes.

rectal columns longitudinal folds of epithelium lining the rectum.

recti- prefix derived from L. *rectus*, straight.

rectigradation *n.* adaptive evolutionary tendency; a structure showing an adaptive trend or sequence in evolution.

rectinerved *a.* with veins straight or parallel, *appl.* leaves.

rectirostral *a.* straight-beaked.

rectiserial *a.* arranged in straight or vertical rows.

rectivenous *a.* with straight veins, *appl.* leaves.

recto- prefix indicating *pert.* or connecting with the rectum.

rectogenital *a. pert.* rectum and genital organs.

rectovesical *a. pert.* rectum and bladder.

recto-uterine *a. appl.* posterior ligaments of uterus.

rectrices *n. plu.* stiff tail feathers of a bird, used in steering. *sing.* rectrix. *a.* rectricial.

rectum *n.* the posterior part of the alimentary canal, leading to the anus, in which water and inorganic ions are absorbed from the gut contents before they are passed out as faeces. *a.* rectal.

rectus *n.* a name for a rectilinear muscle, as rectus femoris.

recurrent *a.* returning or reascending towards origin; reappearing at intervals.

recurrent sensibility sensibility shown by motor roots of spinal cord due to sensory fibres of sensory roots.

recurved *a.* bent backwards.

recurvirostral *a.* with beak bent upwards.

recutite *a.* apparently devoid of epidermis.

red algae common name for the Rhodophyta *q.v.*

red blood cell, red blood corpuscle erythrocyte *q.v.*

red body, gland or spot rete mirabile *q.v.*

red cell ghosts empty red cell membranes obtained by haemolysis and used in investigations of cell membranes.

red drop the phenomenon that the quantum yield of photosynthesis falls sharply when the wavelength of light is greater than 680 nm. This is due to the fact that only photosystem I can be driven by light of longer wavelength.

red fibres *see* red muscle.

red light light of wavelength 620–680 nm.

red muscle type of muscle in fish, containing myoglobin, and with chiefly aerobic respiration, involved in slow swimming. *cf.* white muscle.

red muscle fibres skeletal muscle fibres rich in myoglobin, which give the slow-twitch response and are capable of sustained activity. *cf.* white muscle fibres.

red nucleus area in midbrain involved in

relaying signals from brain to motor neurones in spinal cord.

red pulp erythroid tissue of spleen.

Red Queen hypothesis idea that each evolutionary advance by one species is detrimental to other species so that all species must evolve as fast as possible simply to survive.

red tide a bloom of red dinoflagellates (e.g. *Gonyaulax polyedra*) which colours the sea red. Toxins contained in some of these microorganisms are concentrated in the shellfish that feed on them and can cause sometimes fatal poisoning in humans who eat the shellfish.

redia *n*. a larval stage in the snail host of some endoparasitic flukes. It is produced asexually from the sporocyst, has a mouth and gut and reproduces asexually to produce a further generation of rediae or cercariae.

redifferentiation *n*. in a cell or tissue, a reversal of the differentiated state and then differentiation into another type. *see also* transdifferentiation.

redox *a. pert*. mutual oxidation and reduction.

redox potential (E_0') an electrochemical measure (in volts) which for any substance that can exist in an oxidized or reduced form gives its affinity for electrons relative to hydrogen in standard conditions (the redox potential of the $H+:H_2$ couple is defined as 0 V). A negative redox potential indicates a strong reducing agent, and a positive redox potential a strong oxidizing agent. *alt*. oxidation reduction potential.

reduced *a*. in an anatomical context *appl*. structures that are smaller than in ancestral forms.

reducer organism decomposer *q.v.*

reducing power a general term for the presence in cells of compounds such as $NADH_2$ and NADPH, which are hydrogen and electron donors in metabolic reduction reactions.

reducing sugar sugar with a free aldehyde or ketone group that can act as a reducing agent in solution, as most monosaccharides and disaccharides.

reductase *n*. any enzyme that catalyses reduction of a compound, classified amongst the oxidoreductases in EC class 1, used esp. where hydrogen transfer from the donor is not readily demonstrable.

reduction *n*. the halving of the number of chromosomes that occurs at meiosis; structural and functional development in a species which is less complex than that of its ancestors, *cf*. amplification; *(biochem.)* decreasing the oxygen content or increasing the proportion of hydrogen in a molecule, or adding an electron to an atom or ion, *cf*. oxidation.

reduction division 1st meiotic division, sometimes used for meiosis as a whole.

reductionism *n*. in biology, the idea that complex phenomena such as embryonic development, inheritance, mental processes, etc. are in principle completely explicable in terms of the basic principles of biochemistry, molecular genetics, etc. The opposing anti-reductionist view maintains that at higher levels of organization novel properties emerge that have no direct correspondence to lower-level processes and therefore cannot be wholly explained in their terms.

reductive pentose phosphate pathway Calvin cycle *q.v.*

reduplicate *a. appl*. the arrangement of petals, etc. in flower bud, in which margins of bud sepals or petals turn outwards at points of contact.

reduviid *a. appl*. eggs of certain insects, protected by micropyle apparatus with porches.

reed abomasum *q.v.*

refection *n*. reingestion of incompletely digested food by some animals, as eating faecal pellets, or in rumination.

referred *a. appl*. sensation in a part of the body remote from that to which the stimulus was applied.

reflected *a*. turned or folded back on itself.

reflector layer layer of cells on inner surface of a light-emitting organ, as in fireflies; silvery reflecting plates in skin of fishes above the argenteum.

reflex *a*. involuntary, *appl*. reaction to stimulus; turned or folded back on itself, *alt*. reflected; *n*. involuntary movement (reflex action) or other response elicited by a stimulus at the periphery which is transmitted to the central nervous system and reflected back to a peripheral effector organ.

reflex arc unit of function in nervous system, consisting of a sensory receptor, an effector organ (muscle or gland) and a pathway of nerve cells conveying the sensory information to the central nervous system and carrying motor or other commands from the central nervous system to the effector organ.

reflexed *a.* turned or curved back on itself.

refracted *a.* bent backwards at an acute angle.

refractory *a.* unresponsive; *appl.* period after excitation of neurone during which repetition of stimulus fails to induce a response.

refuge, refugium *n.* an area that has remained unaffected by environmental changes to the surrounding area, such as a mountain area that was not covered with ice during the Pleistocene, and in which, therefore, the previous flora and fauna has survived.

regeneration *n.* renewal of a portion of body which has been injured or lost; reconstitution of a compound after dissociation, as e.g. of rhodopsin.

regma *n.* dry dehiscent fruit whose valves open by elastic movement, as in *Geranium* species.

regosols *n.plu.* soils which are developed on fairly deep unconsolidated parent material such as dune sands or volcanic ash.

regression *n.* reversal in the apparent direction of evolutionary change, with simpler forms appearing; the replacement of a climax ecosystem with a previous stage in the succession, as e.g. the replacement of forest by grassland after felling.

regular *a. appl.* any structure, such as flower, organism, showing radial symmetry. *alt.* actinomorphic *q.v.*

regulation *n.* in embryogenesis, the ability to compensate for the addition or removal of cells.

regulative *appl.* embryos which can compensate for the removal or addition of cells. In such embryos the fate of cells at the beginning of any developmental process is not yet irrevocably determined and they can be influenced by developmental signals to follow another course if required.

regulator *n.* an animal that maintains its internal environment in a state that is largely independent of external conditions.

regulator gene in bacteria, a gene that codes for a protein involved in the control of gene expression, e.g. repressor or apoinducer proteins.

regulatory genes genes that direct the production of proteins that regulate the activity of other genes, or which represent control sites in DNA at which gene expression is regulated.

regulatory protein any protein that acts as a cofactor in enzyme reactions. *see also* gene regulatory protein.

Reichert's membrane connective tissue membrane (basement membrane) covering mural trophectoderm in mouse embryo.

reinforcement *n.* an event that alters an animal's response to a stimulus, positive reinforcement being reward and increasing its response, negative reinforcement being disagreeable or painful and suppressing its response.

reinforcing selection operation of selection pressures on two or more levels of organization, such as population, family and individual, in such a way that certain genes are favoured at all levels and their spread through the population is accelerated.

Reissner's membrane the vestibular membrane of cochlea, stretching from inner to outer wall of cochlea above basilar membrane and separating the scala vestibuli from the scala media.

rejungant *a.* coming together again, *appl.* related but hitherto separate taxa when in the course of time their ranges come to rejoin.

rejuvescence *n.* a renewal of youth; regrowth from injured or old parts.

relationship, coefficient of coefficient of relationship *q.v.*

relative molecular mass (M_r) the ratio of the mass of one molecule of a substance to one-twelfth the mass of an atom of ^{12}C. It is a ratio and therefore dimensionless. *alt.* molecular weight. *see also* molecular mass.

relaxation time the period during which excitation subsides after removal of a stimulus.

relaxed DNA a circular double-helical DNA

molecule without any superhelical turns.

relaxed mutants mutant bacteria that do not show the stringent response in conditions of amino acid starvation.

relaxin *n.* hormone produced by the corpus luteum which produces relaxation of the pelvic ligaments during pregnancy.

relay cell interneurone *q.v.*

release channel calcium channel in sarcoplasmic reticulum which opens on muscle stimulation and releases calcium into the muscle cell.

release factor any of several proteins (e.g. RF 1, RF 2 in *E. coli*) which recognize termination codons in mRNA during translation and are involved in releasing the completed polypeptide chain from the ribosome.

releaser *n.* a stimulus or group of stimuli that activates an inborn tendency or pattern of behaviour, as of species-specific behaviour.

releasing hormone any of a number of small peptides produced by the hypothalamus and which act on the pituitary to induce the release of hormones such as growth hormone, thyroid-stimulating hormone, follicle-stimulating hormone and luteinizing hormone.

relict *a.* not now functional, but originally adaptive, *appl.* structures; surviving in an area isolated from the main area of distribution owing to intervention of environmental events such as glaciation, *appl.* species, populations. *alt.* relic.

rem roentgen equivalent man, the unit dose of ionizing radiation that gives the same biological effect as that due to 1 roentgen of X-rays. The rem has been replaced by the sievert, with 1 sievert = 100 rem.

REM sleep rapid-eye movement sleep, or paradoxical sleep, during which people are dreaming.

remex *sing.* of remiges *q.v.*

remiges *n.plu.* primary wing feathers of bird.

remiped *a.* having feet adapted for rowing motion.

remotor *n.* a retractor muscle.

renal *a. pert.* kidneys.

renal columns cortical tissue between the pyramids of the medulla of kidneys.

renal corpuscle Malpighian corpuscle *q.v.*

renal portal *appl.* a circulation system in which some blood returning to heart passes through kidneys.

renaturation *n.* the return of a denatured macromolecule such as a protein or nucleic acid to its original configuration.

rendzina *n.* any of a group of rich, dark, greyish-brown, limy soils of humid or subhumid grasslands, having a brown upper layer and yellowish-grey lower layers.

reniculus *n.* lobe of kidney comprising papillae, pyramid and surrounding part of cortex.

reniform *a.* kidney-shaped.

renin *n.* proteinase enzyme secreted by kidney and which converts angiotensinogen in the blood to angiotensin I. EC 3.4.99.19.

Renner complex a group of chromosomes that passes from generation to generation as a unit, as in the evening primrose, *Oenothera.*

rennin chymosin *q.v.*

renopericardial *a. appl.* a ciliated canal connecting kidney and pericardium in higher molluscs.

Renshaw cell type of interneurone in spinal cord, involved in regulation of motor neurones supplying a single muscle.

Reoviridae, reoviruses *n., n.plu.* family of icosahedral, non-enveloped, double-stranded RNA viruses of animals and plants, including human rotaviruses that cause diarrhoea in children.

rep protein enzyme from *E. coli* responsible for unwinding the double helix during DNA replication, using ATP as energy source.

repand *a.* with undulated margin, *appl.* leaf; wrinkled, *appl.* bacterial colony.

repandodentate *a.* varying between undulate and toothed, *appl.* margin of leaf.

repandous *a.* curved convexly.

reparative *a.* restorative, *appl.* buds that develop after injury to a leaf.

repeat *n.* duplication or further serial repetition of a stretch of DNA due to unequal crossing-over or other molecular events. *see also* inverted repeats, tandem repeats.

repent *a.* creeping along the ground.

repertoire *n.* in immunology, all the different antigen receptor specificities that can be produced by a single individual.

repetitive DNA repeated DNA sequences

present in most eukaryotes, which may be from tens to thousands of bases long and be present in up to hundreds of thousands of copies. Most repetitive DNA seems to have no function and does not encode protein, although some is transcribed. *see also* selfish DNA, satellite DNA. *cf.* single copy DNA.

replacement-level fertility birth rate that keeps a population constant, exactly replacing deaths by births.

replacement name scientific name adopted as substitute for one found invalid under the rule of the International Codes of Nomenclature.

repletes *n. plu.* workers with distensible crops for storing and regurgitating honeydew and nectar, and constituting a physiological caste of honey ants.

replica *n.* in electron microscopy a thin shell of electron-dense metal atoms following the surface contours of the specimen, which is made by vaporizing metal atoms onto the surface, and which shows up details of the surface structure.

replica plating production of an exact replica of a plate containing bacterial colonies by transfer of bacteria to a new plate by "blotting" with a velvet pad, filter paper, etc. which retains the exact positions of the colonies relative to each other. Destructive identification procedures can be carried out on the replica plate leaving the master plate as an untouched source of bacteria, DNA, etc.

replicase *n.* general term for the enzyme activity involved in the replication of nucleic acids, as in DNA synthesis and the replication of RNA in some viruses.

replicate *a.* doubled back over itself; *v.* to duplicate; to copy itself, as DNA.

replicatile *a. appl.* wings folded back on themselves when at rest.

replication *n.* duplication, as of DNA, by making a new copy of an existing molecule; duplication of organelles such as mitochondria, chloroplasts and nuclei, and of cells.

replication bubble, replication eye structure formed in replicating DNA when replication proceeds in both directions from an origin of replication.

replication-defective virus a virus which

has lost the ability to multiply as the result of e.g. the replacement of part of its genome with host cell genes as in the case of some RNA tumour viruses.

replication fork site of simultaneous unwinding of double-stranded parental DNA and synthesis of new DNA, seen in electron micrographs of replicating DNA as a Y-shaped structure.

replication origin sequence at which DNA replication is started.

replicative form (RF) double-stranded DNA produced by certain single-stranded DNA viruses after infection, and which directs mRNA synthesis.

replicative recombination type of genetic recombination seen in transposition events, in which a DNA sequence is copied and then inserted elsewhere in the genome.

replicative transposition transposition of a DNA sequence to a new site on the chromosome by making a copy of it and integrating the copy into the new site, the old copy remaining at the original site.

replicon *n.* a unit length of DNA that replicates sequentially and which contains an origin of replication, such as a bacterial plasmid or bacterial chromosome, or a region of eukaryotic chromatin.

replisome *n.* multiprotein assembly containing the various activities needed for DNA replication, formed in association with DNA.

replum *n.* a wall, not the carpellary wall, formed from ingrowths from the placenta and dividing a fruit into sections.

reporter gene a marker gene inserted in a recombinant DNA vector, etc. whose activity can be easily tracked and the distribution of the introduced DNA in transgenic animals, etc. assessed.

repressed *a. appl.* generally to any gene that is not being expressed as a result of interaction with a repressor or other regulatory mechanism.

repression *n.* the specific shut-down of gene expression and consequent inhibition of protein synthesis that occurs when, e.g., an enzyme substrate is not available, or a nutrient being synthesized by a microorganism becomes available in the medium; more generally refers to the blocking of transcription of a gene by a repressor protein binding to a control region in DNA, or to the inhibition of translation of mRNA by a repressor protein binding to a specific site on the mRNA.

repressor *n.* in bacteria, a protein which shuts down (represses) transcription of an operon by binding to the operator region on DNA and which is produced by a regulatory gene specific for the operon; a protein which inhibits translation of mRNA by binding to mRNA.

reproduction *n.* the formation of new individuals by sexual or non-sexual means.

reproduction curve a plot giving the relationship between the number of individuals at a particular stage in one generation and the numbers at that stage in a previous generation.

reproductive isolation the inability of two populations to interbreed because they are geographically isolated, or isolated from each other by differences of behaviour, mating time (or in plants maturation times of male and female sex organs), or genital morphology. This is a phase in the development of new species.

reproductive value the relative number of female offspring remaining to be born to each female of age *x,* symbolized by v_x.

reptant *a.* creeping, *appl.* a polyzoan colony with zooecia lying on substrate; *appl.* gastropod molluscs.

Reptilia, reptiles *n., n.plu.* class of amniote, air-breathing, poikilothermic ("cold-blooded") tetrapod vertebrates, mostly terrestrial, having dry horny skin with scales, plates or scutes, functional lungs throughout life, one occipital condyle and a four-chambered heart. Most reptiles lay eggs with a leathery shell but some are ovoviviparous. Reptiles include the tortoises and turtles, the tuatara, lizards and snakes, crocodiles, and many extinct forms, such as dinosaurs, pterosaurs, etc.

reptiloid *a.* having the characteristics of a reptile.

repugnatorial *a.* defensive or offensive, *appl.* glands and other structures.

repulsion *n.* the case when the diploid genotype at two linked loci A and B is *Ab/aB,* where *A* is dominant over *a* and *B* is dominant over *b.*

reserve cellulose cellulose found in plant storage tissue and subsequently used for nutrition after germination.

reservoir *n.* a non-contractile space discharging into gullet of some Mastigophora; host which carries a pathogen but is unharmed and acts as a source of infection to others.

residual bodies anucleate portions of spermatids left after separation of spermatozoa; secondary lysosomes in cytoplasm of eukaryotic cell which contain the indigestible remains of material taken in by phagocytosis.

residual volume volume of air remaining in lungs after strongest possible breathing out.

residue *n.* in chemistry and biochemistry, a compound such as a monosaccharide, nucleotide or amino acid when it is part of a larger molecule.

resilience *n.* ability of a living system to restore itself to its original condition after being disturbed.

resilifer, resiliophore *n.* the projection of shell carrying the flexible hinge in bivalve shells.

resilin *n.* protein present as cross-linked aggregates with rubber-like properties in some insect wing muscles and attachments of wing to thorax, and which is involved in taking up the kinetic energy of the wingbeat at the end of each stroke.

resilium *n.* the flexible horny hinge of a bivalve shell.

resin *n.* any of various high molecular weight substances, including resin acids, esters and terpenes, which are found in mixtures in plants and often exuded from wounds where they may protect against insect and fungal attack, as they harden to glassy amorphous solids.

resin canals ducts in bark, wood, etc., esp. in conifers, lined with glandular epithelium secreting essential oils, e.g. terpenes, which form resin oxidation products.

resiniferous *a.* producing resin.

resinous *a. appl.* a class of odours.

resistance factors R-factors *q.v.*

resistance transfer factor (RTF) that part of a transmissible drug-resistance plasmid that mediates conjugation and plasmid transfer to another bacterium.

resolution *n.* the regeneration of normal duplex DNA molecules from structures formed during e.g. recombination; the regeneration of two replicons from a cointegrate by recombination between two transposons; in microscopy, the size of object that can be viewed clearly, *alt.* resolving power.

resolvase *n.* enzyme activity involved in site-specific recombination between 2 transposons present in a cointegrate, resulting in regeneration of 2 replicons both containing copies of the transposon sequence.

resource *n.* anything provided by the environment to satisfy the requirements of a living organism, e.g. food, living space.

resource-holding potential the ability of an animal to gain and maintain possession of essential resources by fighting.

resource partitioning the division of scarce resources in an ecosystem so that species with similar requirements use the same resources at different times, in different ways, or in different places.

respiration *n.* any or all of the processes used by organisms to generate metabolically usable energy, chiefly in the form of ATP, from the oxidative breakdown of foodstuffs. Refers to processes ranging from the exchange of oxygen and carbon dioxide between organism and environment in aerobic respiration, to the biochemical processes generating ATP at the cellular level. *a.* respiratory, *pert.* or involved in respiration. *see also* aerobic respiration, anaerobic respiration, glycolysis, oxidative phosphorylation.

respiratory centres nuclei in medulla oblongata involved in the involuntary control of breathing.

respiratory chain series of electron carriers (such as quinones, flavoproteins, cytochromes etc.) located in mitochondria and along which electrons derived from respiratory metabolites are transferred in a series of redox reactions resulting in the reduction of oxygen to water and the formation of ATP and NAD. *alt.* electron transport chain, electron transfer chain. *see also* oxidative phosphorylation.

respiratory control regulation of the rate of oxidative phosphorylation by the cellular ADP level, low ADP inhibiting, high ADP stimulating oxygen consumption and ATP synthesis.

respiratory heart a name given to the auricle and ventricle of right side of heart where there is no direct communication between right and left sides. *cf.* systemic heart.

respiratory index the amount of carbon dioxide produced per unit of dry weight per hour.

respiratory movements any movements connected with the supply of oxygen to respiratory surfaces and the removal of carbon dioxide, such as the movements of the thorax and diaphragm in mammals.

respiratory pigments pigments such as haemoglobin and other haem proteins which form an association with oxygen and carry it from the respiratory surfaces to tissue cells; pigments concerned with cellular respiration, components of the respiratory chain such as cytochromes.

respiratory quotient (RQ) ratio of volume of carbon dioxide produced to volume of oxygen used in respiration.

respiratory sac a backward extension of the suprabranchial cavity, in certain air-breathing bony fishes.

respiratory substrate any substance that can be broken down by living organisms during respiration to yield energy.

respiratory surface the surface at which gas exchange occurs between the environment and the body, such as gill lamellae, alveoli of lungs, etc.

respiratory trees in echinoderms, a respiratory system consisting of a series of tubules arising just within the anus into which water can be drawn and expelled.

respondent behaviour animal behaviour performed in response to an obvious stimulus.

response *n.* the activity of a cell or organism in terms of movement, hormone or other secretion, enzyme production, changes in gene expression, etc. as a result of a stimulus; the behaviour of an organism as a result of fluctuations in the environment.

response latency the time interval between a stimulus and the response.

restibrachium *n.* the restiform body or inferior peduncle of the cerebellum.

restiform *a.* having the appearance of a

rope, *appl.* two bodies of nerve fibres on medulla oblongata, the inferior cerebellar peduncles.

resting cell or nucleus one that is not undergoing mitosis or meiosis.

resting metabolic rate (RMR) the metabolic rate, esp. of an animal, measured at rest. *cf.* basal metabolic rate.

resting potential the potential difference across a nerve or muscle cell membrane when it is not being stimulated, and which is usually about -70 millivolts (mV) in a nerve cell. *cf.* action potential.

Restionales *n.* order of xeromorphic tufted or climbing monocots including the family Restionaceae and others.

restitution *n.* the reformation of a tissue or body by union of separated parts; the reunion of breaks in chromosomes.

restriction *n.* in bacteria, the breakdown of incoming DNA of another bacterial strain, or phage or plasmid DNA grown in another strain, by particular types of endonucleases called restriction enzymes, *see also* modification; (*imm.*) MHC restriction *q.v.*

restriction analysis the determination of the identity, internal structure, etc. of a gene or other piece of DNA by restriction mapping and other techniques depending on the use of restriction enzymes to cut DNA into identifiable pieces.

restriction endonuclease any of a large group of endonucleases produced by microorganisms, which recognize short palindromic base sequences in DNA, cutting the double helix at a particular point within the sequence, each different restriction endonuclease recognizing a different DNA sequence. Now used widely in genetic engineering, DNA cloning and gene mapping. *alt.* restriction enzyme, site-specific endonuclease.

restriction enzyme restriction endonuclease *q.v.*

restriction fragment a DNA fragment generated by treatment of DNA with a restriction enzyme or combination of restriction enzymes.

restriction fragment length polymorphism (RFLP) the presence or absence of a cutting site for a particular restriction enzyme in DNA of different individuals, which can be detected by comparison of the DNA fragment lengths generated by that enzyme. It can be used, e.g., to detect defective alleles of a gene, as in antenatal diagnosis of genetic disease.

restriction mapping the procedure of characterizing a region of DNA by the number and relative positions of the sites at which particular restriction enzymes cut the DNA. Treatment of DNA with different combinations of enzymes creates a series of DNA fragments of different lengths, which can be separated and sized by gel electrophoresis and arranged to provide a "restriction map" of the region with distances between each site defined in base pairs (bp) or kilobase pairs (kb).

restriction-modification system in bacteria, a dual function enzymic system which both destroys incoming "foreign" DNA and makes chemical modifications to the bacterium's own DNA which protect it against degradation. Some systems (type I) consist of a single multisubunit enzyme containing both nuclease and methylase (modification) activity, others (type II) employ site-specific restriction endonucleases *(q.v.)* and separate modifying enzymes.

restriction point of eukaryotic cell cycle, the point of no return, occurring in the middle of G_1, and after which the cell is committed to completing its cycle.

restriction site a site of defined sequence in DNA at which a particular restriction enzyme cuts.

restriction-site polymorphism variation in the presence of a specific restriction site between individuals in a population, which can sometimes be used as a convenient marker for an important genetic trait if the restriction site is sufficiently closely linked to it. *alt.* restriction fragment length polymorphism.

resupinate *a.* so twisted that parts are upside down.

resupination *n.* inversion.

rete *n.* mesh of tissue; network of interlaced vessels, nerves or fibres.

rete Malpighii Malpighian layer *q.v.*

rete mirabile small dense network of mainly arterial blood vessels in various organs of some vertebrates.

rete mucosum Malpighian layer *q.v.*

reteform, retiform *a.* in form of a network.

retention signal salvage sequences *q.v.*

retial *a. pert.* a rete.

retiary *a.* making or having a net-like structure; constructing a web; net-like.

reticular *a.* like a meshwork, netted; *pert.* a reticulum.

reticular cells cells of bone marrow from which white blood cells derive.

reticular formation area in brainstem involved in maintaining consciousness and level of arousal. Many sensory pathways from the body have branches that terminate in the reticular formation.

reticulate *a.* like network; *appl.* venation of leaf or insect wing; *appl.* a pattern of lignin deposition in plant cell walls.

reticulocyte *n.* a precursor cell from which mature red blood cell develops, having a reticular appearance when stained.

reticuloendothelial system (RES) the fine fibrillar meshwork of phagocytic cells supported on connective tissue that extends throughout lymphoid organs such as spleen and lymph nodes and also in other organs such as liver and kidneys and which is involved in the uptake and clearance of foreign particulate matter from the blood. Foreign antigens taken up by cells of the reticuloendothelial system in lymphoid organs encounter the T cells and B cells of the immune system which then mount a specific immune response.

reticulopodia *n.plu.* anastomosing thread-like pseudopodia, as of foraminiferans.

reticulose *a.* formed like a network.

reticulospinal *a. appl.* tracts of nerve fibres connecting reticular formation in brain with the spinal cord.

reticulum *n.* network, esp. the framework of fibrous tissue in many organs; in ruminants, the second chamber of the stomach, in which water is stored; (*bot.*) cross-fibres about base of petioles in palms.

retiform *a.* in the form of a network.

retina *n.* the inner light-sensitive layer of the eye, in vertebrates comprising a double layer of epithelium lining the back of the eyeball, composed of a neural retina containing the photoreceptor cells (rods and cones) and neurones, and an inner pigmented epithelium containing melanin granules.

retinaculum *n.* (*bot.*) a small glandular mass to which orchid pollinium adheres on dehiscence; (*zool.*) a fibrous band that holds parts together; a structure linking together fore- and hindwings of some insects. *plu.* retinacula.

retinaculum tendinum ligament around wrist or ankle.

retinal *n.* aldehyde of retinol, part of the visual pigment rhodopsin, from which it is split off by the action of light, *alt.* (formerly) retinene; *a. pert.* the retina.

retinella *n.* neurofibrillary network of phaeosome.

retinene retinal *q.v.*

retinerved *a.* having a network of veins, *appl.* leaf or insect wing.

retinitis pigmentosa genetically determined disorder characterized by night blindness, constriction of the visual fields, and changes in the fundus of the retina.

retinoblast *n.* retinal epithelial cells that give rise to neuroblasts and neuroglial precursor cells.

retinoblastoma *n.* a rare tumour of retinal cells, usually occurring only in very young children. Inheritable familial retinoblastoma is due to a defective allele of the *Rb* gene, a "tumour suppressor gene" (*q.v.*).

retinoic acid vitamin A derivative, used experimentally to induce differentiation in tumour cells. It may also be a natural morphogen in vertebrate development. *see* Appendix 1 (43) for chemical structure.

retinoid *n.* analogue or metabolite of retinol (vitamin A). Retinoids have a role in embryonic development, as shown by the deleterious effects of vitamin A deficiency, the teratogenic effects of very large doses of vitamin A, and the influence of retinoid treatment on various developmental processes, e.g. pattern formation in the vertebrate limb.

retinoid receptor the receptors for retinoic acid and other retinoids, which belong to the steroid receptor superfamily of intracellular receptors.

retinol vitamin A_1, *see* vitamin A.

retinula *n.* group of elongated sensory photoreceptor cells, the innermost element of ommatidium of a compound eye.

retisolution *n.* dissolution of the Golgi apparatus in a cell.

retispersion *n.* peripheral distribution of Golgi bodies.

retractile *a. appl.* a part of an organ that may be drawn inwards, as feelers, claws, etc.

retraction fibres fine cellular processes adhering to the substratum, which extend from the rear of a cell moving over a substrate as it moves forward.

retractor *n.* a muscle which on contraction withdraws that part attached to it, bringing it towards the body.

retrahens *n.* a muscle which draws a part backwards.

retraherence *n.* escape behaviour by which animals survive in conditions of extreme heat or cold, e.g. burrowing.

retral *a.* backward; posterior.

retro- prefix derived from L. *retro,* backwards.

retroarcuate *a.* curving backwards.

retrobulbar *a.* posterior to eyeball; on dorsal side of medulla oblongata.

retrocerebral *a.* situated behind the cerebral ganglion, in invertebrates.

retrocurved recurved *q.v.*

retrofract *a.* bent backwards at an angle.

retrograde *a. appl.* transport or movement of material in axons of neurones towards cell body.

retrogressive *n. appl.* evolutionary trends towards more primitive rather than more complex forms.

retrolingual *a.* behind the tongue, *appl.* a gland.

retromandibular *a. appl.* posterior or temperomaxillary vein.

retroperitoneal *a.* behind peritoneum; *appl.* space between peritoneum and spinal column.

retropharyngeal *a.* behind the pharynx, *appl.* a space, lymph glands.

retroposon retrotransposon *q.v.*

retropubic *a. appl.* a pad or mass of fatty tissue behind pubic symphysis.

retrorse *a.* turned or directed backward.

retroserrulate *a.* with small backward-pointing teeth.

retrosiphonate *a.* with septal necks directed backwards.

retrotransposition *n.* transposition by means of reverse transcription of an RNA and subsequent insertion of the new DNA copies into the genome.

retrotransposon *n.* a transposon that has arisen and is transposed by retrotransposition. *alt.* retroposon.

retro-uterine *a.* behind the uterus.

retroverse retrorse *q.v.*

retroversion *n.* the state of being reversed or turned backwards.

retroviral *a. pert.* or produced by retroviruses; *appl.* vectors, vectors for genetic engineering derived from retroviruses.

Retroviridae, retroviruses *n., n.plu.* family of enveloped, single-stranded RNA viruses including the RNA tumour viruses and HIV (human immunodeficiency virus) as well as many apparently harmless, nononcogenic viruses. They have a unique life history, copying their RNA into DNA by means of the viral enzyme reverse transcriptase. The DNA then enters a host cell chromosome, where it may continue to direct the production of virus particles, or may remain quiescent for many cell generations. The integrated DNA (the provirus) is passed on to all the cell's progeny. Vertebrates appear to carry a number of so-called endogenous proviruses permanently in their genomes without any ill effects.

retuse *a.* obtuse with a broad shallow notch in middle, *appl.* leaves, molluscan shells.

Retzius, striae of *see* striae of Retzius.

revehent *a.* in renal portal system, *appl.* vessels carrying blood back from excretory organs.

reverse mutation back mutation *q.v.*

reverse transcriptase a DNA polymerase found in retroviruses (a class of RNA viruses including the RNA tumour viruses) which synthesizes DNA on a viral RNA template. *alt.* RNA-directed DNA polymerase.

reverse transcription the synthesis of DNA on an RNA template, catalysed by the enzyme reverse transcriptase.

reversed *a.* inverted, *appl.* a spiral shell whose turns are left-handed.

reversion back mutation *q.v.*

revertant *n.* organism or cell in which a reversion has occurred.

revolute *a.* rolled backwards from margin upon undersurface, as some leaves.

reward positive reinforcement *q.v.*

R-factor R plasmid *q.v.*

R form the relaxed form of an allosteric protein, the form having a greater affinity for the substrate.

rhabd-, rhabdo- prefixes derived from Gk. *rhabdos*, a rod.

rhabdacanth *n.* in certain roses, a compound thorn, consisting of small rod-like trabeculae wrapped around with lamellar tissue.

rhabdi *plu.* of rhabdus *q.v.*

rhabdite *n.* short, rod-like body in epidermal cell of turbellarians, which is discharged if the worm is injured and swells up to form a gelatinous covering.

rhabditiform *a. appl.* larvae of nematodes with short straight oesophagus, with double bulb.

rhabditis *n.* larva of certain nematodes.

rhabdocoel *n.* member of the order Rhabdocoela, small turbellarian flatworms whose intestine is simple and sac-like.

rhabdoid *a.* rod-shaped.

rhabdolith *n.* calcareous rod found in some protozoans, strengthening the wall.

rhabdome *n.* pigmented rod formed by microvilli of retinula cells of ommatidium of a compound eye.

rhabdomere *n.* individual element of rhabdome of retinula in ommatidium of a compound eye.

rhabdomyosarcoma *n.* a tumour of muscle.

rhabdopod *n.* element of clasper in some male insects.

rhabdosphere *n.* aggregated rhabdoliths found in deep sea calcareous oozes.

Rhabdoviridae, rhabdoviruses *n., n.plu.* family of bullet-shaped, enveloped, single-stranded RNA viruses including rabies and vesicular stomatitis virus.

rhabdus *n.* a rod-like spicule; the stalk of certain fungi. *plu.* rhabdi.

Rhaetic, Rhaetian *a. appl.* fossils found in marls, shales, and limestone between Trias and Lias.

rhagon *n.* a bun-shaped type of sponge with an apical osculum and large gastral cavity.

Rhamnales *n.* order of trees, shrubs or woody climbers and including Rhamnaceae (buckthorn) and Vitaceae (grape).

rhamnose *n.* 6-carbon (hexose) sugar found in the lipopolysaccharide outer membrane of some Gram-negative bacteria, and in plant cell wall polysaccharides.

rhamphoid *a.* beak-shaped.

rhamphotheca *n.* the horny sheath of a bird's beak.

Rheiformes *n.* an order of flightless birds, including the rheas.

rheobasis *n.* the minimal electrical stimulus that will produce a response.

rheophile, rheophilic *a.* preferring to live in running water. *n.* rheophily.

rheophyte *n.* plant that lives in running water.

rheoplankton *n.* the plankton of running waters.

rheoreceptors *n.plu.* cutaneous sense organs of fishes and some amphibians, receiving stimulus of water current, as lateral line organs, ampullae of Lorenzini etc.

rheotaxis *n.* a taxis in response to the stimulus of a current, usually a water current.

rheotropism *n.* a growth curvature in response to a water or air current. *a.* rheotropic.

rhesus factor (Rh factor) blood group antigen found on red cells of rhesus monkey and a proportion of the human population, determined by a dominant gene. It is of medical importance because a Rh-negative mother carrying a Rh-positive foetus will produce antibodies against the foetal Rh antigen which in subsequent pregnancies will lead to haemolytic disease of the newborn in any Rh-positive offspring (routine preventive treatment is now given to mothers at risk).

rheumatoid factor IgM antibodies found in the serum of individuals with rheumatoid arthritis. They react with IgG antibodies to form immune complexes deposited in joints, etc.

rhexigenous *a.* resulting from rupture or tearing.

rhexilysis *n.* the separation of parts, or production of openings or cavities, by rupture of tissues.

rhexis *n.* fragmentation of chromosomes, caused by physical or chemical agents.

rhigosis *n.* sensation of cold.

rhin-, rhino- prefixes derived from Gk. *rhis*, nose.

rhinarium *n.* the muzzle or external nasal area of mammals; nostril area; part of nasus of some insects.

rhinencephalon *n.* the part of the forebrain forming most of the hemispheres in fishes, amphibians and reptiles, and comprising in humans the olfactory lobe, uncus, the supracallosal, subcallosal and dentate gyri, fornix and hippocampus.

rhinion *n.* the most prominent point at which the nasal bones touch.

rhinocaul *n.* narrowed portion of brain which bears the olfactory lobe.

rhinocoel *n.* cavity in olfactory lobe of brain.

rhinopharynx nasopharynx *q.v.*

rhinophore *n.* in some molluscs, an organ of sensory epithelium, sometimes borne in a pit, usually found on the tentacles and thought to have an olfactory function.

rhinotheca *n.* the sheath of upper jaw of a bird.

rhinoviruses *n. plu.* a numerous group of RNA viruses of the family Picornaviridae, the cause of the common cold and similar minor respiratory ailments in humans.

rhipidate *a.* fan-shaped.

rhipidistians *n.plu.* group of extinct crossopterygian fish existing from Devonian to Permian times, and believed to include the ancestors of land vertebrates.

rhipidium *n.* a fan-shaped cymose inflorescence; a fan-shaped colony of zooids.

rhipidoglossate *a.* having a radula with numerous teeth in a fan-shaped arrangement, as ear shells.

rhipidostichous *a. appl.* fan-shaped fins.

rhiptoglossate *a.* having a long prehensile tongue, as a chameleon.

rhiz-, rhizo- prefixes derived from Gk. *rhiza*, a root.

rhizanthous *a.* having flowers arising so low down on a much reduced stem that they appear as if arising from root.

rhizautoicus *a.* with archegonial and antheridial branches coherent.

rhizine *n.* fine projection from lower fungal cortex of a lichen that attaches it to the substrate.

rhizobacteria *n. plu.* soil bacteria associated with root surfaces.

rhizobia *n.plu.* soil bacteria of the genus *Rhizobium* and related genera, Gram-negative rods that form nodules on the roots of leguminous plants, in which they carry out symbiotic nitrogen fixation.

rhizocarp *n.* a perennial herbaceous plant whose stems die down each winter, so that it persists by underground organs only; a plant producing underground flowers. *a.* rhizocarpic, rhizocarpous.

rhizocaul hydrorhiza *q.v.*

rhizocorm *n.* an underground stem like a single-jointed rhizome; a bulb or corm.

rhizodermis *n.* the outermost layer of tissue in roots, which may be the piliferous layer, or the exodermis in an older root where the piliferous layer has worn away.

rhizogenesis *n.* differentiation and development of roots.

rhizogenic *a.* root-producing; arising from endodermic cells, not developed from pericycle; *pert.*, or stimulating, root formation.

rhizoid *n.* a filamentous outgrowth from prothallus that functions like a root; *a.* rootlike, *appl.* form of a bacterial colony.

rhizomatous *a.* resembling a rhizome, *appl.* fungal mycelium within a substratum or host.

rhizome *n.* thick horizontal stem, usually underground, bearing buds and scale leaves, sending out shoots above and roots below.

rhizomorph *n.* a root-like or bootlace-like structure formed from interwoven hyphae of some basidiomycete fungi.

rhizomorphic rhizomorphous *q.v.*

rhizomorphoid *a.* resembling a rhizomorph; branching like a root.

rhizomorphous *a.* in the form of a root.

rhizomycelium *n.* a many-branched system of hypha-like filaments, usually lacking nuclei, that anchor some chytrids to their substratum.

rhizophagous *a.* root-eating.

rhizophore *n.* a naked outgrowth of thallus, which grows down into soil and develops roots at the apex, as in club mosses.

rhizophorous *a.* root-bearing.

rhizoplane *n.* part of the rhizosphere immediately adjacent to the root surface, comprising a layer approx. 1 μm thick.

rhizoplast *n.* contractile structure in some green algae, connected to the basal bodies of the flagella.

Rhizopoda, rhizopods *n.*, *n.plu.* in protist classification a phylum of mainly free-living unicellular non-photosynthetic microorganisms, the amoebas, found in freshwater and marine habitats, characterized by the formation of pseudopodia and no flagella or cilia. In some groups, the body is surrounded by a casing or test, sometimes calcified. Part of the class Rhizopodea in older classifications of protozoa. *see* Appendix 5.

Rhizopodea, rhizopods *n.* *n.plu.* in older classifications, a class of protozoans the foraminiferans and the amoebas.

rhizosphere *n.* area of soil immediately surrounding and influenced by plant roots.

rhizotaxis *n.* root arrangement.

rho factor (ρ) protein factor regulating the termination of transcription of some bacterial genes.

rhod-, rhodo- prefixes derived from the Gk. *rhodon*, rose, signifying reddish.

rhodamine *n.* fluorescent compound used to visualize cell structures by immuno-fluorescence techniques.

rhodogenesis *n.* formation or reconstitution after bleaching, of rhodopsin.

rhodophane *n.* a red chromophane in retinal cones of some fishes, reptiles and birds.

Rhodophyceae Rhodophyta *q.v.*

Rhodophyta *n.* the red algae, a group of largely multicellular, structurally complex photosynthetic organisms classified either in the plant kingdom or as a phylum of the Protista. They are composed of close-packed filaments. The red colour is due to water-soluble phycobilin pigments. The storage carbohydrate is floridean starch, resembling amylopectin. The red algae are largely marine and there are many tropical species, and they usually grow attached to rocks or other substrates. Their cells have no flagella at any stage.

rhodoplast *n.* a reddish plastid in red algae.

rhodopsin *n.* rose-purple, light-sensitive pigment found in rod cells of the vertebrate retina (and in invertebrates) and which is a conjugate of a protein (opsin) a phospholipid, and the vitamin A aldehyde, retinal, splitting into opsin and retinal on exposure to light. *alt.* visual purple.

rhodosporous *a.* with pink spores.

rhodoxanthin *n.* a carotenoid pigment, found in aril of yew.

rhombencephalon hindbrain *q.v.*

rhombic *a. appl.* lips and grooves of brain at rhomboid fossa.

rhombocoel *n.* dilation of the central canal of medulla spinalis near its terminal end, the terminal vesicle.

rhomboid *a.* having the shape of a rhombus, i.e. a diamond in a pack of playing cards, *appl.* ligament, scales, fossa, etc.

rhomboideum *n.* the rhomboid or costoclavicular ligament.

rhomboideus *n.* major and minor, parallel muscles connecting scapula with thoracic vertebrae.

rhomboid-ovate *a.* between rhomboid and oval in shape, *appl.* leaves.

rhombomeres *n.plu.* the seven distinct segments into which the vertebrate hindbrain is divided at a stage in its development.

Rhombozoa *n.* class of the invertebrate phylum Mesozoa (*q.v.*) that includes the dicyemids and heterocyemids, parasites of the kidneys of marine invertebrates such as cephalopods, other molluscs, flatworms and annelids.

rhopalium *n.* a marginal sense organ in some jellyfish.

Rhynchocephalia, rhynchocephalians *n.*, *n.plu.* order of mainly extinct reptiles, with one living member, the tuatara (*Sphenodon punctatus*), a lizard-like animal confined to a few islands off New Zealand. *alt.* Sphenodonta.

rhynchocoel *n.* cylindrical cavity in body of nemertean worms that houses the proboscis when not extended.

rhynchodont *a.* with a toothed beak.

rhynchoporous *a.* beaked, *appl.* weevils.

rhynchostome *n.* anterior terminal pore through which proboscis is everted in nemertean worms.

α-rhythm spontaneous rhythmic fluctuations of electrical potential of cerebral cortex during mental inactivity.

β-rhythm spontaneous rhythmic fluctuations of electrical potential of cerebral cortex during mental activity.

rhytidome *n.* the outer bark consisting of the periderm and tissues isolated by it.

rib *n.plu.* in tetrapod vertebrates, curved thin bone of the thorax articulating with spine at one end and either free (floating ribs) or fixed to sternum (breast bone) at other; central vein of leaf; any elongated protrusion. *alt.* cota.

ribbon worms Nemertea *q.v.*

rib meristem a meristem in which cells divide perpendicular to the longitudinal axis, producing a complex of parallel files or ribs of cells.

ribitol *n.* sugar alcohol derived from ribose, a constituent of the teichoic acids of bacterial cell walls.

riboflavin *n.* vitamin B_2, consisting of ribose linked to the nitrogenous base dimethylisoalloxazine, synthesized by all green plants and most microorganisms, occurring free in milk and in some tissues of higher organisms and green plants, and in all living cells as a component of the coenzymes flavin adenine dinucleotide (FAD) and flavin mononucleotide (FMN). Liver, yeast and green vegetables are particularly rich sources. Deficiency causes skin cracking and lesions (ariboflavinosis).

riboflavin phosphate flavin mononucleotide *q.v.*

ribonuclease (RNase) *n.* any of various enzymes that cleave RNA into shorter oligonucleotides or degrade it completely into its constituent ribonucleotide subunits. *alt.* nuclease. *see also* RNase H, RNase P, RNase III.

ribonucleic acid *see* RNA in body of text.

ribonucleoprotein (RNP) *n.* any complex of RNA and protein.

ribonucleoside *see* nucleoside.

ribonucleotide *n.* a nucleotide containing the sugar ribose. *see* Appendix 1 (22, 23) for chemical structure.

ribonucleotide reductase either of two enzymes that reduce ribonucleotides to deoxyribonucleotides, the enzyme from animals acting on ribonucleoside diphosphates (*r.n.* ribonucleoside-diphosphate reductase, EC 1.17.4.1), that from some bacteria acting on ribonucleoside triphosphates (*r.n.* nucleoside-triphosphate reductase, EC 1.17.4.2).

ribophorin *n.* protein on the cytoplasmic face of the endoplasmic reticulum, to which ribosomes are anchored.

riboprobe *n.* a probe (*q.v.*) for gene isolation and identification, formed of RNA.

ribose *n.* a pentose sugar, present in RNA

and also an intermediate in the Calvin cycle of photosynthesis. *see* Appendix 1 (4) for chemical structure.

ribosomal DNA (rDNA), ribosome genes the DNA encoding ribosomal RNAs, which in many eukaryotes is present in many copies and clustered in chromosomal regions which form the nucleolar organizers.

ribosomal protein *see* ribosome.

ribosomal RNA (rRNA) major component of ribosomes and the most abundant RNA species in cells. In eukaryotes synthesized in nucleolus from rRNA genes tandemly repeated many times in the chromosomes. Several different types known, denoted by their sedimentation coefficients, e.g. 23S, 16S and 5S RNAs in eukaryotic ribosomes.

ribosome *n.* small particle found in large numbers in all cells both free in cytoplasm and attached to the endoplasmic reticulum, being composed of RNA and protein, and at which translation of messenger RNA and protein synthesis takes place. Composed of 2 subunits which differ in size (50S and 30S in *E. coli*, 60S and 40S in a typical eukaryotic cell) and which associate prior to protein synthesis, each made up of a major RNA species complexed with proteins (52 in the *E. coli* ribosome, approx. 80 in eukaryotes) and also with minor RNA species in the large subunit. *see* Fig.1. *see also* microsome, polysome.

ribosome binding site region in mRNA preceding the coding region, to which the ribosome binds to start translation.

ribothymidylate *n.* unusual nucleotide found in tRNA which is formed by modification of a uridylate residue after transcription and which is the ribonucleotide of thymine.

ribozyme *n.* an RNA molecule with enzymatic acitivity, such as the self- splicing introns of some RNAs, which can excise themselves from the molecule without the agency of protein enzymes.

ribulose (Ru) *n.* 5-carbon ketose sugar which as the phosphate and bisphosphate is involved in carbon dioxide fixation in photosynthesis and in other metabolic pathways such as the pentose phosphate pathway. *see* Appendix 1 (6).

ribulose 1,5-bisphosphate (RuBP) 5-

carbon sugar phosphate which is the primary carbon dioxide acceptor in photosynthesis. *alt.* (formerly) ribulose 1,5-diphosphate (RuDP). *see* Appendix 1 (7).

ribulosebisphosphate carboxylase/ oxygenase (RuBPc/o, Rubisco) enzyme found in chloroplasts of all green plants and in photosynthetic bacteria, and which catalyses the fixation of carbon dioxide into carbohydrate via ribulose-1,5-bisphosphate as acceptor, also having oxygenase activity which is involved in photorespiration. Estimated to be one of the most abundant enzymes on earth. EC 4.1.1.39. *alt.* ribulose-1,5-diphosphate carboxylase.

rickets *n.* inadequate calcification of bone in children, caused by deficiency of vitamin D.

rickettsia-like organism (RLO) any of a group of organisms which resemble rickettsiae but are found in plants.

rickettsiae *n.plu.* small obligate prokaryotic parasites, causing typhus and similar fevers in humans and animals and transmitted by ticks, mites and lice. Their simple cells lack cell walls and are obligate intracellular parasites of mammalian cells. Diseases caused by louse-borne rickettsiae include typhus fever and trench fever, tick-borne rickettsiae cause Rocky Mountain spotted fever and other tick-borne typhuses, scrub typhus is transmitted by mites. *sing.* rickettsia.

rictal *a. pert.* mouth and gape of birds.

rictus *n.* the opening or throat of calyx; the gape of a bird's beak.

rifampicin *n.* semisynthetic derivative of the antibiotic rifamycin.

rifamycin *n.* antibiotic from a *Streptomyces* sp. which specifically inhibits the initiation of RNA synthesis in bacterial cells.

riffle *n.* shallow broken water in a stream running over a stony bed.

rigor *n.* the rigid state of plants when not sensitive to stimuli; contraction and loss of irritability of muscle on heating or after death, or in some states such as shock, fever, etc.

rigor mortis stiffening of body after death due to temporary rigidity of muscles.

rimate *a.* having fissures.

rimiform *n.* in the shape of a narrow fissure.

rimose *a.* having many clefts and fissures.

rimulose *a.* having many small clefts.

ring bark bark peeling off in rings, as in some cherries, etc. *cf.* scale bark.

ring canal circular canal around margin of bell in medusae, connecting via radial canals to body cavity; circular vessel around gullet in echinoderms.

ring cell a thick-walled cell of annulus of sporangium in ferns.

ring centriole disc at end of body or middle portion of sperm, perforated for axial filament of flagellum.

ring chromosomes chromosomes formed by fusion of the ends of a chromosome fragment containing the centromere.

ring-form form assumed by immature trophozoite of the malaria parasite in red blood cells.

ring gland a glandular structure round aorta in insects, composed of various elements such as corpus allatum, corpus cardiacum, pericardial gland and hypocerebral ganglion.

ring-porous *appl.* wood in which the vessels tend to be larger and have thinner walls than in diffuse-porous wood; *appl.* wood in which the vessels formed early in the season are clearly larger than those formed later, producing a clear ring in cross-section.

ring species two species which overlap in range and behave as true species with no interbreeding, but are connected by a series (the ring) of interbreeding subspecies so that no true specific separation can be made.

ring vessel a structure in head of cestodes which unites the four longitudinal excretory trunks.

ringed worms common name for the annelids *q.v.*

ringent *a.* having lips, as of corolla, or valves, separated by a distinct gap; with upper lip arched; gaping.

ripa *n.* a line of ependymal fold over a plexus or tela.

riparian *a.* frequenting, growing on, or living on the banks of streams or rivers.

Ri plasmid a plasmid carried by *Agrobacterium rhizogenes*, the cause of hairy root in various dicotyledonous plants. The plasmid becomes stably integrated into the chromosomes of infected tissue and has been extensively modified by genetic engineering to produce a vector for introducing novel genes into plant cells to generate transgenic plants.

risorius *n*. a cheek muscle stretching from over masseter muscle to corner of mouth.

riverine *a*. living in rivers.

rivinian *a*. *appl*. sublingual glands and ducts; *appl*. notch in ring of bone surrounding tympanic membrane.

rivose *a*. marked with irregularly winding furrows or channels.

rivulose *a*. marked with sinuous narrow lines or furrows.

RNA *n*. ribonucleic acid, large linear molecule of varied composition, made up of a single chain of ribonucleotide (*q.v.*) subunits, containing the bases uracil, guanine, cytosine and adenine. Found in all cells as transfer RNA (*q.v.*), ribosomal RNA (*q.v.*), and messenger RNA (*q.v.*), all cellular RNAs being synthesized by transcription (*q.v.*) of chromosomal DNA acting as a template. It is the primary genetic material in some viruses and in some cases can be synthesized using viral RNA as a template. *see also* small cytoplasmic RNA, small nuclear RNA. *cf*. DNA.

RNA capping *see* cap.

RNA-dependent DNA polymerase reverse transcriptase *q.v*. *alt*. RNA-directed DNA polymerase.

RNA-dependent RNA polymerase RNA replicase *q.v*.

RNA-directed DNA polymerase reverse transcriptase *q.v*.

RNA editing the addition, deletion and conversion of nucleotides in primary RNA transcripts after synthesis to form a functional mRNA, which occurs in some mitochondrial mRNAs of the protozoan *Trypanosoma* and other simple organisms, in some plant chloroplast and mitochondrial mRNAs and in one nuclear transcript in mammals. In trypanosomes it involves the addition and deletion of specific uridine residues, while in plant mitochondria cytosine residues are converted to uridines.

RNA helicase protein that unwinds double-stranded RNAs.

RNA ligase enzyme which catalyses the rejoining of exons in the splicing of certain mRNAs.

RNA maturase protein specified by the second intron in the *box* gene of yeast mitochondria and which is thought to be involved in splicing the RNA transcript of the gene.

RNA plasmid linear or circular RNAs found in some plant mitochondria.

RNA polymerase I eukaryotic RNA polymerase present in nucleolus, responsible for transcribing the rRNA genes.

RNA polymerase II the RNA polymerase in eukaryotic nuclei which is responsible for transcribing protein-coding genes.

RNA polymerase III eukaryotic RNA polymerase present in nucleoplasm, responsible for transcribing tRNAs, 5S RNA, and some other small RNAs.

RNA primase *see* primase.

RNA primer *see* primer.

RNA processing any or all of the processes that result in the generation of a functional tRNA, rRNA or mRNA from a primary RNA transcript, including trimming the ends, the removal of introns (in eukaryotic RNAs), capping (in eukaryotic mRNAs), and cutting out individual rRNAs from their precursor transcripts, and which in eukaryotes occurs in the nucleus.

RNA replicase RNA polymerase which synthesizes RNA using an RNA template, the means by which some RNA viruses replicate their genomes. *alt*. RNA synthetase, replicase.

RNA splicing the process by which introns are removed from the primary RNA transcripts of eukaryotic genes, in which the introns are cut out at precisely defined splice points and the ends of the remaining RNA rejoined to form a continuous mRNA, rRNA or tRNA. *alt*. splicing.

RNA synthetase RNA replicase *q.v*.

RNA tumour viruses members of the retroviruses that can cause tumours in animals. Many have been derived from non-tumorigenic retroviruses by the incorporation of a cellular gene, often in altered or truncated form (an oncogene), which is solely responsible for their tumorigenic capacity. Others are tumorigenic due to the effects of their insertion into the host cell's DNA, leading to the abnormal or excessive expression of cellular genes.

RNA viruses *n. plu.* viruses having RNA as their genetic material, and including the Reoviridae, Togaviridae, Coronaviridae, Picornaviridae, Caliciviridae, Rhabdoviridae, Paramyxoviridae, Orthomyxoviridae, Arenaviridae, Bunyaviridae, and Retroviridae amongst vertebrate viruses, and most plant viruses. *cf.* DNA viruses.

RNA world a proposed stage in the evolution of life, predating the appearance of protein synthesis and DNA, in which self-replicating RNA molecules comprised the genetic material and "enzymes" of very primitive cells. *see also* ribozyme.

Robertsonian translocation type of chromosome abnormality in humans involving exchange between the long and short arms of two nonhomologous acrocentric chromosomes producing one long chromosome and a tiny fragment which is usually lost. *alt.* chromosomal fusion.

robust *a.* heavily-built, *appl.* australopithecines: *Australopithecus robustus*.

rod *n.* rod-shaped, light-sensitive (photoreceptor) sensory cell in retina, containing the light-sensitive pigment rhodopsin, and which is responsible for non-colour vision and vision in poor light; straight or slightly curved cylindrical bacterial cell.

rock mosses granite mosses *q.v.*

Rodentia, rodents *n., n.plu.* the largest order of placental mammals, known from the Paleocene and including rats, mice, voles, hamsters, porcupines, beavers and squirrels. They are omnivorous and/or herbivorous, and have continuously growing chisel-like incisors adapted for gnawing and no canines.

rod fibre nerve process that synapses with rod cell of retina.

rod vision vision in dim light, dark-adapted or "night" vision.

roding *n.* patrolling flight of birds defending territory.

roentgen (r, R) unit of ionizing radiation corresponding to an amount of ionizing radiation sufficient to produce 2 ionizations per cubic μm of water or living tissue.

rolandic *a. appl.* fissure or central sulcus of cerebral hemispheres; *appl.* tubercle of posterior region of medulla oblongata and gelatinous substance of dorsal horn of spinal medulla.

rolling circle mode of replication of some circular DNAs, as in certain phages, in which replication of only one strand is initiated at the origin, the newly synthesized strand displacing the other parental strand, which in some cases is also converted to double-stranded DNA.

root *n.* descending portion of plant, fixing it in soil, and absorbing water and minerals, and having a characteristic arrangement of vascular tissues; radix *q.v.*; embedded portion of tooth, hair, nail, or other structure; pulmonary veins and artery joining lung to heart and trachea; pedicle of vertebra; efferent and afferent fibres of a spinal nerve, leaving or entering the spinal cord.

root cap a protective cap of tissue at tip of root. *alt.* calyptra, pileorhiza.

root cell clear, colourless base of an alga, attaching thallus to substratum.

root climber plant which climbs by roots developed from the stem.

root hairs unicellular outgrowths from epidermal cells of roots, concerned with uptake of water and solutes from soil.

root nodule structure formed on the roots of leguminous and some non-leguminous plants and which contains nitrogen-fixing bacteria.

root parasitism condition shown by semi-parasitic plants, the roots of which penetrate the roots of neighbouring plants and draw from them elaborated food material.

root pocket a sheath containing a root, especially of aquatic plants.

root pressure a positive hydrostatic pressure in xylem, developed when transpiration is low, and created by ion movements and consequent osmosis in root cells which forces water and dissolved ions up the xylem. Can be demonstrated in roots in which the shoot has been cut off, when water exudes from the root stump.

root process a branched structure attaching an algal thallus to the substratum.

root sheath *n. (bot.)* a protective sheath surrounding the developing radicle of some flowering plants such as grasses, *alt.* coleorhiza; velamen of orchid; *(zool.)* that part of hair follicle continuous with epidermis.

root stalk a rootstock or rhizome; root-like horizontal portion of certain hydrozoan colonies.

root tubers swollen food-storing roots of certain plants such as lesser celandine and orchids.

rootlet *n.* an ultimate branch of a root.

rootstock *n.* more-or-less underground part of stem; rhizome *q.v.*

roridous *a.* covered with droplets.

rosaceous *a.* resembling a rose; *pert.* the Rosaceae family of dicot flowers, which includes, as well as roses, cherries, apples, plums, etc., mountain ash, and many other trees and shrubs.

Rosales *n.* order of mainly woody dicots, including the families Chrysobalanaceae (coco plum), Neuradaceae and Rosaceae (rose, apple, etc.)

rosellate *a.* arranged in rosettes.

rosette *n.* (*bot.*) a cluster of leaves arising in close circles from a central axis; a plant disease due to deficiency of boron or zinc; a cluster of crystals as in some plant cells; (*zool.*) a swirl or vortex of hair in pelt of animal; a small cluster of blood cells; group of spiracular channels in exocuticle of some aquatic insects; various other rosette-like structures in other invertebrates.

rosette organ in some ascidians, a ventral complex stolon from which buds are constricted off.

rosin *n.* a resin obtained from pine.

rostel rostellum *q.v.*

rostellar *a. pert.* a rostellum.

rostellate *a.* furnished with a rostellum.

rostelliform *a.* shaped like a small beak.

rostellum *n.* a small beak or beak-like structure, e.g. the tubular mouthparts of some insects, the rounded hooked prominence on scolex of tapeworm.

rostrad *adv.* towards anterior end of body. *cf.* caudad.

rostral *a.* towards the anterior end of the body; *pert.* a beak or rostrum.

rostral gland premaxillary part of labial gland, as in snakes; labral gland of spiders.

rostrate *a.* beaked.

rostriform, rostroid *a.* beak-shaped.

rostro-caudal axis antero-posterior axis, i.e. head to tail axis, of the animal body.

rostrulate *a.* with, or like, a rostrum.

rostrum *n.* beak of birds; beak-like process; prenasal region; upper lip of spiders; modified labium (lower lip) in aphids, with a groove into which the feeding stylets fit; a median ventral plate of shell of barnacles; (*neurobiol.*) the prolongation of the anterior end of corpus callosum which curves under and backwards.

rosula rosette *q.v.*

rosular, rosulate rosellate *q.v.*

rotate *a.* shaped like a wheel.

rotation *n.* turning as on a pivot, as limbs; circulation, as of cell sap.

rotator *n.* a muscle that allows of circular motion.

rotatores spinae paired muscles, one on each side of thoracic vertebra, each arising from transverse process and inserted into vertebra next above.

rotatorium *n.* a trochoid joint or pivot joint.

Rotifera, rotifers *n., n.plu.* phylum of microscopic, multicellular, pseudocoelomate animals, living mostly in fresh water. They are generally cone-shaped with a crown of cilia at the widest end surrounding the mouth. The epidermis is separated from the internal organs by a fluid-filled space, the pseudocoel. Formerly called wheel animals, as the crown of beating cilia looks as though it is rotating.

rotiform rotate *q.v.*

rotula *n.* one of five radially parallel bars bounding circular aperture of oesophagus of sea urchin; patella or knee cap. *a.* rotular.

rotuliform *n.* shaped like a small wheel.

rotundifolious *a.* with rounded leaves.

rough endoplasmic reticulum *see* endoplasmic reticulum.

rouleaux *n.* formations like piles of coins into which red blood cells tend to aggregate.

round dance type of repeated circular dance in bees that alerts other bees to the existence of a food source near the hive.

round window fenestra rotunda *q.v.*

roundworms common name for the Nematoda *q.v.*

Rous sarcoma virus (RSV) an RNA tumour virus of chickens, the source of the oncogene v-*src*, one of the first animal tumour viruses identified, and, unlike many oncogene-containing RNA tumour viruses, is able to replicate without a helper virus.

royal cell in honey bees the large, pitted waxen cell constructed by the workers to rear queen larvae; in termites, the special cell in which the queen is housed.

royal jelly material supplied by workers to female larvae in royal cells, which is necessary for transformation of larvae into queens.

R plasmid plasmid carrying genes for resistance to various commonly used antibiotics, present in many of the enterobacteria. Some R plasmids are transmissible to other bacteria of the same and other related species. *alt.* drug-resistance plasmid, R-factor.

R point restriction point *q.v.*

***r*-selected species** species typical of variable or unpredictable environments, characterized by small body size and rapid rate of increase. *alt.* opportunist.

***r* selection** selection favouring rapid rates of population increase, esp. prominent in species that colonize short-lived environments or undergo large fluctuations in population size. *cf. K* selection.

rubber *n.* the coagulated latex of several trees, mainly *Hevea* sp., being long-chain polymers of isoprene and hydrocarbons.

rubiaceous *a. pert.* a member of the dicot family Rubiaceae (the blackberries, raspberries, etc.).

rubiginose, rubiginous *a.* of a brownish-red tint, rust-coloured.

Rubisco ribulose bisphosphate carboxylase *q.v.*

rubrospinal *a. appl.* a descending tract of nerve fibres from red nucleus, in the ventrolateral column of the spinal cord. *alt.* prepyramidal tract.

ruderal *a.* growing among rubbish or debris; growing by the roadside or in disused fields.

rudiment *n.* an initial, or primordial group of cells which gives rise to a structure.

rudimentary *a.* in an imperfectly developed condition; at an early stage of development; arrested at an early stage.

ruff *n.* fringe of fur or feathers around neck.

Ruffini endings pressure receptors in the dermis of vertebrates.

Ruffini's organs heat receptors in subcutaneous tissue of fingers.

rufine *n.* a reddish pigment in the mucous glands of slugs.

rufinism *n.* red pigmentation due to inhibition of formation of dark pigment. *alt.* rutilism.

ruga *n.* fold or wrinkle, as of skin or mucosal membranes in some animals.

rugate *a.* wrinkled, ridged.

rugose *a.* with many wrinkles or ridges.

ruling reptiles archosaurs *q.v.*

rumen *n.* in ruminants (cud-chewing mammals) the first stomach, in which food is digested by bacteria and from which it can be regurgitated into the mouth for further chewing.

ruminants *n.plu.* herbivorous mammals such as cows, sheep, goats, deer, antelopes and giraffes, that chew the cud and have complex, usually four-chambered, stomachs containing microorganisms that break down the cellulose in plant material.

ruminate *a.* appearing as if chewed, *appl.* endosperm with infolding of testa or of perisperm, appearing mottled in section; *appl.* seeds having such endosperm, as betel nut and nutmeg; *v.* to chew the cud.

rumination *n.* the act of ruminant animals in returning partly digested food from 1st stomach to mouth in small quantities for thorough mastication. *alt.* chewing the cud.

runcinate *a. appl.* pinnatifid leaf when divisions point towards the base, as in dandelions.

runner *n.* a specialized stolon consisting of a prostrate stem rooting at the node and forming a new plant which eventually becomes detached from the parent, as strawberry.

runoff *n.* the drainage of water from waterlogged or impermeable soil.

rupestrine, rupicoline, rupicolous *a.* growing or living among rocks.

ruptile *a.* bursting in an irregular manner.

Ruribulose *q.v.*

rush *see* Juncaceae.

rust fungi common name for a group of basidiomycete fungi, the Uredinales, many of which are serious and widespread plant pathogens of numerous important crops, causing rust disease which appears as small black, orange or brown pustules on the stem or leaf surface. They are characterized by the production of thick-walled teleutospores which germinate to produce a basidium and basidiospores, which give rise to a mycelium and, in many rusts, one or more other kinds of binucleate spore. They have complex life histories on two different hosts, e.g. the black stem rust of cereals (*Puccinia graminis*) uses the barberry (*Berberis vulgaris*) as an intermediate host.

rut *n*. period of sexual heat in male animals, when they often fight for females, defend territory, etc., before mating.

Rutales *n*. order of dicot trees and shrubs, rarely herbs, with leaves often dotted with glands, and including the families Anacardiaceae (cashew), Meliaceae (mahogany), Rutaceae (citrus, rue), Simaroubaceae, and others.

rutilant *a*. of a bright bronze-red colour.

rutilism rufinism *q.v.*

S

σ sigma, symbol for 0.001 seconds; symbol for standard deviation.

S serine *q.v.*; Svedberg unit *q.v.*

S₁ 1st selfing generation, the offspring of a self-cross.

SAM *S-adenosylmethionine q.v.*

SAN sinuatrial node (sinuauricular node) *q.v.*

SBA soybean agglutinin, a lectin isolated from *Glycine max*.

SBMV Southern bean mosaic virus.

SC synaptonemal complex *q.v.*

SCID severe combined immunodeficiency *q.v.*

SCP single-cell protein *q.v.*

scRNA small cytoplasmic RNAs *q.v.*

scRNPs scyrps *q.v.*

SDS sodium dodecyl sulphate *q.v.*

SEM scanning electron microscope *q.v.*

Ser serine *q.v.*

SER smooth endoplasmic reticulum, *see* endoplasmic reticulum.

sis *v-sis*, oncogene carried by simian sarcoma virus and others. *c-sis*, corresponding proto-oncogene in normal cells, which specifies the B chain of platelet-derived growth factor.

SIV simian immunodeficiency virus, a retrovirus closely related to HIV, and which has been found in humans.

SLE systemic lupus erythematosus *q.v.*

snRNA small nuclear RNAs *q.v.*

snRNPs snurps *q.v.*

SOD superoxide dismutase *q.v.*

SP suction pressure *q.v.*

sp. species *q.v.*

SPF specified pathogen-free.

spp. *plu.* of species *q.v.*

src *v-src*, oncogene carried by the RNA tumour virus Rous sarcoma virus. *c-src*, the corresponding proto-oncogene found in normal cells, specifying a protein tyrosine kinase.

SRE serum-responsive element, a control site in various mammalian genes that is responsible for the induction of such genes in response to the stimulation of the cell by growth factors (or serum containing growth factors) and phorbol esters.

SRS-A slow-reacting substance of anaphylaxis, *see* leukotrienes.

SSB *see* single-stranded DNA binding protein.

ssDNA single-stranded DNA.

ssp. subspecies *q.v.*

SSV simian sarcoma virus, an RNA tumour virus, carries oncogene *v-sis*.

STH somatotropin (growth hormone) *q.v.*

STM short-term memory *q.v.*; scanning tunnelling microscope *see* tunnelling electron microscope.

STX saxitoxin *q.v.*

Sv sievert *q.v.*

SV40 simian virus 40 *q.v.*

sabuline *a.* sandy; growing in sand, especially coarse sand.

sabulose, sabulous *a.* sandy.

saccadic *a. appl.* brief movements of the eyes when suddenly looking at a different fixation point.

saccate *n.* pouched.

saccharide *see* monosaccharide, disaccharide, polysaccharide, oligosaccharide.

saccharobiose sucrose *q.v.*

saccharomycetes *n.plu.* the yeasts, unicellular ascomycete fungi of the family Saccharomycetaceae, which include the bread and brewing yeast *Saccharomyces cerevisiae* and the fission yeast *Schizosaccharomyces pombe*, both of which are widely used as simple model organisms in experimental cell biology and genetics and as hosts in recombinant DNA work. In particular, work on the control of the cell cycle in eukaryotic organisms has made much use of yeasts. Yeasts are ubiquitous inhabitants of the soil and plant surfaces esp. on sugary substrates, which they ferment, producing alcohols and carbon dioxide, the basis of their use in wine making, brewing and baking. *see also* false yeasts.

saccharose sucrose *q.v.*

sacciferous *a.* furnished with a sac.

sacciform *a.* like a sac or pouch.

sacculate *a.* provided with sacculi.

sacculation *n.* the formation of sacs or saccules; a series of sacs, as of haustra of colon.

saccule, sacculus *n.* a small sac or pouch; part of the membranous labyrinth of inner ear, together with utricle forms vestibule of inner ear.

sacculus rotundus in rabbits and hares, a dilatation of intestine between ileum and caecum.

sacral *a. pert.* the sacrum.

sacral index one hundred times the breadth of sacrum at base, divided by anterior length.

sacral ribs elements of sacrum joining true sacral vertebrae to pelvis.

sacralization *n.* fusion of sacral and lumbar vertebrae.

sacrocaudal *a. pert.* sacrum and tail region.

sacrococcygeal *a. pert.* sacrum and coccyx.

sacro-iliac *a. pert.* sacrum and ilium (dorsal bone of pelvic girdle).

sacrolumbar *a. pert.* region of loins and termination of vertebral column (sacrum).

sacrospinal *a. pert.* sacral region and spine; *appl.* to ligament between sacrum and spine of ischium (ventral and posterior bone of each half of pelvic girdle).

sacrovertebral *a. pert.* sacrum and vertebrae.

sacrum *n.* bone forming termination of vertebral column, usually composed of several fused vertebrae, *alt.* os sacrum; vertebra or vertebrae to which the pelvic girdle is attached.

saddle clitellum *q.v.*

S-**adenosylmethionine (SAM)** *n.* a compound of adenosine and methionine, a major donor of methyl groups in biosynthetic reactions.

safe concentration the maximum concentration of a toxic substance that has no observable effect on a species after long-term exposure over one or more generations.

sagitta *n.* an elongated otolith in sacculus of teleost fishes; a genus of arrow worms.

sagittae *n.plu.* the inner genital valves in hymenopterans.

sagittal *a.* section or division in median longitudinal plane; *appl.* sinus running between the two hemispheres of the brain; *appl.* the suture between the parietal bones of the skull which forms a ridge or crest on top of skull in some primates.

sago palm *see* Cycadophyta.

Saharo-Arabian Floral Region part of the Palaeotropical Realm comprising Arabia and an area on the eastern side of the Persian Gulf.

salamanders *see* Urodela.

Salicales *n.* order of dicot trees and shrubs comprising the family Salicaceae (willow).

Salientia *n.* in some classifications the name given to the order of amphibians comprising the frogs and toads. *alt.* Anura.

salinization *n.* the deposition of excessive amounts of soluble mineral salts in the soil, making it unfit for cultivation, caused by high surface evaporation often exacerbated by artificial irrigation over long periods.

saliva *n.* secretion produced by salivary glands which open into or near the mouth in many vertebrates and invertebrates, and which in humans contains mucoproteins and the starch-digesting enzyme α-amylase, in insects various digestive enzymes depending on diet, and in blood-sucking invertebrates various anticoagulants.

salivarium *n.* recess of preoral food cavity, with opening of the salivary duct, in insects.

salivary *a. pert.* saliva; *appl.* glands opening into or near the mouth which secrete saliva.

salmonella *n.* bacterium of the genus *Salmonella* which includes spp. causing food poisoning and the causal agent of typhoid fever, *S. typhi. plu.* salmonellae.

Salmonella typhimurium bacterium

causing typhoid fever in mice, commonly used as an experimental organism for genetic studies.

Salmonidae, salmonids *n.*, *n.plu.* a family of the Salmoniformes which includes the genus *Salmo* (salmon, rainbow and brown trout).

Salmoniformes *n.* large order of fairly primitive marine and freshwater teleost fishes, including trout, salmon, pike etc.

salpingian *a. pert.* Eustachian or Fallopian tube.

salpingopalatine *a. pert.* Eustachian tubes and palate.

salpinx *n.* Eustachian tube *q.v.*; Fallopian tube *q.v.*; any of various trumpet-shaped structures.

salps Thaliacea *q.v.*

salsuginous *a.* growing in soil impregnated with salts, as in a salt marsh.

salt bond, salt bridge electrostatic bond *q.v.*

salt gland organ near eye in marine reptiles and birds for excretion of excess sodium chloride; similar structure in gills of fishes; (*bot.*) an epidermal gland exuding salts in certain leaves.

salt linkage electrostatic bond *q.v.*

salt marsh the intertidal area on sandy mud in sheltered coastal areas and in estuaries, supporting characteristic plant and animal communities.

saltation *n.* a jumping movement; (*evol.*) the idea that major evolutionary changes can take place within a single generation through "macromutations" – mutations of large effect. In its most extreme form this idea is now no longer held by most modern evolutionary biologists. *cf.* phyletic gradualism.

saltatorial *a.* adapted for, or used in, leaping, *appl.* limbs of jumping insects.

saltatorians *n.plu.* crickets and grasshoppers, members of the insect order Orthoptera (called Saltatoria in some classifications).

saltatory saltatorial *q.v.*; *appl.* jerky movements of particles and organelles within cells.

saltatory conduction mode of impulse propagation in myelinated nerve fibres, where impulse "jumps" from node to node.

saltigrade *a.* moving by leaps, as some insects and spiders.

salvage reaction metabolic reaction that uses preformed compounds as precursors for new biosyntheses.

salvage sequences sequences present in proteins destined to remain in the endoplasmic reticulum.

samara *n.* winged fruit typical of elm and ash, a single winged achene.

sand *n.* a soil having most particles between 2 mm and 0.02 mm in size, composed usually of silica, and being well drained and aerated.

sand dollars common name for sea urchins of the order Clypeasteroidea, having flattened tests.

sanguicolous *a.* living in blood.

sanguiferous *a.* conveying blood, as arteries, veins.

sanguimotor *a. pert.* circulation of the blood.

sanguivorous *a.* feeding on blood.

Santales *n.* an order of woody dicots, often parasitic on other angiosperms or rarely on gymnosperms and including the families Santalaceae (sandalwood), Viscaceae (mistletoe) and others.

sap *see* cell sap; sugary fluid carried by phloem.

saphena *n.* a conspicuous vein of leg, extending from foot to femur.

saphenous *a. pert.* internal or external saphena, *appl.* a branch of the femoral nerve.

sapients *n.plu.* members of the species *Homo sapiens*, including Neanderthal Man, Cro-Magnon Man and modern man.

Sapindales *n.* order of dicot trees, shrubs, lianas, rarely herbs, and including the Aceraceae (maple), Hippocastanaceae (horse chestnut), Sapindaceae (soapberry) and others.

sapogenin *n.* the non-sugar part of a saponin, usually obtained from saponin by hydrolysis.

saponin *n.* any of various steroid glycosides present in many plants, such as soapwort and soapbark, and which produce a soapy solution in water.

saprobe *n.* a saprobic organism.

saprobic *a.* living on decaying matter; saprophytic or saprozoic.

saprobic classification, Saprobien sys-

tem a biotic index (*q.v.*), used esp. in continental Europe, for assessing the degree of organic pollution of a body of water, which is based on recognition of four stages in the oxidation of organic mattter, each characterized by the presence and relative abundance of certain groupings (saprobic groupings) of indicator species. *see* α-mesosaprobic, β-mesosaprobic, oligosaprobic, polysaprobic.

saprobiont *n.* saprobe *q.v.*

saprogenic *a.* causing decay; resulting from decay.

sapropelic *a.* living among debris of bottom ooze.

saprophage saprobe, *see* saprobic.

saprophyte *n.* plant, fungus or bacterium that gains its nourishment directly from dead or decaying organic matter. *a.* saprophytic. *see also* saprotroph.

saprophyte chain a food chain starting with dead organic matter and passing to saprophytic microrganisms.

saprotroph *n.* any organism that feeds on dead organic matter. *a.* saprotrophic. *alt.* saprophyte, esp. *appl.* fungi and bacteria.

saprozoic *a. appl.* animal that lives on dead or decaying organic matter. *n.* saprozoite.

sapwood *n.* the more superficial, younger, paler, softer wood of trees, which is water-conducting and contains living cells. *alt.* splintwood, alburnum. *cf.* heartwood.

sarcenchyma *n.* parenchyma whose ground substance is granular and not abundant.

sarciniform *a.* arranged in more-or-less cubical clumps, *appl.* cocci.

sarcocarp *n.* the fleshy or pulpy part of a fruit.

sarcocyte *n.* the middle layer of ectoplasm in some protozoans such as gregarines.

sarcode *n.* the body protoplasm of protozoans; protoplasm in general.

sarcoderm *n.* fleshy layer between a seed and its outer covering.

sarcodictyum *n.* the 2nd or network protoplasmic zone of radiolarians.

Sarcodina *n.* class or superclass of Protozoa containing the Rhizopodea (amoebae and foraminiferans) and Actinopodea (heliozoans and radiolarians), which have pseudopodia or actinopodia, little differentiation of the body and no flagella at any stage.

sarcogenic *a.* flesh-producing.

sarcoid *a.* fleshy, as sponge tissue.

sarcolemma *n.* membranous sheath around an individual muscle fibre.

sarcolyte *n.* a non-nucleated muscle fragment undergoing phagocytosis in development of insects.

sarcoma *n.* a tumour of connective tissue, e.g. of fibrous tissue (fibrosarcoma) or bone (osteosarcoma). *cf.* carcinoma.

sarcomatrix *n.* the 4th protoplasmic zone of a radiolarian, the area of digestion and assimilation.

sarcomere *n.* the portion of a striated muscle fibre between two Z-discs, comprising a complete contractile unit.

sarcophagous *a.* flesh-eating.

sarcoplasm *n.* the cytoplasm between fibrils of muscle tissue.

sarcoplasmic reticulum membrane forming a network of fine channels around myofibrils in striated muscle cells, and which acts as an intracellular store of Ca^{2+} which is needed for muscle contraction.

Sarcopterygii, sarcopterygians *n., n.plu.* a group of mostly extinct bony fishes having fleshy fins and nostrils opening into the mouth, comprising the lungfishes (Dipnoi *q.v.*) and the crossopterygians (Crossopterygii *q.v.*).

sarcosoma *n.* the fleshy, as opposed to the skeletal portion of an animal's body.

sarcotesta *n.* softer fleshy outer portion of testa.

sarcotheca *n.* the sheath of a hydrozoan polyp.

sarcotubular system the sarcoplasmic reticulum and t tubules that form a series of membrane-bounded channels around the myofibrils of muscle fibres.

sarcous *a. pert.* flesh or muscle tissue.

sargasterol *n.* a sterol present in some algae.

sarmentaceous, sarmentose, sarmentous *a.* having slender prostrate stems or runners.

sarmentum *n.* the slender stem of a climbing or creeping plant.

Sarraceniales *n.* order of carnivorous herbaceous dicots, typically with pitcher-like leaves for trapping insects. It comprises a single family the Sarraceniaceae (pitcher plants).

sartorius *n*. strap-like muscle in thigh which helps to flex both hip and knee and enables legs to be moved inwards.

satellite *n*. the 2nd of any pair of pseudoconjugant individuals in colonies of gregarines; small piece of chromosomal material attached to the short arm of a chromosome by a slender thread.

satellite cells cells in close physical association with another type of cell, as neuroglial cells with neurones in central nervous system, Schwann cells with peripheral neurones, etc.; small flattened inactive cells lying within the basal lamina of mature skeletal muscle fibres and from which new skeletal muscle cells are produced when required, *alt.* myoblasts.

satellite DNA certain highly repeated short DNA sequences, not transcribed and with no known function, found in eukaryotic chromosomes and confined to centromere regions.

satellite RNA encapsidated small plant pathogenic RNAs which require co-infection with a specific virus, a helper virus, for replication and encapsidation. *alt.* virusoid, satellite virus.

satellite virus satellite RNA *q.v.*

saturated *a. appl.* fatty acids with a fully hydrogenated carbon backbone.

saturnine *a.* forming, having, or *pert.* an equatorial ring.

saurian *a.* appl., *pert.* or resembling a lizard.

Saurischia *n.* a large order of Mesozoic archosaurs, commonly called lizard-hipped dinosaurs, including both bipedal carnivores (the theropods) and very large quadrupedal herbivores (e.g. the sauropods).

saurognathous *a.* with a lizard-like arrangement of jaw-bones.

sauropods *n.plu.* group of gigantic, herbivorous lizard-hipped dinosaurs which included *Diplodocus, Apatosaurus (Brontosaurus)* and *Brachiosaurus*.

savanna *n.* subtropical or tropical dry grassland with drought-resistant vegetation and scattered trees; transitional zone between dry grassland or semi-desert and tropical rain forest. *alt.* savannah.

sawflies *see* Symphyta.

saxatile, saxicoline, saxicolous *a.* living in, on, or among rocks.

Saxifragales *n.* order of dicot trees, shrubs, lianas and herbs, including the families Crassulaceae (orpine), Escalloniaceae (escallonia), Grossulariaceae (gooseberry), Hydrangeaceae (hydrangea), Saxifragaceae (saxifrage) and others.

saxitoxin (STX) a poison produced by marine dinoflagellates and isolated from shellfish that feed on them, selectively blocks the regenerative sodium conductance channel in neurones thereby blocking the generation of nerve impulses, used as a neurotoxin in experimental neurophysiology.

scaberulous *a.* somewhat rough.

scabrate, scabrous *a.* rough, with a covering of stiff hairs, scales or points.

scaffold *n.* of a chromosome, the proteinaceous structure in the shape of a sister chromatid pair which is generated on depletion of chromosomes of histones.

scala *n.* any of the three fluid-filled canals separated by membranes and running the length of the cochlea of inner ear. The scala tympani lies below the basilar membrane, the scala media is delimited by the organ of Corti and Reissner's membrane, and the scala vestibuli lies on the other side of Reissner's membrane. The scala tympani and vestibuli contain perilymph, the scala media, endolymph.

scalariform *a.* ladder-like, *appl.* structures having bars like the rungs of a ladder, such as the walls of some xylem vessels.

scale *n.* a flat, small, plate-like external structure. In plants it is formed from epidermis, in animals it may be made of chitin (in invertebrates), bone or horn (keratin) (in vertebrates).

scale bark bark flaking off in irregular sheets or patches. *cf.* ring bark.

scale insect member of the family Coccidae of the Hemiptera (bugs), feeding on plants, in which the wingless females remain fixed to the food plant and are covered with a waxy covering, the "scale". Some are serious plant parasites, others yield shellac and cochineal. *alt.* mealy bug.

scale leaf a small dry or hard leaf.

scalene *a. pert.* the scalenus muscles; *appl.* tubercle on 1st rib for attachment of scalenus anticus or anterior.

scalenus *n.* one of three neck muscles:

scalenus posticus, scalenus medius, scalenus anticus.

scalids *n.plu.* spines arranged in a series of rings around the mouth in Kinorhyncha.

scaliform *a.* ladder-shaped, *see* scalariform.

scalpella *n.plu.* paired pointed processes, parts of maxillae of dipteran flies.

scalpriform *a.* chisel-shaped, *appl.* incisors of rodents.

scalprum *n.* the cutting edge of an incisor.

scandent *a.* climbing by stem roots or tendrils; trailing, as grasses over shrubs.

scanning electron microscope (SEM) electron microscope that produces a "three-dimensional" image from electrons reflected from the surface of a specimen.

scansorial *a.* adapted for climbing; habitually climbing.

scape *n.* a flower stalk arising at or under ground level; basal part of antenna in some flies.

scapha *n.* narrow curved groove between helix and anthelix of ear.

scaphium *n.* process of 9th (copulatory) segment of male lepidopterans; anterior Weberian ossicle; keel of leguminous flower; a boat-shaped structure.

scaphocephalic *a.* with a narrow elongated skull.

scaphocerite *n.* a boat-shaped exopodite of 2nd antenna of decapod crustaceans.

scaphognathite *n.* process on 2nd maxilla of some decapod crustaceans that pumps water over the gills by a paddle-like action.

scaphoid *a.* shaped like a boat, *appl.* certain bones of wrists and ankles.

scapholunar *a. pert.* scaphoid and lunar carpal bones, or those bones fused. *n.* scapholunatum.

Scaphopoda *n.* a class of marine molluscs, commonly called tusk shells or elephant-tooth shells, which have a tubular shell, a reduced foot and no ctenidia.

scapiform, scapigerous, scapoid, scapose *a.* resembling or consisting of a scape.

scapula *n.* in vertebrates, the shoulder blade, i.e. the dorsal part of pectoral girdle. *a.* scapular.

scapulars *n.plu.* feathers covering the junction of wing with body in bird.

scapulus *n.* modified submarginal region in certain sea anemones.

scapus *n.* scape *q.v.*; stem of feather; hair shaft; part of column below, and including, parapet in sea anemones.

scarab(a)eiform *a. appl.* the C-shaped larval type of certain beetles.

scarce mRNA the class of cellular mRNA consisting of a few copies each of many different sequences. *alt.* complex mRNA.

scarious *a.* thin, dry, scurfy or scaly.

Scarpa's fascia deep layer of superficial abdominal fascia.

Scarpa's foramina two openings, for nasopalatine nerves, in middle line of palatine process of maxilla.

Scarpa's ganglion vestibular ganglion in internal ear.

Scarpa's triangle the femoral triangle formed by the adductor longus, sartorius and inguinal ligament.

Scatchard analysis, Scatchard plot graphical method of analysing the result of equilibrium binding experiments of receptors and ligands, which gives the association constant of binding and the number of binding sites per molecule.

scatophagous *a.* dung-eating.

scatter factor hepatocyte growth factor, a protein expressed in the liver of patients with fulminant hepatic failure, and which causes certain epithelial cells to scatter and to adopt fibroblast-like morphology.

scavenger *n.* an animal feeding on animals that have been killed by other predators, or have died naturally, or on organic refuse.

scavenger receptor receptor on liver cells that binds and removes damaged glycoproteins from blood.

scent scales androconia *q.v.*

Schild plot graphical method of analysing the results of experiments comparing the potency of competitive antagonists on the responses to an agonist drug.

schindylesis *n.* articulation in which a thin plate of bone fits into a cleft or fissure, as that between vomer and palatine bones.

Schisandrales Illicidales *q.v.*

schistocytes *n.* erythrocytes undergoing fragmentation and the resulting hollow fragments.

schistosome *n.* parasitic digenean blood fluke (Trematoda) infesting mammals, e.g. *Schistosoma mansoni*, which causes schistosomiasis (bilharzia) in humans in tropical regions. The larvae develop in certain freshwater snails.

schizocarp *n.* fruit derived from a compound ovary but which splits into two or more one-seeded portions at maturity, e.g. the double "keys" of sycamore and maple.

schizocarpic *a. appl.* dry fruits which split into two or more mericarps, as carcerulus, cremocarp, lomentum, regma, compound samara.

schizocele schizocoel *q.v.*

schizocoel *n.* coelom formed by splitting of mesoderm into layers.

schizogamy *n.* fission into a sexual and non-sexual zooid in some polychaetes.

schizogenesis *n.* reproduction by fission.

schizogenetic *a.* reproducing or formed by fission; *appl.* intercellular spaces or glands in plants formed by separation of cell walls along middle lamella. *alt.* schizogenous. *n.* schizogeny.

schizogony *n.* reproduction by fission into many cells, in protozoans.

schizokinete *a.* a motile worm-like stage in the life history of some sporozoan blood parasites.

schizolysigenous *a.* formed schizogenously, by separation, and enlarged lysigenously, by breakdown, such as glands and cavities in pericarp, e.g. of citrus fruits.

schizolysis *n.* fragmentation.

Schizomycetes obsolete name for the bacteria, from the time when they were classified as "fission fungi".

schizont *n.* in some protozoans, esp. the sporozoan parasites, the stage in the life cycle following the trophozoite and reproducing by multiple fission.

schizontoblast agametoblast *q.v.*

schizontocytes *n.plu.* cells into which a schizont divides and which themselves divide into clusters of merozoites.

schizopelmous *a.* with two separate flexor tendons connected with toes, as some birds.

Schizophyta *n.* in older classifications, a group containing the bacteria and cyanobacteria, when they were considered as plants. Now obsolete. In modern classifications the bacteria, cyanobacteria and other prokaryotic organisms are placed in a separate kingdom, the Monera or Prokaryotae.

schizophytes *n.plu.* now rarely used term to denote organisms formerly considered as plants, such as bacteria, cyanobacteria and yeasts, that reproduce solely by fission.

schizopod stage that stage in development of decapod crustacean larva when it resembles an adult *Mysis* in having an exopodite and endopodite to all thoracic limbs.

schizorhinal *a.* having external narial opening elongated and posterior border angular or slit-like. *cf.* holorhinal.

schizostele *n.* one of a number of strands formed by division of initial apical meristem of stem.

schizostely *n.* condition of stem in which apical meristem gives rise to a number of strands, each composed of one vascular bundle.

schizothecal *a.* having scale-like horny tarsal plates.

schizozoite *n.* in sporozoan parasites, the stage in the life cycle produced by schizogony; small cell produced by multiple fission of a schizont.

school *n.* a group of fish or marine mammals, such as porpoises, that swim together in an organized fashion.

Schwann cell glial cell which forms the fatty sheath around myelinated nerve fibres in the peripheral nervous system.

Schwann sheath neurilemma *q.v.*

sciatic *a. pert.* hip region, *appl.* artery, nerve, veins, etc.

scientific method the rational formulation of hypotheses, collection of data, and testing of hypotheses against observations or experimental results that is the basis of the scientific approach to explaining natural phenomena.

scion *n.* a branch or shoot which is to be grafted on to another plant.

scissile *a.* cleavable; splitting, as into layers.

scissiparity schizogenesis *q.v.*

Scitamineae Zingiberales *q.v.*

sclera *n.* the tough, opaque, fibrous coat of the eyeball, the white of the vertebrate eye.

scleractinians *n.plu.* an order (the Scleractinia or Madreporina) of usually colonial Zooantharia, known as true corals, having a compact calcareous skeleton and polyps with no siphonoglyph.

scleratogenous layer strand of fused sclerotomes formed along the neural tube, later surrounding the notochord.

sclere *n.* a small skeletal structure; a sponge spicule.

sclereid *n.* a type of sclerenchyma cell with a thick lignified wall, making up some seed coats, nutshells, the stone or endocarp of stone fruits, and which gives the flesh of pears its gritty texture.

sclerenchyma *n.* plant tissue with thickened, usually lignified cell walls, which acts as a supporting tissue; hard tissue of coral. *a.* sclerenchymatous.

sclerid sclereid *q.v.*

sclerification *n.* the process of becoming sclerenchyma.

sclerins scleroproteins *q.v.*

sclerite *n.* hard plate or spicule which may be of keratin, calcium carbonate or chitin and is a skeletal or supporting element in invertebrates. *a.* scleritic.

sclerobase *n.* the calcareous axis of alcyonarians.

sclerobasidium *n.* a thick-walled resting body or encysted probasidium of rust and smut fungi.

scleroblast *n.* a sponge cell from which a sclere or spicule develops; an immature sclereid.

scleroblastema *n.* embryonic tissue involved in development of skeleton.

scleroblastic *a. appl.* skeleton-forming tissues.

sclerocarp *n.* the hard seed coat or stone, usually the endocarp, of succulent fruit.

sclerocauly *n.* condition of excessive skeletal structure in a stem.

sclerocorneal *a. pert.* sclera and cornea.

scleroderm *n.* a hard integument; the hard skeletal part of corals.

sclerodermatous *a.* with an external skeletal structure; with horny, bony or calcareous plates in the skin.

sclerodermite *n.* the part of exoskeleton over one arthropod segment.

sclerogen *n.* wood-producing cells, i.e. the vascular cambium.

sclerogenic, sclerogenous *a.* producing lignin.

scleroid, sclerous *a.* hard; skeletal.

scleromeninx dura mater *q.v.*

sclerophyll *n.* plant with tough evergreen leaves; one of the leaves of such a plant.

sclerophyllous *a.* hard-leaved, *appl.* leaves that are resistant to drought through having a thick cuticle, much sclerenchymatous tissue and reduced intercellular air spaces.

sclerophylly *n.* condition of excessive skeletal structure in leaves.

scleroprotein *n.* any of a group of proteins occurring in connective, skeletal and epidermal tissues, such as collagen, chondrin, elastin, keratin etc.

scleroseptum *n.* a radial vertical wall of calcium carbonate in scleractinian corals.

sclerosis *n.* hardening by an increase in the amount of connective tissue (in animals) or lignin (in plants).

sclerospermous *a.* having the seeds covered by a hard coat.

sclerotal sclerotic *q.v.*

sclerotesta *n.* the hard lignified inner layer of testa (seed coat).

sclerotic *a.* hard; containing lignin; *pert.* sclerosis; *pert.* sclera; having undergone sclerosis; *n.* the sclera of eye.

sclerotic ossicles a ring of small bones around sclera of birds; plates surrounding the eye in certain fishes.

sclerotin *n.* a highly resistant, stable, quinone-tanned protein occurring in insect cuticle and amongst structural proteins in many groups of invertebrates and vertebrates.

sclerotioid, sclerotiform *a.* like or *pert.* a sclerotium.

sclerotium *n.* a resting or dormant stage of some fungi when they become a mass of hardened or mummified tissue. *plu.* sclerotia.

sclerotization *n.* the process of hardening and darkening the new exoskeleton which occurs in insects after moulting.

sclerotome *n.* part of somite of vertebrate embryo that develops into cartilage, and in some vertebrates later into bone.

sclerous *a.* hard; sclerotic *q.v.*

scobina *n.* a spikelet of grasses.

scobinate *a.* having a rasp-like surface.

scobiscular, scobisculate, scobiform *a.* granulated; resembling sawdust.

scoleces *plu.* of scolex.

scolecid *a. pert.* to a scolex.

scoleciform *a.* like a scolex.

scolecite *n.* a worm-shaped body branching from mycelium of discomycete fungi; Woronin hypha, a hypha inside coil of perithecial hyphae and giving rise to ascogonia.

scolecoid *a.* like a scolex.

scolespore *n.* a worm-like or thread-like spore.

scolex *n.* region at anterior end of tapeworm containing minute hooks and a sucker, by which it attaches itself to the gut wall.

scolite *n.* a fossil worm burrow.

scolopale *n.* vibratile central peg-like portion of a sensory sensilla.

scolophore, scolpidium chordotonal sensilla *q.v.*

scolus *n.* a thorny process of some insect larvae.

scombrids *n.plu.* fish of the mackerel and tuna family (Scombridae).

scopa pollen brush *q.v.*

scopate, scopiform, scopiferous, scopulate *a.* like a brush.

scopula *n.* a small tuft of hairs; in climbing spiders, an adhesive tuft of club-like hairs on each foot, replacing 3rd claw.

scopuliferous *a.* having a small brush-like structure.

scopuliform *a.* resembling a small brush.

scorpioid *a.* resembling a scorpion; (*bot.*) circinate, *appl.* inflorescence.

scorpioid cyme a cymose inflorescence with one axis at each branching, and in which daughter axes are developed right and left alternately.

scorpion *see* Scorpiones.

Scorpiones, Scorpionoidea *n.* order of arachnids including the scorpions, which have a dorsal carapace on the prosoma, an opisthosoma divided into two regions, with the posterior segments forming a flexible tail bearing a terminal poisonous sting which is used to paralyse prey. They are viviparous.

scorpion flies common name for the Mecoptera *q.v.*

scorteal *a. appl.* or *pert.* a tough cortex, as of certain fungi.

Scotobacteria *n.* in modern classifications the large class of bacteria containing all heterotrophic Gram-negative bacteria and some other heterotrophic groups, the name signifying "bacteria indifferent to light".

scotoma *n.* blind spot, point at which vision is absent within visual field.

scotopia *n.* adaptation of the eye to darkness.

scotopic vision vision at low intensities of light, in shades of grey, involving the rod cells of the vertebrate retina; dark-adapted or "night" vision.

scotopsin opsin *q.v.*

scramble competition situation where a resource is shared equally between competitors.

scrapie *n.* a neurodegenerative disease of sheep, one of the transmissible spongiform encephalopathies, caused by an agent not yet fully characterized.

scrobe *n.* a groove on either side of rostrum of beetles.

scrobicula, scrobicule *n.* the smooth area around the boss of an echinoderm test.

scrobicular *a.* in the region of the scrobicula.

scrobiculate *a.* marked with little pits or depressions.

scrobiculus, scrobicule *n.* a small pit or depression.

Scrophulariales *n.* an order of dicot trees, shrubs, herbs and vines, including the families Acanthaceae (acanthus), Bignoniaceae (jacaranda), Buddleiaceae (buddleiea), Solanaceae (nightshade, potato) and others.

scrotal *a. pert.* or in the region of the scrotum.

scrotum *n.* external sac or sacs containing testicles in mammals; covering of testis in insects.

scrounger *n.* in animal behaviour, an animal that waits for another animal of a different species to catch its prey, and then takes the prey from it. *see also* kleptoparasitism.

scrub *n.* a plant community dominated by shrubs.

scurf *n.* scaly skin; dried outer skin peeling off in scales; scaly epidermal covering of some leaves.

scurvy *n.* deficiency disease caused by a lack of vitamin C (ascorbic acid), which amongst other symptoms prevents formation of effective collagen fibres leading to skin lesions and blood vessel fragility.

scuta *plu.* of scutum *q.v.*

scutate *a.* protected by large scales or horny plates.

scute *n.* an external scale as of reptile, fish or scaly insect; plate of shell of turtles and tortoises. *see also* scutum.

scutella *plu.* of scutellum *q.v.*

scutellar *a. pert.* a scutellum.

scutellate *a.* shaped like a small shield.

scutellation *n.* arrangement of scales, as on leg of bird.

scutelliform *a.* shaped like a small shield.

scutelligerous *a.* furnished with scutella or a scutellum.

scutelliplantar *a.* having tarsus covered with small plates or scutella.

scutellum *n.* (*bot.*) development of part of cotyledon that separates embryo from endosperm in seed of grasses; (*zool.*) any small shield-shaped structure; posterior part of insect notum; scale on tarsus of birds.

scutiferous *a.* having scutella or a scutellum.

scutiform *a.* shaped like a shield.

scutigerous *a.* bearing a shield-like structure.

scutiped *a.* having foot, or part of it, covered by scutella.

scutum *n.* (*bot.*) broad apex of style as in dicots of family Asclepiadaceae; (*zool.*) a shield-like plate, horny, bony or chitinous, formed in the outer covering of an animal (as scales); middle portion of insect notum; dorsal shield of ticks; one of the pair of anterior valves in goose barnacles. *alt.* scute, shield. *plu.* scuta.

scyllitol *n.* a sweet alcohol, $C_6H_6(OH)_6$, related to inositol.

scyphi *plu.* of scyphus *q.v.*

scyphiferous *a.* bearing scyphi, as some lichens.

scyphiform, scyphoid, scyphose *a.* shaped like a cup.

scyphistoma *n.* inconspicuous asexual polyp stage of jellyfish (Scyphozoa).

Scyphozoa, scyphozoans *n., n.plu.* the jellyfish, a class of marine coelenterates of the phylum Cnidaria, with a dominant medusa stage which is free-swimming or attached by an aboral stalk, and no velum.

scyphula scyphistoma *q.v.*

scyphulus *n.* small cup-shaped structure.

scyphus *n.* funnel-shaped corolla of daffodil; a cup-shaped outgrowth bearing apothecium in some lichens.

scyrps colloquial term for small cytoplasmic ribnucleoprotein particles, from the abbreviation scRNPs.

SDS-polyacrylamide gel electrophoresis technique for separating e.g. membrane proteins, in which the membrane complex is first solubilized in the detergent sodium dodecyl sulphate and then subjected to electrophoresis in a polyacrylamide gel to separate the constituent proteins.

sea anemones common name for an order (Actiniaria) of coelenterates of the Zoantharia, which are generally solitary and have no skeleton. They have a hollow cylindrical body, often anchored to rocks, opening at one end in a small mouth surrounded by a ring of tentacles often numbering multiples of six.

sea butterflies pteropods *q.v.*

sea combs common name for the Ctenophora *q.v.*

sea cows the Sirenia, including the dugong and manatee, marine placental mammals highly specialized for an aquatic life with a naked body and front limbs modified as paddles.

sea cucumbers a common name for the Holothuroidea *q.v.*

sea fans gorgonians *q.v.*

sea gooseberries common name for the Ctenophora *q.v.*

sea lilies common name for the Crinoidea *q.v.*

sea mouse common name for the polychaete *Aphrodite*, which has a broad stout oval shape.

sea pens a group of corals of the subclass Alcyonaria, which form stalked colonies markedly resembling quill pens, composed of two different kinds of polyp.

sea slugs shell-less marine molluscs of the class Gastropoda, subclass Opisthobranchia.

sea spiders Pycnogonida *q.v.*

sea squirts common name for the Ascidiacea *q.v.*

sea stars a common name for the Asteroidea *q.v.*

sea urchins a common name for the Echinoidea *q.v.*

sea wasps box-jellies *q.v.*

search(ing) image a transitory filtering of external visual stimuli that enables an animal to focus its attention on finding e.g. a prey item of a particular colour or shape.

seashore *n*. the ground bordering the sea, between the highest high-water and lowest low-water marks, also including the splash zone above high-water mark. *see also* intertidal, littoral, lower shore, middle shore, splash zone, sublittoral, upper shore. *see Fig. 9* for shore zonation.

seaweed *n*. marine multicellular algae belonging to various groups. *see* Chlorophyta, Phaeophyta, Rhodophyta.

sebaceous *a*. secreting or containing oils or fats, *appl*. glands of the skin secreting

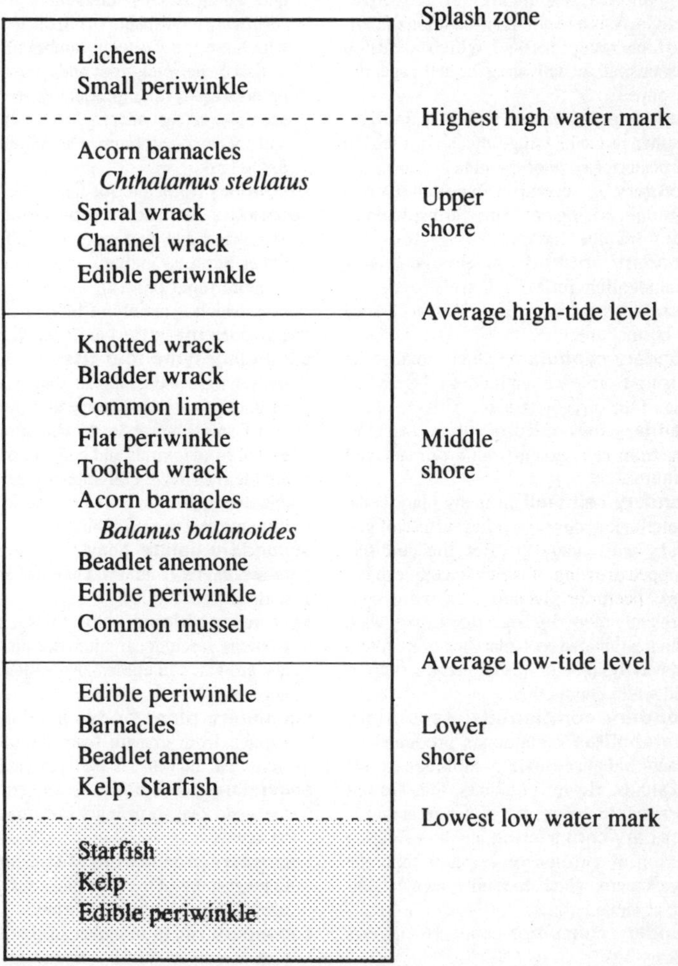

	Splash zone
Lichens Small periwinkle	
- -	Highest high water mark
Acorn barnacles *Chthalamus stellatus* Spiral wrack Channel wrack Edible periwinkle	Upper shore
	Average high-tide level
Knotted wrack Bladder wrack Common limpet Flat periwinkle Toothed wrack Acorn barnacles *Balanus balanoides* Beadlet anemone Edible periwinkle Common mussel	Middle shore
	Average low-tide level
Edible periwinkle Barnacles Beadlet anemone Kelp, Starfish	Lower shore
- -	Lowest low water mark
Starfish Kelp Edible periwinkle	

Fig. 9 Typical zonation for a sheltered Atlantic European rocky shore.

sebum.

sebiferous *a.* conveying fatty material.

sebific *a.* sebaceous *q.v.*; *appl.* gland: colleterium *q.v.*

sebum *n.* material rich in lipids (oils and fats) secreted by sebaceous glands of skin.

secodont *a.* furnished with teeth adapted for cutting.

second messengers compounds such as cyclic AMP, diacylglycerol and inositol phosphates, which are formed intracellularly as a result of stimulation of cell-surface receptors and which are then responsible for activating the cell's specific response.

secondaries *n.plu.* one type of main flight feather in bird's wing, attached in region of posterior edge of the ulna.

secondary *a.* second in importance or in position; arising not from a growing point but from other tissue.

secondary antibody heterologous anti-immunoglobulin *q.v.*

secondary bud an axillary bud, accessory to normal one.

secondary capitulum the six small cells arising from each capitulum of green algae of the order Charales.

secondary cell culture culture originating from cells taken from a primary cell culture.

secondary cell wall in many plant cells, material laid down on inner surface of primary wall, usually after the cell has stopped growing. It is rich in cellulose but lacks pectin or glycoproteins and is consequently more rigid than the primary wall. Thick laminated secondary walls are found especially in cells specialized for support and water conduction.

secondary compounds, secondary metabolites compounds produced by plants and microbes, e.g. antibiotics, plant alkaloids, flower pigments, that are not essential to the growth of the organism.

secondary constriction any non-staining region of chromosome, other than the centromere, which does not attach to spindle at metaphase.

secondary consumer carnivore that eats herbivores.

secondary cortex phelloderm *q.v.*

secondary ecological succession secondary succession *q.v.*

secondary forest, secondary woodland forest or woodland that has developed as a result of secondary succession after complete clearance of pre-existing forest, or which has been planted.

secondary growth in plants, growth bringing about an increase in the thickness of stem and root, as opposed to extension of plant body at the apices of shoots and roots, and which is most marked in trees and shrubs. It is initiated at lateral meristems which are the vascular cambium, a layer of tissue encircling root and stem between phloem and xylem, producing new xylem and phloem (secondary xylem and phloem) and the cork cambium which contributes to the bark.

secondary host intermediate host *q.v.*

secondary immune response immune response made on a second or subsequent exposure to an antigen, usually resulting in more rapid onset of antibody production, which is mainly of IgG.

secondary meristem cork cambium *q.v.*

secondary lymphoid tissues in mammals, lymph nodes, spleen, tonsils, Peyer's patches, adenoids and appendix. They contain T and B lymphocytes which have migrated from thymus and bone marrow, and are the sites at which lymphocytes mature, encounter foreign antigen, and at which immune reactions are initiated.

secondary palate bony plate separating mouth cavity from nasal cavities in mammals and crocodiles.

secondary phloem phloem tissue formed from the vascular cambium during secondary growth, sometimes also called the inner bark.

secondary plant body the plant body formed from growth from lateral meristems, i.e. the vascular and cork cambiums.

secondary production in an ecosystem, the yield due to primary consumers, i.e. herbivores.

secondary prothallium a tissue produced in megaspore of the club moss *Selaginella* after true prothallium is formed.

secondary roots branches of the primary root, arising within its tissue and in turn giving rise to tertiary roots; roots arising at other than normal points of origin.

secondary sexual characteristics features characteristic of a particular sex other than the gonads and genitalia, usually developing under the influence of androgens and oestrogens, and including growth of a beard in men, antlers in stags, and enlarged breasts in women.

secondary structure the two-dimensional configuration of a protein chain or a polynucleotide chain in terms of interactions between amino acids or nucleotides relatively close to one another in the linear sequence.

secondary succession a plant succession following the interruption of the normal or primary succession.

secondary thickening *see* secondary growth.

secondary wood secondary xylem, *see* secondary growth.

secondary xylem *see* secondary growth.

second-order conditioning (*q.v.*) type of classical conditioning (*q.v.*) in which a second stimulus is associated with the conditional stimulus so that the conditioned animal comes to respond to the second conditional stimulus alone.

secrete *v.* to release material or fluid from a cell or tissue.

secretin *n.* polypetide hormone produced by duodenum during digestion and which stimulates pancreas to produce pancreatic juice containing digestive enzymes.

secretion *n.* material or fluid which is produced and released from a cell or gland.

secretitious *a.* produced by secretion.

secretor *n.* person who secretes blood group antigens in saliva and other body fluids, a genetically-determined trait.

secretory *a. appl.* cells and tissues that secrete substances such as digestive enzymes, polypeptide hormones, neuro-transmitters or complex material such as mucus, slime etc.; *appl.* proteins and other material that are secreted.

secretory component, secretory piece small protein component present in IgA found in mucous secretions but not in serum IgA.

secretory granule small membrane-bounded vesicle in cytoplasm of eukaryotic cells, derived from the Golgi apparatus, and containing material to be secreted.

sectile *a.* cut into small partitions or compartments.

section *n.* thin slice of tissue prepared for microscopy; a taxonomic group, often used as a subdivision of a genus, but used in different ways by different authors and never precisely defined.

secular *a.* long term, over a long period of time.

secund *a.* arranged on one side, *appl.* flowers on a stem.

secundiflorous *a.* having flowers on one side of stem only.

secundine *n.* the internal integument of ovule.

secundines *n.plu.* the foetal membranes collectively.

secundly *adv.* on one side of a stem or axis.

sedentaria *n.plu.* sessile or sedentary organisms.

sedentary *a.* not free living, *appl.* animals attached by a base to some substratum.

sedoheptulose *n.* 7-carbon ketose sugar, as the phosphate and bisphosphate involved esp. in carbon dioxide fixation in photosynthesis.

seed *n.* reproductive unit formed from a fertilized ovule, and consisting of an embryo, food store and protective coat. Produced by gymnosperms and angiosperms; semen *q.v.*; *v.* to introduce microorganisms into a culture medium.

seed bank a conservation collection of seeds of wild plant species and local cultivated varieties, usually of important crop plants, kept in long-term storage, usually freeze-dried under liquid nitrogen. Seeds can be germinated as required to provide material for study and for plant breeding programmes.

seed coat thin outer coat of mature seed, which may be dry and papery or hard and highly impermeable to water, which develops from the integuments of ovule. *alt.* testa.

seed ferns the Pteridospermophyta, a division of fossil seed-bearing vascular plants from the late Devonian and Carboniferous, which had fern-like leaves bearing seeds. They included both small shrubby and tall tree-like forms. The large pinnately compound leaves were borne at the top of the trunk.

seed leaf cotyledon *q.v.*

seed plants all seed-bearing plants, i.e. the gymnosperms and angiosperms, sometimes collectively termed Spermatophyta.

seed stalk *see* funicle.

seed storage proteins simple proteins produced in large quantities within seeds where they act as nitrogen storage compounds, and which are broken down and the amino acids utilized during germination and seedling growth.

seed vessel fruit, esp. a dry fruit.

segment n. a division formed by cleavage of an ovum; part of an animal or of a jointed appendage; division of a leaf left nearly to base; portion of a chromosome.

segmental a. of the nature of a segment; *pert.* a segment.

segmental duct an embryonic nephridial duct which gives rise to Wolffian or Müllerian duct.

segmental interchange exchange of non-homologous segments as between two chromosomes; reciprocal translocation.

segmental organ an embryonic excretory organ. *alt.* nephridium.

segmental arteries diverticula from dorsal aorta arising in spaces between successive somites.

segmental papillae conspicuous pigment spots by which true segments can be recognized in leeches.

segmental reflex a reflex involving a single segment of spinal cord, i.e. not involving additional input from brain or other parts of nervous system.

segmentation n. splitting into segments or portions; the repetition of a series of essentially similar segments along the length of the body of the animal, as seen esp. in arthropods and annelids. In annelids, each segment has a similar pattern of blood vessels, nerves, muscles, etc., but in adult arthropods the internal structure is less obviously segmented; the process in the embryonic development of insects and other segmented animals by which the segmented structure is established.

segmentation nucleus body formed by union of male and female pronuclei at fertilization of an ovum.

segmented a. *appl.* the genomes of double-stranded and some single-stranded RNA viruses, which consist of two or more separate molecules each carrying a different gene or genes.

segregation n. separation of parental homologous chromosomes at meiosis, and the consequent separation of alleles at a locus and their distribution to different gametes.

segregation of alleles the first of Mendel's laws, which describes the fact that alleles of a gene segregate unchanged by passing into different gametes at the formation of the next generation.

seiroderm n. dense outer tissue, composed of parallel chains of hyphae, in certain fungi.

seirospore n. one of spores arranged like a chain.

seismathesia n. perception of mechanical vibration.

seismonasty n. plant movements in response to mechanical shock or vibration. *a.* seismonastic.

seismotaxis n. a taxis in response to mechanical vibrations.

sejugate, sejugous *a.* with six pairs of leaflets.

Selachii, selachians *n., n.plu.* a class (or in some classifications an order) of cartilaginous fishes containing the sharks, dogfishes, skates and rays, having claspers and fins with a constricted base, existing from the Devonian to the present day.

selectable marker any characteristic by which a cell with a particular property can be selected during an experiment. One commonly used type of selectable marker is a gene, e.g. for antibiotic resistance, that is placed on a vector so that any cell receiving a recombinant DNA molecule can be selected by survival in medium containing antibiotics.

selectin n. any of a large family of structurally related cell-surface proteins, found on white blood cells and epithelial cells, which cause a weak adhesion of white blood cells to epithelia by binding to ligands on the other cell surface and which are particularly involved in cell–cell interactions during immune responses and inflammation.

selection n. non-random differential reproduction of different genotypes; natural selection *q.v. see also* directional selection, disruptive selection, sexual selection, stabilizing selection.

selection, coefficient of coefficient of selection *q.v.*

selection pressure the effect of any feature of the environment that results in natural selection, e.g. food shortage, predator activity, competition from members of the same or other species, etc.

selective advantage *pert.* any character that gives an organism a greater chance of surviving to reproductive age, breeding and rearing viable offspring.

selector genes genes involved in selecting alternative states in development, as e.g. the homoeotic genes of *Drosophila*.

selenodont *a. appl.* molars lengthened out anteroposteriorly and curved; *appl.* or having molar teeth with crescent-shaped ridges on the grinding surface, as in artiodactyls.

selenoid *a.* shaped like a crescent.

selenophyte *n.* plant tolerating quite high levels of selenium in the soil.

selenotropism *n.* tendency to turn towards the moon's rays.

selenozone *n.* lateral stripe of crescentic growth lines on whorl of shell of a gastropod mollusc.

self-assembly *n.* the capacity of multisubunit proteins and large macromolecular complexes such as a virus particle to assemble without the agency of other proteins. In many cases correct self-assembly may be facilitated *in vivo* by proteins known as molecular chaperones.

self-compatible self-fertile *q.v.*

self-fertile, self-sterile capable, or incapable, respectively, of being fertilized by its own male gametes, *appl.* hermaphrodite animals and flowers.

self-fertilization the fusion of male and female gametes from the same individual.

self-incompatibility self-sterility *q.v.*

self-pollination transfer of pollen grains from anthers to stigma of the same flower.

self-splicing introns introns in some RNAs, such as that from the 16S rRNA from the protozoan *Tetrahymena thermophila*, which are able to excise themselves, leaving the exons correctly joined together, without the agency of protein enzymes.

self-sterility the inability of a hermaphrodite to produce viable offspring by self-fertilization.

self-tolerance the inability to mount an immune response against the body's own antigenic components.

selfing *n.* self-pollination or self-fertilization.

selfish DNA DNA without any apparent function in its host cell, and which maintains itself in the genome by virtue of its own intrinsic characteristics including the ability to integrate and transpose, and in some cases to amplify itself to form multiple repeated copies.

selfish gene the idea that genes are primarily concerned with replicating themselves and passing on more copies of themselves to future generations, the organism in which they are carried being their means for doing this.

selfishness *n.* in sociobiology, behaviour that benefits the individual in terms of genetic fitness at the expense of genetic fitness of other members of the same species.

sella turcica a deep depression in upper surface of sphenoidal bone, lodging the pituitary body.

sellaeform *a.* saddle-shaped.

selliform *a.* saddle-shaped.

selva *n.* the tropical rain forest.

semantide *n.* a molecule that carries information, such as DNA and RNA.

sematectonic *a. appl.* communication by means of objects.

sematic *a.* functioning as a danger signal, as warning colours or odours, *appl.* warning and recognition markings. *see also* episematic, parasematic.

semelparity *n.* production of offspring by an organism all at one time. *a.* semelparous. *cf.* iteroparity.

semen *n.* fluid secreted from testes and accessory glands, and containing sperm.

semi- prefix derived from L. *semi*, half, signifying half or partly.

semiamplexicaul *a.* partly surrounding the stem.

semianatropous *a.* with half-inverted ovule.

semiarid regions dry regions with sufficient rainfall (280–400 mm per annum) to support steppe or savanna grassland and some agriculture.

semiautonomous *a. appl.* organelles such as chloroplasts and mitochondria which contain DNA directing the synthesis of some of their own proteins, and are self-replicating.

semicaudate *a.* with a rudimentary tail.

semicells *n.plu.* the two halves of a cell which are interconnected by an isthmus in some green algae.

semicircular canals the semicircular membranous tubes filled with endolymph in the inner ear, which are concerned with perception of body position, balance etc. In mammals there are three, one vertical and two more-or-less horizontal at right angles to each other, all interconnected at their bases in the sac-like utriculus. Each canal has a swollen base, the ampulla, containing sensory cells that detect the movement of endolymph with body movements. Other sensory cells detect the movement of small calcareous particles under gravity.

semiclasp *n.* one of two apophyses which may combine to form the clasper in certain male insects.

semicomplete *a.* incomplete, *appl.* metamorphosis.

semiconservative replication way in which double-stranded DNA replicates itself, each strand of the double helix serving as a template for synthesis of a new strand.

semicylindrical *a.* round on one side, flat on the other, *appl.* leaves.

semidiscontinuous replication the usual method of DNA replication in which elongation of one new strand proceeds continuously and the other discontinuously because of the opposite polarities of the 2 strands in the DNA duplex and the ability of DNA polymerase to elongate DNA in one direction only.

semidominant codominant *q.v.*

semifloret, semifloscule *n.* a strap-shaped floret in flowers of Compositae.

semiherbaceous *a.* having lower part of stem woody and upper part herbaceous.

semilethal *a.* not wholly lethal; *appl.* genes causing a mortality of more than 50% or permitting survival until reproduction has been effected.

semiligneous *a.* partially lignified, with stem woody only near the base.

semilocular *a. appl.* ovary with incomplete loculi.

semilunar *a.* half-moon shaped, *appl.* branches of carotid artery, fibrocartilages of knee, ganglia, fascia, lobules of cerebellum, valves; *appl.* notch: greater sigmoid cavity between olecranon and coronoid process of ulna; *n.* a carpal bone: lunar bone *q.v.*

semimembranosus *n.* a thigh muscle with flat membrane-like tendon at upper extremity.

semimetamorphosis *n.* partial, or semi-complete metamorphosis.

seminal *a.* (*zool.*) *pert.* semen.

seminal fluid fluid component of semen, secreted by the seminal vesicles and prostate gland in mammals.

seminal funnel internal opening of vasa deferentia in oligochaete worms.

seminal receptacle spermatheca *q.v.*

seminal root the first formed root, developed from the radicle of the seed.

seminal vesicle one of pair of glands of male mammals, which secrete the alkaline fluid component of semen; organ for storing sperm in some invertebrates and lower vertebrates.

semination *n.* dispersal of seeds: discharge of sperm.

seminiferous *a.* secreting or conducting semen; bearing seed.

seminiferous tubules long tightly coiled tubules in the mammalian testis in which the sperm are produced.

seminude *a.* with ovules or seeds exposed.

seminymph *n.* stage in development of insects approaching complete metamorphosis.

semiorbicular *a.* half-rounded, hemispherical.

semiotics *n.* the study of communication.

semiovate *a.* half-oval; somewhat oval.

semioviparous *a.* between oviparous and viviparous, as a marsupial whose young are imperfectly developed when born.

semiovoid *a.* somewhat ovoid in shape.

semipalmate *a.* having toes webbed half-way down.

semiparasite *n.* a plant which only derives part of its nutrients from host and has some photosynthetic capacity.

semipenniform *a. appl.* certain muscles bearing some resemblance to the lateral half of a feather.

semipermeable *a.* partially permeable; *appl.* membrane permeable to a solvent, esp. water, but not to solutes.

semiplacenta *n.* a non-deciduate placenta.

semiplume *n.* a feather with ordinary shaft but with a downy web.

semipupa *n.* coarctate pupa *q.v.*

semirecondite *a.* half-concealed, as insect head by thorax.

semisagittate *a.* shaped like a half arrow-head.

semisaprophyte *n.* a plant partially saprophytic.

semisocial insects those social insects in which there is cooperative care of the brood, a separate sterile worker caste but in which there is no overlap of generations caring for the brood. *cf.* eusocial insects.

semispecies *n.* a taxonomic group intermediate between a species and subspecies esp. as a result of geographical isolation.

semispinalis *n.* a muscle of back and neck, on each side of spinal column.

semistreptostylic *a.* between monimostylic and streptostylic; with slightly moveable quadrate.

semitendinosus *n.* a dorsal muscle of the thigh stretching from the tuberosity of the ischium to the tibia.

semitendinous *a.* half-tendinous.

semituberous *a.* having somewhat tuberous roots.

Sendai virus an enveloped RNA virus of the family Paramyxoviridae, which enters cells by fusion of viral envelope with cell membrane and which also may cause cells to fuse. Inactivated form used to induce cell fusion in the production of somatic cell hybrids.

senescence *n.* advancing age; the complex of ageing processes that eventually lead to death. *a.* senescent.

senile plaque area of abnormal deposition of the complex proteinaceous material amyloid, and destruction of neurones, in brains of patients with Alzheimer's disease.

senility *n.* degeneration due to old age.

sense organ an organ receptive to external stimuli, as eye, ear, etc. *alt.* receptor.

sense strand the DNA strand that is transcribed. *alt.* anticoding strand.

sensile *a.* capable of affecting a sense.

sensilla *n.* a small sense organ. *plu.* sensillae.

sensitive *a.* capable of receiving impressions from external objects; reacting to a stimulus.

sensitization *n.* condition in which an animal or human produces an enhanced immune response, or in some cases an allergic response, on a second encounter with an antigen. *see also* hypersensitivity.

sensorium *n.* the seat of sensation or consciousness; the entire nervous system with sense organs; the sensory, neuromuscular and glandular system. *a.* sensorial.

sensory *a.* involved in the reception, or processing, of stimuli from the external or internal environment; of or concerned with sensation or the senses.

sensory cortex those parts of the cortex in brain devoted to the reception and initial analysis of sensory signals, e.g. visual cortex, auditory cortex, olfactory cortex, somatosensory cortex.

sensory neurone nerve cell concerned with carrying impulses from a sense organ or sensory receptor to the central nervous system.

sensory receptor any cell or part of a cell specialized to respond to stimuli such as light, vibration, mechanical deformation, heat, etc. and to convey signals to the central nervous system, either directly as in sensory neurones whose terminals are modified as receptors or indirectly via connections to sensory neurones.

sensu lato in a broad sense.

sensu stricto in a restricted sense.

sentient *a.* capable of perceiving through the senses; conscious; *appl.* sensory cells.

sepal *n.* modified sterile leaf, often green and with others forming the calyx, or outer series of perianth segments, of a flower, sometimes the same colour as and resembling the petals. *see* Fig. 4.

sepaled *a.* having sepals.

sepaline, sepaloid *a.* like a sepal.

sepalody *n.* conversion of other parts of the flower into sepals.

sepaloid *a.* like a sepal.

sepalous *a.* having sepals.

separation layer abscission layer *q.v.*

sepia *n.* the brown "ink" released by the cuttlefish *Sepia* to distract predators when threatened.

sepiapterin *n.* yellow pteridine pigment, a presumed intermediate in the formation of the eye pigment drosopterin in certain insects.

sepicolous *a.* living in hedges.

sepiment *n.* a partition.

sepion *n.* calcareous shell of cuttlefish.

sepiparous sebaceous *q.v.*

septa *plu.* of septum *q.v.*

septal *a. pert.* a septum; *pert.* hedgerows, *appl.* flora.

septal neck in the cephalopod *Nautilus*, a shelly tube continuous for some distance beyond each septum as support for the siphuncle.

septate a. divided by partitions (septa).

septate junction intercellular junction in insect cells, which has a ladder-like or honeycomb appearance in the electron microscope.

septempartite *a. appl.* leaf with seven divisions extending nearly to the base.

septenate *a.* with parts in sevens.

septibranch *a. appl.* gills of bivalve molluscs which are small and transformed into a transverse muscular pumping partition.

septicidal *a.* dividing through middle of ovary septa; dehiscing at septa, *appl.* fruits.

septiferous *a.* having septa.

septifolious *a.* with seven leaves or leaflets.

septiform *a.* in the shape of a septum.

septifragal *a.* with slits as in septicidal dehiscence, but with septa broken and placentae and seeds left in middle.

septomaxillary *a. pert.* maxilla and nasal septum; *appl.* a small bone in many amphibians and reptiles and in certain birds.

septonasal *a. pert.* the internal partition between the nostrils.

septulum *n.* a small or secondary septum. *a.* septulate.

septum *n.* a partition separating two cavities or masses of tissue, as in fruits, chambered shells, fungal hyphae, corals, heart, nose, tongue, etc. *plu.* septa.

septum lucidum thin inner walls of cerebral hemispheres, between corpus callosum and fornix. *alt.* septum pallucidum.

septum narium the partition between nostrils.

septum transversum foetal diaphragm; ridge within ampulla of semicircular canal.

sequence *see* DNA sequence, protein sequence.

sequenator *n.* machine for the automated sequencing of proteins.

sera *plu.* of serum.

seral *a. pert.* sere; *appl.* a plant community before reaching equilibrium or climax; *pert.* blood serum.

sere *n.* a successional series of plant communities; a stage in a succession.

serial endosymbiosis theory the idea, now generally accepted, that mitochondria and chloroplasts, and possibly some other organelles of eukaryotic cells originated as endosymbiotic microorganisms, mitochondria being acquired first, chloroplasts being acquired later only in the line or lines that led to algae and plants.

seriate *a.* arranged in a row or series.

sericate, sericeous *a.* covered with fine close-pressed hairs; silky.

serine (Ser, S) *n.* β-hydroxyalanine, an amino acid with an aliphatic hydroxyl side chain, non-essential in human diet, constituent of protein, and which is also an important intermediate in phosphatide synthesis.

serine proteinases (proteases) a group of proteinases with a histidine and a serine residue involved in catalysis at the active site, e.g. chymotrypsin, elastase, subtilisin.

serine-threonine protein kinase a protein kinase that phosphorylates serine or threonine side chains on its target proteins. The kinases activated by cyclic AMP are of this type.

serodeme *n.* a deme, e.g. of parasites, differing from others on immunological criteria, i.e. possessing different surface antigens.

serological *n. pert.* serology; *appl.* use of specific antisera to characterize bacteria, viruses, histocompatibility antigens, etc., as opposed to direct biochemical, genetic or molecular biological investigations.

serology *n.* the study of immune sera and the use of antisera to characterize pathogens, antigens, cells, etc.

serosa *n*. lining of peritoneal, pleural and pericardial cavities, also extending as a fine membrane over internal organs; false amnion or outer layer of amniotic fold; outer larval membrane of insects. *a*. serosal.

serosity *n*. watery part of animal fluid; condition of being serous.

serotinal, serotinous *a*. appearing or blooming late in the season; *pert*. later summer; flying late in the evening, *appl*. bats.

serotonin 5-hydroxytryptamine *q.v.*

serotoninergic *a. appl*. neurones releasing the neurotransmitter serotonin (5-hydroxytryptamine).

serotype *n*. a subdivision of a species, esp. of bacteria or viruses, characterized by its antigenic character.

serous *a*. watery; *pert*. serum or other watery fluid.

serous membrane a thin membrane of connective tissue, lining some closed cavity of the body, and reflected over viscera, as a mesentery.

Serpentes *n*. suborder of reptiles comprising the snakes.

serpin *n*. any of a structurally related family of proteins, including α_1-antitrypsin, which inhibit proteases.

serra *n*. any saw-like structure.

serrate *a*. saw-toothed.

serrate-ciliate *a*. with hairs fringing toothed edges, *appl*. leaves.

serrate-dentate *a*. with serrate edges themselves toothed, *appl*. leaves.

serratiform *a*. like a saw.

serration *n*. a series of saw-like notches, on edge of leaf, etc.

serratirostral *a*. with serrated beak, *appl*. birds.

serratodenticulate *a*. with many-toothed serrations.

serratulate serrulate *q.v.*

serrature *n*. a saw-like notch; serration.

serratus anterior, serratus magnus a muscle stretching from upper ribs to scapula.

serratus posterior superior and inferior: two thin muscles of the chest aiding in respiration, spreading respectively backward to anterior ribs and forward to posterior ribs.

serriferous *a*. having a saw-like organ or part.

serriform *a*. like a saw.

serriped *a*. with notched feet.

serrula *n*. a comb-like ridge on chelicerae of some arachnids; a free-swimming larva of alcyonarians, preceding planula.

serrulate *a*. finely notched.

serrulation *n*. fine notches.

Sertoli cell large cell of the epithelium lining in testes, connected with group of developing spermatozoa, acts as a nurse cell.

serule *n*. a minor sere; a succession of minor organisms.

serum *n*. fluid component of blood after removal of cells and fibrinogen; the secretion of a serous membrane; antiserum *q.v.*

serum albumin small protein abundant in blood plasma where it is involved in osmotic regulation and transport of metabolic products.

serum-responsive element *see* SRE.

sesamoid *a. appl*. a bone developed within a tendon and near a joint, as patella, radial or ulnar sesamoid, fabella.

sesamoidal *a. pert*. a sesamoid bone.

sessile *a*. sitting directly on base without a stalk or pedicel, *appl*. flowers, leaves, etc; attached or stationary, as opposed to free living or motile, *appl*. animals, protozoans, etc.

seston *n*. microplankton *q.v.*; all bodies, living and non-living, floating or swimming in water.

set of chromosomes the basic haploid set of chromosomes of any organism.

seta *n*. chitinous hair, bristle, arising from epidermis of many invertebrates, e.g. polychaete and oligochaete annelid worms and insects, *alt*. chaeta; (*bot.*) the stalk bearing the capsule in mosses and liverworts; a bristle-like hair in or on fruiting bodies of some fungi. *plu*. setae.

setaceous *a*. bristle-like; set with bristles.

setiferous, setigerous *a*. bearing setae or bristles.

setiform *a*. bristle-shaped, *appl*. teeth when very fine and closely set.

setigerous *a*. having setae or bristles.

setigerous sac a sac in which is lodged a bundle of chaeta, formed by invagination of epidermis in parapodium of Chaetopoda.

setiparous *a*. producing setae or bristles.

setirostral *a. appl.* birds with bristles on beak.

setobranchia *n.* a tuft of setae attached to gills of certain decapod crustaceans, being coxopodite setae.

setose *a.* bristly.

setula, setule *n.* a thread-like hair or bristle.

setuliform *a.* thread-like, like a seta or fine bristle.

setulose *a.* set with small bristles.

severe combined immunodeficiency (SCID) immunodeficiency, generally genetically based, in which both T and B lymphocytes are absent.

Sewall Wright effect genetic drift *q.v.*

sewage fungus pale slimy growth that often occurs in water with heavy organic pollution, either as a slime or a fluffy fungoid growth with streamers, attached to the bed of the river or pond. It consists of a characteristic community of microorganisms dominated by filamentous and zoogloeal bacteria, but also including fungi and protozoa.

sex *n.* the sum characteristics, structures, features and functions by which a plant or animal is classed as male or female. In some animals sex is entirely genetically determined, in others the sex may change in response to environmental conditions.

sex cell gamete *q.v.*

sex-chromatin body Barr body *q.v.*

sex chromosome chromosomes such as X and Y in humans, which form non-homologous pairs in one of the sexes, and whose presence, absence, or particular form may determine sex, called W and Z in other groups of vertebrates.

sex comb a row of bristles on tarsus of first leg of male in certain insects.

sex-conditioned trait sex-influenced trait *q.v.*

sex-controlled genes genes present and expressed in both sexes, but manifesting themselves differently in males and females.

sex cords proliferations from germinal epithelium of developing gonads which give rise either to seminiferous tubules or to medullary cords of ovary.

sex determination any of various ways in which the sex of an animal is determined.

In many animals sex is determined genetically, i.e. by which combinations of sex chromosomes are carried, or, as in bees, by whether the organism is haploid or diploid. In other animals, however, such as some fishes, sex is environmentally determined by, e.g., temperature.

sex differentiation differentiation of gametes into male and female; differentiation of organisms into kinds with different sexual organs.

sex factors a class of plasmids in bacteria which can induce conjugation and their own and chromosomal gene transfer, including the F-factor, some colicin plasmids and some R-factors.

sex gland gonad *q.v.*

sex hormones chiefly the oestrogens, androgens and gonadotropins.

sex-influenced trait phenotypic character whose expression is influenced by the sex of the individual, such as the simple Mendelian trait of baldness in humans in which the allele is dominant in males but recessive in females. *alt.* sex-conditioned trait.

sex-limited trait phenotypic character expressed only in one sex.

sex-linked *a. appl.* genes carried on only one of the sex chromosomes and which therefore show a different pattern of inheritance in crosses where the male carries the gene from those in which the female is the carrier; *appl.* an inherited trait that is manifest in only one sex.

sex mosaic an intersex *q.v.*; a gynandromorph *q.v.*

sex pilus F pilus *q.v.*

sex ratio the ratio of males to females in a population, may be given as proportion of male births, number of males per 100 females or per 100 births, or as percentage males in the population.

sex reversal, sex transformation a changeover from one sex to the other, natural, pathological or artificially induced.

sexdigitate *a.* having six fingers or toes.

sexduction *n.* in bacteria, the transfer of genes from one bacterium to another mediated by their attachment to a transferable piece of extrachromosomal DNA, the fertility factor (F).

sexfid *a.* cleft into six segments, as calyx.

sexfoil *n.* a group of six leaves or leaflets

around one axis.

sexual *a. pert.* sex; *pert.* reproduction: any kind of reproduction that involves the fusion of gametes to form a zygote.

sexual cell gamete *q.v.*, or cells giving rise to gametes.

sexual coloration colours displayed during the breeding season but not at other times, often different in the two sexes.

sexual cycle menstrual cycle *q.v.*; oestrus cycle *q.v.*

sexual dimorphism marked differences, in shape, size, morphology, colour, etc. between male and female of a species.

sexual imprinting imprinting (*q.v.*) influencing mating preference as adult.

sexual reproduction reproduction involving the formation and fusion of two different kinds of gametes to form a zygote, usually resulting in progeny with a somewhat different genetic constitution from either parental type and from each other.

sexual selection the difference in the ability of individuals of different genetic types to aquire mates, and therefore the differential transmission of certain characteristics to the next generation. It is made up of the choices made between males and females on the basis of some outward characteristic such as bright plumage, length of tail etc. in birds, and competition between members of the same sex.

sexuparous *a.* producing sexual offspring, as after bearing parthenogenetic females in Pterygota.

shadowing *n.* technique for preparing specimens for electron microscopy in which electron-dense metal atoms are evaporated onto the specimen at an angle, throwing surface features into relief.

shaft *n.* stem of feather; stem of hair; scapus *q.v.*; straight cylindrical part of a long bone.

sheath *n.* a protective covering; theca *q.v.*; lower part of leaf enveloping a stem or culm; insect wing cover, especially elytron.

β-sheet secondary structure element common in many proteins, composed of a folding of the polypeptide chain so that strands lie parallel to each other held together by hydrogen bonding.

sheet erosion the removal of topsoil over a wide area after heavy rain.

shell *n.* the hard outer covering of an animal, fruit or some eggs; a calcareous, siliceous, bony, horny or chitinous covering.

shell gland, shell sac organ in which material for forming a shell is secreted.

shield *n.* carapace *q.v.*; clypeus *q.v.*; scutellum *q.v.*; scutum *q.v.*; (*bot.*) disc-like ascocarp or apothecium borne on thallus of lichens.

shikimate (shikimic acid) aromatic carboxylic acid, intermediate in synthesis of aromatic amino acids in microorganisms.

Shine Dalgarno sequence a consensus sequence 5′AGGAGG3′ found in bacterial mRNAs a few bases before the initiation codon, and which is believed to be involved in initiation of translation, possibly by binding to a complementary sequence in the bacterial 16S rRNA of ribosome.

shipworm *n.* bivalve mollusc which is wormlike in form, with a much-reduced shell at the foot and two chalky plates closing the siphon, and which bores into wood.

shivering *n.* mechanism of heat production found in many animals, used by birds and mammals as an emergency protection to maintain body temperature in cold conditions.

shoot *n.* the part of a vascular plant derived from the plumule, being the stem and usually the leaves; a sprouted part, branch or offshoot of a plant.

shore *see* seashore.

short-day plants plants that will only flower if the daily period of light is shorter than some critical length: they usually flower in early spring or autumn. The critical factor is in fact the period of continuous darkness they are exposed to. *cf.* long-day plants, day-neutral plants.

short-grass community type of grassland that develops on poorer soils in dry regions, e.g. short-grass prairie, short-grass steppe, and which consists of grasses no more than 60 cm tall and small herbaceous plants.

short-horned flies Brachycera *q.v.*

short-horned grasshoppers Acrididae *q.v.*

short period interspersion type of sequence arrangement within eukaryotic genomes in which moderately repetitive DNA sequences of average length 300 bp alternate with nonrepetitive sequences ranging from approx. 800 to 1500 bp.

short-term memory (STM) a memory process underlying behaviour, available for seconds or a few minutes after exposure to information; the actual behaviour of re-calling such information. *cf.* long-term memory.

shotgun cloning the random fragmentation of an entire genome and cloning of the in-dividual fragments.

shrub *n.* low-growing woody plant that does not have a main trunk and which branches from the base, often restricted to woody plants less than 6 m high.

shrub layer bush layer *q.v.*

sialaden salivary gland *q.v.*

sialic *a. pert.* saliva.

sialic acid (sialate) any of a class of acidic sugars such as *N*-acetylneuraminic acid, found in many glycoproteins and in gangliosides.

sialoid *a.* like saliva.

sibling species true species which do not interbreed but are difficult to separate on morphological grounds alone.

siblings *n.plu.* offspring of the same parents. *alt.* sibs.

sibmating *n.* mating between siblings.

sibs siblings *q.v.*

sibship *n.* collectively, the siblings of one family.

siccicolous *a.* drought-resistant.

siccous *a.* dry; with little or no juice.

sickle dance a dance of bees where the bee performs a sickle-like semicircular move-ment and the axis of the semicircle indi-cates the direction of the food source.

sickle-cell anaemia the homozygous state of the genetically determined condition in which an abnormal haemoglobin is pro-duced, causing deformation, "sickling", of red blood cells and severe anaemia and other symptoms, leading to premature death if untreated.

sickle-cell haemoglobin (HbS) defective haemoglobin found in the genetically de-termined conditions of sickle-cell trait and sickle-cell anaemia, in which a glutamate has been replaced by a valine in the β chain of the molecule, a change which causes

"sickling" of red blood cells, reducing their capacity for oxygen transport.

sickle-cell trait heterozygous state of the genetically-determined condition which causes sickle-cell anaemia in the homo-zygous state, in heterozygotes causing only a mild anaemia and apparently conferring resistance to falciparum malaria.

sicyoid *a.* shaped like a gourd.

siderocyte *n.* a red blood cell containing free iron not utilized in haemoglobin for-mation.

siderophil(ic) *a.* staining deeply with iron-containing stains; tending to absorb iron; thriving in the presence of iron.

siderophore *n.* compound that chelates iron, found in many bacteria.

sierozem *n.* grey soil, containing little hu-mus, of middle latitude continental desert regions.

sieve area a region in the cell wall of a sieve tube element, sieve cell, or parenchyma cell, having pores (sieve pores) through which cytoplasmic connections pass to adjacent cells. In sieve tube elements these are mostly on end walls between adjoin-ing elements where they form sieve plates.

sieve cell conducting cell of phloem of ferns and gymnosperms. It is tapering and elon-gated and is distinguished from the sieve tube elements of angiosperms by having small sieve pores only and no sieve plate.

sieve disc sieve plate *q.v.*

sieve elements the conducting parts of the phloem, consisting of sieve tube elements.

sieve field sieve area *q.v.*

sieve pit a primary pit giving rise to a sieve pore. *see* sieve area.

sieve plate *see* sieve tubes; madreporite *q.v.*; area of pedipalp of spiders, with open-ings of salivary glands.

sieve tube element *see* sieve tubes.

sieve tubes long tubes in phloem of angiosperms, whose function is the translocation of nutrients, esp. sugars. They are made up of sieve tube elements, elongated cells with no nucleus at matu-rity, joined end to end and communicating through clusters of pores in the walls, the end wall having clusters of larger pores forming a sieve plate.

sievert (Sv) the SI unit of radiation dose equivalent, equal to 1 joule of energy per kilogram of absorbing tissue, and which replaces the rem, with 1 Sv = 100 rem.

sigillate *a.* having seal-like markings, as certain roots and rhizomes.

sigla *n.* name formed from letters or other characters taken from the words in a compound term.

sigma (σ) subunit of bacterial RNA polymerase involved in recognizing and selecting initiation sites (promoters) for transcription and in opening the template DNA double helix; a C-shaped or S-shaped sponge spicule; symbol for 0.001 seconds; symbol for standard deviation.

sigmaspire *a.* sigma-type spicule with an additional twist.

sigmoid *a.* curved like a sigma, in two directions, *appl.* arteries, cavities, valves; *appl.* curve, an S-shaped curve.

sigmoid flexure an S-shaped double curve as in a bird's neck; S-shaped curve of colon; an S-shaped bend.

signal peptidase enzyme resident in the endoplasmic reticulum, which splits off the signal sequence from membrane and secretory proteins.

signal peptide *see* signal sequence.

signal recognition particle nucleoprotein particle comprising six different proteins and a small RNA, which binds to the signal sequence region of prospective secretory and some other proteins as they are being synthesized, and is believed to guide them to the endoplasmic reticulum.

signal sequence, signal peptide a hydrophobic amino-terminal sequence of around 15 amino acids, present in the newly-synthesized forms of many secretory and membrane proteins, and which is involved in the passage of the protein across or into the cell membrane (in bacteria) or into the endoplasmic reticulum (in eukaryotic cells) for transport to its final destination. It is usually subsequently removed.

signal sequence receptor protein in the membranes of the endoplasmic reticulum to which signal sequences bind and which is involved in the transport of proteins from the cytoplasm into the endoplasmic reticulum.

signal transduction the process of converting a biological signal represented by

activation of a cell-surface receptor into an intracellular response.

sign stimulus an environmental stimulus which acts as a releaser of species-specific behaviour.

silencer *n.* a site in DNA which is required for maintaining a neighbouring gene in an inactive state.

silent mutation a mutation which does not affect function or production of gene product and therefore has no effect on phenotype. They are usually point mutations which change one codon into another specifying the same amino acid or a substitute amino acid which does not affect protein function (*alt.* neutral mutations), or mutations in a DNA region which has no genetic function.

silicified *a.* being impregnated with silica, as the walls of diatoms.

silicole *a.* plant thriving in markedly siliceous soil.

silicula, silicule *n.* broad flat capsule divided into two by a false septum and found in members of the Cruciferae, such as honesty. *cf.* siliqua.

siliqua *n.* long thin capsule divided into two by a false septum, found in members of the Cruciferae such as wallflower, *cf.* silicula; a pod-shaped group of fibres around the olive in mammalian brain.

silk *n.* very strong fine protein fibre produced by various insects and spiders. It is extruded as a fluid from specialized glands and hardens on contact with the air. It is composed of the fibrous protein fibroin and other proteins such as sericin, and is used to make e.g. spiders' webs, and egg and pupal cocoons. Commercial silk is produced by the silkworm (*Bombyx mori*).

silk moth the moth *Bombyx mori*, whose larva (the silkworm) spins a cocoon of silk around itself, from which the silk thread can be unwound.

silkworm *n.* the larva of the silk moth *Bombyx mori*, which spins a cocoon of silk around itself when it pupates, from which silk fibre is gathered commercially.

silt *n.* a soil intermediate between sands and clays in size of the particles.

Silurian *a. pert.* or *appl.* geological period lasting from about 438 to 408 million years ago, and in which the first land plants arose.

silva *alt.* spelling of sylva *q.v.*

silverfish common name for many members of the Thysanura *q.v.*

silvicolous *a.* inhabiting or growing in woodlands.

silviculture *n.* the cultivation of trees and management of forests and woodland for timber.

simian *a.* possessing characteristics of, or *pert.* the anthropoid apes.

simian virus 40 (SV40) DNA virus, a papovavirus, originally isolated from monkeys, can cause malignant transformation of susceptible cells in culture and is much studied as a model of carcinogenesis and as a well-understood model of eukaryotic gene structure and expression.

simple eyes ocelli that occur either as well as or without compound eyes in adults of many insects, and usually the only eyes possessed by the larvae; eyes with only one lens.

simple fruits fruits developing from a single carpel or several united carpels, e.g. berries, apples, plums.

simple Mendelian trait trait determined by a single gene.

simple sequence DNA highly repetitive DNA *q.v.*

simple tissues tissues composed of one type of cell.

simplex *a.* having one dominant gene, in polyploidy

simulation n. assumption of features or structures intended to deceive enemies, as forms of leaf and stick insects, and all varieties of protective coloration.

Sinanthropus *see* Pekin man.

sinciput *n.* upper or fore-part of head. *a.* sincipital.

SINEs short interspersed elements, repeated sequences in the mammalian genome.

single-cell protein protein derived from unicellular microorganisms (e.g. bacteria or fungi) grown on a feedstock of crude oil, carbohydrate raw materials, or wastes from food processing as a carbon source, and which may be used for human or animal consumption.

single-copy DNA that portion of the genomic DNA comprising gene sequences present in only one or a few copies per haploid genome, and which is assumed to represent the structural genes. *alt.* unique DNA. *cf.* repetitive DNA.

single-copy plasmid plasmid present in a bacterial cell as a single copy, replicating only once each cell cycle.

singleton *n.* a single offspring.

sinigrin *n.* a glycoside present in plants of the family Cruciferae and their relatives, whose breakdown products give the pungent taste to capers, mustard, etc.

sinistral *a.* on or *pert.* the left.

sinistrorse *a.* growing in a spiral which twines from right to left, anticlockwise, as in most gastropod shells.

sink *n.* any cell, tissue or organism that is a net importer and end-user of a metabolite or other resource. A storage root of a plant is, for example, a sink for sugars synthesized in the leaves, converting them into storage polysaccharides.

Sino-Japanese Floral Region part of the Holarctic Realm comprising Japan, northern and eastern China, and the northern Himalayas.

sinuate, sinuous *a.* having a wavy indented margin as leaves, or gills of an agaric.

sinuatrial node (SAN) group of specialized cells in the wall of right atrium near opening of superior or anterior vena cava, in which the heart beat is initiated. *alt.* pacemaker; sinuauricular node.

sinuatrial or sinuauricular valves valves between sinus venosus and atrium of heart.

sinupalliate *a.* in molluscs, having a well-developed siphon, and so an indented pallial line.

sinus *n.* a cavity, depression, recess or dilatation; a groove or indentation. *alt.* lacuna.

sinus glands endocrine glands in eye-stalks of decapod crustaceans.

sinus rhomboidalis in vertebrate embryos, the posterior incompletely closed part of medullary canal; later a dilatation of canal in sacral region that is formed from it.

sinus venosus posterior chamber of tubular heart of embryo; in lower vertebrates, a corresponding structure receiving venous blood and opening into auricle; cavity of auricle.

sinus venosus sclerae circular canal near sclerocorneal junction and joining with the anterior chamber of eye and anterior ciliary veins.

sinuses of Valsalva dilations of pulmonary artery and of aorta, opposite aortic semilunar and pulmonary valves of heart.

sinusoid *n.* a minute blood-filled space or channel in the tissues of an organ, as in liver; blood-filled space of irregular shape connecting arterial and venous capillaries.

siphon *n.* of aquatic molluscs, funnel shaped structure from mantle cavity to exterior, through which water is drawn in and out, and which in some molluscs can be used as a means of jet propulsion.

siphonaceous, siphoneous *a.* tubular, in algae a form consisting of a more-or-less elaborate multinucleate thallus, not divided into separate filaments.

Siphonaptera *n.* order of wingless insects with large hind legs, commonly called fleas, which as adults are ectoparasites of skin of birds and mammals. They drink blood through piercing mouthparts, have a worm-like larval stage and a pupa.

siphonate *a.* having a siphon.

siphonet *n.* the honey-dew tube of aphids.

siphonium *n.* membranous tube connecting air passages of quadrate bone with air space in mandible.

siphonocladial *a. appl.* filaments with tubular segments, as in green algae.

siphonogamy *n.* fertilization by means of pollen tube through which contents of pollen grain pass to the embryo sac.

siphonoglyph *n.* one of (usually) two longitudinal grooves in pharynx (gullet) of some anthozoans.

siphonophores *n.plu.* a group of pelagic hydrozoans (Siphonophora) which form colonies consisting of both polyps and medusoid forms, some individuals of the colony modified as a float for swimming as in the Portuguese Man o' War, sometimes mistakenly called jellyfish.

siphonoplax *n.* a calcareous plate connected with siphon in certain molluscs.

Siphonopoda Cephalopoda *q.v.*

siphonostele *n.* type of tubular stele in which the vascular tissue surrounds a central core of parenchyma, the pith. The phloem may either be external to the xylem or both internal and external. Present in stems of many ferns and of some gymnosperms and angiosperms.

siphonostomatous *a.* with a tubular mouth; having front margin of shell notched for emission of the siphon.

siphonous *a. appl.* coenocytic green algae of the class Chlorophyceae, which form hollow tubular colonies.

siphonozoid, siphonozooid *n.* small modified polyp without tentacles, serving to draw water through canal system of some soft corals. Possesses reproductive organs and cannot feed.

siphorhinal *a.* with tubular nostrils.

siphuncle *n.* a thin strand of living tissue running in a calcareous tube through all the compartments of a nautiloid shell; the honey dew tube of an aphid.

Siphunculata *n.* the Anoplura (*q.v.*), the sucking lice.

siphunculate *a.* having a siphuncle; having mouth parts modified for sucking, as certain lice.

Sipuncula, sipunculids *n., n.plu.* phylum of marine unsegmented coelomate worms, which have the anterior end of the body introverted and used as a proboscis and tentacles round the mouth.

Sirenia, sirenians *n., n.plu.* an order of placental mammals commonly called sea cows, including the dugong and manatee, which are highly modified for aquatic life, with a naked body and front limbs modified as paddles.

sirenin *n.* substance secreted by some water moulds during development of female gametes and facilitating fertilization.

sirens *n.plu.* family of eel-like amphibians that live in muddy pools in southeast USA and northern Mexico.

sister cell or nucleus one of a pair of cells or nuclei produced by division of an existing cell or nucleus.

sister-chromatid exchange the exchange of segments between sister chromatids during mitosis.

sister chromatids the two copies of a replicated chromosome held together at the centromere, seen in the prophase and metaphase of mitosis and meiosis.

site-directed mutagenesis techniques of mutation *in vitro* in which a mutation is made at a specific predetermined site in DNA.

site-specific endonuclease restriction

endonuclease *q.v.*

site-specific mutation a general term covering various techniques by which specific changes can be introduced into isolated DNA *in vitro*.

site-specific recombination genetic recombination occurring between 2 particular but not necessarily homologous short DNA sequences, as in the integration or excision of phage from a bacterial chromosome or in transposition. *alt.* conservative recombination.

sitology *n.* the science of food, diet and nutrition.

sitophore *n.* trough of hypopharynx between arms of suspensorium.

sitosterol *n.* complex mixtures of sterols in fatty or oily tissue of higher plants, esp. in corn or wheat-germ oil and in certain algae.

skeletal *a. pert.* the skeleton.

skeletal muscle striated muscle *q.v.*

skeletogenous *a. appl.* embryonic structures or parts which later become parts of skeleton; in sponges, the mesogloeal layer of scattered cells and jelly in which the spicules arise and are embedded.

skeleton *n.* hard framework, internal or external, which supports and protects softer parts of plant, animal or unicellular organism, and to which muscles usually attach in animals. *see also* endoskeleton, exoskeleton.

Skene's glands racemose mucous glands of the female urethra. *alt.* para-urethral glands, Guerin's glands.

skin *n.* the outermost covering of an animal, plant, fruit or seed, *see also* epidermis. In mammals the skin is composed of two layers, an outer epidermis derived from ectoderm which becomes cornified at the surface and is continuously being renewed, and an inner dermis, derived from mesoderm, composed of connective, vascular and muscular tissues. *alt.* cutis.

Skinner box apparatus developed by the behaviourist psychologist B. F. Skinner for use in operant conditioning experiments, in which a rat must press a lever to gain a reward or avoid punishment.

skiophyllous *a.* shade-loving, growing in shade.

skiophyte *n.* a shade plant.

skotoplankton *n.* plankton living at depths

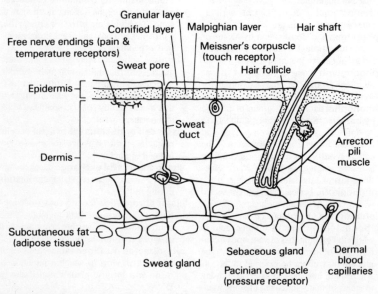

Fig. 10 Schematic cross-section of mammalian skin.

of 500 m.

skototaxis *n*. a positive taxis towards darkness, not a negative phototaxis.

skull *n*. hard cartilaginous or bony part of head of vertebrate, containing the brain, and including the jaws.

slavery dulosis *q.v.*

sleep movements change in position of leaves, petals, etc. at night, which may be brought about by external stimuli of light and temperature changes, or may be an endogenous circadian rhythm.

sliding filament model of muscle contraction, the mechanism whereby the shortening of a muscle fibre is brought about by the sliding between each other of the interdigitating filaments of the myofibril.

sliding growth of plant tissue, when new part of cell wall slides over walls of cells with which it comes in contact.

slime *n*. viscous substance secreted by plants, fungi and animals and rich in glycans (fungal and plant slimes) or glycoproteins (most animal slimes).

slime bacteria myxobacteria *q.v.*

slime layer the proteinaceous or polysaccharide-rich sheath of some bacteria, termed a capsule when thick.

slime moulds *n.plu.* common name for members of the Gymnomycota, the acellular slime moulds *(q.v.)* and cellular slime moulds *(q.v.)*.

slime pits cavities in plant body of hornworts (Anthocerotales) filled with mucilage and sometimes colonized by algae.

slime spore myxospore *q.v.*

slime tubes Cuvierian organs *q.v.*

slow-reacting substance of anaphylaxis (SRS-A) leukotrienes *q.v.*

slow-twitch *appl*. muscle fibres incapable of very rapid contraction, *see* SO fibres.

slow viruses viruses that only show effects a long time after infection.

slug *n*. shell-less terrestrial gastropod mollusc; migrating pseudoplasmodium of a cellular slime mould.

small cytoplasmic RNAs (scRNA) small RNA species ranging in size from 100-300 bases found as ribonucleoprotein particles (scRNP) in the cytoplasm of eukaryotic cells.

small intestine collectively the duodenum, jejunum and ileum.

small nuclear RNAs (snRNA) small RNA species ranging in length from 100-300 bases, found as ribonucleoprotein particles (snRNP) in the nuclei of eukaryotic cells.

small t (small T) protein specified by polyoma virus and produced in infected cells. *see also* large T, middle T.

smegma *n*. secretion of preputial glands or of clitoris glands.

smooth endoplasmic reticulum (SER) *see* endoplasmic reticulum.

smooth muscle non-striated muscle, e.g. of the walls of the alimentary canal, arteries and also found in many other organs, composed of spindle-shaped uninucleate cells and not under voluntary control.

snail *n*. generally refers to a member of the large group of terrestrial and freshwater gastropod molluscs of the subclass Pulmonata, which have a helically coiled shell, no ctenidia, and in which the mantle cavity is used as a lung.

smut fungi common name for members of the Ustilaginales, a group of parasitic heterobasidiomycetes, some of which are serious pathogens of important crops, and which form black, dusty masses of thick-walled teleutospores, resembling soot or smut.

snakes *see* Squamata.

SNAP receptors (SNAREs) membrane proteins of intracellular and plasma membranes which bind to a protein complex (NSF/SNAP) and target the membrane (e.g. of a transport vesicle) to the membrane with which it is to fuse. The targeted fusion is mediated by different SNAP receptors on vesicle and target membranes.

SNAPs soluble NSF attachment proteins found in the membranes of transport vesicles and the membranes of intracellular compartments. They form a complex with the N-ethylmaleimide-sensitive fusion protein (NSF) and bind to intracellular membranes and plasma membrane by means of specific SNAP receptors, thus mediating the specific targeting and fusion of transport vesicles and their target cellular compartments.

SNAREs SNAP receptors *q.v.*

S1 nuclease deoxyribonuclease that degrades single-stranded DNA only.

snurps colloquial term for small nuclear

ribonucleoproteins. *see* small nuclear RNAs.

sobemovirus group plant virus group containing isometric single-stranded RNA viruses, type member southern bean mosaic virus.

sobole *n. see* sucker; an underground creeping stem developing roots and leaves at intervals.

soboliferous *a.* having soboles.

social facilitation the effect of the actions of one animal in a group on the activity of others.

social insects the species of insect that live in organized groups and colonies, with division of labour between different morphological forms and which include many bees, ants and wasps (Hymenoptera) and the termites (Isoptera).

sociation *n.* a minor unit of vegetation, or microassociation.

socies *n.* an association of plants representing a stage in the process of succession.

society *n.* a number of organisms forming a community; in animals, a true society is characterized by cooperative behaviour between individuals, a division of labour and, in the social insects, morphological distinctions between members carrying out different functions; a community of plants other than dominants, within an association or consociation.

sociobiology *n.* the study of the biological and genetic basis of social organization and social behaviour and their evolution in animals, a field of study which has caused controversy when applied to human social behaviour and organization.

sociohormone pheromone *q.v.*

sodium (Na) essential macronutrient, which as the ion Na^+ is involved in generating the membrane potential in all cells and is also required for the generation of nerve impulses in animals.

sodium channel ion channel in cellular membranes that allows the passage of sodium ions (Na^+).

sodium dodecyl sulphate (SDS) detergent widely used in experimental biology for solubilizing membrane protein assemblies for analysis as it disrupts most protein-protein and lipid-protein interactions.

sodium pump Na^+-K^+ ATPase *q.v.*

sodium-potassium pump Na^+-K^+ ATPase *q.v.*

soft corals a group of colonial coelenterates, typified by the sponge-like "dead men's fingers" (*Alcyonarium*), in which the gastric cavities of individual polyps are connected by fine tubes and the bulk of the colony is made up of mesogloea.

SO fibres slow oxidative muscle fibres in muscles of mammalian limbs, red muscle fibres adapted mainly for aerobic metabolism and which are used when the animal is walking or running. *cf.* FG fibres.

soft-rayed *appl.* fish having jointed fin-rays.

softwoods *n.plu.* conifers and their woods.

soil horizon a horizontal zone of soil of distinct composition and texture. A mature soil is made up of several distinct soil horizons. *see* soil profile.

soil profile a vertical section through soil showing the different layers from surface to underlying bedrock.

soil structure the texture of a soil, which is determined by the size and type of the soil particles and how they clump together.

soil water water that fills the spaces between soil particles and pores in rocks above the level of the water table, *cf.* groundwater.

sola *plu.* of solum *q.v.*

solaeus soleus *q.v.*

solar plexus network of sympathetic nerves and ganglia situated behind stomach and supplying abdominal viscera.

solarization *n.* retardation or inhibition of photosynthesis due to prolonged exposure to intense light.

soleaform *a.* slipper-shaped.

solenia *n.plu.* endoderm-lined canals, diverticula from coelenteron of a zooid colony.

solenidion *n.* a modified blunt seta associated with a sensory cell on legs of mites and ticks.

solenocyte *n.* cell resembling flame cell of protonephridial system but with a single flagellum.

solenoid *n.* the 300–500 Å diameter chromatin fibre formed by coiling of the nucleosome fibre.

solenophage *a.* a blood-sucking insect that feeds from the lumen of a blood vessel.

solenostele *n.* a stage after siphonostele in

fern development, having phloem both internal and external to xylem.

soleus *n.* a flat muscle in calf of leg.

solifuga, solifugids *n., n.plu.* order of arachnids commonly called false spiders or sun spiders, having very hairy bodies, large chelicerae and a segmented prosoma and opisthosoma. *alt.* Solpugida, solpugids.

soligenous *a. appl.* wet habitats that are supplied by groundwater rather than by precipitation, e.g. fens.

solitaria phase in locusts, the phase during which they live separately and do not aggregate into swarms.

solitary cells large pyramidal cells of brain, with axons terminating in superior colliculus or in midbrain.

solitary glands or follicles lymphoid nodules occurring singly on intestines and constituting Peyer's patches when aggregated.

solonchak *n.* light-coloured alkali soil with a high salt content found in poorly drained semi-desert regions. It is infertile and low in organic matter.

solonetz *n.* dark-coloured alkali soil found in semi-desert regions where the more soluble salts have been leached out. It supports some vegetation but is relatively infertile.

solpugids Solifuga *q.v.*

soluble NSF attachment proteins SNAPs *q.v.*

solum *n.* floor, as of cavity; soil between source material and topsoil.

soma *n.* the animal or plant body as a whole with the exception of the sex cells; cell body of neurone.

somaclonal variation genetic variation arising from mutations in somatic plant cells undergoing regeneration in culture.

somactids *n.plu.* endoskeletal supports of dermal fin-rays.

somaesthesis n. sensation due to stimuli from skin, muscle and internal organs.

somata *plu.* of soma.

somatic *a. pert.* purely bodily part of plant or animal; *appl.* cell: body cell as opposed to cells of the germ line; *appl.* mutation: mutation occurring in a body cell; *appl.* number: basic number of chromosomes in somatic cells.

somatic cell genetics the study of genetics using cultured somatic cells, involving, e.g., the construction of hybrid cells by cell fusion to enable linkage between various genetic markers to be determined as chromosomes are lost from the hybrids, and also including other techniques such as nuclear transplantation, etc.

somatic cell hybrid a hybrid cell constructed from two somatic cells, often of two different species, by induced cell fusion.

somatic nervous system the parts of the peripheral nervous system that carry sensory information from the periphery to the central nervous system and motor commands from the central nervous system to the skeletal muscles. *alt.* somatomotor system.

somatic polyploidy having polyploid somatic cells.

somatic segregation unequal segregation of maternally and paternally derived organelle genes (as opposed to nuclear genes) in different tissues of the same organism.

somatic sensory fibres nerve fibres conveying impulses from sensory organs (other than those in head) involved in sensing external stimuli. They run from the periphery to the spinal cord and have cell bodies in dorsal root ganglia.

somatocoels *n.plu.* a pair of sacs which bud off from the primary coelomic sacs in the larval stage of echinoderms and later form the main coleom on the adult.

somatocyst *n.* a cavity in pneumatophore of a siphonophore containing air or an oil droplet.

somatogamy *n.* sexual reproduction by fusion of two compatible somatic cells or hyphae, eventually giving rise to a zygote. *alt.* pseudomixis.

somatogenic *a.* developing from somatic cells; *appl.* variation and adaptation arising from external stimuli.

somatoliberin *n.* a small polypeptide produced by hypothalamic neurones which regulates the synthesis of growth hormone in the pituitary.

somatome somite *q.v.*

somatomedin C insulin growth factor I, polypeptide growth factor produced by the liver under the influence of growth hormone, which mediates the effects of growth hormone on skeletal growth. Also a mitogen for some cells *in vitro*.

somatomotor system somatic nervous system *q.v.*

somatopleure *n.* dorsal sheet of mesoderm that forms the outer wall of the coelom, and from which is derived the mesoderm of the limb bud in higher vertebrates (apart from skeletal muscles and dermis).

somatosensory *a. appl.* sensation due to stimuli from skin, muscle or internal organs.

somatosensory cortex area of cerebral cortex in brain concerned with analysis of sensory information from body surface and internal organs.

somatostatin *n.* 14-amino acid peptide hormone, produced by hypothalamus, inhibiting the secretion of growth hormone, insulin and glucagon.

somatotrophic *a.* stimulating nutrition and growth; *appl.* hormone: somatotropin *q.v.* alt. somatotropic.

somatotropin *see* growth hormone.

somatotype *n.* body type or conformation as rated by measurements.

somesthetic system the system of internal receptors that is responsible for bodily sensations in animals.

somite *n.* one of the series of segments into which the trunk mesoderm of a developing animal embryo becomes divided, each somite corresponding to one unit in the final sequence of articulated elements, e.g. vertebra and associated bones and muscles in vertebrates; a segment of a metamerically segmented animal.

somitic *a. pert.* or giving rise to somites, *appl.* mesoderm.

songbirds Passeriformes *q.v.*

sonic *a. pert.* or produced by sound.

Sonoran *a. appl.* or *pert.* a zoogeographical region of southern North America, including northern Mexico, between nearctic and neotropical regions.

soral *a. pert.* a sorus.

soralium *n.* a well-defined group of soredia surrounded by a distinct margin on the thallus of a lichen.

sorbitol *n.* a faintly sweet alcohol isomeric with mannitol.

soredia *plu.* of soredium *q.v.*

soredial *a. pert.* or resembling a soredium.

sorediate *a.* bearing soredia; with patches on the surface.

soredium *n.* a small round or scale-like body on the thallus of some lichens, which contains both algal or cyanobacterial cells and fungal hyphae, and by which the lichen is propagated. *plu.* soredia.

sori *plu.* of sorus *q.v.*

soriferous *a.* bearing sori.

sorocarp *n.* the fruiting body of the cellular slime moulds, composed of a stalk and spore head(s).

sorogen *n.* the cells or tissue that develops into a sorus.

sorophore *n.* thallus or stalk bearing a sorus or sorocarp.

sorosis *n.* a composite fruit formed by fusion of fleshy axis and flowers, as pineapple.

sorotrochous *a.* having a compound ring of cilia or trochal disc, as some rotifers.

sorption *n.* retention of material at surface by adsorption or absorption.

sorus *n.* cluster of sporangia in ferns, borne in large numbers on underside of frond; similar mass of sporangia in other organisms. *plu.* sori.

SOS box short consensus operator DNA sequence found in various genes involved in the SOS response in *E. coli*, and which is recognized by the LexA repressor protein.

SOS response a series of phenotypic changes in *E. coli*, including stimulation of the various repair and retrieval systems, inhibition of cell division and induction of prophages, which are produced by treatments that damage DNA or inhibit replication. *see also* RecA.

South African Floral Region part of the Austral Realm comprising South Africa, Botswana and Namibia.

South-East Asian Floral Region part of the Palaeotropical Realm comprising southern Burma, northern and central Malaysia, and Indochina.

Southern blotting technique in which DNA fragments separated by electrophoresis in an agarose gel are transferred by "blotting" to a nitrocellulose filter for subsequent hybridization with radioactively labelled nucleic acid probes for identification and isolation of sequences of interest.

Southern Realm Austral Realm *q.v.*

South Oceanic Floral Region part of the Austral Realm comprising the islands of the South Atlantic and Indian Oceans south of latitude 50°, e.g. South Georgia and the South Sandwich Islands.

spacer region (spacer) stretches of DNA characteristically found between tandemly repeated genes, and which may be transcribed into RNA and later cut out to leave a separate RNA for each gene (transcribed spacers) or which are not transcribed and therefore separate individual transcription units.

spadiceous *a.* arranged like a spadix.

spadiciform, spadicose *a.* resembling a spadix.

spadix *n.* (*bot.*) a branched inflorescence with an elongated axis, sessile flowers and an enveloping spathe, as of wild arum; a succulent spike; (*zool.*) rudiment of developing manubrium in some coelenterates; amalgamation of internal lobes of tentacles in the cephalopod, *Nautilus*.

spanandry *n.* a scarcity of males; progressive decrease in the number of males, as in some insects.

spangle gall small reddish gall found on the underside of oak leaves and caused by larvae of the gall wasp *Neuroterus quercusbaccarum*.

spanogamy *n.* progressive decrease in the number of females.

spasm *n.* involuntary muscular contraction; spasmodic or spastic contraction of muscle fibres.

spat *n.* the spawn or young of bivalve molluscs.

spathaceous, spathal *a.* resembling or bearing a spathe.

spathe *n.* a large, enveloping leaf-like structure, green or coloured, protecting a spadix.

spathella *n.* a small spathe surrounding division of palm spadix.

Spathiflorae Arales *q.v.*

spathose *a.* with or like a spathe.

spatia *n.plu.* spaces, e.g. intercostal spaces.

spatia zonularia a canal surrounding marginal circumference of lens of eye.

spatula *n.* a "breast-bone" or anchor process of certain dipteran larvae; a spoon-shaped

structure.

spatulate *a.* spoon-shaped.

spawn *n.* collection of eggs deposited by bivalve molluscs, fishes, frogs, etc.; mycelium of certain fungi; *v.* to deposit eggs, as by fishes, etc.

Special Creation *see* Creationism.

specialist, specialist species species which can survive and thrive only within a narrow range of habitat and/or climatic conditions, or which can use only a very limited range of food, and is therefore usually less able to adapt to changing environmental conditions.

specialization *n.* adaptation to a particular mode of life or habitat in the course of evolution.

speciation *n.* the evolution of new species.

species *n.* in sexually-reproducing organisms a group of interbreeding individuals not normally able to interbreed with other such groups, being a taxonomic unit having two names in binomial nomenclature (e.g. *Homo sapiens*), the generic name and specific epithet (italicized in the scientific literature), similar and related species being grouped into genera. Species can be subdivided into subspecies (e.g. *Homo sapiens sapiens* – modern man), geographic races, varieties, and for cultivated plants named cultivars, and for domesticated animals, breeds or strains. Species in asexually reproducing organisms such as bacteria are to a large extent based on morphological, genetic and biochemical characteristics, habitat and host range.

species aggregate a group of very closely related species which have more in common with each other than with other species of the genus.

species composition the different species in a given area or ecosystem.

species diversity the number and abundance of different species within a given area, which is one measure of biological diversity, a diverse environment having relatively small numbers of many different species.

species diversity index the extent to which different species are represented within a community.

species flock group of numerous species, endemic to a particular small area and ecologically diverse, such as the cichlid fish of some African lakes, and which are thought to have evolved from a single ancestor species.

species pair sibling species *q.v.*

species richness the number of different species within a given community or area.

species selection the view that selection can act at the level of the species as well as the individual, which is advanced as an explanation of some long term evolutionary trends.

species-specific behaviour behaviour patterns that are inborn in a species and are performed by all members under the same conditions and which are not modified by learning.

specific *a.* peculiar to; *pert.* a species; *appl.* characteristics distinguishing a species; restricted to interaction with a particular substrate, *appl.* enzymes, antibodies, etc.

specific epithet the second name in a Latin binomial.

specific hunger the preference for certain foods containing an essential mineral or vitamin shown by animals on a diet deficient in the substance.

specification *n.* commitment of cells to form a particular part of a structure, which is independent of their determination.

specificity *n.* the condition of being specific; being limited to a species; restriction of parasites, bacteria and viruses to particular hosts; restriction of enzymes to certain substrates, restriction of antibody to interactions with particular antigens, etc.

spectrin *n.* a fibrous protein located on the cytoplasmic face of the red blood cell membrane and a constituent of the membrane cytoskeleton.

speculum *n.* an ocellus *q.v.*; a wing bar with a metallic sheen, as in mallard drakes.

speleology *n.* the study of caves and cave life.

Spemann organizer the mesoderm of the dorsal lip of the blastopore in amphibian embryos, which is capable when transplanted to another site on the embryo of generating a complete new body axis.

sperm *n.* a spermatozoan; any male gamete produced by an animal; semen.

sperm competition type of sexual competition in animals in which there is competition not for access to females but for fertilization.

sperm nucleus male pronucleus *q.v.*

spermagglutination *n.* agglutination of spermatozoa, such as is brought about by some myxoviruses.

spermangium *n.* an organ producing male spore-like cells, in ascomycetes.

spermaphore *n.* placenta of ovary in plants.

Spermaphyta Spermatophyta *q.v.*

spermaphytic *a.* seed-bearing.

spermary, spermarium *n.* any organ in which male gametes are produced.

spermatangium *n.* male sex organ of red algae, each developing into one non-motile gamete (spermatium).

spermatheca *n.* in female or hermaphrodite invertebrates, a sac for storing spermatozoa received on copulation. *plu.* spermathecae.

spermatia *plu.* of spermatium *q.v.*

spermatic *a. pert.* spermatozoa; *pert.* testis.

spermatid *n.* haploid cell arising by division of secondary spermatocyte in testes and which becomes a sperm.

spermatiferous *a.* bearing spermatia.

spermatiform *a.* like a spermatium.

spermatiophore *n.* a hypha or hyphal outgrowth bearing spermatia.

spermatium *n.* small non-motile cell that functions as a male sex cell in some ascomycete and basidiomycete fungi, and which fuses with a receptive female sex organ; non-motile male gamete produced by red algae. *plu.* spermatia.

spermatoblast *n.* spermatid *q.v.*; Sertoli cell *q.v.*

spermatoblastic *a.* sperm-producing.

spermatocyst *n.* a seminal sac.

spermatocyte *n.* male cell in which meiosis occurs to form sperm. Primary spermatocytes are diploid cells developing from spermatogonia and which undergo a first meiotic division to form two secondary spermatocytes each of which completes meiosis, producing four haploid spermatids from each original diploid spermatocyte.

spermatogenesis *n.* sperm formation, from spermatogonium through primary and secondary spermatocytes and spermatid to spermatozoon.

spermatogenic *a. pert.* sperm formation; sperm-producing.

spermatogenous *a.* giving rise to sperm.

spermatogonium *n.* primordial male germ cell, diploid stem cell precursor of male germ cells, making up the inner layer of the lining of the seminiferous tubules, and which gives rise to spermatocytes. *see also* spermogonium. *plu.* spermatogonia.

spermatoid *a.* like a sperm.

spermatophore *n.* a number of sperms enclosed in a sheath of gelatinous material, the form in which sperm is released by many invertebrates.

Spermatophyta, spermatophytes *n., n.plu.* in some classifications a major division of plants containing all seed-bearing plants, i.e. gymnosperms (e.g. cycads, conifers, *Ginkgo*, *Ephedra* and quillworts) and angiosperms (flowering plants).

spermatoplasm *n.* cytoplasm of sperm.

spermatoplast *n.* a male gamete.

spermatostrate *a.* spread by means of seeds.

spermatozeugma *n.* union by conjugation of two or more spermatozoa, as in the vas deferens of some insects.

spermatozooid antherozooid *q.v.*

spermatozoon *n.* mature motile male gamete in animals, typically consisting of a head containing the nucleus, and a tail consisting of a single flagellum by which it moves. *alt.* sperm. *plu.* spermatozoa.

spermidine *n.* a widely distributed polyamine, $H_2N(CH_2)_3NH(CH_2)_4\ NH(CH_2)_3\ NH_2$, constituent of ribosomes, and acting generally as a membrane-stabilizing agent.

spermiducal *a. appl.* glands into or near which sperm ducts open, in many vertebrates; *appl.* glands associated with male ducts in oligochaete worms.

spermiduct *n.* duct for conveying sperm from testis to exterior.

spermine *n.* a widely distributed polyamine, $H_2N(CH_2)_4NH(CH_2)_3NH_2$, a constituent of ribosomes, and acting generally as a membrane-stabilizing agent.

spermiocyte primary spermatocyte, *see* spermatocyte.

spermiogenesis *n.* development of spermatozoon from spermatid.

spermo- spermato-.

spermocentre *n.* the male centrosome during fertilization.

spermodochium *n.* a group of spermatiophores derived from a single cell and lacking a capsule, in fungi.

spermogenesis spermatogenesis *q.v.*

spermogoniferous *a.* having spermogonia.

spermogonium *n.* the structure containing the sex organs of rust fungi and some lichen fungi, in which non-motile male gametes (spermatia) are formed. *plu.* spermogonia.

spermogonous *a.* like or *pert.* a spermogonium.

spermospore *n.* a male gamete produced in a spermangium.

spermotheca spermatheca *q.v.*

spermozeugma *n.* a mass of regularly aggregated spermatozoa, for delivery into a spermatheca.

sphacelate *a.* decayed; withered; looking as though decayed and withered.

sphacelia *n.* the conidial stage of a fungus that produces a sclerotium or ergot.

sphaeraphides *n.plu.* globular clusters of minute crystals in plant cells.

sphaerenchyma *n.* a tissue composed of spherical cells.

sphaeridia *n.plu.* small rounded bodies, possibly balancing organs or other type of sense organ, on the surface of some echinoderms.

sphaerite *n.* a rounded mass of calcium oxalate or starch forming crystals inside plant cells.

sphaerocyst *n.* large globose or oval cell in trama of *Russula* and *Lactarius* species.

sphaeroid *a.* globular, ellipsoidal, or cylindrical; *appl.* an aggregate of individual protozoans; *appl.* a dilated hyphal cell containing oil droplets, in lichens.

sphagnicolous *a.* inhabiting sphagnum peat moss.

sphagniherbosa *n.* plant community on peat containing large amounts of *Sphagnum* moss.

Sphaeropsida *n.* form class of deuteromycete fungi that reproduce by conidia borne in pycnidia, and which include numerous plant pathogens (e.g. *Phyllosticta, Dendrophoma, Septoria*). *alt.* Sphaeropsidales.

Sphagnidae sphagnum mosses *q.v.*

sphagnum mosses mosses of the class

Sphagnidae, also called peat mosses or bog mosses, having the gametophore with branches in whorls, many dead water-absorbing cells, and comprising one genus *Sphagnum*. They are distinguished from other mosses by leaves lacking midribs and a plant body with no rhizoids when mature. The spore capsule lacks a peristome and the protonema germinating from the spore is plate-like rather than filamentous as in the true mosses.

sphagnous *a. pert.* peat moss.

S phase *see* cell cycle.

sphecology *n.* the study of wasps.

sphenethmoid *n.* single bone replacing the orbitosphenoids in anuran amphibians.

sphenic *a.* like a wedge.

Sphenisciformes *n.* an order of birds including the penguins.

spheno- prefix signifying *pert.* or involving the sphenoid, as sphenomandibular, *appl.* ligament connecting sphenoid and mandible.

spheno-ethmoidal *a. pert.* or in the region of sphenoid and ethmoid, *appl.* a recess above superior nasal concha and a suture.

Sphenodonta *see* Rhynchocephalia.

sphenofrontal *a. pert.* sphenoid and frontal bones, *appl.* a suture.

sphenoid *n.* a basal bone of skull in some vertebrates, composed of several fused bones. *alt.* butterfly bone.

sphenoidal *a.* wedge-shaped; *pert.* or in region of sphenoid.

sphenolateral *n.* one of dorsal pairs of cartilages parallel to the trabeculae embracing the hypophysis in embryo.

sphenomaxillary *a. pert.* sphenoid and maxilla, *appl.* fissure and (pterygopalatine) fossa.

sphenopalatine ganglion an autonomic ganglion on the maxillary nerve in pterygopalatine fossa.

Sphenophyta *n.* one of the four main divisions of extant seedless vascular plants, commonly called horsetails and represented by a single living genus *Equisetum*. They have a sporophyte with roots, jointed stems, and leaves in whorls, and have strobili of reflexed sporangia borne on sporangiophores.

sphenopterygoid *a. pert.* sphenoid and pterygoid; *appl.* mucous pharyngeal glands, near openings of Eustachian tubes, as in birds.

sphenosquamosal *a. appl.* cranial suture between spheroid and squamosal.

sphenotic *a.* postfrontal cranial bone in many fishes.

sphenoturbinal *a.* laminar process of sphenoid.

sphenozygomatic *a. appl.* cranial suture between sphenoid and zygomatic.

spher- sphaer-.

spheraster *n.* a many-rayed globular spicule.

spheroidal *a.* globular, but not perfectly spherical, *appl.* glandular epithelium (from shape of cells).

spheroidocyte *n.* type of haemocyte in insects.

spherome *n.* cell inclusions producing oil or fat globules; intracellular fatty globules as a whole.

spheromere *n.* a segment of a radially symmetrical animal.

spheroplast *n.* bacterial or yeast cell from which the wall has been removed leaving a naked protoplast.

spherosome *n.* organelle derived from endoplasmic reticulum and bounded by a single membrane leaflet, and which synthesizes lipids.

spherula *n.* a small sphere: a small spherical spicule.

spherulate *a.* covered with small spheres.

sphincter *n.* a muscle that closes or contracts an orifice, as that of bladder, anus, etc.

sphingolipid *n.* any of a group of complex lipids, glycolipids and phospholipids, which contain the amino alcohol sphingosine and not glycerol, and including sphingomyelin, cerebrosides, gangliosides and ceramide.

sphingomyelin *n.* any of a group of phospholipids, found esp. in the myelin sheath of nerve cells, and containing the amino alcohol sphingosine, fatty acids, and phosphoryl choline.

sphingosine *n.* an amino alcohol, containing a long unsaturated hydrocarbon chain, and which is found in gangliosides, cerebrosides, sphingomyelin and ceramide.

sphragis *n.* a structure sealing the bursa copulatrix on female abdomen of certain lepidopterans after pairing, and consisting of hardened sphragidal fluid secreted by male.

sphygmic *a*. *pert*. the pulse, *appl*. 2nd phase of systole.

sphygmoid *a*. pulsating; like a pulse.

spica *n*. a spike.

spicate *a*. arranged in spikes, as a flowerhead; bearing spikes; with a spur-like prominence.

spiciform *a*. spike-shaped.

spicigerous spicate *q.v.*

spicose *a*. with spikes or ears, as wheat.

spicula *n*. a small spike; a needle-like body; spicule *q.v.*; *plu*. of spiculum.

spicular *a*. *pert*. or like a spicule.

spiculate *a*. set with spicules; divided into small spikes.

spicule *n*. a minute needle-like body; siliceous or calcareous, found in invertebrates; a minute pointed process. *alt.* spicula.

spiculiferous, spiculigerous, spiculose *a*. furnished with or protected by spicules.

spiculiform *a*. spicule-shaped.

spiculum *n*. the dart of a snail; a spicular structure.

spider cell fibrous astrocyte, *see* astrocyte.

spiders an order of arachnids, the Araneida, having spinning glands on the opisthosoma and poison glands on the chelicerae.

spike *n*. a flower-head with stalkless flowers or secondary small spikes (spikelets) of flowers borne alternately along a single axis.

spikelet *n*. one of the units of the flower-head of grasses, consisting of several florets along a thin stalk, at the base of which are two bracts (glumes) marking the end of the spikelet; any small spike of flowers.

spinal *a*. *pert*. backbone, spinal cord; *appl*. foramen, ganglia, nerves, etc.

spinal canal canal formed by vertebrae which encloses the spinal cord.

spinal column backbone, vertebral column *q.v.*

spinal cord column of nervous tissue running from brain along the back in vertebrates, enclosed in the spinal canal formed by vertebrae. With the brain it forms the central nervous system.

spinalis name given to muscles connecting vertebrae.

spinasternum *n*. an intersegmental sternal sclerite or poststernellum with an internal spine, in certain insects.

spinate *a*. bearing spines; spine-shaped.

spination *n*. the occurrence, development or arrangement of spines.

spindle *n*. the structure formed by microtubules stretching between opposite poles of the cell during mitosis or meiosis and which guides the movement of the chromosomes. *see also* muscle spindle, meiosis, mitosis; a tree of the genus *Euonymus*.

spindle cell a spindle-shaped coelomocyte.

spindle fibre a bundle of microtubules and associated proteins in the mitotic or meiotic spindle that forms a fibre thick enough to be visible in the light microscope. One end may be attached to a pole of the spindle and the other to the kinetochore of a chromosome.

spindle pole body structure found at the poles of the mitotic spindles of many fungi, and which serves the role of a centrosome in organizing the spindle microtubules.

spine *n*. sharp-pointed outgrowth as on leaves, stems, bones, echinoids; sharp pointed modified hair as in hedgehogs, porcupines; the backbone or vertebral column; pointed process of vertebra; scapular ridge; fin-ray; (*neurobiol*.) dendritic: small protuberances stiffened by actin filaments, at the ends of fine branches of dendrites of neurones, where many neurotransmitter receptors are concentrated.

spinescent *a*. tapering; tending to become spiny.

spiniferous, spinigerous *a*. spine-bearing, *appl*. pads on ventral side of leg in the onychophoran *Peripatus*.

spiniform *a*. in the shape of a spine.

spinisternite *n*. a small sternite with spiniform apodeme, between thoracic segments of insects.

spinneret *n*. organ perforated with tubes connecting to the silk glands in spiders, and from which the liquid silk is released to form webs; similar organ from which cocoon is spun in some insects.

spinnerule *n*. a tube discharging silk secretion of spiders.

spinning glands glands that secrete silky material in arthropods, as for webs in spiders and cocoons in insect caterpillars.

spinocaudal *a. pert.* trunk of vertebrates.

spino-occipital *a. appl.* nerves arising in trunk somites which later form part of the skull.

spinose *a.* bearing many spines.

spinothalamic tracts tracts of nerve fibres connecting spinal cord and thalamus in brain, involved in relaying pain and temperature information.

spinous *a.* spiny; spine-like.

spinous process median dorsal spine-like process of vertebra; a process of sphenoid; a process between articular surfaces of proximal end of tibia.

spinulation *n.* a defensive spiny covering; the state of being spinulate.

spinulate *a.* covered with small spines.

spinule *a.* a small spine.

spinulescent *a.* tending to be spinulate.

spinuliferous *a.* bearing small spines.

spinulose, spinulous spinulate *q.v.*

spiny-finned *a.* bearing fins with spiny rays for support.

spiny-headed worms Acanthocephala *q.v.*

spiny-rayed *a. appl.* fins supported by spiny rays.

spiracle *n.* hole in the sides of thoracic and abdominal segments of insects, and of myriapods, through which the tracheal respiratory system connects with the exterior, and which can be opened and closed; small round opening, a vestigial gill slit, immediately anterior to the hyomandibular cartilage in elasmobranch fishes; various other exterior openings connected with breathing or respiration in other animals.

spiraculate, spiraculiferous *a.* having spiracles.

spiraculiform *a.* spiracle-shaped.

spiraculum spiracle *q.v.*

spiral *a.* winding like a screw, *appl.* leaves alternately placed up a stem, *appl.* flowers with spirally inserted parts, *appl.* pattern of lignified thickening in cell wall of xylem vessels and tracheids.

spiral cleavage mode of cleavage in which blastomeres divide obliquely so that at the eight-cell stage the four upper cells are not directly above the four lower cells, as in radial cleavage, but in the grooves between them. Further divisions are also oblique, alternating between right and left. Found in turbellarians, annelids, molluscs, nemertean worms and some other groups, and associated with a determinate type of development.

spiral valve in certain more primitive fishes, such as elasmobranchs, ganoids and dipnoans, a spiral infolding of the intestine wall; of Heister, folds of mucous membrane in neck of gallbladder; of heart, incomplete partition in conus arteriosus in dipnoans, preventing complete mixing of oxygen-rich and oxygen-poor blood.

spiral vessels first xylem elements of a stele, spiral fibres coiled up inside tubes and so adapted for rapid elongation.

spiralia *n.plu.* coiled structures supported by crura, in some brachiopods.

spiranthy *n.* displacement of flower parts through twisting.

spire *n.* the whorled part of a spiral shell.

spiriferous *a.* having a spiral structure.

spirillum *n.* any of a group of bacteria with helically curved or twisted thread-like cells. Some are flagellate. Some are free-living in freshwater or marine environments, others saprophytic or parasitic, including some human pathogens such as *Spirillum minor*, the cause of rat-bite fever. *plu.* spirilla.

spirivalve *n.* a gastropod with spiral shell.

spirochaetes *n.plu.* group of slender, helically coiled bacteria with flexible body and no rigid cell wall. They include free-living, commensal and parasitic forms including some human pathogens such as *Treponema pallidum*, the causal organism of syphilis. *alt.* spirochetes.

spiroid *a.* spirally formed.

spirolophe *n.* a stage in the development of the brachiopod lophophore in which two spirally twisted arms are formed with cirri on one side.

spiroplasm *n.* very small motile helical prokaryote lacking a cell wall, which causes plant disease. *alt.* spiroplasma.

spiroplasms *n.plu.* spiral-shaped organisms found in plants and which resemble mycoplasms.

Spirotrichia, spirotrichans *n., n.plu.* a group of ciliate protozoans having a well-defined gullet surrounded by a ring of composite cilia, the undulating membrane, e.g. *Stentor*.

spite *n*. in animal behaviour, the name given to a behaviour by which an animal reduces its own fitness in the process of harming another animal.

splanchnic *a. pert.* internal organs.

splanchnocoel *n*. the cavity of lateral somites of embryo, persisting as visceral cavity in adults.

splanchnopleure *n*. ventral sheet of mesoderm that forms the inner wall of the coelom.

splash zone zone of a seashore above the high-tide mark but which may be wetted by sea spray at high tide, and which supports some seaweeds (e.g. *Pelvetia caniculata*, channel wrack), small periwinkles and lichens. *see* Fig. 9.

spleen *n*. vascular lymphoid organ in vertebrates in which lymphocytes are produced and red blood cells destroyed.

splenetic splenic *q.v.*

splenial *a. pert.* the splenium of corpus callosum; *pert.* the splenial bone in vertebrate lower jaw; *pert.* the splenius muscle of upper dorsal region and back of neck.

splenic *a. pert.* the spleen. *alt.* splenetic.

splenium *n*. posterior border of corpus callosum in brain.

splenius *n*. muscle of upper dorsal region and back of neck.

splenophrenic *a. pert.* spleen and diaphragm.

spliceosome *n*. multisubunit complex of proteins and RNAs which assembles on RNA and carries out RNA splicing.

splicing *n*. RNA splicing *q.v.*; joining of 2 different pieces of DNA to form a recombinant DNA in genetic engineering.

splicing junctions the junctions between intron and exon in a primary transcript from a eukaryotic "split gene", the points at which introns are excised from the transcript and adjacent exons rejoined, and the immediate nucleotide sequences around these points.

splint bone rudiments of metacarpals and some metatarsals in horses.

splintwood sapwood *q.v.*

split gene typical organization of many eukaryotic (but not prokaryotic) genes, in which the coding region is interrupted by one or more non-coding nucleotide sequences (introns) separating the coding sequences into blocks (exons), both introns and exons being transcribed into a continuous RNA and the introns subsequently removed by RNA splicing to generate a functional RNA. *alt.* interrupted gene.

spondyle, spondylus vertebra *q.v.*

spondylous vertebral *q.v.*

sponges *n.plu.* common name for the Porifera *q.v.*

spongicolus *a.* living in sponges.

spongin *n.* fibrous protein component of the horny sponges, such as the bath sponge.

spongioblast *n.* embryonic epithelial cell that gives rise to neuroglial cells and fibres radiating to periphery of spinal cord.

spongiocoel *n.* the cavity, or system of cavities in sponges.

spongiocyte *n.* a vacuolated cell of the zona fasciculata of the adrenal cortex.

spongiose *a.* of a spongy texture, spongoid; full of small cavities.

spongophyll *n.* a leaf having spongy parenchymatous tissue, without palisade tissue, between upper and lower epidermis, as in some aquatic plants.

spongy *a.* of open texture, containing air spaces, *appl.* parenchyma of mesophyll, *appl.* tissue surrounding embryo sac, as in gymnosperms.

spontaneous generation the idea that organisms could arise spontaneously from non-living material or from unlike living matter, widely held before the 19th century, and which was shown to be untrue by Pasteur, who showed that if air was excluded from a sterilized flask of hay in water, no living organisms materialized.

spontaneous mutations those occurring as a result of normal cellular processes and random interaction with the environment. *cf.* induced mutations.

spools *n.plu.* minute tubes of spinnerets of spiders, from which the silk thread emerges.

spoon *n.* small sclerite at base of balancers in Diptera; pinion of tegula.

spoon worms Echiura *q.v.*

sporabola *n.* the trajectory of a spore discharged from a sterigma.

sporadic *a. appl.* plants confined to limited localities; *appl.* scattered individual cases of a disease; *appl*. cases of spontaneously arising cancers, as opposed to familial cancers of the same type.

sporangia *plu*. of sporangium *q.v*.

sporangial *a. pert*. a sporangium.

sporangiferous *a*. bearing sporangia.

sporangiform, sporangioid *a*. like a sporangium.

sporangiocarp *n*. an enclosed collection of sporangia; a structure of asci and sterile hyphae surrounded by a peridium.

sporangiocyst *n*. a membrane enclosing a sporangium; a thick-walled resistant sporangium. *alt*. sporangiole.

sporangiolum *n*. small sporangium containing only one or a few spores. *plu*. sporangiola.

sporangiophore *n*. a stalk-like structure bearing sporangia.

sporangiosorus *n*. a compact group of sporangia.

sporangiospore *n*. a spore, esp. if non-motile, formed in a sporangium.

sporangium *a*. a cell or multicellular structure in which asexual non-motile spores are produced in fungi, algae, mosses, ferns. *plu*. sporangia.

spore *n*. a small, usually unicellular, reproductive body from which a new organism arises, produced by some plants (ferns and mosses), fungi, bacteria and protozoa.

spore case *see* theca.

spore coat envelope of a bacterial spore, external to cortex and surrounded by exosporium.

spore mother cell a diploid cell which by meiosis gives rise to four haploid cells.

sporidium *n*. an alternative name for basidiospore in smut and rust fungi. *plu*. sporidia.

sporiferous *a*. spore-bearing.

sporification *n*. formation of spores.

sporiparous sporogenous *q.v*.

sporo-, -spore, -sporous word elements from Gk. *sporos*, a seed, *pert*. spores and the structures that produce them.

sporoblast *n*. the meristematic founder cell of a sporangium; spore mother cell *q.v*.

sporocarp *n*. structure inside which spores are produced.

sporocladium *n*. a hyphal branch bearing sporangia or sporangiola, in some fungi.

sporocyst *n*. one of the larval stages in the life cycle of endoparasitic flukes, which develops from the miracidium in the snail host. It has no mouth or gut and reproduces asexually to produce rediae or cercariae; protective envelope of a spore in protozoans; stage in spore formation preceding liberation of spores.

sporocystid *a. appl*. oocyst of sporozoans when the zygote forms sporocysts.

sporocyte *n*. spore mother cell, cell which gives rise to spores.

sporodochium *n*. a mass of conidiophores that erupts from the bark of trees and shrubs infected with *Nectria*, the causative agent of cankers of many hardwood trees and shrubs, and which, together with the brightly coloured fruiting bodies that follow it, is the "coral spot" characteristic of this disease.

sporogenesis *n*. spore formation.

sporogenous *a*. spore-producing.

sporogonium *n*. the sporophyte generation in bryophytes, consisting of capsule and seta, developing from the fertilized ovum in the archegonium, and giving rise to spores. *a*. sporogonial.

sporogony *n*. spore formation; the formation of sporozoites or spores from a sporont in protozoans; the formation of gametes from a sporont, their fusion and subsequent formation of spores and sporozoites from the zygote (sporont), *alt*. gamogony, in protozoans.

sporoid *a*. like a spore.

sporokinete *n*. a motile spore from the oocyst in certain protozoan blood parasites.

sporonine *n*. a terpene-like substance found in the walls of spores and pollen grains.

sporont *n*. the individual or generation which gives rise to a generation of sporozoites.

sporophore *n*. a spore-bearing structure in fungi, which may be a simple sporangiophore or a complex structure such as the fruit-body of a mushroom or toadstool; part of plasmodium producing spores on its free surface, in slime moulds; an inflorescence.

sporophyll *n*. a leaf, or structure derived from a leaf, that bears a sporangium, and which may be much modified, as the stamens and carpels of a flower.

sporophyte *n*. the diploid or asexual phase in the alternation of generations in plants, which in some plants produces spores. *cf.* gametophyte.

sporoplasm *n.* the protoplasm that gives rise to a spore; in some sporozoans, the cell released from cyst and forming an amoebula.

sporopollenin *n.* an alcohol found in the walls of spores and pollen grains, which is related to suberin and cutin but is much more durable, resulting in spores and pollen grains surviving for millions of years.

sporothallus *n.* a thallus that produces spores.

sporotheca *n.* a membrane enclosing sporozoites.

sporosac *n.* an ovoid pouch-like body, consisting of a gonad, a degraded reproductive zooid of a siphonophore.

Sporozoa, sporozoans *n., n.plu.* subphylum of parasitic protozoans containing many that cause disease in humans and domestic animals, and in other vertebrates and invertebrates. They include *Plasmodium*, the causal agent of human malaria, and *Eimeria*, the agent of coccidiosis in cattle, sheep and poultry. They usually have no feeding or locomotory organelles. Sporozoans have a complex life cycle, with asexual and sexual generations, sometimes in two hosts. The stage that infects new cells is a haploid sporozoite.

sporozoid *n.* a motile spore, zoospore *q.v.*

sporozoite *n.* spore released from sporocyst of sporozoan protozoa. In malaria it is the stage in the salivary glands of the mosquito host and is transmitted to humans.

sport *n.* a somatic mutation, producing a plant or part of a plant with altered characteristics, and which can only be propagated vegetatively.

sporula, sporule *n.* a small spore; formerly a spore.

sporulation *n.* the process of spore formation; liberation of spores; in bacteria, the segregation of the DNA to one part of the cell where it is surrounded by a spore coat, forming an endospore; brood formation by multiple cell fission (in some protozoans).

spot desmosome *see* desmosome.

springtails the common name for the Collembola *q.v.*

spumaviruses *n.plu.* a subfamily of nononcogenic retroviruses which produce a characteristic foamy appearance of the cytoplasm in the cells they infect.

spuriae *n.plu.* feathers of bastard wing *q.v.*

spurious *a. appl.* structures that appear to be e.g. teeth, fruit, etc. but are not.

squalene *n.* a C_{30} hydrocarbon consisting of 6 isoprene units, an intermediate in cholesterol biosynthesis.

Squamata *n.* order of reptiles comprising the snakes, lizards and amphisbaenians.

squamate *a.* scaly.

squamation *n.* the arrangement of scales on the surface of a lizard, snake, etc.

squame *n.* flattened cell of the outermost layers of the skin, consisting largely of keratin, and which eventually flakes off; a scale; a part arranged like a scale.

squamella *n.* a small scale or bract.

squamellate, squamelliferous *a.* having small scales or bracts.

squamelliform *n.* resembling a squamella.

squamid *a.* scaly; *n.* any member of the order Squamata, the lizards, amphisbaenians and snakes.

squamiferous, squamigerous *a.* bearing scales.

squamiform *a.* scale-like.

squamose *a.* covered in scales.

squamosal *n.* a membrane bone of vertebrate skull forming part of posterior side wall.

squamous *a.* consisting of scales; *appl.* simple epithelium of flat nucleated cells; squamose *q.v.*

squamula, squamule *n.* a small scale.

squamulate *a.* having minute scales. *alt.* squamulose.

squamulose *a.* squamulate *q.v.*; *appl.* lichens having a foliose growth form with many loosely attached lobes to the thallus.

squarrose *a.* rough with projecting scales or rigid leaves.

squarrulose *a.* tending to become squarrose.

ss-binding protein single-stranded DNA binding protein, which binds to unwound DNA during replication preventing reformation of the original helix. *alt.* helix-destabilizing protein (HD protein).

stabilate *n.* a stable population of an organism similar to a strain.

stability *n.* ability of a community or ecosystem to withstand or recover from changes or stresses imposed from outside. *see* constancy, inertia, resilience.

stabilizing selection selection that operates against the extremes of variation in a population and therefore tends to stabilize the population around the mean.

stachyose *n.* a tetrasaccharide present in certain plant roots and other sources, made up of two galactose residues, one of glucose and one of fructose.

stachysporous *a.* bearing sporangia on the axis.

stade *n.* a stage in development or life history of a plant or animal; interval between two successive moults. *alt.* stadium.

stag-horned *a.* having large branched mandibles, as the stag-beetles.

stagnicolous *a.* living or growing in stagnant water.

stalk cell the barren cell of two into which the antheridial cell of gymnosperms divides; basal cell of crozier in discomycete fungi.

stamen *n.* male reproductive organ (microsporophyll) of a flower, consisting of a stalk or filament bearing an anther in which pollen is produced. *see* Fig. 4.

staminal *a. pert.* or derived from a stamen.

staminate *a. appl.* a flower containing stamens but not carpels.

staminiferous, staminigerous *a.* bearing stamens.

staminode *n.* a leaf-like structure in some flowers, derived from a metamorphosed stamen; a rudimentary, imperfect or sterile stamen.

staminody *n.* the conversion of any floral structures into stamens.

staminose *a. appl.* flowers having very obvious stamens.

stand *n.* aggregation of plants of uniform species and age, distinguishable from the adjacent vegetation.

standard *n.* petal standing up at the back of a papilionaceous flower, such as pea, bean, vetch, which helps to make the flower conspicuous; upstanding petal in flower of iris; a tree or shrub not supported by a wall; a unit of measurement or a material used in calibration.

standard metabolic rate the metabolic rate measured under a set of given conditions.

standard nutritional unit the unit expressing the energy available at a certain trophic level for the next level in the food chain, usually expressed as 10^6 kilocalories per hectare per year.

standing crop the biomass of a particular area, ecosystem etc. at any specified time.

stapedius *n.* the muscle pulling the head of the stapes, one of the ossicles of the middle ear.

stapes *n.* stirrup-shaped innermost bone connecting incus and oval window of middle ear in mammals; bone connecting eardrum and fenestra ovalis of middle ear in amphibians and reptiles.

staphylococci *n.* Gram-positive bacteria of the genus *Staphylococcus*, which are small spherical cells (cocci) often arranged in irregular clusters. Pathogenic strains cause skin lesions and wound infections, and occasionally more disseminated disease. *sing.* staphylococcus.

star cells Kupffer cells *q.v.*

star navigation learned method of navigation apparently used by many migrating songbirds, which have been shown to orient themselves according to the stars.

starch *n.* polysaccharide made up of a long chain of glucose units joined by α-1,4 linkages, either unbranched (amylose) or branched (amylopectin) at a α-1,6 linkage, and which is the storage carbohydrate in plants, occurring as starch granules in amyloplasts, and which is hydrolysed by animals during digestion by amylases, maltase and dextrinases to glucose via dextrins and maltose. *see* Appendix 1 (14).

starch gums dextrins *q.v.*

starch sheath an endodermis with starchy grains.

starch sugar glucose *q.v.*

starfish common name for the Asteroidea *q.v.*

start codon the RNA triplet signalling the start of translation of a polypeptide chain, usually AUG.

startpoint of transcription, the base pair in DNA corresponding to the first nucleotide incorporated into RNA.

stasimorphy *n.* deviation in form due to arrested growth.

stasis *n.* in evolution, the apparent lack of major evolutionary change over long periods of time in any given lineage, as seen in the fossil record, *see* punctuated equilibrium, stabilizing selection; stoppage or retardation in growth or development, or of movement of animal fluids.

static *a. pert.* system at rest or in equilibrium, *appl.* postural reactions; *appl.* proprioceptors, as otoliths and semicircular canals.

stationary phase the third stage in growth of a bacterial colony when multiplication slows down and virtually ceases, due to exhaustion of nutrients.

stato-acoustic *a. pert.* sense of balance and of hearing, *appl.* 8th cranial or cochlear nerves.

statoblast *n.* specialized asexual bud or "winter egg" of some Bryozoa, enclosed in a chitinous shell. It remains dormant during winter and develops into a new colony in spring.

statocone *n.* a minute structure contained in a statocyst.

statocyst *n.* vesicle lined with sensory cells and containing minute calcareous particles, either free in fluid or enclosed in cells, that move under gravity. Present in many invertebrates and concerned with perception of gravity; statocyte *q.v.*

statocyte *n.* a cell containing statoliths and probably acting as a georeceptor, such as a root cap cell containing granules such as starch grains. *alt.* statocyst.

statokinetic *a. pert.* maintenance of equilibrium and associated movements.

statolith *n.* particle of inorganic matter in fluid of statocysts (organs of balance and gravity detection) whose movement under gravity is detected by sensory cells, *alt.* otolith; a cell inclusion, as oil droplet, starch grain, or crystal, which changes its intracellular position under the unfluence of gravity.

staurophyll *n.* a leaf having palisade or other compact tissue throughout.

steapsin *n.* a lipolytic enzyme of the digestive juice of many animals, including the pancreatic juice of vertebrates. EC. 3.1.1.1. *r.n.* triacylglycerol lipase.

stearate (stearic acid) saturated widely distributed, fatty acid, esp. in animals. *see* Appendix 1 (38) for chemical structure.

steatogenesis *n.* production of lipid material.

steganopodous *a.* having feet completely webbed.

stegocarpic, stegocarpous *a.* having a capsule with operculum and periostome, *appl.* mosses.

stelar *a. pert.* stele.

stelar parenchyma pith *q.v.*

stele *n.* column of primary vascular tissues – primary phloem and primary xylem – and pith if present, extending throughout the primary plant body.

stellar stellate *q.v.*

stellate *a.* star-shaped, radiating, *appl.* cells: Kupffer cells *q.v.*; *appl.* hairs, spicules, ganglia of sympathetic nervous system, ligament of rib, etc. *alt.* stellar, asteroid.

stellate cell type of interneurone found in cerebellum with cell body in the molecular layer, acts on Purkinje cells; Kupffer cell *q.v.*

Stelleroidea *n.* class of echinoderms containing the starfish (Asteroidea) and brittle stars (Ophiuroidea).

stelliform stellate *q.v.*

stelocyttarous *a.* building, or *pert.,* stalked combs, as of certain wasps.

stem *n.* the main axis of a vascular plant, bearing buds and leaves or scale leaves, and reproductive structures (e.g. flowers), usually borne above ground (but *see* rhizome), and having a characteristic arrangement of vascular tissue.

stem cell undifferentiated cell in embryo or adult which can undergo unlimited division and can give rise to one or several different cell types. In adults, an undifferentiated cell from which some renewable tissues, e.g. blood and skin, are formed.

stemma *n.* ocellus of arthropods; a tubercle bearing an antenna. *plu.* stemmata, stemmas.

stenobaric *a. appl.* animals adaptable only to small differences in pressure or altitude.

stenobathic *a.* having a narow vertical range of distribution.

stenobenthic *a. pert.,* or living within a narrow range of depth of the sea bottom. *cf.* eurybethic.

stenochoric *a.* having a narrow range of distribution.

stenocyst *n.* one of the auxiliary cells in

leaves of certain mosses.

stenoecious *a.* having a narrow range of habitat selection.

stenohaline *a.* appl. organisms adaptable to a narrow range of salinity only.

stenohygric *a. appl.* organisms adaptable to a narrow variation in atmospheric humidity.

Stenolaemata *n.* class of Bryozoa.

stenomorphic *a.* dwarfed; smaller than typical form, owing to cramped habitat.

stenonatal *a.* with a very small thorax, as worker insect.

stenopetalous *a.* with narrow petals.

stenophagous, stenophagic *a.* subsisting on a limited variety of food.

stenophyllous *a.* narrow-leaved.

stenopodium *n.* a crustacean limb in which the protopodite bears distally both exopodite and endopodite.

stenosepalous *a.* with narrow sepals.

stenosis *n.* narrowing or constriction of a tubular structure, as of a pore, duct or vessel.

stenostomatous *a.* narrow-mouthed.

stenothermic *a. appl.* organisms adaptable to only slight variations in temperature.

stenotopic *a.* having a restricted range of geographical distribution.

stenotropic *a.* having a very limited adaptation to varied conditions.

stephanion *n.* point on skull where superior temporal ridge is crossed by coronal suture.

steppe *n.* dry and treeless grassland covering extensive areas of Asia.

stercome *n.* excreted waste material of Sarcodina, in masses of brown granules.

stercoral *a. appl.* a dorsal pocket or sac of proctodaeum in spiders.

stereid *n.* a lignified parenchyma cell with pit canals; stone cell *q.v.*

steroid bundles bands or bundles of sclerenchymatous fibres.

stereo pair a pair of images of a schematic diagram or model of a protein or other macromolecular structure which together give a three-dimensional image when viewed through special spectacles.

stereocilium *n.* projection on hair cell of cochlea, stiffened by a permanent cytoskeleton and which is involved in sensing vibration caused by sound waves; similar non-motile projections on other cells. *plu.* stereocilia.

stereognosis *n.* the ability to recognize three-dimensional shapes by touch, the sense that appreciates size, weight, shape of an object. *a.* stereognostic.

stereoisomer *n.* any of two or more compounds with the same atomic composition but differing in their structural configuration.

stereokinesis *n.* movement, or inhibition of movement, in response to contact stimuli.

stereom(e) *n.* sclerenchymatous and collenchymatous masses along with hardened parts of vascular bundles forming supporting tissue in plants; the thick-walled elongated cells of the central cylinder in mosses.

stereoscopic vision binocular vision, in which images from both eyes are integrated, allowing the seeing of objects in three dimensions.

stereospecificity *n.* of enzymes, specificity for only one of several possible stereoisomers of a substrate.

stereospondylous *a.* having vertebrae each fused into one piece.

stereotaxis, stereotaxy *n.* the response of an organism to the stimulus of contact with a solid, such as the tendency of some organisms to attach themselves to solid objects or to live in crannies or tunnels.

stereotropism *n.* growth movement in plants associated with contact with a solid object, as the tendency for stems and tendrils of climbers to twine around a support.

stereotyped behaviour *see* fixation.

sterigma *n.* a short outgrowth arising from basidium in basidiomycete fungi and which develops the basidiospore at its tip. *plu.* sterigmata.

sterile *a.* incapable of reproduction; aseptic; axenic *q.v.*

sterilize *v.* to render incapable of reproduction, of animals; to render incapable of conveying infection, of material containing microorganisms.

sternal *a. pert.* sternum, or sternite.

sternebrae *n.plu.* divisions of a segmented breast bone or sternum.

sternellum *n.* a sternal sclerite of insects,

esp. sclerite behind antesternite.

sternite *n.* a ventral plate of an arthropod segment.

sternobranchial *a. appl.* vessel conveying blood to gills, in some crustaceans.

sternoclavicular *a. appl.* and *pert.* articulation between sternum and clavicle.

sternohyoid *a. appl.* a muscle between back of manubrium of sternum and hyoid.

sternokleidomastoid *a. appl.* an oblique neck muscle stretching from sternum to mastoid process.

sternopericardial *a. appl.* ligament connecting dorsal surface of sternum and fibrous pericardium.

sternopleurite *n.* a thoracic sternite formed by union of episternum and sternum in insects.

sternoscapular *a. appl.* a muscle connecting sternum and scapula.

sternothyroid *a. appl.* muscle connecting manubrium of sternum and thyroid cartilage.

sternotribe *a. appl.* flowers with fertilizing elements so placed as to be brushed by sternites of visiting insects.

sternoxiphoid *a. appl.* plane through junction of sternum and xiphoid cartilage.

sternum *n.* breastbone of vertebrates; the ventral plate of exoskeleton of typical arthropod segment; the ventral plates of exoskeleton of a thoracic segment in insects.

steroid *n.* any of a large group of complex polycyclic lipids with a hydrocarbon nucleus and various substituents, synthesized from acetyl CoA via isoprene, squalene and cholesterol, and which include bile acids, sterols, various hormones, cardiac glycosides and saponins. *see* Appendix 1 (45, 47) for some examples of chemical structures.

steroid hormones the oestrogens, testosterone and its derivatives, glucocorticoids, mineralocorticoids, and the insect hormone ecdysterone.

steroid receptor superfamily family of structurally related intracellular receptors for steroid hormones, thyroid hormones, retinoids and some other similar compounds, which when complexed with their ligands become activated transcription factors, and switch on the transcription of selected genes.

sterol *n.* any steroid alcohol, ubiquitous in plants, animals and fungi, as components of the cell membranes, and including ergosterol (a typical fungal sterol), cholesterol (in animal cells), and phytosterol (in plants). *see* Appendix 1 (45) for chemical structure of cholesterol.

stichic *a.* in a row parallel to long axis.

stichidium *n.* a tetraspore receptacle in some red algae.

stick insects common name for many of the Phasmida *q.v.*

sticky ends single-stranded extensions found in double-stranded DNA which has been digested with a restriction enzyme that cuts in a staggered fashion. The ends are complementary and can therefore base-pair; cohesive ends *q.v.*

stigma *n.* (*bot.*) the upper portion of the carpel which receives the pollen and which is usually connected to the ovary by an elongated structure, the style, *see* Fig. 4; (*zool.*) various pigmented spots and markings, such as the coloured wing spot of some butterflies and other insects; a pigmented spot near base of flagellum in photosynthetic euglenoids, involved in photoreception and phototaxis.

stigmasterol *n.* a plant sterol, also present in milk, dietary deficiency causing muscular atrophy and calcium phosphate deposits in muscles and joints.

stigmata *plu.* of stigma *q.v.*

stigmatic *a. pert.* a stigma.

stigmatiferous *a.* bearing stigmas or stigmata.

stigmatiform, stigmatoid *a.* bearing stigmata.

stile(t) style(t) *q.v.*

stilliform guttiform *q.v.*

stilt roots buttress roots *q.v.*

stimulose *a.* having stinging hairs or cells.

stimulus *n.* an agent that causes a reaction or change in an organism or any of its parts.

stimulus–response theory theoretical explanation of classical conditioning which proposes that conditional responses occur because of reinforcement by being followed by a reward.

stimulus substitution theory theory proposed by Pavlov to explain classical conditioning, in which it is held that the animal comes to associate the unconditional and conditional stimuli and the conditional stimulus becomes a substitute for the unconditional stimulus. *cf*. stimulus–response theory.

sting *n*. stinging cell or hair; spine of stingray; offensive or defensive organ for piercing, also for inoculating with poison.

stinging cell nematocyst *q.v.*

stinkhorns common name for fungi of the Phallaes, an order of Gasteromycetes having a foetid odour to the gleba and a mature fruit body resembling a phallus.

stipe *n*. stalk, esp. of mushrooms, toadstools and other stalked fungi; stalk of seaweeds; stem of a fern frond.

stipella *n*. the stipule of a leaflet in a compound leaf.

stipes *n*. the stalk of a stalked eye, as in crustaceans; part of first segment of 1st maxilla of insects; distal portion of embolus in spiders.

stipiform *a*. resembling a stalk or stem.

stipitate *a*. stalked.

stipites *n.plu*. small exoskeletal plates anterior to mentum forming part of maxilla in some arthropods.

stipitiform stipiform *q.v.*

stipular *a*. like, *pert*., or growing in place of stipules.

stipulate *a*. having stipules.

stipule *n*. one of two leaf-like or bract-like outgrowths at the base of a leaf-stalk, sometimes modified as spines or tendrils; paraphyll *q.v.*; pin feather *q.v.*

stipuliferous *a*. having stipules.

stipuliform *a*. in the form of a stipule.

stirps *n*. the sum total of determinants to be found in a fertilized ovum; a group of organisms descended from a common ancestor; a variety of plants with stable characteristics which are retained under cultivation; a group of animals equivalent to a superfamily. *plu*. stirpes.

stochastic *a. appl*. a process in which there is an element of chance or randomness.

stock *n*. one or a group of individuals initiating a line of descent; (*bot*.) stem of tree or bush receiving bud or scion in grafting; the perennial part of a herbaceous plant; (*zool*.) an asexual zooid which produces sexual zooids of one sex by gemmation, as in polychaetes; livestock.

stocking rate the number of a particular kind of grazing animal feeding on a given area of grassland.

stoichiometry *n*. description of a metabolic reaction or cycle in terms of net proportions of molecules of each reactant consumed and produced. *a*. stoichiometric.

stolon *n*. a creeping plant stem or runner capable of developing rootlets and stem and ultimately forming a new individual; hypha connecting two bunches of rhizoids in fungi; similar structures in other organisms.

stolonate *a*. having stolons; resembling a stolon; dveloping from a stolon; *appl*. plants and animals which develop by means of stolons. *alt*. stoloniferous.

stolonial *a. pert*. a stolon or stolons, *appl*. a mesodermal septum in certain tunicates.

stoloniferous *a*. bearing a stolon or stolons.

stoma *n. sing*. of stomata *q.v.*; part of alimentary canal between mouth opening and oesophagus in nematodes.

stomach *n*. the large pouch of the intestine between oesophagus and intestines in vertebrates, and the corresponding part or entire digestive cavity in invertebrates. *a*. stomachic.

stomal, stomatal, stomatic *a. pert*. or like a stoma.

stomata *n.plu*. minute openings in epidermis of aerial parts of plants, esp. on undersides of leaves, through which air and water vapour enters the intercellular spaces, and through which carbon dioxide and water vapour from respiration is released. Stomata can be opened or closed by changes in turgor of the two guard cells that surround the central pore; any small openings or pores in various structures. *a*. stomatal. *sing*. stoma.

stomatal index the ratio between number of stomata and number of epidermal cells per unit area.

stomate *a*. possessing stomata.

stomatic stomal *q.v.*

stomatiferous *a*. bearing stomata.

stomatogastric *a. pert*. mouth and stomach; *appl*. visceral system of nerves supplying anterior part of alimentary canal; *appl*. recurrent nerve from frontal to stomachic ganglion, in insects.

stomatogastric ganglion nerve centre in

crustaceans situated on surface of stomach and controlling the movements of the teeth of the gastric mill.

stomatogenesis *n.* the formation of a mouth, as in ciliates.

stomions *n.plu.* pores or ostia in body wall of developing sponge.

stomium *n.* group of thin-walled cells in fern sporangium where rupture of mature capsule takes place; slit of dehiscing anther.

stomodaeal canal in Ctenophora, a canal given off by each perradial canal and situated parallel to the stomodaeum.

stomodaeum *n.* anterior portion of the alimentary canal, lined with ectodermally derived epithelium. *alt.* fore-gut.

stone canal cylindrical canal extending from madreporite to near mouth border in echinoderms.

stone cells short, isodiametric sclereids, as found in the flesh of a pear, giving it a gritty texture.

stone fruit drupe *q.v.*

stoneflies common name for the Plecoptera *q.v.*

stoneworts *n.plu.* common name for the Charophyta *q.v.*

stony corals a group of colonial coelenterates, typified by the reef- building corals, in which individual polyps are embedded in a matrix of calcium carbonate and connected by living tissue.

stop codon termination codon *q.v.*

stop-transfer sequence amino-acid sequence present in membrane proteins that stops translocation of the polypeptide chain across the membrane of the endoplasmic reticulum.

storey *n.* layer of a given height in a plant community.

storied *a.* arranged in horizontal rows on tangential surfaces, *appl.* axial cells and ray cells of wood cambium; *appl.* cork in monocotyledons with suberized cells in radial rows.

stotting *n.* warning behaviour in some gazelles, which bound away with a stiff-legged gait and tails raised, displaying a white rump.

strain *n.* pure-breeding variant line of a species of domesticated animal or cultivated plant; subspecific group whose members

are not sufficiently different genetically from the rest of the species to form a variety; (*microbiol.*) in bacteria and other microorganisms, a population of genetically identical individuals with some characteristic differentiating them from other strains of the same species.

β-strand one of the polypeptide strands in a β-sheet (*q.v.*).

strangulated *a.* constricted in places; contracted and expanded irregularly.

strata *plu.* of stratum *q.v.*

stratification *n.* arrangement in layers; superimposition of layers of epithelial cells; vertical grouping within a community or ecosystem; differentiation of horizontal layers of soil.

stratified epithelium epithelium several cell layers thick.

stratiform *a.* layered.

stratum *n.* a layer, as of cells, or of tissue; a group of organisms inhabiting a vertical division of an area; vegetation of similar height in a plant community, as trees, shrubs, herbs, mosses; a layer of rock. *plu.* strata.

stratum corneum outer keratinized layer of vertebrate epidermis, consisting of flattened dead cells filled with keratin, which flake off the surface. *alt.* cornified layer.

stratum germinativum basal layer of epidermis, with actively dividing cells.

stratum granulosum layer of small cells developing from the basal layer or stratum germinativum of mammalian epidermis.

stratum lucidum layer of cells becoming keratinized in epidermis of skin.

stratum Malpighii Malpighian layer *q.v.*

stratum spinosum layer of prickle cells in epidermis.

stratum transect a profile of vegetation, drawn to scale and intended to show the heights of plant shoots. *alt.* profile transect.

Strepsiptera *n.* an order of small insects with incomplete metamorphosis, commonly called stylopids, whose larvae and females are parasites of other insects and the males free-living. Forewings are halteres, hindwings are fan-shaped.

streptococci *n.* Gram-positive bacteria of the genus *Streptococcus*, which are small spherical cells (cocci) forming long chains. Many are harmless commensals living in the throat and gut, but some are human and animal pathogens, causing tonsilitis, scarlet fever and tissue destruction. *sing.* streptococcus.

streptolydigins *n.plu.* class of antibiotics, which inhibit bacterial transcription by interacting with the β subunit of RNA polymerase.

streptomycete *n.* member of the family Streptomycetaceae, filamentous prokaryotic microorganisms widespread in soil, characterized by formation of a permanent mycelium and reproduction by means of conidia, some spp. producing antibiotics including streptomycin, chloramphenicol and the tetracyclines.

streptomycin *n.* trisaccharide antibiotic synthesized by the actinomycete *Streptomyces griseus*, which inhibits bacterial protein synthesis by interfering with the binding of formylmethionyl-tRNA to ribosomes, and also by causing misreading of mRNA. It is used to treat tuberculosis in humans, and downy mildew on hops.

Streptoneura prosobranch *q.v.*

streptoneurous *a.* having visceral cord twisted, forming a figure of eight, as certain gastropods.

streptonigrin *n.* antibiotic synthesized by the actinomycete *Streptomyces flocculus,* which causes chromosome breakage.

streptostylic *a.* having quadrate in moveable articulation with squamosal.

stress fibres bundles of actin filaments in cultured cells, lying parallel with the substrate surface.

stress responses the various responses made by cells to heat shock, cold shock and other stresses.

stretch receptor sensory structure that monitors the degree of stretch of a muscle, e.g. muscle spindle in mammals, consisting of a specialized, modified muscle cell innervated by sensory neurones.

stria *n.* a narrow line, band, groove, streak or channel.

striae of Retzius growth markings in enamel of teeth, representing successive surfaces of enamel formation.

striate cortex also known as visual area I or area 17, part of visual cortex concerned with the initial processing of visual information.

striate(d) *a.* marked by narrow parallel lines or grooves.

striated muscle muscle tissue composed of transversely striped (striated) fibres formed from the fusion of many individual muscle cells, and which makes up the muscles attached to the skeleton. It is under the control of the voluntary nervous system.

striatum corpus striatum *q.v.*

stridulating organs special structures on various parts of the body of certain insects such as grasshoppers, crickets and cicadas, which produce the characteristic "song" of these insects. In crickets and grasshoppers the mechanism resembles that of a scraper being run along a toothed file as two parts of the body are rubbed against each other. In cicadas, membranes on either side of the abdomen are made to vibrate by muscular action.

stridulation *n.* the characteristic sound made by grasshoppers, crickets and cicadas.

striga *n.* a band of stiff upright hairs or bristles; a bristle-like scale.

strigate *a.* bearing strigae.

Strigiformes *n.* the owls, an order of mainly nocturnal, short-necked, large-headed birds of prey.

strigilis *n.* a structure for cleaning antennae, at junction of tibia and tarsus of 1st leg of certain insects.

strigillose *a.* minutely strigose.

strigose *a.* covered with stiff hairs; ridged; marked by small furrows.

stringency *n.* in DNA hybridization reactions refers to the degree to which DNAs of differing sequence will form duplexes, conditions of low stringency (e.g. low temperature) allowing duplex formation between nonidentical related DNAs, conditions of high stringency (e.g. high temperature) allowing duplex formation only between identical DNAs.

stringent factor enzyme in bacteria which catalyses the formation of ppGpp and pppGpp in conditions of amino acid starvation and is involved in the stringent response.

stringent response phenomenon seen in bacterial cells under conditions of amino acid starvation, when many cellular functions such as general RNA synthesis are shut down.

striola *n*. a fine narrow line or streak.

striolate *a*. finely striped.

striped muscle striated muscle *q.v.*

strobilaceous *a. pert.* or having strobiles.

strobila strobilus *q.v.*

strobilation, strobilization *n*. reproduction by separating off successive segments of the body, as in some jellyfish and in tapeworms.

strobile strobilus *q.v.*

strobili *plu.* of strobilus *q.v.*

strobiliferous *a*. producing strobiles.

strobiliform, strobiloid *a*. resembling or shaped like a strobilus or cone.

strobilus *n*. (*bot.*) cone-shaped assemblage of sporophylls in horse-tails, club mosses and conifers, *alt.* cone; in flowering plants, a spike formed by persistent membranous bracts, each having a pistillate flower; (*zool.*) stage in development of some jellyfish, a sessile polyp-like form that separates off a succession of disc-like embryos by segmentation; chain of proglottids of tapeworns.

stroma *n*. in ovary, soft vascular framework in which ovarian follicles are embedded; (*bot.*) in chloroplasts, the colourless material enclosed by the inner membrane and in which the grana are embedded, and in which carbon dioxide fixation takes place during photosynthesis; the non-pigmented part of other plastids; (*mycol.*) tissue of hyphae, or of fungal cells and host tissue, in or upon which a spore-bearing structure may be produced, *plu.* stromata.

stromata *plu.* of stroma *q.v.*

stromate *a*. having, or being within or upon, a stroma, *appl.* fruit bodies of fungi.

stromatic *a. pert.,* like, in form or nature of, a stroma.

stromatolites *n.plu.* layered structures, sometimes of considerable size, formed in certain warm shallow waters by mats of cyanobacteria (blue-green algae). Fossils of similar structure have been found in Precambrian rocks, indicating the presence of life at that time.

strombuliferous *a*. having spirally coiled organs or structures.

strombuliform *a*. spirally coiled.

strongyle *n*. a type of nematode larva.

strophanthidin *n*. a digitalis glycoside whose effect on the heart is mediated by inhibition of Na^+-K^+ ATPase.

strophanthin *n*. a glycoside with effects on the nervous system, obtained from various plants of the family Apocynaceae and used as a tropical arrow poison.

strophiolate *a*. having excrescences around hilum, *appl.* seeds.

strophiole *n*. one of the small excrescences arising from various parts of seed testor, arising after fertilization.

strophotaxis *n*. twisting movement or tendency, in response to an external stimulus.

structural colours colours of fish skin, insect wings, etc. that are not due to pigment but to surface structure, e.g. reflecting layers, plates of guanine crystals, etc.

structural gene a gene that codes for an enzyme or other protein required for a cell's structure or metabolism, or for tRNA or rRNA. *cf.* regulatory gene.

θ-structure intermediate structure formed in the replication of circular DNA molecules.

struma *n*. a swelling on a plant organ.

strumiform *a*. cushion-like.

strumose, strumulose *a*. having small cushion-like swellings.

strut roots buttress roots *q.v.*

Struthioniformes *n*. an order of flightless birds including the ostriches.

strychnine *n*. an alkaloid produced from seeds of *Strychnos* species and some other plants, a mammalian poison because of its effects on the nervous system.

Stuart factor Factor X *q.v.*

stupeous, stupose *a*. like tow; having a tuft of matted filaments.

stylar *a. pert.* style.

stylate *a*. having a style.

style *n*. (*bot.*) portion of carpel connecting stigma and ovary, often slender and elongated, *see* Fig. 4; (*zool.*) translucent revolving rod of protein and carbohydrate in stomach of bivalve molluscs which contains digestive enzymes; bristle-like process on clasper of male insect; arista *q.v.*; embolus of spider.

style sac a tubular gland in some molluscs which secretes the crystalline style.

stylet *n*. slender, hollow mouthpart, present in two pairs in aphids, through which they suck sap; any of various small, sharp appendages, used for stinging or piercing prey.

stylifer *n.* portion of clasper that carries the style.

styliferous *a.* bearing a style; having bristly appendages.

styliform *a.* prickle- or bristle-shaped.

styloconic *a.* having terminal peg on conical base, *appl.* type of olfactory sensilla in insects.

styloglossus *n.* a muscle connecting styloid process and side of tongue.

stylohyal *n.* distal part of styloid process of temporal bone; a small interhyal between hyal and hyomandibular.

stylohyoid *a. appl.* a ligament attached to styloid process and lesser cornu of hyoid; *appl.* a muscle; *appl.* a branch of facial nerve.

styloid *a.* pillar-like, *appl.* processes of temporal bone, fibula, radius, ulna; *n.* a columnar crystal.

stylomandibular *a. appl.* ligamentous band extending from styloid process of temporal bone to angle of lower jaw.

stylomastoid *a. appl.* foramen between styloid and mastoid processes, also an artery entering that foramen.

stylopharyngeus *n.* a muscle extending from the base of styloid process downwards along side of pharynx to thyroid cartilage.

stylopids common name for the Strepsiptera *q.v.*

stylospore *n.* a spore borne on a stalk.

stylostegium *n.* the inner corolla of milkweed plants.

stylus *n.* style *q.v.*; stylet *q.v.*; simple pointed spicule; molar cusp; pointed process.

sub- prefix from L. *sub*, under, signifying beneath, below (as in anatomical terms), less than (as in subthreshold), not quite, nearly, somewhat (esp. in descriptions of plant and animal parts, as subdentate, slightly toothed (of leaves), subcarinate, somewhat keel-shaped). In classification it indicates a group just below the status of the taxon following it, as in subclass, etc.

subabdominal *a.* nearly in the abdominal region.

subacuminate *a.* somewhat tapering.

subaduncate *a.* somewhat crooked.

subaerial *a.* growing just above the surface of the ground.

subalpine *a. appl.* zone just below the timber line and to plants and animals that live there.

subalpine *a.* ecological zone immediately below the alpine zone on high mountains.

subalternate *a.* tending to change from alternate to opposite.

subanconeous *n.* small muscle extending from triceps to elbow.

subapical *a.* nearly at the apex.

subarachnoid space cavity filled with cerebrospinal fluid between arachnoid membrane and pia mater surrounding brain and spinal cord.

subarborescent *a.* somewhat like a tree.

subarcuate *a. appl.* a blind fossa which extends backwards under superior semicircular canal in infant skull.

subatrial *a.* below the atrium, *appl.* longitudinal ridges on inner side of metapleural folds, uniting to form ventral part of atrium, in development of amphioxus.

subauricular *a.* below the ear.

subaxillary *a. appl.* outgrowths just beneath the axil.

sub-basal *a.* situated near the base.

sub-branchial *a.* under the gills.

sub-bronchial *a.* below the bronchial tubes.

subcalcareous *a.* somewhat limy.

subcalcarine *a.* under the calcarine fissure, *appl.* lingual gyrus of brain.

subcallosal *a. appl.* a gyrus below corpus callosum.

subcampanulate *a.* somewhat bell-shaped.

subcapsular *a.* under a capsule.

subcardinal *a. appl.* a pair of veins between mesonephroi.

subcarinate *a.* somewhat keel-shaped.

subcartilaginous *a.* not entirely cartilaginous.

subcaudal *a.* beneath or on the ventral side of the tail.

subcaudate *a.* having a small tail-like process.

subcaulescent *a.* borne on a very short stem.

subcellular *a. appl.* functional units or organelles within a cell.

subcentral *a.* nearly central.

subchela *n.* in some arthropods, a prehensile claw of which the last joint folds back on the preceding one.

subchelate *a.* having subchelae; having imperfect chelae.

subcheliceral *a.* beneath the chelicerae.

subchordal *a.* situated under the notochord.

subcingulum *n.* the lower lip part of cingulum or girdle of rotifers.

subclass *n.* taxonomic grouping between class and order, e.g. Theria (mammals that have live-born young) in the class Mammalia.

subclavate *a.* somewhat club-shaped.

subclavian *a.* below the clavicle, *appl.* artery, vein, nerve.

subclavius *n.* a small muscle connecting 1st rib to clavicle.

subclimax *n.* stage in plant succession preceding the climax, which persists because of some arresting factor such as fire, human activity, etc.

subclone *n.* part of a cloned DNA recloned into another vector.

subcoracoid *a.* below the coracoid.

subcordate *a.* tending to be heart-shaped.

subcorneous *a.* under a horny layer; slightly horny.

subcortical *a.* beneath the cortex or cortical layer.

subcosta *n.* an auxiliary vein joining costa of insect wing.

subcostal *a.* beneath ribs.

subcoxa *n.* basal ring of arthropod segment, which articulates with coxa of leg.

subcrenate *a.* tending to have rounded scallops, as a leaf margin.

subcrureus *n.* muscle extending from lower femur to knee.

subcubical *a. appl.* cells not quite so long as broad.

subcutaneous *a.* under the skin; *appl.* parasites living just under skin; *appl.* fat under the skin.

subcuticula *n.* epidermis beneath cuticle, as in nematodes.

subcuticular *a.* under the cuticle, epidermis or outer skin.

subcutis *n.* a loose layer of connective tissue between corium and deeper tissues of dermis; (*mycol.*) inner layer of cutis of agaric fungi, under the epicutis.

subdentate *a.* slightly toothed or notched.

subdermal *a.* beneath the skin; beneath the dermis.

subdorsal *a.* situated almost on the dorsal surface.

subdominant *n.* species that may seem more abundant than the true dominant species in a climax plant community at particular times of the year, or which is more abundant than the dominant species but occurs at a lower frequency.

subdural *a. appl.* the space separating the spinal dura mater from arachnoid.

subepicardial *a. appl.* areolar tissue attaching visceral layer of pericardium to muscular wall of heart.

subepiglottic *a.* beneath the epiglottis.

subepithelial *a.* below epithelium, *appl.* a plexus of cornea; *appl.* endothelium: Débove's membrane *q.v.*

suber *n.* cork tissue.

subereous *a.* of corky texture.

suberic *a. pert.* or derived from cork.

suberiferous *a.* cork-producing.

suberification *n.* conversion into cork tissue.

suberin *n.* waxy substance developed in thickened plant cell wall, characteristic of cork tissues.

suberization *n.* modification of plant cell walls due to suberin deposition.

suberose *a.* with corky waterproof texture; somewhat gnawed.

subesophageal suboesophageal *q.v.*

subfusiform *a.* somewhat spindle-shaped.

subgalea *n.* part of maxilla at base of stipes, of insects.

subgeniculate *a.* somewhat bent.

subgenital *a.* below the reproductive organs.

subgenomic RNA viral RNA produced by transcription of part of a genomic RNA, and which serves as an mRNA.

subgerminal *a.* below the germinal disc, *appl.* cavity.

subglenoid *a.* below the glenoid cavity.

subglossal *a.* beneath the tongue.

subharpal *a. appl.* plate in area below harpe in insects.

subhyaloid *a.* beneath hyaloid membrane or fossa of eye.

subhymenium *n.* layer of small cells between trama and hymenium in gill of agarics.

subhyoid *a.* below hyoid at base of tongue.

subimago *n.* a winged stage between pupa

and full adult (imago) in some insects, e.g. mayflies and other Ephemeroptera, which undergoes a further moult.

subinguinal *a.* situated below a horizontal line at level of great saphenous vein termination, *appl.* lymph glands.

subjugal *a.* below jugal or cheek bone.

subjugular *a. appl.* a ventral fish fin nearly far enough forward to be jugular.

sublanceolate *a.* tending to be narrow and to taper towards both ends.

sublaryngeal *a.* situated below the larynx.

sublenticular *a.* somewhat lens-shaped.

sublethal *a.* not causing death directly, but having cumulative deleterious effects.

subliminal *a. appl.* stimuli that are not strong enough to evoke a sensation.

sublingua *n.* a double projection or fold beneath tongue in some mammals.

sublingual *a.* beneath the tongue, *appl.* gland, artery, etc.; *appl.* ventral pharyngeal gland in hymenopterans.

sublittoral *a.* below littoral, *appl.* the shallow water zone of the sea from the extreme low-tide level to a depth of around 200 m, *see* Fig. 9; zone of a lake too deep for rooted plants to grow.

sublobular *a. appl.* veins at base of lobules in liver.

sublocular *a.* somewhat locular or cellular.

sublocus *n.* a part of a complex genetic locus that acts as an individual locus in some genetic tests.

submalleate *a.* somewhat hammer-shaped, *appl.* trophi of rotifer mastax.

submandibular submaxillary *q.v.*

submarginal *a.* placed nearly at margin.

submarginate *a.* a bordering structure near a margin.

submaxilla mandible *q.v.*

submaxillary *a.* beneath lower jaw, *appl.* duct, ganglion, gland, triangle.

submedian *a. appl.* tooth or vein next median.

submental *a.* beneath the chin, *appl.* artery, glands, triangle, vibrissae; *pert.* submentum.

submentum *n.* part of labium of insects.

submersed *a. appl.* plants growing entirely under water.

submetacentric *a. appl.* chromosomes with the centromere nearer one end than the other giving arms of unequal length;

n. a submetacentric chromosome.

submission *n.* the behaviour of a losing animal in a conflict where it takes up a submissive posture to prevent further attack.

submitochondrial particles small particles produced by sonication of mitochondria, having the inner membrane on the outside, used to study respiratory chain function and arrangement in the membrane.

submucosa *n.* a layer of gut wall between the mucosa and external muscular coat, composed of connective tissue and accommodating blood vessels, nerves, Meissner's plexus, and some glands.

subnasal *a.* beneath the nose.

subneural *a. appl.* blood vessel in annelid worms; *appl.* gland and ganglion of nervous system in tunicates.

subnotochord *n.* a rod ventral to true notochord.

suboccipital *a. appl.* muscles, nerve, triangle, under occipitals of skull.

suboesophageal *a.* below the gullet, *appl.* anterior ganglion of ventral nerve cord in invertebrates.

subopercular *a.* under operculum of fishes, or shell of molluscs.

suboperculum *n.* a membrane bone of operculum in fishes.

suboptic *a.* below the eye.

suboral *a.* below the mouth.

suborbital *a.* below the orbit of the eye.

subovate *a.* somewhat oval or egg-shaped, *appl.* leaves.

subpalmate *a.* tending to become palmate, *appl.* leaves.

subparietal *a.* beneath parietals, *appl.* sulcus which is lower boundary of parietal lobe of brain.

subpectinate *a.* tending to be comb-like in structure.

subpedunculate *a.* resting on a very short stalk.

subpericardial *a.* under the pericardium.

subperitoneal *a. appl.* connective tissue under the peritoneum.

subpessural *a.* below the pessulus of syrinx, *appl.* air sac.

subpetiolar, subpetiolate *a.* within petiole, *appl.* bud so concealed; almost sessile.

subpharyngeal *a.* below the throat; *appl.* gland or endostyle beneath pharynx, with cells containing iodine in ammocoetes.

subphrenic *a.* below the diaphragm.

subphylum *n.* taxonomic grouping between phylum and class, e.g. Vertebrata (in the phylum Chordata).

subpial *a.* under the pia mater.

subpleural *a.* beneath inner lining of thoracic wall.

subpubic *a.* below the pubic region, *appl.* arcuate ligament.

subpulmonary *a.* below the lungs.

subradicate *a.* to have a slight downward extension at base, as of stipe.

subradius *n.* in radially symmetrical animals, a radius of the 4th order, that between adradius and perradius or between adradius and interradius.

subradular *a. appl.* organ containing nerve endings situated at anterior end of odontophore.

subramose *a.* slightly branching.

subreniform *a.* slightly kidney-shaped.

subretinal *a.* beneath the retina.

subrostral *a.* below the beak or rostrum, *appl.* a cerebral fissure.

subsacral *a.* below the sacrum.

subsartorial *a. appl.* plexus under sartorius muscle of thigh.

subscapular *a.* beneath the scapula, *appl.* artery, muscles, nerves, etc.

subsclerotic *a.* beneath sclera; between sclerotic and choroid layers of eye.

subscutal *a.* under a scutum.

subsere *n.* plant succession on a denuded area, secondary succession.

subserous *a.* beneath serous membrane.

subserrate *a.* somewhat notched or saw-toothed.

subsessile *a.* nearly sessile, with almost no stalk.

subsong *n.* the first attempts at song by a young bird, which resembles the adult song but is imprecise and lacks some elements, with ill-defined phrasing and lack of tonal purity.

subspatulate *a.* somewhat spoon-shaped.

subspecies (ssp.) *n.* a taxonomic term usually meaning a group consisting of individuals within a species having certain distinguishing characteristics separating them from other members and forming a

breeding group, but which can still interbreed with other members of the species. *alt.* variety.

subspinous *a.* tending to become spiny.

substance K a neuropeptide related to substance P.

substance P peptide found in gut tissue and in central nervous system, thought to act as a neurotransmitter and be involved in pain pathways.

substantia adamantina enamel of teeth.

substantia alba white matter of brain and spinal cord.

substantia eburnea dentine *q.v.*

substantia gelatinosa the gelatinous neuroglia in spinal cord.

substantia grisea grey matter of spinal cord.

substantia nigra semicircular layer of grey matter in midbrain, contains cell bodies of dopaminergic neurones.

substantia ossa cement of teeth.

substantia reticularis reticular formation *q.v.*

substantia spongiosa cancellous tissue of bone.

substantive variation change in actual constitution or substance of parts. *cf.* meristic variation.

substernal *a.* below the sternum or breast bone.

substitution base substitution *q.v.*

substrate *n.* substance on which an enzyme acts in biochemical reactions; respiratory substrate, substance undergoing oxidation during respiration; any material used by microorganisms as a source of food; inert substance containing or receiving a nutrient solution on which microorganisms grow; the base to which a sedentary animal or plant is fixed. *alt.* substratum.

substrate cycle pair of irreversible metabolic reactions catalysed by 2 different enzymes in which one compound is converted into another and back again, and which may amplify metabolic signals and possibly generate heat as in the flight muscle of bumble bees. *alt.* futile cycle.

substrate-level phosphorylation formation of energy-rich phosphate compounds such as ATP by transfer of phosphate from a metabolic substrate to ADP directly with no respiratory chain involvement, as occurs in glycolysis. *cf.* oxidative phosphorylation, photophosphorylation.

substratose *a.* slightly or indistinctly stratified.

substratum *n.* the base to which a sedentary animal or plant is fixed. *see also* substrate.

subtalar *a. appl.* joint: the articulation between talus and calcaneus.

subtectal *a.* lying under a roof, esp. of skull.

subtegminal *a.* under the tegmen or inner coat of a seed.

subtentacular canals two prolongations of echinoderm coelom.

subterminal *a.* situated near the end.

subthalamus *n.* hypothalamus *q.v.*; part of hypothalamus excluding optic chiasma and region of the mamillary bodies.

subthoracic *a.* not quite so far forward as to be called thoracic, *appl.* certain fish fins.

subtilisin *n.* proteolytic enzyme produced by *Bacillus subtilis.* Included in EC 3.4.21.14.

subtrapezoidal *a.* somewhat trapezoid-shaped.

subtruncate *a.* terminating abruptly.

subtypical *a.* deviating slightly from type.

subulate *a.* awl-shaped, i.e. narrow and tapering from base to a fine point, *appl.* leaves, as of onion.

subumbellate *a.* tending to an umbellate arrangement with peduncles arising from a common centre.

subumbonal *a.* beneath or anterior to the umbo of a bivalve shell.

subumbonate *a.* slightly convex; having a low rounded protuberance.

subumbrella *n.* the concave inner surface of the bell of a medusa.

subuncinate *a.* having a somewhat hooked process; somewhat hooked.

subungual *a.* under a nail, claw or hoof.

subunguis *n.* the ventral scale of a claw or nail.

subunit *see* protein subunit.

subunit vaccine vaccine made from purified protein components of viruses, bacteria and other parasites, rather than the complete organism.

subvaginal *a.* within or under a sheath.

subvertebral *a.* under the vertebral column.

subzonal *a. appl.* layer of cells internal to zona radiata.

subzygomatic *a.* below the cheek bone.

succate *a.* containing juice, juicy.

succession *n.* a geological, ecological, or seasonal sequence of species; the sequence of different communities developing over time in the same area, leading to a dynamic steady state or climax community (used esp. of plant or microbial communities); the occurrence of different species over time in a given area.

successive percentage mortality apparent mortality *q.v.*

succiferous *a.* conveying sap.

succinate (succinic acid) *n.* 4-carbon dicarboxylic acid of the tricarboxylic acid cycle, converted to fumarate by succinate dehydrogenase. *see* Appendix 1 (55) for chemical structure.

succinate dehydrogenase a flavoprotein enzyme catalysing the oxidation of succinate to fumarate in the tricarboxylic acid cycle, providing a direct link with the respiratory chain. EC 1.3.99.1.

succinyl CoA "energy-rich" compound formed from succinate and CoA during the tricarboxylic acid cycle and other metabolic cycles, and which also provides carbon skeleton for porphyrin synthesis.

succise *a.* abrupt; appearing as if part were cut off.

succubous *a.* with each leaf covering part of that under it.

succulent *a.* full of juice or sap; *appl.* fruit having a fleshy pericarp, as berries; *appl.* plants adapted to dry and desert conditions, with swollen water-storing stems and leaves.

succus entericus the secretions of the epithelium lining the small intestine, containing peptidases, maltase, sucrase, lactase, lipase, nucleases, nucleotidases.

sucker *n.* (*bot.*) a branch of stem, at first running underground and then emerging, and which may eventually form an independent plant; haustorium *q.v.*; (*zool.*) an organ adapted to attach to a surface by creating a vacuum, in some animals for the purpose of feeding, in others to assist locomotion or attachment.

sucking disc a disc assisting in attachment, as at end of echinoderm tube-foot.

sucking lice common name for the Anoplura *q.v.*

sucrase *n.* digestive enzyme hydrolysing the disaccharides sucrose and maltose to their component monosaccharides (glucose and fructose, and glucose respectively) by an α-D-glucosidase action, found in intestinal mucosa. EC 3.2.1.48, *r.n.* sucrose-α-D-glucohydrolase.

sucrose *n.* a non-reducing disaccharide present in many green plants and hydrolysed by the enzymes invertase or sucrase or by dilute acids to glucose and fructose. *alt.* cane sugar, invert sugar, beet sugar, saccharose, saccharobiose. *see* Appendix 1 (12) for chemical structure.

suction pressure (SP) the capacity of a plant cell to take up water by osmosis, being the difference between the osmotic pressure of the cell sap causing water to enter and the back pressure exerted by the cell wall (turgor pressure). When a cell is turgid it has no suction pressure.

Suctoria *n.* group of predatory ciliate protozoans which usually lose their cilia as adults and possess one or more suctorial tentacles.

suctorial *a.* adapted for sucking.

Sudanian-Sindian Floral Region part of the Palaeotropical Realm comprising the Sahel region of Africa, Sudan and northwest India.

sudation *n.* discharge of water and other substances in solution, as through pores; sweating.

sudden correction model proposed mechanism for maintenance of fidelity of multiple repeated DNA sequences (such as the rRNA genes) in which the entire gene cluster is replaced from time to time with a set of copies derived from one or a few original sets of genes.

sudoriferous *a.* conveying, producing or secreting sweat, *appl.* glands and their ducts.

sudorific *a.* causing or *pert.* secretion of sweat.

sudoriparous sudoriferous *q.v.*

suffrescent *a.* slightly shrubby, *appl.* plants that are woody at the base but herbaceous above and that do not die back to ground level in winter.

suffrutex *n.* an under-shrub. *plu.* suffrutices.

suffruticose *a.* somewhat shrubby.

sugar *n.* the general name for any mono-, di- or trisaccharide; sucrose *q.v.*

sugar nucleotide a nucleotide covalently linked to a sugar, e.g. UDP-glucose, which is the activated form of sugars for glycan synthesis.

sugent suctorial *q.v.*

suids *n.plu.* members of the mammalian family Suidae: the pigs.

suines *n.plu.* members of the mammalian suborder Suina, which includes the non-ruminant artiodactyls: the hippopotamuses, pigs, peccaries, and a number of extinct groups.

sulcate *a.* grooved or furrowed.

sulcation *n.* fluting; formation of ridges and furrows, as in elytra.

sulci *plu.* of sulcus.

sulcus *n.* a groove; a groove between two convolutions on the surface of the brain. *plu.* sulci.

sulf- sulph-.

sulfanilamide sulphanilamide *q.v.*

sulphanilamide *n.* an antibacterial compound, a sulphur-containing aromatic amide and its derivatives which prevent bacterial growth by inhibiting purine synthesis. *alt.* sulfanilamide.

sulphatase *n.* enzyme catalysing the hydrolysis of sulphuric esters, e.g. arylsulphatase (EC 3.1.6.1), which hydrolyses a phenol sulphate to phenol plus sulphate ion.

sulphatide *n.* any of several sulphur-containing glycolipids, derivatives of ceramide, found in animal brain and other tissues.

sulpholipid *n.* sulphur-containing glycolipid, occurring in chloroplasts in green plants and chromatophores of photosynthetic bacteria.

sulphur bacteria a group of unrelated bacteria which can variously utilize sulphur or sulphide as a respiratory substrate or electron acceptor in photosynthesis or reduce sulphate to sulphide. They comprise the photosynthetic green sulphur bacteria and purple suphur bacteria, which oxidize sulphide to sulphur (and which can also fix nitrogen), the colourless non-photosynthetic sulphur bacteria (e.g. *Thiobacillus*) which oxidize sulphide to sulphur and sulphate, and the heterotrophic sulphate-reducing bacteria which reduce sulphate to sulphide.

sulphur cycle a cycle of biological processes by which sulphur circulates within the biosphere, and which includes assimilation of sulphur by plants as soil sulphate, incorporation into plant and animal protein, putrefaction of dead organic matter by bacteria which releases sulphide, which can be converted to elemental sulphur, to sulphate and back to sulphide by a heterogeneous group of sulphur bacteria.

summation *n.* combined action of either simultaneous or successive nerve impulses, subliminal stimuli or subthreshold potentials which produces an excitatory or inhibitory response.

summer wood late wood *q.v.*

sun-basking *n.* behaviour shown by many poikilothermic animals to control body temperature.

sun spiders common name for the Solifuga *q.v.*

super- prefix derived from L. *super*, over. In classification, a group just above the status of the taxon following it, as in superclass.

superantigen *n.* antigen such as the endogenous Mls antigens of some strains of mice and the staphylococcal enterotoxins, which when used for immunization provoke a massive T-cell response by non-specifically activating all T cells with receptors carrying a particular variable region.

supercarpal *a.* above the carpus; *n.* an upper carpal.

superciliary *a. pert.* eyebrows; above orbit of eye.

superciliary arches two arched elevations on forehead below frontal eminences.

superclass *n.* taxonomic grouping between subphylum and class.

supercoiled DNA a closed circular DNA molecule in which the double helix is further twisted on itself to form a more compact molecule, left-handed turns (negative supercoiling) also leading to a loosening of the strands of the double-helix (underwinding), right-handed turns (positive supercoiling, not found *in vivo*) leading to an overwound helix. *alt.* superhelical DNA, supertwisted DNA.

superdominance overdominance *q.v.*

superfamily *see* gene superfamily; protein superfamily: a diverse family of proteins with different functions but all having related sequences and presumed to be encoded by genes derived from a common ancestral gene.

superfemale metafemale *q.v.*

superfetation *alt.* spelling of superfoetation *q.v.*

superficial *a.* on or near the surface; *appl.* placentation in ovary of flower in which ovules are scattered over inner surface of ovary wall.

superfluent *n.* an animal species of the same importance in an ecosystem as a subdominant plant species in a succession.

superfoetation *n.* fertilization of ovules of an ovary with more than one kind of pollen; successive fertilization of two or more ova of different oestrous periods, in the same uterus.

supergene *n.* region of chromosome which contains a number of genes but in which crossing over does not occur, so that the genes are transmitted together from generation to generation.

supergene family gene superfamily *q.v.*

superglottal *a.* above the glottis.

superhelical DNA supercoiled DNA *q.v.*

superior *a.* upper; higher; anterior; growing or arising above another organ; *appl.* ovary having perianth inserted around the base; *appl.* sepals, petals or stamens attached to the receptacle of a flower below the ovary; *appl.* vena cava: precava *q.v.*

superlinguae *n.plu.* paired lobes of hypopharynx in certain insects.

supermale metamale *q.v.*

supernumerary chromosomes extra heterochromatic chromosomes present in some plants above the normal number for the species, as B chromosomes.

superorganism *n.* any society, such as a colony of a eusocial insect species, possessing features of organization analogous to the properties of a single organism.

superovulation *n.* the production of an unusually large number of eggs at any one time.

superoxide dismutase (SOD) widely distributed enzyme which destroys superoxide (O_2^-) radicals with the formation of hydrogen peroxide and molecular oxygen. *alt.* erythrocuprein, haemo-cuprein, cytocuprein.

superparasite hyperparasite *q.v.*

super-regeneration *n.* the development of additional or superfluous parts in the process of regeneration.

superrepressed *a. appl.* mutant genes that cannot be derepressed. *alt.* uninducible.

supersacral *a.* above the sacrum.

supersecondary structure level of protein structure, such as a β-barrel, in which secondary structure motifs are combined into a more complex discrete structural element.

supersolenoid *n.* the approx. 2000 Å diameter chromatin fibre formed by coiling of the solenoid, and which is thought to be further coiled to form a chromatid.

supersonic ultrasonic *q.v.*

superspecies *n.* a group of closely related species having many morphological resemblances.

supersphenoidal *a.* above sphenoid bone.

supertwisted DNA supercoiled DNA *q.v.*

supervolute *a.* having a plaited and rolled arrangement in the bud.

supinate *a.* inclining or leaning backwards.

supination *n.* movement of arm by which palm of hand is turned upwards.

supinator brevis and longus two forearm muscles used in turning the palm upwards.

supplemental air volume of air that can be expelled from the lungs after normal breathing out.

supporting tissue in plants, tissue made of cells with thickened walls such as collenchyma and sclerenchyma, adding strength to plant body; in animals, skeletal tissue forming endo- or exoskeleton.

suppression *n.* non-development of an organ or part; the cancelling out of the effects of one mutation by another, usually in another gene (if in the same gene it is termed intragenic suppression), the second mutation being called a suppressor.

suppressor mutation mutation that cancels out the effects of another mutation elsewhere in the genome. *see* suppression.

suppressor T lymphocyte, suppressor T cell (T$_s$) T lymphocyte involved in suppressing an immune reaction, probably by interacting in an antigen-specific manner

with helper T lymphocytes and/or B lymphocytes.

suppressor tRNA mutant tRNA with altered codon specificity, which can correct a nonsense mutation elsewhere in the genome by inserting the correct amino acid at the mutant codon.

supra- prefix derived from L. *supra*, above, signifying situated above.

supra-acromial *a.* above the acromion of shoulder blade.

supra-anal *a.* above anus or anal region.

supra-angular surangular *q.v.*

supra-auricular *a.* above the auricle of the ear, *appl.* feathers.

suprabranchial *a.* above the gills.

suprabuccal *a.* above cheek and mouth.

suprabulbar *a. appl.* region between hair bulb and fibrillar region of hair.

supracallosal *a. appl.* a gyrus on the upper surface of corpus callosum of brain.

supracaudal *a.* above the tail or caudal region.

supracellular *a. appl.* structures originating from many cells; *appl.* level of organization above cellular level, as of tissues, organs, etc.

supracerebral *a. appl.* lateral pharyngeal glands, as in hymenopterans.

suprachoroid *a.* over the choroid; between choroid and sclera of eye.

supraclavicle *n.* a bone of pectoral girdle of fishes. *alt.* supracleithrum.

supraclavicular *a.* above or over the clavicle, *appl.* nerves.

supracondylar *a.* above a condyle, *appl.* ridge and process.

supracoracoideus *n.* a flight muscle in birds, running indirectly from breast bone to humerus and responsible for raising the wing.

supracostal *a.* over or external to the ribs.

supracranial *a.* over or above the skull.

supradorsal *a.* on or over the dorsal surface; *appl.* small cartilaginous elements in connection with primitive vertebral column.

supra-episternum *n.* upper sclerite of episternum in some insects.

supraesophageal supraoesophageal *q.v.*

supraglenoid *a.* above the glenoid cavity; *appl.* tuberosity at apex of glenoid cavity.

suprahyoid *a.* over or above the hyoid

bone(s) lying at base of tongue (in mammals).

supralabial *a.* on the lip, *appl.* scutes or scales.

supraliminal *a.* above the threshold of sensation, *appl.* stimuli.

supralittoral *a. pert.* seashore above high-water mark, or spray zone.

supraloral *a.* above the loral region, as in birds, snakes.

supramastoid crest ridge at upper boundary of mastoid region of temporal bone.

supramaxillary *a. pert.* upper jaw.

suprameatal *a. appl.* triangle and spine over external auditory meatus.

supranasal *a.* over nasal bone or nose.

supraoccipital *n.* a large bone in the middle of the back of the skull.

supraocular *a.* over or above the eye.

supraoesophageal *a.* above or over the gullet.

supraorbital *a.* above eye orbits.

suprapatellar *a. appl.* pouch between upper part of patella and femur.

suprapericardial *see* ultimobranchial.

suprapharyngeal *a.* above or over the pharynx.

suprapubic *a.* above the pubic bone.

suprapygal *a.* above the pygal bone.

suprarenal *a.* situated above the kidneys; *appl.* bodies, glands: adrenals *q.v.*

suprarostral *a. appl.* a cartilaginous plate anterior to trabeculae of skull in amphibians.

suprascapula *n.* cartilage of dorsal part of pectoral girdle in certain cartilaginous fishes; incompletely ossified extension of scapula of amphibians and some reptiles.

suprascapular *a.* above the shoulder blade, *appl.* ligament, nerve.

supraseptal *a. appl.* two plates diverging from interorbital septum of skull.

suprasphenoid *n.* membrane bone dorsal to sphenoid cartilage.

suprasphenoidal *a.* above sphenoid bone of skull.

supraspinal *a.* above and over the spinal column; above ventral nerve cord in insects.

supraspinatous *a. appl.* scapular fossa and fascia for origin of supraspinatus.

supraspinatus *n.* shoulder muscle inserted into proximal part of greater tubercle of humerus.

suprastapedial *n.* the part of columella of ear above stapes, homologous with mammalian incus.

suprasternal *a.* over or above breast bone, *appl.* a slit-like space in cervical muscle, *appl.* supernumerary sternal elements in some mammals, *appl.* body plane.

suprastigmal *a.* above a stigma or breathing pore of insects.

supra-temporal *a. pert.* upper temporal region of skull, *appl.* bone, arch, fossa; pterotic of teleosts.

suprathoracic *a.* above thoracic region.

supratidal *a.* above high-water mark, *appl.* the spray zone and organisms living there.

supratonsillar *a. appl.* a small depression in lymphoid mass of palatine tonsil.

supratrochlear *a.* over trochlear surface, *appl.* nerve, foramen, lymph glands.

supratympanic *a.* above the ear drum.

sural *a. pert.* calf of leg.

surangular *n.* a bone of lower jaw of some fishes, reptiles, and birds.

surculose *a.* of plants, bearing suckers.

surcurrent *a.* proceeding or prolonged up a stem.

surface exclusion the inability of a plasmid to enter a bacterial cell already carrying a plasmid of the same type, and which is mediated at the surface of the bacterium.

surface water water from rain or other precipitation that does not sink into the ground or evaporate, and which runs off the land surface, flowing eventually into streams and rivers.

surrogate genetics the study of gene expression by making defined alterations in an isolated DNA template and looking at the effects of these changes on its transcription in *in vitro* or other transcription systems, i.e. bypassing the need to induce or search for suitable mutations in the living organism.

survivorship schedule demographic data giving the number of individuals surviving to each particular age in a population.

suspensor *n.* a chain of cells developing from the zygote in angiosperms, attaching embryo to embryo sac; similar structure in other plants; modified portion of a hypha from which a gametangium or a zygospore is suspended.

suspensorium *n*. skeletal element forming the side wall of mouth cavity in bony fishes and other vertebrates.

suspensory *a. pert.* a suspensorium; serving for suspension, *appl.* various ligaments.

sustainable yield highest rate at which a renewable resource can be used without reducing its supply.

sustentacular *a.* supporting, *appl.* cells; *appl.* connective tissue acting as a supporting framework for an organ.

sutural *a. pert.* a suture; *appl.* dehiscence taking place along a suture.

sutural bones irregular isolated bones occurring along sutures.

suture *n*. line of junction of two parts immovably connected, as between bones of skull, sclerites of exoskeleton covering an arthropod segment, etc.; line of seed capsule, etc. along which dehiscence occurs.

Svedberg unit (S) unit in which the rate of sedimentation of a particle in the ultracentrifuge is expressed ($1\ S = 10^{-13}$ s under standard conditions) and which is an indirect measurement of size and molecular weight.

swamp *n*. wet ground, saturated or periodically flooded, dominated by woody plants and with no surface accumulation of peat.

swarm cells motile zoospores, amoebae, etc. in fungi, algae and slime moulds.

sweat glands specialized glands in the skin of some mammals, through which water and salts are exuded to aid evaporative cooling of the body. They are of two types: apocrine sweat glands and eccrine sweat glands.

sweating *n*. exudation of water from the body surface through sweat glands, which is used as a cooling mechanism by some mammals, evaporation of the water from the surface having a cooling effect.

sweet potato *Ipomoea batatus*, a crop plant originating in Central America, now grown as a staple food worldwide, a member of the family Convolvulaceae (order Passiflores), whose tubers are eaten.

swimbladder *n*. a gas-filled sac in body cavity of most teleost fishes, developed as an outgrowth of the alimentary canal, and which is an aid to buoyancy.

swimmeret *n*. small paired appendage of crustaceans, present on up to five abdominal segments, possibly involved partly in swimming.

swimming bell nectocalyx *q.v.*

swimming funnel the siphon in some cephalopods through which water is expelled from mantle cavity, producing a means of jet propulsion.

swimming ovaries groups of ripe ova in some acanthocephalans, detached from ovary and floating in body cavity.

swimming plates in Ctenophora, ciliated comb-like plates, arranged in eight equidistant bands, and which serve to propel the organism along. *alt.* ctenes, comb-ribes.

switch gene a gene such as the homoeotic genes, which when mutant cause development to switch from one pathway to another, and therefore are believed normally to act as genetic master switches, selecting a particular developmental pathway.

switch plant a xerophyte which produces normal leaves when young, then sheds them, and photosynthesis is taken over by another structure such as a cladode or phyllode.

switch region in immunoglobulin loci, a site found to the 5′ side of each C gene at which recombination occurs during immunoglobulin class switching, resulting in production of antibodies with the same antigen specificity but different class during an immune response.

syconium, syconus *n*. a composite fruit consisting mainly of an enlarged succulent receptacle, as a fig.

sylva *n*. forest of a region; forest trees collectively.

sym-, syn- prefixes from the Gk. *syn*, with.

symbiont *a.* one of the partners in a symbiosis.

symbiosis *n*. close and usually obligatory association of two organisms of different species living together, not necessarily to their mutual benefit; often used exclusively for an association in which both partners benefit, which is more properly called mutualism. *a.* symbiotic.

symmetrodonts *n.plu.* an order of Mesozoic trituberculate mammals having molar teeth with three or more cusps in a triangle.

symmetrical *a. pert.* symmetry.

symmetry *n.* regularity of form, *see* bilateral symmetry, radial symmetry.

symparasitism *n.* the development of several competing species of parasite within or on one host.

sympathetic *a. appl.* components of sympathetic nervous system, as sympathetic neurone; *appl.* segmental nerves supplying spiracles in insects; *appl.* coloration in imitation of surroundings.

sympathetic nervous system part of the autonomic nervous system comprising nerve fibres that leave the spinal cord in the thoracic and lumbar regions and supply viscera and blood vessels by way of a chain of sympathetic ganglia running on each side of the spinal column which communicate with the central nervous system via a branch to a corresponding spinal nerve. The sympathetic nervous system controls movements and secretions from viscera and monitors their physiological state, stimulation of the sympathetic system inducing e.g. the contraction of gut sphincters, heart muscle and the muscle of artery walls, and the relaxation of gut smooth muscle and the circular muscles of the iris. The chief neurotransmitter in the sympathetic system is adrenaline which is liberated in heart, visceral muscle, glands and internal vessels, acetylcholine acting as a neurotransmitter at ganglionic synapses and at sympathetic terminals in skin and skeletal muscle blood vessels. The actions of the sympathetic system tend to be antagonistic to those of the parasympathetic system.

sympathomimetic *a. appl.* substances that produce effects similar to those produced by stimulation of the sympathetic nervous system.

sympatric *a. appl.* species inhabiting the same or overlapping geographic areas. *cf.* allopatric.

sympetalous *a.* having petals joined into a tube, at least at base.

Symphyla *n.* class of arthropods (*q.v.*) allied to the myriapods, which have 14 body segments and 6 pairs of walking legs.

Symphyta *n.* the sawflies, a suborder of hymenopteran insects, having no well-defined "waist", and considered more primitive than members of the suborder Apocrita (bees, wasps, ants, ichneumon flies). The ovipositor is serrated like a saw.

symphily *n.* the situation where one species of insect lives as a guest (symphile) in the nest of a social insect which feeds and protects it in return for its secretions which are used as food, as certain beetles in the nests of ants and termites.

symphoresis *n.* movement or conveyance collectively, as movement of spermatid group to a Sertoli cell.

symphygenesis *n.* the development of an organ from the union of two others.

symphyllodium *n.* a structure formed by coalescence of external coats of two or more ovules; a compound ovuliferous scale.

symphysis *n.* the line of junction at which two bones, e.g. left and right halves of jaw, are fused; slightly moveable articulation of two bones connected by fibrocartilage; the growing together of parts which are separate in early development. *a.* symphyseal, symphysial.

symplast *n.* the interconnected protoplasm of plant tissue, the protoplasm of individual cells connected by plasmodesmata through the cell walls. *a.* symplastic.

symplastic transport in plant tissue, the movement of ions and other material from cell to cell via plasmodesmata (cytoplasmic bridges in the cell wall).

symplesiomorphy *n.* in the cladistic method of classification, the case where a homologous character state shared between two or more taxa is believed to have originated as a novelty in a common ancestor earlier than the most recent common ancestor. *a.* symplesiomorphic, symplesiomorphous. *cf.* homoplasty, synapomorphy.

sympodial *a. pert.* or resembling a sympodium in mode of branching.

sympodite protopodite *q.v.*

sympodium *n.* a plant or part of a plant whose main axis arises not from growth of an apical bud but from that of a lateral branch which also stops growing after a while, growth continuing from a lateral bud near the apex, and so on, e.g. many orchids; a stem vascular bundle and its associated leaf traces. *plu.* sympodia.

symport *n.* membrane protein that transports a solute across the membrane in one direction, transport depending on the sequential or simultaneous transport of another solute in the same direction.

synacme, synacmy *n.* conditions when pistils and stamens mature simultaneously.

synaesthesia *n.* the accompaniment of a sensation due to stimulation of the appropriate receptor, as sound, by a sensation characteristic of another sense, as colour.

synandrium *n.* the cohesion of anthers in male flowers of some aroids (e.g. arum lilies).

synandry *n.* condition where stamens normally separated are united.

synangium *n.* a compound sporangium in which sporangia are coherent, as in some ferns; most anterior portion of the foetal or amphibian heart, through which blood is driven from the ventricle; any arterial trunk from which arteries arise.

Synanthae Cyclanthales *q.v.*

synantherous *a.* having anthers united to form a tube.

synanthous *a.* having flowers and leaves appearing simultaneously; having flowers united together as in Compositae.

synanthropic *a.* associated with humans or their dwellings.

synanthy *n.* adhesion of flowers usually separate.

synapomorphy *n.* in cladistic phylogenetics denotes a homologous character common to two or more taxa and thought to have originated in their most recent common ancestor. *a.* synapomorphous. *cf.* apomorphous.

synaporium *n.* an animal association owing to unfavourable environmental conditions or disease.

synaposematic *a.* having warning colours in common, *appl.* mimcry of a more powerful or dangerous species as a means of defence.

synapse *n.* the point of communication between one nerve cell and another or between nerve cell and a non-neuronal target cell such as muscle. The two main types of synapses are chemical synapses and electrical synapses. Synapses are commonly made between the axon terminals of the transmitting cell and the dendrites or cell body of another, but may also be made by an axon terminal on the axon of another neurone. At chemical synapses the signal is transmitted across a narrow fluid-filled gap between two neurones by chemical messengers (neurotransmitters) whose release from the transmitting (presynaptic) cell is stimulated by the arrival of an electrical impulse at the axon terminal. Neurotransmitter binds to specific receptors on the cell membrane of the receiving (postsynaptic) cell. At excitatory synapses this stimulates the generation of an electrical impulse in the postsynaptic cell. At inhibitory synapses it prevents the generation of an impulse. Electrical synapses are formed by gap junctions between two neurones, through which electrical current can flow directly.

synapse *v.* come into contact with each other, *appl.* nerve endings, homologous chromosomes during meiosis.

synapsid *a.* having skull with one ventral temporal fenestra on each side.

Synapsida, synapsids *n., n.plu.* subclass of reptiles living from the Carboniferous to Triassic, the mammal like reptiles, with synapsid skulls, some forms of which gave rise to the mammals, and including the pelycosaurs and therapsids.

synapsin I protein associated with synaptic vesicles in neurone terminals and which may be involved in mediating their release.

synapsis *n.* association of two homologous chromosomes at the start of meiosis to form a bivalent.

synaptic *a. pert.* or occurring at a synapse.

synaptic cleft narrow fluid-filled gap between two opposing cell membranes at a chemical synapse in nervous tissue or at a neuromuscular junction.

synaptic potential electrical potential difference produced across the postsynaptic membrane by the action of neurotransmitter at a single synapse, individual synaptic potentials being summed by the receiving neurone to produce a grand postsynaptic potential, which may be inhibitory or excitatory.

synaptic vesicles vesicles containing neurotransmitter which aggregate in the cytoplasm at the tip of axon terminals and which liberate their contents into the synaptic cleft. *alt.* presynaptic vesicles.

synaptinemal complex synaptonemal complex *q.v.*

synaptonemal complex proteinaceous ladder-like structure linking two paired homologous chromosomes in meiosis, seen in the electron microscope.

synaptosomes *n.plu.* membrane-bounded structures, representing pinched off nerve terminals and postsynaptic membrane, formed when brain tissue is homogenized.

synaptospermous *a.* having seeds germinating close to the parent plant.

synaptospore *n.* aggregate spore.

synarthrosis *n.* a joint in which bone surfaces are in almost direct contact, fastened together by connective tissue or hyaline cartilage, with no appreciable capability for movement.

syncarp *n.* an aggregate fruit with united carpels.

syncarpy *n.* condition of having carpels united to form a compound gynaecium. *a.* syncarpous.

syncerebrum *n.* a secondary brain formed by union with brain of one or more ventral cord ganglia, in some arthropods.

synchondrosis *n.* a synarthrosis where the connecting medium is cartilage.

synchorology *n.* study of the distribution of plant and animal associations; geographical distribution of communities.

synchronic *a.* contemporary; existing at the same time, *appl.* species, etc. *cf.* allochronic.

synchronizer *n.* some environmental factor, such as light or temperature, that interacts with an endogenous circadian rhythm, causing it to be precisely synchronized (entrained) to a 24-hour cycle rather than free-running. *alt.* zeitgeber.

syncladous *a.* with offshoots or branchlets in tufts, *appl.* some mosses.

synconium, synconus syconium *q.v.*

syncraniate *a.* with vertebral elements fused with skull.

syncranterian *a.* with teeth in a continuous row.

syncryptic *a. appl.* animals which appear alike, although unrelated, due to common protective resemblance to the same surroundings.

syncytium *n.* a multinucleate mass of protoplasm which is not divided into separate cells. *a.* syncytial. *cf.* coenocyte.

syndactyl *a.* with fused digits, as many birds.

syndactylism *n.* whole or part fusion of two or more digits.

syndesis synapsis *q.v.*

syndesmology *n.* that branch of anatomy dealing with ligaments and articulations.

syndesmosis *n.* a slightly moveable joint, with bone surfaces connected by a ligament.

syndetocheilic *a. appl.* type of stomata found in gymnosperms in which the subsidiary cells are derived from the same cell as the guard mother cell.

syndrome *n.* a group of concomitant symptoms, characteristic of a particular condition.

synecology *n.* the ecology of plant or animal communities. *cf.* autecology.

synema synnema *q.v.*

synencephalon *n.* the part of embryonic brain between diencephalon and mesencephalon.

synenchyma *n.* fungal tissue composed of laterally joined hyphae.

syneresis *n.* contraction of a gel with expression of fluid; contraction of clotting blood and expression of serum.

synergic, synergistic *a.* acting together, often to produce an effect greater than the sum of the two agents acting separately. *n.* synergism.

synergid *n.* each of two cells without cell walls lying beside ovum at micropylar end of ovule.

synesthesis synaesthesia *q.v.*

syngametic *a. pert.* union of morphologically similar cells.

syngamodeme *n.* a population unit made up of coenogamodemes whose members can form sterile hybrids with each other.

syngamy *n.* sexual reproduction in unicellular organisms where the two cells that fuse are morphologically similar.

syngeneic *a.* genetically identical.

syngenetic *n.* arising from sexual reproduction; descended from the same ancestors.

syngraft homograft *q.v.*

syngynous epigynous *q.v.*

synkaryon *n.* a zygote nucleus resulting from fusion of gamete nuclei.

synkaryotic *a.* diploid, *appl.* nucleus.

synnema *n.* in some fungi, a group of conidiophores cemented together by their stalks; in some flowers, a bundle of stamens united by their filaments. *plu.* synnemata.

synochreate, synocreate *a.* with stipules united and enclosing stem in a sheath.

synoecious *a.* having antheridia and archegonia on same receptacle, or stamens and pistils on same flower, or male and female flowers on same capitulum.

synonym *n.* in molecular biology any of two or more codons specifying the same amino acid; in classification an alternative Latin name.

synosteosis, synostosis *n.* ossification from two or more centres in the same bone, as from diaphysis and epiphyses in long bones.

synotic tectum in higher vertebrates, a cartilaginous arch between otic capsules representing cartilaginous roof or tegmentum of cranium in lower vertebrates.

synovia *n.* mucopolysaccharide and mucoprotein secretion from cells lining joint cavities, serving to lubricate the joint.

synovial cell epithelial cell lining joint cavities, secreting hyaluronic acid.

synovial fluid fluid contained in joint cavity, secreted by synovial cells, serving to lubricate the joint.

synovial membrane the inner layer of capsule around a moveable joint, made of connective tissue and secreting material to lubricate the joint.

synoviparous *a.* secreting synovia.

synpelmous *a.* having two tendons united before they go to separate digits.

synpolydesmic *a. appl.* scales growing by apposition at margin, and which are made up of fused bony teeth covered with a continuous layer of dentine.

synsacrum *n.* mass of fused vertebra supporting the pelvic girdle of birds.

synsepalous *n.* having a calyx composed of fused sepals. *n.* synsepaly.

synspermous *a.* with seeds united.

synsporous *a.* propagating by cell conjugation, as in algae.

syntagma *see* tagma.

syntelome *n.* a compound telome.

syntenic *a. appl.* genetic loci that lie in the same order on the same chromosome in different species.

syntenosis *n.* articulation of bones by means of tendons.

synthetase *see* ligase.

synthetic theory of evolution *see* Neo-Darwinism.

syntopic *a.* sharing the same habitat within the same geographical range, *appl.* different species or to phenotypic variants within a species.

syntrophy *n.* nutritional interdependence.

syntropic *a.* turning or arranged in the same direction.

syntype *n.* any one specimen of a series used to designate a species when holotype and paratypes have not been selected. *alt.* cotype.

synusia *n.* a plant community of relatively uniform composition, living in a particular environment and forming part of the larger community of that environment.

syringeal *a. pert.* the syrinx.

syringes *plu.* of syrinx *q.v.*

syringium *n.* a syringe-like organ through which some insects eject a disagreeable fluid.

syringograde *a.* jet-propelled, moving by alternate suction and ejection of water through siphons, as squids and salps.

syringyl sinapyl alcohol, *see* lignin.

syrinx *n.* sound-producing organ in birds, situated at junction of windpipe with bronchi. It consists of two patches of thin membrane (tympaniform membrane) in the wall of each bronchus. Sound is produced when the tympanic membranes are pushed inwards partially blocking the bronchi. Air pushed out of the lungs makes the membranes vibrate and produces a sound. *plu.* syringes, syrinxes.

systaltic *a.* contractile; alternately contracting and dilating.

systematics *n.* the study of the identification, taxonomy (*q.v.*) and nomenclature of organisms, including the classification of living things with regard to their natural relationships and the study of variation and the evolution of taxa.

systemic *a.* throughout the body, involving the whole body.

systemic arteries arteries leading from the heart to the aorta in reptiles, carrying blood to the body (as opposed to the lungs).

systemic circulation in vertebrates, the course of blood from ventricle through body to auricle, as opposed to the pulmonary circulation.

systemic heart the heart of invertebrates, and the left auricle and ventricle in higher vertebrates.

systemic lupus erythematosus (SLE) autoimmune disease involving antibodies directed against nucleic acids (amongst other cellular consituents).

systole *n*. contraction of heart causing circulation of blood; contraction of any contractile cavity. *cf.* diastole.

systrophe *n*. aggregation of starch grains in chloroplasts, induced by illumination.

systylous *a*. with coherent styles; with fixed columella lid, as in mosses.

syzygium *n*. a group of associated gregarines.

syzygy *n*. a close suture of two adjacent arms, found in crinoids; a number of individuals, from two to five, adhering in strings in association of gregarines; reunion of chromosome fragments at meiosis.

T

θ- *for all entries with prefix θ- or theta- refer to the headword itself.*

T threonine *q.v.*; thymine *q.v.*

T_C cytotoxic T lymphocyte *q.v.*

T_H helper T lymphocyte *q.v.*

T_m temperature at which DNA is 50% denatured, varies for different DNAs.

T_S suppressor T lymphocyte *q.v.*

T1,2,3,4 etc. *(imm.)* see CD1, 2, 3, 4 etc.

T3, T4, T7 DNA bacteriophages of *E.coli*.

TAP transporter of antigenic peptides, an activity in antigen-processing cells that transports peptides from the cytoplasm into the endoplasmic reticulum for binding to MHC molecules and transport to the cell surface.

TBF, TBP TATA box binding factor/protein *q.v.*

TBSV tomato bushy stunt virus.

TC_50 a measure of infectivity for viruses, the dose at which 50% of tissue cultures become infected and show degeneration.

TCA tricarboxylic acid *q.v.*

TCR T-cell receptor *(q.v.)*, the specific receptor for antigens which is borne on the surface of T lymphocytes.

TDN total digestible nutrients *q.v.*

T-DNA part of the Ti plasmid of *Agrobacterium* that is transferred into the host plant cell genome.

T-DNA *see* Ti plasmid.

TDP, dTDP thymidine diphosphate *q.v.*

TE DNA sequence in *D. melanogaster* flanked by FB elements and which can transpose.

TEA tetraethylammonium *q.v.*

TEM transmission electron microscope *q.v.*

TGF-α transforming growth factor-α *q.v.*

TGF-β transforming growth factor-β *q.v.*

TGMV tomato golden mosaic virus.

Thr threonine *q.v.*

Thy1 cell surface antigen typical of T lymphocytes.

Ti the Ti plasmid *(q.v.)* of *Agrobacterium tumefaciens;* T-cell receptor *q.v.*

TIM triose phosphate isomerase.

TIV tipula iridescent virus, a double stranded DNA virus of insects.

TK thymidine kinase *q.v.*

TMP, dTMP thymidine monophosphate *q.v.*

TMV tobacco mosaic virus, a rod-shaped RNA virus causing mottling (mosaic) disease of tobacco and other plants.

TnA family of bacterial transposons (e.g. *Tn1, Tn3*) carrying antibiotic-resistance genes and which do not rely on insertion sequences for their transposition.

TnA, 9, 10 etc. transposons *(q.v.)* carrying antibiotic resistance or other genes, found in bacteria.

TnC troponin C, the calcium-binding subunit of the troponin complex of striated muscle.

TNF tumour necrosis factor *q.v.*

TnI troponin I, the ATPase-inhibiting subunit of the troponin complex of striated muscle.

TnT troponin T, the tropomyosin-binding subunit of the troponin complex of striated muscle.

TP turgor pressure *q.v.*

t-PA tissue plasminogen activator *q.v.*

TPA tissue plasminogen activator *q.v.*; tetraphorbol acetate, a compound that acts as a tumour promoter *q.v.*

TPP thiamine pyrophosphate *q.v.*

TRH thyroid stimulating hormone releasing hormone *q.v.*

tRNA transfer RNA *q.v.*

tRNA_f^Met tRNA in bacteria which specifically recognizes formylmethionine, the initial amino acid in most newly-synthesized polypeptide chains.

tRNA$_i$^{Met} initiator tRNA in eukaryotic translation, which picks up methionine and is used solely for initiation of translation.

Trp tryptophan *q.v.*

TSH thyroid-stimulating hormone *q.v.*

TTP, dTTP thymidine triphosphate *q.v.*

TTX tetrodotoxin *q.v.*

TYMV turnip yellows mosaic virus.

Tyr tyrosine *q.v.*

tables *n.plu.* outer and inner layers of flat compact bones, esp. of skull.

tabula n. horizontal partitions in the vertical canals in some colonial hydrozoans.

tabular *a.* arranged in a flat surface; flattened, as certain cells.

tabulare *n.* skull bone posterior to parietals in some vertebrates.

tachyauxesis *n.* relatively quick growth; growth of a part at a faster rate than that of the whole.

tachyblastic *a.* with cleavage immediately following oviposition, *appl.* quickly hatching eggs.

tachycardia *n.* excessive heartbeat rate of >100 per minute.

tachygenesis *n.* development with omission of some embryonic stages as in certain crustaceans, or of nymphal stages in some insects, or of tadpole stages in some amphibians.

tachykinins *n.plu.* small group of neuroactive peptides including substance P and substance K.

tachytelic *a.* evolving at a rate faster than the standard rate.

TACTAAC box internal nucleotide sequence in the introns of yeast nuclear genes, thought to be involved in the splicing process.

tactic *a. pert.* a taxis.

tactile *a.* serving the sense of touch, *appl.* sensory hairs, cells, cones, etc.

tactor touch receptor *q.v.*; any sensory receptor receptive to touch.

tactual *a. pert.* sense of touch.

taenia *n.* a band, as of nerve or muscle.

taeniate *a.* ribbon-like; striped.

taenidium *n.* spiral ridge of cuticle strengthening the chitinous layer of insect tracheae and tracheoles.

taenioid *a.* ribbon-shaped: like a tapeworm.

tagma *n.* a segment of the body of a metamerically segmented animal formed by fusion of somites, e.g. head of insects. *plu.* tagmata.

tagmosis *n.* the fusion or grouping of somites to form tagmata in a metamerically segmented animal.

tail fan in decapod crustaceans, the uropods and telson.

taiga *n.* northern coniferous forest zone, esp. in Siberia, adjacent to tundra.

tali *plu.* of talus *q.v.*

talin *n.* a protein associated with the cytoskeleton in many animal cells, providing a connection between cell-surface molecules that bind to the extracellular matrix and the cell's internal cytoskeleton.

tall-grass community type of grassland with grasses up to 2 m or more high that develops in parts of the tropics (tall-grass savanna), and in temperate regions (tall-grass prairie, tall-grass steppe).

talocalcaneal *a. pert.* talus and calcaneus, *appl.* articulation, ligaments.

talocrural *a. pert.* ankle and shank bones; *appl.* articulation: the ankle joint.

talon *n.* posterior heel of upper molar teeth.

talonid *a. appl.* hollow (talonid basin) and cusps, at posterior heel of lower molar teeth of therian mammals.

taloscaphoid *a. pert.* talus and scaphoid bone.

talus *n.* the second largest tarsal bone in man, the astragalus; a tarsal bone in vertebrates.

Tamaricales *n.* order of dicot trees, shrubs or rarely herbs, with small scale-like or heather-like leaves, including the families Fouquieriaceae (ocotilla), Frankeniaceae (sea heath) and Tamaricaceae (tamarisk).

tandem repeat(s) two or more adjacent copies of a gene or other nucleotide sequence, all arranged in the same orientation. *cf.* inverted repeat.

tangential *a.* at right angles to a radius of a cylindrical structure, *appl.* section of e.g. a stem.

tannase *n.* an enzyme which hydrolyses the tannin digallate to gallate, but will also hydrolyse certain ester linkages in other tannins. EC 3.1.1.20.

tannins *n.plu.* complex aromatic compounds some of which are glucosides, occurring in the bark of various trees, possibly giving protection or concerned with pigment formation, and used in tanning and dyeing.

tanyblastic *a.* with a long germ band.

T antigen virus-specified protein produced in cells transformed by the SV40 virus and which is involved in the generation and maintenance of the transformed state, and in inducing virus replication; teratoma antigen.

tapetum *n.* (*zool.*) tapetum lucidum *q.v.*; main body of fibres of corpus callosum in brain; (*bot.*) special nutritive layer investing sporogenous tissue of a sporangium; layer of cells lining the cavity of anther and absorbed as pollen grains mature.

tapetum lucidum light-reflecting layer in vertebrate eye, the pigmented layer of the retina.

tapeworms common name for the Cestoda *q.v.*

taphrophyte *n.* a ditch-dwelling plant.

taproot *n.* long straight main root formed from radicle of embryo in gymnosperms and dicotyledons. In plants such as carrot and turnip it forms a swollen food-storage organ.

taraxanthin *n.* a xanthophyll carotenoid pigment found in red algae.

Tardigrada, tardigrades *n., n.plu.* a phylum of small animals, commonly called water bears, that have some features similiar to arthropods and are sometimes included in the Arachnida; an infraorder of Edentata, including the tree sloths.

target organ the end-organ upon which a hormone or nerve acts.

tarsal *a. pert.* tarsus, of foot and eyelid; *n.* ankle bone, *plu.* tarsalia.

tarsal glands modified sebaceous glands of the eyelids, the ducts opening on the free margins. *alt.* Meibomian glands.

tarsi *plu.* of tarsus; two elongated plates of dense connective tissue helping to support the eyelid.

tarsomeres *n.plu.* the two parts of the dactylopodite in spiders, basitarsus and telotarsus.

tarsometatarsal *a. pert.* an articulation of the tarsus with the metatarsus.

tarsometatarsus *n.* a short straight bone of bird's leg formed by fusion of distal row of tarsals with second to fifth metatarsals.

tarsophalangeal *a. pert.* tarsus and phalanges.

tarsus *n.* ankle bones; segment of insect leg beyond the tibia, which bears the terminal claw and pulvillus; fibrous connective tissue plate of eyelid.

tartareous *a.* having a rough and crumbling surface.

tartaric acid organic acid that gives acid taste to grapes.

tassel *n.* male inflorescence of maize plant.

tassel-finned fishes a common name for the crossopterygians *q.v.*

taste bud a sense organ consisting of a small flask-shaped group of cells, found chiefly on the upper surface of the tongue, and by which the sensations of sweetness, sourness, bitterness and saltiness are perceived.

taste pore orifice in epithelium leading to terminal hairs of sensory cells in a taste bud.

TATA box conserved 7-base sequence preceding the startpoint of transcription in many eukaryotic genes, usually including the sequence TATA, involved in the binding of the complex of transcription factor proteins and RNA polymerase required for initiation of transcription. *alt.* Hogness box.

TATA box binding factor/protein (TBF, TBP) protein subunit of the transcription factor TFIID that binds to the TATA box in the promoters of many eukaryotic genes, and on which the transcription initiation complex is formed.

tau *n.* a microtubule-associated protein present in normal cells, and also in the abnormal paired helical filaments in neurones in brains of patients with Alzheimer's disease.

taurine *n.* an amino acid containing sulphonic acid, and which is a possible neurotransmitter in some invertebrate nervous systems.

taurocholic acid bile acid hydrolysed to taurine and cholic acid.

tautomerism *n.* form of isomerism consisting of movement of hydrogen atoms from one atom to another, e.g. in the rings of compounds such as purines and pyrimidines, resulting in rearrangement of double bonds, the alternative forms of the molecules being called tautomers. *a.* tautomeric.

tautonyn *n*. the same name given to a genus and one of its species or subspecies.

taxa *plu*. of taxon *q.v.*

Taxales *n*. an order of gymnosperms, being evergreen shrubs or small trees with small linear leaves, and ovules solitary and surrounded by an aril, such as the yews (Taxaceae).

taxeopodous *a*. having proximal and distal tarsal bone in straight lines parallel to limb axis.

taxes *plu*. of taxis *q.v.*

taxis *n*. a movement of a freely motile cell, such as a bacterium or protozoan, or of part of an organism, towards (positive) or away from (negative) a source of stimulation, such as light (phototaxis) or chemicals (chemotaxis); orientation behaviour related to a directional stimulus. *a*. tactic. *plu*. taxes.

taxol anticancer drug obtained from the Pacific yew tree, which inhibits cell division by binding to microtubules and preventing their normal growth and disassociation.

taxon *n*. the members of any particular taxonomic group e.g. a particular species, genus, family. *plu*. taxa.

taxonometrics numerical taxonomy *q.v.*

taxonomic category a category used in the classification of living organisms, e.g. Phylum, Class, Order, Family, Genus or Species.

taxonomy *n*. the science of classification (*q.v.*), esp. of living organisms, which arranges organisms into hierarchical groupings.

taxonomy *n*. the analysis of an organism's characteristics for the purpose of classification. *see also* systematics.

taxospecies, taxonomic species species defined by similarities in morphological characters only, and which do not necessarily correspond to biological species.

taxy taxis *q.v.*

Tay-Sachs disease an inherited condition characterized by blindness, seizures, degeneration of mental and motor function and early death in childhood in homozygotes, caused by a recessive defect in the gene for hexosaminidase A leading to an abnormal accumulation of ganglioside GM_2 in the central nervous system in homozygotes.

TCA cycle tricarboxylic acid cycle *q.v.*

T cell T lymphocyte *q.v.*; sensory cell responding to touch in leech segmental ganglion.

T-cell differentiation antigens proteins borne on the surface of different types of T lymphocytes at different stages in their development, and which can be used to distinguished the various classes.

T-cell growth factors various protein factors which stimulate division and differentiation of T lymphocytes.

T-cell receptor (TCR) the antigen receptor of T lymphocytes, a highly variable antigen-specific heterodimeric glycoprotein on the surface of T lymphocytes, which recognizes and binds antigen in the form of a peptide bound to an MHC molecule on the surface of another cell. Like antibodies, the receptor consists of an antigen-specific variable portion and a constant region. In conjunction with other signals, activation of the T-cell receptor leads to activation and proliferation of the T lymphocyte. Two types of T-cell receptor have been identified in mammals: $\alpha\beta$ made up of an α subunit and a β subunit, and $\gamma\delta$, made up of a γ subunit and a δ subunit. Like antibody genes, T-cell receptor genes undergo rearrangement before transcription and protein synthesis. *see also* B lymphocytes, cytotoxic T lymphocytes, helper T lymphocytes, immune response, T lymphocytes.

tear pit dacryocyst *q.v.*

tectal *a*. of or *pert*. tectum.

tectorial membrane a shelf of tissue running the length of the cohlea above the hair cells, which have their stereocilia (hairs) embedded in it. Movement of the tectorial membrane in response to vibrations in the fluid of the inner ear and movements of the cochlear partitions stimulates the hair cells.

tectorium *n*. tectorial membrane *q.v.*; coverts of birds.

tectospondylic *a*. having vertebrae with several concentric rings of calcification, as in some elasmobranchs.

tectostracum *n*. thin waxy outer covering of exoskeleton in ticks and mites.

tectrices *n.plu*. small feathers covering bases of remiges. *alt*. wing coverts.

tectum *n.* a roof-like structure, as the corpora quadrigemina in brain which forms the roof of the mesencephalon (midbrain).

teeth *n.plu.* hard, bony, outgrowths from jaws of mammals, each tooth composed of a core of soft tissue (pulp) supplied with nerves and blood vessels, surrounded by a layer of dentine and covered with an outer layer of enamel. The adult set generally comprises molars or grinding teeth, premolars, canines and incisors or biting teeth; similar structures in jaws, throat etc. of reptiles and fish; any similarly shaped structures in invertebrates used for rasping, seizing or grinding food, and which are generally composed of chitin or keratin; (*bot.*) the pointed outgrowths on edges of leaves, calyx etc.

tegmen *n.* (*bot.*) the integument or inner seed coat; (*zool.*) thin, hardened forewing of grasshoppers, stick insects and cockroaches. *plu.* tegmina.

tegmentum *n.* tegmen *q.v.*; (*bot.*) a protective bud scale; (*neurobiol.*) dorsal part of cerebral peduncles. *plu.* tegmenta.

tegmina *plu.* of tegmen *q.v.*

tegula *n.* a tile-shaped structure.

tegument *n.* integument *q.v.*; type of epidermis present in flukes and tapeworms, consisting of an outer syncytial layer overlying a layer of muscle fibres and an inner zone of cell bodies containing nuclei and connected by fine cytoplasmic strands to the outer layer.

teichoic acid polymer of glycerol or ribitol, often also containing *N*-acetylglucosamine, *N*-acetylgalactosamine, and alanine, forming the surface coat of Gram-positive bacteria.

tela *n.* a web-like tissue; folds of the pia mater forming roof of 3rd and 4th ventricles in brain; interlacing fibrillar or hyphal tissue in fungi.

telarian *a.* web-spinning.

teleceptor distance receptor *q.v.*

telegamic *a.* attracting females from a distance, *appl.* scent apparatus of lepidopterans.

teleianthous *a.* having both gynaecium and androecium.

telemetacarpal *a. appl.* condition of retaining distal elements of metacarpals, as in some cervids (deer).

telencephalon *n.* the anterior part of forebrain, including the cerebral hemispheres, lateral ventricles, optic part of hypothalamus and anterior portion of third ventricle. *a.* telencephalic.

teleodont *a. appl.* forms of stag beetle with the largest mandible development.

teleological *a. appl.* explanations for the evolution of particular functions or structures that suppose a purpose or design to evolution.

teleology *n.* the doctrine of final causes, the invalid view that evolutionary developments are due to the purpose or design that is served by them; similar type of explanation applied to biological or cellular process, or animal behaviour, which presupposes an impossible awareness of a particular goal.

teleonomy *n.* the idea that if a structure or process exists in an organism it must have conferred an evolutionary advantage. *a.* teleonomic.

teleoptile *n.* a feather of definitive plumage of adult bird; a pennaceous feather.

Teleostei, teleosts *n., n.plu.* group of fish including all modern bony fishes except lungfishes, holosteans and crossopterygians, which have thin bony scales covered by an epidermis, a homocercal tail, a hydrostatic air bladder (swimbladder), no spiracle and no spiral valve in the gut.

telereceptor distance receptor *q.v.*

telescopiform *a.* having joints that telescope into each other.

teleutosorus *n.* in rust fungi, a group of developing teleutospores forming a pustule on the host. *alt.* telium, teliosorus.

teleutospore *n.* thick-walled spore in rust and smut fungi, known as smut spores in the latter.

teleutostage *n.* the stage in life cycle of a rust fungus when teleutospores are produced. *alt.* telial stage, teliostage, teleutoform stage.

telia *plu.* of telium *q.v.*

telial *a. pert.* or having telia.

telic *a.* purposive; *pert.* teleosis.

Teliomycetes *n.* group of heterobasidiomycete fungi containing the rusts (Uredinales) and smuts (Ustilaginales).

teliosis *n.* purposive development or evolution.

telium *n.* group of cells that produce teleutospores in rust fungi. *plu.* telia.

telmophage *n.* a blood-sucking insect which feeds from a blood pool produced by laceration of blood vessels.

teloblast pole cell *q.v.*

telocentric *a.* with a terminal centromere; *n.* a telocentric chromosome.

telocoel *n.* 1st or 2nd ventricle in brain.

telofemur *n.* segment of femur between basifemur and genu, in ticks and mites.

telogen *n.* resting stage in the hair growth cycle.

teloglia *n.plu.* glial cells around endings of axon at a neuromuscular junction.

telokinesis *n.* telophase *q.v.*; changes in cell after telophase, where this is considered only as a cytoplasmic division, resulting in the reconstitution of daughter nuclei.

telolecithal *a.* having yolk accumulated in one hemisphere, as eggs of amphibians.

telolemma *n.* capsule containing a nerve fibre termination, in neuromuscular spindles. *alt.* end-sheath.

telome *n.* a morphological unit in a vascular plant which is either a terminal branch bearing a sporangium and a vascular supply, or the simplest part of the plant body, whether terminal or not; the sterile and fertile axes of certain fossil leafless primitive vascular plants.

telomerase *n.* enzyme involved in forming the telomeres, the structures that seal the ends of chromosomes.

telomere *n.* end of each chromosome arm distal to the centromere, and which consists of special structural DNA elements responsible for "sealing" the ends of the chromosomes. They comprise inverted repeated DNA sequences and are replicated by a special mechanism distinct from that of the main body of the chromosomal DNA, involving a telomerase enzyme.

telophase *n.* the stage of mitosis and meiosis in which sister chromatids or homologous chromosomes (in 1st meiotic division) reach the poles of the spindle, cytoplasmic division takes place and nuclei are reformed.

telopod *n.* male copulatory appendage in some millipedes.

Telosporea *n.* in older classifications, a class of sporozoan protozoa including the gregarines and coccidia. Part of the Apicomplexa in protist classification. *see* Appendices 4 and 5.

telotarsus *n.* distal segment of dactylopodite of spiders.

telotaxis *n.* movement along line between animal and source of stimulus.

telotroch *n.* preanal tuft or circlet of cilia of a trochophore.

telson *n.* the unpaired terminal abdominal segment of crustaceans and the king crab *Limulus*; the 12th abdominal segment in some insect larvae, and in Protura.

telum *n.* last abdominal segment of insects.

Temnospondyli, temnospondyls *n.*, *n.plu.* extinct order of labyrinthodont amphibians, found in the Carboniferous.

temnospondylous *a.* with vertebrae not fused but in articulated pieces. *n.* temnospondyly.

temperate *a. appl.* phage capable of integrating its DNA into the bacterial chromosome where it lies dormant for many generations, instead of causing a lytic infection; *appl.* climate, moderate, having long warm summers and short cold winters.

temperate rain forest type of forest that develops in some temperate coastal areas with high rainfall or continual moisture from ocean fogs, as the coastal forest of western North America.

temperature coefficient (Q_{10}) quotient of two reaction rates at temperatures differing by 10 °C.

temperature-sensitive mutations mutations that only produce an effect when the cell or organism carrying them is kept at a certain temperature. They are generally the result of a small change in the amino acid sequence of the protein product which does not affect the protein's function at one temperature, but is sufficient to disrupt its normal structure and function at another (usually higher) temperature.

template *n.* a blueprint or pattern from which something can be made, *appl.* DNA and RNA acting as templates in transcription and translation respectively; in animal behavioural development, a hypothetical internal representation of the final version of a particular behaviour against which the animal measures its current performance at any point during development.

temporal *a*. in region of temples; *appl*. compound bone on side of mammalian skull whose formation includes fusion of petrosal and squamosal; *pert*. time.

temporal isolation prevention of interbreeding between species of plants and animals as a result of time differences, such as shedding of pollen or mating.

temporal lobe of brain, lateral lobe of cerebral hemispheres behind the main sulcus.

temporomalar *a*. *appl*. branch of maxillary nerve supplying the cheek.

temporomandibular *a*. *appl*. articulation: the hinge of the jaws; *appl*. external lateral ligament between zygomatic process of temporal bone and neck of mandible.

temporalis *n*. broad radiating muscle arising from whole temporal fossa and extending to coronoid process of mandible.

tenacle, tenaculum *n*. holdfast of algae; ectodermal area modified for adhesion of sand grains, in certain sea anemones; in Collembola, paired appendages of 3rd abdominal segment modified to retain furcula; in teleost fishes, a fibrous band stretching from the eyeball to skull.

tendines *plu*. of tendo.

tendinous *a*. of the nature of a tendon; having tendons.

tendo tendon *q.v.*

tendon *n*. a band or cord of white fibrous connective tissue connecting a muscle with a moveable structure such as a bone.

tendon cells fibroblasts in white connective tissue, with wing-like processes extending between bundles of collagen fibres.

tendon organ type of sensory receptor within muscle, found near tendon-muscle junctions, sensitive to contraction of nearby muscle fibres. *alt*. Golgi tendon organ.

tendon reflex contraction of muscles in a state of slight tension by a tap on their tendons.

tendril *n*. a specialized twining stem, leaf, petiole, or inflorescence by which climbing plants support themselves.

tendril fibres cerebellar nerve fibres with branches communicating with dendrites of Purkinje neurones.

tendrillar *a*. acting as a tendril; twining.

tenent *a*. holding; *appl*. tubular hairs with

expanded tips, of arolium; *appl*. hairs secreting an adhesive fluid, on tarsus of spiders.

teneral *a*. immature; *appl*. stage of some insects on emergence from the nymphal covering.

Tenericutes *n*. one of the main divisions of the prokaryotic kingdom in some classifications, which includes the various wall-less prokaryotes such as rickettsiae, mycoplasmas and chlamydiae.

tenia *alt*. spelling of taenia *q.v.*

Tenon, capsule of the fibroelastic membrane surrounding the eyeball from optic nerve to ciliary region.

tenoreceptor *n*. a proprioceptor in tendon, reacting to contraction.

ten per cent law the generalization that 90% of energy at one trophic level in a food chain is lost as respiration when being transformed into the energy of the next trophic level.

tension wood reaction wood of dicotyledons, having little lignification and many gelatinous fibres, and produced on the upper side of bent branches.

tensor *a*. *appl*. muscles that stretch parts of the body.

tentacles *n.plu*. slender flexible organs on head of many invertebrates, used for feeling, exploration, grasping or attachment; adhesive structures of insectivorous plants, as of sundew. *alt*. tentacula.

tentacular *a*. *pert*. tentacles; *appl*. canal branching from perradial canal to tentacle base in ctenophores.

tentaculiferous *a*. bearing tentacles.

tentaculiform *a*. like a tentacle in shape or structure.

tentaculocyst *n*. in some coelenterate medusae, a sense organ consisting of a modified club-shaped tentacle on the margin of the bell, containing one or more lithites.

tentaculozooids *n.plu*. long, slender tentacular individuals at outskirts of hydrozoan colony.

tentaculum tentacle *q.v*. *plu*. tentacula.

tentilla, tentillum *n*. a branch of a tentacle.

tentorium *n*. chitinous framework supporting brain in insects; a transverse fold of dura mater, ossified in some mammals, that separates cerebellum and occipital lobes of brain.

tenuiissimus *a. appl.* a slender muscle beneath biceps femoris in some mammals.

tenuivirus group group of single-stranded RNA plant viruses with filamentous particles, occasionally branched.

teosinte *n. Zea mexicana,* a member of the Gramineae which occurs in Central America and is thought to be an ancestor of cultivated maize.

tepal *n.* a perianth segment which is not differentiated into petal or sepal.

teratocarcinoma *n.* a cancer originating from a teratoma.

teratogen *n.* any agent that can cause malformations during embryonic development, e.g. the drug thalidomide. *a.* teratogenic.

teratogenesis *n.* abnormal development, resulting in a foetus with congenital deformities.

teratology *n.* the study of abnormal embryonic development and congenital malformations.

teratoma *n.* abnormal growth of an oocyte or testis germ cell *in situ,* resulting in a disorganized mass of cells of many different types and undifferentiated stem cells that continue to divide.

terebra *n.* ovipositor modified for boring, sawing or stinging as in certain bees and wasps.

terebrate *a.* having a boring organ; adapted for boring.

terebrator *n.* a boring organ, a trichogyne (*q.v.*), of lichens.

teres *n.* the round ligament of liver; two muscles, teres major and minor, extending from scapula to humerus.

terete, teretial *a.* nearly cylindrical in section, as stems.

terga *plu.* of tergum *q.v.*

tergal *a. pert.* tergum; situated at the back.

tergite *n.* dorsal plate of exoskeleton of a typical arthropod skeleton.

tergosternal *a.* connecting tergite and corresponding sternite, *appl.* muscles in insects.

tergum *n.* the back; dorsal part of typical arthropod segment; the notum (*q.v.*) in insects; dorsal plate of barnacles. *plu.* terga.

terminal *a. pert.,* or situated at the end, as terminal bud at end of twig; final, *appl.* the last stage of cell differentiation, after which a cell cannot differentiate further.

terminal deoxynucleotidyl transferase (TdT) enzyme that adds nucleotides to the 3′ ends of DNA, found e.g. in T cells and B cells, where it adds non-templated nucleotides at the junctions between V, J and D gene segments during the rearrangement of immunoglobulin heavy chain genes and T-cell α and β receptor genes. It is also used to add oligonucleotide linkers to the ends of plasmid and insert DNAs in the construction of recombinant DNAs. *alt.* terminal transferase.

terminal oxidase cytochrome *c* oxidase *q.v.*; an oxidase which reacts with oxygen to form water at the end of an electron transfer chain, as cytochrome *c* oxidase, ascorbic oxidase, polyphenol oxidase.

terminalia *n.plu.* external genitalia in dipterans.

terminalization *n.* movement of chiasmata towards ends of chromosomes during diplotene and telekinesis stages of meiosis.

termination codon any of three codons signalling the end of a protein-coding sequence in DNA or RNA, and at which polypeptide synthesis stops, and which are (in mRNA) UAA, UAG or UGA. *alt.* nonsense codon, stop codon.

terminator *n.* region of DNA involved in and necessary for correct termination of transcription and release of the RNA transcript.

termitarium *n.* elaborately constructed nest of a termite colony.

termites common name for the Isoptera *q.v.*

termiticole *n.* organism that lives in a termite nest, e.g. some fungi and insects.

ternary, ternate *a.* arranged in threes; having three leaflets to a leaf; trilateral, *appl.* symmetry; *appl.* crosses involving three genera.

ternatopinnate *a.* having three pinnate leaflets to each compound leaf.

terpene *n.* any of a class of volatile aromatic compounds built up from 5-carbon isoprene units, components of plant essential oils, and including menthol, limonene, myrcene (from bay leaves), geraniol.

terpene, terpenoid *n.* any of a large class of natural plant products based on the isoprenoid unit, and including β-carotene, lycopene, violaxanthin and other xanthophylls, and gibberellins.

terraneous *a. appl.* land vegetation.

terrestrial *a.* living or found on land, as opposed to in rivers, lakes or oceans or in the atmosphere.

terricolous *a.* inhabiting the soil.

terriherbosa *n.* terrestrial herbaceous vegetation.

territoriality *n.* a social system in which an animal establishes a territory which it defends against other members of the same species.

territory *n.* area defended by an animal or group of animals, mainly against other members of the same species; an area sufficient for food requirements of an animal or aggregation of animals; foraging area.

Tertiary *a. pert.* or *appl.* geological period lasting from about 65 million years to 2 million years ago.

tertiary consumer a carnivore that eats other carnivores.

tertiary feathers third row of flight feathers in bird's wing, attached to humerus. *alt.* scapulars; tertials.

tertiary parasite an organism parasitic on a hyperparasite.

tertiary roots roots produced by secondary roots.

tertiary structure the overall conformation of a polypeptide chain in a protein comprising the relationships of elements of primary and secondary structure to each other.

tessellate(d) *a.* chequered, *appl.* markings or colours arranged in squares; *appl.* epithelium.

test, testa *n.* shell or hard outer covering; (*bot.*) seed coat.

test cross the mating of an organism to a double recessive in order to determine whether it is homozygous or heterozygous for a character under consideration.

testaceous *a.* protected by a shell-like outer covering; made of shell or shell-like material; red-brick in colour.

testes *plu.* of testis *q.v.*

testicle testis *q.v.*

testicular *a. pert.* testis; testicle-shaped; (*bot.*) having two oblong tubercles, as in some orchids.

testicular-feminization syndrome inherited X-linked condition in humans (and other mammals) in which genetic (XY) males show an immature female human phenotype as a result of the unresponsiveness of their tissues to the male sex hormone testosterone.

testis *n.* male reproductive gland producing spermatozoa. *alt.* testicle. *plu.* testes.

testis-determining region small region of the mammalian Y chromosome that is required for development as a male.

testosterone *n.* an androgen produced chiefly by the testis in vertebrates, the main male sex hormone in mammals, a steroid hormone secreted from an early stage in mammalian development by the gonads of male embryos and which stimulates the differentiation of male sexual organs and later, secondary sexual characteristics, and maintains male sexual function in adult males.

testudinate *a.* having a hard protective shell, as of tortoise.

tetanic *a. pert.* tetanus.

tetaniform, tetanoid *a.* like tetanus.

tetanus *n.* state of a muscle undergoing a continuous series of contractions due to electrical stimulation, *a.* tetanic; disease characterized by progressive paralysis, caused by infection with the bacterium *Clostridium tetani*, which produces a powerful protein neurotoxin, tetanus toxin.

Tethys (Tethyian sea) the sea between Laurasia and Gondwanaland.

tetra- prefix derived from Gk. *tetras*, four, signifying having four of, arranged in fours, divided into four parts.

tetra-allelic *a. appl.* a polyploid with four different alleles at a locus.

tetrabranchiate *a.* having four gills.

tetracarpellary *a.* having four carpels.

tetracerous *a.* four-horned.

tetrachaenium *n.* fruit composed of four adherent achenes, as in members of the mint family (Labiatae).

tetrachotomous *a.* divided up into fours.

tetracoccus *n.* any microorganism found in groups of four.

tetracotyledonous *a.* having four cotyledons.

tetract *n.* a four-rayed spicule.

tetractinal *a.* with four rays.

tetractine *a.* a spicule of four equal and

similar rays meeeting at equal angles.

tetracyclic *a.* with four whorls, *appl.* flowers.

tetracycline *n.* antibiotic produced by the actinomycete *Streptomyces,* which inhibits the binding of tRNA to bacterial ribosomes, and which can be used against both Gram-negative and Gram-positive bacterial infections.

tetracytic *a. appl.* plant stomata accompanied by four subsidiary cells.

tetrad *n.* a group of four; *appl.* four spores formed by 1st and 2nd meiotic divisions of spore mother cell, esp. in fungi; a group of four chromatids at meiosis; a pair of homologous chromosomes separating at mitosis, forming a quadrangular shape.

tetradactyl *a.* having four digits.

tetradidymous *a.* having or *pert.* four pairs.

tetradymous *a.* having four cells, *appl.* spores.

tetradynamous *a.* having four long stamens and two short.

tetraethylammonium (TEA) quaternary ammonium compound, selectively blocks potassium conductance channels in neurones, thereby blocking the generation of nerve impulses.

tetragenic *a.* controlled by four genes, *appl.* characters.

tetragonal *a.* having four angles.

tetragynous *a.* having four pistils; having four carpels to a gynaecium.

tetrahedral *a.* having four triangular sides; *appl.* apical cell in plants having a unicellular growing point.

tetrahydrofolate *n.* tetrahydropteroylglutamate, a compound containing a substituted pteridine, *p*-aminobenzoate and glutamate, and which is a donor of 1-carbon units in many biosyntheses, e.g., in the synthesis of dTMP and other nucleotides and methionine, and an acceptor of 1-carbon units in degradative reactions. It is required in the diet of mammals, e.g. as the vitamin folic acid.

tetraiodothyronine thyroxine *q.v.*

tetralophodont *a. appl.* molar teeth with four ridges.

tetralophous *a. appl.* spicule with four rays branched or crested.

tetramerous *a.* composed of four parts; in multiples of four.

tetramorphic *a.* having four forms.

tetrandrous *a.* having four stamens.

tetrapetalous *a.* having four petals.

tetraphyllous quadrifoliate *q.v.*

tetraploid *a.* having four times the haploid number of chromosomes, *appl.* cells, nuclei, organisms; *n.* an organism with four sets of chromosomes per somatic cell.

tetrapod *n.* a four-footed animal.

tetrapterous *a.* having four wings or wing-like processes.

tetrapyrenous *a.* having four fruit stones.

tetraquetrous *a.* having four angles, as some stems.

tetraradiate *a. appl.* pelvic girdle consisting of pubis, prepubis, ilium, and ischium.

tetrasaccharide *n.* any of a group of carbohydrates, such as stachyose, made up of four monosaccharide units.

tetrarch *n.* refers to roots, stems in which the primary xylem of the stele forms a four-lobed cylinder (in cross-section somewhat like a maltese cross), with four alternating bundles of phloem.

tetraselenodont *a.* having four crescentic ridges on molar teeth.

tetrasepalous *a.* having four sepals.

tetraseriate *a.* arranged in four rows.

tetrasome *n.* association of four homologous chromosomes in meiosis, *see* quadrivalent.

tetrasomic *a. pert.* or having four homologous chromosomes; *n.* an organism with four chromosomes of one type.

tetraspermous *a.* having four seeds.

tetrasporangium *n.* sporangium produced on tetrasporophyte phase of red algae, each producing four haploid tetraspores from which the male and female gametophytes develop.

tetraspore *n.* one of a group of four non-motile spores produced by sporangium of red algae; one of four basidial spores, in certain fungi.

tetrasporic, tetrasporous *a.* four-spored.

tetrasporocystid *a. appl.* oocyst of sporozoans when four sporocysts are present.

tetrasporophyte *n.* asexual phase in life cycle of red algae which bears tetrasporangia.

tetraster *n.* a mitotic figure having four astral poles rather than two, found in zygotes produced by polyspermy.

tetrastichous *a*. arranged in four rows.

tetrathecal *a*. having four compartments or chambers.

tetraxon tetractine *q.v.*

Tetraviridae *n*. family of insect RNA viruses.

tetrazoic *a*. having four sporozoites.

tetrazooid *n*. zooid developed from each of four parts constricted from stolon process of embryonic ascidians.

tetrodotoxin (TTX) *n*. a poison obtained from puffer fish which selectively blocks the regenerative sodium conductance channel in neurones and muscle fibres thereby blocking the generation of nerve impulses and muscle contraction.

tetrose *n*. any monosaccharide having the formula $(CH_2O)_4$, such as erythrose.

textura tissue *q.v.*

T form the "tense" form of an allosteric protein, the form having a lower affinity for the substrate.

thalamencephalon *n*. the part of the forebrain comprising thalamus, corpora geniculata and pineal body; diencephalon *q.v.*

thalamomamillary *a. appl.* fasciculus: Vicq-d'Azyr, bundles of *q.v.*

thalamus *n*. large paired egg-shaped masses of grey matter in the diencephalon of vertebrate forebrain, below lateral ventricles. Many sensory pathways converge on the various regions of the thalamus before being passed on to the cerebral cortex; (*bot.*) the receptacle of a flower.

thalassaemia *n*. any of a group of hereditary anaemias in which various defects in haemoglobin production are present as a result of defective globin genes. α-thalassaemia is characterized by non-production of α-globin, β-thalassaemia by the non-production of functional β-globin. Thalassaemia minor is the mild or asymptomatic condition in heterozygotes for a defective gene, thalassaemia major the severe and potentially fatal anaemia that develops in some homozygotes.

thalassemia thalassaemia *q.v.*

thalassin *n*. a toxin of sea anemone tentacles.

thalassoid *a. pert.* freshwater organisms resembling, or originally, marine forms.

thalassoplankton *n*. marine plankton.

thalliform thalloid *q.v.*

thalline *a*. consisting of a thallus.

Thallobacteria *n*. a class of prokaryotes containing the actinomycetes and related forms.

thalloid, thallose *n*. growing as a thallus; resembling a thallus.

thallome *n*. a thallus-like structure; thallus *q.v.*

thallophyte *n*. member of the plant kingdom in which the plant body is not divided into root, stem and leaves, e.g. an alga.

thallus *n*. a simple plant body not differentiated into leaf and stem, as of lichens, multicellular algae and some liverworts.

thamnium *n*. branched and shrub-like thallus of certain lichens.

thanatoid *a*. deadly poisonous; resembling death.

thanatosis *n*. habit or act of feigning death; death of a part.

thaumatin protein with an intensely sweet taste obtained from the tropical plant *Thaumatococcus*.

Theales *n*. order of dicot trees, shrubs and woody climbers, rarely herbaceous, with evergreen leaves. Includes many families e.g. Dipterocarpaceae, Hypericaceae (St. John's Wort) and Theaceae (tea).

thebesian *a. appl.* valve of coronary sinus.

theca *n*. protective covering encapsulating various structures, organs or organisms, as spore- or pollen-case, pupal case, the tube of tube animals, the covering of the spinal cord, the coat around an ovarian follicle, the calcareous shell (test) of echinoderms, a fungal ascus, a sporangium.

theca interna internal layer of covering of ovarian follicle which contains oestrogen-secreting cells.

thecacyst *n*. sperm envelope or spermatophore formed by spermatheca.

thecal *a*. surrounded by a protective membrane or tissue; *pert.* a theca; *pert.* an ascus.

thecate *a*. covered or protected by a theca.

thecial *a*. within or *pert.* a thecium.

thecium *n*. that part of fungus or lichen bearing spores; ascus hymenium.

thecodont *a*. having teeth in sockets.

thecodonts *n.plu*. group of primitive reptiles of Permian to Triassic age, including bipedal or crocodile-like forms, and thought to be ancestral to several groups.

thelygenic *a*. producing offspring preponderantly or entirely female.

thelyotoky *n*. parthenogenesis where females only are produced.

thenal *a. pert*. the palm of the hand.

thenar *n*. the muscular mass forming ball of thumb.

theobromine *n*. 3,7-dimethylxanthine, a purine similar to caffeine, found in coffee, tea and chocolate, which is a stimulant of the central nervous sytem and a diuretic.

theophylline *n*. plant alkaloid closely related to theobromine and obtained originally from tea leaves which inhibits the hydrolysis of cyclic AMP by phosphodiesterase, is also a diuretic and heart stimulant.

Therapsida, therapsids *n., n.plu*. order of extinct mammal-like reptiles, living from the Permian to the Triassic, believed to be ancestral to mammals.

Theria, therians *n.plu*. a subclass of mammals, including all living mammals except monotremes, with molar teeth bearing a triangle of cusps (tribosphenoid), cervical ribs fused to vertebrae, and a spiral cochlea.

thermaesthesia *n*. sensitivity to temperature stimuli.

thermal panting method of body cooling employed by some birds and mammals, and a few reptiles, in which heat is lost from the body by panting, which increases the evaporation of water from the moist surface tissues of the respiratory tract.

thermal pollution discharge of heat into the environment from industrial processes (e.g. the release of cooling water from a power station into a river), raising the temperature of the environment. The rise in temperature may affect living organisms directly, and the warmer water also holds less dissolved oxygen.

thermium *n*. plant community in warm or hot springs.

thermoacidophilic bacteria group of archaebacteria (e.g. *Sulfolobus, Thermococcus, Thermoproteus*) adapted for life in acidic hot springs, thriving at temperatures of 70–75 °C and pH range 1–3. *alt*. thermoacidophiles. *see also* Appendix 6.

thermobiology *n*. study of the effects of thermal energy on all types of living organisms and biological molecules.

thermocleistogamy *n*. self-pollination of flowers when unopened owing to unfavourable temperature.

thermocline *n*. the upper layer of water of rapidly changing temperature in lakes and seas in summer; zone between warm surface water and colder deep water, in which the temperature gradually decreases, in lake, reservoir or ocean.

thermoduric *a*. resistant to relatively high temperatures (70–80 °C), *appl*. microorganisms.

thermogenesis *n*. the generation of heat by metabolism, carried out by some animals (e.g. mammals) and some plants (e.g. the *Arum* spadix).

thermolysis *n*. loss of body heat; chemical dissociation owing to heat.

thermonasty *n*. nastic movement in plants in response to variations of temperature. *a*. thermonastic.

thermoneutral zone for an endothermic animal, the temperature range it can tolerate without change in its metabolic rate.

thermoperiodicity, thermoperiodism *n*. the response of living organisms to regular changes of temperature, either with day or night or season to season.

thermophase *n*. first developmental stage in some annual and perennial plants, and which can be partly or entirely completed during seed ripening if temperature and humidity are favourable. *alt*. vernalization phase.

thermophil(ic) *a*. thriving at relatively high temperatures, above 45 °C, *appl*. certain bacteria. *n*. thermophil.

thermophilous *a*. heat-loving, *appl*. plants.

thermophobic *a*. able to live or thrive only at relatively low temperatures.

thermophylactic *a*. heat-resistant; tolerating heat, as certain bacteria.

thermophyte *n*. a heat-tolerant plant.

thermoreceptor *n*. a sense organ that responds to temperature stimuli.

thermoregulation *n*. control of body temperature, either by metabolic or behavioural means, so that it maintains a more-or-less constant temperature.

thermoscopic *a*. adapted for recognizing changes of temperature, as special sense organs of certain cephalopods.

thermotactic *a. pert.* thermotaxis; *appl.* optimum: the range of temperature preferred by an organism.

thermotaxis *n.* movement in response to temperature stimulus.

thermotropism *n.* curvature in plants in response to temperature stimulus.

theromorphs *n.plu.* an order of aberrant and primitive mammal-like reptiles of the Carboniferous to Permian, having a primitive sprawling gait and including the sail-back lizards.

therophyllous *a.* having leaves in summer; with deciduous leaves.

therophyte *n.* a plant which completes its life cycle within a single season, being dormant as seed during unfavourable period, i.e. an annual.

theropod *a. appl.* a group of small carnivorous bipedal dinosaurs, a suborder of the Saurischia, the lizard-hipped dinosaurs.

thiamin(e) a water-soluble vitamin, a member of the vitamin B complex, found especially in seed embryos and yeast, its absence causing beriberi in humans, or polyneuritis, and which is a precursor of the coenzyme thiamine pyrophosphate required for carbohydrate metabolism. *alt.* vitamin B$_1$.

thiamine pyrophosphate (TPP) important prosthetic group/coenzyme for key enzymes in the pentose phosphate pathway, Calvin cycle of photosynthesis and some decarboxylation reactions. *see* Appendix 1 (58) for chemical structure.

thick filament filament made of the protein myosin, which is one of the two types of interdigitating protein filaments in muscle cell myofibrils. *cf.* thin filament.

thigmaesthesia *n.* the sense of touch.

thigmokinesis stereokinesis *q.v.*

thigmotaxis stereotaxis *q.v.*

thigmotropism *n.* growth curvature in reponse to a contact stimulus found in clinging plant organs such as stems, tendrils; response of sessile organisms to stimulus of contact. *a.* thigmotropic.

thin filament one of two types of interdigitating protein flaments in muscle cell myofibrils, containing actin, tropomyosin and troponin. *cf.* thick filament.

thinophyte *n.* plant of sand dunes.

thiobacilli *n.plu.* non-filamentous, chemo-autotrophic bacteria of the genus *Thiobacillus*, characterized by their use of elemental sulphur or other inorganic sulphur compounds.

thioester *n.* an ester with sulphur instead of oxygen.

thiogenic *a.* sulphur-producing, *appl.* bacteria utilizing sulphur compounds.

thiolytic *a. appl.* reactions or enzymes that break S–C bonds.

thiophil(ic) *a.* an organism thriving in the presence of sulphur compounds, as certain bacteria.

Thiopneutes *n.* in some classifications the name for the sulphate-reducing bacteria (e.g. *Desulfovibrio, Desulfuromonas*). *see also* Appendix 6.

thioredoxin *n.* protein cofactor in some enzyme reactions, undergoes redox reactions and is a carrier of reducing power.

thoracic *a. pert.* or in region of, thorax.

thoracic duct vessel conveying lymph from abdomen to left subclavian vein.

thoracic index one hundred times depth of thorax at nipple level divided by breadth.

thoracolumbar *a. pert.* thoracic and lumbar region of spine; *appl.* nerves: the sympathetic system.

thoracopod *n.* any thoracic leg of crabs, lobsters, etc.

thorax *n.* in higher vertebrates that part of trunk between neck and abdomen, the chest in humans, containing heart, lungs etc.; the region behind head in other animals; in insects, first three segments behind head, bearing legs and wings.

thorny-headed worms Acanthocephala *q.v.*

thread cells stinging cells of coelenterates; in the skin of hagfishes, cells whose long threads form a network in which mucous secretions of ordinary gland cells is entangled.

threadworms Nematomorpha *q.v.*; small nematode worms, commonly also called pin worms, common inhabitant of human bowel.

threatened *a. appl.* wild species that is still abundant in its natural range but is likely to become endangered because of declining numbers. *see also* endangered, rare, vulnerable.

threonine (Thr, T) *n.* an amino acid, aminohydroxybutyric acid, constituent of protein, essential in human diet. *see* table in Appendix 1 for chemical formula.

threshold *n*. critical value of membrane potential at which an impulse is initiated in a neurone; minimal stimulus required for producing sensation; level or value which must be reached before an event occurs.

threshold effect the harmful effect of a small change in environment which exceeds the limit of tolerance of an organism or population, and which becomes evident e.g. as a sudden and dramatic decrease in population size.

threshold trait in genetics, a phenotypic character that cannot be measured quantitatively, such as a predisposition to certain illnesses with a genetic component.

thrips common name for the insect order Thysanoptera *q.v.*

thrombin *n*. a proteinase produced from prothrombin at a wound, which converts the soluble protein fibrinogen to insoluble fibrin in blood clot formation. EC 3.4.21.5.

thrombinogen prothrombin *q.v.*

thrombocyte *n*. platelet *q.v.*, in mammals; in non-mammalian vertebrates, spindle-shaped nucleated cells involved in blood clotting.

thrombocytopenia *n*. lack of platelets in the blood.

thrombokinase Factor X *q.v.*

thrombokinesis *n*. the process of blood clotting.

thrombomodulin *n*. protein produced by endothelial cells which is involved in limiting the extent of blood clotting. It forms a complex with thrombin, which interacts with one component of the coagulation pathway – protein C – to downregulate its clotting activity.

thromboplastid *n*. former name for a blood platelet.

thromboplastin Factor X *q.v.*

thrombosis *n*. blood clot formation, esp. when it causes blockage of a blood vessel.

thromboxane *n*. member of a class of lipid-derived chemical signal molecules produced mainly in platelets by the action of thromboxane synthetase on prostaglandin H_2, in response to injury, during inflammation, etc. Thromboxanes are involved in blood clotting through their ability to activate platelets.

thromboxanes *n.plu.* compounds derived from arachidonic acid, and which are involved in inflammation, acting to constrict blood vessels and aggregate platelets.

thrum-eyed short-styled, with long stamens extending to mouth of tubular corolla. *cf.* pin-eyed.

thylacogens *n. plu.* substances produced by parasites and causing reactive hypertrophy of the host's tissue at the site of infection.

thylakoid *n*. membrane vesicle, numbers of which are stacked up to form the grana of chloroplasts, collectively enclosing the thylakoid space and bearing chlorophyll, photosystems I and II, the photosynthetic electron transport chain, and the ATP-synthesizing complex in the membrane; invaginations of the cell membrane in cyanobacteria forming lamellae on which the photosynthetic pigments are borne.

thylosis tylosis *q.v.*

thymectomy *n*. removal of the thymus.

Thymelaeales *n*. order of dicot shrubs comprising the family Thymelaeaceae (mezereon).

thymic *a. pert.* thymus.

thymic humoral factor polypeptide produced by thymus, involved in maturation of T cells.

thymidine *n*. a nucleoside made up of the pyrimidine base thymine linked to deoxyribose. *alt.* deoxythymidine.

thymidine diphosphate (TDP, dTDP) thymidine nucleotide containing a diphosphate group, forms part of some activated sugars.

thymidine kinase (TK) enzyme catalysing the conversion of thymidine to thymidine-5'-monophosphate in the minor pathway of DNA biosynthesis. EC 2.7.1.21.

thymidine monophosphate (TMP, dTMP) nucleotide composed of thymine, deoxyribose and a phosphate group, a product of partial hydrolysis of DNA, synthesized *in vivo* by reduction of UMP to dUMP and methylation of the uracil ring. *alt.* deoxythymidine monophosphate, deoxythymidylate, thymidylate, thymidylic acid, thymidine 5'-phosphate. *cf.* ribothymidylate.

thymidine triphosphate (TTP, dTTP) thymidine nucleotide containing a triphosphate group, one of the four deoxyribonucleotides needed for DNA synthesis, also acts in some metabolic reactions in a manner analogous to ATP.

thymidylate thymidine monophosphate *q.v.*

thymidylyltransferase *see* nucleotidyltransferase.

thymine (T) *n.* a pyrimidine base, constituent of DNA, and which is the base in the nucleoside thymidine. See Appendix 1(19) for chemical structure.

thymine dimer *see* pyrimidine dimer.

thymopoietins *n.plu.* class of polypeptides produced by thymus, involved in maturation of T lymphocytes.

thymosins *n.plu.* class of polypeptides produced by thymus, involved in maturation of T lymphocytes.

thymovidin *n.* a hormone produced by thymus in birds, which influences egg albumen and shell formation.

thymulin *n.* nonapeptide produced by thymus, involved in maturation of T lymphocytes.

thymus *n.* a primary lymphoid organ, located in humans at base of neck, in which precursor T lymphocytes proliferate, acquire antigen specificity and differentiate into the different types of T cell. It also produces various hormones or factors involved in maturation of T lymphocytes. In humans the thymus begins to atrophy after puberty.

thyridium *n.* hairless whitish area on certain insect wings.

thyro-arytaenoid *n.* a muscle of larynx.

thyrocalcitonin calcitonin *q.v.*

thyroepiglottic *a. appl.* ligament connecting epiglottis stem and angle of thyroid cartilage.

thyroglobulin *n.* a glycoprotein containing iodine, and mannose or *N*-acetylglucosamine, the iodine of the thyroid gland being stored chiefly as this protein.

thyroglossal *a. pert.* thyroid and tongue; *appl.* an embryonic duct from which the thyroid gland develops.

thyrohyals *n.plu.* greater cornua of hyoid bone.

thyrohyoid *a. appl.* muscle extending from thyroid cartilage to hyoid cornu.

thyroid *a.* shield-shaped, peltate; *pert.* or produced by the thyroid gland.

thyroid gland, thyroid shield-shaped gland in neck of vertebrates which secretes the iodine-containing hormones thyroxine and tri-iodothyronine which regulate oxidative metabolism in the body, and in tadpoles initiate metamorphosis. In mammals it also secretes the hormone calcitonin. Its function is essential for proper growth.

thyroid hormone receptor receptor protein for thyroid hormones (e.g. thyroxine) which is an intracellular protein of the steroid receptor superfamily.

thyroid hormones the hormones secreted by the thyroid gland - chiefly thyroxine and tri-iodothyronine.

thyroid-stimulating hormone (TSH) glycoprotein hormone secreted by the anterior pituitary which regulates the growth of the thyroid gland and the synthesis and secretion of its hormones. *alt.* thyrotropin.

thyroid-stimulating hormone releasing hormone (TRH) tripeptide secreted by hypothalamus which causes the secretion and release of thyroid-stimulating hormone from the anterior pituitary.

thyrotropic, thyrotrophic *a.* influencing the activity of the thyroid gland, *appl.* hormone: thyroid-stimulating hormone or thyrotropin.

thyrotropin thyroid-stimulating hormone *q.v.*

thyroxin(e) *n.* iodine-containing thyroid hormone, tetraiodothyronine, derived from the amino acid tyrosine, initiates metamorphosis in tadpoles, and is essential for normal growth and development in mammals. *see* Appendix 1(34) for chemical structure.

thyrsoid *a.* resembling a thyrsus in shape.

thyrsus *n.* a mixed inflorescence with main axis racemose, later axes cymose, with cluster almost double-cone shaped. *alt.* thyrse.

Thysanoptera *n.* order of small slender sap-sucking insects, commonly called thrips, with piercing mouthparts and no or very narrow wings fringed with long setae, which can become pests. They have an incomplete metamorphosis.

Thysanura, thysanurans *n., n.plu.* order of insects containing the silverfish and the firebrat, small primitive wingless insects, with small leg-like appendages on the abdomen and three long processes at the posterior end.

tibia *n.* shin bone, inner and larger of leg bones between knee and ankle; 4th segment of legs of insects, spiders and some myriapods.

tibial *a. pert.*, or in region of tibia.

tibiale *n.* a sesamoid bone in tendon of posterior tibial muscle.

tibialis *n.* anterior and posterior, tibial muscles acting on ankles and intertarsal joints.

tibiofibula *n.* bone formed of fused tibia and fibula.

tibiofibular *a. pert.* tibia and fibula, or tibiofibula.

tibiotarsal *a. pert.* tibia or tarsus; *pert.* or in region of tibiotarsus.

tibiotarsus *n.* tibial bone to which proximal tarsals (ankle-bones) are fused, as in birds.

ticks common name for many of the Acari *q.v.*

tidal *a. pert.* tides; ebbing and flowing; *appl.* air: volume of air normally inhaled and exhaled at each breath; *appl.* wave: main flow of blood during systole.

tidal rhythms endogenous physiological or behavioural rhythms governed by the bimonthly tidal cycle, seen in many marine and shore animals which use it to synchronize their daily behavioural rhythms with the changes in high and low-tide levels that occur throughout the cycle.

tigellum *n.* the central embryonic axis of a plant, consisting of radicle and plumule.

tight junction area of closely apposed plasma membrane of two adjacent animal cells. Tight junctions prevent lateral diffusion of proteins within the lipid bilayer and are found especially in epithelial cells where they "seal" the sheet of epithelium, preventing the movement of large and small molecules between the cells.

tigrolysis *n.* breakdown of ribosomes in nerve cells.

tiller *n.* flowering stem of a grass.

timbal *n.* sound-producing organ in cicadas.

timberline *n.* the altitude (in mountains) and latitude above which trees are unable to grow.

time-energy budget the amounts of time and energy allotted by animals to various activities.

time/energy budget in behavioural ecology, the quantitative evaluation of the amount of time and energy an animal expends on various tasks.

Timofeev's corpuscles specialized sensory nerve endings in submucosa of urethra and in prostatic capsule.

Tinamiformes *n.* an order of birds including the tinamous.

tinctorial *a.* producing dyestuff.

tinsel *a. appl.* flagella which are branched and feathery. *cf.* whiplash.

tip cell the uninucleate ultimate cell of a hyphal crozier, distal to the dome cell, and directed towards the basal cell.

Ti plasmid a plasmid carried by the crown-gall bacterium *Agrobacterium tumefaciens*, part of which (T-DNA) becomes integrated into the chromosomes of infected tissue. Normally encodes unusual amino acids that are used by the bacterium as a food source. Foreign genes artificially spliced into T-DNA are also stably integrated into the plant cell chromosomes and *Agrobacterium* and its T-DNA is one of the most important vectors in present use for introducing foreign DNA into plant cells to produce transgenic plants. Modified T-DNA that retains its capacity for integration but does not cause tumours is used.

tipophyte *n.* a pond plant.

tissue *n.* an organized aggregate of cells of a particular type or types, e.g. nervous tissue, connective tissue.

tissue culture the culture *in vitro* of cells taken from an animal or plant. *see also* cell culture.

tissue factor non-enzyme lipoprotein released from damaged blood vessels, which together with Factor VIIa initiates the blood clotting process.

tissue fluid interstitial fluid *q.v.*

tissue plasminogen activator (t-PA, TPA) a serine proteinase that converts plasminogen to plasmin by limited proteolytic cleavage.

tissue type the set of histocompatibility antigens (*q.v.*) displayed by any individual on the surface of their body cells.

titer *alt.* spelling of titre *q.v.*

titre *n.* the concentration of specific antibodies, antigens, virus particles, etc. in a sample, often presented in terms of the maximum dilution at which the sample still gives a reaction in an immunological test with a standard preparation of an appropriate antigen or antibody or in other biological or biochemical assays.

TL antigens antigens found specifically on thymus cells in early stages of differentiation.

T lymphocyte, T cell a type of small antigen-specific lymphocyte originating in thymus (in mammals) and present in secondary lymphoid tissues (e.g. lymph nodes, spleen) and blood, and which is involved in cellular immune reactions and aiding the production of antibodies. T lymphocytes bear antigen-specific receptors on their surface and only react to foreign antigen presented to them on the surface of a cell. The main types of T lymphocytes are: cytotoxic T cells which recognize and kill body cells that have become antigenically altered in some way (e.g. by virus infection), helper T cells which are activated by foreign antigen on the surface of antigen-presenting cells and in turn activate the appropriate B cells, suppressor T cells which are involved in suppressing immune responses and the general regulation of the immune system. *see also* T-cell receptor and individual entries for type of lymphocyte.

toad *n.* common name for those members of the amphibian order Anura that are stout-bodied, with a warty skin, and live in damp places, returning only to the water to spawn. There is no biological or evolutionary distinction between frogs and toads and many anuran families contain species of both types. The term frog is generally used to cover all anurans.

toadstool *n.* common name for agaric fungi other than edible mushrooms of the genus *Agaricus*, but which has no biological significance.

tobamovirus group group of single-stranded RNA plant viruses with rigid rod-shaped particles, type member tobacco mosaic virus, which has been the subject of much work on the molecular biology of virus structure.

tobravirus group group of rigid rod-shaped single-stranded RNA viruses, type member tobacco rattle virus. They are multicomponent viruses in which two genomic RNAs are encapsidated in two different virus particles.

tocopherol vitamin E *q.v.*

Togaviridae *n.* family of small, enveloped, single-stranded RNA viruses that includes Sindbis virus and rubella (German measles). Many are arthropod-borne viruses (arboviruses) such as Semliki Forest virus, Eastern encephalomyelitis virus and Ross River virus.

token stimulus a stimulus which operates indirectly by having become linked through experience with an action or object, e.g. colour is the token stimulus attracting bees to some flowers, although they actually want to eat the nectar.

tokology *n.* the study of reproductive biology.

tolerance *n.* the ability to survive and grow in the presence of a normally toxic substance, e.g. heavy metals; (*imm.*) the situation in which an animal does not mount an immune response against an antigen, as e.g. against the antigens carried on its own cells.

tomato spotted wilt virus group group of single-stranded RNA plant viruses with spherical particles comprising a lipid envelope surrounding a ribonucleoprotein core. The genome is made up of three different linear RNAs.

tombusvirus group group of single-stranded RNA plant viruses with isometric particles, type member, tomato bushy stunt virus.

tomentose *a.* covered closely with matted hairs or fibres.

tomentum *n.* the closely matted hair on leaves or stems; or filaments on cap and stem in fungi; other matted structures.

tomite *n.* free-swimming, non-feeding stage following protomite stage in life cycle of certain ciliates.

tomium *n.* the sharp edge of a bird's beak.

tomont *n.* stage in life cycle of holotrichan ciliates when body divides, usually inside a cyst.

tone *n.* sound, especially with reference to pitch, quality and strength; tonus *q.v.;* the condition of tension found in living animal tissue, especially muscle, *alt.* tonicity.

tongue *n.* usually moveable and protrusible

organ on floor of mouth; any tongue-like structure, as radula, ligula; hypopharynx, in some insects.

tongue worms Pentastomida *q.v.*

tonic *a.* producing tension; *see also* tonus.

tonicity *see* tone, tonus.

tonofilament *n.* keratin filament found in and connecting epithelial cells through spot desmosomes.

tonoplast *n.* the membrane delimiting the central vacuole of plant cells.

tonotaxis *n.* a taxis in response to a change in density.

tonsilla *n.* tonsil *q.v.*; posterior lobule at side of cerebellar hemisphere.

tonsillar ring ring of lymphoid tissue formed by the palatine, pharyngeal and lingual tonsils. *alt.* Waldeyer's tonsillar ring.

tonsils *n.plu.* paired lymphoid tissues in pharynx or near base of tongue.

tonus *n.* condition of persistent partial excitation, which in muscles results in a state of partial contraction; in certain nerve centres, the state in which motor impulses are given out continuously without any sensory impulses from the receptors. *alt.* tonicity, tone. *a.* tonic.

tool use the manipulation of an object by an animal to achieve some end which it could not achieve without it.

tooth *see* teeth.

tooth shells Scaphopoda *q.v.*

toothed whales Odontoceti *q.v.*

toothplate *n.* a flexible plate of cartilage carrying horny teeth in hagfishes, used to tear pieces of prey off and pull them into the mouth.

topaestheia, topaesthesis *n.* appreciation of the location of a tactile sensation.

topochemical *a. appl.* sense: the perception of odours in relation to track or place, as in ants.

topocline *n.* a geographical variation, not always related to an ecological gradient, but to other factors such as topography and climate.

topodeme *n.* a deme occupying a particular geographical area.

topogamodeme *n.* individuals occupying a precise locality which form a reproductive or breeding unit.

topogenous mire a mire that develops in places where there is a permanently high water table.

topographical maps in cerebral cortex, the point to point mapping of points in the sensory field to points in the cortex, resulting in an internal map. The relationship between neighbouring points in the sensory field is maintained in the cortex, albeit often in a distorted, fragmented and duplicated form.

topoinhibition *n.* inhibition of a cellular process as a result of the proximity of other cells.

topoisomerase *n.* any enzyme that untwists supertwisted DNA by nicking one of the DNA strands, producing relaxed DNA.

topoisomerase I enzyme that removes negative supercoils from DNA.

toponym *n.* the name of a place or region; a name designating the place of origin of a plant or animal.

toral *a. pert.* a torus *q.v.*

topotaxis *n.* movement induced by spatial differences in stimulation intensity, and orientation in relation to sources of stimuli.

topotropism *n.* orientation towards the source of a stimulus. *a.* topotropic.

topotype *n.* a specimen from locality of original type.

torcular *n.* occipital junction of venous sinuses with dura mater.

tori *plu.* of torus *q.v.*

tornaria *n.* the free larval stage in development of some Enteropneusta *q.v.*

tornate *a.* with blunt extremities, as a spicule; rounded off.

Toroviridae *n.* family of enveloped single-stranded RNA viruses that cause enteric infections in humans and other mammals. The elongated virus particles sometimes bend into the shape of a torus.

torose *a.* having fleshy swellings.

torpor *n.* state of complete inactivity accompanied by decreased body temperature and greatly reduced metabolic rate shown by some animals which may occur on a daily basis, or seasonally, when it is more usually known as hibernation.

torques *n.* a necklace-like arrangement of fur, feathers, or the like.

torsion *n.* spiral bending; the twisting around of a gastropod body as it develops. *a.* torsive.

torticone *n.* a turreted, spirally twisted shell.

tortoises Chelonia *q.v.*

torula *n.* a small torus; a small round protuberance; a yeast cell.

torulose *a.* with small swellings; beaded.

torulus antennifer *q.v.*

torus *n.* receptacle of a flower; firm prominence, or marginal fold or ridge; any of the pads on feet of various animals, as of cat. *plu.* tori.

total digestible nutrients (TDN) a measure of the nutrient value of a food which combines chemical analysis with estimates of digestibility obtained from analysis of the composition of the faeces.

total range the entire area covered by an individual animal in its lifetime.

totipotent *a. appl.* blastomeres (*q.v.*) that can develop into complete individuals when separated; *appl.* cells capable of forming any cell type.

touch receptors swellings of the terminals of sensory cutaneous nerves in epidermis of skin which transmit sensation of touch.

toxa toxon *q.v.*

toxic *a. pert.*, caused by, or of the nature of a poison; poisonous.

toxicity *n.* the nature of a poison; the virulence of a poison.

toxicology *n.* the study of poisons and their effects on living organisms.

toxicogenic toxigenic *q.v.*

toxicophorous, toxiferous *a.* holding or carrying poison.

toxigenic *a.* producing a poison.

toxiglossate *a.* having hollow radula teeth conveying poisonous secretion of salivary glands, as certain carnivorous marine gastropods.

toxin *n.* any poison derived from a plant, animal or microorganism. *see also* phytotoxin, zootoxin, mycotoxin.

toxognaths *n.plu.* first pair of limbs, with opening of poison duct, in centipedes.

toxoid *n.* a toxin which has been inactivated but which can still stimulate the production of antibodies effective against the original toxin.

toxon *n.* a bow-shaped spicule. *alt.* toxa.

toxophore *n.* the part of a molecule responsible for its toxic properties.

trabecula *n.* a columnar structure bridging a space, such as: row of cells spanning a cavity, small fibrous band forming part of imperfect septa or framework of organs, muscular column projecting from inner surface of ventricles of heart; (*bot.*) rod-like part of cell wall extending across lumen of cell, plate of sterile cells extending across sporangium of pteridophytes, primordial gill of agarics. *plu.* trabeculae.

trabecular *a.* having a cross-barred framework; *pert.* or formed of trabeculae. *alt.* trabeculate.

trace elements elements such as Zn, Cu, Mn, occurring in minute quantities in living tissue and required in very small amounts in food. *alt.* micronutrients.

tracer *n.* a molecule or atom that has been labelled chemically or radioactively and can thus be followed in a biochemical reaction, during metabolism, during passage through the body, etc.

trachea *n.* (*zool.*) the windpipe in vertebrates; in insects and other arthropods one of the air-filled tubules of the respiratory system, which opens to the exterior through openings (spiracles) in the sides of thorax and abdomen; (*bot.*) element of xylem tissue of plants with spiral or annular thickenings of the wall. *plu.* tracheae. *a.* tracheal.

tracheal *a. pert.*, resembling or having tracheae.

tracheal gills small wing-like respiratory outgrowths from the abdomen of some aquatic insect larvae.

Tracheophyta, tracheophytes vascular plants *q.v.*

tracheary tracheal *q.v.*; tracheate *q.v.*

tracheary elements the water-conducting cells of the xylem.

tracheate *a.* having tracheae.

tracheid *n.* a type of water-conducting xylem cell present in all vascular plants, with lignified secondary cell wall usually containing spiral thickening or bordered pits, but which does not contain perforations through cell wall. *cf.* vessel element.

tracheidal cells cells of the pericycle resembling tracheids.

trachelate *a.* narrowed as in neck formation.

trachelomastoid *a. pert.* neck region and mastoid process; *appl.* muscle, longissimus capitis.

trachenchyma *n.* tracheal vascular tissue.

tracheobronchial *a. pert.* trachea and bronchi, *appl.* lymph glands; *appl.* a syrinx formed of lower end of trachea and upper bronchi.

tracheole *n.* fine branch of the tracheal or respiratory system in insects and other arthropods, directly supplying the tissues.

trachyglossate *a.* with rasping or toothed tongue.

tract *n.* a region or area or system considered as a whole, as alimentary tract; a band, a bundle, or system of nerve fibres.

tractellum *n.* flagellum at anterior end of Mastigophora, or of zoospores, which causes rotatory motion.

tragus *n.* a small pointed eminence in front of concha of ear, well-developed in many bats.

trait *n.* a distinct phenotypic character, which may be either heritable or environmentally determined or both. Examples of heritable traits include white and other flower colours, white or red eyes in *Drosophila*, common baldness in humans, and sickle-cell anaemia and other genetic diseases.

trama *n.* the inner core of interwoven hyphae of the gill of agaric fungi.

trans in genetics, two different mutations at the same locus on a homologous pair of chromosomes are said to be in *trans* if one is on one chromosome and one on the other, *cf. cis*; *appl.* molecular configuration, one of two configurations of a molecule caused by the limitation of rotation around a double bond, the alternative configuration being the *cis*-configuration.

trans-acting *a. appl.* genes or their protein products produced by one chromosome and cooperating with or acting on genes elsewhere.

trans-splicing the generation of a mature RNA by the joining of two separate transcripts, as opposed to the more usual process of RNA splicing (*q.v.*).

transacetylase acetyltransferase *q.v.*

transacylase acyltransferase *q.v.*

transad *n.* closely related species that have become separated by a geographical or environmental barrier; *adv. appl.* organisms of the same or closely related species which have become separated by an environmental barrier, as European and American reindeer.

transaldolase *n.* widely distributed enzyme catalysing the transfer of a 3-carbon (aldehyde) unit from a ketose to an aldose, e.g. the conversion of sedoheptulose 7-phosphate + glyceraldehyde 3-phosphate into erythrose 4-phosphate + fructose 6-phosphate in the pentose phosphate pathway. EC 2.2.1.2.

transalpine *a.* situated to the north of the Alps.

transaminase aminotransferase *q.v.*

transamination *n.* transfer of amino (NH_2) groups from one molecule to another.

transcript *n.* the RNA that is synthesized by RNA polymerase on a DNA or RNA template.

transcriptase *n.* any enzyme catalysing transcription. *alt.* RNA polymerase, DNA-dependent RNA polymerase.

transcription *n.* the copying of any DNA strand nucleotide by nucleotide, following the base-pairing rules, by an RNA polymerase to produce a complementary RNA copy, in eukaryotes occurring in the nucleus. *see* Fig. 11.

transcription complex a complex of transcription factors and RNA polymerase that is formed on the promoter of a gene and initiates transcription.

transcription factor any protein that is directly involved in regulating the initiation of transcription.

transcription unit stretch of chromosome (or DNA *in vitro*) transcribed into a continuous length of RNA which may comprise one or more genes.

transcriptional control control of gene expression exerted at the level of initiation of transcription.

transcriptional regulators regulatory proteins, esp. in eukaryotes, that bind to DNA at specific control sites to initiate or prevent transcription and gene expression. Eukaryotic genes typically have complex control sites to which a combination of regulatory proteins must bind to attain a maximum level of transcription.

transcriptional terminator short sequence in a gene that signals the end of transcription.

transcytosis *n.* the passage of materials or solutes across a cell layer by endocytosis at one face, transport in vesicles and release at the other face.

transdetermination *n.* the switch of a cell from one state of determination to another as seen in some cultured *Drosophila* imaginal discs.

transdifferentiation *n.* the change of a cell from one fully differentiated state to another. *alt.* metaplasia.

transduce *v. see* transduction.

transducin (G$_t$) *n.* G protein involved in transducing the signal from activated rhodopsin to the biochemical machinery of rod and cone cells to produce an electrical signal for transmission to the brain.

transducing phage a bacteriophage that carries genes picked up from one bacterium and which can transfer them to a new host bacterium.

transduction *n.* the process of conveying or carrying over, as sensory transduction, the transmission of a stimulus from a sensory receptor to the central nervous system, or as signal transduction, the process whereby a hormonal or other extracellular signal is transmitted across the plasma membrane of a cell to activate the intercellular biochemical pathways that lead to the cell's response; (*genet.*) the transfer of genes from one bacterium to another by means of carriage within a bacteriophage. *v.* transduce.

transect *n.* a line, strip, or profile, as of vegetation, chosen for studying and charting.

transection *n.* section across a longitudinal axis. *alt.* cross-section, transverse section.

transeptate *a.* having transverse septa.

transfection *n.* the genetic modification of cultured eukaryotic cells by DNA added to the culture medium which enters the cells and in some cases is stably incorporated into the genome. *v.* transfect.

transfer cell type of parenchyma cell in many plant tissues, with ingrowths of cell wall greatly increasing surface area of plasma membrane, probably involved in transport of solutes over short distances.

transfer RNA (tRNA) type of small RNA found in all cells which carries amino acids to the ribosomes for protein synthesis, each different tRNA being specific for one kind of amino acid only (e.g. tRNA^{Cys}, tRNA^{Lys} etc.), and which lines up amino acids in the correct sequence by specific base pairing of a 3-base anticodon on each tRNA with complementary codons on messenger RNA.

transferase *n.* any enzyme which catalyses the transfer of a radical or group of atoms from one molecule to another. EC 2.

transferrin *n.* iron-transporting protein found in blood plasma.

transformation *n.* the genetic modification of a bacterium by DNA added to the culture and which enters the bacterial cell. It can occur naturally, during mixed infections, and is used in experimental bacterial genetics and to produce bacteria containing artificially produced recombinant DNA; the changes that occur in cultured cells after infection with tumour viruses, treatment with carcinogens etc., in particular the ability to divide indefinitely, and which sometimes result in a cell that can form a malignant tumour if transplanted into an animal, *alt.* malignant or neoplastic transformation; metamorphosis, or a change in form.

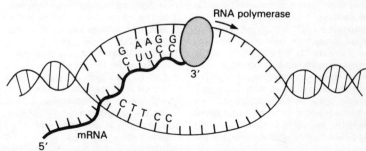

Fig. 11 Transcription of RNA from DNA.

transformation series *see* evolutionary transformation series.

transformed cell a cultured cell which has undergone certain changes, usually as a result of infection by certain tumour viruses or treatment with carcinogens, which result in its becoming able to divide indefinitely, losing normal contact inhibition and showing other characteristics of a cancer cell. In some cases a transformed cell has become fully malignant and can produce a tumour if transplanted into an animal.

transforming *a. appl.* viruses, proteins, genes, etc. that can induce neoplastic transformation if introduced into susceptible cultured cells.

transforming growth factor-α (TGF-α) a polypeptide produced by cancerous and embryonic cells, which inhibits the growth of some cells and stimulates division in others, depending on what other growth factors are present.

transforming growth factor-β (TGF-β) glycoprotein structurally unrelated to TGF-α which promotes division in some cells and inhibits it in others. Produced by many cells. A related protein in amphibian eggs is a morphological inducer.

transforming principle name originally given to the substance responsible for the transformation of avirulent (R) pneumococci to the virulent (S) form, and which was later identified as DNA.

transfusion tissue tissue of gymnosperm leaves consisting of parenchymatous and tracheidal cells.

transgene *n.* any gene introduced into an animal or plant artificially by the techniques of genetic engineering, such organisms being known as transgenic animals or plants.

transgenic *a. appl.* animals and plants into which genes from another species have been deliberately introduced by genetic engineering. In mammals this involves introducing the gene into an ovum, fertilized egg or blastocyst taken from the animal, fertilizing it *in vitro* if necessary and then replacing it in a pseudopregnant foster mother. In plants, the bacterium *Agrobacterium tumefaciens* carrying the required gene is often used to infect and transmit the gene to cultured plant tissue

from which a transgenic plant can then be regenerated.

transgressive *a. appl.* a species that overlaps two adjacent communities.

transhydrogenase *n.* formerly, one of a group of enzymes which catalyse the transfer of hydrogen from one molecule to another, now placed in EC group 1.

transient *a.* passing; of short duration.

transient polymorphism the existence of two or more distinct types of individuals in the same breeding population only for a short while, one type then replacing the other.

transilient *a. appl.* nerve fibres connecting non-adjacent parts of the cerebral cortex.

transit sequence a sequence of amino acids, generally at the N-terminal end, in newly-synthesized proteins destined for chloroplasts, and which is required for their transport across the chloroplast membranes from their site of synthesis in the cytoplasm. It is believed to be recognized by specific receptors and transport proteins in the chloroplast membrane.

transition *n.* base substitution mutation in DNA in which a pyrimidine or purine is substituted for another pyrimidine or purine respectively.

transition state high-energy intermediate in a chemical reaction that is the form of the substrate recognized and bound by an enzyme.

transition state analogue compound similar in structure to the transition state of a chemical reaction, but which inhibits an enzyme-catalysed reaction as it binds to the enzyme but does not react further.

transitional *a. appl.* epithelium occurring in ureters and urinary bladder renewing itself by mitotic division of 3rd and inner layer; *appl.* inflorescence intermediate between racemose and cymose.

transketolase *n.* widely distributed enzyme catalysing the transfer of a 2-carbon unit from a ketose to an aldose, e.g. the conversion of ribose 5-phosphate + xylulose 5-phosphate into sedoheptulose 7-phosphate + glyceraldehyde 3-phosphate in photosynthesis and in the pentose phosphate pathway. EC 2.2.1.1.

translation *n.* process by which the genetic information encoded in messenger RNA directs the synthesis of specific proteins, being the orderly synthesis of polypeptide chains at the ribosomes using messenger RNA as a template, by matching codons in mRNA with complementary anticodons on transfer RNAs carrying individual amino acids. *see* Figs. 5, 12. *see also* genetic code.

translational control control of gene expression at the level of translation, when mRNAs are produced but not translated, sometimes being stored for future use, as in eggs and early embryos.

translocase *n.* any enzyme mediating the facilitated diffusion of a substance across a membrane permeability barrier; the elongation factor (*q.v.*) EF-G.

translocation *n.* movement or removal to a different place or habitat; movement of material in solution within an organism, esp. in phloem of plant; chromosomal rearrangement in which part of a chromosome breaks off and is rejoined to a non-homologous chromosome, *see also* reciprocal translocation; *appl.* protein, movement of a protein across a membrane.

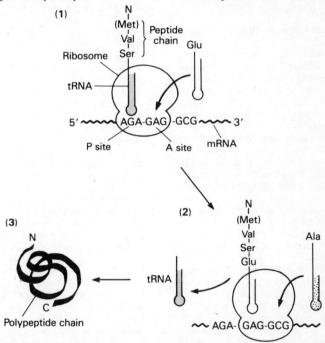

Fig.12 Stages in translation of messenger RNA to produce a protein. (1) A transfer RNA bearing the part of the polypeptide chain that has already been synthesized occupies the P site on a ribosome. The appropriate aminoacyl-tRNA lines up beside it at the A site, matching its anticodon to the codon in the mRNA. (2) The peptide chain is transferred to the amino acid carried by the incoming tRNA by the formation of a peptide bond. In this way a new amino acid is added to the growing polypeptide. At the same time, the ribosome moves one place along the mRNA, leaving the A site free for the next aminoacyl-tRNA to bind. Having transferred its peptide chain, a tRNA exits from the ribosome. (3) The outcome of translation is a complete protein chain. After synthesis it folds up into a three-dimensional conformation characteristic of each protein.

translocation factor elongation factor (*q.v.*) EF-G.

translocation quotient ratio of content of a particular substance in shoot to that in root, a measure of a substance's mobility or relative translocation.

translocator *n.* membrane protein mediating the transfer of a substance across a membrane.

transmedian *a. pert.* or crossing the median line.

transmembrane *a.* across the membrane, as in transmembrane potential; *appl.* proteins spanning a biological membrane. *see* Fig. 7.

transmembrane potential membrane potential *q.v.*

transmethylase alt. methyltransferase.

transmissible *a. appl.* diseases that can be transmitted from one individual to another by contact, or by air, water, food or insect vectors, i.e. diseases caused by bacteria, viruses, fungi and other parasites. *cf.* non-transmissible.

transmissible spongiform encephalopathies group of neurodegenerative diseases of humans and animals in which the brain has a characteristic sponge-like appearance after death, and which include scrapie in sheep, bovine spongiform encephalopathy, and Kreutzfeld–Jakob disease in humans. The transmissible agent, sometimes called a prion, and the nature of the transmission, are not yet fully identified.

transmission electron microscope (TEM) electron microscope that produces an image from the diffraction of electrons after passage through the specimen.

transmission genetics genetics studied from the point of view of heredity, the inheritance of characteristics, etc. as opposed to molecular genetics which covers the study of the physical struc- ture of genes and their context and the molecular basis of gene expression.

transmitter neurotransmitter *q.v.*

transpalatine *n.* a cranial bone of crocodiles, connecting pterygoid with jugal and maxilla.

transpinalis *n.* a muscle connecting transverse processes of vertebrae.

transpiration *n.* the evaporation of water through stomata of plant leaves and stem.

transpiration stream the movement of water and inorganic solutes upwards from roots to leaves through the xylem.

transpirational pull the continuous loss of water through leaves by transpiration, causing the flow of water through xylem from roots to leaves.

transplant *v.* to transfer tissue from one part to another part of the body of the same individual, or to transfer tissue from one individual to another; *n.* tissue transferred in this way, *alt.* graft.

transplantation antigen histocompatibility antigen *q.v.*

transport *see* active transport, facilitated diffusion, passive transport, translocation.

transport protein any protein that binds and transports a particular substance within the body or within a cell; protein that transports a substance or ion across a biological membrane. *alt.* transporters. *see also* membrane transport proteins.

transport vesicle small membrane-bounded vesicle which carries material from one cellular compartment to another within eukaryotic cells, budding off one compartment and fusing with the membrane of another.

transporter *see* transport proteins.

transposable element *n.* any of a wide variety of genetic elements (DNA sequences) present in both eukaryotic and prokaryotic genomes that possess the property of inserting themselves by transposition (*q.v.*) at various nonhomologous regions on the chromosomes and other DNAs in the bacterial cell or eukaryotic nucleus. Such transpositions can result in the inactivation of genes into which the transposable elements insert or changes in the expression of nearby genes. Permanent insertions into genes in germline cells result in heritable mutations. *alt.* transposon. *see also* controlling element, copia, insertion sequence, transposition.

transposase *n.* protein(s) specified by a transposon and responsible for its transposition.

transposition *n.* the movement of a DNA sequence to another position on the same or a different DNA molecule by replication of the sequence and insertion of the copy either at random or at some preferred but not homologous site (usually without loss of the original sequence), which occurs in both bacteria and eukaryotic cells, the transposable sequences being known generally as transposable elements or transposons, and which variously results in deletions, duplications, and changes in gene expression at the site of insertion.

transposition immunity the inability of a plasmid carrying transposon Tn3 to accept further insertions of Tn3 from another DNA molecule.

transposon *n.* transposable element (*q.v.*) esp. those found in bacteria. The simplest types of transposon are bacterial insertion sequences containing only the genetic functions necessary for their own transposition, larger bacterial transposons also carry various genes such as those for antibiotic resistance (e.g. in some bacterial plasmids). *see also* drug-resistance plasmid, insertion sequence, transposition.

transpyloric plane upper of imaginary horizontal planes dividing abdomen into artificial regions.

transtubercular *a. appl.* plane of body through tubercles of iliac crests.

transudate *n.* any substance which has oozed out through a membrane or pores.

transvection *n.* the case where a chromosomal rearrangement that prevents synapsis between homologous regions changes the phenotype although the overall genotype remains unchanged, and which is due to the ability of a mutation at or near one allele to influence the other allele only when they are synapsed.

transversal *a.* lying across or between, as a transversal wall.

transverse *a.* lying across or between, *appl.* section, a cross-section.

transverse process projection from each side of the neural arch of vertebra, with which a rib articulates.

transverse tubules ingrowths of plasma membrane which form interconnected tubules surounding each myofibril in a skeletal muscle fibre.

transversion *n.* in nucleic acids, the substitution of a purine for a pyrimidine base, or vice versa.

transversum *n.* in many reptiles, a cranial

bone extending from pterygoid to maxilla.

transversus *n.* a transverse muscle, as of abdomen, thorax, tongue, foot, etc.

trapeziform *a.* trapezium-shaped.

trapezium *n.* the 1st carpal bone at base of 1st metacarpal; a portion of the pons Varolii in brain.

trapezius *n.* a broad, flat, triangular muscle of neck and shoulders.

trapezoid *a.* shaped like a trapezium, *appl.* various muscles, bones.

traumatic *a. pert.* or caused by, a wound or other injury.

traumatin *n.* substance obtained from wounded bean pods and considered to be a wound hormone, which is effective in inducing cell division.

traumatonasty *n.* a movement in response to wounding.

traumatotropism *n.* a growth curvature in plants in response to wounds.

tree *n.* a woody perennial plant which has a single main trunk at least 7.5 cm in diameter at 1.3 m height, a definitely formed crown of foliage, and a height of at least 4 m.

tree ferns a group of tropical and subtropical ferns (*Cyathea*) with aerial stems several metres high.

tree line line marking the northern, southern, or altitude limit beyond which trees do not grow.

tree layer the highest horizontal layer in a plant community, comprising the tree canopy.

tree rings growth rings *q.v.*

tree-ring dating dendrochronology *q.v.*

trefoil *n.* a flower or leaf with three lobes.

trehalose *n.* a disaccharide composed of two glucose units, esp. abundant in some lichens and in insects, in which it is one of the principal fuels for the flight muscles.

Trematoda, trematodes *n., n.plu.* class of parasitic flatworms, including the digenean gut, liver and blood flukes, such as *Fasciola*, the liver fluke of sheep and cattle, and *Schistosoma* spp. which cause the serious disease schistosomiasis, in humans, with tissue damage and inflammation.

tremelloid, tremellose *a.* gelatinous in substance or appearance.

triactinal *a.* three-rayed.

triacylglycerol *n.* any of a class of lipids

common in living organisms, uncharged fatty acid esters of the 3-carbon sugar alcohol glycerol, storage form of fatty acids and important components of plant and animal fats and oils. alt. neutral fat, triglyceride. |see Appendix 1 (36) for chemical structure.

triadelphous a. with stamens united by their filaments into three bundles.

triaene n. a somewhat trident-shaped spicule.

trial-and-error conditioning or learning a kind of learning in which a random and spontaneous response becomes associated with a particular stimulus, because that response has always produced a reward whereas other responses have not done so. alt. instrumental conditioning, operant conditioning.

triallelic a. appl. polyploid with three different alleles at a locus.

triandrous a. having three stamens.

triangularis n. muscle from mandible to lower lip, which pulls down corner of mouth; muscle and tendinous fibres between dorsal surface of sternum and costal cartilages, transversus thoracis, which assists breathing out.

trianthous a. having three flowers.

triarch n. refers to roots, stems in which the primary xylem of the stele forms a three-lobed cylinder, with three alternating bundles of phloem.

triarticulate a. three-jointed.

Triassic a. pert. or appl. geological period lasting from about 250 to 213 million years ago. n. Trias.

triaster n. three chromatin masses resulting from abnormal tripolar mitosis, as seen in some cancer cells.

triaxon n. a sponge spicule with three axes.

tribe n. subdivision of a family in plant taxonomy, differing in minor characters from other tribes. Tribal names generally have the ending -eae.

triboloid a. like a burr; prickly.

triboluminescence n. luminescence produced by friction.

tribosphenic a. trituberculate, appl. teeth of therian mammals.

tribracteate a. with three bracts.

tricarboxylic acid organic acid bearing three COOH groups, as citric acid and aconitic acid.

tricarboxylic acid cycle (TCA cycle) a key series of metabolic reactions in aerobic cellular respiration, occurring in the mitochondria of animals and plants, and in which acetyl CoA formed from pyruvate produced during glycolysis is completely oxidized to CO_2 via interconversions of various carboxylic acids (oxaloacetate, citrate, ketoglutarate, succinate, fumarate, malate). It results in the reduction of NAD and FAD to NADH and $FADH_2$, whose reducing power is then used indirectly in the synthesis of ATP by oxidative phosphorylation. The TCA cycle also produces substrates for many other metabolic pathways. alt. citric acid cycle, Krebs cycle.

tricarpellary a. having three carpels.

tricentric a. having three centromeres.

triceps n. appl. a muscle with three heads or insertions.

trichilium n. a pad of matted hairs at base of certain leaf petioles.

trichoblast n. plant epidermal cell that develops into a root hair.

trichobothrium n. a conical protuberance with sensory hair, on either side of anal segment in myriapods; a vibratory sensory hair or thread-like bristle in spiders.

trichocarpous a. with hairy fruits.

trichocyst n. oval or spindle-shaped protrusible body in ectoplasm of ciliates and dinoflagellates.

trichoderm n. filamentous outer layer of cap and stipe in agamic fungi.

trichogen n. hair-producing cell.

trichogyne n. elongated hair-like receptive cell at end of female sex organ in some fungi, lichens, red or green algae, which may receive the male gamete.

trichoid a. hair-like.

trichome n. any of various outgrowths of the epidermis in plants, including branched and unbranched hairs, vesicles, hooks, spines and stinging hairs; a hair tuft; (bact.) strand of vegetative cells ensheathed in a hollow tubular structure forming the filament of filamentous cyanobacteria and some filamentous sulphur bacteria.

Trichomycetes n. class of endo- and exoparasitic fungi living in or on arthropods, having a simple or branched mycelium, and reproducing by non-motile spores.

trichophore *n.* a group of cells bearing a trichogyne; chaetigerous sac of annelid worms.

trichopore *n.* opening for emerging hair or bristle, as in spiders.

Trichoptera *n.* order of insects known commonly as caddis flies, whose aquatic larvae (caddis worms) often build elaborate protective cases incorporating pieces of sand, leaf fragments, etc., or nets of silk in which food is trapped. They have a complete metamorphosis. The adults have two pairs of long slender wings and much reduced mouthparts and rarely feed.

trichosclereid *n.* a sclereid with thin hair-like branches extending into intercellular spaces.

trichosiderin *n.* iron-containing red pigment isolated from human hair.

trichosis *n.* distribution of hair; abnormal hair growth.

trichothallic *a.* having a filamentous thallus; *appl.* growth of a filamentous thallus in certain algae by division of intercalary meristematic cells at the base of a terminal hair.

trichotomous *a.* divided into three branches. *n.* trichotomy, a three-way branch.

trichroic *a.* showing three different colours.

trichromatic *a. pert.* or able to perceive, the three primary colours.

trichromatic theory theory developed in 19th century that three different pigments must be present in retinal cones to account for data on colour vision, since proved to be true. *alt.* Young-Helmholtz trichromatic theory.

tricipital *a.* having three heads or insertions, as triceps muscle; *pert.* triceps.

triclads *n.plu.* order of elongated turbellarians, the Tricladida, commonly called planarians, having an intestine with three main branches and well-developed sense organs.

Tricoccae Euphorbiales *q.v.*

tricoccus *a. appl.* a fruit consisting of three carpels.

triconodont *a. appl.* tooth with three crown prominences in a line parallel to jaw axis.

Triconodonta, triconodonts *n., n.plu.* or-der containing the earliest known mammals, living from the late Triassic to the Jurassic, somewhat resembling shrews, and with molar teeth with three cusps in a straight line.

tricostate *a.* with three ribs.

tricotyledonous *a.* with three cotyledons.

tricrotic *a.* having a triple beat in the arterial pulse.

tricrural *a.* with three branches.

tricuspid *a.* with three cusps, *appl.* triangular valve of heart.

tricuspidate *a.* having three points, *appl.* leaf.

tridactyl *a.* having three digits.

tridentate *a.* having three tooth-like divisions.

tridynamous *a.* with three long and three short stamens.

trifacial *a. appl.* 5th cranial nerve, the trigeminal.

trifarious *a.* in groups of three; of three kinds; in three rows; having three surfaces.

trifid *a.* cleft into three lobes.

triflagellate *a.* having three flagella.

trifoliate *a.* having three leaves growing from the same point.

trifoliolate *a.* with three leaflets growing from the same point.

trifurcate *a.* with three forks or branches.

trigamous *a. appl.* flowerhead with staminate, pistillate and hermaphrodite flowers.

trigeminal *a.* consisting of, or *pert.*, three structures; *appl.* nerve: 5th cranial nerve whose root is divided into three branches conveying sensation from eye orbit, forehead and front of scalp (ophthalmic), upper jaw, teeth and overlying skin (maxillary), and lower jaw, teeth and overlying skin and sensation other than taste from tongue and mouth (mandibular). *alt.* facial nerve.

trigeneric *a. pert.* or derived from three genera.

trigenetic *a.* requiring three different hosts in the course of a life cycle.

trigenic *a. pert.* or controlled by three genes.

triglyceride triacylglycerol *q.v.*

trigon *n.* the triangle of cusps on molar teeth of upper jaw.

trigonal *a.* ternary or triangular when *appl.* symmetry with three parts to a floral whorl;

triangular in cross-section, *appl.* stems.

trigone *n.* a small triangular space.

trigonelline *n.* a methyl derivative of nicotinic acid (niacin) found in plants, esp. legume seeds, potatoes and dahlia tubers, has no activity in animals.

trigoneutic *a.* producing three broods in each breeding season.

trigonid *n.* triangle of cusps of lower molar teeth.

trigonum *n.* posterior process of talus forming a separate ossicle.

trigynous *a.* having three pistils or styles; having three carpels to the gynaecium.

triheterozygote *n.* an organism heterozygous for three genes.

trihybrid *n.* a cross whose parents differ in three distinct characters; *a.* heterozygous for three pairs of alleles.

tri-iodothyronine *n.* iodine-containing hormone derived from tyrosine produced by the thyroid gland.

trijugate *a.* having three pairs of leaflets.

trilabiate *a.* having three lips, with lip of corolla divided into three.

trilobate *a.* having three lobes.

Trilobita, Trilobitomorpha, trilobites *n., n., n.plu.* group of fossil arthropods (*q.v.*) found throughout the Palaeozoic era. They had a body divided into a head and a posterior region of numerous segments, a single pair of antennae, and numerous similar biramous appendages.

trilocular *a.* having three compartments or loculi.

trilophodont *a.* having teeth with three crests.

Trimerophyta *n.* group of primitive vascular plants, known from the mid-Devonian and now extinct, thought to represent the ancestors of the ferns and progymnosperms. The main axis was branched, with some branches bearing sporangia, and they lacked leaves.

trimerous *a.* composed of three or multiples of three, as parts of flower.

trimonoecious *a.* with male, female and hermaphrodite flowers on the same plant.

trimorphic *a.* having three different forms, *appl.* species; having three different types of individual; with stamens and pistils of three different length.

trimorphism *n.* occurrence of three distinct

forms or forms of organs in one life cycle or one species.

trinervate *a.* having three veins or ribs running from base to margin, as leaf.

trinomial *a.* consisting of three names, as names of subspecies.

trinucleotide repeats consecutive repeats of a three-nucleotide sequence, which are found in some genes. In Huntington's disease and some other inherited diseases, affected individuals have an increased number of trinucleotide repeats in the disease gene.

trioecious *a.* producing male, female and hermaphrodite flowers on different plants.

trionym trinomial name.

triose *n.* any monosaccharide having the formula $(CH_2O)_3$, such as glyceraldehyde, dihydroxyacetone phosphate.

triose phosphate isomerase enzyme (EC 5.3.1.1) that catalyses the isomerization of dihydroxyacetone phosphate to glyceraldehyde-3-phosphate for use in glycolysis.

triosseum *a. appl.* foramen, the opening between coracoid, clavicle and scapula.

triovulate *a.* having three ovules.

tripartite *a.* divided into three parts, as leaf.

tripetalous *a.* having three petals.

triphyllous *a.* having three leaves.

tripinnate *a.* divided pinnately three times.

tripinnatifid *a.* divided three times in a pinnatifid manner.

tripinnatisect *a.* three times lobed, with divisions nearly to midrib, *appl.* leaves.

triplechinoid *a.* of sea urchins, having two primary pore plates with one or more secondaries between.

triplet *n.* three consecutive bases in DNA or RNA, encoding an amino acid. *see* genetic code.

triplex *a.* having three dominant genes, in polyploidy.

triplicostate *a.* having three ribs.

triplinerved *a. appl.* leaves with three prominent veins. *alt.* triple-nerved.

triploblastic *a. appl.* embryos with three primary germ layers.

triplocaulescent *a.* having axes of the third order, i.e. having a main stem with branches which are branched.

triploid *a.* having three times the haploid number of chromosomes, *appl.* cells, nuclei, organisms; *n.* an organism with three sets of chromosomes per somatic cell.

triplostichous *a*. arranged in three rows, as of cortical cells on small branches of green algae of the genus *Chara*; *appl*. eyes with preretinal, retinal and postretinal layers, as of larval scorpion.

tripolite diatomaceous earth *q.v.*

tripton *n*. non-living material difting in plankton.

triquetral, triquetrous *a. appl*. stem with three angles and three concave faces; *appl*. three-cornered or wedge-shaped bone: one of the carpal bones.

triquetrum *n*. the cuneiform carpal bone.

triquinate *a*. divided into three, with each lobe again divided into five.

triradial *a*. having three branches arising from one centre.

triradiate *a. appl*. pelvic girdle consisting of pubis, ilium and ischium.

triramous, triramose *a*. divided into three branches.

trisaccharide *n*. a carbohydrate made up of three monosaccharide units, e.g. raffinose.

trisepalous *a*. having three sepals.

triseptate *a*. having three septa.

triserial *a*. arranged in three rows; having three whorls.

triskelion *n*. three-legged protein assembly.

trisomic *a. pert*. or having three homologous chromosomes or genetic loci. *n*. trisomy.

trisomy *n*. the trisomic condition.

trispermous *a*. having three seeds.

trisporic, trisporous *a*. having three spores.

tristachyous *a*. with three spikes.

tristichous *a*. arranged in three vertical rows.

tristyly *n*. the condition of having short, medium-length and long styles. *a*. tristylic.

triternate *n*. thrice ternately divided.

tritibial *n*. compound ankle bone formed when centrale unites with talus.

tritium *n*. radioactive isotope of hydrogen, ^{3}H, widely used in experimental biology to label tracer compounds and biological molecules.

tritocerebrum *n*. the third lobe of insect brain indicated during development; part of brain of higher crustaceans, consisting of antennal nerve centres.

tritocone *n*. a cusp of premolar tooth.

tritor *n*. grinding surface of a tooth.

tritubercular *a. appl*. molar teeth with three cusps.

trituberculate *a*. having three cusps to the molar teeth.

trituberculates *n.plu*. a group of Mesozoic therian mammals known mainly from remains of jaws and teeth, the molar teeth having the characteristic triangle of cusps, and which are forerunners of the living therians. *alt*. tribosphaenids.

Triuridales *n*. an order of saprophytic monocots with scale leaves and comprising the family Triuridaceae.

trivalent *n*. association of three chromosomes held together by chiasmata between diplotene and metaphase of 1st division in meiosis; *a. appl*. antigen that can bind three antibody molecules.

trivoltine *a*. having three generations of broods in a year.

trixenous *a*. of a parasite, having three hosts.

trizoic *a. appl*. protozoan spore containing three sporozoites.

troch *n*. a circlet or segmental band of cilia of a trochophore.

trochal *a*. wheel-shaped.

trochantellus *n*. segment of leg between trochanter and femur, in some insects.

trochanter *n*. prominences at upper end of thigh bone; small 2nd segment of leg between coxa and femur in insects and spiders.

trochanteric fossa deep depression on medial surface of neck of femur.

trochate *a*. having a wheel-like structure; wheel-shaped.

trochlea *n*. a pulley-shaped structure, esp. one through which a tendon passes, as of femur, humerus.

trochlear *a*. shaped like a pulley; *pert*. trochlea; *appl*. nerve: 4th cranial nerve to superior oblique muscle of eye.

Trochodendrales *n*. order of dicot shrubs and trees including the families Tetracentraceae and Trochodendraceae.

trochoid *a*. wheel-shaped; capable of rotating motion, as a pivot joint.

trochophore *n*. free-swimming top-shaped pelagic larval stage of annelids, bryozoans, and some molluscs, forming part of the zooplankton. It has a ring of cilia around the rim and a terminal ring or tuft of cilia in front of the mouth. *alt.* trochosphere.

trochus *n.* the inner, anterior coarser ciliary zone of a rotifer disc.

troglobiont *n.* any organism living only in caves.

Trogoniformes *n.* an order of very soft-plumaged birds including the trogons.

tropeic *a.* keel-shaped.

trophallaxis *n.* in social insects, the exchange of alimentary liquid among colony members and guest organisms, either mutually or unilaterally.

trophamnion *n.* sheath around developing egg of some insects, and passing nourishment to the embryo.

trophectoderm *n.* outer layer of mammalian blastocyst, which will form chorion. *alt.* trophoblast.

trophi *n.plu.* hard jaw-like structures in the pharynx of rotifers, used to grind food; the mouthparts of arthropods, esp. insects.

trophic *a. pert.* or connected with nutrition and feeding; *appl.* hormones influencing the activity of endocrine glands and growth, such as those secreted by the anterior lobe of the pituitary.

trophic factor protein that is required for the survival and growth, and sometimes repair, of a particular tissue or cell type, e.g. proteins such as nerve growth factor, brain-derived neurotrophic factor, glial growth factors.

trophic level a level in a food chain defined by the method of obtaining food, and in which all organisms are the same number of energy transfers away from the original source of the energy (e.g. photosynthesis) entering the ecosystem. *see* autotroph, herbivore, secondary consumer, tertiary consumer; the nutrient status of a body of water, *see* eutrophic, oligotrophic.

trophidium *n.* 1st larval stage of certain ants.

trophifer, trophiger *n.* posterolateral region of insect head with which mouthparts articulate.

trophoblast *n.* the outer layer of cells of epiblast or of morula. *alt.* trophectoderm.

trophobiont *n.* organism that lives in a symbiosis where each partner feeds the other, as ants and aphids. *see* trophobiosis.

trophobiosis *n.* the life of ants in relation to their nutritive organisms, as to fungi and insects.

trophocytes *n.plu.* cells providing nourishment for other cells, as nurse cells to oocyte.

trophoderm *n.* outer layer of chorion; trophectoderm with a mesodermal cell layer.

trophogenic *a.* due to food or feeding, *appl.* characters in social hymenopterans.

trophogone *n.* a nutritive organ in ascomycetes.

trophonemata *n.plu.* uterine villi or hair-like projections which transfer nourishment to embryo.

trophont *n.* stage of vegetative growth in holotrichan ciliates.

trophophore *n.* in sponges, an internal bud or group of cells destined to become a gemmule.

trophophyll *n.* a sterile or foliage leaf of certain pteridophytes.

trophosome *n.* polyp involved in feeding in a hydroid colony; vascularized organ in vestimentiferan tubeworms, which contains symbiotic sulphur-oxidizing CO_2-fixing chemoautotrophic bacteria which supply fixed carbon to their host.

trophospongia *n.* spongy vascular layer of mucous membrane between uterine wall and trophoblast.

trophotaxis *n.* response to stimulation by an agent which may serve as food.

trophotropism *n.* tendency of a plant organ to turn towards food, or of an organism to turn towards food supply.

trophozoite *n.* the adult stage of a sporozoan.

trophozooid *n.* in some tunicates, lateral buds which collect food for the colony.

tropic *a. pert.* tropism, *appl.* movement or curvature in response to a directional or unilateral stimulus; having or *pert.* a directive influence, *appl.* hormones, as -tropins; tropical, *appl.* regions.

tropical *a. appl.* climate characterized by high temperature, humidity and rainfall, found in a belt on both sides of the equator.

tropical rain forest evergreen broadleaf forest that develops in areas near the Equator with a climate of high temperature, humidity and rainfall and no marked seasons, and which is characterized by a high biological diversity and productivity. Tropical rain forest is found in the Amazon basin, parts of Central America, central West Africa, parts of the south-eastern African coast and Madagascar, South-east Asia and Indonesia, New Guinea, and the northern tip of Australia. *see also* rain forest.

tropine *n.* the base, $C_8H_{15}NO$, of atropine.

tropism *n.* a plant or sessile animal growth movement, usually curvature towards (positive) or away from (negative) the source of stimulus.

tropocollagen *n.* basic structural unit of the fibrous protein collagen, consisting of three long polypeptide chains wound round each other in a helical conformation.

tropomyosin *n.* protein in the thin filaments of striated muscle, which in the absence of Ca^{2+} prevents the interaction of actin and myosin, and thus prevents contraction.

troponin *n.* protein complex in thin filaments of striated muscle, consisting of three components, one of which (TnC) binds Ca^{2+}. In this state it interacts with tropomyosin, allowing interaction of actin and myosin and muscle contraction.

tropophil(ous) *a.* tolerating alternating periods of cold and warmth, or of moisture or dryness; adapted to seasonal changes.

tropophyte *n.* a plant which adapts to the changing seasons, being more or less hygrophilous in summer and xerophilous in winter; a plant growing in the tropics.

tropotaxis *n.* a taxis in which an animal orients itself in relation to source of stimulus by simultaneously comparing the amount of stimulus on either side of it by symmetrically placed sense organs.

true-breeding *a. appl.* organisms that are homozygous for any given genotype and therefore pass it on to all their progeny in a cross with a similar homozygote.

true corals common name for the Scleractinia, an order of mainly colonial hydrozoans, having a compact calcareous skeleton and no siphonoglyph.

true flies Diptera *q.v.*

true mosses the Bryophyta *q.v.*

true ribs ribs connected directly with the sternum (breast bone).

trumpet hyphae large trumpet-shaped sieve cells in some brown algae.

truncate *a.* terminating abruptly, as if tapering end were cut off. *alt.* abrupt.

truncus arteriosus the most anterior region of foetal, or of amphibian, heart, through which blood is driven to ventricles.

trunk *n.* the main stem of tree; the body, exclusive of head and limbs; main stem of a vessel or nerve; elongated proboscis, as of elephant.

tryma *n.* a drupe with separable rind and two-halved endocarp with spurious dissepiment, as walnut.

trypanomonad *a. appl.* phase of development of trypanosome while in its invertebrate host.

trypanosome *n.* member of a genus of parasitic flagellate protozoans, *Trypanosoma*, which includes the organisms causing trypanosomiasis, or sleeping sickness, in Africa (*T. brucei*) and Chagas' disease in South America (*T. cruzi*), both transmitted by blood-sucking insects.

trypsin *n.* proteolytic digestive enzyme found in pancreatic juice of mammals (and similar enzymes found in other animals and in plants), formed from an inactive precursor, trypsinogen, by enzymatic cleavage in small intestine by enteropeptidase. EC 3.4.21.4.

trypsin inhibitors various proteins that inhibit the enzyme trypsin by binding tightly to the active site. Some belong to the serpin family (e.g. α_1-antitrypsin, a deficiency of which is responsible for lung damage due to unrestrained proteinase activity, resulting in the clinical condition emphysema) and some to another family of proteinase inhibitors (e.g. pancreatic trypsin inhibitor).

trypsinogen *see* trypsin.

tryptic *a.* produced by or *pert.* trypsin.

tryptophan (Trp, W) *n.* β-indolealanine, an amino acid with an aromatic side chain, constituent of protein, essential in human diet and a precursor of the auxin indoleacetic acid in plants. *see* table in Appendix 1 for chemical formula.

tryptophan synthase enzyme that catalyses the synthesis of tryptophan from indoleglycerol phosphate and serine. EC 4.2.1.20. *alt.* tryptophan synthetase.

tryptophanase *n.* enzyme catalysing the breakdown of tryptophan to ammonia, pyruvic acid and indole, present esp. in some colon bacteria. EC 4.1.99.1.

T tubules transverse tubules *q.v.*

tuatara *see* Rhyncocephalia.

tuba *n.* a salpinx or tube, as tuba acustica or auditiva, the Eustachian tube; tuba uterina: Fallopian tube.

tubal *a. pert.* the Eustachian tube or Fallopian tubes.

tubar *a.* consisting of an arrangement of tubes, or forming a tube, as *appl.* system and skeleton of sponges.

tubate *a.* tube-shaped, tubular.

tube feet projections from the body wall in echinoderms which are connected to the water vascular system and generally used for locomotion. They may also be modified to serve respiratory, food-catching and sensory functions.

tubenoses Procellariiformes *q.v.*

tuber *n.* thickened, fleshy, food-storing underground root, or similar underground stem with surface buds.

tuber vermis part of superior vermis of cerebellum, continuous laterally with inferior semilunar lobules.

tubercle *n.* a small rounded protuberance; root swelling or nodule; a dorsal articular knob on a rib; a cusp of a tooth.

tubercle bacillus *Mycobacterium tuberculosis*, the cause of tuberculosis.

tubercular, tuberculate *a. pert.,* resembling or having tubercles.

tuberculose *a.* having many tubercles.

tuberiferous *a.* bearing or producing tubers.

tuberoid *a.* shaped like a tuber.

tuberose, tuberous *a.* covered with or having many tubers.

tuberosity *n.* a rounded eminence on a bone, usually for muscle attachment.

tubeworms *see* Annelida, Echiura, Phoronida, Pogonophora, Pterobranchia, Vestimentifera.

tubicolous *a.* inhabiting a tube.

tubicorn *a.* with hollow horns.

tubifacient *a.* tube-making, as some polychaete worms.

tubificid worms freshwater worms of the genus *Tubifex*, which are tolerant of heavy organic pollution.

tubiflorous tubuliflorous *q.v.*

tubiform *a.* having the form of a tube or tubule.

tubilingual *a.* having a hollow tongue, adapted for sucking.

tubiparous *a.* secreting tube-forming material, *appl.* glands.

tubo-ovarian *a. pert.* oviduct and ovary.

tubotympanic *a. appl.* recess between 1st and 3rd visceral arches, from which are derived the tympanic cavity and the Eustachian tubes.

tubular *a.* having the form of a tube; containing tubules.

tubulate *a.* in the form of a tube, having tubes, consisting of tubes.

tubule *n.* a small tube.

Tubulidentata *n.* order of placental mammals known from the Miocene, or possible the Eocene, whose only living member is the African ant-eater or aardvark (*Orycteropus*), which possesses unique peg-like teeth with tubular canals in the dentine, is ant-eating and has powerful digging forelimbs.

tubuliferous *a.* having tubules.

tubuliflorous *a.* having florets with a *tubular corolla*.

tubuliform *a.* tube-shaped; *appl.* type of spinning glands in spiders.

tubulin *n.* polypeptide subunit of microtubules, present in two types, α-tubulin and β-tubulin.

tubulose *a.* having, or composed of tubular structures; hollow and cylindrical.

tubulus *n.* any of various small tubular structures; tubule.

tumescence *n.* swelling.

tumid *a.* swollen; turgid.

tumor *alt.* spelling of tumour *q.v.*

tumorigenesis *n.* the development of a tumour.

tumorigenic *a. appl.* any agent that can cause a tumour, such as a tumour virus, certain chemicals, etc.; *appl.* a cell that can give rise to a tumour.

tumour *n.* a growth resulting from the abnormal proliferation of cells, and which may be self-limiting or non-invasive, when it is called a benign tumour, or continue proliferating indefinitely and invade underlying tissues and metastasize, when it is called a malignant tumour or cancer.

tumour angiogenesis factor substance released from malignant tumours which induces the formation of a capillary network invading the tumour.

tumour antigen novel cell-surface protein that appears on tumour cells, and which is not present on the normal cell.

tumour necrosis factor (TNF) a lymphokine produced by activated macrophages. It also has direct antitumour activity. *alt.* cachectin.

tumour promoters compounds that are not carcinogenic in themselves, but which hasten the effects of a carcinogen.

tumour suppressor genes genes believed to be involved in the development of some familial cancers. Heterozygotes inheriting a defect in one copy of such a gene show a much greater predisposition to develop various types of cancer. The development of the cancer seems to be due in many cases to subsequent loss or inactivation of the "good" copy of the gene in certain somatic cells, from which the cancer then develops.

tumour viruses *see* DNA tumour viruses, RNA tumour viruses.

tundra *n.* treeless Polar region with permanently frozen subsoil, bare of vegetation or may support mosses, lichens, herbaceous plants and dwarf shrubs.

tunic *n.* membrane or tissue enclosing or surrounding a structure; the body wall or test of a tunicate.

tunica *n.* body wall or outer covering; investing membrane or tissue or outer wall of organ.

tunica albuginea perididymis *q.v.*

tunica corpus type of cellular organization, e.g. of plant apical meristem, in which the region is differentiated into two parts distinguished by their plane of cell division, the outer or tunica layer having mainly anticlinal divisions and the inner corpus having divisions in various planes.

tunica intima innermost layer of a blood vessel wall.

tunicamycin *n.* antibiotic that inhibits the glycosylation of glycoproteins in eukaryotic cells.

Tunicata, tunicates *n.*, *n.plu.* subphylum of chordates containing the classes Ascidiacea (the sessile sea squirts), the free-swimming tadpole-like Larvacea, and the Thaliacea (free-swimming salps). Chordate features (i.e. notochord and nerve cord) are found only in the larva and are generally lost in the adult. The adult secretes a tough cellulose sac (tunic) in which the animal is embedded. *alt.* urochordates.

tunicate *a.* provided with a tunic or test.

tunicin *n.* a polysaccharide related to cellulose, found in the tunic of ascidians.

tunicle *n.* a natural covering or integument.

tunnelling electron microscope type of electron microscope that gives atomic level resolution of certain types of structure or surfaces, including biological macro molecules.

Turbellaria, turbellarians *n.*, *n.plu.* class of free-living flatworms with a leaf-like shape and a ciliated epithelium.

turbid plaque type of plaque produced by the growth of certain bacteriophages on a lawn of bacteria, which has a cloudy appearance due to the survival and growth of some bacteria within the area.

turbinal *a.* spirally rolled or coiled, as bone or cartilage; *n.* one of the nasal bones in vertebrates, supporting the olfactory tissues, *alt.* turbinate bone.

turbinate *a.* top-shaped, spirally rolled or coiled, *appl.* shells.

turbinate bones fragile scrolled bones forming the side walls of nasal cavities.

turbinulate *a.* shaped like a small top; *appl.* certain apothecia.

turgescence, turgidity, turgor *n.* the swelling of a plant cell due to the internal pressure of vacuolar contents; distention of any living tissue due to internal pressures. *a.* turgid, turgescent.

turgor pressure (TP) the pressure set up inside a plant cell due to the hydrostatic pressure of the vacuole contents pressing on the rigid cell wall. It provides mechanical support to non-woody plant stems, and changes in turgor pressure due to osmosis are responsible for the opening and closing of stomata and some seismonastic movements.

turio *n.* young scaly shoot budded off from underground stem; detachable winter bud used for perennation in many water plants. *alt.* turion.

Turkish saddle sella turcica *q.v.*

β-turn structure common in proteins whereby polypeptide chains make a sharp bend by hydrogen bonding between the CO group of one amino acid residue and the

NH group of the next third residue. *alt.* β-bend.

turnover *n.* in an ecosystem, the ratio of productive energy flow to the biomass; the fraction of a population which is exchanged per unit time through loss by death or emigration and replacement by reproduction and immigration.

turnover number in enzyme reactions, the number of substrate molecules converted into product per second when the enzyme is fully saturated with substrate.

turnover time time taken to complete a biological cycle; time from birth to death of an organism.

turtles *see* Chelonia.

tusk shells common name for the Scaphopoda *q.v.*

tutamen *n.* a protective structure, as eyelid.

two-dimensional gel electrophoresis technique for the separation of proteins, in which the mixture is first separated by isoelectric focusing or electrophoresis in one direction, and then subjected to electrophoresis in a direction at right-angles to the first.

Ty element any of a family of transposon-like DNA sequences found in yeast *Saccharomyces cerevisiae*.

tycholimnetic *a.* temporarily attached to bed of lake and at other times floating.

tychoplankton *n.* drifting or floating organisms which have been detached from their previous habitat, as in plankton of the Sargasso Sea; inshore plankton.

tychopotamic *a.* thriving only in backwaters.

tylopods *n.plu.* members of the mammalian suborder Tylopoda, of the Artiodactyla, which include the camel family and a number of extinct groups.

tylosis *n.* development of irregular cells in a cavity; intrusion of parenchyma cells into a xylem vessel, esp. of secondary xylem, through pits; a callosity; callus formation.

tylosoid *a.* resembling a tylosis, as a resin duct filled with parenchymatous cells.

tylotate *a.* with a knob at each end.

tylotic *a.* affected by tylosis.

tymbal *alt.* spelling of timbal *q.v.*

tymovirus group group of single-stranded RNA plant viruses with isometric particles, type member turnip yellow mosaic virus.

tympanic *a. pert.* tympanum.

tympanic membrane eardrum, thin membrane at the internal end of the external channel of the ear via which sound vibrations are transmitted to the ossicles of the middle ear.

tympaniform membrane *see* syrinx.

tympanohyal *a. pert.* tympanum and hyoid; *n.* part of hyoid arch embedded in petromastoid.

tympanoid *a.* shaped like a flat drum, *appl.* certain diatoms.

tympanum *n.* the drum-like cavity constituting the middle ear; the eardrum or tympanic membrane; membrane of auditory organ borne on tibia, metathorax or abdomen of insect; an inflatable air sac on the neck of birds of the grouse family; (*bot.*) the membrane closing capsule in some mosses. *a.* tympanic.

type *n.* sum of characteristics common to a large number of individuals, e.g. of a species, and serving as the basis for classification; a primary model, the actual specimen described as the original of a new genus or species. *alt.* holotype.

type locality locality in which the holotype or other type used for designation of a species was found.

type number the most frequently occurring chromosome number in a taxonomic group.

type specimen the single specimen chosen for the designation and description of a new species.

Typhales *n.* order of marsh or aquatic monocots with rhizomes and linear leaves, comprising the families Sparaginiaceae (bur reed) and Typhaceae (cat-tail).

typhlosole *n.* median longitudinal fold of the intestine projecting into lumen of gut in some vertebrates and in cyclostomes.

typical *a. appl.* specimen conforming to type or primary example; exhibiting in marked degree the characteristics of species or genus.

typogenesis *n.* phase of rapid type formation in phylogenesis; quantitative or explosive evolution.

typonym *n.* a name designating or based on type specimen or type species.

tyramine *n.* a phenolic amine formed by decarboxylation of tyrosine, produced in small amounts in animal liver, also produced by bacterial action on tyrosine-rich substrates, secreted by cephalopods, and found in various plants such as mistletoe and ergot of rye, and which causes a rise in arterial blood pressure.

tyrocidin *n.* cyclic decapeptide antibiotic produced by *Bacillus brevis*.

tyrosinase *n.* general name for a group of copper-containing enzymes catalysing the oxidation of tyrosine with the formation of dopa (dihydroxyphenylalanine). *r.n.* monophenol monooxygenase, EC 1.14.18.1; catechol oxidase, EC 1.10.3.1.

tyrosine (Tyr, Y) *n.* amino acid with aromatic side chain, constituent of protein, also important as a precursor of adrenaline and noradrenaline, and of thyroxine and melanin. *see* table in Appendix 1 for chemical formula.

tyrosine hydroxylase tyrosinase *q.v.*

tyrosine kinase *see* tyrosine protein kinase.

tyrosine protein kinase enzyme that specifically phosphorylates a target protein at a tyrosine residue. Tyrosine protein kinases are involved in the transmission of signals from cell-surface receptors, and some receptors for growth factors have an intrinsic tyrosine kinase activity. Several potential oncogenes specify proteins with tyrosine kinase activity. *alt.* tyrosine kinase, protein tyrosine kinase.

tyvelose *n.* a 3,6-dideoxyhexose sugar found in the lipopolysaccharide outer membrane of some enteric bacteria.

U

U uracil *q.v.*

ψU pseudouridine *q.v.*

UCR unconditional response or reflex *q.v.*

UCS unconditional stimulus *q.v.*

UDP uridine diphosphate *q.v.*

UDPG UDP-glucose *q.v.*

UDP-(sugar) activated form of monosaccharide attached to the nucleotide UDP, formed from UTP and the sugar, donors of monosaccharide residues in polysacchar-ide synthesis, interconversion of sugars etc., and including UDP-glucose, UDP-galactose.

UH2A a temporarily modified form of histone H2A in which the histone is linked to the protein ubiquitin by an isopeptide bond, found in appreciable amounts in nucleosomes in euchromatin of interphase nuclei. Previously called A24.

UMP uridine monophosphate *q.v.*

UTP uridine triphosphate *q.v.*

u-PA urokinase *q.v.*

UV ultraviolet light *q.v.*

ubiquinone *n.* quinone derivative with a tail of isoprenoid units (the number varying with species), a mobile electron carrier between flavoproteins and cytochromes of the respiratory electron transport chain.

ubiquitin *n.* small acidic protein found in a wide range of organisms from bacteria to mammals and which is involved in the targeting and degradation of abnormal intracellular proteins in the cytoplasm, where it binds to the protein to be degraded. It has also been found bound to histone H2A in chromatin.

ula *n.* the gums (of teeth).

uletic *a. pert.* the gums.

uliginose, uliginous *a.* swampy; growing in mud or swampy soil.

ulna *n.* one of the long bones of vertebrate forearm, parallel with radius and in some vertebrates combined with it to form a single bone. *a.* ulnar.

ulnar nervure radiating or cross vein in insect wing.

ulnare *n.* one of the wrist bones, lying at the far end of the ulna.

ulnocarpal *a. pert.* ulna and carpus.

ulnoradial *a. pert.* ulna and radius.

uloid *a.* resembling a scar.

ulotrichous *a.* having woolly or curly hair.

ultimate *a. appl.* factor thought to be the fundamental cause of some biological phenomenon. *cf.* proximate.

ultimobranchial bodies pair of gland rudiments derived from the 5th pharyngeal pouches, which secrete calcitonin and later degenerate and disappear.

ultra-abyssal hadral *q.v.*

ultracentrifuge *n.* instrument in which extracts of broken cells can be separated into their different components by spinning at various speeds (up to 150,000 *g*), the different organelles sedimenting at different speeds, and which is also used to separate large molecules of different molecular weights.

ultradian rhythm biological rhythm with a periodicity greater than 24 hours.

ultramicroscopic *a. appl.* structures or organisms too small to be visible under the light microscope but which can be seen in the electron microscope.

ultramicrotome *n.* machine with fine glass or diamond knife-blade for slicing ultrathin tissue sections for electron microscopy.

ultrastructure *n.* the fine structure of cells as seen in the electron microscope.

ultraviolet light electromagnetic radiation of wavelengths between those of the violet end of the visible light spectrum and X-rays. Although invisible to the human eye, it can be captured on photographic film. The ultraviolet (UV) spectrum is subdivided by wavelength into A (400–320 nm), B (320–280 nm) and C (280–10 nm) bands, of which UV-B is most harmful to living organisms. Much of the solar UV radiation, esp. the shorter wavelengths, is absorbed by the stratospheric ozone layer before reaching the Earth's surface.

umbel *n*. a flower-head in which each flower or cluster of flowers arises from a common centre, forming a flat-topped or rounded cluster; various structures in other organisms resembling an umbel.

Umbellales Cornales *q.v.*

umbella an umbel *q.v.*; umbrella of jellyfish.

umbellate *a*. arranged in umbels.

umbellifer *n*. member of the large dicot family Umbelliferae, typified by cow parsley, carrot, etc., having small flowers borne in umbels and much-divided leaves. They include many plants grown for food, but also highly poisonous species (e.g. *Conium maculatum*, called hemlock in the UK).

umbelliferous *a*. having flower-heads in umbels, as the umbellifers (e.g. cowparsley).

umbelliform *a*. shaped like an umbel.

umbelligerous *a*. bearing flowers or polyps in umbellate clusters.

umbellula *n*. a large cluster of polyps at tip of elongated stalk. *alt*. umbellule.

umbellulate *a*. arranged in umbels and umbellules.

umbellule *n*. small secondary umbel; *q.v.* umbellula.

umber opal *q.v.*

umbilical *a. pert*. navel or umbilical cord.

umbilical cord *n*. cord of tissue connecting embryo with placenta.

umbilicate *a*. having a central depression; like a navel.

umbilicus *n*. the navel, place of attachment of the umbilical cord; hilum *q.v.*; basal depression of certain spiral shells; an opening near base of feather.

umbo *n*. a protuberance like a boss on a shield; (*bot.*) swollen part of cone scale; (*zool.*) beak or older part of shells of bivalve molluscs. *plu*. umbones.

umbonal *a. pert*. an umbo.

umbonate *a*. having, *pert.*, or resembling an umbo.

umbones *plu*. of umbo *q.v.*

umbraculiform *a*. shaped like an expanded umbrella.

umbraculum *n*. any umbrella-like structure; the pigmented fringe of iris, in some ungulates.

umbraticolous *a*. growing in a shaded habitat.

umbrella *n*. the upper surface of the bell of a medusa.

Umwelt *n*. the total sensory input of an animal, distinctive for each species.

unbalanced *a*. in genetics, *appl*. aneuploid cells or gametes arising as the result of mitosis or meiosis in cells bearing certain types of chromosomal rearrangements, or to chromosomal rearrangements themselves that result in duplications and deficiencies.

uncate *a*. hooked.

unciferous *a*. bearing hooks.

unciform *a*. shaped like a hook or barb; *n*. unciform bone of wrist.

uncinate *a*. hook-like; *appl*. apex, as of a leaf.

uncinate fasciculus band of fibres connecting temporal and frontal lobes of brain.

uncinus *n*. small hooked, or hook-like, structure; crotchet *q.v.*; a marginal tooth of radula in gastropods.

unconditional reflex (UCR) an inborn reflex, produced involuntarily in response to a stimulus. *alt*. unconditioned reflex.

unconditional response (UCR) *see* classical conditioning. *alt*. unconditioned response.

unconditional stimulus (UCS) a stimulus that produces a simple reflex response. *alt*. unconditioned stimulus.

uncoupling agent any chemical that allows electron and proton transport along the respiratory chain of mitochondria or the electron transport chain of chloroplasts without the generation of ATP.

uncus *n*. hook-shaped extremity of hippocampus in brain.

undate *a*. wavy, undulating.

undecaprenyl phosphate, undecaprenyl pyrophosphate phosphate or pyrophosphate derivatives of the very long chain isoprenyl lipid undecaprenol, carrier of activated oligosaccharides in the biosynthesis of glycans in eubacteria.

understorey *n.* vegetation layer between tree canopy and the ground cover in a forest or wood, composed of shrubs and small trees.

under-wing one of the posterior wings of any insect.

undifferentiated *a.* not differentiated, in immature state, *appl.* embryonic cells that have not yet acquired a specialized structure and function; *appl.* meristematic and stem cells.

undose *a.* having undulating and nearly parallel depressions which run into each other and resemble ripple marks.

undulate *a.* having wave-like undulations, *appl.* leaves.

undulating membrane a cytoplasmic membrane in some flagellates which attaches part of the flagellum along the length of the cell; similar structure in tail of spermatozoon.

undulipodium *n.* flagellum or cilium of a eukaryotic cell. *plu.* undulipodia.

unequal crossing-over type of recombination that sometimes occurs between chromosomes carrying clusters of identical or related genes, in which misalignment occurs and recombinants are produced of unequal length, one containing fewer and the other more copies of the gene than normal. *alt.* non-reciprocal recombination.

ungual *a. pert.,* or having a nail or claw.

unguiculate *a.* clawed; (*bot.*) *appl.* petals with a claw-like stalk.

unguiculus *a.* a small nail or claw.

unguis *n.* nail or claw; the fang of an arachnid chelicera, through which the poison gland opens; (*bot.*) narrow stalk of some petals.

ungula *n.* hoof; (*bot.*) claw-like stalk of some petals.

ungulate *a.* hoofed; hoof-like.

ungulates *n.plu.* hoofed mammals. *see* Artiodactyla, Perissodactyla.

unguliform *a.* hoof-shaped.

unguligrade *a.* walking upon hoofs, which are formed from the tips of the digits.

uni- prefix from L. *unus*, one, generally meaning having one of.

uniascal *a.* containing a single ascus, *appl.* locules.

uniaxial *a.* with one axis; *appl.* movement only in one plane, as a hinge joint.

unibranchiate *a.* having one gill.

unicamerate *a.* one-chambered.

unicapsular *a.* having only one capsule.

unicell *n.* single-celled organism.

unicellular *a.* having only one cell, or consisting of one cell.

uniciliate *a.* having one cilium or flagellum.

unicolour *a.* having only one colour, of the same colour throughout.

unicorn *a.* having a single horn.

unicostate *a.* having a single prominent midrib, as certain leaves.

unicuspid *a.* having one tapering point, as tooth.

unidactyl *a.* having one digit.

uniembryonate *a.* having one embryo only.

unifacial *a.* having one face or chief surface; having similar structure on both sides.

unifactorial *a. pert.* or controlled by a single gene.

uniflagellate *a.* having only one flagellum.

uniflorous *a.* having one flower.

unifoliate *a.* with one leaf; with a single layer of zooecia, *appl.* polyzoan colony.

unifoliolate *a.* having one leaflet only.

uniforate *a.* having only one opening.

uniformity *n.* in ecology, the tendency of the component species of an association to be uniformly distributed within it.

unijugate *a. appl.* a pinnate leaf having one pair of leaflets.

unilabiate *a.* with one lip or labium.

unilacunar *a.* with one lacuna; having one leaf gap, *appl.* nodes.

unilaminate *a.* having one layer only.

unilateral *a.* arranged on one side only.

unilocular *a.* having a single compartment or cell; containing a single oil droplet, as cells in white fat.

unimodal *a.* having only one mode, *appl.* frequency distribution with a single maximum.

unimucronate *a.* having a single sharp point at tip.

uninducible *a. appl.* mutant, normally inducible genes in a permanent state of repression and which cannot be induced. *alt.* superrepressed.

uninemal *a.* single-stranded.

uninuclear, uninucleate *a.* having one nucleus.

uniovular *a. pert.* a single ovum; *appl.* twinning, twins produced from a single egg.

uniparental *a.* arising from a single parent.

uniparental disomy condition in which an embryo inherits two copies of a chromosome or part of a chromosome from one parent.

uniparous *a.* producing one offspring at a birth; having a cymose inflorescence with one axis of branching.

unipennate *a. appl.* muscle having its tendon of insertion extending along one side only.

unipetalous *a.* having one petal.

unipolar *a.* having one pole only, *appl.* some nerve cells with a single process.

uniport *n.* membrane protein that transports a solute across the membrane in one direction only.

unipotent(ial) *a. appl.* cells that can develop into only one type of cell. *cf.* totipotent.

unique sequences single-copy DNA *q.v.*

uniradiate *a.* one-rayed.

Uniramia *n.* in some classifications a group of arthropods containing the insects and myriapods. *alt.* Atelocerata.

uniramous *a.* having one branch; *appl.* crustacean appendage lacking an exopodite; *appl.* antennule.

unisegmental *a. pert.* or involving a single segment.

unisepalous *a.* having a single sepal.

uniseptate *a.* having one septum or dividing partition.

uniserial *a.* arranged in one row or series; *appl.* certain ascospores; *appl.* fins with radials on one side of basilia; *appl.* medullary rays.

uniseriate *a.* occurring in a single row, or layer.

uniserrate *a.* with one row of serrations along edge.

uniserrulate *a.* with one row of fine serrations along edge.

unisetose *a.* having one bristle.

unisexual *a.* of one or other sex, distinctly male or female; (*zool.*) sometimes *appl.* animal producing both sperm and eggs; (*bot.*) *appl.* plants and flowers having stamens and carpels in separate flowers.

unispiral *a.* having one spiral only.

unistrate, unistratose *a.* having one layer.

unit membrane any cell membrane with a lipid bilayer structure.

unitubercular *a.* having a single small prominence, tubercle or cusp.

unitunicate *a. appl.* asci in which both inner and outer wall are rigid and inelastic.

univalent *a. appl.* a single unpaired chromosome at meiosis.

univalve *n.* a shell in one piece, as of a gastropod mollusc.

universal donor *see* blood groups.

universal genetic code *see* genetic code.

universal recipient *see* blood groups.

universal veil tissue that completely encloses the developing toadstool in some fungi such as *Amanita*, and which tears as the toadstool grows to leave a cup-shaped remnant (volva) around base of stalk.

univoltine *a.* producing only one brood in a season.

unken reflex defensive posture adopted by newts in which they hold themselves rigidly immobile, tail upheld, showing brightly coloured underparts.

unsaturated *a. appl.* fatty acids with one or more double bonds, C=C, in their hydrocarbon chain.

unpaired *a.* situated in median line of body, consequently single; *appl.* nucleotide in DNA lacking a complementary nucleotide on opposite strand.

unscheduled DNA synthesis DNA synthesis occurring outside of the S phase during the cell cycle.

unstriped muscle smooth muscle *q.v.*

up mutation mutation in which transcription of a particular gene(s) is much enhanced, usually due to a mutation in the promoter controlling that gene(s).

upper shore zone of seashore between the average high-tide level and the highest high-water mark, which supports only a few species.

upregulation *n.* increase in number, as of receptors on a cell surface, or in rate of production.

upstream *a. appl.* DNA sequences, control sites, etc. on the proximal side of any given point in relation to the direction of transcription. Generally refers to points before the startpoint of transcription. *cf.* downstream.

urachus *n.* the median umbilical ligament; fibrous cord extending from apex of bladder to umbilicus.

uracil (U) *n.* a pyrimidine base, constituent of RNA, and which is the base in the nucleoside uridine. *see* Appendix 1 (18) for chemical structure.

uracil-DNA glycosidase an enzyme that removes uracil from uracil-containing nucleotides in DNA (which are then excised and replaced by other enzymes) and thus prevents mutations caused by the spontaneous deamination of cytosine to uracil and its subsequent mispairing with adenine. *alt.* DNA-uracil glycosidase.

urate (uric acid) *n.* 2,6,8-trioxypurine, an almost insoluble end-product of purine metabolism in certain mammals, the product of the breakdown of nucleic acids and proteins, excreted in urine in primates, main nitrogenous excretion product in birds, reptiles and some invertebrates esp. insects, also produced in plants.

urate oxidase enzyme found in liver and kidney and in some fungi, responsible for the oxidation of urate to allantoin which occurs in animals other than primates. EC 1.7.3.3. *alt.* uricase.

urceolate *a.* urn- or pitcher-shaped.

urceolus *n.* any pitcher-shaped structure.

urea *n.* carbamide, NH_2CONH_2, soluble waste product of the breakdown of proteins and amino acids in mammals and some other animals, chief nitrogenous constituent of the urine, and also found in some fungi and higher plants.

urea cycle metabolic cycle principally involving arginine, citrulline and ornithine, and found in all terrestrial vertebrates except reptiles and birds, in which ammonium ion formed during amino acid breakdown is converted to urea for excretion. *alt.* arginine-urea cycle, Krebs-Henseleit cycle, ornithine cycle.

urease *n.* enzyme which catalyses hydrolysis of urea into ammonia and carbon dioxide. EC 3.5.1.5.

uredia *sing.* of uredium *q.v.*

uredial *a. appl.* or *pert.* the summer stage of rust fungi.

uredinium, uredium *n.* in macrocyclic rust fungi a structure resembling an acervulus, in which binucleate uredospores, or summer spores, are produced. *plu.* uredinia, uredia.

uredo *n.* summer stage of rust fungi.

uredospore *n.* in macrocyclic rust fungi, the summer spore, the main propagative phase of the rust life cycle, a binucleate spore which germinates to form a mycelium from which new uredospores are produced, and so on.

ureide *n.* nitrogen-containing compound formed during amino acid metabolism (e.g. citrulline) or by the oxidation of purines (e.g. allantoin and allantoic acid), and which are used as nitrogen transport compounds in some plants, and as excretion products in some animals.

ureotelic *a.* excreting nitrogen as urea, as adult amphibia, elasmobranch fishes, mammals. *cf.* ammonotelic, uricotelic.

ureter *n.* duct conveying urine from kidney to bladder or cloaca.

ureteric *a. appl.* bud: embryonic diverticulum of metanephros giving rise ultimately to ureters.

urethra *n.* duct leading off urine from bladder, and in male also conveying semen.

uric acid *see* urate.

uricase urate oxidase *q.v.*

uricotelic *a.* excreting nitrogen as uric acid, as insects, birds, reptiles. *cf.* ureotelic.

uridine *n.* a nucleoside made up of the pyrimidine base uracil linked to ribose.

uridine diphosphate (UDP) uridine nucleotide containing a diphosphate group, forms part of activated sugars in many metabolic reactions.

uridine monophosphate (UMP) nucleotide consisting of uracil, ribose and a phosphate group, product of the partial hydrolysis of RNA, synthesized *in vivo* from orotidylate by decarboxylation. *alt.* uridylate, uridylic acid, uridine 5′-phosphate.

uridine triphosphate (UTP) uridine nucleotide containing a triphosphate group, one of the four ribonucleotides needed for synthesis of RNA, takes part in many metabolic reactions in a manner analogous to ATP, esp. forming activated UDP-sugars.

uridylate (uridylic acid) uridine monophosphate *q.v.*

uridylyltransferase *see* nucleotidyltransferase.

urinary *a. pert.* urine, *appl.* organs including kidneys, ureters, bladder and urethra.

urine *n.* fluid excretion from kidney in mammals, containing waste nitrogen chiefly as urea; solid or semisolid excretion in birds and reptiles, containing waste nitrogen chiefly as uric acid.

uriniferous *a.* urine-producing; *appl.* tubules of nephron leading from Bowman's capsules to collecting ducts.

urinogenital *a. pert.* urinary and genital systems.

urinogenital ridge a paired ridge in embryo from which urinary and genital systems are developed.

urinogenital sinus bladder or pouch in connection with the urinary and genital systems in many animals.

urkingdoms *n.plu.* three kingdoms proposed for the classification of living organisms, i.e. eubacteria, archaebacteria, and urkaryotes (eukaryotes).

urn *n.* theca or capsule of mosses; an urn-shaped structure; one of the ciliated bodies floating in coelomic fluid of annulates.

urobilin *n.* a brown pigment in urine.

urobilinogen *n.* a colourless compound derived from bilirubin, oxidized to urobilin and excreted in urine.

urocardiac ossicle stout short bar forming part of gastric mill in certain crustaceans.

urochord *n.* the notochord when confined to tail region, as in tunicates.

urochrome *n.* a yellowish pigment to which ordinary colour of urine is due.

urocoel *n.* an excretory organ in molluscs.

urocyst *n.* the urinary bladder.

urodaeum *alt.* spelling of urodeum *q.v.*

Urodela, urodeles *n., n.plu.* one of the three orders of extant amphibians, containing the newts and salamanders, amphibians with well-developed tails and two pairs of more-or-less equal legs. Called the Caudata in some classifications. *cf.* Anura.

urodelous *a.* with a persistent tail.

urodeum *n.* the part or chamber of the cloaca into which ureters and genital ducts open. *alt.* urodaeum.

urogastric *a. pert.* posterior part of gastric region in some crustaceans.

urogastrone epidermal growth factor *q.v.*

urogenital urinogenital *q.v.*

urohyal *n.* a median bony element in the hyoid arch.

uroid *n.* in some amoebae, a region of posterior gel-like cytoplasm.

urokinase (u-PA) *n.* proteolytic enzyme present in mammalian urine, closely related to tissue plasminogen activator and also known as urinary plasminogen activator, and which can convert plasminogen to plasmin.

uromere *n.* an abdominal segment in arthropods.

uromorphic *a.* like a tail.

uroneme *n.* tail-like structure of some ciliate protozoans.

uronic acid any of a group of acids formed as oxidation products of sugars, and found in urine and as subunits of some polysaccharides.

uropatagium *n.* membrane stretching from one femur to the other in bats; part of patagium extending between hind-feet and tail in flying lemurs.

urophan *n.* any ingested substance found chemically unchanged in urine. *a.* urophanic.

urophysis *n.* in fishes, a concentration of neurosecretory nerve endings at the end of the spinal cord, resembling the neurohypophysis of mammals.

uropod *n.* fan-shaped paired appendage on the penultimate segment of crustaceans, used for swimming.

uropore *n.* opening of excretory duct in certain invertebrates.

uroporphyrin *n.* a brownish-red product of haem metabolism, lacking iron, a pigment of urine.

uropygium *n.* the hump at the end of a bird's body, containing vertebrae and supporting tail feathers. *a.* uropygial. *alt.* uropyge.

uropyloric *a. pert.* posterior region of crustacean stomach.

urorectal *a. appl.* embryonic septum which ultimately divides intestine into anal and urinogenital parts.

urorubin *n.* red pigment of urine.

urosacral *a. pert.* caudal and sacral regions of the vertebral column.

urosome *n.* the tail region of fish; abdomen of some arthropods.

urostege, urostegite *n.* ventral tail plate of snakes.

urosternite *n.* ventral plate of arthropod

abdominal segment.

urosthenic *a.* having the tail strongly developed for propulsion.

urostyle *n.* an unsegmented bone forming the posterior part of vertebral column of frogs and toads.

Urticales *n.* an order of dicots including herbs, vines, shrubs and trees, and including the families Cannabaceae (hemp), Moraceae (mulberry), Ulmaceae (elm) and Urticaceae (nettle).

urticant *a.* stinging, irritating.

urticaria *n.* local inflammatory swellings on skin as a result of allergic reactions. *alt.* hives.

urticarial *a.* producing a rash as of stinging nettle, *appl.* hairs of some caterpillars.

uterine *a. pert.* uterus.

uterine cervix the neck of the uterus where it adjoins the vagina.

uterine crypts depressions in uterine mucosa, for accommodation of chorionic villi.

uteroabdominal *a. pert.* uterus and abdominal region.

uterosacral *a. appl.* two ligaments of sacrogenital folds attached to sacrum.

uterovaginal *a. pert.* uterus and vagina.

uterovesical *a. pert.* uterus and urinary bladder.

uterus *n.* in female mammals, the organ in which the embryo develops and is nourished before birth; in other animals, an enlarged portion of oviduct modified to serve as a place for development of young or of eggs.

utricle utriculus *q.v.*

utricular, utriculate *a. pert.* utricle; shaped like a utricle; containing vessels like small bags.

utriculiform *a.* shaped like a utricle or small bladder.

utriculus *n.* part of the membranous labyrinth of inner ear, which together with saccule (sacculus rotundus) forms vestibule of inner ear.

utriform *a.* bladder-shaped, with a shallow constriction.

uva *n.* a berry formed from a superior ovary and with a central placenta, such as a grape.

uvea *n.* pigmented epithelium covering posterior surface of iris.

uvula *n.* conical flap of soft tissue hanging from soft palate.

V

V valine *q.v.*

V the symbol for the element vanadium; the SI symbol for volt.

V$_H$ variable region of antibody heavy chain. *see* V region.

V$_L$ variable region (*q.v.*) of antibody light chain. *see* V region.

V$_m$ax *see* Michaelis-Menten kinetics *q.v.*

Val valine *q.v.*

VIP vasoactive intestinal peptide *q.v.*

VLDL very low density lipoprotein *q.v.*

v-*onc* general designation for viral oncogenes (e.g. v-*src*, v-*fes*), the genes carried by RNA viruses which are responsible for viral transformation of the infected cell and tumorigenesis. *see also* c-*onc*, oncogene.

VOR vestibulo-ocular reflex *q.v.*

VPL ventroposterolateral nucleus of thalamus.

VSV vesicular stomatitis virus, a rhabdovirus, a mild pathogen of cattle.

v–a mycorrhiza vesicular arbuscular mycorrhiza *q.v.*

vaccine *n.* a preparation of microorganisms or their antigenic components which can induce protective immunity against the appropriate pathogenic bacterium or virus but which does not itself cause disease. Vaccines may be composed of killed pathogenic microorganisms (killed vaccines), or live non-pathogenic strains of virus or bacterium (live vaccines), or may be composed of isolated protein antigens (subunit vaccines).

vacuolar *a. pert.* or resembling a vacuole.

vacuolated *a.* containing vacuoles.

vacuole *n.* large fluid-filled membrane-bounded cavity in the cytoplasm of eukaryotic cells, found esp. in plant and algal cells (where it is often known as the central vacuole and takes up a large part of the cellular volume), and protozoa, and which contains cell sap (in plant and algal cells) or water or partially digested food (in protozoans). *see also* contractile vacuole.

vacuolization *n.* formation of vacuoles; appearance or formation of drops of clear fluid in growing or ageing cells. *alt.* vacuolation.

vacuum activity a fixed-action pattern of activity carried out by an animal in the absence of any external stimulus.

vagal *pert.* the vagus *q.v.*

vagiform *a.* having an indeterminate form; amorphous.

vagile *a.* freely motile; able to migrate. *n.* vagility.

vagina *n.* canal leading from uterus to external opening of genital canal; sheath or sheath-like tube.

vaginal *a. pert.* or supplying the vagina, as arteries, nerves, etc.

vaginal process projecting lamina on inferior surface of petrous portion of temporal bone; a lamina on sphenoid.

vaginate *a.* sheathed.

vaginervose *a.* with irregularly arranged veins.

vaginicolous *a.* building and inhabiting a sheath or case.

vaginipennate *a.* having wings protected by a sheath.

vagus *n.* 10th cranial nerve, supplies many internal organs including stomach and duodenum, liver, spleen and kidneys, forms cardiac and pulmonary plexuses, supplies muscles of oesophagus and bronchi and their glands, and pharynx and larynx; visceral accessory nervous system in insects.

valency *n.* of an antibody, the number of antigen-binding sites per molecule.

valine (Val, V) *n.* amino acid with a hydrocarbon side chain, constituent of protein, essential in the diet of man and some animals. *see* table in Appendix 1 for chemical formula.

valinomycin *n.* antibiotic which increases the permeability of cell membranes to potassium ions.

Valium trade name for one of the commonly used benzodiazepines *q.v.*

vallate *a.* with a rim surrounding a depression, *appl.* papillae with taste buds on back of tongue.

vallecula *n.* a depression or groove.

valleculate *a.* grooved.

Valsalva's sinuses sinuses of Valsalva *q.v.*

valvate *a.* hinged at margin only; meeting at edges; opening or furnished with valves.

valve *n.* any of various structures permitting flow of blood, etc. in only one direction, and preventing backward flow; each half of the hinged two-part shells of, e.g., clams and cockles; each part of the silica case of diatoms; any of sections formed by a seed capsule on dehiscence; lid-like structure of certain anthers.

valve of Thebesius valve in the coronary sinus in right atrium.

valvula *n.* a small valve in the coronary sinus in right atrium.

valvulae conniventes circular, spiral or bifurcated folds of mucous membrane found in alimentary canal from duodenum to ileum, affording an increased area for secretion and absorption.

valvular *a. pert.* or like a valve, *appl.* dehiscence of certain capsules and anthers which split into several sections.

valvule *n.* upper palea of grasses; valvula *q.v.*

vanadium (V) A component of some bacterial nitrogenases.

vanadocyte *n.* blood cell containing a vanadium compound, in certain ascidians.

vancomycin *n.* an antibiotic which blocks cell wall formation in bacteria.

van der Waals forces weak non-covalent interatomic attractive forces, of importance in forming and maintaining the three-dimensional structure of proteins and other biological macromolecules.

vane *n.* the web of a feather, consisting of the thread-like barbs.

vannus *n.* fan-like posterior lobe of hindwing of some insects. *a.* vannal.

variable expressivity of a genetic trait, expression to a different degree in different individuals.

variable region *see* V region.

variance *n.* the condition of being varied; the mean of the squares of individual deviations from the mean.

variant *n.* an individual or species deviating in some character or characters from type.

variate *n.* the variable quantity; a character that is variable in quality or magnitude.

variation *n.* divergence from type in certain characteristics; the phenotypic differences that exist between individuals of the same species (other than those due to age) and which reflect both genetic differences and the influence of the environment. *see also* continuous variation, discontinuous variation, genetic variation; (*stat.*) deviation from the mean.

varicella-zoster virus the virus that causes chickenpox and shingles, a member of the Herpesviridae.

varicellate *a. appl.* shells with small and indistinct ridges.

varices *n.plu.* prominent ridges across whorls of various univalve shells, showing previous position of the outer lip. *sing.* varix.

variegation *n.* patchiness of pigmentation of leaves or other plant organs caused by viral interference or various genetic factors such as non-development of chloroplasts in some cell lineages, somatic mutations leading to a change of phenotype during development etc.; any discontinuous phenotype caused by a change in genotype in somatic cells during development.

variety *n.* a taxonomic group below the species level and used in different senses by different specialists.

variola major, smallpox; minor, cowpox.

variole *n.* small pit-like marking on various parts, in insects.

varix *sing.* of varices *q.v.*

vas *n.* a small vessel, duct, canal or blind tube. *plu.* vasa.

vasa afferentia lymphatic vessels entering

lymph nodes. *sing.* vas afferens.

vasa vasorum blood vessels supplying the larger arteries and veins and found in their coats.

vasal *a. pert.* or connected with a vessel.

vascular *a. pert.*, consisting of, or containing vessels adapted for the carriage or circulation of fluid, in animals refers to the blood and lymphatic systems, in plants to the xylem and phloem.

vascular bundle discrete strand of xylem and phloem cells, sometimes separated by a strip of cambium, in stems of some plants.

vascular cambium *see* cambium.

vascular cylinder the xylem and phloem of stem and root.

vascular plants common name for all plants containing xylem and phloem and which include the club mosses, ferns, cycads, gymnosperms and angiosperms.

vascular strand fine strand of primary xylem and primary phloem in stems of some plants, leaves and flowers.

vascular system (*bot.*) the plant tissues that carry water and solutes around the plant, i.e. the xylem and phloem; (*zool.*) the blood and/or lymphatic systems.

vascular tissue in plants, xylem and phloem.

vascularized *a.* infiltrated with blood capillaries.

vasculogenesis *n.* the formation of blood vessels.

vasculum *n.* a pitcher-shaped leaf; a small blood vessel.

vasifactive *a.* producing new blood vessels.

vasoactive *a.* affecting blood vessels, causing them either to dilate or contract.

vasoactive intestinal peptide (VIP) a biologically active peptide occurring in both gut and nervous tissue.

vasoconstrictor *a.* causing constriction of blood vessels (by contraction of smooth muscle in their walls).

vasodentine *n.* a variety of dentine permeated with blood vessels.

vasodilator *a.* causing relaxation or enlargement of blood vessels.

vasoganglion *n.* dense network of blood vessels such as a rete mirabile.

vasoinhibitory vasodilator *q.v.*

vasomotor *a. appl.* nerves supplying muscles of walls of blood vessels and regulating the calibre of the vessel by vasodilation and vasoconstriction.

vasopressin *n.* peptide hormone produced by the neurohypophysis (the posterior lobe of the pituitary). It stimulates smooth muscle contraction, causes constriction of small blood vessels and raises blood pressure, and has an antidiuretic effect by causing water resorption in kidney tubules. *alt.* antidiuretic hormone.

vasotocin *n.* a peptide hormone having similar properties to oxytocin and vasopressin, secreted by the posterior lobe of the pituitary gland in lower vertebrates, and by foetus and pineal in mammals.

vastus *n.* a division of the quadriceps muscle of thigh.

vector *n.* any agent (living or inanimate) that acts as an intermediate carrier or alternative host for a pathogenic organism and transmits it to a susceptible host; in genetic engineering, phage, plasmid or virus DNA into which another DNA is inserted for introduction into bacterial or other cells for amplification (DNA cloning) or studies of gene expression.

vegetable starch starch, as compared with glycogen.

vegetal pole that part of a yolky egg that cleaves more slowly than the rest, due to the presence of the yolk, *cf.* animal pole; the end of a blastula at which the larger cleavage products (megameres) collect.

vegetation *n.* the plant cover of an area, considered generally, and not taxonomically.

vegetative *a. appl.* stage of growth in plants when reproduction does not occur; *appl.* foliage shoots on which flowers are not formed; *appl.* the assimilative phase in fungi, when mycelium is being produced; *appl.* reproduction by bud formation or other asexual method in plants and animals; *appl.* nervous system: the autonomic nervous system.

vegetative apomixis asexual reproduction in plants by e.g. rhizomes, stolons and bulbils.

vegetative nucleus macronucleus *q.v.*; pollen tube nucleus.

veil *n.* in fungi, sheet of fine tissue stretching from stipe over cap in some basidiomycete fungi, and which is reptured as fruit body develops, remaining as the ring on stalk and sometimes patches on cap. *alt.* velum.

veins *n.plu.* branched vessels that convey blood to heart; of insect wing, fine extensions of the tracheal system that support the wing; of leaves, branching strands of vascular tissue.

vela *plu.* of velum *q.v.*

velamen *n.* the multiple-layered epidermis of aerial orchid roots, providing mechanical protection, reducing water loss, and possibly specialized for water absorption.

velaminous *a.* having a velamen.

velar *a. pert.* or situated near a velum.

velate *a.* veiled; covered by a velum.

veld *n.* the open temperate grasslands of southern Africa.

veliger *n.* second larval stage in some molluscs, developing from the trochophore.

velum *n.* a membrane or structure similar to a veil; in Hydrozoa and some jellyfishes the ring of tissue projecting inwards from margin of bell; in lampreys and some other vertebrates, a flap of muscular tissue in the buccal cavity; (*bot.*) veil, a sheet of tissue stretching from stipe over top of cap in some basidiomycete fungi. *plu.* vela.

velvet *n.* soft vascular skin which covers the antlers of deer during their growth and is rubbed off as the antlers mature.

velvet worms Onychophora *q.v.*

vena *n.* a vein, esp. large blood vessel carrying blood to heart.

vena cava one of the main veins that carries blood to the right auricle of heart.

vena comitantes vein accompanying or alongside an artery or nerve.

venation *n.* the arrangement of veins of leaf or insect wing.

venin *n.* a toxic substance of snake venom.

venomous *a.* having poison glands and able to inflict a poisonous wound.

venose *a.* having many and prominent veins.

venous *a. pert.* veins; *appl.* blood returning to heart after circulation in body.

vent *n.* the anus; cloacal or anal aperture in lower vertebrates.

vent community *see* hydrothermal vent community.

venter *n.* (*zool.*) abdomen; lower abdominal surface; (*bot.*) the swollen lower portion of archegonium of bryophytes, containing a single egg.

ventrad *adv.* towards lower or abdominal surface.

ventral *a. pert.* or nearer the belly or underside of an animal or under surface of leaf, wing, etc.; *pert.* or designating that surface of a petal, etc. that faces centre or axis of flower. *cf.* dorsal.

ventral aorta large artery in fish and in amniote embryos running forward from ventricle of heart.

ventral root of cranial nerve, a nerve root with some sensory fibres; of spinal nerve, a nerve root with some motor fibres.

ventricle *n.* cavity or chamber; one of several fluid-filled cavities in centre of brain; one of pair of lower chambers in heart.

ventricose *a.* swelling out towards the middle, or unequally.

ventricular *a. pert.* a ventricle, *appl.* ligaments and folds of larynx, *appl.* septum and valves in heart.

ventricular membrane basement membrane underlying epithelial tissues of retina.

ventrobronchus *n.* one of a number of tubes in lungs of birds which branch off the bronchi and are connected with the anterior air sacs.

ventrodorsal *a.* extending from ventral to dorsal.

ventrolateral *a.* at the side of the ventral region; central and lateral.

venule *n.* small vessel conducting venous blood from capillaries to a vein; small vein of leaf or insect wing.

veratridine *n.* alkaloid poison acting on nervous system.

vermicular *a.* resembling a worm in appearance or movement.

vermiculate *a.* marked with numerous sinuate fine lines or bands of colour or by irregular depressed lines.

vermiform *a.* shaped like a worm.

vermiform appendix a remnant of the caecum present in some mammals, in humans being a worm-like blind tube extending from the gut.

vermis *n.* the median portion of the cerebellum, distinguished from the cerebellar cortex.

vernal *a. pert.* or appearing in mid or late

spring.

vernalization *n.* the exposure of certain plants or their seeds to a period of cold which is necessary either to cause them to flower at all or to make them flower earlier than usual, and is used esp. on cereals such as winter varieties of wheat, oats and rye.

vernation *n.* the arrangement of leaves within a bud.

vermicose *a.* having a varnished appearance; glossy.

verruca *n.* a wart-like projection.

verruciform *a.* wart-shaped.

verrucose *a.* covered with wart-like projections.

versatile *a.* swinging freely. *appl.* anthers; capable of turning backwards and forwards, *appl.* bird's toe.

versical *a. pert.* or in relation to bladder.

versicoloured *a.* variegated in colour; capable of changing colour.

versiform *a.* changing shape; having different forms.

vertebra *n.* any of the bony or cartilaginous segments that make up a backbone, having a central hole for the passage of the spinal cord. *plu.* vertebrae.

vertebral *a. pert.* backbone; *appl.* various structures situated near or connected with backbone, or with any structure like a backbone; *appl.* artery supplying the hind parts of cerebral hemispheres.

vertebral column the series of vertebrae running from head to tail along the back of vertebrates, and which encloses the spinal cord. *alt.* backbone, spinal column.

vertebrarterial canal canal formed by foramina in transverse process of cervical vertebrae or between cervical rib and vertebra.

Vertebrata, vertebrates *n., n.plu.* subphylum of the Chordata, animals characterized by the possession of a brain enclosed in a skull, ears, kidneys and other organs, and a well-formed bony or cartilaginous vertebral column or backbone enclosing the spinal cord. The Vertebrata includes the classes Agnatha (lampreys and hagfish), Holocephalii (rabbit fish), Selachii (sharks, dogfishes and rays), Osteichthyes (bony fish), Amphibia, Reptilia, Aves (birds) and Mammalia.

vertebration *n.* division into segments or parts resembling vertebra.

vertex *n.* top of the head; region between compound eyes of insect.

vertical *a.* standing upright; lengthwise; in direction of axis; *pert.* vertex.

verticil whorl *q.v.*

verticillaster *n.* a much condensed cyme with appearance of whorl but really arising in axils of opposite leaves.

Verticillatae Casuarinales *q.v.*

verticillate *a.* arranged in whorls or verticils.

very low density lipoprotein (VLDL) a group of plasma lipoproteins, synthesized by the liver, in which form triacylglycerols are transported from liver to adipose tissue.

vesica *n.* bladder, esp. urinary bladder.

vesical *a. pert.* or in relation to bladder, *appl.* arteries, etc.

vesicle *n.* general term for small membrane-bounded sacs in eukaryotic cells, which are derived mainly from plasma membrane, endoplasmic reticulum and Golgi apparatus and carry materials from one cellular compartment to another or for secretion. *see* secretory granule, synaptic vesicles, transport vesicle; small spherical air space in tissues; small cavity or sac usually containing fluid; one of three primary cavities in the human brain; (*mycol.*) hyphal swelling as in mycorrhiza; (*zool.*) hollow prominence on shell or coral; base of postanal segment in scorpions. *alt.* vesicula.

vesicle-mediated transport the intracellular transport of material enclosed in a membrane vesicle from one cellular compartment to another. *see also* transport vesicle.

vesicula *n.* a small bladder-like cyst or sac; vesicle *q.v.*

vesicula seminalis seminal vesicle *q.v.*

vesicular *a.* composed of or marked by the presence of vesicle-like cavities; bladder-like.

vesicular arbuscular mycorrhiza (v–a mycorrhiza) common type of endomycorrhiza characterized by the occurrence of vesicles (swellings on invading hyphae) and arbuscules (discrete masses of branched hyphae) in infected tissues, the lack of a fungal sheath around the roots, and hyphae ramifying within and between the cells of the root cortex.

vesicular gland gland in tissue underlying epidermis in plants and containing essential oils.

vespertine *a.* blossoming or active in the evening.

vespoid *a.* wasp-like.

vessel *n.* any tube with properly defined walls in which fluids such as blood, lymph, sap, etc. move. *see also* xylem vessel.

vessel element one type of water-conducting cell in xylem of angiosperms, with heavily lignified secondary cell walls and large perforations through the cell wall, especially the end walls. Joined end to end with similar cells to form a long hollow tube or xylem vessel.

vestibular *a. pert.* a vestibule.

vestibular apparatus sensory apparatus in inner ear concerned with maintaining balance.

vestibular nerve branch of auditory nerve.

vestibulate *a.* in the form of a passage between two channels; resembling, or having, a vestibule.

vestibule, vestibulum *n.* a cavity leading into another cavity or passage; portion of ventricle directly below opening of aortic arch in heart; cavity leading to larynx; nasal cavity; posterior chamber of bird's cloaca; (*bot.*) pit leading to stoma of leaf.

vestibulocochlear *a. appl.* nerve: the auditory nerve which innervates the inner ear.

vestibulo-ocular reflex (VOR) the reflex eye movements that ensure that the eyes remain stably pointing in one direction when the head turns, so that the image of the visual field on the retina does not become blurred.

vestige *n.* a small degenerate or imperfectly developed organ or part which may have been complete and functional in some ancestor.

vestigial *a.* of smaller and simpler structure than corresponding part in an ancestral species; small and imperfectly developed.

Vestimentifera, vestimentiferans *n.*, *n.plu.* a phylum proposed to include certain genera of the Pogonophora (e.g. *Riftia, Lamellibrachia*), sessile deep-sea worms that produce fixed chitin tubes in bottom sediments or on decaying wood on the sea floor, which carry symbiotic sulphide-oxidizing chemoautotrophic bacteria contained within a structure called a trophosome, and whose haemoglobin can carry hydrogen sulphide as well as oxygen.

vestiture *n.* a body covering, as of scales, hair, feathers, etc.

vexilla *plu.* of vexillum *q.v.*

vexillar(y) *a. pert.* a vexillum; *appl.* type of imbricate aestivation in which upper petal is folded over others.

vexillate *a.* bearing a vexillum.

vexillum *n.* the upper petal standing at the back of a papilionaceous flower, such as pea, bean, etc., which helps to make the flower conspicuous, *alt.* banner, standard; the vane of a feather, *alt.* web.

V gene any of the gene segments coding for part of the variable regions of immunoglobulin molecules, several hundred different forms of which exist in the typical mammalian genome. To produce the variable region of an immunoglobulin gene one V gene segment is joined at random to a J segment, and in heavy chain genes, a D segment also, this process occurring only in the precursors of B lymphocytes.

via *n.* a way or passage.

viable *a.* capable of living; capable of developing and surviving parturition, *appl.* foetus.

viatical *a. appl.* plants growing by the roadside.

vibraculum *n.* whip-like cell modified for defence in Ectoprocta.

vibratile *a.* oscillating; *appl.* antennae of insects.

vibratile corpuscles cells closely resembling sperms present in coelomic fluid of starfish.

vibratile membrane structure formed by fused cilia for wafting food to mouth in ciliate protozoans.

vibrio *n.* any of a group of bacteria with short curved cells, appearing comma-shaped under the microscope, esp. the cholera bacillus, *Vibrio cholerae* and related organisms.

vibrissa *n.* stiff hair growing on nostril or face of animal, as whiskers of cat or mouse, often acting as a tactile organ; feather at base of bill or around eye; one of paired bristles near upper angles of mouth cavity in Diptera; one of the sensitive hairs of an insectivorous plant, as of *Dionaea*, Venus' fly trap. *plu.* vibrissae.

vicariation *n.* the separate occurrence of corresponding species, as reindeer and caribou, in corresponding but separate environments, divided by a natural barrier.

vicarious *a. appl.* species belonging to a closely related group that are equivalent in ecological terms but which live in separate regions divided by environmental barriers.

vicilin *n.* a seed storage protein of legumes.

vicinism *n.* tendency to variation due to proximity of related forms.

Vicq-d'Azyr, bundles of a bundle of nerve fibres running from corpora mamillaria to the thalamus. *alt.* mamillothalamic tract, thalamomamillary fasciculus.

villi *plu.* of villus. *q.v.*

villiform *a.* having form and appearance of velvet, *appl.* dentition.

villin *n.* protein component of the cytoskeleton of intestinal microvilli.

villose, villous *a.* shaggy; having villi or covered with villi.

villus *n.* one of the small vascularized projections on the lining of the small intestine; one of the processes on chorion through which nourishment passes to the embryo; invagination of a synovial membrane into joint cavity; a fine straight process on epidermis of plants. *plu.* villi.

vimen *n.* long slender shoot or branch. *plu.* vimina.

vimentin *n.* protein component of intermediate filaments in many cell types.

vinblastine, vincristine plant alkaloids derived from *Vinca* spp. and used as anticancer drugs. They inhibit microtubule formation and kill rapidly dividing cells by disrupting the mitotic spindle.

vinculin *n.* a protein associated with the cytoskeleton in some mammalian cells.

vinculum *n.* slender band of tendon; band uniting two main tendons in bird's foot; part of sternum bearing claspers in male insects. *plu.* vincula.

Violales *n.* order of dicot shrubs or small trees, less often herbs, and including the families Cistaceae (rock rose), Violaceae (violet) and others.

violaxanthin *see* xanthophyll.

viral *a. pert.*, belonging to, consisting of, or due to, a virus.

viral interference the case where the multiplication of one virus in a cell is inhibited when the cell is also infected by another type of virus.

virescence *n.* production of green colouring in petals instead of usual colour.

virescent *a.* turning greenish or green.

virgalium *n.* a series of rod-like elements forming petaloid rays of an ambulacral plate, as in some Asteroidea.

virgate *a.* rod-shaped; striped.

virgin lymphocytes immature lymphocytes that have acquired antigen specificity but have not yet encountered antigen.

virgula *n.* a small rod; a paired or bilobed structure or organ at oral sucker in some trematodes.

virgulate *a.* with or like a small rod or twig; having minute strips.

viridant *a.* becoming or being green.

virilization *n.* masculinization of genetic females caused by disturbances of sex hormone metabolism, due to various causes; precocious sexual development in genetic males.

virino *n.* hypothetical infective agent composed of a host protein and a host-independent nucleic acid which has been suggested as the type of agent responsible for the transmissible spongiform encephalopathies such as scrapie.

virion *n.* mature virus particle consisting of nucleic acid core and protein coat, and in some types an outer lipid envelope.

viroids *n.plu.* small circles of RNA which cause various diseases in plants, replicated entirely by host cell enzymes and not coding for any proteins. Transmitted from plant to plant by insect vectors.

virology *n.* the study of viruses.

virose, virous *a.* containing a virus.

virulence *n.* the ability to cause disease. *a.* virulent, *appl.* bacteria, viruses, etc.

virulence gene gene in a pathogenic microorganism which is responsible for its ability to cause disease.

virulent bacteriophage (phage) a bacteriophage which can only enter the lytic cycle of infection, causing lysis of the host. *cf.* temperate phage.

virus *n*. minute, intracellular obligate parasite, visible only under the electron microscope. A virus particle consists of a core of nucleic acid, which may be DNA or RNA, surrounded by a protein coat, and in some viruses a further lipid/glycoprotein envelope. It is unable to multiply or express its genes outside a host cell as it requires host cell enzymes to aid DNA replication, transcription and translation. Viruses cause many diseases of man, animals and plants. Viruses infecting bacteria are called bacteriophages. *see also* DNA viruses, RNA viruses.

virusoid satellite RNA *q.v.*

virus receptor protein on surface of host cell to which virus particles bind and which aids their entry into the cell.

viscera *n.plu.* the internal organs collectively. *a.* visceral.

visceral afferent fibres nerve fibres conveying impulses from sensory receptors in internal organs to the spinal cord. Their cell bodies are in dorsal root ganglia.

visceral arches a series of skeletal arches and associated tissue developed in connection with mouth and pharynx and including the gill arches and gill bars.

visceral clefts gill slits *q.v.*

visceral efferent fibres nerve fibres of the sympathetic and parasympathetic systems of the autonomic (involuntary) nervous system, conveying impulses from brain and spinal cord to smooth muscle and glandular tissue of internal organs.

visceral hump or mass in molluscs a central concentration of viscera covered by a soft skin, the mantle.

viscerocranium *n*. jaws and visceral arches.

visceromotor *a. appl.* nerves carrying motor impulses to internal organs.

viscid *a.* sticky.

viscid silk a highly extensible and sticky type of silk produced by spiders, which forms the spiral of a typical orb web. *see also* frame silk.

viscidium *n*. in orchid flower, a sticky disc at end of the stalk of pollinium, by which it is attached to an insect's head.

viscin *n*. sticky substance obtained from various plants, esp. from berries of mistletoe.

visible light light between the wavelengths 380–780 nm, which is perceptible by the human eye.

visual axis the straight line between the point to which the focused eye is directed and the fovea.

visual cortex that part of cerebral cortex in brain concerned with visual perception.

visual pigments rhodopsin and the cone pigments, which are the photoreceptor pigments of the vertebrate eye; other pigments in other species used for vision.

visual purple rhodopsin *q.v.*

visual violet iodopsin *q.v.*

visual transduction the conversion of the light signal received by the photoreceptor pigments in the eye into a nerve impulse.

vital capacity of lungs, the sum of complemental, tidal and supplemental air.

vital staining the staining of living cells or tissues with non-toxic dyes.

vitalism *n*. a belief that phenomena exhibited in living organisms are due to a special force distinct from physical and chemical forces.

vitamin *n*. any of various organic compounds needed in minute amounts for various metabolic processes and synthesized by plants and some lower animals, but which must be supplied in the diet of higher animals. The lack of the appropriate vitamin causes a deficiency disease. *see individual entries.*

vitamin A a fat-soluble vitamin derived from carotenes, found in liver oils of certain fish and in milk and eggs. It is a precursor of retinal, the light-sensitive pigment of the rods and cones of the eye. Its deficiency retards growth, causes night blindness and keratinization of epithelia. *see* Appendix 1 (43) for chemical structure.

vitamin B complex a group of water-soluble vitamins obtained from yeast, wheat germ, liver, and given separate B numbers, now usually replaced by specific names, *see below.*

vitamin B$_c$ folic acid *q.v.*

vitamin B$_{12}$ *see* cobalamine.

vitamin B$_{17}$ laetrile *q.v.*

vitamin B$_1$ thiamine *q.v.*

vitamin B$_2$ riboflavin *q.v.*

vitamin B$_3$ pantothenic acid *q.v.*

vitamin B$_4$ biotin *q.v.*

vitamin B$_5$ pantothenic acid *q.v.*

vitamin B₆ any or all of three interconvertible compounds, pyridoxine, pyridoxal and pyridoxamine, found in eggs, milk, meat, whole grains, fresh vegetables, and yeast. *see individual entries.*

vitamin B₇ niacin *q.v.*; nicotinamide or nicotinic acid *q.v.*

vitamin B₈ *see* adenosine monophosphate.

vitamin C ascorbic acid, a water-soluble vitamin found in fresh fruit and vegetables, esp. citrus fruit and blackcurrants, required in the diet of primates and some other animals, its deficiency causing scurvy. Its precise role in metabolism is not yet known but it is believed to act as a cofactor in oxidation-reduction reactions, and is involved in the synthesis of bone, cartilage and dentine. *see* Appendix 1 (11).

vitamin D any or all of several fat-soluble steroids, found esp. in fish liver oils, egg yolk and milk, or formed from precursors in skin exposed to ultraviolet light (sunlight). A deficiency in children causes rickets due to deficient and abnormal bone growth. They are necessary for normal bone and tooth structure, increasing calcium and phosphate absorption from the gut. Vitamin D_2 is calciferol, vitamin D_3 is cholecalciferol, and vitamin D_4 is dihydrotachysterol.

vitamin E any of several fat-soluble vitamins, the most common being α-tocopherol (vitamin E or E_1), β-tocopherol (vitamin E_2), and γ-tocopherol (vitamin E_3), which occur in leaves of various plants and in oils of some seed germs. Their absence leads to sterility in some animals, and they are possibly necessary for reproduction in all mammals. They have strong antioxidant properties and may be necessary for stabilizing membranes and preventing oxidation in cells.

vitamin G riboflavin *q.v.*

vitamin H biotin *q.v.*

vitamin K a group of vitamins necessary for blood clotting as they are concerned in the production of prothrombin and other co-agulation factors in the liver. They are obtained from green leaves, putrefying fish, or are synthesized by bacteria in the gut. Vitamin K_1 is α-phylloquinone or phytonadione, vitamin K_2 is β-phyllo-quinone (farnoquinone), vitamin K_3 is menadione.

vitamin P citrin (*q.v.*) with its active constituent hesperidin, which affects the permeability and fragility of blood capillaries.

vitamin P P pellagra-preventive factor *q.v.*

vitellarium yolk gland *q.v.*

vitellin *n.* abundant phosphoprotein in egg yolk; similar or related substance in seeds.

vitelline *a. pert.* yolk or yolk-producing organ; *appl.* membrane: zona pellucida *q.v.*

vitelline duct duct conveying yolk from yolk gland to oviduct.

vitelline layer thick transparent layer surrounding plasma membrane of ovum of vertebrates and invertebrates.

vitellogenesis *n.* yolk formation.

vitellogenic hormone juvenile hormone *q.v.*

vitellogenin *n.* protein produced in the liver of certain female amphibians, and which is converted into yolk protein.

vitellus *n.* yolk of ovum or egg.

vitrella *n.* cell of an ommatidium which secretes the crystalline cone.

vitreodentine *n.* a very hard variety of dentine.

vitreous *a.* hyaline; transparent.

vitreous body, vitreous humor clear jelly-like substance in inner chamber of eye.

vitreous membrane innermost layer of dermal coat of hair follicle; innermost layer of cornea.

vitreum *n.* the vitreous humor of eye.

vitronectin *n.* a glycoprotein of the extracellular matrix.

vitta *n.* one of the resinous canals in pericarp of dicots of the Umbelliferae and some other families; a longitudinal ridge on diatoms; a band of colour. *plu.* vittae.

vittate *a.* having lengthwise ridges, bands or stripes.

viverrids *n.plu.* members of the Viverridae, a family of carnivores including the genet.

viviparous *a.* producing young alive rather than laying eggs, as all mammals except monotremes, and some animals in other groups; *appl.* plants, having seeds that germinate while still attached to the parent plant, e.g. mangrove. *n.* viviparity.

vocal chords folds of mucous membrane that project into larynx and whose vibration produces sound.

vocal sac extension of mouth cavity in some amphibians that is involved in sound production.

voice box larynx *q.v.*

volant *a.* adapted for flying or gliding.

volar *a. pert.* palm of hand or sole of foot.

Volkmann's canals *see* Haversian canals.

voltage clamp apparatus for studying membrane conductance changes in response to changes in membrane potential, but which prevents the conductance changes from influencing the membrane potential by "clamping" membrane potential at a level determined by the experimenter.

voltage-gated *appl.* ion channels in cell membranes whose opening (or closing) depends on a certain threshold membrane potential being reached in the membrane. *alt.* voltage-sensitive.

voltine *a. pert.* number of broods in a year.

voltinism *n.* a polymorphism in some insect species, where some individuals enter diapause and some do not.

voluble *a.* twining spirally.

voluntary *a.* subject to or regulated by the will.

voluntary muscle striated muscle *q.v.*

volute *a.* rolled up; spirally twisted.

volutin granules polyphosphate granules found in some microorganisms.

volva *n.* cup-shaped remnant of the universal veil that remains around base of stalk in some fungi such as *Amanita. plu.* volvae.

volvate *a.* having a volva.

vomer *n.* paired bone forming floor of nasal cavity. *alt.* ploughshare bone.

vomerine *a. pert.* vomer; *appl.* teeth borne on vomers.

vomeronasal organ Jacobson's organ *q.v.*

vomeropalatine *n.* fused vomer and palatine in some fishes and amphibians.

von Baer's law *see* recapitulation theory.

von Ebner's gland gland in tongue that secretes a watery fluid.

von Willebrand factor glycoprotein involved in blood clotting which is required for the adhesion of platelets to damaged regions of blood vessels, and stabilization of Factor VIII. A significant deficiency (von Willebrand disease) causes a bleeding disorder.

V region the variable region of an immunoglobulin molecule, which differs in amino acid sequence in different antibodies and which is the part of the molecule involved in forming the antigen-binding site, each light chain and each heavy chain having one variable region of equal length at the carboxy end of the polypeptide chain. *see also* V gene, C region, C gene.

V-type ATPase type of proton-transporting ATPase, found e.g. in the membranes of plant vacuoles and lysosomes.

vulnerable *a.* IUCN definition *appl.* species or larger taxa (1) thought likely to move into the endangered category in the near future if circumstances do not change, as most or all of their populations are decreasing through e.g. loss of habitat or over-exploitation, (2) whose populations are seriously depleted and whose security is not assured, and (3) which are at present abundant but are threatened by major adverse factors throughout their range. *see also* endangered, rare, rarity.

vulva *n.* external female genitalia in mammals; opening of ovary to exterior in nematodes.

W

W tryptophan *q.v.*
WGA wheat-germ agglutinin *q.v.*
WP wall pressure, *see* turgor pressure; water potential, *see* suction pressure.
WTV wound tumour virus, a segmented, double-stranded RNA plant virus transmitted by insect vectors.

waggle dance the sequence of movements by which honeybees communicate the location and distance of food sources and new nest sites. It comprises a repeated figure of eight movement, made up of a straight run, a loop back to right (or left), then another straight run, then a loop back in the opposite direction and so on, the straight run containing information on the direction and distance away of the target.
wall pressure (WP) turgor pressure *q.v.*
Wallace's line an imaginary line separating the Australian and Oriental zoogeographical regions, between Bali and Lombok, between Celebes and Borneo, and then eastwards of the Philippines.
Wallerian degeneration degeneration of nerve fibres following injury, produced distally to the injury.
wandering cells amoeboid cells of mesogloea; leukocytes that enter tissues.
Warburg–Dickens pathway pentose phosphate pathway *q.v.*
Warburg's factor former name for cytochrome oxidase *q.v.*
warm-blooded homoiothermal *q.v.*
warning coloration bright and distinctive coloration, such as the yellow and black stripes of a wasp, that warns a potential predator that it is unpleasant-tasting or dangerous. *alt.* aposematic coloration.
wart *n.* a small dry benign growth on skin, *alt.* papilloma; glandular protuberance.

wasp *n.* insect of the superfamily Vespoidea (true wasps) of the order Hymenoptera, generally with a smooth shiny body and a well-defined waist between thorax and abdomen. Some species are social and some solitary. All feed their young on small insects, larvae, etc. Social wasps live in colonies with a queen, males and workers, in a cellular nest made from paper produced from chewed wood pulp. *see also* digger wasps.
water balance the balance between the water intake of an organism directly, in food, and as metabolic water (*q.v.*), and the water lost by excretion and evaporation.
water bears the common name for the tardigrades *q.v.*
water cycle the processes by which the Earth's supply of water is converted from one physical state to another by evaporation and condensation, and is moved between the land, oceans and atmosphere, and is passed through the bodies of plants and animals.
water fleas *see* Branchiopoda.
water gland a structure in leaf mesophyll that regulates exudation of water through pores (hydathodes) in the leaf.
water meadow grassland bordering a river which regularly floods in winter.
water pore in various invertebrates, a pore by which water tubes connect to exterior, esp. a pore in echinoderms from which the madreporite is derived; hydathode *q.v.*
water potential (WP) suction pressure *q.v.*
water slater small freshwater isopod crustacean, resembling a terrestrial woodlouse, with no carapace, and carrying young in a brood pouch under the hind part of the body.
water table upper surface of the zone of

saturation, below which all available pores in soil and rock are filled with water.

water vascular system system of vessels through which water circulates, characteristic of echinoderms, extending into the arms and opening to the exterior through the madreporite.

Watson–Crick base pairing the normal A to T (or U) and C to G pairing that occurs in double-helical DNA or RNA.

Watson Crick helix *see* DNA.

wattle *n.* fleshy excrescence under throat of cock or turkey or of some reptiles.

wax cells modified leukocytes charged with wax, as in certain insects.

wax hair a filament of wax extruded through pore of a wax gland, as in certain scale insects.

wax pocket one of the paired wax-secreting glands on abdomen of honeybee.

waxes *n.plu.* esters of fatty acids with long chain monohydric alcohols, insoluble in water and difficult to hydrolyse, found as protective waterproof coatings on leaves, stems, fruits, animal fur and integument of insects, etc., and including beeswax and lanolin.

W chromosome the X chromosome in animals in which the female is the heterogametic sex.

weathering *n.* the action of external factors such as rain, frost, snow, sun or wind on rocks, altering their texture and composition and converting them to soil.

web *n.* membrane stretching from toe to toe as in frog and swimming birds; network of threads spun by spiders, *see* orb web.

Weberian apparatus Weberian ossicles *q.v.*

Weberian ossicles chain of four small bones that connect the swimbladder to the ear in teleost fish of the superorder Ostariophysi (cyprinid fish and their relatives).

Weber's line imaginary line separating islands with a preponderant Indo-Malayan fauna from those with a preponderant Papuan fauna.

wedge bones small infravertebral ossifications at junction of two vertebrae, often present in lizards.

weed *n.* a plant growing where it is not wanted.

Wernicke's area region in temporal lobe of left cerebral cortex involved in language comprehension. *alt.* area 22.

Weismannism *n.* the concepts of evolution and heredity put forward by the German biologist A. Weismann in the late 19th century, and which deal chiefly with the continuity of germ plasm and the non-transmissibility of acquired characteristics.

Welwitschiales *n.* an order of Gnetophyta including the single living genus *Welwitschia*, having a mainly subterranean steam and two thick long leaves surviving throughout the plant's life.

West African Floral Region part of the Palaeotropical Realm comprising the southern coastal region of West Africa and Africa east to Lake Tanganyika and south to central Angola.

Western blotting the transfer by a blotting technique of proteins separated by electrophoresis from the gel to a medium on which they can be further analysed by treatment with specific antibodies.

wetland *n.* area habitually saturated with water, and which may be partly or wholly covered permanently, occasionally or periodically by fresh or salt water up to a depth of 6 metres, and which includes bogs, fens, flood meadows, marshland and salt marshes, shallow ponds, river estuaries, and intertidal mud flats, but excludes rivers, streams, lakes and oceans.

wetlands *n.plu.* areas of shallow water containing much vegetation.

whales *see* cetaceans.

Wharton's duct the duct of the submaxillary gland.

Wharton's jelly the gelatinous core of the umbilical cord.

wheat-germ agglutinin (WGA) protein from wheat-germ, a lectin which binds specifically to terminal N-acetylgalactosamine residues in carbohydrate chains.

wheel animals *n.plu.* common name for the Rotifera *q.v. alt.* wheel animalcules.

whip scorpions common name for members of the Uropygi, an order of arachnids having the last segment bearing a long jointed flagellum.

whiplash *a. appl.* flagella that are unbranched and not feathery.

whisk ferns common name for the

Psilophyta *q.v.*

white blood cell leukocyte *q.v.*

white body so-called optic gland of molluscs, a large soft body of unknown function.

white commissure a transverse band of white fibres forming floor of median ventral tissue of spinal cord.

white fat type of adipose tissue containing cells with large lipid globules of triacylglycerols, almost filling the cytoplasm. *alt.* white adipose tissue.

white fibres myelinated nerve fibres; white muscle fibres *q.v.*; unbranched inelastic fibres of connective tissue, made of collagen and occurring in wavy bundles.

white matter regions of brain and spinal cord that appear white, consisting chiefly of myelinated axons (nerve fibres).

white muscle type of muscle in fish, with chiefly anaerobic respiration, involved in bursts of fast swimming. *cf.* red muscle.

white muscle fibres skeletal muscle fibres which give the fast-twitch response and are involved in rapid intermittent movement. *cf.* red muscle fibres.

white pulp lymphoid tissue surrounding arterioles in spleen.

whorl *n.* (*bot.*) circle of flowers, parts of a flower, or leaves arising from one point; (*zool.*) the spiral turn of a univalve shell; the concentric arrangement of ridges of skin on fingers.

wilderness *n.* land that has never been permanently occupied by humans, or exploited by mining, agriculture, logging, etc. in any way.

wildlife corridor in urban and suburban areas, narrow continuous areas of favourable habitat that connect built-up areas with the country and allow the movement of animals, birds and plants along them.

wild type *n.* the organism carrying the normal (unaltered) form of a gene or genes. *cf.* mutant. *a.* wild-type: may refer to genotype or phenotype.

wilting *n.* loss of turgidity in plant cells, due to inadequate water absorption

wilting coefficient percentage of moisture in soil when wilting takes place.

wing *n.* forelimb modified for flying, in pterodactyls, birds, bats; epidermal structure modified for flying, in insects; large

lateral process on sphenoid bone; (*bot.*) one of two lateral petals in a papilionaceous flower; lateral expansion on many fruits and seeds; any broad membranous expansion.

wing cells distally rounded polyhedral cells in epithelium of cornea, proximally with extensions between heads of basal cells.

wing coverts tectrices *q.v.*

wing disc small undifferentiated sac of epithelium in dipteran larva that develops into wing at metamorphosis.

wing pad undeveloped wing of insect pupa.

wing quill remex *q.v.*

wing sheath elytron *q.v.*

winged insects common name for the Pterygota *q.v.*

winged stem stem having expansions of photosynthetic tissue, as some vetches.

wingless insects common name for the Apterygota *q.v.*

Winslow's foramen opening of bursa omentalis and large sac of peritoneum.

winter bud dormant bud, protected by hard scales during winter.

winter egg egg of many freshwater animals, provided with a thick shell which preserves it as it lies quiescent over the winter, and which hatches in spring.

wisdom teeth four back molar teeth which complete the permanent set in humans, erupting late.

witches' broom twiggy growth occurring on some trees, caused by infection with fungi or mites.

wobble hypothesis explanation for the recognition by some anticodons of a wider range of codons than strictly specified by the genetic code, by relaxation of the base-pairing rules between the 3rd base of the codon and 1st base of the anticodon.

Wolffian ducts a pair of ducts developing in the early mammalian embryo which represent a primitive kidney and which will give rise to the internal sex organs in males.

Wolffian ridges ridges which appear on either side of the middle line of early embryo, and upon which limb buds are formed.

wood *n.* secondary xylem *q.v.*; the hard, generally non-living part of a tree.

wood vessel xylem vessel *q.v.*

woodlice *see* Isopoda.

woody plant any perennial plant, e.g. tree or shrub, having secondary lignified xylem in stem.

worker *n.* a member of the labouring, non-reproductive caste of semisocial and eusocial insect species.

Wormian bones sutural bones *q.v.*

Woronin hypha a hypha inside coil of perithecial hyphae and giving rise to ascogonia.

wound cambium cambium forming protective tissue at site of injury in plants.

wound hormones substances produced by wounded plant cells, which stimulate renewed growth of tissue near wound.

wound response *see* plant defence response.

wyosine (Y) *n.* unusual nucleoside found only in tRNA, an extensively modified form of guanosine which can pair with U as well as C.

writhe(*W*) number of turns of superhelix in a supercoiled DNA molecule. *alt.* writhing number.

XYZ

φX174 single-stranded DNA phage of *E. coli.*
XMP xanthine monophosphate *q.v.*
XP xeroderma pigmentosum *q.v.*

xanth- prefix from the Gk. *xanthos,* yellow, indicating yellow colouring.
xanthein *n.* a water-soluble yellow pigment in cell sap.
xanthin *n.* a yellow carotenoid pigment in flowers.
xanthine *n.* a purine, 2,6-dioxypurine, found especially in animal tissues such as muscle, liver, pancreas, spleen, and in urine, and also in certain plants, and which is a breakdown product of AMP and guanine, and is oxidized to urate (uric acid).
xanthine monophosphate (xanthylate, xanthylic acid) (XMP) *n.* a ribonucleotide containing the purine base xanthine, a biosynthetic precursor of guanosine monophosphate (GMP).
xanthine oxidase enzyme which oxidizes hypoxanthine to xanthine and then to urate (uric acid), using molecular oxygen. EC 1.2.3.2.
xanthism *n.* colour variation in which the normal colour is replaced almost entirely by yellow.
xanthocarpous *a.* having yellow fruits.
xanthochroic *a.* having a yellow or yellowish skin.
xanthodermic *a.* having a yellowish skin.
xanthommatin *n.* brown ommatochrome pigment of eyes, ocelli, larval Malpighian tubules in certain insects.
xanthophore *n.* a yellow chromatophore.
xanthoplast *n.* a yellow plastid or chromatophore.
xanthophyll *n.* any of a group of widely distributed yellow or brown carotenoid pigments, oxygenated derivatives of carotenes, and including lutein, violaxanthin, neoxanthin, cryptoxanthin. *alt.* phylloxanthin.
Xanthophyta *n.* phylum of yellow-green photosynthetic protists, formerly considered as algae, which have two unequal flagella, the longer hairy and the shorter smooth. They contain the chlorophylls *a, c, c$_2$* and *e* and have xanthin pigments. Storage products are oils. Common in fresh water, many xanthophytes have multicellular and syncytial forms.
xanthopous *a.* having a yellow stem.
xanthopterin *n.* a yellow pigment, a pterin, found esp. in wings of yellow butterflies and the yellow bands of wasps and other insects, and also in mammalian urine, and which can be oxidized to leucopterin and converted to folic acid by microorganisms.
xanthospermous *a.* having yellow seeds.
X cell type of retinal ganglion cell that responds to stationary spots and lines of light, and, in animals with good colour vision, to particular wavelengths of light.
X chromosome the female sex chromosome in mammals, two copies of which are present in each somatic cell of females with one copy being permanently inactivated, and one (active) copy being present in males; in general, the sex chromosome present in two copies in the homogametic sex, and in one copy in the heterogametic sex.
X-chromosome inactivation the inactivation of all but one copy of the X chromosome in the somatic cells of female mammals. *alt.* X inactivation. *see also* Lyon hypothesis, dosage compensation.
xenarthral *a.* having additional articular facets on dorsolumbar vertebrae.
xenic *n. pert.* a culture containing one or

more unidentified microorganisms.

xeno- prefix derived from Gk. *xenos,* strange.

xenobiosis *n.* the condition where colonies of one species of social insect live in the nests of another species and move freely among them, obtaining food from them by various means but keeping their broods separate.

xenobiotic *a.* foreign to a living organism; *appl.* foreign substances such as drugs etc.

xenodeme *n.* a deme of parasites differing from others in host specificity.

xenoecic *a.* living in the empty shell of another organism.

xenogamy cross-fertilization *q.v.*

xenogeneic *a. appl.* immunization of one animal with antibodies of another of a different species, *alt.* heterologous; *appl.* grafting of tissue between animals of different species.

xenogenous *a.* originating outside the organism; caused by external stimuli.

xenograft *n.* a graft of tissue from one species to another.

xenomorphosis heteromorphosis *q.v.*

xenoplastic *a. appl.* graft established in a different host.

Xenopus laevis a species of frog, commonly known as the African clawed toad, widely used in research in developmental biology because of the size and robustness of its eggs and their amenability to surgical manipulation. Its oocytes are also used as "living test-tubes" to study the expression and function of isolated foreign genes and RNAs injected into them.

xerad xerophyte *q.v.*

xerantic *a.* drying up; withering, parched.

xerarch *a. appl.* seres progressing from xeric towards mesic conditions.

xeric *a.* dry, arid; tolerating, or adapted to, arid conditions.

xero- prefix derived from the Gk. *xeros,* dry.

xerochasy *n.* dehiscence of fruits when induced by drying. *a.* xerochastic.

xeroderma pigmentosum rare inheritable skin disease in humans, in which skin is abnormally sensitive to ultraviolet or sunlight, producing parchment skin, ulceration of the cornea and a predisposition to skin cancer, and which is caused by a defect in a DNA repair enzyme responsible for excising pyrimidine dimers formed in DNA on exposure to ultraviolet or sunlight.

xeromorphic *a.* structurally modified so as to withstand drought, *appl.* desert plants such as cacti, etc. *n.* xeromorphy.

xerophilous *a.* able to withstand drought, *appl.* plants adapted to a limited water supply. *n.* xerophil.

xerophyte *n.* a plant adapted to arid conditions, either having xeromorphic characteristics, or being a mesophyte growing only in a wet period.

xerosere *n.* stages in a plant succession that begins on a dry site and develops towards moister conditions.

xerothermic *a. appl.* organisms thriving in hot dry conditions.

X inactivation X-chromosome inactivation *q.v.*

xiphihumeralis *n.* a muscle extending from xiphoid cartilage to humerus.

xiphisternum *n.* the posterior portion of the sternum, usually cartilaginous.

xiphoid *a.* sword-shaped.

xiphoid process xiphisterum *q.v.*; tail or telson of the king crab, *Limulus.*

Xiphosura *n.* order of aquatic arthropods, commonly called king or horseshoe crabs, in the class Merostomata, having a heavily chitinized body with the prosoma covered with a horseshoe-shaped carapace, *Limulus* being a living example, but the group being known from the Paleozoic.

X-linked gene any gene carried on the X chromosome. *alt.* sex-linked gene.

X-organ small compact sac-like neurosecretory organ in eye-stalk of some crustaceans.

X-ray crystallography technique for determining the three-dimensional atomic structures of molecules that can be crystallized, from the diffraction patterns of X rays passed through the crystal.

xylan *n.* any of a group of polysaccharides composed of a central chain of linked xylose residues with other monosaccharides attached as single units or side chains, found in the cell walls of many angiosperms.

xylary *a. pert.* xylem, *appl.* fibres, etc; *appl.* procambium which gives rise to xylem.

xylem *n.* the main water-conducting tissue

in vascular plants which extends throughout the body of the plant and is also involved in transport of minerals, food storage and support. Primary xylem is derived from the procambium, secondary xylem (e.g. the wood of trees and shrubs) from the vascular cambium. Xylem is composed of tracheary elements: tracheids and (in angiosperms) vessel elements. Both are elongated hollow cells, with thickened, usually heavily lignified walls, and lacking protoplasts when mature. They are joined end to end to form a continuous conducting tube. *see also* vascular bundle, phloem.

xylem canal narrow tubular space replacing central xylem in demersed stem of some aquatic plants.

xylem parenchyma short lignified cells surrounding conducting elements or produced with other xylem cells towards the end of the growing season.

xylem ray ray or plate of xylem between two medullary rays; part of a ray of parenchyma found in secondary xylem.

xylocarp *n.* hard woody fruit.

xylochrome *n.* a pigment of tannin, produced before death of xylem cells and giving colour to heartwood.

xylogen *n.* the forming xylem in a vascular bundle.

xylogenesis *n.* the formation of xylem.

xyloglucan *n.* a polysaccharide found in the cell walls of most dicotyledonous plants, composed of a central chain of linked glucose residues with xylose units attached as single units or as part of a complex side chain.

xyloic xylary *q.v.*

xyloid ligneous *q.v.*

xyloma *n.* hardened mass of mycelium which gives rise to spore-bearing structures in certain fungi; a tumour of woody plants.

xylophagous *a.* wood-eating, as certain termites and beetles.

xylophilous *a.* preferring wood; growing on wood.

xylophyte *n.* a woody plant.

xylose *n.* a 5-carbon aldose sugar, a constituent of polysaccharides, esp. in the cell walls of some plants.

xylostroma *n.* the felt-like mycelium of

some wood-destroying fungi.

xylotomous *a.* able to bore or cut wood.

xylulose *n.* 5-carbon ketose sugar, as xylulose 5-phosphate involved esp. in carbon dioxide fixation in photosynthesis.

Y tyrosine *q.v.;* wyosine *q.v.;* male chromosome in mammals.

YAC yeast artificial chromosome, an artificial chromosome constructed from any DNA and telomeres and centromeres from yeast chromosomes, and which can replicate in yeast cells, used as a recombinant DNA vector for large fragments of DNA.

Y-cartilage cartilage joining ilium, ischium, and pubis in the acetabulum.

Y cell type of retinal ganglion cell that responds to changes in illumination or moving stimuli.

Y chromosome the male sex chromosome in mammals, smaller than and non-homologous with the X chromosome, being absent from cells of females and present in one copy in the somatic cells of males; in general, the sex chromosome that pairs with the X chromosome in the heterogametic sex.

yeast artificial chromosome *see* YAC.

yarovization vernalization *q.v.*

yeast mating type locus *see* mating type locus.

yeasts *see* saccharomycetes, and *see also* false yeasts.

yellow body corpus luteum *q.v.*

yellow cartilage a cartilage with matrix pervaded by yellow or elastic connective tissue fibres.

yellow cells chloragogen cells (*q.v.*) surrounding gut of annelid worms.

yellow fibres elastic connective tissue fibres composed largely of elastin.

yellow spot macula lutea *q.v.*

yellow-green algae *see* Chrysophyta.

Y-ligament iliofemoral ligament *q.v.*

Y-linked gene any gene carried on the Y chromosome.

yolk nutrient material rich in protein and fats, forming a large part of the ova of many egg-laying animals (e.g. amphibians, reptiles and birds) and which nourishes the developing embryo.

yolk duct vitelline duct *q.v.*

yolk gland gland associated with the reproductive system in most animals and which produces yolk cells filled with nutrients, which in most animals become incorporated into the egg as the yolk.

yolk plug mass of yolky cells filling up the blastopore in some gastrulating amphibian eggs.

yolk sac membranous sac rich in blood vessels which develops around the yolk in the eggs of vertebrates and which is attached to the embryo and through which nutrients pass from the yolk. In some viviparous lower vertebrates the yolk sac forms a placenta with the uterine wall. In mammals the yolk sac is empty of yolk and in the marsupials forms the main, and short-lived, placenta.

yolk sac placenta placenta formed by embryonic yolk sac and chorion, the only type of placenta in most marsupials. Viviparous selachians, lizards and snakes also develop a yolk sac placenta.

Y-organs in crabs and lobsters, a pair of glands in the antennary or maxillary segment, resembling the prothoracic glands of insects, and which secrete ecdysone.

ypsiloid *a*. U-shaped, *appl.* cartilage anterior to pubis in salamanders for attachment of muscles used in breathing.

Z either glutamine or glutamic acid in the single-letter code for amino acids.

Z-DNA *see* deoxyribonucleic acid.

ZPA zone of polarizing activity *q.v.*

zalambdodont *a*. *appl*. insectivores with very narrow molar teeth with V-shaped transverse ridges.

Z chromosome the Y chromosome when female is the heterogametic sex.

Z-disc, Z-line a dark line seen separating sarcomeres of muscle myofibrils under the microscope and which represents the membrane to which the actin filaments of each sarcomere are anchored.

zeatin *n*. a natural cytokinin isolated from maize (*Zea mays*), a derivative of adenine.

zeaxanthin *n*. yellow carotenoid pigment found in many plant cells including maize, in some classes of algae and in egg yolk and which is an isomer of lutein.

zebrafish *Danio rerio*, a small subtropical freshwater fish widely used for studies in developmental biology.

zein *n*. a simple protein in seeds of maize, lacks tryptophan and lysine.

zeitgeber *n*. a synchronizing agent, as environmental cues responsible for keeping circadian rhythms of plants in tune with the daily 24-hour light–dark cycle.

zeugopodium *n*. forearm; shank.

zidovudine AZT *q.v.*

zinc (Zn) an essential micronutrient.

zinc finger a structural feature shared by various proteins that bind to DNA and act as transcriptional regulators, and which is believed to be involved in DNA binding. Each zinc finger is a hairpin fold of amino acid chain held together by a Zn atom.

Zingiberales *n*. an order of tropical monocot herbs with rhizomes, and including Cannaceae (canna), Musaceae (banana), Strelitziaceae (bird of paradise flower), Zingiberaceae (ginger) and others.

Zinjanthropus *see* australopithecines.

Zoantharia *n*. subclass of Anthozoa, including the stony corals and sea anemones, which are solitary or colonial with paired mesenteries usually in multiples of six, and having the skeleton, if present, external and not made of spicules.

zoarium *n*. all the individuals of a polyzoan colony.

zoëa *n*. early larval form of certain decapod crustaceans.

zoecial, zoecium zooecial, zooecium *q.v.*

zoic *a*. containing remains of organisms and their products; *pert*. animals or animal life.

zoid zoospore *q.v.*

zona *n*. a zone, band or area. *a*. zonal.

zona fasciculata radially arranged columnar cells in adrenal cortex below zona glomerulosa, secreting glucocorticoids.

zona glomerulosa groups of cells forming external layer of adrenal cortex beneath capsule, secreting glucocorticoids and mineralocorticoids.

zona granulosa granular zone in a Graafian follicle, the mass of cells in which the ovum is embedded.

zona orbicularis circular fibres of capsule of hip joint, around the neck of femur.

zona pellucida thick transparent layer surrounding plasma membrane of mammalian ovum.

zona reticularis or reticulata inner layer of adrenal cortex.

zona striata zona pellucida *q.v.*

zonal *a.* of or *pert.* a zone.

zonary *a. appl.* placenta with villi arranged in a band or girdle.

zonate *a.* zoned or marked with rings; arranged in a single row, as some tetraspores.

zonation *n.* arrangement or distribution in zones.

zone *n.* an area characterized by similar fauna or flora; a belt or area to which certain species are limited; stratum or set of beds characterized by typical fossil or set of fossils; an area or region of the body. *a.* zonal.

zone fossil index fossil *q.v.*

zone of polarizing activity (ZPA) small area of tissue in vertebrate limb bud which directs development along the dorso-ventral axis (thumb-little finger axis).

zonite *n.* a body segment of Diplopoda.

zonoid *a.* like a zone.

zonolimnetic *a.* of or *pert.* a certain zone in depth.

zonula zonule *q.v.*

zonula adh(a)erens belt desmosome, intermediate junction. *see* desmosome.

zonula ciliaris the hyaloid membrane forming suspensory ligament of lens of eye.

zonula occludens tight junction *q.v.*

zonule *n.* a small zone, belt, or girdle.

zoo-, -zooid, -zoite word elements derived from Gk. *zōos*, animal.

zooanthellae *n. plu.* cryptomonads symbiotic with certain marine protozoans.

zoobenthos *n.* the fauna of the sea bottom, or of the bottom of inland waters.

zoo blot comparison of DNAs from different animals, by digestion of DNA with the same restriction enzymes and electrophoresis of the resulting DNA fragments.

zoobiotic *a.* parasitic on, or living on an animal.

zoochlorellae *n.plu.* symbiotic green algae living in the cells of various animals.

zoochoric *a. appl.* plants dispersed by animals. *n.* zoochory.

zoocoenocyte *n.* a coenocyte bearing cilia, in certain algae.

zoocoenosis *n.* an animal community.

zoocyst sporocyst *q.v.*

zooecial *a. pert.* or resembling a zooecium.

zooecium *n.* chamber or sac enclosing zooid in bryozoan colony.

zooerythrin *n.* red pigment found in plumage of various birds.

zoofulvin *n.* yellow pigment found in plumage of various birds.

zoogamete *n.* a motile gamete.

zoogamous *a.* having motile gametes.

zoogamy *n.* sexual reproduction in animals.

zoogenesis *n.* the origin of animals, *appl.* usually to phylogenetic origin.

zoogenetics *n.* animal genetics.

zoogenic *a.* arising from the activity of animals.

zoogenous *a.* produced or caused by animals.

zoogeographical regions large areas of the world with distinct natural faunas. At the highest level are three large regions or zoogeographic realms: Arctogaea, Neogaea and Notogaea. Arctogaea comprises the Palaeartic (Europe, North Africa, Asia south to the Himalayas), the Nearctic (Greenland, North America south to central Mexico), the Ethiopian (Africa south of the Sahara), and Oriental (India, Indochina, Malaysia, the Philippines and Indonesian islands west of Wallace's Line) Regions. The Palaearctic and Nearctic Regions are sometimes considered together as the Holarctic Region. Neogaea or the Neotropical Region comprises South America, Central America and southern Mexico, and the West Indies. Notogaea comprises Australia, New Zealand, most of the Pacific islands and the Indonesian islands east of Wallace's Line.

zoogeography *n.* the geographical distribution of animal species.

zoogloea *n.* a mass of bacteria embedded in a mucilaginous matrix and frequently forming an iridescent film on surface of water.

zooid *n.* an individual in a colonial animal, such as an individual polyp in a coral.

zoology *n.* the science dealing with the structure, functions, behaviour, history, classification, and distribution of animals.

Zoomastigina *n.* a heterogeneous phylum of non-photosynthetic, mainly unicellular, protists bearing from one to many thousands of flagella, commonly known as the animal flagellates, zooflagellates or zoomastigotes. Some (e.g. *Naegleria*) can change from a flagellated to an amoeboid form. They include intestinal parasites of fish and amphibians (e.g. the opalinids), the kinetoplastids, typified by the genus *Trypanosoma*, which causes sleeping sickness in humans, wood-digesting protozoa that live in the gut of termites (e.g. *Staurojenia*), the choanoflagellates, which resemble the body cells of sponges, and several other groups. They correspond to the Zoomastigophora or Protomonadina in older classifications.

Zoomastigophora *n.* in older classifications, a class of protozoans including the non-photosynthetic flagellates. *see* Appendices 4 and 5.

zoomorphic *a.* having the form of an animal.

zoonomy *n.* physiology, esp. animal physiology.

zoonosis *n.* a disease of animals that can be transmitted to man. *plu.* zoonoses.

Zoopagales *n.* an order (in some classifications considered as a class) of zygomycete fungi parasitic on small soil animals. They are known as the predaceous fungi or animal traps, and capture and parasitize soil amoebae, rhizopods and nematodes by attaching to them and growing within or on them.

zooparasite *n.* any parasitic animal.

zoophilous *a. appl.* plants adapted for pollination by animals other than insects.

zoophobic *a. appl.* plants shunned by animals because they are protected by spines, hairs, etc.

zoophyte *n.* an animal resembling a plant in appearance, as some colonial hydrozoans.

zooplankton *n.* animal plankton.

zoosis *n.* any disease caused by animals.

zoosphere *n.* biciliate zoospore of algae.

zoosporangiophore *n.* structure bearing zoosporangia.

zoosporangium *n.* a sporangium in which zoospores are produced.

zoospore *n.* motile, flagellated asexual reproductive cell in protozoans, algae and fungi.

zoothecium *n.* in certain ciliates, the common gelatinous and often branched matrix.

zootic climax any stable climax community dependent for its maintenance on animal activity such as grazing.

zootoxin *n.* any toxin produced by animals.

zootrophic heterotrophic *q.v.*

zootype *n.* representative type of animal.

zooxanthellae *n.plu.* parasitic or symbiotic yellow or brown algae living in various marine invertebrates.

zooxanthin *n.* yellow pigment found in the plumage of certain birds.

Zuckerkandl's bodies aortic bodies *q.v.*

zwitterion *n.* an ion with both positive and negative charges, as all amino acids.

zygantrum *n.* fossa on posterior surfaces of neural arch of vertebrae of snakes and certain lizards.

zygapophysis *n.* one of the processes of a vertebra by which it articulates with adjacent vertebrae.

zygobranchiate *a.* having gills symmetrically placed and renal organs paired, *appl.* certain gastropods.

zygodactyl(ous) *a.* having two toes pointing forward, two backwards, as in parrots.

zygodont *a.* having molar teeth in which the four cusps are united in pairs.

zygogenetic, zygogenic *a.* produced by fertilization.

zygoid *a.* diploid, *appl.* parthenogenesis.

zygoma *n.* the bony arch of the cheek.

zygomatic *a. pert.* or in the area of the cheekbone.

zygomatic arch the bony arch of the cheek, formed by the jugal and squamosal bones, that runs along the lower edge of the temporal fenestra from the end of the jaw to the orbit of the eye.

zygomatic gland the infraorbital salivary gland.

zygomaticofacial *a. appl.* foramen on malar surface of zygomatic bone for passage of nerve and vessels; *appl.* branch of zygomatic or temporomalar nerve.

zygomaticotemporal *a. appl.* suture, foramen, nerve, etc., at temporal surface of zygomatic bone.

zygomaticus *n.* muscle from zygomatic bone to angle of mouth.

zygomelous *a.* having paired appendages, *appl.* fins.

zygomorphic *a. appl.* flowers, etc. that are bilaterally, rather than radially, symmetrical.

Zygomycota, Zygomycotina, zygo-

mycetes *n.*, *n.*, *n.plu.* diverse class, or in some classifications a division (Zygomycota), of terrestrial fungi including the bread moulds, fly fungi and animal traps (predacious fungi), characterized by sexual reproduction by fusion of gametangia, the production of a resting sexual spore (zygospore) and asexual reproduction by non-motile spores.

zygoneury *n.* in certain gastropods, having a connective between pleural ganglion and ganglion on visceral branch of opposite side.

zygophore *n.* hyphal branch bearing zygospores.

zygophyte *n.* plant with two similar reproductive cells which unite in fertilization.

zygopleural *a.* bilaterally symmetrical.

zygopodium *n.* forearm, shank.

zygopterans *n.plu.* damsel flies, members of the suborder Zygoptera, of the order Odonata.

zygosis *n.* conjugation; union of gametes.

zygosphene *n.* an articular process on anterior surface of neural arch of vertebrae of snakes and certain lizards, which fits into zygantrum.

zygosphere *n.* gamete which unites with a similar one to form a zygospore.

zygospore *n.* thick-walled sexual spore resulting from fusion of gametangia in zygomycete fungi.

zygotaxis *n.* mutual attraction between male and female gametes.

zygote *n.* cell formed from the union of two gametes or reproductive cells. *alt.* fertilized ovum or egg.

zygotene *n.* stage of prophase of meiosis at which homologous chromosomes pair.

zygotic *a. pert.* a zygote; *appl.* a mutation occurring immediately after fertilization; *appl.* number: the diploid or somatic number of chromosomes; *appl.* activity of an embryo's own genes, as opposed to the activity of gene products laid down in the egg by the mother.

zymogen *n.* functionally inactive precursor of certain enzymes, the active form being produced by specific cleavage of the polypeptide chain. *alt.* proenzyme.

zymogen granules in pancreatic cells secreting digestive enzymes, small dense vesicles containing enzyme precursors (zymogens) for secretion from the cell.

zymogenic *a. pert.* or causing fermentation.

zymogenous *a. appl.* microflora in soil normally present in the resting state and only becoming active when a fresh supply of organic material is added.

Appendix 1

SELECTED STRUCTURAL FORMULAE

A. Carbohydrates and related compounds

(1) D-glucose

        (a) open chain   (b) pyranose ring form

       (2) D-galactose   (3) D-fructose

Chemical structures

(4) D-ribose

CH₂OH
C=O
HCOH
HCOH
CH₂OH

(5) D-deoxyribose

(6) D-ribulose

(7) D-ribulose 1,5-bisphosphate

CH₂O (P)
C=O
CHOH
CHOH
CH₂O (P)

(8) inositol 1,4,5-trisphosphate

(9) glucosamine

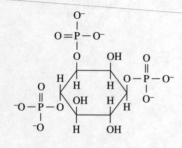

(10) *N*-acetylgalactosamine

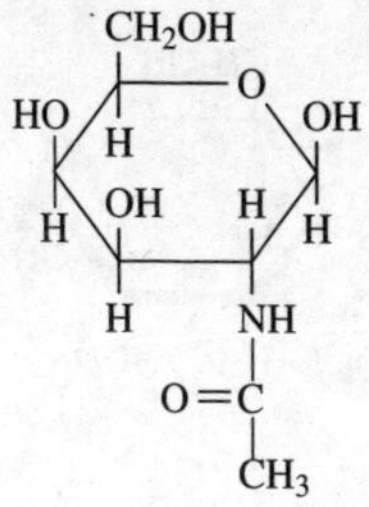

(11) ascorbic acid (vitamin C)

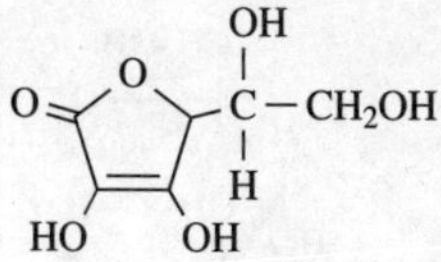

(12) sucrose

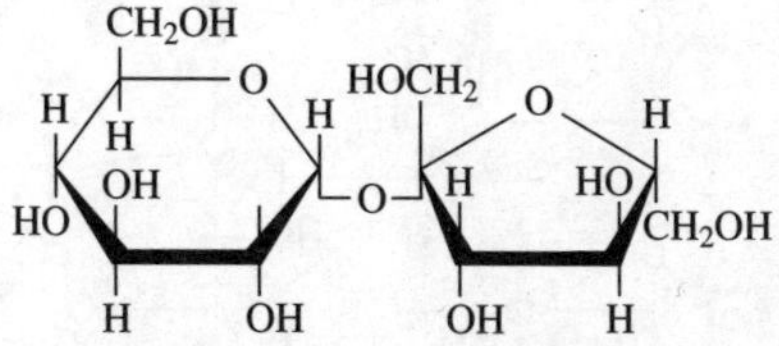

Glucose   Fructose

(13) raffinose

Galactose   Glucose   Fructose

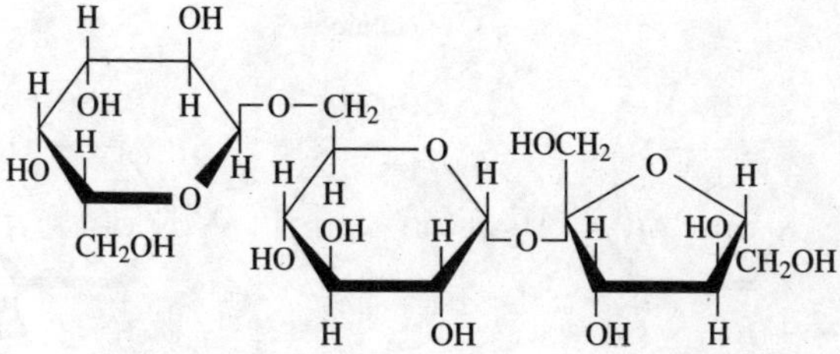

Chemical structures

(14) starch (amylose)

(15) glycogen (branch point)

(16) cellulose

B. Bases, nucleotides and their derivatives

(17) cytosine

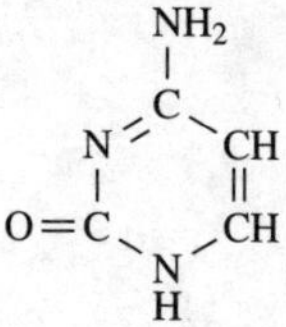

$$
\begin{array}{c}
NH_2 \\
|\\
C \\
N \quad\quad CH \\
\| \\
O=C \quad\quad CH \\
N \\
H
\end{array}
$$

(18) uracil

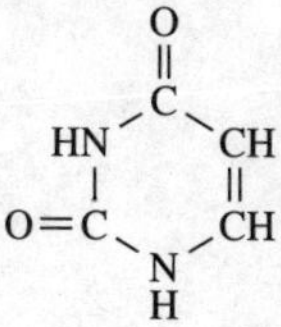

$$
\begin{array}{c}
O \\
\|\\
C \\
HN \quad\quad CH \\
\| \\
O=C \quad\quad CH \\
N \\
H
\end{array}
$$

(19) thymine

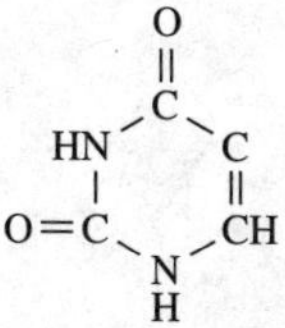

$$
\begin{array}{c}
O \\
\|\\
C \\
HN \quad\quad C \\
\| \\
O=C \quad\quad CH \\
N \\
H
\end{array}
$$

(20) guanine

$$
\begin{array}{c}
O \\
\|\\
C \\
HN \quad C \quad N \\
\| \quad\quad CH \\
H_2N \quad C \quad C \quad N \\
N \quad\quad H
\end{array}
$$

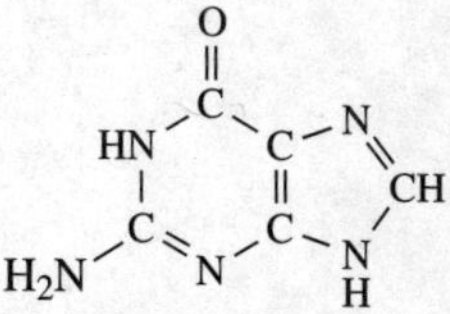

(21) adenine

$$
\begin{array}{c}
NH_2 \\
|\\
C \\
N \quad C \quad N \\
\quad\quad\quad CH \\
HC \quad C \quad N \\
N \quad\quad H
\end{array}
$$

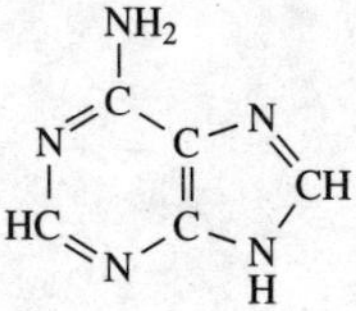

Chemical structures

(22) 5'-CMP (cytidine monophosphate)

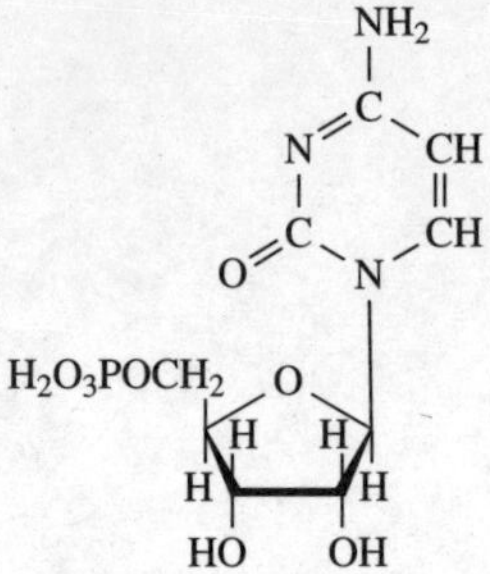

(23) ATP (adenosine triphosphate)

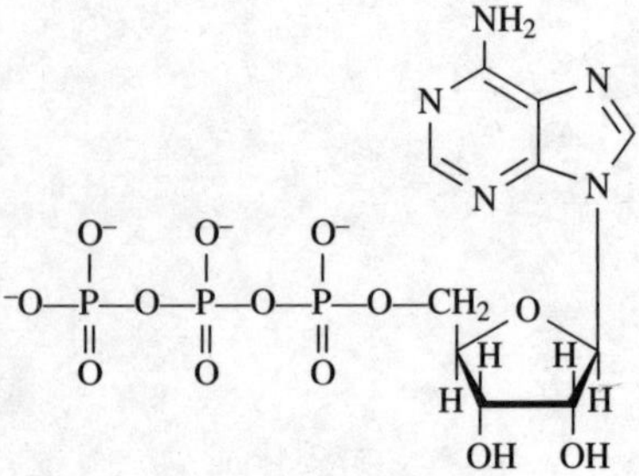

(24) cyclic AMP

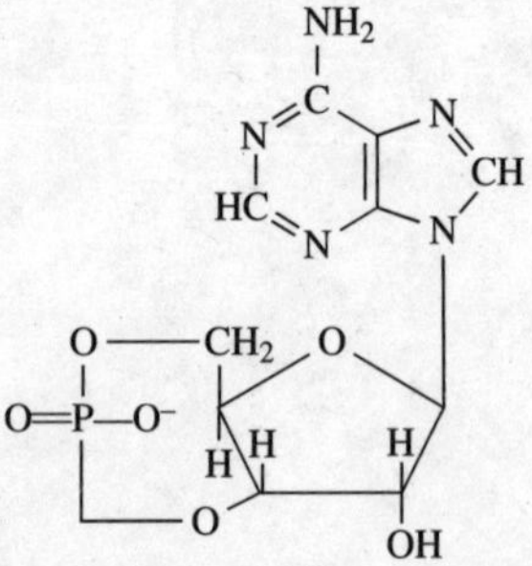

(25) a polynucleotide

Chemical structures

(26) NAD (nicotinamide adenine dinucleotide)/
NADP (nicotinamide adenine dinucleotide phosphate)

(27) FMN (flavin mononucleotide)

(28) coenzyme A

C. Amino acids and compounds derived from them

(29) general formula of amino acid

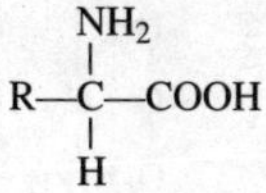

$$\underset{\underset{H}{|}}{\overset{\overset{NH_2}{|}}{R-C-COOH}}$$

(30) a dipeptide

$$R_1-\underset{\underset{NH_2}{|}}{\overset{\overset{H}{|}}{C}}-\underset{\underset{O}{\|}}{C}-N-\underset{\underset{R_2}{|}}{\overset{\overset{H}{|}}{C}}-COOH$$

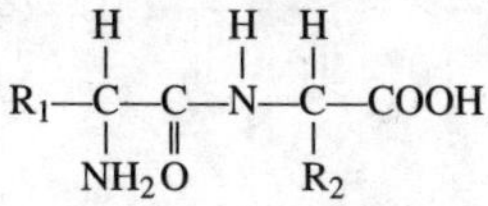

See table at end of Appendix 1 for individual amino acids

(31) dopamine

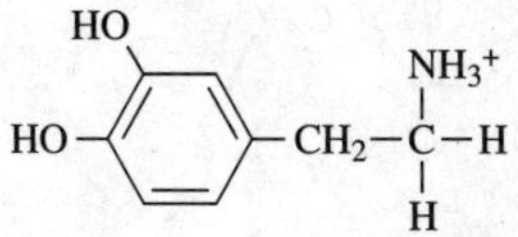

$$\text{HO} \\ \text{HO} \end{array} \right\rangle - CH_2 - \underset{\underset{H}{|}}{\overset{\overset{NH_3{}^+}{|}}{C}} - H$$

Chemical structures

(32) adrenaline/noradrenaline
(epinephrine/norepinephrine)

(33) 5-hydroxytryptamine (serotonin)

(34) thyroxine
(tetraiodothyronine)

D. Lipids

(35) glycerol

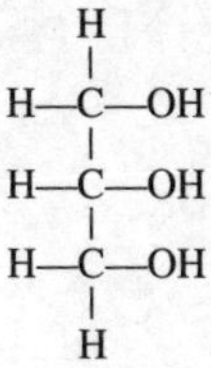

$$
\begin{array}{c}
H \\
| \\
H-C-OH \\
| \\
H-C-OH \\
| \\
H-C-OH \\
| \\
H
\end{array}
$$

(36) triacylglycerol/
diacylglycerol

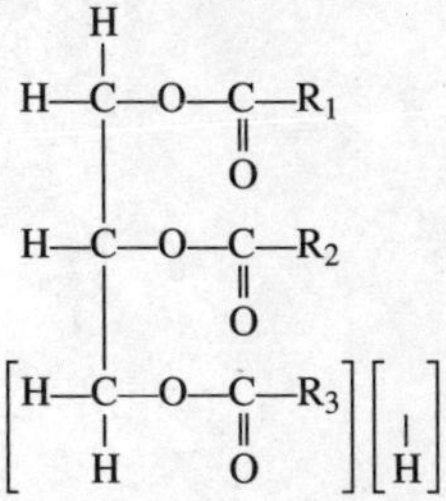

$$
\begin{array}{c}
H \\
| \\
H-C-O-C-R_1 \\
\qquad \quad \| \\
\qquad \quad O \\
| \\
H-C-O-C-R_2 \\
\qquad \quad \| \\
\qquad \quad O \\
\left[H-C-O-C-R_3 \right] \left[\; | \; \right] \\
\quad \; | \quad \; \| \qquad \quad H \\
\quad \; H \quad \; O
\end{array}
$$

(37) oleic acid

$$CH_3(CH_2)_7CH=CH(CH_2)_7COOH$$

(38) stearic acid

$$CH_3(CH_2)_{16}COOH$$

(39) linolenic acid

$$CH_3CH_2CH=CH\ CH_2CH=CH\ CH_2CH\ (CH_2)_7\ COOH$$

(40) phosphatidylcholine

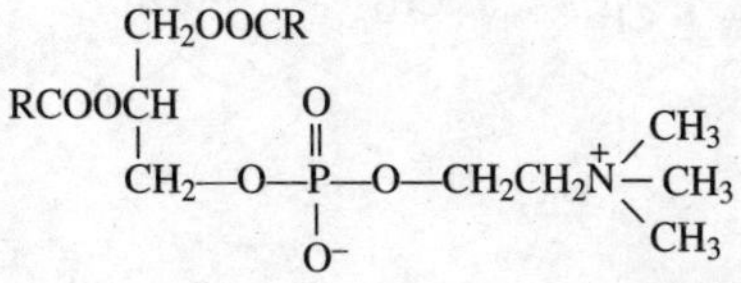

$$
\begin{array}{c}
CH_2OOCR \\
| \\
RCOOCH \qquad O \\
\qquad \quad | \qquad \quad \| \qquad \qquad \qquad CH_3 \\
\qquad \quad CH_2-O-P-O-CH_2CH_2\overset{+}{N}-CH_3 \\
\qquad \qquad \qquad \; | \qquad \qquad \qquad \qquad CH_3 \\
\qquad \qquad \qquad \; O^-
\end{array}
$$

Chemical structures

(41) phosphatidylinositol phosphate

(42) a ganglioside (GM$_2$)

(R, = fatty acid, Glc = glucose,
Gal = galactose, GalNAc = N-acetylgalactosamine,
NAN = N-acetylneuraminic acid)

(43) retinol (vitamin A)/ retinoic acid

(44) gibberellic acid (GA₃)

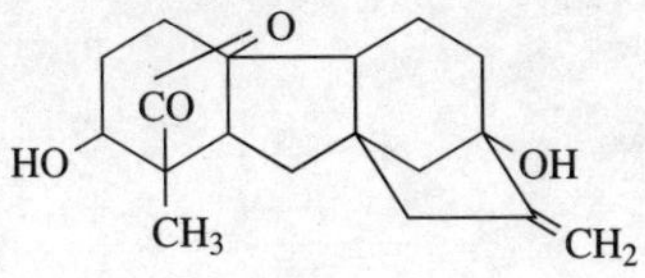

(45) cholesterol

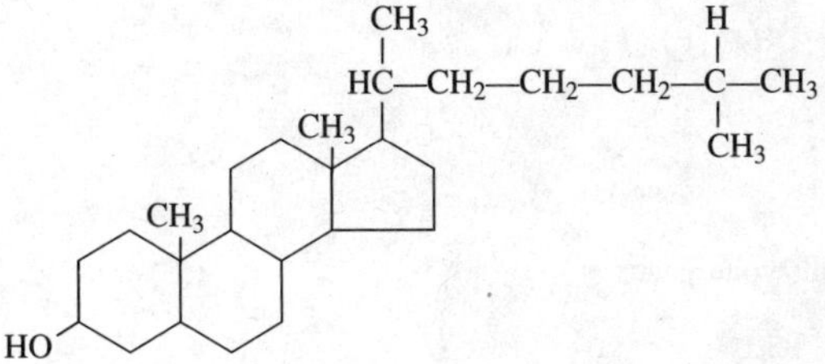

(46) corticosterone

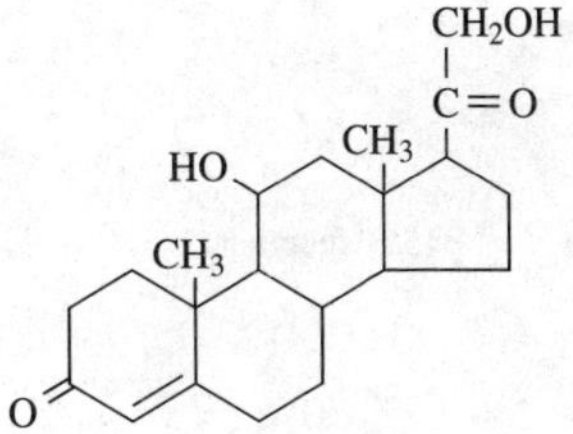

(47) oestradiol

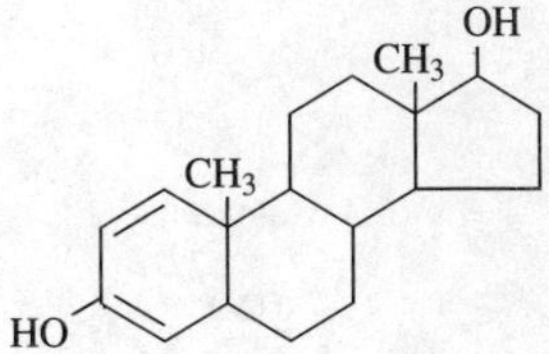

Chemical structures

E. Metabolites and other small molecules

(48) pyruvate

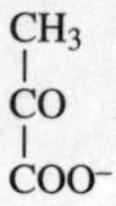

$$CH_3$$
$$|$$
$$CO$$
$$|$$
$$COO^-$$

(49) glyceraldehyde-3-phosphate

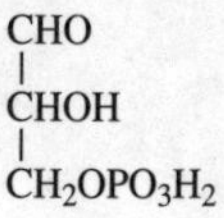

$$CHO$$
$$|$$
$$CHOH$$
$$|$$
$$CH_2OPO_3H_2$$

(50) 3-phosphoglycerate

$$COO^-$$
$$|$$
$$HOCH$$
$$|$$
$$CH_2O \, \textcircled{P}$$

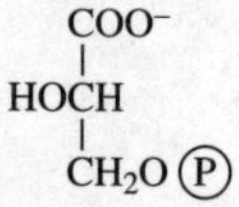

(51) phosphoenolpyruvate

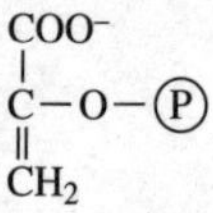

$$COO^-$$
$$|$$
$$C-O-\textcircled{P}$$
$$||$$
$$CH_2$$

(52) creatine phosphate

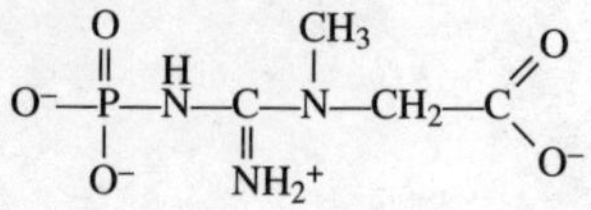

$$
\begin{array}{ccccc}
O & & CH_3 & & O \\
|| & H & | & & \diagup \\
O^--P-N-C-N-CH_2-C & & & \\
| & & || & & \diagdown \\
O^- & & NH_2^+ & & O^-
\end{array}
$$

(53) citrate

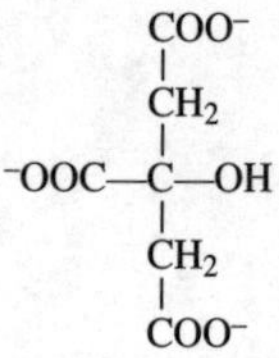

$$COO^-$$
$$|$$
$$CH_2$$
$$|$$
$$^-OOC-C-OH$$
$$|$$
$$CH_2$$
$$|$$
$$COO^-$$

(54) oxaloacetate

$$COO^-$$
$$|$$
$$C=O$$
$$|$$
$$CH_2$$
$$|$$
$$COO^-$$

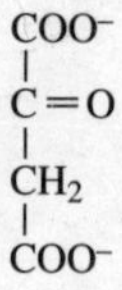

(55) succinate

$$COO^-$$
$$|$$
$$CH_2$$
$$|$$
$$CH_2$$
$$|$$
$$COO^-$$

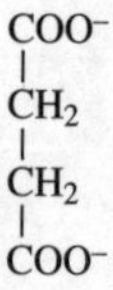

(56) malate

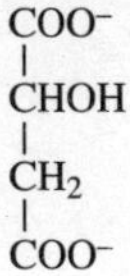

```
COO⁻
|
CHOH
|
CH₂
|
COO⁻
```

(57) acetylcholine

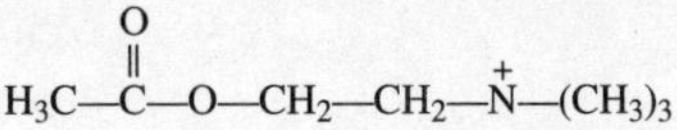

(58) thiamine pyrophosphate

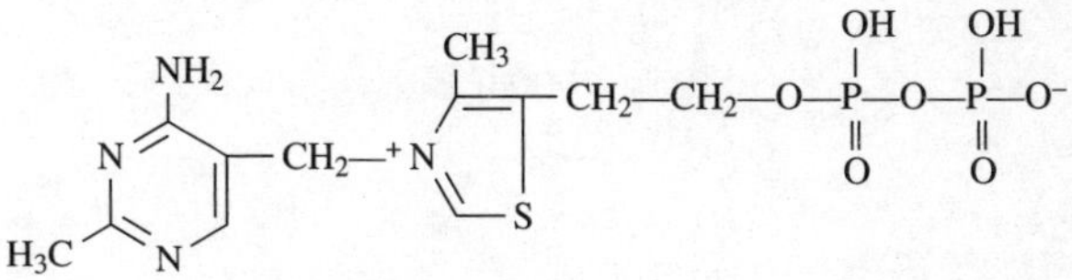

F. Others

(59) haem (as ferroprotoporphyrin IX)

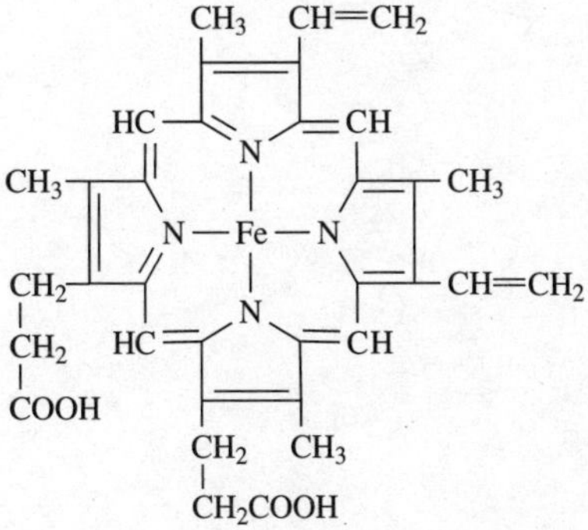

Chemical structures

(60) chlorophyll A

(61) β-carotene

The amino acids commonly found in proteins

Amino acid	Abbreviations		R group (side chain)
Glycine	Gly	G	H
Alanine	Ala	A	CH_3
Valine	Val	V	$CH(CH_3)_2$
Leucine	Leu	L	$CH_2CH(CH3)_2$
Isoleucine	Ile	I	$CH(CH_3)CH_2CH_3$
Serine	Ser	S	CH_2OH
Threonine	Thr	T	$CH(OH)CH_3$
Lysine	Lys	K	$(CH_2)_4NH_2$
Arginine	Arg	R	$(CH_2)_3NHCNHNH_2$
Histidine	His	H	
Aspartic acid	Asp	D	CH_2COOH
Asparagine	Asn	N	CH_2CONH_2
Glutamic acid	Glu	E	$(CH_2)CONH_2$
Proline	Pro	P	
Tryptophan	Trp	W	
Phenylalanine	Phe	F	
Tyrosine	Tyr	Y	
Methionine	Met	M	$(CH_2)_2SCH_3$
Cysteine	Cys	C	CH_2SH

Two cysteine side groups can form a covalent disulphide bond S=S. Disulphide bonds between cysteines are widespread in proteins and are important in determining three-dimensional conformation and in holding together multisubunit proteins.

Appendix 2

AN OUTLINE OF THE PLANT KINGDOM

The following outline is intended as a brief summary of the different types of plants that exist and does not represent a rigorous taxonomic classification. The names of divisions, classes and orders, and the rank assigned to the various groups, differ considerably from authority to authority, and no attempt has been made to give all alternatives. The nomenclature of the main groups largely follows that used in P.H. Raven, R.F. Evert and H. Curtis, *Biology of Plants*, 5th edn (Worth, New York, 1992) and L. Margulis & K.V. Schwartz, *Five Kingdoms*, 2nd edn (Freeman, 1988). The nomenclature and arrangement of the flowering plants follows that proposed by Takhatajan and Cronquist rather than the older systems of Engler and Bentham and Hooker (see S. Holmes, *An Outline of Plant Classification*, Longman, Harlow, 1983). Fungi and lichens have in the past been classified within the plant kingdom, but the quite different nature and evolutionary origins of the fungi are now recognized by placing them in a separate kingdom, and they are considered as such in Appendix 3. The lichens are also included there. The blue-green algae (cyanobacteria) are also sometimes considered as honorary plants for historical reasons, but are prokaryotes and are included in the Kingdom Prokaryotae (Appendix 6). The algae, both unicellular and multicellular, are considered as members of the Protista (Appendix 5), but those divisions traditionally included in the plant kingdom are also listed here.

All divisions and classes named here, and some orders and common names, have an entry in the body of the dictionary.

Algae

DIVISION CHRYSOPHYTA (diatoms and golden algae)
DIVISION PYRROPHYTA (dinoflagellates)
DIVISION EUGLENOPHYTA (euglenoids)
DIVISION RHODOPHYTA (red algae)
DIVISION PHAEOPHYTA (brown algae)
DIVISION CHLOROPHYTA (green algae)

Bryophytes

DIVISION HEPATOPHYTA (liverworts)
DIVISION ANTHOCEROPHYTA (hornworts)
DIVISION BRYOPHYTA
      class Sphagnidae (sphagnum or peat mosses)
      class Andreaeidae (granite or rock mosses)
      class Bryidae (true mosses)

Early vascular plants, now extinct

DIVISION RHYNIOPHYTA
DIVISION ZOSTEROPHYLLOPHYTA
DIVISION TRIMEROPHYTA (possibly the progenitor of the ferns, horsetails and progymnophytes)

Pteridophytes: seedless vascular plants

DIVISION PSILOPHYTA (PSILOTOPHYTA) (whisk ferns, only two living
  genera, *Psilotum* and *Tmesipteris*)
DIVISION LYCOPHYTA (lycophytes)
            order Lycopodiales (lycopods or club mosses)
                  Lepidodendrales (tree lycophytes, extinct)
                  Selaginellales (one living genus, *Selaginella*)
                  Isoetales (quillworts)
                  and other orders (extinct)
DIVISION SPHENOPHYTA (horsetails)
            order Equisetales (one living genus, *Equisetum*)
                  and other orders (extinct)
DIVISION PTEROPHYTA (FILICOPHYTA) (ferns)
            order Marattiales (giant ferns)
                  Ophioglossales (e.g. *Ophioglossum*, adder's tongue)
                  Osmundales (e.g. *Osmunda*)
                  Filicales (most living ferns, e.g. maidenhair fern,
                        filmy fern, hart's tongue, bracken)
                  Marsileales (water ferns, e.g. *Pilularia*, pillwort)
                  Salviniales (water ferns, e.g. *Azolla*)
DIVISION PROGYMNOPHYTA (progymnosperms, extinct, e.g. *Archaepteris*)

Spermatophytes: seed plants

Gymnosperms

DIVISION PTERIDOSPERMOPHYTA (seed ferns, extinct)
DIVISION CYCADEOIDOPHYTA (BENNETTITALES) (cycadeoids, extinct)
DIVISION CYCADOPHYTA (cycads)
DIVISION GINKGOPHYTA (one species, *Ginkgo*)
DIVISION CONIFEROPHYTA (conifers and their allies)
            order Coniferales (most living conifers, e.g., firs, monkey puzzle,
                        cedars, cypresses, junipers, pines, redwoods)
                Cordaitales (extinct)
                Voltziales (extinct)
                Taxales (yews, *Torreya*)
DIVISION GNETOPHYTA
            order Welwitschiales (one species, *Welwitschia*)
                Ephedrales (one genus, *Ephedra*)
                Gnetales (one genus, *Gnetum*)

Angiosperms

DIVISION ANTHOPHYTA (MAGNOLIOPHYTA) (the common names in
  brackets after orders refer to families)
      class Dicotyledones (Magnoliopsida)
        subclass Magnoliidae
            order Magnoliales (e.g. magnolia, nutmeg)
                Laurales (e.g. laurel, calycanthus)

                Piperales (e.g. pepper (the spice))
                Aristolochiales (birthwort)
                Rafflesiales (rafflesia)
                Nymphaeales (e.g. water lily)
    subclass Ranunculida
        order Illiciales (e.g. star anise)
                Nelumbonales (Indian lotus)
                Ranunculales (e.g. buttercup, barberry)
                Papaverales (poppy, fumitory)
                Sarraceniales (pitcher plant (fam. Sarraceniaceae))
    subclass Hamamelididae
        order Trochodendrales
                Hamamelidales (e.g. witch hazel, plane)
                Eucommiales
                Urticales (e.g. elm, nettle, mulberry)
                Casuarinales (she oak)
                Fagales (beech (incl. oaks), birch, hazel)
                Myricales (sweet gale)
                Leitneriales
                Juglandales (e.g. walnut)
    subclass Caryophyllidae
        order Caryophyllales (e.g. pink, amaranth, goosefoot)
                Cactaceae (cacti)
                Polygonales (polygonum)
                and other orders
    subclass Dilleniidae
        order Paeoniales (peonies)
                Theales (e.g. tea, St John's Wort)
                Violales (e.g. violet, rock rose)
                Passiflores (passion flower, pawpaw)
                Cucurbitales (cucurbits: e.g. marrow, squash, gourd)
                Datiscales (e.g. begonia)
                Capparales (e.g. crucifers (brassicas), caper, mignonette)
                Tamaricales (e.g. tamarisk)
                Salicales (willow)
                Ericales (e.g. heather, wintergreen)
                Primulales (e.g. primrose)
                Malvales (e.g. mallow, cocoa, lime-tree)
                Euphorbiales (e.g. spurge, box-tree, jojoba)
                Thymelaeales (daphne (mezereon))
    subclass Rosidae
        order Saxifragales (e.g. gooseberry, saxifrage, hydrangea)
                Rosales (e.g. rose (rose, apple, hawthorn, etc.), coco plum)
                Fabales (Leguminosae) (e.g. peas, beans, etc., mimosa)
                Nepenthales (sundews, pitcher plant (fam. Nepenthaceae))
                Myrtiflorae (e.g. myrtle, mangrove, pomegranate, evening
                                primrose, loosestrife)
                Hippuridales (mare's tail, gunnera)

Rutales (e.g. rue, citrus fruits, mahogany)
Sapindales (Acerales) (e.g. maple, horse chestnut)
Geraniales (e.g. balsam, geranium, nasturtium)
Cornales (e.g. dogwood, umbellifers, *Davidia*, ginseng)
Celastrales (e.g. holly)
Rhamnales (e.g. grape, buckthorn)
Oleales (Ligustrales) (olive, privet)
Elaeagnales (oleaster)
and other orders
subclass Asteridae
order Dipsacales (e.g. honeysuckles, valerian)
Gentianales (e.g. gentian, bog bean)
Polemoniales (e.g. borage, convolvulus, dodder, phlox)
Scrophulariales (personatae) (e.g. acanthus, buddleia,
African violet, nightshade)
Lamiales (e.g. verbena, mint)
Campanulales (e.g. bellflower, lobelia)
Asterales (composites (e.g. daisy, cornflower, thistle,
dandelion))

class Monocotyledones (Liliopsida, Liliatae)
subclass Alismidae
order Alismales (e.g. flowering rush)
Potamogetonales (e.g. pondweed, eel-grass)
subclass Liliidae
order Liliales (e.g. agave, lily, daffodil, onion, yam)
Iridales (e.g. iris)
Zingiberales (e.g. canna, banana, ginger)
Orchidales (orchid)
subclass Commelinidae
order Juncales (e.g. rush)
Cyperales (sedge)
Bromeliales (pineapple)
Commelinales (e.g. tradescantia, xyris)
Restionales
Poales (grasses)
subclass Arecidae
order Arecales (palms)
Arales (e.g. arum)
Pandanales (screw pine)
Typhales (bur reed, cat-tail)

Appendix 3

AN OUTLINE OF THE FUNGI (KINGDOM MYCETAE)

This outline follows that given in L. Margulis & K.V. Schwartz, *Five Kingdoms*, 2nd edn (Freeman, 1988) and C.J. Alexopolous and C.W. Mims, *Introductory Mycology*, 3rd edn (Wiley, New York, 1979) but includes only those groups regarded as the 'true fungi' or Eumycota. The slime moulds, chytrids, hypochytrids, oomycetes and hyphomycetes are now considered as protists and are included in Appendix 5.

True fungi (Eumycota)

DIVISION ZYGOMYCOTA
      class Mucorales (e.g. *Mucor, Rhizopus, Pilobolus*)
      class Entomophthorales (fly fungi)
      class Zoopagales (predaceous fungi or animal traps)
DIVISION ASCOMYCOTA
      class Hemiascomycetae (yeasts (e.g. *Saccharomyces cerevisiae*), leaf-curl
                        fungi)
      class Euascomycetae (black moulds, blue moulds, pyrenomycetes (flask
                        fungi, e.g. powdery mildews), discomycetes (cup
                        fungi, e.g. *Monilinia* brown rot of peach, *Rhytisma*
                        acerinum tar spot of maples), morels, truffles)
      class Laboulbeniomycetae (parasitic on insects)
      class Loculoascomycetae (e.g. *Elsinoe* spp. (citrus scab, grape anthracnose,
                        raspberry anthracnose))
DIVISION BASIDIOMYCOTA
      class Homobasidiomycetae
        subclass Hymenomycetes
          order Aphyllophorales (chanterelles, coral fungi)
              Stereales
              Thelephorales
              Polyporales (bracket fungi)
              Corticales
              Boletales (ceps and other boletes)
              Russulales (russulas and lactarias)
              Trichlomatales (e.g. tricholomas)
              Pluteales (e.g. *Pluteus*)
              Cortinariales (cortinarias)
              Agaricales (e.g. *Agaricus, Amanita*)
              and other orders
        subclass Gasteromycetes
          order Hymenogastrales
              Lycoperdales (puffballs, earthstars)
              Sclerodermales (earthballs)
              Phallales (stinkhorns)
              Nidulariales (bird's nest fungi)

class Heterobasidiomycetae
subclass Teliomycetes (rusts and smuts)

DIVISION DEUTEROMYCOTA (FUNGI IMPERFECTI) (an artificial grouping of fungi with no known sexual stage)
form class Sphaeropsida (pycnidial fungi)
form class Melanconia (conidia)
form class Monilia (e.g. *Penicillium*)
form class Mycelia Sterilia (e.g. *Rhizoctonia*)

Lichens

DIVISION MYCOPHYCOPHYTA (Lichens)
class Ascolichenes (lichens in which an ascomycete is the fungal partner)
class Basidiolichenes (lichens in which a basidiomycete is the fungal partner)
class Deuterolichenes (lichens in which a deuteromycete is the fungal partner)

Appendix 4

AN OUTLINE OF THE ANIMAL KINGDOM

This outline is intended simply to give an overall view of the different types of animals that exist, and is not a rigorous or comprehensive taxonomic classification. Only extant phyla are included, but some extinct groups within phyla are indicated. The names of phyla, classes and orders and the rank given to the various groups differ from authority to authority, and no attempt has been made to give all the alternatives. The arrangement and nomenclature of phyla used here generally follows L. Margulis & K.V. Schwartz, *Five Kingdoms*, 2nd edn (Freeman, 1988). A traditional zoological classification of the protozoans is included here for historical reasons; a more modern treatment of the protozoa, in which the group is divided into numerous separate phyla, is included in Appendix 5.

See entries in the body of the dictionary for more information.

PHYLUM PROTOZOA (unicellular organisms)
  subphylum Sarcomastigophora
    superclass Mastigophora
      class Phytomastigophora (photosynthetic flagellates e.g.
                    Chlamydomonas, Euglena, Gymnodinium)
      class Zoomastigophora (non-photosynthetic flagellates, e.g.
                    Trypanosoma (parasitic), choanoflagellates)
    superclass Opalinata (multiflagellate protozoans inhabiting the gut of
                some amphibians)
    superclass Sarcodina
      class Rhizopodea (amoebas (including the parasitic amoebas) and
                foraminiferans)
      class Actinopodea (radiolarians and heliozoans)
  subphylum Ciliophora (the ciliates)
    class Ciliatea
      subclass Holotrichia (e.g. *Paramecium, Tetrahyena*)
      subclass Peritrichia (e.g. *Vorticella*)
      subclass Suctoria (stalked sessile cells, lacking cilia when mature,
                e.g. *Podophyra*)
      subclass Spirotrichia (e.g. *Stentor*)
  subphylum Sporozoa (exclusively parasitic protozoa)
    class Telosporea
      subclass Gregarinia (gregarines: endoparasites of invertebrates)
      subclass Coccidia (e.g. *Eimeria*, the cause of coccidioidosis, and
                Plasmodium, the agent of malaria)
    class Piroplasmea (e.g. *Babesia*)
  subphylum Cnidospora (exclusively parasitic protozoa)
    class Myxosporidea (e.g. Myxobolus, the cause of a skin disease in
                cyprinid fish)
    class Microsporidea

In modern classifications protozoa are included in the separate kingdom Protista along with other groups of eukaryotic microorganisms of uncertain affinity (see Appendix 5).

Invertebrates

PHYLUM PLACOZOA one species known, *Trichoplax adhaerens*

PHYLUM PORIFERA (sponges)
     class Calcarea (calcareous sponges)
     class Hexactinellidea (glass sponges, e.g. *Euplectella*, Venus's flower basket)
     class Demospongia (e.g. *Spongia*, the bath sponge)

PHYLUM CNIDARIA
     class Anthozoa (corals and sea anemones)
      subclass Alcyonaria (soft corals, sea fans, sea pens)
      subclass Zoantharia (sea anemones, stony corals)
     class Hydrozoa (milleporine corals, solitary hydroids, e.g. *Hydra*, and colonial hydroids, e.g. *Bouganvillea*, and the siphonophores, e.g. *Physalia*, the Portuguese Man o' War)
     class Scyphozoa (true jellyfishes)
     class Cubozoa (box-jellies)

PHYLUM CTENOPHORA (sea gooseberries or comb jellies)
     class Tentacula
     class Nuda

The Cnidaria and Ctenophora collectively are the coelenterates.

PHYLUM MESOZOA
     class Rhombozoa (dicyemids, heterocyemids)
     class Orthonectidea (orthonectids)

PHYLUM PLATYHELMINTHES (flatworms)
     class Turbellaria (free-living flatworms)
      order Acoela
         Rhabdocoela
         Tricladida
         Polycladida
         and other orders
     class Monogenea (flukes)
      order Monopisthocotylea (skin flukes)
         Polyopisthocotylea (gill flukes)
     class Trematoda (flukes)
      order Aspidobothrea
         Digenea (gut, liver and blood flukes)
     class Cestoda (tapeworms)

PHYLUM NEMERTINA (NEMERTEA, RHYNCHOCOELA) (nemertine
worms, ribbon worms)
    class Anopla
    class Enopla

PHYLUM GNATHOSTOMULIDA

PHYLUM GASTROTRICHA

PHYLUM ROTIFERA (rotifers)
    class Seisonacea
    class Bdelloidea
    class Monogononta

PHYLUM KINORHYNCHA

PHYLUM LORICIFERA (only 10 species known)

PHYLUM ACANTHOCEPHALA (thorny-headed worms)

PHYLUM ENTOPROCTA (moss animals)

PHYLUM NEMATODA (nematodes or roundworms)
    class Aphasmidia (Adenophorea)
    class Phasmidia (Secernentea)

PHYLUM NEMATOMORPHA (gordian worms, horse-hair worms)

PHYLUM ECTOPROCTA (moss animals)

(The Entoprocta and Ectoprocta were formerly grouped together as the Bryozoa.)

PHYLUM PHORONIDA (tubeworms, *ca.* 15 species known)

PHYLUM BRACHIOPODA (lamp shells)
    class Inarticulata
    class Articulata

PHYLUM MOLLUSCA (molluscs)
    class Monoplacophora (*Vema, Neopilina*)
    class Aplacophora (solenogasters)
    class Caudofoveata
    class Polyplacophora (chitons)
    class Pelecypoda (Bivalvia) (clams, etc.)
      subclass Protobranchia (*Nucula*)
      subclass Lamellibranchia (most other genera)
    class Gastropoda
      subclass Prosobranchia (winkles, etc.)
      subclass Opisthobranchia (sea slugs, sea hares)
      subclass Pulmonata (whelks, snails and land slugs)
    class Scaphopoda (tusk shells)
    class Cephalopoda
      subclass Nautiloidea (pearly nautilus)
      subclass Ammonoidea (ammonites, extinct)
      subclass Coleoidea (octopus, squid, cuttlefish)

PHYLUM PRIAPULIDA

PHYLUM SIPUNCULA (peanut worms)

PHYLUM ECHIURA (spoon worms)

PHYLUM ANNELIDA (ringed worms)
      class Polychaeta (e.g. ragworms, lugworms, myzostomarians)
      class Oligochaeta (earthworms, etc.)
      class Hirudinea (leeches)

PHYLUM TARDIGRADA (water bears)

PHYLUM PENTASTOMIDA (PENTASTOMA) (tongue worms)

PHYLUM ONYCHOPHORA (velvet worms)

PHYLUM ARTHROPODA (arthropods)
  subphylum Trilobitomorpha (extinct)
      class Trilobita
  subphylum Chelicerata
      class Merostomata (horseshoe crabs)
      class Arachnida
        order Scorpiones (scorpions)
            Pseudoscorpiones (false scorpions)
            Araneae (Araneida) (spiders)
            Palpigrada (palpigrades)
            Solifuga (Solpugida) (solifugids)
            Opiliones (harvestmen)
            Acari (Acarina) (ticks and mites)
      class Pycnogonida (sea spiders)
  subphylum Crustacea
      class Cephalocarida
      class Branchiopoda (water fleas, etc.)
      class Ostracoda (ostracods)
      class Copepoda (copepods)
      class Mystacocarida
      class Branchiura (fish lice)
      class Cirripedia (barnacles)
      class Malacostraca (crabs, lobsters, shrimps, woodlice)
  subphylum Atelocerata
      class Diplopoda (millipedes)
      class Chilopoda (centipedes)
      class Pauropoda
      class Symphyla
      class Insecta
        order Diplura
            Thysanura (silverfish)
            Collembola (springtails)
            Protura (bark lice)
            Odonata (dragonflies)
            Ephemeroptera (mayflies)

Plecoptera (stoneflies)
Dictyoptera (cockroaches)
Dermaptera (earwigs)
Embioptera (web spinners)
Isoptera (termites)
Psocoptera (book-lice and their allies)
Anoplura (biting and sucking lice)
Orthoptera (locusts, grasshoppers and crickets)
Thysanoptera (thrips)
Hemiptera (bugs)
Neuroptera (lace-wings, ant-lions)
Mecoptera (scorpion flies)
Trichoptera (caddis-flies)
Lepidoptera (moths and butterflies)
Coleoptera (beetles)
Diptera (house flies, mosquito, tsetse fly, crane-flies)
Siphonaptera (fleas)
Strepsiptera (stylopids)
Hymenoptera (ants, wasps, bees, saw-flies, ichneumon flies, gall-wasps)

PHYLUM POGONOPHORA (beard worms, tubeworms)

PHYLUM VESTIMENTIFERA (beard worms, tubeworms)

PHYLUM ECHINODERMATA (echinoderms)
  subphylum Pelmatozoa
      class Crinoidea (sea lilies and feather stars)
  subphylum Eleutherozoa
      class Stelleroidea (starfish, brittle stars) (Asteroidea (starfish) and Ophiuroidea (brittle stars))
      class Echinoidea (sea urchins, sand dollars)
      class Holothuroidea (sea cucumbers)
      and many extinct classes

PHYLUM CHAETOGNATHA (arrow worms)

PHYLUM HEMICHORDATA (hemichordates)

      class Enteropneusta (acorn worms, tongue worms)
      class Pterobranchia (pterobranchs)
      class Planktosphaeroidea (known from a larval form only)
      class Graptolita (graptolites, extinct)

PHYLUM CHORDATA (the chordates, including tunicates (urochordates), cephalochordates and vertebrates)
  subphylum Tunicata (Urochordata)
      class Ascidiacea (sea squirts)
      class Larvacea
      class Thaliacea (salps)
  subphylum Cephalocordata (lancelets)
      class Leptocardii

Vertebrates

subphylum Agnatha (vertebrates without jaws)
    class Cyclostomata (lampreys, hagfishes, slime eels)
and the extinct class Ostracodermi (ostracoderms: cephalaspids, anaspids,
pteraspids and thelodonts)

subphylum Gnathostomata (jawed vertebrates)
  superclass Pisces
    class Chondrichthyes (cartilaginous fishes)
      subclass Elasmobranchii
                order Heterodontiformes (the Port Jackson shark)
                       Hexanchiformes
                       Lamniformes (most dogfishes and sharks)
                       Raiiformes (rays and skates)
                       Torpediniformes (electric rays)
                       and extinct orders
    class Osteichthyes (bony fishes)
      subclass Actinopterygii
        infraclass Palaeoniscoidei (extinct except *Polypterus* and
                                                  Erpetoichthys)
        infraclass Chondrostei (sturgeons)
        infraclass Holostei (extinct except *Amia* and *Lepisosteus*)
        infraclass Teleostei (most living bony fishes)
          superorder Elopomorpha (eels, tarpons, etc.)
          superorder Clupeomorpha (herrings, etc.)
          superorder Osteoglossomorpha (mooneyes, knife-fish and bony
                                                  tongues)
          superorder Protacanthopterygii
              order Salmoniformes (salmon)
              order Gonorhynchiformes (millifishes and deep-sea lantern
                                                  fishes)
                    and extinct orders
          superorder Ostariophysi
              order Cypriniformes (characins, American knife-fishes,
                                                  carps and minnows)
                    Siluriformes (catfishes)
          superorder Paracanthopterygii
              order Gadiformes (cod, etc.)
              and other orders
          superorder Acanthopterygii
              order Atheriniformes (tooth-carps, etc.)
                    Perciformes (perches, mackerel, tuna, etc.)
                    Pleuronectiformes (flatfishes)
                    and other orders

```
            class  Sarcopterygii
              subclass Crossopterygii
                        order  Rhipidistia (extinct)
                               Actinistia (extinct except for the coelacanth
                                                      Latimeria)
              subclass  Dipnoi (lungfishes)
              and extinct classes acanthodians and placoderms
         superclass  Tetrapoda
            class  Amphibia
              subclass  Labyrinthodontia (extinct)
              subclass  Lepospondyli (extinct)
              subclass  Lissamphibia (includes all living amphibians)
                        order  Anura (Salientia) (frogs and toads)
                               Urodela (newts and salamanders)
                               Apoda (caecilians)
            class  Reptilia
              subclass Anapsida
                        order  Cotylosauria (Captorhinida) (extinct)
                               Mesosauria (extinct)
              subclass  Testudinata
                        order  Chelonia (tortoises and turtles)
              subclass  Lepidosauria
                        order  Sphenodonta (extinct except the tuatara,
                                                      Sphenodon)
                               Eosuchia (extinct)
                               Squamata
                        suborder  Lacertilia (lizards)
                                  Amphisbaenia (amphisbaenids)
                                  Serpentes (snakes)
              subclass  Archosauria
                        order  Thecodontia (extinct)
                               Saurischia (lizard-hipped dinosaurs, extinct)
                               Ornithischia (bird-hipped dinosaurs, extinct)
                               Pterosauria (pterosaurs, extinct)
                               Crocodylia (crocodiles and alligators)
                               and other extinct orders
              subclass uncertain
                        order  Nothosauria (extinct)
                               Placodontia (extinct)
                               Plesiosauria (extinct)
                               Ichthyosauria (extinct)
                               and other extinct orders
              subclass  Synapsida (mammal-like reptiles, extinct)
                        order  Pelycosauria
                               Therapsida
```

class Aves (birds)
    subclass Archaeornithes (*Archaeopteryx* only, extinct)
    subclass Odontornithes (extinct)
        order Hesperornithiformes
            Ichthyorniformes
    subclass Neornithes (all other birds)
        order Tinamiformes (tinamous)
            Rheiformes (rheas)
            Struthiorniformes (ostriches)
            Casuariiformes (cassowaries and emus)
            Aepyornithiformes (e.g. *Aepyornish*, extinct)
            Dinornithiformes (moas and kiwis)
            Podicipediformes (Colymbiformes) (grebes)
            Procellariformes (albatrosses, shearwaters and
                    petrels
            Sphenisciformes (penguins)
            Pelecaniformes (pelicans and allies)
            Anseriformes (waterfowl, ducks, geese, swans)
            Phoenicopteriformes (flamingoes and allies)
            Ciconiiformes (herons and allies)
            Falconiformes (falcons, hawks, eagles, buzzards and
            other birds of prey)
            Galliformes (grouse, pheasants, partridges, etc.)
            Gruiformes (hemipodes, cranes, bustards, rails,
                    coots, etc.)
            Charadriiformes (shorebirds, waders, gulls, auks)
            Gaviiformes (divers)
            Columbiformes (doves, pigeons, sandgrouse)
            Psittaciformes (parrots, etc.)
            Cuculiformes (cuckoos and others)
            Strigiformes (owls)
            Caprimulgiformes (nightjars)
            Apodiformes (swifts)
            Coliiformes (colies)
            Trogoniformes (trogons)
            Coraciiformes (kingfishers, bee-eaters, hoopoes)
            Piciformes (woodpeckers)
            Passeriformes (perching birds and songbirds, a very
            large order, including finches, crows, warblers,
            sparrows and weavers, etc.)
class Mammalia
   subclassPrototheria
            order Monotremata (duck-billed platypus)
            and several extinct orders

subclass Theria
  infraclass   Metatheria (marsupials)
      order Didelphiformes (opossums)
            Peramelina (bandicoots)
            Diprotodonta (kangaroos, etc.)
  infraclass   Eutheria (placental mammals)
      order Insectivora (shrews, moles, etc.)
            Chiroptera (bats)
            Dermoptera (flying lemurs)
            Fissipedia (dogs, cats, bears, mustelids and other specialized carnivores)
            Pinnipedia (seals)
            Cetacea (whales and dolphins)
         suborder   Odontoceti (toothed whales and dolphins)
                 Mysticeti (baleen whales)
      order Rodentia (rodents)
            Lagomorpha (rabbits, hares, etc.)
            Artiodactyla (even-toed ungulates)
         suborder   Suina (pigs, hippopotamus, etc.)
                 Tylopoda (camels, etc.)
                 Ruminantia (deer, antelope, etc.)
      order Perissodactyla (odd-toed ungulates)
         suborder   Hippomorpha (horses)
                 Ceratomorpha (rhinos and tapirs)
      order Proboscidea (elephants)
             Hyracoidea (hyraxes)
             Tubulidentata (anteaters)
             Sirenia (dugongs, manatees)
             Pholidota (pangolins)
             Primates
         suborder   Prosimii (tree shrews, lemurs, etc.)
                 Anthropoidea (monkeys, apes and man)
                 and many extinct groups

Appendix 5

AN OUTLINE OF THE KINGDOM PROTISTA

The Kingdom Protista (or Protoctista) comprises a diverse assemblage of unicellular and simple multicellular eukaryotic organisms, which form groups each representing a separate evolutionary line that split off early in eukaryote evolution, and which do not sit happily in the animal, plant or fungal kingdoms. The classification here follows L. Margulis & K.V. Schwartz, *Five Kingdoms*, 2nd edn (Freeman, 1988). See entries in the body of the dictionary for more information.

Commonly known as algae
PHYLUM DINOFLAGELLATA (DINOMASTIGOTA or PYRROPHYTA) (dinoflagellates, e.g. *Gymnodinium, Gonyaulax*))
PHYLUM CHRYSOPHYTA (golden algae, e.g. *Ochromonas*)
PHYLUM HAPTOPHYTA (PRYMNESIOPHYTA) (haptomonads, coccolithophoroids, e.g. *Prymnesium*)
PHYLUM EUGLENOPHYTA (euglenoids, e.g. *Euglena*)
PHYLUM CRYPTOPHYTA (cryptomonads)
PHYLUM XANTHOPHYTA (yellow-green algae) (e.g. *Vaucheria*)
PHYLUM EUSTIGMATOPHYTA (e.g. *Pleurochloris*)
PHYLUM BACILLARIOPHYTA (diatoms, e.g. *Asterionella, Navicula, Thalassiosira*)
PHYLUM PHAEOPHYTA (brown algae, e.g. *Fucus, Laminaria*)
PHYLUM RHODOPHYTA (red algae, e.g. *Porphyra, Corallina, Chondrus*)
PHYLUM GAMOPHYTA (conjugating green algae, e.g. *Mougeotia, Spirogyra*)
PHYLUM CHLOROPHYTA (green algae, e.g. *Chlamydomonas, Chlorococcus, Chlorella, Volvox, Ulva, Oedogonium, Stigeoclonium, Chaetomorpha, Acetabularia, Nitella, Platymonas*)

Commonly known as protozoa
PHYLUM CARYOBLASTEA (one species, *Pelomyxa palustris*)
PHYLUM RHIZOPODA (amoebas, e.g. *Acanthamoeba, Amoeba, Difflugia*)
PHYLUM ZOOMASTIGINA (zooflagellates, e.g. *Crithidia, Giardia, Naegleria, Trypanosoma*)
PHYLUM ACTINOPODA (radiolarians, acantharians)
PHYLUM FORAMINIFERA (e.g. *Globigerina*)
PHYLUM CILIOPHORA (ciliates, e.g. *Colpoda, Didinium, Stentor, Vorticella*)
PHYLUM APICOMPLEXA (sporozoans: gregarines, coccidians, e.g. *Eimeria*, hemosporidians, e.g. *Plasmodium*, piroplasms, e.g. *Babesia*)
PHYLUMJ CNIDOSPORIDIA (microsporidians or myxosporidians, e.g. *Glugea*)

Formerly classified in the Fungi

PHYLUM LABYRINTHULOMYCOTA (slime nets, e.g. *Labyrinthula*)

PHYLUM ACRASIOMYCOTA (acrasiomycetes or cellular slime moulds, e.g. *Dictyostelium*)

PHYLUM MYXOMYCOTA (myxomycetes or plasmodial slime moulds, e.g. *Echinostelium, Physarum, Stemonitis*)

PHYLUM PLASMODIOPHOROMYCOTA (plasmodiophorans, e.g. *Plasmodiophora brassicae*, club-root)

PHYLUM HYPHOCHYTRIDIOMYCOTA (water moulds or hyphochytrids, e.g. *Rhizidiomyces*)

PHYLUM CHYTRIDIOMYCOTA (water moulds or chytrids, e.g. *Allomyces, Blastocladiella*)

PHYLUM OOMYCOTA (oomycetes: e.g. *Phytophthora infestans,* potato blight; downy mildews, e.g. *Peronospora*; parasitic water moulds)

Appendix 6

AN OUTLINE OF THE PROKARYOTES (KINGDOM PROKARYOTAE)

Until recent years there have been few means to classify prokaryotes in a way that reflects their evolutionary relationships. The bacteria have traditionally been divided into some dozen large groups on the basis of shape, Gram-staining (which reflects the structure and composition of the cell wall) and other biochemical and physiological properties. The brief outline given below largely follows that proposed in *Bergey's Manual of Systematic Bacteriology*.

Eubacteria
DIVISION I GRACILICUTES

    class Scotophobia ('bacteria indifferent to light': includes all heterotrophic Gram-negative bacteria and some other groups, e.g.:

        Gliding bacteria (e.g. *Beggiatoa*)
        Myxobacteria (e.g. *Stigmatella*)
        Sheathed bacteria (e.g. *Sphaerotilus*)
        Budding bacteria (e.g. *Hyphomicrobium*)
        Stalked bacteria (e.g. *Caulobacteria*)
        Spirochaetes (e.g. *Treponema, Borrelia*)
        Spiral and curved bacteria (e.g. *Spirillum*)
        Sulphate-reducing bacteria (e.g. *Desulfovibrio*)
        Gram-negative aerobic rods and cocci (e.g. *Pseudomonas* and the nitrogen-fixing bacteria *Rhizobium, Azotobacter*)
        Gram-negative facultative anaerobic rods (e.g. the enterobacteria *Escherichia, Shigella, Salmonella, Klebsiella,* and the vibrios e.g. *Vibrio cholerae*)
        Gram-negative cocci and coccobacilli (e.g. *Neisseria*)
        Gram-negative anaerobic cocci (e.g. *Veillonella*)
        Gram-negative anaerobic rods (e.g. *Bacteroides*)
        Chemoautotrophic bacteria (the nitrifying and denitrifying bacteria, e.g. *Nitrobacter, Nitrosomonas*; the sulphur-oxidizing bacteria, e.g. *Thiobacillus*; and the methane-oxidizing bacteria, e.g. *Methylomonas*)

    class Anoxyphotobacteria (anaerobic phototrophic bacteria: the green and purple photosynthetic sulphur bacteria, and the purple non-sulphur bacteria, e.g. *Chlorobium, Rhodospirillum, Rhodomicrobium*)

683

    class Oxyphotobacteria (the cyanobacteria, e.g. *Anabaena,
        Nostoc, Microcystis*)
    class Prochlorophyta (the prochlorophytes, *Prochloron,
        Prochlorothrix*)

DIVISION II FIRMICUTES

    class Firmibacteria (Gram-positive rods and cocci)
        Aerobic and facultatively anaerobic cocci (e.g.
            Micrococcus, Staphylococcus, Streptococcus)
        Aerobic endospore-forming rods (e.g. *Bacillus,
            Clostridium*)
        Asporogenous rods (e.g. *Lactobacillus*)
    class Thallobacteria
        Actinobacteria (the actinomycetes e.g. *Streptomyces* and
            Actinomyces, and the mycobacteria, e.g.
            Mycobacterium tuberculosis)
        Corynebacteria (e.g. *Corynebacterium*)

DIVISION III TENERICUTES

    class Mollicutes (diverse wall-less prokaryotes of doubtful
        origins e.g. rickettsiae, chlamydiae, mycoplasmas)

Archaebacteria
DIVISION IV MENDOSICUTES

    class Archaebacteria (prokaryotes with cell walls lacking
        peptidoglycan, ether-linked membrane lipids, and
        distinctive ribosomes and rRNA sequences)
        Methanogens (e.g. *Methanobacterium, Methanococcus*)
        Extreme halophiles (e.g. *Halobacterium*)
        Thermoacidophiles (e.g. *Thermococcus, Thermoproteus*)

Appendix 7

VIRUS FAMILIES

Nucleic acid	*Bacteriophages*	*Plant virus groups*	*Animal viruses*
a. DNA virus families			
double-stranded DNA	Corticoviridae (e.g. PM2) Lipothrixviridae (e.g. TTV1) Myoviridae (T4 and the T-even phages) Plasmaviridae (e.g. MVL2) Podoviridae (T7 and the T-odd phages) Siphoviridae (e.g. lambda and P22) SSVI group Tectiviridae (e.g. PRD1)	Caulimoviruses	Adenoviridae (adenovirus) Baculoviridae (e.g. insect baculovirus) Hepadnaviridae (e.g. Hepatitis B) Herpesviridae (e.g. herpesviruses, chickenpox (varicella), Epstein-Barr virus) Iridoviridae (e.g. insect iridescent viruses) Papovaviridae (e.g. papilloma-viruses, SV40, polyomaviruses) Polydnaviridae Poxviridae (e.g. vaccinia, smallpox)
single-stranded DNA	Microviridae (e.g. ΦX174, G4) Inoviridae (fd)	Geminiviruses	Parvoviridae (e.g. canine distemper virus)
b. RNA virus families			
double-stranded RNA	Cystoviridae (f6)	Cryptoviruses Plant reoviruses (fujivirus and phytoreovirus groups)	Birnaviridae Reoviridae (e.g. human rotaviruses)

Virus families

Nucleic acid	Bacteriophages	Plant virus groups	Animal viruses
single-stranded RNA	Leviviridae (MS2)	Alfalfa mosaic virus group Bromoviruses Carlaviruses Carmoviruses Closteroviruses Comoviruses Cucumoviruses Dianthoviruses Fabaviruses Furoviruses Hordeiviruses Ilarviruses Luteoviruses Maize chlorotic dwarf virus group Marafiviruses Necroviruses Parsnip yellow fleck virus group Pea enation mosaic virus group Plant rhabdoviruses Potexviruses Potyviruses Sobemoviruses Tenuiviruses Tobraviruses Tobamoviruses Tomato spotted wilt virus group Tombusviruses Tymoviruses	Arenaviridae Bunyaviridae (e.g. Rift Valley fever) Caliciviridae (e.g. swine vesicular exanthema) Coronaviridae (e.g. human coronaviruses) Filoviridae Flaviviridae Nodaviridae Paramyxoviridae (e.g. influenza) Picornaviridae (e.g. poliovirus, foot-and-mouth disease) Retroviridae (e.g. HIV, animal tumour viruses) Rhabdoviridae (e.g. rabies) Tetraviridae Togaviridae (e.g. rubella, yellow fever) Toroviridae

Appendix 8

ETYMOLOGICAL ORIGINS OF SOME COMMON WORD ELEMENTS IN BIOLOGY

The word element as it appears in English is in bold type. The Greek (Gk) or Latin (L.) word from which it is derived is shown in italics and is followed by its original meaning.

a- *a* (Gk) not.

ab- *ab* (L.), from.

absci- *abscidere* (L.), to cut off.

abyss- *abyssos* (Gk), unfathomed.

acanac- *akanos* (Gk), thistle.

acanth- *akantha* (Gk), thorn.

acer- *acer* (L.), sharp.

acid- *acidus* (L.), sour.

acra- *akros* (Gk), tip.

actin- *aktis* (Gk), ray.

ad- *ad* (L.), to, towards.

adeno- *aden* (Gk), gland.

adipo- *adeps* (L.), fat.

aeolian *Aeolus* (Gk), god of the winds.

aer- *aer* (L.), *aēr* (Gk), air.

-aesthesia, -aesthetic *aisthēsis* (Gk), sensation.

agrost- *agrōstis* (Gk), grass.

albo-, albu-, albino *albus* (L.), white.

alga-, algo- *alga* (Gk), seaweed.

allele, allelo- *allēlōn* (Gk), one another.

allo- *allos* (Gk), other.

ambi- *ambo* (L.), both.

amoeba *amoibē* (Gk), change.

amphi- *amphi* (Gk), both.

amylo- *amylum* (Gk), starch.

an- *an* (Gk), not.

ana- *ana* (Gk), up, again.

andr-, andro- *anēr* (Gk), male, *andrikos* (Gk), masculine.

anemo- *anemos* (Gk), wind.

angio-, -angium *anggeion* (Gk), vessel.

aniso- *anisos* (Gk), unequal.

ankylo- *agkylos* (Gk), crooked.

anlage *Anlage* (Ger.) predisposition.

annelid, annulate *annulus* (L.), a ring.

ano- *anus* (L.), anus.

anomalo- *anomalos* (Gk), uneven.

anomo- *anomos* (Gk), lawless, irregular.

ante- *ante* (L.), before.

antha-, antho-, -anthous, -anthy *anthos* (Gk), flower.

anthero- *anthēros* (Gk), flowering.

anthropo- *anthrōpos* (Gk), man.

anti- *anti* (Gk), against, opposite.

-apl'oid *aploos* (Gk), onefold, and *eidos* (Gk), form.

apo- *apo* (Gk), from.

arachni-, arachno- *arachnē* (Gk), spider, cobweb.

arbor- *arbor* (L.), tree.

archaeo- *archaios* (Gk), primitive, ancient.

arche- *archē* (Gk), beginning.

archi- *archi* (Gk), first.

archo- *archon* (Gk), ruler.

arci- *arcus* (L.), bow.

argent- *argentum* (L.), silver.

argyro- *argyros* (Gk), silver.

arthro- *arthron* (Gk), a joint.

artio- *artios* (Gk), even (numbered).

-asci, -ascus, asco- *askos, askidion* (Gk), bag, little bag.

astra-, astro-, -aster *astra* (Gk), star.

-atomy, -otomy *tomē* (Gk), cutting.

auri-, auricul- *auris, auricula* (L.), ear, small ear.

auto- *autos* (Gk), self.

auxi-, auxo- *auxein* (Gk), to increase.

avi- *avis* (L.), bird.

axill- *axilla* (L.), armpit.

axis, axial *axis* (L.), axle.

axo- *axōn* (Gk), axis.

barb-, barba *barba* (L.), beard.

baro- *baros* (Gk), pressure, weight.

bathy- *bathys* (Gk), deep.

batrach- *batrachos* (Gk), frog.

benthos, benthic *benthos* (Gk), depths of sea.

bi- *bis* (L.), twice.

bio-, -biotic *bios* (Gk), life, *biosis*, living, *biōtikos*, pert. life

blast- *blastos* (Gk), bud.

botany *botanē* (Gk), pasture.

bothr- *bothros* (Gk), pit.

botry- *botrys* (Gk), bunch of grapes.

brachia- *brachium* (L.), arm.
brachy- *brachys* (Gk), short.
brady- *bradys* (Gk), slow.
branchi- *branchiae* (L.), gills, or
  brangchia (Gk), gills.
brevi- *brevis* (L.), short.
bryo- *bryon*(Gk), moss.
bucco- *bucca* (L.), cheek.
caeno- *kainos* (Gk), recent.
calci- *calx* (Gk), lime.
calyptr- *kalyptra* (Gk), covering.
cambium, cambio- *cambium* (L.),
  change.
capit- *caput* (L.), head, *capitellum*, small
  head.
capsid, capso-, capsul- *capsa* (L.), box,
  capsula, little box.
carbo- *carbo* (L.), coal.
carcino- *karkinos* (Gk), a crab.
cardia-, cardio- *kardia* (Gk), heart,
  stomach.
-carp , -carpous *karpos* (Gk), fruit.
carpa- *carpal* (L.), wrist.
cata- *katalysis* (Gk), dissolving.
cata- *kata* (Gk), down.
cauda- *cauda* (L.), tail.
caul-, cauli- *caulis* (L.), stalk or *kaulos*
  (Gk), stalk.
cell, cellular *cellula* (L.), small room.
centro-, -centric *kentron* (Gk), centre.
cephal- *kephalē* (Gk), head.
-ceptor, -ceptive *capere* (L.), to take.
cerca-, cerco- *kerkos* (Gk), tail.
cerebr- *cerebrum* (L.), brain.
-cerous *keras* (Gk), horn.
cervic- *cervix* (L.), neck.
ceta-, ceto- *cetus* (L.), whale.
-chaene-, -chene *chainein* (Gk), to gape.
chaet- *chaitē* (Gk), hair.
chela- *chēlē* (Gk), claw.
chemi-, chemistry *chēmeia* (Gk),
  transmutation.
chiasm- *chiasma* (Gk), cross.
chitin *chiton* (Gk), tunic.
chlamy- *chlamys* (Gk), cloak, mantle.
chloro- *chlōros* (Gk), yellow, green, pale.
chondro- *chondros* (Gk), cartilage.
-chord, chorda-, chordo- *chordē* (Gk),
  string.
-chore *chōros* (Gk), place.
chroma-, chromo-, -chrome *chrōma*
  (Gk), colour.
chrono- *chronos* (Gk), time.

-cidal *caedere* (L.), to kill.
clade, -cladous *klados* (Gk), branch.
clav- *clava* (L.), a club.
cleisto- *kleistos* (Gk), closed.
clino-, -cline, -clinous *klinē* (Gk), bed.
-clinous *klinein* (Gk), to bend.
clype- *clypeus* (L.), shield.
cocc-, cocco- *koccos* (Gk), berry.
cochli- *kochlias* (Gk), a snail.
coel- *koilos* (Gk), hollow.
coen-, -coenosis *koinos* (Gk), shared in
  common.
coleo- *koleos* (Gk), sheath.
conch- *concha* (L.), shell, *kongchē* (Gk),
  shell.
cono-, -cone *kōnos* (Gk), a cone.
copro- *kopros* (Gk), dung.
-corn, corne-, corni- *cornus* (L.), horn.
corona- *corona* (L.), crown.
corp-, corpor- *corpus* (L.), body.
cortex, cortic- *cortex* (L.), bark.
costa- *costa* (L.), rib.
cotyl- *kotylē* (Gk), cup.
coxa-, coxo- *coxa* (L.), hip.
crani-, crania- *kranion* (Gk), skull.
-crine *krinein* (Gk), to separate.
cruci- *crux* (L.), cross.
cryo- *kryos* (Gk), frost.
crypto- *kryptos* (Gk), hidden.
cyano- *kyanos* (Gk), dark blue.
cyath- *kyathus* (Gk), a cup.
cyclo-, -cyclic *kyklos* (Gk), circle.
cyst- *kystis* (Gk), bladder.
-cyte, cyto- *kytos* (Gk), hollow.
-dactyl *daktylos* (Gk), finger.
de- *de* (L.), away.
deme, demo- *dēmos* (Gk), people.
demi- *dimidius* (L.), half.
dendr- *dendron* (Gk), tree.
dent- *dens* (L.), tooth.
derma-, dermo-, -derm *derma* (Gk),
  skin.
desm- *desmos* (Gk), bond.
di- *dis* (Gk), twice.
dia- *dia* (Gk), asunder.
dicho- *dicha* (Gk), in two.
dictyo- *dictyon* (Gk), net.
digito- *digitus* (L.), finger.
dino- *dinos* (Gk), rotation.
dino- *deinos* (Gk), terrible.
diplo- *diploos* (Gk), double.
dors- *dorsum* (L.), back.
-drome, -dromic, -dromous *dramein*

(Gk), to run, *drōmos* (Gk), running

-duct *ducere* (L.), to lead.

duplico- *duplex* (L.), double.

dynamo- *dynamis* (Gk), power.

dys- *dys* (Gk), mis-.

e- *ex* (L.), out of.

ec- *ek* (Gk), out of.

echino- *echinos* (Gk), spine.

eco- *oikos* (Gk), house, household.

ect- *ektos* (Gk), without, outside.

-ectomy *ektomē* (Gk), a cutting out.

elaeo-, elaio- *elaion* (Gk), oil.

electro- *elektron* (Gk), amber (a fossil resin which produces static electricity when rubbed).

embryo- *embryon* (Gk), embryo.

endo- *endon* (Gk), within, inside.

-ennial *annus* (L.), year.

entero- *enteron* (Gk), gut.

entomo- *entomon* (Gk), insect.

epi- *epi* (Gk), upon.

equi- *aequus* (L.), equal.

erg-, -ergic, -ergy *ergon* (Gk), activity, work.

erythro- *erythros* (Gk), red.

eu- *eu* (Gk), well.

eury- *eurys* (Gk), wide.

exo- *exō* (Gk), outside.

extra- *extra* (L.), beyond.

-farious *fariam* (L.), in rows.

fauna- *faunus* (L.), god of woods.

ferre-, ferri-, ferro- *ferrum* (L.), iron.

fibrino- *fibra* (L.), a band.

-fid *findere* (L.), to split.

fili- *filum* (L.), a thread.

flavo- *flavus* (L.), yellow.

flor- *flos* (L.), flower.

folia- *folium* (L.), leaf.

fronto- *frons* (L.), forehead.

fuco- *fucus* (L.), seaweed.

-fugal, -fuge *fugere* (L.), to flee.

galacto- *gala* (Gk), milk.

gamete, gameto- *gametes* (Gk), spouse.

gamo-, -gamy, -gamous *gamos* (Gk), marriage.

ganglio- *ganglion* (Gk), swelling.

gastro- *gaster* (Gk), stomach.

-geminal *geminus* (L.), double.

-gen, -genous *genos* (Gk), descent.

gene-, -genetic *genesis* (Gk), birth, descent, origin.

-genic, -genous *gennaein* (Gk), to produce.

genito- *gignere* (L.), to beget.

geno- *genos* (Gk), race.

geo- *gē*, or *gaia* (Gk), earth.

germ-, germin- *germen* (L.), , a bud.

geronto- *gerōn* (Gk), old man.

glia-, -gloea *gloia* (Gk), glue.

-globin, -globulin *globus* (L.), a sphere.

glosso- *glossa* (L.), tongue.

gluco-, glyco- *glykys* (Gk), sweet.

gnatho-, -gnath, -gnathous *gnathos* (Gk), jaw.

gonad-, -gone, -gonic *gonē* (Gk), seed.

-gone, goni- *gonos* (Gk), offspring.

-grade *gradus* (L.), step.

-gram, -graphy *graphein* (Gk), to write.

gyn-, -gynous *gynē* (Gk), female.

haem-, haema- *haima* (Gk), blood.

halo- *hals* (Gk), sea, salt.

haplo- *haploos* (Gk), simple.

hapto- *haptos* (Gk), touch.

helio- *helios* (Gk), sun.

hemi- *hēmi* (Gk), half.

hepa-, hepatico- *hepar* (Gk), liver.

hepta- *hepta* (Gk), seven.

hetero- *heteros* (Gk), other.

hex- *hex* (Gk), six.

histio- *histion* (Gk), tissue.

histo- *histos* (Gk), tissue.

holo- *holos* (Gk), whole.

homeo- *homoios* (Gk), alike.

homo- *homos* (Gk), the same.

hormone *hormaein* (Gk), to excite.

hyalo- *hyalos* (Gk), glass.

hydr-, hydro- *hydor* (Gk), water.

hygro- *hygros* (Gk), wet.

hyper- *hyper* (Gk), above.

hypo- *hypo* (Gk), under.

ichthy- *ichthys* (Gk), fish.

-icole, -icolous *colere* (L.), to dwell.

-iferous *ferre* (L.), to carry.

-ific, -ification *facere* (L.), to make.

-igen *generare* (L.), to beget.

-igerous *gerere* (L.), to bear.

im-, in- *in* (L.), not.

immuno- *immunis* (Gk), free.

in- *in* (L.), into.

infero- *inferus* (L.), beneath.

infra- *infra* (L.), below.

inter- *inter* (L.), between.

intra- *intra* (L.), within.

iso- *isos* (Gk), equal.

-jugate *jugare* (L.), to join.

jugo- *jugum* (L.), yoke.

juxta- *juxta* (L.), close to.
kary- *karyon* (Gk), nucleus, nut.
kera- *keras* (Gk), horn.
-kinesis, -kinetic *kinesis* (Gk), movement, *kinein* (Gk), to move.
labia-, labio- *labium* (L.), lip.
lacto- *lac* (L.), milk.
lati-, latero- *latus* (L.), wide.
lepido- *lepidotos* (Gk), scaly.
lepto- *leptos* (Gk), slender.
leuco-, leuko- *leukos* (Gk), white.
limn- *limne* (Gk), marsh.
lipo- *lipos* (Gk), fat.
litho-, -lith *lithos* (Gk), stone.
lopho- *lophos* (Gk), crest.
luci- *lux* (L.), light.
luteo- *luteus* (L.), orange-yellow.
-lysin, -lysis, lyso-, -lytic *lysis* (Gk), loosing, *lyein* (Gk), to dissolve.
macro- *makros* (Gk), large.
masto- *mastos* (Gk), breast.
matro- *mater* (L.), mother.
medi- *medius* (L.), middle.
mega- *megas* (Gk), large.
megalo- *megalon* (Gk), great.
meio- *meion* (Gk), smaller.
meiosis- *meiosis* (Gk), diminution.
-mere, mero- *meros* (Gk), a part.
meso- *mesos* (Gk), middle.
meta- *meta* (Gk), after.
metabolism, metabolic *metabole* (Gk), change.
-metric, -metry *metron* (Gk), measure, *metreo* (Gk), to count.
micro- *mikros* (Gk), small.
mito-, mitosis *miton* (Gk), a thread.
mono- *monos* (Gk), alone, single.
morpho-, -morph, -morphism, -morphy *morphe* (Gk), shape, form, *morphosis* (Gk), form.
multi- *multus* (L.), many.
mutate, muta- *mutare* (L.), to change.
myco-, -mycin *mykes* (Gk), fungus.
myelo- *myelos* (Gk), marrow.
myo- *mys* (Gk), muscle.
myrme- *myrmēx* (Gk), ant.
myxo- *myxa* (Gk), slime.
nano- *nanos* (Gk), dwarf.
necro- *nekros* (Gk), dead.
nema-, nemato-, -neme *nēma* (Gk), a thread.
neo- *neos* (Gk), new.
nephr- *nephros* (Gk), kidney.

neuro- *neuron* (Gk), nerve.
nexus, -nexed *nectare* (L.), to bind.
nigro- *niger* (L.), black.
nitro- *nitron* (Gk), soda.
noci- *nocere* (L.), to hurt.
-nomics, -nomy *nomos* (Gk), law.
nomin-, -nomial *nomen* (L.), name.
noto- *noton* (Gk), back.
nucleus *nucleus* (Gk), kernel.
ob- *ob* (L.), against, reversely.
occipi- *occiput* (L.), back of head.
octa-, octo- *okta* (Gk), *octo* (L.), eight.
-odont, odonto- *odous* (Gk), tooth.
-oecious, -oecium *oikos* (Gk), house.
oestro-, oestrus *oistros* (Gk), gadfly.
-ogen *genos* (Gk), birth.
-ogony *gonos* (Gk), generation.
oi-, -oo- *ōon* (Gk), egg.
-oid *eidos* (Gk), form.
olei-, oleo- *oleum* (L.), oil.
oligo- *oligos* (Gk), few.
-ology *logos* (Gk), discourse.
onco- *onkos* (Gk), bulk, mass.
onto- *on* (Gk), being.
-oo- *ōon* (Gk), egg.
ophthal- *ophthalmos* (Gk), eye.
opsi-, -opsin, -opsy, opto- *opsis* (Gk), eye.
ora-, oro- *os, oris* (L.), mouth.
organ *organon* (Gk), instrument.
orni- *ornis* (Gk), bird.
ortho- *orthos* (Gk), straight.
osmo- *ōsmos* (Gk), impulse.
ost-, osteo- *osteon* (Gk), bone.
ostraco- *ostrakon* (Gk), shell.
oto- *ous* (Gk), ear.
-otomy *temnein* (Gk), to cut.
ova-, ovi-, ovo- *ovum* (L.), egg.
oxy- *oxys* (Gk), sharp.
pachy- *pachys* (Gk), thick.
palaeo- *palaios* (Gk), ancient.
palpi- *palpare* (L.), to stroke.
pan-, panto- *pan* (Gk), all.
para- *para* (Gk), beside.
-parous *parere* (L.), to produce.
patho-, -pathy *pathos* (Gk), suffering.
patri- *pater* (L.), father.
-patric *patria* (L.), native land.
ped-, -pedal *pes* (L.), foot.
-pelagic *pelagos* (Gk), sea.
penta-, pento- *pente* (Gk), five.
per- *per* (L.), through.
peri- *peri* (Gk), around.

perisso- *perissos* (Gk), odd (numbered).
petalo- *petalon* (Gk), leaf.
petro- *petros* (Gk), stone.
-phage, phago, -phagous *phagein* (Gk), to eat.
-phase *phasis* (Gk), aspect, appearance.
-phene, pheno- *phainein* (Gk), to appear.
-phil, -phile, -phily *philein* (Gk), to love.
philo- *philos* (Gk), loving.
-phobe, -phobic *phobos* (Gk), fear.
phono- *phōnē* (Gk), sound.
-phore *phorein* (Gk), to carry.
phospho- *phosphoros* (Gk), bringing light.
photo- *phos* (Gk), light.
phragmo- *phragmos* (Gk), fence.
-phylactic *phylaktikos* (Gk), fit for preserving.
-phyll, -phyllous *phyllon* (Gk), leaf.
physi-, -physis *physis* (Gk), growth.
physics *physis* (Gk), nature.
phyt-, -phyte *phyton* (Gk), plant.
pinna-, penna- *penna* (L.), feather.
pisci- *piscis* (L.), fish.
placo- *plax* (Gk), plate.
plana- *planatus* (L.), flattened.
-planetic, planeto- *planetes* (Gk), wanderer.
plano- *planos* (Gk), wandering.
-plasia *plasis* (Gk), a moulding, *plassein* (Gk), to form.
plasma-, -plasm, *plasma* (Gk), form.
-plast, -plastic, plastid *plastos* (Gk), formed.
plasti-, plasto- *plastos* (Gk), formed.
platy- *platys* (Gk), flat.
pleio- *pleion* (Gk), more.
plero- *plērēs* (Gk), full.
plesio- *plēsios* (Gk), near.
pleura-, -pleurite *pleuros* (Gk), side.
-plicate *plicare* (L.), to fold.
pluri- *plus* (L.), more.
pneu- *pnein* (Gk), to breathe.
-pod, -podite *pous* (Gk), foot.
-poiesis, -poietic *poiesis* (Gk), making.
poly- *polys* (Gk), many.
poro-, -pore *poros* (Gk), channel.
porphyr- *porphyra* (Gk), purple.
post- *post* (L.), after.
pre- *prae* (L.), before.
primo- *primus* (L.), first.
pro- *pro* (Gk), before.
proto- *prōtos* (Gk), first.

pseudo- *pseudes* (Gk), false.
psycho- *psychē* (Gk), mind.
-pter- *pteron* (Gk), wing.
pterido- *pteris* (Gk), fern.
ptero- *pteron* (Gk), wing.
pteryg- *pterygion* (Gk), little wing.
pteryg- *pterygion* (Gk), fin.
-ptile *ptilon* (Gk), feather.
pulmo- *pulmo* (L.), lung.
pycno- *pyknos* (Gk), dense.
pygo- *pygē* (Gk), rump.
-pyle *pyle* (Gk), gate.
pyreno- *pyrēn* (Gk), fruit stone.
pyri- *pyrum* (L.), pear.
pyrro- *pyrrhos* (Gk), tawny-red.
quadrato- *quadratus* (L.), squared.
quadri-, quadru- *quattuor* (L.), four.
quin-, quinque- *quinque* (L.), five.
racem- *racemus* (L.), bunch.
rachi- *rachis* (Gk), spine.
radic- *radix* (L.), a root.
radio- *radius* (L.), a ray.
-ramous *ramus* (L.), branch.
rani- *rana* (L.), frog.
re- *re* (L.), back.
rena-, reni- *renes* (Gk), kidney.
reti- *rete* (L.), net.
reticulo- *reticulum* (L.), small net.
retro- *retro* (L.), backwards.
rhabdo- *rhabdos* (Gk), rod.
rheo- *rheein* (Gk), to flow.
rhin- *rhis* (Gk), nose.
rhiza-, rhizo- *rhiza* (Gk), root.
rhodo- *rhodon* (Gk), rose.
rhyncho- *rhyngchos* (Gk), snout.
ribo- *ribes* (L.), currant.
rostra- *rostrum* (L.), beak.
rubi- *ruber* (L.), red.
rubro- *ruber* (L.), red.
sacchar- *sakchar* (Gk), sugar.
salpingo- *salpingx* (Gk), trumpet.
sangui- *sanguis* (L.), blood.
sapo- *sapo* (L.), soap.
sapro- *sapros* (Gk), decayed.
-sarc, sarco- *sarx, sarkōdēs* (Gk), flesh, fleshy.
-saur- *sauros* (Gk), lizard.
schisto-, -schist *schistos* (Gk), split.
schizo- *schizein* (Gk), to cleave.
-scopic *skopein* (Gk), to view.
scoto- *skotos* (Gk), dark.
scute, scutum *scutum* (L.), shield.
seismo- *seismos* (Gk), a shaking.

seleno- *selēnē* (Gk), moon.
-sematic *sema* (Gk), signal.
semi- *semi* (L.), half.
septa-, septi-, septo- *septum* (L.), partition.
septi- *septum* (L.), seven.
-sere *serere* (L.), to put in a row.
serum *serum* (L.), whey.
seta-, seti-, seto- *seta* (L.), bristle.
siali-, sialo- *sialon* (Gk), saliva.
sidero- *sidēros* (Gk), iron.
soma-, somato-, -some *sōma* (Gk), body.
sora-, sori-, soro- *sōros* (Gk), heap.
speleo- *spelaion* (Gk), cave.
sperma-, -sperm *sperma* (Gk), seed.
sphaero- *sphaira* (Gk), globe.
spheno- *sphen* (Gk), a wedge.
sphero- *sphaira* (Gk), globe.
spondyl- *sphondylos* (Gk), vertebra.
spor-, -spore *sporos* (Gk), seed.
squame, squama- *squama* (L.), scale.
stachy- *stachys* (Gk), ear of corn.
stamin-, -stemonous *stemon* (Gk), spun thread.
-stat, -static *stare* (L.), to stand.
stato- *statos* (Gk), stationary, standing.
stega-, stegi- *stega* (Gk), roof.
-stelic, -stely *stele* (Gk), pillar.
stereo-, -steric *stereos* (Gk), solid.
stern- *sternum* (L.), breast.
steroid, -sterone *stear* (Gk), suet.
stoma-, stomato-, -stome *stoma* (Gk), mouth.
-strate *stratum* (L.), layer.
strepto- *streptos* (Gk), twisted, pliant.
strobilus *strobilos* (Gk), fir-cone.
strom-, stroma *stroma* (Gk), bedding.
-stylic *stylos* (Gk), pillar.
sub- *sub* (L.), under.
super- *super* (L.), over.
supra- *supra* (L.), above.
sylv- *sylva* (L.), forest.
sym- *syn* (Gk), with.
syn- *syn* (Gk), with.
synaps-, synapto- *synapsis, synaptos* (Gk), union, joined.
synthesis *synthesis* (Gk), composition.
tachy- *tachys* (Gk), quick.
talo- *talus* (L.), ankle.
tarso- *tarsos* (Gk), sole of foot.
tauto- *tautos* (Gk), the same.
-taxis, -taxy *taxis* (Gk), arrangement.

taxo- *taxis* (Gk), arrangement.
tect- *tectum* (L.), roof.
tele- *tēle* (Gk), far.
teleo- *teleos* (Gk), complete.
telo-, telio- *telos* (Gk), end.
tempor- *tempora* (L.), temples.
-tene *tainia* (Gk), a band.
tensin- *tonos* (Gk), tension.
terga- *tergum* (L.), back.
ternato- *terni* (L.), three.
terr- *terra* (L.), earth.
tetra- *tetras* (Gk), four.
thallo- *thallos* (Gk), branch.
-theca, -thecium *theke* (Gk), box.
-theria *therion* (Gk), small animal.
thermo- *thermē* (Gk), heat.
thero- *theros* (Gk), summer.
thigmo- *thigēma* (Gk), touch.
thio- *theion* (Gk), sulphur.
thrombo- *thrombos* (Gk), clot.
thylako- *thylakos* (Gk), pouch.
thyro- *thyra* (Gk), door.
-tocin *tokos* (Gk), birth.
-tope, -topic, topo- *topos* (Gk), place.
toti- *totus* (L.), all.
toxico-, *-toxin* toxikon *(Gk), poison.*
tracheo- *trachia* (L.), windpipe.
trachy- *trachys* (Gk), rough.
trans- *trans* (L.), across.
trauma- *trauma* (Gk), wound.
tri- *tria* (Gk), three, *tres* (L.), three.
trich- *thrix* (Gk), hair.
trocho- *trochos* (Gk), hoop.
-troph, -trophic, tropho-, -trophy *trophē* (Gk), maintenance, nourishment.
-tropic, -tropism trope *(Gk), turn.*
tropo- *tropos* (Gk), turn.
tubi-, tubo- *tubus* (L.), pipe.
tympano- *tympanon* (Gk), drum.
-type *typos* (Gk), pattern.
ulna-, ulno- *ulna* (L.), elbow.
ultra- *ultra* (L.), beyond.
umbell- *umbella* (L.), a sunshade.
umbona- *umbo* (L.), shield boss.
unci- *uncus* (L.), hook.
uni- *unus* (L.), one.
uredo- *urēdo* (L.), blight.
uro- *ouron* (Gk), urine or *oura* (Gk), tail.
vacuol- *vacuus* (L.), empty.
vagini- *vagina* (L.), sheath.
valv- *valvae* (L.), folding doors.
vasa-, vaso- *vas* (L.), vessel.
ventr- *venter* (L.), belly.

vermi- *vermis* (L.), worm.
versi- *versare* (L.), to turn.
vesicul- *vesicula* (L.), small bladder.
viro- *virus* (L.), poison.
vitello- *vitellus* (L.), yolk.
vitreo- *vitreus* (L.), glassy.
vivi- *vivus* (L.), living.
-vorous *vorare* (L.), to devour.
xantho- *xanthos* (Gk), yellow.

xeno-, -xenous *xenos* (Gk), host or strange.
xero- *xeros* (Gk), dry.
zo-, zoo- *zōon* (Gk), animal.
-zoic *zoikos* (Gk), pertaining to life.
zygo-, zygote *zygon* (Gk), yoke, *zygotos* (Gk), yoked.
zymo-, -zyme *zymē* (Gk), leaven.